The Kaleidoscope of Gender

Sixth Edition

The Kaleidoscope of Gender

Prisms, Patterns, and Possibilities

Sixth Edition

Catherine G. Valentine

Nazareth College

Mary Nell Trautner

University at Buffalo, State University of New York

with

Joan Z. Spade

The College at Brockport, State University of New York

Los Angeles | London | New Delhi
Singapore | Washington DC | Melbourne

FOR INFORMATION:

SAGE Publications, Inc.
2455 Teller Road
Thousand Oaks, California 91320
E-mail: order@sagepub.com

SAGE Publications Ltd.
1 Oliver's Yard
55 City Road
London, EC1Y 1SP
United Kingdom

SAGE Publications India Pvt. Ltd.
B 1/I 1 Mohan Cooperative Industrial Area
Mathura Road, New Delhi 110 044
India

SAGE Publications Asia-Pacific Pte. Ltd.
18 Cross Street #10-10/11/12
China Square Central
Singapore 048423

Acquisitions Editor: Joshua Perigo
Editorial Assistant: Noelle Cumberbatch
Production Editor: Jane Martinez
Copy Editor: Tammy Giesmann
Typesetter: Hurix Digital
Proofreader: Jeff Bryant
Cover Designer: Candice Harman
Marketing Manager: Jennifer Jones

Printed in the United States of America

Library of Congress Cataloging-in-Publication Data

Names: Valentine, Catherine G., editor. | Spade, Joan Z., editor. | Trautner, Mary Nell, editor.

Title: The kaleidoscope of gender : prisms, patterns, and possibilities/ Catherine G. Valentine, Nazareth College, Mary Nell Trautner, University at Buffalo, State University of New York, Joan Z. Spade, The College at Brockport, State University of New York.

Description: Sixth Edition. | Thousand Oaks, CA : SAGE Publications, [2019] | Includes bibliographical references and index. | Revised edition of The kaleidoscope of gender, [2017]

Identifiers: LCCN 2018061129 | ISBN 9781506389103 (pbk. : alk. paper)

Subjects: LCSH: Sex role. | Sex differences (Psychology) | Gender identity. | Man-woman relationships. | Interpersonal relations.

Classification: LCC HQ1075 .K35 2019 | DDC 305.3—dc23 LC record available at https://lccn.loc.gov/2018061129

This book is printed on acid-free paper.

SFI label applies to text stock

19 20 21 22 23 10 9 8 7 6 5 4 3 2 1

Contents

PART II: PATTERNS

PREFACE

This sixth edition of *The Kaleidoscope of Gender: Prisms, Patterns, and Possibilities* provides an overview of the cutting-edge literature and theoretical frameworks in the sociology of gender and related fields for understanding the social construction of gender. Although not ignoring classical contributions to gender theory and research, this book focuses on where the field is moving and the changing paradigms and approaches to gender studies. *The Kaleidoscope of Gender* uses the metaphor of a kaleidoscope and three themes—prisms, patterns, and possibilities—to unify topic areas. It focuses on the prisms through which gender is shaped, the patterns gender takes, and the possibilities for social change through a deeper understanding of ourselves and our relationships with others, both locally and globally.

The book begins, in the first part, by looking at gender and other social prisms that define gendered experiences across the spectrum of daily lives. We conceptualize prisms as social categories of difference and inequality that shape the way gender is defined and practiced, including culture, race/ethnicity, social class, sexuality, age, and ability/disability. Different as individuals' lives might be, there are patterns to gendered experiences. The second part of the book follows this premise and examines these patterns across a multitude of arenas of daily life. From here, the last part of the book takes a proactive stance, exploring possibilities for change. Basic to the view of gender as a social construction is the potential for social change. Students will learn that gender transformation has occurred and can occur and, consequently, that it is possible to alter the genderscape. Because prisms, patterns, and possibilities themselves intersect, the framework for this book is fluid, interweaving topics and emphasizing the complexity and ever-changing nature of gender.

We had multiple goals in mind as we first developed this book, and the sixth edition reaffirms these goals:

1. Creating a book of readings that is accessible, timely, and stimulating in a text whose structure and content incorporate a fluid framework, with gender presented as an emergent, evolving, complex pattern—not one fixed in traditional categories and topics;
2. Selecting articles that creatively and clearly explicate what gender is and is not and what it means to say that gender is socially constructed by incorporating provocative illustrations and solid scientific evidence of the malleability of gender and the role of individuals, groups, and social institutions in the daily performance and transformation of gender practices and patterns;
3. Including readings that untangle and clarify the intricate ways gender is embedded in, intersects with, and is defined by the prisms of culture/nation, race/ethnicity, class, sexuality, age, ability/disability, and other patterns of identities, groups, and institutions;
4. Integrating articles with cross-cultural and global foci to illustrate that gender is a continuum of categories, patterns, and expressions whose relevance is contextual and continuously shifting, and that gender inequality is not a universal and natural social pattern, but at the same time, emphasizing how patriarchal social systems result in similar patterns of experiences and inequalities;
5. Assembling articles that offer students useful cognitive and emotional tools for making sense of the shifting and contradictory genderscape they inhabit, its personal relevance, its implications for relationships both locally and globally, and possibilities for change.

These goals shaped the revisions in the sixth edition of *The Kaleidoscope of Gender*. New selections in this edition emphasize sex and gender diversity, including the experiences of transgender and intersex people. Global and intersectional analyses as well as new contemporary social movements for gender justice are incorporated throughout the book. We continue to explore the role of institutions in maintaining gender difference and inequality. Across the chapters, readings examine the individual, situational, and institutional bases for gendered patterns in relationships, behaviors, and beliefs. Additionally, many readings illustrate how multiple prisms of difference and inequality, such as race, age, and social class, create an array of patterns of gender—distinct but sometimes similar to the idealized patterns in a culture.

As in the fifth edition, reading selections include theoretical and review articles; however, the emphasis continues to be on contemporary contributions to the field. The introduction to the book provides an overview of theories in the field, particularly theories based on a social constructionist perspective. In addition, the introduction to the book develops the kaleidoscope metaphor as a tool for viewing gender and a guide for studying gender. Revised chapter introductions contextualize the literature in each part of the book, introduce the readings, and illustrate how they relate to analyses of gender. Introductions and questions for consideration precede each reading to help students focus on and grasp the key points of the selections. Additionally, each chapter ends with questions for students to consider and topics for students to explore.

It is possible to use this book alone, as a supplement to a text, or in combination with other articles or monographs. It is designed for undergraduate audiences, and the readings are appropriate for a variety of courses focusing on the study of gender, such as sociology of gender, gender and social change, and women's studies. The book may be used in departments of sociology, anthropology, psychology, women's studies, and gender studies.

We would like to thank those reviewers whose valuable suggestions and comments helped us develop the book throughout five editions, including the following:

Sixth edition reviewers:

Hortencia Jimenez, Hartnell College; Pamela McMullin-Messier, Central Washington University; Michael Ramirez, Texas A&M University–Corpus Christi; Regina Davis-Sowers, Middle Tennessee State University; Natalie Jolly, University of Washington; Amanda Miller, University of Indianapolis; Kristi Brownfield, Northern State University.

Fifth edition reviewers:

Kathryn Feltey, University of Akron; Tennille Allen, Lewis University; Michelle Deming, University of South Carolina; Andrea Collins, University of St Mark & St John Plymouth; Kimberly Hoang, Boston College; Pamela Danker, Blackburn College; Amanda Miller, University of Indianapolis; Regina Davis-Sowers, Santa Clara University.

Fourth edition reviewers:

Nancy Ashton; Allison Alexy, Lafayette College; John Bartkowski, University of Texas at San Antonio; Beth Berila, St Cloud State University, Women's Studies Program; Ted Cohen, Ohio Wesleyan University; Francoise Cromer, Stony Brook University; Pamela J. Forman, University of Wisconsin–Eau Claire; Ann Fuehrer, Miami University; Katja Guenther, University of California, Riverside; William Hewitt, West Chester University of PA; Bianca Isaki, University of Hawai'i at Manoa; Kristin J. Jacobson, The Richard Stockton College of New Jersey; Brian Kassar, Montana State University; Julia Mason, Grand Valley State University; Janice McCabe, Florida State University; Kristen McHenry, University of Massachusetts Dartmouth; Elizabeth Markovits, Mount Holyoke College; Jennifer Pearson, Wichita State University; Sara Skiles-duToit, University of Texas, Arlington; Mary Nell Trautner, University at Buffalo, SUNY; Julianne Weinzimmer, Wright State University; and Lori Wiebold, Bradley University.

Third edition reviewers:

ChaeRan Freeze, Brandeis University; Patti Giuffre, Texas State University; Linda Grant, University of Georgia; Todd Migliaccio, California State University, Sacramento; J. Michael Ryan, University of Maryland, College Park; and Diane Kholos Wysocki, University of Nebraska at Kearney.

Second edition reviewers:

Patti Giuffre, Texas State University, San Marcos; Linda Grant, University of Georgia; Minjeong Kim, University at Albany, SUNY; Laura Kramer, Montclair State University; Heather Laube, University of Michigan, Flint; Todd Migliaccio, California State University, Sacramento; Kristen Myers, Northern Illinois University; Wendy Simonds, Georgia State University; Debbie Storrs, University of Idaho; and Elroi Waszkiewicz, Georgia State University.

Finally, we would like to thank students in our sociology of gender courses for challenging us to think about new ways to teach our courses and making us aware of arenas of gender that are not typically the focus of gender studies books.

Open Access Teaching and Learning Resources

The SAGE Gender and Sexuality Resource Site is an open access site meant to enhance the teaching and learning environments in gender and sexuality courses. Access the site by visiting **study.sagepub.com/socgsrc**.

Video, podcasts, web links, and articles are provided for the following topic areas:

- Theories of Gender and Sexuality
- Learning and "Doing Gender"
- Sexual Minorities
- Sexual Violence and Commodification
- Crime, Social Control, and the Legal System
- Religion
- Politics and Power
- Families, Intimate Relationships, and Reproduction
- The Workplace
- Health and Medicine
- Education
- Sports
- Media and Popular Culture
- Social Movements and Activism
- Gender and Sexuality Across Cultures

INTRODUCTION

CATHERINE G. VALENTINE AND MARY NELL TRAUTNER, WITH JOAN Z. SPADE

This book is an invitation to you, the reader, to enter the fascinating and challenging world of gender studies. Gender is briefly defined as the meanings, practices, and relations of femininities and masculinities that people create as we go about our daily lives in different social settings in the contemporary United States. Although we discuss gender throughout this book, it is a very complex term to understand and the reality of gender goes far beyond this simple definition. While a more detailed discussion of what gender is and how it is related to biological maleness and femaleness is provided in Chapter 1, we find the metaphor of a kaleidoscope useful in thinking about the complexity of the meaning of gender from a sociological viewpoint.

THE KALEIDOSCOPE OF GENDER

A real kaleidoscope is a tube containing an arrangement of mirrors or prisms that produces different images and patterns. When you look through the eyepiece of a kaleidoscope, light is typically reflected by the mirrors or prisms through cells containing objects such as glass pieces, seashells, and the like to create ever-changing patterns of design and color (Baker, 1999). In this book, we use the kaleidoscope metaphor to help us grasp the complex and dynamic meaning and practice of gender as it interacts with other social prisms—such as race, ethnicity, age, sexuality, and social class—to create complex patterns of identities and relationships. Three themes then emerge from the metaphor of the kaleidoscope: prisms, patterns, and possibilities.

Part I of the book focuses on prisms. A prism in a kaleidoscope is an arrangement of mirrors that refracts or disperses light into a spectrum of patterns (Baker, 1999). We use the term *social prism* to refer to socially constructed categories of difference and inequality through which our lives are reflected or shaped into patterns of daily experiences. In addition to gender, when we discuss social prisms, we consider other socially constructed categories such as race, ethnicity, age, social class, and sexuality. Culture is also conceptualized as a social prism in this book, as we examine how gender is shaped across groups and societies. The concept of social prisms helps us understand that gender is not a universal or static entity but, rather, is continuously created within the parameters of individual and group life. Looking at the interactions of the prism of gender with other social prisms helps us see the bigger picture—gender practices and meanings are a montage of intertwined social divisions and connections that both pull us apart and bring us together.

Part II of the book examines the patterns of gendered expressions and experiences created by the interaction of multiple prisms of difference and inequality. Patterns are regularized, prepackaged ways of thinking, feeling, and acting in society, and gendered patterns are present in almost all aspects of daily life. In the United States, examples of gendered patterns include the association of the color pink with girls and blue with boys (Paoletti, 2012). However, these patterns of gender are experienced and expressed in different ways depending on the other social prisms that shape our identities and life chances. Furthermore, these patterns are not static, as Paoletti illustrates.

Before the 1900s, children were dressed similarly until around the age of 7, with boys just as likely as girls to wear pink—but both more likely to be dressed in white. In addition, dresses were once considered appropriate for both genders in Europe and America. It wasn't until decades later, in the 1980s, that color became rigidly gendered in children's clothing, in the pink-and-blue schema. You will find that gendered patterns restrict choices, even the colors we wear—often without our even recognizing it is happening.

Another example of a gendered pattern is the disproportionate numbers of female educators and male engineers (see Table 7.1 in this book). If you take a closer look at Table 7.1, you will note that architects and engineers are predominately White men and educational occupations are predominantly White women (Bureau of Labor Statistics, 2018a). These patterns of gender are a result of the complex interaction of multiple social prisms across time and space.

Part III of the book concerns possibilities for gender change. Just as the wonder of the kaleidoscope lies in the ever-evolving patterns it creates, gendered patterns are always in flux. Each life and the world we live in can be understood as a kaleidoscope of unfolding growth and continual change (Baker, 1999). This dynamic aspect of the kaleidoscope metaphor represents the opportunity we have, individually and collectively, to transform gendered patterns that are harmful to women and men. Although the theme of gender change is prominent throughout this book, it is addressed specifically in Chapter 10. One caveat must be presented before we take you through the kaleidoscope of gender. A metaphor is a figure of speech in which a word ordinarily used to refer to one thing is applied to better understand another thing. A metaphor should not be taken literally. It does not directly represent reality. We use the metaphor of the kaleidoscope as an analytical tool to aid us in grasping the complexity, ambiguity, and fluidity of gender. However, unlike the prisms in a real kaleidoscope, the meaning and experience of social prisms (e.g., gender, race, ethnicity, social class, sexuality, and culture) are socially constructed and change in response to patterns in the larger society. Thus, although the prisms of a real kaleidoscope are static, the prisms of the gender kaleidoscope are fluid and shaped by the patterns of society.

As you step into the world of gender studies, you'll need to develop a capacity to see what is hidden by the cultural blinders we all wear at least some of the time. This capacity to see into the complexities of human relationships and group life has been called the sociological imagination or, to be hip, a "sociological radar." It is a capacity that is finely honed by practice and training both inside and outside the classroom. A sociological perspective enables us to see through the cultural smokescreens that conceal the patterns, meanings, and dynamics of our relationships.

Gender Stereotypes

The sociological perspective will help you think about gender in ways you might never have considered. It will, for example, help you debunk *gender stereotypes,* which are rigid, oversimplified, exaggerated beliefs about femininity and masculinity that misrepresent most women and men (Walters, 1999). To illustrate, let's analyze one gender stereotype that many people in American society believe—women talk more than men (Anderson & Leaper, 1998; Swaminathan, 2007; Wood, 1999).

Social scientific research is helpful in documenting whether women actually talk more than men, or whether this belief is just another gender stereotype. To arrive at a conclusion, social scientists study the interactions of men and women in an array of settings and count how often men speak compared with women. They almost always find that, on average, men talk more in mixed-gender groups (Brescoli, 2011; Wood, 1999). Researchers also find that men interrupt more and tend to ignore topics brought up by women (Anderson & Leaper, 1998; Wood, 1999). In and of themselves, these are important findings—the stereotype turns reality on its head.

So why does the stereotype continue to exist? First, we might ask how people believe something to be real—such as the stereotype that women talk more than men—when, in general, it isn't true. Part of the answer lies in the fact that culture, briefly defined as the way of life of a group of people, shapes what we experience as reality (see Chapter 3 for a more detailed discussion). As Allan Johnson (1997) aptly puts it, "Living in a culture is somewhat like participating in the magician's magic because all the while we think we're paying attention to what's 'really' happening, alternative realities unfold without even occurring to us" (p. 55). In other words, we don't usually reflect on our own culture; we are mystified by it without much awareness of its bewildering effect on us. The power of beliefs, including gender beliefs, is quite awesome. Gender stereotypes shape our perceptions, and these beliefs shape our reality.

A second question we need to ask about gender stereotypes is: What is their purpose? For example, do they set men against women and contribute to the persistence of a system of inequality that disadvantages women and advantages men? Certainly, the stereotype that many Americans hold of women as nonstop talkers is not a positive one. The stereotype does not assume that women are assertive, articulate, or captivating speakers. Instead, it tends to depict women's talk as trivial gossip or irritating nagging. In other words, the stereotype devalues women's talk while, at the same time, elevating men's talk as thoughtful and worthy of our attention. One of the consequences of this stereotype is that both men and women take men's talk more seriously (Brescoli, 2011; Wood, 1999). This pattern is reflected in the fact that the voice of authority in many areas of American culture, such as television and politics, is almost always a male voice (Brescoli, 2011). The message communicated is clear—women are less important than men. In other words, gender stereotypes help legitimize status and power differences between men and women (Brescoli, 2011).

However, stereotypical images of men and women are not universal in their application, because they are complicated by the kaleidoscopic nature of people's lives. Prisms, or social categories, such as race/ethnicity, social class, and age, intersect with gender to produce stereotypes that differ in symbolic meaning and functioning. For example, the prisms of gender, race, and age interact for African American and Hispanic men, who are stereotyped as dangerous (as noted in Adia Harvey Wingfield's reading in Chapter 7 and Dawn Marie Dow's reading in Chapter 8). These variations in gender stereotypes act as controlling images that maintain complex systems of domination and subordination in which some individuals and groups are dehumanized and disadvantaged in relationship to others (see readings in Chapter 2).

Development of the Concept of Gender

Just a few decades ago, social scientists described gender as two discrete categories called sex roles—masculine/men and feminine/women. These sex roles were conceptualized in a biological "essentialist" framework to be either an automatic response to innate personality characteristics and/or biological sex characteristics such as hormones and reproductive functions (Kimmel, 2004; Tavris, 1992) or a mix of biological imperatives and learning reinforced by social pressure to conform to one or the other sex role (Connell, 2010). For example, women were thought to be naturally more nurturing because of their capacity to bear children, and men were seen as prewired to take on leadership positions in major societal institutions such as family, politics, and business. This "sex roles" model of women and men was one-dimensional, relatively static, and ethnocentric, and it is *not* supported by biological, psychological, historical, sociological, or anthropological research.

The concept of gender developed as social scientists conducted research that questioned the simplicity and accuracy of the "sex roles" perspective. One example of this research is that social scientists have debunked the notion that biological sex characteristics cause differences in men's and women's behaviors (Tavris, 1992). Research on hormones illustrates this point. Testosterone, which women as well as men produce, does not cause aggression in men (Sapolsky, 1997), and the menstrual cycle does not cause women to be more "emotional" than men (Tavris, 1992; see L. Ayu Saraswati's reading in Chapter 6).

Another example is that social scientific research demonstrated that men and women are far more physically, cognitively, and emotionally alike than different. What were assumed to be natural differences and inequalities between women and men were clearly shown to be the consequence of the asymmetrical and unequal life experiences, resources, and power of women compared with men (Connell, 2010; Tavris, 1992). Consider the arena of athletics. It is a common and long-held belief that biological sex is related to physical ability and, in particular, that women are athletically inferior to men. These beliefs have been challenged by the outcomes of a recent series of legal interventions that opened the world of competitive sports to girls and women. Once legislation such as Title IX was implemented in 1972, the expectation that women could not be athletes began to change as girls and young women received the same training and support for athletic pursuits as did men. Not surprisingly, the gap in physical strength and skills between women and men decreased dramatically. Today, women athletes regularly break records and perform physical feats thought impossible for women just a few decades ago.

Yet another example of how the "sex roles" model was discredited was the documentation of inequality as a human-created social system. Social scientists highlighted the social origins of patterns of gender inequality within the economy, family, religion, and other social institutions that benefit men as a group and maintain patriarchy as a social structure. To illustrate, in the 1970s, when researchers began studying gender inequality, they found that women made between 60 and 70 cents for every dollar men made. Things are not much better today. In 2017, the median weekly salary for women was 81.8% of men's median salary (Bureau of Labor Statistics, 2018b).

The intellectual weaknesses of "sex roles" theory (Connell, 2010), buttressed by considerable contradictory evidence, led social scientists to more sophisticated theories and modes of studying gender that could address the complexities and malleability of sex (femaleness and maleness) and gender (femininities and masculinities). In short, social science documented the fact that we are made and make ourselves into gendered people through social interaction in everyday life (Connell, 2010). It is not natural or normal to be a feminine woman or a masculine man. Gender is a socially constructed system of social relations that can be understood only by studying the social processes by which gender is defined into existence and maintained or changed by human actions and interactions (Schwalbe, 2001). This theory of gender social construction will be discussed throughout the book.

One of the most important sources of evidence in support of the idea that gender is socially constructed is derived from cross-cultural and historical studies as described in the earlier discussion of the gendering of pink and blue. The variations and fluidity in the definitions and expressions of gender across cultures and over time illustrate that the American gender system is not universal. For example, people in some cultures have created more than two genders (see Serena Nanda's reading in Chapter 1). Other cultures define men and women as similar, not different (see Christine Helliwell's reading in Chapter 3). Still others view gender as flowing and changing across the life span (Herdt, 1997).

As social scientists examined gender patterns through the prism of culture and throughout history, their research challenged the notion that masculinity and femininity are defined and experienced in the same way by all people. For example, the meaning and practice of femininity in orthodox, American religious subcultures is not the same as femininity outside those communities (Rose, 2001). The differences are expressed in a variety of ways, including women's clothing. Typically, orthodox religious women adhere to modesty rules in dress, covering their heads, arms, and legs.

Elaborating on the idea of multiple or plural masculinities and femininities, Australian sociologist Raewyn Connell coined the terms *hegemonic masculinity* and *emphasized femininity* to understand the relations between and among masculinities and femininities in patriarchal societies. Patriarchal societies are dominated by privileged men (e.g., upper-class White men in the United States), but they also typically benefit less privileged men in their relationships with women. According to Connell (1987), hegemonic masculinity is the idealized pattern of masculinity in patriarchal societies, while emphasized femininity is the vision of femininity held up as the model of womanhood in those societies. In Connell's definition, hegemonic masculinity is "the pattern of practice (i.e., things done, not just a set of role expectations or an identity) that allowed men's dominance over women to continue" (Connell & Messerschmidt, 2005, p. 832). Key features of hegemonic masculinity include the subordination of women, the exclusion and debasement of gay men, and the celebration of toughness and competitiveness (Connell, 2000). However, hegemony does not mean violence per se. It refers to "ascendancy achieved through culture, institutions, and persuasion" (Connell & Messerschmidt, 2005, p. 832). Emphasized femininity, in contrast, is about women's subordination, with its key features being sociability, compliance with men's sexual and ego desires, and acceptance of marriage and child care (Connell, 1987). Both hegemonic masculinity and emphasized femininity patterns are "embedded in specific social environments" and are, therefore, dynamic as opposed to fixed (Connell & Messerschmidt, 2005, p. 846).

According to Connell, hegemonic masculinity and emphasized femininity are not necessarily the most common gender patterns. They are, however, the versions of manhood and womanhood against which other patterns of masculinity and femininity are measured and found wanting (Connell & Messerschmidt, 2005; Kimmel, 2004). For example, hegemonic masculinity produces marginalized masculinities, which, according to Connell (2000), are characteristic of exploited groups such as racial and ethnic minorities. These marginalized forms of masculinity may share features with hegemonic masculinity, such as "toughness," but are socially debased (see Adia Harvey Wingfield's reading in Chapter 7).

In patriarchal societies, the culturally idealized form of femininity, emphasized femininity, is produced in relation to male dominance. Emphasized femininity insists on compliance, nurturance, and empathy as ideals of womanhood to which all women should subscribe (Connell, 1987). Connell does not use the term *hegemonic* to refer to emphasized femininity, because, she argues, emphasized femininity is always subordinated to masculinity. James Messerschmidt (2012) adds to our understanding of femininities by arguing that the construction of hegemonic masculinity requires some kind of "buy-in" from women and that, under certain circumstances and in certain contexts, there are women who create emphasized femininities. By doing so, they contribute to the perpetuation of coercive gender relations and identities. Think of circumstances and situations—such as within work, romantic, or family settings—when women are complicit in maintaining oppressive gender relations and identities. Why would some women participate in the production of masculinities and femininities that are oppressive? The reading by Karen D. Pyke and Denise L. Johnson in Chapter 2 is helpful in answering these questions, employing the term *hegemonic femininity* rather than *emphasized femininity*. They describe the lives of young, second-generation Asian women and their attempts to balance two cultural patterns of gender in which White femininity, they argue, is hegemonic, or the dominant form of femininity.

Another major source of gender complexity is the interaction of gender with other social categories of difference and inequality. Allan Johnson (2001) points out,

> Categories that define privilege exist all at once and in relation to one another. People never see me solely in terms of my race, for example, or my gender. Like everyone else's, my place in the social world is a package deal—white, male, heterosexual, middle-aged, married . . .—and that's the way it is all the time. . . . It makes no sense to talk about the effect of being in one of these categories—say, white—without also looking at the others and how they're related to it. (p. 53)

Seeing gender through multiple social prisms is critical, but it is not a simple task, as you will discover in the readings throughout this book. Social scientists commonly refer to this type of analysis as intersectionality, but other terms are used as well (see Chapter 2 for a discussion of this). We need to be aware of how other social prisms alter life experiences and chances. For example, although an upper-class African American woman is privileged by her social class category, she will face obstacles related to her race and gender. Or consider the situation of a middle-class White man who is gay; he might lose some of the privilege attached to his class and race because of his sexual orientation.

Finally, gender is now considered a social construct shaped at individual, interactional, and institutional levels. If we focus on only one of these levels, we provide only a partial explanation of how gender operates in our lives. This idea of gender being shaped at these three different levels is elaborated in Barbara J. Risman's article in Chapter 1 and throughout the book. Consider these three different ways of approaching gender and how they interact or influence one another. At the *individual* level, sociologists study the social categories and stereotypes we use to identify ourselves and label others (see Chapter 4). At the *interactional* level, sociologists study gender as an ongoing activity carried out in interaction with other people, and how people vary their gender presentations as they move from situation to situation (see Carla A. Pfeffer's reading in Chapter 1). At the *institutional* level, sociologists study how "gender is present in the processes, practices, images and ideologies, and distributions of power in the various sectors of social life," such as religion, health care, language, and so forth (Acker, 1992, p. 567; see also Joan Acker's reading in Chapter 7).

Theoretical Approaches for Understanding Gender

Historically, conflict and functionalist theories explained gender at a macro level of analysis, with these theories having gone through many transformations since first proposed around the turn of the 20th century. Scholars at that time were trying to sort out massive changes in society resulting from the industrial and democratic revolutions. However, a range of theories—for example, feminist, postmodernist, and queer theories—provide more nuanced explanations of gender. Many of these more recent theories frame their understanding of gender in the lived experiences of individuals, what sociologists call micro level theories, rather than focusing solely on a macro level analysis of society, wherein gender does not vary in form or function across groups or contexts.

Functionalism

Functionalism attempts to understand how all parts of a society (e.g., institutions such as family, education, economy, and the polity or state) fit together to form a smoothly running social system. According to this theoretical paradigm, parts of society tend to complement each other to create social stability (Durkheim, 1933). Translated into separate sex role relationships, Talcott Parsons and Robert Bales (1955), writing after World War II, saw distinct and separate gender roles in the heterosexual nuclear family as a functional and logical adaptation to a modern, complex society. Women were thought to be more "functional" if they were socialized and aspired to raise children. And men were thought to be more "functional" if they were socialized and aspired to support their children and wives. However, as Michael Kimmel (2004) notes, this "sex-based division of labor is functionally anachronistic," and if there ever was any biological basis for specific tasks being assigned to men or women, it has been eroded (p. 55). The functionalist viewpoint has largely been discredited in sociology, although it persists as part of common culture in various discourses and ideologies, especially conservative religious and political thought.

Evolutionary Psychology and Neuroscience

Functionalist thinking is also replicated in the realms of neuroscience and evolutionary psychology. In brief, the former tries to explain gender inequality by searching for neurological differences in human females and males assumed to be caused by hormonally induced differences in the brain. The hypothesized behavioral outcomes, according to neuroscientists such as Simon Baron-Cohen (2003), are emotionally tuned in, verbal women in contrast to men who are inclined to superior performance in areas such as math and music (Bouton, 2010). The latter, evolutionary psychology, focuses on "sex differences" (e.g., high-risk-taking male behaviors) between human females and males that are hypothesized to have their origins in psychological adaptations to early human, intrasexual competition. Both approaches, which assume there are essential differences between males and females embedded in their bodies or psyches, have been roundly critiqued by researchers (e.g., Fine, 2010; Fine, Jordan-Young, Kaiser, & Rippon, 2013) who uncovered a range of problems, including research design flaws, no significant differences between female and male subjects, overgeneralization of findings, and ethnocentrism. Feminist neuroscientists have carefully set out the serious, negative consequences of the tendency of evolutionary psychology and neuroscience to produce "untested stereotype-based speculation" that reinforces popular misconceptions about women and men (Fine, 2013). In a short essay titled "Plasticity, Plasticity, Plasticity . . . and the Rigid Problem of Sex," Cordelia Fine and colleagues (2013) discuss the ways in which behavioral neuroendocrinology "has been transformed by an increasingly large body of research demonstrating the power of an individual's behavior, the behavior of others, and aspects of the environment to influence behavior through reciprocal modulation of the endocrine system" (p. 551). Simply put, we are not hardwired. Our brains are "adaptively plastic," interacting and changing with individual life experiences and social contexts (p. 551).

Conflict Theories

Karl Marx and later conflict theorists, however, did not see social systems as functional or benign. Instead, Marx and his colleague Friedrich Engels described industrial societies as systems of oppression in which one group, the dominant social class, uses its control of economic resources to oppress the working class. The economic resources of those in control are obtained through profits gained from exploiting the labor of subordinate groups. Marx and Engels predicted that the tension between the "haves" and the "have-nots" would result in an underlying conflict between these two groups. Most early Marxist theories focused on class oppression; however, Engels (1942/1970) wrote an important essay on the oppression of women as the earliest example of oppression of one group by another. Marx and Engels inspired socialist feminists, discussed later in this introduction under "Feminist Theories."

Current theorists, while recognizing Marx and Engels's recognition of the exploitation of workers in capitalist economies, criticize early conflict theory for ignoring women's reproductive labor and unpaid work (Federici, 2012). They focus on the exploitation of women by global capitalism (see article by Bandana Purkayastha in Chapter 2). Conflict theories today call for social action relating to the oppression of women and other marginalized groups, particularly within this global framework.

Social Constructionist Theories

Social constructionist theories offer a strong antidote to biological essentialism and psychological reductionism in understanding the social worlds (e.g., institutions, ideologies, identities) constructed by people. This theory, as discussed earlier, emphasizes the social or collective processes by which people actively shape reality (e.g., ideas, inequalities, social movements) as we go about daily life in different contexts and situations. The underpinnings of social constructionist theory are in sociological thought (e.g., symbolic interactionism, dramaturgy, and ethnomethodology), as well as in anthropology, social psychology, and related disciplinary arenas.

Social constructionism has had a major impact on gender analysis, invigorating both gender research and theoretical approaches (e.g., discussions of doing gender theory, relational theory, and intersectional analysis). From a social constructionist viewpoint, we must learn and do gender (masculinities and femininities) in order for gender differences and inequalities to exist. We also build these differences and inequalities into the patterns of large social arrangements such as social institutions. Take education. Men predominate in higher education and school administration, while women are found at the elementary and preschool levels (Connell, 2010). Theories rooted in the fundamental principles of gender social construction follow.

"Doing Gender" Theory

Drawing on the work of symbolic interactionism, specifically dramaturgy (Goffman) and ethnomethodology (Garfinkel), Candace West and Don H. Zimmerman published an article in 1987 simply titled "Doing Gender." In this article, they challenged assumptions of the two previous decades of research that examined "sex differences" or "sex roles." They argued that gender is a *master identity,* which is a product of social interactions and "doing," not simply the acting out of a role on a social stage. They saw gender as a complicated process by which we categorize individuals into two sex categories based on what we assume to be their sex (male or female). Interaction in contemporary Western societies is based on "knowing the sex" of the individual we are interacting with. However, we have no way of actually knowing an individual's sex (genitalia or hormones); therefore, we infer sex categories based on outward characteristics such as hairstyle, clothing, etc. Because we infer sex categories of the individuals we meet, West and Zimmerman argue that we are likely to question those who break from expected gendered behaviors for the sex categories we assign to them. We are also accountable for our own gender-appropriate behavior. Interaction in most societies becomes particularly difficult if one's sex category or gender is ambiguous, as you will read in Betsy Lucal's article in Chapter 1.

Thus, this process of being accountable makes it important for individuals to display appropriate gendered behavior at all times in all situations. As such, "doing gender" becomes a salient part of social interactions and embedded in social institutions. As they note, "Insofar as a society is partitioned by 'essential' differences between women and men and placement in a sex category is both relevant and enforced, doing gender is unavoidable" (West & Zimmerman, 1987, p. 137).

Of course, they recognize that not everyone has the same resources (such as time, money, and/or expertise) to "do gender" and that gender accomplishment varies across social situations. In considering the discussion of who talks more, "doing gender" might explain why men talk more in work groups, as they attempt to portray their gendered masculinity while women may be doing more gender-appropriate emotion work such as asking questions and filling in silences. As such, when men and women accomplish gender as expected for the sex categories they display and are assigned to by others, they are socially constructing gender.

This concept of "doing gender" is used in many articles included in this book, but the use of the concept is not always consistent with the way the authors originally presented it (West & Zimmerman, 2009). Doing gender is a concept that helped move the discussions of sex/gender to a different level where interactions (micro) and institutions (macro) can be studied simultaneously and gender becomes a more lived experience, rather than a "role."

Postmodern Theories

Postmodernism focuses on the way knowledge about gender is constructed, not on explaining gender relationships themselves. To postmodernists, knowledge is never absolute—it is always situated in a social reality that is specific to a historical time period. Postmodernism is based on the idea that it is impossible for anyone to see the world without presuppositions. From a postmodernist perspective, then, gender is socially constructed through discourses, which are the "series of stories" we use to explain our world (Andersen, 2004). Postmodernists attempt to "deconstruct" the discourses or stories used to support a group's beliefs about gender (Andersen, 2004; Lorber, 2001). For example, Jane Flax argues that to fully understand gender in Western cultures, we must deconstruct the meanings in Western religious, scientific, and other discourses relative to "biology/sex/gender/nature" (cited in Lorber, 2001, p. 199). As you will come to understand from the readings in Chapters 1 and 3 (e.g., Nanda and Christine Helliwell), the association between sex and gender in Western scientific (e.g., theories and texts) and nonscientific (e.g., films, newspapers, media) discourses is not shared in other cultural contexts. Thus, for postmodernists, gender is a product of the discourses within particular social contexts that define and explain gender.

Queer Theories

Queer theories borrow from the original meaning of the word *queer* to refer to that which is "outside ordinary and narrow interpretations" (Plante, 2006, p. 62). Queer theorists are most concerned with understanding sexualities in terms of the idea that (sexual) identities are flexible, fluid, and changing, rather than fixed. In addition, queer theorists argue that identity and behavior must be separated. Thus, we cannot assume that people are what they do. From the vantage point of this theory, gender categories, much like sexual categories, are simplistic and problematic. Real people cannot be lumped together and understood in relationship to big cultural categories such as men and women, heterosexual and homosexual (Plante, 2006). Carla A. Pfeffer's reading in Chapter 1 sets out the premises and impact of queer theory on gender studies in considerable detail. She argues that the discipline of sociology is well positioned to examine the lives of queer social actors, and her work is an excellent example of the application of queer theory.

Relational Theory

The relational theory of gender was developed in response to the problems of the "sex roles" model and other limited views of gender (e.g., categoricalism, as critiqued by queer theory). Connell (2000) states that a gender relations approach opens up an understanding of "the different dimensions of gender, the relation between bodies and society, and the patterning of gender" (pp. 23–24). Specifically, from a relational viewpoint, (1) gender is a way of organizing social practice (e.g., child care and household labor) at the personal, interactional, and institutional levels of life; (2) gender is a social practice related to bodies and what bodies do but *cannot* be reduced to bodies or biology; and (3) masculinities and femininities can be understood as *gender projects* that produce the *gender order* of a society and interact with other social structures such as race and class (pp. 24–28).

Feminist Theories

Feminist theorists expanded on the ideas of theorists such as Marx and Engels, turning attention to the causes of women's oppression. There are many schools of feminist thought. Here, we briefly introduce you to those typically covered in overviews. One group, socialist feminists, continued to emphasize the role of capitalism in interaction with a patriarchal social structures as the basis for the exploitation of women. These theorists argue that economic and power benefits accrue to men who dominate women in capitalist societies. Another group, radical feminists, argues that patriarchy—the domination of men over women—is the fundamental form of oppression of women. Both socialist and radical feminists call for far-reaching changes in all institutional arrangements and cultural forms, including the dismantling of systems of oppression such as sexism, racism, and classism; replacing capitalism with socialism; developing more egalitarian family systems; and making other structural changes (e.g., Bart & Moran, 1993; Daly, 1978; Dworkin, 1987; MacKinnon, 1989).

Not all feminist theorists call for deep, structural, and cultural changes. Liberal feminists are inclined to work toward a more equitable form of democratic capitalism. They argue that policies such as Title IX and affirmative action laws opened up opportunities for women in education and increased the number of women professionals, such as physicians. These feminists strive to achieve gender equality by removing barriers to women's freedom of choice and equal participation in all realms of life, eradicating sexist stereotypes, and guaranteeing equal access and treatment for women in both public and private arenas (e.g., Reskin & Roos, 1990; Schwartz, 1994; Steinberg, 1982; Vannoy-Hiller & Philliber, 1989; Weitzman, 1985).

Although the liberal feminist stance may seem to be the most pragmatic form of feminism, many of the changes brought about by liberal varieties of feminism have "served the interests of only the most privileged women" (Motta, Fominaya, Eschle, & Cox, 2011, p. 5). Additionally, liberal feminist approaches that work with the state or attempt to gain formal equal rights within a fundamentally exploitive labor market fail to challenge the growth of neo-liberal globalism and the worsening situation of many people in the face of unfettered markets, privatization, and imperialism (Motta et al., 2011; e.g., see discussion of the Great Recession in Chapter 5). In response to these kinds of issues and problems, 21st-century feminists are revisiting and reinventing feminist thinking and practice to create a "more emancipatory feminism" that can lead to "post-patriarchal, anti-neoliberal politics" (Motta et al., 2011, p. 2; see readings in Chapter 10).

Intersectional or Prismatic Theories

A major shortcoming with many of the theoretical perspectives just described is their failure to recognize how gender interacts with other social categories or prisms of difference and inequality within societies, including race/ethnicity, social class, sexuality, age, and ability/disability (see Chapter 2). A growing number of social scientists are responding to the problem of incorporating multiple social categories or social positions in their research by developing a new form of analysis, often described as intersectional analysis, which we also refer to as prismatic analysis in this book. Chapter 2 explores these theories of how gender interacts with other prisms of difference and inequality to create complex patterns. Without an appreciation of the interactions of socially constructed categories of difference and inequality, or what we call prisms, we end up with not only an incomplete but also an inaccurate explanation of gender.

As you read through the articles in this book, consider the basis for the authors' arguments in each reading. How do the authors apply the theories just described? What observations, data, or works of other social science researchers do these authors use to support their claims? Use a critical eye to examine the evidence as you reconsider the assumptions about gender that guide your life.

The Kaleidoscope of Gender: Prisms, Patterns, and Possibilities

Before beginning the readings that take us through the kaleidoscope of gender, let us briefly review the three themes that shape the book's structure: prisms, patterns, and possibilities.

Part I: Prisms

Understanding the prisms that shape our experiences provides an essential basis for the book. Chapter 1 explores the meanings of the pivotal prism—gender—and its relationship to biological sex and sexuality. Chapter 2 presents an array of prisms or socially constructed categories that interact with gender in many human societies, such as race/ethnicity, social class, sexuality, age, and ability/disability. Chapter 3 focuses on the prism of culture/nation, which alters the meaning and practice of gender in surprising ways.

Part II: Patterns

The prisms of the kaleidoscope create an array of patterned expressions and experiences of femininity and masculinity. Part II of this book examines some of these patterns. We look at how people learn, internalize, and "do" gender (Chapter 4); how gender is exploited by corporate capitalism (Chapter 5); how gender engages bodies, sexualities, and emotions (Chapter 6); how gendered patterns are reproduced and modified in work (Chapter 7); how gender is created and transformed in our intimate relationships (Chapter 8); and how conformity to patterns of gender is enforced and maintained (Chapter 9).

Part III: Possibilities

In much the same way as the colors and patterns of kaleidoscopic images flow, gendered patterns and meanings are inherently changeable. Chapter 10 examines the shifting sands of the genderscape and reminds us of the many possibilities for change.

We use the metaphor of the gender kaleidoscope to discover what is going on under the surface of a society whose way of life we don't often penetrate in a nondefensive, disciplined, and deep fashion. In doing so, we will expose a reality that is astonishing in its complexity, ambiguity, and fluidity. With the kaleidoscope, you never know what's coming next. Come along with us as we begin the adventure of looking through the kaleidoscope of gender.

References

Acker, J. (1992). From sex roles to gendered institutions. *Contemporary Sociology*, *21*(5), 565–570.

Anderson, K. J., & Leaper, C. (1998). Meta-analysis of gender effects on conversational interruption: Who, what, when, where, and how. *Sex Roles*, *39*(3–4), 225–252.

Andersen, M. L. (2004). *Thinking about women: Sociological perspectives on sex and gender* (6th ed.). Boston: Allyn & Bacon.

Baker, C. (1999). *Kaleidoscopes: Wonders of wonder.* Lafayette, CA: C&T.

Baron-Cohen, S. (2003). *The essential difference: The truth about the male and female brain.* New York: Basic Books.

Bart, P. B., & Moran, E. G. (Eds.). (1993). *Violence against women: The bloody footprints.* Thousand Oaks, CA: Sage.

Bouton, K. (2010, August 23). Peeling away theories on gender and the brain. *New York Times.* Retrieved from http://www.nytimes.com/2010/08/24/science/24scibks.html

Brescoli, V. (2011). Who takes the floor and why: Gender, power, and volubility in organizations. *Administrative Science Quarterly*, *56*(4), 622–641.

Bureau of Labor Statistics. (2018a). Table A-2: 2017 Median weekly earnings of full-time wage and salary workers by detailed occupation, sex, race and Hispanic or Latino ethnicity and Non-Hispanic ethnicity. Annual Average 2017. Source: *Current Population Survey.* [Unpublished data - sent by request].

Bureau of Labor Statistics. (2018b). Table 39: Usual weekly earnings of employed full-time wage and salary workers by intermediate occupation and sex. Annual Average 2017. Source: *Current Population Survey.*

Connell, R. W. (1987). *Gender and power: Society, the person and sexual politics.* Stanford, CA: University of Stanford Press.

Connell, R. W. (2000). *The men and the boys.* Berkeley: University of California Press.

Connell, R. W. (2010). Retrieved from www.raewyn-connell.net/

Connell, R. W., & Messerschmidt, J. W. (2005). Hegemonic masculinity: Rethinking the concept. *Gender & Society*, *19*(6), 829–859.

Daly, M. (1978). *Gyn/ecology: The metaethics of radical feminism.* Boston: Beacon Press.

Durkheim, E. (1933). *The division of labor in society.* Glencoe, IL: Free Press.

Dworkin, A. (1987). *Intercourse.* New York: Free Press.

Engels, F. (1970). *Origin of the family, private property, and the state.* New York: International Publishers. (Original work published in 1942)

Federici, S. (2012). *Revolution at point zero: Housework, reproduction, and feminist struggle.* Oakland, CA: PM Press.

Fine, C. (2010). *Delusions of gender: How our minds, society, and neurosexism create difference.* New York: Norton.

Fine, C. (2013). New insights into gendered brain wiring, or a perfect case study in neurosexism? *The Conversation.* Retrieved from http://theconversation.com/new-insights-into-gendered-brain-wiring-or-a-perfect-case-study-in-neurosexism-21083

Fine, C., Jordan-Young, R., Kaiser, A., & Rippon, G. (2013). Plasticity, plasticity, plasticity . . . and the rigid problem of sex. *Trends in cognitive sciences*, *17*(11), 550–551.

Herdt, G. (1997). *Same sex, different cultures.* Boulder, CO: Westview Press.

Johnson, A. (1997). *The gender knot: Unraveling our patriarchal legacy.* Philadelphia: Temple University Press.

Johnson, A. (2001). *Privilege, power, and difference.* Mountain View, CA: Mayfield.

Kimmel, M. S. (2004). *The gendered society* (2nd ed.). New York: Oxford University Press.

Lorber, J. (2001). *Gender inequality: Feminist theories and politics.* Los Angeles: Roxbury.

MacKinnon, C. A. (1989). *Toward a feminist theory of the state.* Cambridge, MA: Harvard University Press.

Messerschmidt, J. (2012). Engendering gendered knowledge: Assessing the academic appropriation of hegemonic masculinity. *Men and Masculinities*, *15*(1), 56–76.

Motta, S., Fominaya, C. F., Eschle, C., & Cox, L. (2011). Feminism, women's movements and women in movement. *Interface*, *3*(2), 1–32.

Paoletti, J. B. (2012). *Pink and blue: Telling the boys from the girls in America.* Bloomington: Indiana University Press.

Parsons, T., & Bales, R. F. (1955). *Family, socialization, and interaction process.* Glencoe, IL: Free Press.

Plante, R. F. (2006). *Sexualities in context: A social perspective.* Boulder, CO: Westview Press.

Reskin, B. F., & Roos, P. A. (1990). *Job queues, gender queues: Explaining women's inroads into male occupations.* Philadelphia: Temple University Press.

Rose, D. R. (2001). Gender and Judaism. In D. Vannoy (Ed.), *Gender mosaics: Social perspectives* (pp. 415–424). Los Angeles: Roxbury.

Sapolsky, R. (1997). *The trouble with testosterone: And other essays on the biology of the human predicament.* New York: Scribner, a Division of Simon and Schuster, Inc.

Schwalbe, M. (2001). *The sociologically examined life: Pieces of the conversation* (2nd ed.). Mountain View, CA: Mayfield.

Schwartz, P. (1994). *Love between equals: How peer marriage really works.* New York: Free Press.

Steinberg, R. J. (1982). *Wages and hours: Labor and reform in twentieth century America.* New Brunswick, NJ: Rutgers University Press.

Swaminathan, N. (2007, July 6). Gender jabber: Do women talk more than men? *Scientific American.* Retrieved from http://www.scientificamerican.com/article.cfm?id=women-talk-more-than-men&print=true

Tavris, C. (1992). *The mismeasure of woman.* New York: Simon & Schuster.

Vannoy-Hiller, D., & Philliber, W. W. (1989). *Equal partners: Successful women in marriage.* Thousand Oaks, CA: Sage.

Walters, S. D. (1999). Sex, text, and context: (In) between feminism and cultural studies. In M. M. Ferree, J. Lorber, & B. B. Hess (Eds.), *Revisioning gender* (pp. 193–257). Thousand Oaks, CA: Sage.

Weitzman, L. J. (1985). *The divorce revolution: The unexpected social and economic consequences for women and children in America.* New York: Free Press.

West, C., & Zimmerman, D. H. (1987). Doing gender. *Gender & Society*, *1*(2), 125–151.

West, C., & Zimmerman, D. H. (2009). Accounting for doing gender. *Gender & Society*, *23*(1), 112–122.

Wood, J. T. (1999). *Gendered lives: Communication, gender, and culture* (3rd ed.). Belmont, CA: Wadsworth.

PART I

PRISMS

1

The Prism of Gender

Catherine G. Valentine

In the metaphorical kaleidoscope of this book, gender is the pivotal prism. It is central to the intricate patterning of social life and encompasses power relations, the division of labor, symbolic forms, and emotional relations (Connell, 2000). The shape and texture of people's lives are affected in profound ways by the prism of gender as it operates in their social worlds. Indeed, our ways of thinking about and experiencing gender, and the related categories of sex and sexuality, originate in our society. As we noted in the introduction to this book, gender is very complex. In part, the complexity of the prism of gender in North American culture derives from the fact that it is characterized by a marked contradiction between people's beliefs about gender and real behavior. Our real behavior is far more flexible, adaptable, and malleable than our beliefs would have it. To put it another way, contrary to the stereotypes of masculinity and femininity, there are no gender certainties or absolutes. Real people behave in feminine, masculine, and nongendered ways as they respond to situational demands and contingencies (Glick & Fiske, 1999; Pfeffer, 2014; Tavris, 1992).

To help us think more clearly about the complexity of gender, two questions are addressed in this chapter: (1) How does Western, i.e., Euro-American, culture condition us to think about gender, especially in relation to sex and sexuality? (2) How does social scientific research challenge Western beliefs about gender, sex, and sexuality?

Western Beliefs About Gender, Sex, and Sexuality

Most people in contemporary Western cultures, such as the United States, grow up learning that there are two and only two sexes, male and female; two and only two genders, feminine and masculine; and two and only two sexualities, heterosexual and homosexual (Bem, 1993; Budgeon, 2014; Lucal, 2008; Pfeffer, 2014; Wharton, 2005). We are taught that a real woman is female-bodied, feminine, and heterosexual; a real man is male-bodied, masculine, and heterosexual; and any deviation or variation is strange, unnatural, and potentially dangerous. Most people also learn that femininity and masculinity flow from biological sex characteristics (e.g., hormones, secondary sex characteristics, external and internal genitalia). We are taught that testosterone, a beard, big muscles, and a penis make a man, while estrogen, breasts, hairless legs, and a vagina make a woman. Many of us never question what we have learned about sex and gender, so we go through life assuming that gender is a relatively simple matter: A person who wears lipstick, high-heeled shoes, and a skirt is a feminine female, while a person who plays rugby, belches in public, and walks with a swagger is a masculine male (Lorber, 1994; Ridgeway & Correll, 2004).

The readings we have selected for this chapter reflect a growing body of social scientific research that challenges and alters the Western view of sex, gender, and sexuality. Overall, the readings are critical of the American tendency to explain virtually every human behavior in individual and biological terms. Americans overemphasize biology and underestimate the power of social facts to explain sex, sexuality, and gender (Connell, n.d.; O'Brien, 1999). For instance, Americans tend to equate aggression with biological maleness and vulnerability with femaleness; natural facility in physics with masculinity and natural facility in child care with femininity; lace and ribbons with girlness and rough-and-tumble play with boyness (Glick & Fiske, 1999; Ridgeway & Correll, 2004). These notions of natural sex, gender, and sexuality difference, opposition, and inequality (i.e., a consistently higher valuation of masculinity than femininity) permeate our thinking, color our labeling of people and things in our environment, and affect our practical actions (Bem, 1993; Haines, Deaux, & Lofaro, 2016; Schilt & Westbrook, 2009; Wharton, 2005).

We refer to the American two-and-only-two sex/gender/sexuality system as the "pink and blue syndrome" (Schilt & Westbrook, 2009). This syndrome is deeply lodged in our minds and feelings and is reinforced through everyday talk, performance, and experience. It's everywhere. Any place, object, discourse, or practice can be gendered. Children's birthday cards come in pink and blue. Authors of popular books assert that men and women are from different planets. People love PMS and alpha-male jokes. In "The Pink Dragon Is Female" (see Chapter 5), Adie Nelson's research reveals that even children's fantasy costumes are predictably gendered as masculine or feminine. The "pink and blue syndrome" is so embedded within our culture and, consequently, within individual patterns of thinking and feeling that most of us cannot remember when we learned gender stereotypes and expectations or came to think about sex, gender, and sexuality as natural, immutable, and fixed. It all seems so simple and natural. But is it?

What is gender? What is sex? What is sexuality? How are gender, sex, and sexuality related? Why do most people in our society believe in the "pink and blue syndrome"? Why do so many of us attribute one set of talents, temperaments, skills, and behaviors to women and another, opposing set to men? These are the kinds of questions social scientists have been asking and researching for well over 50 years. Thanks to the good work of an array of scientists, we now understand that gender, sex, and sexuality are not so simple. Social scientists have discovered that the gender landscape is complicated, shifting, and contradictory. Among the beliefs called into question by research are

- the notion that there are two and only two sexes, two and only two genders, and two and only two sexualities;
- the assumption that the two-and-only-two system is universal; and
- the belief that nature, rather than nurture, causes the "pink and blue syndrome."

Using Our Sociological Radar

Before we look at how social scientists answer questions such as, "What is gender?" let's do a little research of our own. Try the following: Relax, turn on your sociological radar, and examine yourself and the people you know carefully. Do all the men you know fit the ideal of masculinity all the time, in all relationships, and in all situations? Do all the women in your life consistently behave in stereotypical feminine fashion? Do you always fit into one as opposed to the other culturally approved gender category? Or are most of the people you know capable of "doing" both masculinity and femininity, depending on the interactional context? If we allow ourselves to think and see outside the contemporary American cultural framework, we will observe that none of the people we know are aggressive all the time, nurturing all the time, sweet and submissive all the time, or strong and silent all the time. Thankfully, we are complex and creative. We stretch and grow and develop as we meet the challenges, constraints, and opportunities of different and new situations and life circumstances. Men can do mothering; women can "take care of business." Real people are not stereotypes.

Yet even in the face of real gender fluidity, variation, and complexity, the belief in sex/gender/sexuality dichotomy, opposition, and inequality continues to dominate almost every aspect of the social worlds we inhabit. For example, recent research shows that even though men's and women's roles have changed and blended, the tendency of Americans to categorize and stereotype people based on the simple male/female dichotomy persists (Glick & Fiske, 1999; Haines, Deaux, & Lofaro, 2016; Miller, Eagly, & Linn, 2014; Shields, Garner, Di Leone, & Hadley, 2006; Snyder, 2014). As Peter Glick and Susan Tufts Fiske (1999) put it, "We typically categorize people by sex effortlessly, even nonconsciously, with diverse and profound effects on social interactions" (p. 368). To reiterate, many Americans perceive humankind as divided into mutually exclusive, nonoverlapping groups: males/masculine/men and females/feminine/women (Bem, 1993; Lucal, 2008; Wharton, 2005). This perception is shored up by the belief that heterosexuality or sexual attraction between the two, and only two, sexes/genders is natural. *Heteronormativity* (see Chapter 6 for detailed discussion) is now the term commonly used by sociologists to refer to the "cultural, legal, and institutional practices" that maintain a binary and unequal system (Schilt & Westbrook, 2009, p. 441). The culturally created model of gender, as well as sex and sexuality, then, is nonkaleidoscopic: no spontaneity, no ambiguity, no complexity, no diversity, no surprises, no elasticity, and no unfolding growth.

Social Scientific Understandings of Sex, Gender, and Sexuality

Modern social science offers a rich and complex understanding of gender, sex, and sexuality. It opens the door to the diversity of human experience and rejects the tendency to reduce human behavior to simple, single-factor explanations. Research shows that the behavior of people, no matter who they are, depends on time and place, context and situation—not on fixed sex/gender/sexuality differences (Lorber, 1994; Tavris, 1992; Vespa, 2009). For example, just a few decades ago in the United States, cheerleading was a men's sport because it was considered too rigorous for women (Dowling, 2000), women were thought to lack the cognitive and emotional "stuff" to pilot flights into space, and medicine and law were viewed as too intellectually demanding for women. As Carol Tavris (1992) says, research demonstrates that perceived gender differences turn out to be a matter of "now you see them, now you don't" (p. 288).

If we expand our sociological examination of sex/gender/sexuality to include cross-cultural studies, the real-life fluidity of human experience comes fully alive (see Chapter 3 for a detailed discussion). In some cultures (e.g., the Aka hunter-gatherers), fathers as well as mothers suckle infants (Hewlett, 2001). In other cultures, such as the Agta Negritos, women as well as men are hunters (Estioko-Griffin & Griffin, 2001). Among the Tharus of India and Nepal, marriage is "woman-friendly" and women readily divorce husbands because each woman "enjoys a more dominant position and can find another husband more easily" (Verma, 2009, para. 14). As Serena Nanda discusses in depth in her reading in this chapter, extraordinary gender diversity was expressed in complex, more-than-two sex/gender/sexuality systems in many precontact Native American societies.

In addition, the complex nature of sex/gender/sexuality is underscored by scholarship on multiple masculinities and femininities, as discussed in the introduction to this book. There is no single pattern of masculinity or femininity. Masculinities and femininities are constantly in flux (Coles, 2009). Recall that Raewyn Connell (2000), in her analysis of masculinities, argued that hegemonic masculinity produces complicit, marginalized, and subordinated masculinities. Similarly, there is no femininity, singular. Instead, the ideal and practice of femininity vary by class, race, sexuality, historical period, nation, and other social factors. In her reading in this chapter, Connell extends analysis of masculinities by critiquing Eurocentric assumptions about gender relations with a focus on the relation between hegemony and masculinity through eras of decolonization, postcolonial development, and neoliberal globalization. Let's use sociological radar again and call on the work of social scientists to help us think more precisely and "objectively" about what gender, sex, and sexuality are. It has become somewhat commonplace to distinguish between gender and sex by viewing sex, femaleness and maleness, as a biological fact unaffected by culture and thus unchanging and unproblematic, while viewing gender as a cultural phenomenon, a means by which people are taught who they are (e.g., girl or boy), how to behave (e.g., ladylike or tough), and what their roles will be (e.g., mother or father) (Sørensen, 2000). However, this mode of distinguishing between sex and gender has come under criticism, largely because new studies have revealed the cultural dimensions of sex itself (Schilt, 2010). That is, the physical characteristics of sex cannot be separated from the cultural milieu in which they are labeled and given meaning. In other words, the relationship between biology and behavior is reciprocal, both inseparable and intertwined (Sapolsky, 1997; Yoder, 2003).

Sex, as it turns out, is not a clear-cut matter of DNA, chromosomes, external genitalia, and the like, factors that produce two and only two sexes—female and male. First, there is considerable biological variation. Sex is not fixed in two categories. Biologist Anne Fausto-Sterling (1993) suggests that sex is more like a continuum than a dichotomy. For example, all humans have estrogen, prolactin, and testosterone but in varying and changing levels (Abrams, 2002). Think about this: In American society, people tend to associate breasts and related phenomena, such as breast cancer and lactation, with women. However, men have breasts. Indeed, some men have bigger breasts than some women, some men lactate, and some men get breast cancer. Also, in our society, people associate facial hair with men. What's the real story? All women have facial hair, and some have more of it than do some men. Indeed, recent hormonal and genetic studies (e.g., Abrams, 2002; Beale, 2001) are revealing that, biologically, women and men, female and male bodies are far more similar than different. In a short article, Vanessa Heggie (2015), an historian of science, notes that as early as the 1930s, scientists (e.g., geneticists) were aware of the non-binary nature of sex and gender. She emphasizes that "there has never been scientific (or philosophical, or sociological) consensus that there are simply two human sexes, that they are easily (and objectively) distinguished, and that there is no overlap between the groups. Nor have they agreed that all of us are 'really' one sex or the other. . . . You can examine someone's genitals, their blood, their genes, their taste in movies, the length of their hair, and make a judgement, but none of these constitute a universal or objective test for sex, let alone gender."

Second, not only do femaleness and maleness share much in common, but variations in and complexities of sex development produce *intersex* people whose bodies do not fit either of the two traditionally understood sex categories (Fausto-Sterling, 2000; Fujimora, 2006). Until recently in the United States, intersex was kept a secret and treated as a medical emergency (Grabham, 2007). Now that activists and researchers are challenging the marginalization and medicalization of intersex people, we understand that intersex is not a rarity. Scientists estimate that up to 2% of live births are intersex. Among intersex births are babies born with both male and female characteristics and babies born with "larger-than-average" clitorises or "smaller-than-average" penises (Lucal, 2008). Joan H. Fujimora (2006) examined recent research on sex genes and concluded that "there is no single pathway through which sex is genetically determined" and we might consider sex variations, such as intersex, as resulting from "multiple developmental pathways that involve genetic, protein, hormonal, environmental, and other agents, actions, and interactions" (p. 71). Judith Lorber and Lisa Jean Moore (2007) argue that intersex people are akin to multiracial people. They point out that just as scientists have demonstrated through DNA testing that almost all of us are genetically interracial, similarly, "if many people were genetically sex-typed, we'd also find a variety of chromosomal, hormonal, and anatomical patterns unrecognized" in our rigid, two-sex system (p. 138). In their chapter reading, Georgiann Davis and Sharon Preves examine the harmful consequences of the medicalization of intersex in the United States. They also discuss in detail the emergence of the intersex rights movement both as a response to medically unnecessary "normalization" surgeries and as a challenge to the two-and-only-two sex/gender/sexuality system. Biology is a complicated business, and that should come as no surprise. The more we learn about biology, the more elusive and complex sex becomes. What seemed so obvious—two opposite sexes—turns out to be a gross oversimplification.

Then, what is gender? As discussed in the introduction to this book, gender is a human invention, a means by which people are sorted (in our society, into two gender categories), a basic aspect of how our society organizes itself and allocates resources (e.g., certain tasks assigned to people called women and other tasks to those termed men), and a fundamental ingredient in how individuals understand themselves and others ("I feel feminine"; "He's manly"; "You're androgynous").

One of the fascinating aspects of gender is the extent to which it is negotiable and dynamic. In effect, masculinity and femininity exist because people believe that women and men are distinct groups and, most important, because people "do gender," day in and day out, and enforce gender conformity. It is now common for gender scholars to refer to gender as a performance or a masquerade, emphasizing that it is through the ways we present ourselves in our daily encounters with others that gender is created and recreated. The chapter reading by Betsy Lucal illustrates vividly how gender is a matter of attribution and enactment.

We even do gender by ourselves, and sometimes quite self-consciously. Have you ever tried to make yourself look and act more masculine or feminine? What is involved in "putting on" femininity or masculinity? Consider *transvestism,* or cross-gender dressing. "Cross-dressers know that successfully being a man or a woman simply means convincing others that you are what you appear to be" (Kimmel, 2000, p. 104). Think about the emerging communities of *transgender* people who are "challenging, questioning, or changing gender from that assigned at birth to a chosen gender" (Lorber & Moore, 2007, p. 139). Although most people have deeply learned gender and view the gender category they inhabit as natural or normal, intersex and transgender activists attack the boundaries of "normal" by refusing to choose a traditional sex, gender, or sexual identity (Lorber & Moore, 2007). In so doing, cultural definitions of sex and gender are destabilized and expanded. Carla A. Pfeffer's chapter reading illustrates this process by exploring transgender identities and relationships, demonstrating how the experiences of "queer" social actors have the potential to shake the foundations of normative binaries of sex, gender, and sexuality.

You may be wondering why we have not used the term *role,* as in *gender role,* to describe "doing gender." The problem with the concept of roles is that many social roles, such as those of teacher, student, doctor, or nurse, are situation specific. However, gender, like race, is a status and identity that cuts across situations and institutional arenas. In other words, gender does not "appear and disappear from one situation to another" (Van Ausdale & Feagin, 2001, p. 32). In part, this is a consequence of the pressures that other people exert on us to "do gender" no matter the social location in which we find ourselves. Even if an individual would like to "give up gender," others will work hard to define and interact with that individual in gendered terms. If you were an accountant, you could "leave your professional role behind you" when you left the office and went shopping or vacationing. Gender is a different story. Could you leave gender at the office? What would that look like, and what would it take to make it happen?

So far, we have explored gender as a product of our interactions with others. It is something we do, not something we inherit. Gender is also built into the larger world we inhabit in the United States, including its institutions, images and symbols, organizations, and material objects. For example, jobs, wages, and hierarchies of dominance and subordination in workplaces are gendered. Even after decades of substantial increase in women's workforce participation, occupations continue to be allocated by gender (e.g., secretaries are overwhelmingly women; men dominate construction work) and a wage gap between men and women persists (Bose & Whaley, 2001; Steinberg, 2001; see also the introduction to this book and the introduction to Chapter 7). In addition, men are still more likely to be bosses and women to be bossed. The symbols and images that surround us and by which we communicate are another part of our society's gender story. Our language speaks of difference and opposition in phrases such as "the opposite sex" and in the absence of any words except awkward medical terms (e.g., hermaphrodite) or epithets (e.g., fag) to refer to sex/sexual/gender variants. In addition, the swirl of standardized gendered images in the media is almost overwhelming. Blatant gender stereotypes still dominate TV, film, magazines, and billboards (Lont, 2001). Gender is also articulated, reinforced, and transformed through material objects and locales (Sørensen, 2000). Shoes are gendered, body adornments are gendered, public restrooms are gendered, ships are gendered, wrapping paper is gendered, and deodorants are gendered. The list is endless. The point is that these locales and objects are transformed into a medium for gender to operate within (Sørensen, 2000). They make gender seem "real," and they give it material consequences (p. 82).

Just as culture spawns the binary and oppositional sex and gender template (Grabham, 2007), sexuality, too, is socially constructed (see discussion in Chapter 6). It is not "a natural occurrence derived from biological sex" (Schilt & Westbrook, 2009, p. 443). But in the United States, the imperative to do heterosexuality dominates and is bound to privilege and power. Kristen Schilt and Laurel Westbrook state that our gender system "must be conceived of as heterosexist, as power is allocated via positioning in the gender and sexual hierarchies" (p. 443). Masculinity and heterosexuality are privileged, while femininity and homosexuality are denigrated. Other sexualities (e.g., bisexuality and pansexuality) are relegated to the margins.

In short, social scientific research underscores the complexity of the prism of gender and demonstrates how gender/sex/sexuality are constructed at multiple, interacting levels of society. The first reading in this chapter, by Barbara J. Risman, is a detailed examination of the ways our gender structure is embedded in the individual, interactional, and institutional dimensions of our society, emphasizing that gender cannot be reduced to one level or dimension: individual, interactional, or institutional. We are literally and figuratively immersed in a gendered world—a world in which difference, opposition, and inequality are the culturally defined themes. And yet, that world is kaleidoscopic in nature. The lesson of the kaleidoscope is that "nothing in life is immune to change" (Baker, 1999, p. 29). Reality is in flux; you never know what's coming next. The metaphor of the kaleidoscope reminds us to keep seeking the shifting meanings as well as the recurring patterns of gender (Baker, 1999).

We live in an interesting time of kaleido scopic change. Old patterns of sex/gender/sexuality difference and inequality keep reappearing, often in new guises, while new patterns of convergence, equality, and self-realization have emerged. Social science research is vital in helping us stay focused on understanding the prism of gender as changeable and helping us respond to its context—as a social dialogue about societal membership and conventions and "as the outcome of how individuals are made to understand their differences and similarities" (Sørensen, 2000, pp. 203–204). With that focus in mind, we can more clearly and critically explore our gendered society.

References

Abrams, D. C. (2002, March–April). Father nature: The making of a modern dad. *Psychology Today,* pp. 38–42.

Baker, C. (1999). *Kaleidoscopes: Wonders of wonder.* Lafayette, CA: C&T.

Beale, B. (2001). The sexes: New insights into the X and Y chromosomes. *The Scientist, 15*(15), 18. Retrieved from https://www.the-scientist.com/research/the-sexes-new-insights-into-the-x-and-y-chromosomes-54434

Bem, S. L. (1993). *The lenses of gender.* New Haven, CT: Yale University Press.

Bose, C. E., & Whaley, R. B. (2001). Sex segregation in the U.S. labor force. In D. Vannoy (Ed.), *Gender mosaics* (pp. 228–239). Los Angeles: Roxbury.

Budgeon, S. (2014). The dynamics of gender hegemony: Femininities, masculinities and social change. *Sociology*, *48*(2), 317–334.

Coles, T. (2009). Negotiating the field of masculinity: The production and reproduction of multiple dominant masculinities. *Men and Masculinities, 12*(1), 30–44.

Connell, R. (n.d.). www.raewynconnell.net/2013/03/feminisms-challenge-to-biological.html

Connell, R. W. (2000). *The men and the boys.* Berkeley: University of California Press.

Dowling, C. (2000). *The frailty myth.* New York: Random House.

Estioko-Griffin, A., & Griffin, P. B. (2001). Woman the hunter: The Agta. In C. Brettell & C. Sargent (Eds.), *Gender in cross-cultural perspective* (3rd ed., pp. 238–239). Upper Saddle River, NJ: Prentice Hall.

Fausto-Sterling, A. (1993, March–April). The five sexes: Why male and female are not enough. *The Sciences,* pp. 20–24.

Fausto-Sterling, A. (2000). *Sexing the body: Gender politics and the construction of sexuality.* New York: Basic Books.

Fujimora, J. H. (2006). Sex genes: A critical sociomaterial approach to the politics and molecular genetics of sex determination. *Signs, 32*(1), 49–81.

Glick, P., & Fiske, S. T. (1999). Gender, power dynamics, and social interaction. In M. Ferree, J. Lorber, & B. Hess (Eds.), *Revisioning gender* (pp. 365–398). Thousand Oaks, CA: Sage.

Grabham, E. (2007). Citizen bodies, intersex citizenship. *Sexualities, 10*(29), 28–29.

Haines, E. L., Deaux, K., & Lofaro, N. (2016). The times they are a-changing . . . or are they not? A comparison of gender stereotypes, 1983-2014. *Psychology of Women Quarterly*, *40*(3), 353–363.

Heggie, V. (2015). "Nature and sex redefined – we have never been binary." *The Guardian*. February 19. Retrieved January 17, 2018 (https://www.theguardian.com/science/the-h-word/2015/feb/19/nature-sex-redefined-we-have-never-been-binary).

Hewlett, B. S. (2001). The cultural nexus of Aka father-infant bonding. In C. Brettell & C. Sargent (Eds.), *Gender in cross-cultural perspective* (3rd ed., pp. 45–46). Upper Saddle River, NJ: Prentice Hall.

Kimmel, M. S. (2000). *The gendered society.* New York: Oxford University Press.

Lont, C. M. (2001). The influence of the media on gender images. In D. Vannoy (Ed.), *Gender mosaics* (pp. 114–122). Los Angeles: Roxbury.

Lorber, J. (1994). *Paradoxes of gender.* New Haven, CT: Yale University Press.

Lorber, J., & Moore, L. J. (2007). *Gendered bodies.* Los Angeles: Roxbury.

Lucal, B. (2008). Building boxes and policing boundaries: (De)constructing intersexuality, transgender and bisexuality. *Sociology Compass, 2*(2), 519–536.

Miller, D. I., Eagly, A. H., & Linn, M. C. (2014, October 20). Women's representation in science predicts national gender-science stereotypes: Evidence from 66 nations. *Journal of Educational Psychology*. Advance online publication. Retrieved from http://dx.doi.org/10.1037/edu0000005

O'Brien, J. (1999). *Social prisms: Reflections on the everyday myths and paradoxes.* Thousand Oaks, CA: Pine Forge Press.

Pfeffer, C. (2014). "I don't like passing as a straight woman": Queer negotiations of identity and social group membership. *American Journal of Sociology*, *120*(1), 1–44.

Ridgeway, C. L., & Correll, S. J. (2004). Unpacking the gender system: A theoretical perspective on gender beliefs and social relations. *Gender & Society, 18*(4), 510–531.

Sapolsky, R. M. (1997). *The trouble with testosterone and other essays on the biology of the human predicament.* New York: Simon & Schuster, Inc.

Schilt, K. (2010). *Just one of the guys? Transgender men and the persistence of gender inequality.* Chicago: University of Chicago Press.

Schilt, K., & Westbrook, L. (2009). Doing gender, doing heteronormativity: "Gender normals," transgender people, and the social maintenance of heterosexuality. *Gender & Society, 23*(4), 440–464.

Shields, S., Garner, D., Di Leone, B., & Hadley, A. (2006). Gender and emotion. In J. Stets & J. Turner (Eds.), *Handbook of the sociology of emotions* (pp. 63–83). New York: Springer.

Snyder, K. (2014). The abrasiveness trap: High-achieving men and women are described differently in reviews. *Fortune*. Retrieved from http://fortune.com/2014/08/26/performance-review-gender-bias/

Sørensen, M. L. S. (2000). *Gender archaeology.* Cambridge, UK: Polity Press.

Steinberg, R. J. (2001). How sex gets into your paycheck and how to get it out: The gender gap in pay and comparable worth. In D. Vannoy (Ed.), *Gender mosaics.* Los Angeles: Roxbury.

Tavris, C. (1992). *The mismeasure of woman.* New York: Simon & Schuster.

Van Ausdale, D., & Feagin, J. R. (2001). *The first R: How children learn race and racism.* Lanham, MD: Rowman & Littlefield.

Verma, S. C. (2009, October). Amazing Tharu women: Empowered and in control. *Intersections: Gender and Sexuality in Asia and the Pacific, 22.* Retrieved from http://intersections.anu.edu.au/issue22/verma.htm

Vespa, J. (2009). Gender ideology construction: A life course and intersectional approach. *Gender & Society, 23*(3), 363–387.

Wharton, A. S. (2005). *The sociology of gender.* Malden, MA: Blackwell.

Yoder, J. D. (2003). *Women and gender: Transforming psychology.* Upper Saddle River, NJ: Prentice Hall.

Introduction to Reading 1

Barbara Risman is a sociologist who has made significant contributions to research and writing on gender in heterosexual American families. In this article, she argues that we need to conceptualize gender as a social structure so we can better analyze the ways gender is embedded in the individual, interactional, and institutional dimensions of social life. You will want to pay special attention to Table 1.1, in which Risman summarizes social processes that create gender in each dimension.

1. Why does Risman include the individual dimension of social life in her theory of gender as a social structure?
2. What are the benefits of a multidimensional structural model of gender?
3. Define the concept "trading power for patronage," and discuss at least two examples from your experience or observations of heterosexual relationships.

Gender as a Social Structure

Theory Wrestling With Activism

Barbara J. Risman

In this article, I briefly summarize my . . . argument that gender should be conceptualized as a social structure (Risman 1998) and extend it with an attempt to classify the mechanisms that help produce gendered outcomes within each dimension of the social structure.

Gender as Social Structure

With this theory of *gender as a social structure,* I offer a conceptual framework, a scheme to organize the confusing, almost limitless, ways in which gender has come to be defined in contemporary social science. Four distinct social scientific theoretical traditions have developed to explain gender. The first tradition focuses on how individual sex differences originate, whether biological (Udry 2000) or social in origin (Bem 1993). The second tradition . . . emerged as a reaction to the first and focuses on how the social structure (as opposed to biology or individual learning) creates gendered behavior. The third tradition, also a reaction to the individualist thinking of the first, emphasizes social interaction and accountability to others' expectations, with a focus on how "doing gender" creates and reproduces inequality (West and Zimmerman 1987). The sex-differences literature, the doing gender interactional analyses, and the structural perspectives have been portrayed as incompatible in my own early writings as well as in that of others (Epstein 1988; Ferree 1990; Kanter 1977; Risman 1987; Risman and Schwartz 1989). England and

Risman, B. J. (2004). Gender as a social structure. *Gender & Society*, *18*(4). Reprinted by permission of SAGE Publications Inc., on behalf of Sociologists for Women in Society.

Browne (1992) argued persuasively that this incompatibility is an illusion: All structural theories must make assumptions about individuals, and individualist theories must make presumptions about external social control. While we do gender in every social interaction, it seems naive to ignore the gendered selves and cognitive schemas that children develop as they become cultural natives in a patriarchal world (Bem 1993). The more recent integrative approaches (Connell 2002; Ferree, Lorber, and Hess 1999; Lorber 1994; Risman 1998) treat gender as a socially constructed stratification system. This article fits squarely in the current integrative tradition.

Lorber (1994) argued that gender is an institution that is embedded in all the social processes of everyday life and social organizations. She further argued that gender difference is primarily a means to justify sexual stratification. Gender is so endemic because unless we see difference, we cannot justify inequality. I share this presumption that the creation of difference is the very foundation on which inequality rests.

I build on this notion of gender as an institution but find the institutional language distracting. The word "institution" is too commonly used to refer to particular aspects of society, for example, the family as an institution or corporations as institutions. My notion of gender structure meets the criteria offered by Martin (forthcoming). . . . While the language we use may differ, our goals are complementary, as we seek to situate gender as embedded not only in individuals but throughout social life (Patricia Martin, personal communication).

I prefer to define gender as a social structure because this brings gender to the same analytic plane as politics and economics, where the focus has long been on political and economic structures. While the language of structure suits my purposes, it is not ideal because despite ubiquitous usage in sociological discourse, no definition of the term "structure" is widely shared. Smelser (1988) suggested that all structuralists share the presumption that social structures exist outside individual desires or motives and that social structures at least partially explain human action. Beyond that, consensus dissipates. Blau (1977) focused solely on the constraint collective life imposes on the individual. Structure must be conceptualized, in his view, as a force opposing individual motivation. Structural concepts must be observable, external to the individual, and independent of individual motivation. This definition of "structure" imposes a clear dualism between structure and action, with structure as constraint and action as choice.

Constraint is, of course, an important function of structure, but to focus only on structure as constraint minimizes its importance. Not only are women and men coerced into differential social roles; they often choose their gendered paths. A social structural analysis must help us understand how and why actors choose one alternative over another. A structural theory of action (e.g., Burt 1982) suggests that actors compare themselves and their options to those in structurally similar positions. From this viewpoint, actors are purposive, rationally seeking to maximize their self-perceived well-being under social-structural constraints. As Burt (1982) suggested, one can assume that actors choose the best alternatives without presuming they have either enough information to do it well or the options available to make choices that effectively serve their own interests. For example, married women may choose to do considerably more than their equitable share of child care rather than have their children do without whatever "good enough" parenting means to them if they see no likely alternative that the children's father will pick up the slack.

While actions are a function of interests, the ability to choose is patterned by the social structure. Burt (1982) suggested that norms develop when actors occupy similar network positions in the social structure and evaluate their own options vis-à-vis the alternatives of similarly situated others. From such comparisons, both norms and feelings of relative deprivation or advantage evolve. The social structure as the context of daily life creates action indirectly by shaping actors' perceptions of their interests and directly by constraining choice. Notice the phrase "similarly situated others" above. As long as women and men see themselves as different kinds of people, then women will be unlikely to compare their life options to those of men. Therein lies the power of gender. In a world where sexual anatomy is used to dichotomize human beings into types, the differentiation itself diffuses both claims to and expectations for gender equality. The social structure is not experienced as oppressive if men and women do not see themselves as similarly situated.

While structural perspectives have been applied to gender in the past (Epstein 1988; Kanter 1977), there has been a fundamental flaw in these applications. Generic structural theories applied to gender presume that if women and men were to experience identical structural conditions and role expectations, empirically observable gender differences would disappear. But this ignores not only internalized gender at the individual level . . . but the cultural interactional expectations that remain attached to women and men because of their gender category. A structural perspective on gender is accurate only if we realize that gender itself is a structure deeply embedded in society.

Giddens's (1984) structuration theory adds considerably more depth to this analysis of gender as a social structure with his emphasis on the recursive relationship between social structure and individuals. That is, social structures shape individuals, but simultaneously, individuals shape the social structure. Giddens embraced the transformative power of human action. He insisted that any structural theory must be concerned with reflexivity and actors' interpretations of their own lives. Social structures not only act on people; people act on social structures. Indeed, social structures are created not by mysterious forces but by human action. When people act on structure, they do so for their own reasons. We must, therefore, be concerned with why actors choose their acts. Giddens insisted that concern with meaning must go beyond the verbal justification easily available from actors because so much of social life is routine and so taken for granted that actors will not articulate, or even consider, why they act.

This nonreflexive habituated action is what I refer to as the cultural component of the social structure: The taken for granted or cognitive image rules that belong to the situational context (not only or necessarily to the actor's personality). The cultural component of the social structure includes the interactional expectations that each of us meet in every social encounter. My aims are to bring women and men back into a structural theory where gender is the structure under analysis and to identify when behavior is habit (an enactment of taken for granted gendered cultural norms) and when we do gender consciously, with intent, rebellion, or even with irony. When are we doing gender and re-creating inequality without intent? And what happens to interactional dynamics and male-dominated institutions when we rebel? Can we refuse to do gender or is rebellion simply doing gender differently, forging alternative masculinities and femininities?

Connell (1987) applied Giddens's (1984) concern with social structure as both constraint and created by action in his treatise on gender and power (see particularly chapter 5). In his analysis, structure constrains action, yet "since human action involves free invention . . . and is reflexive, practice can be turned against what constrains it; so structure can deliberately be the object of practice" (Connell 1987, 95). Action may turn against structure but can never escape it.

A theory of gender as a social structure must integrate this notion of causality as recursive with attention to gender consequences at multiple levels of analysis. Gender is deeply embedded as a basis for stratification not just in our personalities, our cultural rules, or institutions but in all these, and in complicated ways. The gender structure differentiates opportunities and constraints based on sex category and thus has consequences on three dimensions: (1) at the individual level, for the development of gendered selves; (2) during interaction as men and women face different cultural expectations even when they fill the identical structural positions; and (3) in institutional domains where explicit regulations regarding resource distribution and material goods are gender specific.

Advantages to Gender Structure Theory

This schema advances our understanding of gender in several ways. First, this theoretical model imposes some order on the encyclopedic research findings that have developed to explain gender inequality. Thinking of each research question as one piece of a jigsaw puzzle, being able to identify how one set of findings coordinates with others even when the dependent variables or contexts of interest are distinct, furthers our ability to build a cumulative science. Gender as a social structure is enormously complex. Full attention to the web of interconnection between gendered selves, the cultural expectations that help explain interactional patterns, and institutional regulations allows each research tradition to explore the growth of their own trees while remaining cognizant of the forest.

A second contribution of this approach is that it leaves behind the modernist warfare version of science, wherein theories are pitted against one another, with a winner and a loser in every contest. In the past, much energy . . . was devoted to testing which theory best explained gender inequality and by implication to discounting every alternative possibility.[1] Theory building that depends on theory slaying presumes parsimony is always desirable, as if this complicated world of ours were best described with simplistic monocausal explanations. While parsimony and theory testing were the model for the twentieth-century science, a more postmodern science should attempt to find complicated and integrative theories (Collins 1998). The conceptualization of gender as a social structure is my contribution to complicating, but hopefully enriching, social theory about gender.

A third benefit to this multidimensional structural model is that it allows us to seriously investigate the direction and strength of causal relationships between gendered phenomena on each dimension. We can try to identify the site where change occurs and at which level of analysis the ability of agentic women and men seem able at this, historical moment, to effectively reject habitualized gender routines. For example, we can empirically investigate the relationship between gendered selves and doing gender without accepting simplistic unidirectional arguments for inequality presumed to be either about identities or cultural ideology. It is quite possible, indeed likely, that socialized femininity does help explain why we do gender, but doing gender to meet others' expectations, surely, over time, helps construct our gendered selves. Furthermore, gendered institutions depend on our willingness to do gender, and when we rebel, we can sometimes change the institutions themselves. I have used the language of dimensions interchangeably with the language of levels because when we think of gender as a social structure, we must move away from privileging any particular dimension as higher than another. How social change occurs is an empirical question, not an a priori theoretical assumption. It may be that individuals struggling to change their own identities (as in consciousness-raising groups of the early second-wave women's movement) eventually bring their new selves to social interaction and create new cultural expectations. For example, as women come to see themselves (or are socialized to see themselves) as sexual actors, the expectations that men must work to provide orgasms for their female partners becomes part of the cultural norm. But this is surely not the only way social change can happen. When social movement activists name as inequality what has heretofore been considered natural (e.g., women's segregation into low-paying jobs), they can create organizational changes such as career ladders between women's quasi-administrative jobs and actual management, opening up opportunities that otherwise would have remained closed, thus creating change on the institutional dimension. Girls raised in the next generation, who know opportunities exist in these workplaces, may have an altered sense of possibilities and therefore of themselves. We need, however, to also study change and equality when it occurs rather than only documenting inequality.

Perhaps the most important feature of this conceptual schema is its dynamism. No one dimension determines the other. Change is fluid and reverberates throughout the structure dynamically. Changes in individual identities and moral accountability may change interactional expectations, but the opposite is possible as well. Change cultural expectations, and individual identities are shaped differently. Institutional changes must result from individuals or group action, yet such change is difficult, as institutions exist across time and space. Once institutional changes occur, they reverberate at the level of cultural expectations and perhaps even on identities. And the cycle of change continues. No mechanistic predictions are possible because human beings sometimes reject the structure itself and, by doing so, change it.

Social Processes Located by Dimension in the Gender Structure

When we conceptualize gender as a social structure, we can begin to identify under what conditions and how gender inequality is being produced within each dimension. The "how" is important because without knowing the mechanisms, we cannot intervene. If indeed gender inequality in the division of household labor at this historical moment were primarily explained (and I do not suggest that it is) by gendered selves, then we would do well to consider the most effective socialization mechanisms to create fewer gender-schematic children and resocialization for adults. If, however, the gendered division of household labor is primarily constrained today by cultural expectations and moral accountability, it is those cultural images we must work to alter. But then again, if the reason many men do not equitably do their share of family labor is that men's jobs are organized so they cannot succeed at work and do their share at home, it is the contemporary American workplace that must change (Williams 2000). We may never find a universal theoretical explanation for the gendered division of household labor because universal social laws may be an illusion of twentieth-century empiricism. But in any given moment for any particular setting, the causal processes should be identifiable empirically. Gender complexity goes beyond historical specificity, as the particular causal processes that constrain men and women to do gender may be strong in one institutional setting (e.g., at home) and weaker in another (e.g., at work).

The forces that create gender traditionalism for men and women may vary across space as well as time. Conceptualizing gender as a social structure contributes to a more postmodern, contextually specific social science. We can use this schema to begin to organize thinking about the causal processes that are most likely to be effective on each dimension. When we are concerned with the means by which individuals come to have a preference to do gender, we should focus on how identities are constructed through early childhood development, explicit socialization, modeling, and adult

Table 1.1 Dimensions of Gender Structure, by Illustrative Social Processes[a]

	Dimensions of the Gender Structure		
	Individual Level	*Interactional Cultural Expectations*	*Institutional Domain*
Social Processes	Socialization Internalization Identity work Construction of selves	Status expectations Cognitive bias Othering Trading power for patronage Altercasting	Organizational practices Legal regulations Distribution of resources Ideology

[a]These are examples of social processes that may help explain the gender structure on each dimension. They are meant to be illustrative and not a complete list of all possible social processes or causal mechanisms.

experiences, paying close attention to the internalization of social mores. To the extent that women and men choose to do gender-typical behavior cross-situationally and over time, we must focus on such individual explanations. Indeed, much attention has already been given to gender socialization and the individualist presumptions for gender. The earliest and perhaps most commonly referred to explanations in popular culture depend on sex-role training, teaching boys and girls their culturally appropriate roles. But when trying to understand gender on the interactional/cultural dimension, the means by which status differences shape expectations and the ways in which in-group and out-group membership influences behavior need to be at the center of attention. Too little attention has been paid to how inequality is shaped by such cultural expectations during interaction. I return to this in the section below. On the institutional dimension, we look to law, organizational practices, and formal regulations that distinguish by sex category. Much progress has been made in the post–civil rights era with rewriting formal laws and organizational practices to ensure gender neutrality. Unfortunately, we have often found that despite changes in gender socialization and gender neutrality on the institutional dimension, gender stratification remains.

What I have attempted to do here is to offer a conceptual organizing scheme for the study of gender that can help us to understand gender in all its complexity and try to isolate the social processes that create gender in each dimension. Table 1.1 provides a schematic outline of this argument.[2]

Cultural Expectations During Interaction and the Stalled Revolution

In *Gender Vertigo* (Risman 1998), I suggested that at this moment in history, gender inequality between partners in American heterosexual couples could be attributed particularly to the interactional expectations at the cultural level: the differential expectations attached to being a mother and father, a husband and wife. Here, I extend this argument in two ways. First, I propose that the stalled gender revolution in other settings can similarly be traced to the interactional/cultural dimension of the social structure. Even when women and men with feminist identities work in organizations with formally gender-neutral rules, gender inequality is reproduced during everyday interaction. The cultural expectations attached to our sex category, simply being identified as a woman or man, has remained relatively impervious to the feminist forces that have problematized sexist socialization practices and legal discrimination. I discuss some of those processes that can help explain why social interaction continues to reproduce inequality, even in settings that seem ripe for social change.

Contemporary social psychological writings offer us a glimpse of possibilities for understanding how inequality is reconstituted in daily interaction. Ridgeway and her colleagues (Ridgeway 1991, 1997, 2001; Ridgeway and Correll 2000; Ridgeway and Smith-Lovin 1999) showed that the status expectations attached to gender and race categories are cross-situational. These expectations can be thought of as one of the engines that re-create inequality even in new settings where there is no other reason to expect male privilege to otherwise emerge. In a sexist and racist society, women and all persons of color are expected to have less to contribute to task performances than are white men, unless they have some other externally validated source of prestige. Status expectations create a cognitive bias toward privileging those of already high status. What produces status distinction, however, is culturally and historically variable. Thus, cognitive bias is one of the causal mechanisms that help to explain the reproduction of gender and race inequality in everyday life. It may also be an important explanation for the reproduction of class and heterosexist inequality in everyday life as well, but that is an empirical question.

Schwalbe and his colleagues (2000, 419) suggested that there are other "generic interactive processes through which inequalities are created and reproduced in everyday life." Some of these processes include othering, subordinate adaptation, boundary maintenance, and emotion management. Schwalbe and his colleagues suggested that subordinates' adaptation plays an essential role in their own disadvantage. Subordinate adaptation helps to explain women's strategy to adapt to the gender structure. Perhaps the most common adaptation of women to subordination is "trading power for patronage" (Schwalbe et al. 2000, 426). Women, as wives and daughters, often derive significant compensatory benefits from relationships with the men in their families. Stombler and Martin (1994) similarly showed how little sisters in a fraternity trade affiliation for secondary status. In yet another setting, elite country clubs, Sherwood (2004) showed how women accept subordinate status as "B" members of clubs, in exchange for men's approval, and how when a few wives challenge men's privilege, they are threatened with social ostracism, as are their husbands. Women often gain the economic benefits of patronage for themselves and their children in exchange for their subordinate status.

One can hardly analyze the cultural expectations and interactional processes that construct gender inequality without attention to the actions of members of the dominant group. We must pay close attention to what men do to preserve their power and privilege. Schwalbe and his colleagues (2000) suggested that one process involved is when superordinate groups effectively "other" those who they want to define as subordinate, creating devalued statuses and expectations for them. Men effectively do this in subversive ways through "politeness" norms, which construct women as "others" in need of special favors, such as protection. By opening doors and walking closer to the dirty street, men construct women as an "other" category, different and less than independent autonomous men. The cultural significance attached to male bodies signifies the capacity to dominate, to control, and to elicit deference, and such expectations are perhaps at the core of what it means for men to do gender (Michael Schwalbe, personal communication).

These are only some of the processes that might be identified for understanding how we create gender inequality based on embodied cultural expectations. None are determinative causal predictors, but instead, these are possible leads to reasonable and testable hypotheses about the production of gender. . . .

Notes

1. See Scott (1997) for a critique of feminists who adopt a strategy where theories have to be simplified, compared, and defeated. She too suggested a model where feminists build on the complexity of each other's ideas.

2. I thank my colleague Donald Tomaskovic-Devey for suggesting the visual representation of these ideas as well as his usual advice on my ideas as they develop.

References

Bem, Sandra. 1993. *The lenses of gender.* New Haven, CT: Yale University Press.

Blau, Peter. 1977. *Inequality and heterogeneity.* New York: Free Press.

Burt, Ronald S. 1982. *Toward a structural theory of action.* New York: Academic Press.

Collins, Patricia Hill. 1998. *Fighting words: Black women and the search for justice.* Minneapolis: University of Minnesota Press.

Connell, R. W. 1987. *Gender and power: Society, the person, and sexual politics.* Stanford, CA: Stanford University Press.

———. 2002. *Gender: Short introductions.* Malden, MA: Blackwell.

England, Paula, and Irene Browne. 1992. Internalization and constraint in women's subordination. *Current Perspectives in Social Theory* 12:97–123.

Epstein, Cynthia Fuchs. 1988. *Deceptive distinctions: Sex, gender, and the social order.* New Haven, CT: Yale University Press.

Ferree, Myra Marx. 1990. Beyond separate spheres: Feminism and family research. *Journal of Marriage and the Family 53*(4): 866–84.

Ferree, Myra Marx, Judith Lorber, and Beth Hess. 1999. *Revisioning gender*. Thousand Oaks, CA: Sage.

Giddens, Anthony. 1984. *The constitution of society: Outline of the theory of structuration.* Berkeley: University of California Press.

Kanter, Rosabeth. 1977. *Men and women of the corporation.* New York: Basic Books.

Lorber, Judith. 1994. *Paradoxes of gender.* New Haven, CT: Yale University Press.

Martin, Patricia. Forthcoming. Gender as a social institution. In *Social Forces,* edited by Kristen A. Myers, Cynthia D. Anderson, and Barbara J. Risman, 1998. *Feminist foundations: Toward transforming society.* Thousand Oaks, CA: Sage.

Ridgeway, Cecilia L. 1991. The social construction of status value: Gender and other nominal characteristics. *Social Forces 70*(2): 367–86.

———. 1997. Interaction and the conservation of gender inequality: Considering employment. *American Sociological Review 62*(2): 218–35.

———. 2001. Gender, status, and leadership. *Journal of Social Issues 57*(4): 637–55.

Ridgeway, Cecilia L., and Shelley J. Correll. 2000. Limiting inequality through interaction: The end(s) of gender. *Contemporary Sociology* 29:110–20.

Ridgeway, Cecilia L., and Lynn Smith-Lovin. 1999. The gender system and interaction. *Annual Review of Sociology* 25:191–216.

Risman, Barbara J. 1983. Necessity and the invention of mothering. Ph.D. diss., University of Washington.

———. 1987. Intimate relationships from a microstructural perspective: Mothering men. *Gender & Society* 1: 6–32.

———. 1998. *Gender vertigo: American families in transition.* New Haven, CT: Yale University Press.

Risman, Barbara J., and Pepper Schwartz. 1989. *Gender in intimate relationships.* Belmont, CA: Wadsworth.

Schwalbe, Michael, Sandra Godwin, Daphne Holden, Douglas Schrock, Shealy Thompson, and Michele Wolkomir. 2000. Generic processes in the reproduction of inequality: An interactionist analysis. *Social Forces 79*(2): 419–52.

Scott, Joan Wallach. 1997. Comment on Hawkesworth's "Confounding Gender." *Signs: Journal of Women in Culture and Society 22*(3): 697–702.

Sherwood, Jessica. 2004. Talk about country clubs: Ideology and the reproduction of privilege. Ph.D. diss., North Carolina State University.

Smelser, Neil J. 1988. Social structure. In *Handbook of sociology,* edited by Neil J. Smelser. Beverly Hills, CA: Sage.

Stombler, Mindy, and Patricia Yancey Martin. 1994. Bring women in, keeping women down: Fraternity "little sister" organizations. *Journal of Contemporary Ethnography* 23:150–84.

Udry, J. Richard. 2000. Biological limits of gender construction. *American Sociological Review* 65: 443–57.

West, Candace, and Don Zimmerman. 1987. Doing gender. *Gender & Society* 1:125–51.

Williams, Joan. 2000. *Unbending gender: Why family and work conflict and what to do about it.* New York: Oxford University Press.

Introduction to Reading 2

By analyzing the challenges she faces in the course of her daily experience of negotiating the boundaries of our gendered society, sociologist Betsy Lucal describes the rigidity of the American binary gender system and the consequences for people who do not fit. Since her physical appearance does not clearly define her as a woman, she must navigate a world in which some people interact with her as though she is a man. Through analysis of her own story, Lucal demonstrates how gender is something we do, rather than something we are.

1. Why does Lucal argue that we cannot escape "doing gender"?
2. How does Lucal negotiate "not fitting" into the American two-and-only-two gender structure?
3. Have you ever experienced a mismatch between your gender identity and the gender that others perceive you to be? If so, how did you feel and respond?

What It Means to Be Gendered Me

Betsy Lucal

I understood the concept of "doing gender" (West and Zimmerman 1987) long before I became a sociologist. I have been living with the consequences of inappropriate "gender display" (Goffman 1976; West and Zimmerman 1987) for as long as I can remember. My daily experiences are a testament to the rigidity of gender in our society, to the real implications of "two and only two" when it comes to sex and gender categories (Garfinkel 1967; Kessler and McKenna 1978). Each day, I experience the

Lucal, B. (1999). What it means to be gendered me. *Gender & Society, 13*(6). Reprinted by permission of SAGE Publications Inc., on behalf of Sociologists for Women in Society.

consequences that our gender system has for my identity and interactions. I am a woman who has been called "Sir" so many times that I no longer even hesitate to assume that it is being directed at me. I am a woman whose use of public rest rooms regularly causes reactions ranging from confused stares to confrontations over what a man is doing in the women's room. I regularly enact a variety of practices either to minimize the need for others to know my gender or to deal with their misattributions.

I am the embodiment of Lorber's (1994) ostensibly paradoxical assertion that the "gender bending" I engage in actually might serve to preserve and perpetuate gender categories. As a feminist who sees gender rebellion as a significant part of her contribution to the dismantling of sexism, I find this possibility disheartening.

In this article, I examine how my experiences both support and contradict Lorber's (1994) argument using my own experiences to illustrate and reflect on the social construction of gender. My analysis offers a discussion of the consequences of gender for people who do not follow the rules as well as an examination of the possible implications of the existence of people like me for the gender system itself. Ultimately, I show how life on the boundaries of gender affects me and how my life, and the lives of others who make similar decisions about their participation in the gender system, has the potential to subvert gender.

Because this article analyzes my experiences as a woman who often is mistaken for a man, my focus is on the social construction of gender for women. My assumption is that, given the gendered nature of the gendering process itself, men's experiences of this phenomenon might well be different from women's.

The Social Construction of Gender

It is now widely accepted that gender is a social construction, that sex and gender are distinct, and that gender is something all of us "do." This conceptualization of gender can be traced to Garfinkel's (1967) ethnomethodological study of "Agnes."[1] In this analysis, Garfinkel examined the issues facing a male who wished to pass as, and eventually become, a woman. Unlike individuals who perform gender in culturally expected ways, Agnes could not take her gender for granted and always was in danger of failing to pass as a woman (Zimmerman 1992).

This approach was extended by Kessler and McKenna (1978) and codified in the classic "Doing Gender" by West and Zimmerman (1987). The social constructionist approach has been developed most notably by Lorber (1994, 1996). Similar theoretical strains have developed outside of sociology, such as work by Butler (1990) and Weston (1996). . . .

Given our cultural rules for identifying gender (i.e., that there are only two and that masculinity is assumed in the absence of evidence to the contrary), a person who does not do gender appropriately is placed not into a third category but rather into the one with which her or his gender display seems most closely to fit; that is, if a man appears to be a woman, then he will be categorized as "woman," not as something else. Even if a person does not want to do gender or would like to do a gender other than the two recognized by our society, other people will, in effect, do gender for that person by placing her or him in one and only one of the two available categories. We cannot escape doing gender or, more specifically, doing one of two genders. (There are exceptions in limited contexts such as people doing "drag" [Butler 1990; Lorber 1994].)

People who follow the norms of gender can take their genders for granted. Kessler and McKenna asserted, "Few people besides transsexuals think of their gender as anything other than 'naturally' obvious"; they believe that the risks of not being taken for the gender intended "are minimal for nontranssexuals" (1978, 126). However, such an assertion overlooks the experiences of people such as those women Devor (1989) calls "gender blenders" and those people Lorber (1994) refers to as "gender benders." As West and Zimmerman (1987) pointed out, we all are held accountable for, and might be called on to account for, our genders.

People who, for whatever reasons, do not adhere to the rules, risk gender misattribution and any interactional consequences that might result from this misidentification. What are the consequences of misattribution for social interaction? When must misattribution be minimized? What will one do to minimize such mistakes? In this article, I explore these and related questions using my biography.

For me, the social processes and structures of gender mean that, in the context of our culture, my appearance will be read as masculine. Given the common conflation of sex and gender, I will be assumed to be a male. Because of the two-and-only-two genders rule, I will be classified, perhaps more often than not, as a man—not as an atypical woman, not as a genderless person. I must be one gender or the other; I cannot be neither, nor can I be both. This norm has a variety of mundane and serious consequences for my everyday existence. Like Myhre (1995), I have found that the choice not to participate in femininity is not one made frivolously.

My experiences as a woman who does not do femininity illustrate a paradox of our two-and-only-two gender system. Lorber argued that "bending gender rules and passing between genders does not erode but rather preserves gender boundaries" (1994, 21). Although people who engage in these behaviors and appearances do "demonstrate the social constructedness of sex, sexuality, and gender" (Lorber 1994, 96), they do not actually disrupt gender. Devor made a similar point: "When gender blending females refused to mark themselves by publicly displaying sufficient femininity to be recognized as women, they were in no way challenging patriarchal gender assumptions" (1989, 142). As the following discussion shows, I have found that my own experiences both support and challenge this argument. Before detailing these experiences, I explain my use of my self as data.

My Self as Data

This analysis is based on my experiences as a person whose appearance and gender/sex are not, in the eyes of many people, congruent. How did my experiences become my data? I began my research "unwittingly" (Krieger 1991). This article is a product of "opportunistic research" in that I am using my "unique biography, life experiences, and/or situational familiarity to understand and explain social life" (Riemer 1988, 121; see also Riemer 1977). It is an analysis of "unplanned personal experience," that is, experiences that were not part of a research project but instead are part of my daily encounters (Reinharz 1992).

This work also is, at least to some extent, an example of Richardson's (1994) notion of writing as a method of inquiry. As a sociologist who specializes in gender, the more I learned, the more I realized that my life could serve as a case study. As I examined my experiences, I found out things—about my experiences and about theory—that I did not know when I started (Richardson 1994).

It also is useful, I think, to consider my analysis an application of Mills's (1959) "sociological imagination." Mills (1959) and Berger (1963) wrote about the importance of seeing the general in the particular. This means that general social patterns can be discerned in the behaviors of particular individuals. In this article, I am examining portions of my biography, situated in U.S. society during the 1990s, to understand the "personal troubles" my gender produces in the context of a two-and-only-two gender system. I am not attempting to generalize my experiences; rather, I am trying to use them to examine and reflect on the processes and structure of gender in our society.

Because my analysis is based on my memories and perceptions of events, it is limited by my ability to recall events and by my interpretation of those events. However, I am not claiming that my experiences provide the truth about gender and how it works. I am claiming that the biography of a person who lives on the margins of our gender system can provide theoretical insights into the processes and social structure of gender. Therefore, after describing my experiences, I examine how they illustrate and extend, as well as contradict, other work on the social construction of gender.

Gendered Me

Each day, I negotiate the boundaries of gender. Each day, I face the possibility that someone will attribute the "wrong" gender to me based on my physical appearance. I am six feet tall and large-boned. I have had short hair for most of my life. For the past several years, I have worn a crew cut or flat top. I do not shave or otherwise remove hair from my body (e.g., no eyebrow plucking). I do not wear dresses, skirts, high heels, or makeup. My only jewelry is a class ring, a "men's" watch (my wrists are too large for a "women's" watch), two small earrings (gold hoops, both in my left ear), and (occasionally) a necklace. I wear jeans or shorts, T-shirts, sweaters, polo/golf shirts, button-down collar shirts, and tennis shoes or boots. The jeans are "women's" (I do have hips) but do not look particularly "feminine." The rest of the outer garments are from men's departments. I prefer baggy clothes, so the fact that I have "womanly" breasts often is not obvious (I do not wear a bra).

Sometimes, I wear a baseball cap or some other type of hat. I also am white and relatively young (30 years old).[2] My gender display—what others interpret as my presented identity—regularly leads to the misattribution of my gender. An incongruity exists between my gender self-identity and the gender that others perceive. In my encounters with people I do not know, I sometimes conclude, based on our interactions, that they think I am a man. This does not mean that other people do not think I am a man, just that I have no way of knowing what they think without interacting with them.

Living With It

I have no illusions or delusions about my appearance. I know that my appearance is likely to be read as "masculine" (and male) and that how I see myself is socially irrelevant. Given our two-and-only-two gender structure, I must live with the consequences of my appearance. These consequences fall into two categories: issues of identity and issues of interaction.

My most common experience is being called "Sir" or being referred to by some other masculine linguistic marker (e.g., "he," "man"). This has happened for years, for as long as I can remember, when having encounters with people I do not know.[3] Once, in fact, the same worker at a fast-food restaurant called me "Ma'am" when she took my order and "Sir" when she gave it to me.

Using my credit cards sometimes is a challenge. Some clerks subtly indicate their disbelief, looking from the card to me and back at the card and checking my signature carefully. Others challenge my use of the card, asking whose it is or demanding identification. One cashier asked to see my driver's license and then asked me whether I was the son of the cardholder. Another clerk told me that my signature on the receipt "had better match" the one on the card. Presumably, this was her way of letting me know that she was not convinced it was my credit card.

My identity as a woman also is called into question when I try to use women-only spaces. Encounters in public rest rooms are an adventure. I have been told countless times that "This is the ladies' room." Other women say nothing to me, but their stares and conversations with others let me know what they think. I will hear them say, for example, "There was a man in there." I also get stares when I enter a locker room. However, it seems that women are less concerned about my presence, there, perhaps because, given that it is a space for changing clothes, showering, and so forth, they will be able to make sure that I am really a woman. Dressing rooms in department stores also are problematic spaces. I remember shopping with my sister once and being offered a chair outside the room when I began to accompany her into the dressing room. Women who believe that I am a man do not want me in women-only spaces. For example, one woman would not enter the rest room until I came out, and others have told me that I am in the wrong place. They also might not want to encounter me while they are alone. For example, seeing me walking at night when they are alone might be scary.[4]

I, on the other hand, am not afraid to walk alone, day or night. I do not worry that I will be subjected to the public harassment that many women endure (Gardner 1995). I am not a clear target for a potential rapist. I rely on the fact that a potential attacker would not want to attack a big man by mistake. This is not to say that men never are attacked, just that they are not viewed, and often do not view themselves, as being vulnerable to attack.

Being perceived as a man has made me privy to male-male interactional styles of which most women are not aware. I found out, quite by accident, that many men greet, or acknowledge, people (mostly other men) who make eye contact with them with a single nod. For example, I found that when I walked down the halls of my brother's all-male dormitory making eye contact, men nodded their greetings at me. Oddly enough, these same men did not greet my brother.

I had to tell him about making eye contact and nodding as a greeting ritual. Apparently, in this case I was doing masculinity better than he was! I also believe that I am treated differently, for example, in auto parts stores (staffed almost exclusively by men in most cases) because of the assumption that I am a man. Workers there assume that I know what I need and that my questions are legitimate requests for information.

I suspect that I am treated more fairly than a feminine-appearing woman would be. I have not been able to test this proposition. However, Devor's participants did report "being treated more respectfully" (1989, 132) in such situations. There is, however, a negative side to being assumed to be a man by other men. Once, a friend and I were driving in her car when a man failed to stop at an intersection and nearly crashed into us. As we drove away, I mouthed "stop sign" to him. When we both stopped our cars at the next intersection, he got out of his car and came up to the passenger side of the car, where I was sitting. He yelled obscenities at us and pounded and spit on the car window. Luckily, the windows were closed. I do not think he would have done that if he thought I was a woman. This was the first time I realized that one of the implications of being seen as a man was that I might be called on to defend myself from physical aggression from other men who felt challenged by me. This was a sobering and somewhat frightening thought.

Recently, I was verbally accosted by an older man who did not like where I had parked my car. As I walked down the street to work, he shouted that I should park at the university rather than on a side street nearby. I responded that it was a public street and that I could park there if I chose. He continued to yell, but the only thing I caught was the last part of what he said: "Your tires are going to get cut!" Based on my appearance that day—I was dressed casually and carrying a backpack, and I had my hat on backward—I believe he thought that I was a young male student rather than a female professor. I do not think he would have yelled at a person he thought to be a woman—and perhaps especially not a woman professor.

Given the presumption of heterosexuality that is part of our system of gender, my interactions with women who assume that I am a man also can be viewed from that perspective. For example, once my brother and I were shopping when we were "hit on" by two young women. The encounter ended before I realized what had happened. It was only when we walked away that I told him that I was pretty certain that they had thought both of us were men. A more common experience is realizing that when I am seen in public with one of my women friends, we are likely to be read as a heterosexual dyad. It is likely that if I were to walk through a shopping mall holding hands with a woman, no one would look twice, not because of their open-mindedness toward lesbian couples but rather because of their assumption that I was the male half of a straight couple. Recently, when walking through a mall with a friend and her infant, my observations of others' responses to us led me to believe that many of them assumed that we were a family on an outing, that is, that I was her partner and the father of the child.

Dealing With It

Although I now accept that being mistaken for a man will be a part of my life so long as I choose not to participate in femininity, there have been times when I consciously have tried to appear more feminine. I did this for a while when I was an undergraduate and again recently when I was on the academic job market. The first time, I let my hair grow nearly down to my shoulders and had it permed. I also grew long fingernails and wore nail polish. Much to my chagrin, even then one of my professors, who did not know my name, insistently referred to me in his kinship examples as "the son." Perhaps my first act on the way to my current stance was to point out to this man, politely and after class, that I was a woman.

More recently, I again let my hair grow out for several months, although I did not alter other aspects of my appearance. Once my hair was about two and a half inches long (from its original quarter inch), I realized, based on my encounters with strangers, that I had more or less passed back into the category of "woman." Then, when I returned to wearing a flat top, people again responded to me as if I were a man.

Because of my appearance, much of my negotiation of interactions with strangers involves attempts to anticipate their reactions to me. I need to assess whether they will be likely to assume that I am a man and whether that actually matters in the context of our encounters. Many times, my gender really is irrelevant, and it is just annoying to be misidentified. Other times, particularly when my appearance is coupled with something that identifies me by name (e.g., a check or credit card) without a photo, I might need to do something to ensure that my identity is not questioned. As a result of my experiences, I have developed some techniques to deal with gender misattribution.

In general, in unfamiliar public places, I avoid using the rest room because I know that it is a place where there is a high likelihood of misattribution and where misattribution is socially important. If I must use a public rest room, I try to make myself look as nonthreatening as possible. I do not wear a hat, and I try to rearrange my clothing to make my breasts more obvious. Here, I am trying to use my secondary sex characteristics to make my gender more obvious rather than the usual use of gender to make sex obvious. While in the rest room, I never make eye contact, and I get in and out as quickly as possible. Going in with a woman friend also is helpful; her presence legitimizes my own. People are less likely to think I am entering a space where I do not belong when I am with someone who looks like she does belong.[5]

To those women who verbally challenge my presence in the rest room, I reply, "I know," usually in an annoyed tone. When they stare or talk about me to the women they are with, I simply get out as quickly as possible. In general, I do not wait for someone I am with because there is too much chance of an unpleasant encounter.

I stopped trying on clothes before purchasing them a few years ago because my presence in the changing areas was met with stares and whispers. Exceptions are stores where the dressing rooms are completely private, where there are individual stalls rather than a room with stalls separated by curtains, or where business is slow and no one else is trying on clothes. If I am trying on a garment clearly intended for a woman, then I usually can do so without hassle. I guess the attendants assume that I must be a woman if I have, for example, a women's bathing suit in my hand. But usually, I think it is easier for me to try the clothes on at home and return them, if necessary, rather than risk creating a scene. Similarly, when I am with another woman who is trying on clothes, I just wait outside.

My strategy with credit cards and checks is to anticipate wariness on a clerk's part. When I sense that there is some doubt or when they challenge me, I say, "It's my card." I generally respond courteously to requests for photo ID, realizing that these might be routine checks because of concerns about increasingly widespread fraud. But for the clerk who asked for ID and still did not think it was my card, I had a stronger reaction. When she said that she was sorry for embarrassing me, I told her that I was not embarrassed but that she should be. I also am particularly careful to make sure that my signature is consistent with the back of the card. Faced with such situations, I feel somewhat nervous about signing my name—which, of course, makes me worry that my signature will look different from how it should.

Another strategy I have been experimenting with is wearing nail polish in the dark bright colors currently fashionable. I try to do this when I travel by plane. Given more stringent travel regulations, one always must present a photo ID. But my experiences have shown that my driver's license is not necessarily convincing. Nail polish might be. I also flash my polished nails when I enter airport rest rooms, hoping that they will provide a clue that I am indeed in the right place.

There are other cases in which the issues are less those of identity than of all the norms of interaction that, in our society, are gendered. My most common response to misattribution actually is to appear to ignore it, that is, to go on with the interaction as if nothing out of the ordinary has happened. Unless I feel that there is a good reason to establish my correct gender, I assume the identity others impose on me for the sake of smooth interaction. For example, if someone is selling me a movie ticket, then there is no reason to make sure that the person has accurately discerned my gender. Similarly, if it is clear that the person using "Sir" is talking to me, then I simply respond as appropriate. I accept the designation because it is irrelevant to the situation. It takes enough effort to be alert for misattributions and to decide which of them matter; responding to each one would take more energy than it is worth.

Sometimes, if our interaction involves conversation, my first verbal response is enough to let the other person know that I am actually a woman and not a man. My voice apparently is "feminine" enough to shift people's attributions to the other category. I know when this has happened by the apologies that usually accompany the mistake. I usually respond to the apologies by saying something like "No problem" and/or "It happens all the time." Sometimes, a misattributor will offer an account for the mistake, for example, saying that it was my hair or that they were not being very observant.

These experiences with gender and misattribution provide some theoretical insights into contemporary Western understandings of gender and into the social structure of gender in contemporary society. Although there are a number of ways in which my experiences confirm the work of others, there also are some ways in which my experiences suggest other interpretations and conclusions.

What Does It Mean?

Gender is pervasive in our society. I cannot choose not to participate in it. Even if I try not to do gender, other people will do it for me. That is, given our two-and-only-two rule, they must attribute one of two genders to me. Still, although I cannot choose not to participate in gender, I can choose not to participate in femininity (as I have), at least with respect to physical appearance. That is where the problems begin. Without the decorations of femininity, I do not look like a woman. That is, I do not look like what many people's commonsense understanding of gender tells them a woman looks like. How I see myself, even how I might wish others would see me, is socially irrelevant. It is the gender that I appear to be (my "perceived gender") that is most relevant to my social identity and interactions with others. The major consequence of this fact is that I must be continually aware of which gender I "give off" as well as which gender I "give" (Goffman 1959).

Because my gender self-identity is "not displayed obviously, immediately, and consistently" (Devor 1989, 58), I am somewhat of a failure in social terms with respect to gender. Causing people to be uncertain or wrong about one's gender is a violation of taken-for-granted rules that leads to embarrassment and discomfort; it means that something has gone wrong with the interaction (Garfinkel 1967; Kessler and McKenna 1978). This means that my non-response to misattribution is the more socially appropriate response; I am allowing others to maintain face (Goffman 1959, 1967). By not calling attention to their mistakes, I uphold their images of themselves as competent social actors. I also maintain my own image as competent by letting them assume that I am the gender I appear to them to be.

But I still have discreditable status; I carry a stigma (Goffman 1963). Because I have failed to participate appropriately in the creation of meaning with respect to gender (Devor 1989), I can be called on to account for my appearance. If discredited, I show myself to be an incompetent social actor. I am the one not following the rules, and I will pay the price for not providing people with the appropriate cues for placing me in the gender category to which I really belong.

I do think that it is, in many cases, safer to be read as a man than as some sort of deviant woman. "Man" is an acceptable category; it fits properly into people's gender worldview. Passing as a man often is "the path of least resistance" (Devor 1989; Johnson 1997). For example, in situations where gender does not matter, letting people take me as a man is easier than correcting them.

Conversely, as Butler noted, "We regularly punish those who fail to do their gender right" (1990, 140). Feinberg maintained, "Masculine girls and women face terrible condemnation and brutality—including sexual violence—for crossing the boundary of what is 'acceptable' female expression" (1996, 114). People are more likely to harass me when they perceive me to be a woman who looks like a man. For example, when a group of teenagers realized that I was not a man because one of their mothers identified me correctly, they began to make derogatory comments when I passed them. One asked, for example, "Does she have a penis?"

Because of the assumption that a "masculine" woman is a lesbian, there is the risk of homophobic reactions (Gardner 1995; Lucal 1997). Perhaps surprisingly, I find that I am much more likely to be taken for a man than for a lesbian, at least based on my interactions with people and their reactions to me. This might be because people are less likely to reveal that they have taken me for a lesbian because it is less relevant to an encounter or because they believe this would be unacceptable. But I think it is more likely a product of the strength of our two-and-only-two system. I give enough masculine cues that I am seen not as a deviant woman but rather as a man, at least in most cases. The problem seems not to be that people are uncertain about my gender, which might lead them to conclude that I was a lesbian once they realized I was a woman. Rather, I seem to fit easily into a gender category—just not the one with which I identify. In fact, because men represent the dominant gender in our society, being mistaken for a man can protect me from other types of gendered harassment. Because men can move around in public spaces safely (at least relative to women), a "masculine" woman also can enjoy this freedom (Devor 1989).

On the other hand, my use of particular spaces—those designated as for women only—may be challenged. Feinberg provided an intriguing analysis of the public rest room experience. She characterized women's reactions to a masculine person in a public rest room as "an example of genderphobia" (1996, 117), viewing such women as policing gender boundaries rather than believing that there really is a man in the women's rest room. She argued that women who truly believed that there was a man in their midst would react differently. Although this is an interesting perspective on her experiences, my experiences do not lead to the same conclusion.[6]

Enough people have said to me that "This is the ladies' room" or have said to their companions that "There was a man in there" that I take their reactions at face value. Still, if the two-and-only-two gender system is to be maintained, participants must be involved in policing the categories and their attendant identities and spaces. Even if policing boundaries is not explicitly intended, boundary maintenance is the effect of such responses to people's gender displays.

Boundaries and margins are an important component of both my experiences of gender and our theoretical understanding of gendering processes. I am in effect both woman and not woman. As a woman who often is a social man but who also is a woman living in a patriarchal society, I am in a unique position to see and act.

I sometimes receive privileges usually limited to men, and I sometimes am oppressed by my status as a deviant woman. I am, in a sense, an outsider within (Collins 1991). Positioned on the boundaries of gender categories, I have developed a consciousness that I hope will prove transformative (Anzaldúa 1987). In fact, one of the reasons why I decided to continue my non-participation in femininity was that my sociological training suggested that this could be one of my contributions to the eventual dismantling of patriarchal gender constructs. It would be my way of making the personal political. I accepted being taken for a man as the price I would pay to help subvert patriarchy. I believed that all of the inconveniences I was enduring meant that I actually was doing something to bring down the gender structures that entangled all of us.

Then, I read Lorber's (1994) *Paradoxes of Gender* and found out, much to my dismay, that I might not actually be challenging gender after all. Because of the way in which doing gender works in our two-and-only-two system, gender displays are simply read as evidence of one of the two categories. Therefore, gender bending, blending, and passing between the categories do not question the categories themselves. If one's social gender and personal (true) gender do not correspond, then this is irrelevant unless someone notices the lack of congruence.

This reality brings me to a paradox of my experiences. First, not only do others assume that I am one gender or the other, but I also insist that I *really am* a member of one of the two gender categories. That is, I am female; I self-identify as a woman. I do not claim to be some other gender or to have no gender at all. I simply place myself in the wrong category according to stereotypes and cultural standards; the gender I present, or that some people perceive me to be presenting, is inconsistent with the gender with which I identify myself as well as with the gender I could be "proven" to be. Socially, I display the wrong gender; personally, I identify as the proper gender.

Second, although I ultimately would like to see the destruction of our current gender structure, I am not to the point of personally abandoning gender. Right now, I do not want people to see me as genderless as much as I want them to see me as a woman. That is, I would like to expand the category of "woman" to include people like me. I, too, am deeply embedded in our gender system, even though I do not play by many of its rules. For me, as for most people in our society, gender is a substantial part of my personal identity (Howard and Hollander 1997). Socially, the problem is that I do not present a gender display that is consistently read as feminine. In fact, I consciously do not participate in the trappings of femininity. However, I do identify myself as a woman, not as a man or as someone outside of the two-and-only-two categories.

Yet, I do believe, as Lorber (1994) does, that the purpose of gender, as it currently is constructed, is to oppress women. Lorber analyzed gender as a "process of creating distinguishable social statuses for the assignment of rights and responsibilities" that ends up putting women in a devalued and oppressed position (1994, 32). As Martin put it, "Bodies that clearly delineate gender status facilitate the maintenance of the gender hierarchy" (1998, 495).

For society, gender means difference (Lorber 1994). The erosion of the boundaries would problematize that structure. Therefore, for gender to operate as it currently does, the category "woman" is expanded to include people like me. The maintenance of the gender structure is dependent on the creation of a few categories that are mutually exclusive, the members of which are as different as possible (Lorber 1994). It is the clarity of the boundaries between the categories that allows gender to be used to assign rights and responsibilities as well as resources and rewards.

It is that part of gender—what it is used for—that is most problematic. Indeed, is it not *patriarchal*—or, even more specifically, *heteropatriarchal*—constructions of gender that are actually the problem? It is not the differences between men and women, or the categories themselves, so much as the meanings ascribed to the categories and, even more important, the hierarchical nature of gender under patriarchy that is the problem (Johnson 1997). Therefore, I am rebelling not against my femaleness or even my womanhood; instead, I am protesting contemporary constructions of femininity and, at least indirectly, masculinity under patriarchy. We do not, in fact, know what gender would look like if it were not constructed around heterosexuality in the context of patriarchy. Although it is possible that the end of patriarchy would mean the end of gender, it is at least conceivable that something like what we now call gender could exist in a postpatriarchal future. The two-and-only-two categorization might well disappear, there being no hierarchy for it to justify. But I do not think that we should make the assumption that gender and patriarchy are synonymous.

Theoretically, this analysis points to some similarities and differences between the work of Lorber (1994) and the works of Butler (1990), Goffman (1976, 1977), and West and Zimmerman (1987). Lorber (1994) conceptualized gender as social structure, whereas the others focused more on the interactive and processual nature of gender. Butler (1990) and Goffman (1976, 1977) view gender as a performance, and West and Zimmerman (1987) examined it as something all of us do. One result of this difference in approach is that in Lorber's (1994) work, gender comes across as something that we are caught in, something that, despite any attempts to the contrary, we cannot break out of. This conclusion is particularly apparent in Lorber's argument that gender rebellion, in the context of our two-and-only-two system, ends up supporting what it purports to subvert. Yet, my own experiences suggest an alternative possibility that is more in line with the view of gender offered by West and Zimmerman (1987): If gender is a product of interaction, and if it is produced in a particular context, then it can be changed if we change our performances. However, the effects of a performance linger, and gender ends up being institutionalized. It is institutionalized, in our society, in a way that perpetuates inequality, as Lorber's (1994) work shows. So, it seems that a combination of these two approaches is needed.

In fact, Lorber's (1994) work seems to suggest that effective gender rebellion requires a more blatant approach—bearded men in dresses, perhaps, or more active responses to misattribution. For example, if

I corrected every person who called me "Sir," and if I insisted on my right to be addressed appropriately and granted access to women-only spaces, then perhaps I could start to break down gender norms. If I asserted my right to use public facilities without being harassed, and if I challenged each person who gave me "the look," then perhaps I would be contributing to the demise of gender as we know it. It seems that the key would be to provide visible evidence of the nonmutual exclusivity of the categories. Would *this* break down the patriarchal components of gender? Perhaps it would, but it also would be exhausting.

Perhaps there is another possibility. In a recent book, *The Gender Knot,* Johnson (1997) argued that when it comes to gender and patriarchy, most of us follow the paths of least resistance; we "go along to get along," allowing our actions to be shaped by the gender system. Collectively, our actions help patriarchy maintain and perpetuate a system of oppression and privilege. Thus, by withdrawing our support from this system by choosing paths of greater resistance, we can start to chip away at it. Many people participate in gender because they cannot imagine any alternatives. In my classroom, and in my interactions and encounters with strangers, my presence can make it difficult for people not to see that there *are* other paths. In other words, following from West and Zimmerman (1987), I can subvert gender by doing it differently.

For example, I think it is true that my existence does not have an effect on strangers who assume that I am a man and never learn otherwise. For them, I do uphold the two-and-only-two system. But there are other cases in which my existence can have an effect. For example, when people initially take me for a man but then find out that I actually am a woman, at least for that moment, the naturalness of gender may be called into question. In these cases, my presence can provoke a "category crisis" (Garber 1992, 16) because it challenges the sex/gender binary system.

The subversive potential of my gender might be strongest in my classrooms. When I teach about the sociology of gender, my students can see me as the embodiment of the social construction of gender. Not all of my students have transformative experiences as a result of taking a course with me; there is the chance that some of them see me as a "freak" or as an exception. Still, after listening to stories about my experiences with gender and reading literature on the subject, many students begin to see how and why gender is a social product. I can disentangle sex, gender, and sexuality in the contemporary United States for them. Students can begin to see the connection between biographical experiences and the structure of society. As one of my students noted, I clearly live the material I am teaching. If that helps me to get my point across, then perhaps I am subverting the binary gender system after all. Although my gendered presence and my way of doing gender might make others—and sometimes even me—uncomfortable, no one ever said that dismantling patriarchy was going to be easy.

Notes

1. Ethnomethodology has been described as "the study of commonsense practical reasoning" (Collins 1988, 274). It examines how people make sense of their everyday experiences. Ethnomethodology is particularly useful in studying gender because it helps to uncover the assumptions on which our understandings of sex and gender are based.

2. I obviously have left much out by not examining my gendered experiences in the context of race, age, class, sexuality, region, and so forth. Such a project clearly is more complex. As Weston pointed out, gender presentations are complicated by other statuses of their presenters: "What it takes to kick a person over into another gendered category can differ with race, class, religion, and time" (1996, 168). Furthermore, I am well aware that my whiteness allows me to assume that my experiences are simply a product of gender (see, e.g., hooks 1981; Lucal 1996; Spelman 1988; West and Fenstermaker 1995). For now, suffice it to say that it is my privileged position on some of these axes and my more disadvantaged position on others that combine to delineate my overall experience.

3. In fact, such experiences are not always limited to encounters with strangers. My grandmother, who does not see me often, twice has mistaken me for either my brother-in-law or some unknown man.

4. My experiences in rest rooms and other public spaces might be very different if I were, say, African American rather than white. Given the stereotypes of African American men, I think that white women would react very differently to encountering me (see, e.g., Staples [1986] 1993).

5. I also have noticed that there are certain types of rest rooms in which I will not be verbally challenged; the higher the social status of the place, the less likely I will be harassed. For example, when I go to the theater, I might get stared at, but my presence never has been challenged.

6. An anonymous reviewer offered one possible explanation for this. Women see women's rest rooms as their space; they feel safe, and even empowered, there. Instead of fearing men in such space, they might instead pose a threat to any man who might intrude. Their invulnerability in this situation is, of course, not physically based but rather socially constructed. I thank the reviewer for this suggestion.

References

Anzaldúa, G. 1987. *Borderlands/La Frontera.* San Francisco: Aunt Lute Books.

Berger, P. 1963. *Invitation to sociology.* New York: Anchor.

Butler, J. 1990. *Gender trouble.* New York: Routledge.

Collins, P. H. 1991. *Black feminist thought.* New York: Routledge.

Collins, R. 1988. *Theoretical sociology.* San Diego: Harcourt Brace Jovanovich.

Devor, H. 1989. *Gender blending: Confronting the limits of duality.* Bloomington: Indiana University Press.

Feinberg, L. 1996. *Transgender warriors.* Boston: Beacon.

Garber, M. 1992. *Vested interests: Cross-dressing and cultural anxiety.* New York: HarperPerennial.

Gardner, C. B. 1995. *Passing by: Gender and public harassment.* Berkeley: University of California.

Garfinkel, H. 1967. *Studies in ethnomethodology.* Englewood Cliffs, NJ: Prentice Hall.

Goffman, E. 1959. *The presentation of self in everyday life.* Garden City, NY: Doubleday.

———. 1963. *Stigma.* Englewood Cliffs, NJ: Prentice Hall.

———. 1967. *Interaction ritual.* New York: Anchor/Doubleday.

———. 1976. Gender display. *Studies in the Anthropology of Visual Communication* 3:69–77.

———. 1977. The arrangement between the sexes. *Theory and Society* 4:301–31.

hooks, b. 1981. *Ain't I a woman: Black women and feminism.* Boston: South End Press.

Howard, J. A., and J. Hollander. 1997. *Gendered situations, gendered selves.* Thousand Oaks, CA: Sage.

Johnson, A. G. 1997. *The gender knot: Unraveling our patriarchal legacy.* Philadelphia: Temple University Press.

Kessler, S. J., and W. McKenna. 1978. *Gender: An ethnomethodological approach.* New York: John Wiley.

Krieger, S. 1991. *Social science and the self.* New Brunswick, NJ: Rutgers University Press.

Lorber, J. 1994. *Paradoxes of gender.* New Haven, CT. Yale University Press.

———. 1996. Beyond the binaries: Depolarizing the categories of sex, sexuality, and gender. *Sociological Inquiry* 66:143–59.

Lucal, B. 1996. Oppression and privilege: Toward a relational conceptualization of race. *Teaching Sociology* 24:245–55.

———. 1997. "Hey, this is the ladies' room!": Gender misattribution and public harassment. *Perspectives on Social Problems* 9:43–57.

Martin, K. A. 1998. Becoming a gendered body: Practices of preschools. *American Sociological Review* 63:494–511.

Mills, C. W. 1959. *The sociological imagination.* London: Oxford University Press.

Myhre, J. R. M. 1995. One bad hair day too many, or the hairstory of an androgynous young feminist. In *Listen up: Voices from the next feminist generation,* edited by B. Findlen. Seattle, WA: Seal Press.

Reinharz, S. 1992. *Feminist methods in social research.* New York: Oxford University Press.

Richardson, L. 1994. Writing: A method of inquiry. In *Handbook of qualitative research,* edited by N. K. Denzin and Y. S. Lincoln. Thousand Oaks, CA: Sage.

Riemer, J. W. 1977. Varieties of opportunistic research. *Urban Life* 5:467–77.

———. 1988. Work and self. In *Personal sociology,* edited by P. C. Higgins and J. M. Johnson. New York: Praeger.

Spelman, E. V. 1988. *Inessential woman: Problems of exclusion in feminist thought.* Boston: Beacon.

Staples, B. 1993. Just walk on by. In *Experiencing race, class, and gender in the United States,* edited by V. Cyrus. Mountain View, CA: Mayfield. (Originally published 1986)

West, C., and S. Fenstermaker. 1995. Doing difference. *Gender & Society* 9:8–37.

West, C., and D. H. Zimmerman. 1987. Doing gender. *Gender & Society* 1:125–51.

Weston, K. 1996. *Render me, gender me.* New York: Columbia University Press.

Zimmerman, D. H. 1992. They were all doing gender, but they weren't all passing: Comment on Rogers. *Gender & Society* 6:192–98.

Introduction to Reading 3

Sociologists Georgiann Davis and Sharon Preves are at the cutting-edge of intersex theory and activism. In this reading, they bring their deep understanding together to explore what intersex is and how intersex advocacy emerged and developed in the United States. Intersex is a natural physical variation occurring in approximately 1 of every 2,000 births worldwide. The majority of intersex traits are not harmful. However, in the United States, intersex has been medicalized and intersex people have commonly been subjected to dangerous "normalization" surgeries and treatments in an effort on the part of medical providers to fit intersex bodies into the two-and-only-two sexes (female or male) binary. The intersex rights movement began in the late 1980s to challenge the medical establishment and has rapidly grown into a global movement. Davis and Preves detail the struggles of intersex advocates to challenge the ethics of normalization surgeries and, on a broader scale, to unsettle the sex binary itself.

1. How does the reality of intersex demonstrate the flaws of binary thinking about sex?
2. What is the terminology debate, and why is the language of intersex important?
3. What is the relationship between the intersex rights movement and other movements for gender and sexual equality?

Reflecting on Intersex

25 Years of Activism, Mobilization, and Change

Georgiann Davis and Sharon Preves

Introduction: The Social Construction of Intersex as a Medical Problem

"A pregnancy test?" I, Georgiann Davis, was so confused. Before the medical scheduler would even agree to arrange the endocrinology consultation that my primary care provider had requested, she insisted that I needed a slew of lab work—eleven orders to be exact: progesterone, leutinizing hormone, prolactin, testosterone, free T4, vitamin D 1,25-dihydroxy, phosphorus, estradiol, glycohemoglobin, TSH ultrasensitive, and serum qualitative pregnancy. I asked the medical scheduler again, but this time with obvious frustration: "Why a pregnancy test? That makes no sense. I can't get pregnant." Apologetically the medical scheduler explained that the endocrinologist required the results of my pregnancy test before even allowing her to schedule a consultation. While a pregnancy test might seem like a harmless and routine test for a medical provider to require of a thirty-four-year-old woman seeking an endocrinology consultation, I'm not your average woman. You might be thinking that I am trans*, but I'm not. I'm an intersex queer woman and a sociologist who studies intersex. I'm also the 2014–2015 president of the AIS-DSD Support Group, one of the largest intersex support groups in the world.[1] I was born with complete androgen insensitivity syndrome (CAIS), an intersex trait that was diagnosed in the mid-1990s. I later learned the mid-1990s was also the same point in history when the intersex rights movement was in its infancy in the United States. I have XY chromosomes and a vagina but no uterus. I had testes, but they were removed when I was a teenager. My parents agreed to this medically unnecessary surgery because my medical providers suggested that doing so would minimize my risk of cancer—a claim that is not empirically supported (Nakhal et al. 2013). Pregnancy is biologically impossible in my body, so the pregnancy test made no sense. I find my experiences with medical care, then and now, unnecessarily frustrating and humiliating, which leaves me asking, with a mentor, colleague, and friend, sociologist Sharon Preves, how much has intersex medical care, and the advocacy that seeks to critically examine and disrupt it, changed over the past twenty-five years, and how much has it stayed the same?[2]

Intersex is a natural physical variation occurring in approximately 1 of every 2,000 births worldwide. The term *intersex* represents the "I" in the acronym LGBTI and refers to the diversity in physical sex development that differs from typical female or male anatomy. The "LGB" in the acronym LGBTI refers to lesbian, gay, or bisexual sexual identities, and the "T" stands for transgender or transsexual (often abbreviated as "T*"), which relates specifically to one's sense of gender identity and gender expression as feminine or masculine in a way that is not congruent with their biological female or male sex at birth. The current medical model of surgically and hormonally "correcting" intersex variations emerged primarily from the work of Johns Hopkins University psychologist John Money in the mid-1950s.

Intersex terminology emerged in the late nineteenth century and was used not only when referring to hermaphrodites, the more popular pre-twentieth-century term for intersex people, but to homosexuals as well (Epstein 1990). Today, the term *hermaphrodite* is considered derogatory by many, although not all, people with intersex traits. The term *intersex*, and its derivatives, including intersex traits, intersex conditions, and the like, is still widely used and accepted by intersex people and their families. However, as we explain later, *intersex* was renamed a *disorder of sex development* throughout the medical community at the beginning of the twenty-first century, which has caused terminological tensions in the intersex community.

In contemporary Western societies, it is commonly understood that biological sex, which comprises chromosomes, hormones, gonads, external genitalia, and internal reproductive structures, is a simple two-category phenomenon that is naturally correlated with our gender identity. Men have penises, testes, and XY chromosomes while women have vaginas, ovaries, uteruses, and XX chromosomes. However, sex is anything but simple and one's biological sex isn't always correlated with their gender or sexual identity. For example, many people with CAIS, Georgiann included, are born with an outward female appearance, and most live their lives as women. They have vaginas, yet they also have undescended testes and XY chromosomes. Women with CAIS do not have a uterus. None of this, however, would be obvious without invasive exploratory surgeries or the power of medical technologies, such as imaging and chromosome testing, which reveal such complexities of biological sex. CAIS is only one example of an intersex trait. In fact, there are more than twenty different documented types of intersex traits. Hypospadias, for example, is an intersex trait in which the urethral opening of the penis is located along the base or shaft of the penis rather than at the tip. Some intersex traits result in externally ambiguous genitalia, but others, like CAIS or minor hypospadias, do not.

Hypospadias is quite common and has been increasing in frequency in recent decades, occurring in an estimated 1 of every 250 male births (Baskin 2012; Holmes 2011). Surgery to "correct" the position of the urethral opening to facilitate standing during urination is very common, as many medical providers view the ability to stand while urinating as central to masculine identity and social acceptance by one's peers. Note that many men with hypospadias do not identify as intersex and that historically men lived full lives with hypospadias prior to the invention of surgical "repair." Men with hypospadias often experience ongoing problems following hypospadias "repair" surgery, such as frequent urinary tract infections, narrowing of the urethral canal due to the buildup of scar tissue, and painful urination. Chronic complications resulting from surgeries to "correct" the position of the urethra are common—so common, in fact, that doctors coined the term *hypospadias cripple* to describe patients who experience ongoing and debilitating surgically induced complications (Craig et al. 2014).

Although the majority of intersex traits are not physiologically harmful, the birth of an intersex baby is often viewed as a medical emergency (see Davis and Murphy 2013; Preves 2003), a rather predictable response given that childbirth is medicalized throughout the Western world, especially in the United States where, more often than not, babies are born in hospitals under the care of medical doctors and nurses whose task is to ensure the safe delivery of a healthy baby. The issue here is that intersex traits rarely pose health concerns. Yet, because intersex bodies are viewed as unhealthy because they deviate from social expectations of what male and female bodies, especially genitalia, ought to look like, medical providers are quick to recommend and perform urgent surgical and hormonal "correction" (Davis and Murphy 2013; Preves 2003). Because childbirth occurs in a medical setting, the response to any "deviance" in a newborn's body is medical. Intersex "deviance" is medically "normalized" by surgical and hormonal interventions to create cosmetically typical female or male bodies.

Prior to the twentieth century, medical providers did not have the tools, for example surgical expertise and chromosomal testing, that they have now to "fix" intersex bodies. As Geertje Mak (2012) notes in a study of nineteenth-century hermaphrodite case histories, rather than attempt to biologically capture or prove an individual's sex, medical providers understood sex as embedded within the social, moral, and legal fabric of the individual's community through the type of occupation one held (or eventually held), the clothes one wore (or chose to wear when the individual was able to independently make such choices), and the social relationships one maintained. Sex was regarded as a social location and not a physical bodily phenomenon.

Medical advances of the twentieth century offered providers the tools to subject intersex people to "normalization" surgeries (Reis 2009; also see Warren 2014 for a discussion of an eighteenth-century surgery). These procedures are designed to "normalize" intersex bodies by erasing evidence of any sex difference that challenges a sex/gender binary. For instance, medical providers often recommend that people with CAIS undergo a gonadectomy, like Georgiann did, to remove their internal testes. Although providers justify these recommendations by claiming that removal of internal testes reduces the risk of testicular cancer, these claims are not empirically supported (Nakhal et al. 2013). Instead, as we and others have argued elsewhere, such "normalization" surgeries are not medically necessary but rather are recommended by medical practitioners in order to uphold a sex/gender binary that insists, for example, that women should not have testes (Davis 2015; Feder 2014; Holmes 2008; Karkazis 2008; Preves 2003; Reis 2009). This insistence on enforcing a sex/gender binary in the face of obvious and consistent sex/gender diversity is no doubt related to an overarching social system in which heterosexuality is deemed normative. If sexual identity were not a concern, diversity of sex development (in the case of intersex) or of gender identity (in the case of trans*) would be of far less concern to medical providers and others.

In the late 1980s and early 1990s, people with intersex traits began organizing to challenge the medically unnecessary interventions providers were performing on intersex babies and children to shoehorn intersex people into the male/female sex binary, planting the seeds of a global intersex rights movement (see Preves 2005). Such "social surgeries" were first conducted on intersex infants and children as early as the nineteenth century, if not before (see Warren 2014). Initially, as we describe in detail later, intersex activists engaged in confrontational mobilization strategies that involved public protests at medical conferences and media appearances where they shared their horrific experiences of medical trauma, notably stories about their diagnosis and the medically unnecessary and irreversible interventions they were subjected to as children. Today, intersex advocacy has shifted to a more collaborative model to promote social change; that is, a mobilization strategy where at least some intersex activists are collaborating with medical allies to bring about change in intersex medical care. This strategy of working within medicine to promote change occurs more frequently in the United States than in other countries, where it is more often than not contested as a viable strategy for changing intersex medical care.

Georgiann and Sharon have come together to write this piece as a critical reflection on intersex that explores the past, present, and potential future of U.S. intersex advocacy. We focus specifically on intersex advocacy in the United States because that is where our expertise resides. The questions we explore in this reflection are why and how did intersex advocacy come to be? In what ways is intersex advocacy different today than it was in the past? In what ways is it similar? And how might the visibility of intersex in mainstream youth media affect the lives of the next generation of intersex people?

The Rise of the Intersex Rights Movement: Challenging the Medical Treatment of Intersex, 1993–2003

Intersex is a relatively new area of sociocultural inquiry. Outside of medicine, relatively few people have studied intersex, in part due to the fact intersex people were rather invisible until the global intersex rights movement was formed toward the end of the twentieth century. One reason for this invisibility is that when providers told people that they were intersexed, and they often did not, they also typically informed them that their anatomical differences were extremely rare and that they were unlikely to ever meet another person with a similar anatomical trait. Providers commonly withheld the intersex diagnosis from their patients, lying to them to allegedly protect their gender identity development (i.e., how young children develop a sense of self as feminine or masculine). Medical providers encouraged their patients' parents to do the same, an experience Georgiann knows firsthand. When Georgiann was a teenager, she had surgery to remove what she was told by her providers and parents were precancerous, underdeveloped ovaries. In actuality, as mentioned earlier, providers removed her internal testes for a medically unnecessary reason: to ensure that a girl didn't have testes. Georgiann's testes were producing the majority of her body's sex hormones. By removing them, providers left her dependent on synthetic hormone replacement therapy for the rest of her life to replace what her testes were already producing naturally. These hormones are essential to prevent people from developing osteoporosis or other potentially debilitating physical ailments.

Intersex medicalization gained the attention of feminist scholars in the early 1990s. For example, in a 1993 article titled "The Five Sexes: Why Male and Female Are Not Enough," biologist Anne Fausto-Sterling refuted the widely accepted assumption that sex was a simple two-category phenomenon consisting only of "females" and "males." If we are going to categorize people into sex categories, Fausto-Sterling maintained in a tongue-to-cheek tone, then we must expand the sex binary to include true-hermaphrodites, male pseudo-hermaphrodites, and female pseudo-hermaphrodites. Social psychologist Suzanne Kessler (1998) further warned that the expansion of biological sex to five categories wouldn't suffice, for it rested on the assumption that people's sex could indeed be categorized. Rather than expand the available sex categories, Kessler argued for the recognition of the diversity of sex development. Later, in 2000, Fausto-Sterling accepted Kessler's critique in a piece she titled "The Five Sexes, Revisited."

By the early 2000s, Sharon was well on her way to documenting how intersex people were treated by medical providers and, more generally, how they live with their intersex traits. It was 1993 when Sharon was a first-year medical sociology doctoral student at the University of California, San Francisco, when she was assigned Fausto-Sterling's "Five Sexes" article in a Gender and Science seminar. She was simultaneously enrolled in a seminar on Medicine and the Family that semester for which she began a literature review to explore how parents made sex assignment and gender rearing decisions when their children were born intersex. What she found was a complete lack of discussion of this topic, or of intersex at all, in the sociology, social work, and psychology literature. When she extended her research to the medical literature, Sharon was shocked to find numerous reports of surgical sex assignment on seemingly healthy infants and children. These reports focused on the physical, rather than the psychosocial, outcomes of medical intervention, and many of them contained disturbing, grainy black and white photos of children's genitals or full naked bodies with their eyes blocked out (in an apparent attempt to protect their identities). Curiously, the majority of these publications didn't report long-term longitudinal follow-up with the patients about their gender and sexual identities or any quality-of-life measures; they were primarily limited to preadolescent reports. It was these alienating photographs coupled with a complete lack of quality-of-life outcomes that compelled Sharon to search further for the voices and stories behind these photos. She decided to document the experiences of intersex adults, including their long-term quality of life and psychosocial health. As a result of her systematic sociological analysis, Sharon produced a number of publications, including her book *Intersex and Identity: The Contested Self* (2003). This book provided the very first in-depth account of intersex experiences. It was in *Intersex and Identity* that we learned that contemporary intersex people felt isolated and stigmatized by medical providers—feelings that were minimized when these same people were able to connect with others who were intersex to offer peer support. We also learned that intersex people felt physically and emotionally harmed by the irreversible intersex "normalization" interventions of early surgery, ongoing examinations, and hormone treatments.

Although a handful of intersex people and their parents were connecting through support groups in the 1980s, the U.S. intersex community truly emerged in the early 1990s after Bo Laurent, using the pseudonym Cheryl Chase to protect her identity, founded the Intersex Society of North America (ISNA). Chase founded ISNA by publishing a letter to the editor of the journal *The Sciences* (Chase 1993). She wrote this letter in direct response to Fausto-Sterling's article "The Five Sexes." In her letter, Chase critiqued intersex medical sex assignment as destructive, raising concerns about the ethics and effectiveness of surgical procedures that impair sexual and psychological function. In the last line of her letter, Chase noted her affiliation with ISNA, an organization she fabricated in that very letter to increase her legitimacy. In her signature line, Chase listed a mailing address for ISNA at a San Francisco post office box. Much to her surprise, she soon began receiving mail from intersex people around the world and decided to form the Intersex Society of North America in earnest.

ISNA published the first issue of its newsletter, cleverly titled *Hermaphrodites with Attitude,* in the winter of 1994 (Intersex Society of North America 1994). By the time this first issue was published, ISNA had already established a mailing list that included recipients in fourteen of the United States and five countries. The political content of the publication, and the organization itself, worked to transform intersex, including the word *hermaphrodite,* from being a source of shame into a source of pride and empowerment. In other words, intersex activists were reclaiming intersex and hermaphrodite terminologies. The newsletter consisted primarily of personal stories, essays, poetry, and humor, providing formerly isolated individuals with the means to connect with others who had similar experiences. *Hermaphrodites with Attitude* was published from 1994 to 1999.

In addition to its newsletter, ISNA also provided support groups, a popular website, and annual retreats. Early on, ISNA's mission was divided between providing peer support to its members and its objective of medical reform. While other intersex organizations chose to address the mission of support as their primary focus, ISNA ultimately decided to pursue social change. The political action of ISNA members alienated them from some other intersex people and groups.

ISNA made deliberate appeals to queer activists, press outlets, and medical organizations, framing intersex as an issue of gender and sexuality. Lesbian, gay, bisexual, and transgender activist organizations, both in and outside of medicine, could easily relate to intersex grievances of stigma, shame, and alienation. At the same time, aligning intersex issues with sexual or gender minorities compromised intersex activists' ability to establish credibility with the non-LGBT medical mainstream, who viewed heterosexual normalcy as one of the primary objectives of intersex medical sex assignment.

In September 1996, former U.S. House Representative Patricia Schroeder's (D. Colorado) anti–female genital mutilation (FGM) bill became law. This law banned genital cutting on girls under the age of 18 in the United States except in cases where "health" demands its necessity, thus allowing for intersex "emergencies" to be exempt. Press coverage of this law included a front-page article in the *New York Times*. Chase and other members of ISNA were outraged by the law's complicit endorsement of intersex genital surgeries. They began to stage protests to draw attention not only to this law's loophole but to "intersex genital mutilation" (IGM) as well (Preves 2003, 2005).

In addition to lobbying members of Congress to extend the anti-FGM bill to include IGM, ISNA staged protests at medical conferences. ISNA's first major protest was at the 1996 American Academy of Pediatrics meeting in Boston. Members of ISNA joined with noted trans* activist Riki Anne Wilchins and members of Transsexual Menace for this event, collectively calling themselves "Hermaphrodites with Attitude" (HWA). They picketed the conference after intersex activists were denied floor time to address the doctors in attendance. ISNA representatives used the name HWA frequently during the 1990s when they engaged in protests (Preves 2003, 2005). This historic 1996 protest in Boston propelled the American Academy of Pediatrics to create a position statement on infant and childhood genital surgery (Committee on Genetics 2000). By 1997, the broader medical community began engaging in a debate about best practices for intersex infants and children, largely in response to the first reports of David Reimer's unsuccessful sex/gender reassignment, which served to discredit the validity of what is now known as the "optimum gender of rearing" (OGR) model (Money, Hampson, and Hampson 1957). The OGR model "held that *all* sexually ambiguous children should—indeed *must*—be made into unambiguous-looking boys or girls to ensure unambiguous gender identities" (Dreger and Herndon 2009:202).

David Reimer and his identical twin brother were born in 1965, as typical, non-intersex boys. During a circumcision accident at the age of eight months, David's penis was tragically burned off by electrocautery. His devastated parents worked with psychologist John Money, the primary clinician who developed the OGR model, to help their child live a healthy life. Dr. Money suggested bringing about optimal gender identity development through a surgical castration and social reassignment of David as female when he was twenty-two months old. For decades following his reassignment, the medical intervention on intersex children relied on the apparent successful outcome of this case until David spoke out against his sex reassignment in 1997 (Colapinto 1997, 2000). David had rejected the female-feminine gender that he had been assigned and had been living as a boy since the age of fourteen. He reported that the treatments that were intended to bring about a feminine gender identity were, in fact, a cause of great stigma, isolation, and shame. Despite being a very private person, he was motivated to speak out publicly after learning that other children were being subjected to the same treatments he received and that his case had been lauded as evidence of the success of sex reassignment in early childhood. Many intersex adults also decry their childhood medical sex assignment when they grow up to identify as a gender different than their surgical sex. Many of these intersex adults choose to physically transition their sex, as David Reimer did and many trans*-identified individuals do, so that their sex is congruent with their gender identity. The rate of intersex adults that are also trans* isn't well known. In Sharon's 2003 study, nearly 25% of her interviewees were living in a gender different from their medical sex assignment.

Medical debates about the efficacy of surgical and hormonal sex assignment of intersex children quickly followed the headlines of Reimer's male identity and the apparent failure of Dr. Money's optimal gender rearing model (Preves 2005). These debates were quite polarized and framed the issue at hand as whether to perform *immediate* or *delayed* medical treatment; that is, these discussions focused on *when* and not *whether* to intervene, and many physicians felt that they were being put on the defensive. In more recent years, some physicians have begun to advocate watchful waiting rather than emergency medical intervention in an appeal for additional and more systematic longitudinal research on intersex children and adults.

This debate came to a head in 2000 and was described as a crisis in medicine by physicians who had formerly considered this treatment to be in the best interest of intersex children and their families. The North American Task Force on Intersex was formed in 2000 with the intention of open and interdisciplinary collaboration and an aim to reach some consensus on best practices in intersex care. The membership of the task force included key players in the American Academy of Pediatrics and ISNA, as well as scholars and clinicians in many related fields (Preves 2005). While the task force was not long-lived, some of the conversations were, ultimately leading to the National Institutes of Health (NIH) issuing a program announcement in 2001 for funding dedicated to new and continued research on intersex. Well over a decade later, the NIH continues to dedicate resources to and requests for research on culturally competent care for intersex people and their families.

As ISNA sought credibility in medical circles by shedding its former confrontational "Hermaphrodites with Attitude" activism, it retooled itself to put forth an image more conducive to collaboration with medical providers. This included the publication of its new newsletter, *ISNA News,* in 2001, in place of its more radical *Hermaphrodites with Attitude* publication (Intersex Society of North America 2001). In addition to the newsletter's change in title, *ISNA News* moved away from the personal stories and humor that were commonplace in *Hermaphrodites with Attitude* to more professional and organizational concerns such as financial reports, profiles of board members, and continued coverage of medical conferences and research. This shift mirrors an overarching change within the intersex movement at the beginning of the twenty-first century when at least some intersex activists and doctors began working alongside one another for change rather than against each other as political adversaries. A mere four years after picketing outside of such conventions, Cheryl Chase began to be featured as an invited keynote speaker at prominent medical conventions (Preves 2005).

ISNA distanced itself even further from a narrative of personal medical trauma when Chase stepped down as the executive director and a non-intersex medical sociologist, Monica Casper, took the helm for one year, from 2003 to 2004. Chase stepped back in to serve as ISNA's executive director in 2004 until ISNA closed down in 2008. During her time at ISNA, Casper helped connect the intersex movement's concerns to other movements and communities, including women's health, disability rights, children's rights, sexual rights, and reproductive rights. She also helped expand ISNA's Medical Advisory Board, on which Sharon served from 2005 to 2008.

How Intersex Became a "Disorder of Sex Development": Terminological Tensions, 2004–2014

In October 2005, a few years before ISNA ceased operations, two medical providers convened a meeting in Chicago of fifty experts on intersex from around the world. This international group of experts consisted of medical specialists from various fields and two intersex activists, including Cheryl Chase. This meeting produced the very first consensus statement on the medical management of intersex conditions, which was published in various scholarly medical outlets (see Houk et al. 2006; Lee et al. 2006). According to meeting attendees, the consensus statement, which was a revision of the earlier American Academy of Pediatrics statement in 2000 (Committee on Genetics 2000), was necessary due to medical advances in intersex care and the recognition of the value of psychosocial support and patient advocacy to overall quality of life (Lee et al. 2006). This new statement made a number of recommendations, including avoiding unnecessary surgical intervention, especially cosmetic genital surgery. The authors also questioned the claim that early surgical intervention "relieves parental distress and improves attachment between the child and the parents" (Lee et al. 2006:491). Although this statement was promising, there was still no guarantee that medical professionals would follow its recommendations (and indeed, few have).

A second recommendation of the 2006 "Consensus Statement on Management of Intersex Disorders" was the call for an interdisciplinary team approach to treating individuals with intersex traits (Lee et al. 2006). This approach calls for various pediatric specialists, including endocrinology, surgery, psychiatry, and others, to collaborate when making medical recommendations and providing intersex medical care. While this team model seems like a step in the right direction away from Dr. John Money's OGR model that dominated much of the second half of the twentieth century, in *Contesting Intersex,* Georgiann questions the ability of this team model to account for the voices of intersex people and/or their parents (Davis 2015). Although the goal of this concentrated expertise is to provide a multidisciplinary approach to intersex medical care, it may work to intimidate intersex people and their parents through the illusion that every concern has been addressed by a diverse group of medical experts.

Perhaps the most controversial component of the consensus statement is the recommended shift away from intersex language and all uses of hermaphrodite terminology. The authors of the consensus statement claim that patients disapprove of such terms, and they also allege providers and parents find such language "confusing" (Lee et al. 2006:488). In place of intersex language and hermaphrodite terminology, the authors advocate for disorders of sex development (DSD) nomenclature. The introduction of DSD language created new conflict in the intersex community, which compelled Georgiann to bridge her personal and professional interests in intersex by conducting a sociological analysis of intersex in contemporary U.S. society during her doctoral studies at the University of Illinois at Chicago. As she first argued in a 2014 paper titled "The Power in a Name: Diagnostic Terminology and Diverse Experiences," many intersex people are adamantly against the DSD label due to the pathologization the word *disorder* implies (Davis 2014). A recent study that appeared in the *International Journal of Pediatric Endocrinology* shows that parents of intersex children also express dissatisfaction with DSD terminology (Lin-Su et al. 2015). However, there are intersex people, and their parents, who prefer DSD terminology because of an internalized belief, problematic or not, that the word *disorder* is an accurate medical description of the intersex body, referring to a disruption in typical gestational development (Davis 2014, 2015). A minority of intersex people feel that individuals should use whatever term makes them feel most comfortable, be it *intersex* or *DSD* (Dreger and Herndon 2009). Indeed, as we have illustrated earlier in discussing the Intersex Society's use of the phrase "hermaphrodites with attitude," language can be an immensely powerful tool used deliberately to affect emotions and one's sense, or lack of, social control.

Having flexibility around terminology might be the most strategic approach, for it allows intersex people to benefit from all that each term provides (Davis 2015). For instance, when evoking DSD terminology, intersex people could benefit from insurance access to requested medical resources, government protection against discriminatory actions in the workplace, and more positive relationships with providers, parents, and society at large due to a societal norm of empathy for those with medical abnormalities. Although it certainly is the case that people with disabilities are subjected to discrimination (Green et al. 2005; Kumar et al. 2014), they are a legally protected class of citizens. Intersex people, on the other hand, are not legally protected in any capacity, but by medicalizing their difference, this could change.[3] There are, of course, serious problems with embracing DSD language. Intersex people who prefer DSD terminology express feelings of abnormality, and, more specifically, serious doubts about their gender authenticity (Davis 2015). For example, some women with CAIS wonder if they are "really" women given they were born with XY chromosomes and testes. Indeed, it is easy to see why accepting the label of being "disordered" could have negative consequences on one's sense of self. If, however, intersex people can be flexible with diagnostic terminology, acknowledging that there may even be power embedded in seemingly pejorative labels such as DSD (e.g., in seeking health insurance coverage for medically recommended hormone replacement therapy, protection from employment discrimination, and the like), they may be empowered to use such language to their advantage. This is precisely why Georgiann argues for flexibility around terminology in her book *Contesting Intersex: The Dubious Diagnosis* (2015). To effectively view and use diagnostic terminology in this flexible manner, one must be able to see medical diagnoses as a socially constructed phenomenon (see, e.g., Jutel 2011).

Other scholars across disciplines have offered different interpretations of the "disorders of sex development" terminology. Some have openly criticized it (see, e.g., Davidson 2009; Holmes 2009; Karkazis 2008; Reis 2009; Topp 2013). Historian Elizabeth Reis (2009) has, for example, offered "divergence of sex development" as an alternative to "disorders of sex development," and communication scholar Sarah Topp (2013) has supported "differences of sex development." Philosopher Ellen Feder (2009b) has argued that "the [nomenclature] change should be understood as normalizing in a positive sense" (134). She has also more directly stated that DSD "could be understood as progressive" (Feder 2009a:226). Sociologist Alyson Spurgas (2009) warns that "the DSD/intersex debate and its associated contest over treatment protocol has consequences for embodied (and thus sexed, gendered and desiring) individuals everywhere" (118). Sharon Preves suggests adopting the phrase "diversity of sex development" when using the DSD acronym.

Medical professionals, on the other hand, have widely embraced the disorders of sex development terminology, although a minority of providers have very recently started using the term *differences* rather than *disorders* of sex development, recommended by Topp (2013). Research demonstrates that providers may have embraced the DSD language because it allowed them to reclaim their medical authority and jurisdiction over intersex, which was in jeopardy as a result of the intersex activism of the 1990s and 2000s (Davis 2015). By referring to *intersex* as a DSD, providers escape criticism from intersex activists and their allies who call for the end of intersex "normalization" surgeries—indeed, in a game of semantics, providers can now claim that they treat *disorders of sex development*, not *intersex*. In 2010, only four years after the consensus statement was published, several medical providers noted that DSD nomenclature had successfully replaced intersex language and hermaphrodite terminology (Aaronson and Aaronson 2010; Hughes 2010a, 2010b; Pasterski, Prentice, and Hughes 2010). But this was never Chase's goal when she participated in the 2005 Chicago "consensus meeting." Rather, by adopting DSD terminology, she had hoped that "disorders of sex development" language would replace only terminology that used or incorporated forms of the word *hermaphrodite*, such as *male pseudo-hermaphrodite*, not all uses of the term *intersex*. While Hughes and other providers frame this renaming as an all-encompassing and victimless victory, we must not forget that they are speaking from a medical perspective rather than from the perspective of those personally affected by intersex traits. This shift in diagnostic language has had major implications throughout the global intersex community, marking a shift in intersex advocacy from collective confrontation against the medical profession to contested collaboration with the medical community (Davis 2015).

This change in nomenclature spawned the death of ISNA and the birth of a new organization in 2008: Accord Alliance. Accord Alliance was also formed by Cheryl Chase in collaboration with other former ISNA leaders and allies. When she co-founded this new organization, Chase used her legal name, Bo Laurent, rather than her pseudonym and formally retired the activist pseudonym "Cheryl Chase." Whereas Accord Alliance embraces the disorders of sex development terminology, many other intersex organizations, activists, scholars, and even some clinicians do not. This debate over terminology is currently a very heated issue among people concerned with this topic. Today Accord Alliance continues to work alongside medical professionals to help educate and build bridges between parents with intersex children and medical providers who specialize in this field (www.accordalliance.org).

A second intersex organization that formed to fill ISNA's void is Advocates for Informed Choice (AIC), with its youth advocacy program Inter/Act (http://aiclegal.org/). AIC was formed in 2006 by legal advocate Anne Tamar-Mattis, the partner of Dr. Suegee Tamar-Mattis, a physician who happens to be intersex. AIC is first and foremost a legal advocacy organization fighting for the human rights of intersex children. AIC also sponsors several programs, including the Interface Project, which is an advocacy campaign curated by Jim Ambrose, a long-time intersex advocate who was involved with the ISNA in the 1990s. Ambrose's Interface Project features brief first-person video accounts of people with intersex traits discussing their experiences and belief that "No Body Is Shameful." (These videos are reminiscent of the "It Gets Better" video campaign.) With representation from around the world, fourteen intersex people, including Georgiann, have contributed their voices via these brief autobiographical video accounts (www. interfaceproject.org).

Since ISNA's closure in 2008, the AIS-DSD Support Group has grown into one of the largest intersex support groups in the world, with membership now extending to those with intersex traits other than androgen insensitivity syndrome. Sherri Groveman Morris, an intersex woman, founded the AIS-DSD Support Group in 1995 so that women with androgen insensitivity syndrome wouldn't have to face their diagnosis alone. Organizational membership was initially only open to women with androgen insensitivity syndrome. The organization was, at the time it was formed, named the Androgen Insensitivity Syndrome Support Group-USA (AISSG-USA), after its sister intersex support group in the United Kingdom, AISSG-UK. In 2010, the then-named AISSG-USA started hosting a Continuing Medical Education (CME) event the day before its annual conference, during which medical experts on intersex would share their knowledge with interested parties, including medical residents, other physicians, and even some intersex activists. AISSG-USA became the AIS-DSD Support Group in the summer of 2011, when it launched its new website (www.aisdsd.org). This organizational name change made the group more inclusive by extending its membership to people with intersex traits besides AIS.

On January 1, 2014, Georgiann started a two-year term as president of the AIS-DSD Support Group. Since her presidency, Georgiann has worked with the AIS-DSD Support Group Board to continue to diversify the organization and has since witnessed several transformations. First, in late 2014, the AIS-DSD Support Group voted to open membership to anyone personally affected by intersex, regardless of gender identity or expression—that is, opening up group membership to men with intersex traits and to people with intersex traits who identify as genderqueer (those who reject conventional gender roles and expectations) or who reject gender labels altogether. This was a remarkable development as the organization previously had a strict women-only policy, with the exception of male parents of children with AIS. This change was inspired by a keynote address that Bo Laurent (the activist formerly known as Cheryl Chase) delivered at the 2012 AIS-DSD Support Group annual meeting that challenged the organization to include men born with intersex traits. This challenge became personal in 2013, when a teen member of the AIS-DSD Support Group decided to gender transition. Without changing the existing policy, this young man would not have been able to attend the annual meeting the following year. Many members considered it unethical and inappropriate to deny support to this young man and his family because he had gender transitioned, a sentiment that played a significant role in expanding organizational membership to include all intersex people, not just those who identify as women. Second, also in 2014, the AIS-DSD Support Group took a terminological stance and replaced "disorders of sex development" with "differences of sex development" language across their website and other publications. This change was made to prevent the problematic pathologization inherent within the "disorders of sex development" terminology. Third, in 2015, the AIS-DSD Support Group Board convened a Diversity Committee, which is committed to increasing the racial, ethnic, and socioeconomic diversity of its membership. Currently, the Diversity Committee is in the early stages of formulating concrete action plans.

Accord Alliance, AIC, and the AIS-DSD Support Group are just three of many intersex support and advocacy groups that exist around the world. We have chosen to focus on these three intersex advocacy organizations because they are among the most visible, especially in the United States. (With its online biographical video Interface Project, AIC also has international representation, and the AIS-DSD Support Group conference has attendees from outside the United States.) Although each of these organizations serves a unique purpose in the intersex rights movement, they share the goal of improving the lives of intersex people and their families. AIC and the AIS-DSD Support Group often collaborate with one another. For example, AIC and the AIS-DSD Support Group have overlapping leadership, and AIC has sponsored the AIS-DSD Support Group's youth program for the past few years. Each of these three organizations has experienced tremendous growth in recent years, in part due to the void created by ISNA's closure in 2008. Despite their collaboration and shared goals, they remain independent organizations with varying perceptions about the effectiveness of partnering with medical professionals to promote change in intersex medical care.

The Future of the Intersex Movement and Medical Care: 2015 and Beyond

Intersex individuals came together to form the U.S. intersex rights movement in the late 1980s and early 1990s. They mobilized in order to change intersex medical care that was based on John Money's "optimum gender of rearing" model. These activists represented the first generation of adults who had been subjected to intersex "normalization" surgeries and the secrecy and shame that surround such treatment. They demanded that medical providers stop performing "normalization" surgeries on intersex babies and young children. They also wanted providers to stop lying to intersex people about their diagnoses and encouraging parents to do the same. These activists wanted to raise public awareness about intersex through numerous venues including media appearances, talks at universities, and LGBT centers.

Although intersex babies and children are still subjected to "normalization" surgeries (see Davis 2015), it appears that now it is far less common for intersex people to be lied to about their diagnosis. Today, it seems that providers no longer instruct parents of intersex children to withhold an intersex diagnosis from their child. This is evidence of change because it likely minimizes at least some of the shame and secrecy tied to the lack of complete and honest diagnosis disclosure. However, many intersex people still struggle with their diagnosis, with some questioning their gender authenticity (Davis 2015).

The ethics of intersex "normalization" surgery, including medical liability for performing such interventions, is currently being determined in the courtroom. In 2013, Pamela and John Mark Crawford filed a lawsuit, in both federal and state courts, against the South Carolina Department of Social Services, the Greenville Hospital System, the Medical University of South Carolina, and specifically named individual employees on behalf of their adopted eight-year-old son, M.C., who was born with an intersex trait (Project Integrity 2013). Before the Crawfords adopted M.C., he was in the South Carolina foster care system, where medical providers, with the support of social service employees, performed surgery on him at the age of sixteen months to address his intersex trait. According to the lawsuit, the surgery removed "healthy genital tissue" with the result of feminizing his body, "potentially sterilizing him and greatly reducing if not eliminating his sexual function" (Project Integrity 2013). Despite undergoing this infant medical sex assignment to make him appear outwardly female, M.C. clearly and strongly identifies as male. While the federal portion of this lawsuit has been dismissed in court, the state suit against specific state entities and individuals is still in litigation.

Another significant recent development has been the unprecedented increase in the number of intersex people, and their parents, who are able to find and connect with each other, seek support, and organize to make up the second generation of intersex activists. This can almost exclusively be attributed to the expansion of various social media outlets, notably Facebook, Tumblr, and Twitter. For example, every active intersex social movement organization has a Facebook page that is easily located on the Internet. As another example, an intersex man created a Facebook page titled "The Commons" that allows intersex people, their parents, medical allies, and sociocultural scholars to connect with each other outside of a formal intersex social movement organization. While "The Commons" is a private Facebook group that one must be invited to join by its founder or one of its moderators, it is a virtual space for intersex people, and their allies from all around the world, to connect with each one another.

The third generation of intersex activists comprises youth born near or after 1990. These youth are coming of age at a time when there is growing societal acceptance of diverse gender and sexual identities, and also greater visibility of intersex in mainstream youth media. Many of these young people are connected through AIC's youth program titled Inter/Act, an advocacy and support program for intersex people between the ages of fourteen and twenty-five (http://interactyouth.org/). Members of Inter/Act, and its program coordinators, were heavily involved as consultants on the second (2014–2015) season of MTV's *Faking It*, a popular television show among youth. In the second season of the show, it was revealed that Lauren, one of the show's main characters, was intersex. This marked the first time that a major television program featured an intersex character. In addition, second-generation intersex activists who were, directly or indirectly, affiliated with Inter/Act, appeared in an immensely popular BuzzFeed video titled "What It's Like to Be Intersex" (BuzzFeed 2015). Within six weeks of its release on YouTube, in March 2015, this video had already amassed more than one million views. Just eight weeks later, at the time of this publication, the views number more than four million. It is likely that this visibility of intersex in mainstream youth media has profoundly impacted how younger intersex people see themselves and how their peers perceive and interact with them.

Since the formation of the U.S. intersex rights movement in the late 1980s and early 1990s, when Sharon was just beginning to shed light on the lives of intersex people and their struggles, there has been considerable change. Unfortunately, this progress has been limited primarily to the expansion of peer support and means of advocacy, notably the use of social media. Georgiann's recent work reveals that intersex people continue to be subjected to "normalization" surgeries, struggle to receive quality medical care, and experience stigma surrounding their bodily difference.

The intersex social movement is situated within the larger context of social change with regard to gender and sexual diversity. In the United States, for example, thirty-seven states legalized same-sex marriage within an eleven-year period (from 2004 to 2015), and in June 2015, the U.S. Supreme Court made same-sex marriage legal throughout the entire country. On the gender diversity front, trans* visibility and acceptance has increased in the 2010s. This social change was propelled in part by the wildly popular Netflix show *Orange Is the New Black* featuring a leading trans* character portrayed by a trans* actor (Laverne Cox). The show premiered in 2013 and Ms. Cox became a household name just a year later when she appeared on the cover of *Time Magazine* in a feature article titled "The Transgender Tipping Point" (Steinmetz 2014). In 2015, major media gave attention to other trans* celebrities, including Caitlyn (formerly Bruce) Jenner's televised coming out on *20/20* and her appearance on the cover of *Vanity Fair* magazine.

The increasing visibility of gender and sexual diversity has been coupled with recent public attention to intersex issues. This has occurred not only through the MTV show *Faking It* but also through the global attention to intersex raised by the International Olympics Committee reinstating gender verification of female athletes in 2012 and Germany becoming the first country to allow an indeterminate gender on a newborn's birth certificate. The future of the intersex movement is likely to be more global than it has in the past, given the vast power of the Internet and social media to connect people throughout the world. Our hope is that the stigma intersex people face will be diminished for future generations as they come of age at a time when there is, for the first time ever, substantial intersex visibility in the mainstream media.

Notes

1. AIS-DSD Support Group is the official name of this organization. *AIS* stands for androgen insensitivity syndrome, whereas *DSD* stands for differences (not disorders) of sex development. www.aisdsd.org.

2. We would like to thank the editors of this volume, Kay Valentine and Joan Spade, for the opportunity to write

this article. Georgiann and Sharon represent two generations of scholars who have been working separately on intersex advocacy and (de)medicalization for many years. It is an honor and joy for us to come together to write this historical narrative.

3. While intersex is not a protected legal class, the Human Rights Commission of the City and County of San Francisco issued a human rights–based investigation into the medical "normalization" of intersex people in 2005 (Arana 2005). This investigation was largely compelled by the activism of Cheryl Chase and other members of ISNA.

References

Aaronson, Ian A., and Alistair J. Aaronson. 2010. "How Should We Classify Intersex Disorders?" *Journal of Pediatric Urology* 6:443–46.

Arana, Marcus de María. 2005. *A Human Rights Investigation into the Medical "Normalization" of Intersex People: A Report of a Public Hearing by the Human Rights Commission of the City & County of San Francisco.* Retrieved from http://www.isna.org/files/SFHRC_Intersex_Report.pdf

Baskin, Laurence S., ed. 2012. *Hypospadias and Genital Development.* New York: Springer Science and Business Media.

BuzzFeed. 2015. "What It's Like to Be Intersex" [Video]. March 28. Retrieved from https://www.youtube.com/watch? v=cAUDKEI4QKI

Chase, Cheryl. 1993. "Letters from Readers." *The Sciences* (July/August):3.

Colapinto, John. 1997. "The True Story of John/Joan." *Rolling Stone,* December 11, pp. 54–73, 92–97.

_____. 2000. *As Nature Made Him: The Boy Who Was Raised as a Girl.* New York: HarperCollins.

Committee on Genetics: Section on Endocrinology and Section on Urology. 2000. "Evaluation of the Newborn with Developmental Anomalies of the External Genitalia." *Pediatrics* 106(1):138–42.

Craig, James R., Chad Wallis, William O. Brant, James M. Hotaling, and Jeremy B. Myers. 2014. "Management of Adults with Prior Failed Hypospadias Surgery." *Translational Andrology and Urology* 3(2):196–204.

Davidson, Robert J. 2009. "DSD Debates: Social Movement Organizations' Framing Disputes Surrounding the Term 'Disorders of Sex Development.'" *Liminalis—Journal for Sex/Gender Emancipation and Resistance.* Retrieved from http://www.liminalis.de/2009_03/Aitikel_Essay/Liminalis-2009-Davidson.pdf

Davis, Georgiann. 2014. "The Power in a Name: Diagnostic Terminology and Diverse Experiences. *Psychology & Sexuality* 5(1):15–27.

_____. 2015. *Contesting Intersex: The Dubious Diagnosis.* New York: New York University Press.

Davis, Georgiann, and Erin L. Murphy. 2013. "Intersex Bodies as States of Exception: An Empirical Explanation for Unnecessary Surgical Modification." *Feminist Formations* 25(2):129–52.

Dreger, Alice D., and April M, Herndon. 2009. "Progress and Politics in the Intersex Rights Movement: Feminist Theory in Action." *GLQ: A Journal of Lesbian and Gay Studies* 15(2):199–224.

Epstein, Julia. 1990. "Either/or–Neither/Both: Sexual Ambiguity and the Ideology of Gender." *Gender* 7(Spring):99–142.

Fausto-Sterling, Anne. 1993. "The Five Sexes: Why Male and Female Are Not Enough." *The Sciences* (March/April):20–25.

_____. 2000. "The Five Sexes, Revisited." *The Sciences* (July/August):18–23.

Feder, Ellen. 2009a. "Imperatives of Normality: From 'Intersex' to 'Disorders of Sex Development.'" *GLQ: A Journal of Lesbian and Gay Studies* 15(2):225–47.

_____. 2009b. "Normalizing Medicine: Between 'Intersexuals' and Individuals with 'Disorders' of Sex Development.'" *Health Care Analysis: Journal of Health Philosophy and Policy* 17(2):134–43.

_____. 2014. *Making Sense of Intersex: Changing Ethical Perspectives in Biomedicine.* Bloomington: Indiana University Press.

Green, Sara, Christine Davis, Elana Karshmer, Pete Marsh, and Benjamin Straight. 2005. "Living Stigma: The Impact of Labeling, Stereotyping, Separation, Status Loss, and Discrimination in the Lives of Individuals with Disabilities and Their Families." *Sociological Inquiry* 75(2):197–215.

Holmes, Lewis B. 2011. *Common Malformations.* New York: Oxford University Press.

Holmes, Morgan. 2008. *Intersex: A Perilous Difference.* Selinsgrove, PA: Susquehanna University Press.

_____, ed. 2009. *Critical Intersex.* Farnham, England: Ashgate.

Houk, Christopher P., Ieuan A. Hughes, S. Faisal Ahmed, Peter A. Lee, and Writing Committee for the International Intersex Consensus Conference Participants. 2006. "Summary of Consensus Statement on Intersex Disorders and Their Management." *Pediatrics* 118(2):753–57.

Hughes, Ieuan A. 2010a. "How Should We Classify Intersex Disorders?" *Journal of Pediatric Urology* 6:447–48.

_____. 2010b. "The Quiet Revolution: Disorders of Sex Development." *Best Practice & Research Clinical Endocrinology & Metabolism* 24(2):159–62.

Intersex Society of North America. 1994. *Hermaphrodites with Attitude* 1(1). Retrieved from http://www.isna.org/files/hwa/winter1995.pdf

_____. 2001. *ISNA News.* February. Retrieved from http://www.isna.org/files/hwa/feb2001.pdf

Jutel, Annemarie. 2011. *Putting a Name to It: Diagnosis in Contemporary Society.* Baltimore: Johns Hopkins University Press.

Karkazis, Katrina. 2008. *Fixing Sex: Intersex, Medical Authority, and Lived Experience.* Durham, NC: Duke University Press.

Kessler, Suzanne. 1998. *Lessons from the Intersexed.* New Brunswick, NJ: Rutgers University Press.

Kumar, Arun, Nivedita Kothiyal, Vanmala Hiranandani, and Deepa Sonpal. 2014. "Ableing Work, Disabling Workers?" *Development in Practice* 24(1):81–90.

Lee, Peter A., Christopher P. Houk, S. Faisal Ahmed, and Ieuan A. Hughes. 2006. "Consensus Statement on Management of Intersex Disorders." *Pediatrics* 118(2):488–500.

Lin-Su, Karen, Oksana Lekarev, Dix P. Poppas, and Maria G. Vogiatzi. 2015. "Congenital Adrenal Hyperplasia Patient Perception of 'Disorders of Sex Development' Nomenclature." *International Journal of Pediatric Endocrinology* 9:1–7.

Mak, Geertje. 2012. *Doubting Sex: Inscriptions, Bodies, and Selves in Nineteenth-Century Hermaphrodite Case Histories.* Manchester, England: Manchester University Press.

Money, John, Joan G. Hampson, and John L. Hampson. 1957. "Imprinting and the Establishment of Gender Role." *Archives of Neurology and Psychiatry* 77: 333–36.

Nakhal, Rola S., Margaret Hall-Craggs, Alex Freeman, Alex Kirkham, Gerard S. Conway, Rupali Arora, Christopher R. J. Woodhouse, Dan N. Wood, and Sarah M. Creighton. 2013. "Evaluation of Retained Testes in Adolescent Girls and Women with Complete Androgen Insensitivity Syndrome." *Radiology* 268(1):153–60.

Pasterski, Vickie, P. Prentice, and Ieuan A. Hughes. 2010. "Impact of the Consensus Statement and the New DSD Classification System." *Best Practice & Research Clinical Endocrinology & Metabolism* 24:187–95.

Preves, Sharon E. 2003. *Intersex and Identity: The Contested Self.* New Brunswick, NJ: Rutgers University Press.

_____. 2005. "Out of the O.R. and into the Streets: Exploring the Impact of Intersex Media Activism." *Cardozo Journal of Law & Gender* 12(1):247–88. (Reprinted from *Research in Political Sociology*.)

Project Integrity. *Advocates for Informed Choice.* Retrieved from http://aiclegal.org/programs/project-integrity/

Reis, Elizabeth. 2009. *Bodies in Doubt: An American History of Intersex.* Baltimore: Johns Hopkins University Press.

Spurgas, Alyson K. 2009. "(Un)Queering Identity: The Biosocial Production of Intersex/DSD." Pp. 97–122 in *Critical Intersex*, edited by Morgan Holmes. Farnham, England: Ashgate.

Steinmetz, Katy. 2014. "The Transgender Tipping Point." *Time Magazine* May 29.

Topp, Sarah S. 2013. "Against the Quiet Revolution: The Rhetorical Construction of Intersex Individuals as Disordered." *Sexualities* 16(1-2):180–94.

Warren, Carol A.B. 2014. "Gender Reassignment Surgery in the 18th Century: A Case Study." *Sexualities* 17(7): 872–84.

Introduction to Reading 4

Carla A. Pfeffer's research is based on in-depth interviews with nontransgender (cis) women partners of transgender men from the United States, Canada, and Australia. The results of her data analysis are reported in detail in this reading. Pfeffer's findings demonstrate the powerful potential of the relationships and identities constructed by the people she interviewed to challenge the sex/gender/sexuality binaries that have been the foundation of binary thinking and acting in Western societies such as the United States. Pfeffer applies the idea of "recognition" to show how we "do" both gender and sexuality. Her study poses a direct challenge to the belief that sexual, as well as gender, identities are fixed and natural.

1. Why does Pfeffer refer to the people she interviewed as "queer social actors"?
2. Why does the author use the concept of "recognition" instead of "passing" in her analysis?
3. How do cis women and their trans partners work to (re)define their identities in ways that challenge linguistic and social categories?
4. What is queer theory, and how is it used in this reading?

"I Don't Like Passing as a Straight Woman"

Queer Negotiations of Identity and Social Group Membership

Carla A. Pfeffer

Despite broader social acknowledgment of gender and sexual diversity, transgender individuals and their significant others remain relatively unrecognized in both mainstream and academic discourse and are often subsumed under the limited theoretical frame of social "passing" when they do appear. Building a sociological critique against overly simplified biological frameworks for understanding complex gender and sexual identities, I analyze in-depth interviews with nontransgender women partners of transgender men. The personal identifications and experiences of this group of "queer" social actors are proposed as sociopolitically distinguishable from those of other more commonly recognized sexual minority groups. Data reveal the interactive social processes that often determine "rightful" social inclusion and exclusion across gender and sexual identity categories as well as their capacities to generate and limit possibilities for social movements and political solidarity.

. . . The present study proposes cis women partners of trans men as queer social actors,[1] arguing that a more developed understanding of this understudied group may fruitfully extend sociological knowledge on contemporary sexual identity groups and communities.[2] The present work broadens the notion of "queer," as a politics established against identity, considering the ways in which "queer," as a relational subjectivity, usefully complicates our understanding of social identities and social group–based membership. In this way, the present study is a move toward theorizing particular queer social actors, identities, social embodiments, and families as embedded within intersecting normative and regulatory social systems, structures, and institutions.

An exploration of the identities and experiences of cis women partners of trans men also provokes consideration of the complex management processes involved in negotiating both individual identity and social group–based memberships. A critical aspect of these social processes is being seen or not seen, recognized or not recognized, as a rightful member of particular social identity groups with which one identifies. For trans men and their cis women partners, these meaningful social recognition processes often include (sometimes unintentional or even undesired) social "passing" with regard to gender and sexual orientation.

A problematic aspect of many of the sociological studies employing this notion of "passing" is their tendency to reinforce the presumed essentiality of sex and gender binaries by assuming that some social actors hold authentic proprietary claims over particular social identity–group membership (e.g., only those categorized as "male" at birth can be "authentic" or "real" men), while others can stake only inauthentic or false claims. Indeed, it is only under such a framework that it makes sense that some individuals might be recognized as authentically (and therefore unremarkably) "belonging" as members, while others may only hope to "pass" into relatively inauthentic membership as wannabes. Notions of "passing," therefore, tend to be predicated upon assumptions of essentialized and naturalized group difference.

. . . In this article, I draw upon Connell's (2009) notion of "recognition" (in lieu of "passing") to argue that social rights, privileges, and group membership connected to categories of sex, gender, and sexuality depend largely upon social interpellation. More specifically, I will demonstrate how gender and sexual identities are interactional accomplishments that often reveal more about the workings of normative social privilege than they reveal about the social actors whose gender and sexual identities are being (mis)-recognized. This study considers queer social actors' often strategic and pragmatic management of these (mis)recognition processes to gain access to particular social and material benefits of social group membership, offering theoretical and empirical insights on identity negotiations, and moments of "trouble" in these negotiations, across contested and regulated social categories and groups more broadly. As such, this work provides insights that actively respond to Irvine's (1994, p. 245) still-relevant

Pfeffer, C. (2014). "I don't like passing as a straight woman": Queer negotiations of identity and social group membership. *American Journal of Sociology*, *120*(1): 1–44. Reprinted with permission from The University of Chicago Press.

call to sociologists nearly two decades earlier: "Sociological theory must . . . [place] social categories such as sexuality and race in the foreground in the context of power and difference." Finally, this work proposes a sociological queer analytic framework that compels solidarity-based approaches to social movement organizing around identity-based rights.

Toward a Sociology of Queer Social Actors and Identities: Extending Theoretical and Analytical Frameworks

Emerging as a late 20th-century outgrowth of post-structuralist thought, a central analytic across much queer theory is its critique of notions of normativity, deviance, and stable/coherent identities. The interface between queer theory and sociology has been slow to develop.[3] Michael Warner, one of the key figures in the development and popularization of academic queer theory, describes social science disciplines' reticence to adopt queer theoretical frames as paradoxical given that "the analysis of normativity . . . should have become central to such disciplines" (2012, p. 8). Epstein (1994, p. 197) writes that displacement of sexual minorities to the periphery rather than the center of social inquiry has had critical limiting effects on the discipline of sociology and that "the challenge that queer theory poses to sociological investigation is precisely in the strong claim that no facet of social life is fully comprehensible without an examination of how sexual meanings intersect with it."[4]

. . . Seidman (1994) argues that one of queer theory's central and most defining contributions is the way in which it challenges taken-for-granted assumptions about the existence of a relatively stable homosexual subject and identity. Queer theory and politics embraced (rather than attempted to reconcile) the messiness and fluidity of sexual acts, boundaries, and identities. Indeed, queer politics galvanized those who shared a burgeoning sense of disenfranchisement from (and reaction against) mainstream lesbian and gay politics of "normalization," generating expressly oppositional politics informed by postmodern and deconstructionist theorizing (Seidman 2001; Bernstein 2005).

Identities as Social Process: Sociological Queer Analysis

. . . As sociologists, rather than ignoring or sidelining critical social analyses of queer social subjects, we might query: What are some of the meaningful social and political processes that regulate queer social actors' membership within, or passage through, various identity and social-membership groups? How might sociologists contribute to a project that expands beyond the textual to consider the everyday lives of queer social actors?

Valocchi (2005, p. 766) offers one possible pathway for sociology, defining "sociological queer analysis" as that which blends "a queer sensibility about the performative nature of identity with sociological sensibility about how these performances are constrained, hierarchical, and rooted in social inequality."[5] As such, one of the primary goals of the present work is to develop an expressly sociological queer analysis that focuses upon fissures and moments of trouble in culture and identity, articulating the social process through which individuals come to embrace and resist subject identities as "queer" even as these identities are (mis)recognized by social others. The discipline of sociology is perhaps uniquely well positioned to seriously consider the daily lives of queer social actors and to begin to theorize the processes through which these lives and identities are constituted, (mis)recognized, resisted, and embraced. Namaste (1994) urges sociologists to consider the social constructedness of genders and sexualities and the ways in which some are normalized (or left unmarked, as nonqueer), as well as how all social actors negotiate various identities and subject positions (and limits to these identities and subject positions).

. . . Queer social actors, like everyone else, lead lives simultaneously produced through and against normative structures of sex, gender, and sexuality. These normative structuring forces of sex, gender, and sexuality operate primarily along presumably "natural," biological, and essentialized binaries of male/female, man/woman, and heterosexual/homosexual. The lives and experiences of cis women partners of trans men, however, call these normative structuring binaries into even greater question in their failure to adequately articulate and encapsulate these queer social actors' identities and social group memberships. The experiences of queer social actors, therefore, hold the potential to rattle the very foundations upon which normative binaries rest, highlighting the increasingly blurry intersections, tensions, and overlaps between sex, gender, and sexual orientation in the 21st century (Pfeffer 2012).

Theorizing Social "(Mis)recognition" Rather Than "Passing"

The incoherence of these normative binaries becomes clearer through focus on interactional processes by which social actors are granted insider/outsider social status. When individuals refer to someone "passing" as a man or "passing" as a woman, the social meaning making that is taking place lies at the thorny intersections of sex and gender categorization, expression, attribution, and identity (for further discussion of these and other concepts, language, and terminology related to transgender identity and experience, see Wentling et al. [2008]; Pfeffer [2010]). Studies of "passing," and the social accomplishments of sex and gender, have a long, revered, and contentious history in sociology, particularly among symbolic interactionists and ethnomethodologists (see Garfinkel 1967; Goffman 1976; Kessler and McKenna 1978; West and Zimmerman 1987; Denzin 1990; Rogers 1992; Zimmerman 1992). "Passing" carries the assumption that certain individuals somehow naturally embody particular identities to which others can stake only inauthentic membership claims. In a sense, some individuals are understood as rightful "owners" of membership to particular social identity groups—most notably, those groups holding disproportionate social power and authority (Harris 1993; Calavita 2000).

The concept of passing also relies on juxtaposed notions of conscious, intentional, deceptive "dupers" and presumably natural, authentic, deceived "dupes" (Serano 2007). Nevertheless, "passing" is often held as the gold standard of "successful" transsexualism—particularly by medical establishments; as such, "passing" is often conceptualized as emblematic of normativity or a desire to *be* normative (as reviewed by Connell [2009]). Analyses of "passing" in racial and class contexts (see Harris 1993; Calavita 2000; Kennedy 2002; Ong 2005), however, adopt a more nuanced lens that views "passing" as a potentially pragmatic (though fraught) interactional strategy for accessing and attaining regulated social, material, and legal resources, and consider the personal, interpersonal, and sociopolitical effects and consequences that the use of such strategies may involve.

While "passing" may grant reprieve from the social stigma and potential danger of ambiguous gender expression, as well as access to social and material resources granted only to particular group members, this access and these reprieves are often tenuous, context specific, and revocable. Trans men who most always "pass" in ordinary social situations may live in fear about the consequences of being involved in a serious accident during which the removal of clothing (or, in some cases, the accessing of identification records indicating legal sex or gender status) would seriously impair their ability to be unambiguously recognized in accordance with their gender identity. Employing a sociological queer analysis, the concept of "passing" may be further illuminated by focusing on those ordinarily granted "natural" and unquestioned status within particular identity categories. Elson (2004, p. 172), for example, presents a compelling exploration into cis women's experience of identity posthysterectomy and whether or not those who undergo this surgical procedure are still considered (and consider themselves) "women" or not—reaching the equivocal conclusion of yes, no, maybe. As such, Elson (2004) probes and destabilizes the supposedly "natural" and essential links between biology, gender identity, and social perceptions of which bodies rightfully constitute "woman."

Connell (2009) usefully troubles the notion of "passing" to consider how "recognition" may be a more precise conceptual framework for thinking about the juxtapositions between one's body, subjective identity, social group memberships, and social appraisals of all of these. Accordingly, we would do better to supplant our biologically essentialist notions of "passing" with a more sociological notion of "recognition." By doing so, we might come to consider and recognize that trans people's efforts to "pass" occur not when living in accordance with their subjective gender identities, but as they attempt to live within gender identities normatively corresponding to their sex assignment or sex categorization (West and Zimmerman 1987, p. 133).[6] In other words, many trans men do, indeed, "pass" for much of their lives—as girls or as women. They often report struggling, within bodies and social identities that do not feel like "home," until these efforts become untenable and they take further steps to bring their bodies and social embodiments in line with their gender identity.

As this study will show, sexual identity is also a relationally formed construct, depending upon a constellation of dynamic, shifting, socially informed understandings that individuals hold about themselves and others. As Vidal-Ortiz (2002, p. 192) writes: "One interactional way in which gender and sexuality collide is as people interpret each others' attractions based on their gender presentations or expressions." Sexuality is about more than personal identities, autonomous desires, and sexual object choice alone. Rather, we "do sexuality"; our sexualities are interpellated every day, arising from social others' (mis)recognition of the ways in which we see and understand ourselves and our partners. I argue that we must further extend Connell's notion of "recognition" to attend not only to the ways in which we may come to see individuals in accordance with how they see themselves but also to the ways in which making any attribution of identity is part of the process of bringing identities into social being. In other words, by focusing on how we recognize and misrecognize others' self-identities, we come to better understand these identities not as individual and predetermined fixed entities, but as dynamic social processes. . . .

Study

Participant Recruitment and Sample

This work represents the largest and most comprehensive study conducted, to date, with cis women partners of trans men (for additional information about the size and growth of this emergent social group, see Pfeffer [2010]). Research participants were recruited using online and paper-flyer postings targeting the significant others, friends, families, and allies of trans men. Most study participants were recruited via Internet-facilitated social network ("snowball") sampling, the primary method of purposeful sampling when targeting sexual minorities and their partners (Patton 1990; Mustanski 2001; Shapiro 2004; Rosser et al. 2007). I also enlisted key informants across the United States and Canada to distribute materials to potential participants in their local regions.

I conducted interviews with 50 cis women partners of trans men for this study. Participants discussed their experiences in 61 individual relationships with trans men (several participants reported multiple relationships with trans men). Participants resided across 13 states in the United States, three Canadian provinces, and one Australian state, expanding existing work on sex and gender minorities that focuses almost exclusively on one or two states, with large urban centers, in the United States. This sample consists of participants from most of the U.S. geographic regions with the highest reported proportions of trans men (see Rosser et al. 2007), including two much underresearched regions with regard to studies of sex and gender minorities—the midwestern United States and Canada. The most frequent sexual orientation self-identification label, used by 50% of participants in this sample, was "queer." Participants' trans partners (according to participant reports) were also most likely to identify as "queer" (48%), with "heterosexual" as second most common (33%). When asked to describe how they would define or label their relationship(s) with their trans partner(s), study participants described their relationships as "queer" 65% of the time among those providing information for this question.

Despite aiming for racial and age diversity, only variation on age was successfully achieved. Interviewees' ages ranged from 18 to 51 years, with an average of 29 years, and, on average, cis women's trans partners were slightly younger. Participants largely self-identified as white. When considering the race/ethnicity of the trans partners of participants, the sample begins to reflect somewhat greater racial/ethnic variation, with 18% identified as "multiracial." Participants and their partners were highly educated (with 24% and 11%, respectively, holding postgraduate degrees) yet reported household incomes that were quite low among participants providing these data. Trans men partners of participants were at various stages of sex or gender transition—with most being just a bit over two years into the process. Most were taking testosterone, a considerable minority had had "top" surgery, while a very slim minority had had "bottom" surgery of any kind. . . .

Findings and Discussion: Doing Gender and Sexuality Through (Mis)recognition Process

Just as trans men have their own transition experiences to manage on multiple levels, so, too, do their cis women partners (see Nyamora 2004; Pfeffer 2008; Brown 2009; Joslin-Roher and Wheeler 2009; Ward 2010). Study participants relayed, in great detail, the various struggles they experienced as they sought to maintain, transform, understand, proclaim, and refute various personal and social identities in the context of their lives. The following sections present narrative data, using pseudonyms to protect participant confidentiality, illustrating the ways in which queer social actors negotiate intersecting and sometimes conflicting social identities, relationships, politics, and social groups. These narratives prompt consideration of the ways in which gender and sexual identity are interactive social accomplishments involving boundary negotiations and (mis)recognition processes that carry tangible personal and social consequences.

Language and Social "Reading"

"Queer" as a distinct social identity category.—Cis women partners of trans men frequently wondered aloud, when I asked them about their own shifting and contingent sexual identities in relation to their trans partners, "What does that make me?" Martha (25 years, Massachusetts) described the

challenge of personally struggling with issues connected to identity in the context of her relationship with her trans partner:

> I thought of myself as a dyke and then now I'm with someone who identifies as a man and I'm thinking—how do I identify now? I'm not a lesbian. . . . I'm not really perceived as queer by many other people right now. And it really messed with me for awhile—what am I? Who am I? Not that I didn't know who I was, but what identity should I give to people? A lot of times I'd try to adopt my identity as my own and it doesn't matter what other people think. But it's hard not to judge myself by other people's judgments.

Having difficulty figuring out how to self-identify was described often by participants in my sample as not only an internal struggle, but one that emerges from various social and cultural imperatives and in social interactions with others. . . .

Another participant, Linda (22 years, Sydney, Australia), explicitly rejected the social imperative to identify her relationship with her partner using particular identity labels: "All these people would go, 'Oh, what does that make you now?' And I would say, 'Happy and in love. That's all.' I didn't see why anything else has to matter." Current and former lesbian-identified respondents reported facing particular challenges in terms of identity and social/community membership and the attributions others made about their personal motivations, desires, and emotional health. As Polly (40 years, New York) noted: "If you're a lesbian, everybody works so hard to accept it. They accept it, then you fuck them up by being with a trans guy. And then they're like, 'Okay, next she's going to go to men.' That it's just this form of evolution . . . and you're just graduating in this progressive chain of eventually getting to the pinnacle of the 'real' man. I sort of feel like people see it as this progressive growth into being fully, Freudianly, 'correctly' socialized to heterosexuality." Cis women partners of trans men described facing persistent challenges in actively negotiating their own (and their partner's) shifting identities across a variety of personal, interpersonal, and social contexts. One of the ways in which this negotiation manifested for many participants was through language and determining how they would self-identify, with regard to sexuality, in the context of their unique relationships.

Just over half of the cis women participants in this study self-identified their sexual orientation as "queer" at the point of interview and about 65% described their relationship with their trans partner as "queer." According to these cis women's accounts, over 60% of their trans men partners were perceived as men in social spaces "always" or "almost always." When in public together, therefore, many cis women in this sample reported being frequently (mis)recognized as part of a heterosexual couple. Verbal evidence participants provided in their accounts of these social encounters included social others using the words "sir," "bro," "boyfriend," "husband," "dad," and "father," as well as pronouns such as "he" and "him" when referring to participants' trans partners, and use of words/pronouns such as "Miss," "Mrs.," "Ms.," "ma'am," "girl," "girlfriend," "wife," "mom," "mother," "she," and "her" when referring to the participants themselves. Several participants also described instances in which clerks "corrected" sex designators from "female/f" to "male/m" on their trans partner's paperwork or in computer records systems, remarking about how there must have been an "error in the system," upon seeing the man in front of them. This was an important example of the way in which being misrecognized (according to medical or legal systems, which serve as gatekeepers for sex marker designation changes on personal identification documents) and recognized (in accordance with one's gender identity) may go hand in hand, providing or preventing access to regulated social and material institutions (such as a marriage license).

Nonverbal indicators that trans partners were being socially "read" as men or that the couple was being "read" as heterosexual included the check being consistently handled to one's trans partner at restaurants and other service establishments, other men giving a head "nod" when passing one's trans partner on the street, being smiled at by older persons when holding hands with one's trans partner in public,[7] and not being scrutinized when in sex-segregated public spaces (such as restrooms). In these instances, (mis)recognition processes often conferred social advantage, privilege, and mainstream acceptance. Yet being (mis)-recognized as heterosexual was described as personally and socially problematic by many participants—particularly insofar as they feared being (mis)recognized as "heteronormative" by social others. Participants described their understandings of heteronormativity as fulfilling stereotypically gendered "roles" in their relationships, endorsing majoritarian politics, and not being seen as queer or politically radical.[8]

Self-identifying as "queer," among study participants, was described as a fraught (though sometimes powerfully political) solution to the inadequacy of other currently existing language choices for expressing sexual identity in the context of one's relationship with a trans partner:

> Before my ex-partner . . . I had been sort of actively claiming that I wasn't straight . . . and I was very comfortable telling people that. But I also come from a small town and the options there were very much "gay," "lesbian," "bisexual" or "straight." I didn't feel that any of those fit me. So I started saying to my friends and to whoever else, "Well, I'm not straight." But that's as far as it went . . . I hadn't had any other partners that would actually complicate that at that point . . . But [once I met my trans ex partner], it just made sense for me to think about identifying as "queer" and that felt comfortable. (Sage, 21 years, Ontario)

Sage's narrative walks us through a process of queer identity consolidation. Sage considers sexual orientation self-identification labels in the context of her own life, coming to the conclusion that none of the existing labels accurately "fit." She first chooses a new identity category rooted in disidentification with an existing identity category ("not straight"). Later, a new relational context (partnering with a trans man) serves as the impetus for self-identifying in yet a new way—adopting an identity label ("queer") that was not part of the original range of self-identification choices of which she was aware or that were available to her. . . .

As this narrative illustrates, for some, choosing to self-identify as "queer" also serves as a conscious and intentional social indicator of a political stance that explicitly resists or rejects normativity in order to imagine a different or transformed social landscape. When asked what identifying as "queer" meant to her, Ani (21 years, Ohio) stated: "I needed a language for *not* being heteronormative." These experiences stand in stark contrast to calls for a "post-queer study of sexuality" (Green 2002, p. 537) in sociology or claims that the term "queer" exists primarily to symbolize a departure from sexual identity categories (Green 2002, 2007). Rather, these participants assert "queer" as one of the few (if not the only) sexual identity categories that does not overly constrain or threaten the relationships they have with their trans partners. Participants told me that self-identifying as "lesbian" in the social world carried the possibility of invalidating their trans partner's identity as a man.

It is possible to connect some of the identity and (mis)recognition struggles of these participants with those of bisexual-identified respondents from other sociological empirical work (Burrill 2001; Wolkomir 2009; Tabatabai and Linders 2011).[9] Specifically, women across each group described being (mis)recognized by social others in ways inconsonant with their own sexual self-identifications and in ways that often shifted based upon social assessments of their partner's gender identity in relation to their own. Empirical comparisons between this sample and earlier work on bisexual-identified cis women (Blumstein and Schwartz 1974, 1977, 1990; Richardson and Hart 1981; Ault 1996a, 1996b; Rust 2000) also attest to the fluidity and dynamic potential of sexual identifications. While many participants in my study reported moving from self-identifications as "lesbian" prior to a partner's transition (or partnering with a trans man) to self-identifications as "queer," women in these earlier studies often reported self-identification as lesbian when partnered with another woman and self-identification as heterosexual when partnered with a man, discussing the ways in which shifts in the sex of one's partner resulted in shifts of group membership, community, and sense of belonging. In other words, sexual identity was understood as largely situational and context/partner/community-dependent, rather than individual, inherent, or fixed and immutable.

One primary point of difference between these groups is that among the group of cis women partners of trans men I interviewed, identification as "bisexual" was reportedly an untenable choice for many as it could introduce identity and relationship insecurity through trans partners wondering whether participants were attracted to them as a man or as a woman. Further, very few of the participants in this study self-identified as "heterosexual" (n = 2), with most participants expressly rejecting such self-identification and discussing how much they valued their connection to (and membership within) LGBTQ communities.

"Queer" as an empty signifier.—Paradoxically, another dominant theme that emerged among participants who self-identified as "queer" was the sentiment that "queer" can become so all-encompassing, as a catchall identity, that it may be in peril of becoming an empty social category. Gamson (1995) describes this tendency as the "queer dilemma." While the lack of boundedness associated with "queer," as an identity, can make it particularly appealing to those for whom other categories feel overly restrictive or inappropriate, for others this very unboundedness can feel quite confining:

> I could say I'm queer but I also am not so sure I want to signal that identity either because I feel sometimes queerness is a little irresponsible because it's just so overused that it becomes sort of meaningless. I don't even know what people [are] trying to indicate to me when they say that. So I don't know if I feel comfortable saying it. . . . I think my sexual identity doesn't have a particular proclivity or erotic choice that has anything to do with a

pre-existing terminology. . . . So I feel like in my life I slide myself into the term that worked mostly to make other people understand me—not necessarily because I feel like it really is an adequate description of who I am. (Polly, 40 years, New York)

For Polly, therefore, "queer" serves as a social identity category in which she reluctantly places herself for the purposes of becoming socially intelligible to others rather than from a sense of its personal resonance. Polly's narrative thus highlights the critical importance and paradox of social recognition with regard to queer identities. Polly adopts a label that makes her socially recognizable and interpretable to social others. This label, however, fails to fully encapsulate or accurately describe the specificity of her particular partner choices and desires.

Amber (19 years, Ontario) offered another example of the limitations of "queer" as an identity signifier: "'Queer' is such a vague term. If you say you're queer then people will often just assume that, if you're a girl, then you're a lesbian. . . . But I date men so I don't want to . . . be just kind of lost in the queer umbrella. If you're going to look at me and want to know what box I go in, put me in the right one." For Amber, then, "queer" is a category that renders her attractions to cis men invisible. Rather than being overly all-encompassing, she finds it overly restrictive and exclusionary in the context of her own attractions and desires. Both Polly and Amber articulate such as "queer." Some of these struggles, once again, echo those of expressly bisexual-identified women who often report being (mis)recognized as heterosexual when partnered with men and as lesbian when partnered with a woman, rendering their bisexual self-identifications invisible (Burrill 2001; Wolkomir 2009; Tabatabai and Linders 2011).

Cis women and their trans partners must often work to (re)define their identities—as individuals and in relationship to one another—in ways that both challenge and extend existing linguistic and social categories. Furthermore, the rising visibility and media presence of partnerships between cis women and trans men, particularly via the medium of the Internet, contributes to the emergence of queer cultural communities through which language and support may be continuously developed, challenged, and shared (see Shapiro 2004). The Internet emergence of a new linguistic identity term, "queer-straight" (which two participants in this study used to describe their relationship with their trans partner), may be one way in which sociolinguistic innovation is developing out of existing frustrations over lack of specificity and meaning with "queer."

In addition to negotiating language and identity-classificatory systems, study participants reported marked and sometimes painful discrepancies between how they see and understand themselves and how they are seen and understood (or not) by others in their social communities and contexts. Two themes that frequently emerged for cis women partners of trans men were actually flip sides of the same "(mis)recognition coin"—being (mis)recognized (or "passing") as unremarkably straight in both queer and nonqueer social spaces and becoming invisibly queer (i.e., no longer being recognized as a rightful member of the queer community) within queer social spaces. Clearly, (mis)recognition—or being "seen" and "not seen"—by various communities is a powerful social process that critically informs, validates, and invalidates personal identities and group memberships. The following sections detail these flip sides of this same coin of social group (mis)recognition and membership processes as well as describe how the cis women in this study negotiated these processes.

Identity and Social Norm Resisting and Affiliating

"I don't want to be a housewife!"—Participants often spoke explicitly about not wanting to fall into relational patterns with their partner that might be interpreted as normative. Some cis women voiced this intention directly to their trans partner—as in the case of Emma (22 years, Ontario), who spoke of a conversation during which she reportedly told him: "I am a feminist and I don't want to be a housewife. . . . That's not who I am and that's not who you're going to be in a relationship with." Some cis women and their trans partners shared in the desire to reject and resist normativity. According to Sage (21 years, Ontario): "It sort of is a little disturbing to both of us—as individuals and together—to think that we might fall into sort of a heterosexuality, a heteronormative pattern. Being queer, interacting as queer, presenting as queer, and being queer in the world is something that's really important to both of us." In a similar vein, Belinda (24 years, Ontario) explained: "We both say that it's a queer relationship. Neither of us are interested in passing as a straight couple or having people believe that we're a straight couple."

Recall that the majority of cis women's accounts include discussion of being (mis)recognized as heterosexual by social others. As such, these cis women's vocal and instrumental resistance to being socially (mis)recognized as anything but "queer" offers possibilities for destabilizing normativity insofar as it challenges social others' notions of what a "heterosexual couple" is like. Further, it reveals the ways in which participants position themselves explicitly against habituated, iterative enactments of normativity—which they explicitly counterpose to feminist and queer identities. Of course, their resistance may be limited given that opportunities to correct the social (mis)recognition of others do not always readily present themselves, may be unsuccessful, may be resisted by one's partner, or may be unsafe in certain social contexts. . . .

Axial coding of the data revealed that cis women participants more often judged themselves to more strongly reject or resist normative practices and politics than their trans partners, particularly when they self-identified as "queer." This finding might be expected when we consider that being recognized by others as male is often socially accomplished through relational enactments of normative or hegemonic masculinity (Connell 1987). In other words, trans men (like cis men) often gain social recognition of their gender identity as men when engaging in stereotypical social behaviors associated with "being a man" (see Connell 1987; Brown 2009; Pfeffer 2010; Ward 2010). While there was no difference in self-reports of enacting traditional versus nontraditional gender performances in relationships across age or sexual identity of participants, younger cis women (those under 35 years of age) more frequently worried that their relationships would be (mis)recognized as heterosexual than older cis women (those 35 years of age and older). These patterns likely reflect the influence of Third Wave feminist and queer politics in the lives of cis women under 35 years of age in this sample (see Pfeffer 2010).

"We're just another straight couple with an extra set of tits!"—Despite the fact that participants most frequently identified themselves (and their relationships) as "queer" and distanced themselves and their relationships from characterization as "heteronormative," a vocal minority made statements that could be interpreted as reflective of heteronormativity. These statements ranged from the seemingly blatant—such as that from Lily (26 years, Florida), which opens this section—to those couched in the feminist post-structuralist language of gender performativity (see Butler 1990, 1993). Axial coding of the data revealed that cis women ages 35 and older reported desires for heteronormativity more often than those younger than 35 years of age. Those cis women who reported that their trans partners were perceived socially as male "always" or "almost always" were most likely to report performing traditional enactments of gender in their relationships and to report that their trans partner embraced normativity. Cis women were also more likely to report performing traditional enactments of gender in their relationships when their partners transitioned over the course of the relationship and were trans identified when the relationship began (as opposed to those whose relationship began as lesbian or those who were with partners who had already completed most of their transition by the time the relationship began).

When Ellia (24 years, New Mexico) was asked how she would describe the type of relationship that she has with her partner, she responded: "We're just a straight couple. He's my fiancé, we're getting married, we're just a straight couple." While Ellia's description is laden with unremarkable, normative descriptors (e.g., "straight," "he," "fiancé," "married," "straight," "couple"), her invocation of the phrase, "We're *just* [my emphasis] a straight couple," twice, may be interpreted as awareness that, without defending the normativity of her partnership, her relationship may be quite unlikely to be understood by others as "just a straight couple." Margaret (29 years, Massachusetts) offered another perspective on distancing her family from counternormativity: "One of the first conversations we ever had was about kids, how many we wanted, and what the time frame was and we aligned completely. . . . Sometimes, when you're *super* radical, you get to *not* be radical! And I want our kids to have one set of parents with one last name." Margaret's conceptualization is an interesting and provocative one—it suggests that privately held queer identities (which may be socially invisible or hidden, particularly in the context of family life) remain socially radical. Furthermore, it suggests that, based on this internally held queer identity, it is possible (and perhaps even acceptable) to access certain privileges and normative institutions that do not challenge or erode the "queerness" of these privately held queer identities. Margaret acknowledges and resists normative understandings of family as she casts herself in the part of "*super* radical" and relays the negotiations and deliberations in which she and her partner have engaged with regard to having and naming children. This vision of a possible future that Margaret envisions allows her to transform normative ("*not* radical") practices of having and naming children into a "*super* radical" enterprise of queer family building.

Cis women participants also articulated their experiences enacting what some may interpret as habituated and stereotypically gendered relational structures in ways they explicitly linked to conscious gender performativity and normative resistance (Pfeffer 2012). According to Rachel (27 years, Ohio): "I think he had this fantasy . . . which I don't think exists for anybody anymore. But, in his head, part of becoming a man was becoming a *Leave It to Beaver* dad—like coming home and mom has dinner on the table and whatever else is happening. But it turns out he cleans house more than I do and he cooks more than I do. So I think, at this point, our relationship is undefinable by present terms; so I would just say, 'queer.' It's just different. It's different than anything available." Eliza (24 years, Nova Scotia) offered another example that paralleled Rachel's but also explicitly considered the importance of others' social perceptions of her relationship structure:

> We're both very sort of intrigued by 50s décor and roles and all that sort of stuff. . . . I will take on the role of housewife and, a lot of the time, it's this tongue-in-cheek sort of thing. He'll be like, "Get me a beer!" and I'll put on an apron and run off into the other room, "Here ya go, dear!" It's very sort of playful. Again, it's the performance of gender instead of *really* taking it all that seriously. But, at the same time . . . the kitchen is *my* kitchen and all this sort of stuff that's very gendered. . . . Sometimes I'm concerned that other people might not quite get it and that they might think that we're really espousing these very traditional roles. . . . I don't want to be the passive wife. . . . I'd much rather be the tough wife.

For these participants, performing normativity is a reportedly conscious dynamic that holds the potential to be simultaneously nostalgic, flexible, ironic, and difficult to define. Cis women and their trans men partners clearly engage in dynamic, relational processes that produce and validate enactments of gender in ways that may be simultaneously normative and counternormative, despite the commonly voiced concern to not be (mis)recognized as traditional or unremarkably heterosexual (for more on this, see Brown [2009]; Ward [2010]; Pfeffer [2010], [2012]).

A sociological queer analysis might also usefully trouble assertions that those in relationships with trans people must have relationships that are somehow more transgressive or counternormative than other types of relationships. As Kessler and McKenna (2003) note, the prefix "trans" in "transgender" does not necessarily refer to the "transcendence" or "transformation" of gender or gender normativity, and to assume that it does is to minimize decades of sociological work testifying to the rigidity and recalcitrance of the socially structuring gender binary in our society. These assertions also fail to consider the ways in which identity choices are socially embedded, strategic, and constrained. From a queer sociological analytic perspective, we might approach questions about whether the relationships between cis women and trans men reflect a radical subversion of cultural normativity or merely mirror and repackage cultural normativity with some degree of critical suspicion. Such questions implicitly suggest that the onus of responsibility for radically reconfiguring gendered power relations ultimately lies with a numerical and marginalized social minority. Indeed, we might usefully redirect such questions toward whether or not relationships between cis women and cis men—the numerical majority in our culture—currently reflect radical subversion of cultural normativity. Doing so reminds us of the powerful structuring forces of inequality for all social actors and also points to potentially fruitful alliances between social actors working toward equality aims. Building these communities of political and social alliance and resistance was described as an area of particular struggle for the cis women in this study.

Community Belonging, Vanishing, and Outcasting

"A normal, boring couple" and "I definitely don't miss being scared."—Brown (2009) describes "sexual identity renegotiation" as a central challenge faced by cis women partners of trans men. When providing accounts of their experiences in social spaces, cis women sometimes discussed how being (mis)recognized as unremarkably heterosexual was a social phenomenon highly desired by one's trans partner, while their own feelings remained more ambivalent or even conflicted. As Frieda (28 years, Ontario) discussed:

> [My partner] definitely was into the whole idea of us passing as a straight couple, so nothing queer really fit into our everyday lives or relationship because his main priority was passing as a man and that I should look like a woman so we can pass as a straight couple and he can blend in. So he encouraged me to look more feminine and to have my hair long and things like that . . . [but] I wanted to shave my head and . . . pierce things and . . . do things that normal, boring, feminine, straight women didn't usually do and they didn't fit in with what he wanted. . . . I kind of felt guilty or selfish if I tried to dress the way that I wanted. . . . When we were going out together, I tried to look as feminine and as boring as I could so we could pass as a normal, boring couple.

Frieda's narrative speaks to the way in which her partner's accomplishment of recognition as a man depends, at least in part, on social others' recognition of her as normatively feminine. This makes sense if we consider that the accomplishment of social recognition as a "normal" man depends, centrally, upon being perceived by others as not a woman and not gay (Connell 1987). In other words, social recognition of Frieda's partner as a man is facilitated through social assumptions linking manhood and heteronormativity. This assumed connection to heteronormativity was both troubling and strange to many participants—particularly those for whom social recognition as lesbian and counternormative had become a critical aspect of their sense of self.

Polly (40 years, New York) discussed challenges connected to reinterpreting her own identity, the social perceptions of others, and social group memberships:

> I think I'm still trying to sort out what it means *not* to be a lesbian. There is a nice *recognition* [author's emphasis] when you're walking down the street with your girlfriend and you're holding hands and see another lesbian and they see you as a lesbian and it's like you feel like you're all in the same club. So I miss that. . . . I just sort of feel like this level of boringness. I guess I have to say I definitely got off on the transgression of having men look at me and then kissing my girlfriend. And now it's like I have men look at me and then I kiss him and it's like, "Big whoop." . . . It's just not the same charge. So I think I miss that. I miss some of that transgressive sort of fucking with people's heteronormative assumptions and now I'm just like basically following the script and it feels a little weird. It's not quite as fun. [I miss] the performativity of being gay. . . . Sometimes it's scary and you don't do it. So I definitely don't miss being scared.

For both Frieda and Polly, social experiences wherein they believed their partner was recognized by others as a man elided their own queer visibility, creating the paradoxical situation of gaining access to heteronormative social privilege while simultaneously losing access to (or recognition by) sexual minority communities with which they strongly identify/identified. Furthermore, both describe "passing" or being (mis)recognized as heterosexual as "boring," highlighting the power of visibly queer social identities to provoke and dynamically elicit sexually charged, emotional responses based upon their connection to transgressiveness. Polly's concluding remark, alluding to the danger associated with public expressions of intimacy that are recognized as lesbian, highlights a pragmatic aspect of being (mis)recognized as heterosexual: reduced threat of physical and sexual violence directed toward those who are more visibly queer.

Most cis women who reported being (mis)recognized as part of a heterosexual couple, by family, friends, or strangers, acknowledged the privilege that such (mis)recognition entails, while simultaneously expressing discomfort with this privilege and bemoaning the inevitable trade-off of losing social recognition as queer. Margaret (29 years, Massachusetts) stated: "I have mixed feelings about it Sometimes I really like passing. There's a real social benefit to it; it makes it a lot easier." Veronica (21 years, New York) told me: "It makes me feel safe in the world," but she also commented on the flip side: "It makes me feel really invisible and that's something he and I both deal with a lot. We don't like the invisibility factor. We're always looking for ways to be visible and to educate others. So maybe that's the only way because I don't really know how much we can walk down the street wearing shirts that say, 'We're not so straight!'" When Maya (30 years, California), who had just had a baby, was asked to discuss how she felt she and her partner are perceived by others, she responded: "It's annoying because we get such privilege everywhere we go. . . . My mother's like, 'Thank God!' And I provided her a grandchild, so I'm 'normal.' In some respects it's good and in other respects I wish *everyone* had that." Eliza (25 years, Nova Scotia), who is legally married to her trans partner, stated: "With family . . . there's a thing in the back of my head that wonders if it's so easy for them because now we're a 'straight couple.' It's almost less explaining for them to do in the future. Sometimes it's a mixed blessing." As Eliza reveals, family members' potential investments in processes of doing sexuality for their relatives further highlight sexuality as an interactive social accomplishment. These narratives also reveal a keenly developed consciousness of the way queer people experience the sometimes-marginalizing gaze of nonqueer people, poignantly highlighting the disjunctive between self-identification and social (mis)recognition.

"Another breeder couple invading."—Participants in this study also described the experience of losing access to (and social recognition within) queer communities as they became "invisibly queer."[10] Margaret (29 years, Massachusetts) said: "When I see lesbian couples with a baby, I smile at them and have this moment of like, 'What a cute couple with a baby.' And [my partner] and I have this experience together because, at one point, he had been externally identified by others as a lesbian. So we have this moment of, 'Oh, another queer couple with a baby!' But [lesbian couples] . . . don't see that we're having this moment of camaraderie like, 'Yay, you did it, we're going to do it!' They see us as like, 'Oh, those straight people are looking at us.'" Maya (30 years, California) offered a

similar story: "We can go anywhere and not have people looking at us except when we're in [a gay neighborhood] and then it's like, 'Oh, another breeder couple invading.' And I just want to wear rainbow flags everywhere I go so I can prove that I belong in this community." Lilia (22 years, California) also articulated the not-uncommon experience of having her queer identity elided by others within the queer community: "My lesbian friends . . . [are] like, basically, 'Oh, so you turned straight.' . . . [But] I don't consider this a straight relationship since he's very queer. . . . I can see how it's straight in some context. But it's queer. His experiences of growing up as a woman [are] what makes it queer." In each of these narratives, participants describe experiencing the elision of their queerness—disappearing into the background of queer communities within which they often previously found community and recognition as queer. Many cis women participants described being (mis)recognized as heterosexual as not only personally invalidating but as alienating from queer communities of social support and belonging. Once again, these experiences echo those of bisexual-identified women who often report being ostracized from lesbian communities when partnered with men and from heterosexual communities when partnered with women (Burrill 2001; Wolkomir 2009; Tabatabai and Linders 2011).

Cis women partners of trans men face challenges of marginalization not only from social distancing, exclusion, and (mis)recognition by others within LGBTQ communities, but sometimes as a result of their trans partner's wish to disassociate from these communities to reinforce their own social recognition as a man. Belinda (24 years, Ontario) spoke about losing her connection to the lesbian community when her partner disengaged from it:

> It was tough for me as someone who had just kind of come out as a lesbian. I remember wanting to do lesbian things and go to lesbian bars and that kind of stuff. And I remember a switch in him where he was like, "No, I'm a straight guy." And I think that was hard because there was this community that I was trying to get involved with that suddenly didn't work with his identity. . . . I didn't really know that there was the option of him saying, "I'm queer." I just figured that's what happened when someone became trans—you were a lesbian and now you're straight.

Belinda articulates the limited (and often limiting) nature of social models of identity in the context of transition. Belinda was unaware that there were other ways (than "straight male") for trans men in relationships with cis women to identify and that these different identifications (if embraced by her partner) might generate alternate possibilities for her own identity and membership to social communities. Narratives like Belinda's also highlight how the accomplishment of social recognition as a man often necessitates social distancing from LGBTQ communities and spaces.

"The people that I dated would make me visibly queer."—When considering the personal and social identities and group memberships of cis women partners of trans men, it is also important to consider the often temporal-relational and contingent aspects of these ways of being and belonging in the world. Susan (23 years, Tennessee) articulated two distinct dilemmas she faced as a formerly lesbian-identified cis woman and as the former partner of a trans man: "I lost my community. . . . You lose the lesbian community and you really don't get anything else. . . . And the partners' [of trans men] community—you're only a valid member of that as long as you're in your relationship, which has nothing to do with *you* and everything to do with *him*." For Susan, carving out a space in the queer community along with other partners of trans men reflected both a contingent and tenuous subject position within such communities. Susan's experiences of being pushed out of lesbian community spaces upon partnering with a trans man was not uncommon. Rather than operating along explicit cut-and-dried practices of inclusion and exclusion, many cis women described more subtle social practices in which their rightful membership within lesbian community spaces was challenged or brought into question once they began relationships with trans men or once a previously lesbian-identified partner began to move away from that identity and transitioned to living as a trans man.

Ani (21 years, Ohio) discussed another challenge in her relationship with a partner who socially identified as a "man" rather than as a "transgender man": "It's a lot easier to be able to [say]: 'Yes, I'm queer, I'm dating a *trans man*,' as opposed to, 'Yes, I'm queer, I'm dating a man.' People won't ask you to justify yourself in the same way. . . . Your sexuality clearly relies on your partner." Ani's partner's gender identity and recognition by social others as a man meant that her own queer identity was frequently made invisible—rendering her unremarkably heterosexual in the eyes of social others, including queer social others.

Nearly 30% of the participant sample self-identified, unprompted, as "femme"—meaning that the actual composition of femme-identifying or feminine-appearing cis women in the sample is likely higher than 30%. Nyamora (2004) and Brown (2009) both describe the ways in which cis femme-identified women partners of trans men frequently experience a grieving process in connection to the perceived loss of their queer femme visibility. Further, many of the participants in my study discussed how others' recognition of their queerness often relies upon their connection to a partner who embodies female masculinity in a visible and culturally intelligible way. For example, Teresa (24 years, Maine) told me:

> I think as a femme. . . . I don't feel like I've ever been seen as queer when I've been by myself. I think so often in my history of dating people that the people that I dated would make me visibly queer. So it's really interesting when the person I'm dating makes me *invisible*. And so I don't gain any visibility as a lesbian or as someone who is queer when being out in public with [my trans partner] the way I would with past partners. So that's really, really hard. However, in a way it sort of feels almost liberating because now I and only I am responsible for my queer visibility. . . . I think that it's sexism, honestly, that femmes are seen as invisible beings when really we're radically queer in our own right and we're just never given that credit.

As Teresa articulates, femme-appearing/identifying cis women partnered with trans men, therefore, may face particular barriers with regard to being recognized as a member of the communities to which they belong and with which they identify (see also Nyamora 2004; Brown 2009; Joslin-Roher and Wheeler 2009; Ward 2010).[11]

These narratives reveal the extent to which queer visibility remains culturally synonymous with social perceptions of female masculinity and male femininity (Hutson 2010), often rendering those who embody cis femininity within queer communities invisible as queer. These narratives also echo earlier writings on lesbian butch and femme genders as socially intelligible identities around which communities materialized and organized (cf., Ponse 1978; Krieger 1983; Taylor and Whittier 1992; Kennedy and Davis 1993). Queer invisibility was of particular concern and consideration to many of the femme-identified cis women I interviewed. This articulated invisibility serves as a marked empirical contrast to theorizing around femme identity (e.g., Hollibaugh 1997; Munt 1998; Levitt, Gerrish, and Hiestand 2003), which marks it as politically transgressive (and even "transgender") in its own right. Such fissures between personal experience and political potential further highlight the need to examine the processes by which gender and sexual identities are produced through social interaction.

"You're not really gay" and "Take your pants off and show me."—Participants spoke about the ways in which queer femininities may not only be rendered invisible within queer and nonqueer cultural spaces, but how they may also be explicitly devalued within some queer communities relative to queer androgynies and queer masculinities (Kennedy and Davis 1993; Cogan 1999; Levitt et al. 2003). As Belinda (24 years, Ontario) told me:

> Basically within the lesbian community I was like completely made fun of. I used to have people make fun of me for carrying a purse and looking "too girly" and, "Oh you're not really gay." Just those kinds of comments. So that was really hard for me when I was coming out because I just wanted to be taken seriously you know. . . . So my response to that [when I first came out] was to kind of change to become *less* feminine, change my body posturing and the way that I dress and cut off all my hair and that kind of stuff.

Narratives like Belinda's exemplified some queer cis women's experience of living in the liminal space of insider/outsider with regard to both queer and nonqueer communities.

Ward (2010) suggests that sidelining of the power and transgressive potential of femme identity among cis women partners of trans men may be an artifact of their primary social status within trans communities as allies and supporters of their partners—one of the forms of "gender labor" in which they engage. Some of the strategies self-identified femme participants described for rendering their queer identities more recognizable included adopting unique and unconventional hairstyles and hair colors, wearing rainbow jewelry and other LGBTQ pride symbols, dressing in vintage clothing, and obtaining visible tattoos and piercings, embodying counternormative embodiment practices with the intention of visually signifying their queer identities (see also Pitts 2000). Participants' narratives revealed the impact of being rendered invisible or an outsider not only in terms of one's own queer identity and relationship but also in determining the parameters of in-group/out-group social membership itself.

While some trans men and their cis women partners described being (mis)recognized as heterosexual and becoming invisible as queer within LGBTQ communities, other participants reported that their partners were (mis)recognized as trans men or as cis women, rather than cis men, more often in gay and lesbian social spaces than in mainstream or non-LGBTQ social spaces. The tensions between these (mis)recognition processes carried striking social consequences. One set of trends that emerged in participants' accounts involved (1) explicit exclusion of trans people and their partners from primarily gay and lesbian social spaces and (2) intimidating and even violent interactions aimed toward "finding out" the "real" sex of those who are trans as they interact within primarily gay and lesbian social spaces. Seventeen (34%) participants described instances of being told by leaders of gay and lesbian organizations (or hearing through the grapevine) that their or their partner's presence was no longer welcome since their partner's transition. Martha (25 years, Massachusetts) described making reservations at a lesbian bed and breakfast only to be told that she and her partner were no longer welcome upon the innkeeper's learning of her partner's transition. Lynne (35 years, California) described the exclusion of trans men from the yearly "dyke march" in her town.

June (21 years, Ontario), Kendra (21 years, Ohio), and Samantha (20 years, Michigan) each relayed harrowing and eerily similar experiences their trans partner had in gay and lesbian bars. According to June: "He went out to a . . . lesbian bar . . . and they wanted him to prove that he was actually male. So there was a lot of, 'Take your pants off and show me,' type of thing. They followed him into the bathroom and it was about an hour of harassment like that." Samantha told me: "He was going to the bathroom . . . and he was waiting for the stalls and . . . this old lesbian got up in [his] face and was like, 'Go use the other bathroom, we need this one more than you do. . . .' And she got really up in his face about it and he was like, 'I'm trans. I have to sit to pee.' And she was like, 'No you're not. . . .' She actually ripped his shirt off to see." In the context of a gay bar, Kendra relayed the following description:

> He almost got beat up that night . . . He went to the women's restroom because he wasn't fully male and he didn't want gay guys to find out that he didn't have a penis; so he chose to use the women's restroom that night. He was still fairly early into his transition and a guy followed him in there and watched him urinate and said, 'Take off that binder. I don't know why you want to be a guy. . . .' Later, the guy lunged across the dance floor at my partner and, luckily, one of our friends pushed him out of the way.

In each of these instances, trans men were held accountable for others' recognition of them as men—social processes that could have frightening and even dangerous consequences, even within communities that had formerly served as relatively safe havens from exclusion and discrimination.[12]

These narratives attest to the permeability and instability of membership and recognition within various identity-based communities. In a social context that continues to affirm fixed and naturalized binaries (male/female, man/woman, heterosexual/homosexual) despite increasing evidence documenting the fluidity and diversity of sex, gender, and sexual identifications, we find herein evidence for these identities as interactive social accomplishments. Perhaps even more important, we are urged to reconsider just who should be held accountable when it comes to recognizing the sex, gender, and sexual identities of others.

Conclusion: Possibilities for Social Solidarity and Broader Application

In this study, I draw from Connell's (2009) notion of "recognition" to demonstrate the myriad ways in which we "do" not only gender, but sexuality as well, revealing sexual identities as interactional social accomplishments through which status, rights, and group membership may be stripped or conveyed. By challenging the essentialist notion that sexual identities are largely fixed and natural/biological, we are better poised to consider what is at stake when social actors recognize and misrecognize their peers' sexual self-identifications. The cis women I interviewed often vocally asserted their self-identification as queer. Yet in many instances, these cis women's accounts focused on being (mis)recognized by both queer and nonqueer social others as unremarkably heterosexual. Which of these accounts of their sexual identity is "true"? These findings prompt consideration of how the social effects of (mis)recognition processes (e.g., being able to access regulated social institutions and social membership within particular groups) are powerfully structuring—perhaps even largely determinant—of social group membership.

. . . Extending Connell's (2009) "recognition" framework, this study highlights what is at stake in social (mis)recognition processes not only for queer social actors but also for everyone, as these processes reveal the ways in which access to regulated social groups and institutions is often mediated largely through interactional and perceptual social processes rather than static or essential aspects of individuals.

Namaste, writing about queerness and queer theory, states: "We cannot assert ourselves to be outside of heterosexuality, nor entirely inside, because each of these terms achieves its meaning in relation to the other. . . . We can think about the *how* of these boundaries . . . how they are created, regulated and contested" (1994, p. 224). This analysis offers further insight into that *how*—detailing the ways in which heterosexual, gay, lesbian, and queer identities and social identity group memberships overlap and are messily embraced, resisted, and (mis)recognized in the context of cis women's relationships with trans men. How might we make sense of the following narrative from a cis woman partner of a trans man that inspired the title for this article and was emblematic of many of the responses that I received? "I don't like passing as a straight woman. I would feel like I wasn't visible at times—and same with him, that he wasn't visible. . . . Both of our identities were very blurred; and that's a tough thing when so much of who we are is about other people perceiving us. . . . I like my queer identity and that's what I want people to see. So it was tough when I knew that wasn't being seen" (Martha, 25 years, Massachusetts). Much of the thrust of the mainstream lesbian and gay social movement over the past two decades has focused on protesting and bringing greater public awareness to discrimination against lesbians and gay men as well as their exclusion from various social institutions and privileges, such as legally recognized marriage.[13]

In calling for expanded rights and inclusion, mainstream lesbian and gay social movements have largely centered upon crafting a politics of sameness and respectability that stands in stark contrast to the oppositional politics of activist groups of the late 1960s through the early 1990s—such as the Gay Liberation Front, ACT UP, and Queer Nation (Duggan 2002; Ward 2008). Further, many of these more recent efforts depend largely upon appeals to the biological/genetic etiology of sexual orientation and gender identity (e.g., Lady Gaga's aforementioned pop culture anthem, "Born This Way"). Couching demands for inclusion, equality, and freedom from discrimination within a framework of biological determinism consistently compels the following presumably rhetorical defense of these demands when they face social opposition: "In the context of historical and contemporary social discrimination and exclusion, why would anyone *choose* this?" Yet narratives and self-identifications like Martha's provide evidence against the counterfactual claim that no one would choose queerness if given such an option, just as they simultaneously recognize and explicitly value queer identities and queer cultures per se. They also reframe the issue of "choice" to consider that choosing to self-identify as queer is not synonymous with choosing social (mis)recognition, exclusion, and discrimination. In other words, many of the women I interviewed refused to be held accountable for other people's (mis)recognition of their or their partner's sex, gender, and sexual identities. . . .

Choosing queer self-identification and alliance as a form of normative resistance (see Pfeffer 2012) is not limited by the contours of one's own body in relation to those of one's partner(s). Normative social structures inscribe the parameters within which all social actors must live their daily lives. As such, all social actors desiring social change (perhaps especially those with normative privilege) are accountable for, and have a vested interest in, resisting and pushing against these parameters, as well as supporting others engaged in similar or parallel forms of resistance. Reframing and reorienting sociological analyses to the normative center, therefore, highlights the accountability and responsibility that those with relative privilege hold with regard to enacting social change, resisting stultifying normativity, and reconfiguring relationships of power. In doing so, we might further shift our inquiries to consider how and why anyone might develop and nurture their own and others' queer identities and relationships for the purposes of greater gender and sexual equality. . . .

Notes

1. As Schilt and Westbrook, drawing from Serano (2007), note, "Cis is the Latin prefix for 'on the same side.' It complements trans, the prefix for 'across' or 'over' . . . to refer to individuals who have a match between the gender they were assigned at birth, their bodies, and their personal identity" (Schilt and Westbrook 2009, p. 461). Use of the phrase "cis women" throughout this manuscript is intended to mark the identities of women in my sample, just as the identities of men who are their partners are also marked. To not do so, as rightfully noted by an *AJS* reviewer, "reproduces the 'otherness' of trans by not marking the unmarked category."

2. The phrase "trans men" is used throughout for sake of consistency and simplicity. It should be noted, however, that gender identity labels and categories are often far from consistent or simple. The cis women in this sample identified their trans partners using various terms—transgender, transsexual, trans, female-to-male (ftm), man, boi, etc. The "trans men" referred to in this study are individuals who were assigned, by sex, as "female" at birth and whose gender identity does not directly correspond with this sex assignment or their sex categorization. Some trans men partners of the cis women I interviewed have pursued hormonal or surgical realignment surgeries to bring their bodies in closer alignment with their gender identities, while others have not. . . . For additional background information on the language, concepts, and terms related to transgender identity and

experience, please see Wentling et al. (2008) and Pfeffer (2010).

3. Epstein (1994) offers the provocative claim that much queer theory is rooted in and dependent upon sociological theoretical precedents, particularly across the areas of symbolic interactionism and labeling theory. These critiques are later echoed by Dunn (1977) and Green (2007), who highlight the particular theoretical and empirical contributions of pragmatists, symbolic interactionists, and ethnomethodologists to the development of poststructuralist and queer theory produced by scholars such as Judith Butler. As Green (2007, pp. 26–27) writes: "With regard to gender and sexuality . . . sociology has been doing a kind of queer theory long before the first queer theorist set pen to paper."

4. As Sedgwick writes (1990, p. 1) in a foundational text of queer, *Epistemology of the Closet:* "An understanding of virtually any aspect of modern Western culture must be, not merely incomplete, but damaged in its central substance to the degree that it does not incorporate a critical analysis of modern homo/heterosexual definition."

5. Judith Butler's (1990) theorization of "gender performativity" draws from Foucault's ([1976] 1990) theorizing around power, repression, and generativity. According to Butler (1990), being a "man" or a "woman" (or "male" or "female") is not a fixed, biological, or immutable human characteristic but, rather, is (re)produced through a system of power and social relations. While these operations of power may compel social relations that (re)produce the normative as ideal, and discipline deviations from normative ideals, these same repressive forces ultimately suggest and generate the potential for disobedience and alternate social relations—producing "gender trouble."

6. "Sex assignment" refers to the assignment of a person, at birth, to "male" or "female" based on bodily signifiers such as presence of a penis or vagina. "Sex categorization" refers to the everyday, iterative placement of a person into social categories such as "girl," "women." "boy," "man."

7. Some participants, who had been with the same partner prior to this transition, found this form of social exchange particularly salient as they noticed very different reactions from older persons when engaging in public handholding with the very same partner. Prior to transition, when their partner was reportedly "read" as female and the couple was "read" as lesbian, they recalled older individuals starting at them while not smiling, whispering, avoiding eye contact, and not returning smiles.

8. Participants themselves used the term "roles" (e.g., "1950s housewife role") to describe the enactments of traditional wife/husband, and mother/father family dynamics as they understood them.

9. See also Pfeffer (2012) for further discussion of the overlaps between this sample and those focusing on bisexual-identified cis women.

10. See Brown (2009) for a discussion of similar experiences among another sample of cis women partners of trans men.

11. It must also be noted that the gender presentation of trans men is of critical importance here in others' constructions of the couple's sexual identity. Women who told me that their trans partner was often perceived as a gay man by social others were often misrecognized as "friends" rather than romantic partners. Some women described instances of their partner being hit on by other men in their presence.

12. Of course it is important to consider that lesbian and gay communities, while often providing shelter from homophobia and heterosexism, still struggle with issues of inclusion and discrimination not only with regard to those who are trans or bisexual but with regard to racism, classism, ableism, and sizeism (to name just a few areas) as well.

13. For an overview of the public response to these efforts, see Stone (2012).

References

Ault, Amber. 1996a. "Ambiguous Identity in an Unambiguous Sex/Gender Structure: The Case of Bisexual Women." *Sociological Quarterly* 37(3):449–63.

_____. 1996b. "The Dilemma of Identity: Bi Women's Negotiations." In *Queer Theory/Sociology,* edited by Steven Seidman. London: Blackwell.

Bernstein, Mary. 2005. "Identity Politics." *Annual Review of Sociology* 31:47–74.

Blumstein, Phillip, and Pepper Schwartz. 1974. "Lesbianism and Bisexuality." Pp. 278–95 in *Sexual Deviance and Sexual Deviants*, edited by Erich Goode and Richard T. Troiden. New York: Morrow.

_____. 1977. "Bisexuality: Some Social Psychological Issues." *Journal of Social Issues* 33:30–45.

_____. 1990. "Intimate Relationships and the Creation of Sexuality." Pp. 307–20 in *Homosexuality/Heterosexuality: Concepts of Sexual Orientation*, edited by David P. McWhirter, Stephanie Sanders, and June Machover Reinisch. New York: Oxford University Press.

Brown, Nicola. 2009. "'I'm in Transition Too': Sexual Identity Renegotiation in Sexual-Minority Women's Relationships with Transsexual Men." *International Journal of Sexual Health* 21:61–77.

Burrill, Katkryn G. 2001. "Queering Bisexuality." *Journal of Bisexuality* 2(2/3):95–105.

Butler, Judith. 1990. *Gender Trouble: Feminism and the Subversion of Identity*. New York: Routledge.

_____. 1993. *Bodies That Matter: On the Discursive Limits of "Sex."* New York: Routledge.

Calavita, Kitty. 2000. "The Paradoxes of Race, Class, Identity, and 'Passing': Enforcing the Chinese Exclusion Acts, 1882–1910." *Law and Social Inquiry* 25:1–40.

Cogan, Jeannie C. 1999. "Lesbians Walk the Tightrope of Beauty: Thin Is In but Femme Is Out." Pp. 77–90 in *Lesbians, Levis, and Lipstick: The Meaning of Beauty in Our Lives*, edited by Jeannie C. Cogan and Joanie M. Erickson New York: Haworth.

Connell, R.W. 1987. *Gender and Power*. Sydney: Allen & Unwin.

Connell, Raewyn. 2009. "Accountable Conduct: 'Doing Gender' in Transsexual and Political Retrospect." *Gender and Society* 23:104–11.

Denzin, Norman K. 1990. "Harold and Agnes: A Feminist Narrative Undoing." *Sociological Theory* 8:198–216.

Duggan, Lisa. 2002. "The New Homonormativity: The Sexual Politics of Neoliberalism." In *Materializing Democracy: Towards a Revitalized Cultural Politics*, edited by Russ Castronovo and Dana D. Nelson. Durham, N.C.: Duke University Press.

Dunn, Robert G. 1997. "Self, Identity and Difference: Mead and the Poststructuralists." *Sociological Quarterly* 38:687–705.

Elson, Jean. 2004. *Am I Still a Woman? Hysterectomy and Gender Identity*. Philadelphia: Temple University Press.

Epstein, Steven. 1994. "A Queer Encounter: Sociology and the Study of Sexuality." *Sociological Theory* 12:188–102.

Foucault, Michel. (1976) 1990. *The History of Sexuality*. Vol. 1, *An Introduction*. New York: Vintage Books.

Gamson, Joshua. 1995. "Must Identity Movements Self-Destruct? A Queer Dilemma." *Social Problems* 42: 390–407.

Garfinkel, Harold. 1967. *Studies in Ethnomethodology*. Englewood Cliffs, N.J.: Prentice Hall.

Goffman, Erving. 1976. "Gender Advertisements." *Studies in the Anthropology of Visual Communication* 3:69–154.

Green, Adam I. 2002. "Gay but Not Queer: Toward a Post-Queer Study of Sexuality." *Theory and Society* 31: 521–45.

_____. 2007. "Queer Theory and Sociology: Locating the Subject and the Self in Sexuality Studies." *Sociological Theory* 25:26–45.

Harris, Cheryl I. 1993. "On Passing: Whiteness as Property." *Harvard Law Review* 106:1707–91.

Hollibaugh, Amber. 1997. "Gender Warriors: An Interview with Amber Hollibaugh." In *Femme: Feminists, Lesbians, Bad Girls*, edited by Laura Harris and Elizabeth Crocker. New York: Routledge.

Hutson, David. 2010. "Standing OUT/Fitting IN: Identity, Appearance, and Authenticity in Gay and Lesbian Communities." *Symbolic Interaction* 33(2): 213–33.

Irvine, Janice M. 1994. "A Place in the Rainbow: Theorizing Lesbian and Gay Culture." *Sociological Theory* 12: 232–48.

Joslin-Roher, Emily, and Darrell Wheeler. 2009. "Partners in Transition: The Transition Experience of Lesbian, Bisexual, and Queer Identified Partners of Transgender Men." *Journal of Gay and Lesbian Social Services* 21:30–48.

Kennedy, Elizabeth L., and Madeline Davis. 1993. *Boots of Leather, Slippers of Gold: The History of a Lesbian Community.* New York: Penguin.

Kennedy, Randall. 2002. *Interracial Intimacies: Sex, Marriage, Identity, and Adoption*. New York: Random House.

Kessler, Suzanne J., and Wendy McKenna. 1978. *Gender: An Ethnomethodological Approach*. New York: Wiley.

_____. 2003. "Who Put the 'Trans' in Transgender? Gender Theory and Everyday Life." In *Constructing Sexualities: Readings in Sexuality, Gender, and Culture,* edited by Suzanne LaFont. Upper Saddle River, N.J.: Prentice Hall.

Krieger, Susan. 1983, *The Mirror Dance: Identity in a Women's Community.* Philadelphia: Temple University Press.

Levitt, Heidi M., Elisabeth A. Gerrish, and Katherine R. Hiestand. 2003. "The Misunderstood Gender: A Model of Modern Femme Identity." *Sex Roles* 3:99–113.

Munt, Sally R., ed. 1988. *Butch/Femme: Inside Lesbian Gender.* London: Cassell.

Mustanski, Brian S. 2001. "Getting Wired: Exploiting the Internet for the Collection of Valid Sexuality Data." *Journal of Sex Research* 38:292–301.

Namaste, Kai. 1994. "The Politics of Inside/Out: Queer Theory, Poststructuralism, and a Sociological Approach to Sexuality." *Sociological Theory* 12:220–31.

Nyamora, Cory M. 2004. "Femme Lesbian Identity Development and the Impact of Partnering with Female-to-Male Transsexuals." Psy.D. dissertation, Alliant International University, California School of Professional Psychology.

Ong, Maria. 2005. "Body Projects of Young Women of Color in Physics: Intersections of Gender, Race and Science." *Social Problems* 52:593–617.

Patton, Michael Q. 1990. *Qualitative Evaluation and Research Methods.* Newbury Park, Calif.: Sage Publications.

Pfeffer, Carla A. 2008 "Bodies in Relation—Bodies in Transition: Lesbian Partners of Trans Men and Body Image." *Journal of Lesbian Studies* 12:325–45.

_____. 2010. "'Women's Work?' Women Partners of Transgender Men Doing Housework and Emotion Work." *Journal of Marriage and Family* 72: 165–83.

_____. 2012. "Normative Resistance and Inventive Pragmatism: Negotiating Structure and Agency in Transgender Families." *Gender and Society* 26:574–602.

Pitts, Victoria. 2000. "Visibly Queer: Body Technologies and Sexual Politics." *Sociological Quarterly* 41(3): 443–63.

Ponse, Barbara 1978. *The Social Construction of Identity and Its Meanings within the Lesbian Subculture.* Westport, Conn.: Greenwood Press.

Richardson, Diane, and John Hart. 1981. "The Development and Maintenance of a Homosexual Identity." Pp. 73–92 in *The Theory and Practice of Homeosexuality,* edited by John Hart and Diane Richardson. London: Routledge & Kegan Paul.

Rogers, Mary. 1992. "They All Were Passing; Agnes, Garfinkel, and Company." *Gender and Society* 6:169–91.

Rosser, B., R. Simon, Michael J. Oakes, Walter O. Bockting, and Michael Miner. 2007. "Capturing the Social Demographics of Hidden Sexual Minorities: An Internet Study of the Transgender Population in the United States." *Sexuality Research and Social Policy: Journal of NSRC* 4:50–64.

Rust, Paula C. 2000. "Bisexuality: A Contemporary Paradox for Women." *Journal of Social Issues* 56(2): 205–21.

Schilt, Kristen, and Laurel Westbrook, 2009. "Doing Gender Heteronormativity: 'Gender Normals,' Transgender People, and the Social Maintenance of Heterosexuality." *Gender and Society* 23:440–64.

Sedgwick, Eve Kosofsky. 1990. *Epistemology of the Closet.* Berkeley and Los Angeles: University of California Press.

Seidman, Steven. 1994. "Queer-ing Sociology, Sociologizing Queer Theory: An Introduction." *Sociological Theory* 12:166–77.

_____. 2001. "From Identity to Queer Politics: Shifts in Normative Heterosexuality and the Meaning of Citizenship." *Citizenship Studies* 5:321–28.

Serano, Julia. 2007. *Whipping Girl: A Transsexual Woman on Sexism and the Scapegoating of Femininity.* Emeryville, Calif.: Seal Press.

Shapiro, Eve. 2004. "'Trans'cending Barriers: Transgender Organizing on the Internet." *Journal of Gay and Lesbian Social Services* 16:165–79.

Stone, Amy L. 2012. *Gay Rights at the Ballot Box.* Minneapolis: University of Minnesota Press.

Tabatabai, Ahoo, and Annulla Linders. 2011. "Vanishing Act: Non-Straight Identity Narratives of Women in Relationships with Women and Men." *Qualitative Sociology* 34:583–99.

Taylor, Verta, and Nancy E. Whittier. 1992. "Collective Identity in Social Movement Communities: Lesbian Feminist Mobilization." In *Frontiers in Social Movement Theory,* edited by Aldon D. Morris and Carol M. Mueller. New Haven, Conn.: Yale University Press.

Valocchi, Stephen. 2005. "Not Yet Queer Enough: The Lessons of Queer Theory for the Sociology of Gender and Sexuality." *Gender and Society* 19:750–70.

Vidal-Ortiz, Salvador. 2002. "Queering Sexuality and Doing Gender: Transgender Men's Identification with Gender and Sexuality." *Gender Sexualities* 6:181–233.

Ward, Jane. 2008. *Respectably Queer: Diversity Culture in LGBT Activist Organizations.* Nashville: Vanderbilt University Press.

_____. 2010. "Gender Labor: Transmen, Femmes, and Collective Work of Transgression." *Sexualities* 13:236–54.

Warner, Michael. 2012. "Queer and Then?" *Chronicle Review*, January 1, 1–14. http://chronicle.com/article/QueerThen-/130161/.

Wentling, Tre, Kristen Schilt, Elroi Windsor, and Betsy Lucal. 2008. "Teaching Trans-gender." *Teaching Sociology* 36:49–57.

West, Candace, and Don H. Zimmerman. 1987. "Doing Gender." *Gender and Society* 1:125–51.

Wolkomir, Michelle. 2009. "Making Heteronormative Reconciliations: The Story of Romantic Love, Sexuality, and Gender in Mixed-Orientation Marriages." *Gender and Society* 23:494–519.

Zimmerman, Don H. 1992. "They Were All Doing Gender, but They Weren't All-Passing: Comment on Rogers." *Gender and Society* 6:192–98.

Introduction to Reading 5

Raewyn Connell's contributions to the sociological understanding of masculinities and gender relations generated major changes in gender theory, research, and policy-making. Connell, as discussed in the Introduction to the book and the Introduction to this chapter, has argued that masculinities are multiple, that masculinities change over time, and that women as well as men have a role in creating masculinities. One of the most powerful concepts to emerge from Connell's work is "hegemonic masculinity," a culturally exalted form of masculinity which tops a hierarchy of masculinities and is a key mechanism in maintaining gender inequality. In this reading, Connell turns her attention to the importance of decolonizing the study of masculinities and, consequently, unsettling dominant views of gender relations produced in Western Europe and North America. As Connell says, gender discourse has been Eurocentric and thus the assumptions underlying much research on gender relations and policy-making are biased. For example, Connell notes in the reading that "Anglophone categories such as 'MSM,' 'identity,' 'heterosexuality,' and even 'men' may misrepresent" the realities of people's lives in the periphery. Connell calls for a critique of the relation between hegemony and masculinity and an historical understanding of masculinities in relationship to structures of imperialism and neoliberal global power structures. As she states at the close of the reading, we need accurate knowledge and insight into hegemonic masculinity that incorporates Southern perspectives on gender relations and power.

1. What is the global South, and why does Connell argue for including Southern research on masculinities?

2. How and why did colonial powers, such as the Christian societies of Europe, employ gendered violence in their strategies for hegemony?

3. What is transnational corporate masculinity?
4. Describe some of the counter-hegemonic projects among men that have emerged alongside the new world gender order.

Masculinities in Global Perspective

Hegemony, Contestation, and Changing Structures of Power

Raewyn Connell

Thirty years ago, three Australian authors proposed "a new sociology of masculinity' (Carrigan et al. 1985). They criticized the popular concept of a "male sex role," offering instead a combination of feminist, gay liberation, and psychoanalytic ideas. Their most influential idea was that multiple masculinities existed, that there was hierarchy among them, and that a hegemonic version, at the top of the hierarchy, connected the subordination of women to the subordination of marginalized groups of men. The term "hegemonic masculinity" named a key mechanism sustaining an oppressive society and implied that contesting this mechanism was an important strategy of change.

In the following decades, as research grew alongside public debates about men and masculinity, the concept of hegemonic masculinity was widely used. The concept has played a role in reform agendas and has guided empirical investigations. It has also been vigorously criticized and re-formulated (Connell and Messerschmidt 2005).

Questions have been raised about the idea of masculinity, the use of Gramscian ideas in understanding gender relations, the location of the concept in modern or postmodern thought, and the relation of hegemonic masculinity to identity, power, and violence (Howson 2006; Meuser 2010; Zhan 2015; Pascoe and Tristan 2016).

Most of this research and debate has occurred within the global North. It is increasingly recognized that the resulting geopolitics of knowledge is a problem. For a deeper understanding of the issues raised in the debates about hegemonic masculinity, we need to learn not only from Western Europe and North America but also from the majority world. We need, in short, to decolonize the study of masculinities.

Decolonizing the Discussion

Modern knowledge production has a global structure (Hountondji 1997). A worldwide division of labor, with its origins in colonial conquest, locates the production of theory in the global metropole and treats the periphery essentially as a data source. Intellectual workers in the periphery normally follow the intellectual authority of the North and seek recognition there. (Our 1985 article "Toward a New Sociology of Masculinity" was a good example of this pattern.) Over the last two decades, however, there has been a sustained critique of Northern dominance in the social sciences, proposing globally inclusive agendas of theory (Connell 2007; Go 2012; Bhambra 2014; Rosa 2014). The same kind of discussion has developed in gender studies (Bulbeck 1998; Lugones 2007; Connell 2015).

These concerns have recently emerged in research on masculinity. Robert Morrell and Sandra Swart in South Africa (2005) pose the situation of the poorest part of the world's population as a key issue for masculinity studies. Margaret Jolly (2008, p. l), introducing research on masculinities in the post-colonial Pacific, emphasises "the crucial importance of colonialism in the construction of indigenous masculinities in both past and present." Paul Amar (2011), in a critical review of Middle East masculinity studies, vigorously argues for a decolonial perspective. Ford and Lyons (2012), introducing research on masculinities from Southeast Asia, question universalized concepts and emphasize the need for local knowledge.

Kopano Ratele (2013), on the basis of experience in southern Africa, questions the assumption that "traditional" masculinity automatically means patriarchal dominance. The Brasilian scholar Diego Santos Vieira de Jesus (2011) shows how a post-colonial approach to

Raewyn, C. (2016). Masculinities in global perspective: Hegemony, contestation, and changing structures of power. *Theory and Society, 45*: 303–318. Reprinted with permission from Springer Nature.

masculinities yields a broad historical framework that throws light on the colonizers and the imperial center as well as the colonized. In a recent article I pointed to the global archive on masculinities and argued for the importance of ideas, as well as data, from the global South (Connell 2014).

The idea of hegemonic masculinity has to be considered in the light of these changes; and the question arises whether the idea of hegemony applies in the colonial world at all. The Indian historian Ranajit Guha, founder of Subaltern Studies, questioned this in an article called "Dominance without Hegemony and its Historiography" (1989). The imperial power, he argued, never achieved hegemony in colonial India. The British persuaded themselves that they operated by the rule of law; but this was self-deception, in a colony actually controlled by autocratic decrees and military force. The truth was revealed by the widespread acts of resistance that British rule continually encountered.

The problem with the Eurocentrism of global gender discourse is that it projects into gender analysis everywhere the image that the society of the global North holds of itself. Specifically, it presumes coherence and a self-sustaining logic for any gender order. This is implicit in the concepts of "patriarchy," "sex/gender system," "gender norms," "gender regime," and "heteronormativity." Eurocentric gender research and policy-making assume that gender has a system-like character, a logical homogeneity and, though it may change, that it does so with continuity in time.

With these assumptions in the background, the concept of hegemony tends to become ahistorical, concerned with the social reproduction of a system. Hence the prevalence, in research on hegemonic masculinity, of ideas of identity formation, socialization, habitus, and the internalization of social norms—which are actually black-box concepts produced by assuming a mechanism of social reproduction (Connell 1983). Hence the familiar slippage between notions of hegemony and notions of domination (Connell and Messerschmidt 2005), which are easily blurred when the reproduction of a hierarchical system is assumed.

Research in postcolonial contexts, however, calls exactly these assumptions into question. Historical discontinuity is the core of colonial conquest. Margrethe Silberschmidt (2004), researching HIV transmission in East Africa, rejects the idea that men's dangerous assertion of sexual privilege reflected the continuity of "traditional" masculinity. She argues that gender violence resulted from the breakdown of traditional gender orders, under the pressures of colonialism and post-colonial economic change.

In a similar vein, when talking about feminist sexuality research in Africa, Jane Bennett (2008, p. 7) observes that mainstream methods textbooks tacitly assume a stable social environment. But a stable environment cannot be assumed for research in postcolonial conditions where "relative chaos, gross economic disparities, displacement, uncertainty and surprise" are the norm not the exception. Discussing the "water wars" in Cochabamba (Bolivia), Nina Laurie (2005) traces the clash of masculinities in this defeated neoliberal privatization attempt. She too makes a strong argument that research in the global South cannot presume a consolidated gender order.

To discard global-North assumptions about social reproduction does not imply that gender concepts such as hegemonic masculinity must be abandoned. Rather, it requires that gender concepts should always be understood historically, as concepts that concern the making and transformation of gender orders through time. Hegemony is a historical possibility, a state of gender relations being struggled for, and struggled against, by different social forces. Since the accomplishment of hegemony is never guaranteed, the most useful way to conceptualize hegemonic masculinity is to treat it as a collective project for realizing gender hierarchy. And that, in the light of the postcolonial critique outlined above, is a process we now have to understand on a world scale.

Hegemony and Empire in the History of Masculinities

Constitutive Violence and the Making of Colonial Societies

Colonialism, as Guha said, involved massive violence. Some conquests destroyed a pre-colonial regime and so established rule over a subject population. The classic case was the invasion led by Cortés in México, smashing the Aztec empire and reducing indigenous men and women to a new kind of serfdom. In other cases, from Hispaniola to Australia, colonizing violence swept over a whole population and directly or indirectly destroyed most of it.

In a powerful argument Amina Mama (1997, p. 48) shows that to understand violence against women in postcolonial Africa we must understand the violence of colonialism; and to understand that, we must start with "gender relations and gender violence at the imperial source." The Christian societies of Europe that launched the global conquests of the last five-hundred years were already patriarchal and organized for war. Until the machine gun and the aeroplane appeared, the only overwhelming weapon they had was the broadside-firing warship. It was their military organization and ruthlessness that enabled conquest on land.

This social technology involved constructions of masculinity. The masculinities of empire were necessarily bound up with the enabling of violence—violence sufficient to overcome the considerable military capabilities of colonized societies. When the colonizers sorted men into categories of "manly" and "effeminate," as they often did (Sinha 1995), it was groups perceived as warriors—Sikh, Pathan, Zulu, Cheyenne—who were admired, though not trusted.

In Northern research on "gender-based violence," violence is usually understood as a consequence of gender arrangements, i.e., as a dependent variable. In postcolonial analyses like Mama's, violence is constitutive for gender relations. In an essay in the journal Feminist Africa, Jane Bennett (2010, p. 35) considers homophobic violence. She muses that, seen in the light of Southern experience, the connection between gender and violence changes shape: "gender, as practiced conventionally despite diversity of contexts, is violence."

The double movement of disrupting indigenous gender orders and creating new ones was a fundamental and persisting feature of colonialism. Memory of the disruption is the driving force in one of the most famous postcolonial documents, Chinua Achebe's (1958) great novel about masculinity in West Africa at the time of conquest, *Things Fall Apart*.

The dis-ordering of gender relations occurred in multiple ways, including rape, which was endemic in conquest and disrupted indigenous kinship and communal relations with the land; forced migrations, up to the huge scale of the Atlantic slave trade; the loss of women's land rights, a feature of colonialism in the Pacific (Stauffer 2004); and the suppression of gender groupings such as the two-souled people of indigenous North America (Williams 1986). Imperial expansion also disrupted gender relations among the colonizers. The early history of the British settlements in Australia is full of debate about sexual anarchy and gender imbalance (Reid 2007). In the 1840s and 1850s there was a celebrated attempt to import a supply of women from England—a distance of twenty-thousand kilometres—to become respectable servants and wives (Kiddle 1992).

The making of colonial societies deeply concerned gender. It required the management of reproductive bodies through relationships that organized sexuality, birth and childrearing, domestic work, and the broad division of labor. Colonial economies required continuing workforces, and colonizing elites required family and inheritance structures.

In trying to stabilize the turbulent situations created by colonizing violence and the resistance of the colonized, the colonizing power brought into play mechanisms that can be seen as the initial hegemonic projects of colonialism. Establishing hegemony was a principal task of missionary religion, as Valentine Mudimbe (1994, p. 140) notes in his powerful analysis of Belgian colonization in the Congo. All over the colonized world, missionary religion concerned itself with imposing a new order on gender relations and especially sexuality.

Hegemony is a matter of institutions as well as beliefs. Where schools were introduced by colonial governments or churches, they were typically gender-segregated. Systems of law regulated indigenous marriage, women's rights, and inheritance. Gender relations were a significant concern in colonial legal codes such as those written by the French for Cambodia (Haque 2012). Colonized men were recruited in considerable numbers into imperial armed forces, especially the British and French. Patriarchal households organized labor forces and allowed white men sexual access to slave and indigenous women (Saffioti 1969).

A spectrum of hegemonic mechanisms also developed among the colonizers. They were sketched in J. O. C. Phillips's (1987) pioneering study of settler-colonial masculinities in New Zealand, and can be seen very clearly in Robert Morrell's (2001) classic study of colonial Natal. Morrell traces the institutionalization of a hegemonic form of masculinity in the schools, military forces, and civil society of the British settlers. It was specifically a harsh and insistent masculinity adapted to the need to dominate a colonized population.

Nothing guaranteed that colonial strategies for hegemony would succeed. Indeed, the project was inherently contradictory. The dynamics of colonialism both created the need and continually disrupted the results achieved. Colonialism disrupted gender order by continuing violence and dispossession; by the turbulence of the global capitalist economy; by continuing resistance, from Tupac Amaru to Abd el-Krim. There is every reason to think gender hegemony in colonial contexts was patchy, contested, and varied greatly from one part of the colonial world to another.

Out of Colonialism: Hegemonic Projects in Resistance and Development

Raymond Suttner (2005), in an illuminating study of the armed struggle against apartheid in South Africa, notes that colonial and apartheid authorities typically denied the manhood of African men. Indigenous men were treated as children in need of control—"boy" was an everyday term. Resistance by men, not surprisingly, took the shape of an assertion of manhood. The ANC mobilized stories of heroic resistance from the past and young men often interpreted joining the struggle as a form of initiation into manhood.

Such collective masculinity projects are widespread among resistance movements, giving prestige to young men on the front line, such as Palestinian youth in the intifada (Peteet 1994). Post-conflict, this can lead to severe problems, with continuing community violence in South Africa (Xaba 2001), Timor Leste (Myrttinen 2012), and other cases.

It is important to note, therefore, the other dimensions of gender in resistance movements. Suttner (2005) carefully documents emotionality, confronting of pain, and desire for the presence of children, dimensions given legitimacy by leaders such as Chris Hani. Very similar conclusions are reached by Ortega and Maria (2012) in an impressive study of militants in Latin American guerrilla movements. These movements had multiple forms of masculinity, a significant place for emotion, and an ideology of social equality that often gave women a prominent role. An oral-history study of gender issues in the Vietnamese wars for independence against the French and Americans is called Even the Women Must Fight (Turner and Hao 1998). Marnia Lazreg (1990, p. 768) says of the Algerian independence struggle that "the very fact that women entered the war willingly was in and of itself a radical break in gender relations."

Yet Lazreg and Maria Mies (1986) have remarked how often national liberation movements mobilized women for struggle, but on gaining independence installed patriarchal regimes. A striking example is the anticolonial rising in Ireland.

Across Dublin, women were in combat in all the insurgent battalions except one. That one was commanded by Eamon de Valera, who sent the women home (Townshend 2005). After independence, with a conservative Catholicism ascendant, women were thoroughly marginalized in Irish public life. After de Valera himself became head of government, he brought in a Constitution that defined woman's place as "within the home" (1937 Constitution, Article 41).

Postcolonial societies have often shown a "shifting terrain of gender relations," as Linden Lewis (2004) puts it in his study of Caribbean masculinity. Fatima Mernissi's pioneering fieldwork on masculinity (1975) found evidence of "sexual anomie" among young men in Morocco and great uncertainty in the transition between generations. In Egypt a couple of decades later, Mai Ghoussoub (2000) found evidence of a great cultural disturbance in gender relations, and "a chaotic quest for a definition of modern masculinity."

Discussing Iran under the neocolonial regime before the Islamic revolution, Al-e Ahmad (1962) described a thin and rootless masculinity among the middle class, "a donkey in a lion's skin."

Not all changes in gender relations, however, were chaotic. Some were driven by the policies of developmental states. In the Turkish successor state to the Ottoman Empire, a military regime under the war hero Mustafa Kernal created a paradigm of secular development in which modernizing the position of women was a central, almost iconic, feature. But military masculinities remained hegemonic, with conscription as a rite of passage into manhood for generations of Turkish men (Sinclair-Webb 2000).

Economic development was another important arena of gender formation. Where steelmaking, machine manufacturing, and large-scale extractive industry were launched, industrial labor provided a central definition of working-class masculinity. Mike Donaldson (1991) showed this for Wollongong in settler-colonial Australia, noting how the gradual destruction of working bodies became part of the enactment of masculinity, demonstrating toughness and endurance. Dunbar Moodie's (1994) study of gold mining in South Africa presents another striking case, with industrial militancy growing while older constructions of masculinity were displaced and family connections with pastoral homelands weakened.

Elite masculinities could change too, as they did in Japan. Starting in the Meiji era, a strong developmental state and a small group of powerful conglomerates, the zaibatsu, launched heavy industry and constructed finance, education, and weapon systems. For a time Japan became an imperial power. It is not surprising that a strongly-marked hegemonic model of masculinity on a national scale was produced, the corporate sarariiman (Ito 1993; Dasgupta 2003). The model was based on stable long-term employment as a manager, a sharp gender division of labor in the home, and a steep hierarchy of authority in the workplace. These conditions eroded in the late twentieth century, accompanied by public debates about salaryman masculinity and greater recognition of diversity (Roberson and Suzuki 2003; Taga et al. 2011).

The social transformations in development, then, involved new waves of gender disordering, and new hegemonic projects. In Turkey and Japan this produced historically original constructions of masculinity that achieved hegemony at the national level. In other situations it seems that a sustainable hegemony was not achieved in the era of decolonization and development, though social dominance for groups of powerful men usually was.

Masculinities in Neoliberal Development

Since the 1970s, development strategies in the periphery have diverged. Many countries under neoliberal regimes abandoned import replacement industrialization and turned to mining and agriculture to find "comparative advantage" in world markets. Others used low wages as their comparative advantage in manufacturing for export (Connell and Dados 2014). On both pathways, states and ruling classes in the periphery used the removal of social protections and the privatization of public assets to bolster their position in global markets.

Neoliberalism almost everywhere has been introduced by male elites, who have rolled back institutional protections and cultural gains by women, while promoting women's labor force participation and a notionally de-gendered ideology of individual advancement. Gender segregation and gendered exploitation flourish in new forms in the factories of the "south China miracle," the maquilas of the Mexican borderlands, the huge expatriate workforce of the oil industry in the Persian Gulf states, and among migrant domestic workers such as the baomu of capitalist China (Yan 2008).

Neoliberalism has had contradictory effects for masculinity formation. For large numbers of men, "structural adjustment" meant unemployment or casualization. Mara Viveros (2001) notes the impact across Latin America, especially the growing difficulty for working-class men in sustaining a breadwinner model of masculinity. South Africa has a similar experience (Hunter 2004), where the transition from apartheid to neoliberalism led to the collapse of secondary industry, with mass unemployment and an increasingly desperate situation for young Black men.

On the other hand, as neoliberal regimes concentrated the profits of development, they created conditions for the growth of entrepreneurial masculinities. The most spectacular examples are in China and India, where elite businessmen now control fortunes comparable to the great fortunes in the US/European metropole.

Yet the money and power of these new elites may not easily translate into achieved hegemony. Writing from post-communist Serbia, Marina Blagojevic (2013) notes how the pressures of the neoliberal era divide masculinities in the eastern European "semi periphery." The dismantling of the state-centered economy, and dependence on Western Europe, threatened men who were bearers of old forms of hegemonic masculinity. Others, who have marketable assets or skills, position themselves in the neoliberal economy and attempt to develop an international-style entrepreneurial masculinity. This split between hegemonic projects is not easily resolved locally.

Neoliberal development may also create, unexpectedly, local conditions for more egalitarian gender relations rather than more hierarchical ones. The violent neoliberal turn in Chile, for instance, created an export fruit industry that drew many women for the first time into wage-earning labor, and eroded patriarchal relations in rural families (Tinsman 2000).

Neoliberal development strategies in the periphery depend on the growth of global markets, global finance, and global communications. The consequence has been the creation of new social arenas in transnational rather than local space. These are powerfully gendered, though in new ways. Transnational manufacturing involves, as Juanita Elias (2008) has shown, a structure of relations between the professionalized masculinities of global corporate managements, local patriarchies in state and factory management, and gender-divided, often feminized, local workforces.

Global Hegemony and Contestation Today

The Offshore Metropole and Masculinity

The growth of European empire in past centuries depended on certain social conditions in the metropole: strong states organized for sustained warfare; ideologies of supremacy, first religious and then racial; population growth able to sustain a flow of bodies to the colonies; and a mercantile capitalism searching for unlimited profits. I will call the complex of institutions and cultural patterns and practices that enabled metropolitan societies to sustain empire the metropole-apparatus. The historical continuity of the metropole-apparatus underlies the coloniality of power and its persistence in the postcolonial world.

In the neoliberal era of globalization, the metropole-apparatus has, to a certain extent, broken free from the territorial states where it was originally based. The capacity to exercise global power is still connected to the wealth of Europe and North America and the military power of the United States. But metropolitan power increasingly operates offshore, through transnational institutions and spaces of a historically new kind: transnational corporations; global markets (especially finance markets, symbolized by the 24-hour operation of stock exchanges); international electronic media, including television and the Internet; and an international state, including both the United Nations complex and the linked-up military, intelligence, and security apparatuses of NATO and other alliances.

The gender research in the metropole most relevant to understanding the contemporary metropole-apparatus concerns managerial masculinities. There is persuasive empirical work documenting power-oriented gender practices in both states and companies (Mulholland 1996; Wajcman 1999). Michael Roper's (1994) excellent history of managerial masculinities in British engineering firms traced changes in a local hegemonic masculinity, as managers' concerns shifted from the workforce and the production process to a neoliberal focus on finance and short-term profit Richard Collier's (2010) careful study of corporate lawyers in Britain shows professional masculinities close to the patterns of corporate management—with possibilities of change, especially in the younger generation, held back by competitive pressures and the conservatism of their seniors.

Some recent studies have traced the gendered character of markets themselves as social institutions. An aggressive, misogynist occupational culture appears in arenas such as commodity and currency trading and financial manipulation generally (Levin 2001; McDowell 2010; Connell 2010a).

Top corporate management in the global economy is overwhelmingly the business of men. Of the five-hundred biggest international corporations listed in *Fortune* magazine's "Global 500" in 2014, 95.2% had a man as CEO. In many ways the social world of these men resembles the managerial masculinities documented in the old metropole—competitive and power-oriented. Elite managers persistently construct hierarchical relationships with women, whether wives, employees, or sex workers. A striking confirmation emerges from an international bank merger in Scandinavia, a region whose gender orders are among the most egalitarian on earth. Janne Tienari et al. (2005) conducted interviews with the top executives after the merger. The senior managers were almost all men, and did not want to hear about gender equality problems. They took management to be naturally men's business, "constructed according to the core family and male-breadwinner model."

But transnational business masculinity cannot simply reproduce historic bourgeois masculinity. The labor of TNC management is secularized, mobile, and highly technologized, being closely integrated with corporate intranets and high-technology communications (Connell 2010b). This is not a "geek" masculinity but it requires interaction with the changing masculinities of the ICT industries (Poster 2013). Because TNC management involves negotiations with local patriarchies (Elias 2008), it requires a degree of tolerance for differences in culture; and there are indications this also applies to sexuality. A professionalization of management has been attempted through the US-style MBA, and elite business schools in the metropole take pride in having an international intake of students. Firms from relatively affluent countries in the periphery, such as Chile and Australia, mostly follow transnational managerial practice though they participate in global business on unequal terms (Olavarría 2009; Connell 2010a).

There seems, then, to be a changed hegemonic project of masculinity formation within the global corporate economy. This is not producing a kinder, more inclusive, or more feminized capitalism; a closer look at the masculinities of the main power-holding elites in the contemporary world shows the huge task still ahead for the project of gender equality (Connell 2016). But we do see hegemonic projects responding to the turbulence faced by global management and the impossibility of imposing any single gender template.

Contemporary Hegemonic Projects

As metropolitan power moves offshore into the complex of transnational institutions, the need for mechanisms of consent that produce hegemony at a local or national level declines. Are we now producing, on the scale of global society, the situation that Guha diagnosed in colonial India: hegemony an illusion and coercion the reality? Only in a few parts of the world do state or economic elites now rely on custom or claim oldestablished authority, and even where they do (e.g., in Thailand) the claim is fragile. Gone too is the old-style paternalism of improving public services or guaranteeing welfare to subaltern groups. The opposition to "Obamacare" from the political right in the United States is a striking example.

But a more limited and complex form of hegemony may be found on the world scale. Three conditions would be sufficient to sustain the position of transnational corporate masculinity:

1. The institutional complex—private property and state authority—currently delivering control of the global economy, remains socially accepted within the most powerful states. Although there is widespread discontent, seen in the 2016 Brexit vote and the Sanders campaign, no organized alternative has much traction in the United States or the European Union. Police-state repression in China and populist conservatism in India are currently well entrenched.

2. The self-selecting masculine elites now in power retain their legitimacy and organizational control within the new metropole-apparatus. The corporate

recovery from the 2007–2008 global financial crisis suggests the capacity for continuing control is there.

3. The metropole-apparatus connects well enough with national power structures in the periphery to allow continued extraction of raw materials, overseas trade and corporate operations, and to sustain compliant states, in the periphery (see Mbeki 2009, for diagnosis of Africa along these lines; Messerschmidt 2010 for the symbolic projection of masculinities by the US political elite).

Yet these conditions have to be worked on. The incessant busy-ness of corporate and political management, with its penumbra of bribery and intimidation and its sponsorship of violent interventions, show there is no automatic global control. The gender dynamics outlined in this article show many examples of tension and dis-articulation. The extension of the neoliberal human rights regime to issues of reproduction and sexuality, to take just one example, has been repeatedly opposed by the most patriarchal governments in the periphery (in UN population debates as recently as 2014). The Islamist insurgencies of recent years, from Afghanistan to Nigeria, are if anything more patriarchal than the regimes they confront.

The emerging world gender order is far from being a smoothly-running machine. Rather, it is a scene of conflicting hegemonic projects. It has multiple tiers, where different configurations of masculinity are at work, and come into conflict. Major gains for gender equality have been made in the last half century, notably in state provision of education for girls and the rising participation of women in wage work. Up to now, however, these changes have yielded only a little ground for democratic projects of change in masculinity.

Counter-Hegemony

Movements for change in masculinity, nevertheless, keep welling up. South Africa, for instance, remains a violent and unequal society, where gender inequalities are deeply implicated in the world's heaviest burden of HIV/AIDS (Epstein et al. 2004). But South Africa has also seen intense debates about changing masculinities, accompanied by local projects of change (Sideris 2005; Shefer et al. 2007; Ratele 2014). India too is a highly unequal society, yet has multiple sources of change among men, revealed in Radhika Chopra's books *From Violence to Supportive Practice* (2002) and *Reframing Masculinities* (2007).

Programs concerned with the reduction of violence or the prevention of AIDS are now widespread. They are found in Latin America (Zingoni 1998), in Africa (e.g., Sonke Gender Justice, www.genderjustice.org.za), and in Southeast Asia and other regions (Lang et al. 2008; United Nations 2013). They have recently been linked internationally through the MenEngage network (www.menengage.org), which has sponsored two international conferences of activists, the most recent producing the "Delhi Declaration" of 2014. These projects represent a historic change, mobilizing men internationally for gender justice.

But to be realistic, they remain relatively small; and mostly follow concepts developed in the global North. As Dowsett (2003) noted in a study of AIDS prevention in Bangladesh, Anglophone categories such as "MSM," "identity," "heterosexuality," and even "men" may misrepresent local social realities. Over time, a greater concern with distinctive local experience and strategy has been developing. Melissa Meyer and Helen Struthers's media project *[Un]covering Men: Rewriting Masculinity and Health in South Africa* (2012) is an example of the creative work that results.

Das and Singh (2014) offer something even more striking. From more than a decade of NGO-based programmes in India, including the well-known Men's Action for Stopping Violence Against Women (MASVAW), they have generated a seven-point theory of change. This theory emphasizes the different starting-points in gender reform for men and women, the inevitability of resistance, and the strategies most likely to overcome it.

Do local gender-equality projects among men represent a counter-hegemonic strategy at the societal level? That question was raised by a team leading workshops about masculinity and violence in the very difficult environment of El Salvador, after a brutal civil war (Bird et al. 2007). Part of their answer is that local interventions bring out alternative practices and desires for peace that already exist in the society. Such possibilities can be seen in other places too (Haque 2013; Myrttinen 2012).

However, the NGO format of social action has been problematic for feminists, because of the way it is integrated into neoliberal politics (Alvarez 1999). The NGOs specifically concerned with gender-based violence overwhelmingly depend on corporate charity, international aid programs, or national states. Published research reveals few connections between masculinity reform efforts and union activism, landless people's movements, environmental activism, or other movements that offer a significant challenge to corporate or state power. They seem, so far, no threat to the corporate masculinity of the new metropole. For such a challenge to develop would require a different structure of politics.

Conclusion

As Rachel Jewkes et al. (2015) show in a valuable current review, the concept of hegemonic masculinity informs much anti-violence activism and when carefully used can illuminate problems of strategy. Their argument is consistent with the approach adopted here of analyzing masculinities in terms of collective hegemonic projects, local, societal, and global.

I am arguing that the changing structures of imperialism and neoliberal global power are a vital part of our understanding of masculinities. They represent both the structural conditions of hegemonic projects now and the sedimented consequences of gender projects in the past. Hegemony cannot be presumed in the violent and exploitative social relations that constitute imperial and transnational gender orders. But hegemony is constantly under construction, renovation, and contestation.

In this contestation, intellectual struggle is required. Knowledge produced in the majority world and Southern perspectives on social relations and power are increasingly important for global gender politics. A notable example is provided by AMEGH, the Mexican Association for the Study of the Gender of Men, and the Colegio de la Frontera Norte. Their work has recently produced a powerful volume on the gendered violence in northern México, Salvador Cruz Sierra's Vida, muerte y resistencia en Ciudad Juárez (2013).

Knowledge is not a substitute for action. But accurate knowledge and theoretical insight are priceless assets for action, when action is concerned with contesting power and achieving social justice. That was our hope in formulating the concept of hegemonic masculinity and remains the reason to build on it today.

References

Achebe, C. (1958). *Things fall apart*. London: Heinemann.

Al-e Ahmad, J. (1982 [1962]). *Weststruckness*. Translated by John Green and Ahmad Alizadeh. Lexington, KY: Mazda Publishers.

Alvarez, S. E. (1999). Advocating feminism: The Latin American feminist NGO "boom." *International Feminist Journal of Politics*, *1*(2), 181–209.

Amar, P. (2011). Middle East masculinity studies: Discourses of "men in crisis," industries of gender in revolution. *Journal of Middle East Women's Studies*, *7*(3), 36–70.

Bennett, J. (2008). Editorial: Researching for life: Paradigms and power. *Feminist Africa*, *11*, 1–12.

Bennett, J. (2010). "Circles and circles": Notes on African feminist debates around gender and violence in the C21. *Feminist Africa*, *14*, 21–47.

Bhambra, G. K. (2014). *Connected Sociologies*. London: Bloomsbury Academic.

Bird, S., Rutilio D., Madrigal, L., Ochoa, J. B., & Tejeda, W. (2007). Constructing an alternative masculine identity: The experience of the Centro Bartolomé de las casas and Oxfam America in El Salvador. *Gender and Development*, *15*(1), 111–121.

Blagojevic, M, (2013). Transnationalization and its absence: The Balkan semiperipheral perspective on masculinities. In J. Hearn, M. Blagojevic, & K. Harrison (Eds.), *Rethinking transnational men: Beyond, between and within nations* (pp. 163–184). New York: Routledge.

Bulbeck, C. (1998). *Re-orienting western feminisms: Women's diversity in a postcolonial world*. Cambridge: Cambridge University Press.

Carrigan, T., Connell, R., & Lee, J. (1985). Toward a new sociology of masculinity. *Theory and Society*, *14*(5), 551–604.

Chopra, R. (Ed.). (2002). *From violence to supportive practice: Family, gender and masculinities in India*. New Delhi, India: United Nations Development Fund for Women.

Chopra, R. (Ed.) (2007). *Reframing masculinities: Narrating the supportive practices of men*. New Delhi, India: Orient Longman Private.

Collier, R. (2010). *Men, law and gender: Essays on the "man" of law*. Abingdon, VA: Routledge.

Connell, R. (1983). The black box of habit on the wings of history: Critical reflections on the theory of social reproduction, with suggestions on how to do it better. In R. Connell (Ed.), *Which way is up? Essays on sex, class and culture* (pp. 140–161). Sydney: Allen & Unwin.

Connell, R. (2007). *Southern theory: The global dynamics of knowledge in social science*. Cambridge: Polity.

Connell, R. (2010a). Im Innern des gläsernen Turms: Die Konstruktion von Männlichkeiten im Finanzkapital. *Feministische Studien*, *28*(1), 8–24.

Connell, R. (2010b). Building the neoliberal world: Managers as intellectuals in a peripheral economy. *Critical Sociology*, *36*(6), 777–792.

Connell, R. (2014). Margin becoming centre: For a world-centred rethinking of masculinities. *NORMA: International Journal for Masculinity Studies*, *9*(4), 217–231.

Connell, R. (2015). Meeting at the edge of fear: Theory on a world scale. *Feminist Theory*, *16*(1), 49–66.

Connell, R. (2016). 100 million Kalashnikovs: Gendered power on a world scale. *Debate Feminista*, *51*, 3–17.

Connell, R., & Dados, N. (2014). Where in the world does neoliberalism come from? The market agenda in southern perspective. *Theory and Society*, *43*(2), 117–138.

Connell, R., & Messerschmidt, J. W. (2005). Hegemonic masculinity: Rethinking the concept. *Gender and Society*, *19*(6), 829–859.

Cruz Sierra, S. (Ed.). (2013). *Vida, muerte y resistencia en Ciudad Juárez: Una aproximación desde la violencia, el género y la cultura*. Tijuana, Mexico: El Colegio de la Frontera Norte.

Das, A., & Singh, S. K. (2014). Changing men: Challenging stereotypes. Reflections on working with men on gender issues in India. *IDS Bulletin*, *45*(1), 69–79.

Dasgupta, R. (2003). Creating corporate warriors: The "salaryman" and masculinity in Japan. In K. Louie & M. Low (Eds.), *Asian Masculinities* (pp. 118–134). London: RoutledgeCurzon.

Donaldson, M. (1991). *Time of our lives: Labour and love in the working class*. Sydney: Allen & Unwin.

Dowsett, G. W. (2003). HIV/AIDS and homophobia: Subtle hatreds, severe consequences and the question of origins. *Culture, Health & Sexuality*, *5*(2), 121–136.

Elias, J. (2008). Hegemonic masculinities, the multinational corporation, and the developmental state: Constructing gender in "progressive" firms. *Men and Masculinities*, *10*(4), 405–421.

Epstein, D., Morrell, R., Moletsane, R., & Unterhalter, E. (2004). Gender and HIV/AIDS in Africa south of the Sahara: Interventions, activism, identities. *Transformation*, *54*, 1–16.

Ford, M., & Lyons, L. (Eds.). (2012). *Men and masculinities in Southeast Asia*. London: Routledge.

Ghoussoub, M. (2000). Chewing gum, insatiable women and foreign enemies: Male fears and the Arab media. In M. Ghoussoub & E. Sinclair-Webb (Eds.), *Imagined Masculinities: Male Identity and Culture in the Middle East* (pp. 227–235). London: Saqi Books.

Go, Julian. (2012). For a postcolonial sociology. *Theory and Society*, *42*(1), 25–55. Retrieved from http://link.springer.com/article/10.1007/s11186-012-9184-6#page-2

Guha, R. (1989). Dominance without hegemony and its historiography. *Subaltern Studies*, *6*, 210–309.

Haque, M. M. (2012). Men, masculinities and social change: Exploring Khmer masculinities and their implications for domestic violence. PhD thesis, University of Sydney.

Haque, M. M. (2013). Hope for gender equality? A pattern of post-conflict transition in masculinity. *Gender, Technology and Development*, *17*(1), 55–71.

Hountondji, P. J. (1997). Introduction: Recentring Africa. In Paulin J. Hountondji (Ed.), *Endogenous Knowledge: Research Trails* (pp. 1–39). Dakar, Senegal: CODESRIA.

Howson, R. (2006). *Challenging Hegemonic Masculinity*. London: Routledge.

Hunter, M. (2004). Masculinities, multiple-sexual-partners, and AIDS: The making and unmaking of Isoka in KwaZulu-Natal. *Transformation*, *54*, 123–153.

Ito, K. (1993). *Otokorashisa-n-yukue [Directions for masculinities: Cultural sociology of manliness]*. Tokyo, Japan: Shiny sha.

Jewkes, R., Morrell, R., Hearn, J., Lundqvist, E., Blackbeard, D., Lindegger, G., Quayle, M., Sikweyiya, Y., & Gotlzén, L. (2015). Hegemonic masculinity: Combining theory and practice in gender interventions. *Culture, Health & Sexuality*, *17*(sup2), 96–111.

Jolly, M. (2008). Moving masculinities: Memories and bodies across Oceania. *The Contemporary Pacific*, *20*(1), 1–24.

Kiddle, M. (1992 [1950]). *Caroline Chisholm*. Melbourne, Australia: Melbourne University Press.

Lang, J., Greig, A., & Connell, R., in collaboration with the Division for the Advancement of Women. (2008). *The role of men and boys in achieving gender equality*. New York: United Nations Division for the Advancement of Women. Retrieved from http://www.un.org/womenwatch/daw/egm/men-boys2003/Connell-bp.pdf

Laurie, N. (2005). Establishing development orthodoxy: Negotiating masculinities in the water sector. *Development and Change*, *36*(3), 527–549.

Lazreg, M. (1990). Gender and politics in Algeria: Unraveling the religious paradigm. *Signs*, *15*(4), 755–780.

Levin, P. (2001). Gendering the market: Temporality, work, and gender on a national futures exchange. *Work and Occupations*, *28*, 112–130.

Lewis, L. (2004). Caribbean masculinity at the Fin de Siècle. In R. E. Reddock (Ed.), *Interrogating Caribbean masculinities: Theoretical and empirical analyses* (pp. 244–266). Jamaica, Barbados, Trinidad and Tobago: University of the West Indies Press.

Lugones, M. (2007). Heterosexism and the colonial/modern gender system. *Hypatia*, *22*(1), 186–219.

Mama, A. (1997). Sheroes and villains: Conceptualizing colonial and contemporary violence against women in Africa. In M. J. Alexander & C. T. Mohanty (Eds.), *Feminist Genealogies, Colonial Legacies, Democratic Futures* (pp. 46–62). New York: Routledge.

Mbeki, M. (2009). *Architects of poverty: Why African capitalism needs changing*. Johannesburg, South Africa: Picador Africa.

Mbembe, A. (2001). *On the postcolony*. Berkeley: University of California Press.

McDowell, L. (2010). Capital culture revisited: Sex, testosterone and the city. *International Journal of Urban and Regional Research*, *34*(3), 652–658.

Memissi, F. (1985 [1975]). *Beyond the veil: Male-female dynamics in modem Muslim society* (Rev. ed.). London: Saqi Books.

Messerschmidt, J. W. (2010). *Hegemonic masculinities and camouflaged politics: Unmasking the Bush dynasty and its war against Iraq*. Boulder, CO: Paradigm Publishers.

Meuser, M. (Ed.). (2010). Erwägen, Wissen, Ethik, 2010, *21*(3): special issue on hegemonic masculinity.

Meyer, M., & Struthers, H. (Eds.). (2012). *[Un]covering men: Rewriting masculinity and health in South Africa*. Auckland Park, South Africa: Fanele.

Mies, M. (1986). *Patriarchy and Accumulation on a World Scale*. London: Zed.

Moodie, T. D., & Ndatshe, V. (1994). *Going for gold: Men, mines and migration*. Johannesburg, South Africa: Witwatersrand University Press.

Morrell, R. (2001). *From boys to gentlemen: Settler masculinity in colonial Natal 1880–1920*. Pretoria: University of South Africa.

Morrell, R., & Swart, S. (2005). Men in the Third World: Postcolonial perspectives on masculinity. In M. S. Kimmel, J. Hearn, & R. Connell (Eds.), *Handbook of*

Studies on Men and Masculinities (pp. 90–113). Thousand Oaks: Sage.

Mudimbe, V. Y. (1994). *The idea of Africa*. Bloomington: Indiana University Press.

Mulholland, K. (1996). Entrepreneurialism, masculinities and the self-made man. In D. L. Collinson & J. Hearn (Ed.), *Men as managers, managers as men* (pp. 123–149). London: Sage.

Myrttinen, H. (2012). Violence, masculinities and patriarchy in post-conflict Timor-Leste. In M. Ford & L. Lyons (Eds.), *Men and Masculinities in Southeast Asia* (pp. 103–120). London: Routledge.

Olavarria, J. (Ed.). (2009). Masculinidades y globalización: Trabajo y vida privada, familias y sexualidades. Santiago, Red de Masculinidiad/es Chile, Universidad Academia de Humanismo Cristiano and CEDEM.

Ortega, D., & Maria, L. (2012). Looking beyond violent militarized masculinities: Guerrilla gender regimes in Latin America. *International Feminist Journal of Politics*, *14*(4), 489–507.

Pascoe, C. J., & Tristan, B. (2016). Introduction: Exploring masculinities: History, reproduction, hegemony, and dislocation. In C. J. Pascoe & T. Bridges (Eds.), *Exploring masculinities: Identity, inequality, continuity, and change* (pp. 1–34). New York: Oxford University Press.

Peteet, J. (1994). Male gender and rituals of resistance in the Palestinian intifada: A cultural politics of violence. *American Ethnologist*, *21*(1), 31–49.

Phillips, J. (1987). *A man's country? The image of the Pakeha male—A history.* Auckland: Penguin.

Poster, W. R. (2013). Subversions of techno-masculinity: Indian ICT professionals in the global economy. In J. Beam, M. Blagojevic, & K. Harrison (Eds.), *Rethinking transnational men: Beyond, between and within nations* (pp. 113–133). New York: Routledge.

Ratele, K. (2013). Masculinities without tradition. *Politikon: South African Journal of Political Studies*, *40*(1), 133–156.

Ratele, K. (2014). Currents against gender transformation of South African men: Relocating marginality to the centre of research and theory of masculinities. *NORMA: International Journal for Masculinity Studies*, *9*(1), 30–44,

Reid, K. (2007). *Gender, clime and empire: Convicts, settlers and the state in early colonial Australia.* Manchester: Manchester University Press.

Roberson, J. E., & Suzuki, N. (Eds.). (2003). *Men and masculinities in contemporary Japan: Dislocating the salaryman doxa*. London: RoutledgeCurzon.

Roper, M. (1994). *Masculinity and the British organization man since 1945*. Oxford: Oxford University Press.

Rosa, M. C. (2014). Theories of the south: Limits and perspectives of an emergent movement in social sciences. *Current Sociology*, 62(6). doi:10.1177/0011392114522171

Saffioti, H. I. B. (1978 [1969]). *Women in class society [A mulher na sociedade de classes]*. New York: Monthly Review Press.

Shefer, T., Ratele, K., Strebel, A., Shabalala, N., & Buikema, R. (Eds.). (2007). *From boys to men: Social constructions of men in contemporary society*. Lansdowne, South Africa: UCT Press.

Sideris, T. (2005). "You have to change and you don't know how!": Contesting what it means to be a man in rural area of South Africa. In G. Reid & L. Walker (Eds.), *Men Behaving Differently* (pp. 111–137). Cape Town: Double Storey Books.

Silberschmidt, M. (2004). Men, male sexuality and HIV/ ATDS: Reflections from studies in rural and urban East Africa. *Transformation*, *54*, 42–58.

Sinclair-Webb, E. (2000). 'Our Bülent is now a commando': military service and manhood in Turkey. In M. Ghoussoub & E. Sinclair-Webb (Eds.), *Imagined masculinities: Male identity and culture in the Middle East* (pp. 65–92). London: Saqi Books.

Sinha, M. (1995). *Colonial masculinity: The 'manly Englishman' and the 'effeminate Bengali' in the late nineteenth century.* Manchester: Manchester University Press.

Stauffer, R. H. (2004). *Kahana: How the land was lost*. Honolulu: University of Hawai'i Press.

Suttner, R. (2005). Masculinities in the African National Congress-led liberation movement: The underground period. *Kleio*, *37*, 71–106.

Taga, F., Higashino, M., Sasaki, M., & Yohei, M. (2011). *Changing lives of salarymen*. Kyoto, Japan: Minerva Shobo.

Tienari, J., Søderberg, A.-M., Holgersson, C., & Vaara, E. (2005). Gender and national identity constructions in the cross-border merger context. *Gender, Work and Organization*, *12*(3), 217–241.

Tinsman, H. (2000). Reviving feminist materialism: Gender and neoliberalism in Pinochet's Chile. *Signs*, *26*(1), 145–188.

Townshend, C. (2005). *Easter 1916: The Irish Rebellion*. London: Allen Lane.

Turner, K. G., & Hao, P. T. (1998). *Even the women must fight memories of war from North Vietnam*. New York: Wiley.

United Nations. (2013). *Why do men use violence and how can we stop it? Quantitative findings from the UN multi-country study on men and violence*. Bangkok: UNDP, UNFPA, UN Women and UNV.

Vieira de Jesus, D. S. (2011). Bravos novos mundos: Uma leitura pós-colonialista sobre masculinidades ocidentais. *Estudos Feministas*, 19(1), 125–139.

Viveros, V. M. (2001). Contemporary Latin American perspectives on masculinity. *Men and Masculinities*, *3*(3), 237–260.

Wajcman, J. (1999). *Managing like a man: Women and men in corporate management*. Cambridge: Polity and Sydney: Allen & Unwin.

Williams, W. L. (1986). *The spirit and the flesh: Sexual diversity in American Indian culture.* Boston: Beacon Press.

Xaba, T. (2001). Masculinity and its malcontents: The confrontation between 'struggle masculinity' and 'post-struggle masculinity' (1990–1997). In R. Morrell (Ed.), *Changing Men in Southern Africa* (pp. 105–124). Pietermaritzburg, South Africa: University of Natal Press.

Yan, H. (2008). *New masters, new servants: Migration, development and women workers in China.* Durham, NC: Duke University Press.

Zhan, J. (2015). *Xing bie zhi lu: Rui wen kang nai er de nan xing qi zhi li lun tan suo [The road of gender: An exploration of Raewyn Connell's theories of masculinity].* Guilin, China: Guangxi Normal University Press.

Zingoni, E. L. (1998). Masculinidades y violencia desde un programa de acción en México. In T. Valdés & J. Olavarría (Eds.), *Masculinidades y equidad de género en América Latina* (pp. 130–136). Santiago: FLACSO/UNFPA.

Introduction to Reading 6

The anthropologist Serena Nanda is widely known for her ethnography of India's Hijaras, titled *Neither Man nor Woman.* The article included here is from her more recent book on multiple sex/gender systems around the world. Nanda's analysis of multiple genders among Native North Americans is rich and detailed. As you read this piece, consider the long-term consequences of the failure of European colonists and early anthropologists to get beyond their ethnocentric assumptions so they could understand and respect the gender diversity of North American Indian cultures.

1. Why does Serena Nanda use the term *gender variants* instead of *two-spirit* and *berdache*?
2. What was the relationship between sexual orientation and gender status among American Indians whose cultures included more than two sex/gender categories? How about hermaphroditism and gender status?
3. Why was there often an association between spiritual power and gender variance in Native American cultures?

Multiple Genders Among Native Americans

Serena Nanda

The early encounters between Europeans and Native Americans in the fifteenth through the seventeenth centuries brought together cultures with very different sex/gender systems. The Spanish explorers, coming from a Catholic society where sodomy was a heinous crime, were filled with contempt and outrage when they recorded the presence of men in Native North American societies who performed the work of women, dressed like women, and had sexual relations with men (Lang 1996; Roscoe 1995).

Europeans labeled these men ***berdache***, a term originally meaning male prostitute. The term was both insulting and inaccurate, derived from the European view that these roles centered on the "unnatural" and sinful practice of sodomy as defined in their own societies. This European ethnocentrism also caused early observers to overlook the specialized and spiritual functions of many of these roles and the positive value attached to them in many Native American societies.

By the late-nineteenth and early-twentieth centuries, some anthropologists included accounts of Native American sex/gender diversity in their ethnographies, attempting to explain the contributions alternative sex/gender roles made to social structure or culture. These accounts, though less contemptuous than earlier ones, nevertheless largely retained the ethnocentric

Nanda, S. (2014). Multiple genders among Native Americans. *Gender diversity: Crosscultural variations*, 2nd ed. Long Grove, IL: Waveland Press, Inc.

emphasis on berdache sexuality, defining it as a form of "institutionalized homosexuality." Influenced by functionalist theory, anthropologists viewed these sex/gender roles as functional because they provided a social niche for male individuals whose personality and sexual orientation did not match the definition of masculinity in the anthropologists' societies, or because the roles provided a "way out" of the masculine or warrior role for "cowardly" or "failed" men (see Callender and Kochems 1983).

Increasingly, however, anthropological accounts paid more attention to the association of Native American sex/gender diversity with shamanism and spiritual powers; they also noted that mixed gender roles were often central and highly valued, rather than with European concepts of homosexuality (erotic feelings for a person of the same sex), transvestism (cross-dressing), or hermaphroditism (the presence of both male and female sexual organs in an individual) continued to distort their indigenous meanings. marginal and deviant within some Native American societies. Still, the identification of Native American sex/gender diversity

In Native American societies, the European homosexual/heterosexual dichotomy was not culturally relevant as a central or defining aspect of gender. While mixed sex/gender individuals in many Native American societies did engage in sexual relations and even married persons of the same sex, this was not central to their alternative gender role. Europeans also overemphasized the function of cross-dressing in these roles, labeling such individuals as ***transvestites***; although mixed gender roles often did involve cross-dressing, this varied both within and among Native American societies. The label "hermaphrodite" was also inaccurate as a general category, although some societies did recognize biological intersexuality as the basis of sex/gender variation.

Given the great variation in Native North American societies, it is perhaps most useful to define their non-normative sex/gender roles as referring to people who partly or completely adopted aspects of the culturally defined role of the other sex or gender and who were classified as neither woman nor man but as mixed, alternative genders; these roles did not involve a complete crossing over to an opposite sex/gender role (see Callender and Kochems 1983:443).

Both Native American sex/gender diversity and anthropological understandings of these roles have shifted in the past 30 years (Jacobs, Thomas, and Lang 1997: Introduction). Most current research rejects institutionalized homosexuality as an adequate explanation of Native American sex/gender diversity: It emphasizes occupation rather than sexuality as its central feature; considers multiple sex/gender roles as normal, indeed often integrated into and highly valued in Native American sex/gender systems (Albers 1989:134; Jacobs et al. 1997; Lang 1998); notes the variation in such roles across indigenous North (and South) America (Callender and Kochems 1983; Jacobs et al. 1997; Lang 1998; Roscoe 1998); and calls attention to the association of such roles with spiritual power (Roscoe 1996; Williams 1992).

Consistent with these new perspectives, the term "berdache" is somewhat out of fashion, though there is no unanimous agreement on what should replace it. One widely accepted suggestion is the term ***two-spirit*** (Jacobs et al. 1997; Lang 1998), a term coined in 1990 by urban Native American gays and lesbians. Two-spirit has the advantage of conveying the spiritual nature of gender variance in both traditional and contemporary Native American societies, although it emphasizes the Euro-American binary sex/gender construction of male and female/man and woman, which did not characterize all Native American groups.

Distribution and Characteristics of Variant Sex/Gender Roles

Multiple sex/gender systems were found in many, though not all, Native American societies. Variant male sex/gender roles are documented for 110 to 150 societies, occurring most frequently in the region extending from California to the Mississippi Valley and the upper-Great Lakes, the Plains and the Prairies, the Southwest, and to a lesser extent along the Northwest Coast. With few exceptions, gender variance is not historically documented for eastern North America, though it may have existed prior to the European invasion and disappeared before it could be recorded historically (Callender and Kochems 1983; Fulton and Anderson 1992).

There were many variations in Native American sex/gender diversity. Some cultures included three or four genders: men, women, male variants, and female variants (e.g., biological females who, by engaging in male activities, were reclassified as to gender). Gender variant roles also differed in the criteria by which they were defined; the degree of their integration into the society; the norms governing their behavior; the way the role was publicly acknowledged or sanctioned; how others were expected to behave toward gender variant persons; the degree to which a gender changer was expected to adopt the role of the opposite sex or was limited in doing so; the power, sacred or secular, that was attributed to them; and the path to recruitment.

In spite of this variety, however, there were also some widespread similarities: transvestism, cross-gender occupation, same-sex (but different gender) sexuality, a special process or ritual surrounding recruitment, special language and ritual roles, and associations with spiritual power.

Transvestism

Transvestism was often associated with gender variance but was not equally important in all societies. Male gender variants frequently adopted women's dress and hairstyles partially or completely, and female gender variants partially adopted the clothing of men; in some societies, however, transvestism was prohibited. The choice of clothing was sometimes an individual matter, and gender variants might mix their clothing and their accessories. For example, a female gender variant might wear a woman's dress but carry (male) weapons. Dress was also sometimes situationally determined: a male gender variant would have to wear men's clothing while engaging in warfare but might wear women's clothing at other times. Similarly, female gender variants might wear women's clothing when gathering (women's work) but male clothing when hunting (men's work) (Callender and Kochems 1983:447). Among the Navajo, a male gender variant, ***nàdleeh***, would adopt almost all aspects of a woman's dress, work, language, and behavior; the Mohave male gender variant, called ***alyha***, was at the extreme end of the cross-gender continuum in imitating female physiology as well as transvestism. The repression and ultimately the almost total decline of transvestism was a direct result of U.S. prohibitions against it.

Occupation

The occupational aspects of Native American gender variance was central in most societies. Most frequently a boy's interest in the tools and activities of women and a girl's interest in the tools of male occupations signaled an individual's wish to undertake a gender variant role (Callender and Kochems 1983:447; Whitehead 1981). In hunting societies, for example, female gender variance was signaled by a girl rejecting the domestic activities associated with women and participating in playing and hunting with boys. In the Arctic and sub-Arctic this might be encouraged by a girl's parents if there were not enough boys to provide the family with food (Lang 1998). Male gender variants were frequently considered especially skilled and industrious in women's crafts and domestic work (though not in agriculture, where this was a man's task) (Roscoe 1991; 1996). Female gender crossers sometimes won the reputation of superior hunters and warriors.

The households of male gender variants were often more prosperous than others, sometimes because they were hired by whites. In their own societies the excellence of male gender variants' craftwork was sometimes ascribed to a supernatural sanction for their gender transformation (Callender and Kochems 1983:448). Female gender variants opted out of motherhood, so they were not encumbered by caring for children, which may explain their success as hunters or warriors. In Borne societies, gender variants could engage in both men's and women's work, and this, too, accounted for their increased wealth. Another source of income was payment for the special social activities due to gender variants' intermediate gender status, such as acting as go-betweens in marriage. Through their diverse occupations, then, gender variants were often central rather than marginal in their societies.

The explanation of male gender variant roles as a niche for "failed" or cowardly men who wished to avoid warfare or other aspects of the masculine role is no longer widely accepted. To begin with, masculinity was not associated with warrior status in all Native American cultures. In some societies, male gender variants were warriors, and in many others, males who rejected the warrior role did not become gender variants. Sometimes male gender variants did not go to war because of cultural prohibitions against their using symbols of maleness, for example, the prohibition against their using the bow among the Illinois. Where male gender variants did not fight, they sometimes had other important roles in warfare, like treating the wounded, carrying supplies for the war party, or directing postbattle ceremonials (Callender and Kochems 1983:449). In a few societies male gender variants became outstanding warriors, such as Finds Them and Kills Them, a Crow Indian who performed daring feats of bravery while fighting with the U.S. Army against the Crow's traditional enemies, the Lakota Sioux (Roscoe 1998:23).

Gender Variance and Sexuality

While generally sexuality was not central in defining gender status among Native Americans, in some Native American societies, same-sex sexual desire or practices were significant in the definition of gender variant roles (Callender and Kochems 1983:449). Some early reports noted specifically that male gender variants lived with and/or had sexual relations with women as well as with men; in other societies they were reported as having sexual relations only with men, and in still other societies, of having no sexual relationships at all (Lang 1998:189–95).

The bisexual orientation of some gender variant persons may have been a culturally accepted expression of their gender variance. It may have resulted from an individual's life experiences, such as the age at which he or she entered the gender variant role, and/or it may have been one aspect of the general freedom of sexual expression in many Native American societies. While male and female gender variants most frequently had sexual relations with, or married, persons of the same biological sex as themselves, these relationships were not considered homosexual in the contemporary Western understanding of that term. In a multiple gender system the partners would be of the same sex but different genders, and homogender, rather than homosexual, practices bore the brunt of negative cultural sanctions (as is true today, for example, in contemporary Indonesia). The sexual partners of gender variants were never considered gender variants themselves.

Among the Navajo there were four genders; man, woman, and two gender variants: the masculine female-bodied nàdleeh and the feminine male-bodied nàdleeh (Thomas 1997). A sexual relationship between a female-bodied nàdleeh and a woman or a sexual relationship between a male-bodied nàdleeh and a man were not stigmatized because these persons were of different genders, although they were of the same biological sex. A sexual relationship between two women, two men, two female-bodied nàdleeh, or two male-bodied nàdleeh, however, was considered homosexual, and even incestual, and was strongly disapproved of.

The relation of sexuality to variant sex/gender roles across North America suggests that sexual relations between gender variants and persons of the same biological sex were a result rather than a cause of gender variance. Sexual relationships between a man and a male gender variant were accepted in most Native American societies, though not in all, and appear to have been negatively sanctioned only when it interfered with child-producing heterosexual marriages. Gender variants' sexual relationships might be casual and wide-ranging (Europeans used the term "promiscuous"), or stable, and sometimes involved life-long marriages. In some societies, however, male gender variants were not permitted to engage in long-term relationships with men, either in or out of wedlock, and many male gender variants were reported as living alone.

A man might desire sexual relations with a (male) gender variant for different reasons: In some societies taboos on sexual relations with menstruating or pregnant women restricted opportunities for sexual intercourse; in other societies sexual relations with a gender variant person were exempt from punishment for extramarital affairs; in still other societies, for example among the Navajo, some gender variants were considered especially lucky, and a man might hope to have this luck transferred to himself though sexual relations (Lang 1998:349).

Biological Sex and Gender Transformations

European observers often confused gender variants with ***hermaphrodites*** (biologically intersexed persons). Some Native American societies explicitly distinguished hermaphrodites from gender variants and treated them differently; others assigned gender variant persons and hermaphrodites to the same alternative gender status. In most Native American societies biological sex (or the intersexed condition of the hermaphrodite) was not the criterion for a gender variant role, nor were the individuals who occupied gender variant roles anatomically abnormal. The Navajo were an exception: They distinguished between the intersexed and the alternatively gendered but treated them similarly, though not exactly the same (Hill 1935; Thomas 1997).

Even as the traditional Navajo sex/gender system had biological sex as its starting point, the Navajo nàdleeh were also distinguished by gender-linked behaviors, such as body language, clothing, ceremonial roles, speech style, and occupation. Feminine, male-bodied nàdleeh might engage in women's activities such as cooking, weaving, household tasks, and making pottery. Masculine, female-bodied nàdleeh, unlike other female-bodied persons, avoided childbirth; today they are associated with male occupational roles such as construction or fire-fighting (although ordinary women also sometimes engage in these occupations). Traditionally, female-bodied nàdleeh had specific roles in Navajo ceremonials (Thomas 1997).

Thus, even where hermaphrodites occupied a special gender variant role, Native American gender variance was defined more by cultural than biological criteria. In the recorded case of a physical examination of a gender variant male, the previously mentioned

Finds Them and Kills Them, his genitals were found to be completely normal (Roscoe 1998).

Native American gender variants were not generally conceptualized as hermaphrodites, but neither were they conceptualized as transsexuals (people who change from their original sex to the opposite sex). Gender transformations among gender variants were recognized as only a partial transformation, and the gender variant was not thought of as having become a person of the opposite sex/gender. Rather, gender variant roles were autonomous gender roles that combined the characteristics of men and women and had some unique features of their own. For example, among the Zuni a male gender variant was buried in women's dress but also in men's trousers on the men's side of the graveyard (Parsons, cited in Callender and Kochems 1983:454; Roscoe 1991:124, 145). Male gender variants were neither men—by virtue of their chosen occupations, dress, demeanor, and possibly sexuality—nor women, because of their anatomy and their inability to bear children. Only among the Mohave do we find the extreme imitation of women's physiological processes related to reproduction and the claims to have female sexual organs—both of which were ridiculed within Mohave society. Even here, however, where informants reported that female gender variants did not menstruate, this did not make them culturally men. Rather, it was the mixed quality of gender variant status that was culturally elaborated in Native North America and was the source of supernatural powers sometimes attributed to them.

Sacred Power

The association between the spiritual power and gender variance occurred in most, if not all, Native American societies. Even where, as previously noted, recruitment to the role was occasioned by a child's interest in occupational activities of the opposite sex, supernatural sanction, frequently appearing in visions or dreams, was also involved, as among Prairie and Plains societies. These visions involved female supernatural figures, often the moon. Among the Omaha, the moon appeared in a dream holding a burden strap—a symbol of female work—in one hand, and a bow—a symbol of male work—in the other. When the male dreamer reached for the bow, the moon forced him to take the burden strap (Whitehead 1981). Among the Mohave, a child's choice of male or female implements heralding gender variant status was sometimes prefigured by a dream that was believed to come to an embryo in the womb (Devereux 1937).

In some but not all societies, sex/gender variants had sacred ritual roles and curing functions (Callender and Kochems 1983:453; Lang 1998). Where feminine qualities were associated with these roles, male gender variants might become spiritual leaders or healers, but where these roles were associated with masculine qualities they were not entered into by male gender variants. The Plains Indians, who emphasized a vision as a source of supernatural power, regarded male gender variants as holy persons, but California Indian societies did not. Moreover, in some Native American societies gender variants were specifically excluded from religious roles (Lang 1998:167). Nevertheless, sacred power was so widely associated with sex/gender diversity in Native North America that scholars generally agree that it is an important explanation of why such roles were so widespread.

In spite of cultural differences among Native American societies, some of their general characteristics are consistent with the positive value placed on sex/gender diversity and the widespread existence of multigender systems (Lang 1996). One cultural similarity is a cosmology (system of religious beliefs) in which transformation and ambiguity are recurring themes, applying to humans, animals, and objects in the natural environment. In many of these cultures, sex/gender ambiguity, lack of sexual differentiation, and sex/gender transformations are central in creation stories (Lang 1996:187). Native American cosmology may not be "the cause" of sex/gender diversity but it certainly (as in India) provides a hospitable context for it.

Female Gender Variants

Female gender variants probably occurred more frequently among Native Americans than in other cultures, a point largely overlooked in the historic and ethnographic record (see Blackwood 1984; Jacobs et al. 1997; Lang 1998; Medicine 1983).

Although the generally egalitarian social structures of many Native American societies provided a hospitable context for female gender variance, it occurred in perhaps only one-quarter to one-half of the societies with male variant roles (Callender and Kochems 1983:446; see also Lang 1998:262–265). This may be explained partly by the fact that in many Native American societies women could—and did—adopt aspects of the male gender role, such as engaging in warfare or hunting, and sometimes dressed in male clothing, without being reclassified into a different gender (Blackwood 1984; Lang 1998:261ff; Medicine 1983). . . .

While most often Native American women who crossed genders occupationally . . . were not reclassified into a gender variant role, several isolated cases of female gender transformations have been documented historically. One of these is Ququnak Patke, a "manlike woman" from the Kutenai (Schaeffer 1965). Ququnak Patke had married a white fur trader, and when she returned to her tribe, she claimed that her husband had transformed her into a man. She wore men's clothes, lived as a man, married a woman, and claimed supernatural sanction for her role change and her supernatural powers. Although whites often mistook her for a man in her various roles as warrior, explorer's guide, and trader, such transformations were not considered a possibility among the Kutenai, and many thought Ququnak Patke was mad. She died attempting to mediate a quarrel between two hostile Indian groups.

Because sexual relations between women in Native American societies were rarely historically documented, it is hard to know how far we can generalize about the relation of sexuality to female gender variance in precontact Native American cultures. The few descriptions (and those for males, as well) are mainly based on ethnographic accounts that relied on twentieth-century informants whose memories were already shaped by white hostility toward gender diversity and same-sex sexuality. Nevertheless, it seems clear that although Native American female gender variants clearly had sexual relationships with women, sexual object choice was not their defining characteristic. In some cases, they were described "as women who never marry"; this does not say anything definitive about their sexuality and it may be that the sexuality of female gender variants was more variable than that of men.

Contact with whites opened up opportunities for gender divergent individuals, males as well as females (see Roscoe 1998; 1991). Overall, however, as a result of Euro-American repression and the growing assimilation of Euro-American sex/gender ideologies, Native American female and male gender variant roles largely disappeared by the 1930s, as the reservation system was well under way. Yet, their echoes may remain, both in the anthropological interest in this subject and in the activism of contemporary two-spirit individuals.

References

Albers, Patricia C. 1989. "From Illusion to Illumination: Anthropological Studies of American Indian Women." In *Gender and Anthropology: Critical Reviews for Research and Teaching*, edited by Sandra Morgan. Washington, DC: American Anthropological Association.

Blackwood, Evelyn. 1984. "Sexuality and Gender in Certain Native American Tribes: The Case of Cross-Gender Females." *Signs: Journal of Women in Culture and Society* 10: 1–42.

Callender, Charles, and Lee M. Kochems. 1983. "The North American Berdache." *Current Anthropology* 24 (4): 443–56 (Commentary, pp. 456–70).

Devereux, George. 1937. "Institutionalized Homosexuality of the Mohave Indians." *Human Biology* 9: 498–587.

Fulton, Robert, and Steven W. Anderson. 1992. "The Amerindian 'Man-Woman': Gender, Liminality, and Cultural Continuity." *Current Anthropology* 33(5): 603–10.

Hill, Willard W. 1935. "The Status of the Hermaphrodite and Transvestite in Navaho Culture." *American Anthropologist* 37: 273–79.

Jacobs, Sue-Ellen, Wesley Thomas, and Sabine Lang, eds. 1997. *Two-Spirit People: Native American Gender Identity, Sexuality and Spirituality*. Urbana and Chicago: University of Illinois Press.

Lang, Sabine. 1996. "There Is More than Just Men and Women: Gender Variance in North America." In *Gender Reversals and Gender Culture*, edited by Sabrina Petra Ramet, pp. 183–86. London and New York: Routledge.

_____. 1998. *Men as Women, Women as Men: Changing Gender in Native American Cultures*. Trans. from the German by John L. Vantine. Austin: University of Texas Press.

Medicine, Beatrice. 1983. "Warrior Women: Sex Role Alternatives for Plains Indian Women." In *The Hidden Half: Studies of Plains Indian Women*, edited by P. Albers and B. Medicine, pp. 267–80. Lanham Park, MD: University Press of America.

Roscoe, Will. 1991. *The Zuni Man-Woman*. Albuquerque: University of New Mexico Press.

_____. 1995. "Cultural Anesthesia and Lesbian and Gay Studies." *American Anthropologist* 97(3): 448–52.

_____. 1996. "How to Become a Berdache: Toward a Unified Analysis of Gender Diversity." In *Third Sex, Third Gender: Beyond Sexual Dimorphism in Culture and History*, edited by Gilbert Herdt, pp. 329–72. New York: Zone (MIT).

_____. 1998. *Changing Ones: Third and Fourth Genders in Native North America*. London: Macmillan.

Schaeffer, Claude E. 1965. "The Kutenai Female Berdache: Courier, Guide, Prophetess, and Warrior." *Ethnohistory: The Bulletin of the Ohio Valley Historic Indian Conference* 12(3): 173–236.

Thomas, Wesley. 1997. "Navajo Cultural Constructions of Gender and Sexuality." In *Two-Spirit People: Native American Gender Identity, Sexuality, and Spirituality*, edited by Sue-Ellen Jacobs, Wesley Thomas, and Sabine Lang, pp. 156–73. Urbana and Chicago: University of Illinois Press.

Whitehead, Harriet. 1981. "The Bow and the Burden Strap: A New Look at Institutionalized Homosexuality in

Native North America." In *Sexual Meanings: The Cultural Construction of Gender and Sexuality,*" edited by Sherry B. Ortner and Harriet Whitehead, pp. 80–115. Cambridge University Press.

Williams, Walter. 1992. *The Spirit and the Flesh: Sexual Diversity in American Indian Culture.* Boston: Beacon.

Topics for Further Examination

- Visit the websites of aisdsd.org, interACT: Advocates for Intersex Youth, and www.interfaceproject.org to learn more about intersex and intersex activism. In addition, see coverage of sex, gender, intersex, and transgender in popular magazines such as National Geographic (https://www.nationalgeographic.com/magazine/2017/01/) and Scientific American (https://www.scientificamerican.com/article/the-new-science-of-sex-and-gender/).
- Google *The Society Pages* (https://thesocietypages.org) and *Sociological Images* at *The Society Pages* (https://thesocietypages.org/socimages/) for short and informative articles on topics such as "gendering intelligence" and gendering objects/products. Gender bending in the arts has a long tradition.
- What can we learn about gender and sexual identity from the work of fine arts photographers such as Samuel Fosso, Marie Hoag & Bolette Berg, JJ Levine, and Yijun Liao? Look at gender bending work in other domains of the arts, including music videos. How does gender bending art work contribute to our understanding of gendered roles, relations, and structures?

2

The Interaction of Gender With Other Socially Constructed Prisms

Catherine G. Valentine with Joan Z. Spade

After considering what gender is and isn't, we are going to complicate things a bit by looking at how other socially constructed categories of difference and inequality, such as race, ethnicity, social class, religion, age, ability/disability, culture/nation, and sexuality, shape gender. As is the case with prisms in a kaleidoscope, the interaction of gender with other social prisms creates complex patterns of identity and relationships for people across groups and situations. Because there are so many different social prisms that interact with gender in daily life, we can discuss only a few in this chapter; however, other social categories are explored throughout this book. The articles we have selected for this chapter illustrate three key arguments. First, gender is a complex and multifaceted array of experiences and meanings that cannot be understood without considering the social context within which they are situated. Second, variations in gender meanings and relationships are connected to different levels of prestige, privilege, and power associated with membership in other socially constructed categories of difference and inequality. Third, gender intersects with other socially constructed categories of difference and inequality at all levels discussed in the introduction to this book—individual, interactional, and institutional.

Privilege

In our daily lives, there usually isn't enough time or opportunity to consider how the interaction of multiple social categories to which we belong affects beliefs, behaviors, and life chances. In particular, we are discouraged from critically examining our culture, as will be discussed in Chapter 3. People who occupy positions of privilege often do not notice how their privileged social positions influence them. In the United States, privilege is associated with white skin color, masculinity, wealth, heterosexuality, youth, able-bodiedness, and so on. It seems normal to those of us who occupy privileged positions and those we interact with that our positions of privilege be deferred to, allowing us to move more freely in society. Peggy McIntosh (1998) is a pioneer in examining these hidden and unearned benefits of privilege. She argues that there are implicit benefits of privilege and that persons with "unearned advantages" often do not understand how their privilege is a function of the disempowerment of others. For example, male privilege seems "normal" and White privilege seems "natural" to those who are male and/or White.

The struggle for women's rights in the United States also has seen the effects of privilege, with White women historically dominating this movement. The privilege of race and social class created a view of woman as a universal category, which essentially represented White women's interests. While many women of color stood up for women's rights, they did so in response to a universal definition of womanhood derived from White privilege (hooks, 1981). This pattern of White dominance is not a recent phenomenon and has been recognized within the African American community for some time, as discussed in the first reading in this chapter, by Bonnie Thornton Dill and Marla H. Kohlman. For example, in 1867, former slave Sojourner Truth responding to a White man who felt women were more delicate than men, described the exertion required of her work as a slave and asked, "Ain't I a woman?" (Guy-Scheftall, 1995). One hundred years later, women of color, including Audre Lorde (1982), bell hooks (1981), Angela Davis (1981), Gloria Anzaldúa (1987), and others, continued to speak out against White privilege within the women's movement. These women, recognizing that the issues facing women of color were not always the same as those of White women, carried on a battle to make African American women visible in the second wave of feminism. For example, while White feminists were fighting for the right to abortion, African American women were fighting other laws and sterilization practices that denied them the right to control their own fertility and bear their own babies. Women of color, including Thornton Dill (with Kohlman, this chapter) and others, some of whom are listed in this chapter, continue to challenge the White-dominated definitions of gender and fight to include in the analysis of gender an understanding of the experience of domination and privilege of all women.

Understanding the Interaction of Gender With Other Categories of Difference and Inequality

Throughout this chapter, you will read about social scientists and social activists who attempt to understand the interactions between "interlocking oppressions." Social scientists develop theories, with their primary focus on explanation, whereas social activists explore the topic of interlocking oppressions from the perspective of initiating social change. Although the goals of explanation and social/political change are rarely separated in feminist research, they reflect different emphases that must be addressed (Collins, 1990; Walby, Armstrong, & Strid, 2012). As such, these two different agendas shape attempts to understand the interaction of gender with multiple social prisms of difference and inequality. Much of this research on intersectionality is written by women, with a focus on women, because it came out of conflicts and challenging issues within the women's movement. As you read through this chapter and book, it is important to remember that the socially defined gender categories in Western culture intersect with other prisms of difference and inequality. And, as Patricia Hill Collins (1998, 2001) has argued, consider how the meanings of gender and other categories of difference and inequality are embedded in the structure of social relations within and between nations (Bose, 2012).

The effort to include the perspectives and experiences of *all* women in understanding gender is complicated. Previous theories had to be expanded to include the interaction of gender with other social categories of difference and privilege. These efforts to refine or redefine the concept of intersectionality continue. Yet, as Kathy Davis (2008) argues, the ambiguity and open-endedness of the concept makes it successful as a feminist theory because it is more accessible and, thus, applied more meaningfully. To better understand the development of the concept of intersectionality, we group these efforts into three different approaches. The earliest approach is to treat each social category of difference and inequality as if it were separate and not overlapping. A second approach is to add up the different social categories that an individual belongs to and summarize the effects of the social categories of privilege and power. The third and newest approach attempts to understand the simultaneous interaction of gender with all other categories of difference and inequality. These three approaches are described in more detail in the following paragraphs because they help us understand the complexity of applying an intersectional approach.

Separate-and-Different Approach

Deborah King (1988) describes the earliest approach as the "race–sex analogy." She characterizes this approach as one in which oppressions related to race are compared with those related to gender, but each is seen as a separate influence. King quotes Elizabeth Cady Stanton, who in 1860 stated, "Prejudice against color, of which we hear so much, is no stronger than that against sex" (p. 43). This approach, the race–sex analogy, continued well into the late 20th century and can thwart a deeper understanding of the complexity of gender. For example, in the race–sex analogy, gender is assumed to have the same effect for African American and White women, while race is defined as the same experience for African American men and women. However, the reading by Karen D. Pyke and Denise L. Johnson in this chapter, which explores Asian American women's definitions of "femininity," illustrates how we cannot assume that gender will mean the same thing to different women from the same society. Although these Vietnamese American and Korean American women were born and raised in the United States, they live with two different cultural definitions of femininity, an idea we will consider in more depth in the next chapter. Although they may try, the reality is that Asian American women cannot draw a line down their bodies separating gender from ethnicity and culture. As these women's words suggest, the effects of prisms of difference on daily experience are inextricably intertwined. For example, African American girls and women cannot always be certain that the discrimination they face is due to race or gender, or both. As individuals, we are complex combinations of multiple social identities. Separating the effects of multiple social prisms theoretically does not always make sense and is almost impossible to do on an individual level.

Looking at these individual challenges, King (1988) argues that attempting to determine which "-ism" (e.g., sexism, racism, or classism) is most oppressive and most important to overcome does not address real-life situations. This approach pits the interests of each subordinated group against the others and asks individuals to choose one group identity over others. For example, must poor, African American women decide which group will best address their situations in society: groups fighting racial inequality, gender inequality, or class inequality? The situations of poor, African American women are more complex than this single-issue approach can address.

The race–sex analogy of treating one "-ism" at a time has been criticized because, although some needs and experiences of oppressed people are included, others are ignored. For instance, Collins (1990) describes the position of African American women as that of "outsiders within" the feminist movement and in relations with women in general. She and others have criticized the women's movement for leaving out of its agenda an awareness of the experiences and needs of African American women (e.g., King, 1988). This focus on one "-ism" or another—the formation of social action groups around one category of difference and inequality—is called identity politics.

Additive Approach

The second approach used by theorists to understand how multiple social prisms interact at the level of individual life examines the effects of multiple social categories in an additive model. In this approach, the effects of race, ethnicity, class, and other social prisms are added together as static, equal parts of a whole (King, 1988). Returning to the earlier example of poor, African American women, the strategy is one of adding up the effects of racism, sexism, and classism to equal what is termed *triple jeopardy*. If that same woman was also a lesbian, her situation would be that of *quadruple jeopardy*, according to the additive model.

Although this approach takes into account multiple social identities in understanding oppression, King and others reject it as too simplistic. We cannot simply add up the complex inequalities across social categories of difference and inequality, because the weight of each social category varies based on individual situations. McIntosh (1998) argues that privileges associated with membership in particular social categories interact to create "interlocking oppressions" whose implications and meanings shift across time and situations. For example, for African American women in some situations, their gender will be more salient, while in other situations, their race will be more salient.

Interaction Approach

The third approach to understanding the social and personal consequences of membership in multiple socially constructed categories is called multiracial feminist theory. Various terms are used by multiracial feminist theorists to describe "interlocking oppressions," including *intersectional analysis* (Baca Zinn & Thornton Dill, 1996), *interrelated* (Weber, 2001), *simultaneous* (Collins, 1990; Weber, 2001), *multiplicative* or *multiple jeopardy* (King, 1988), *matrix of domination* (Collins, 1990), and *relational* (Baca Zinn, Hondagneu-Sotelo, & Messner, 2001; Baca Zinn & Thornton Dill, 1996). The first article in this chapter, by Thornton Dill and Kohlman, provides an overview of the meaning of intersectional analysis. For the purposes of this book, we describe this approach as consisting of "prismatic" or intersectional interactions, which occur when socially constructed categories of difference and inequality interact with other categories in individuals' lives. A brief discussion of some of these different models for explaining "interlocking oppressions" is useful in deepening our understanding of the complex interactions of membership in multiple social prisms.

King (1988) brings these interactions to light, discussing the concept of multiple jeopardy, in which she refers not only "to several, simultaneous oppressions, but to the multiplicative relationships among them as well" (p. 47). As a result, socially constructed categories of difference and inequality fold into individual identities, not in an additive way but in a way in which the total construction of an individual's identity incorporates the relationship of the identities to one another. King's model includes both multiple social identities and situational factors to understand individual differences. For example, being a submissive woman might matter more in certain religious groups where women have more restricted roles, while race or class may be less salient in that situation because the latter social prisms are likely to be similar across the religious group.

Maxine Baca Zinn, Pierrette Hondagneu-Sotelo, and Michael Messner (2001) emphasize that gender is relational. Focusing on gender as a process (Connell, 1987), they argue that "the meaning of *woman* is defined by the existence of women of different races and classes" (Baca Zinn et al., 2001, p. 174). As Pyke and Johnson's reading in this chapter illustrates, the fact that the Asian American women they interviewed were confused about whether to accept Asian standards of gender or American standards "normalizes" and makes White femininity dominant.

Collins (1990), on the other hand, conceptualizes oppressions as existing in a "matrix of domination" in which individuals not only experience but also resist multiple inequalities. Collins argues that domination and resistance can be found at three levels: personal, cultural, and institutional. Individuals with the most privilege and power—White, upper-class men—control dominant definitions of gender in this model. Collins discusses how "White skin privilege" has limited White feminists' understandings of gender oppression to their own experiences and created considerable tension in the women's movement. Tensions occur on all three levels—personal, within and between groups, and at the level of institutions such as the women's movement itself—all of which maintain power differentials. Furthermore, as she has since argued, these intersections of inequality are embedded in national identity systems of gendered social organizations (Collins, 1998, 2001).

In more recent work (2015), Collins addresses the dilemma of defining intersectionality. She notes that there is general agreement that the term "references the critical insight that race, class, gender, sexuality, ethnicity, nation, ability, and age operate not as unitary, mutually exclusive entities, but rather as reciprocally constructing phenomena" (p. 1). But, exactly what counts as intersectionality remains unclear. Collins brings some clarity to the definitional problem by arguing that intersectionality can be understood as a "knowledge project" that is simultaneously a field of study, an analytical strategy, and a form of social justice practice (p. 5). In her reading in this chapter, Collins develops this argument by looking at violence, especially in the lives of African American women, as a key to grasping the theoretical and political shape of intersectionality. As you can see, this third approach does not treat interlocking oppressions as strictly additive. The unequal power of groups created by systems of inequality affects interpersonal power in all relationships, both within and across gender categories. Thus, there is a clear political or activist orientation to intersectional analyses, which many scholars, including Thornton Dill and Kohlman, feel must be addressed (see also Walby et al., 2012). The articles in this chapter illustrate how multiple socially constructed prisms interact to shape both the identities and opportunities of individuals. They also show how interpersonal relationships are intricately tied to the larger structures of society or nations, as Collins (2001) argues, and how gender is maintained across groups in society.

These efforts to understand gender through the lens of multiple social prisms of difference and inequality can be problematic. One concern is that gender could be reduced to what has been described as a continually changing quilt of life experiences (Baca Zinn et al., 2001; Connell, 1992). That is, if the third approach is taken to the extreme, gender is seen as a series of individual experiences and the approach can no longer be used as a tool for explaining patterns across groups, which is meaningless in generating social action. Thus, the current challenge for researchers and theorists is to forge an explanation of the interaction of gender with other socially constructed prisms that both recognizes and reflects the experiences of individuals, while at the same time highlighting the patterns that occur across groups of individuals.

The application of intersectional analyses continues to be riddled with debates over how it should be done. An article in this chapter by Bandana Purkayastha adds to this discussion by elaborating on the ways intersectionality can be applied in transnational or global spaces. This piece clearly points to the fact that intersectional analysis is a dynamic and lively addition to the study of gender difference and inequality. Kaveri Thara's reading in this chapter offers an example of what we can learn about axes of domination and marginalization by looking at the dynamics of intersectionality outside of Euro-American societies. She examines the ways in which gender, caste, and poverty status are mobilized by South Indian fisherwomen to protect their caste-based occupation against the threats from fish shops in a capitalist economy. Another current debate addresses the methods or the way we do research on intersectionality (Choo & Ferree, 2010). As you will see, much of the research applying intersectional analysis uses qualitative research. However, Leslie McCall (2005) carefully laid out an argument that considered why this was so. She argued that the methodology of intersectionality research is complicated by the need to define "categories" of difference and inequality. But when defining categories, such as gender or race, to individuals you are studying, you take away the meanings they bring to those categories and the way they experience them. McCall argues for using categories such as gender, race, and social class, even though they are imperfect, in quantitative research to study the complexity associated with examining multiple categories of difference and inequality.

Since we live in a world that includes many socially constructed categories of difference and inequality, understanding the ways social prisms come together is critical for understanding gender. How we visualize the effects of multiple social prisms depends on whether we seek social justice, theoretical understanding, or both. How can we use these experiences to better understand the lived experiences of individuals and work to make those lives better? If multiple social categories are linked, then what are the mechanisms by which difference is created, supported, and changed? Are multiple identities multiplied, as King (1988) suggests; added, as others suggest; or combined into a matrix, as Collins (1990) suggests? Or, as Purkayastha suggests in this chapter, must we also consider nation in our analysis as we examine the interactions of gender with other categories of difference and inequality (see also Collins, 2001)? We raise these questions not to confuse you, but rather, to challenge you to try to understand the complexity of gender relations.

Prismatic Interactions

We return now to the metaphor of the kaleidoscope to help us sort out this question of how to deal with multiple social identities in explaining gender. Understanding the interaction of several socially constructed identities can be compared to the ray of light passing through the prisms of a kaleidoscope. Socially constructed categories serve as prisms that create life experiences. Just as the kaleidoscope produces a flowing and constantly changing array of patterns, we find individual life experiences to be unique and flowing. However, similar colors and patterns often reoccur in slightly different forms. Sometimes, when we look through a real kaleidoscope, we find a beautiful image in which blue is dominant. Although we may not be able to replicate the specific image, it would not be unusual for us to see another blue-dominant pattern. Gender differences emerge in a similar form—not as a single, fixed pattern but as a dominant, broad pattern that encompasses many unique but similar patterns. However, introducing the notion of category of gender adds questions relating to how these categories are defined. Are categories of gender based in individual identities, symbolic representations, or social structures (Choo & Ferree, 2010; Winker & Degele, 2011)?

The prism metaphor offers an avenue for systematically envisioning the complexity of gender relations. We would argue that to fully understand this interaction of social influences, one must focus on power. The distribution of privilege and oppression is a function of power relations (Baca Zinn et al., 2001). All the articles in this chapter examine differences defined by power relations, and power operates at every level of life, from the intimate and familial (see the reading by Pyke and Johnson in this chapter) to the global (see the reading by Purkayastha in this chapter). It is difficult to explain the combined effects of multiple social prisms without focusing on power. We argue that the power one accrues from a combination of socially constructed categories explains the patterns created by these categories. However, one cannot add up the effects from each category one belongs to, as in the additive approach described earlier. Instead, one must understand that, like the prisms in the kaleidoscope, the power of any single socially constructed identity is related to all other categories. The final patterns that appear are based on the combinations of power that shape the patterns. Individuals' life experiences, then, take unique forms as race, class, ethnicity, religion, age, ability/disability, body type, and other socially constructed characteristics are combined to create patterns that emerge across contexts and daily life experiences.

Consider your own social identities and the social categories to which you belong. How do they mold you at this time, and how did they, or might they, frame your experience of gender at other times and under different circumstances? Consider other social prisms such as age, ability/disability, religion, and national identity. If you were to build your own kaleidoscope of gender, what prisms would you include? What prisms interact with gender to shape your life? Do these prisms create privilege or disadvantage for you? Think about how these socially constructed categories combine to create your life experiences and how they are supported by the social structure in which you live. Keep your answers in mind as you read these articles to gain a better understanding of the role prisms play in shaping gender and affecting your life.

References

Anzaldúa, G. (1987). *Borderlands/La Frontera: The new mestiza.* San Francisco: Aunt Lute Books.

Baca Zinn, M., Hondagneu-Sotelo, P., & Messner, M. A. (2001). Gender through the prism of difference. In M. L. Andersen & P. H. Collins (Eds.), *Race, class, and gender: An anthology* (4th ed., pp. 168–176). Belmont, CA: Wadsworth.

Baca Zinn, M., & Thornton Dill, B. (1996). Theorizing difference from multiracial feminism. *Feminist Studies, 22*(2), 321–327.

Bose, C. (2012). Intersectionality and global gender inequality. *Gender & Society*, *26*(1), 67–72.

Choo, H. Y., & Ferree, M. M. (2010). Practicing intersectionality in sociological research: A critical analysis of inclusions, interactions, and institutions in the study of inequalities. *Sociological Theory*, *28*(2), 129–149.

Collins, P. H. (1990). *Black feminist thought: Knowledge, consciousness, and the politics of empowerment.* New York: Routledge.

Collins, P. H. (1998). Intersections of race, class, gender, and nation: Some implications for Black family studies. *Journal of Comparative Family Studies, 29*(1), 27–36.

Collins, P. H. (2001). It's all in the family: Intersections of gender, race, and nation. *Hypatia*, *13*(3), 62–82.

Collins, P. H. (2015). Intersectionality's Definitional Dilemmas. *Annual Review of Sociology*, 41, 1–20.

Connell, R. W. (1987). *Gender and power: Society, the person, and sexual politics.* Stanford, CA: Stanford University Press.

Connell, R. W. (1992). A very straight gay: Masculinity, homosexual experience, and the dynamics of gender. *American Sociological Review, 57,* 735–751.

Davis, A. Y. (1981). *Women, race, and class.* New York: Random House.

Davis, K. (2008). Intersectionality as buzzword: A sociology of science perspective on what makes a feminist theory successful. *Feminist Theory*, *9*(1), 67–85.

Guy-Scheftall, B. (Ed.). (1995). *Words of fire: An anthology of African-American feminist thought.* New York: New Press.

hooks, b. (1981). *Ain't I a woman: Black women and feminism.* Boston: South End Press.

King, D. (1988). Multiple jeopardy: The context of a Black feminist ideology. *Signs, 14*(1), 42–72.

Lorde, A. (1982). *Zami: A new spelling of my name.* Trumansburg, NY: Crossing Press.

McCall, L. (2005). The complexity of intersectionality. *Signs, 30*(3), 1771–1800.

McIntosh, P. (1998). *White privilege and male privilege: Unpacking the invisible knapsack.* Wellesley, MA: Wellesley College Center for Research on Women.

Walby, S., Armstrong, J., and Strid, S. (2012). Intersectionality: Multiple inequalities in social theory. *Sociology, 46*(2), 224–240.

Weber, L. (2001). *Understanding race, class, gender, and sexuality: A conceptual framework.* Boston: McGraw-Hill.

Winker, G., & Degele, N. (2011). Intersectionality as multilevel analysis: Dealing with social inequality. *European Journal of Women's Studies*, *18*(1), 51–66.

Introduction to Reading 7

Intersectional theory is not an easy concept to define for many of the reasons described in the introduction to this chapter. In this piece, Bonnie Thornton Dill and Marla H. Kohlman discuss some of the key issues relating to intersectional theory today. They look at the roots and history of intersectionality from the perspective of Black feminists and also review some debates about methodological issues surrounding intersectionality in terms of qualitative versus quantitative analyses. While their focus is also on social justice, we are not able to include that entire discussion in this excerpt, but we encourage you to be sensitive to issues of social justice and intersectionality that they weave throughout this piece, particularly in their conclusion to this reading. They end by contrasting "strong intersectionality" with "weak intersectionality," further illustrating the various ways this paradigm has been applied. This reading helps us understand the value of thinking deeply about differences and the need for intersectional analyses.

1. How have Blacks, particularly Black women, helped shape the development of intersectional analyses?

2. What are the problems of studying intersectionality using quantitative methodology, and what are the advantages of doing so?
3. Explain what the authors mean by the difference between "strong intersectionality" and "weak intersectionality."

Intersectionality

A Transformative Paradigm in Feminist Theory and Social Justice

Bonnie Thornton Dill and Marla H. Kohlman

As a Black scholar writing about women's issues in the mid-1980s, Thornton Dill joined with several colleagues in calling for a feminist theoretical paradigm that would expose the disconnect between experience and theory in the often untold stories of women of color and those without economic privilege (Baca Zinn, Cannon, Dill, & Higginbotham, 1986). They understood that feminist theory was quite limited without the purposeful integration of the notion of difference, beginning with race, ethnicity, class, and culture. They also understood that the integration of race and class into the gendered discourses extant at that time would change the nature of feminist discourse in important and powerful ways.

More than two decades later, one of the first things students learn in women's studies classes is how to look at women's and men's lives through multiple lenses. The concept of *intersectionality* has been a key factor in this transition. This conceptual tool has become integral to both theory and research endeavors, as it emphasizes the interlocking effects of race, class, gender, and sexuality, highlighting the ways in which categories of identity and structures of inequality are mutually constituted and defy separation into discrete categories of analysis. Intersectionality provides a unique lens of study that does not question difference; rather, it assumes that differential experiences of common events are to be expected.

As scholars producing intersectional work began to apply their insights to institutional dynamics, they began to speak and write about the challenges and opportunities that exist within and through the academy and the labor market, and in law and public policy. Thus, intersectional scholarship is engaged in transforming both theory and practice across disciplinary divides, offering a wide range of methodological approaches to the study of multiple, complex social relations. In her widely cited 2005 article, "The Complexity of Intersectionality," sociologist Leslie McCall states that "intersectionality is the most important theoretical contribution that women's studies, in conjunction with related fields, has made so far" (p. 1771).

In this chapter, we map the developments in intersectional theorizing and institutional transformation in the past decade while also offering our views on the future of intersectionality for feminist theory and methodology.

Roots and History

Intersectional scholarship emerged as an amalgamation of aspects of women's studies and race and ethnic studies. Its foundations are in the scholarly tradition that began in the 19th century with Black women such as Sojourner Truth, Maria Stewart, and Anna Julia Cooper and men like W. E. B. Du Bois—intellectuals who first articulated the unique challenges of Black women facing the multiple and simultaneous effects of race, gender, and class. What distinguished this early work on Black women was that it argued forcefully and passionately that the lives of African American women could not be understood through a unidimensional analysis focusing exclusively on either race or gender.

Thornton Dill, B., & Kohlman, M. H. (2011). Intersectionality: A transformative paradigm in feminist theory and social justice. In S. Hesse-Biber (Ed.), *Handbook of Feminist Research*, 2nd ed. (pp. 154–174). Reprinted with permission from SAGE Publications, Inc.

Intersectional scholarship, as we know it today, fused this knowledge from race and ethnic studies with aspects of women's studies and refined it in the debates and discourse that informed the civil and women's rights activism of the 1960s and 1970s. Before that time, women's studies emphasized the importance of gender and sexism while Black and Latino studies focused on race and racism as experienced within these respective communities. Each field sought to interrogate historical patterns of subordination and domination, asserting that we live in a society that is organized around complex and layered sets of inequalities. . . .

Categories of race/ethnicity, class, and gender were defined as major markers and controllers of oppression in the earliest discussions of intersectionality, with limited attention given to other categories such as sexuality, nation, age, disability, and religion, which have been discussed in more recent years. One result of this historical trajectory is a perspective asserting that individuals and groups can simultaneously experience oppression and privilege. Mere recognition of this history is not sufficient to form a complete understanding of the extensive ranges of "structures and experiences produced by intersecting forms of race and gender"; neither does it ensure proper acknowledgment of the "interlocking inequalities, what Patricia Hill Collins calls the 'matrix of domination'" (Baca Zinn & Dill, 1996, p. 326).

Collins explains that "the matrix of domination is structured on several levels. People experience and resist oppression on three levels: the level of personal biography; the group or community level of cultural context created by race, class, and gender; and the systemic level of social institutions" (Collins, 1990, p. 227). Collins distinguishes her conceptualization of interlocking theories, or oppressions, from the traditional additive models of oppression found in much traditional feminist theory. . . .

By noting the ways in which men and women occupy variant positions of power and privilege across race, space, and time, intersectionality has refashioned several of the basic premises that have guided feminist theory as it evolved following the 1950s. Many have explicitly recognized that the prototypical model for feminist theory post-1950s was based on the lives of White women whose experiences as wives, daughters, and mothers were to be strictly differentiated from the experience of Black women as informed by historical precedent. "Judged by the evolving nineteenth century ideology of femininity, which emphasized women's roles as nurturing mothers and gentle companions and housekeepers for their husbands, Black women were practically anomalies" (Davis, 1981, p. 5). Indeed, "one cannot assume, as have many feminist theorists and activists, that all women have had the same experience of gender oppression—or that they will be on the same side of a struggle, not even when some women define that struggle as 'feminist' . . . [F] or peoples of color, having children and maintaining families have been an essential part of the struggle against racist oppression. [Thus, it is not surprising that] many women of color have rejected the white women's movement's view of the family as the center of 'women's oppression'" (Amott & Matthaei, 1991, pp. 16–17), which many find to be the decisive message of books regarded as pivotal foundations of the second wave of the feminist movement such as *The Feminine Mystique* by Betty Friedan or *Of Woman Born* by Adrienne Rich. We do not mean to imply, either, that all women of color would renounce the accounts offered in the pages of these books. We offer these textual examples as evidence that intersectional scholarship has been able to highlight the myriad ways in which the experiences of some White women and women of color differ in the nuances of the maintenance of family. More specifically, these differences are to be found within the inextricable lines of racial ethnicity and gender that were influential in the fomentation of a feminist consciousness for some women that was both distinct from and dependent upon the experiences of other women.

Indeed, the family has been for many women of color a sort of "haven in a heartless world" of racism that provides the needed support to fight against oppression of many types (Dill, 1979; hooks, 1984). Drawing on the work of a number of the pioneering Black feminist intersectional scholars, Landry (2000) argues in his book on Black middle-class women that this support enabled Black women to produce a new ideology of womanhood that permitted the formation of the modern dual-career and dual-earner family. This model of womanhood rejected the notion that "outside work was detrimental to [Black women's roles] as wives and mothers" (Landry, 2000, p. 73). Indeed, Black women of the late 19th and early 20th centuries realized that "their membership in the paid labor force was critical to achieving true equality with men" (p. 74) in the larger U.S. society in a way that was not available to White women under the cult of true womanhood that constrained them to the exclusive domains of hearth and home. Women of color, having always been regarded as a source of labor in the United States, were never the beneficiaries of this ideology of protectionism and were not, therefore, hampered from developing an ideology that saw beyond the dictates of traditional feminist principles based in the experience of gender subordination perceived as endemic to all women (see, for example, Davis, 1972).

Winifred Breines (2006) provides an interesting reflection on the role and relationship of early intersectional thinking that promoted an understanding of mutually constituted structures of difference and inequality in feminist experiences. As she argues in *The Trouble Between Us: An Uneasy History of White and Black Women in the Feminist Movement*,

> In the development of the feminist movement, one of the most dramatic political shifts was from a desire to overcome difference to its promotion. Integration or interracialism as a goal migrated toward difference and an embrace of identity that precluded togetherness. This was a disturbing process but, in retrospect, probably inevitable. Postwar young people, especially whites, knew very little about racism and sexism. They had to separate to learn who they were in the race, class, and gender terms constructed by American society. . . . Just as identity politics divided the society that created such politics in the first place, they divided the movements. (Breines, 2006, p. 16)

Breines (2006) concludes her text on the differences that emerged between Black and White women in the feminist movement with words from several young feminists, one of whom contends that "unlike second wave feminism, which has operated from a monolithic center, multiplicity offers the power of existing insidiously and simultaneously everywhere. 'Women' as a primary identity category has ceased to be the entry point for much young activist work" (p. 196). Breines (2006) follows this with the admonition that young feminists have come to this knowledge having read the experiences of those who struggled before them: "They may not be aware of it, but the racial learning curve that began in the early 1960s continues among younger—and older—feminists in the twenty-first century" (p. 199). Similar to the project embarked upon by Breines, the research and writing of feminist scholars of color continues the tradition of theorizing the experience of women of color who have been ignored in the scholarship on both race and gender. These scholars have produced landmark studies based on lived experience at the intersections of race, gender, ethnicity, class, and sexuality.[1]

Growth and Dissemination: Emerging Inquiries and Controversies

As an approach to creating knowledge that has its roots in analyses of the lived experiences of women of color—women whose scholarly and social justice work reveals how aspects of identity and social relations are shaped by the simultaneous operation of multiple systems of power—intersectional scholarship is interdisciplinary in nature and focuses on how structures of difference combine to create a feminist praxis that is new and distinct from the social, cultural, and artistic forms emphasized in traditional feminist paradigms that focus primarily upon contrasting the experiences of women in society to those of men. Intersectionality is intellectually transformative not only because it centers the experiences of people of color and locates its analysis within systems of ideological, political, institutional, and economic power as they are shaped by historical patterns of race, class, gender, sexuality, nation, ethnicity, and age but also because it provides a platform for uniting different kinds of praxis in the pursuit of social justice: analysis, theorizing, education, advocacy, and policy development.

The people who engage with this work do so out of strong commitments to diversity, multiculturalism, and human rights, combined with a desire to create a more equitable society that recognizes, validates, and celebrates difference. The social justice agenda of this scholarship is crucial to its utility in fomenting theory and praxis specifically designed for analyzing inequalities of power and privilege, and, consequently, intersectionality is of interest to persons outside the academy who share concerns that underlie this scholarship. As Catharine MacKinnon contends, "What is important about intersectionality is what it is doing in our world, how it is traveling around the world and being used in defense of human rights, not just what it says" (MacKinnon, 2010).

The intellectual vibrancy within and around intersectional theory is yielding new frontiers of knowledge production that include, but are not limited to, scholarship on identity and the applicability of intersectionality to groups in other social locations or in multiple social locations simultaneously (Browne & Misra, 2003; Henderson & Tickamyer, 2009; Kohlman, 2010). Discussions about methodologies, language, and images most accurately convey the complexities of these interrelationships. For example, the development of the queer of color critique (Ferguson, 2003; Johnson & Henderson, 2005) as an intervention into sexuality studies establishes race and ethnicity as critical dimensions of queer studies, scholarship on globalization and international human rights moves intersectionality beyond the U.S. context (Davis, 2008; Knapp, 2005; Mohanty, 2003; Yuval-Davis, 2006), and work that continues explicitly to link theory and practice provides an analytical foundation for social justice and critical resistance. Within each of these topics, there are disagreements about approach and perspective, . . . and the debates and discussions contribute to the vibrancy of the topic and thus to advancing this scholarship and producing knowledge that illuminates the many factors that shape processes of experiencing multiple identities and social locations.

Because the contemporary growth of intersectionality as a theoretical approach is relatively recent and has developed in a number of different fields, future growth is largely defined by the trajectory of current debates and inquiries.

* * *

Methodological Concerns

Debates around methodologies center on the concern of remaining grounded in the questions, struggles, and experiences of particular communities that generate an intersectional perspective. At the same time, methodological debates about intersectionality often extend this approach to identify common themes and points of connection between specific social locations and broader social patterns. Stated somewhat differently: How do we benefit from comparisons and interrelationships without negating or undermining the complex and particular character of each group, system of oppression, or culture? Answers to this question are embedded in discussions about the language and metaphors that most effectively convey the concept of intersectionality as well as in debates about the use of qualitative and historical versus quantitative research methodologies.

Central to the discussion of language have been disputes about the adequacies and limitations of the term "intersectionality" and the metaphors associated with it. Scholars working with these ideas continue to seek ways to overcome an image that suggests that these dimensions of inequality, such as race, class, and gender, are separable and distinct and that it is only at certain points that they overlap or intersect with one another. This concern was specifically articulated by Deborah King in 1988 when she called for a model of analysis permitting recognition of the "multiple jeopardy" constituted by the interactive oppressions that circumscribe the lives of Black women and defy separation into discrete categories of analysis. The modifier "multiple" refers not only to several simultaneous oppressions but to the multiplicative relationships among them as well (King, 1988, p. 47). It is now widely recognized that intersectionality is more than a car crash at the nexus of a set of separate roads (Crenshaw, 1989). Instead, it is well understood that these systems of power are mutually constituted (Weber, 2009) such that there is no point at which race is not simultaneously classed and gendered or gender is not simultaneously raced and classed. How to capture this complexity in a single term or image has been an ongoing conversation.

Recent work by Ivy Ken (2007) provides a useful overview of a number of the conceptual images in use: that is, the notion of "intersecting versus interlocking" inequalities, which has been expressed in metaphors such as crossing roads or a matrix or the importance of locating oppressions within "systems versus structures versus institutions." She then moves on to analyze the limitations of these analytical approaches and suggests aspects of intersectionality that remain unexplained. She further proposes an innovative and promising approach to thinking about these ideas using the processes of producing, using, experiencing, and digesting sugar as a metaphor for describing, discussing, and theorizing intersectionality (Ken, 2008). For example, she addresses the importance of context-specific relationships in understanding how race, class, and gender oppression is "produced, what people and institutions do with it once they have it in their hands, what it feels like to experience it, and how it then comes to shape us" (Ken, 2008, p. 154). Ken argues that the relationships among sources of oppression, like race, class, and gender, start with production—every aspect of race, class, and gender has been and is produced under particular social, historical, political, cultural, and economic conditions.

Given the intersectional argument that race, class, gender, and other axes of inequality are always intertwined, co-constructed, and simultaneous (Weber, 2009), questions and debates have arisen about how quantitative approaches that rely on the analysis of separate and distinct variables can account for such interactivity. Two issues frame this debate. The first is the idea that these axes of inequality are not simply characteristics of individuals to be used as variables, isolated from the particular histories, social relations, and institutional contexts that produced them (Amott & Matthaei, 1996; Stacey & Thorne, cited in Harnois, 2009). The second is the task of developing quantitative approaches that address and reveal the overlapping differences present in intersectional analyses in a way that will yield important, generalizable results.

The work of Leslie McCall (2001, 2005) with regard to race, class, and gender in different types of labor markets has been particularly important in efforts to rethink the use of quantitative tools so that they can reveal the differential ways race, class, and gender interact within different social contexts. Her *Signs* article (2005) has been an important tool in efforts to provide empirical evidence of the value of intersectional analysis in the quantitative social sciences. Specifically, McCall applies an intersectional approach to an examination of the impact of economic restructuring on wage inequalities (see also Hancock, 2008; Simien, 2007; Valentine, 2007). To do this, she studies the effect of multiple factors on different racial/ethnic, class, and gender groups and on the relationships both within and between those groups in different regional economies. What she finds is that the patterns are not the same: that a single economic environment may create advantages for some in a group and disadvantages for others in the same group relative to other groups, thus making some environments more appropriate for one set of social policies while a different set may be more appropriate for another. She states: "different contexts reveal different configurations of inequality [and] no single dimension of overall inequality can adequately describe the full structure of multiple, intersecting, and conflicting dimensions of inequality" (McCall, 2005, p. 1791).

Kohlman (2006, 2010) has utilized intersection theory to illustrate how the experiences of men and women who report having experienced sexual harassment in the U.S. labor market differ because of the interaction of several forces of oppression that influence behavior simultaneously. She employs quantitative methods to illustrate successfully that it is both possible and imperative to deconstruct commonly used additive models of analysis, which mask the intersectional effects shaping the experiences of those embedded within them. Catherine Harnois (2005, 2009), by applying multiracial feminist theory in the design of a quantitative analysis of women's paths to feminism, has both revealed and offered meaningful explanations for variations among women by race and ethnicity, differences within racial-ethnic categories that had not been thought to exist.

Because empirical findings from quantitative analyses dominate the social sciences, are seen as authoritative and generalizable, and often provide the documentation upon which social policy is built, quantitative research that demonstrates the importance of intersectionality offers the opportunity to expand the framework's applicability and impact. The danger, however, is that these axes of inequality may be read as a reductive analysis of the interaction of a set of individual characteristics, thus diluting the power and full meaning of the experiential theory these interactions have been constructed to illustrate. Quantitative methodologies, when read in conjunction with findings produced from the qualitative studies that continue to dominate research within the intersectional paradigm, provide analytic frames that complement, apply, and extend the impact and understanding of intersectionality.

It is also important to note that debates about methodology are not limited to a quantitative versus qualitative discussion, but, rather, embrace the idea that we must continue to explore and expand the approaches we use to address an even broader range of questions that can be generated by this scholarship. These should include applied and theoretical and interdisciplinary and transversal modes of inquiry, among others. Interdisciplinary research must embrace multiple methodological approaches to capture the complexities and nuances in the lives of individuals and the experiences of groups of people (see also Hancock, 2007). A key criterion is to avoid essentializing people's experiences by burying intragroup diversity within isolated analytical categories.

* * *

Conclusion

In 1979, Audre Lorde stood before an audience at a conference devoted to Simone de Beauvoir's book *The Second Sex* in New York City and spoke these words:

> Those of us who stand outside the circle of this society's definition of acceptable women; those of us who have been forged in the crucibles of difference—those of us who are poor, who are lesbians, who are Black, who are older—know that *survival is not an academic skill.* It is learning how to stand alone, unpopular and sometimes reviled, and how to make common cause with those others identified as outside the structures in order to define and seek a world in which we all can flourish. It is learning how to take our differences and make them strengths. (Lorde, 1984, p. 112)

More than three decades later, Lorde's clear mandate for social justice resonates at the very core of what intersectionality was, is, and must continue to be in order to serve the aims of those individuals who have found themselves to be situated in different social locations around the margins of feminist debates and inquiries. Just as Lorde called for feminists to turn difference into strengths, the theoretical paradigm of intersectionality has provided a voice and a vision to scholars seeking to make visible the interlocking structures of inequality to be found within the academic and everyday concerns that shape both our livelihoods and our experiences of the world.

Intersectionality has traveled a long distance, and, indeed, we recognize that it has taken many forms across academic disciplines and life histories. Intersectionality has now reached the point where it may be regarded as a member of the theoretical cannon taught in courses on law, social sciences, and the humanities. Intersectionality has, thus, increased in strength and, perhaps alternatively, suffered significant dilution in application as often happens when any theoretical tool is either misinterpreted or misapplied. As to the evidence of intersection theory's increasing strength, we know that intersectionality has been the practice of Black feminist scholars for generations; in that respect, this paradigm is not at all new. It has just been a long time coming into its own, and now, having been newly embraced as a powerful tool of social justice, social thought, and social activism by a larger population of feminist researchers, it has become more visible than ever before.

As to the evidence of its misapplication, we caution scholars to be mindful of what we consider to be "strong intersectionality" and "weak intersectionality." "Strong intersectionality" may be found in theoretical and methodological rubrics that seek to analyze institutions and identities *in relation to one another.* That is, "strong intersectionality" seeks to ascertain how phenomena are mutually constituted and interdependent, how we must understand one phenomenon in deference to understanding another. On the other hand, "weak intersectionality" explores differences without any true analysis. That is to say, "weak intersectionality" ignores the very mandate called for by Audre Lorde and seeks to explore no more than how we are different. "Weak intersectionality" eschews the difficult dialogue(s) of how our differences have come to be—or how our differences might become axes of strength, fortification, and a renewed vision of how our world has been—and continues, instead, to be socially constructed by a theory and methodology that seeks only occasionally to question difference, without arriving at a deep and abiding understanding of how our differences are continuing to evolve.

Part of the proliferation of "weak intersectionality" may be found in the interdisciplinary narratives advancing the argument that this paradigmatic tool operates in different ways in different institutional spaces. This argument is also reminiscent of the contention that intersectional theory has been individualized within separate fields of knowledge. We can only reply to such arguments by, first, acknowledging that intersectionality has developed disparately within different spheres of knowledge because of the way in which intersectional theorizing is applied across disciplinary fields. For example, some might encounter intersectionality as a concrete reality that hinders effective litigation under the law because whole people are literally required to split their identity(ies) in order to be properly recognized in a court of law. But when this same legal phenomenon is read and discussed in the social sciences or humanities, the scholars at issue might study it as a structural impediment or analytic frame that defies discrete analysis. That being said, this dilemma has the very real potential of diverting one's attention away from the theoretical imperative of intersectionality as a source of illumination and understanding to one that distorts and misrepresents lived experiences of the law, social norms, and social justice.

But we also recognize, as a second proposition, that intersectionality has developed differently across spheres of knowledge because of the differing experiences and privileges we enjoy as scholars and everyday citizens of the world. We noted previously, for example, that intersectionality has benefited tremendously from differing methodological applications and transnational discourses, even as we remain steadfast in our contention that this paradigm was born of the experiences of Black women in the United States that could not be properly understood using the unidimensional lens of race or of gender in academic and legal discourses.

Having established the foregoing premises of "strong intersectionality" and "weak intersectionality," we now see intersectional theorizing developing into a paradigm of analysis that defies separation into distinct fields of knowledge because of its explanatory power as a theoretical tool that does not require tweaking "to make it fit," so to speak. This is because the primacy of the basic core principles of intersectionality—that is, mutually constituted interdependence; interlocking oppressions and privileges; multiple experiences of race, gender, sexuality, and so forth—are now more widely recognized as such and scholars are more apt to hold one another to these basic rules of application, whatever methodology is employed. In fact, the debates occurring within intersectional scholarship today reflect the growth and maturation of this approach and provide the opportunity to begin, as Lynn Weber says, to "harvest lessons learned" (as cited in Dill & Zambrana, 2009, p. 287).

Among the lessons learned and knowledge produced is a broader and more in-depth understanding of the notion of race, racial formation, and racial projects. Another is a broader understanding of the concept of nation and of notions of citizenship both in the United States and globally. Concepts such as situated knowledge (Lorde), oppositional consciousness (Sandoval), and strategic essentialism (Hurtado) offer ways to theorize about difference and diversity. A third lesson is the knowledge that there is no single category (race, class, ethnicity, gender, nation, or sexuality) that can explain human experience without reference to other categories. Thus we have and will need to continue to develop more nuanced and complex understandings of identity and more fluid notions of gender, race, sexuality, and class. The work relies heavily on a more expanded sense of the concept of social construction and rests much of its analysis on the principle of the social construction of difference. Organista (2007) contests the dominant culture's imperialism and resistance to discussion of human differences within and across cultures and calls for a discussion of difference "beyond the kind of defensive and superficial hyperbole that leaves social oppression unchallenged" (p. 101). And, although the scholarship still struggles with the pull to establish either a hierarchy of difference or a list that includes all forms of social differentiation, both of which are antithetical to the specific objectives of intersectionality, a body of knowledge is being produced that provides a basis for understanding the various histories and organizations of these categories of inequality. This evolving body of knowledge is helping us better understand what differences render inequalities and how to resist reductionist impulses. As a theoretical paradigm,[2] intersectionality is unique in its versatility and ability to produce new knowledge. We remain optimistic about the future of intersectionality, particularly if this scholarship respects its crucial commitments to laying bare the roots of power and inequality, while continuing to pursue an activist agenda of social justice.

Notes

1. This list is not meant to be exclusive or exhaustive but a reflection of the breadth of early intersectional scholarship: Patricia Hill Collins (1990), Kimberlé Crenshaw (1989), Gloria Anzaldúa (1987), Maxine Baca Zinn and Bonnie Dill (1996), Audre Lorde (1984), Angela Y. Davis (1981), Cherrie Moraga (1983), Chela Sandoval (1991), Chandra Talpade Mohanty (1988), and bell hooks (1984).

2. In 1998, Collins referred to intersectionality as an "emerging paradigm." In 2007, Hancock argues it has become a normative and empirical paradigm.

References

Amott, T. L., & Matthaei, J. A. (1991). *Race, gender and work.* Boston: South End Press.

Amott, T. L., & Matthaei, J. A. (1996). *Race, gender and work* (New ed.). Boston: South End Press.

Anzaldúa, G. (1987). *La Frontera/Borderlands: The new Mestiza.* San Francisco: Aunt Lute Books.

Baca Zinn, M., Cannon, L. W., Dill, B. T., & Higginbotham, E. (1986). The costs of exclusionary practices in women's studies. *Signs: Journal of Women in Culture and Society, 11*, 290–303.

Baca Zinn, M., & Dill, B. T. (1996). Theorizing difference from multi-racial feminism. *Feminist Studies, 22*(2), 321–331.

Breines, W. (2006). *The trouble between us: An uneasy history of white and black women in the feminist movement.* New York: Oxford University Press.

Browne, I., & Misra, J. (2003). The intersection of gender and race in the labor market. *Annual Review of Sociology, 29*, 487–513.

Collins, P. H. (1990). *Black feminist thought: Knowledge, consciousness, and the politics of empowerment.* Boston: Unwin Hyman.

Crenshaw, K. (1989). Demarginalizing the intersection of race and sex: A black feminist of antidiscrimination doctrine, feminist theory, and antiracist politics. *University of Chicago Legal Forum,* 139–167.

Davis, A. Y. (1972). Reflections on the black woman's role in the community of slaves. *The Massachusetts Review, 13*(1/2), 81–100.

Davis, A. Y. (1981). *Women, race, and class.* New York: Random House.

Davis, K. (2008). Intersectionality as buzzword: A sociology of science perspective on what makes a feminist theory successful. *Feminist Theory, 9*(1), 67–85.

Dill, B. T. (1979). The dialectics of black womanhood. *Signs: Journal of Women in Culture and Society, 4*(3), 543–555.

Dill, B. T., & Zambrana, R. E. (2009). *Emerging intersections: Race, class, and gender in theory, policy, and practice.* New Brunswick, NJ: Rutgers University Press.

Ferguson, R. A. (2003). *Aberrations in black: Toward a queer of color critique.* Minneapolis: University of Minnesota Press.

Hancock, A. M. (2007). When multiplication doesn't equal quick addition: Examining intersectionality as a research paradigm. *Perspectives on Politics, 5*(1), 63–79.

Hancock, A. M. (2008). Intersectionality as a normative and empirical paradigm. *Politics & Gender, 3*(2), 248–254.

Harnois, C. E. (2005). Different paths to different feminisms? Bridging multiracial feminist theory and quantitative sociological gender research. *Gender & Society, 19*(6), 809–828.

Harnois, C. E. (2009). Imagining a "feminist revolution": Can multiracial feminism revolutionize quantitative social science research? In M. T. Berger & K. Guidroz (Eds.), *The intersectional approach: Transforming the academy through race, class, and gender* (pp. 157–172). Chapel Hill, NC: UNC Press.

Henderson, D., & Tickamyer, A. (2009). Staggered inequalities in access to higher education by gender, race, and ethnicity. In B. T. Dill & R. E. Zambrana (Eds.), *Emerging intersections: Race, class, and gender in theory, policy, and practice* (pp. 50–72). New Brunswick, NJ: Rutgers University Press.

hooks, b. (1984). *Feminist theory: From margin to center.* Cambridge, MA: South End Press.

Johnson, E. P., & Henderson, M. G. (Eds.). (2005). *Black queer studies: A critical anthology.* Durham, NC: Duke University Press.

Ken, I. (2007). Race-class-gender theory: An image(ry) problem. *Gender Issues, 24*, 1–20.

Ken, I. (2008). Beyond the intersection: A new culinary metaphor for race-class-gender studies. *Sociological Theory, 26*(2), 152–172.

King, D. (1988). Multiple jeopardy, multiple consciousness: The context of a black feminist ideology. *Signs: Journal of Women in Culture and Society, 14*(1), 42–72.

Knapp, G. A. (2005). Race, class, gender: Reclaiming baggage in fast-travelling theories. *European Journal of Women's Studies, 12*(3), 249–265.

Kohlman, M. H. (2006). Intersection theory: A more elucidating paradigm of quantitative analysis. *Race, Gender & Class, 13*, 42–59.

Kohlman, M. H. (2010). Race, rank and gender: The determinants of sexual harassment for men and women of color in the military. In V. Demos & M. Segal (Series Eds.), *Advances in Gender Research: Vol. 14. Interactions and intersections of gendered bodies at work, at home, and at play* (pp. 65–94). Boston: Elsevier.

Landry, B. (2000). *Black working wives: Pioneers of the American family revolution.* Berkeley: University of California Press.

Lorde, A. (1984). *Sister outsider.* Freedom, CA: Crossing Press.

MacKinnon, C. (2010, March 11). *Panelist remarks on "Rounding intersectionality: Critical foundations and contested trajectories."* Paper at the 4th Annual Critical Race Studies Symposium—Intersectionality: Challenging Theory, Reframing Politics, Transforming Movements, UCLA School of Law, Los Angeles, CA.

McCall, L. (2001). *Complex inequality: Gender, class, and race in the new economy.* New York: Routledge.

McCall, L. (2005). The complexity of intersectionality. *Signs: Journal of Women in Culture and Society, 30*(3), 1771–1800.

Mohanty, C. T. (1988). Under Western eyes: Feminist scholarship and colonial discourses. *Feminist Review, 30*, 61–88.

Mohanty, C. T. (2003). *Feminism without borders.* Durham, NC: Duke University Press.

Moraga, C. (1983). *Loving in the war years.* Boston: South End Press.

Organista, C. K. (2007). *Solving Latino psychosocial and health problems: Theory, practice, and populations.* Hoboken, NJ: John Wiley & Sons, Inc.

Sandoval, C. (1991). U.S. third world feminism: The theory and method of oppositional consciousness in the postmodern world. *Genders, 10*, 1–24.

Simien, E. (2007). Doing intersectionality research: From conceptual issues to practical examples. *Politics and Gender, 3*(2), 264–271.

Valentine, G. (2007). Theorizing and researching intersectionality: A challenge for feminist geography. *Professional Geographer, 59*, 10–21.

Weber, L. (2009). *Understanding race, class, gender, and sexuality: An intersectional framework* (2nd ed.). New York: Oxford University Press.

Yuval-Davis, N. (2006). Intersectionality and feminist politics. *European Journal of Women's Studies, 13*(3), 193–210.

Introduction to Reading 8

Patricia Hill Collins (see Introduction to Chapter) is a celebrated social theorist who developed ideas of multiple, intersecting oppressions (i.e., the matrix of domination) and the "outsider-within" in her now classic book, Black Feminist Thought (1990). In this reading, Collins extends her analysis of the matrix of domination by looking at the connections among violence, power relationships, and political resistance in the United States with a focus on African American women's lives.

1. What is hate speech and, according to Collins, how and why has it been reconfigured in recent years?
2. How does Collins use violence as a navigational tool for understanding power relations?
3. Why is Collins critical of shorthand terms such as "race, class, and gender"?
4. How and why have African American women employed flexible solidarity as a form of political action?

On Violence, Intersectionality and Transversal Politics

Patricia Hill Collins

The persistence of violence over the past several decades, coupled with the explosive growth of intersectionality as a form of critical inquiry and praxis (see e.g. Collins and Bilge 2016), suggest that the connections between violence, intersecting power relations and political resistance remain highly salient. In this commentary, I ask, in what ways might continuing to focus on violence illuminate the connections between intersecting systems of power and on the contours of political resistance?

Violence, Intersecting Power Relations and Domination

In "The Tie That Binds" (1998), I presented a set of conceptual tools for exploring how violence is shaped by and helps structure intersecting power relations. In that essay, I aimed to unsettle prevailing definitions of violence by embedding definitions of violence within the specific race and gender formations of the U.S. Influenced by the intellectual and political climate of the 1990s, I explored three ways that hierarchical power relations in the U.S. framed systemic violence, namely, (1) the power to legitimate what counts as violence; (2) the symbiotic relationship linking actions and speech; and (3) how routinized violence across a range of social institutions normalized the violence targeted toward specific groups as well as the culpability of some groups in causing it.

Revisiting my argument twenty years later, I see how the ideas in "The Tie That Binds" speak to contemporary power relations. My treatment of race and gender constituted starting points and not end points of intersectional analysis, which has grown to encompass hierarchies of class, ethnicity, nation, sexuality, age, ability and religion. Whereas my overall framework still resonates within this expanding intersectional universe, social trends over the past several decades have tempered my perceptions of its main ideas. For example, the speech climate that shaped my initial essay has changed dramatically. In "The Tie That Binds," I took inspiration from the then prominent debates about hate speech that were concerned not just with stereotypes, but also with efforts to prove how negative representations harmed people (see e.g. Matsuda et al. 1993). One of the victories of mid-twentieth century social movements lay in supporting African Americans, women, Latino/as, sexual minorities and similarly subordinated people who had long argued that hate speech harmed them because it often catalysed violent acts. By the 1990s, everyday racial slurs, gender epithets and off-colour jokes became increasingly stigmatized as hate speech. Because hate speech ostensibly violated norms of so-called political correctness, such speech seemed to be on the wane. Yet far from disappearing, the 1990s signaled the temporary replacement of overt hate speech with a seemingly more genteel, covert version. The kind of overt hate speech against Blacks, Latinos, women and sexual minorities that permeated U.S. culture prior to mid-twentieth century social movements became muted in public. Covert micro-aggressions that eschewed crass stereotypes but that invoked derogatory meanings concerning race, gender, class and sexuality filled the void left by the retreat of hate speech.

Social institutions increasingly failed to support the efforts of people who challenged this residual, covert hate speech and the behaviour that it engendered. Under the guise of protecting free speech, institutional pushback consisted of dismissing the claims of those who were targeted by hate speech and trivializing efforts to regulate it. For me, the disappearance of hate speech coupled with its persisting effects became visible during a visit to a prominent liberal arts campus. In a private conversation, a white undergraduate student described the dilemma she faced as a woman living in a co-ed dormitory. Each evening, a group of boys gathered in her dormitory's lounge and insisted on watching what she experienced as porn on the dormitory's television. Ignoring her claims about the hostile dormitory situation in which she was living, her college denied her request to make her male dorm dwellers stop, citing institutional commitments to

Collins, P. H. (2017). On violence, intersectionality and transversal politics. *Ethnic and Racial Studies*, *40*(9): 1460–1473. Reprinted with permission from Taylor & Francis.

protecting free speech for all students. As a result, the student had to walk through the lounge every evening to get to her room. But was this a protection of free speech, or a case illustrating how sexual harassment encompasses both speech and action?

The growth of the Internet, social media and the changing contours of journalism means that the very definition of public speech has shifted dramatically, as have the power dynamics associated with it. The 2016 U.S. Presidential campaign and its aftermath promises to be a watershed marking tectonic shifts concerning what constitutes acceptable and unacceptable public speech. The most recent "Hate Map" published by the Southern Poverty Law Center, a non-profit organization founded in 1971 that aims to combat hate, intolerance and discrimination through education and litigation, documents the rise in racist, sexist and anti-immigrant actions after the election (Southern-Poverty-Law-Center https://www.splcenter.org/). The increasing incidents of hate speech during campaign rallies and in the aftermath of the election seemingly contributed to the reemergence of hate speech in public venues as well as an increase in accompanying violent acts.

These events suggest that existing social hierarchies in the U.S. are sustained by reconfigured hate speech and the actions such speech engenders. The lies, half-truths and bluster of fake news that makes it difficult to recognize truth from propaganda seemingly weaken free speech as an antidote to hate speech. Does this resurgence in hate speech and its related actions signal a new slippery slope back into the kinds of violence that sustain power hierarchies through the threat of violence and force? Or might U.S. society be settling into a new form of disciplinary power, one where the technology of interconnected global communications has dramatically changed the stakes of the game?

The routinized nature of violence as an important dimension of disciplinary power has also undergone substantial change in the U.S. Despite important and widespread intersectional scholarship on violence during the past two decades, not only has systemic violence not diminished, the myriad forms that it now takes are staggering to consider. Is it that violence itself has become more entrenched, or rather that we now possess more ways to witness it? The shifting political context in Western democracies that normalizes right-wing, nationalistic politics, as well as the coming-of age of an entire generation that consumes violence as entertainment through television, movies, social media and music, point to a qualitative difference in perceptions of violence and the willingness to endorse, use or at minimum, ignore it.

Certainly the U.S. is awash with individuals who perpetrate violent acts that, in the absence of intersectional analyses, can seem random, individualized and senseless. Yet what exactly is it about being a young white male in the U.S. that leads some individuals to engage in extreme acts of violence? How are we to make sense of the actions of young white men who murder LGBTQ Latino youth enjoying a night out at a local nightclub, shoot Black worshippers after a prayer service in an African American church, take a machine gun to movie-goers in a suburban Colorado theater, and kill first-grade students as they huddled together in their elementary school classroom? Certainly many other individual Americans perpetrate violent acts. Yet these acts caught the public's attention because they were so heinous. If young white American male citizens engage in racist, sexist and homophobic violence, what good will closing borders against imagined Muslim "terrorists" or illegal Mexican "rapists" achieve? Contemporary events such as these make me wonder whether I may have written "The Tie That Binds" during an unusual decade, a transition period marked by the fall of one empire (the Berlin wall symbolized what many hoped would be the end of such walls), and the growth of state-sanctioned, homegrown terrorism that masquerades as normal politics.

Within intersecting power relations, the statistically insignificant actions of individual young white men who engage in domestic terrorism may mask the broader contours of deeply entrenched, routinized violence. Without human oversight or agency, social institutions routinely replicate power hierarchies where violence is vested less in speech but rather in bureaucratic action and custom. Racism, sexism, class exploitation and homophobia become hegemonic when they become uncritically embedded in the rules and regulations of normal society. Take for example, how policies of mass incarceration in the U.S. took decades to put into place, mainly through the decision-making of ordinary governance. There was no coup that installed a dictator who proclaimed that the new policy was to lock up Blacks, Latinos and poor people. Instead, the convergence of disparate practices that by themselves seemed innocuous, coalesced to build a prison industry that will take decades to dismantle. But maybe dismantling it is not the point. The brilliance of mass incarnation is that, because it does not overtly target Blacks and Latinos, it cannot be accused of being racially discriminatory. Individuals who implement the rules or fail to challenge them are free to hate one another, yet neither the state or major corporations should interfere. Rather, ostensibly colourblind rules and regulations reinscribe social inequality as firmly as the use of force. In this context, violence did not disappear. Instead, it became embedded in the rules, and became even more routinized via a system of seemingly non-discriminatory ideas and practices. State-sanctioned violence that is not defined as violence at all, yet that is essential in sustaining racial inequality persists, seemingly hidden in plain sight.

Violence as a Navigational Tool

Focusing on violence provides an important navigational tool for building theoretical arguments about political domination, especially the centrality of violence itself within and across distinctive systems of power. Different systems of power each rely on distinctive forms of violence, the case for example, of the interpersonal violence of domestic abuse that women experience (sexism), hate speech targeted toward Muslims, Jews and religious minorities (religious intolerance), the routinized violence targeted against racial and ethnic minorities (racism), and the state-sanctioned violence of warfare (nationalism). Collectively, these seemingly disparate expressions of violence constitute a malleable conceptual glue that both structures the forms that violence takes within distinctive systems of power and that facilitates their smooth interaction. In this sense, violence constitutes a saturated site of intersectionality where intersecting power relations are especially visible. Saturated sites of intersecting power relations are intensified points of convergence, or crossroads for intersectional power relations that facilitate the naturalization and normalization of political domination. Violence constitutes the conceptual glue that binds intersecting systems of power together (see e.g. Collins 1998b).

The significance of violence as a tie that permeates intersecting oppressions raises questions about the treatment of power within intersectionality. On the one hand, within some segments of intersectional scholarship, references to power appear to be everywhere; power is constantly mentioned, referenced and cited. Yet merely mentioning power may do more harm than good. Within intersectional discourse, conventions that substitute "race" for racism, "sex" for sexism and "class" for capitalism foster abstract references to power that neglect how political domination operates. Relying on a series of shorthand terms to invoke intersecting power hierarchies, much as "race, class and gender" becomes reduced to a slogan through overuse, the phrase "intersecting systems of power," itself a replacement for intersecting oppressions, may be headed for a similar fate. Phrases such as intersecting systems of power that circulate as hyper-visible signifiers render power as a descriptive, placeholder term with ostensibly minimal political impact. The hypervisibility granted abstract power-talk simultaneously limits the kind of politics that become possible within these abstractions.

On the other hand, for scholars and activists who see the links joining intersectionality's inquiry and praxis, power and politics take on a different demeanour. Social actors within social movement contexts often use intersectionality as a touchstone for political action. Frontline social actors within bureaucracies as well as those working in grassroots organizations often look to intersectionality to help solve thorny social problems such as homelessness, health disparities, mass incarceration, educational disparities and ever-present violence. Social workers, teachers, lawyers, nurses and similar practitioners engage intersectionality to help solve social problems. For front line political actors, the power hierarchies that create social inequalities and their concomitant social problems seem evident. Within bureaucratic contexts, social actors who claim intersectionality seek guidance for how it might inform their problem-solving strategies. Blacks, women, Latinos/as, indigenous people, women, undocumented people and other similarly subordinated groups who are most affected by social problems often see intersectionality as essential for their political projects (Roberts and Jesudason 2013; Terriquez 2015).

In this context, because the literature on political domination directly engages questions of power, it raises a mirror to intersectionality, asking whether the tendency to replace terms such as *oppression* with the term *power* offers an empty signifier that minimizes the effects of political domination. Racism, sexism and class exploitation constitute distinctive and intersecting systems of oppression that rely both on political domination and on violence as an important tool of organization. The expansive literature on political domination, which has largely developed uninformed by intersectional frameworks, provides important clues concerning power relations and politics as well as the centrality of violence to social hierarchy.

Colonialism, imperialism, heteropatriarchy, capitalism, nationalism and racism contain distinctive arguments about the nature of power that resonate with intersectionality. For example, by distinguishing racisms of extermination or elimination (exclusive racisms, such as Nazi Genocide), from racisms of oppression or exploitation (internal racisms, such as racial segregation in the U.S., racial apartheid in South Africa and colonial racisms), Etienne Balibar provides a crucial intervention in critical racial theory (Balibar 1991). Balibar argues that these ideal types are rarely found in isolation, and that connections among these types is more common. Zygmunt Bauman's classic book *Modernity and the Holocaust*, develops this thesis of a racism of extermination, extending Balibar's argument beyond nationalism to link racisms of extermination to modernity itself (Bauman 1989).

Placing this literature on political domination in dialogue with intersectionality provides new avenues for thinking through how violence works as a saturated site of intersectionality. I see two promising dimensions of expanding intersectionality's analysis of power relations by explicitly thinking about domination. First, using violence as a navigational tool provides an entry point for the broader question of how domination is organized across multiple systems of power. Racism, sexism, heterosexism, capitalism, nationalism and similar systems of power all rely on violence to sustain domination. All systems of oppression rely on violence, yet the forms that violence takes varies tremendously.

Second, attending to political domination captures the complexities and instabilities that characterize how oppression and resistance coexist. Whether racism or sexism, resistance is always present, even if it seems to be invisible. Just as violence is routinized within social institutions, immanent and/or tangible political resistance also persists.

Towards Transversal Politics: Flexible Solidarity and Coalition Building

When I wrote "The Tie That Binds," I hoped that developing intersectionality's theoretical contours might contribute to ameliorating violence as a social problem. Yet intersectional analyses, on their own, are unlikely to yield more effective political solutions to violence. Analysis is important, yet action also matters. Because "thinking" one's way out of domination is unrealistic, I now ask, how might more sophisticated analyses of power that take into account the ties linking violence, intersecting oppressions and domination facilitate more robust analyses of political resistance?

In "The Tie That Binds," I discussed transversal politics as a form of political engagement that had important implications for understanding organized political resistance. Here I return to that argument via a brief discussion of solidarity and coalition-building. The responses of African American women and similarly historically subordinated groups to intersecting oppressions illuminates the nature of political domination as well the transversal politics it might engender. For example, African American women's intellectual and political traditions raise some provocative questions concerning the nature of political solidarity. The trajectory of Black feminism within African American communities suggests that Black women's responses to racial violence moved beyond racial solidarity that was centred exclusively on racism. Black feminist agendas regarding gender and sexuality existed much earlier than when they became visible to a broader public. The story of how Black feminism's analyses of gender and sexuality advanced intersectional arguments is well-known. Yet the ways in which Black feminist understandings of political solidarity may have been shaped by intersectional analyses remains less familiar.

In my recent work, I have returned to issues of political solidarity and coalition politics within African American women's history (see e.g. Collins forthcoming 2017). Black women's experiences with violence provide guidance, not as a universal case for oppressed groups, but rather as a catalyst for theoretical insight concerning the interconnections of domination and resistance. Because contemporary forms of violence visited upon African American women have become so routinized in U.S. social institutions and normalized within public hate speech, it is easy to overlook the centrality of violence to the origins and history of African Americans as a U.S. population group. African Americans became Black people in the context of a forced migration within global capitalist expansion, the differential exploitation of productive and reproductive labour of men and women during slavery, and the subsequent structural disadvantages have shaped African American life. Ghettoization and racial segregation were key to African American domination in the U.S. context. Persistent high levels of residential, educational and employment segregation in the U.S. constitute fundamental structural features that contribute to racial hierarchy. Violence was essential to forming Black people as a population as well as the shared meanings that were associated with this political Blackness.

For African American women, these social conditions catalysed a particular form of politics, one characterized on the one hand by a stance of dissemblance from the external world whereby Black women hid the harm they experienced from rape, abuse and forms sexual violence (Hine 1989); and on the other hand, a distinctive Black feminist politics that reflected Black women's analyses and actions in response to the ever-present threat of violence. As a collective, Black women in the U.S. could not ignore how anti-Black hate speech and routinized racial violence took gender-specific forms. In this social context, Black women developed a more strategic, dynamic and sophisticated approach to solidarity that refutes understandings of solidarity as ideological uniformity that, within African American communities, took patriarchal and homophobic forms. Yet many Black feminist intellectual-activists never fully accepted this kind of group-think that defines solidarity through the ideological lens of a homogeneous blackness that privileged masculinity and heterosexuality. Instead, Black women were more likely to see the ways in which they were simultaneously in solidarity with Black men regarding racism as well as the ways in which such solidarity was problematic regarding sexism and homophobia.

Historically, Black women intellectual-activists developed forms of political action that were characterized by a *flexible solidarity*, one where alliances within African American communities have been grounded in ongoing relationships of compromise and contestation (Collins forthcoming 2017). Black women's community work in particular fostered a commitment to Black solidarity as a core feature of African American women's political engagement both *within* and *on behalf* of Black communities (Collins 2006, 123–160). Without solidarity among African Americans, political struggles to upend racial domination were doomed. Yet for Black women, an unquestioned solidarity could be neither inherently desirable nor effective when it rested on male-dominated, intergenerational gender hierarchies. Such solidarity was hierarchical, rigid, often backed up by religious theology or tradition, and created roadblocks for effective political action. Black women saw the need for solidarity, yet calibrated their ideas and actions to hone critical understandings of solidarity that were better suited for specific political projects, for example, opposing both lynching and rape because they were interconnected practices of violence. Solidarity was not an essentialist category, a bundle of rules that was blindly applied across time and space. Instead, a flexible understanding of solidarity enabled African American women to work with the concept, moulding it to the particular challenges at hand.

Working within African American organizations often sensitized Black women to inequalities of gender and sexuality within African American communities as well as within broader society. This awareness catalysed a deepening analysis of intersectionality during the nineteenth and twentieth centuries. Moreover, these intellectual and political understandings of solidarity were also worked out over time, primarily through everyday, organized political behaviour within African American communities. Stated differently, sustaining political vigilance in the face of racism required being attuned to the political implications both of ideology and strategy.

This idea of flexible solidarity *within* Black feminism lays a foundation for the kind of elasticity that Nira Yuval-Davis assigns to transversal politics (Yuval-Davis 1997, 125–132). Drawing on the work of Italian feminists, Yuval-Davis concludes her book *Gender and Nation* with a section sketching out several political and intellectual projects that seem to point toward a transversal politics. Rereading Yuval-Davis's arguments, especially in light of the shifting interpretive climate of hate speech, the more visible connections between hate speech and actions, and the scope of routinized violence, suggests that revisiting the main ideas of transversal politics may be especially constructive.

Several points stand out that merit review. For one, Yuval-Davis eschews understandings of groups that are based solely on self-chosen identities or identifications. Instead, she focuses on the authority of nation-states in creating and reproducing historically constituted, socially stratified population groups. She notes, "the boundaries of the groupings were determined not by an essentialist notion of difference, but by a concrete and material political reality" (Yuval-Davis 1997, 129). In the 1980s and into the 1990s, an emerging feminist literature on intersectionality engaged nationalism, examining topics such as how the public policies of nation-states were inherently intersectional, how the national identities of various nation-states relied on intersecting systems of power, and how differential citizenship rights underlay social inequalities (see e.g. Anthias and Yuval-Davis 1992). Scholars in the 1990s seemingly moved away from the literature on nation-states and nationalism, especially its emphasis on the structures of state power. Rereading Yuval Davis in the aftermath of this discursive turn shows how she places far more emphasis on historically constituted groups and the opportunities and constraints they bring to coalition politics than contemporary emphases on individuals and their rights.

Bringing groups back into analysis creates space to analyse inter-group politics. Yuval-Davis describes the structured yet dynamic sense of coalitions as being "rooted" in a particular social context but also "shifting" in order to engage in transversal dialogues and politics. Yuval-Davis's depiction of transversal politics requires processes of shifting that do not mean losing one's own rooting within historically situated communities and the intellectual and political sensibilities that rooting engenders. In this sense, ideas about intersectionality and flexibility that Black women develop within African American communities need not be jettisoned when shifting toward transversal politics. Far from one of subordinating one's issues into some greater good, as suggested within prevailing understandings of solidarity, remaining rooted while shifting constitutes a viable if not essential political option.

Another dimension of shifting is equally significant: the process of shifting must maintain the multiplicity of perspectives both within a group and across groups. This is the difficult challenge, one that recognizes that some coalitions may not be possible. My reading of the historical trajectory of Black feminism in the U.S. is that the flexible solidarity that Black women exhibit across many historical periods, and that informs intersectionality, constitute a missing dimension of transversal politics. Instead, the flexible solidarity by Black women within African American communities, when coupled with Yuval Davis's framework of the rooting and shifting of transversal politics, potentially facilitates thinking through coalition politics within a context of intersecting power relations.

How might the concept of flexible solidarity honed through Black women's politics within African American communities and the idea of transversal politics as a framework for coalitions among groups inform anti-violence initiatives? Flexible solidarity and transversal politics remain abstract, and some might argue, unrealistic aspirational constructs. One construct seems wedded to past practices (flexible solidarity) whereas the other points toward an as yet unrealized future (transversal politics). Perhaps, however, both constructs inform contemporary anti-violence initiatives.

Take, for example, the effective political mobilization of the Black Lives Matter movement against state-sanctioned violence (Cobb 2016).This movement illustrates how the ideas of intersectionality and flexible solidarity honed within Black feminism suggest a move toward transversal politics as a way to resist violence. Initially led by three queer African American women who created the hashtag #BlackLivesMatter, the stellar growth of Black Lives Matter from 2012 to 2016 illustrates how the legacy of Black feminism has been brought to bear on the contemporary social problem of state-sanctioned racial violence. The deaths of several young African American men, widely shared on social media, was the spark that catalysed the movement. Yet Black women have been visible within the Black Lives Matter movement, from the initial leadership of the movement, to the large numbers Black women participating in the protests, demonstrations and urban rebellions that sustained the vitality of the movement. In essence, Black women who catalysed this movement drew upon the legacy of a Black feminism's long history of resisting violence targeted toward Black people.

At its inception, the Black Lives Matter movement also invoked the idea of intersectionality to expand the categories of Black people who should be respected by the movement. Historically, Black women themselves had used the idea of flexible solidarity to choose strategic moments to broaden Black solidarity to address issues of gender. The initial hashtag #BlackLivesMatter expressed a similar deepening of an intersectional analysis of Blackness, now expanded to highlight the issues of groups that were historically subordinated within Black communities. The web site of the Black Lives Matter movement has undergone substantial updating as the organization has grown, yet the initial intersectional description of their mission has remained constant:

> Rooted in the experiences of Black people in this country who actively resist our dehumanization, #BlackLivesMatter is a call to action and a response to the virulent anti-Black racism that permeates our society. Black Lives Matter is a unique contribution that goes beyond extrajudicial killings of Black people by police and vigilantes . . . Black Lives Matter affirms the lives of Black queer and trans folks, disabled folks, black-undocumented folks, folks with records, women and all Black lives along the gender spectrum. It centers those that have been marginalized within Black liberation movement. (blacklivesmatter.com)

The movement as laid out by the founders of #BlacklivesMatter is clearly intersectional by highlighting how all Black individuals within Black communities were worthy of political protection. Their intersectional mandate deepens analysis of how different sub-groups within Black communities experience racial domination. It is rooted in a collective Black past, yet not one that is uncritically celebrated or that mandates knee-jerk adherence to solidarity.

Significantly, as the movement has grown, its organizational practices also illustrate the goal of drawing upon flexible solidarity to strengthen both its own organizational capacities as a political community as well as those of other Black political communities. As the movement has evolved, it rejected the hierarchical bureaucracies of traditional civil rights organizations in favour of a more fluid decentralized organizational structure that allows it to draw upon the flexibility of networks. This focus on flexible coalitions within a

Black movement sets the stage for potential coalitions with external groups. In this sense, Black Lives Matter remains rooted in its anti-violence project, yet embraces a form of flexible solidarity within its practices that sees coalition as always under construction and not as ideologically fixed. This orientation positions it to remain rooted in the needs of its own praxis. Black Lives Matter points to the necessary interconnectedness of intersectionality and flexible solidarity within its own praxis as well as the continued challenges of using these ideas within broader social movements. This example also signals the challenges of future coalitionbuilding with other groups that have been inspired by this movement, yet must find ways to bring more sophisticated understandings of their own group histories to the transversal politics that might ensue.

Developing more complex analyses of intersectionality as a form of critical inquiry and praxis that resists violence promises to be a long-term intellectual and political project. Because violence is so deeply embedded into the fabric of society, it is unlikely to yield to the efforts of any one theory or group of social actors. Yet just as intersecting oppressions are far from static, forms of political resistance that are similarly flexible are well-positioned for such sustained intellectual and political struggle. In this endeavour, continuing to focus on violence should illuminate new connections between intersecting systems of power and on new possibilities for political resistance.

References

Anthias, Floya, and Nira Yuval-Davis. 1992. *Racialized Boundaries: Race, Nation, Gender, Colour and Class and the Anti-Racist Struggle*. New York: Routledge.

Balibar, Etienne. 1991. "Racism and Nationalism." In *Race, Nation, Class: Ambiguous Identities,* edited by Etienne Balibar and Immanuel Wallerstein, 37–67. New York: Verso.

Bauman, Zygmunt. 1989. *Modernity and the Holocaust*. Ithaca, NY: Cornell University Press.

blacklivesmatter.com. "Black Lives Matter." Accessed February 15, 2017.

Cobb, Jelani. 2016. "The Matter of Black Lives Matter." *The New Yorker*, March 14.

Collins, Patricia Hill. 1998. "The Tie That Binds: Race, Gender and U.S. Violence." *Ethnic and Racial Studies*, 21(5), 918–938.

Collins, Patricia Hill. 2001. "Like One of the Family: Race, Ethnicity, and the Paradox of US National Identity." *Ethnic and Racial Studies*, 24(1), 3–28.

Collins, Patricia Hill. 2006. *From Black Power to Hip Hop: Essays on Racism, Nationalism and Feminism*. Philadelphia, PA: Temple University Press.

Collins, Patricia Hill. forthcoming. "The Difference That Power Makes: Intersectionality and Participatory Democracy." Investigaciones Feministas.

Collins, Patricia Hill, and Sirma Bilge. 2016. *Intersectionality*. Cambridge: Polity.

Hine, Darlene Clark. 1989. "Rape and the Inner Lives of Black Women in the Middle West: Preliminary Thoughts on the Culture of Dissemblance." *Signs: Journal of Women in Culture and Society*, 14(4), 912–920.

Matsuda, Mari J., Charles Lawrence III, Richard Delgado, and Kimberle Crenshaw. 1993. *Words that Wound: Critical Race Theory, Assaultive Speech, and the First Amendment*. Boulder: Westview Press.

Roberts, Dorothy, and Sujatha Jesudason. 2013. "Movement Intersectionality: The Case of Race, Gender, Disability and Genetic Technologies." *Du Bois Review: Social Science Research on Race*, 10, 313–328.

Southern-Poverty-Law-Center. https://www.splcenter.org/.

Terriquez, Veronica. 2015. "Intersectional Mobilization, Social Movement Spillover, and Queer Youth Leadership in the Immigrant Rights Movement." Social Problems, 62, 343–362.

Yuval-Davis, Nira. 1997. *Gender and Nation*. Thousand Oaks, CA: Sage.

Introduction to Reading 9

In this article, Karen D. Pyke and Denise L. Johnson use both the social construction of gender and intersectional analysis to examine the experiences of second-generation Asian American women. They interviewed 100 daughters of Korean American (KA) and Vietnamese American (VA) immigrants to better understand how gender, ethnicity, and culture influenced the meaning respondents gave to their experiences. By living in two worlds, the Asian American women were acutely aware of the social construction of gender within culture, as they had to move between two cultural constructions of femininity. Thus, as in the previous reading by Nikki Jones, race and gender interact in ways that made the women conscious of their decision to "do gender" based on the women's strategic use of culturally defined, situational expectations for femininity.

1. Using this article as an example, explain what it means to "do gender."
2. Why don't these women just be "who they are" across situations?
3. How do these women's struggles between cultural definitions of femininity reinforce, and make dominant, White femininity?

Asian American Women and Racialized Femininities

"Doing" Gender Across Cultural Worlds

Karen D. Pyke and Denise L. Johnson

The study of gender in recent years has been largely guided by two orienting approaches: (1) a social constructionist emphasis on the day-to-day production or doing of gender (Coltrane 1989; West and Zimmerman 1987), and (2) attention to the interlocking systems of race, class, and gender (Espiritu 1997; Collins 2000). Despite the prominence of these approaches, little empirical work has been done that integrates the doing of gender with the study of race. A contributing factor is the more expansive incorporation of social constructionism in the study of gender than in race scholarship where biological markers are still given importance despite widespread acknowledgment that racial oppression is rooted in social arrangements and not biology (Glenn 1999). In addition, attempts to theoretically integrate the doing of gender, race, and class around the concept of "doing difference" (West and Fenstermaker 1995) tended to downplay historical macrostructures of power and domination and to privilege gender over race and class (Collins et al. 1995). Work is still needed that integrates systems of oppression in a social constructionist framework without granting primacy to any one form of inequality or ignoring larger structures of domination.

The integration of gender and race within a social constructionist approach directs attention to issues that have been overlooked. Little research has examined how racially and ethnically subordinated women, especially Asian American women, mediate cross-pressures in the production of femininity as they move between mainstream and ethnic arenas, such as family, work, and school, and whether distinct and even contradictory gender displays and strategies are enacted across different arenas. Many, if not most, individuals move in social worlds that do not require dramatic inversions of their gender performances, thereby enabling them to maintain stable and seemingly unified gender strategies. However, members of communities that are racially and ethnically marginalized and who regularly traverse interactional arenas with conflicting gender expectations might engage different gender performances depending on the local context in which they are interacting. Examining the ways that such individuals mediate conflicting expectations would address several unanswered questions. Do marginalized women shift their gender performances across mainstream and subcultural settings in response to different gender norms? If so, how do they experience and negotiate such transitions? What meaning do they assign to the different forms of femininities that they engage across settings? Do racially subordinated women experience their production of femininity as inferior to those forms engaged by privileged white women and glorified in the dominant culture?

We address these issues by examining how second-generation Asian American women experience and think about the shifting dynamics involved in the doing of femininity in Asian ethnic and mainstream cultural worlds. We look specifically at their assumptions about gender dynamics in the Eurocentric mainstream and Asian ethnic social settings, the way they

Pyke, K. D., & Johnson, D. L. (2003). Asian American women and racialized femininities: "Doing" gender across cultural worlds. *Gender & Society, 17*(1): 33–53.

think about their gendered selves, and their strategies in doing gender. Our analysis draws on and elaborates the theoretical literature concerning the construction of femininities across race, paying particular attention to how controlling images and ideologies shape the subjective experiences of women of color. This is the first study to our knowledge that examines how intersecting racial and gender hierarchies affect the everyday construction of gender among Asian American women.

Constructing Femininities

Current theorizing emphasizes gender as a socially constructed phenomenon rather than an innate and stable attribute (Lorber 1994; Lucal 1999; West and Zimmerman 1987). Informed by symbolic interactionism and ethnomethodology, gender is regarded as something people do in social interaction. Gender is manufactured out of the fabric of culture and social structure and has little, if any, causal relationship to biology (Kessler and McKenna 1978; Lorber 1994). Gender displays are "culturally established sets of behaviors, appearances, mannerisms, and other cues that we have learned to associate with members of a particular gender" (Lucal 1999, 784). These displays "cast particular pursuits as expressions of masculine and feminine 'natures'" (West and Zimmerman 1987, 126). The doing of gender involves its display as a seemingly innate component of an individual.

The social construction of gender provides a theoretical backdrop for notions of multiple masculinities put forth in the masculinities literature (Coltrane 1994; Connell 1987, 1995; Pyke 1996). We draw on this notion in conceptualizing a plurality of femininities in the social production of women. According to this work, gender is not a unitary process. Rather, it is splintered by overlapping layers of inequality into multiple forms of masculinities (and femininities) that are both internally and externally relational and hierarchical. The concepts of hegemonic and subordinated masculinities are a major contribution of this literature. . . .

The concept of femininities has served mostly as a placeholder in the theory of masculinities where it remains undertheorized and unexamined. Connell (1987, 1995) has written extensively about hegemonic masculinity but offers only a fleeting discussion of the role of femininities. He suggested that the traits of femininity in a patriarchal society are tremendously diverse, with no one form emerging as hegemonic. Hegemonic masculinity is centered on men's global domination of women, and because there is no configuration of femininity organized around women's domination of men, Connell (1987, 183) suggested the notion of a hegemonic femininity is inappropriate. He further argued that women have few opportunities for institutionalized power relations over other women. However, this discounts how other axes of domination, such as race, class, sexuality, and age, mold a hegemonic femininity that is venerated and extolled in the dominant culture, and that emphasizes the superiority of some women over others, thereby privileging white upper-class women. To conceptualize forms of femininities that are subordinated as "problematic" and "abnormal," it is necessary to refer to an oppositional category of femininity that is dominant, ascendant, and "normal" (Glenn 1999, 10). We use the notion of hegemonic and subordinated femininities in framing our analysis.

Ideas of hegemonic and subordinated femininities resonate in the work of feminist scholars of color who emphasize the multiplicity of women's experiences. Much of this research has focused on racial and class variations in the material and (re)productive conditions of women's lives. More recently, scholarship that draws on cultural studies, race and ethnic studies, and women's studies centers the cultural as well as material processes by which gender and race are constructed, although this work has been mostly theoretical (Espiritu 1997; Collins 2000; St. Jean and Feagin 1998). Collins (2000) discussed "controlling images" that denigrate and objectify women of color and justify their racial and gender subordination. Controlling images are part of the process of "othering," whereby a dominant group defines into existence a subordinate group through the creation of categories and ideas that mark the group as inferior (Schwalbe et al. 2000, 422). Controlling images reaffirm whiteness as normal and privilege white women by casting them as superior.

White society uses the image of the Black matriarch to objectify Black women as overly aggressive, domineering, and unfeminine. This imagery serves to blame Black women for the emasculation of Black men, low marriage rates, and poverty and to control their social behavior by undermining their assertiveness (Collins 2000). While Black women are masculinized as aggressive and overpowering, Asian women are rendered hyperfeminine: passive, weak, quiet, excessively submissive, slavishly dutiful, sexually exotic, and available for white men (Espiritu 1997; Tajima 1989). This Lotus Blossom imagery obscures the internal variation of Asian American femininity and sexuality, making it difficult, for example, for others to "see" Asian lesbians and bisexuals (Lee 1996).

Controlling images of Asian women also make them especially vulnerable to mistreatment from men who view them as easy targets. By casting Black women as not feminine enough and Asian women as too feminine, white forms of gender are racialized as normal and superior. In this way, white women are accorded racial privilege.

The dominant culture's dissemination of controlling imagery that derogates nonwhite forms of femininity (and masculinity) is part of a complex ideological system of "psychosocial dominance" (Baker 1983, 37) that imposes elite definitions of subordinates, denying them the power of self-identification. In this way, subordinates internalize "commonsense" notions of their inferiority to whites (Espiritu 1997; Collins 2000). Once internalized, controlling images provide the template by which subordinates make meaning of their everyday lives (Pyke 2000), develop a sense of self, form racial and gender identities, and organize social relations (Osajima 1993; Pyke and Dang 2003). For example, Chen (1998) found that Asian American women who joined predominately white sororities often did so to distance themselves from images of Asian femininity.

In contrast, those who joined Asian sororities were often surprised to find their ideas of Asian women as passive and childlike challenged by the assertive, independent women they met. By internalizing the racial and gendered myth making that circumscribes their social existence, subordinates do not pose a threat to the dominant order. As Audre Lorde (1984, 123) described, "the true focus of revolutionary change is never merely the oppressive situations which we seek to escape, but that piece of the oppressor which is planted deep within us."

Hegemonies are rarely without sites of resistance (Espiritu 2001; Gramsci 1971; Collins 2000). Espiritu (1997) described Asian American writers and filmmakers whose portraits of Asians defy the gender caricatures disseminated in the white-dominated society. However, such images are often forged around the contours of the one-dimensional stereotypes against which the struggle emerges. Thus, controlling images penetrate all aspects of the experience of subordinates, whether in a relationship of compliance or in one of resistance (Osajima 1993; Pyke and Dang 2003).

The work concerning the effects of controlling images and the relational construction of subordinated and hegemonic femininities has mostly been theoretical. The little research that has examined how Asian American women do gender in the context of racialized images and ideologies that construct their gender as "naturally" inferior to white femininity provides only a brief look at these issues (Chen 1998; Lee 1996). Many of the Asian American women whom we study here do not construct their gender in one cultural field but are constantly moving between sites that are guided by ethnic immigrant cultural norms and those of the Eurocentric mainstream. A comparison of how gender is enacted and understood across such sites brings the construction of racialized gender and the dynamics of hegemonic and subordinated femininities into bold relief. We examine how respondents employ cultural symbols, controlling images, and gender and racial ideologies in giving meanings to their experiences.

Gender in Ethnic and Mainstream Cultural Worlds

We study Korean and Vietnamese Americans, who form two of the largest Asian ethnic groups in southern California, the site of this research. We focus on the daughters of immigrants as they are more involved in both ethnic and mainstream cultures than are members of the first generation. . . . The second generation, who are still mostly children and young adults, must juggle the cross-pressures of ethnic and mainstream cultures without the groundwork that a long-standing ethnic enclave might provide. This is not easy. Disparities between ethnic and mainstream worlds can generate substantial conflict for children of immigrants, including conflict around issues of gender (Kibria 1993; Zhou and Bankston 1998).

Respondents dichotomized the interactional settings they occupy as ethnic, involving their immigrant family and other coethnics, and mainstream, involving non–Asian Americans in peer groups and at work and school. They grew up juggling different cultural expectations as they moved from home to school and often felt a pressure to behave differently when among Asian Americans and non–Asian Americans. Although there is no set of monolithic, stable norms in either setting, there are certain pressures, expectations, and structural arrangements that can affect different gender displays (Lee 1996). Definitions of gender and the constraints that patriarchy imposes on women's gender production can vary from culture to culture. The Confucian moral code, which accords male superiority, authority, and power over women in family and social relations, has influenced the patriarchal systems of Korea and Vietnam (Kibria 1993; Min 1998). Women are granted little decision-making power and are not accorded an individual identity apart from their family role, which emphasizes their service to male members. A woman who violates her role brings shame to herself and her family. Despite Western observers' tendency to regard Asian families as uniformly and rigidly patriarchal, variations exist (Ishii-Kuntz 2000). Women's resistance strategies, like the exchange of information in informal social groups, provide pockets of power (Kibria 1990). Women's growing educational and economic opportunities and the rise of women's rights groups in Korea and Vietnam challenge gender inequality (Palley 1994). Thus, actual gender dynamics are not in strict compliance with the prescribed moral code.

As they immigrate to the United States, Koreans and Vietnamese experience a shift in gender arrangements centering on men's loss of economic power and increased dependency on their wives' wages (Kibria 1993; Lim 1997; Min 1998). Immigrant women find their labor in demand by employers who regard them as a cheap labor source. With their employment, immigrant women experience more decision-making power, autonomy, and assistance with domestic chores from their husbands. However, such shifts are not total, and male dominance remains a common feature of family life (Kibria 1993; Min 1998). Furthermore, immigrant women tend to stay committed to the ethnic patriarchal structure as it provides resources for maintaining their parental authority and resisting the economic insecurities, racism, and cultural impositions of the new society (Kibria 1990, 1993; Lim 1997). The gender hierarchy is evident in parenting practices. Daughters are typically required to be home and performing household chores when not in school, while sons are given greater freedom.

Native-born American women, on the other hand, are perceived as having more equality, power, and independence than women in Asian societies, reflecting differences in gender attitudes. A recent study of Korean and American women found that 82 percent of Korean women agreed that "women should have only a family-oriented life, devoted to bringing up the children and looking after the husband," compared to 19 percent of U.S. women (Kim 1994). However, the fit between egalitarian gender attitudes and actual behavior in the United States is rather poor. Patriarchal arrangements that accord higher status to men at home and work are still the norm, with women experiencing lower job status and pay, greater responsibility for family work even when employed, and high rates of male violence. Indeed, the belief that gender equality is the norm in U.S. society obscures the day-to-day materiality of American patriarchy. Despite cultural differences in the ideological justification of patriarchy, gender inequality is the reality in both Asian and mainstream cultural worlds.

* * *

Gender Across Cultural Terrains: "I'm Like a Chameleon. I Change My Personality"

The 44 respondents who were aware of modifying their gender displays or being treated differently across cultural settings framed their accounts in terms of an oppressive ethnic world and an egalitarian mainstream. They reaffirmed the ideological constructions of the white-dominated society by casting ethnic and mainstream worlds as monolithic opposites, with internal variations largely ignored. Controlling images that denigrate Asian femininity and glorify white femininity were reiterated in many of the narratives. Women's behavior in ethnic realms was described as submissive and controlled, and that in white-dominated settings as freer and more self-expressive.

Some respondents suggested they made complete personality reversals as they moved across realms. They used the behavior of the mainstream as the standard by which they judged their behavior in ethnic settings. As Elizabeth (19, VA) said,

> I feel like when I'm amongst other Asians . . . I'm much more reserved and I hold back what I think. . . . But when I'm among other people like at school, I'm much more outspoken. I'll say whatever's on my mind. It's like a diametric character altogether. . . . I feel like when I'm with other Asians that I'm the *typical* passive [Asian] person and I feel like that's what's expected of me and if I do say something and if I'm the *normal* person that I am, I'd stick out like a sore thumb. So I just blend in with the situation. (emphasis added)

Elizabeth juxtaposes the "typical passive [Asian] person" and the "normal," outspoken person of the mainstream culture, whom she claims to be. In so doing, she reaffirms the stereotypical image of Asians as passive while glorifying Americanized behavior, such as verbal expressiveness, as "normal." This implies that Asian ethnic behavior is aberrant and inferior compared to white behavior, which is rendered normal. This juxtaposition was a recurring theme in these data (Pyke 2000). It contributed to respondents' attempts to distance themselves from racialized notions of the typical Asian woman who is hyperfeminine and submissive by claiming to possess those traits associated with white femininity, such as assertiveness, self-possession, confidence, and independence. Respondents often described a pressure to blend in and conform with the form of gender that they felt was expected in ethnic settings and that conflicted with the white standard of femininity. Thus, they often described such behavior with disgust and self-loathing. For example, Min-Jung (24, KA) said she feels "like an idiot" when talking with Korean adults:

> With Korean adults, I act more shy and more timid. I don't talk until spoken to and just act shy. I kind of speak in a higher tone of voice than I usually do. But then when I'm with white people and white adults, I joke around, I laugh, I talk, and I communicate about how I feel. And then my voice gets stronger. But then when I'm with Korean adults, my voice gets really high. . . . I just sound like an idiot and sometimes when I catch myself I'm like, "Why can't you just make conversation like you normally do?"

Many respondents distanced themselves from the compliant femininity associated with their Asianness by casting their behavior in ethnic realms as a mere act not reflective of their true nature. Repeatedly, they said they cannot be who they really are in ethnic settings and the enactment of an authentic self takes place only in mainstream settings. . . .

Wilma (21, VA) states, "Like some Asian guys expect me to be passive and let them decide on everything. Non-Asians don't expect anything from me. They just expect me *to be me*" (emphasis added). Gendered behavior engaged in Asian ethnic settings was largely described as performative, fake, and unnatural, while that in white-dominated settings was cast as a reflection of one's true self. The femininity of the white mainstream is glorified as authentic, natural, and normal, and Asian ethnic femininity is denigrated as coerced, contrived, and artificial. The "white is right" mantra is reiterated in this view of white femininity as the right way of doing gender.

The glorification of white femininity and controlling images of Asian women can lead Asian American women to believe that freedom and equity can be acquired only in the white-dominated world. For not only is white behavior glorified as superior and more authentic, but gender relations among whites are constructed as more egalitarian. . . .

Controlling images of Asian men as hypermasculine further feed presumptions that whites are more egalitarian. Asian males were often cast as uniformly domineering in these accounts. Racialized images and the construction of hegemonic (white) and subordinated (Asian) forms of gender set up a situation where Asian American women feel they must choose between white worlds of gender equity and Asian worlds of gender oppression. Such images encourage them to reject their ethnic culture and Asian men and embrace the white world and white men so as to enhance their power (Espiritu 1997). . . .

In these accounts, we can see the construction of ethnic and mainstream cultural worlds—and Asians and whites—as diametrically opposed. The perception that whites are more egalitarian than Asian-origin individuals and thus preferred partners in social interaction further reinforces anti-Asian racism and white superiority. The cultural dominance of whiteness is reaffirmed through the co-construction of race and gender in these narratives. The perception that the production of gender in the mainstream is more authentic and superior to that in Asian ethnic arenas further reinforces the racialized categories of gender that define white forms of femininity as ascendant. In the next section, we describe variations in gender performances within ethnic and mainstream settings that respondents typically overlooked or discounted as atypical.

Gender Variations Within Cultural Worlds

Several respondents described variations in gender dynamics within mainstream and ethnic settings that challenge notions of Asian and American worlds as monolithic opposites. Some talked of mothers who make all the decisions or fathers who do the cooking. These accounts were framed as exceptions to Asian male dominance. For example, after Vietnamese women were described in a group interview as confined to domesticity, Ngâ (22, VA), who immigrated at 14 and spoke in Vietnamese-accented English, defined her family as gender egalitarian. She related,

> I guess I grow[sic] up in a *different* family. All my sisters don't have to cook, her husbands[sic] cooking all the time. Even my oldest sister. Even my mom—my dad is cooking. . . . My sisters and brothers are all very strong. (emphasis added)

Ngâ does not try to challenge stereotypical notions of Vietnamese families but rather reinforces such notions by suggesting that her family is different. Similarly, Heidi (21, KA) said, "Our family was kind of *different* because . . . my dad cooks and cleans and does dishes. He cleans house" (emphasis added). Respondents often framed accounts of gender egalitarianism in their families by stating they do not belong to the typical Asian family, with "typical" understood to mean male dominated. This variation in gender dynamics within the ethnic community was largely unconsidered in these accounts.

Other respondents described how they enacted widely disparate forms of gender across sites within ethnic realms, suggesting that gender behavior is more variable than generally framed. Take, for example, the case of Gin (29, KA), a law student married to a Korean American from a more traditional family than her own. When she is with her husband's kin, Gin assumes the traditional obligations of a daughter-in-law and does all the cooking, cleaning, and serving. The role exhausts her and she resents having to perform it. When Gin and her husband return home, the gender hierarchy is reversed. . . .

Controlling images of Asian men as hyperdomineering in their relations with women obscures how they can be called on to compensate for the subservience exacted from their female partners in some settings. Although respondents typically offered such stories as evidence of the patriarchy of ethnic arenas, these examples reveal that ethnic worlds are far more variable than generally described. Viewing Asian ethnic worlds through a lens of racialized gender stereotypes renders such variation invisible or, when acknowledged, atypical.

Gender expectations in the white-dominated mainstream also varied, with respondents sometimes expected to assume a subservient stance as Asian women. These examples reveal that the mainstream is not a site of unwavering gender equality as often depicted in these accounts and made less so for Asian American women by racial images that construct them as compliant. Many respondents described encounters with non-Asians, usually whites, who expected them to be passive, quiet, and yielding. Several described non-Asian (mostly white) men who brought such expectations to their dating relationships. Indeed, the servile Lotus Blossom image bolsters white men's preference for Asian women (Espiritu 1997). As Thanh (22, VA) recounted,

> Like the white guy that I dated, he expected me to be the submissive one—the one that was dependent on the guy. Kind of like the "Asian persuasion," that's what he'd call it when he was dating me. And when he found out that I had a spirit, kind of a wild side to me, he didn't like it at all. Period. And when I spoke up—my opinions—he got kind of scared.

So racialized images can cause Asian American women to believe they will find greater gender equality with white men and can cause white men to believe they will find greater subservience with Asian women. This dynamic promotes Asian American women's availability to white men and makes them particularly vulnerable to mistreatment.

There were other sites in the mainstream, besides dating relationships, where Asian American women encountered racialized gender expectations. Several described white employers and coworkers who expected them to be more passive and deferential than other employees and were surprised when they spoke up and resisted unfair treatment. Some described similar assumptions among non-Asian teachers and professors. Diane (26, KA) related,

> At first one of my teachers told me it was okay if I didn't want to talk in front of the class. I think she thought I was quiet or shy because I'm Asian. . . . [Laughing.] I am very outspoken, but that semester I just kept my mouth shut. I figured she won't make me talk anyway, so why try. I kind of went along with her.

Diane's example illustrates how racialized expectations can exert a pressure to display stereotyped behavior in mainstream interactions. Such expectations can subtly coerce behavioral displays that confirm the stereotypes, suggesting a kind of self-fulfilling prophecy. Furthermore, as submissiveness and passivity are denigrated traits in the mainstream, and often judged to be indicators of incompetence, compliance with such expectations can deny Asian American women personal opportunities and success. Not only is passivity unrewarded in the mainstream; it is also subordinated. The association of extreme passivity with Asian women serves to emphasize their otherness. Some respondents resist this subordination by enacting a more assertive femininity associated with whiteness. Lisa (18, KA) described being quiet with her relatives out of respect, but in mainstream scenes, she consciously resists the stereotype of Asian women as passive by adjusting her behavior. . . .

To act Asian by being reserved and quiet would be to "stand out in a negative way" and to be regarded as "not cool." It means one will be denigrated and cast aside. Katie (21, KA) consciously engages loud and gregarious behavior to prove she is not the typical Asian and to be welcomed by white friends. Whereas many respondents describe their behavior in mainstream settings as an authentic reflection of their personality, these examples suggest otherwise. Racial expectations exert pressure on these women's gender performances among whites. Some go to great lengths to defy racial assumptions and be accepted into white-dominated social groups by engaging a white standard of femininity. As they are forced to work against racial stereotypes, they must exert extra effort at being outspoken and socially gregarious. Contrary to the claim of respondents, gender production in the mainstream is also coerced and contrived. The failure of some respondents to recognize variations in gender behavior within mainstream and ethnic settings probably has much to do with the essentialization of gender and race. That is, as we discuss next, the racialization of gender renders variations in behavior within racial groups invisible.

The Racialization of Gender: Believing Is Seeing

In this section, we discuss how respondents differentiated femininity by race rather than shifting situational contexts, even when they were consciously aware of altering their own gender performance to conform with shifting expectations. Racialized gender was discursively constructed as natural and essential. Gender and race were essentialized as interrelated biological facts that determine social behavior.

Among our 100 respondents, there was a tendency to rely on binary categories of American (code for white) and Asian femininity in describing a wide range of topics, including gender identities, personality traits, and orientations toward domesticity or career. Racialized gender categories were deployed as an interpretive template in giving meaning to experiences and organizing a worldview. Internal variation was again ignored, downplayed, or regarded as exceptional. White femininity, which was glorified in accounts of gender behavior across cultural settings, was also accorded superiority in the more general discussions of gender.

Respondents' narratives were structured by assumptions about Asian women as submissive, quiet, and diffident and of American women as independent, self-assured, outspoken, and powerful. That is, specific behaviors and traits were racialized. As Ha (19, VA) explained, "sometimes I'm quiet and passive and shy. That's a Vietnamese part of me." Similarly, domesticity was linked with Asian femininity and domestic incompetence or disinterest, along with success in the work world, with American femininity. Several women framed their internal struggles between career and domesticity in racialized terms. Min-Jung said,

> I kind of think my Korean side wants to stay home and do the cooking and cleaning and take care of the kids whereas my American side would want to go out and make a difference and become a strong woman and become head of companies and stuff like that.

This racialized dichotomy was central to respondents' self-identities. Amy (21, VA) said, "I'm not Vietnamese in the way I act. I'm American because I'm not a good cook and I'm not totally ladylike." In fact, one's ethnic identity could be challenged if one did not comply with notions of racialized gender. In a group interview, Kimberly (21, VA) described "joking around" with coethnic dates who asked if she cooked by responding that she did not. . . .

Similarly, coethnic friends tell Hien (21, VA), "You should be able to cook, you are Vietnamese, you are a girl." To be submissive and oriented toward family and domesticity marks Asian ethnicity. Conformity to stereotypes of Asian femininity serves to symbolically construct and affirm an Asian ethnic identity. Herein lies the pressure that some respondents feel to comply with racialized expectations in ethnic settings, as Lisa (18, KA) illustrates in explaining why she would feel uncomfortable speaking up in a class that has a lot of Asians:

> I think they would think that I'm not really Asian. Like I'm whitewashed . . . like I'm forgetting my race. I'm going against my roots and adapting to the American way. And I'm just neglecting my race.

American (white) women and Asian American women are constructed as diametric opposites. Although many respondents were aware that they contradicted racialized notions of gender in their day-to-day lives, they nonetheless view gender as an essential component of race. Variation is ignored or recategorized so that an Asian American woman who does not comply is no longer Asian. This was also evident among respondents who regard themselves as egalitarian or engage the behavioral traits associated with white femininity. There was the presumption that one cannot be Asian and have gender-egalitarian attitudes. Asian American women can engage those traits associated with ascendant femininity to enhance their status in the mainstream, but this requires a rejection of their racial/ethnic identity. This is evident by the use of words such as "American," "whitewashed," or "white"—but not Asian—to describe such women. Star (22, KA) explained, "I look Korean but I don't act Korean. I'm whitewashed. [Interviewer asks, 'How do you mean you don't act Korean?'] I'm loud. I'm not quiet and reserved."

As a result, struggles about gender identity and women's work/family trajectories become superimposed over racial/ethnic identity. The question is not simply whether Asian American women like Min-Jung want to be outspoken and career oriented or quiet and family oriented but whether they want to be American (whitewashed) or Asian. Those who do not conform to racialized expectations risk challenges to their racial identity and charges that they are not really Asian, as occurs with Lisa when she interacts with her non-Asian peers. She said,

> They think I'm really different from other Asian girls because I'm so outgoing. They feel that Asian girls have to be the shy type who is very passive and sometimes I'm not like that so they think, "Lisa, are you Asian?"

These data illustrate how the line drawn in the struggle for gender equality is superimposed over the cultural and racial boundaries dividing whites and Asians. At play is the presumption that the only path to gender equality and assertive womanhood is via assimilation to the white mainstream. This assumption was shared by Asian American research assistants who referred to respondents' gender egalitarian viewpoints as evidence of assimilation. The assumption is that Asian American women can be advocates of gender equality or strong and assertive in their interactions only as a result of assimilation, evident by the display of traits associated with hegemonic femininity, and a rejection of their ethnic culture and identity. This construction obscures gender inequality in mainstream U.S. society and constructs that sphere as the only place where Asian American women can be free. Hence, the diversity of gender arrangements practiced among those of Asian origin, as well as the potential for social change within Asian cultures, is ignored. Indeed, there were no references in these accounts to the rise in recent years of women's movements in Korea and Vietnam. Rather, Asian ethnic worlds are regarded as unchanging sites of male dominance and female submissiveness.

Discussion and Summary

Our analysis reveals dynamics of internalized oppression and the reproduction of inequality that revolve around the relational construction of hegemonic and subordinated femininities. Respondents' descriptions of gender performances in ethnic settings were marked by self-disgust and referred to as a mere act not reflective of one's true gendered nature. In mainstream settings, on the other hand, respondents often felt a pressure to comply with caricatured notions of Asian femininity or, conversely, to distance one's self from derogatory images of Asian femininity to be accepted. In both cases, the subordination of Asian femininity is reproduced.

In general, respondents depicted women of Asian descent as uniformly engaged in subordinated femininity marked by submissiveness and white women as universally assertive and gender egalitarian. Race, rather than culture, situational dynamics, or individual personalities, emerged as the primary basis by which respondents gave meaning to variations in femininity. That is, despite their own situational variation in doing gender, they treat gender as a racialized feature of bodies rather than a sociocultural product. Specific gender displays, such as a submissive demeanor, are required to confirm an Asian identity. Several respondents face challenges to their ethnic identity when they behave in ways that do not conform with racialized images. Indeed, some claimed that because they are assertive or career oriented, they are not really Asian. That is, because they do not conform to the racialized stereotypes of Asian women but identify with a hegemonic femininity that is the white standard, they are different from other women of Asian origin. In this way, they manipulate the racialized categories of gender in attempting to craft identities that are empowering. However, this is accomplished by denying their ethnicity and connections to other Asian American women and through the adoption and replication of controlling images of Asian women.

Respondents who claim that they are not really Asian because they do not conform with essentialized notions of Asian femininity suggest similarities to transgendered individuals who feel that underneath, they really belong to the gender category that is opposite from the one to which they are assigned. The notion that deep down they are really white implies a kind of transracialized gender identity. In claiming that they are not innately Asian, they reaffirm racialized categories of gender just as transgendered individuals reaffirm the gender dichotomy (Kessler and McKenna 1978; Lorber 1994).

However, there are limitations to notions of a transracialized identity as racial barriers do not permit these women to socially pass into the white world, even though they might feel themselves to be more white than Asian. Due to such barriers, they use terms that are suggestive of a racial crossover, such as "whitewashed" or "American" rather than "white" in describing themselves. Such terms are frequently used among Asian Americans to describe those who are regarded as assimilated to the white world and no longer ethnic, further underscoring how racial categories are essentialized (Pyke and Dang 2003). Blocked from a white identity, these terms capture a marginalized space that is neither truly white nor Asian. As racial categories are dynamic, it remains to be seen whether these marginalized identities are the site for new identities marked by hybridity (Lowe 1991) or whether

Asian Americans will eventually be incorporated into whiteness. This process may be hastened by outmarriage to whites and high rates of biracial Asian Americans who can more easily pass into the white world, thereby leading the way for other Asian Americans. While we cannot ascertain the direction of such changes, our data highlight the contradictions that strain the existing racial and gender order as it applies to second-generation Asian American women.

While respondents construct a world in which Asian American women can experience a kind of transracial gender identity, they do not consider the same possibility for women of other races. A white woman who is submissive does not become Asian. In fact, there was no reference in these accounts to submissive white women who are rendered invisible by racialized categories of gender. Instead, white women are constructed as monolithically self-confident, independent, assertive, and successful—characteristics of white hegemonic femininity. That these are the same ruling traits associated with hegemonic masculinity, albeit in a less exaggerated, feminine form, underscores the imitative structure of hegemonic femininity. That is, the supremacy of white femininity over Asian femininity mimics hegemonic masculinity. We are not arguing that hegemonic femininity and masculinity are equivalent structures. They are not. Whereas hegemonic masculinity is a superstructure of domination, hegemonic femininity is confined to power relations among women. However, the two structures are interrelated with hegemonic femininity constructed to serve hegemonic masculinity, from which it is granted legitimacy.

Our findings illustrate the powerful interplay of controlling images and hegemonic femininity in promoting internalized oppression. Respondents draw on racial images and assumptions in their narrative construction of Asian cultures as innately oppressive of women and fully resistant to change against which the white-dominated mainstream is framed as a paradigm of gender equality. This serves a proassimilation function by suggesting that Asian American women will find gender equality in exchange for rejecting their ethnicity and adopting white standards of gender. The construction of a hegemonic femininity not only (re)-creates a hierarchy that privileges white women over Asian American women but also makes Asian American women available for white men. In this way, hegemonic femininity serves as a handmaiden to hegemonic masculinity.

By constructing ethnic culture as impervious to social change and as a site where resistance to gender oppression is impossible, our respondents accommodate and reinforce rather than resist the gender hierarchal arrangements of such locales. This can contribute to a self-fulfilling prophecy as Asian American women who hold gender egalitarian views feel compelled to retreat from interactions in ethnic settings, thus (re)creating Asian ethnic cultures as strongholds of patriarchy and reinforcing the maintenance of a rigid gender hierarchy as a primary mechanism by which ethnicity and ethnic identity are constructed. This marking of ethnic culture as a symbolic repository of patriarchy obscures variations in ethnic gender practices as well as the gender inequality in the mainstream. Thus, compliance with the dominant order is secured.

Our study attempts to bring a racialized examination of gender to a constructionist framework without decentering either race or gender. By examining the racialized meaning systems that inform the construction of gender, our findings illustrate how the resistance of gender oppression among our respondents draws ideologically on the denigration and rejection of ethnic Asian culture, thereby reinforcing white dominance. Conversely, we found that mechanisms used to construct ethnic identity in resistance to the pro-assimilation forces of the white-dominated mainstream rest on narrow definitions of Asian women that emphasize gender subordination. These findings underscore the crosscutting ways that gender and racial oppression operates such that strategies and ideologies focused on the resistance of one form of domination can reproduce another form. A social constructionist approach that examines the simultaneous production of gender and race within the matrix of oppression, and considers the relational construction of hegemonic and subordinated femininities, holds much promise in uncovering the micro-level structures and complicated features of oppression, including the processes by which oppression infiltrates the meanings individuals give to their experiences.

References

Baker, Donald G. 1983. *Race, ethnicity and power.* Boston: Routledge Kegan Paul.

Chen, Edith Wen-Chu. 1998. The continuing significance of race: A case study of Asian American women in white, Asian American, and African American sororities. Ph.D. diss., University of California, Los Angeles.

Collins, Patricia Hill. 2000. *Black feminist thought.* New York: Routledge.

Collins, Patricia Hill, Lionel A. Maldonado, Dana Y. Takagi, Barrie Thorne, Lynn Weber, and Howard Winant. 1995. Symposium: On West and Fenstermaker's "Doing difference." *Gender & Society* 9:491–513.

Coltrane, Scott. 1989. Household labor and the routine production of gender. *Social Problems* 36:473–90.

———. 1994. Theorizing masculinities in contemporary social science. In *Theorizing masculinities,* edited by Harry Brod and Michael Kaufman. Thousand Oaks, CA: Sage.

Connell, R. W. 1987. *Gender and power.* Stanford, CA: Stanford University Press.

———. 1995. *Masculinities.* Los Angeles: University of California Press.

Espiritu, Yen L. 1997. *Asian American women and men.* Thousand Oaks, CA: Sage.

———. 2001. "We don't sleep around like white girls do": Family, culture, and gender in Filipina American life. *Signs: Journal of Women in Culture and Society* 26:415–40.

Glenn, Evelyn Nakano. 1999. The social construction and institutionalization of gender and race. In *Revisioning gender,* edited by Myra Marx Ferree, Judith Lorber, and Beth B. Hess. Thousand Oaks, CA: Sage.

Gramsci, Antonio. 1971. *Selections from the prison notebooks of Antonio Gramsci,* edited and translated by Quintin Hoare and Geoffrey Nowell Smith. New York: International.

Ishii-Kuntz, Masako. 2000. Diversity within Asian American families. In *Handbook of family diversity,* edited by David H. Demo, Katherine Allen, and Mark A. Fine. New York: Oxford University Press.

Kessler, Suzanne, and Wendy McKenna. 1978. *Gender: An ethnomethodological approach.* Chicago: University of Chicago Press.

Kibria, Nazli. 1990. Power, patriarchy, and gender conflict in the Vietnamese immigrant community. *Gender & Society* 4:9–24.

———. 1993. *Family tightrope: The changing lives of Vietnamese Americans.* Princeton, NJ: Princeton University Press.

Kim, Byong-suh. 1994. Value orientations and sex-gender role attitudes on the comparability of Koreans and Americans. In *Gender division of labor in Korea,* edited by Hyoung Cho and Pil-wha Chang. Seoul, Korea: Ewha Womans University Press.

Lee, Jee Yeun. 1996. Why Suzie Wong is not a lesbian: Asian and Asian American lesbian and bisexual women and femme/butch/gender identities. In *Queer studies,* edited by Brett Beemyn and Mickey Eliason. New York: New York University Press.

Lim, In-Sook. 1997. Korean immigrant women's challenge to gender inequality at home: The interplay of economic resources, gender, and family. *Gender & Society* 11:31–51.

Lorber, Judith. 1994. *Paradoxes of gender.* New Haven, CT: Yale University Press.

Lorde, Audre. 1984. *Sister outsider.* Trumansburg, NY: Crossing Press.

Lowe, Lisa. 1991. Heterogeneity, hybridity, multiplicity: Marking Asian American differences. *Diaspora* 1:24–44.

Lucal, Betsy. 1999. What it means to be gendered me: Life on the boundaries of a dichotomous gender system. *Gender & Society* 13:781–97.

Min, Pyong Gap. 1998. *Changes and conflicts.* Boston: Allyn & Bacon.

Osajima, Keith. 1993. The hidden injuries of race. In *Bearing dreams, shaping visions: Asian Pacific American perspectives,* edited by Linda Revilla, Gail Nomura, Shawn Wong, and Shirley Hune. Pullman: Washington State University Press.

Palley, Marian Lief. 1994. Feminism in a Confucian society: The women's movement in Korea. In *Women of Japan and Korea,* edited by Joyce Gelb and Marian Lief Palley. Philadelphia: Temple University Press.

Pyke, Karen. 1996. Class-based masculinities: The interdependence of gender, class, and interpersonal power. *Gender & Society* 10:527–49.

———. 2000. "The normal American family" as an interpretive structure of family life among grown children of Korean and Vietnamese immigrants. *Journal of Marriage and the Family* 62:240–55.

Pyke, Karen, and Tran Dang. 2003. "FOB" and "whitewashed": Intra-ethnic identities and internalized oppression among second generation Asian Americans. *Qualitative Sociology* 26(2).

Schwalbe, Michael, Sandra Godwin, Daphne Holden, Douglas Schrock, Shealy Thompson, and Michele Wolkomir. 2000. Generic processes in the reproduction of inequality: An interactionist analysis. *Social Forces* 79:419–52.

St. Jean, Yanick, and Joe R. Feagin. 1998. *Double burden: Black women and everyday racism.* Armonk, NY: M. E. Sharpe.

Tajima, Renee E. 1989. Lotus blossoms don't bleed: Images of Asian women. In *Making waves,* edited by Asian Women United of California. Boston: Beacon.

West, Candace, and Sarah Fenstermaker. 1995. Doing difference. *Gender & Society* 9:8–37.

West, Candace, and Don H. Zimmerman. 1987. Doing gender. *Gender & Society* 1:125–51.

Zhou, Min, and Carl L. Bankston III. 1998. *Growing up American.* New York: Russell Sage.

Introduction to Reading 10

Kaveri Thara is an Indian social scientist who studies poverty, caste, gender, and international development. In this reading, she discusses preliminary findings from her ethnographic research with a fisherwomen's association in Udupi City in South India. Thara uses intersectional analysis to understand how the fisherwomen, who are often in direct competition with each other, unite in struggle against the threat that large fish shops pose to their occupation. Thara finds that the fisherwomen not only mobilize gender but also occupational caste position and poverty status to attain state and local political support for their livelihoods.

1. In Reading 10, Collins discusses flexible solidarity as a strategy employed by African American women to build and maintain solidarity. How do the fisherwomen in Thara's study utilize flexible solidarity?
2. Discuss the role of breadwinning in the construction of Mogaveera fisherwomen's identity.
3. How do the fisherwomen use appeals to gendered poverty to argue for state protection and benefits?
4. Why and how do Mogaveera fisherwomen include Mogaveera men in meetings of the Association?

Protecting Caste Livelihoods on the Western Coast of India

An Intersectional Analysis of Udupi's Fisherwomen

Kaveri Thara

Introduction

This article is based on preliminary research conducted with a fisherwomen's association in Udupi, south India, and is part of a larger project aimed at a feminist analysis of solidarity economy practices in Latin America and Asia. This research in Udupi seeks to understand how members of an association, called *Udupi Hasi Meenu Marathagarara Sangha* or Udupi Fresh Fish Sellers Association, have been able to protect their livelihoods in a capitalist context, in which fish shops with refrigeration and home delivery services have taken over fish sales in other parts of Mangaluru, where Udupi is located. (Since women fish sellers are referred to as fisherwomen in this region, I use this term in the paper.) With increasing urbanization of the city of Udupi, the Association supports solidarity amongst fisherwomen, enabling them to negotiate with the state for the protection of their livelihoods. The project seeks to broadly answer questions of women's solidarity economy practices and social reproduction, engaging critically with these concepts. However, the specific questions this article poses are: How do women protect their livelihoods in a capitalist context? What political and personal strategies do they employ in this regard? How do they view and perceive their livelihoods in terms of their socially reproductive roles?

As Kabeer et al. argue, the biggest challenge in organizing workers in the informal economy is constructing a shared identity. Many of these women are in direct competition with each other and are located at different intersections of inequality in terms of class, race, caste and legal status. Their working conditions do not enable their recognition as workers. Given this context, organizations have to adopt a wide array of strategies to build and maintain solidarity between women, including the dissemination of information vital to their work, the provision of credit and other facilities to improve the conditions of their work, engagement in politics and policies, and mobilization of public and political consciousness of the importance of their work.[1]

In this instance, the fact that women sell in markets facilitates organization, and yet the fact that they compete with each other as sellers brings about tensions within the group. Maintaining a unified front thus requires broader strategies such as the provision of low-interest credit, enabling access to welfare services from the government and the negotiation of a place to sell fish. However, even for these material resources to be mobilized, a case has to be made in favour of the fisherwomen in the larger context of a capitalist economy. Since larger revenues can be made from bigger businesses selling fresh fish, which would then augment state revenues in the form of taxes, the sale of

Thara, K. (2016). Protecting caste livelihoods on the western coast of India: An intersectional analysis of Udupi's fisherwomen. *Environment and Urbanization, 28*(2): 423–436. Reprinted with permission from SAGE Publications, Inc.

fish in the traditional way through fisherwomen is less attractive to the state. Why then would the state support these women? How has the Association garnered support?

It is here that an analysis of these fisherwomen's identities and the intersection of their identities is important. In the act of mobilizing for resources, women's groups tend to use a variety of strategies and political resources, drawing attention to their gender and frequently to other identities such as class, race and, in the Indian context, caste. While analysis of women's organizations often focuses on their gendered experiences, an intersectional analysis can reveal how women's experiences are in constant co-construction with other experiences of caste, class, etc.

An intersectional analysis can thus provide a deeper understanding of how identities intersect in specific spaces and times to produce effects that may not be the same in other similar contexts. Using an intersectional approach to women's organizations allows us to ask why certain women face greater oppression than others, and what differences there are between women, beyond their gender, that produce certain constellations of marginality or dominance. It allows us to see gender as differentially experienced by different women and allows us to uncover specific historical trajectories and relations that put certain groups of women at certain social axes.

In the Indian context the state has had to deal with intersectionality in its policies, accommodating groups suffering plural marginal identities. Laura Dudley Jenkins looks at this engagement with plural social categories by the Indian state, examining how affirmative action policies (called reservations or quotas in India) use an intersectional lens, addressing plural marginal identities, such as caste, class and gender.[2] The issue of difference amongst women has also surfaced within Indian feminist debates. The exclusion of Dalit women's concerns from within the dominant feminist movement in India has led to academics calling for an analysis of difference.[3] Given this context of difference, an analysis of feminism in India has to account for differences between women of religion, caste and class.[4] This analysis has been further enriched by studies of caste and religion in the contemporary context, in interaction with modern democratic structures and institutions. Rajni Kothari points to how caste and religion are rearticulated within secular frameworks of democracy and citizenship to produce new constellations of identity that emerge in interaction.[5] Within this framework, it becomes important to draw from intersectionality not only in terms of the intersection of identities but also in terms of the intersection of social categories that draw from frameworks of social organization, traditional and modern, that inform each other and play out in local politics in India.

In the case of the Udupi Fresh Fish Sellers Association, intersectionality enables a more complex analysis, uncovering elements that cannot be explained purely by gender. First of all, fishing is the traditional occupation of the Mogaveera community in the region and Mogaveera women have traditionally bartered and sold fish locally to sustain their families. In their interviews, fisherwomen continue to lay claim to their caste occupation: often speaking of their right to caste work, calling it *jati kasubu* in the Kannada language. Secondly, the Mogaveera community, being the second most numerically dominant community in Udupi, is politically powerful. This provides Mogaveera women with the political resources to muster support. Third, the Mogaveera community was traditionally matrilineal, like many other Tuluva (Tulu-speaking[6]) communities in the region,[7] making women central to the family. According to the interviews conducted so far with older fisherwomen, Mogaveera families have in the past depended primarily on women's earnings. Women's work of selling fish has been traditionally and historically encouraged. Fourth, as a considerable number of these women lack formal education and belong to the lower classes, poverty emerges as a concern around which secular policies tackling women's poverty are mobilized.

Mogaveera fisherwomen, positioned within these axes of caste, gender, poverty and political status, discursively enlist these experiences, strategically choosing from them depending on the context and the position of those they seek to convince. The success of this initiative lies in the fact that these women have been able to muster both the marginal and dominant identities they inhabit, discursively constructing the terrain of fish sales as a woman's right.[8] I argue that the way in which the gender identity of these fisherwomen is intertwined with their caste identity results in a specific understanding of "womanhood" that links Mogaveera women with the work of selling fish. This caste acceptance of fisherwomen, intertwined with their lower-class status and their status as primary breadwinners, enables political support from local political elites and state functionaries, allowing them to protect their livelihoods despite a larger capitalist context.

Methods

This article relies on interviews, observation and participant observation conducted over a period of four months from January to May 2016 in ongoing research. A total of 16 interviews have been conducted with fisherwomen who are members of this Association, in addition to interviews with the head of the Association, the administrative assistant of the Association, the head of a credit cooperative society that supports this organization, and a founding member of a cooperative bank founded by men from the Mogaveera community.[9] Observation and participant observation were carried out at three meetings of the fisherwomen,[10] as well as observation of their everyday activities to understand time use and the distribution of domestic tasks. I followed fisherwomen from their homes in the early morning as they went to the harbour to purchase fish and transported the fish to the market, observed their market activities, and finally followed them home in the evening. Apart from the field research, a considerable amount of locally written and published material on the Mogaveera community is also being reviewed, part of which is used here. While this research is in the beginning stages, with fieldwork meant to be completed by January 2017, these preliminary findings already allow for insights into the strategies that enable low-income women fish sellers to protect their livelihoods.

The Context

Udupi City is administratively classified as a town within the larger Udupi Taluk, which includes 15 other towns and 86 other villages.[11] The Udupi district is home to a majority Hindu population with minority Muslim and Christian populations. The Billava caste is numerically dominant in the region, followed by the Mogaveera caste. The Mogaveeras in the region were traditionally involved in fishing and related occupations, and are listed under the Other Backward Classes (OBC) category.[12]

> Fisheries on the western coast of India make use primarily of mechanized and motorized fishing vessels, differing in this regard from those on the eastern coast, where two-thirds of fishing vessels are still manually propelled.[13] The western coast also has a long history of fish exports. Udupi District is particularly distinct as it is a temple town (home to the udupi Krishna Temple) that attracts tourists from all over India and the world, and also houses the Manipal University, one of the largest private education institutions in India. Thus the demand for fish locally is also quite high, driven not only by the local inhabitants but also by the many students and visitors from outside the region who throng the restaurants in the area.

The occupation of fishing and allied activities continues to be dominated by Mogaveeras today, even if some Mogaveeras have transitioned into non-fishing activities. While a considerable number of migrants work as head loaders, the sale of fish is largely dominated by Mogaveera women, who have traditionally sold fish that in the past were caught by male family members. Prior to the establishment of local markets, women bartered fish for rice through what was termed the *Keka* system, dealing primarily with agricultural caste groups that they were linked to through generations of bartering women. Older fisherwomen describe the harsh conditions in the past when they had to walk miles with a heavy basket of fish on their heads to the market or to *Keka* households to sell fish.[14] During these long Journeys, Mogaveera fisherwomen often relied on other caste groups—such as the Bunts in the region—who accepted fish from them in exchange for rice and provided them with shelter. Women were able to sell fish because other women—older women, mothers, mothers-in-law, sisters or sisters-in law—took over domestic chores at home. Thus women selling or exchanging fish for other necessities is something that is not just accepted, but encouraged.

The conditions of selling fish have changed. The majority of the fisherwomen sell directly to clients in the local markets. While these women may not be "poor" in the sense of earning less than two dollars a day, our research indicates that most of them could be considered close to poor. Even if they have food security, a considerable number of these women face crises during periods of disease or ill health and most of them send their children to public schools, being unable to pay for the private education that so many in India, even among manual labourers, consider essential to their children's success.[15]

Today, even though the traditional matrilineal organization of the group has been replaced by patrilineal family organization through changes in laws,[16] Mogaveera fisherwomen continue to be economically active. However, fishing, as well as fish sales, has changed considerably. In the 1980s the construction of roads and highways and the improvement in travel, particularly the introduction of the auto rickshaws that many of the older women in this research referred to, allowed them to more easily transport the fish to markets. At around the same time, the mechanization of fishing and the introduction of ice factories resulted in a larger supply of fish that could stay fresh for longer periods of time. With ice and storage facilities at the harbour, women today experience less pressure to dispose of fish quickly. They are also able to sell fish in markets for better prices, without having to rely on barter under the Keka system, which was at times quite exploitative, as upper-caste groups o en exercised power over these women.[17] However, as one fisherwoman told me, *"we continue to be stressed about whether we will be able to sell our fish for a decent price, that worry remains."*

With mechanization of fishing and fewer Mogaveera men in fishing, a substantial number of fishermen, especially those who are employed on boats for fishing, are migrants from North Karnataka (primarily Bijapur), Orissa, Goa and Kerala. Interestingly, of the 16 fisherwomen who were interviewed, only three spoke of spouses who were fishermen, another three spoke of spouses working in fishing-related activities, and the remaining 10 spoke of husbands working in non-fishing activities.[18] Fish purchased by Mogaveera women at the harbour is loaded by migrant female day workers (also primarily from north Karnataka-Bijapur), called head loaders, into vehicles that transport both the fish and the women to the fish markets where they sell their wares. Women from the Kharvi community from neighbouring Goa and members of scheduled castes in the area often clean fish purchased by Mogaveera women. Both head loaders and women who clean fish are paid wages by Mogaveera women, who then sell the fish either directly to customers or to restaurants in their areas. Caste distinctions are thus clear among the fish sellers, the head loaders and the cleaners. Even if the fisherwomen's Association insists that many non-Mogaveera women also sell fish, the dominant majority of its members are Mogaveeras.[19] It is undeniable that selling fish provides more flexibility than many other jobs available to women. Women can sell fish when they want and take a holiday when they want. While it may provide the same earnings as many other jobs available to them, such as in construction or in factories, women value their independence as fish sellers. It is a livelihood that fisherwomen consider viable and it enables them to educate their children, who then do not have to sell fish. Amongst the 16 women I interviewed, all spoke of wanting their children to be well educated and in more comfortable jobs.

The Struggle Against Shops

While transportation and refrigeration have considerably improved the livelihood of selling fish, these facilities also make it possible for big fish shops to sell fresh fish, thus threatening the fisherwomen's livelihoods. While the fisherwomen have no issue with smaller shops, arguing that these will not be able to make a profit after paying rent and expenses, they fear the large shopping mall type of establishments that can clearly threaten their livelihoods.

The Udupi Fresh Fish Seller's Association was founded in 2010 to unite fisherwomen in their struggle against such large fish shops. This cooperative comprises about 1,631 women selling fresh fish, of which the large majority belong to the Mogaveera caste while a very small minority are women from the Kharvi community, also a traditional fishing community from neighbouring Goa.[20] The Association brings together fisherwomen from 36 fish markets in Udupi District.[21] It was founded by Baby Salian, a fisherwoman who has been selling fish for over 30 years now. In 2010, when a fish shop was set up in Saligrama, a village in Udupi Taluk, Baby Sailan founded the Association to bring fisherwomen together in order to make demands of the state.[22] The Association made an appeal to the then home minister of Karnataka, requesting that he refuse licences to any new fresh fish outlets in Udupi District, as this would affect the livelihoods of over 10,000 fisherwomen directly selling fish and about 30,000 women indirectly associated with the sales of fresh fish—such as the head loaders who are involved in carrying fish in the harbour, and women who sort, clean and prepare seafood for consumption. Interestingly, only women selling fish are members of the Association, and head loaders and cleaners are not. As Baby Salian explained, because the fish sellers share the same work space—the fish market—as well as similar conditions of work that head loaders and cleaners working in the harbour do not, the Association only covers those who sell fish. Even though these other women are not members, the Association refers to the larger numbers of women working in fish sales-related activities in its advocacy claims. In response to their protest, the minister promised the Association that he would ensure that no new outlets for sale if fresh fish would be permitted.[23]

This informal arrangement made it possible for Udupi Taluk, to a considerable extent, to keep fish sales with the women, a trend that has died away in other parts of Mangaluru. Such informal political arrangements are common in cities in India and have been examined by Benjamin, who speaks of "informal connections" distinguishing connections at the local level with the local bureaucracy from those at the state and national levels.[24] In Mangaluru City, for instance, fisherwomen face heavy competition from fish shops.[25] This is explained in part by the greater size of Mangaluru, which makes a single fish market difficult to access for residents living far away. Udupi City is small enough for a single fish market. In other small towns and villages that are part of Udupi Taluk, women are also able to take advantage of the proximity of the market to all residents, and to continue selling fresh fish directly.[26] Restaurants in Manipal (where the university is) source fish from fisherwomen in Udupi City who supply both restaurants and the local population. Thus women have been able to largely protect their direct sales of fish to both individuals and larger establishments without any intermediaries. Since the activities of sorting, cleaning and sales are all done by women, their solidarity within the larger women's group, extending beyond the members of the Association, ensures that intermediaries are largely kept out.

However, market forces are quite strong in the region, and with upper-class men who have access to capital vying to enter this profitable trade, women fish sellers are increasingly under pressure from fresh fish shops that provide competition. Despite the agreement with the home minister, fish shops resorted to obtaining permits from other administrative offices. For example, in 2014, following the issuance of permits by gram panchayaths (local village councils) to shops in Udupi, Brahmavar, Sastan, Kota, Saligrama, Saibarakatte, Hebri and Kundapura, the Association once again had to launch a protest before the deputy commissioner's Office in Udupi. As a result of this protest the then Udupi-Chikmaglur member of parliament, Shobha Karandlaje, once again assured the Association that she would discuss this with the deputy commissioner of Udupi, urging him to provide instructions to the gram panchayaths.[27] While gram panchayaths are authorized to issue permits and licences to shops, the agreement at a higher level with the home minister and the deputy commissioner would override their authority, and this overriding authority was disregarded in these cases. With their frequent protests, fisherwomen continue to apply pressure on the local administration to ensure the protection of their occupations. Within a larger neoliberal capitalist context, in which the state is largely allied to capital interests and enterprises, this fear of competition from larger vendors continues to run high amongst the fisherwomen.

Apart from resisting the establishment of fish shops in Udupi, the Association also organizes the fish markets that come under it. Women who want to sell fish in the affiliated markets are required to become members of the Association and are expected to adhere to the rules and regulations set forth for the peaceful management of fish markets. The Association has never refused any application for membership. In a meeting when one fisherwoman representing one of the markets responded negatively to an application by a woman for membership in the Association and a space to sell fish, the fisherwomen from the other markets supported the new entrant and argued that the Association is meant to help women in need. The Association organizes the space allotment of members, rotating women's sites in the market every few months through a picking of lots. It also engages in other issues not related to fishing. For example, in a meeting held on February 2016, a fisherwoman suffering from cancer and undergoing chemotherapy was given a sum of INR 10,000 (approx. US$ 150) toward her treatment, collected on a voluntary basis from women from all the fish markets in Udupi. Thus the Association plays the role of a women's solidarity network, in which women support each other not only professionally but also personally.

The Association maintains close ties with other fish federations and credit organizations. It is closely linked to the S.K. & Udupi District Co-operative Fish Marketing Federation, a government federation of fishermen and women, which provides financial and other forms of support to both men and women undertaking fish sales, fishing or related activities. The Federation channels credit through MMVSSN—herein referred to as the Society), which was formed in 2011, a year after the women's Association was founded. Through membership in the Society fishermen and fisherwomen have access to credit at bank lending/preferential rates, as well as assistance in marketing seafood, access to subsidized fuel and ice for storage of fish, access to state welfare schemes including microcredit loans, state-sponsored housing, and supply of safety equipment to fishermen at sea.[28] The head of the fisherwomen's Association is also a member of both the Federation and the Society. Her presence on the boards of the several other fishing-related organizations in the area and links to key functionaries of these organizations provide for a deeply entrenched network that the Association depends on. These links and networks enable the organizations' access to government schemes and benefits, which they use to improve the lives of fisherwomen. For example, one such scheme that is presently being channelled through the Society is the Masthyashraya housing scheme, under which several fisherwomen have been able to access free housing provided by the state.[29]

Enlisting Support Through Intersectional Identities

In our interviews, the director of the Association, Baby Salian, stressed the importance of support from diverse groups. The support of Mogaveera men is crucial and the Association strategically invites powerful and resourceful Mogaveera men to its events and meetings. In an Association meeting in which I participated, the foll: 6 women office holders were accompanied by two Mogaveera men, who provided advice to the women at different points in time. Baby Salian argues that Mogaveera men are important to the organization as they "guide" and "support" the Association in organizing the women. And members argue that the presence of Mogaveera men in meetings is important as they bring in broader knowledge of fishing. Yet the nature of discussions and the issues brought up and decided during these meetings have nothing to do with fishing in itself, but more to do with the organization of markets, allotment of spaces, threats to their livelihoods, etc. The role of men as "advisors" to the Association therefore seems to be more strategic than essential. For example, in a meeting held in February 2016, one of the men present was the chief executive officer of the Society.

Upper-class Mogaveera men have moved out of working in boats to owning boats or factories processing fish. While a considerable number of Mogaveera men have moved out of fishing, the large majority remain linked to fishing and fishing-related activities. For example, Mogaveera men work in cooperative banks that largely service fishermen in the region and in associations or organizations that provide services to fisherman.[30] As such, they become important supporters for the women's Association, without taking over decision-making powers. Decision making within the Association is largely dependent on majority votes taken in monthly meeting, and while men can provide advice, women take decisions after discussing and debating the issue at hand.[31]

Involving Mogaveera men in the activities of the Association in a certain sense marks the territory of fishing and fish sales as a Mogaveera domain and reinforces caste ties and links between men and women within the community. The sense of community that women build and maintain with their men through the symbolic presence of men in their activities, enables them to obtain political support from their fellow Mogaveeras. In the last 13 elections of the Udupi Assembly Constituency, Mogaveera elected representatives (MLAs) were successful in seven instances, and represented the constituency. Amongst these seven instances, in five cases the same Mogaveera woman - Manorama Madhavaraj (from the Mogaveera community but not a fisherwoman herself)—was elected. With Mogaveeras dominating the political landscape in the area, the caste identity of fisherwomen clearly plays a role in their ability to keep fish sales to themselves.

The fisherwomen have also been able to capitalize on their gender identity, recruiting politically strong women from other castes to support their Association. For example, since the establishment of the Association in 2010, they have enlisted the support of Shobha Karandlaje, a woman politician from the Gowda caste and the present MLA for the UdupiChikmagalur constituency. Drawing on a shared gender identity, the Association frequently invites her to its programmes and enlists her support. The appeal to gender was evident in an address to the Association in which she stated, "*women today play a significant role in religious, political and social arena. She surpasses men in all spheres*."[32]

Along with gender, a common appeal is also made to women's poverty while lobbying the state. In interviews with members of the Association and the Society, the sale of fish comes forth as critical to survival. In the statements made before the government, fisherwomen point to the inability of the government to provide alternatives to the thousands of fisherwomen engaged in this activity, and necessity and poverty are frequently invoked in arguments for state support. The lack of other viable alternatives is expressed in the interviews and the press reports on the demands made by the Association to the state.

The work and livelihoods of fisherwomen are key to lifting families out of poverty, and this fact too is consistently put forth before the state. The sale of fish is no longer preferred by younger Mogaveera women with higher education. Both the Association's and the Society's members speak of Mogaveera women transitioning out of fish sales, and other women—scheduled caste women and Kharvi women—moving into their place. And yet the sale of fish by these women is what has ensured their daughters' transition out of the occupation and out of conditions of poverty. Poverty is thus crucial to their appeals and intersects with gender in the effects it has on women and their families.

Gender, poverty and caste intertwine in the manner in which the sale of fish as an occupation is constructed. That women sell fish and that only women should sell fish is asserted by the Association but also by all the fisherwomen interviewed. That men should not occupy this domain, that it is a woman's space, emerges from all the interviews. That it has to be women selling fish is claimed on the ground that they are largely, and in many cases solely, responsible for their households.

This reliance on women's work is not new. Most women spoke of their mothers working to take care of them. During the interviews, all but one woman spoke of their mothers selling fish, and 10 out of 16 women spoke of their mothers being the primary breadwinners. Within the matrilineal system, women were primarily responsible within households, and according to these interviewees, the role of men seemed to be marginal in sustaining their families. While the younger fisherwomen (two) spoke often of husbands contributing substantially to household incomes, the older fisherwomen (14) were more likely to be primarily responsible for managing their homes, with spouses only supplementing household incomes. In interviews with three Mogaveera men, women were described as primarily responsible for running the home. Men were frequently referred to as less responsible, often alcoholic or frequently sick. The Mogaveera fisherwoman is thus constructed as essentially hardworking, nurturing and caring, but also breadwinning, independent and responsible. There were no references made to men's responsibility—to the contrary, women's work was constructed as essential to support the larger good of the community. Thus, enabling women's work also in a certain way ensures that women continue to be primarily responsible as breadwinners. Women also adopt this discourse and often refer to their work as essential to food security and education for their children, a project in which men seem to be less implicated.[33] This is less the case with the two younger women interviewed, who spoke of equally shared financial and domestic responsibilities at home.

But in all the interviews, what emerges is fish selling as an activity to sustain oneself and one's family. The sale of fish for profit is rarely discussed. But here again this is because profits are rarely possible, not because women resist capital modes of production and exchange. For these women the task of selling fish encapsulates both activities of production and reproduction, with no necessity to make explicit what is implicit. Women speak of working to feed and educate their children and while referring to profits also refer to losses. Because of the unpredictability of sales, women speak of being able to make enough money to sustain their families. While some of the women do sell fish to larger buyers such as restaurants and tourist enterprises, these are few and far apart.

Thus, a complex and nuanced argument is made that is acceptable to local elites who are also located within local relations of caste, while at the same time acceptable to the policy discourse on poverty. The historical specificity of Mogaveera women as poor fisherwomen struggling to rise above poverty is interwoven with references to women's primary responsibility towards their children, in a certain way fixing both caste and gendered meanings to the livelihood of selling fish. The occupation is in turn also constructed around this rhetoric of poverty, as a resource for poor women, as an occupation of necessity and thus as an occupation of the poor. While caste livelihoods are not explicitly claimed, it remains a central organizing element in the local politics and informal arrangements that sustain the Association.

Conclusions

Even at this preliminary stage of research, this case offers good potential for an intersectional analysis as it shows how gender is intertwined with caste, political status and class. A purely gendered analysis without attention to these intersections would lack the historical, political and economic specificity that reveals the complex social relations that sustain this livelihood. While intersectionality as a concept emerged out of the fact that multiple marginal identities could render some women more vulnerable than others,[34] this example reveals how both marginal and dominant identities/statuses have been used by Mogaveera fisherwomen to enlist local political and state support in the protection of their livelihoods.

Caste continues to structure social relations, and the Mogaveera identity surfaces in conversations, interviews and exchanges even if it remains absent from the name of the Association as well as in its political rhetoric. Castes and livelihoods continue to be closely linked in India, and in the particular case of Udupi's fisherwomen, this explains the solidarity that the Association has been able to maintain between fisherwomen as well as between Mogaveera men and women. However, the secular context in which they are located requires them to accept not only Mogaveera women but also others selling fish, however small and insignificant their numbers are.

But gender is also closely intertwined with caste to enable this success. Matrilineal organization and the gendered distribution of work and responsibilities within the family make Mogaveera women central to fish selling. Fisherwomen thus continue to be synonymous with Mogaveera women, even if many upper-class Mogaveeras are no longer engaged in fishing or fishing-related activities. Even if a so-called "right to a livelihood" cannot be claimed in a modern open economy, the right to a caste-based occupation emerges significantly in the interviews. The right to a caste occupation to fall back on in times of need is reframed in a secular context through appeals to poverty. Here, women ask what alternatives the state can provide poor women, if they lose their livelihoods. Drawing from the secular, rational discourses of poverty and gender, the Association advances the cause of "poor fisherwomen" and the "protection of livelihoods." In a broader political context of ending poverty and ensuring gender equality, women as primary breadwinners—the new development agents—emerge as more responsible and thus more deserving of state protection and benefits. Both caste and secular resources are thus mobilized in protecting their livelihoods. From the demands made of the state, the political rhetoric of the organizations, and the everyday narratives that women engage in to describe their work, the terrain of fish sales thus emerges as primarily a poor Mogaveera woman's domain—to be protected for the purpose of survival itself.

Notes

1. Kabeer, N, K Milward and R Sudarshan (2013), "Organising women workers in the informal economy," *Gender & Development* Vol 21, No2, pages 249–263.

2. Jenkins, L D (2003), *Identity and Identification in India: Defining the Disadvantaged*, Routledge, London and New York.

3. Guru, Gopal (1995), "Dalit Women Talk Differently," *Economic and Political Weekly* Vol 30, No 41, pages 2548–2550; also Kannabiran, Kalpana (2006), "A Cartography of Resistance: The National Federation of Dalit Women," in Nira Yuval-Davis, Kalpana Kannabiran and Ulrike Vieten (editors). *The situated politics of belonging,* SAGE, pages 54–72; and Rege, Sharimila (1998), "Dalit Women Talk Differently: A Critique of 'Difference' and Towards a Dalit Feminist Standpoint Position," *Economic and Political Weekly* Vol 33, No 44, pages 39–46.

4. Chaudhuri, Maitrayee (2004), *Feminism in India,* Kali for Women, New Delhi.

5. Kothari, R (editor) (1970), *Caste in Indian Politics,* Orient Blackswan, Hyderabad.

6. Tulu is a southern language spoken in this region and the communities located in this region are commonly called Tuluvas, meaning Tulu-speaking communities.

7. Ishii, M (2014), "Traces of reflexive imagination: Matriliny, modern law, and spirit worship in South India," *Asian Anthropology* Vol 13, No 2, pages 106–123.

8. While these women muster caste identity informally, they argue for poor women's rights to livelihoods, including not only Mogaveera women but also women of other castes who support their work. I examine this in Section IV in detail.

9. The head of the Association, Baby Salian; the Association's administrative assistant, Ashwini; the Chief Executive Officer of *Meenu Maratagarara Vividoddesha Souharda. Sahakara Niyamita, Udupi* (MMVSSN, meaning Fish Sellers Varied Interests Assistance Society), Prakash Suvarna; and one of the founder members of the Mahalakshmi Cooperative Bank, Udupi, Anand Putrana.

10. I observed the first meeting, but in the second meeting, when the fisherwomen discussed health issues and the lack of medical insurance coverage for the group, I was roped into the discussion, I was asked for advice on how to negotiate for and obtain low-cost medical insurance because of my work at the Manipal University, which runs one of the best hospitals in the region—the Kasturba Medical College Hospital. During the third meeting, after gathering some information from the university, I discussed some of the options available with the university health care scheme, called Arogyaand. This was particularly relevant given my job at the Manipal University, which is linked to the Manipal Kasturba Medical Hospital.

11. Udupi City, which is administratively called Udupi City Municipal Council, is also the administrative capital of Udupi District. See http://www.census2011.co.in/data/subdistrict/5523-udupi-udupi-karnataka.html, accessed 22 February 2016.

12. http://www.mogaveeracommunity.com/mogaveera-community.html, accessed 10 March 2016. As per the information provided on this website, while a majority of the Mogaveera community is classified as OBC, a few Mogaveera groups are classified as Scheduled Castes and and Scheduled Tribes.

13. http://igssf.icsf.net/en/page/1061-India.html, accessed 10 March 2016.

14. Interviews with Baby Salian on 18 January 2016, relating how her mother sold fish, and Prakash Suvarna on 25 January 2016, on how his mother carried fish to surrounding villages.

15. Since the quality of public education in India is poor, particularly at the primary school level, families that can afford it prefer to send their children to private schools, many of which charge quite low fees.

16. Suchitra, J Y and H Swaminathan (2010), *Asset Acquisition among Matrilineal and Patrilineal Communities: A Case Study of Coastal Karnataka* No 3, Bangalore.

17. Interview with Prakash Suvarna, who related what his mother told him about the Keka system, in which fisherwomen could sell only to those families their mothers had sold to. At times the prices fixed for fish were exploitative and fisherwomen were also asked to Work in the fields during their visits to Keka households to sell fish.

18. Interviews conducted in the Beedinagudde market and Maple Bandar, both of which are located in Udupi City.

19. While exact numbers of non-Mogaveera women selling fish are not available with the Association, in an

interview with a staff member of the Association's office, she revealed that of the total 1,631 members, about 20–30 women may be non-Mogaveeras. From my preliminary research in three fishing markets in and around Udupi—Udupi City, Parkala, Maple Bandar and Manipal—I have yet to meet a non-Mogaveera woman selling fish.

20. In Kundapura about 500 fisherwomen have their own association, which works closely with the Udupi Association to keep fresh fish shops at bay. Interview with Ashwini, administrative assistant at the Association, on 12 March 2016.

21. The fish markets with active members of the Association include fish markets in Karkala, Kapu and Brahmavara. It also includes fish markets within Udupi Taluk. In all it covers 36 fish markets that take active part in the Association's activities. Interview with Ashwini, administrative assistant at the Association, on 12 March 2016.

22. Alva, A (2010), "Fisherwomen's association inaugurated in south Indian State," Women in Fisheries, International Collective in Support of Fishworkers, 2 February, accessed 31 May 2016 at http://wif.icsf.net/en/samudra-news-alert/articledetail/42985-Fisherwomen's.html?language=EN.

23. The then home minister, V S Acharya, made a promise to the Association that licences would not be granted to fish shops in Udupi. See Katapadi, P (2010), "Fresh Fish Sellers' Association submits appeal to the Home Minister-Dr. V.S. Acharya," *Bellevision,* 23 July, accessed 10 March 2016 at http://www.bellevision.com/belle/index.php?action=topnews&type=593.

24. Benjamin, S (2000), "Governance, economic settings and poverty in Bangalore," *Environment and Urbanization* Vol 12, No 1, pages 35–56.

25. Interviews with Baby Salian and Prakash Survarna.

26. Interviews with Prakash Suvarna on 10 March 2016 and Anand Putrana on 23 March 2016.

27. Katpadi, P (2014), "Udupi: Women fish sellers urge MP Shobha not to allow fish shops," *Utupi Today,* 22 October, accessed 10 March 2016 at http://www.udupitoday.com/udtoday/news_Udupi-Women-fish-sellers-urge-MP-Shobha-not-to-allow-fish-shop_4706.html.

28. http://www.fishmark.in/activities.html, accessed 10 March 2016; also interviews with Prakash Suvarna, Chief Executive Officer of MMVSSN, on 25 January 2016, 26 January 2016, 18 February 2016, 10 March 2016 and 20 May 2016, and Baby Salian, Director of the Association, on 18 January 2016.

29. Interview with Prakash Suvarna, Chief Executive Officer of MMVSSN, on 10 March 2016; also http://www.fishmark.in/activities/html, accessed 10 March 2016.

30. For example, a group of Mogaveera men founded the Mahalakshmi Cooperative Bank Limited. Interview with Anand Putrana (on 23 May 2016), one of the founder members of the bank.

31. Prior to the delimitation of constituencies in 2008. In the elections held from 2009 onwards, the Udupi constituency was replaced by the Udupi-Chikmaglur constituency. After 2009 all three terms were occupied by candidates from the Chikmaglur region.

32. *Deccan Herald* (2010), "Unity among women yield tremendous benefits: Shobha," 1 February, accessed 10 March 2016 at http://www.deccanherald.com/content/50235/unity-amount-women-yield-tremendous.html.

33. This is of course with respect to fisherwomen only as my interviews are only with these women. As my research does not extend to upper-class Mogaveera women, I am unable to say anything about them at the moment.

34. Benjamin, S (2000), "Governance, economic settings and poverty in Bangalore," *Environment and Urbanization* Vol 12, No 1, pages 35–56.

Introduction to Reading 11

One of the most recent applications of intersectionality is in global/transnational research, as mentioned by Bonnie Thornton Dill and Marla H. Kohlman in the first reading in this chapter. In this article, Bandana Purkayastha develops the meaning of intersectionality using a transnational framework. As Nikki Jones and Karen D. Pyke and Denise L. Johnson did in their analyses, Purkayastha also considers race in this context. As you will see from this piece, scholars are still struggling with what intersectional analysis means in a changing world, but the process of doing so is exciting. This article gives good examples of how we could and should study the intersections of multiple identities while also considering a transglobal context.

1. What is meant by "transnational spaces"?
2. How does Purkayastha recommend doing an intersectional analysis in transnational spaces?
3. What does she mean by "axes of domination"? What other "axes" can you name in addition to race?

Intersectionality in a Transnational World

Bandana Purkayastha

As a late entrant into the world of sociology, I read Patricia Hill Collins's *Black Feminist Thought: Knowledge, Consciousness, and the Politics of Empowerment* (1990) in a graduate seminar on gender. . . . Over the years, insights of many other scholars have helped me to understand intersectionality. As Hae Yeon Choo and Myra Marx Ferree (2010) have recently pointed out, these diverse works on intersectionality have highlighted the importance of "including the perspectives of multiply marginalized people, especially women of color; an analytic shift from addition of multiple independent strands of inequality toward a multiplication and thus transformation of their main effects into interactions; and a focus on seeing multiple institutions as overlapping in their co-determination of inequalities to produce complex configurations" (2010, 4). Nonetheless, I remain grateful to Professor Collins for moving the conversation on intersectionality, in the early 1990s, to a more visible level.

Professor Collins's work has remained dynamic, expanding far beyond the original idea of power relations organized through intersecting axes of race/class/gender (Collins 1990) to her recent articulation of the "intersecting power relations of race, class, gender, ethnicity, sexuality, age, ability, nation" (Collins 2010, 8). In this brief essay, I touch upon two related aspects of intersectionality that require additional clarification as we study social lives in the twenty-first century. My observations are focused on the ways in which we understand "race" even as our lives expand onto transnational—including virtual—social spaces. . . .

Social Lives in Transnational Spaces

Over the past decade, a rapidly growing literature has described how individuals and groups maintain connections across countries so that social lives are constructed, not only in single countries, but in transnational spaces. Transnational spaces are composed of tangible geographic spaces that exist across multiple nation-states *and* virtual spaces. With improvements in personal and media communication and travel technology, the ability to move money easily across the globe, and the marketing and ease of consuming "cultural" products—including fashions, cosmetics, music, foods, and art—have made it easier for many groups to create lives that extend far beyond the boundaries of single nation-states. We now know about first-generation immigrant "transnational villagers" who build lives in more than one country by traveling back and forth regularly, organizing family lives across countries, and remitting and investing money, as well as engaging in politics in "homelands" (e.g., Guevarra 2009; Hondagneu-Sotelo and Avila 1997; Levitt 2001; Purkayastha 2009). We also know about post-immigrant generations who actively maintain links with their parents' homelands (e.g., Purkayastha 2005); cyber migrants who work for Northern employers but are geographically based in the South (e.g., Abraham 2010); and participants in web-based communities, some of whom seek community, while others try on less-essentialist, choice-driven, multiple, fragmented, and hybrid identities on the web and thus dilute the consequences of gendering, racialization, class, and other social hierarchies to which they are subjected in their tangible lives (e.g., Diamandaki 2003; Ignacio 2006; Lee 2003; Mitra 1997; Mitra and Gajjala 2008; Narayan and Purkayastha 2011). Other scholars have analyzed web-based transnational linkages that enable geographically dispersed groups to form close-knit political networks (e.g., Earl 2006; Narayan, Purkayastha, and Banerji, forthcoming; Pudrovska and Ferree 2004). As a rapidly growing number of people are tied to transnational spaces—that is, they build lives that combine intersecting local, regional, national, and transnational spaces—single nation-states no longer wholly contain their lives.

At the same time, nation-states have responded in a variety of ways to control social lives in transnational spaces. For instance, the literature on immigration shows how nation-states are creating gendered categories of "overseas citizens" in order to attract

Purkayastha, B. (2012). Intersectionality in a transnational world. *Gender & Society, 26*(1): 55–66. Reprinted by permission of SAGE Publications Inc., on behalf of Sociologists for Women in Society.

remittances from migrants (e.g., Guevarra 2009) or to draw on the expertise or lobbying power of people settled in other countries (Purkayastha 2009). Equally important, nation-states have attempted to expand their ability to control people across transnational spaces; ideologies, interactions, and institutions that have sustained raced/gendered/classed and other hierarchies *within nations,* have expanded in new ways across nations.[1] For most of the twentieth century, nation-states maintained separate apparatus for controlling groups within nations (e.g., police, prisons) from the apparatus used to dominate and control groups/states outside nations (e.g., the military, foreign intelligence agencies, facilities to house prisoners of wars). Now, these tools of control are increasingly blurred within nations; for instance, policies such as the PATRIOT Act and organizations such as Homeland Security have blurred the distinction between foreign surveillance and national surveillance in the United States. Security agreements *across* nations have created transnational security regions, where profiles developed in one powerful country are likely to be rapidly disseminated and acted on in other countries within the transnational security regime (Purkayastha 2009; Vertovec 2001). A *suspect* in a terrorism case in Scotland or Spain can, almost immediately, be arrested in Australia or the United States.[2] The profile of a "turbaned terrorist" or the suspicions against "Muslim-looking" people have generated contemporary racial profiles so that Sikh men—who wear turbans to comply with their religious tenets—and a range of people of Middle Eastern and South Asian origin are profiled as potential security threats. They are searched more stringently at airports, subject to extra questioning at national borders, subject to surveillance for communicating with people in "enemy countries," frequently visiting these countries, or sending money to "suspect" organizations through institutional arrangements as they travel through security regime (see Iwata and Purkayastha 2011; Purkayastha 2009). These new global security arrangements intersect with other processes for controlling racially marked populations within nations.

Overall, then, transnational spaces are composed of tangible and virtual social spaces that exist through and beyond single nation-states. Individuals, groups, corporations, and nation-states continue to expand their purview into such spaces.[3] At the same time, those who cannot access transnational spaces—for a variety of reasons, including the digital divide and stringent government control over travel and internet access—are marginalized in new ways within this expanded context. Contemporary discussions of marginalization and privilege have to take these new developments into account.

Intersectionality in Transnational Spaces

While her earlier work on Black feminist thought was focused on the United States, in her recent writings Professor Collins recognizes the expansion of social life beyond the nation-state (e.g., Collins 2010). She discusses the dispersion and consumption of cultural products, such as hip-hop, around the globe and the cultural familiarity this engenders among consumers, and the possibilities for "creat(ing) shifting patterns of face-to-face and mediated interactions . . . (as) new technologies create organizational opportunities for new sorts of political communities" (2010, 18). She discusses the ways in which people imagine local and far-flung political communities using new technologies, and the ways in which these multiple communities somewhat dilute concentrated power of the privileged. Despite this recognition of other worlds of experience, Professor Collins has not, as far as I am aware, discussed the structures of domination and control in transnational spaces. As a result, it is not clear how our current conceptualization of intersectionality—including the expanded version of race/class/gender/age/ability/sexuality/ethnicity/nation—might change if we incorporate social life in transnational space.

Professor Collins has offered a powerful critique of the "race-neutral" scholarship on gender, and her discussions of racism (and the ways this racism interacts with other axes of domination) led to the visibility of concepts such as "controlling images" and "women of color." Her references to Black women and other people of color in the United States continue to serve as a reminder of their continuing marginalization and open up some space to include their experiences in developing theory. But she does not discuss the deviations from Euro-American organizations of racial hierarchies in different countries around the globe, which coexist with global-level Euro-American racialization processes. As a result, it is not always clear when and how we are to conceptualize "race" within the intersectionality matrix if we study transnational social lives.

I will begin with a simple example that focuses on women of color. A Ugandan Black immigrant and a Ugandan Indian immigrant—whose family lived for many generations in Uganda before being forcibly evicted by Idi Amin—are both racially marginalized, though in different ways, in the United States. While both share the effects of gendered/racialized migration policies that would prohibit or slow the process through which they might form families in the United States, their experiences differ in other ways. The Ugandan Black migrant is likely to experience the gamut of racisms experienced by African Americans, while the Indian Ugandan is likely to experience the racisms faced by Muslims and "Muslim-looking" people in the United States, and they may share other structural discriminations experienced by Asian Americans (Narayan and Purkayastha 2009). These similarities and differences are consistent with racist ideologies, interactions, and institutional arrangements in the United States. But if both return to their home country Uganda, they would encounter a different set of privileges and marginalization in this Black-majority country; the Black Ugandan migrant is advantaged here (though the other intersecting factors would together shape her exact social location). If both visit or temporarily live in India, the Indian-origin Uganda-born person may experience the privileges associated with the dominant group in the country. However, if she is a Muslim or a low-caste Hindu, she might experience a different set of social hierarchies. Similarly, Japanese-origin people from Brazil who returned to Japan, or Japanese-origin Americans who were forced to return to Japan, encountered different sets of social hierarchies. A broadly similar argument could be made about the relative position of Blacks and Indians under the different historical circumstances, for instance, during the apartheid regime and after apartheid in South Africa (see Govinden 2008).

There are variations of who is part of the privileged majority versus the marginalized minority *within* a country, and these hierarchies do not always fit the white-yellow/brown-Black hierarchy extant in Western Europe and North America. Thus concepts such as "women of color"—which act as an effective framework for indicating the social location of these women in Western Europe and North America, and continuing global hierarchies *between* countries in the global North and South—do not work as well if we wish to track the array of the axes of power and domination within countries *along with* existing global-level hierarchies. Yet considering these multiple levels is important if intersectionality is to retain its explanatory power in an increasingly transnational world where within-country *and* between-country structures shape people's experiences.

The possibility of forming community on virtual spaces and using the web to maintain meaningful connections with people in other countries also emphasizes the need to consider transnational spaces. A South African Black female immigrant in the United States who is able to maintain active connections with her friends, family, and political networks in her home country (via phone, email, and a variety of web-based media) may be able to minimize some of the toll of racism she experiences by making her South African relationships most salient in her life. The Indian-origin post-immigrant-generation American who regularly participates in a religio-social Hindu online community and visits India regularly is also able to position herself as a member of the majority group in India (and the Indian diaspora) even as she experiences the deleterious effects of structural racism in the United States. In other words, people who can access transnational social spaces attempt to balance their lack of privilege in one country (their raced/classed/gendered/ability/sexual/age/nationality status in one nation-state) by actively seeking out privilege and power in another place and/or in virtual spaces.

While many of the axes Professor Collins identified—race, gender, class, sexuality, age, ability, and nation—remain relevant, they may not work in the same way as "women of color" constructs suggest. Being able to build transnational lives—the ability of groups to live within and beyond single nation-states—suggests that it is quite possible for groups to be part of *the racial majority and minority simultaneously* (Purkayastha 2010). Indeed, in places where caste and religious or ethnic hierarchies—with their own set of ideologies, interactions, and institutional structures—are more salient, we should consider the relative importance of these axes of domination within those countries (and the extent to which these structure transnational social lives) as we use intersectional frameworks.

I do not intend to suggest that we stop considering racial hierarchies. Along with variations in who makes up the racial majority or minority *within* different countries, hierarchies among nation-states continue to promote Western hierarchies of race and whiteness, yellowness, and Blackness across the world in ways that are broadly similar to the period of colonialism (see, e.g., Gilroy 1989; Kim 2008; Kim-Puri 2007; Nandy 2006; Sardar et al. 1993). Such racial hierarchies are maintained through ideologies, actions, and institutional arrangements associated with political and economic control. As Evelyn Nakano Glenn and her colleagues have documented, color-based hierarchies continue to structure people's lives in many countries around the globe, especially as "fairness as beauty" is marketed to places where the majority of the people are nonwhite (Glenn 2008).

Since Professor Collins's discussions focus on the United States and minority groups within the boundaries of this nation-state, the ideas about race she discusses are built upon the structures that are particularly relevant to the United States and Western Europe. While intersectionality remains an important framework, we need to encapsulate marginalization structures that are salient in other locales and the ways in which these hierarchies play out in transnational spaces.

* * *

Last Remarks

As a scholar who continues to use intersectionality as my primary theoretical framework, I can enumerate many ways in which the work of Professor Collins and others who have developed this framework have improved our ability to study social lives. The framework remains important, but we have to pay attention to and elucidate the complexities of using this framework beyond Euro-American societies. Understanding and attending to the complexities of transnationalism—composed of structures within, between, and across nation-states, and virtual spaces—alerts us to look for other axes of domination and the limits of using "women of color" concepts, as we use them now, to look across *and* within nation-states to understand the impact of transnationalism. My examples here were focused on those who can access transnational spaces. A focus on transnational intersectionality should alert us to the position of those who are unable to afford access to technology to build virtual communities, to participate in a medium because they are not proficient in English, which has become the dominant language in virtual spaces, or to build transnational social lives because of active government surveillance and control of their lives or because they are too poor and isolated to access transnational tangible and virtual spaces.

While I focused on "race" in this brief essay, the other axes of domination are likely to show some variations if we analyze multiple and simultaneous social locations to develop a better understanding of intersectionality. The organization of power and processes of marginalization has continued to change in this century. We need to further elucidate the theoretical implications and methodologies for adequately capturing different mechanisms of domination and how these meld with the ones with which we are most familiar within Euro-American scholarship.

Notes

1. The existence of global matrices of domination are not new phenomena—colonialism sustained the power and privilege of white Euro-America over Africa, Asia, and Latin America for centuries—but the contemporary organization of economic/social/political power, privilege, and marginalization reflect the development of transnational social spaces.

2. I deliberately picked Scotland and Spain because an Indian doctor in Australia was charged with complicity in the Glasgow bombings, while an American lawyer was charged with the Spanish bombings. Both were proven innocent (see Armaline, Glasberg, and Purkayastha 2011 for further details).

3. Individuals and groups need not participate in two or more countries (or virtual communities) equally. Indeed, their node of experience is often their country of residence. Other countries or virtual communities are often part of a larger field of experience, and the salience of these other spaces is likely to vary. My point here is that we need to seriously consider the node *and* the field, as these contribute to the experiences of privilege and marginalization that shape our lives in more complex ways than the model of intersectionality, based on single nations, suggests.

References

Abraham, Margaret. 2010. Globalization, work and citizenship: The call centre industry in India. In *Contours of citizenship: Women, diversity and practices of citizenship*, edited by Margaret Abraham, Esther Ngan-ling Chow, Laura Maratou-Alipranti, and Evangelia Tastsoglou. Aldershot, UK: Ashgate.

Choo, Hae Yeon, and Myra Marx Ferree. 2010. Practicing intersectionality in sociological research: A critical analysis of inclusions, interactions, and institutions in the study of inequalities. *Sociological Theory* 28:129–49.

Collins, Patricia Hill. 1990. *Black feminist thought: Knowledge, consciousness, and the politics of empowerment.* Boston: Unwin Hyman.

Collins, Patricia Hill. 2010. The new politics of community. *American Sociological Review* 75:7–30.

Diamandaki, K. 2003. Virtual ethnicity and digital diasporas: Identity construction in cyber-space. *Global Media Journal.* http://lass.calumet.purdue.edu/cca/gmj/sp03/graduatesp03/gmj-sp03grad-diamandaki.htm (accessed June 15, 2010).

Earl, Jennifer. 2006. Pursuing social change online: The use of four protest tactics on the Internet. *Social Science Computer Review* 24:362–77.

Gilroy, Paul. 1989. *There ain't no Black in the Union Jack.* Chicago: University of Chicago Press.

Glenn, Evelyn Nakano. 2008. *Shades of citizenship.* Stanford, CA: Stanford University Press.

Govinden, Devarakhsnam. 2008. *Sister outsiders: The representation of identity and difference in selected writings by South African Indian women*. Pretoria: University of South Africa Press.

Guevarra, Anna. 2009. *Marketing dreams, manufacturing heroes: The transnational labor brokering of Filipino workers*. New Brunswick, NJ: Rutgers University Press.

Hondagneu-Sotelo, Pierette, and Ernestine Avila. 1997. I'm here, but I'm there: The meanings of Latina transnational motherhood. *Gender & Society* 11:548–71.

Ignacio, Emily. 2006. *Building diaspora: Filipino community formation on the Internet.* New Brunswick, NJ: Rutgers University Press.

Iwata, Miho, and Bandana Purkayastha. 2011. Cultural human rights. In *Human rights in our own backyard: Social justice and resistance in the U.S.,* edited by William Armaline, Davita Glasberg, and Bandana Purkayastha. Philadelphia: University of Pennsylvania Press.

Kim, Nadia. 2008. *Imperial citizens: Koreans and race from Seoul to LA.* Stanford, CA: Stanford University Press.

Kim-Puri, H.-J. 2007. Conceptualizing gender/sexuality/state/nation: An introduction. *Gender & Society* 19:137–59.

Lee, Rachel. 2003. *Asian America.net: Ethnicity, nationalism, and cyberspace.* New York: Routledge.

Levitt, Peggy. 2001. *The transnational villagers.* Berkeley: University of California Press.

Mitra, Ananda. 1997. Virtual commonality: Looking for India on the Internet. In *Virtual culture: Identity and communication in cybersociety,* edited by S. Jones. Thousand Oaks, CA: Sage.

Mitra, Rahul, and Radhika Gajjala. 2008. Queer blogging in Indian digital diasporas: A dialogic encounter. *Journal of Communication Inquiry,* originally published online July 16, 2008.

Nandy, Ashis. 2006. *The intimate enemy: Loss and recovery of self under colonialism.* New Delhi, India: Oxford University Press.

Narayan, Anjana, and Bandana Purkayastha. 2009. *Living our religions: South Asian Hindu and Muslim women narrate their experiences.* Stirling, VA: Kumarian Press.

Narayan, Anjana, and Bandana Purkayastha. 2011. Talking gender superiority in virtual spaces. *Journal of South Asian Diasporas* 3:53–69.

Narayan, Anjana, Bandana Purkayastha, and Sudipto Banerji. Forthcoming. Constructing virtual, transnational identities on the web: The case of Hindu student groups in the U.S. and U.K. Special Issue on Virtual Ethnicities. *Journal of Intercultural Studies.*

Pudrovska, T., and M. M. Ferree. 2004. Global activism in "virtual space": The European women's lobby in the network of transnational women's NGOs on the web. *Social Politics: International Studies in Gender State and Society* 11:117–43.

Purkayastha, Bandana. 2005. *Negotiating ethnicity: Second-generation South Asian Americans traverse a transnational world.* New Brunswick, NJ: Rutgers University Press.

Purkayastha Bandana. 2009. Another word of experience? South Asian diasporic groups and the transnational context. *Journal of South Asian Diasporas* 1:85–99.

Purkayastha, Bandana. 2010. Interrogating intersectionality: Contemporary globalization and racialized gendering in the lives of highly educated South Asian Americans and their children. *Journal of Intercultural Studies* 31:29–47.

Sardar, Ziauddin, Ashis Nandy, Merryl Wyn Davies, and Claude Alvares. 1993. *The blinded eye: 500 years of Christopher Columbus.* New York: Apex Press; Goa, India: The Other India Press.

Vertovec, Steven. 2001. Transnational challenges to the "new" multiculturalism. www.transcomm.ox.ac.uk/working papers.

Topics for Further Examination

- The term *intersectionality* is now widely used and debated in an array of arenas in the United States. A simple Google search will call up an almost overwhelming number of scholarly and popular articles addressing intersectionality. Focus on the 2016 presidential race by googling "Hillary Clinton and intersectionality" and "Bernie Sanders and intersectionality." How was the concept of intersectionality employed in these two campaigns for political purposes? And, how was the meaning of intersectionality compromised by its use in that political arena?
- The #MeToo movement has generated lively discussion and debate about intersectionality. Community activist, Tarana Burke, identified as the founder of the movement, has argued that the movement needs to be more intersectional. Her criticisms are echoed by scholars as well. For example, Google the article entitled "Ambiguities and dilemmas around #MeToo" by Dubravka Zarkov and Kathy Davis (European Journal of Women's Studies: 2018). What is the #MeToo movement? How could it benefit from intersectional practice?
- View the Kimberlé Crenshaw TED Talk on "The urgency of intersectionality" at https://www.ted.com/talks/kimberle_crenshaw_the_urgency_of_intersectionality. What is the central message of Crenshaw's talk with respect to intersectionality?

3

Gender and the Prism of Culture

Catherine G. Valentine

Now that we have introduced you to the ways the contemporary U.S. gender system interacts with, and is modified by, a complex set of categories of difference and inequality, we turn to an exploration of the ways the prism of culture intersects with gender definitions and arrangements. Generations of researchers in the social sciences have opened our eyes to the array of "genderscapes" around the globe. When we look through our kaleidoscope at the interaction between the prisms of gender and culture, we see different patterns that blur, blend, and are cast into a variety of culturally gendered configurations (Baker, 1999).

What Is Culture?

Culture consists of the "implicit and explicit patterns of representations, actions, and artifacts" that are created and shared by people in their networks of interaction and social environments (DiMaggio & Markus, 2010, p. 348). The knowledge structures of human cultures provide predictability and meaning in human life, but they also vary significantly across history and place. Consider one simple example. We would not know what, how, when, where, and why to eat without cultural knowledge. Note, too, that the how/what/when/where/why of eating differs across cultures and over time. Additionally, in highly heterogeneous contemporary societies such as the United States, people shift between multiple, overlapping cultural configurations of norms, values, and the like (Patterson, 2014). The social scientific view of culture makes it clear that, without culture, human experience would have little, if any, shape or meaning (Schwalbe, 2005). That is, culture provides people with the assumptions and expectations on which their social interactions are built and in which their identities, behaviors, feelings, and thoughts are forged.

The term *cultural frame* is helpful for analyzing the socially created, unevenly shared schemas or knowledge structures by which people organize their social relations and coordinate behavior in the contexts in which they interact (Patterson, 2014; Ridgeway, 2009). Cultural frames rely on the human capacity to create categories, such as age-based categories (e.g., children and adults) and gender-based categories (e.g., women and men) (Patterson, 2014; Ridgeway, 2009). In the modern United States, the dominant gender frame is rooted in the Western history of oppositional binary thinking that molds people, in all their variety and complexity, into opposed categories: Black versus White; masculine versus feminine. This primary gender frame is reductionistic. It defines people as belonging to one of two, and only two, sex/gender/sexuality categories, as discussed in Chapter 1. This frame is the lens through which many, if not most, Americans perceive and label others and themselves. Central to the frame are sex/gender/sexuality stereotypes, the pink and blue syndrome presented in Chapter 1, of women and men as opposite types of people, and of men, in general, as having higher status and greater value than women

(Ridgeway, 2009). The U.S. gender frame may seem fixed or absolute from the viewpoint of many Americans who have learned to perceive and interpret people in binary terms, but it is not. There is nothing universal or natural about oppositional binaries. As you move through the readings in this chapter, compare and contrast gender frames. For example, the articles by Christine Helliwell, Dredge Byung'chu Käng, and Maria Alexandra Lepowsky examine cultural worlds in which people's beliefs about sex, sexuality, and gender—and thus their representations, identities, and institutions—depart dramatically from American dualistic beliefs and patterns of interaction.

To repeat, cultures are created by people in the different social worlds in which they live, day-to-day and over time. Consequently, cultures can be strikingly different (DiMaggio & Markus, 2010, p. 349), as expressed in the extraordinary sex/gender/sexuality variation found within and across networks and groups. For example, in some cultures, people do not perceive or categorize humans as homosexual (vs. heterosexual), and yet young male members of such cultures regularly and ritually engage in same-sex "sexual encounters" (Herdt, 1997). The work of Gilbert Herdt (1997, 2011) covers several decades of life among the Sambia of New Guinea. Until quite recently, the Sambia were patriarchal and their way of life was characterized by warfare, male privilege, secret male cults, strong antagonisms between men and women, female pollution rituals, and male insemination rites (Miller, 1993). Believing that boys had to be made into men through the ingestion of semen, defined as a potent substance, initiation rituals involved an extended period in which older boys orally inseminated younger boys. Upon marriage, heterosexuality was enforced. Herdt is very careful to distinguish between sexual behavior and identity in his analysis of these rituals, although he notes that the same-sex rituals involved expressions of pleasure. We suggest that it is valuable to think through what might seem to be a vast cultural gulf between traditional Sambian gender/sex/sexuality beliefs and practices as compared to those in the United States. Are there contexts in the United States, both past and present, in which homoerotic encounters between "straight-identified" boys and men take place and are ritualized? Jane Ward (2015), a sociologist, studies the fluidity of sexuality in U.S. society. Her research shows that homoerotic encounters between heterosexual men are commonplace in the United States and, in some settings, these relations are ritualized. For example, hazing rites in college fraternities and the military often involve sexual contact. She points to the "elephant walk" in which fraternity pledges "strip naked and stand in a circle, with one thumb in their mouth and the other in the anus of the pledge in front of them" (p. 4). Instead of interpreting such behaviors in terms of gendered sexuality stereotypes, Ward, much as Herdt in his work on Sambian culture, proposes that we understand male heterosexuality as "a fluid set of desires," constrained by gender norms, not by biological imperatives. Not only do different groups of people produce different cultures, but the cultures they produce are dynamic. That is, people continually generate and alter culture, including gender systems, both as individuals and as members of particular networks and groups; as a result, all cultures undergo change as their members evaluate, resist, and challenge beliefs and practices (DiMaggio & Markus, 2010; Stone & McKee, 1999). To illustrate, in the United States today, transgender activists are pushing back against the boundaries and discipline of the gender binary, and the outcomes are encouraging as, for example, all-women's colleges are moving to accept trans women, gender-neutral bathrooms are becoming more common, and workplace rights for trans people are slowly being instituted. Similarly, intersex activists have organized to seek change in how biological sex is understood and how natural, intersex variations are regarded and treated (see Davis and Preves, Chapter 1).

The prisms of gender and culture are inextricably intertwined. That is, people construct specific gender beliefs and practices in relation to particular cultural traditions and societal conditions. Cultures are gendered in distinctive ways, and gender systems, in turn, shape both material and symbolic cultural products (see, e.g., Chapter 5). As you will discover, the cross-cultural analyses of gender presented in this chapter provide critical support for the social constructionist argument that gender is a situated, negotiated, contested, and changing set of practices and understandings.

Let's begin with observations about gender in different cultures that add to those discussed above. Do you know there are cultures in which individuals can move from one gender category to another without being stigmatized? If you traveled from country to country around the world, you would find cultures in which men are gentle, soft-spoken, and modest, and cultures in which women are viewed as strong and take on roles labeled masculine in the United States. Although you might hear news about extreme forms of oppression of women in some places in the world (e.g., bride burning or dowry death in parts of India), there are other places where women and men live in relative harmony and equality. Also, there are cultures in which the social prisms of difference and inequality that operate in the United States (e.g., social class, race, sexual orientation) are minimal, inconsequential, or nonexistent (see the chapter reading by Helliwell).

The Problem of Ethnocentrism in Cross-Cultural Research

If you find any of these observations unsettling or even shocking, then you have probably tapped into the problem of bias in cross-cultural studies. One of the great challenges of cross-cultural research is learning to transcend one's own cultural assumptions of what is normal or natural to be able to understand cultural differences. It takes practice, conscious commitment, and self-awareness to get outside one's own cultural frame. After all, seeing what our culture wants us to see is precisely what socialization is about (see Chapter 4). Not only do cultural blinders make it difficult for us to comprehend the fiction of the gender binary that characterizes cultural beliefs in the United States today, but they can make it even more challenging to grasp the profoundly different ways people in other cultures think about and organize human relations.

We tend to "see what we believe," which means we are likely to deny gender patterns that vary from our own cultural experience and/or to misinterpret patterns that are different from our own. For example, the Europeans who first explored and colonized Africa were horrified by the ways African forms of gender and sexuality diverged from their own. They had no framework in which to understand warrior women, such as Nzinga of the Ndongo kingdom of the Mbundu (Murray & Roscoe, 1998). Nzinga was king of her people, dressed as a man, and was surrounded by young men who dressed as women and were her "wives" (Murray & Roscoe, 1998). However, her behavior made sense in the context of her culture, one in which people defined gender as situational and symbolic, thus allowing for alternative genders (Murray & Roscoe, 1998). In his chapter reading, Dredge Byung'chu Käng analyzes the history of the emergence of five gender categories in Thailand, a culture in which gender-variant people have long been a part of the genderscape. The role of language in structuring and reflecting cultural frames, including gender schemas, is highlighted by Käng in his discussion of everyday Thai language which, unlike English, does not distinguish among sex, gender, and sexuality and does not collapse differences into two and only two categories. It is a challenge to resist the tendency toward ethnocentrism (i.e., the belief that the ideas and practices of one's own culture are the standard and that divergent cultures are substandard or inferior). However, the rewards for bracketing the ethnocentric attitude are extremely valuable because one is then able to understand how and why gender operates in cultures that are different from one's own experience and, thus, from one's cultural frame. Thanks to the wide-ranging research of sociologists and anthropologists, we are increasingly able to grasp the peculiarities of our gender system and understand more deeply lifeways, including genderscapes, in other places in the world.

The readings in this chapter will introduce you to some of the variety in gender beliefs and practices across cultures and illustrate three of the most important findings of cross-cultural research on gender: (1) There is no universal definition or experience of gender; indeed, gender is not constituted as oppositional and binary in all societies. (2) Gender inequality, specifically the dominance of men over women, is not the rule everywhere in the world. (3) Gender arrangements, whatever they may be, are socially constructed and, thus, ever evolving.

There Is No Universal Definition or Experience of Gender

Although people in many contemporary cultures perceive at least some differences between women and men, and assign different tasks and responsibilities to people based on gender categories, these differences vary both from culture to culture and within cultures. There is no unified ideal or definition of masculinity or femininity across cultures. In some cultures, such as the Ju/'hoansi of Namibia and Botswana, women and men alike can become powerful and respected healers, whereas in others, such as the United States, powerful healing roles have been dominated by men (Bonvillain, 2001). Among the seminomadic, pastoral Tuareg of the Sahara and the Sahel, women have considerable economic independence as livestock owners, herders, gardeners, and leathersmiths, whereas in other cultural groups, such as the Taliban of Afghanistan, women are restricted to household labor and economic dependence on men (Rasmussen, 2001).

The readings in this chapter highlight some of the extraordinary cross-cultural differences in beliefs about men and women and in the tasks and rights assigned to them. They offer insights into how gender is shaped across cultures by a number of factors, including ideology, participation in economic production, and control over sexuality and reproduction. For example, in the chapter reading with the titillating title "It's Only a Penis," Christine Helliwell provides an account of the Gerai of Borneo, a cultural group in which rape does not exist. Helliwell argues that the Gerai belief in the biological sameness of women and men is a key to understanding their rape-free society. Her research offers an important report of how assumptions about human biology, in this case femaleness and maleness, are culturally shaped and have profound consequences for gender relations.

In addition, the two-sex (male or female), two-gender (feminine or masculine), and two-sexual-orientation (homosexual or heterosexual) system of Western, U.S. culture is not a universal mode of categorization and organization. As you know from reading Serena Nanda's article on gender variants in Native North America (Chapter 1), the two-spirit role was widespread and accepted in many American Indian tribes. Gilbert Herdt (1997; see discussion of Sambia, earlier), an expert on the anthropology of sexual orientation and gender, points out that the two-spirit role reached a high point in its cultural elaboration among the Mojave Indians, who "sanctioned both male (alyha) and female (hwame) two-spirit roles, each of which had its own distinctive social positions and worldviews" (p. 92). The chapter reading by Sabine Lang addresses multiple gender systems in indigenous North American cultures with a focus on the roles of female-bodied people who took up activities and tasks typically defined as masculine. She also examines indigenous concepts of sexuality and contemporary two-spirit women's lives.

Gender Inequality Is Not the Rule Everywhere

Gender and power go together but not in only one way. The relationship of power to gender in human groups varies from radical male dominance to relative equality between women and men (see the chapter reading by Lepowsky). At the extreme are intensely patriarchal societies in which women are dominated by men in multiple contexts and relationships. In traditional China, for example, sons were preferred, female infanticide was common, divorce could be initiated only by husbands, restrictions on girls and women were embodied in the mutilating practice of foot binding, and the suicide rate among young wives—who typically endured extreme isolation and hardship—was higher than in any other age and gender category (Bonvillain, 2001; Wong, 2016).

The United States has a complex history of gender relations in which White men as a group have had power over women as a group (and over men of color). This history was forged by European settlers who "seized and established property rights over land and resources and systematically removed Native Americans" through various strategies (e.g., cultural erasure through assimilation), including militarized genocide. Evelyn Nakano Glenn (2015) uses the concept of "settler colonialism" to analyze how the European White settler undertaking involved the creation of racialized gender and gendered racial dualisms (p. 71). "White settler society understood extreme gender differentiation as a mark of civilization," and white settlers racialized certain groups as sub-human (e.g., Mexicans, Chinese, Native Americans as well as Blacks). "Masculine whiteness . . . became central to settler identity," and it enshrined heteropatriarchal nuclear family arrangements as the model for other social arrangements, including the state (p. 60). For many decades, White men's power was overt and legal. For example, in the 19th century, husbands were legally empowered to beat their wives, women did not have voting rights, and women were legally excluded from many occupations (Nakano Glenn, 2015; Stone & McKee, 1999). Today, gender inequality takes more covert and subtle forms. For example, women earn less, on average, than do men of equal educational and occupational level; women are far more likely to be sexually objectified; and women are more likely to shoulder the burden of a double workday inside and outside the home (Coltrane & Adams, 2001; Chapters 7 and 8, this volume).

Understanding the relationship between power and gender requires us to use our sharpest sociological radar. To start, it is important to understand that power does not reside in individuals per se. For example, neither presidents nor bosses have power in a vacuum. They require the support of personnel and special resources such as media, weapons, and money. Power is a group phenomenon, and it exists only so long as a powerful group, its ruling principles, and its control over resources are sustained (Kimmel, 2000).

In addition, not all members of an empowered group have the same amount of power. In the United States and similar societies, male power benefits some men more than others. In fact, many individual men do not hold formal positions of power, and many do not feel powerful in their everyday lives (Johnson, 2001; Kimmel, 2000). Yet major institutions and organizations (e.g., government, big business, the military, the mass media) in the United States are gendered masculine, with controlling positions in those arenas dominated by men, but not just any men (Johnson, 2001). Controlling positions continue to be overwhelmingly held by White men who come from privileged backgrounds and whose lives appear to conform to a straight and narrow path (Johnson, 2001; Kimmel, 2000). As we learned in Chapter 2, the relationship of gender to power in a nation such as the United States is complicated by interactions among structures of domination and subordination such as race, social class, and sexual orientation. Not all societies are as highly and intricately stratified by gender, race, social class, and other social categories of privilege and power as is the United States. Many cultural groups organize relationships in ways that give most or all adults access to similar rights, prestige, decision-making authority, and autonomy in their households and communities. Traditional Agta, Ju/'hoansi, and Iroquois societies are good examples. In other cultural groups, such as the precontact Tlingit and Haida of the Canadian Pacific coastal region, relations among people were based on their position in economic and status hierarchies, yet egalitarian valuation of women and men prevailed (Bonvillain, 2001). In a widely reported new study (Migliano & Vinicius, 2015), anthropologists present their findings from research among the BaYaka from Congo and the Agta from the Philippines. Both are contemporary hunter-gatherer societies whose way of life is suggestive of how our human ancestors lived for most of human history. They are strikingly egalitarian. In fact, anthropologists such as Barry Hewlitt (2005), who has studied the Aka Pygmy people of central Africa, argue that they are fiercely egalitarian. Hierarchical structures are nonexistent. For example, among the Aka, while women hunt, men tend to babies and vice versa. Flexibility is the rule. Big egos are discouraged. Play is encouraged. Caring and fairness are the norm.

The point is that humans do not inevitably create inequalities out of perceived differences. Thus, even though there is generally some type of division of labor by gender across cultures today, differences in men's and women's work do not inexorably lead to patriarchal relations in which men monopolize high-status positions in important institutions and women are relegated to a restricted world of low-status activities and tasks. To help illustrate, Maria Alexandra Lepowsky's ethnography of Vanatinai social relations in the 1970s provides us with a model of a society in which the principles of personal autonomy and freedom of choice prevailed. The gender ideology of the Vanatinai was egalitarian, and their belief in equality manifested itself in daily life. For example, women as well as men owned and inherited land and other valuables. Women chose their own marriage partners and lovers, and they divorced at will. Any individual on Vanatinai might try to become a leader by demonstrating superior knowledge and skill.

Gender inequality is not the rule everywhere. Male dominance, patriarchy, gender inequality—whatever term one uses—is not the inevitable state of human relations. Additionally, patriarchy itself is not unitary. Patriarchy does not assume a particular shape, and it does not mean that women have no control or influence in their communities. Even in the midst of patriarchy, women and men may create identities and relationships that allow for autonomy and independence. In her reading in this chapter, Charlotte Wrigley-Asante details the empowering benefits that young women migrant workers in Accra, Ghana, accrue from economic independence. Using trading as a resource, the young women are able to save and invest in housing and personal items. They are also able to supplement household budgets and, as a consequence, their status is enhanced, as is their ability to control intimate relations with men.

Gender Arrangements Are Ever Evolving

The cross-cultural story of gender takes us back to the metaphor of the kaleidoscope. Life is an ongoing process of change from one pattern to another. We can never go back to "the way things were" at some earlier moment in time, nor can we predict exactly how the future will unfold. This is, of course, the story of gender around the world.

Two of the major sources of change in gender meanings and practices across cultures are culture contact and diffusion of beliefs and practices around the globe (Ritzer, 2004; Sørensen, 2000). Among the best documented accounts of such change have been those that demonstrate how Western gender systems were imposed on people whose gender beliefs and arrangements varied from Western assumptions and practices. For example, Native American multiple-gender systems were actively, and sometimes violently, discouraged by European colonists (Herdt, 1997; Nakano Glenn, 2015; see Nanda's reading in Chapter 1). Today, globalization—a complex process of worldwide diffusion of practices, images, and ideas (Ritzer, 2004)—raises the problem of the development of a world order, including a gender order, that may be increasingly dominated by Western cultural values and patterns (Held, McGrew, Goldblatt, & Perraton, 1999; see Chapter 10 for further discussion). Cenk Özbay, in his chapter reading, examines the emergence of exaggerated masculinity among rent boys in Istanbul. As part of his analysis, he offers insights into the process of diffusion as it plays out in the interpenetration of Western gay culture and the gender/sexuality frame of the rent boys and gay men of Istanbul.

Culture contact and diffusion via globalization are by no means the only source of changing gender arrangements (see Chapter 10 for detailed discussion of gender change). The forces of change are many and complicated, and they have resulted in a mix of tendencies toward rigid, hierarchical gender relations and toward gender flexibility and equality, depending on the specific cultural context and forces of change experienced by particular groups of people. In all this, there is one fact: People are not bound by any set of gender beliefs and practices. Culture change is inevitable, and so is change in the genderscape.

References

Baker, C. (1999). *Kaleidoscopes: Wonders of wonder.* Lafayette, CA: C&T.

Bonvillain, N. (2001). *Women and men: Cultural constructs of gender* (3rd ed.). Upper Saddle River, NJ: Prentice Hall.

Coltrane, S., & Adams, M. (2001). Men, women, and housework. In D. Vannoy (Ed.), *Gender mosaics* (pp. 145–154). Los Angeles: Roxbury.

DiMaggio, P., & Markus, H. R. (2010). Culture and social psychology: Converging perspectives. *Social Psychology Quarterly*, *73*(4), 347–352.

Held, D., McGrew, A., Goldblatt, D., & Perraton, J. (1999). *Global transformations.* Stanford, CA: Stanford University Press.

Herdt, G. (1997). *Same sex, different cultures.* Boulder, CO: Westview Press.

Herdt, G. (2011). Talking about sex: On the relationships between discourse, secrecy and sexual subjectivity in Melanesia. In D. Lipset & P. Roscoe (Eds.), *Echoes of the Tambaran: Masculinity, history and the subject in the work of Donald F. Tuzin* (pp. 259–273). Canberra: Australian National University Press.

Hewlitt, B. (2005, June 15). Are the men of the African Aka tribe the best fathers in the world? *The Guardian.* Retrieved from http://www.theguardian.com/society/2005/jun/15/childrensservices.familyandrelationships

Johnson, A. G. (2001). *Privilege, power, and difference.* Mountain View, CA: Mayfield.

Kimmel, M. (2000). *The gendered society.* New York: Oxford University Press.

Migliano, A., & Vinicius, L. (2015, May 18). Why our ancestors were more gender equal than us. *The Conversation.* Retrieved from http://theconversation.com/why-our-ancestors-were-more-gender-equal-than-us-41902

Miller, B. D. (1993). Anthropology of sex and gender. In B. D. Miller (Ed.), *Sex and gender hierarchies* (pp. 3–31). Boston: Cambridge University Press.

Murray, S. O., & Roscoe, W. (1998). *Boy-wives and female-husbands: Studies of African homosexualities.* New York: St. Martin's Press.

Nakano Glenn, E. (2015). Settler colonialism as structure: A framework for comparative studies of U.S. race and gender formation. *Sociology of Race and Ethnicity*, *1*(1), 54–74.

Patterson, O. (2014). Making sense of culture. *Annual Reviews of Sociology*, *40*(1), 1–30.

Rasmussen, S. (2001). Pastoral nomadism and gender. In C. B. Brettell & C. F. Sargent (Eds.), *Gender in cross-cultural perspective* (pp. 280–293). Upper Saddle River, NJ: Prentice Hall.

Ridgeway, C. (2009). Framed before we know it: How gender shapes social relations. *Gender & Society*, *23*(2), 145–160.

Ritzer, G. (2004). *The globalization of nothing.* Thousand Oaks, CA: Pine Forge Press.

Schwalbe, M. (2005). *The sociologically examined life.* New York: McGraw-Hill.

Sørensen, M. L. S. (2000). *Gender archeology.* Cambridge, UK: Polity Press.

Stone, L., & McKee, N. P. (1999). *Gender and culture in America.* Upper Saddle River, NJ: Prentice Hall.

Ward, J. (2015, February 2). It turns out that male sexuality is just as fluid as female sexuality. *The Conversation.* Retrieved from http://theconversation.com/it-turns-out-male-sexuality-is-just-as-fluid-as-female-sexuality-36189

Wong, O. M. H. (2016). The changing relationship of women with their natal families. *Journal of Sociology*, *52*(1), 53-67.

Introduction to Reading 12

Anthropologist Christine Helliwell provides a challenging account of a cultural group, the Gerai of Indonesia, in which rape does not exist. She links the freedom from rape among the Gerai people to the relatively egalitarian nature of their gender relations. Helliwell's research questions many gender beliefs held by members of Western cultures today.

1. How are men's and women's sexual organs conceptualized among the Gerai, and what are the consequences for Gerai understandings of sexual intercourse?
2. Genitalia do not determine identity in Gerai. What does?
3. Why is Gerai culture rape-free, and are there lessons we can learn from the Gerai about how to diminish or even eliminate rape in the United States?

"It's Only a Penis"

Rape, Feminism, and Difference

Christine Helliwell

In 1985 and 1986 I carried out anthropological fieldwork in the Dayak community of Gerai in Indonesian Borneo. One night in September 1985, a man of the village climbed through a window into the freestanding house where a widow lived with her elderly mother, younger (unmarried) sister, and young children. The widow awoke, in darkness, to feel the man inside her mosquito net, gripping her shoulder while he climbed under the blanket that covered her and her youngest child as they slept (her older children slept on mattresses nearby). He was whispering, "be quiet, be quiet!" She responded by sitting up in bed and pushing him violently, so that he stumbled backward, became entangled with her mosquito net, and then, finally free, moved across the floor toward the window. In the meantime, the woman climbed from her bed and pursued him, shouting his name several times as she did so. His hurried exit through the window, with his clothes now in considerable disarray, was accompanied by a stream of abuse from the woman and by excited interrogations from wakened neighbors in adjoining houses.

I awoke the following morning to raucous laughter on the longhouse verandah outside my apartment where a group of elderly women gathered regularly to thresh, winnow, and pound rice. They were recounting this tale loudly, and with enormous enjoyment, to all in the immediate vicinity. As I came out of my door, one was engaged in mimicking the man climbing out the window, sarong falling down, genitals askew. Those others working or lounging near her on the verandah—both men and women—shrieked with laughter.

When told the story, I was shocked and appalled. An unknown man had tried to climb into the bed of a woman in the dead, dark of night? I knew what this was called: attempted rape. The woman had seen the man and recognized him (so had others in the village, wakened by her shouting). I knew what he deserved: the full weight of the law. My own fears about being a single woman alone in a strange place, sleeping in a dwelling that could not be secured at night, bubbled to the surface. My feminist sentiments poured out. "How can you laugh?" I asked my women friends; "this is a very bad thing that he has tried to do." But my outrage simply served to fuel the hilarity. "No, not bad," said one of the old women (a particular friend of mine), "simply stupid."

I felt vindicated in my response when, two hours later, the woman herself came onto the verandah to

Helliwell, C. (2000). "It's only a penis": Rape, feminism, and difference. *Signs*, *25*(3): 789–816. Reprinted with permission of University of Chicago Press.

share betel nut and tobacco and to broadcast the story. Her anger was palpable, and she shouted for all to hear her determination to exact a compensation payment from the man. Thinking to obtain information about local women's responses to rape, I began to question her. Had she been frightened? I asked. Of course she had—Wouldn't I feel frightened if I awoke in the dark to find an unknown person inside my mosquito net? Wouldn't I be angry? Why then, I asked, hadn't she taken the opportunity, while he was entangled in her mosquito net, to kick him hard or to hit him with one of the many wooden implements near at hand? She looked shocked. Why would she do that? she asked—after all, he hadn't hurt her. No, but he had wanted to, I replied. She looked at me with puzzlement. Not able to find a local word for rape in my vocabulary, I scrabbled to explain myself: "He was trying to have sex with you." I said, "although you didn't want to. He was trying to hurt you." She looked at me, more with pity than with puzzlement now, although both were mixed in her expression. "Tin [Christine], it's only a penis," she said. "How can a penis hurt anyone?"

Rape, Feminism, and Difference

A central feature of many feminist writings about rape in the past twenty years is their concern to eschew the view of rape as a natural function of male biology and to stress instead its bases in society and culture. It is curious, then, that so much of this work talks of rape in terms that suggest—either implicitly or explicitly—that it is a universal practice. To take only several examples: Pauline Bart and Patricia O'Brien tell us that "every female from nine months to ninety years is at risk" (1985, 1); Anna Clark argues that "all women know the paralyzing fear of walking down a dark street at night. . . . It seems to be a fact of life that the fear of rape imposes a curfew on our movements" (1987, 1); Catharine MacKinnon claims that "sexuality is central to women's definition and forced sex is central to sexuality," so "rape is indigenous, not exceptional, to women's social condition" (1989b, 172) and "all women live all the time under the shadow of the threat of sexual abuse" (1989a, 340); Lee Madigan and Nancy Gamble write of "the global terrorism of rape" (1991, 21–22); and Susan Brison asserts that "the fact that all women's lives are restricted by sexual violence is indisputable" (1993, 17). . . . This is particularly puzzling given that Peggy Reeves Sanday, for one, long ago demonstrated that while rape occurs widely throughout the world, it is by no means a human universal: some societies can indeed be classified as rape free (1981).

There are two general reasons for this universalization of rape among Western feminists. The first of these has to do with the understanding of the practice as horrific by most women in Western societies. In these settings, rape is seen as "a fate worse than, or tantamount to, death" (S. Marcus 1992, 387): a shattering of identity that, for instance, left one North American survivor feeling "not quite sure whether I had died and the world went on without me, or whether I was alive in a totally alien world" (Brison 1993, 10). . . .

A second, equally deep-seated reason for the feminist tendency to universalize rape stems from Western feminism's emphasis on difference between men and women and from its consequent linking of rape and difference. Two types of difference are involved here. The first of these is difference in social status and power; thus rape is linked quite explicitly, in contemporary feminist accounts, to patriarchal social forms. Indeed, this focus on rape as stemming from difference in social position is what distinguishes feminist from other kinds of accounts of rape (see Ellis 1989, 10). In this view, inequality between men and women is linked to men's desire to possess, subjugate, and control women, with rape constituting a central means by which the freedom of women is limited and their continued submission to men ensured. Since many feminists continue to believe that patriarchy is universal—or, at the very least, to feel deeply ambivalent on this point—there is a tendency among us to believe that rape, too, is universal.[1]

However, the view of women as everywhere oppressed by men has been extensively critiqued within the anthropological literature. A number of anthropologists have argued that in some societies, while men and women may perform different roles and occupy different spaces, they are nevertheless equal in value, status, and power.[2] . . .

But there is a second type of difference between men and women that also, albeit largely implicitly, underlies the assumption that rape is universal, and it is the linkage between this type of difference and the treatment of rape in feminist accounts with which I am largely concerned in this article. I refer to the assumption by most Western feminists writing on rape that men and women have different bodies and, more specifically, different genitalia: that they are, in other words differently sexed. Furthermore, it is taken for granted in most feminist accounts that these differences render the former biologically, or "naturally," capable of penetrating and therefore brutalizing the latter and render the latter "naturally" able to be brutalized. . . . Rape of women by men is thus assumed to be universal because the same "biological" bodily differences between men and women are believed to exist everywhere.

Unfortunately, the assumption that preexisting bodily difference between men and women underlies rape has blinded feminists writing on the subject to the ways the practice of rape itself creates and inscribes such difference. This seems particularly true in contemporary Western societies where the relationship between rape and bodily/genital dimorphism appears to be an extremely intimate one. Judith Butler (1990, 1993) has argued (following Foucault 1978) that the Western emphasis on sexual difference is a product of the heterosexualization of desire within Western societies over the past few centuries, which "requires and institutes the production of discrete and asymmetrical oppositions between 'feminine' and 'masculine' where these are understood as expressive attributes of 'male' and 'female'" (1990, 17).[3] The practice of rape in Western contexts can only properly be understood with reference to this heterosexual matrix, to the division of humankind into two distinct—and in many respects opposed—types of body (and hence types of person).[4] While it is certainly the case that rape is linked in contemporary Western societies to disparities of power and status between men and women, it is the particular discursive form that those disparities take—their elaboration in terms of the discourse of sex—that gives rape its particular meaning and power in these contexts.

Sharon Marcus has already argued convincingly that the act of rape "feminizes" women in Western settings, so that "the entire female body comes to be symbolized by the vagina, itself conceived of as a delicate, perhaps inevitably damaged and pained inner space" (1992, 398). I would argue further that the practice of rape in these settings—both its possibility and its actualization—not only feminizes women but masculinizes men as well.[5] This masculinizing character of rape is very clear in, for instance, Sanday's ethnography of fraternity gang rape in North American universities (1990b) and, in particular, in material on rape among male prison inmates. In the eyes of these rapists the act of rape marks them as "real men" and marks their victims as not men, that is, as feminine.[6] In this iconography, the "masculine" body (along with the "masculine" psyche) is viewed as hard, penetrative, and aggressive, in contrast to the soft, vulnerable, and violable "feminine" sexuality and psyche. Rape both reproduces and marks the pronounced sexual polarity found in these societies.

Western understandings of gender difference have almost invariably started from the presumption of a presocial bodily difference between men and women ("male" and "female") that is then somehow acted on by society to produce gender. In particular, the possession of either male genitals or female genitals is understood by most Westerners to be not only the primary marker of gender identity but, indeed, the underlying cause of that identity. . . .

I seek to do two things in this article. First, in providing an account of a community in which rape does not occur, I aim to give the lie to the widespread assumption that rape is universal and thus to invite Western feminists to interrogate the basis of our own tendency to take its universality for granted.[7] The fundamental question is this: Why does a woman of Gerai see a penis as lacking the power to harm her, while I, a white Australian/New Zealand woman, am so ready to see it as having the capacity to defile, to humiliate, to subjugate and, ultimately, to destroy me?

Second, by exploring understandings of sex and gender in a community that stresses identity, rather than difference, between men and women (including men's and women's bodies), I aim to demonstrate that Western beliefs in the "sexed" character of bodies are not "natural" in basis but, rather, are a component of specifically Western gendering and sexual regimes. And since the practice of rape in Western societies is profoundly linked to these beliefs, I will suggest that it is an inseparable part of such regimes. This is not to say that the practice of rape is always linked to the kind of heterosexual regime found in the West; even the most cursory glance at any list of societies in which the practice occurs indicates that this is not so.[8] But it is to point out that we will be able to understand rape only ever in a purely localized sense, in the context of the local discourses and practices that are both constitutive of and constituted by it. In drawing out the implications of the Gerai stress on identity between men and women for Gerai gender and sexual relations, I hope to point out some of the possible implications of the Western emphasis on gender difference for Western gender and sexual relations—including the practice of rape.

Gender, Sex, and Procreation in Gerai

Gerai is a Dayak community of some seven hundred people in the Indonesian province of Kalimantan Barat (West Borneo).[9] In the twenty months I spent in the community, I heard of no cases of either sexual assault or attempted sexual assault (and since this is a community in which privacy as we understand it in the West is almost nonexistent—in which surveillance by neighbors is at a very high level [see Helliwell 1996]—I would certainly have heard of any such cases had they occurred). In addition, when I questioned men and women about sexual assault, responses ranged from puzzlement to outright incredulity to horror.

While relations between men and women in Gerai can be classified as relatively egalitarian in many respects, both men and women nevertheless say that men are "higher" than women (Helliwell 1995, 364). This greater status and authority does not, however, find expression in the practice of rape, as many feminist writings on the subject seem to suggest that it should. This is because the Gerai view of men as "higher" than women, although equated with certain kinds of increased potency vis-à-vis the world at large, does not translate into a conception of that potency as attached to and manifest through the penis—of men's genitals as able to brutalize women's genitals.

Shelly Errington has pointed out that a feature of many of the societies of insular Southeast Asia is a stress on sameness, even identity, between men and women (1990, 35, 39), in contrast to the Western stress on difference between the passive "feminine" object and the active, aggressive "masculine" subject.[10] Gerai understandings of gender fit Errington's model very well. In Gerai, men and women are not understood as fundamentally different types of persons: there is no sense of a dichotomized masculinity and femininity. Rather, men and women are seen to have the same kinds of capacities and proclivities, but with respect to some, men are seen as "more so" and with respect to others, women are seen as "more so." Men are said to be braver and more knowledgeable about local law (*adat*), while women are said to be more persistent and more enduring. All of these qualities are valued. Crucially, in terms of the central quality of nurturance (perhaps the most valued quality in Gerai), which is very strongly marked as feminine among Westerners, Gerai people see no difference between men and women. As one (female) member of the community put it to me: "We all must nurture because we all need."[11] The capacity both to nurture and to need, particularly as expressed through the cultivation of rice as a member of a rice group, is central to Gerai conceptions of personhood: rice is the source of life, and its (shared) production humanizes and socializes individuals (Helliwell, forthcoming). Women and men have identical claims to personhood based on their equal contributions to rice production (there is no notion that women are somehow diminished as persons even though they may be seen as less "high"). As in Strathern's account of Hagen (1988), the perceived mutuality of rice-field work in Gerai renders inoperable any notion of either men or women as autonomous individual subjects.

It is also important to note that while men's bravery is linked to a notion of their greater physical strength, it is not equated with aggression—aggression is not valued in most Gerai contexts.[12] As a Gerai man put it to me, the wise man is the one "who fights when he has to, and runs away when he can"; such avoidance of violence does not mark a man as lacking in bravery. . . . While it is recognized that a man will sometimes need to fight—and skill and courage in fighting are valued—aggression and hotheadedness are ridiculed as the hallmarks of a lazy and incompetent man. In fact, physical violence between adults is uncommon in Gerai, and all of the cases that I did witness or hear about were extremely mild.[13] Doubtless the absence of rape in the community is linked to this devaluing of aggression in general. However, unlike a range of other forms of violence (slapping, beating with a fist, beating with an implement, knifing, premeditated killing, etc.), rape is not named as an offense and accorded a set punishment under traditional Gerai law. In addition, unlike these other forms of violence, rape is something that people in the community find almost impossible to comprehend ("How would he be able to do such a thing?" one woman asked when I struggled to explain the concept of a man attempting to put his penis into her against her will). Clearly, then, more is involved in the absence of rape in Gerai than a simple absence of violence in general.

Central to all of the narratives that Gerai people tell about themselves and their community is the notion of a "comfortable life": the achievement of this kind of life marks the person and the household as being of value and constitutes the norm to which all Gerai people aspire. Significantly, the content of such a life is seen as identical for both men and women: it is marked by the production of bountiful rice harvests each year and the successful raising of a number of healthy children to maturity. The core values and aspirations of men and women are thus identical; of the many life histories that I collected while in the community—all of which are organized around this central image—it is virtually impossible to tell those of men from those of women. Two points are significant in this respect. First, a "comfortable life" is predicated on the notion of a partnership between a man and a woman (a conjugal pair). This is because while men and women are seen to have the same basic skills and capacities, men are seen to be "better" at certain kinds of work and women to be "better" at other kinds. Second, and closely related to this, the Gerai notion of men's and women's work does not constitute a rigid division of labor: both men and women say that theoretically women can perform all of the work routinely carried out by men, and men can perform all of the work routinely carried out by women. However, men are much better at men's work, and women are much better at women's work. Again, what we have here is a stress on identity between men and women at the expense of radical difference.

This stress on identity extends into Gerai bodily and sexual discourses. A number of people (both men and women) assured me that men sometimes menstruate; in addition, menstrual blood is not understood to be polluting, in contrast to how it is seen in many societies that stress more strongly the difference between men and women. While pregnancy and childbirth are spoken of as "women's work," many Gerai people claim that under certain circumstances men are also able to carry out this work—but, they say, women are "better" at it and so normally undertake it. In line with this claim, I collected a Gerai myth concerning a lazy woman who was reluctant to take on the work of pregnancy and childbirth. Her husband instead made for himself a lidded container out of bark, wood, and rattan ("like a betel nut container"), which he attached around his waist beneath his loincloth and in which he carried the growing fetus until it was ready to be born. On one occasion when I was watching a group of Gerai men cut up a boar, one, remembering an earlier conversation about the capacity of men to give birth, pointed to a growth in the boar's body cavity and said with much disapproving shaking of the head: "Look at this. He wants to carry his child. He's stupid." In addition, several times I saw fathers push their nipples into the mouths of young children to quiet them; while none of these fathers claimed to be able to produce milk, people nevertheless claimed that some men in the community were able to lactate, a phenomenon also attested to in myth. Men and women are thought to produce the same genital fluid, and this is linked in complex ways to the capacity of both to menstruate. All of these examples demonstrate the community's stress on bodily identity between men and women.

Furthermore, in Gerai, men's and women's sexual organs are explicitly conceptualized as the same. This sexual identity became particularly clear when I asked several people who had been to school (and hence were used to putting pencil to paper) to draw men's and women's respective organs for me: in all cases, the basic structure and form of each were the same. One informant, endeavoring to convince me of this sameness, likened both to wooden and bark containers for holding valuables (these vary in size but have the same basic conical shape, narrower at the base and wider at the top). In all of these discussions, it was reiterated that the major difference between men's and women's organs is their location: inside the body (women) and outside the body (men).[14] In fact, when I pressed people on this point, they invariably explained that it makes no sense to distinguish between men's and women's genitalia themselves; rather, it is location that distinguishes between penis and vulva.[15]

Heterosexuality constitutes the normative sexual activity in the community and, indeed, I was unable to obtain any information about homosexual practices during my time there. In line with the stress on sameness, sexual intercourse between a man and a woman in Gerai is understood as an equal coming together of fluids, pleasures, and life forces. The same stress also underlies beliefs about conception. Gerai people believe that repeated acts of intercourse between the same two people are necessary for conception, since this "prepares" the womb for pregnancy. The fetus is deemed to be created through the mingling of equal quantities of fluids and forces from both partners. Again, what is seen as important here is not the fusion of two different types of bodies (male and female) as in Western understandings; rather, Gerai people say, it is the similarity of the two bodies that allows procreation to occur. As someone put it to me bluntly: "If they were not the same, how could the fluids blend? It's like coconut oil and water: they can't mix!"

What needs to be stressed here is that both sexual intercourse and conception are viewed as involving a mingling of similar bodily fluids, forces, and so on, rather than as the penetration of one body by another with a parallel propulsion of substances from one (male) body only into the other, very different (female) one. What Gerai accounts of both sexual intercourse and conception stress are tropes of identity, mingling, balance, and reciprocity. In this context it is worth noting that many Gerai people were puzzled by the idea of gender-specific "medicine" to prevent contraception—such as the injectable or oral contraceptives promoted by state-run health clinics in the area. Many believed that, because both partners play the same role in conception, it should not matter whether husband or wife received such medicine (and indeed, I knew of cases where husbands had taken oral contraceptives meant for their wives). This suggests that such contraceptive regimes also serve (like the practice of rape) to reinscribe sex difference between men and women (see also Tsing 1993, 104–20). . . .

While Gerai people stress sameness over difference between men and women, they do, nevertheless, see them as being different in one important respect: their life forces are, they say, oriented differently ("they face different ways," it was explained to me). This different orientation means that women are "better" at certain kinds of work and men are "better" at other kinds of work—particularly with respect to rice-field work. Gerai people conceive of the work of clearing large trees for a new rice field as the definitive man's work and regard the work of selecting and storing the rice seed for the following year's planting—which is correlated in fundamental ways with the process of giving birth—as the definitive woman's work. Because women are perceived to lack appropriate skills with respect to the first, and men are perceived to lack appropriate skills with respect to the second, Gerai people say that to be viable a household must contain both adult males and adult females. And since a "comfortable life" is marked by success in production not only of rice but also of children, the truly viable household must contain at least one conjugal pair. The work of both husband and wife is seen as necessary for the adequate nurturance of the child and successful rearing to adulthood (both of which depend on the successful cultivation of rice). Two women or two men would not be able to produce adequately for a child since they would not be able to produce consistently successful rice harvests; while such a household might be able to select seed, clear a rice field, and so grow rice in some rudimentary fashion, its lack of expertise at one of these tasks would render it perennially poor and its children perennially unhealthy, Gerai people say. . . .

Gender difference in Gerai, then, is not predicated on the character of one's body, and especially of one's genitalia as in many Western contexts. Rather, it is understood as constituted in the differential capacity to perform certain kinds of work, a capacity assigned long before one's bodily being takes shape.[16] In this respect it is important to note that Gerai ontology rests on a belief in predestination, in things being as they should (see Helliwell 1995). In this understanding, any individual's *semongan* is linked in multifarious and unknowable ways to the cosmic order, to the "life" of the universe as a whole. Thus the new fetus is predestined to become someone "fitted" to carry out either men's work or women's work as part of the maintenance of a universal balance. Bodies with the appropriate characteristics—internal or external genitalia, presence or absence of breasts, and so on—then develop in line with this prior destiny. At first sight this may not seem enormously different from Western conceptions of gender, but the difference is in fact profound. While, for Westerners, genitalia, as significant of one's role in the procreative process, are absolutely fundamental in determining one's identity, in Gerai the work that one performs is seen as fundamental, and genitalia, along with other bodily characteristics, are relegated to a kind of secondary, derivative function.

Gerai understandings of gender were made quite clear through circumstances surrounding my own gender classification while in the community. Gerai people remained very uncertain about my gender for some time after I arrived in the community because (as they later told me) "I did not . . . walk like a woman, with arms held out from the body and hips slightly swaying; I was 'brave' trekking from village to village through the jungle on my own; I had bony kneecaps; I did not know how to tie a sarong in the appropriate way for women; I could not distinguish different varieties of rice from one another; I did not wear earrings; I had short hair; I was tall" (Helliwell 1993, 260). This was despite the fact that people in the community knew from my first few days with them both that I had breasts (this was obvious when the sarong that I wore clung to my body while I bathed in the river) and that I had a vulva rather than a penis and testicles (this was obvious from my trips to defecate or urinate in the small stream used for that purpose, when literally dozens of people would line the banks to observe whether I performed these functions differently from them). As someone said to me at a later point, "Yes, I saw that you had a vulva, but I thought that Western men might be different." My eventual, more definitive classification as a woman occurred [later]. . . . As I learned to distinguish types of rice and their uses, I became more and more of a woman (as I realized later), since this knowledge—including the magic that goes with it—is understood by Gerai people as foundational to femininity. . . .

Gerai people talk of two kinds of work as defining a woman: the selection and storage of rice seed and the bearing of children.[17] But the first of these is viewed as prior, logically as well as chronologically. People are quite clear that in the womb either "someone who can cut down the large trees for a ricefield is made, or someone who can select and store rice." When I asked if it was not more important whether or not someone could bear a child, it was pointed out to me that many women do not bear children (there is a high rate of infertility in the community), but all women have the knowledge to select and store rice seed. In fact, at the level of the rice group the two activities of "growing" rice and "growing" children are inseparable: A rice group produces rice in order to raise healthy children, and it produces children so that they can in turn produce the rice that will sustain the group once their parents are old and frail (Helliwell, forthcoming). For this reason, any Gerai couple unable to give birth to a child of their own will adopt one, usually from a group related by kinship. The two activities of growing rice and growing children are constantly talked about together, and the same imagery is used to describe the development of a woman's pregnancy and the development of rice grains on the plant. . . .

Gerai, then, lacks the stress on bodily—and especially genital—dimorphism that most feminist accounts of rape assume. Indeed, the reproductive organs themselves are not seen as "sexed." In a sense it is problematic even to use the English categories *woman* and *man* when writing of this community, since these terms are saturated with assumptions concerning the priority of biological (read, bodily) difference. In the Gerai context, it would be more accurate to deal with the categories of, on the one hand, "those responsible for rice selection and storage" and, on the other, "those responsible for cutting down the large trees to make a ricefield." There is no discursive space in Gerai for the distinction between an active, aggressive, penetrating male sexual organ (and sexuality) and a passive, vulnerable, female one. Indeed, sexual intercourse in Gerai is understood by both men and women to stem from mutual "need" on the part of the two partners; without such need, people say, sexual intercourse cannot occur, because the requisite balance is lacking. . . . [T]he sexual act is understood as preeminently mutual in its character, including in its initiation. The idea of having sex with someone who does not need you to have sex with them—and so the idea of coercing someone into sex—is thus almost unthinkable to Gerai people. In addition, informants asserted that any such action would destroy the individual's spiritual balance and that of his or her rice group and bring calamity to the group as a whole.[18]

In this context, a Gerai man's astonished and horrified question "How can a penis be taken into a vagina if a woman doesn't want it?" has a meaning very different from that of the same statement uttered by a man in the West. In the West, notions of radical difference between men and women—incorporating representations of normative male sexuality as active and aggressive, normative female sexuality as passive and vulnerable, and human relationships (including acts of sexual intercourse) as occurring between independent, potentially hostile, agents—would render such a statement at best naive, at worst misogynist. In Gerai, however, the stress on identity between men and women and on the sexual act as predicated on mutuality validates such a statement as one of straightforward incomprehension (and it should be noted that I heard similar statements from women). In the Gerai context, the penis, or male genitalia in general, is not admired, feared, or envied. . . . In fact, Gerai people see men's sexual organs as more vulnerable than women's for the simple reason that they are outside the body, while women's are inside. This reflects Gerai understandings of "inside" as representing safety and belonging, while "outside" is a place of strangers and danger, and it is linked to the notion of men as braver than women.[19] In addition, Gerai people say, because the penis is "taken into" another body, it is theoretically at greater risk during the sexual act than the vagina. This contrasts, again, quite markedly with Western understandings, where women's sexual organs are constantly depicted as more vulnerable during the sexual act—as liable to be hurt, despoiled, and so on (some men's anxieties about *vagina dentata* notwithstanding). In Gerai a penis is "only a penis": neither a marker of dimorphism between men and women in general nor, in its essence, any different from a vagina.

Conclusions

. . . With this background, I return now to the case with which I began this article—and, particularly, to the great differences between my response to this case and that of the Gerai woman concerned. On the basis of my own cultural assumptions concerning the differences—and particularly the different sexual characters—of men and women, I am inclined (as this case showed me) to read any attempt by a man to climb into a woman's bed in the night without her explicit consent as necessarily carrying the threat of sexual coercion and brutalization. The Gerai woman, in contrast, has no fear of coerced sexual intercourse when awakened in the dark by a man. She has no such fear because in the Gerai context . . . women's sexuality and bodies are no less aggressive and no more vulnerable than men's.

In fact, in the case in question, the intruding man did expect to have intercourse with the woman.[20] He claimed that the woman had already agreed to this through her acceptance of his initiatory gifts of soap.[21] The woman, however, while privately agreeing that she had accepted such gifts, claimed that no formal agreement had yet been reached. Her anger, then, did not stem from any belief that the man had attempted to sexually coerce her ("How would he be able to do such a thing?"). Because the term "to be quiet" is often used as a euphemism for sexual intercourse in Gerai, she saw the man's exhortation that she "be quiet" as simply an invitation to engage in sex with him, rather than the implicit threat that I read it to be.[22] Instead, her anger stemmed from her conviction that the correct protocols had not been followed, that the man ought to have spoken with her rather than taking her acceptance of the soap as an unequivocal expression of assent. She was, as she put it, letting him know that "you have sexual relations together when you talk together. Sexual relations cannot be quiet."[23]

Yet, this should not be taken to mean that the practice of rape is simply a product of discourse: that brutality toward women is restricted to societies containing particular, dimorphic representations of male and female sexuality and that we simply need to change the discourse in order to eradicate such practices.[24] Nor is it to suggest that a society in which rape is unthinkable is for that reason to be preferred to Western societies. To adopt such a position would be still to view the entire world through a sexualized Western lens.

In order to understand the practice of rape in countries like Australia and the United States, then—and so to work effectively for its eradication there—feminists in these countries must begin to relinquish some of our most ingrained presumptions concerning differences between men and women and, particularly, concerning men's genitalia and sexuality as inherently brutalizing and penetrative and women's genitalia and sexuality as inherently vulnerable and subject to brutalization. Instead, we must begin to explore the ways rape itself produces such experiences of masculinity and femininity and so inscribes sexual difference onto our bodies.

Notes

1. Among "radical" feminists such as Andrea Dworkin and Catharine MacKinnon this belief reaches its most extreme version, in which all sexual intercourse between a man and a woman is viewed as akin to rape (Dworkin 1987; MacKinnon 1989a, 1989b).

2. Leacock 1978 and Bell 1983 are well-known examples. Sanday 1990a and Marcus 1992 are more recent examples, on Minangkabau and Turkish society, respectively.

3. See Laqueur 1990 for a historical account of this process.

4. On the equation of body and person within Western (especially feminist) thought, see Moore 1994.

5. See Plaza 1980: "[Rape] is very sexual in the sense that [it] is frequently a sexual activity, but especially in the sense that it opposes men and women: it is social sexing which is latent in rape. . . . Rape is sexual essentially because it rests on the very social difference between the sexes" (31).

6. The material on male prison inmates is particularly revealing in this respect. As an article by Stephen Donaldson, a former prisoner and the president of the U.S. advocacy group Stop Prisoner Rape, makes clear, "hooking up" with another prisoner is the best way for a prisoner to avoid sexual assaults, particularly gang rapes. Hooking up involves entering a sexual liaison with a senior partner ("jocker," "man," "pitcher," "daddy") in exchange for protection. In this arrangement, the rules are clear: the junior partner gives up his autonomy and comes under the authority of the senior partner; he is often expected by the senior partner to be as feminine in appearance and behavior as possible, including shaving his legs, growing long hair, using a feminine nickname, and performing work perceived as feminine (laundry, cell cleaning, giving backrubs, etc.) (Donaldson 1996, 17, 20). See also the extract from Jack Abbott's prison letters in Halperin 1993 (424–25).

7. While I am primarily concerned here with the feminist literature (believing that it contains by far the most useful and insightful work on rape), it needs to be noted that many other (nonfeminist) writers also believe rape to be universal. See, e.g., Ellis 1989; Palmer 1989.

8. For listings of "rape-prone" societies, see Minturn, Grosse, and Haider 1969; Sanday 1981.

9. I carried out anthropological fieldwork in Gerai from March 1985 to February 1986 and from June 1986 to January 1987. The fieldwork was funded by an Australian National University Ph.D. scholarship and carried out under the sponsorship of Lembaga Ilmu Pengetahuan Indonesia. At the time that I was conducting my research a number of phenomena were beginning to have an impact on the community—these had the potential to effect massive changes in the areas of life discussed in this article. These phenomena included the arrival of a Malaysian timber company in the Gerai region and the increasing frequency of visits by Malay, Bugis, Chinese, and Batak timber workers to the community; the arrival of two American fundamentalist Protestant missionary families to live and proselytize in the community; and the establishment of a Catholic primary school in Gerai, resulting in a growing tendency among parents to send their children (both male and female) to attend Catholic secondary school in a large coastal town several days' journey away.

10. The Wana, as described by Jane Atkinson (1990), provide an excellent example of a society that emphasizes sameness. Emily Martin points out that the explicit Western opposition between the "natures" of men and women is assumed to occur even at the level of the cell, with biologists commonly speaking of the egg as passive and immobile and

the sperm as active and aggressive even though recent research indicates that these descriptions are erroneous and that they have led biologists to misunderstand the fertilization process (1991). See also Lloyd 1984 for an excellent account of how (often latent) conceptions of men and women as having opposed characteristics are entrenched in the history of Western philosophical thought.

11. The nurture-need dynamic (that I elsewhere refer to as the "need-share dynamic") is central to Gerai sociality. Need for others is expressed through nurturing them; such expression is the primary mark of a "good" as opposed to a "bad" person. See Helliwell (forthcoming) for a detailed discussion.

12. In this respect, Gerai is very different from, e.g., Australia or the United States, where, as Michelle Rosaldo has pointed out, aggression is linked to success, and women's constitution as lacking aggression is thus an important element of their subordination (1980, 416; see also Myers 1988, 600).

13. See Helliwell 1996, 142–43, for an example of a "violent" altercation between husband and wife.

14. I have noted elsewhere that the inside-outside distinction is a central one within this culture (Helliwell 1996).

15. While the Gerai stress on the sameness of men's and women's sexual organs seems, on the face of it, to be very similar to the situation in Renaissance Europe as described by Laqueur 1990, it is profoundly different in at least one respect: in Gerai, women's organs are not seen as emasculated versions of men's—"female penises"—as they were in Renaissance Europe. This is clearly linked to the fact that, in Gerai, as we have already seen, *people* is not synonymous with *men,* and women are not relegated to positions of emasculation or abjection, as was the case in Renaissance Europe.

16. In this respect Gerai is similar to a number of other peoples in this region (e.g., Wana, Ilongot), for whom difference between men and women is also seen as primarily a matter of the different kinds of work that each performs.

17. In Gerai, pregnancy and birth are seen not as semi-mystical "natural" processes, as they are for many Westerners, but simply as forms of work, linked very closely to the work of rice production.

18. Sanday 1986 makes a similar point about the absence of rape among the Minangkabau. See Helliwell (forthcoming) for a discussion of the different kinds of bad fate that can afflict a group through the actions of its individual members.

19. In Gerai, as in nearby Minangkabau (Sanday 1986), vulnerability is respected and valued rather than despised.

20. The man left the community on the night that this event occurred and went to stay for several months at a nearby timber camp. Community consensus—including the view of the woman concerned—was that he left because he was ashamed and distressed, not only as a result of having been sexually rejected by someone with whom he thought he had established a relationship but also because his adulterous behavior had become public, and he wished to avoid an airing of the details in a community moot. Consequently, I was unable to speak to him about the case. However, I did speak to several of his close male kin (including his married son), who put his point of view to me.

21. The woman in this particular case was considerably younger than the man (in fact, a member of the next generation). In such cases of considerable age disparity between sexual partners, the older partner (whether male or female) is expected to pay a fine in the form of small gifts to the younger partner, both to initiate the liaison and to enable its continuance. Such a fine rectifies any spiritual imbalance that may result from the age imbalance and hence makes it safe for the relationship to proceed. Contrary to standard Western assumptions, older women appear to pay such fines to younger men as often as older men pay them to younger women (although it was very difficult to obtain reliable data on this question, since most such liaisons are adulterous and therefore highly secretive). While not significant in terms of value (women usually receive such things as soap and shampoo, while men receive tobacco or cigarettes), these gifts are crucial in their role of "rebalancing" the relationship. It would be entirely erroneous to subsume this practice under the rubric of "prostitution."

22. Because Gerai adults usually sleep surrounded by their children, and with other adults less than a meter or two away (although the latter are usually inside different mosquito nets), sexual intercourse is almost always carried out very quietly.

23. In claiming that "sexual relations cannot be quiet," the woman was playing on the expression "be quiet" (meaning to have sexual intercourse) to make the point that while adulterous sex may need to be even "quieter" than legitimate sex, it should not be so "quiet" as to preclude dialogue between the two partners. Implicit here is the notion that in the absence of such dialogue, sex will lack the requisite mutuality.

24. Foucault, e.g., once suggested (in a debate in French reprinted in *La Folie Encerclée* [see Plaza 1980]) that an effective way to deal with rape would be to decriminalize it in order to "desexualize" it. For feminist critiques of his suggestion, see Plaza 1980; de Lauretis 1987; Woodhull 1988.

References

Atkinson, Jane Monnig. 1990. "How Gender Makes a Difference in Wana Society." In *Power and Difference: Gender in Island Southeast Asia,* ed. Jane Monnig Atkinson and Shelly Errington, 59–93. Stanford, Calif.: Stanford University Press.

Bart, Pauline B., and Patricia H. O'Brien. 1985. *Stopping Rape: Successful Survival Strategies.* New York: Pergamon.

Bell, Diane. 1983. *Daughters of the Dreaming.* Melbourne: McPhee Gribble.

Brison, Susan J. 1993. "Surviving Sexual Violence: A Philosophical Perspective." *Journal of Social Philosophy 24*(1): 5–22.

Butler, Judith. 1990. *Gender Trouble: Feminism and the Subversion of Identity.* New York and London: Routledge.

———. 1993. *Bodies That Matter: On the Discursive Limits of "Sex."* New York and London: Routledge.

Clark, Anna. 1987. *Women's Silence, Men's Violence: Sexual Assault in England, 1770–1845.* London and New York: Pandora.

de Lauretis, Teresa. 1987. "The Violence of Rhetoric: Considerations on Representation and Gender." In *Technologies of Gender: Essays on Theory, Film and Fiction,* 31–50. Bloomington and Indianapolis: Indiana University Press.

Donaldson, Stephen. 1996. "The Deal behind Bars." *Harper's* (August): 17–20.

Dworkin, Andrea. 1987. *Intercourse.* London: Secker & Warburg.

Ellis, Lee. 1989. *Theories of Rape: Inquiries into the Causes of Sexual Aggression.* New York: Hemisphere.

Errington, Shelly. 1990. "Recasting Sex, Gender, and Power: A Theoretical and Regional Overview." In *Power and Difference: Gender in Island Southeast Asia,* ed. Jane Monnig Atkinson and Shelly Errington, 1–58. Stanford, Calif.: Stanford University Press.

Foucault, Michel. 1978. *The History of Sexuality.* Vol. 1, *An Introduction.* Harmondsworth: Penguin.

Halperin, David M. 1993. "Is There a History of Sexuality?" In *The Lesbian and Gay Studies Reader,* ed. Henry Abelove, Michele Barale, and David M. Halperin, 416–31. New York and London: Routledge.

Helliwell, Christine 1993. "Women in Asia: Anthropology and the Study of Women." In *Asia's Culture Mosaic,* ed. Grant Evans, 260–86. Singapore: Prentice Hall.

———. 1995. "Autonomy as Natural Equality: Inequality in 'Egalitarian' Societies." *Journal of the Royal Anthropological Institute* 1(2): 359–75.

———. 1996. "Space and Sociality in a Dayak Longhouse." In *Things as They Are: New Directions in Phenomenological Anthropology,* ed. Michael Jackson, 128–48. Bloomington and Indianapolis: Indiana University Press.

———. Forthcoming. *"Never Stand Alone": A Study of Borneo Sociality.* Williamsburg: Borneo Research Council.

Laqueur, Thomas. 1990. *Making Sex: Body and Gender from the Greeks to Freud.* Cambridge, Mass., and London: Harvard University Press.

Leacock, Eleanor. 1978. "Women's Status in Egalitarian Society: Implications for Social Evolution." *Current Anthropology 19*(2): 247–75.

Lloyd, Genevieve. 1984. *The Man of Reason: "Male" and "Female" in Western Philosophy.* London: Methuen.

MacKinnon, Catharine A. 1989a. "Sexuality, Pornography, and Method: 'Pleasure under Patriarchy.'" *Ethics* 99: 314–46.

———. 1989b. *Toward a Feminist Theory of the State.* Cambridge, Mass., and London: Harvard University Press.

Madigan, Lee, and Nancy C. Gamble. 1991. *The Second Rape: Society's Continued Betrayal of the Victim.* New York: Lexington.

Marcus, Julie. 1992. *A World of Difference: Islam and Gender Hierarchy in Turkey.* Sydney: Allen & Unwin.

Marcus, Sharon. 1992. "Fighting Bodies, Fighting Words: A Theory and Politics of Rape Prevention." In *Feminists Theorize the Political,* ed. Judith Butler and Joan W. Scott, 385–403. New York and London: Routledge.

Martin, Emily 1991. "The Egg and the Sperm: How Science Has Constructed a Romance Based on Stereotypical Male-Female Roles." *Signs: Journal of Women in Culture and Society 16*(3): 485–501.

Minturn, Leigh, Martin Grosse, and Santoah Haider. 1969. "Cultural Patterning of Sexual Beliefs and Behaviour." *Ethnology 8*(3): 301–18.

Moore, Henrietta L. 1994. *A Passion for Difference: Essays in Anthropology and Gender.* Cambridge and Oxford: Polity.

Myers, Fred R. 1988. "The Logic and Meaning of Anger among Pintupi Aborigines." *Man 23*(4): 589–610.

Palmer, Craig. 1989. "Is Rape a Cultural Universal? A Re-Examination of the Ethnographic Data." *Ethnology 28*(1): 1–16.

Plaza, Monique. 1980. "Our Costs and Their Benefits." *m/f* 4: 28–39.

Rosaldo, Michelle Z. 1980. "The Use and Abuse of Anthropology: Reflections on Feminism and Cross-Cultural Understanding." *Signs 5*(3): 389–417.

Sanday, Peggy Reeves. 1981. "The Socio-Cultural Context of Rape: A Cross-Cultural Study." *Journal of Social Issues 37*(4): 5–27.

———. 1986. "Rape and the Silencing of the Feminine." In *Rape,* ed. Sylvana Tomaselli and Roy Porter, 84–101. Oxford: Blackwell.

———. 1990a. "Androcentric and Matrifocal Gender Representations in Minangkabau Ideology." In *Beyond the Second Sex: New Directions in the Anthropology of Gender,* ed. Peggy Reeves Sanday and Ruth Gallagher Goodenough, 141–68. Philadelphia: University of Pennsylvania Press.

———. 1990b. *Fraternity Gang Rape: Sex, Brotherhood, and Privilege on Campus.* New York and London: New York University Press.

Strathern, Marilyn. 1988. *The Gender of the Gift: Problems with Women and Problems with Society in Melanesia.* Berkeley: University of California Press.

Tsing, Anna Lowenhaupt. 1993. *In the Realm of the Diamond Queen: Marginality in an Out-of-the-Way Place.* Princeton, N.J.: Princeton University Press.

Woodhull, Winifred. 1988. "Sexuality, Power, and the Question of Rape." In *Feminism and Foucault: Reflections on Resistance,* ed. Irene Diamond and Lee Quinby, 167–76. Boston: Northeastern University Press.

Introduction to Reading 13

Dredge Byung'chu Käng is an anthropologist who conducted extensive fieldwork in Thailand between 2004 and 2011. His informants were *kathoey/gay* Thais as well as their family and friends. The traditional Thai gender/sexuality system consisted of three genders/sexes, including *kathoey*, a general term referring to all three-gender categories. However, the autocolonial, modern Thai state has promoted a Western heteronormative, two-and-only-two gender system and monogamous marriage. The efforts of the state have failed. Käng's research shows that a complex set of historical and contemporary forces have instead produced five key gender categories. Käng uses the term *genderscapes* to emphasize the possibilities for various genders, and he explores the multifaceted nature of the Thai *phet* system.

1. Why does Käng emphasize the public performance of gender and nonnormative gender presentation rather than sexuality?
2. Käng states that the categories of woman and man are not natural, nor do they require heterosexuality. How does this factor play out in the Thai genderscape?
3. What is the relationship between *kathoey* and *gay* in Thailand?
4. Discuss the conditions that may contribute to opportunities for female same-sex relationships in Thailand.

Conceptualizing Thai Genderscapes

Transformation and Continuity in the Thai Sex/Gender System

Dredge Byung'chu Käng

Historical Background and Theoretical Framework

Transformations in the Thai Sex/Gender System

Thailand is the only country in Southeast Asia that circumvented direct colonization. Without a colonial power, in either direct or indirect senses, Thailand is characterized by its semicolonial status (Jackson 2010). Yet, colonial history is important in understanding Thai gender/sexuality because the threat of foreign encroachment and subsequent nationalist modernization projects reformulated gender and produced unexpected consequences in gender relations. In part, Siam avoided external rule by proffering and employing autocolonial practices to demonstrate its *siwilai* (civilized) status (Winichakul 1994, 2000, 2010). While it would seem that these effects are more indirect since Thailand was never colonized, the consolidation of monarchical power and colonial impulse in the pursuit of *siwilai* actually heightened the ability of the state to recast cultural conventions. Arguably, autocolonial governmentality in Thailand was more direct and stronger a force of change than that imposed in its colonized neighbors by European rule. An intact absolute monarchy followed by a constitutional monarchy provided greater ability of the state to model and legislate gender norms than was possible in the colonies. For Thailand to remain free from Western domination, the population was subjected to new forms of rule (Connors 2007). . . .

The project of modernization contrasts the traditional Thai system of three sexes with a system that promotes standardization into two gender-normative sexes ideally engaged in monogamous marriage (Loos 2006). Semicolonialism imposed sexual dimorphism

---

Käng, D. B. (2014). Conceptualizing Thai genderscapes. In P. Liamputtong (Ed.), *Contemporary socio-cultural and political perspectives in Thailand*. Reprinted with permission from Springer Nature.

and heterosexual matrices that become refashioned in the Thai case of *siwilai*—but this state process of de-androgynizing local genders is a Foucauldian (1990) disciplinary mechanism that is productive of new kinds of thirds, which may or may not be ambiguous, since polarizing binary sex makes their transgressions more apparent. The early twentieth-century interventions, particularly those during the Field Marshal Plaek Phibunsongkhram era, were particularly virulent, legislating gender-specific hair, dress, and behavior (e.g., long hair and pantyhose for women, husbands kissing wives) that would be recognizable to Westerners (van Esterik 2000; Barmé 2002). These changes were not uniformly adopted, with urban elites being most influenced. Van Esterik (2000) notes how working-class women avoided and resisted Thai bureaucratic pressures to conform to ideals of Western femininity. Besides European encroachment and subsequent attempts at modernization, rapid pulses of gender and sexual transformation also followed other key historical, economic, and technological events such as the Vietnam/American War, AIDS, the Asian financial crisis, and the availability of birth control, hormones, and sexual reassignment and cosmetic surgeries. Furthermore, cultural influence from the migration of Chinese immigrants has affected gender relations, particularly among the merchant class (Bao 2005).

A new era of development post-WWII increasingly drove capitalist modes of production and consumption in Thailand. These came to be inhabited under the skin and combined with new discourses around economic restructuring, consumerism, development, modern identity, and cosmopolitanism that produced a post-Fordist system of flexible gender identities that modified historical forms but whose contours vary by subject position and moral valence. Involvement with Vietnam as an American ally greatly expanded sex tourism in Thailand and helped develop Thailand's infrastructure and economy (Bishop and Robinson 1998; Jeffrey 2002). Ironically, the expansion of the middle class made possible by prostitution has increased stigma for sex workers among those eager to make novel class distinctions. More recently, the Asian financial crisis and the subsequent recovery have both produced anti-Western sentiment and greater political, economic, and cultural integration with other Southeast and East Asian nations.

Thai biopolitical intervention in the production of modern sex is thus the basis for the production of the current genderscapes. The autocolonial intervention rearticulates a newly localized sex/gender system. Discursive governmentality is the catalyst to create this shift, yet cannot fully account for the variation present. Its erasure of former gender conceptualizations and reconstruction of a system in line with modern Western sexual dimorphism remains incomplete. In fact, hybrid resistances, cosmopolitan engagements, and everyday practices exceed the limitations placed on sex forms. The microstructural adjustments in the performance of gender also provide a degree of fluidity that cannot be accounted for in taxonomic models of the Thai sex/gender system. Nevertheless, these mundane performances are enabled and constrained via socio-moral understandings of appearance. Here, I turn to a theoretical elaboration of these additional processes.

Globalization and Localization of Gender Transformations

. . . I refer to genderscapes as a sex/gender system, a culturally elaborated mode of inhabiting reproductive differences in relation to subsistence, kinship, politics, and so forth (Rubin 1975). While I agree with Rubin (1984) that gender and sexuality should be separated analytically, they are emically intertwined and mutually constituted in Thailand and elsewhere. With the exception of masculine gay men, Thais typically do not distinguish gender and sexuality in their own lives. Sexuality is generally neither necessary nor sufficient to transform one's gender. Thus, following Jackson (2003) and Sinnott (2004), I conceptualize gender and sexuality within the purview of genderscapes as autonomous but related domains of a sex/gender order. Eroticism is corollary to but not necessarily derivative of gender, requiring relational rather than independent analysis.

Conceptualization as a -scape reveals greater possibilities for various genders. Modeling gender in a three-dimensional space rather than as a dualism opens up crosses, unities, mixes, alternatives, and liminalities beyond male-to-female and female-to-male transgenders. Incorporating time, the nodes of gender can shift and need not be fully formed to be recognized. I thus benefit from Herdt's conceptualization of third sexes, although I do not go as far as arguing that the ontology of third sexes requires a "stable social role, that can be inhabited—marking off a clear social status position, rights and duties, with indications for the transmission of corporeal and incorporeal property and rights" (Herdt 1996: 60). Instead, I embrace dynamism within the realm of habituated practice (Bourdieu 1990), as key nodes in the system help to anchor genderscapes. The performance or practice of gender allows for their ongoing rearticulation; yet the possibilities are always shaped through ritualization and regulatory discourses (Butler 2004; Morris 1995).

Currently, I argue for five major gender nodes, which exist in relation to one another based on anatomical differences, presentation of self, socioeconomic roles, desire for certain others, and so on. Each of these nodes has spin-offs or subdivisions, creating a large number of possible categories. However, they coalesce around five forms with significant ontological fixity in public everyday life. These are realized through the repetition of symbolic processes and everyday practices which make them appear to be real and part of the natural and hierarchical order of things (Bourdieu 1990). These processes include the presentation of a gendered self in daily life as well as the dissemination and reinstantiation of these practices via media representations. Face and appearance are key to Thai social norms and the regulation of behavior and propriety (Mulder 1997), especially gender performance and its formation (Barmé 2002). Surfaces (van Esterik 2000) and the gaze (Morris 1997) are technologies for the regulation of Thai gender and sexual relations. Appearances act as a mechanism to discipline idealized social forms, irrespective of interiority, through the fear of losing face. Within this public "regime of images" (Jackson 2004), private sexual practices are sequestered while gender presentation is brought to the fore. Therefore, public enactments of gender variance become highlighted for their difference at the same time sexuality can, and "should," remain unknown. In sum, the images one projects about the self are more important than identity in public interactions. Such exteriorities are not expected to access an essential truth. Furthermore, the Thai concept of *phet* frames sexuality as an extension of gender. Thus, my focus in constructing Thai genderscapes emphasizes the public performance of gender, and nonnormative gender presentation, rather than sexuality.

Thai Genderscapes

Kathoey:	a male-to-female transgender person, typically engaging in or desiring relationships with heterosexual men, irrespective of operative status
Gay:	a male, masculine or feminine, who engages in or desires same-sex relationships with other males
Tom:	a masculine woman who engages in or desires same-sex relationships with a woman
Dee:	a feminine woman who engages in same-sex relationships with *tom*

Contemporary Thai Gender and Sexuality

. . . Every day Thai does not distinguish between sex, gender, and sexuality (*phet*). Jackson and Sullivan (1999: 5) suggest that "within Thai discourse, gay and *kathoey* are not distinguished as a sexuality and a gender, respectively. Rather, gay, *kathoey*, together with 'man' (*phu-chai*), 'woman' (*phu-ying*), and the lesbian identities *tom* and *dee*, are collectively labelled as different varieties of *phet*." Academics use specialized terms to differentiate sex, gender, sexuality, and other aspects of *phet*, but these are not commonly understood. Activists also have developed specialized terms. In the last decade, the development of HIV services for males who have sex with males and transgender women has also elaborated new specialized terms, often derived from international NGO, public health, and human rights discourses. In the general operations of *phet*, sex, gender, and sexuality are bound up in metaphorical packages which discourage dissonance between gender presentation and a presumably gendered desire rather than between sex and gender.

The multifaceted nature of *phet* has been conceptualized in two primary ways in the literature: a Westernization model and an indigenization model. In the former, modern Western understandings of gender/sexuality usurp and supplant Thai ones. In the latter, new sexual identities are ensconced within the Thai *phet* system. Morris (1994) contrasts the traditional three-sex system with a modern four-sexuality system. She argues that the Thai ternary of man, woman, and *kathoey* is increasingly being replaced by four modern Western sexualities based on the two binaries of male:female and homosexual:heterosexual, which create the four positions of female-heterosexual, female-homosexual, male-heterosexual, and male-homosexual. According to Morris (1994), these systems coexist but are incommensurable, and thus the "modern" system is replacing the "traditional" one. Similarly, van Esterik (2000: 218–219) states that "the influence of international gay culture including new media (magazines and videos) may be increasing the numbers of category labels, but is breaking down the diversity in the Thai gender system to stress identity based on object choice, heterosexual or homosexual."

In contrast, Jackson (2000) and Sinnott (2004) argue that Western sexual identities are indigenized through local conceptualizations of gender, thereby multiplying gender categories. Jackson (2003) uses the term "eroticized genders" and Sinnott (2004) uses the term "gendered sexualities." They emphasize that sexual desire is an extension of gender identification rather than separate domains of gender and sexual orientation. Based on the historical analysis of the Thai press and academic texts, Jackson (2000: 412) asserts that there are at least ten gender terms commonly used in contemporary Thai discourse. He charts how the three categories of (1) man, (2) *kathoey* (transgender), and (3) woman have proliferated into ten categories from the 1960s to the 1980s: (1a) man, (1b) *seua bai* (male bisexual); (2a1) *gay king*, (2a2) *gay queen*, (2b1) *kathoey*, (2b2) *kathoey plaeng phet* (transsexual), (2b3) *khon sorng phet* (hermaphrodite), (2c) *tom*; (3a) *dee* and (3b) woman, respectively. These terms refer to seven common *phet* categories: "man," *gay king*, *gay queen*, *kathoey*, *tom*, *dee*, and "woman" (414).

Yet, the categories of salience that I documented in everyday talk are those that are visibly distinguishable by outward appearance: man, effeminate *gay*, *kathoey*, *tom*, and woman. I want to note that the categories of woman and man are not natural, preexisting forms. Nor do they require heterosexuality. For instance, a woman can be a *dee* or a man can have a female, *kathoey*, or male partners while maintaining a gender-normative status. Additionally, the range of acceptable demeanor, dress, and other gender markers is quite wide and has historically shifted. All Thai genders are modern formations that have undergone tremendous transformation over the last century. However, for lack of space, here I focus on individuals who would be considered gender variant and update the contemporary literature.

In Thai, *kathoey* is a general term encompassing all third-gender categories, theoretically referencing all nonnormative gender presentations and sexualities beyond heterosexual male and female. But in practice, *kathoey* seldom refers to female-bodied individuals, regardless of their gender expression. In cosmopolitan Bangkok, among the middle classes, *kathoey* only refers to male to female transgender persons, that is, transgender women (Jackson 2000; Sinnott 2004). *Gay* are typically offended when others refer to them as *kathoey*, although the term is used for in-group joking and accepted when outside Bangkok, as the locals are considered not to know better. People identified as *kathoey* may also be offended by the term as it can be used as a slur. There are numerous words that are considered more polite or respectful. Thus, if a person who is not *kathoey* is in the presence of one, she might use a term like *sao-praphet-sorng* (second category woman). Thai academics often refer to gender "fluidity," as identities follow a developmental trajectory and situational positioning. Witchayanee (2008) differentiates between "half and half" (those who have either breast implants or neo-vaginas but not both) and "fully transformed" transgender sex workers. Pramoj Na Ayuttaya (2008) identifies five types of *kathoey*: postoperative transgender, preoperative transgender, drag queen, penetrating girl (active in sexual intercourse), and those who live part time as transgender and part time as men.

In February 2010, the term *phu-ying-kham-phet* (transsexual woman) was introduced by Nok Yollada to differentiate a transsexual (who desires or has had gender surgery) from a transvestite. Around the same time, the Rainbow Sky Association of Thailand, a sexual diversity NGO primarily funded to provide HIV prevention services, started using the term *sao-thi-ji* (transgender woman), borrowing from the English abbreviation "TG." However, most *kathoey* use the term among themselves or simply use *sao* (young woman). "Ladyboy" refers to *kathoey* who are cabaret performers, beauty pageant contestants, and bar-based sex workers. Some *kathoey* consider "ladyboy" distasteful as it upholds the stereotype that they are prostitutes and thus inherently indecent and criminal. *Kathoey* are also differentiated by their operative status. But, many *kathoey* consider it offensive to be asked whether they have had sexual reassignment or gender confirmation surgery because they feel their identity does not need to follow their genitalia. While there has been a proliferation of terms around male-to-female transgenderism, I suggest that they coalesce around one commonly understood gender form: *kathoey*.

For *kathoey*, transgenderism is made visible via sartorial practice, cosmetic use, bodily comportment, and language (Thai uses gendered particles that mark the speaker as male or female). Bisexual men, who are labeled based on sexual behavior rather than desire or gender presentation, are generally said to be gay, but ashamed (*ap-ai*) to identify themselves. There is no equivalent term for a bisexual woman. *Dee* and masculine *gay* express gender normativity. Thus, *dee* are only discernible when with their *tom* partners and masculine *gay* are said not to "show" (*sadaeng-ok*). Importantly, as public display of affection is considered impolite, nonnormative sexuality is generally not apparent while nonnormative gender presentation is. Sexuality is understood as private and thus not subject to social condemnation. However, for gender nonnormative individuals, sexuality is presumed as an extension of gender presentation. Thus, only effeminate *gay*, *kathoey*, and *tom* noticeably do not conform to gender norms; and among them, *kathoey* are the most stigmatized. Though, with the growing use of surgery and other medical technologies, *kathoey* visibility is decreasing, as they increasingly pass as women.

Gay are further characterized by age and effeminacy (*tut*, sissy or queen; *taeo*, sissy; *sao-sao,* girly; *i-aep*, a feminine person who presents masculinely in public). These terms are usually not labels of self-identity but are used as insults or for in-group joking. Use of the terms "*king*" and "*queen*" in relation to *gay* is now considered passé (although the terms have been taken up by lesbians in *lesking* and *lesqueen*), perhaps because gender presentation has become more independent of preferred sexual positioning (*baep*), and *baep* is often flexible depending on the partner. *Gay* rather matter-of-factly disclose their *baep* (*ruk*, penetrate; *rap*, receive; *bot*, versatile; *salap*, reciprocate or alternate) as Thais would their age. These are among the first questions one might be asked upon meeting a stranger in a gay venue or online. However, *baep* does not constitute a public identity.

Along the continuum of *kathoey-gay*, distinction making occurs at both ends (see also Jackson 1997). Masculine *gay* refer to effeminate *gay* as *tut*, *kathoey*, or *i-aep*, individuals who would be *kathoey* given the opportunity. Postoperative transsexuals differentiate themselves from those who have not had surgery. They say they have already become women and often assume they pass (even when they do not), confident in the alignment of their essentially female mind and body. At the same time, there is fluidity between *gay* and *kathoey* categories, both in identity and sartorial practice. *Kathoey-noi* (little *kathoey* or just a little transgendered), for instance, use makeup like women but dress in men's clothing. *Kathoey-noi* are not transgender; they are not *gay;* they are an in-between category. *Kathoey-noi* are not uncommon. They are generally young, around 16–26 and often said to be transitioning into a *kathoey* lifestyle. There are, however, adult males who fit into this pattern. Some *kathoey* become *gay* and vice versa, although the former conversion is more prevalent. As *kathoey-kathoey*, *gay-kathoey*, and effeminate *gay* pairings become more common, the disgust associated with similar-gender coupling is diminishing. New terms such as *sao-siap* (penetrating girl, referring to *kathoey* who are active in anal intercourse) and *tom-gay* (a *tom*, masculine female, in a relationship with another *tom*) describe variations that incorporate putatively discordant gender expression and sexual practice. These changes point to the breakdown of the heterogender sexual matrix (Peletz 2006), in which only sex between individuals of "opposite" genders is socially acceptable.

In English, we say "gay man," with "gay" sexuality (an adjective) modifying "man" (a noun). But in Thai, "*gay*" is already a noun so that one is either "*gay*" or "man." Thus, one can say "I am a *gay*" or "I am a man." However, these are not exclusive categories and there is a recognition that sexual desire does not have to follow gender identification. For example, one can say the two in sequence, as in *phom pen ke; phom pen maen* to emphasize that one is a masculine gay man. Bee, who is *gay*, once told me: *phom mai pen ke; phom pen maen* (literally "I am not gay; I am a man") to stress that he does not see himself as effeminate at all, since the term "*gay*" can invoke effeminacy, as in *ke map koen pay* (too *gay*). In the statement *mai pen ruk, mai pen rap; pen ke* (He is not top/penetrator or bottom/receiver; he is *gay*), the term is referencing sexual versatility. Thus, the term "*gay*" is polysemic in everyday use, contextually referencing effeminacy, masculinity, or sexual versatility. These examples show that while gender and sexuality are linked, they can be distinguished from one another. At the same time, they are conflated in everyday life.

Phet terms are not isomorphic with identities. Neologisms and variants do not necessarily constitute new forms; they can be situationally employed or used to label others and make fine distinctions. Masculine *gay* often refer to themselves as *maen-maen* (very manly, like a man), although this is more a descriptor of gender presentation than an identity. Similarly, the term *me-tho sek-chuan* (metrosexual), while defined in Thai Wikipedia and in many lifestyle magazines as a heterosexual man who appears gay based on his interest in fitness, fashion, and grooming, is typically used in speech to refer to someone who is *gay* or closeted. Indeed, this *phu-chai saiphan mai* (new breed of man) category was popularized by the film *Metrosexual* (2006, directed by Yongyoot Thongkongtoon), in which a group of women tries to prove to their friend that her fiancé, who is too perfect to be heterosexual, must be *gay*. The Thai title is *Kaeng chani kap i-aep* (Gang of Girls and the Closet Case). Metrosexuality is marked visibility by being too perfect a man to be straight, just as *kathoey* beauty contestants are often said to possess *owoe-bioti* (over beauty) in comparison to women. Many new gender expressions are derived from films, songs, and Internet sites, although their usage is typically short-lived and the terms do not constitute identities, though they may be variations upon them.

Female-bodied gender forms are highly visible but have less social presence in Thai society and media

(Käng 2011). Like female same-sex sexuality in other parts of the world, it is limited by income differentials, safety concerns, and social proscriptions around propriety. Additionally, female same-sex relationships can also be less challenging to the sex/gender order. Many Thai women are able to circumvent the issue of respectability as being in a same-sex relationship can be morally less damaging than a heterosexual one (Sinnott 2004). In particular, one can avoid unwanted pregnancy and the contamination associated with the loss of virginity. This, however, is more applicable to young women and those from wealthier backgrounds. Compared to women in other regions of the world, Thai women also have relatively high employment and financial independence, allowing many of them to live independently of family and forestall marriage (Mills 1999; Muecke 1984). Thailand has the highest rate of single women in Asia and one of the lowest birthrates in the world. These factors perhaps provide more opportunity for female same-sex relationships than in other geographic areas.

Tom-dee (the terms are derived from the English "tomboy" and "lady") couples are ubiquitous and easily identifiable in mainstream commercial venues such as shopping malls (Wilson 2007). Suburban malls have stores targeted to *tom*, selling men's clothes in smaller sizes (in contrast to stores targeting *kathoey* that provide women's clothes in larger sizes). I suggest that *tom-dee* couples are actually more visible than gay men in Thai public space, where they become legible by learning their social codes. Female couples can be seen holding hands, one sporting short hair and men's clothes, the other with long hair and women's clothes. Sinnott (2004: 142) has referred to the public conspicuousness of these relationships in terms of "visual explicitness and verbal silence." The couples are discernible but unremarkable. Women are also highly active in Thai activism around sexual diversity, with organizations such as Anjaree (founded in 1986) among the most vocal advocates for sexuality rights. However, their work is often overshadowed by an infusion of international HIV prevention funding for *gay* and *kathoey* in the mid-2000s, which has been mobilized to promote the acceptance of sexual minorities and their human rights more broadly.

In *tom-dee* relationships, it is only the former who can be considered misgendered. *Toms* are masculine women. They are biological females attracted to women, but this attraction to women is an extension of their masculinity. Their gender does not match their sex, though their sexuality can be seen as an extension of their gender. *Dees* are their female partners. But, as gender normative or "ordinary women," *dee* identity becomes relevant in relation to *tom*. That is, *dee* are labeled as counterparts of *tom* rather than as another category of women, often being temporary members of the *tom-dee* community (Sinnott 2004). Thus, while *tom* represent a more or less stable *phet* identity in contemporary Thailand, *dee* have a more liminal position. They exist primarily in association with *tom* rather than as an independent category. *Dee* remain peripheral to the nodes of *tom* and woman in Thai genderscapes.

Dee inhabit a relational and situational identity (Sinnott 2007). Sexuality can be tied to temporal gender positions in the life cycle: what is appropriate in youth may not be so in adulthood. Engaging in sexual acts inappropriate for one's status can modify ascribed gender positions. Yet, being *dee* is considered relatively unproblematic. Many women, often married, have told me, without remorse or shame: "I was a *dee* before." From a Western perspective, *dee* is a lesbian sexual identity, women who desire or engage in sexual relations with other women. However, gender difference is more important to *tom-dee* couplings than sexual identity (Sinnott 2004). From the perspective of Thai *phet*, as a feminine woman attracted to the masculinity of another woman, their desire is homosexual but heterogender (Peletz 2006) and thus normatively paired.

In the alternate genders of *tom* and *kathoey*, blending of sex/gender attributes is a more typical feature than the disavowal of a biological sex and the recreation of a newly contrasting gendered persona. *Tom* maintain many characteristics of women (e.g., being good caretakers) as *kathoey* maintain many characteristics of men (e.g., being sexually assertive). Some *kathoey* are more like transsexuals in the classic sense of seeing themselves as an "opposite" sex, and this is essential to the new conceptualization of *phu-ying-kham-phet* (transsexual woman), literally a woman who has crossed sex. But, this is generally not the case for *tom* who see themselves as masculine women rather than as men. Sinnott (2004: 22) refers to *tom* "in the feminine form ('she,' 'her') to reflect the common understanding among Thais that *toms* are female, and although they are masculine, they are distinct from males."

Five Genders?

The proposal of *five phet* categories is reminiscent of Davies (2010) five genders among the Bugis of Sulawesi, Indonesia. Yet, the sex/gender system in Thailand does not follow the same pattern of crossings (i.e., female, female to male, male/female androgyny, male to female, and male). Nor are the spiritual aspects of *bissu* unity at play.[1] The five gender forms I propose are similar to the seven *phet* categories Jackson (2000) states are commonly described in Thai literature. However, rather than representing the categories like a phylogenetic tree, which emphasizes the historical transformations of the categories, I propose a multidimensional -scape. On the one hand, this formation provides greater flexibility in the enactments of gender as they are practiced in everyday life at a point in time. On the other, a -scape model shows the layered relationships between nodes and can shift based on historical transformations, one's social positioning, or ideological perspective.

With the proposed five gender nodes, I have not included *dee*. As Sinnott (2004: 9) notes, "masculine women have long been evident in the Thai system of sex and gender, but the linguistic and social marking of feminine women who are partners of masculine women creates a new and precarious field of identity." Sinnott likens *dee* to the gender-normative partners of *kathoey*, who are simply categorized as men. As women who are feminine in dress, comportment, and speech, they are not marked as gender different. Most *dee* do not refer to themselves as *dee* but as "women." *Dee* is more frequently a label that *tom* use for their feminine partners. Ploy, a neighbor of mine who ran a noodle shop, talking to me about her brother's transition from *kathoey* to *gay,* stated matter-of-factly: "Well, I was a *dee* before I was with him," nodding her head in the direction of the unmarried father of their daughter. Many other women told me that they had been in relationships with women when younger, but typically did not refer to the relationship as *tom-dee*. Instead, the relationship was with another *phu-ying* (girl/woman). The expectation among *tom* is that *dee* will eventually leave *tom* for a man. Similarly, a "real" man is anticipated to leave a *kathoey* for a "real" woman in order to have a family, or keep the *kathoey* as a mistress. A *dee* is thus generally considered *tham-mada* (ordinary or normal) but could shift positions within genderscapes, for example, if one emphasizes sexual object choice over gender presentation.

Additionally, I have not differentiated nodes for masculine and feminine *gay,* which can be thought of as two different categories based on different gender presentation, public visibility, and sense of self. Like *dee*, masculine *gay* are gender normative in presentation or often hypermasculine. However, unlike *dee*, *gay* have a strong sense of identity based on their sexual desires. Sinnott (2004: 29) suggests that "the category 'gay' has introduced a third possible kind of masculinity positioned between normative 'men' and *kathoeys*, in that gay men are masculine yet desire other masculine men as sexual partners. . . . [Masculine gay men] are not highly visible, in that they do not match the perception that gayness equals effeminacy—they simply fall off the radar for many Thais." Masculine *gay* break the rules of the heterogender matrix; they represent a singularity where the traditional three-sex system and the modern four-sexuality systems intersect in Thai genderscapes. Of course, feminine *gay* would provide a fourth male position. Yet, while only the feminine *gay* is visibly identifiable, both masculine and feminine *gay* have a sexual identity that is often an important component of self-concept and lifestyle. They also conceive of themselves as a community using shared social space and having a common identity and transnational ties with other gay communities around the world. The terms *king* and *queen* are rarely used and tend to reference sexual position more than an identity. Thus, while focusing on the appearance of gender-variant performance, which would exclude masculine *gay,* who are relatively invisible, I combine them here as a single node. Indeed, no genderscape nodes are homogenous or consistent. Yet, the gay node has a particularly two-faced characteristic. As Sinnott (2004: 30) notes, "an important difference between gay men and *dees* is that *dees* are only *dees* in relation to a *tom*. The masculine female gives the *dee* her identity, whereas gay men take pains to distinguish their community and identities from *kathoey* identity." *Gay*, *tom,* and *kathoey* are more like what Foucault (1990) might refer to as a "species" with discursive and ontological fixity. They all have elaborated subcultures with specific social spaces and organization. At the same time, elaborating gender categories as nodes in the Thai genderscapes acknowledges the variation within each in relation to the others.

Social Position and Ideological Stance

Class, education, geography, and *phet* identification also affect how people conceptualize gender/sexuality categories. In Thailand, geographical regions and urban/rural distinctions are very important and highly associated with class differences. Central Thais, who are culturally and linguistically dominant, conceive of Northerners as "soft." The women are thought of as more gracile, polite, and lighter skinned. The men are similarly effeminized. Southern men, by contrast, are considered rough, hard, and dark. They are also portrayed as more patriarchal as a result of Islamic conversion or interaction. Northeasterners are often said not to be Thai and, instead, referred to as Lao, and therefore, less developed. For many Central Thais, Bangkok is the *only city*. All other areas are considered *tang-changwat* (provincial, upcountry) or, less politely, *ban-nok* (the boonies). While rural populations consume the same national media as produced in the capital, interpretations are filtered through their daily experience, which includes the out-migration of many young people, including the gender variant. There is also less access to the consumer products that allow for greater gender differentiation.

One of my best friends, Wan, who is a *kathoey* in her mid-30s from *Isan* (the Northeast region), often makes the statement: "I was the first *kathoey* in my village." This assertion struck me as strange because I imagined there must have been *kathoey* who preceded her since there is a history of *kathoey* in Thailand. I have traveled with her to her home village three times, and people have confirmed that she was the first *kathoey* that they can remember. Before, villagers were quite hostile to *kathoey*, until she and a few others showed themselves as *kathoey*, dressing and living as transgender women.[2] Now there are approximately ten adult *kathoey* in her village, one of whom was married to a man in 2009 in a day-long celebration attended by several hundred guests. Both sets of parents gave speeches about their happiness during the wedding. There are also several *kathoey* children. When I asked a mother of a 10-year-old boy about when the child started expressing herself as a girl, the mother replied: "Since birth. When she/he started talking, she/he used *kha*" (the female polite particle).[3] But, this increased acceptance has only occurred since around 2000. Wan only began living as a *kathoey* after the death of her father, who was a respected village leader. Others soon followed and transgenderism became a visible part of village life. Thus, Wan feels that the situation for *kathoey* is improving rapidly.

Wan believes that while there is a strong inclination toward being *kathoey*, one can choose to be *gay* instead. When we met with a 16-year-old at a temple fair who was cross-dressing, Wan immediately went to speak with her. "What are you?" she asked. "Are you like me or him [nodding toward me]?" She went on to ask whether her father accepted her cross-dressing, which the young *kathoey* affirmed. She then said: "You have to choose whether you will be like me or him. If you choose to be like me, life will be harder." Wan counseled the youth, acknowledging the fluidity between *kathoey* and *gay* as well as the different opportunities associated with these life trajectories.

Wan notes that the number of *gay* in her village is increasing, although their normative masculinity renders them relatively invisible and uncontroversial. Villagers simply refer to men and women; the distinction between the two is based on outward presentation. However, when someone is verbally identified as *gay*, the reaction is often: "I didn't know she/he was a woman." Such comments show that villagers perceive sexuality in terms of a gendered desire that should be an extension of their gender presentation. For most villagers, a *gay* is someone who appears like a man, but is actually a woman based on their *jit-jai* (mind/heart or inner being), their desire for male partners. One of Wan's friends, who used to be *kathoey* when living in the provinces but has since become *gay* after moving to Bangkok, overheard our discussion. He commented: "Before, things were bad for us, but now it is getting much better." The repetition of an improved situation for the "third gender" has become a refrain among Thais of all genders. But, the increased acceptability of gender variance is neither embraced nor uncontested (Jackson 1995, 1999b; Sinnott 2000).

In particular, moral stance can override other classificatory schemes, as was evident from a pile sort exercise I conducted to develop a conceptual map of Thai genders among a diverse group of Bangkok residents.[4] Respondents were asked to think aloud while making their taxonomic decisions. Individuals used a variety of factors in creating groups: anatomy, gender expression on a male-female continuum, romantic attraction, common/normal/natural status, and personal experience. I was not surprised when an early free list by a man in his 50s returned two items: man and woman. I was, however, taken aback when, after elaborating a wide number of gender categories, he created two piles: man and woman in one called "normal" and the rest in another called "abnormal" (*phit-pukati*). I had erroneously assumed that man and woman were counterparts and would remain in separate piles because I failed to account for the moral valence attached to *phet* categories.

Gender classification is not an amoral process. *Phet* are defined by factors which are variously invoked by different people, situationally dependent, and experientially based. Instead of seeing Morris's and Jackson's and Sinnott's interpretations as orthogonal, I suggest the three are complementary. Class, generation, rural/urban upbringing, moral stance, personal experience, and context mediate how the local repertoire of gender/sexuality is practiced, interpreted, and labeled in relation to differential exposure to market mechanisms, bureaucratic institutions, and cultural forms. That is, social stances and life opportunities condition how Thais inhabit and interpret *phet*. Furthermore, I argue that gender forms are interpolated by the moral valence attached to their normativity. These concerns not only expand the terrain of gender/sexuality but also force a reconsideration of their topography. I suggest that *phet* should not be enumerated individually but conceptualized in nodes and clusters. That is, gender/sexuality categories are not fixed to four modern sexual positions. Nor are they proliferating with each new addition of a term. Rather, *phet*, which may or may not be publicly visible, cluster around several key nodes (man, woman, *kathoey*, *gay*, and *tom*), which are renewed through everyday experience. These forms shift over time, often proliferating in punctuated bursts that retreat into refined forms. Furthermore, as Thais use different criteria to assess *phet* (e.g., anatomy, sartorial presentation, desired partner, normality, personal experience), their classifications vary widely and the boundaries between groups overlap. For example, *phu-ying-kham-phet* (transsexual women) are variously grouped with men (based on anatomy at birth), women (based on postoperative anatomy, social presentation, or desired partner), or *kathoey* (based on their being transgender or "not normal"). The framework Thais use to think about these differences is conditioned by social experience. There are multiple stances and layers to the evaluation and categorization of gender/sexuality. Thus, I argue that the multidimensional nature of Thai *phet* are best conceptualized as a localized genderscape, a terrain of archetypes in which fields of power, morality, and experience shape its continually shifting boundaries over time.

Discussion and Conclusion

Conceptualizing Thai Genderscapes

I argue that the gender-inflected sexualities of Thailand were produced by the very forces of autocolonial governmentality, modernization, and globalization (including the institutionalization of sexual dimorphism, restructuring of kin relations, and the construction of tradition) that tried to erase them. While masculine *gay* represent a gender node where a "modern" sexuality intrudes into a system of multiple genders, *gay* is itself a product of modern *phet* formation that developed concomitantly with the West and in relation to *kathoey* and other local gender configurations. The addition of gay identity forms in Thailand cannot be described as a "rupture" as transgenderal homosexualities both continue unabated and modernize alongside new forms of homosexuality that, at least on the surface, appear modern and Western. That is, gay identity as Altman (2001) describes did not diffuse to Thailand from the West, but developed in parallel dialogue at approximately the same time (Jackson 1999a). While capitalist modes of consumption facilitate such new identities, the conditions are not the same, particularly in relation to a break in kinship relations.

. . . Thai genderscapes are a localized production of gender and sexual differences that negotiate the tensions between local and global gender/sexuality forms. As such, they are hybrid: indigenizing the global and recasting the local for international audiences. The disjuncture or tension arising from the differences of autochthonous and global forms produces local distinction, distinguishing Thailand from many of its Asian neighbors. Furthermore, regardless of whether a gender form is considered "traditionally" Thai or not, they are all constituted via historical transformations interacting within the forces of human, capital, technological, ideological, and other flows. Thai genderscapes are already completely hybrid and globalized. But, they maintain local character as gender/sexuality forms are, for the most part, indigenized through local conceptualizations of *phet*.

I have argued for the conceptualization of contemporary Thai gender and sexuality as genderscapes grounded in five major gender/sexuality categories: *kathoey*, *tom*, *gay*, woman, and man. These categories are not the only ones that exist. Rather, they possess an ontological fixity which comes from their reproduction in the repetition of everyday performances and symbolic practices, which can shift norms over time. Moreover, key historical incidents produce punctuated expansions of possibilities, which may or may not shift the terrain of gender forms. Finally, the social position and ideological stance of an individual shapes how she/he inhabits and interprets *phet*, making genderscapes a perspectival endeavor.

Notes

1. The *bissu* are transgender ritual specialists whose mixture of male and female characteristics identify and represent the undifferentiated nature of the universe. *Bissu* gender unity allows them to access spiritual powers unattainable by males or females. Peletz (2006) notes that this pattern exists throughout Southeast Asia. However, the situation in Thailand is more complex. Transgender and gay ritual specialists are currently increasing in popularity in both the North and Central regions, yet the lack of an historical record makes it unclear whether this is a resurgence of prior

practices. Of course, transgender ritual specialists exist in other world areas, with the *hijras* of South Asia perhaps the most well-known case.

2. As in other parts of Asia, there is not an emphasis on "coming out" in Thailand. However, unlike Confucian Asian societies, there is less emphasis on hiding one's gender/sexual nonconformity. Effeminate *gay* will often state that people know about their sexual orientation, even if they have not been told, because they "show" themselves.

3. The third person singular pronoun in Thai is gender neutral.

4. Pile sorts are a cognitive mapping procedure to understand how community members think about and attach meaning to different items within a conceptual domain. I began the exercise with a free list to identify the *phet* respondents conceived of as most salient. Up to 22 terms were then sorted based on similarity. If there were more than three initial piles, I asked participants to subsequently sort into three piles and then two piles as I wanted to see if the three-sex system would be reproduced and how genders in the third category, especially *kathoey,* would be categorized as males or females. There were 37 participants.

References

Altman, D. (2001). *Global sex*. Chicago: University of Chicago Press.

Bao, J. (2005). *Marital acts: Gender, sexuality, and identity among the Chinese Thai Diaspora*. Honolulu: University of Hawaii Press.

Barmé, S. (2002). *Woman, man, Bangkok: Love, sex, and popular culture in Thailand*. Lanham, MD: Rowman & Littlefield.

Bourdieu, P. (1990). *The logic of practice*. Stanford: Stanford University Press.

Butler, J. (2004). *Undoing gender*. New York: Routledge.

Connors, M. K. (2007). *Democracy and national identity in Thailand.* London, UK: NIAS Press.

Foucault, M. (1990). *The history of sexuality:* Vol. 1. *An introduction.* New York: Vintage.

Herdt, G. (Ed.). (1996). *Third sex, third gender: Beyond sexual dimorphism in culture and history*. New York: Zone Books.

Jackson, P. (1997). *Kathoey* >< gay >< man: The historical emergence of gay male identity in Thailand. In L. Manderson & M. Jolly (Eds.), *Sites of desire, economies of pleasure: Sexualities in Asia and the Pacific* (pp. 166–190). Chicago: University of Chicago Press.

Jackson, P. (1999a). An American death in Bangkok: The murder of Darrell Berrigan and the hybrid origins of gay identity in 1960s Thailand. *GLQ: A Journal of Lesbian and Gay Studies, 5*, 361–411.

Jackson, P. (1999b). Tolerant but unaccepting: The myth of a Thai "gay paradise." In P. Jackson & N. Cook (Eds.), *Genders and sexualities in modern Thailand* (pp. 226–242). Bangkok, Thailand: Silkworm Books.

Jackson, P. (2000). An explosion of Thai identities: Global queering and re-imagining queer theory. *Culture, Health & Sexuality, 40*(4), 405–424.

Jackson, P. (2003). Performative genders, perverse desires: A bio-history of Thailand's same-sex and transgender cultures. *Intersections: Gender, History and Culture in the Asian Context, 9*. Retrieved June 6, 2004, from http://intersections.anu.edu.au/issue9/jackson.html

Jackson, P. (2010). The ambiguities of semicolonial power in Thailand. In R. Harrison & P. Jackson (Eds.), *The ambiguous allure of the west: Traces of the colonial in Thailand* (pp. 37–56). Hong Kong: Hong Kong University Press.

Jackson, P. A., & Sullivan, G. (Eds.). (1999). *Lady boys, tom boys, rent boys: Male and female homosexualities in contemporary Thailand*. New York: Routledge.

Käng, D. B. (2011). Paradise lost and found in translation: Queer media loci in Bangkok. *GLQ: A Journal of Lesbian and Gay Studies, 17,* 169–191.

Käng, D. B. (2012). Kathoey "in trend": Emergent genderscapes, national anxieties, and the re-signification of male-bodied effeminacy in Thailand. *Asian Studies Review, 36*, 475–494.

Loos, T. (2006). *Subject Siam: Family, law, and colonial modernity in Thailand.* Ithaca, NY: Cornell University Press.

Milk, M. B. (1999). *Thai women in the global labor force: Consuming desires, contested selves.* New Brunswick, NJ: Rutgers University Press.

Morris, R. (1994). Three sexes and four sexualities: Redressing the discourses of gender and sexuality in contemporary Thailand. *Positions: East Asian Cultures Critique, 2,* 215–243.

Morris. R. (1995). All made up: Performance theory and the new anthropology of sex and gender. *Annual Review of Anthropology, 24*(1), 567–592.

Morris, R. (1997). Educating desire: Thailand, transnationalism, and transgression. *Social Text, 15*, 53–79.

Muecke, M. (1984). Make money not babies: Changing status markers of Northern Thai women. *Asian Survey, 24*, 459–470.

Mulder, N. (1997). *Thai images: The culture of the public world*. Chiang Mai, Thailand: Silkworm Books.

Peletz, M. (2006). Transgenderism and gender pluralism in Southeast Asia since early modern times. *Current Anthropology, 47*(2), 309–340.

Pramoj Na Ayuttaya, P. (2008, September 14). *The variety of kathoey in Thai society*. Unpublished paper presented at "*Kathoey* are not women and don't have to be as beautiful as a pageant contestant" seminar, Chiang Mai.

Rubin, G. (1975). The traffic in women: Notes on the political economy of sex. In R. R. Reiter (Ed.), *Toward an anthropology of women* (pp. 157–210). New York: Monthly Review Press.

Rubin, G. (1984). Thinking sex: Notes for a radical politics of sexuality. In C. S. Vance (Ed.), *Pleasure and danger: Exploring female sexuality* (pp. 267–319). Boston: Routledge & Kegan Paul.

Sinnott, M. (2000). The semiotics of transgendered sexual identity in the Thai print media: Imagery and discourse of the sexual other. *Culture, Health & Sexuality, 2*(4), 425–440.

Sinnott, M. (2004). *Toms and Dees: Transgender identity and female same-sex relationships in Thailand.* Honolulu: University of Hawaii Press.

Sinnott, M. (2007). Gender subjectivity: Dees and Toms in Thailand. In S. Wieringa, E. Blackwood, & A. Bhaiya (Eds.), *Women's sexualities and masculinities in a globalizing Asia* (pp. 119–138). New York: Palgrave Macmillan.

van Esterik, P. (2000). *Materializing Thailand.* Oxford, England: Berg.

Wilson, A. (2007). *The intimate economies of Bangkok: Tomboys, tycoons, and Avon ladies in the global city.* Berkeley: University of California Press.

Winichakul, T. (1994). *Siam mapped: A history of the geo-body of a nation.* Honolulu: University of Hawaii Press.

Winichakul, T. (2000). The quest for "siwilai": A geographical discourse of civilizational thinking in the late nineteenth- and early twentieth-century Siam. *Journal of Asian Studies, 59*(3), 528–549.

Winichakul, T. (2010). Coming to terms with the West: Intellectual strategies of bifurcation and post-Westernism in Siam. In R. Harrison & P. Jackson (Eds.), *The ambiguous allure of the west: Traces of the colonial in Thailand* (pp. 135–151). Hong Kong: Hong Kong University Press.

Introduction to Reading 14

Sociologist Cenk Özbay used participant observation and interviews to understand the world of "rent boys" in contemporary Istanbul, Turkey. Özbay speaks Turkish and created a "careful ethnographic plan" to move with relative ease in the bars where he observed interactions among rent boys, gay men, and transvestites. Rent boys come from poor neighborhoods in the outlying areas of Istanbul, neighborhoods called *varoş*. Ranging in age from 16 to 25, rent boys are heterosexually identified and engage in compensated sex with gay men. The key concept in this reading is "exaggerated masculinity." Özbay examines the ways rent boys perform exaggerated masculinity as a strategy for dealing with their sexual interactions with gay men. Be sure to read the footnotes at the end of this article, because they provide valuable details about rent boys.

1. What are the tactics used by rent boys to maintain their masculine identities vis-à-vis gay men? How does *varoş* play a role in these tactics?
2. Discuss the risks faced by rent boys in their construction and maintenance of exaggerated masculinity.
3. What is global gay culture, and how does it shape the enactment of exaggerated masculinity by rent boys?

Nocturnal Queers

Rent Boys' Masculinity in Istanbul

Cenk Özbay

Recently, 'rent boys'[1] have become increasingly visible in the queer social spaces of Istanbul. Rent boys engage in different forms of compensated sex (Agustin, 2005) with other men. They construct their masculine identities through their clandestine homoerotic involvements. They invent and practice an embodied style that I call 'exaggerated masculinity'[2] in order to mark their manly stance and

Özbay, C. (2010). Nocturnal queers: Rent boys' masculinity in Istanbul. *Sexualities, 13*(5). Reprinted with permission of SAGE Publications, Inc.

deal with the risks that same-sex sexual activities pose for the reproduction of their masculine selves. In this article, I examine how these heterosexually identified rent boys assemble and perform exaggerated masculinity in order to negotiate the tensions between their local socially excluded environments and a burgeoning Western-style gay culture[3] while they conduct their 'risky' sexual interactions with other men.

Male prostitution takes place in different social settings around the world across a wide diversity of class, racial, cultural, and organizational arrangements (see, for example, Aggleton, 1999; Dorais, 2005; Fernández-Dávila et al., 2008; Hall, 2007; Jackson and Sullivan, 1999; McNamara, 1994; Minichiello et al., 2001; Mujtaba, 1997; Schifter, 1998; West, 1993). . . . By studying male prostitution we can gain insight into the social dynamics behind how dissident sexualities are experienced and interpreted in the margins of hegemonic masculinities. In this article, I aim to make a contribution to the gap in the field of compensated sex between men of different social classes who embody distinct masculinities in . . . non-Western sexual geographies by using the Istanbul case in which a number of southern sexual cultures such as the Mediterranean, East European and Islamic meet and interface (Bereket and Adam, 2008; Tapinc, 1992). . . .

Rent boys come from lower-class neighborhoods in the outskirts of the Istanbul metropolitan region called *varoş*—a term . . . similar to the Brazilian 'favela' (Goldstein, 2003) and the French 'banlieue' (Wacquant, 2008). Rent boys (aged between 16 and 25) are mostly sons of the recently migrated large families that have coped with dislocation, poverty, and cultural exclusion. They speak Turkish with different regional accents, which show their symbolic marginalization and lack of cultural capital. Rent boys self-fashion their masculinity to produce a niche for themselves within a highly stratified, increasingly hegemonic gay culture in Istanbul. This self-fashioning via the embodied, stylized, continuously refined exaggerated masculinity operates through an 'outsider within' (Collins, 1986) position amongst self-identified gay men in Istanbul.

Varoş boys narrate a story . . . of their 'real' selves while they strive to become rent boys, which they claim is a temporary and transitory position. Exaggerated masculinity is a critical part of this construction in the context of male prostitution. *Varoş* boys transform themselves to achieve the rent boy identity through a discursive process, in which they reiterate the rules and characteristics of being a rent boy, and through a bodily process in which they learn and do exaggerated masculinity. . . .

In . . . Istanbul . . . *varoş* is a highly marginalized social identity regarding the mainstream culture of the middle classes. When they attempt to enter the spaces of the Western-style gay venues in Istanbul, *varoş* boys are discriminated against and rejected in terms of their alterity to the . . . modern, urban, and liberal lifestyles that middle classes have long adopted.

'Rent boy' emerges in the liminal space between the *varoş* identity and the local reflection of the global gay culture: A rent boy neither becomes gay nor stays as *varoş*. Rent boys animate a dynamic process of cultural hybridization and theatrical displays of exaggerated masculinity as a response to double marginalization. While they strategically use their *varoş* backgrounds to underline their masculinity and consolidate their authenticity in order to attract gay men who are supposed to have a fantasy of having sex with heterosexual men, they concomitantly take advantage of their encounters with middle-class gay men and empower themselves in their *varoş* environments. In this sense, the agility of the identity of rent boy permits its subjects to be enriched and strengthened in the symbolic hierarchies that they face in both *varoş* and gay cultures. Masculine embodiment and its deliberate and nuanced uses become crucial in rent boys' symbolic and material culture. . . .

Varoş as Culture and Identity

After the 1980 military coup, neoliberal reforms in Turkey transformed both Istanbul's position within the country as well as its own socio-spatial organization. The population in Istanbul has multiplied almost four times and recently approached 12 million people. Urban segregation and social fragmentation escalated and reshaped Istanbul as a space of contestation in which previously silenced social groups including Islamists, Kurds, and queers claimed legitimacy and public visibility (Kandiyoti, 2002; Keyder, 1999, 2005).

Varoş was one of the names given by the middle-class . . . citizens of Istanbul . . . to the illegal squatter settlement neighborhoods around the city and to the migrant people who built houses and worked in the temporary jobs in the informal sector (White, 2004). . . . *Varoş* became synonymous with a regressive, 'pre-modern' subjectivity that is abject and disenfranchised.

In the 1990s, the term *varoş* started to designate urban poverty instead of backwardness and rurality while people living in *varoş* areas were increasingly identified as the 'threatening Other' (Demirtas and

Sen, 2007; Erman, 2004). *Varoş* was constructed as a space where fundamental Islamism, Kurdish separatism, illegality, criminality, and violence met. Through media representations, *varoş* was otherized in terms of culture, economy, ethnicity, and politics. . . . At the same time, inhabitants of *varoş* reclaimed and appropriated the word as a way to identify their own cultural position distinctly from the Istanbulite. . . .

Rent boys are the children of *varoş*. They tactically constitute their identities as *varoş* to underline their differences from their gay clients not only in terms of sexuality but also in terms of class position. In this sense, being *varoş* refers to an embodied cultural difference as well as certain gendered meanings regarding masculinity. Rent boys repetitively state that they are 'real' men because they are coming from *varoş*. In this way, *varoş* is naturalized and linked to an inherent masculinity that gay men do not (and cannot) have. In other words, *varoş* becomes a sign of an uncontaminated, natural, physical, and authentic masculinity, while gay stands for feminine values and norms such as culture, refinement, and cleanness. In a symbolic order of masculinity, *varoş* boys turn to be 'naturally' and unchangeably masculine while gay men's bodies represent a modern, inauthentic, and imperfect masculinity.

Tactics of Masculinity

In addition to the symbolic significance of *varoş* in creating a 'naturally' virile character, rent boys also employ tactics to maintain their masculine identities vis-à-vis gay men. The most important strategy is being 'top only.' Thus, rent boys claim that they engage sexually with other men only when they play the top (active) role.[4] Protecting their bodies from penetration and becoming sexually available only as tops allow rent boys to reclaim their incontestably masculine identities. The gender of their sexual partner does not make a real difference either for their sexual repertoire or for their erotic subjectivities (for a similar situation among Brazilian male prostitutes, see Parker, 1998).

Another way that rent boys secure their masculinity is their heterosexualizing discourse. When they talk, rent boys position themselves in relation to an imagined girlfriend, fiancé, or long-term lover to-be-married with whom they have ongoing emotional and sexual affairs. When challenged, this discursive heterosexuality enables rent boys to prove their 'real' heterosexual identities. In order to distinguish themselves from gay men and to buttress their masculinity, rent boys also humiliate and denigrate gay men. It is important to note that rent boys' homophobia is, in most cases, a performative 'utterance' (Butler, 1993) to help maintain their masculine identities. It does not really prevent them from mingling, negotiating, and having sex with gay men in other situations.

Masculinity has always been a contested subject in the construction of queer sexualities in Turkey (Bereket and Adam, 2006; Hocaoglu, 2002; Özbay and Soydan, 2003; Tapinc, 1992; Yuzgun, 1986). However, rent boys' 'top only' positions and homophobic utterances are only one aspect of the exaggerated performances of masculinity. Different than the archetypical macho sexual pose of Latin America (Lancaster, 1994) rent boys do not brag about their sexual escapades with gay men. Instead, they have an evasive manner about their queer sexual practices. In addition to homophobia, the silence of rent boys about their homosexual involvements coincides with the tradition of the strict separation of intimate affairs from public sphere in some Muslim societies as Murray calls it 'the will not to know' (Murray, 1997). . . .

Within the framework of interpenetrating western gay culture and local constellations of gender and sexuality, masculinity matters for rent boys and gay men on another level: the appeal of passing and acting straight (Clarkson, 2006). Gay men in Istanbul have an increasing obsession with the 'straight-acting' and 'straight-looking' self-presentation, which demands a certain degree of heterosexual masculinity for erotic engagement. This fetishism for the 'more masculine' attributes and bodily gestures contributes to a hierarchy in which feminine qualities, as in effeminate men, are deemed inferior and unwanted, while masculine traits are presented as rare, desired, and superior. The negative attitude toward effeminacy and the desire for more masculine attributes contribute to an exaggerated masculinity to prevail as the 'most masculine,' and thus craved, in the gay culture in Istanbul. Rent boys take advantage of this erotic climate and relocate themselves in the eyes of their potential clients. Put in other words, rent boys convert their erotic and sexual positionalities into social and economic capital through their use of the encounter and desire between different masculinities.

The Interplay of Multiple Masculinities

Since gender is conceptualized as a continual 'doing' rather than as a natural 'being' (Butler, 1999: 25; West and Zimmerman, 1987), gendered subjectivities are constituted through 'the repeated stylization of the body, a set of repeated acts within a highly rigid regulatory frame that congeal over time to produce the appearance of substance, of a natural sort of being' (Butler, 1999: 33). Gendered subjectivity comes into being via the constellation of bodily performances within the 'regulatory frame' of the heterosexual matrix. Rent boys subvert their regular and 'normal' heterosexual script with male prostitution while they simultaneously try to re-stabilize it by enacting exaggerated masculinity—a style that requires well-defined gendered performances before different audiences. The omnipresent sense of risk inaugurates the possibility for the exaggeratedly masculine identity to be questioned and imperiled. In this sense, the rent boy's masculinity is a[n] . . . insecure subject position that needs to be repetitively asserted and proven while it continuously introduces new risks to be contemplated by rent boys in order to achieve their heterosexual and masculine status.

In her seminal works, Raewyn Connell (1987, 1995) demonstrated that multiple masculinities coexist and interact in a society at any given time. The encounter and dialogue between a *varoş* boy and a middle-aged upper-class gay man might be seen as a manifestation of what Connell terms the relations between divergent masculinities. These relations ought to be seen through the prism of power. In this sense, the culturally exalted hegemonic masculinity brings complicity, subordination, intimidation, and exploitation into relations between different masculinities. The exclusion of same-sex desire is critical for the constitution of hegemonic masculinity (Connell, 2000: 83). As a model, an ideal, or a reference point, hegemonic masculinity—in relation to the heterosexual matrix—affects all other ways of being a man including its imitations (as in rent boys) as well as the resistant or alternative versions (as in queer masculinity).

In the eastern Mediterranean region, configurations of masculinities take shape between the westernizing influences of modernity and the history of Islamic culture and tradition (Bereket and Adam, 2008; Ghoussoub and Sinclair-Webb, 2000; Ouzgane, 2006). The case of rent boys in Istanbul is not an exception. . . .

Locating Rent Boys

Place: Taksim Square.

Time: Any evening, especially after 10 p.m.

The crowded Istiklal Street, which is a major promenade connected to . . . Taksim Square, is full of intermeshing people from all classes, ages, genders, ethnicities, religions, sexualities, and cultures representing Istanbul's social diversity. Among the carnivalesque crowd an attentive eye can notice some young men walking or leaning against walls, checking the passerby. It is obvious for these attentive eyes that these young men, who carefully prepared themselves for the peak hours, reciprocate with curious gazes that can speak the same language of the looker. Around midnight these young men suddenly disappear from the street. Now, it is . . . bar time.

After paying the entry fee (around $10) I enter Bientot, the most famous and much-frequented club of rent boys in Istanbul. Bientot is very close to the vivid Istiklal Street, near a well-known transgender dance club and the only gay bathhouse of the city. Bientot, like two other similar bars, is a 'limitative and disciplining' (Hammers, 2008) space in the sense that types of people (i.e., rent boys, transvestites, clients) are set, their roles are prescribed (i.e., who dances, who looks, who buys drinks), and interactions between visitors are stabilized (i.e., negotiations, flirting, cruising, kissing). . . . Gay men (whether clients of rent boys or not) told me that they do not 'have fun' in Bientot as they do in other gay bars and they come here just to see or talk to the *varoş* boys only in the predefined ways that are available to them.

Bientot is full of its . . . frequenters: Several single gay men from all ages, some mixed-friend groups, several transvestites, and more than 70 rent boys. In general, everyone seems to know each other. Everybody except rent boys drinks and rests on the walls surrounding the dance floor enjoying music (popular Turkish pop songs of the day) while most of the rent boys dance in a unique style without drinking unless a client is generous enough to buy them one. . . .

Here is a quotation from my field notes immediately after arriving home from Bientot:

> A shocking place . . . High volume of music, really bad ventilation, the smell of alcohol, the smell of sweat, the smell of cologne, the smoke from cigarettes . . . You can't escape from the piercing looks into your eyes. These looks are masculine, you can tell, but they are also very inviting and flirtatious, which contradicts . . . the assertive masculinity. The dancing bodies are very close to each other. They are very straight looking like the ordinary boys at the street; but, on the other hand, the male-to-male intimacy of the dance destroys the desired heterosexual ambiance. It seems like they are straight boys in a gay club, dancing together passionately. . . .

Playing With Fire: Elements of Rent Boys' Style

A weekly TV show filmed the gay bathhouse near Bientot with hidden cameras in early 2005. After recording each possible 'proof' of male prostitution (including negotiations for prices and actions) the programmers tried to talk with the manager of the bathhouse about the organization . . . while he kept refuting that he hired rent boys. During the interview the camera focuses on a young rent boy, half naked in his towel, arguing angrily with another one about the recruitment of new rent boys that they already knew. He said 'I told you don't bring everyone here from your neighborhood. Look at me. I only bring my brother. You may have a fight with one of them in the future and he can go and tell people, including your father, what you do here. You are playing with fire. I told you this before. Don't play with fire.'

. . . This warning against 'playing with fire' is neither unique to this rent boy nor restricted within the walls of the bathhouse. It offers a useful framework to better comprehend a rent boy's unceasing physical and social negotiations with other rent boys, gay men, and transvestites. Rent boy is a conditional and fragile identity. It surfaces between the contradictory discursive and sexual practices, which subvert the line between homo and heterosexuality. It is a contingent performance that links the *varoş* culture of Istanbul and the ostensibly global gay life-style. It is an interplay of competing working- and upper-middle class meanings and signifiers. Through the incessant play of risk taking, a rent boy invests his heterosexuality as well as his social position and kinship networks which are likely to be harmed by an undesired disclosure, as the rent boy quoted earlier fears.

Here, I follow Agustin's (2005: 619) proposition to define and study prostitution, sex work, and compensated sex as a 'culture' to expose the previously under-researched links with systems of inequality and the production of social meaning. Wright (2005: 243) also highlights the 'percolation of queer theoretical concerns' and 'an array of cultural studies interventions' into the sociology of masculinities in order to pose new questions on masculine performances, cultural practices, and 'engenderment' processes that men undertake through the routes of non-hegemonic masculinities in diverse settings. Hence, I frame exaggerated masculinity as a product of the culture of rent boys in Istanbul. Rent boys learn, practice, and transform exaggerated masculinity through the mechanisms of social control and self-governance. The process of the construction and reconstruction of exaggerated masculinity is constantly under risk of disappointment and failure.

. . . Risk appears three-fold in animating exaggerated masculinity by rent boys. First and foremost, rent boys' involvement with male prostitution should not be revealed to their friends, family, and extended relatives. Otherwise, they cannot sustain their ordinary lives as young, decent, and respected members of their community. On the other hand, while the rent boy reproduces *varoş* culture as corroboration to his 'natural' masculinity, he should also play with and transmute it symbolically in order to have a subject position within gay culture instead of being abjected. So, the second risk is . . . a nuanced middle space between the two unwanted identities that a rent boy must navigate carefully: staying as an unmodified *varoş* or becoming (too) gay. While connecting closely with gay men, rent boys' third risk is about protecting their heterosexuality. Although rent boys have sex with gay men, they are not supposed to have a gay identity. In sum, a rent boy has to control meticulously and manage risk regarding his bodily acts, behaviors, and relations with other people in order not to be exposed while balancing between the discrepant meanings of *varoş* and gay positions. In this framework, I will now outline the elements of how rent boys sustain exaggerated masculinity through their risk-taking activities and their entanglements with different segments of the culture of male sex work in Istanbul.

The Body

The first point of risk that rent boys take into consideration focuses on their bodies. Almost all rent boys have athletic or skinny bodies. . . . They think that their gay clients like their bodies as skinny, fatless, and 'toned' and not over-muscular and 'hung.' They also believe that they look younger this way. . . . Hakan (aged 22) says, 'body is everything we have in this job, of course we need to take care of it.' Rent boys have a certain tension around their bodies in order to keep them in good condition, to seem young(er), and not to lose their virility through developing an over-muscled look.

Appearance

Another significant issue in the material culture of rent boys is about what they wear and how they look.

Most rent boys wear denim jeans. They almost never wear shorts even when it is unbearably hot and humid in Istanbul. Burak (aged 18) states that 'real men never wear shorts, jeans are the best.' For their upper parts they commonly opt for white. 'White is better because it looks more attractive when you are tanned. Also, it shines in the dark bar and makes you more visible,' says Arda (aged 23). Black tops are also very popular for their taste because it is deemed to be more masculine and mature. They also wear some bright and lively colors like red and yellow to be seen in the bar. . . .

Rent boys do not wear earrings as Okan (aged 18) told me, 'Earrings would harm masculinity.' They are more tolerant toward wristbands, chains, and rings, but earrings are identified with gays and/or foreigners. . . . Rent boys in Istanbul insist on wearing sport shoes and sneakers even on snowy days. . . .

Perfume

Perfumes and colognes are significant manifestations of rent boys' risky relations with their gay clientele. It is always good for a rent boy to have the fragrance of a charming perfume because it increases his attractiveness when his client has to whisper into his ear in the noisy bar. Perfumes are very expensive for rent boys' budgets but sometimes they receive perfumes after satisfying a client with their sexual performance. . . . As Burak told me, 'if you smell [of] perfume, it shows that you recently got some work done.'

The risky point is . . . the gendered quality of the fragrance. The fragrance must smell masculine because otherwise it cannot contribute to the exaggerated masculinity of the rent boy. . . . On the other hand, a client who uses a very masculine perfume for himself endangers a rent boy's masculinity because it implies that the gay client was not a feminine man and he could turn active in sexual penetration. . . .

Most of the rent boys that I talked to said that they were against stealing or any other kind of criminal activity. On the other hand, they also revealed that they were not against asking for or even stealing perfumes from gays' houses after they have sex. Perfumes clearly are the exception for rent boys' moral stance against stealing. Hasan (aged 24) says 'when I see a nice perfume I ask for it. Honestly, if he does not want to give it to me I will try to take it anyway. I don't think this is stealing.' Hakan also noted that 'I am not interested in anything else, but if he has a nice perfume I will take it . . . He can buy another bottle easily and I will smell nice. Good deal.' Murat (aged 23) states that 'perfume is a connection between the Rich's life and mine. I can take it, I can use it and when I smell it I remember what I did and I enjoy about it. It makes my life more beautiful.'

Dance

Dance is another risky subject in the context of male prostitution in Istanbul. Rent boys have to dance in the bar in order to be seen by clients. The particular motions and gyrations of the boys' dancing give the impression that they are carrying out a predefined script of performing a task, but not reflecting pleasure by moving in a relaxed manner with the music and the rhythm. In other words, when rent boys dance, they perform another requirement of their work. Their dance is never visibly homoerotic although their bodies are pretty close and sometimes touch each other. It has its own sense of humor: If a rent boy puts himself at the back of another, the one in the front bounces in sudden panic—in an anxiety to save his back (his bottom). It manifests a rigid top-bottom code concerning the control and defense of your own back and a constant search to attack the others' backs.

If a rent boy oversteps the boundary of touching another's back or exhibits signs of pleasure, other rent boys explicitly disapprove the act and call him 'pervert,' *ergo*, humorous pleasure that comes from sodomizing others should be limited to activity with gays and not with other rent boys. In the bar, this is the main reason behind physical fights amongst rent boys. Thus, bodily humor is dangerous to play with, although avoiding it brings social exclusion because a rent boy ought to dance. He needs to 'show' in order to charm his audience. A motionless rent boy renders himself invisible, which seriously reduces his chances of finding a client. Anil (aged 20) says, 'dancing is the moment where we get the gays. We attract them when we dance. They love to watch us.'

Most importantly, a rent boy has to dance without looking feminine. Okan says, 'it is better not to do it [dance] if you do it like a girl.' Riza (aged 24) told me 'you should not shake your ass like a belly dancer. Arms and legs must be straight. Gaze is also important.' There are strict performative codes that most rent boys obey to protect the masculine image during the dance: The body should not be curved or shaken too much and it must repeat the same rough movements without flexibility. It must show strength. Shoulders and arms should be kept wide open, the waist should move only back and forth, imitating the sexual act of penetration. Dance is controlled and regulated by the surveillance of other rent boys. As long as they can perform it according to the unwritten rules of exaggerated masculinity, dance guarantees rent boys' masculine identities and makes them the center of attraction before potential clients.

Friends

As I recounted earlier . . . taking part in male prostitution or being seen while cruising is very risky. . . . Concealment paralyzes friendship mechanisms amongst rent boys. Most of the time, they come to the bars or other cruising places alone or at most in the company of one other rent boy, who is supposed to be trustworthy (mostly one's kin, for example a cousin). They usually know other rent boys personally and they have an intimate network of gossip and information exchange. They also spend time together chatting and dancing in the bars, but they always wind up alone while working or cruising. The solitude of rent boys might be seen as a tactic to increase their chance of negotiation for higher prices or as a part of the tradition of mendacity about what they do for how much. It actually protects them from unwanted rumors and from the dangers of unexpected disclosure. Can (aged 21) elaborates that 'I know some people in the bar, some other "rents" but I never see them out of the bar. Nobody knows that I am coming here in my neighborhood. I must be very careful. When my regular friends ask I tell them I will hang out with my cousins.' Mert (aged 24) adds, 'If you go out together he [a friend] can say that Mert let the guy fuck him, Mert turned bottom, etc. If he won't say it today, he will say it tomorrow. This is how it works. So, it is better to be alone instead of dealing with gossip and lies.'

Another point that poses a risk to the exaggerated masculinity is about emotions and sexual attraction between rent boys. In order to sustain fraternal heterosexuality, homoeroticism must be tamed and eliminated (Connell, 2000; Sedgwick, 1986). In male prostitution, who is feminine (gay) and who is masculine (rent boy) is rigidly defined. For rent boys, intimate relations are allowed only between these distinct gendered groups and not within them. Therefore, the possibility or manifestation of any kind of affect, eroticism, or sexuality between rent boys subverts their masculine positions as well as their 'natural' heterosexuality. Just like the uneasiness when they dance together, the risk of emotional and bodily intimacy as well as the ways it might be talked about create a certain tension and prevent rent boys from becoming further attached to each other.

Drinks and Drugs

. . . Drinking alcoholic beverages in the bar is a vital chance to look like an adult and demonstrate toughness. Soft drinks and soda are not preferred because they look juvenile and gentle. Beer is the drink that rent boys consume mostly because it is the cheapest and the most masculine beverage. . . . Beer is easy to drink while dancing, and more importantly, it does not make one drunk easily. Alcohol is a very risky issue just like drugs. Mixing different beverages, drinking tequila shots fast, or taking drugs can make a rent boy dizzy—sometimes almost unconscious. Emre (aged 25) notes 'gays try to make you drunk by buying you many drinks. They want to use you when you are drunk. If you are new here they can easily entrap you. You can have sex for no money, or worse things can happen.' These 'worse things' that Emre notes may lead to losing the masculine pose and roughness, which was carefully constructed. . . .

Transvestites

My framing of risk for rent boys' exaggerated masculinity includes their multifaceted relations with transvestite sex workers. Almost all the rent boys that I talked with have had sexual experiences with transvestites. A rent boy and a transvestite can become friends, sexual partners, and even lovers. The stories told about rent boys and transvestites range from scandals such as a drunken rent boy who was raped by a transvestite to some poignant love stories. Despite the fact that they are in two different sides of sex work, neither rent boys nor transvestites pay to have sex with each other. As Aykut (aged 25) says 'we are free for them, they are free for us. For all the rest, only money talks.'

While transvestites enjoy the young virility and 'real' masculinity of rent boys, the latter are happy to show how masculine and sexually active they are by having sex with the 'girls.' In most cases, a transvestite mentors an inexperienced rent boy and she teaches him how to have good sex. Although it seems a mutually satisfying relationship, these escapades with transvestites are indeed very risky for rent boys. Transvestites can easily ridicule a rent boy for not having a sufficiently large penis or for not achieving a fulfilling sexual performance. Emir (aged 20) said, 'I saw many guys like this. Everybody knows that they ejaculate really fast or it [the penis] is really small because one of the girls talked about it. They can still convince some clients, especially tourists, but it is more difficult to find a client for them.' Such a public display of physical or sexual insufficiency would permanently destroy a rent boy's sources of masculine respectability and reputation.

Safe Sex

The last component of what I conceptualize as risk for rent boys' construction of the exaggerated masculinity, is about 'sexual risk' (Fernández-Dávila et al., 2008) and bodily health. All the rent boys that I conducted interviews with had knowledge about STDs, HIV, condoms and how to use them. Nevertheless, my conversations with both rent boys and their clients testify that rent boys have a certain disinclination and resistance to concede their vulnerability and to use a condom during sexual intercourse. They prefer to have *doğal* (natural) or çiplak (naked, without a condom) sex especially when the client asks for or pays more. Ilker (aged 19) told me 'I use it [a condom] sometimes. It does not really bother me. I prefer cleaner gays so it is not a big threat for me. I know many rents do it without condoms with tourists because they pay more. It is crazy because there is a higher chance for a foreigner to be sick.' Their negative attitude might originate from the practical difficulty to use condoms, or as more likely, the construction of their masculine self-identities rejects expressions of fear and protection while it promotes courage and adventure. Rent boys interpret the sexual encounter as an opportunity to challenge and prove their manhood; as Ozgur (aged 22) says, 'little boys might get scared of it, but for me, it is not the case. I know how to fuck a guy without a condom in a safe way. It is not necessary for me to put one on. I can protect myself.' Also, some clients opt for unprotected sex with younger rent boys whom they believe do not have a long sexual history and are thus 'cleaner'. On the other hand, Burak mentioned, 'Probably because I am younger they ask my age and how many times I did it [having sex]. Then, they say "it is OK with you, you are clean" and I don't put a condom on. That's what they want.' Therefore, rent boys' desire to demonstrate their courage and fearlessness operates along with some clients' demands for unprotected sex and produces a risky and dangerous encounter for both sides.

Concluding Remarks

In this article, I have explained how rent boys in Istanbul have developed cultural, bodily, symbolic, and material strategies both to challenge tacitly and to negotiate inventively with the social norms of hegemonic male sexuality (Plummer, 2005) and hegemonic masculinity (Connell, 1995). The 'top only' sexual positions whereby they make themselves sexually available, the protection of their bodies from penetration, and the distance they place between themselves and feminine connotations by the way they dance, smell, or dress, can be seen as attempts to save the penis-and-penetration-centered hegemonic virile sexuality. On the other hand, the enactment of exaggerated masculinity and the production of a story of authentic manhood via *varoş* culture are manifestations of their complicity in the hegemonic forms of masculinity in Istanbul despite their dissident sexual practices that contradict these narratives.

Is it acceptable for the embodiments of hegemonic masculinity, or its imitations, to operate alongside queer sex? Is it possible for one to reclaim his privileged heterosexual status while he engages in compensated sex with other men? Gary W. Dowsett and his colleagues note that the definitions and conceptualizations in which masculinities have been theorized are in need of reconsideration and recalibration since 'the prevailing formulation of masculinity represents a failure to engage with the creative meanings and embodied experiences evident in non-hegemonic sexual cultures, and with the effects these meanings and experiences may generate beyond their boundaries' (2008:124). In this sense, rent boys and their ambivalent sexual acts and identifications provide an excellent case for such inquiries regarding their involvement with the active meaning-making process of sexuality and masculinity. As a response to the possible inquiries and challenges toward their heterosexual and masculine self-identities, they use exaggerated masculinity in order to be able to continue their everyday lives as heterosexual members of their families and kinship networks. In other words, exaggerated masculinity repairs and masks the subverting effects of compensated sex for rent boys' heterosexuality and makes them closer to the hegemonic ideal of masculinity. They perform an assiduous self-governance through symbols and implicit meanings vis-à-vis different and contradictory class positions, gender identities, and sexual acts.

Rent boys constitute exaggerated masculinity relationally and strategically at the nexus of contradictory contexts of the local *varoş* culture and the impact of the global gay culture. Risk is central in understanding the mechanisms of exaggerated masculinity since it is a fragile, insecure, playful combination of various bodily acts, gestures, and symbols. . . .

Notes

1. Rent boy (as in English) is the term my informants use for defining themselves. They say '*ben bir rent bovum'* (I am a rent boy) or just '*rentim'* (I am rent). Sometimes they prefer to say, '*parayla veya ücretli çikiyorum'* (I am going for money). I never encountered any other terms, either in the English versions or in Turkish translations, such as *erkek fahişe* (male prostitute), *seks işçisi* (sex worker) or *jigolo* (gigolo) used by my informants, their clients, or in the mass media. The subject of this article, rent boys, who are from the *varoş* segments of the city, is the only group of men who engage in compensated sex (receiving money or gifts) in the gay scene in Istanbul (Hocaoglu, 2002).

2. I prefer to describe the rent boys' stylized embodiment as 'exaggerated masculinity' in order to underline its theatrical, playful, performative, and decontexualizing characteristics. It is a constellation of learnt, imitated, calculated, and socially regulated displays of doing masculinity. There are other similar terms for such excessive masculine performances like hypermasculinity (Healey, 1996) or machismo (Gutmann, 1996) that are conceptualized in different webs of relations.

3. With the Western-style gay culture, I basically mean the emergence of men who call themselves gay (as in English) or sometimes *gey* in Turkish (Bereket and Adam, 2006) because they engage in sexual, erotic, and emotional relations with other men. There are many components of this culture including enclosed spaces called gay bars or gay clubs, and access to foreign or local websites with gay content for various purposes such as online dating. Before the emergence of the modern gay identity in Turkey, there were various sorts of same-sex sexual relations going on under different identifications and social organizations (Tapinc, 1992; Yuzgun, 1986).

4. Rent boys claim that they are 'top only' in order to insist they do not let their clients penetrate their bodies, while they can insert their penises into their clients' bodies through oral and anal sex. Rent boys also claim that they never touch their client's penises and they never let their clients caress their bodies. In addition to the 'top only' rule, some of the rent boys I talked to stated that they never kiss their clients from their mouths and some told me that they do not 'make out' with clients and delimit their sexual activities with oral and anal penetration (Özbay, 2005).

References

Aggleton, P. (Ed.). (1999). *Men Who Sell Sex: International Perspectives on Male Prostitution and HIV/AIDS.* Philadelphia, PA: Temple University Press.

Agustin, L. M. (2005). The cultural study of commercial sex. *Sexualities, 8*(5), 618–631.

Bereket, T., & Adam, B. D. (2006). The emergence of gay identities in Turkey. *Sexualities, 9*(2), 131–151.

Bereket, T., & Adam, B. D. (2008). Navigating Islam and same-sex liaisons among men in Turkey. *Journal of Homosexuality, 55*(2), 204–222.

Butler, J. (1993). *Bodies That Matter: On the Discursive Limits of 'Sex'.* New York: Routledge.

Butler, J. (1999). *Gender Trouble: Feminism and the Subversion of Identity.* New York: Routledge.

Clarkson, J. (2006). 'Everyday Joe' versus 'pissy, bitchy, queens': Gay masculinity on StraightActing.com. *The Journal of Men's Studies, 14*(2), 191–207.

Collins, P. H. (1986). Learning from the outsider within. *Social Problems, 33*(6), 14–32.

Connell, R. W. (1987). *Gender and Power: Society, the Person and Sexual Politics.* Stanford, CA: Stanford University Press.

Connell, R. W. (1995). *Masculinities.* Berkeley: University of California Press.

Connell, R. W. (2000). *The Men and the Boys.* Berkeley: University of California Press.

Demirtas, N., & Sen, S. (2007). *Varoş* identity: The redefinition of low income settlements in Turkey. *Middle Eastern Studies,* 43(1), 87–106.

Dorais, M. (2005). *Rent Boys: The World of Male Sex Workers.* Montreal: McGill-Queen's University Press.

Dowsett, G. W., Williams, H., Ventuneac, A., et al. (2008). Taking it like a man: Masculinity and barebacking online. *Sexualities, 11*(1/2), 121–141.

Erman, T. (2004). Gecekondu Çalişmalarinda 'Öteki' Olarak Gecekondulu Kurgulari. *European Journal of Turkish Studies* 1. URL (accessed 15 June 2010): http://www.ejts.org/document85.html.

Ghoussoub, M., & Sinclair-Webb, E. (Eds.). (2000). *Imagined Masculinities: Male Identity and Culture in the Modern Middle East.* London: Saqi.

Goldstein, D. M. (2003). *Laughter Out of Place: Race, Class, Violence, and Sexuality in a Rio Shantytown.* Berkeley: University of California Press.

Gutmann, M. C. (1996). *The Meanings of Macho: Being a Man in Mexico City.* Berkeley: University of California Press.

Hall, T. (2007). Rent-boys, barflies, and kept men: Men involved in sex with men for compensation in Prague. *Sexualities, 10*(4), 457–472.

Hammers, C. J. (2008). Making space for an agentic sexuality? The examination of a lesbian/queer bathhouse. *Sexualities, 11*(5), 547–572.

Healey, M. (1996). *Gay Skins: Class, Masculinity, and Queer Appropriation.* London: Cassell.

Hocaoglu, M. (2002). *Escinsel Erkekler.* Istanbul: Metis.

Jackson, P., & Sullivan, G. (Eds.). (1999). *Lady Boys, Tom Boys, Rent Boys: Male and Female Homosexualities in Contemporary Thailand.* New York: Haworth.

Kandiyoti, D. (2002). Introduction: Reading the fragments. In D. Kandiyoti & A. Saktanber (Eds.), *Fragments of Culture: The Everyday of Modern Turkey.* London: I.B. Tauris, 1–21.

Keyder, C. (Ed.). (1999). *Istanbul: Between the Global and the Local.* Boston, MA: Rowman & Littlefield Publishers.

Keyder, C. (2005). Globalization and social exclusion in Istanbul. *International Journal of Urban and Regional Research, 29*(1), 124–134.

Lancaster, R. (1994). *Life is Hard: Machismo, Danger, and the Intimacy of Power in Nicaragua.* Berkeley: University of California Press.

McNamara, R. P. (1994). *The Times Square Hustler: Male Prostitution in New York City.* New York: Praeger.

Minichiello, V., Mariño, R., Browne, J., et al. (2001). Male sex workers in three Australian cities: Socio-demographic and sex work characteristics. *Journal of Homosexuality, 42*(1), 29–51.

Mujtaba, H. (1997). The other side of midnight: Pakistani male prostitutes. In S. Murray & W. Roscoe (Eds.), *Islamic Homosexualities: Culture, History, and Literature.* New York: New York University Press, 267–274.

Murray, S. (1997). The will not to know: Islamic accommodations of male homosexuality. In S. Murray & W. Roscoe (Eds.), *Islamic Homosexualities: Culture, History, and Literature.* New York: New York University Press, 14–54.

Ouzgane, L. (Ed.). (2006). *Islamic Masculinities.* London: Zed.

Özbay, C. (2005). Virilities for rent: Navigating masculinity, sexuality and class in Istanbul. Unpublished MA thesis, Bogazici University.

Özbay, C., & Soydan, S. (2003). *Escinsel Kadinlar.* Istanbul: Metis.

Parker, R. (1998). *Beneath the Equator: Cultures of Desire, Male Homosexuality, and Emerging Gay Communities in Brazil.* New York: Routledge.

Plummer, K. (2005). Male sexualities. In M. Kimmel, J. Hearn, & R. W. Connell (Eds.), *Handbook of Studies on Men and Masculinities.* Thousand Oaks, CA: Sage, 178–195.

Schifter, J. (1998). *Lila's House: Male Prostitution in Latin America.* New York: Haworth Press.

Sedgwick, E. K. (1986). *Between Men: English Literature and Male Homosexual Desire.* New York: Columbia University Press.

Tapinc, H. (1992). Masculinity, femininity, and Turkish male homosexuality. In K. Plummer (Ed.), *Modern Homosexualities: Fragments of Lesbian and Gay Experience.* London: Routledge, 39–50.

Wacquant, L. (2008). *Urban Outcasts: A Comparative Sociology of Advanced Marginality.* Cambridge: Polity.

West, C., & Zimmerman, D. (1987). Doing gender. *Gender & Society, 1*(2), 125–151.

Wright, L. (2005). Introduction to 'queer' masculinities. *Men & Masculinities, 7*(3), 243–247.

Yuzgun, A. (1986). *Escinsellik.* Istanbul: Huryuz.

Introduction to Reading 15

Sabine Lang is a German cultural anthropologist who conducts fieldwork on gender and sexuality among the Dine (Navajo) in New Mexico, the Shoshone-Bannock in Idaho, and urban Native Americans. Her reading addresses the critical importance of a nuanced understanding of Native American men-women, lesbians, and two-spirit people, situating them within the broader historical, cultural, and political context of Native tribal communities and paying attention to how Native American men-women, lesbians, and two-spirits create and maintain their identities and the worldviews of Native communities. The focus of Lang's reading on female-bodied people who take up a third gender status is a valuable contribution to the literature on non-binary sex/sexuality/gender systems.

1. Why does Lang prefer to use the terms woman-man and man-woman?
2. How do Shoshone men-women of the past and present express masculine characteristics and habits?
3. Why is it problematic to apply European American or Western definitions of homosexuality and heterosexuality to sexual relationships among Native Americans?
4. What roles do caretaking and motherhood play in the lives of Native lesbians/twospirit women?

Native American Men-Women, Lesbians, Two-Spirits

Contemporary and Historical Perspectives

Sabine Lang

Introduction

I came to the United States in the early 1990s as a young German ethnographer to study "third and fourth genders" in contemporary Native American communities. As I talked to members of the first generation of Native gay and lesbian activists (e.g., Randy Burns and Erna Pahe who had been co-founders of Gay American Indians), however, they frankly told me that rather than asking questions about "third-gender" people who have been dead for many decades, they expected me to listen to what they had to say about their present-day lives and concerns. In that context, they addressed not only issues of the ongoing historical trauma of colonization, violence, racism, HIV/AIDS, and homophobia both in U.S. society at large and Native communities, but also issues of decolonization and sovereignty, in much the same way more recent activists and scholars do (e.g., Driskill et al. 2011a, 2011b; Justice et al. 2010). While being suspicious of members of the anthropological profession, they nevertheless appreciated the work of individual non-Native scholars such as Williams and Roscoe, which they found useful in establishing a continuum between the "old-time" gender diversity and contemporary lesbian, gay, bisexual, transgender, and queer (LGBTQ) Native Americans. This continuum is important in two respects. On the one hand, it is part of the definition of a contemporary indigenous LGBTQ identity as "two-spirit people" combining feminine and masculine qualities, though not in a physical but rather in a spiritual sense. On the other, references to the once respected status of people who were not heterosexual and/or did not fit into the categories of "woman" or "man" can be helpful in gaining acceptance in contemporary Native communities.

In the most recent academic approaches to the two-spirit, many of whose proponents have Native roots, the term and concept undergoes still another change. "Two-spirit" becomes a key concept, a liberating icon in a larger political project: a radical vision of decolonizing not only indigenous communities but also the settler states/societies they live in, including the academy and research.[1] As is stated by Driskill et al. in their introduction to *Queer Indigenous Studies,* this project is about designing "practices and futures—both inside and outside the academy—that can both remember and create radical, decolonial LGBTQ2 [the "2" denotes two-spirit people, S.L.] indigenous communities. (. . .) The book invites looking to Indigenous genders and sexualities—both 'traditional' and contemporary—for their potential to disrupt colonial projects and to rebalance Indigenous communities. Beginning to articulate and practice specific LGBTQ2 Indigenous communities is a way of continuing radical movements and scholarship that work for collective decolonial futures" (2011b: 18). The unifying power of two-spirit as a collective, inclusive LGBTQ Native identify, as well as its potential in developing an indigenist approach to sexuality and gender, are also emphasized by authors who discuss two-spirit issues from the perspective of social work and health (e.g., Walters, Evans-Campbell, Simoni, Ronquillo, and Bhuyan 2006).

In the following, I will first briefly outline multiple gender systems in indigenous North American cultures, then the roles of female-bodied persons who took up the ways of men. This is followed by a discussion of definitions of homosexuality and heterosexuality in Native American cultures, and of the factors that shape the two-spirit identity. In the concluding section, I will give some glimpses of contemporary two-spirit women's lives, roles, and concerns. The data on which the present contribution are based include conversations I had with lesbian, gay, and bisexual two-spirit people, as well as with contemporary people who were raised to fulfill the special gender roles described in anthropological literature. Quite a few of the latter prefer to identify in terms of specific indigenous gender classifications rather than as "gay/lesbian," "queer," "transgender," or even "two-spirit." While anthropologists are criticized for focusing on "gender constructions unavailable to [contemporary, S.L.] Indigenous GLBTQ people who used the term Two-Spirit" (Driskill et al. 20llb: 14), such gender

Lang, S. (2016). Native American men-women, lesbians, two-spirits: Contemporary and historical perspectives. *Journal of Lesbian Studies, 20*(3-4): 299–323. Reprinted with permission from Taylor & Francis.

constructions were still available to some people who grew up on reservations in the 1950s and 1960s. In some local contexts on the reservations they are still viable today (cf. House, this issue). And still today, people who have grown up to be nádleeh, tainna wa'ippe or another gender other than woman and man are looked upon with utmost respect, and viewed as teachers and role models, in two-spirit communities. The same was true when I did my fieldwork in the early to mid-1990s. While there are not many of them, they have by no means vanished, which is why the present contribution begins with a discussion of their roles and the cultural contexts in which these roles were, and sometimes still are, embedded.

Men-Women and Women-Men: Some General Remarks

Greater honor was paid to her than to the Great chief, for she occupied the 1st place in all the Councils, and when she walked about, was always preceded by four young men, who sang and danced the Calumet to her. She was dressed as an Amazon; she painted her face and Wore her Hair like the men. (Gravier 1959 [1700]: 147–148; orthography as in the original.)

The account by the French Jesuit Missionary Jacques Gravier, from which the above quote is taken, is arguably the first description of a Native American femalebodied person in a masculine role. Being a member of the Houma tribe in what today is Louisiana, the "femme chef" (woman chief) had so distinguished herself in war that she was awarded the highest honors. She participated in the council meetings of the men and, like members of the nobility, was laid to rest in the temple of the Howna (Gravier 1959: 147). This "woman chief" exemplifies one among several categories of female-bodied people who assumed a status resembling that of men in indigenous North American cultures. Among the Houma and other tribes across the Subcontinent, it was not unusual for women to go to war. In rare individual cases remembered for generations, they would perform such extraordinary feats as warriors that they were assigned a quasi-male status in their respective society (cf. Lang 1998: 303–308; cf. also Hemmilä, this issue). Far more common and much less spectacular, however, were institutionalized statuses, usually based on occupational preferences, within gender systems that acknowledged gender variability or gender diversity, that is, "cultural expressions of multiple genders (i.e., more than two) and the opportunity for individuals to change gender roles and identities over the course of their lifetimes" (Jacobs and Cromwell 1992: 63–64). Such special statuses for people who were neither women nor men but genders of their own existed in a considerable number of pre-reservation and early reservation Native American tribal societies both for male and female individuals who chose to live—completely or partially—in the culturally defined role of the "other sex." Due to the massive impact of colonization, missionary work, and forced acculturation, these traditions of gender diversity, which often included same-sex relationships for reasons to be discussed below, have now largely disappeared in indigenous North American communities, to the extent that people will even deny their former existence (cf. Jacobs 1997: 25).

The first to give accounts of indigenous North American males in a woman's role and females in a woman's role were travelers, traders, and missionaries (cf. Hemmilä, this issue), followed by representatives of the then budding discipline of U.S. anthropology from the early twentieth century onward. In anthropological literature, the traditions of what is now referred to as gender variability in Native North America were formerly subsumed under the term "berdache." However, that word has fallen into disuse due to its original Arab/French meaning of male prostitute or "kept boys" (Angelino and Shedd 1955; Williams 1986a: 9–10), which has come to be viewed as inappropriate not only by contemporary LGBTQ Native Americans, but also by Native and Non-native anthropologists (cf. Jacobs, Thomas, and Lang 1997a). In addition, the term becomes downright absurd when applied to females in a man's role. In academic literature, it has been largely replaced by the term "two-spirit" from the mid-1990s onward as a result of two conferences on "The 'North American Berdache' Revisited," whose participants included Native gay and lesbian activists as well as non-Native anthropologists (Jacobs, Thomas, and Lang 1997a). In addition, the term has gained almost universal currency in LGBTQ Native American contexts and activism.

Like "berdache," however, "two-spirit," which originated at an intertribal gathering of LGBT Native Americans/First Nations people in the late 1980s, encompasses an entire host of identities and sexualities across history, ranging from the pre-reservation and early reservation systems of multiple genders to contemporary LGBTQ indigenous people. While the inclusiveness of the term has great advantages when it comes to uniting LGBTQ Native Americans/First Nations in supratribal political and HIV/AIDS organizing, I nevertheless prefer to use the terms-womanman (male in a woman's role) and man-woman (female in a man's role) to refer to people past and present who were, and are, neither women nor men.[2] As I learned in conversations with contemporary women-men and men-women who were raised on their people's reservation to fulfill their people's reservation to fulfill their special role, this is very much in line with how they view themselves, as has already been alluded to above. While they will agree to be subsumed under the label of "two-spirit" in intertribal contexts (e.g., at meetings of LGBTQ indigenous people from all over the United States and Canada), they will usually make a clear distinction between themselves and LGBTQ2 Native Americans/members of First Nations. They do neither identify as gay, lesbian, bisexual, or queer, nor as transgender, but will refer to themselves by the term appropriate for their specific gender status in their respective culture, such as *nádleeh* (Navajo), *tainna wa'ippe* (Shoshone), *dubads* (Paiute), and so on.

While there is a considerable body of anthropological literature discussing Native male-bodied people in a culturally defined woman's role (e.g., Callender and Kochems 1983; Jacobs 1968; Jacobs and Cromwell 1992; Lang 1990, 1998; Roscoe 1987, 1991; Williams 1986a, b), much less attention has been paid to females who took up, and in some cases still take up, the ways of men, with some exceptions (e.g., Blackwood 1984; Callender and Kochems 1983; Lang 1990, 1998, 1999; Roscoe 1988a, 1998; Whitehead 1981; Williams 1986a). An institutionalized, socially sanctioned specific gender status for men-women has been reported from at least 61 tribes as compared to 131 tribes with a special status for women-men (Roscoe 1988b: 217–222).[3] It is difficult to determine whether there were really twice as many special gender statuses for women-men than there were for menwomen on the North American Subcontinent, or whether the latter just have often not been recorded by ethnographers and other writers on the subject. If manwoman gender statuses in fact did only exist in 61 tribes, an explanation still has to be found.[4]

While gender diversity at least traditionally was a widespread trait of indigenous North American cultures, its individual expressions varied. Women-men and men-women would sometimes adopt the role of the "other" sex completely, sometimes only partially (cf. Lang 1998: 59–90, 261–267). As a rule, however, and regardless of the degree to which they mixed or did not mix gender role components, they were classified as neither men nor women, but as genders of their own within systems of three or four genders, as becomes apparent from the words used refer them. These differ from the terms for man and woman, and often indicate that men-women and women-men are viewed as combining the masculine and the feminine (cf. Roscoe 1988b; Lange 1998: 248–251, 263–265). The Shoshone, for example, call both men-women and women-men *tainna wa'ippe*, "man-woman" (Michael Owlfeather, personal communication). Among the Diné (Navajo), there likewise exists a common term for both, *nádleeh* ("someone who is in a constant process of change," Wesley Thomas, personal communication).[5] The Sduk-al-bixw (Quinault) world for men-women is "man acting" (Olson 1936: 99), and the Akimel O'odham (Pima) call women-men *wik'ovat,* "like a girl" (Hill 1938: 339).

As I have argued elsewhere (Lang 2011), the gender statuses of women-men and men-women can be characterized as being "hermaphroditic," that is, a mixture of the masculine and feminine, and this dual nature—rather than a complete transition from the masculine to the feminine gender role or vice versa—was also emphasized particularly by traditionally raised women-men I talked to. Their ambiguity reflects general patterns found in Native American world views which appreciate and emphasize transformation, ambiguity, and change. For example, humans can transform into animals and vice versa, or beings can be both humans and animals at the same time. Individuals combining the masculine and the feminine are therefore just another aspect of the transformations and ambiguities that are a central feature of indigenous North American systems of thought. Within world views that recognize that the act of creation did not necessarily establish everything in binary categories, human beings who are ambiguous in terms of their sexual features, or display a discrepancy between their physical sex and their occupational preferences, are accepted as part of the natural order of things. In other words, they are not what in Western culture is termed "deviant." In some cases they are even welcomed as people who have been blessed, or touched, by the supernatural (cf. Lang 2011).

Men-Women in Tribal Societies

When the systems of multiple genders were still intact in Native American cultures, the classification of a person as a woman-man or man-woman was usually based on her/his occupational preferences, which often became manifest as early as in childhood. A gendered division of labor, which assigned specific work tasks to either the feminine or the masculine sphere, existed in all tribal societies on the Subcontinent, and while there was some flexibility in gender roles, activities within that division of labor were decisive in defining an individual's gender. Prospective men-women usually showed a profound interest in masculine occupations in childhood. As became apparent from conversations I had with contemporary women-men and men-women, this is still the case today. "Men pretenders" among the Ingalik (autonym: Deg Xit'an) in Alaska, for example, refused to learn women's skills when they were children, so their fathers would take them under their wings and teach them men's work. They joined the boys and men in the kashim (men's house), and are said to have concealed their female physical characteristics during the sweat bath. As adults they assumed the complete social role of a man but only rarely married women (Osgood 1958: 262–263). Female-bodied nádleeh also began to behave like boys in childhood (Hill 1935: 273). Young Cocopa warrhameh played with the boys, made arrows and bows for themselves, and went hunting for rabbits and birds (Gifford 1933: 294).

In some cultures, girls' entrance into the role and status of men-women was not due to their personal inclinations; parents would decide to raise daughters as sons in the interest of the respective community. This was the case in the Arctic and Subarctic, where the diet was largely composed of meat obtained by the men by hunting, and vegetable food gathered by women played a less important role. If there were not enough boys in a community or family, some fathers would teach their daughters hunting skills and raise them to fulfill a hunter's role. Among some Inuit groups, for example, such girls learned to hunt seals from a kayak, acquired a quasi-masculine status, and wore men's garb (Mirsky 1937: 84; Kjellström 1973: 180). In addition, the role and status of men-women among the Inuit is sometimes related to the reincarnation beliefs that are omnipresent in Inuit cultures, as has been described by Bernard Saladin d'Anglure. Some children, defined by him as a "troisième sexe social," are dressed like members of the opposite sex by their family until they reach adolescence; this is done when a deceased relative of the opposite sex has become reincarnated in a child, or—in the case of females—when no boys are born into a family.

From time to time, parents among the Canadian Anishinaabe (Ojibwa) whose only child was a girl decided to raise her as a boy, and sometimes fathers would choose their favorite daughter to fulfill such a man-woman role. Such girls were treated like boys in every respect and, like boys, were sent on a vision quest to summon a guardian spirit, or supernatural helper (Landes 1937: 119, 121). Masculine occupations of men-women in Native American cultures included hunting, fishing, trapping, trade, and participation in warfare. Sometimes they practiced specialized occupations in their tribal societies. In some California groups such as the Achomawi, Klamath, and Tolowa, as well as among the Mohave living in the Mojave Desert, they were traditionally curers or "shamans" (Lang 1998: 286). This might or might not be related to their gender status. In some indigenous North American cultures, curers were predominantly men, and men-women chose that specialization as part of their masculine role. In others, there is no predominance of either women or men in healing, and men-women probably became healers if they were gifted for it but were not expected to pursue that profession if they did not possess that gift (cf. Lang 1998: 286–289). Among some tribes, men-women and women-men were believed to possess special powers due to their specific nature.

In addition, men-women are said to display generally "masculine" characteristics and habits which are, unfortunately, not usually described in detail. For example, the sources describe them as "behaving like members of the opposite sex" (Lillooet, Teit 1906: 267), "acting, talking and expressing themselves like members of the opposite sex" (Pima, Hill 1938: 339), or acting "like a man" (Paiute, Kelly 1932: 158). Some glimpses of Shoshone men-women's roles and demeanor past and present can be gained from the literature and from conversations I had with Michael Owlfeather. Female-bodied *tainna wa'ippe* traditionally went hunting and did other culturally defined men's work; some of them married women while others stayed "single" (Steward 1941: 312, 353); again others maintained—and still maintain—relationships with both other genders and give birth to children. Like the men, they formerly used to be warriors. Contemporary Shoshone men-women still go hunting and fishing, and tend to choose "modern" occupations that are defined as belonging to the masculine sphere; for example, they work in road construction or as firemen fighting the forest fires that regularly erupt in the region. In addition, their behavior and habitus is generally characterized as "masculine," both in earlier sources (cf. Stewart 1941: 385) and today:

> [They are] very butch, you know, and they' re more men than men sometimes. (. . .) And we Shoshones, we've always had manly women, as you want to call it. Like L. C. [one of the men-women on the reservation, S.L.], she has always dressed as a man, always acted like a man, and the rest of the [manly] women do, too, they're very masculine. (Michael Owlfeather, personal communication)

Besides masculine occupations, men-women also adopted other components of masculinity. Among the Cocopa and Maidu, for example, they had their nasal septum pierced like the men, and among the latter they were initiated into the men's secret society (Gifford 1933: 294; Loeb 1933: 183). Among the Mohave, where both sexes were tattooed but with different patterns, *hwame* were tattooed like men (Devereux 1937: 501)

The sources suggest that whenever an institutionalized gender status existed for men-women, a girl's choice to adopt a masculine role was usually not opposed by his/her parents and community (cf. Lang 1998: 311–322). In some cases there was even supernatural support for such a choice: in Native American cultures of the Plains, as well as in parts of the Plateau, Northeast, and California, men-women (and women-men) were, and sometimes still are, viewed as acting upon a dream or vision. This both explained and legitimized their inclination to be a gender other than man or woman, and was another reason why prospective men-women and women-men would usually get their way: their families and communities did not want to incur the wrath of the supernatural. In addition, as has already been pointed out above, the genders of women-men and men-women are situated within world views that put an emphasis on ambiguity and transformation.

Indigenous Definitions of Homosexuality and Heterosexuality

Both women-men and men-women often entered into sexual or even marital relationships with members of the same sex. Hence, their roles have long been interpreted as institutionalized homosexuality in anthropological writings. This "homosexuality" was usually considered innate, and sometimes classified according to Western psychiatric concepts of perversion and deviance,[6] even though some authors early on pointed to the fact that definitions of " deviance" vary widely between cultures: behavior considered deviant in one culture may be part of social norms in another (cf. Benedict 1934: 60; Mead 1961: 1454). In addition, this interpretation of gender variability as a way to culturally integrate, or accommodate, "homosexual" people ignores the fact that by no means all men-women and women-men entered into same-sex relationships. The sources, as well as conversations I had with people who are familiar with their tribe's traditions of gender variability, reveal that women-men had sex, or partner relationships, with women in a number of tribes, and men-women had relationships with men (Lang 1998: 189–195, 190–291). In some tribes women-men and men-women were, and still are, apparently free to have relationships with both women and men. With regard to Shoshone *tainna wa'ippe,* for example, Michael Owlfeather (personal communication) commented: "Some of those [male-bodied] two-spirited people were married people. They dressed in feminine clothes, they dressed in women's things, but they were married. Some of them even had children." The same is true of female-bodied *tainna wa'ippe.*

In the 1970s, feminist approaches in anthropology facilitated a new understanding of ways in which gender and sex are conceptualized and constructed differently in different cultures. This resulted in a shift away from the earlier essentialist views, toward new interpretations of men-women's and women-men's statuses and roles in terms of gender rather than sexual partner choice (cf. Callender and Kochems 1983; Jacobs and Cromwell 1992; Jacobs, Thomas, and Lang 1997; Kessler and McKenna 1977; Lang 1990, 1998, 20ll; Martin and Voorhies 1975; Whitehead 1981; Williams 1986a,b). Kessler and McKenna were the first to define the North American male- and female-bodied "berdaches" as part of cultural constructions of gender that recognize additional genders besides woman and man (1977: 24–36). The 1980s and 1990s witnessed a surge of anthropological literature exploring systems of multiple genders not only in North America, but also in other parts of the world such as Southeast Asia, India, and Polynesia (e.g., Blackwood 1984; Blackwood and Wieringa 1999; Herdt 1994; Nanda 1990).

What, then, is the implication of cultural systems of more than two genders for possible concepts of heterosexuality and homosexuality in indigenous North American cultures? Given the presence of multiple genders, sexual relationships between a woman-man and a man or a man-woman with a woman are "homosexual" on the level of physical sex, but not on the level of gender. If a man-woman has sex with either a man or a woman, he/she is having sex with someone whose gender differs from his/her own; the same is true for sexual relationships between women-men and either women and men. Regardless of whether they are of the same sex or not—the partners in such relationships are never of the same gender. Within such systems of multiple genders, Native classifications of sexual partner choice are based on the gender rather than the physical sex of those involved in a relationship.

In most cases, little is known about the details of these classifications, but thanks to the work of Wesley Thomas (Diné) and information generously shared by Michael Owlfeather (Shoshone/Métis) in personal conversations it is possible to compare Diné and Shoshone concepts of appropriate and non-appropriate sexual relationships. Among the Diné (Navajo), there traditionally existed four genders—women, men, women-men, and men-women. As has been mentioned above, the latter two genders were called nádleeh. The same term is, by the way, also applied to intersex people; the latter's supernatural prototypes feature in the Diné Origin Story, and the human nádleeh (whether physically ambiguous or not) are viewed as the worldly likenesses of these mythical hermaphrodite beings. Within the Diné classification, the equivalent to a homosexual relationship is a sexual relationship between people of the same gender (two women, two men, two female-bodied nádleeh, two male-bodied nádleeh), or of closely related genders (woman and male-bodied nádleeh, man and female-bodied nádleeh).

Among the Shoshone, *tainne wa'ippe* of both sexes are viewed as acting on a vision that demands them to adopt the occupations, ways of behavior, and garb of the "other sex." However, their gender status does not limit their partner choice. In contrast to *nádleeh,* they are free to have relationships with both men and women. The same is true for women-men and men-women in some other Native cultures, as has been pointed out above; we can thus assume that the Shoshone classification of sexual relationships is representative of other indigenous North American cultures as well. The only type of relationship that was at least traditionally forbidden—not only among the Shoshone and Diné, but universally in Native American cultures—was between two women-men and two men-women. Where multiple genders continue to exist, women-men, for example, are still believed to be connected by a bond of kinship, sometimes referring to each other as "sisters." Hence, a sexual relationship between them is considered incestuous (Michael Owlfeather, personal communication; Thomas 1997 and personal communication; Williams 1986b; 93–94).

As a matter of fact, very little is known about homosexual relationships as defined within Western categories in North American tribal societies, not least because ethnographers tended to equate women-men and men-women with "homosexuals." Hence, many of them did not even inquire about same-sex relationships outside the so-called "berdache institution." Some sources indicate that such homosexual behavior occurred and was sometimes tolerated. Among the Yuma, for example, "casual secret homosexuality" was said to be well known in the early 1930s, and "not considered objectionable" (Forde 1931:157). The Yuma themselves made a dear distinction between such same-sex relationships and those of women-men *(elxa')* and men-women *(kwe'rhame),* as becomes apparent from Forde's statement that those involved in discreet homosexual relationships "would resent being called elxa" or kwe'rhame" (Forde 1931: 157). In other cases, those involved were probably wedded in heterosexual marriages and led lives that were inconspicuous, above all for ethnographers' eyes (cf. Lang 1998: 323–330).

A distinction between themselves and gay/lesbian Native Americans is also made by contemporary women-men and men-women. This distinction is based on varying criteria. Among the Shoshone, a "gay" Native person as opposed to a *tainna wa'ippe* is defined as lacking the spiritual component that is so essential to the *tainna wa'ippe's* role and status:

> One person told me, "Well, those old-time 'berdaches,' they were nothing but drag queens, weren't they?" And I said, "No, they weren't," because they didn't dress in women's clothes just because of personal preference. It was because of the manifestation of Spirit. They *had* to do it. You know, (. . .) they had to act on this vision in order to be a complete person. (Michael Owlfeather, personal communication; emphasis his)

Among the Diné, the distinction between *nádleeh* and gay/lesbian is based on the fact that the *nádleeh* role and status is based on occupational preferences while "gay" is viewed as being based on sexual preference. According to Wesley Thomas (personal communication), "a true *nádleeh* or traditional *nádleeh* is somebody who is one hundred percent—[he pauses to think:] a woman, who was born as a man but is a woman in Navajo society, not in their sexual preferences or sexual persuasion, but as an occupational [preference]."

Two-Spirit People

The old-time statuses and roles of women-men and men-women have disappeared on many reservations and in urban Native communities. When I talked to members of the first generation of Native LGBT activists in the early 1990s, they told me that the last "true" winkte, tainna wa'ippe, and so on, had lived on the reservations in the 1930s and 1940s. When one of my friends grew up as a male-bodied nádleeh on the Navajo Reservation in the 1950s, there was no role model around. However, her/his grandmother, who had been born in the early 1920s, recognized her/his special nature. She remembered the nádleeh, their rootedness in Diné religion and philosophy, and raised her grandchild to fulfill that special role. The same was true for other people I met; they, too, had been raised by grandparents who still remembered and appreciated the traditions of gender diversity in their respective cultures.

On the other hand, there were a growing number of Native lesbians, gays, and bisexuals, the large majority of them urban, who sought to gain acceptance by their Native communities, began to make their voices heard, and researched into the history of same-sex relationships in their indigenous cultures. From the mid-1970s onward, and largely unnoticed by non-Native anthropologists and historians who began to explore "old-time" gender systems, Native American gay and lesbian writers and activists made their voices heard as well, addressing their concerns and experiences, including the impact of colonialism on gay/lesbian and "straight" indigenous North Americans alike.

At the same time, the first Native gay and lesbian organization, Gay American Indians (GAI), was founded in San Francisco in 1975. What was to become the two-spirit identity has been originally emerging in such urban settings as well as in "pan-tribal" contexts such as the annual Two-Spirit Gatherings, not on the reservations. The concept and term of two-spirit are the result of urban gay and lesbian Native Americans' search for and rediscovery of the cultural roots of homosexual behavior in indigenous North American cultures. Initially, these activists did research into the written historical and anthropological sources, as the traditions of gender diversity were often forgotten or repressed in their communities. While anthropological approaches to multiple genders in Native American cultures are now criticized for ignoring indigenous knowledge and the concerns of contemporary LGBTQ2 indigenous North Americans (cf. Driskill et al. 2011a, b), the emergent two-spirit identity initially reflected, ironically, the very way non-Native researchers interpreted the roles and statuses of men-women and womenmen: as ways to culturally integrate "homosexual"/LGBT individuals who, in addition, often held revered and highly respected statuses in their respective communities.

While "two-spirit" has come to be widely used in LGBTQ indigenous contexts and organizations, it needs to be mentioned that not all LGBTQ Native Americans/First Nations people identify as two-spirit. The gay poet, teacher, and social worker Gregory Scofield (Cree/Métis), for example, stated: "First of all, in relation to the ideology of Two-Spirited theory, I always back away from that three-hundred fold. I mean I don't consider myself Two-Spirited. (. . .) Not that I'm disparaging of it. It's just that I think it's very multi-layered insofar as the politicization of the word and how it's come about and its interpretation and its reinvention and the reinterpretation of things" (McKegney 2014: 218; cf. also Gilley 2006: 90–95, 113–121).

A key feature of the concept, and identity, of "two-spirit" is that unlike for many "white" gays and lesbians, people's (homo)sexuality is not a prime marker of their identity even though sexuality and desire are, of course, important in their lives. Their primary identity is as Native Americans and /or members of a specific ethnic group, as was stressed again and again by people I talked to. Erna Pahe (Navajo), for example, commented with regard to GAI: "One of our major emphases is that we are Indian first, we're Navajo, we're Pima, we're Apaches. And we do not divide our group and say that we're gay, and making us different. We're all Indians, and that's the way we portray our feelings, and that's the priority in terms of our organization." This was echoed by Randy Burns (Paiute), who told me: "I used to say I was gay first and I was Indian second, and then I, of course, through peer pressure got convinced that, 'No, you're Indian! Remember who you are!' And, you know, 'When you're born into this world, you're born Native, you're not born queer.'" Another recurrent aspect of being two-spirit is spirituality; the very term/concept of "*two-spirit*" expresses a connection of contemporary Native American LGBTQ to indigenous spirituality in the broadest sense. As has been observed by Lüder Tietz with regard to Canada, "For some *two-spirited people,* the word *spirit* seems to be at the core of the use of the term 'Two-Spirited.' (. . .) The concept of 'two-spirited' can be seen as the attempt to transcendentally substantiate a modem lesbian/gay identity that is specifically First Nations" (Tietz 1996: 207, emphasis his, translation mine).

This priority in identity needs to be seen in the context of two-spirit people's life experiences and concerns, which have much more in common with those of "straight" Native Americans than with those of white lesbians and gays, or even other LGBTQ people of color (cf. Driskill 2010: 78, 80). With heterosexual Native Americans, they share the experience and impact of five hundred years of colonization, cultural genocide, and forced acculturation. Just like other Native Americans, middle-aged two-spirits raised on reservations still vividly remember the physical and emotional violence they suffered in boarding schools run under the motto of "Kill the Indian, save the Man," which was given out in 1879 by Richard E. Pratt, the founder of Carlisle Indian School. Given these shared experiences of ongoing historical trauma, which have left a deep, lasting impact on heterosexual and LGBTQ Native Americans alike, and the Native-oriented priority in their identity, most two-spirit people do not wish to set themselves apart from their Native communities. Many of them feel that they are innately imbued with special spiritual and/or practical gifts that they would like to use for the benefit of their ethnic group and/or the Native American community at large.

This desire to be an integral part of Native communities implies that twospirit people are not prone to separatism on the grounds of their sexual orientation. For example, they do not strive for any type of new religious community shared exclusively by two-spirit people; for them, Native religions are not a matter to be eclectically trifled with. If they engage in religious practices of their respective ethnic group, they attach importance to doing so in the customary and appropriate manner of their people. An example of this is a women-only Sun Dance founded by a Lakota woman in the mid-1980s (cf. Lang 2013; see also below). Though specifically addressing Native focus is not on "lesbianism" but on the proper performance of a sacred ceremony.

Two-Spirit Women/Native Lesbians

This primary identity as Native Americans influences the way two-spirit women see themselves, and the way they relate to Native gays in particular and their communities in general. For example, the concept of lesbian separatism seemed to hold little appeal for most Native lesbians/two-spirit women I talked to. While they acknowledge that some of their experiences, identities, and goals differ from those of male two-spirits, there is an emphasis on solidarity due to shared experiences of colonization, racism, and social marginalization. In addition, a very common concept in North American indigenous cultures is that the feminine and masculine qualities are complementary, and that it is advisable to combine them in both worldly and spiritual matters for efficiency. Similar observations were made by Walters et al. (2006), and summed up by one of their Native lesbian respondents:

> A lot of times in the White community, lesbians will say, you know, "I just don't like men." Actually, I think that identifying as two-spirit, I have more of an alliance with Native men (. . .) because they're Native men and they have experienced a lot of similar racist attitudes as well as homophobic attitudes on the reservation that I have. Um, we seem to bond together better, the male and the female sides sort of complement one another. I have difficulty explaining it to White lesbians who would say, "Well, why would you want gay men at an event?" Because Native gay men are not gay men, they're my two-spirit brothers. (132)

Another feature of both two-spirit women and men is that they do not wish to set themselves apart from their Native communities on the grounds of their sexuality, as has already been pointed out above. There are specific roles they assume, or wish to assume, for the benefit of these communities; these roles are culturally attributed to two-spirit people in general, so it makes no sense to treat women and men separately in the present context.

One of the specific gifts attributed to two-spirit people is caretaking. The caretaker role was already associated with the traditional women-men and men-women, and continues to be stressed both by contemporary two-spirit people and their communities (Evans-Campbell, Fredriksen-Goldsen, Walters, and Stately 2007; Jacobs, Thomas, and Lang 1997; Lang 1998). Michael Owlfeather told me: "We have always been the caretakers, we have always been the cultural keepers," and went on to elaborate on other culturally ascribed roles of two-spirits/menwomen/women-men: "We are the ones that people go to when they want to know how to do things right, be it funerals, be it give-aways, be it a way of dressing, a style of headwork, a song. We are the people that keep those things. We always have been, and we always will be." The role and contribution of two-spirit people as caregivers for the elderly and children, and positive reactions to these activities by families and Native communities, have been stressed by the respondents in EvansCampbell et al. (2007). The authors conclude:

> Caregiving is perceived as an important and integral role of two-spirit people, and it is clear that many two-spirit people already engage in caregiving or expect to provide care for others at some point during their lifetime. (. . .) Based on culturally proscribed roles, two-spirit people may be asked to care for the elderly, relatives, or children. In some communities, they may be asked to care for the community as a whole as they take on specific social or ceremonial responsibilities. (88)

This role was also emphasized by people I talked to. A Nez Percé woman pointed out in a conversation that elders will pass on their knowledge to the two-spirit people who take care of them, and who thus become keepers of their tribe's traditions. Being caretakers of children in particular was an important aspect of the role of women-men and men-woman. In Native American communities, it is not unusual for children to be raised by people other than their parents. Navajo *nádleeh* of both sexes are traditionally expected to adopt and raise children (Wesley Thomas, personal communication). George (pseudonym), a Colville *wenatcha* (woman-man) I talked to, had raised no less than twenty-six children from her/his extended family, and Michael Owlfeather raised three of his nieces because their parents had alcohol abuse problems and could not properly take care of the girls.

Another recurrent aspect in the lives of Native lesbians/two-spirit women lives is motherhood. Many of those I met had become pregnant "the natural way," as one of them said with a wink, that is, the children were not conceived by artificial insemination but by intercourse with men. Quite a few of the Native lesbians I talked to had started out having sexual relationships with men, or had had such relationships at some time in their lives. Within their Native American cultural frames of reference they would stress motherhood as an essential aspect of womanhood, regardless whether a woman identifies as lesbian or straight. George told me that the female-bodied *wenatcha* of his tribe usually give birth to and raise children, explaining that this is just part of their nature, which requires a balance of the feminine and the masculine rather than switching completely from one to the other.

Other two-spirit women, rather than subscribing to rigid European categories of "gay," "lesbian," or "straight," will rather self-label as bisexual due to their recognition of the fact that sexual preferences may change in the course of an individual's lifetime. In other cases, they will use the term "lesbian" in a sense that includes the possibility of having sexual relationships with both women and men; in still others, they will use "lesbian" synonymous with "man-woman," referring to contemporary manly women on the reservations including their relationships that are by no means exclusively homosexual.

The role of two-spirit women as parents, and the Native definition of "lesbian" as not excluding sexual relationships with men, was also emphasized by Michael Owlfeather in one of our conversations. Among the Shoshone, there were manly women of his grandmother's generation who would have male husbands and female lovers: "My grandmother was a two-spirited person herself, and it's known among the lesbian people here. (. . .) [Women like her] took pride in doing anything a man could do, and doing it better, you know? But these women also had children, got married. But they had their 'women friends' and were always respected for that and everything else. It was nothing overt, they didn't hold it out to the community, but everybody knew what was going on."

Most Native lesbians/two-spirit women I talked to were acutely aware of the triple discrimination they are exposed to as women, lesbians, and Native Americans. Recent studies address the impact of historical trauma, discrimination, and other stressors on the health and wellness of Native American lesbian, gay, bisexual, transgender, and two-spirit men and women (see, e.g., Burks, Robbins, and Durtschi 2010; Walters et al. 2006; Walters 2010). The health status of LGBTQ2 Native Americans/First Nations people reflects, in many ways, that of the indigenous population of the United States and Canada in general. Issues related to physical and mental health include a high prevalence of chronic illnesses and premature mortality related to these, disproportionately high levels of poverty and socioeconomic deprivation, exposure to racism, and the persistent impact of colonization.

Given the discrimination and racism they encounter even within the "white" lesbian subculture, some lesbians of Native American or mixed descent have come to prefer partners who share the same, or at least a related, ethnic background. One of the reasons for this is doubtlessly the fact that, no matter how sensitive the white partner of a Native woman may be as far as matters of racism are concerned, racism often seems to become an issue in a relationship between a Native American and a white woman. White lesbians may also have a yearning for Native spirituality and a romanticized image of their Native partners, which is likewise annoying to the latter. Native lesbians may feel that they do not want to cope with these issues in their relationships anymore, having to deal with them already in all spheres of their lives. They may reach the conclusion that it will further their personal growth if they are in a relationship with a woman where no strength and energy is consumed by suffering from and educating their partner on racism and related issues.

In some cases, Native lesbians/two-spirit women have become so disenchanted with the culture and values of the colonizers that they have established small communities restricted to Native women or women of color. And in spite of the fact that many two-spirit women emphasize and appreciate cooperation with Native gays in projects and organizations, they feel that their specific experiences and needs sometimes require women-only gatherings. One example is the Women's Sun Dance founded by a Lakota woman in the mid-1980s (Lang 2013). While white women and non-Native women of color are welcome to attend as long as they behave respectfully toward the ceremony and its participants, only Native women may pledge to actually be Sun Dancers. The Women's Sun Dance has meanwhile become integrated into a larger project called Kunsi Keya Tamakoce (the Lakota term for "Grandmother Turtle Land") located in a remote rural mountain area. Its mission is very clearly outlined in terms of indigenous knowledge, a decided stance against heteronormativity and racism, and a focus on ecology. According to its homepage, it "preserves indigenous religious heritage and fights against racism, sexism, classim, homophobia, ageism and ecological violence experienced in tribal populations and the world at large" (http://www.kunsikeya.org/about.html). The general emphasis of the project, which was recognized by the United Nations as an Indigenous People's Organization in 2013, seems to have shifted from lesbians to Native women in general and has also come to include Native men (http://www.kunsikeya.org/); the Women's Sun Dance, however, is still women-only.

Notes

1. With regard to its research agenda and desiderata, this approach is situated within a larger current of efforts towards decolonization in the academy and beyond as outlined, for example, in the pioneering book Decolonizing Methodologies by Linda Tuhiwai Smith (2012 [1999]). On decolonization strategies in research with/on contemporary two-spirit people see, for example. Walters and simony (2009).

2. This differs from the way these terms are used in other anthropological writings (e.g., Roscoe 1991; Fulton and Anderson 1992), where "man-woman" usually refers to male-bodied persons in a woman's role. However, in Native American cultures more importance is attached to an individual's gender status than to his/her physical sex. Hence, it seems logical to follow Bleibtreu-Ehrenberg (1984), who titled her monograph on make gender diversity in various cultures Der Weibmann ("The Woman-Man"), and to put the gender to which the chosen gender role belongs first and the physical sex second.

3. A few of these references may be doubtful, such as koskalaka as a term referring to female alternative gender among the Yankton Dakota (Medicine 1997: 146–147).

4. Whitehead has argued that Native American cultures tend to lack such statuses because of the feminine physiological functions, most prominently menstrucation, which are dangerous to masculine persuits and thus prevent women's access to a masculine role (Whitehead 1981: 91–92). Morever, she assumes a universal gender hierarchy in Native American cultures where "the man was everywhere considered superior to the woman. As in most hierarchical systems, downward mobility was more easily achieved than upward mobility" (Whitehead 1981: 86). Yet such a supposed universal gender hierarchy favoring men in Native American cultures has been challenged, as well as the generalizing statement hat menstruation keeps women from doing men's work (cf. Callender and Kochems 1983: 455–456; for a critique of the gender hierarchy model see also Powers 1986: 1–19).

5. Michael Owlfeather (Shoshone/Métis) and Prof. Wesley Thomas (Diné) are experts in the traditions of gender diversity in their respective cultures. On the nádleeh, see also Hill (1935); Thomas (1997); Epple (1997). In Shoshoni, tainkwa or tainna means "man" and wa'ippe "woman" (Miller 1972: 136, 172).

6. For a discussion of anthropological interpretations of gender diversity, see, for example, Lane (1998: 17–56).

References

Angelino, H., and C. L. Shedd. 1955. A note on berdache. *American Anthropologist* 57(1): 121–126.

Benedict, R. F. 1934. Anthropology and the abnormal. *Journal of General Psychology* 10: 59–82.

Blackwood, E. 1984. Sexuality and gender in certain Native American tribes: The case of cross-gender females. *Signs: Journal of Women in Culture and Society* 10: 1–42.

Blackwood, E., and S. E. Wieringa (eds.). 1999. Female desires: *Same-sex relations and transgender practices across cultures*. New York: Oxford University Press.

Bleibtreu Ehrenberg, G. 1984. *Der Weibmcmn: Kultischer Geschlechtswechsel im Schamanismus*. Frankfurt am Main: Fischer.

Burks, D. J., R. Robbins, and J.P. Durtschi. 2011. American Indian gay, bisexual and two spirit men: A rapid assessment of HIV/AIDS risk factors, barriers to prevention and culturally sensitive intervention. Culture, Health and Sexuality: *An International Journal for Research, Intervention and Care* 13(3): 283–298.

Carpenter, E. 1914. *Intermediate types among primitive folk*. London: George Allen.

Callender, C., and L. Kochems. 1983. The North American berdache. *Current Anthropology* 24: 443–470.

D'Anglure, B. S. 1992. Le "troisieme sexe." *La Recherche* 245: 836–844.

———. 1988. Du foetus au chamane: La construction d'un "troisiéme sexe" inuit. Études/*lnuit/Studies* 10(1/2): 25–113.

Devereux, G. 1937. Homosexuality among the Mohave Indians. *Human Biology* 9:498–527.

Driskill, Q.-L. 2010. Doubleweaving two-spirit critiques: Building alliances between native and queer studies. GLQ: *A Journal of Lesbian and Gay Studies* 16(1-2): 69–92.

———. 2011b. Introduction. In Q.L. Driskill, C. Finley, B. J. Gilley, and S. L. Morgensen (eds.), *Queer indigenous studies: Critical interventions in theory, politics, and literature*. Tucson: University of Arizona Press, pp. 1–28.

Epple, C. 1997. A Navajo worldview and nádleehí: Implications for western categories. In S.E. Jacobs, W. Thomas, and S. Lang (eds.), *Two-spirit people: Native American gender identity, sexuality, and spirituality*. Urbana and Chicago: University of Illinois Press, pp. 174–191.

Evans-Campbell, T., K. I. Fredriksen-Goldsen, K. L. Walters, and A. Stately. 2007. Caregiving experiences among American Indian two-spirit men and women: Contemporary and historical roles. *Journal of Gay and Lesbian Social Services* 18(3/4): 75–92.

Forde, C. D. 1931. Ethnography of the Yuma Indians. *University of California Publications in American Archaeology and Ethnology* 28(4):83–278.

Fulton, R., and S. W. Anderson. 1992. The Amerindian "man-woman": Gender, liminality, and cultural continuity. *Current Anthropology* 33(5): 603–609.

Gay American Indians and W. Roscoe (eds.). 1988. *Living the spirit: A gay American Indian anthology*. New York: St. Martin's Press.

Gifford, E. W.!933. The Cocopa. *University of California Publications in American Archaeology and Ethnology* 31(5): 257–333.

Gilley, B. J. 2006. *Becoming two-spirit: Gay identity and social acceptance in Indian country*. Lincoln and London: University of Nebraska Press.

Goldenweiser, A. 1929. Sex and primitive society. In S. D. Calverton and V. F. Schmalhausen (eds.), *Sex in civilization*. Garden City, NY: Garden City Publishing, pp. 53–66.

Gravier, J. 1959. Relation ou Journal du voyage du Père Gravier de Ia Compagnie de Jesus en 1700. In R. G. Thwaites (ed.), *The Jesuit relations and allied documents*, vol. 65. New York: Pageant Book Company, pp. 100–179. [Reprint.]

Guerra, F. 1971 *The pre-Columbian mind*. New York: Seminar Press.

Herdt, G. (ed.). 1994. *Third sex, third gender: Beyond sexual dimorphism in culture and history*. New York: Zone Books.

Hill, W. W. 1938. Note on the Pima berdache. *American Anthropologist* 40: 338–340.

———.1935. The status of the hermaphrodite and transvestite in Navaho culture. *American Anthropologist* 37: 273–279.

Jacobs, S.-E. 1997. Is the "North American berdache" merely a phantom in the imagination of western social scientists? In S.-E. Jacobs, W. Thomas, and S. Lang (eds.), *Two-spirit people: Native American gender identity, sexuality, and spirituality*. Urbana and Chicago: University of Illinois Press, pp. 21–44.

———. 1968. Berdache: A brief review of the literature. *Colorado Anthropologist* 1: 25–40.

Jacobs, S.-E., W. Thomas, and S. Lang (eds.). 1997a. *Two-spirit people: Native American gender identity, sexuality, and spirituality*. Urbana and Chicago: University of Îllinois Press.

———. 1997b. Introduction. In S.-E. Jacobs, W. Thomas, and S. Lang (eds.), *Two-spirit people: Native American gender identity, sexuality, and spirituality*. Urbana and Chicago: University of Illinois Press, pp. l–19.

Jacobs. S.-E., and J. Cromwell. 1992. Visions and revisions of reality: Reflections on sex, sexuality, gender, and gender variance. *Journal of Homosexuality* 23(4): 43–69.

Kelly, L 1932. Ethnography of the Surprise Valley Paiute. *University of California Publications in American Archaeology and Ethnology* 31(3): 67–210.

Kessler, S. J., and W. McKenna. 1977. Gender: *An ethnomethodological approach*. New York: Wiley.

Kjellström, R 1973. *Eskimo marriage: An account of traditional Eskimo courtship and marriage*. Stockholm: Nordiska museets handlingar 80.

Landes, R 1937. The Ojibwa of Canada. In M. Mead (ed.), *Cooperation and competition among primitive people*. New York: McGraw-Hill, pp. 87–126.

Lang, S. 2013. Re-gendering sacred space: An all-women's sun dance. In A. Blätter, and S. Lang (eds.), "Contemporary Native American studies" Special issue, *Ethnoscripts* 15(1): 124–139.

———. 2011. Transformations of gender in Native American cu1tures. In R. Potter-Deimel, and K. Kolinská (eds.), "Transformation, translation, transgression: Indigenous American cultures in contact and context." Special issue, *Litteraria Pragensia* 21(4): 70–81.

———. 1999. Lesbians, men-women, and two-spirits: Homosexuality and gender in Native American cultures. In E. Blackwood and S. E. Wieringa (eds.), *Female desires: Same-sex relations and transgender practices across cultures*. New York: Columbia University Press, pp. 91–116.

———. 1998. *Men as women, women as men: Changing gender in Native American cultures*. Austin: University of Texas Press.

———. 1990. *Männer als Frauen –Frauen als Männer: Geschlechtsrollenwechsel bei den Indianern Nordamerikas*.Hamburg: Wayasbah.

Loeb, E. M. 1933. The eastern Kuksu cult *University of California Publications in American Archaeology and Ethnology* 33(2): 139–231.

Martin, M. K, and B. Voorhies. 1975. *Female of the species*. New York: Columbia University Press.

McKegney, S. 2014. *Masculindians: Conversations about indigenous manhood*. East Lansing: Michigan State University Press.

Mead, M. 1961. Cultural determinants of sexual behavior. In E. C. Young (ed.), *Sex and internal secretions*. Baltimore: Williams and Wilkins, pp. 1433–1479.

———. 1932. *The changing culture of an Indian tribe*. New York: Columbia University Press.

Medicine, B. 1997. Changing Native American roles in an urban context and changing Native American sex: roles in an urban context. In S.-E. Jacobs, W. Thomas, and S. Lang (eds.), *Two-spirit people: Native American gender identity, sexuality, and spirituality*. Urbana and d1icago: University of Illinois Press, pp. 145–155.

Miller, W. R. 1972. *Newe natekwinappeh: Shoshoni stories and dictionary*. Salt Lake City: University of Utah Press.

Mirsky, J. 1937. The Eskimo of Greenland. In M. Mead (eel), *Cooperation and competition among primitive people*. New York: McGraw-Hill, pp. 51–86

Nanda, S. 1990. *Neither man nor woman: The hijras of India*. Belmont, CA: Wadsworth.

Olson, R L 1936. The Quinault Indians. *University of Washington Publications in Anthropology* 6(1): 1–194.

Osgood, C. 1958. Ingalik social culture. *Yale University Publications in Anthropology* 53: 1–289.

Powers, M. N. 1986. *Oglala women: Myth, ritual and reality*. Chicago and London: University of Chicago Press.

———. 1993. How to become a berdache: Toward a unified analysis of gender diversity. In G. Herdt (ed.), *Third sex, third gender: Beyond sexual dimorphism in culture and history*. New York: Zone Books, pp. 329–372.

———. 1991. The *Zuni man-woman*. Albuquerque: University of New Mexico Press.

———. 1988a. Strange country this: Images of berdaches and warrior-women. In Gay American Indians and W. Roscoe (eds.), *Living the spirit: A gay American Indian anthology*. New York: St. Martin's Press, pp. 48–76.

———. 1988b. North American tribes with berdache and alternative gender roles. In Gay American Indians, and W. Roscoe (eds.), *Living the spirit: A gay American Indian anthology*. New York: St. Martin's Press, pp. 217–222.

———. 1987. Bibliography of berdache and alternative gender roles among North American Indians. *Journal of Homosexuality* 14(3–4):81–171.

Steward, J. H. 1941. Culture element distributions 13: Nevada Shoshone. *Anthropological Records* 4(2): 209–359.

Stewart, O. C. 1941. Culture element distributions 14: Northern Paiute. *Anthropological Records* 4(3): 361–446.

Teit, J. 1906. The Lillooet Indians. *Memoirs of the American Museum of National History* 4(5) (= Publications of the Jesup North Pacific Expedition II, part 5). New York: American Museum of Natural History.

Thomas, W. 1997. Navajo cultural constructions of gender and sexuality. In S.-E. Jacobs, W. Thomas, and S. Lang (eds.), *Two-spirit people: Native American gender identity, sexuality, and spirituality*. Urbana and Chicago: University of Illinois Press, pp. 156–173.

Tietz, L 1996. *Moderne Rückbezüge auf Geschlechtsrollen indianischer Kulturen*. Unpublished Master's thesis. University of Hamburg, Germany, Institut für Ethnologie.

Tuhiwai Smith, L. 2012. *Decolonizing methodologies: Research and indigenous peoples* (second edition). New York: Zed Books.

Walters, K. L. 2010. Critical issues and LGBT-two spirit populations: Highlights from the HONOR project study. IOM Presentation, March 2010. http://www.iom.edu/~/media/Files/Activity%20Files/SelectPops/LGBTHealthIssues/Walters%20presentationl.pdf.

Walters, K. L., and J. M. Simoni. 2009. Decolonizing strategies for mentoring American Indians and Alaska Natives in HIV and mental health research. *American Journal of Public Health* 99(1): S71–S76.

Walters, K. L., T. Evans-Campbell, J. M. Simoni, T. Ronquillo, and R. Bhuyan. 2006. "My spirit in my heart": Identity experiences and challenges among American Indian two-spirit women. *Journal of Lesbian Studies* 10(1/2): 125–149.

Whitehead, H. 1981. The bow and the burden-strap: A new look at institutionalized homosexuality. In S. B. Ortner and H. Whitehead (eds.), *Sexual meanings: The cultural construction of gender and sexuality*. London: Cambridge University Press, pp. 80–115.

Williams, W. L. 1986a. *The spirit and the flesh: Sexual diversity in American Indian culture*. Boston: Beacon Press.

———. 1986b. Persistence and change in the berdache traditions among contemporary Lakota Indians. In E. Blackwood (ed.), *The many faces of homosexuality. Anthropological approaches to homosexual behavior*. New York and London: Harrington Park Press, pp. 191–200.

Introduction to Reading 16

Maria Alexandra Lepowsky is an anthropologist who lived among the Melanesian people of Vanatinai, a small, remote island near New Guinea, from 1977 to 1979, for 2 months in 1981, and again for 3 months in 1987. She chose Vanatinai, which literally means "motherland," because she wanted to do research in a place where "the status of women" is high. The egalitarianism of the Vanatinai challenges the Western belief in the universality of male dominance and female subordination.

1. What is the foundation of women's high status and gender equality among the people of Vanatinai?
2. What does gender equality mean on Vanatinai? Does it mean that women and men split everything fifty-fifty? Are men and women interchangeable?
3. What are the similarities and differences between the egalitarianism of the Gerai people (depicted in Christine Helliwell's article in this chapter) and that of the people of Vanatinai?

Gender and Power

Maria Alexandra Lepowsky

Vanatinai customs are generally egalitarian in both philosophy and practice. Women and men have equivalent rights to and control of the means of production, the products of their own labor, and the products of others. Both sexes have access to the symbolic capital of prestige, most visibly through participation in ceremonial exchange and mortuary ritual. Ideologies of male superiority or right of authority over women are notably absent, and ideologies of gender equivalence are clearly articulated. Multiple levels of gender ideologies are largely, but not entirely, congruent. Ideologies in turn are largely congruent with practice and individual actions in expressing gender equivalence, complementarity, and overlap.

There are nevertheless significant differences in social influence and prestige among persons. These are mutable, and they fluctuate over the lifetime of the individual. But Vanatinai social relations are egalitarian overall, and sexually egalitarian in particular, in that at each stage in the life cycle all persons, female and male, have equivalent autonomy and control over their own actions, opportunity to achieve both publicly and privately acknowledged influence and power over the actions of others, and access to valued goods, wealth, and prestige. The quality of generosity, highly valued in both sexes, is explicitly modeled after parental nurture. Women are not viewed as polluting or dangerous to themselves or others in their persons, bodily fluids, or sexuality.

Vanatinai sociality is organized around the principle of personal autonomy. There are no chiefs, and nobody has the right to tell another adult what to do. This philosophy also results in some extremely permissive childrearing and a strong degree of tolerance for the idiosyncrasies of other people's behavior. While working together, sharing, and generosity are admirable, they are strictly voluntary. The selfish and antisocial person might be ostracized, and others will not give to him or her. If kinfolk, in-laws, or neighbors disagree, even with a powerful and influential big man or big woman, they have the option, frequently taken, of moving to another hamlet where they have ties and can expect access to land for gardening and foraging. Land is communally held by matrilineages, but each person has multiple rights to request and be given space to make a garden on land held by others, such as the mother's father's matrilineage. Respect and tolerance for the will and idiosyncrasies of individuals is reinforced by fear of their potential knowledge of witchcraft or sorcery.

Anthropological discussions of women, men, and society over the last one hundred years have been framed largely in terms of "the status of women," presumably unvarying and shared by all women in all social situations. Male dominance and female subordination have thus until recently been perceived as easily identified and often as human universals. If women are indeed universally subordinate, this implies a universal primary cause: hence the search for a single underlying reason for male dominance and female subordination, either material or ideological.

More recent writings in feminist anthropology have stressed multiple and contested gender statuses and ideologies and the impacts of historical forces, variable and changing social contexts, and conflicting gender ideologies. Ambiguity and contradiction, both within and between levels of ideology and social practice, give both women and men room to assert their value and exercise power. Unlike in many cultures where men stress women's innate inferiority, gender relations on Vanatinai are not contested, or antagonistic: There are no male versus female ideologies which vary markedly or directly contradict each other. Vanatinai mythological motifs, beliefs about supernatural power, cultural ideals of the sexual division of labor and of the qualities inherent to men and women, and the customary freedoms and restrictions upon each sex at different points in the life course all provide ideological underpinnings of sexual equality.

Since the 1970s writings on the anthropology of women, in evaluating degrees of female power and influence, have frequently focused on the disparity between the "ideal" sex role pattern of a culture, often based on an ideology of male dominance, publicly proclaimed or enacted by men, and often by women as well, and the "real" one, manifested by the actual behavior of individuals. This approach seeks to uncover

Lepowsky, M. A. (1993). Gender and power. *Fruit of the motherland*. Copyright © 1993. Reprinted with permission of Columbia University Press.

female social participation, overt or covert, official or unofficial, in key events and decisions and to learn how women negotiate their social positions. The focus on social and individual "action" or "practice" is prominent more generally in cultural anthropological theory of recent years. Feminist analyses of contradictions between gender ideologies of female inferiority and the realities of women's and men's daily lives—the actual balance of power in household and community—have helped to make this focus on the actual behavior of individuals a wider theoretical concern.[1]

In the Vanatinai case gender ideologies in their multiple levels and contexts emphasize the value of women and provide a mythological charter for the degree of personal autonomy and freedom of choice manifested in real women's lives. Gender ideologies are remarkably similar (though not completely, as I discuss below) as they are manifested situationally, in philosophical statements by women and men, in the ideal pattern of the sexual division of labor, in taboos and proscriptions, myth, cosmology, magic, ritual, the supernatural balance of power, and in the codifications of custom. Women are not characterized as weak or inferior. Women and men are valorized for the same qualities of strength, wisdom, and generosity. If possessed of these qualities an individual woman or man will act in ways which bring prestige not only to the actor but to the kin and residence groups to which she or he belongs.

Nevertheless, there is no single relationship between the sexes on Vanatinai. Power relations and relative influence vary with the individuals, sets of roles, situations, and historical moments involved. Gender ideologies embodied in myths, beliefs, prescriptions for role-appropriate behavior, and personal statements sometimes contradict each other or are contradicted by the behavior of individuals.

* * *

Material and Ideological Bases of Equality

Does equality or inequality, including between men and women, result from material or ideological causes? We cannot say whether an idea preceded or followed specific economic and social circumstances. Does the idea give rise to the act, or does the act generate an ideology that justifies it or mystifies it?

If they are congruent, ideology and practice reinforce one another. And if multiple levels of ideology are in accord, social forms are more likely to remain unchallenged and fundamentally unchanged. Where levels of ideology, or ideology and practice, are at odds, the circumstances of social life are more likely to be challenged by those who seek a reordering of social privileges justified according to an alternative interpretation of ideology. When social life embodies these kinds of contradictions, the categories of people in power—aristocrats, the rich, men—spend a great deal of energy maintaining their power. They protect their material resources, subdue the disenfranchised with public or private violence, coercion, and repression, and try to control public and private expressions of ideologies of political and religious power.

On Vanatinai, where there is no ideology of male dominance, the material conditions for gender equality are present. Women—and their brothers—control the means of production. Women own land, and they inherit land, pigs, and valuables from their mothers, their mothers' brothers, and sometimes from their fathers equally with men. They have the ultimate decision-making power over the distribution of staple foods that belong jointly to their kinsmen and that their kinsmen or husbands have helped labor to grow. They are integrated into the prestige economy, the ritualized exchanges of ceremonial valuables. Ideological expressions, such as the common saying that the woman is the owner of the garden, or the well-known myth of the first exchange between two female beings, validate material conditions.

I do not believe it would be possible to have a gender egalitarian society, where prevailing expressions of gender ideology were egalitarian or valorized both sexes to the same degree, without material control by women of land, means of subsistence, or wealth equivalent to that of men. This control would encompass anything from foraging rights, skills, tools, and practical and sacred knowledge to access to high-paying, prestigious jobs and the knowledge and connections it takes to get them. Equal control of the means of production, then, is one necessary precondition of gender equality. Vanatinai women's major disadvantage is their lack of access to a key tool instrumental in gaining power and prestige, the spear. Control of the means of production is potentially greater in a matrilineal society.

* * *

Gender Ideologies and Practice in Daily Life

In Melanesian societies the power of knowing is privately owned and transmitted, often through ties of kinship, to heirs or younger supporters. It comes not simply from acquiring skills or the experience and the wisdom of mature years but is fundamentally a spiritual power that derives from ancestors and other spirit forces.

In gender-segregated societies, such as those that characterize most of Melanesia, this spiritual knowledge power is segregated as well into a male domain through male initiations or the institutions of men's houses or male religious cults. Most esoteric knowledge—and the power over others that derives from it—is available to Vanatinai women if they can find a kinsperson or someone else willing to teach it to them. There are neither exclusively male nor female collectivities on Vanatinai nor characteristically male versus female domains or patterns of sociality (cf. Strathern 1987:76).

Decisions taken collectively by Vanatinai women and men within one household, hamlet, or lineage are political ones that reverberate well beyond the local group, sometimes literally hundreds of miles beyond. A hundred years ago they included decisions of war and peace. Today they include the ritualized work of kinship, more particularly of the matrilineage, in mortuary ritual. Mortuary feasts, and the inter-island and inter-hamlet exchanges of ceremonial valuables that support them, memorialize the marriages that tied three matrilineages together, that of the deceased, the deceased's father, and the widowed spouse. Honoring these ties of alliance, contracted by individuals but supported by their kin, and threatened by the dissolution of death, is the major work of island politics. . . .

The small scale, fluidity (cf. Collier and Rosaldo 1981), and mobility of social life on Vanatinai, especially in combination with matriliny, are conducive of egalitarian social relations between men and women and old and young. They promote an ethic of respect for the individual, which must be integrated with the ethic of cooperation essential for survival in a subsistence economy. People must work out conflict through face-to-face negotiation, or existing social ties will be broken by migration, divorce, or death through sorcery or witchcraft.

Women on Vanatinai are physically mobile, traveling with their families to live with their own kin and then the kin of their spouse, making journeys in quest of valuables, and attending mortuary feasts. They are said to have traveled for these reasons even in precolonial times when the threat of attack was a constant danger. The generally greater physical mobility of men in human societies is a significant factor in sexual asymmetries of power, as it is men who generally negotiate and regulate relationships with outside groups (cf. Ardener 1975:6).

Vanatinai women's mobility is not restricted by ideology or by taboo, and women build their own far-ranging personal networks of social relationships. Links in these networks may be activated as needed by the woman to the benefit of her kin or hamlet group. Women are confined little by taboos or community pressures. They travel, choose their own marriage partners or lovers, divorce at will, or develop reputations as wealthy and generous individuals active in exchange.

Big Men, Big Women, and Chiefs

Vanatinai giagia, male and female, match Sahlins's (1989) classic description of the Melanesian big man, except that the role of gia is gender-blind. There has been renewed interest among anthropologists in recent years in the big man form of political authority.[2] The Vanatinai case of the female and male giagia offers an intriguing perspective. . . .

Any individual on Vanatinai, male or female, may try to become known as a gia by choosing to exert the extra effort to go beyond the minimum contributions to the mortuary feasts expected of every adult. He or she accumulates ceremonial valuables and other goods both in order to give them away in acts of public generosity and to honor obligations to exchange partners from the local area as well as distant islands. There may be more than one gia in a particular hamlet, or even household, or there may be none. A woman may have considerably more prestige and influence than her husband because of her reputation for acquiring and redistributing valuables. While there are more men than women who are extremely active in exchange, there are some women who are far more active than the majority of men.

Giagia of either sex are only leaders in temporary circumstances and if others wish to follow, as when they host a feast, lead an exchange expedition, or organize the planting of a communal yam garden. Decisions are made by consensus, and the giagia of both sexes influence others through their powers of persuasion, their reputations for ability, and their knowledge, both of beneficial magic and ritual and of sorcery or witchcraft. . . .

On Vanatinai power and influence over the actions of others are gained by achievement and demonstrated superior knowledge and skill, whether in the realm of gardening, exchange, healing, or sorcery. Those who accumulate a surplus of resources are expected to be generous and share with their neighbors or face the threat of the sorcery or witchcraft of the envious. Both women and men are free to build their careers through exchange. On the other hand both women and men are free not to strive toward renown as giagia but to work for their own families or simply to mind their own business. They can also achieve the respect of their peers, if they seek it at all, as loving parents, responsible and hardworking lineage mates and affines, good gardeners, hunters, or fishers, or skilled healers, carvers, or weavers.

Mead (1935) observes that societies vary in the degree to which "temperament types" or "approved social personalities" considered suitable for each sex or a particular age category differ from each other. On Vanatinai there is wide variation in temperament and behavior among islanders of the same sex and age. The large amount of overlap between the roles of men and women on Vanatinai leads to a great deal of role flexibility, allowing both individual men and women the freedom to specialize in the activities they personally enjoy, value, are good at performing, or feel like doing at a particular time. There is considerable freedom of choice in shaping individual lifestyles.

An ethic of personal autonomy, one not restricted to the powerful, is a key precondition of social equality. Every individual on Vanatinai from the smallest child to an aged man or woman possesses a large degree of autonomy. Idiosyncrasies of personality and character are generally tolerated and respected. When you ask why someone does or does not do something, your friends will say, emphatically and expressively, "We [inclusive we: you and I both] don't know," "It is something of theirs" [their way], or "She doesn't want to."

Islanders say that it is not possible to know why a person behaves a certain way or what thoughts generate an action. Persisting in a demand to "know" publicly the thoughts of others is dangerous, threatening, and invasive. Vanatinai people share, in part, the perspectives identified with postmodern discussions of the limits of ethnographic representation: It is impossible to know another person's thoughts or feelings. If you try they are likely to deceive you to protect their own privacy or their own interests. Your knowing is unique to you. It is your private property that you transmit only at your own volition, as when you teach magical spells to a daughter or sister's son.[3]

The prevailing social sanction is also individualistic: the threat of somebody else's sorcery or witchcraft if you do not do what they want or if you arouse envy or jealousy. But Vanatinai cultural ideologies stress the strength of individual will in the face of the coercive pressures of custom, threat of sorcery, and demands to share. This leads to a Melanesian paradox: The ethic of personal autonomy is in direct conflict to the ethic of giving and sharing so highly valued on Vanatinai, as in most Melanesian cultures. Nobody can make you share, short of stealing from you or killing you if you refuse them. You have to want to give: your nurture, your labor, your valuables, and your person. This is where persuasion comes in. It comes from the pressure of other people, the force of shame, and magical seduction made potent by supernatural agency. Vanatinai custom supplies a final, persuasive argument to resolve this paradox: By giving, you not only strengthen your lineage and build its good name, you make yourself richer and more powerful by placing others in your debt.

What can people in other parts of the world learn from the principles of sexual equality in Vanatinai custom and philosophy? Small scale facilitates Vanatinai people's emphasis on face-to-face negotiations of interpersonal conflicts without the delegation of political authority to a small group of middle-aged male elites. It also leaves room for an ethic of respect for the will of the individual regardless of age or sex. A culture that is egalitarian and nonhierarchical overall is more likely to have egalitarian relations between men and women.

Males and females on Vanatinai have equivalent autonomy at each life cycle stage. As adults they have similar opportunities to influence the actions of others. There is a large amount of overlap between the roles and activities of women and men, with women occupying public, prestige-generating roles. Women share control of the production and the distribution of valued goods, and they inherit property. Women as well as men participate in the exchange of valuables, they organize feasts, they officiate at important rituals such as those for yam planting or healing, they counsel their kinfolk, they speak out and are listened to in public meetings, they possess valuable magical knowledge, and they work side by side in most subsistence activities. Women's role as nurturing parent is highly valued and is the dominant metaphor for the generous men and women who gain renown and influence over others by accumulating and then giving away valuable goods.

But these same characteristics of respect for individual autonomy, role overlap, and public participation of women in key subsistence and prestige domains of social life are also possible in large-scale industrial and agricultural societies. The Vanatinai example suggests that sexual equality is facilitated by an overall ethic of respect for and equal treatment of all categories of individuals, the decentralization of political power, and inclusion of all categories of persons (for example, women and ethnic minorities) in public positions of authority and influence. It requires greater role overlap through increased integration of the workforce, increased control by women and minorities of valued goods—property, income, and educational credentials—and increased recognition of the social value of parental care. The example of Vanatinai shows that the subjugation of women by men is not a human universal, and it is not inevitable. Sex role patterns and gender ideologies are closely related to overall social systems of power and prestige. Where these systems stress personal autonomy and egalitarian social relations among all adults, minimizing the formal authority of one person over another, gender equality is possible.

Notes

1. See, for example, Rogers (1975) and Collier and Rosaldo (1981) on ideal versus real gender relations. Ortner (1984) summarizes approaches to practice; cf. Bourdieu (1977).

2. The appropriateness of using the big man institution to define Melanesia versus a Polynesia characterized by chiefdoms, the relationship of big men to social equality, rank, and stratification, and the interactions of this form of leadership with colonialism and modernization are central issues in recent anthropological writings on big men (e.g., Brown 1987; Godelier 1986; Sahlins 1989; Strathern 1987; Thomas 1989; Lederman 1991). I discuss the implications of the Vanatinai case of the giagia at greater length in Lepowsky (1990).

3. See, for example, Clifford (1983); Clifford and Marcus (1986); and Marcus and Fischer (1986) on representations. In this book I have followed my own cultural premises and not those of Vanatinai by publicly attributing thoughts, motives, and feelings to others and by trying to find the shapes in a mass of chaotic and sometimes contradictory statements and actions. But my Vanatinai friends say, characteristically, that my writing is "something of mine"—my business.

References

Ardener, Edwin. 1975. "Belief and the Problem of Women." In Shirley Ardener, ed., *Perceiving Women*. London: Malaby.

Bourdieu, Pierre. 1977. *Outline of a Theory of Practice*. T. R. Nice. Cambridge: Cambridge University Press.

Brown, Paula. 1987. "New Men and Big Men: Emerging Social Stratification in the Third World, A Case Study from the New Guinea Highlands." *Ethnology* 26:87–106.

Clifford, James. 1983. "On Ethnographic Authority." *Representations* 1:118–146.

Clifford, James, and George Marcus, eds. 1986. *Writing Culture: The Poetics and Politics of Ethnography*. Berkeley: University of California Press.

Collier, Jane, and Michelle Rosaldo. 1981. "Politics and Gender in Simple Societies." In Sherry Ortner and Harriet Whitehead, eds., *Sexual Meanings: The Cultural Construction of Gender and Sexuality*. Cambridge: Cambridge University Press.

Godelier, Maurice. 1986. *The Making of Great Men: Male Domination and Power Among the New Guinea Baruya*. Cambridge: Cambridge University Press.

Lederman, Rena. 1991. "'Interests' in Exchange: Increment, Equivalence, and the Limits of Bigmanship." In Maurice Godelier and Marilyn Strathern, eds., *Big Men and Great Men: Personifications of Power in Melanesia*. Cambridge: Cambridge University Press.

Lepowsky, Maria. 1990. "Big Men, Big Women, and Cultural Autonomy." *Ethnology* 29(10):35–50.

Marcus, George, and Michael Fischer, eds. 1986. *Anthropology as Cultural Critique: An Experimental Moment in the Human Sciences*. Chicago: University of Chicago Press.

Mead, Margaret. 1935. *Sex and the Temperament in Three Primitive Societies*. New York: William Morrow.

Ortner, Sherry. 1984. "Theory in Anthropology Since the Sixties." *Comparative Studies in Society and History* 26(1):126–166.

Rogers, Susan Carol. 1975. "Female Forms of Power and the Myth of Male Dominance: A Model of Female/Male Interaction in Peasant Society." *American Ethnologist* 2:727–756.

Sahlins, Marshall. 1989. "Comment: The Force of Ethnology: Origins and Significance of the Melanesia/Polynesia Division." *Current Anthropology* 30:36–37.

Strathern, Marilyn. 1987. "Introduction." In Marilyn Strathern, ed., *Dealing with Inequality: Analysing Gender Relations in Melanesia and Beyond*. Cambridge: Cambridge University Press.

Thomas, Nicholas. 1989. "The Force of Ethnology: Origins and Significance of the Melanesia/Polynesia Division." *Current Anthropology* 30:27–34.

Topics for Further Examination

- Expand your understanding of non-binary gender systems by reading scholarly articles and articles in reliable popular journals such as *National Geographic* on the Hijras of India, the Fa'afafines of Samoa, and the Xanith of Oman.

- The challenges posed by ethnocentrism, including Eurocentrism, are vast, and they impact every field of study. For example, archeology, a subfield of anthropology, seeks to understand human history through a variety of physical remains (e.g., artifacts, architecture, and biofacts). Feminist archeology arose as a critique of the heteronormative patriarchal biases in Western archeology. That critique is on-going. Read various accounts of the incorrect "sexing" of the famous mummy discovered in a Viking warrior grave in Sweden (see e.g., "Viking 'Warrior' Presumed to Be a Man Is Actually a Woman," Live Science, https://www.livescience.com › History) Why was the mummy assumed to be male? How does the new "sex" information about the mummy stand to change how we view human history in the Viking Age period in Northern Europe?

- Challenge Western/Euro-American beliefs that women who practice Islam are antifeminist by reading investigative journalist and scholarly accounts of the rise of feminism among Muslim women in countries such as Indonesia, Pakistan, and Egypt.

PART II

PATTERNS

4

LEARNING AND DOING GENDER

MARY NELL TRAUTNER WITH JOAN Z. SPADE

We began this book by discussing the shaping of gender in Western and non-Western cultures. Part II expands on the idea of prisms by examining the *patterns* of gendered experiences that emerge from the practices of daily life and the interaction of gender with other socially constructed prisms. Patterns of individuals' lives are influenced by gender and other social prisms, just as multiple patterns are created by the refraction of light as it travels through a kaleidoscope containing prisms.

GENDERED PATTERNS

Social patterns are at the center of social scientists' work. Michael Schwalbe (1998), a sociologist, defines social patterns as "a regularity in the way the world works" (p. 101). For example, driving down the "right" side of the street is a regularity American people appreciate. You will read about different gendered patterns in Part II, many of which are regularities you will find problematic because they deny the individuality of women and men. Clearly, there are exceptions to social patterns; however, these exceptions are in the details, not in the regularity of social behavior itself (Schwalbe, 1998). Patterns in society are not simple and are even contradicted by other patterns. We have rigid gender expectations for things such as which colors are appropriate for children, teens, and even adults. At the same time, we practice resistance to these patterns and fluidity in the way gender is displayed in daily life. For example, an upper-class man might feel comfortable wearing a pink polo shirt to a golf tournament but not so comfortable putting a ruffled pink shirt on his 2-year-old son.

A deeper understanding of how and why particular social patterns and practices exist helps us interpret our own behavior and the world around us. Gender, as we discussed in Part I, is not a singular pattern of masculinity or femininity that carries from one situation to another. Instead, it is complex, multifaceted, and ever changing, depending on the social context, whom we are with, and where we are, as illustrated in the reading by Michela Musto in this chapter. Gender is also interpreted differently based on the community or group we associate with. That is, African American women are much less likely to adhere to idealized forms of gender or, as Karen D. Pyke and Denise L. Johnson labeled it in Chapter 2, hegemonic femininity—White, middle-class femininity. Our behavior in almost all situations is framed within our knowledge of idealized gender—hegemonic masculinity and emphasized femininity. Whether we resist or ridicule gender practices, we are almost always aware of them.

Keep the concepts of hegemonic masculinity and emphasized femininity in mind as we examine social patterns of gender. To illustrate this, let's return to the stereotype discussed in the introduction to this book—that women talk more than men. We know from research that the real social pattern in mixed-gender groups is that men talk more, interrupt more, and change the topic more often than do women (Anderson & Leaper, 1998; Brescoli, 2011;

Wood, 1999). The stereotype, while trivializing women's talk and ignoring the dominance of men in mixed-gender groups, maintains the patterns of dominance and subordination associated with hegemonic masculinity and emphasized femininity, influencing women's as well as men's behaviors. Girls—particularly White, middle-class girls—are encouraged to use a pleasant voice and not talk too much. Later, as they grow older and join mixed-gender groups at work or in play, women's voices are often ignored and they are subordinated as they monitor what they say and how often they talk, checking to make sure they are not dominating the conversation. And since gender is relational, others learn that girls talk too much and should either shut up or speak in a "nice voice." Gender is an ever-present force in defining daily behavior and is used in marketing to entice us to "buy into" gender as we purchase all kinds of products (Chapter 5). By examining how these idealized versions of masculinity and femininity pattern daily practices, we can better understand the patterns and meanings of our behavior and the behaviors of others.

Gendered patterns of belief and behavior influence us throughout our lives, from birth until old age, in almost every activity in which we engage. Readings in Part II examine the process and consequences of learning to do gender (Chapters 4 and 5) and then describe gendered patterns in work (Chapter 7) and in daily intimate relationships (Chapter 8). We also explore how gendered patterns affect our bodies, sexualities, and emotions (Chapter 6), and how patterns of dominance, control, and violence enforce gender patterns (Chapter 9).

The patterns that emerge from the gender kaleidoscope are not unique experiences in individual lives; they are regularities that occur in many people's lives. They are not static patterns that remain the same across lifetimes or history, nor are they singular patterns with one and only one way of doing gender. Gendered patterns are many and fluid across time and space. If you don't pay attention, gendered patterns may seem as though they are individual choices. Institutions and groups enforce gendered patterns and practices in the home, workplace, and daily life, as described in the readings throughout Part II. These patterns overlap and reinforce gender differences and inequalities. For example, gender discrimination in wages and promotions affects families' decisions about parenting roles and relationships. Since most men still earn more than most women, the choices of families who wish to break away from idealized gender patterns and practices are limited by decisions surrounding household income. However, these patterns are complicated by intersections with race and social class, as you will see from the readings in Chapters 7 and 8.

Learning and Doing Gender

The readings in Chapter 4 examine the processes by which we acquire self-perceptions and behaviors and learn our culture's expectations for idealized patterns of masculinities and femininities. These readings emphasize that, regardless of our inability or unwillingness to attain idealized femininity and masculinity, almost everyone in a culture learns what idealized gender is and organizes their lives around those expectations, even if in resistance to them, as is the case of parents raising gender-variant children in the reading by Elizabeth P. Rahilly in this chapter. Of course, the genderscape is complex. While some people resist idealized gender and others try to ignore these signals, some communities develop alternatives to idealized gender, such as some lesbian co-parenting relationships (Padavic & Butterfield, 2011).

Socializing Children

There are many explanations for why children gravitate toward idealized gender-appropriate behavior. The term sociologists use to describe how we learn gender is *gender socialization*, and sociologists approach it from a variety of different perspectives (Coltrane, 1998). Socialization is the process of teaching members of a society the values, expectations, and practices of the larger culture. Socialization takes place in all interactions and situations, with families and schools typically having primary responsibility for socializing infants and children in Western societies. Early attempts to explain gender socialization gave little attention to the response of individuals to agents of socialization, such as parents, peers, and teachers, and to the influence of mass media and a consumer culture. There was an underlying assumption in this early perspective that individuals were blank tablets (*tabulae rasae*) on which the cultural definitions of idealized gender and other appropriate behaviors were written. This perspective assumed that, as individuals developed, they took on a gender identity appropriate to their assigned biological sex category (Howard & Alamilla, 2001). Accepting the gender that is associated with your sex assignment is referred to as *cisgender*.

Social scientists now realize that individuals are not blank tablets, that sex categories are not easily determined (see reading by Georgiann Davis and Sharon Preves in Chapter 1), and that gender socialization is not just something that is "done" to us. Theorists now describe socialization into gender as a series of complex and dynamic processes. Individuals create, as well as respond to, social stimuli in their environments (Carlton-Ford & Houston, 2001; Howard & Alamilla, 2001). Moreover, socialization doesn't simply end after childhood. Socialization is a process that lasts across one's lifetime, from birth to death (Lutfey & Mortimer, 2003), and occurs continually with everyone we interact with—friends, peers, coworkers, and acquaintances—as well as the environment around us, including mass media, as discussed in Chapters 4, 5, and 6. Furthermore, there is a fluidity to gender ideology, with changes occurring across the life course, across race–gender categories (Vespa, 2009), and even across social contexts. For example, beauty means something different for a young child, teenager, or older person. Throughout our lives, we assess cues around us and behave as situations dictate. All socialization is, of course, reinforced by social institutions, as Barbara J. Risman discussed in the first reading in this book, and which we will discuss later in this introduction. Thus, whether we want it or not, idealized gender is a key factor in determining what is appropriate throughout our lives—even though few of us actually attain an idealized form of gender.

The dominant pattern of gender expectations, the pink and blue schema described in the introduction to this book (Paoletti, 2012), begins at birth. Once external genital identification takes place, immediate expectations for masculine and feminine behavior follow. Exclamations of "He's going to be a great baseball (or football or soccer) player" and "She's so cute" are accompanied by gifts of little sleepers in pink or blue with gender-appropriate decorations. Try as we might, it is very difficult to find gender-neutral clothing for children (see Adie Nelson's article in Chapter 5). These expectations, and the way we treat young children, reinforce idealized gender constructions of dominance and subordination and illustrate how influential the role of marketing and consumer culture is in defining idealized gender.

It is not long then, before most children come to understand that they should be "boys" and "girls," and segregate themselves accordingly. These children learn their appropriate gender behavior (defined as cisgender in many studies). Family members are not alone in teaching children to behave as "good boys" or "good girls." Almost every person a child comes into contact with and virtually all aspects of a child's material world reinforce gender. In effect, children are taught that males and females are different and that they are expected to behave accordingly. In Chapter 5, you will read more about how capitalist societies reinforce and maintain gender difference and inequality for children and adults. Television, music, books, clothing, and toys differentiate and prescribe idealized gender behavior for girls and boys. For example, studies of children's books find some distinctive patterns that reinforce idealized forms of gender. As Janice McCabe, Emily Fairchild, Liz Grauerholz, Bernice A. Pescosolido, and Daniel Tope discuss in their reading in this chapter, boys outnumber girls in books (titles and main characters) published across the 20th century. Although books continue to depict traditional gender patterns, on the plus side, researchers find that girls and women are more likely to be portrayed in gender-atypical roles in many recent children's books (Gooden & Gooden, 2001).

Most children quickly understand the idealized gender-appropriate message directed toward them and try to behave accordingly. Although not all boys are dominant and not all girls are subordinate, studies in a variety of areas find that most White boys tend toward active and aggressive behaviors, while most White girls tend to be quieter and more focused on relationships. The patterning for African American boys is similar, and African American boys who do not act in gender-appropriate ways are seen as "soft" or feminine (Carter, 2005). These patterns for boys and girls have been documented in school and in play (e.g., Sadker & Sadker, 1994; Thorne, 1993).

It is important to note that the consequences for gender-appropriate behavior are not entirely positive. Gender-appropriate behavior is related to lower self-confidence and self-esteem for girls (e.g., Eder, 1995; Orenstein, 1994; Spade, 2001), whereas boys are taught to "mask" their feelings and compete with everyone for control, thus isolating themselves and ignoring their own feelings (e.g., Connell, 2000; Messner, 1992; Pollack, 2000).

Social Institutions and Socialization

Many of our social institutions segregate children and adults by gender as well, thus creating gendered identities. All adults, not just parents, play a major role in teaching gender. Teachers also teach gender, and when they separate children into gender-segregated spaces in lunch lines or playground areas, they reinforce gender differences (Sadker & Sadker, 1994; Thorne, 1993).

However, teachers are becoming more aware of their role in gendering children and, in some contexts, such as the swimming team described in the reading by Musto in this chapter, the importance of gender can be made irrelevant by focusing on the task at hand. Yet, as shown in Musto's reading, these lessons may not carry over to other contexts.

Schools, however, typically reinforce separate and unequal spheres for boys and girls (Orenstein, 1994; Sadker & Sadker, 1994; Thorne, 1993). Considerable research by the American Association of University Women (1992, 1998, 1999) documents how schools "shortchange" girls. Schools are social institutions that maintain patterns of power and dominance. Indeed, we teach dominance in schools in patterns of teacher–student interactions such as respecting the responses of boys while encouraging girls to be helpers in the classroom (Grant, 1985; Sadker & Sadker, 1994). A study using data collected from individuals during their high school years (2002, 2004, and 2006) showed that gender socialization in schools varies based on the race of the student. For example, math teachers tend to hold a biased perception of girls' abilities, particularly when comparing the abilities of White girls with those of White boys (Riegle-Crumb & Humphries, 2012). The reading by Maria Charles in this chapter describes how schools reinforce choices related to science, technology, engineering, and mathematics (STEM), eventually solidifying the gender segregation of STEM careers. Unfortunately, gender socialization and expectations continue well into the STEM careers, with some women scientists enforcing gender norms and expectations by distancing themselves from femininity and "typical feminine practices" as other women fight gender discrimination (Rhoton, 2011). Thus, according to Laura A. Rhoton, women scientists prefer to associate with women who act the role of "scientist" rather than with women who practice femininity, further reinforcing the perception of appropriate gender in science. And, unfortunately, this climate of gender difference and inequality persists in the workplace, as you will read about in the readings Chapter 7.

However, not all boys and men are allowed to be dominant across settings (Eder, 1995), and few girls come close to achieving idealized femininity. Ann Arnett Ferguson (2000) describes how schools discourage African American boys from claiming their Blackness and masculinity. Although White boys may be allowed to be "rambunctious" and disrespectful, African American boys are punished more severely than their White peers when they "act out," and there is less tolerance for African American boys who try to dominate. Girls also exist within a hierarchy of relationships (Eder, 1995). Girls from racial, ethnic, economically disadvantaged, or other subordinated groups must fight even harder to succeed under multiple systems of domination and inequality in schools. To illustrate this point, Julie Bettie (2002) compared the paths to success for upwardly mobile White and Mexican American high school girls and found some similarities in gender experiences at home and school, such as participation in sports, which facilitated mobility for both groups of girls. But, there were also differences in their experiences because race was always salient and was a barrier for the Mexican American girls. However, Bettie believes that achieving upward mobility may have been easier for these Mexican American girls than for their brothers because it is easier for girls to transgress gender boundaries. Their brothers, on the other hand, felt pressure to "engage in the rituals of proving masculinity" even though this behavior was rejected by those in control at school (p. 419).

Bettie's (2002) study emphasizes the fact that multiple social prisms of difference and inequality create an array of patterns, which would not be possible if gender socialization practices were singular or universal. Individuals' lives are constructed around many factors, including gender. Cultural values and expectations influence, and frequently contradict, the maintenance of hegemonic masculinity and emphasized femininity in Western societies. Pyke and Johnson's reading in Chapter 2 and other readings throughout this book illustrate how the practice of gender is strongly influenced by culture. The process of gender socialization is rooted in the principle that girls/women and boys/men are not equal and that the socially constructed categories of difference and inequality (gender, race, ethnicity, class, religion, age, culture, etc.) are legitimate.

Sports, particularly organized sports, provide other examples of how institutionalized activities reinforce the gender identities children learn. Boys learn the meaning of competition and success, including the idea that winning is everything (e.g., Messner, 1992). Girls, on the other hand, are more often found on the edges of the playing field, or on the sides of the playground, watching the boys (Thorne, 1993). And, as you will read in Chapter 9, moms are typically relegated to the sidelines as well, while men coach. Even though girls are more involved in playing sports than in previous generations, many scholars note that girls and women are still expected to maintain some level of femininity during athletic competitions. Yet not all children play in the same ways. Marjorie Harness Goodwin (1990) finds that children from urban, lower-class, high-density neighborhoods—where households are closer together—are more likely to play in mixed-gender and mixed-age groups. In suburban, middle-class households, which are farther apart than urban households, parents are more likely to drive their children to sporting activities or houses to play with same-sex/gender, same-age peers. The consequences of social class and place of residence are that lower-class children are more comfortable with their sexuality as they enter preadolescence and are less likely to gender segregate in school (Goodwin, 1990).

Sports and play continue to segregate us and define gender throughout our lives. Although women are increasingly participating in traditionally "male" sports activities, the gendered nature of these sports still exists in the institutions supporting the activities (Buysse & Embser-Herbert, 2004), the game rules, and the minds of the participants, even in traditionally "male" sports such as basketball (Berlage, 2004), ice hockey (Theberge, 2000), and body building (Wesely, 2001). Socialization into gender does not stop at any particular age but occurs throughout our lifetimes and throughout our activities.

Gender Transgressions

Change in social expectations comes slowly, but today there is more acceptance of individuals who do not accept or feel comfortable in the gender appropriate for their assigned sex, such as Bruce Jenner, the former Olympian and reality show star who now identifies as a woman, Caitlyn. Rahilly addresses this issue in her piece in this chapter, which focuses on parents who are raising children who refuse to align their genders with expectations for behavior appropriate to their assigned sex. While all this attention to transgender might seem like a breakthrough for creating more flexibility in the hold gender has on our life choices, the acceptance of transgender may not be breaking down the gender binary. Instead, Eleanor Burkett (2015) asks in a *New York Times* editorial whether Bruce Jenner's coming out as a woman, Caitlyn, further reinforces the stereotypes and a gender binary. In fact, Jenner's public transgender event actually made those "tidy boxes" of boys/men and girls/women much more rigid as Bruce went from a man who was a star athlete to Caitlyn, a voluptuous woman laid out neatly and passively on the cover of *Vanity Fair*.

Although there is an increasing acceptance of transgendered people, beginning in the early 2000s, there still occurs what Laurel Westbrook and Kristen Schilt (2014) call "gender panics," when people's assumptions about biology-based gender ideology are disrupted. Using newspaper articles to determine how the press and public responded to such disruptions, they examined press reports of several events, including the move by New York City to allow individuals to change the sex listed on their birth certificates without requiring proof of genital reconstruction (a biological change that corresponded with their new sex identification). They also examined articles relating to competitive sports and workplace discrimination. They concluded that people were more likely to require biology-based criteria for gender if the activity was sex-segregated, particularly if the requirements protected females in female spaces from males or trans women who may still have some male biological sex characteristics. In sex-segregated competitive sports, however, identity-based definitions of gender were less likely to be accepted and, trans women who want to compete in women's competitions are more likely to have biological markers that indicate they are female, such as genital reconstruction and/or testosterone levels similar to females. However, in workplace discrimination, a space that is not sex-segregated, they found there is more acceptance of an identity-based definition of gender. These gender panics, as Westbrook and Schilt (2014) call them, are not new; rather, they are just becoming more apparent and a bit more complicated by the acceptance of identity-based markers of gender as opposed to biological markers.

Until recently, most persons who went through sex change operations were referred to as transsexuals. Yet, even with the appropriate biological markers, the transition from one gender to another was not necessarily easy. Renée Richards, who, in 1975 underwent sex reassignment surgery at the age of 40, was initially denied a spot in the U.S. Open Tennis Tournament in 1977 because she wasn't "really" a woman. That decision was reversed and she was allowed to compete in professional women's tennis. This change in terminology over time reflects an acknowledgment of the powerful force of social factors in determining our gender identities and moves away from the idea that gender is an essential part of our biological nature.

While most children learn to display idealized gender behaviors, at times we all step out of gender-appropriate zones in our daily lives. Girls and women are more likely to transgress and do masculine things than boys and men are to participate in feminine activities. C. Shawn McGuffey and B. Lindsay Rich (1999) find that girls who transgress into the "boys' zone" may eventually be respected by their male playmates if they are good at conventionally male activities, such as playing baseball. Boys, however, are harassed and teased when they try to participate in any activity associated with girls (McGuffey & Rich, 1999). By denying boys access to girls' activities, the dominance of masculinity is reinforced as when boys are ridiculed because they are not sufficiently dominant or because they "throw like a girl." Therefore, boys reinforce and maintain masculinity by goading one another to perform "manhood acts" (Schrock & Schwalbe, 2009), and we all end up devaluing feminine acts, a point underscored in Kristen Myers' article in this chapter about non-hegemonic representations of masculinity in popular television shows aimed at children. She finds that male characters in these shows who are not hegemonically masculine actually reestablish the dominance of hegemonic masculinity and ultimately reinforce traditional ideas about gender.

As you can see, learning gender is complicated. Clearly, gender is something that we "do" as much as learn, and in doing gender, we are responding to structured expectations from institutions in society as well as interpersonal cues from those we are interacting with. Throughout our lives, every time we enter a new social situation, we look around for cues and guides to determine how to behave in an appropriate manner. In some situations we might interpret gender cues as calling for a high degree of idealized gender difference and inequality, while in other situations the clues allow us to be more flexible. Thus, we create gender as well as respond to expectations for it. And we change gender when we resist it!

Doing Masculinity and Femininity Throughout Our Lives

Most men have learned to "do" the behaviors that maintain hegemonic masculinity, while at the same time suppressing feelings and behaviors that might make them seem feminine (Connell, 1987). As a result, being a "man" or a "woman" requires an awareness of and responses to the other gender. Our cues and behaviors change whether we are responding to someone we identify as being of the same gender or of a different gender. That is, masculinities or femininities are enacted based on how those we are interacting with are displaying femininities or masculinities (Connell & Messerschmidt, 2005).

As argued, hegemonic masculinity is maintained in a hierarchy in which only a few men achieve close-to-idealized masculinity, with everyone else subordinated to them—women, poor White men, men of color, gay men, and men from devalued ethnic and religious groups. Furthermore, this domination is not always one-on-one but, rather, can be institutionalized in the structure of the situation. As you read the articles in this chapter, you will see that gender is not something we learn once, in one setting. Instead, we learn to do gender over time in virtually everything we undertake. And, although we do gender throughout our lives, we rarely achieve idealized gender; yet, by doing gender, we continue to maintain a system of gender difference and inequality. Also, remember that learning to "do" gender is complicated by the other prisms that interact in our lives. Recall the lessons from Chapter 2 and remember that gender does not stand alone but, rather, is reflected in other social identities.

It is not easy to separate the learning and doing of gender from other patterns. As you read selections in other chapters in Part II of this book, you will see the influence of social processes and institutions on how we learn and do gender across all aspects of our lives. Before you start to read, ask yourself how you learned gender and how well you do it. Not succeeding at doing gender is normal. That is, if we all felt comfortable with ourselves, no one would be striving for idealized forms of gender—hegemonic masculinity or emphasized femininity. Imagine a world in which we all feel comfortable just the way we are! As you read through the rest of this book, ask yourself why that world doesn't exist.

References

American Association of University Women. (1992). *How schools shortchange girls.* Washington, DC: Author.

American Association of University Women. (1998). *Gender gaps: Where schools still fail our children.* Washington, DC: Author.

American Association of University Women. (1999). *Voices of a generation: Teenage girls on sex, school, and self.* Washington, DC: Author.

Anderson, K. J., & Leaper, C. (1998). Meta-analysis of gender effects on conversational interruption: Who, what, when, where, and how. *Sex Roles, 39*(3–4), 225–252.

Berlage, G. I. (2004). Marketing and the publicity images of women's professional basketball players from 1997 to 2001. In J. Z. Spade & C. G. Valentine (Eds.), *The kaleidoscope of gender: Prisms, patterns, and possibilities* (pp. 377–386). Belmont, CA: Wadsworth.

Bettie, J. (2002). Exceptions to the rule: Upwardly mobile White and Mexican American high school girls. *Gender & Society, 16*(3), 403–422.

Brescoli, V. (2011). Who takes the floor and why: Gender, power, and volubility in organizations. *Administrative Science Quarterly, 56*(4), 622–641.

Burkett, E. (2015, June 6). "What makes a woman?" *The New York Times*. Retrieved from http://www.nytimes.com/2015/06/07/opinion/sunday/what-makes-a-woman.html?emc=edit_th_20150607&nl=todaysheadlines&nlid=44913438

Buysse, J. M., & Embser-Herbert, M. S. (2004). Constructions of gender in sport: An analysis of intercollegiate media guide cover photographs. *Gender & Society, 18*(1), 66–81.

Carlton-Ford, S., & Houston, P. V. (2001). Children's experience of gender: Habitus and field. In D. Vannoy (Ed.), *Gender mosaics: Societal perspectives* (pp. 65–74). Los Angeles: Roxbury.

Carter, P. L. (2005). *Keepin' it real: School success beyond Black and White*. New York: Oxford University Press.

Coltrane, S. (1998). *Gender and families*. Thousand Oaks, CA: Pine Forge Press.

Connell, R. W. (1987). *Gender and power: Society, the person, and sexual politics*. Stanford, CA: Stanford University Press.

Connell, R. W. (2000). *The men and the boys*. Berkeley: University of California Press.

Connell, R. W., & Messerschmidt, J. W. (2005). Hegemonic masculinity: Rethinking the concept. *Gender & Society, 19*(6), 829–859.

Eder, D. (1995). *School talk: Gender and adolescent culture*. New Brunswick, NJ: Rutgers University Press.

Ferguson, A. A. (2000). *Bad boys: Public schools in the making of Black masculinity*. Ann Arbor: University of Michigan Press.

Gooden, A. M., & Gooden, M. A. (2001). Gender representation in notable children's picture books: 1995–1999. *Sex Roles, 45*(1–2), 89–101.

Goodwin, M. H. (1990). *He-said-she-said: Talk as social organization among Black children*. Bloomington: Indiana University Press.

Grant, L. (1985). Race-gender status, system attachment, and children's socialization in desegregated classrooms. In L. C. Wilkinson & C. Bagley Marret (Eds.), *Gender influences in classroom interaction* (pp. 57–77). New York: Academic Press.

Howard, J. A., & Alamilla, R. M. (2001). Gender and identity. In D. Vannoy (Ed.), *Gender mosaics: Societal perspectives* (pp. 54–64). Los Angeles: Roxbury.

Lutfey, K., & Mortimer, J. C. (2003). Development and socialization through the adult life course. In J. Delamater (Ed.), *Handbook of social psychology* (pp. 183–204). New York: Kluwer/Plenum.

McGuffey, C. S., & Rich, B. L. (1999). Playing in the gender transgression zone. *Gender & Society, 13*(5), 608–627.

Messner, M. A. (1992). *Power at play: Sports and masculinity*. Boston: Beacon Press.

Orenstein, P. (1994). *School girls: Young women, self-esteem, and the confidence gap*. New York: Anchor Books.

Padavic, I., & Butterfield, J. (2011). Mothers, fathers, and "mathers" negotiating a lesbian co-parental identity. *Gender & Society*, *25*(2), 176-196.

Paoletti, J. B. (2012). *Pink and blue: Telling the boys from the girls in America*. Bloomington: Indiana University Press.

Pollack, W. S. (2000). *Real boys' voices*. New York: Random House.

Rhoton, L. A. (2011). Distancing as a gendered barrier: Understanding women scientists' gender practices. *Gender & Society, 25*(6), 696–716.

Riegle-Crumb, C., & Humphries, M. (2012). Exploring bias in math teachers' perceptions of students' ability by gender and race/ethnicity. *Gender & Society, 26*(2), 290–322.

Sadker, D., & Sadker, M. (1994). *Failing at fairness: How our schools cheat girls*. New York: Simon & Schuster.

Schrock, D., & Schwalbe, M. (2009). Men, masculinity, and manhood acts. *Annual Review of Sociology, 35*, 277–295.

Schwalbe, M. (1998). *The sociologically examined life: Pieces of the conversation*. Mountain View, CA: Mayfield.

Spade, J. Z. (2001). Gender and education in the United States. In D. Vannoy (Ed.), *Gender mosaics: Societal perspectives* (pp. 85–93). Los Angeles: Roxbury.

Theberge, N. (2000). *Higher goals: Women's ice hockey and the politics of gender*. Albany: State University of New York Press.

Thorne, B. (1993). *Gender play: Girls and boys in school*. New Brunswick, NJ: Rutgers University Press.

Vespa, J. (2009). Gender ideology construction: A life course and intersectional approach. *Gender & Society, 23*(3), 363–387.

Wesely, J. K. (2001). Negotiating gender: Bodybuilding and the natural/unnatural continuum. *Sociology of Sport Journal, 18*, 162–180.

Westbrook, L., & Schilt, K. (2014). Doing gender, determining gender: Transgender people, gender panics, and the maintenance of the sex/gender/sexuality system. *Gender & Society, 28*(1), 32-57.

Wood, J. T. (1999). *Gendered lives: Communication, gender, and culture* (3rd ed.). Belmont, CA: Wadsworth.

Introduction to Reading 17

Parents play a major role in socializing young children into appropriate genders, the genders that match the sexes assigned to them at birth. Males become boys and females become girls, by learning to present themselves in appropriate ways such that others recognize their gender. An individual whose gender and sex align is referred to as cisgender. However, not every child wants to be socialized into their "appropriate" gender. Elizabeth P. Rahilly addresses this issue, focusing on parents who are raising children who refuse to align their genders with expectations for behavior appropriate to their assigned sex. She conducted interviews with 24 parents of 16 gender-variant children. These parents are predominantly White, middle class, and well educated. In this reading, she describes the strategies parents developed to deal with their children's gender variance, with most of their attention focused on the people their children interacted with. She recruited these parents at a support conference for parents of gender-variant children and also from an Internet blog that one of the parents in her study authored. This research provides an interesting glimpse at parenting against the norm.

1. How did the parents respond to their children's desire to be gender variant?
2. What role did outsiders play in the way parents managed the gender variance of their children?
3. In accepting the gender variance of their children, are these parents still upholding the "truth regime" of a gender binary?

The Gender Binary Meets the Gender-Variant Child

Parents' Negotiations With Childhood Gender Variance

Elizabeth P. Rahilly

Tristan's just everything, he's not limited, and I think part of it is that gender thing, there's no boxes for him . . . I just want to keep it that way, I don't want the world to crush him.

—Shella[1]

Transgender identity has long been significant in sociocultural analyses of gender (Bornstein 1995; Halberstam 2005). Gender variance exposes the limits of the gender binary and the overly deterministic role it ascribes to assigned sex, in turn signaling possibilities for social change against dominant ideologies and practices. Pursuant to West and Zimmerman's (1987) canonical distinctions between sex, sex category, and gender, several empirical studies have addressed trans persons' experiences to illuminate the logics of the gender binary, both when it prevails and when it is troubled (Connell 2010; Gagne and Tewksbury 1998; Jenness and Fenstermaker 2014; Schilt and Westbrook 2009). As crucial as these studies are to a sociology of gender, their principal substrate for analysis has been *adult* experiences and perspectives. Save Tey Meadow's research (2011, 2013), childhood gender variance is largely absent from the empirical repertoire. Only recently has the prospect of raising children as categorically "gender-variant" or "transgender"[2] surfaced on the cultural landscape.

Over the last decade, preadolescent gender-variant children have garnered widespread visibility, beyond the walls of the "medicopsychological" clinic, where much of the research on, and management of, childhood gender variance traditionally has occurred (Bryant 2006). These children's behaviors consistently and significantly stray from the expectations of their assigned sex—from the clothes, toys, and play groups they prefer to their repeated articulations about their sense of self (e.g., "I'm your son, not your daughter!"). This visibility is due in no small part to the parents who raise these children and reject traditional

Rahilly, E. P. (2015). The gender binary meets the gender-variant child. *Gender & Society, 29*(3): 338–61. Reprinted by permission of SAGE Publications Inc., on behalf of Sociologists for Women in Society.

reparative interventions (e.g., Green 1987; Zucker 2008). An increasing number of mental health practitioners reject reparative approaches as well (Ehrensaft 2011; Lev 2004).

This article draws on interview data with 24 parents of gender-variant children, who represent 16 childhood cases altogether and are part of a larger longitudinal project on parents of gender-variant and transgender children. I examine three practices—"gender hedging," "gender literacy," and "playing along"—to illuminate the ways in which these parents come to an awareness of the gender binary as a limited cultural ideology, or a "truth regime" (Foucault 2000), and in turn devise various practical and discursive strategies to navigate that regime and accommodate their children's nonconformity. These parents widen the options their children have, not only regarding interests and activities, as conventional "gender-neutral" parenting would advocate, but also with regard to a potentially transgender sense of self. They also adhere to essentialist understandings of gender identity and expression, in ways that expand, rather than limit, the range of gendered possibilities. Altogether, these families are inventing a new mode of social response to a problem that would, in previous decades, be the province of psychotherapeutic intervention and exposing new challenges to the gender binary during early childhood development.

Parenting and Gender: The Gender "Truth Regime"

Following her work on "transgender families," Pfeffer (2012) called for more concerted research into "the increasingly diverse family forms of the twenty-first century," whose members expose new strategies for negotiating and resisting gender norms (Pfeffer 2012, 596). Meadow's (2011, 2013) ethnography offered some of the first insights into a new generation of parents who are raising transgender children. Meadow found that parents drew on traditional explanatory tropes—including biomedical, psychiatric, and spiritual—to explain their child's gender-variant "self" to others, thereby "assimilat[ing] their children's atypical identities into familiar knowledge and belief systems" (Meadow 2011, 728). Meadow argues that these traditional frameworks bear as much potential for embracing non-normative genders as they do for constricting them. I build on this budding sociological terrain, turning my focus to specific methods and strategies parents develop in everyday interactions to navigate the gender binary, starting with their initial reckonings with the gender binary as faulty cultural ideology.

Of course, parents' potential to challenge the gender binary is not new. Attendant with ideological aspects of second-wave feminism, many scholars have been interested in parenting practices that resist stereotyping male and female children, often referred to as "gender-neutral" or "feminist" parenting (Bem 1983; Pogrebin 1980; Risman and Meyers 1997; Statham 1986). Both Kane (2006) and Martin (2005), however, have noted the limited legacy of such parenting ideals, which they attribute to negative cultural associations between childhood gender nonconformity and adult homosexuality, fostering parents' maintenance of compulsory heterosexuality and hegemonic masculinity. More recently, Kane (2012) revealed a range of tendencies among contemporary parents, from those who presume stereotypical gender behaviors in their children to those who consciously seek to widen their children's social options. Nonetheless, Kane observed that almost all parents succumb to the "gender trap," or social expectations that limit parents' best intentions against the binary (Kane 2012, 3). Even the most progressively minded parents in Kane's sample still felt accountable to a modicum of gender normativity in public, especially with their sons. And few, if any of them, seemed cognizant of the prospect of a transgender child. Indeed, one of the parents in Kane's sample—who chased down a store clerk when the clerk assumed her boys would not use glitter in a crafts project—easily dismissed the notion that her three-year-old son would grow up to be a "girl": "So I said, 'Eli, you'll never be a girl, but if you want that Barbie pool you can have it'" (Kane 2012, 150). As traditionally conceived, gender-progressive parenting encouraged boys and girls to be whatever they wanted to be, regardless of stereotypes—but they were ever and always (cisgender[3]) boys and girls, respectively.[4]

In this article, the "gender binary" refers to a dominant cultural presumption about sex and gender: namely, that there is an expected "congruent" relationship between one's sexed body and their gender identity and expression—that is, babies assigned "male" grow up to be "boys" and babies assigned "female" grow up to be "girls," and without many options in between. I use "male" and "female" to refer to the sexual anatomy that is coded at birth, and "boy" and "girl" to refer to the gender identities that are presumed of bodies assigned as such. Many parents no longer expect stereotypically "masculine" and "feminine" behaviors from their children—and often laud a child for stepping outside these in certain respects (e.g., boys who exhibit sensitivity, girls who prefer sports to Barbies). However, the presumption that a child's assigned sex will predict and circumscribe their gendered sensibilities and identities ("boy" or "girl") still holds force in our culture. The first question that is asked after a child is born is the first of many iterations of this belief system, around which myriad institutions and practices are arranged. . . .

I employ the concept of "truth regime" to analyze the practical, discursive, and intellectual strategies these parents engage in to navigate the gender binary and legitimize childhood gender-variant subjectivities. The power of the gender truth regime lies in its erasure of childhood transgender possibilities; children assigned male who present as "girls" and children assigned female who present as "boys"[5] are culturally unintelligible. And the parents who permit such possibilities are implicated negatively by others, including neighbors, doctors, teachers, and extended family, who might question the apparent "mismatch" (especially if they were aware of the assigned sex). I draw on the concept of "truth regime" to examine these parents' newfound negotiations with, and increasing resistance to, the gender binary in the face of its regulatory effects, particularly during everyday discursive interactions.

The "truth regime" framework intersects with the "doing gender" approach. The dictates of the gender truth regime powerfully inform interactional practices, to which parents at first feel accountable. However, parents' growing awareness of the falsehoods of the gender binary enables "redoing gender" (West and Zimmerman 2009), or "doing gender differently" (Dalton and Fenstermaker 2002), through which alternative gender practices become possible. Rather than "undoing gender" altogether, parents still attribute gendered meaning to their children's preferences—they are atypically masculine or feminine, but masculine and feminine nonetheless. I use the term "truth regime" to emphasize the discursive and ideological foundations of the gender binary that parents work to resist, through which changes to the system of normative gender accountability might transpire.

As I demonstrate, parents' strategies emerge *in response* to children's demands and preferences, and not necessarily due to a "gender-neutral" agenda of their own. This child-directed dynamic speaks to more general "bidirectional" or "reciprocal influence" theories of childhood socialization (Coltrane and Adams 1997; Peterson and Rollins 1987), in which both parents and children are seen as active agents in the progress.

* * *

Challenging the Gender Truth Regime

By the time they enter parenthood, many adults have internalized a dominant cultural ideology that presumes a deterministic relationship between sex and gender; "males" are boys and "females" are girls. But the parents in this study confront the limits of these "certainties" in the face of their children's persistent gender-variant preferences and expressions. In this section, I describe three major practices that surfaced in parents' narratives: "gender hedging," "gender literacy," and "playing along." Through these practices, parents come to an awareness of the gender binary as a restrictive truth regime, and work to carve out more inclusive understandings of, and practices around, gender nonconformity, despite a world that is largely ignorant of childhood transgender possibilities.

Gender Hedging: "Walking the Fine Line" of the Gender Binary

When referring to the early stages of their parenting careers, before they grew cognizant of the prospect of a "gender-variant" or "transgender" child, almost all parents described engaging in a kind of boundary work with their children's "atypical" behaviors, especially as the child approached school age. I refer to this work as "gender hedging," or parents' creative efforts to curb their child's nonconformity and stay within gender-normative constraints. A parent purchases pink socks for their "son," for example, but not a skirt. I introduce gender hedging as parents' first strategic negotiations with the gender truth regime, as it marks a crucial phase in their developing consciousness about this dominant belief system: gender proves as much a set of cultural dictates to which parents feel beholden as it does a given "truth" about their child's sex, which offers little reference for their child's persistent preferences and behaviors. While gender hedging largely upholds the gender truth regime, as parents work to fashion an overall front of normativity (e.g., no dresses at the store), it also permits small concessions to a child's gender-variant interests (e.g., a pink shirt is okay), and stirs parents' questioning about how much of these they should regulate and restrain, if at all.

When I asked parents to orient me to their child's gender nonconformity, they listed a variety of activities their child engaged in, often starting around the age of two, that were the stereotypical stock of the "other sex's" interests and preferences, including toys, clothing, types of play, and friend groups.[6] Tim, for example, adored playing dress-up in an Ariel mermaid costume (which Beth purchased for him after he repeatedly begged for it at the store), which came complete with jewelry and high heels. However, the outfit proved "too much" for Beth and Barry to accept, and lines were drawn regarding the extent to which Tim could wear it: The dress was allowed, but not the accessories, and only indoors. Tim also wanted to carry a purse in public. Beth offered him a "substitute"; she gave him a small boutique shopping bag with handles instead of a woman's handbag, so as to be less conspicuous. Beth described such efforts as a "daily tightrope walk" and "a fine line that [they] walk."

Katy also remembered trying to accommodate her child's preferences for girly clothes in public: "He started wearing some feminine stuff, [at the store] I'd pick out, okay, it's not pink but it's got Hello Kitty on it, that'll be okay, you know." Theresa recalled her efforts to "soft pedal" around Lisa's girliness in one emblematic move: when Lisa started kindergarten, Theresa made an interactive chart with popsicle sticks designating the kinds of daily attire Lisa could wear to school. On some days, Lisa could wear more feminine clothes (skorts—half shorts, half skirt); on others, she had to wear boy clothes (pants). Now, Theresa cringes at the thought of it, but at the time, she felt she had to "enforce a balance . . . not to go all the way into girly-girly land." . . .

Parents, in both male and female cases, also expressed fear about their child's risk of bullying and exclusion, which largely compelled their early efforts to keep the nonconformity at bay and indoors.

Notably, Kane (2006, 2012) described similar kinds of "boundary maintenance" among the parents she studied, who allowed gender-atypical play indoors but ensured gender-normative presentations in public, especially with male children. However, the parents in this study eventually allowed children assigned male access to proverbial "icons of femininity" (Kane 2006), including frilly skirts and dresses outside of the house, and long hair. With children assigned female, parents obliged more and more clothing from the boys' department and short haircuts. Moreover, these parents mentioned what their children *said* (i.e., "I'm your son, not your daughter!" or "I feel more like a girl than a boy") as much as what they *did or liked*, and the significance attributed to these verbal declarations cannot be underestimated. These parents would argue that their child's repeated self-identifications are what set them apart from other children who "just" prefer occasional gender-atypical activities (and whose parents permit this).[7]

Interestingly enough, in a quarter of the cases, parents confessed to cloaking their regulation of certain behaviors in excuses that did not have to do with gender: Molly told Gil that his clothing preferences were too "sloppy," versus too masculine for a little girl, which she now recognizes was her "ulterior motive." Beth gave Tim's favorite dress-up heels to the dog so she didn't have to tell him they didn't want him wearing them. Theresa routinely framed pants as more comfortable for playtime with peers, versus more appropriate for boys. Parents' rhetorical moves to hide the true motives of their gender hedging are perhaps the most intriguing element of the practice: While parents felt bound to conform, they sought to avoid teaching that conformity explicitly to their children.

Parents' strategic work in gender hedging makes them increasingly frustrated with the regulatory forces of the gender truth regime, which presumes certain behaviors and dispositions relative to particular sex categories but which do not align with those of their children, time and again. In attempting to comply with the regime and not "bother other people"—including, fundamentally, protecting their children from negative attention—parents devise a variety of crafty maneuvers to satisfy their child's preferences while staying just within binary limits, but these continue to belie what their children really want. As Carl relayed, "We saw him when he was being pushed into, because of our own ignorance, a gender that wasn't his to accept . . . he would push back and [say], 'I'm not doing that.'" These tiresome negotiations ultimately catalyze their search for insights online, where they encounter a body of trans-affirming discourses that radically shift their perceptions about gender.

Gender Literacy: Talking Back to the Gender Binary

Parents' encounters on the Internet usher in a new stage of consciousness about childhood gender nonconformity, which challenges their attempts to curb it and breeds a new set of strategies. These strategies manifest in the form of explicit dialogues and discourse, with their children and with others, about more expansive (trans)gender possibilities than the gender truth regime allows. Through their online searches, parents find a flurry of talk among other parents, professionals, advocates, blogs, listservs, and advocacy organizations about gender-variant and transgender children. In these virtual forums—which often lead to live support groups and conferences with other parents—gender variance is affirmed as a natural, normal part of human diversity. Longstanding cultural beliefs rooted in the gender binary are the problem, as represented in the following excerpt from one prominent advocacy organization: "When a child is born, a quick glance between the legs determines the gender label that the child will carry for life. But . . . a binary concept still fails to capture [that] . . . biological gender occurs across a continuum of possibilities" (Gender Spectrum, n.d.). This discursive community also asserts that gender and sexuality are "separate, distinct parts of our overall identity" and that "gender expression should not be viewed as an indication of sexual orientation" (Gender Spectrum, n.d.). This distinction reverberated, often passionately, in my interviews. Tellingly, a striking majority of parents also volunteered awareness that their child could be both "trans and gay" as adults (the two adolescents in the sample, for example, transgender boys, identified as "gay" and "bi" at the time of our interviews). Parents' affirmation of their children's nonconformity as a matter of *gender*, and not (homo)sexuality, surfaced as a key component of the newfound transgender-aware principles they espoused.

During our interviews, it became apparent that parents sought to reiterate these discourses within their homes. Parents frequently recounted conversations with their children in which they aimed to pass on a more inclusive, less binary understanding of gender. I refer to these efforts as "gender literacy," which I adapt from France Winndance Twine's (2010) work on "racial literacy." . . .

One aspect of gender literacy entails parents' efforts to equip their children with a simple vocabulary for explaining their nonconformity to peers. Laurie said, "We would have to coach him on the kinds of responses to have to other kids . . . [he says] he's a boy who likes feminine things." Similarly, Heather claimed, "We kind of say together . . . You're always gonna be a girl in your heart."' Both Molly and Lynne said that prior to their children's transitions they used the phrase "boy with a girl's body." Katy actually tried defining "transgender," "gender-variant," and "intersex" for Liam, because she thought these might resonate with how he feels. While Katy worried that these terms were too complicated for Liam, they signify her enduring attempts to provide a language in her home that normalizes gender variance. In contrast, Becca and Sara preferred using labels their children derived themselves. Becca, who adopted her child's coinage "boygir," exemplified this philosophy: "One of the things I've really had to struggle with . . . is the labeling . . . We're just trying to put our own experiences around it . . . [but] I want him to define himself." Here, Becca testifies not only to the child-directed nature of this process (parents defer to their children's self-conceptions) but to the intellectual work she does to deconstruct conventional "truths" about sex and gender, including their categorical referents, that she has internalized.

Another facet of this strategy is parents' warning their children about prejudice toward gender nonconformity, similar to the "preparation for bias" that racial socialization scholars have observed among parents and children of color (Hughes et al. 2006). Ally, for example, believes that she has to be candid with Ray about potential harassment from peers: "I think that was how I explained it to him early on was, there are some people . . . who are gonna be really mean, 'cause they don't understand that . . . boys can wear girly clothes, play with girly toys." Tracy compared the importance of these lessons to dialoguing about racism: "I still think that we have to talk openly about what society is gonna expect because I think, just like with racism . . . ignoring race and pretending it doesn't exist is . . . not helpful to children."

Parents also strive to articulate trans-inclusive understandings of bodies and gender. Sam, for example, recalled making the following "edits" for Jamie when the topic of bodies appeared in a children's book: "I'd say, 'Nearly all girls' bodies are like this and nearly all boys' bodies are like that' . . . I [told him] that there are some people whose bodies don't match up with the genital parts that you traditionally associate." Tracy said that when her children use public restrooms, she will ask them which bathroom someone would use who does not identify as man or woman, "just to kind of plant the seed [that] it doesn't have to be one or the other." Moreover, in half of the preadolescent cases, parents indicated that they made their children aware, in the simplest terms possible, that there are "drugs," "medication," and/or "surgeries" that can help with body change in the future, when such questions surfaced (Liam, for example, expressed interest in having breasts "like his Mommy's"). These are striking examples of parents' attempts to actively affirm transgender and transsexual subjectivities during early childhood, versus regurgitating the body logics of the gender truth regime (i.e., "You can't have breasts like your Mommy's, you're a boy").

Parents engage their gender-normative children in gender literacy as well. For example, when their younger son, Eddy, asked them if Liam identifies as a boy or girl, Katy and Brian responded, "Well, sometimes Liam doesn't know, and sometimes Liam feels like a girl, and sometimes Liam feels like a boy, and that's okay . . . how do you feel on the inside?" As a testament to the gender-progressive potential of these strategies, Eddy wore a skirt to school over his shorts so that he could tell his friends, "Boys can wear skirts [too]." Clarise described her youngest child, who is six, as the one who "gets it" the best: "[She] gets that there's all kinds of varieties of gender . . . because it's always been that way for her." The gender literacy in which parents engage *all* of their children is indicative of how the presumptions of the gender truth regime are being radically resisted and retold in these families.

Parents practice gender literacy in more public institutions, too, including their children's schools. Parents work with teachers and administrators to coordinate gender-inclusiveness training, as well as to draft school policies that specifically protect "gender identity and/or expression." Carl joined an organization that teaches LGBT awareness to religious bodies in his community: "I wouldn't [have] done it if it weren't for Mark . . . I don't want him growing up in an environment that doesn't accept him." Several parents also launched online blogs detailing their experiences. Alicia reflected, "[Parents are] starting to move into an advocacy role, so they're wanting to include the general public in these discussions . . . parents are looking to have a voice." . . .

"Playing Along" (or Not): "Head Games" With the Gender Binary

While parents enact multiple forms of "gender literacy" to challenge the gender truth regime, they also feel that not every instance is appropriate for, or receptive to, such explicit deconstructionist efforts. This proved particularly true for interactions with strangers, who often attribute the wrong gender to a child (for example, at the grocery store, someone refers to a gender-variant male child as a "beautiful little girl"). . . . During such interactional routines, parents confront the power effects of the gender truth regime and must manage others' normative assumptions about sex and gender vis-á-vis their gender-variant children. Becca described these encounters as a "head game": "Up until this point, I have a little boy, and I know what's going on with my little boy, but [then] . . . suddenly, it's like I have to think of this as like having a little girl, which is its own . . . head game." . . . Parents advised that "playing along," as it was often described to me, was the most appropriate strategy with people whom they were unlikely to see again, when candid lessons about gender variance felt inapt: "In the interest of just keeping . . . the social construct together, I went with it, and I was just, like, whatever, I'm not in a space to educate" (Becca). Moreover, most children ask their parents *not* to correct strangers in these instances (such early requests are often regarded as indicative of a transgender identity later on). While "playing along" may not rupture the "social construct" for parents' interlocutors—and I seek to emphasize parents' heightened awareness that the construct exists—this strategy permits gender-variant expressions in public in a way that the norms of a child's assigned sex would disallow. Parents' decisions to honor their children's requests and "play along" with strangers thus affords their children safety and privacy that more explicit kinds of "gender literacy" might make uncomfortable. Indeed, many parents adopt the perspective that what their child has "between their legs" is nobody's business and irrelevant to their preferred gender presentation. Theresa reflected on these early negotiations: "She did start saying, 'Don't tell anybody I'm a boy'. . . I realize now that I was very anxious to take care of [other people], how do I help people to understand. . . . What I'm really trying to figure out [is] how to protect her privacy and still run interference." In short, playing along and not saying anything, versus effectively revealing their child's assigned sex to strangers, proved an important discursive practice in its own right to accommodate their children's most comfortable self-expressions, particularly for gender-variant children who had not claimed a binary identity.

When it comes to people parents see more frequently, "playing along" feels less viable. Beth claimed, "I felt the need to explain it to acquaintances and friends . . . you see the parents every day at drop-off . . . so I did feel the need to say, 'He prefers girl stuff.'" Here, Beth seeks to mediate between her child's apparent nonconformity (the boy who likes the girly toys and dress-up at school) and others' potential scrutiny, in turn signaling her own allowances of these preferences. Parents' discursive interventions in more familiar contexts, versus staying silent, work to carve out space for gender nonconformity where it might be otherwise inhibited. Disclosure is also important when parents fear their child's safety and well-being is at stake. Theresa, for instance, advised the host parent of a girls-only slumber party that her daughter was transgender, just in case her status was revealed by a potential "wild card" from her old school. Parents' use of terms like "playing along," "head game," and "wild card" are duly reflective of the strategic awareness they have developed to navigate the gender truth regime in everyday life, protecting and accommodating their children in the most appropriate ways they see fit with different audiences.

In contrast to the logics of the gender truth regime, parents adopt new ideologies that imagine a wider "spectrum" of (trans)gender possibilities, which are not moored to two, static sex categories. Shella reflected, "It's amazing to watch somebody really be strong in who they are to try and tackle something huge, because okay, you're born with a penis, okay, you're a boy, boom, done—*NO, not necessarily.*" Ally reiterated this perspective: "You wanna call somebody with a penis 'male'? Yeah, talk to the hyenas."[8] Alongside this de-linking of sex and gender, parents discussed more fluid, nonbinary identities, which Ally mused about as "a whole 'nother space that doesn't have to be just girl, just boy." Katy also mentioned a desire to "[go] beyond the binary" and advocate for the "boy in the skirt." This kind of intellectual work is particularly important for parents whose children had not articulated a binary identity (boy or girl), but were more fluid or switched their expressions day-to-day.

As parents reject traditional binary beliefs for a more spectrum-oriented perspective, they also embrace the idea that we are "born" with our gender, that it is an innate, "immutable" part of us. Several of the mothers in my study—self-proclaimed feminists who came of age during the 1970s—advised that having a gender-variant child has made them rethink the constructionist beliefs they adopted during second-wave feminism. Laurie typified this attitude:

> Having grown up with this sort of . . . feminist attitude . . . I grew up in the '70s, the free-to-be-you-and-me generation . . . and I always thought that we could choose our gender expression, and I didn't realize until I had a kid that gender expression or gender identity is just this immutable part of you, like the color of your skin, or any other fixed part of you.

Similarly, Brian asserted that gender variance stems "from the first duplication of those cells . . . this is how they're made." Joe raised the possibility of genetic or hormonal factors: "It's got to be either in vitro [utero] or hormonal . . . or may be there's some gene . . . there seems to be a gene that causes everything else." For these parents, only an innate hard-wiring during fetal development, or a "core biology," could explain a child's gender (variance) that resisted all cues to normative socialization. Evidently, these parents reconceive of gender in ways that harness both essentialist and constructionist frameworks. They reject the conventional sex-based assumptions of the gender truth regime, which they now see as hardly representative of the various ways masculinity and femininity manifest in the human population. Simultaneously, they embrace gender variance as a matter of "natural" human variation, often literally at the genomic or cellular level. In imagining "beyond the binary," these parents do not abandon the essentialist underpinnings of normative gender ideology.

* * *

Notes

1. All names are pseudonyms.
2. "Gender-variant" serves as an umbrella term for all the children represented in this study, whose behaviors are considered significantly more masculine or more feminine relative to their assigned sex. When referring to individual cases, I use "transgender" to refer to a child who had a "cross-gender" identity (i.e., transgender girls who were assigned male at birth and transgender boys who were assigned female). For children who did not identify with one specific gender, I refer to their assigned sex to signal their gender variance (i.e., "gender-variant male") and use the pronouns parents used at the time of the interviews.
3. I use "cisgender" to mean not transgender, or identifying with the gender presumed at birth.
4. In contrast, Jane Ward (2011) advocates "cultivating children's gender-queerness."
5. "Tomboys," of course, complicate this simple symmetry, as "masculine" girls are often given more latitude than "feminine" boys. However, parents of transgender boys in this sample advised that their children's persistent desires to be addressed as "boys" (including requesting boys' haircuts and male pronouns) made the "tomboy" category feel nonviable early into grade school, signaling the potential limits of this category for significant female masculinity.
6. The children represented here come from a variety of family contexts that would impact the availability of gender-atypical items (i.e., older/younger siblings, only children). However, these children often expressed their preferences through objects found at the store.
7. I do not give these observations to suggest objective distinctions between these children and their potentially more normative counterparts. Rather, I aim to highlight specific actions and interpretations these parents cited for coming to conceive of, and embrace, their child as categorically gender-variant or transgender.
8. The female spotted hyena has an enlarged clitoris that becomes erect, which scientists name a "pseudo-penis."

References

Bern, Sandra. 1983. Gender schema theory and its implications for child development. *Signs* 8: 598–616.

Bornstein, Kate. 1995. *Gender outlaw.* New York: Routledge.

Bryant, Karl. 2006. Making gender identity disorder of childhood. *Sexuality Research and Social Policy* 3:23–39.

Charmaz, Kathy. 2006. *Constructing grounded theory.* Thousand Oaks, CA: Sage.

Coltrane, Scott, and Michelle Adams. 1997. Children and gender. In *Contemporary parenting*, edited by Terry Arendell. Thousand Oaks, CA: Sage.

Connell, Catherine. 2010. Doing, undoing, or redoing gender? Learning from the workplace experiences of transpeople. *Gender & Society* 24:31–55.

Dalton, Susan, and Sarah Fenstermaker. 2002. "Doing gender" differently. In *Doing gender, doing difference.* New York: Routledge.

Ehrensaft, Diane. 2011. *Gender born, gender made.* New York: Experiment, LLC.

Foucault, Michel. 2000. Truth and power. In *Essential works of Foucault 1954-1984,* Vol. 3, edited by Paul Rabinow. New York: The New Press.

Gagne, Patricia, and Richard Tewksbury. 1998. Conformity pressures and gender resistance among transgendered individuals. *Social Problems* 45:81–102.

Gender Spectrum. n.d. Understanding gender, https://www.genderspectrum.org/ understanding-gender.

Green, Richard. 1987. *The sissy boy syndrome and the development of homosexuality.* New Haven, CT: Yale University Press.

Halberstam, Judith. 2005. *In a queer time and place.* New York: New York University Press.

Hughes, Diane, James Rodriguez, Emilie P. Smith, Deborah J. Johnson, Howard C. Stevenson, and Paul Spicer. 2006. Parents' ethnic-racial socialization practices. *Developmental Psychology* 42:747–70.

Jenness, Valerie, and Sarah Fenstermaker. 2014. Agnes goes to prison. *Gender & Society* 28:5–31.

Kane, Emily. 2006. "No way my boys are going to be like that!" *Gender & Society* 20:149–76.

Kane, Emily. 2012. *The gender trap.* New York: New York University Press.

Lev, Arlene Istar. 2004. *Transgender emergence.* New York: Berkley Books.

Martin, Karin. 2005. William wants a doll. Can he have one? *Gender & Society* 19:456–79.

Meadow, Tey. 2011. "Deep down where the music plays." *Sexualities* 14:725–47.

Meadow, Tey. 2013. Studying each other. *Journal of Contemporary Ethnography* 42:466–81.

Peterson, Gary, and Boyd Rollins. 1987. Parent-child socialization. *In Handbook of marriage and the family*, edited by Marvin B. Sussman and Suzanne K. Steinmetz. New York: Plenum.

Pfeffer, Carla. 2012. Normative resistance and inventive pragmatism. *Gender & Society* 19:456–79.

Pogrebin, Letty Cottin. 1980. *Growing up free.* New York: McGraw-Hill.

Risman, Barbara J., and Kristen Myers. 1997. As the twig is bent. *Qualitative Sociology* 20:229–52.

Introduction to Reading 18

This reading by Michela Musto is about gender socialization, how males and females learn to be boys and girls, emphasizing the role of adult authority figures other than parents in this process. In this reading, Musto describes peer influence on gender socialization, but she also describes something we have been emphasizing throughout this book—the influence of context on the way we display gender. Over the course of one season, she observed children in the Shark swimming group, the fastest swimmers in their age bracket. She watched these children disregard gender during swim practice but continue to practice gender "borderwork" during free time, albeit in a modified form. Her research helps us understand how gender is not an essential or inborn characteristic of people but rather something that is reinforced in the structural and contextual settings in which we enact gender.

1. What makes the children in the Shark group less likely to do gender than children in the other swimming groups with the same coach?
2. Think about ways you can use this reading to argue that gender is not an inborn, biological trait of children.
3. Imagine other settings in which children are less likely to engage in gender "borderwork." What characteristics of these settings are important in undoing gender?

Athletes in the Pool, Girls and Boys on Deck

The Contextual Construction of Gender in Coed Youth Swimming

Michela Musto

Although it is only eight o'clock in the morning, the swimming pool at the Sun Valley Aquatics Center is bustling with activity.[1] It is a warm, sunny day in southern California, and 300 kids are participating in a Sun Valley Swim Team (SVST) swim meet. Girls and boys as young as five years old rummage through their swim bags, grabbing goggles and swim caps as they walk toward the starting blocks. Between races, swimmers slather their arms with waterproof sunblock, laugh with their friends, and offer each other bites of half-eaten bagels. To my right, three 11-year-old boys, Alex, Kevin, and Andrew, are sitting in a semicircle, scrutinizing a "heat sheet" that lists the names of other boys and girls they are racing against in their upcoming events.[2] Alex notices he is the only boy in his race, sparking the following conversation:

Alex: They're all girls! That's sad.

Kevin: That must suck.

Andrew: I know her [points to a name on the paper]. You're the only male! Have fun! You have the second-fastest time—she's first, you're second.

Alex: What's her time?

Andrew: [Sophia's] really fast. She was in Sharks.

Andrew flips the page, and the boys continue looking at their other events.

Throughout their conversation, Alex, Kevin, and Andrew draw upon multiple and contradictory meanings of gender. Although they agree that it "sucks" to be the "only male" in an "all girls' event, the boys then discuss Sophia's athleticism in a relatively unremarkable manner. Instead of teasing Alex for being slower than a girl, Andrew nonchalantly informs Alex that Sophia is "really fast," something neither Alex nor Kevin contests. How was it possible for gender to simultaneously be of minimal and significant interest to the swimmers?

Because gender is a social structure that is embedded within individual, interactional, and institutional relations, social change toward gender equality is uneven across the gender order (Connell 1987, 2009; Lorber 1994; Martin 2004; Risman 2004). The salience of gender varies across contexts, allowing some contexts to support more equitable patterns of gender relations than others (Britton 2000; Connell 1987; Deutsch 2007; Schippers 2002; Thorne 1993). Within a context, both structural mechanisms and hegemonic beliefs play an important role in determining whether individuals draw on and affirm group boundaries between the genders—what Thorne (1993) calls "borderwork" (see also Messner 2000; Morgan and Martin 2006; Ridgeway 2009; Ridgeway and Correll 2004). Although scholars have theorized that alternative patterns of gender relations may shape social relations when gender is less sailent (Britton 2000; Connell 1987; Deutsch 2007; Ridgeway 2009; Schippers 2007), few empirical studies have followed a group of individuals across different contexts to understand how gender relations and meanings may change. As the dialogue among the boys on the swim team suggest, because individuals negotiate different systems of accountability as they move from one setting to the next, gender can take on multiple meanings as a result of contextually specific, group-based interactions.

In what follows, I analyze nine months of participant observation research and 15 semi-structured interviews conducted with 8- to 10-year-old swimmers at SVST. . . .

The Variable Salience of Gender Across Contexts

Existing scholarship has identified specific structural mechanisms, such as formal and informal policies and practices, within an array of institutions that help

Musto, M. (2014). Athletes in the pool, girls and boys on deck: The contextual construction of gender in coed youth swimming. *Gender & Society, 28*(3): 359–380. Reprinted by permission of SAGE Publications Inc., on behalf of Sociologists for Women in Society.

explain how gender becomes a salient organizing principle in interactions (Messner 2000; Thorne 1993). In schools, the formal age separation and large number of students encourage boys and girls to engage in borderwork (Thorne 1993). At the same time, teachers can implement rules and seating charts that allow children to interact in "relaxed and non-gender-marked ways" (Thorne 1993, 64; see also Moore 2001). Similarly, bureaucratic policies reduce the amount of discrimination women experience within office workplaces (Morgan and Martin 2006; Ridgeway and Correll 2000). Yet the organization of many out-of-office business settings—such as having different tees for men and women on golf courses—continues to hold women professionals accountable to normative conceptualizations of gender (Morgan and Martin 2006).

In addition to structural mechanisms, hegemonic cultural beliefs also impact the salience of gender within interactions (Ridgeway and Correll 2000, 2004; Ridgeway 2009, 2011). Although the "default expectation" (Ridgeway and Correll 2004, 513) may be to treat individuals in accordance with hegemonic beliefs, these beliefs can be less salient within interactions depending on a context's gender composition, gender-typing, and institutional frame (Ridgeway and Correll 2004; Ridgeway 2009, 2011). However, even when structural mechanisms allow for less oppressive gender relations within some contexts, individuals often "implicitly fall back on cultural beliefs about gender" in new and unscripted settings (Ridgeway 2009, 156), thus reinscribing hegemonic patterns of gender relations.

Although individuals are often framed by hegemonic patterns of gender relations within interactions, interactions can also be framed by less oppressive patterns of gender relations and meanings (Deutsch 2007; Hollander 2013; Ridgeway 2009; Ridgeway and Correll 2000, 2004; Schippers 2007). However, the processes that allow individuals to enact alternative patterns of gender relations remain undertheorized within existing scholarship (for exceptions, see Finley 2010; Hollander 2013; Schippers 2002; Wilkins 2008). As sociologists have argued, there is not always a direct relationship between the cultural order and the meanings individuals associate with cultural representations (Connell 1987; Eliasoph and Lichterman 2003; Fine 1979; Swidler 1986). Instead, hegemonic meanings are negotiated and contested within group-based interactions (Eliasoph and Lichterman 2003). When applied to gender theory, the meanings people associate with gender may vary, perhaps dramatically, across contexts depending on whether gender is a salient organizing principle within group-based relations. Furthermore, if a context allows for nonhegemonic patterns of gender relations, perhaps aspects of the more egalitarian patterns of social relations can transfer across contexts (Hollander 2013).

Competitive youth swimming is an ideal setting to examine how gender boundaries and meanings are constructed and negotiated within everyday life (McGuffey and Rich 1999; Messner 2000). Within the United States, sport has historically played a visible role in naturalizing hierarchical, categorical, and essentialist differences between the genders (Kimmel 1996; Lorber 1994; Messner 2011). Because the institutional "center" of sport often affirms hegemonic masculinity (Messner 2002), girls' and boys' interactions within athletic contexts often help strengthen hierarchical and categorical group boundaries between the genders, thus maintaining the power, prestige, and resources boys have over girls (McGuffey and Rich 1999; Messner 2000; Thorne 1993). Yet at the same time, research finds that girls' and women's athleticism is becoming normalized (Ezzell 2009; Heywood and Dworkin 2003; Messner 2011), potentially calling into question hegemonic gender meanings pertaining to athleticism (Kane 1995; Messner 2002). Since it may be easier for individuals to enact alternative patterns of gender relations within contexts that are considered feminine (Finley 2010), the enactment of alternative patterns of gender relations may be especially apparent within competitive youth swimming, a sport that has historically been considered acceptable for white, middle-class girls to participate in (Bier 2011; Cahn 1994).

In this article, I follow a group of 8- to 10-year-old swimmers across different contexts at swim practices, asking: Do the meanings swimmers associate with gender vary as a result of their contextually specific, group-based interactions? If so, what are the conditions that allow swimmers to associate alternative cultural meanings with gender? To answer these questions, I outline the "gender geography" of swimmers' gender relations within two main contexts, arguing that when gender was less salient and children could "see" athletic similarity between the genders, children interacted in ways that undermined hegemonic beliefs about gender. Yet when the salience of gender was high and structural mechanisms encouraged kids to engage in borderwork, swimmers affirmed beliefs in essentialist and categorical—but nonhierarchical—differences between the genders. By paying attention to structural mechanisms and the variable salience of gender, we can thus see whether and how children associate different meanings with gender across contexts. Furthermore, because the swimmers enacted nonhierarchical gender relations in both contexts, this article contributes to gender theory by introducing the concept of "spillover," theorizing that aspects of less oppressive gender relations may transfer across contexts.

* * *

THE "GENDER GEOGRAPHY" OF SUN VALLEY SWIM TEAM

On my first day of research with SVST, Coach Elizabeth started Sharks swim practice with a team meeting. The day before, she explained, she had to "excuse" the athletes from practice early for misbehaving—something she had not done to a group of swimmers in more than three years. While solemnly addressing the swimmers, Elizabeth reminded the athletes that they were the fastest swimmers in their age category; she thus expected more from them than if they were in the Dolphins or Piranhas groups. Elizabeth told the swimmers that while they were at swim practices, "There is a time to listen and a time for fun." When it was "time to listen," Elizabeth stressed that the swimmers should pay attention, remain focused, and follow her instructions. By doing so, they would achieve their goals of becoming faster swimmers.

As Elizabeth's speech suggests, there were two main contexts that organized swimmers' relations at the pool: focused athletic contexts in which swimmers were expected to follow their coach's instructions, and unfocused free time in which swimmers had fun with their friends. . . . [T]he variable salience of gender at the pool played an important role in shaping the different meanings swimmers associated with gender within and across these contexts. As a result of the structural mechanisms instituted by Coach Elizabeth during focused aspects of practice, gender was less salient in this context, and the swimmers interacted in nonantagonistic ways. While doing so, the swimmers regularly witnessed athletic parity between the genders and associated alternative, nonhegemonic meanings pertaining to athleticism. Because the gender meanings changed across contexts at the pool, however, gender was highly salient during the swimmers' free time. Structural mechanisms instead encouraged the kids to engage in borderwork in this context. Because swimmers tended to interact in antagonistic ways in their free time, similarities between the genders were less visible, ultimately encouraging the swimmers to associate gender with categoricalism and essentialism.

RACING "FOR TIMES": FOCUSED ASPECTS OF PRACTICE

The most focused aspects of Sharks swim practices occurred when athletes raced "for times." While racing "for times," athletes swam a distance, such as 50 or 100 yards, as fast as they could—like they did during formal competitions. Afterwards, the athletes calculated how fast they swam the interval and reported their times to Coach Elizabeth. In interviews, athletes described racing "for times" as having "to sprint and go as fast as you can" while trying "to get the same time [as] in the [swim] meet." Zoe, a nine-year-old Asian girl, told me that during these workouts, she compares herself to Olympians like Michael Phelps and Natalie Coughlin, reminding herself, "If you were one of them, you wouldn't be able to stop, so just try to push through it and work hard and think of something else besides how hard it is." As Zoe's strategy suggests, racing "for times" was not a time to goof around. Instead, in this context, swimmers were expected to work hard, swim fast, and push themselves when tired.

During these workouts, Elizabeth often organized swimmers into groups according to their athletic ability. While assigning the swimmers to lanes, Elizabeth either instructed the fastest athletes to share a lane or assigned several fast swimmers to each lane. To motivate the athletes, Elizabeth often encouraged the swimmers to race the swimmers next to them, catch the swimmers in front of them, and compare times with other swimmers in their lanes. Because the girls and boys had relatively equal athletic abilities, racing for times was a context where swimmers of both genders regularly trained and raced together. . . .

[D]uring focused aspects of practice, Coach Elizabeth organized lanes based on athletes' fastest times, not gender. While following Elizabeth's instructions to race other swimmers, Nick and Jon compared themselves to Sophia—a girl. After hearing Lesley's time on the first 200-yard Individual Medley, Elizabeth instructed Lesley to swim with faster swimmers—both were boys. Instances where the girls and boys compared times and raced each other occurred regularly in this context.

As previous scholarship has argued, gender is often highly salient when kids engage in mixed-gender competitions—especially within athletic contexts (McGuffey and Rich 1999; Messner 2000; Moore 2001; Thorne 1993). While racing "for times," Sophia could have teased Nick and Jon for losing to "a girl," or Jon could have told Lesley that "girls suck" at swimming. However, when coaches or athletes directly compared girls' and boys' performances during SVST practices, I never heard athletes use these comparisons as an opportunity to evoke antagonistic interactions. Instead, similar to how teachers can encourage boys and girls to interact in relaxed, nonantagonistic ways by dividing students by reading abilities instead of gender (Thorne 1993), the informal policies instituted by Coach Elizabeth minimized the salience of gender within this context. The swimmers were instructed to complete tasks with specific goals (Moore 2001; Ridgeway 2011), allowing the girls and boys to interact in ways that did not affirm group boundaries between the genders.

"It's Just, Like, the Same Thing": Alternative Meanings of Gender

Because gender was less salient during focused aspects of Sharks swim practices, the swimmers interacted in ways that allowed them to associate alternative meanings with gender. This became clear when the athletes discussed instances they raced against swimmers of the other gender. Without nervously giggling or averting his eyes, Jon talked about getting "killed" by Sophia when they swam breaststroke. Cody leaned back and shrugged as he told me, "It doesn't matter . . . It's just, like, the same thing" if he loses to a girl or a boy. When asked who he races during practices, Nick spontaneously compared his times to Sophia's:

> When Brady (11, white) was in the group I always raced against him. Now that he's gone the only one left is Sophia. Which, 200 IMs, no question, she's gonna win because my breaststroke sucks. Butterfly . . . I'll usually [win]—well, most of the time. Backstroke, it's a 50-50 game, and freestyle, 50-50. Breaststroke, no doubt she's in front.

Without a hint of defensiveness about losing to a girl, Nick made detailed comparisons between himself and Sophia. Even in butterfly, his fastest stroke, Nick recognized that he wins only "most of the time." Although boys often have much at stake in maintaining hierarchical and categorical differences between the genders (McGuffey and Rich 1999; Thorne 1993), at SVST the boys instead associated alternative meanings with gender while talking about racing "for times," where athleticism was not associated with hierarchy or difference.

The girls also talked about racing "for times" in ways that suggested they were not inferior or fundamentally different athletes because of their gender. Chelsea, a 10-year-old Asian girl, told me that boys are "not always faster [in swimming], sometimes they can be slower." Similarly, Anna, a 10-year-old white girl, discussed a race she lost to Elijah, a 10-year-old white boy. Instead of justifying defeat by saying that boys are always faster than girls, Anna identified a specific reason why she lost. She explained that when she dove into the water, "I [dove] to the side. It was not a good dive." Even Wendy, a nine-year-old white girl—one of the slowest athletes in the group—told me that because Sophia is as fast as Nick, there was "not really" a difference between the girls' and boys' swimming abilities.

There are two reasons Sharks swim practices were an ideal context for swimmers to enact nonhegemonic patterns of gender relations pertaining to athleticism. First, the Sharks swimmers were the fastest group of "ten and under" swimmers on the team, and highly committed to athletics. Many of the Sharks swimmers told me they attended practice to "get better times" or to "get better" at swimming. Several of the boys and girls expressed a desire to swim in the Olympics one day, and Grace, a 10-year-old white girl, even chose to attend swim practice instead of her best friend's birthday party. Because of the athletes' commitment, the majority of swimmers willingly followed Elizabeth's instructions—even if it meant sharing lanes with swimmers of the other gender. Swimmers in the other "10 and under" groups on the team, however, did not always follow their coaches' instructions as readily. I occasionally noticed girls and boys in the Dolphins and Piranhas groups make faces and shriek when instructed to share lanes with swimmers of the other gender—something Sharks swimmers never did while racing "for times." As a result of the Sharks swimmers' commitment to athleticism, acting in accordance with the structural mechanisms instituted by Elizabeth likely mattered more than it did to other "10 and under" athletes.

Additionally, while following Elizabeth's instructions to share lanes and race one another, the swimmers compared times, a relatively transparent measure of ability. If the athletes were playing a team sport like basketball or soccer, where athleticism is assessed through less quantifiable skills, such as dribbling or blocking ability, it may have been easier for the boys to marginalize or masculinize the girls' abilities (McGuffey and Rich 1999; Thorne 1993). Indeed, during interviews, several of the girls and boys discussed instances during recess and physical education classes when boys invoked hierarchical and categorical notions of gender while playing team sports, such as when they refused to play with girls or became upset after losing to girls. During Sharks swim practices, however, the swimmers were frequently provided with specific, quantifiable instances of girls beating boys and boys beating girls. Through these time-based comparisons, it became clear that the girls' and boys' abilities overlapped (Kane 1995). As a result, within a context where swimmers willingly interacted in ways that illuminated similarities between the genders, girls and boys associated nonhegemonic meanings with gender.

Having Fun With Friends: Unsupervised Free Time

The least focused aspects of swimming occurred during the swimmers' free time. Sharks swimmers were never completely unsupervised on the pool deck, but there were times—such as before swim practices or between races at swim meets—when SVST coaches were busy coaching other swimmers. As opposed to focused aspects of practice, which were "hard" and "tiring," unsupervised free time was a chance for the kids to have fun with their friends. David, a 10-year-old Latino boy, explained that before and after swim practice, he and his friends had "lots of fun together." Grace told me, "It's always fun to come here and see [my friends]," and Chelsea similarly said that she had "fun" while "hanging out" with her friends before practices.

The unsupervised aspect of the swimmers' free time played an important role in shaping kids' social interactions with their friends. At the pool, I did not observe patterns of age and racial separation that other scholars have observed among children in schools and summer camps (Lewis 2003; Moore 2001; Perry 2002). In interviews, furthermore, most of the swimmers had a difficult time naming their closest friends on the team, explaining that they were close friends with "everyone" and had "a lot of good friends [on the team]." Despite the ostensive unity among the swimmers, none of the swimmers reported being friends with kids of the other gender. For example, Nick, a multiracial nine-year-old, named every male swimmer and male coach he could think of when describing his "good friends." . . .

Nick's "good friends" range from boys in the Sharks group to one of the men who worked at the pool's café. Like other swimmers in the Sharks group, moreover, Nick developed friendships across racial and age categories. Although the requirements for being Nick's friend are not particularly stringent—you simply needed to "chat a lot" or give him a discount on food—the only girl he mentions is his sister, who can be "very annoying." This is striking because Nick's parents were good friends with the parents of Chelsea, a girl in the Sharks group. On several different occasions, Nick talked about going fishing with Chelsea's family and having her family over for barbecues. Once, he even told me he dreamed about raiding her family's food pantry. Based on Nick's criteria, Chelsea should count as a friend. However, when I asked Nick if he ever "hangs out" with Chelsea, he simply responded, "No." When asked to elaborate, he explained, "I don't hang out with girls."

As Nick's comments suggest, gender was a highly salient category that structured kids' friendships during their free time. Among swimmers in the Sharks group, this gender separation was marked with extensive physical separation. After changing into their swim suits in sex-segregated locker rooms, the girls would set their swim bags near the right end of the bleachers that lined the length of the pool. The boys would walk past the girls, often without even glancing in their direction, to the far end of the benches, placing their bags almost 50 meters from the girls' space.

There were three reasons why gender became a highly salient organizing principle within the kids' group-based relations during unsupervised free time. First, as opposed to when Coach Elizabeth instructed the girls and boys to share lanes and compare times, in unsupervised aspects of practices, no policies encouraged the boys and girls to interact. Because the swimmers' unsupervised free time was not formally scripted, the kids relied on gender as a highly salient criterion when developing friendships (Ridgeway 2011). Furthermore, similar to how formal gender segregation on soccer teams and golf courses can increase the salience of gender within interactions (Messner 2000; Morgan and Martin 2006), the policy of physically separating the swimmers into gender-segregated locker rooms before and after practice formally marked the boys and girls as different when they entered and exited the pool deck. And finally, the crowded nature of the pool deck may have contributed to the salience of gender in this context (Thorne 1993). Because there were often between 50 and 100 kids on the pool deck during SVST practices, there were plenty of witnesses who could tease kids for having "crushes" on kids of the other gender, making it risky for the girls and boys to socialize with each other. Thus, in the swimmers' free time, rather than developing friendships based on similar interests or athletic ability, the lack of rules, the threat of heterosexual teasing, and gender-segregated locker rooms helped create a context where gender was highly salient.

"Boys Are Always Wild" and "Girls Are Very Nice and Sweet": Hegemonic Meanings of Gender

Given the high salience of gender boundaries during swimmers' unsupervised free time, the girls' and boys' interactions often strengthened gender-based group boundaries during unsupervised aspects of practice (McGuffey and Rich 1999; Messner 2000; Thorne 1993). Once, before practice, Nick shouted his last name while jumping toward Katie. While mimicking his motion, Katie shouted, "Weirdo!" back at him. Several times, I watched Nick, Brady, and Sophia dump cold water on one another's heads after practice. After a swimming fundraiser, Katie, Jon, and Cody spent 10 minutes hitting and splashing one another with foam swimming "noodles" in the pool. Toward the end of a swim meet, several boys filled their swim caps with water and tried splashing Lesley and Grace. After wrestling the swim caps out of the boys' hands, Grace came over to me and told me that Elijah gave her "cooties."

Although the swimmers tended to interact in antagonistic ways during their free time, borderwork at the pool did not seem to be based on beliefs in male supremacy. Unlike existing research has suggested (McGuffey and Rich 1999; Messner 2000; Thorne 1993), boys did not provoke antagonistic relations more frequently than girls, nor did the boys control more space on the pool deck. Furthermore, the girls never tried to avoid confrontations with the boys, and instead seemed confident in their ability to interact as equals. Once, for example, I was talking with Katie when Amy, an 11-year-old Asian girl, walked over to us. Elijah and Jon were standing several feet away, wearing swimming flippers on their hands. Katie warned Amy that the boys would "smack you with that fin" if Amy got too close. Amy, however, rolled her eyes and told Katie, "I'm not scared." She then punched Katie's arm a couple of times, demonstrating how she would fight if provoked. If the swimmers had believed that boys on the pool deck were stronger than the girls, Amy may have been more cautious about fighting Jon and Elijah. Instead, she confidently proclaimed that she was "not scared" and demonstrated how she would punch them.

Other girls in the Sharks group also seemed confident in their ability to engage in borderwork as equals with the boys. Once, much to the girls' excitement, Katie "pantsed" Elijah at a swim meet.[3] Another time, after Nick dumped what he described as "ice cold" water on Sophia's head, she got "revenge" by pouring red Gatorade on him. If Katie or Sophia had believed the boys were stronger or more powerful than the girls, they may have been afraid to instigate such interactions. Fear of the boys' reactions, however, did not stop Katie from "pantsing" Elijah, or Sophia from dumping a red, sticky drink on Nick's head. Although hegemonic cultural beliefs about gender often become activated when gender is a salient aspect of social interactions (Ridgeway 2011), the swimmers' antagonistic interactions in this context did not appear to be based upon a sense of male supremacy. Instead, they were transformative in the sense that they allowed the girls to occupy space and express agency when interacting as equals with the boys. However, because these interactions continued to affirm categorical and essentialist differences between the genders, they simultaneously undermined and reproduced aspects of hegemonic gender relations.

Furthermore, all the swimmers talked about sharing close physical space with kids of the other gender in ways that were markedly different from how they talked about racing one another. When talking about racing "for times," the swimmers willingly recognized and discussed the overlap between girls' and boys' abilities. However, on a social level, the meanings kids associated with gender were firmly grounded in categorical differences. Perhaps because of the risk of heterosexual teasing (Thorne 1993), boys and girls told me that spending time with athletes of the other gender was "not fun," "awkward," "annoying," "awful," "super uncomfortable," "gross," "kinda weird," and "really bad and really messed up." Furthermore, many of the kids articulated essentialist understandings of gender within these narratives, explaining that "boys are always wild," "girls are very nice and sweet," "girls are more limber," and "boys are more competitive." Notably, however, the swimmers did not include assumptions about male supremacy within these explanations. Instead, as suggested by their patterns of borderwork, the swimmers associated categorical and essentialist—but nonhierarchical—meanings with gender.

As an observer who spent an equal amount of time with the girls and boys, it was puzzling to hear girls and boys make categorical and essentialist distinctions between the genders. If girls were always "more limber" than boys, then how could the swimmers account for the boy from the Sharks group who frequently did the splits before swim practice? If "girls are very nice and sweet," then how could they explain the times when the girls screamed at and hit one another? Although it was easy for me to think of exceptions to the kids' generalizations, whenever I asked kids about these exceptions, my questions were met with shrugs and surprise.

Despite being quite knowledgeable about one another's swimming abilities, the girls and boys were relatively unaware of the other group's social experiences. Because the swimmers tended to provoke antagonistic interactions with one another, similarities between the genders were obscured. Unlike focused aspects of practice, structural mechanisms did not illuminate the similarities between the girls and boys (Kane 1995). In this less scripted context, the kids instead drew upon and enacted aspects of hegemonic patterns of gender relations (Morgan and Martin 2006; Ridgeway 2009, 2011). The swimmers, however, did not default to enacting all aspects of hegemonic gender relations. The swimmers' group-based interactions led the swimmers to associate gender with categorical and essentialist meanings, but the assumption that boys are superior to girls was notably absent from swimmers' interactions during unsupervised free time.

Discussion and Conclusion

Gender is a social structure embedded within individual, interactional, and institutional relations (Connell 2009; Lorber 1994; Martin 2004; Messner 2000; Risman 2004). At the institutional level, femininities and masculinities are ranked in societal-wide, historically based hierarchies that are created and re-created through laws, policies, practices, collective representations, symbols, and hegemonic meanings of gender (Connell 1987; Lorber 1994), but the impact of structural mechanisms and hegemonic cultural beliefs within interactions varies based on the context (Britton 2000; Connell 1987; Deutsch 2007; Messner 2000; Morgan and Martin 2006; Ridgeway 2009, 2011; Ridgeway and Correll 2004; Thorne 1993).

My research contributes to existing literature by exploring how gender meanings and relations change across contexts. By following the same group of individuals across different contexts, I found that the meanings kids associated with a social category such as "gender" did not always align with hegemonic beliefs. Instead, the swimmers' understandings of gender were filtered through group-based interactions and thus varied dramatically depending on the context (Fine 1979; Eliasoph and Lichterman 2003; Swidler 1986).

* * *

Notes

1. The swim team name and names of all participants are pseudonyms.
2. SVST coaches occasionally organized intrasquad swim meets, where boys and girls of all ages race against one another in heats arranged from slowest to fastest.
3. Because the kids often wore pants and T-shirts over their swimsuits during their free time, swimmers occasionally tried to pull down other swimmers' pants, revealing their swimsuits in the process.

References

Anderson, Eric. 2008. "I used to think women were weak": Orthodox masculinity, gender segregation, and sport. *Sociological Forum* 23:257–80.

Bier, Lisa. 2011. *Fighting the current: The rise of American women's swimming: 1870–1926.* Jefferson, NC: McFarland.

Britton, Dana M. 2000. The epistemology of the gendered organization. *Gender & Society* 14:418–34.

Cahn, Susan K. 1994. *Coming on strong: Gender and sexuality in twentieth-century women's sport.* Cambridge, MA: Harvard University Press.

Connell, Raewyn. 1987. *Gender and power: Society, the person and sexual politics.* Stanford, CA: Stanford University Press.

Connell, Raewyn. 2009. *Short introductions: Gender.* Maiden, MA: Polity Press.

Deutsch, Francine M. 2007. Undoing gender. *Gender & Society* 21:106–27.

Eliasoph, Nina, and Paul Lichterman. 2003. Culture in interaction. *American Journal of Sociology* 108:735–94.

Ezzell, Matthew B. 2009. "Barbie dolls" on the pitch: Identity work, defensive othering, and inequality in women's rugby. *Social Problems* 56:111–31.

Fine, Gary Alan. 1979. Small groups and culture creation: The idioculture of little league baseball teams. *American Sociological Review* 44: 733–45.

Heywood, Leslie, and Shari L. Dworkin. 2003. *Built to win: The female athlete as cultural icon.* Minneapolis: University of Minnesota Press.

Kane, Mary Jo. 1995. Resistance/transformation of the oppositional binary: Exposing sport as a continuum. *Journal of Sport and Social Issues* 19:191–218.

Kimmel, Michael S. 1996. *Manhood in America*. New York: Free Press.

Lewis, Amanda. 2003. *Race in the schoolyard: Negotiating the color line in classrooms and communities*. New Brunswick, NJ: Rutgers University Press.

Lorber, Judith. 1994. *Paradoxes of gender*. New Haven: Yale University Press.

Martin, Patricia Yancey. 2004. Gender as social institution. *Social Forces* 82:1249–73.

McGuffey, C. Shawn, and B. Lindsay Rich. 1999. Playing in the gender transgression zone: Race, class, and hegemonic masculinity in middle childhood, *Gender & Society* 13:608–27.

Messner, Michael A. 1992. *Power at play: Sports and the problem of masculinity*. Boston: Beacon Press.

Messner, Michael A. 2000. Barbie girls versus sea monsters: Children constructing gender. *Gender & Society* 14:765–84.

Messner, Michael A. 2002. *Taking the field: Men, women and sports*. Minneapolis: University of Minnesota Press.

Messner, Michael A. 2009. *It's all for the kids: Gender, families, and youth sports*. Berkeley: University of California Press.

Messner, Michael A. 2011. Gender ideologies, youth sports, and the production of soft essentialism. *Sociology of Sport Journal* 28:151–70.

Moore, Valerie Ann. 2001. "Doing" racialized and gendered age to organize peer relations: Observing kids in summer camp. *Gender & Society* 15:835–58.

Morgan, Laurie A. and Karin A. Martin. 2006. Taking women professionals out of the office: The case of women in sales. *Gender & Society* 20:108–28.

Ridgeway, Cecilia L. 2009. Framed before we know it: How gender shapes social relations. *Gender & Society* 23:145–60.

Ridgeway, Cecilia L. 2011. *Framed by gender: How gender inequality persists in the modern world*. Oxford, UK: Oxford University Press.

Ridgeway, Cecilia L., and Shelley J. Correll. 2000. Limiting inequality through interaction: The end(s) of gender. *Contemporary Sociology* 29:110–20.

Ridgeway, Cecilia L., and Shelley J. Correll. 2004. Unpacking the gender system: A theoretical perspective on gender beliefs and social relations. *Gender & Society* 18:510–31.

Risman, Barbara J. 2004. Gender as a social structure: Theory wrestling with activism. *Gender & Society* 18:429–50.

Schippers, Mimi. 2002. *Rockin'out of the box: Gender maneuvering in alternative hard rock*. New Brunswick, NJ: Rutgers University Press.

Schippers, Mimi. 2007. Recovering the feminine other: Masculinity, femininity, and gender hegemony. *Theory and Society* 36:85–102.

Swidler, Ann. 1986. Culture in action: Symbols and strategies. *American Sociological Review* 51:273–86.

Thorne, Barrie. 1993. *Gender play: Girls and boys in school*. New Brunswick, NJ: Rutgers University Press.

Wilkins, Amy. 2008. *Wannabes, goths, and Christians: The boundaries of sex, style and status*. Chicago: University of Chicago Press.

Introduction to Reading 19

In this piece, Janice McCabe, Emily Fairchild, Liz Grauerholz, Bernice A. Pescosolido, and Daniel Tope conducted a quantitative analysis of Little Golden Books (1942–1993), Caldecott award winners (1938–2000), and *Children's Catalog* (1900–2000). This sample covers a wide range of children's books through the 20th century. Little Golden Books are relatively inexpensive and available widely, including in grocery stores; the Caldecott award is given annually by the Association for Library Service to Children for "the most distinguished picture book for children" (books receiving honorable mention were also included in the sample); and *Children's Catalog* is an extensive listing of all books for children. The authors coded gender information from titles and main characters—including animals if they were the subject of the story—for 5,618 books. Their findings of the distribution of female and male characters in children's books throughout the 20th century might surprise you.

1. Were males or females more likely to be included in the titles of books and as main characters?
2. In which type of book were the differences in gender representation more extreme?
3. Why does it matter that both sexes are not equally represented in the books studied?

Gender in Twentieth-Century Children's Books

Patterns of Disparity in Titles and Central Characters

Janice McCabe, Emily Fairchild, Liz Grauerholz, Bernice A. Pescosolido, and Daniel Tope

Research on gender representation in children's literature has revealed persistent patterns of gender inequality, despite some signs of improvement since Weitzman et al.'s (1972) classic study more than 35 years ago. Recent studies continue to show a relative absence of women and girls in titles and as central characters (e.g., Clark, Lennon, and Morris 1993; Hamilton et al. 2006), findings that mirror those from other sources of children's media, including cartoons and coloring books (e.g., Fitzpatrick and McPherson 2010; Klein and Shiffman 2009). Theoretically, this absence reflects a "symbolic annihilation" because it denies existence to women and girls by ignoring or underrepresenting them in cultural products (Tuchman 1978). As such, children's books reinforce, legitimate, and reproduce a patriarchal gender system.

Because children's literature provides valuable insights into popular culture, children's worlds, stratification, and socialization, gender representation in children's literature has been researched extensively. Yet most studies provide snapshots of a small set of books during a particular time period while making sweeping claims about change (or lack thereof) and generalizing to all other books. . . . While examining particular books during limited time periods may reveal important insights about these periods and books, we know little about representation of males and females in the broad range of books available to children throughout the twentieth century. . . .

Children's Understandings of Gender: Schemas, Reader Response, and Symbolic Annihilation

No medium has been more extensively studied than children's literature. This is no doubt due, in part, to the cultural importance of children's books as a powerful means through which children learn their cultural heritage (Bettelheim 1977). Children's books provide messages about right and wrong, the beautiful and the hideous, what is attainable and what is out of bounds—in sum, a society's ideals and directions. Simply put, children's books are a celebration, reaffirmation, and dominant blueprint of shared cultural values, meanings, and expectations.

Childhood is central to the development of gender identity and schemas. By preschool, children have learned to categorize themselves and others into one of two gender identity categories, and parents, teachers, and peers behave toward children based on these categories. The development of a gender identity and understandings of the expectations associated with it continue throughout childhood. Along with parents, teachers, and peers, books contribute to how children understand what is expected of women and men and shape how they think of their place in the social structure: Through stories, "children learn to constitute them selves [*sic*] as bipolar males or females with the appropriate patterns of power and desire" (Davies 2003, 49). Books are one piece of a socialization and identity formation process that is colored by children's prior understandings of gender, or gender schemas. Because schemas are broad cognitive structures that organize and guide perception, they are often reinforced and difficult to change. It takes consistent effort to combat dominant cultural messages (Bem 1983), including those sent by the majority of books.

The extensive body of research (often referred to as "reader response") examining the role of the reader in constructing meanings of literature (e.g., Applebee 1978; Cullingford 1998) comes to a similar conclusion. We interpret stories through the filter of our prior knowledge about other stories and everyday experiences; in other

McCabe, J., Fairchild, E., Grauerholz, L., Pescosolido, B. A., & Tope, D. (2011). Gender in twentieth-century children's books: Patterns of disparity in titles and central characters. *Gender & Society, 25*(2): 197–226.

words, schemas shape our interpretations. Reading egalitarian books to children over a sustained period of time shapes children's gender attitudes and beliefs (e.g., Barclay 1974; Trepanier-Street and Romatowski 1999). However, one book is unlikely to drastically change a child's gender schema.

The effects of gender schemas can be seen in children's preferences for male characters. Boys and, to a lesser extent, girls prefer stories about boys and men (e.g., Bleakley, Westerberg, and Hopkins 1988; Connor and Serbin 1978). This research suggests that children see girls and women as less important and interesting. Even seeming exceptions to the pattern of male preference support the underlying premise: When boys identify with a girl as a central character, they redefine her as a secondary character (Segel 1986) and they identify male secondary characters as central characters when retelling stories (Davies 2003). Patterns of gender representation in children's books, therefore, work with children's existing schemas and beliefs about their own gender identity. A consistently unequal pattern of males and females in children's books thus contributes to and reinforces children's gender schemas and identities.

While representation in the media conveys social existence, exclusion (or underrepresentation) signifies nonexistence or "symbolic annihilation" (Tuchman 1978). Not showing a particular group or showing them less frequently than their proportion in the population conveys that the group is not socially valued. This phenomenon has been documented in a range of outlets—from television (Tuchman 1978) to introductory sociology textbooks (Ferree and Hall 1990) to animated cartoons (Klein and Shiffman 2009). Yet, research on "symbolic annihilation" has neglected children's books and failed to tie representations to broader historical changes.

Historical Change: Gender Throughout the Twentieth Century

Inequitable gender representations may have diminished over time in the United States, corresponding with women gaining rights throughout the century (e.g., voting and reproductive rights) and entrance into the public sphere via the workplace, politics, and media. However, it seems more likely that there will be periods of greater disparity and periods of greater parity, corresponding with upsurges in feminist activism and backlash against progressive gender reforms. For instance, Cancian and Ross (1981) identified a curvilinear pattern in newspapers' and magazines' coverage of women, finding that coverage peaked during the first wave of feminist activism (1908–1920) and dipped until the second wave was well underway in 1970, when it began to rise again.

Thus, we have reason to believe that representations during midcentury—after the 19th Amendment gave women the right to vote but before the second-wave women's movement—may differ from other parts of the century. Historians have identified the 1930s as a time of backlash against the changes in gender expectations and sexual freedom of the 1920s (Cott 1987; Scharf 1980). While resistance to these changes existed in the first two decades of the century (Kimmel 1987), the tide shifted with the Great Depression. Women were scorned for taking "male jobs" (Evans 1997; Scharf 1980), the increase in the number of women in the professions "came to a halt" (Scharf 1980, 85), and the media asked "Is Feminism Dead?" in 1935 (Scharf 1980, 110). Even when women's employment skyrocketed during WWII, traditional notions of gender persisted through the valuation of the "domestic ideology" (Evans 1997; Friedan 1963; Rupp and Taylor 1987) and women were "criticized for failing to raise their sons properly" (Evans 1997, 234). This gender traditionalism and antifeminism persisted into the 1960s, although feminist challenges to gender expectations began to swell again with President Kennedy's Commission on the Status of Women, the Equal Pay Act, the publication of *The Feminine Mystique*, and the founding of the National Organization for Women (NOW) (Rupp and Taylor 1987). The cumulative effects of these events were apparent in the 1970s as feminism rapidly expanded in a second wave of activism (Cancian and Ross 1981; Evans 1997). Although there was some resistance to feminism during the 1980s (Evans 1997; Faludi 1991), this latter part of the century saw a more consistent presence of activism; by the mid-1990s, feminist solidarity was growing among younger women (Evans 1997) identified as feminism's "third wave."

Based on these patterns of feminist activism and backlash, we expect representation of women and girls to be closer to parity during activist periods (1900–1929 and 1970–2000) and more absent during greater gender traditionalism (1930–1969). We link the theoretical concept of symbolic annihilation to gender representation throughout the century.

* * *

Findings

Twentieth-Century Representations

We first provide general yearly trends of the percentage of books featuring males and females in titles, as well as among central characters. Here, the unit of analysis is year rather than book. With all book series combined, there are 101 cases (representing 5,618 books across 101 years).

Because we are interested primarily in (dis)parity between representations of male and female characters, we focus on the *presence* of males or females. However, it is noteworthy that male or female characters are not present in many titles: 55 to 57 percent of Caldecott award winners and *Children's Catalog*; 43 percent in Little Golden Books. There were also some instances in which it was not possible to determine whether a character was male or female: 4 percent of Goldens, 8 percent of Caldecotts, and 19 percent of *Catalogs* had at least one such character.

[There are] three interesting patterns in representations. First, there is a clear disparity across all measures: Males are represented more frequently than females in titles and as central characters. For instance, on average, 36.5 percent of books each year include a male in the title compared to 17.5 percent that include a female. By no measure are females present more frequently than males. In fact, the mode for males in titles is 33, meaning that the most common distribution is that one-third of the books published that year include a male in the title, whereas the mode for females is 0, meaning that the most common distribution is that no book titles include females. Similarly, the mode for male central characters (overall) is 50, but 0 for females. . . . For instance, 13 years had no male animal characters while 24 years had no female animals. Examining each variable's range shows that males are present in up to 100 percent of the books, but females never exceed 75 percent. More striking, no more than 33 percent of books published in a year contain central characters who are adult women or female animals, whereas adult men and male animals appear in up to 100 percent.

Second, [there are] important variations by type of character. The greatest parity exists for child central characters; the greatest disparity exists for animal characters. Boys appear as central characters in 26.4 percent of books and girls in 19 percent, but male animals are central characters in 23.2 percent of books while female animals are in only 7.5 percent. The data show one instance of a higher range of books including female characters than male: that for children, where up to 75 percent of books in a year contain girl central characters while a maximum of 50 percent contain boys. It should be noted, however, that only one year has 75 percent girls and that most years have higher ranges for boys than for girls.

Third, there are differences across book series, but—as with variations by type of character—these differences are by degree, not direction. Regardless of book series, males are always represented more often than females in titles and as central characters; however, the *extent* of the disparities differs. Golden Books tend to have the most unbalanced representations; Goldens have the highest mean and mode of males in the titles of any of the book types and the highest mean value of male central characters, followed by Caldecotts and the *Catalog*. The greatest disparity—animal characters—and the smallest—child characters—are also consistent across book types.

. . . All of the male to female comparisons . . . are statistically significant; in other words, for each variable in each book series, males are present in significantly more books than are females. When all books are combined, we find 1,857 (out of 5,618) books where males appear in the titles, compared to 966 books with females; a ratio of 1.9:1. For central characters, 3,418 books featured any male and 2,098 featured any female (1.6:1). Once again, the greatest disparity is for animal characters (2.6:1) and the least for child characters (1.3:1). . . .

A closer look at the types of characters with the greatest disparity reveals that only one Caldecott winner has a female animal as a central character without any male central characters. The 1985 Honor book *Have You Seen My Duckling?* . . . follows Mother Duck asking other pond animals this question as she searches for a missing duckling. One other Caldecott has a female animal without a male animal also in a central role; however, in *Officer Buckle and Gloria*, the female dog is present alongside a male police officer. Although female animal characters do exist, books with male animals, such as *Barkley* . . . and *The Poky Little Puppy* . . . were more than two-and-a-half times more common across the century than those with female animals.

The greatest disparity in titles and overall characters occurs among the Little Golden Books and Caldecott award winners and the least disparity in the *Catalog* books. . . . Regardless of type of character (i.e., child or adult, human or animal), books in the *Catalog* are significantly more equal than the Goldens. . . .

Trends by Historical Period

Data presented thus far provide a general picture of disparity in children's books. However, we expect historical and social factors to affect representation. . . . Books published during the 1930s–1960s are more likely than earlier or later decades to feature males in the titles and, with one exception (1900s), as central characters. Books in early and later years are more likely to feature females, such as Harriet and Mirette while midcentury books, like *The Poky Little Puppy* feature more males. In rare cases, there are actually more females than males in both the early and later parts of the twentieth century. . . . The most equitable category is child central characters. In contrast, animal characters are the least equitable. Although the most recently published books come quite close to parity for human characters (ratios of 0.9:1 [children] to 1.2:1 [adults] for the 1990s), a significant disparity remains for animals (1.9:1). All of [our analyses] show a nonlinear pattern, with greatest inequality midcentury.

* * *

Discussion

Gender is a social creation; cultural representation, including that in children's literature, is a key source in reproducing and legitimating gender systems and gender inequality. The messages conveyed through representation of males and females in books contribute to children's ideas of what it means to be a boy, girl, man, or woman. The disparities we find point to the symbolic annihilation of women and girls, and particularly female animals, in twentieth-century children's literature, suggesting to children that these characters are less important than their male counterparts.

We provide a comprehensive picture of children's books and demonstrate disparities on multiple measures. Still, there may be reason to believe that our findings are conservative regarding the unequal representation children actually experience. This is due in part to how gender schemas and developing gender ideologies are compounded. Reader response research suggests that as children read books with male characters, their preferences for male characters are reinforced, and they will continue reaching for books that feature boys, men, and male animals. Children's exposure, moreover, is likely narrower than the range of books we studied.

Adults also play important roles as they select books for their own children and make purchasing decisions for schools and libraries. Because boys prefer male central characters while girls' preferences are less strong, textbooks in the 1980s advised: "the ratio of 'boy books' should be about two to one in the classroom library collection" (Segel 1986, 180). Given this advice, disparities in actual libraries and classrooms could be even larger than what we found. Although feminist stories have circulated since at least the 1970s, "neither feminist versions of old stories nor new feminist stories are readily available in bookshops and libraries, and schools show almost no sign of this development" (Davies 2003, 49). Therefore, combating the patterns we found with "feminist stories" requires parents' conscious efforts. While some parents do this, most do not. A study of parents' reasons for selecting books finds most choices are based on parents' personal childhood favorites—indicating the continued impact of books from generations ago—and rarely on concern for stereotypes, particularly gender stereotypes (Peterson and Lach 1990).

Our historical lens allowed us to see change over time, but not consistent improvement. Rather, our findings support what other studies of media have shown: that coverage of social groups corresponds to changes in access to political influence (Burstein 1979; Cancian and Ross 1981). We found that the period of greatest disparity between males and females in children's books was the 1930s–1960s—precisely the period following the first-wave women's movement. Historians have noted, "No question, feminism came under heavy scrutiny—and fire—by the end of the 1920s" (Cott 1987, 271), coinciding with the beginning of this midcentury period. And, "'women's lib' was on everyone's lips" by 1970 (Evans 1997, 287), coinciding with the end of this period. Certainly, shifts in gender politics affect representation. . . .

Why is there a persistence of inequality among animal characters? There is some indication that publishers, under pressure to publish books that are more balanced regarding gender, used animal characters in an attempt to avoid the problem of gender representation. . . . As one book editor in Turow's (1978) study of children's book publishing remarked about the predominant use of animal central characters: "It's easier. You don't have to determine if it's a girl or boy—right? That's such a problem today. And if it's a girl, God forbid you put her in a pink dress" (p. 89). However, our findings show that most animal characters are sexed and that inequality among animals is greater—not less—than that among humans. The tendency of readers to interpret even gender-neutral animal characters as male exaggerates the pattern of female underrepresentation. For example, mothers (even those scoring high on the Sex Role Egalitarianism Questionnaire) frequently label gender-neutral animal characters as male when reading or discussing books with their children (DeLoache, Cassidy, and Carpenter 1987) and children assign gender to gender-neutral animal characters (Arthur and White 1996). Together with research on reader interpretations, our findings regarding imbalanced representations among animal characters suggest that these characters could be particularly powerful, and potentially overlooked, conduits for gendered messages. The persistent pattern of disparity among animal characters may reveal a subtle kind of symbolic annihilation of women disguised through animal imagery—a strategy noted by others (Adams 2004; Grauerholz 2007; Irvine 2007).

Although children's books have provided a steady stream of characters privileging boys and men over girls and women, examining representation across the long range illuminates areas where such messages are being challenged. Clearly, children's book publishing has been responsive to social change, and girls are more likely to see characters and books about individuals like themselves today than midcentury. Feminist activism during the 1970s specifically targeted children's books. For example, the publication of Weitzman et al.'s (1972) study appears to have influenced the publishing industry in important ways. Weitzman received funding from the NOW Legal Defense and Education Fund to reproduce children's book illustrations for a slide show to parents, educators, and publishers. This presentation made its way around the world in an effort to promote social change (Tobias 1997). Some argue that Weitzman et al.'s study profoundly shaped the children's book industry as a "rallying point for feminist activism," including the creation of "nonsexist" book lists and feminist publishing companies and the "raising of consciousness among more conventional publishers, award committees, authors, parents, and teachers" (Clark, Kulkin, and Clancy 1999, 71). The linear change we found since 1970 for most measures suggests this second-wave push for gender equity in children's books may have had a lasting impact.

Nonetheless, disparities remain in recent years, and our findings suggest ways that children's books are less amenable to change, especially in the case of animals. Although we do not know the complete impact of unequal representation on children, these data, in conjunction with previous research on the development and maintenance of gender schemas and gender identities, reinforce the importance of continued attention to symbolic annihilation in children's books. While children do not always interpret messages in books in ways adults intend (see, e.g., Davies 2003), the messages from the disparities we find are reinforced by similar—or even more unequal—ones among characters in G-rated films (Smith et al. 2010), cartoons (Klein and Shiffman 2009), video games (Downs and Smith 2010), and even coloring books (Fitzpatrick and McPherson 2010). This widespread pattern of underrepresentation of females may contribute to a sense of unimportance among girls and privilege among boys. Gender is a structure deeply embedded in our society, including in children's literature. This research highlights patterns that give us hope for the success of feminist attention to issues of disparity and remind us that continued disparities have important effects on our understandings of gender and ourselves.

References

Adams, Carol. 2004. *The sexual politics of meat*. New York: Continuum.

Applebee, Arthur. 1978. *The child's concept of a story*. Chicago: University of Chicago Press.

Arthur, April G., and Hedy White. 1996. Children's assignment of gender to animal characters in pictures. *Journal of Genetic Psychology* 157:297–301.

Barclay, Lisa K. 1974. The emergence of vocational expectations in preschool children. *Journal of Vocational Behavior* 4:1–14.

Bem, Sandra Lipsitz. 1983. Gender schema theory and its implications for child development: Raising gender-aschematic children in a gender-schematic society. *Signs* 8:598–616.

Bettelheim, Bruno. 1977. *Uses of enchantment*. New York: Vintage.

Bleakley, Mary Ellen, Virginia Westerberg, and Kenneth D. Hopkins. 1988. The effect of character sex on story interest and comprehension in children. *American Education Research Journal* 25:145–55.

Burstein, Paul. 1979. Public opinion, demonstrations, and the passage of anti-discrimination legislation. *Public Opinion Quarterly* 43:157–72.

Cancian, Francesca M., and Bonnie L. Ross. 1981. Mass media and the women's movement: 1900–1977. *Journal of Applied Behavioral Science* 17:9–26.

Clark, Roger, Heidi Kulkin, and Liam Clancy. 1999. The liberal bias in feminist social science research on children's books. In *Girls, boys, books, toys: Gender in children's literature and culture*, edited by B. L. Clark and M. R. Higonnet. Baltimore: Johns Hopkins University Press.

Clark, Roger, Rachel Lennon, and Leanna Morris. 1993. Of Caldecotts and Kings: Gendered images in recent American children's books by black and non-black illustrators. *Gender & Society* 7:227–45.

Connor, Jane Marantz, and Lisa A. Serbin. 1978. Children's responses to stories with male and female characters. *Sex Roles* 4:637–45.

Cott, Nancy F. 1987. *The grounding of modern feminism*. New Haven, CT: Yale University Press.

Cullingford, Cedric. 1998. *Children's literature and its effects: The formative years*. London: Continuum.

Davies, Bronwyn. 2003. *Frogs and snails and feminist tales: Preschool children and gender.* Rev. ed. Cresskill, NY: Hampton Press.

DeLoache, Judy S., Deborah J. Cassidy, and C. Jan Carpenter. 1987. The three bears are all boys: Mothers' gender labeling of neutral picture book characters. *Sex Roles* 17:163–78.

Downs, Edward, and Stacy L. Smith. 2010. Keeping abreast of hypersexuality: A video game character content analysis. *Sex Roles* 62:721–33.

Evans, Sara. M. 1997. *Born for liberty: A history of women in America.* New York: Free Press.

Faludi, Susan. 1991. *Backlash: The undeclared war against American women.* New York: Doubleday.

Ferree, Myra Marx, and Elaine J. Hall. 1990. Visual images of American society: Gender and race in introductory sociology textbooks. *Gender & Society* 4:500–33.

Fitzpatrick, Maureen L., and Barbara J. McPherson. 2010. Coloring within the lines: Gender stereotypes in contemporary coloring books. *Sex Roles* 62:127–37.

Friedan, Betty. 1963. *The feminine mystique*. New York: W. W. Norton.

Grauerholz, Liz. 2007. Cute enough to eat: The transformation of animals into meat for human consumption in commercialized images. *Humanity & Society* 31: 334–54.

Hamilton, Mykol C., David Anderson, Michelle Broaddus, and Kate Young. 2006. Gender stereotyping and underrepresentation of female characters in 200 popular children's picture books: A twenty-first century update. *Sex Roles* 55:757–65.

Irvine, Leslie. 2007. Introduction: Social justice and the animal question. *Humanity & Society* 31: 299–304.

Kimmel, Michael. 1987. Men's responses to feminism at the turn of the century. *Gender & Society* 1:261–83.

Klein, Hugh, and Kenneth S. Shiffman. 2009. Underrepresentation and symbolic annihilation of socially disenfranchised groups ("out groups") in animated cartoons. *Howard Journal of Communications* 20:55–72.

Peterson, Sharyl Bender, and Mary Alyce Lach. 1990. Gender stereotypes in children's books: Their prevalence and impact on cognitive and affective development. *Gender & Education* 2:185–97.

Rupp, Leila J., and Verta Taylor. 1987. *Survival in the doldrums: The American women's rights movement, 1945 to the 1960s*. New York: Oxford University Press.

Scharf, Lois. 1980. *To work and to wed: Female employment, feminism, and the Great Depression*. Westport, CT: Greenwood Press.

Segel, Elizabeth. 1986. "As the twig is bent . . .": Gender and childhood reading. In *Gender and reading: Essays on readers, texts, and contexts*, edited by E. A. Flynn and P. P. Schweickart. Baltimore: Johns Hopkins University Press.

Smith, Stacy L., Katherine M. Pieper, Amy Granados, and Marc Choueiti. 2010. Assessing gender-related portrayals in top-grossing G-rated films. *Sex Roles* 62:774–86.

Tobias, Shelia. 1997. *Faces of feminism: An activist's reflections on the women's movement*. Boulder, CO: Westview.

Trepanier-Street, Clary A., and Kimberly Wright Romatowski. 1999. The influence of children's literature on gender role perceptions: A reexamination. *Early Childhood Education Journal* 26:155–59.

Tuchman, Gaye. 1978. The symbolic annihilation of women by the mass media. In *Hearth and home: Images of women in the mass media*, edited by G. Tuchman, A. K. Daniels, and J. Benét. New York: Oxford University Press.

Turow, Joseph. 1978. *Getting books to children: An exploration of publisher-market relations*. Chicago: American Library Association.

Weitzman, Lenore, Deborah Eifler, Elizabeth Hokada, and Catherine Ross. 1972. Sex-role socialization in picture books for preschool children. *American Journal of Sociology* 77:1125–150.

Introduction to Reading 20

Kristen Myers analyzes the depiction of male characters in four television shows that were popular among elementary school children in the mid-2000s: *Hannah Montana, Suite Life on Deck, Wizards of Waverly Place,* and *iCarly*. Unlike many shows that promote ideas that masculinity is the polar opposite of femininity, these shows all featured male characters who were non-hegemonically masculine—that is, they were emotional, not athletic, or soft-spoken. She asks whether such portrayals promote progressive masculinities, or whether they ultimately reinforce hegemonic masculinity and a traditional gender order.

1. What can be learned from analyzing television shows aimed at children compared to a study about real-life children and how they interpret messages?
2. What is "regional-level" masculinity and how does this concept help us understand socialization?
3. What does Myers mean when she refers to "mask-ulinity"?
4. How is hegemonic masculinity a resource that could be deployed to regulate gender enactment?

"Cowboy Up!"

Non-Hegemonic Representations of Masculinity in Children's Television Programming

Kristen Myers*

From 2006–2011, Disney Channel featured a wildly popular show called *Hannah Montana,* aimed at an audience of pre-adolescent children. The show was about a teenaged girl named Miley (played by Miley Cyrus), who had a secret life as a pop-star named Hannah Montana. She lived with her father, a country singer named Robby Ray (played by Billy Ray Cyrus), and her brother Jackson (Jason Earles). On one episode (Season 2, Episode 22), a teenaged boy named Rico (Moises Arias) asked Robby Ray to teach him to line-dance so that he could impress a girl. When Rico danced, he moved his hips in big, fluid arcs, rather than stiffly shuffling about. He shimmied his shoulders. Robby Ray said, "A good ole boy ain't going to be wanting to do all that kind of stuff. I mean, your girl's wanting a championship line dancer, not the spin cycle on a washing machine!" Although Robby Ray didn't accuse Rico of being gay, he did problematize his feminine motion. To correct the problem, Robby Ray decided that Rico should "get in touch with his inner cowboy." He sat Rico on a deck railing, to pretend to be a cowboy riding a horse. After three hours of riding the deck railing, Rico was in so much pain that he walked bowlegged, like a real cowboy. He danced like one too—awkward, halting, and stiff. Rico was cured of his girly swishing moves—he had "cowboyed up."

Hannah and other television programs proffer complex messages about masculinity for young American children to consume and perhaps emulate (Baker-Sperry, 2007; Corsaro, 1997). In introducing young children to cultural conceptualizations of masculinity, television series help produce "regional masculinities" (Connell & Messerschmidt, 2005) that shape the ways that masculinity plays out in actual children's lives. Because they live in "media-rich worlds," (Martin & Kayzak, 2009, 317), children easily absorb these messages. Television has an especially great impact on children, who consume it while their identities are being formed (Kelley et al., 1999; McAllister & Giglio, 2005; Baker-Sperry, 2007; and Corsaro, 1997). Television helps construct a rhetorical

Myers, K. (2012). "Cowboy up!": Non-hegemonic representations of masculinity in children's television programming. *The Journal of Men's Studies, 20*(2): 125–143.

"frame" (Goffman, 1974; Ridgeway, 2011) that shapes people's perceptions of the world (Kuypers, 2009), despite the fact that the characters are fictional and viewers may never actually meet the actors in the series (Ferris, 2001).

This article focuses on the contradictory versions of masculinity that were presented in four television series aimed at children. On the one hand, these shows featured protagonist boys who were soft-spoken, unathletic, emotional, and thoughtful—antitheses to the hyper-masculine heroes of years past (Bereska, 2003). Popularizing images of non-traditional masculinity could help shift the patriarchal gender order in a feminist direction (Butler, 1999; Connell, 1987; Renold, 2004; Walsh et al., 2008). On the other hand, these boys were almost always the butt of jokes. They were consistently feminized, with femininity signifying weakness and failure. Traditionally masculine characters often lurked in the background, reminding viewers what a "real man" looked like. Here, I explore the extent to which these programs promote progressive masculinities, or if traditional gender orthodoxy prevails.

Masculinities: Theory and Practice

In the popular imagination, masculinity is the polar opposite of femininity (Fausto-Sterling, 2000). In recent years, however, researchers have argued that masculinity and femininity are complexly interconnected social constructions (West & Zimmerman, 1987), and there is not simply one masculinity juxtaposed to one femininity (Connell, 1987). Connell has shown that societies construct multiple masculinities and multiple femininities, with one form of masculinity dominating all others: "hegemonic masculinity." All boys and men are measured by hegemonic masculinity, even though most boys and men will never accomplish it. Connell and Messerschmidt (2005, p. 844) have elaborated on the concept of hegemonic masculinity, explaining that "To sustain a given pattern of hegemony requires the policing of men as well as the exclusion or discrediting of women." Women, girls, men, and boys all engage in this policing. Masculinity is embodied and enacted through displays of strength, athleticism, risk-taking, and heterosexual prowess.

Boys recognize hegemonic masculinity at an early age. For example, Messner (1990) researched elementary-school boys as they learned that playing basketball was not just about having fun with your friends—it was about being evaluated by older, higher status boys and men. Renold's (2007) work on 10 and 11 year old boys showed that even elementary school-aged boys adopted hegemonically masculine personae—what she calls a "hyperheterosexual" identity (see also Haywood, 1996). Boys in her study distanced themselves from femininity by avoiding girls, but they knew that being desired by girls was a good thing. As one girl commented about a hyperheterosexual boy: "Todd likes having girlfriends, but he dislikes girls" (p. 286).

Older boys participate in the construction and reinforcement of hegemonic masculinity within their own peer groups. Oranksy and Maracek (2009) found that high school boys hid their feelings from their male friends. If a friend expressed emotion, subjects said that they would ignore it or tell him to "just suck it up." By eschewing emotion, a boy avoided "being a girl." Rather than complaining about being teased, the boys were grateful to be distracted from their emotions. Oransky and Maracek write, "By disallowing interactions that would be labeled as girly or gay, boys protected one another's manliness" (p. 232).

Because heterosexuality is a major component of successful masculinity, boys spend a lot of energy addressing it. As Korobov (2005, p. 228) writes, "adolescence is a time when young men in particular begin to routinely practice forms of heteronormative masculinity that may implicitly or explicitly sanction sexism, homophobia, and 'compulsory heterosexuality.'" Pascoe (2005) calls this discourse "fag talk." In Pascoe's study, adolescents used "fag" to mean weak and unmanly. "Fag talk" was central to boys' joking discourse. At the same time, however, "fag talk" was a potent threat—boys could be targeted at any time by anyone. Pascoe writes,

> Fag talk and fag imitations serve as a discourse with which boys discipline themselves and each other through joking relationships. . . . The fluidity of the fag identity is what makes the specter of the fag such a powerful disciplinary mechanism. (p. 330)

Calling someone "fag" was also a clever way to announce to other boys, "Not it!" Ramlow (2003, p. 108) says that homophobic comments are effective because they ultimately demasculinize men: "Being called a 'faggot,' a 'pussy,' or 'gay,' then, is not always or overtly about the material fact of sexual difference or same-sex relations; it is about the failures of heteronormative masculinity." In name-calling, many boys use "gay" and "girl" interchangeably (Oranksy & Maracek, 2009). Indeed, Epstein (1997) argued that, in primary or elementary school, the worst thing a boy could be called is a girl. As Butler (1999) explains, the "heterosexual matrix" complexly interconnects masculinity and heterosexuality, rejecting femininity.

Taken together, this literature shows that the masculinity, femininity, heteronormativity, and homophobia are intertwined so that markers of one become imbricated as evidence of another. Boys reject femininity in order to establish their dominance, and they must continually degrade girls and feminize other boys so as to maintain their status—even as they pursue girls sexually.

Because of the harmful effects of hegemonic masculinity on boys and girls, some scholars have insisted that we begin to cultivate non-hegemonic masculinities (Pollack, 1998; Kimmel, 2006). Risman and Seale (2010), for example, urge adults to interrupt boyhood as we know it. Butler (1999) theorizes that we could "queer" the gendered order by refusing to conform to hegemonic proscriptions. Despite persuasive literature suggesting that we raise less hegemonic boys, there is little actual research on what it is like to grow up as a non-hegemonic boy, as Renold (2004) points out. Is being a non-hegemonic boy a panacea for patriarchy and its ill effects? Renold found that, instead of subverting the dominant gender order, "othered" boys' management strategies actually reinforced dominant masculinities by treating hegemonic boys as the standard. Rather than embracing their counter-hegemonic potential, these boys longed to be "normal." They adopted the misogynist practices of their bullying classmates, rejecting all things feminine, including girls. Renold says that, ironically, "they appeared not to make the connection between the devaluing of femininity more widely and the subordination of non-hegemonic masculinities" (p. 261). Rather than altering the gender regime, non-hegemonic boys actually helped reinforce the traditional order (see also Connell & Messerschmidt, 2005; Kimmel, 2006; Pascoe, 2003).

Culture and Media

Hyperheterosexual hegemonic masculinity is a sociocultural product. Connell and Messerschmidt (2005) explained that the "geography of masculinity" occurs at three levels: 1) Local level of face-to-face interaction; 2) Regional level of culture or nation-state; and 3) Global or transnational level. Regional-level masculinity is pertinent to this paper:

> Hegemonic masculinity at the regional level is symbolically represented through the interplay of specific local masculine practices that have regional significance, such as those constructed by feature film actors, professional athletes, and politicians. The exact content of these practices varies over time and across societies. Yet regional hegemonic masculinity shapes a society-wide sense of masculine reality and, therefore, operates in the cultural domain as on-hand material to be actualized, altered, or challenged through practice in a range of different local circumstances. A regional hegemonic masculinity, then, provides a cultural framework that may be materialized in daily practices and interactions. (pp. 849–850)

This article focuses on the production of regional masculinity by American children's television programming, which affects the ways that children conceptualize masculinity in their localized lives. . . .

Children easily absorb "the rules of play" communicated by popular media, because "young children are immersed in media-rich worlds" (Martin & Kayzak, 2009, p. 317). McAllister and Giglio (2005) explain that media have an especially great impact on children, who consume it while their identities are being formed (see also Baker-Sperry, 2007; and Corsaro, 1997). Television media have been shown to impact children's early gendered behavior (Powell & Abels, 2002). Indeed, the media might have a greater impact on children today given the nearly ubiquitous nature of television programming directed at them (McAllister and Giglio 2005). McAllister and Giglio (2005) explain that cable television has produced entire networks—or "kidnets"—aimed at children. Recognized for their buying potential, and for their influence over parents' buying practices, children have become an important audience for advertisers. Children's programming is broadcast 24 hours a day, daily, and the number of adolescent and pre-adolescent' channels grows each year. The two largest kidnets are Disney and Nickelodeon. On March 14, 2011, Disney channel ranked as the number one network among tweens for the 16th week in a row, according to Tvbythenumbers.com. Even during the NBA playoffs in spring 2011, Disney ranked as the 3rd most watched network during prime-time, among all viewers. When rating all-day programming between November 2010 and May 2011, Nickelodeon consistently topped the charts at about 2.1 million viewers, with Disney coming in second at 1.6 million. *Hannah's* final episode garnered a monster 6+ million viewers. Adults and children watch these networks. They also buy t-shirts, dolls, cds, and dvds. These shows are an entire industry (McAllister & Giglio, 2005).

Nickelodeon has several channels: Nick Jr., Nick, Teennick, Nick at Nite, and Nicktoons. Programming on these channels often overlaps. Similarly, Disney broadcasts on three channels: Disney Channel, Disney XD, and Playhouse Disney. To fill air time, episodes from a single series are rebroadcast five to six times daily, with a single episode often airing twice a day. Thus, the opportunity for children to consume the same program repeatedly is substantial. And, as Crawley, et al (1999) have shown, repeated viewing leads to greater comprehension by child viewers.

Even though children consume a great deal of television on a weekly basis, they do not passively absorb larger cultural messages. Children are actors in their own right, negotiating meanings among themselves (Myers & Raymond, 2010; Thorne, 1993; Van Ausdale & Feagin, 2002), with adults (Baraldi, 2008; DeMol & Buysee, 2008; Ludvigsen & Scott, 2009), and with the media itself (Bragg & Buckingham, 2004; Fingerson, 1999). However, children are affected by and grapple with cultural frames (Myers & Raymond, 2010; Neitzel & Chafel, 2010). Martin and Kayzak (2009) have argued that it is therefore important to understand the messages that are available to the children who consume them.

In this paper, I analyze the content of four television programs on Disney and Nickelodeon that were aimed at young children. What real-life children do with these messages is not addressed in this project. Instead, I explore the contradictory messages about masculinity communicated to children, examining the implications of these messages for children's conceptualization of masculinity at the local level.

Methods and Analysis

To explore these issues, I conducted qualitative textual analysis of the content of four popular children's television programs: Disney's *Suite Life on Deck* (broadcast from 2008–2011); *Hannah Montana* (2006–2011); and *Wizards of Waverly Place* (2007–2011); and Nickelodeon's *iCarly* (2007–2011). I chose these shows based on focus groups with elementary school girls, aged 5–11 (N = 63) (see Myers & Raymond, 2010). These were their most-watched programs: they quoted them, gushed over characters' exploits, gossiped about their sexual liaisons, and even wore clothes that featured the characters' likenesses. Because these shows informed most of the girls' cultural references, I decided to systematically study them so as to uncover patterns in their content.

These children's shows were not made for girls only. They were watched by millions of viewers of all ages nationwide, as discussed above. *Hannah* and *iCarly* were hugely popular shows, rated in the top 10 most viewed cable shows week after week in the summer of 2010, during the period when I coded data. Each series was well-established, in at least its 3rd season, and each season had 25–30 half-hour episodes (about 22 minutes long without commercials).

In selecting episodes to watch and code, I relied heavily on the programming schedules of each network. I did not randomly select episodes, as many were not easily available. I recorded the shows, watched them on Youtube.com, and/or ordered them through cable's "OnDemand." This paper includes systematic analysis of 45 episodes: 13 episodes of *Suite Life*, 10 episodes of *Hannah*, 10 episodes of *Wizards*, and 12 episodes of *iCarly*. Table 4.1 describes each program for non-viewers.

I coded each episode in this manner: 1) I watched each episode entirely without taking notes to make sure I caught the physical and verbal content. 2) I watched each episode again, taking copious notes. I stopped each program numerous times to transcribe who said what to whom and under which circumstances. Dialogue was rich, bombarding the viewer with messages about boys, girls, romance, and bodies at a rapid-fire pace. It often took me 10 minutes to get through less than two minutes of one episode. A typical episode took at least an hour and a half to code. 3) I trained a graduate assistant to independently code 25% of the episodes, randomly selected from the total list of episodes included in this study. Our percent agreement was .83, meaning we coded data the same way 83% of the time, a good rate of agreement (Kurasaki, 2000). . . .

Findings

I find that these television programs highlighted non-hegemonic masculinities routinely. However, rather than undermining hegemonic masculinity, non-hegemonic characters actually re-valorized hegemonic masculinity (Walsh et al., 2008). I argue that the mutually-reinforcing dynamics suggested by Connell and Messerschmidt (2007) and by Renold (2004)—in which subordinated masculinities unwittingly conspire to recreate a hegemonic gendered order—"frame" masculinity for the audience (Goffman, 1974; Ferris, 2001; Kuypers, 2009) in a traditional manner.

Table 4.1 Description of Television Programs Analyzed

Program	*Premise*	*Main characters*	*Gender*	*Race*
Hannah Montana (Disney)	Tennessee native, Miley, moves with her family to LA, where she lives a double life: she's secretly a pop-star (Hannah Montana).	Miley	Girl	White
		Robbie Ray	Man	White
		Jackson	Boy	White
		Oliver	Boy	White
		Lilly	Girl	White
		Rico	Boy	Latino
Suite Life on Deck (Disney)	In this sequel to *The Suite Life of Zack and Cody,* twins Zack and Cody go to high school on a cruise ship.	Zack	Boy	White
		Cody	Boy	White
		London Tipton	Girl	Asian
		Woody	Boy	White
		Bailey	Girl	White
		Marcus	Boy	Black
		Mr. Moseby	Man	Black
Wizards of Waverly Place (Disney)	A family of wizards live on Waverly Place, in NYC. The kids learn to be wizards from their father, and they interact with magical creatures from the wizarding world.	Alex	Girl	Latina
		Justin	Boy	Latino
		Max	Boy	Latino
		Theresa	Woman	Latina
		Jerry	Man	White
		Harper	Girl	White
iCarly (Nickelodeon)	Carly lives with her brother, Spencer, in an apartment in Seattle. Carly and her best friend, Sam, star in a weekly web show, *iCarly,* with friend, Freddy.	Carly	Girl	White
		Spencer	Man	White
		Sam	Girl	White
		Freddy	Boy	White
		Mrs. Benson	Woman	White
		Gibby	Boy	White

* indicates a hegemonically masculine character.

Non-Hegemonic Masculinities

Almost all of the male characters on these shows embodied non-hegemonic masculinities: 14 out of 16 males (88%) listed in Table 1 were non-hegemonic. They were not domineering, competitive, or sexually predatory. Instead, these males were gentle and emotional. Rather than shunning femininity, they often marked themselves in feminine ways. Most were heterosexual failures, although none was overtly gay. I describe them here.

Suite Life's Cody (played by Cole Sprouse) was smart, polite, non-athletic, romantic, clean, goal-oriented, and cautious. Much of the show's humor centered on his being a nerd. Cody's roommate, Woody (Matthew Timmons), was overweight, wore braces, and had a head of bouncy, unkempt curly hair. Food-obsessed and gassy, Woody was the butt of many jokes. *Suite Life's* Mr. Moseby (Phill Lewis) was one of the few adults on the show. He directed the ship's activities and acting as the students' guardian. Moseby was short, black, and effeminate. His uniform consisted of a blazer with a pocket handkerchief, shorts, and knee socks. One day, Moseby was excited to receive a package: "It's here! My pocket hanky of the month!" He took it out of the box and exclaimed, "Oooo! *Stripes!*" (Season 1, Episode 19). The kids teased him about his diminutive size and his failure with women.

On *Hannah*, Miley's older brother, Jackson, was short, lazy, and an academic failure. He had trouble getting dates. Jackson and his father, Robby Ray—also a non-hegemonic character—had a close relationship. For example, Robby Ray insisted that Jackson register at the local community college (S3, E14). Jackson resisted, telling his father, "Just accept that I'm a slacker." Robby Ray said, "Son, it's totally natural to be scared to go to college." Jackson protested that he was not scared, but he quickly began sputtering, "I'm completely terrified." Robby Ray put his arm around him and said, "It's ok son."

Robby Ray was a widower, raising the children alone. Although he was a "good ole boy," he was also non-hegemonic: he wore pink and spent a lot of time on his hair. Robby Ray took great pride in his housework. In the same episode discussed above, Jackson pulled a towel from the laundry basket and held it to his face, saying "That [towel] smells great! Are you doing something different?" Robby Ray grinned and said, "I started adding fabric softener halfway through the rinse cycle. It's made all the difference in the world." Both men valued what is traditionally considered "women's work."

On *iCarly*, Carly's (played by Miranda Cosgrove) older brother, Spencer (Jerry Trainor), was her legal guardian. Spencer was a sculptor who worked at home. He was impulsive, playful, and clownish. For example, Spencer ordered a personalized credit card with a bunny hologram on it (S2, E12). He proudly showed the kids how the bunny changed when you moved it: "Happy bunny. Sad bunny. My bunny has conflicting emotions." *iCarly's* Freddy (Nathan Kress) was Carly's neighbor and the producer of the iCarly webshow. Freddy lived alone with his over-protective mother, who infantilized him. She followed him around with ointment and anti-bacterial spray, rubbing and spraying him whenever he stood still. Freddy had "Galaxy War" sheets on his bed, like a little boy. He was a "techy," unathletic, smart, and cautious. Their classmate, Gibby (Noah Munck), was doughy and quiet. He removed his shirt frequently, for no apparent reason—in the middle of the halls at school, at restaurants, parties, etc. His flabby body was a source of humor. Sam (Jennette McCurdy)—a girl—beat him up and stole his lunches regularly. Gibby vomited when he was scared or in trouble. The kids treated Gibby as a cute pet.

On *Wizards*, Alex's (played by Selena Gomez) older brother, Justin (David Henrie), was the most successful wizard in the family. He was studious, obedient, and risk-averse. For example, Justin refused to sing in the shower because he thought it was dangerous: "singing leads to dancing and dancing leads to slipping." A consummate "geek," he was a member of quiz bowl, the alien language club, chess club, and the Captain Jim Bob Sherwood Space Farmer fan club. He was openly emotional. On one episode, the "wizards" reunited their friend with his birth parents (S2, E18). Touched, Justin said, "I promised myself I wouldn't cry, I promised myself I wouldn't cry. . . . I'm not afraid of my emotions."

All of these males possessed at least some of the characteristics of non-hegemonic masculinity. In other words, all of them were feminized in some way. Nevertheless, they did not make up a cohesive trope of non-hegemonic masculinity. Some (like Robby Ray, Cody, and Justin) were heterosexually successful, smart, body-conscious, and ambitious, while others (like Woody and Jackson) were lazy, dirty, and heterosexual failures. Clearly, however, none of them was hegemonically masculine. These shows promoted non-hegemonic protagonists, who were adored by viewers. On the surface, one might assume that these masculinities could undermine gender orthodoxy, given the huge audience who consumes them. If so, the gender order could be transformed (Butler, 1999; Lotz & Ross, 2004). In the next section, however, we see that non-hegemonic masculinity was a comedic tool rather than a transformative archetype.

Hegemonic Masculinity

On these programs, characters co-constructed hegemonic masculinity at the expense of non-hegemonic characters. They constructed hierarchies, emphasized ideal masculine bodies, and celebrated hyperheterosexuality. Although few of the male characters actually achieved hegemonic status, they all were affected by it. As Connell and Messerschmidt (2005) posited, the non-hegemonic characters served a larger purpose: reinforcing hegemonic masculinity and the traditional gendered order.

Hyperheterosexuality

Renold (2007) argued that hyperheterosexuality is an important marker of hegemonic masculinity, even among children. Some of the boys on these shows were hyperheterosexual, and others aspired to be. For example, when the *Suite Life* ship docked in Greece, the students went to a museum (S1, E7). Cody watched helplessly while Bailey (played by Debby Ryan), his crush, flirted with their tour guide: "Look at her, drooling over him like he's some kind of Greek Adonis." The tour guide introduced himself to everyone: "Hello, my name is Adonis." Cody groaned, and said to his twin, Zack, "This guy might mess up my 6 month plan to win Bailey. This month, we have to get to at least hand-holding." Zack (Dylan Sprouse) told Cody, "Dude, while you're working on your 6 month plan, this guy is just *working it*." Sure enough, Bailey went to the roof with Adonis to "see the view," leaving Cody crushed. Cody wanted to be a smooth operator like Adonis, but he lacked hyperheterosexual seduction skills.

Zack—Cody's opposite—was a hyperheterosexual predator. Non-hegemonic boys studied Zack's techniques. Zack told Woody, "There is nothing, *nothing* better in this world than an unhappy hot girl." On one episode, he and Woody posed as janitors to scope out new girls as they arrived on the ship (S2, E13). Woody pointed out one girl he thought was pretty. Zack showed him that the girl had a lock on her luggage: "That means she's suspicious and cautious. I'm looking for naïve and vulnerable." Woody grinned, "There's so much to *learn* from you!" Zack said, "Now focus. These girls are only here for a week." Woody tried again: "How about that one: cute, blonde, *nice legs*, and carrying a text book—repressed book worm badly in need of a good time!" The "girl" whom Woody was talking about turned around, and Zack exclaimed, "That's *Cody*!" This scene secured Zack's superiority above both Woody—who failed his hyperheterosexual training—and Cody—who looked like a cute girl from the back.

Zack was surrounded by lesser boys, whom he mentored in the art of scoring. For example, Zack concocted a scheme to meet gullible girls, a fake beauty pageant:

Zack: "All the girls will fill out an application and send in a photo, which makes it easy to weed out the ones not worth pursuing. Then we just cancel the whole thing and they'll never know we were involved."

Woody: "Come on—no girl's stupid enough to fall for that."

London (played by Brenda Song) walked by, saw the sign, and exclaimed: "Ooo! A beauty contest! And I'm beauty-*ful!*"

The boys grinned at each other.

Zack wanted to "weed out" girls who were ugly and/or cautious. Dumb girls were easier to manipulate. For example, a pretty blonde girl named Capri (Brittany Ross) carried her pageant application to Zack and said, "Excuse me, I don't understand Question 4. What is your ick?" Zack said: "That's *IQ*." Capri said, "Ohhhhh!." Zack grinned at Woody: "This is gonna be *awesome*!"

Zack used the language of predator versus prey as he coached Woody in hyperheterosexuality. On another episode (S1, E19), Zack explained:

> When the lions are out hunting gazelles, they don't attack the strong healthy ones. Oh no. They attack the weak ones. The ones crying and eating ice cream.

Woody grinned and said, "I worship this man." Hyperheterosexual bad-boy characters, like Zack, sat at the top of the hierarchy, reminding non-hegemonic boys how much they had to learn. They teased "others" about their failures with women. For example, when Mr. Moseby learned about the fake pageant, he chastised the boys. Zack said, "It's ok! We've canceled it." Moseby said, "It's too late. Now I have a bunch of disappointed girls on my hands." Zack smirked: "Oh come, it can't be the *first* time." Woody and Marcus laughed. Zack's hyperhetersexuality entitled him to ridicule his elders and his peers with impunity.

Cowboy-Up: "Fag Talk" and Feminization

Hegemonic masculinity is not easily attainable—indeed, most males will never accomplish it. But it serves as a ready-made tool that can be used by anyone to police a male's masculinity. Questioning a male's manhood serves a similar purpose to Pascoe's "fag talk:" it is a disciplinary mechanism to regulate boys' doing gender. Of course, characters on these G-rated children's' shows did not actually call each other "fags." Instead, they feminized each other, which paralleled "fag talk." One way that characters on the series indicted someone's masculinity was to tell him to "man up." For example, on *Hannah*, Robby Ray nagged Jackson for sleeping until 2 pm: "It's time for you to cowboy up and act like a man." On *Wizards*, Justin announced that he was going to perform the "Thin Man" spell (S2, E3). Alex laughed: "Great! It's a spell that will make Justin thin *and* a man." Everyone laughed, including their father. On *iCarly*, Sam called Freddy, Spencer and Carly "a bunch of prancies" when they criticized her. On *Suite Life*, Marcus complained that Woody got him a cranberry hat instead of a magenta hat, to wear while judging the beauty pageant. Woody said, "You are such a diva! Other divas look at you and say, 'Wow, what a diva!'" Feminizing boys was a central part of these shows' humor, and it functioned like "fag talk."

On one episode of *Suite Life,* Woody arm-wrestled a female classmate (S2, E21). She slammed his arm down so hard that he fell out of his desk. He laid on the floor in a daze. Victorious, she raised her fists and yelled, "In your face! Eighty-three pounds of pure power!" Cody said, "Dude, you just got beat by a girl who can fit in a keyhole." Their teacher asked Woody what he was doing on the floor. Zack said, "Looking for his pride." Woody called out, "Can't find it! Oo, but I found a piece of gum!" He peeled the gum from beneath his desk and popped into his mouth, grinning goofily at the onlookers. Woody's repeated failings at masculinity provided dependable comic fodder for this series.

Like Oranksy and Marecek's (2009) subjects, some of the boys actually seemed to appreciate being policed. As described earlier, Rico asked Robby Ray to police and correct his fluid dance style. In another example, on *Suite Life*, Cody recreated a country fair to cheer up Bailey, who was homesick (S1, E19). Bailey's ex-boyfriend, Moose (played by Hutch Dano), came to visit during the fair. Moose—a hegemonic character, as implied by his name—beat Cody at every single game, including arm-wrestling, bobbing for corncobs, and "even something cerebral, like chess." Cody confided in Zack that he was worried Moose would win Bailey too. Zack sneered, "If the next competition is whining like a girl, you're gonna win by a mile!" Cody said, "Ok slightly harsh, but I guess I needed to hear that. I'm going to show Bailey that I'm ten times the man that Moose is. Right after I floss." He flossed his teeth and said, "Ow. Floss burn!" The audience laughed.

Gender Crossing: Boys in Drag

On the shows, a common comedic ploy involved dressing male characters in women's clothing. Sometimes boys would just appear in women's clothes momentarily. For example, on Suite Life, Cody tripped and fell into a rack of clothes in a boutique (S2, E5). When he stood back up, he wore a pink feather boa. He immediately took it off, but not before the audience laughed. On another episode, Woody opened a package from his mom to find a muumuu, which she accidentally sent to him instead of his sister (S1, E19). Zack saw him holding it up and quipped, "Nice dress Woody. Really brings out your eyes." Neither incident was central to the story line. They were just gratuitous episodes of unintentional gender crossing.

Other times, drag was integral to the plot line. For example, on *iCarly*, the kids asked Spencer to help them solve a problem, and they needed him to wear a disguise (S2, E12). Carly asked Spencer, "What size dress do you wear?" Without missing a beat, Spencer replied, "10, why?" His quick response implied that Spencer wore dresses frequently. Spencer dressed as an old woman and went out to try to help the kids. While he was out, an old man mistook Spencer for a woman, asking "Hey, you come alone?" Spencer dismissed him, but the old man persisted: "Hey baby. Let's go get some chicken pot pie!" Spencer said, "I don't care for pot pie. Run along." Someone recognized Spencer, who protested, "I'm just a busty old woman!" The old man reappeared, leering, "You're not too old for me!" He grabbed hold of Spencer. Spencer angrily removed his wig, revealing his gender. Shocked, the old man ran for the police. Spencer replaced his wig, but the old man returned, dragging the police behind him saying, "That's the one. That's the man lady." Spencer said, "No, I'm a simple old lady." He was escorted from the building.

On *Hannah*, Jackson avoided his chores by flirting with a girl on the beach (S2, E27). When he saw his father looking for him, Jackson tried to hide from him. He borrowed the girl's sunglasses and hat:

Girl: "What are you doing?"

Jackson: "Having fun, being my own man." He raised his voice a couple of octaves and said, "Let's talk like girls! Oh those are nice shoes where'd you get 'em, they're so cute!"

Robby Ray: "Excuse me, ladies?"

Jackson: "Not interested! Taken!"

Robby Ray: "Uh, Jacksina, can I have a word with you?"

Jackson, in his own voice: "Hey Rob-o! [To the girl:] That's my roommate. Probably needs to borrow some money. Why don't you wait over there, I don't want to embarrass him [to Robby Ray:] like he's embarrassing me!"

Robby Ray: "Oh yeah, said the guy in the floppy sunflower hat with the girly glasses!"

Taylor and Rupp (2003) argued that the spectacle of drag challenges the binary conceptualization of men and women as polar opposites. Drag operates outside of masculinity and femininity, "troubling" gender and sexuality by undermining people's perceptions of what is natural. In this way, drag is a form of social protest. On these television shows, drag did call attention to the gender binary, but *not* so as to undermine the naturalness of gender. Drag was not counter-hegemonic, but a comedic tool for ultimately re-inscribing the binary. Drag made a spectacle out of crossers not to celebrate them, but to punish them. Spencer's women's clothes put him in the position of being objectified by a strange man. Amusing because it bent heterogendered boundaries, this incident also communicated that even old women must fend off unwanted advances by men in public, and that men can be sexually predatory at any age. The scene changed once Spencer took off his wig, and the old man called the police. Spencer was no longer desirable, but became a threatening "man lady." Spencer was escorted away, and the message communicated was that crossing is deviant if not illegal. Equilibrium was restored, and the binary was re-naturalized.

Homoeroticism

Occasionally, homoerotic elements appeared in plot-lines. These included bawdy phrases or propositions, and physical intimacy among male characters. For example, Spencer's encounter with the leering old man did not end when he was escorted from the building. Back at home, Spencer answered his door to find the man standing there. The man said, "Look, why don't you just put the wig back on and we can start over?" Spencer slammed the door in his face. This moment transformed from being merely sexist to being homoerotic. The man *knew* Spencer was a man, followed him home, and proposed that Spencer "put the wig back on" to consummate the fantasy that he had concocted.

On *Suite Life*, Cody and Moose's country-fair competition was tinged with homoeroticism (S1, E19). As already mentioned, one of the contests was bobbing for corn cobs. The boys went under water and came out with cobs—sometimes two at a time—sticking lengthwise from their mouths. The sight of the phalluses protruding from their mouths was borderline pornographic, underscored by their bawdy dialogue: Zack told Cody, "You really want to impress Bailey. Get in there and bob for a cob." Cody nodded and said to Moose: "You better kiss your kernels goodbye, pal, because you're going down." Moose said, "We both are going down. That's how you pick up the corn." Here, Cody likened Moose's testicles to corn kernels, underscoring the impression that the cobs were meant to represent phalluses. And the boys' use of the phrase "going down"—which often refers to oral sex—further punctuated the homoerotic undertones of the interchange. Later, Cody lost to Moose in chess and said, "Well, kiss my bishop!" Again, because of the phallic shape of the bishop and the larger context of their competition, this comment renewed the homoerotic tension between Moose—the strong, hegemonic boy—and Cody—the non-hegemonic, feminized boy.

Homoeroticism also manifested as physical intimacy between males. Intimacy was more than sporty physical contact between tough guys. It breached the normative, expected level of touching in a homophobic culture. Intimate moments were remarkable and remarked upon by other characters. For example, on *Suite Life*, Cody was guest lecturing in a home economics class, due to his culinary expertise (S2, E4). Woody was the only male student in the class. Note that Woody's very name is slang for an erection. Cody told the students to gather around to feel his muscles move as he whisked, teaching them the proper technique. The girls flocked around him, massaging him, ooing and ah-ing over his flexing arms. Woody joined the group and touched Cody's tricep. Woody grinned lasciviously into the camera and said, "Impressive!" Cody whipped around and said, "Woody!" Woody let go, looked down, and mumbled, "Sorry Mr. Martin." Everyone in the class clearly enjoyed touching Cody's arm. Yet only Woody's pleasure was sanctioned.

In a bizarre incident on *iCarly*, Spencer hired a surly old cowboy, Bucky, to teach him to ride a mechanical bull (S3, E5). Bucky insisted that the only way Spencer would learn to ride the bull was to learn to *act* like the bull:

Bucky: "Get down on your hands and knees."

Spencer: "I'm sorry?"

Bucky: "GET DOWN!"

Spencer got on all fours.

Bucky: "Now when I climb on you, I want you to try to throw me off."

Spencer, rising up: "I don't think this is a good idea."

Bucky forced him back down: "Do you want to learn to ride that tin can or don't ya?"

Spencer: "I'm not sure any more!"

Spencer tried to throw Bucky off.

Bucky: "Come on boy! You can do better than that!"

Spencer: "I don't know! This is so new to me!"

This scene felt more like a sexual assault than homoerotic, as Bucky forced Spencer into unwanted, unequal physical intimacy. It embodied the masculine hierarchy, with a hegemonic male mounted upon a non-hegemonic male. Spencer was clearly bothered by the experience. When he ran into Carly at the smoothie shop, she said, "What are you doing here? I thought you had a bull riding lesson." Spencer said, "I did. And my teacher put his *butt* on me. I never want to think about him or that stupid bull ever again. That thing's an instrument of torture." He felt victimized and gave away his mechanical bull.

By allowing male characters to interact in intimate ways, these shows flirted with homoeroticism, which is definitely counter-hegemonic in a homophobic culture. These shows had the opportunity to introduce viewers to alternative ways for boys to interact, beyond competition and violence. Yet, for the most part, homoerotic moments were not presented as serious possibilities. They were humorous, disruptive, and in Spencer's case, harmful. Thus, homoeroticism was a tool for underscoring homophobia and competition among males.

Mask-ulinity Play

As mentioned above, the literature on non-hegemonic boys shows that, rather than undermining the hegemonic gendered order, non-hegemonic boys actually reinforce hegemony by valorizing hegemonic masculinity themselves (Pascoe, 2003; Renold, 2004). Non-hegemonic characters on these shows did the same thing: they temporarily crossed into hegemonic territory, and then crossed back (see also Thorne, 1993). Occasionally, these otherwise non-hegemonic males donned a "mask" of masculinity (Pollack, 1998), and asserted themselves over others in hegemonic, domineering ways.

Boys crossed to take charge of a situation. On *Hannah*, Oliver (played by Mitchell Musso) transformed from Lilly's (Emily Osment) compliant, "whipped" boyfriend to pimp in a matter of seconds (S3, E15). Sitting on the beach, Oliver and Lilly told Miley that they wanted her to have a boyfriend, instead of being lonely. Miley insisted that she did not need a boyfriend. Lilly gave Oliver a knowing look, and he said, "I'm on it." Oliver stood up and announced to the people on the beach: "Yo! Single dudes! Listen up! I got a perfectly good girl here, good height-to-weight ratio, not hard to look at. Let's start the bidding at uh. . ." Miley tackled him. From the ground, Oliver said, "Got a temperament issue, but the right guy could tame her." She grabbed him by the nose. He said, "You're blowing the sales pitch." In the guise of friendship, Oliver turned Miley into a piece of meat.

On one episode of *iCarly,* Gibby transformed from passive heterosexual failure to hyperheterosexual brute. The friends were surprised to hear that Gibby had a girlfriend, but they were shocked when they met her: Tasha (Emily Ratajkowski) was a tall, beautiful brunette. Freddy stared at her, mouth agape, and said, "Why can't *I* have one of those?" Gibby, usually demonstrative and goofy, began speaking in monotonal monosyllables, grunting orders at Tasha. In a series of misunderstandings, Gibby thought he saw Freddy kiss Tasha. He flew into a jealous rage, broke up with Tasha and promised retribution. Gibby kicked a beanbag chair across the room and told him Freddy would "beat him down." Gibby told Freddy to "bring a mop for your blood!" Sam intervened before Gibby and Freddy actually fought, clearing up the misunderstanding. Gibby apologized but remained aloof: he shook hands with Freddy, and said, "Bros?" Freddy agreed, "Bros." Tasha whined, "What about me?" Gibby nodded and said, "You're back in." She squealed, "Oh thank you!" In this episode, Gibby acted like a completely different boy than in any other episode. He intimidated and impressed his peers. Hegemonic masculinity was an effective, temporary tool for him to manage a particular situation.

Nevel (played by Reed Alexander), a recurring guest character on *iCarly* (featured in 7 episodes), was simultaneously feminine and domineering. On one episode, Nevel refused to help Carly solve a problem unless she agreed to kiss him (S2, E12). He said, "A kiss, a kiss is what I seek. Upon your lips, not on your cheek." Carly snarled, "Nobody likes a rhymer." Nevel wanted Carly to kiss him because he wanted to control her, not because he was hyperheterosexual. To his surprise, she agreed. He said, "Bring your sweet lips. Get ready for a real kiss from a real man." Then he turned and yelled, "Mother! Run a bath!" Unlike the other boys who transformed temporarily, Nevel's feminine markers did not fade as he enacted hegemonic masculinity. He straddled both realms simultaneously.

Carly and Nevel agreed to meet in an alley behind her apartment building. Nevel brought big, thuggish men to stand at each entrance to the alley, preventing Carly from escaping. He said, "You look wonderful in low light." He prepared to kiss her: "First, this:" applying lip balm with his pinky. Carly outsmarted him and avoided the kiss. Infuriated, Nevel waved his fist and yelled, "I declare that you'll rue this day! You'll rue it!" Nevel was unable to conquer Carly, and his enactment of hegemonic masculinity failed as well.

Discussion and Conclusion

These television shows offered a variety of masculinities for audiences of children and adults. Hegemonic boy characters did exist—and girl characters found them irresistible. But most (88%) male characters tended toward a non-hegemonic, even feminized masculinity: they were sensitive, non-athletic, and unsuccessful with girls. Casual viewers might conclude that these shows have rewritten male characters in a way that undermines the traditional gender order, allowing for a broader, more feminist conceptualization of masculinity. Given the popularity of these shows, these counter-hegemonic masculinities could help alter gendered expectations among a group of young people, who might become more tolerant of non-conformists (Butler, 1999).

Although the potential for reconceptualizing gender existed—through the use of non-hegemonic protagonists, drag, and homoeroticism—these incidents were largely comedic, rather than serious challenges to the gender order. Systematic analysis of the portrayal of masculinity on these programs reveals a *leger de main*: non-hegemonic boys were not heroic, but clowns, serving as foils for hegemonic masculinity. Comedy centered on the ways that these boys failed at masculinity. Humor was used throughout the programs, disguising hegemonic messages as benign. Walsh et al (2008, p. 132) write, "As the male protagonists remain likable characters, sexism is reduced to only a momentary digression, easily laughed off, as opposed to part of a systemic repressive ideology."

Pascoe (2003) and Renold (2004) argue that non-hegemonic boys participate in the hierarchy by approximating hegemonic masculinity, by donning the mask of hegemonic masculinity (Pollack 1998). In these shows, boys' masculinity-play suggests that hegemonic masculinity is a resource for all boys, a "patriarchal dividend" (Connell, 1995) regardless of where they fall in the hierarchy. They can use hegemonic masculinity to correct power imbalances—particularly over girls, as seen in these examples—and secure their relative status, even if they are lower status males. Hegemonic boy characters used non-hegemonic males to underscore their own dominance (Connell & Messerschmidt, 2005). Even though most characters were not hegemonic, hegemonic masculinity was a resource to be deployed by anyone at any time, to regulate a boy's gender enactment.

Rather than truly celebrating non-hegemonic masculinity, hegemonic masculinity remained the standard. There were consequences for challenging the gendered binary. Characters who flirted with non-hegemonic masculinities were policed—sometimes literally—for their transgressions. In post-feminist fashion (McRobbie, 2004; Ringrose, 2007; Walsh et al., 2004), the masculine hierarchy was highlighted, reinscribing a hierarchy in which hegemonic boys ruled over girls and boys of lesser status. Because hegemonic masculinity is an especially damaging incarnation of gender, these findings are troubling.

The effects of this gender frame (Ridgeway, 2011) are widespread due to the massive audience who consume these messages daily. The television programs are part of the regional construction of masculinity that reflects the larger American culture. If Connell and Messerschmidt (2005) are correct in their supposition that people's local, face-to-face interactions are shaped by culture, then everyday children will police each other in ways that they see on television (see also Corsaro, 1997). My own research reveals that children invoke these images and use them to police each other (Myers & Raymond, 2010). By valorizing hegemonic masculinity, children's' television programming missed an opportunity to transform and expand—in a positive light—cultural representations of boyhood and masculinity. Rather than contributing a feminist portrayal of gender, these shows could be characterized a "postfeminist" (McRobbie, 2004). These negative cultural meanings propagated by television have real-life consequences for everyday young people, limiting their imaginations (Butler, 2004) to "patriarchal constellations" (Walsh et al., 2008, p. 125).

References

Baker-Sperry, L. (2007). The production of meaning through peer interaction. *Sex Roles*, *56*, 717–727.

Baraldi, C. (2008). Promoting self-expression in classroom interactions. *Childhood*, *15*, 238–257.

Bereska, T. (2003). The changing boys' world in the 20th century. *The Journal of Men's Studies, 11*, 157–174.

Bragg, S., & Buckingham, D. (2004). Embarrassment, education, and erotics. *European Journal of Cultural Studies*, *7*, 441–459.

Butler, J. (1999). *Gender trouble*. New York: Routledge.

Butler, J. (2004). *Undoing gender*. New York: Routledge.

Connell, R.W. (1987). *Gender and power*. Stanford University Press.

Connell, R.W. (1995). *Masculinities*. Cambridge: Polity Press.

Connell, R.W., & Messerschmidt, J.W. (2005). Hegemonic masculinity. *Gender & Society, 19,* 829–859.

Corsaro, W. (1997). *The sociology of childhood*. Berkeley: Pine Forge Press.

DeMol, J., & Buysse, A. (2008). Understanding children's influence in parent-child relationships. *Journal of Social and Personal Relationships*, *25*, 359–379.

Epstein, D. (1997). Boyz' own stories. *Gender and Education, 9*, 105–115.

Fausto-Sterling, A. (2000). *Sexing the body*. New York: Basic Books.

Ferris, K. (2001). Through a glass darkly. *Symbolic Interaction, 24*, 25–47.

Fingerson, L. (1999). Active viewing. *Journal of Contemporary Ethnography*, *28*, 389–418.,

Goffman, E. (1974). *Frame analysis*. New York: Northeastern University Press.

Haywood, C. (1996). Out of the curriculum. *Curriculum Studies, 4*, 229–251.

Keddie, A. (2003). Little boys: Tomorrow's macho lads. *Discourse: Studies in the Cultural Politics of Education, 24*, 289–306.

Kelley, P., Buckingham, D., & Davies, H. (1999). Talking dirty: Sexual knowledge and television. *Childhood, 6*, 221–242.
Kimmel, M. (2006). A war against boys? *Dissent, 53*, 65–70.
Korobov, N. (2005). Ironizing masculinity. *The Journal of Men's Studies, 13*, 225–246.
Kurasaki, K. (2000). Intercoder reliability for validating conclusions drawn from open-ended interview data. *Field Methods, 12*, 179–194.
Kuypers, J. (2009). *Rhetorical criticism: Perspectives in action.* New York: Lexington Press.
Lotz, A., & Ross, S. (2004). Bridging media-specific approaches: The value of feminist television criticism's synthetic approach. *Feminist Media Studies, 4*, 185–202.
Ludvigsen, A., & Scott, S. (2009). Real kids don't eat quiche. *Food, Culture & Society, 12*, 417–436.
Martin, K., & Kazyak, E. (2009). Hetero-romantic love and heterosexiness in children's G-rated films. *Gender & Society, 23*, 315–336.
McAllister, M., & Giglio, M. (2005). The commodity flow of U.S. children's television. *Critical Studies in Media Communication, 22*, 26–44.
McRobbie, A. (2004). Notes on postfeminism and popular culture. In A. Harris (Ed.), *All about the girl* (pp. 3–14). New York: Routledge.
Messner, M. (1990). Boyhood, organized sports, and the construction of masculinities. *Journal of Contemporary Ethnography, 18*, 416–444.
Myers, K., & Raymond, L. (2010). Elementary school girls and heteronormativity: The girl project. *Gender & Society, 24*, 167–188.
Neitzel, C., & Chafel, J. (2010). "And no flowers grow there and stuff:" Young children's social representations of poverty. *Sociological Studies of Children and Youth, 13*, 33–59.
Oransky, M., & Marecek, J. (2009). I'm not going to be a girl. *Journal of Adolescent Research, 24*, 218–241.
Pascoe, C.J. (2003). Multiple masculinities? *American Behavioral Scientist, 46*, 1423–1438.
Pascoe, C.J. (2005). *Dude, you're a fag. Sexualities, 8*, 329–346.
Pollack, W.S. (1998). *Real boys*. New York: Henry Holt.
Powell, K., & Abels, L. (2002). Sex role stereotypes in tv programs aimed at the preschool audience. *Women and Language, 25*, 14–22.
Ramlow, T. (2003). Bad boy. *GLQ, 9*, 107–132.
Renold, E. (2004). 'Other' boys: negotiating non-hegemonic masculinities in the primary school. *Gender and Education, 16*, 247–266.
Renold, E. (2007). Primary school "studs:" (De)constructing young boys' heterosexual masculinities. *Men and Masculinities, 9*, 275–297.
Ridgeway, C. (2011). Framed by gender. New York: Oxford.
Ringrose, J. (2007). Successful girls? *Gender and Education, 19*, 471–489.
Risman, B., & Seale, E. (2010). Be twixt and between. In B. Risman (Ed.), *Families as they really are* (pp. 340–361). New York: Norton.
Thorne, B. (1993). *Gender play*. Rutgers University Press.
Van Ausdale, D., & Feagin, J. (2002). *The first r*. New York: Rowman & Littlefield.
Walsh, K., Fursich E., & Jefferson, B. (2008). Beauty and the patriarchal beast. *Journal of Popular Film and Television*, 36, 123–132.
West, C., & Zimmerman, D. (1987). Doing gender. *Gender & Society, 1*, 125–151.

Introduction to Reading 21

Many people erroneously believe that the reason women do not pursue careers in STEM (science, technology, engineering, and mathematics) fields is because of innate differences between females and males. This article takes a different direction in trying to understand why these fields are so sex segregated. Maria Charles pursues cultural, economic, social, and institutional explanations for sex segregation in these areas. She compares gender across culture to give a deeper explanation for why there are fewer women than men in STEM careers in the United States.

1. Are interests in STEM subjects in school innate? Why or why not?
2. As you read this article, consider three arguments you could make for why more women do not pursue STEM careers in the United States.
3. What does Charles mean when she says "believing in difference can actually produce difference"? How does this fit with a socialization explanation?

What Gender Is Science?

Maria Charles

Gender equality crops up in surprising places. This is nowhere more evident than in science, technology, engineering, and mathematics (STEM) fields. The United States should be a world leader in the integration of prestigious male-dominated occupations and fields of study. After all, laws prohibiting discrimination on the basis of sex have been in place for more than half a century, and the idea that men and women should have equal rights and opportunities is practically uncontested (at least in public) in the U.S. today.

This egalitarian legal and cultural context has coincided with a longstanding shortage of STEM workers that has spurred countless initiatives by government agencies, activists, and industry to attract women into these fields. But far from leading the world, American universities and firms lag considerably behind those in many other countries with respect to women among STEM students and workers. Moreover, the countries where women are best represented in these fields aren't those typically viewed as modern or "gender-progressive." Far from it.

Sex segregation describes the uneven distributions of women and men across occupations, industries, or fields of study. While other types of gender inequality have declined dramatically since the 1960s (for example, in legal rights, labor force participation rates, and educational attainment), some forms of sex segregation are remarkably resilient in the industrial world.

In labor markets, one well-known cause of sex segregation is discrimination, which can occur openly and directly or through more subtle, systemic processes. Not so long ago, American employers' job advertisements and recruitment efforts were targeted explicitly toward either men or women depending on the job. Although these gender-specific ads were prohibited under Title VII of the 1964 Civil Rights Act, less blatant forms of discrimination persist. Even if employers base hiring and promotion solely on performance-based criteria, their taken-for-granted beliefs about average gender differences may bias their judgments of qualification and performance. (See Chapter 7 for a fuller discussion of sex segregation and discrimination in the labor force.)

Discrimination isn't the whole story. It's well-established that girls and young women often avoid mathematically intensive fields in favor of pursuits regarded as more human-centered. Analyses of gender-differentiated choices are controversial among scholars because this line of inquiry seems to divert attention away from structural and cultural causes of inequalities in pay and status. Acknowledging gender-differentiated educational and career preferences, though, doesn't "blame the victim" unless preferences and choices are considered in isolation from the social contexts in which they emerge. A sociological analysis of sex segregation considers how the economic, social, and cultural environments influence preferences, choices, and outcomes. Among other things, we may ask what types of social context are associated with larger or smaller gender differences in aspirations. Viewed through this lens, preferences become much more than just individuals' intrinsic qualities.

An excellent way to assess contextual effects is by investigating how career aspirations and patterns of sex segregation vary across countries. Recent studies show international differences in the gender composition of STEM fields, in beliefs about the masculinity of STEM, and in girls' and women's reported affinity for STEM-related activities. These differences follow unexpected patterns.

STEM Around the World

Many might assume women in more economically and culturally modern societies enjoy greater equality on all measures, since countries generally "evolve" in an egalitarian direction as they modernize. This isn't the case for scientific and technical fields, though.

Statistics on male and female college graduates and their fields of study are available from the United Nations Educational, Scientific, and Cultural Organization (UNESCO) for 84 countries covering the period between 2005 and 2008. Sixty-five of those countries have educational systems large enough to offer a full range of majors and programs (at least 10,000 graduates per year).

One way of ranking countries on the sex segregation of science education is to compare the (female-to-male) gender ratio among science graduates to the gender ratio among graduates in all other fields. By this measure, the rich and highly industrialized U.S. falls in about the middle of the distribution (in close

Charles, M. (2011). What gender is science? *Contexts, 10*: 22–28. Reprinted with permission from the author and the American Sociological Association.

proximity to Ecuador, Mongolia, Germany, and Ireland—a heterogeneous group on most conventional measures of "women's status"). Female representation in science programs is weakest in the Netherlands and strongest in Iran, Uzbekistan, Azerbaijan, Saudi Arabia, and Oman, where science is disproportionately female. Although the Netherlands has long been considered a gender-traditional society in the European context, most people would still be intrigued to learn that women's representation among science graduates is nearly 50 percentage points lower there than in many Muslim countries. . . . The most gender-integrated science programs are found in Malaysia, where women's 57-percent share of science degree recipients precisely matches their share of all college and university graduates.

"Science" is a big, heterogeneous category, and life science, physical science, mathematics, and computing are fields with very different gender compositions. For example, women made up 60 percent of American biology graduates, but only about 19 percent of computing graduates, in 2008, according to the National Center for Educational Statistics.

But even when fields are defined more precisely, countries differ in some unexpected ways. A case in point is computer science in Malaysia and the U.S. While American computer scientists are depicted as male hackers and geeks, computer science in Malaysia is deemed well-suited for women because it's seen as theoretical (not physical) and it takes place almost exclusively in offices (thought to be woman-friendly spaces). In interviews with sociologist Vivian Lagesen, female computer science students in Malaysia reported taking up computing because they like computers and because they and their parents think the field has good job prospects. The students also referenced government efforts to promote economic development by training workers, both male and female, for the expanding information technology field. About half of Malaysian computer science degrees go to women.

Engineering is the most strongly and consistently male-typed field of study worldwide, but its gender composition still varies widely across countries. Female representation is generally weaker in advanced industrial societies than in developing ones. In our 2009 article in the *American Journal of Sociology*, Karen Bradley and I found this pattern using international data from the mid-1990s; it was confirmed by more recent statistics assembled by UNESCO. Between 2005 and 2008, countries with the most male-dominated engineering programs include the world's leading industrial democracies (Japan, Switzerland, Germany, and the U.S.) along with some of the same oil-rich Middle Eastern countries in which women are so well-represented among science graduates (Saudi Arabia, Jordan, and the United Arab Emirates). Although women do not reach the 50-percent mark in any country, they come very close in Indonesia, where 48 percent of engineering graduates are female (compared to a 49-percent share of all Indonesian college and university graduates). Women comprise about a third of recent engineering graduates in a diverse group of countries including Mongolia, Greece, Serbia, Panama, Denmark, Bulgaria, and Malaysia.

While engineering is uniformly male-typed in the West, Lagesen's interviews suggest Malaysians draw gender distinctions among engineering *subfields*. One female student reported, ". . . In chemical engineering, most of the time you work in labs. . . . So I think it's quite suitable for females also. But for civil engineering . . . we have to go to the site and check out the constructions."

Girl Geeks in America

Women's relatively weak presence in STEM fields in the U.S. is partly attributable to some economic, institutional, and cultural features that are common to affluent Western democracies. One such feature is a great diversity of educational and occupational pathways. As school systems grew and democratized in the industrial West, educators, policymakers, and nongovernmental activists sought to accommodate women's purportedly "human-centered" nature by developing educational programs that were seen to align functionally and culturally with female domestic and social roles. Among other things, this involved expansion of liberal arts programs and development of vocationally oriented programs in home economics, nursing, and early-childhood education. Subsequent efforts to incorporate women, *as women,* into higher education have contributed to expansion in humanities programs, and, more recently, the creation of new fields like women's studies and human development. These initiatives have been supported by a rapid expansion of service-sector jobs in these societies.

In countries with developing and transitional economies, though, policies have been driven more by concerns about advancing economic development than by interests in accommodating women's presumed affinities. Acute shortages of educated workers prompted early efforts by governments and development agencies to increase the supply of STEM workers. These efforts often commenced during these fields' initial growth periods—arguably before they had acquired strong masculine images in the local context.

Another reason for stronger sex segregation of STEM in affluent countries may be that more people (girls and women in particular) can afford to indulge tastes for less lucrative care and social service work in these contexts. Because personal economic security and national development are such central concerns to young people and their parents in developing societies, there is less latitude and support for the realization of gender-specific preferences.

Again, the argument that women's preferences and choices are partly responsible for sex segregation doesn't require that preferences are innate. Career aspirations are influenced by beliefs about ourselves (What am I good at and what will I enjoy doing?), beliefs about others (What will they think of me and how will they respond to my choices?), and beliefs about the purpose of educational and occupational activities (How do I decide what field to pursue?). And these beliefs are part of our cultural heritage. Sex segregation is an especially resilient form of inequality because people so ardently believe in, enact, and celebrate cultural stereotypes about gender difference.

Believing Stereotypes

Relationship counselor John Gray has produced a wildly successful series of self-help products in which he depicts men and women as so fundamentally different that they might as well come from different planets. While the vast majority of Americans today believe women should have equal social and legal rights, they also believe men and women are very different, and they believe innate differences cause them to *freely choose* distinctly masculine or feminine life paths. For instance, women and men are expected to choose careers that allow them to utilize their hardwired interests in working with people and things, respectively.

Believing in difference can actually produce difference. Recent sociological research provides strong evidence that cultural stereotypes about gender difference shape individuals' beliefs about their own competencies ("self-assessments") and influence behavior in stereotype-consistent directions. Ubiquitous cultural depictions of STEM as intrinsically male reduce girls' interest in technical fields by defining related tasks as beyond most women's competency and as generally unenjoyable for them. STEM avoidance is a likely outcome.

Shelley Correll's social psychological experiment demonstrates the self-fulfilling effects of gender beliefs on self-assessments and career preferences. Correll administered questions purported to test "contrast sensitivity" to undergraduates. Although the test had no objectively right or wrong answers, all participants were given identical personal "scores" of approximately 60 percent correct. Before the test, subjects were exposed to one of two beliefs: that men on average do better, or that men and women perform equally well. In the first group, male students rated their performance more highly than did female students, and male students were more likely to report aspiring to work in a job that requires contrast sensitivity. No gender differences were observed among subjects in the second group. Correll's findings suggest that *beliefs about difference* can produce gender gaps in mathematical self-confidence even in the absence of actual differences in ability or performance. If these beliefs lead girls to avoid math courses, a stereotype-confirming performance deficit may emerge. . . .

Enacting Stereotypes

Whatever one believes about innate gender difference, it's difficult to deny that men and women often behave differently and make different choices. Partly, this reflects inculcation of gender-typed preferences and abilities during early childhood. This "gender socialization" occurs through direct observation of same-sex role models, through repeated positive or negative sanctioning of gender-conforming or nonconforming behavior, and through assimilation of diffuse cultural messages about what males and females like and are good at. During much of the 20th century, math was one thing that girls have purportedly not liked or been good at. Even Barbie said so. Feminists and educators have long voiced concerns about the potentially damaging effects of such messages on the minds of impressionable young girls.

But even girls who don't believe STEM activities are inherently masculine realize others do. It's likely to influence their everyday interactions and may affect their life choices. For example, some may seek to affirm their femininity by avoiding math and science classes or by avowing a dislike for related activities. Sociologists who study the operation of gender in social interactions have argued that people expect to be judged according to prevailing standards of masculinity or femininity. This expectation often leads them to engage in behavior that reproduces the gender order. This "doing gender" framework goes beyond socialization because it doesn't require that gender-conforming dispositions are internalized at an early age, just that people know others will likely hold them accountable to conventional beliefs about hard-wired gender differences.

The male-labeling of math and science in the industrial West means that girls and women may expect to incur social sanctions for pursuing these fields. Effects can be cumulative: taking fewer math classes will negatively affect achievement in math and attitudes toward math, creating a powerful positive feedback system.

Celebrating Stereotypes

Aspirations are also influenced by general societal beliefs about the nature and purpose of educational and occupational pursuits. Modern education does more than bestow knowledge; it's seen as a vehicle for individual self-expression and self-realization. Parents and educators exhort young people, perhaps girls in particular, to "follow their passions" and realize their "true selves." Because gender is such a central axis of individual identity, American girls who aim to "study what they love" are unlikely to consider male-labeled science, engineering, or technical fields, despite the material security provided by such degrees.

Although the so-called "postmaterialist" values of individualism and self-expression are spreading globally, they are most prominent in affluent late-modern societies. Curricular and career choices become more than practical economic decisions in these contexts; they also represent acts of identity construction and self-affirmation. Modern systems of higher education make the incursion of gender stereotypes even easier, by allowing wide latitude in course choices.

The ideological discordance between female gender identities and STEM pursuits may even generate attitudinal aversion among girls. Preferences can evolve to align with the gender composition of fields, rather than vice versa. Consistent with these arguments is new evidence showing that career-related aspirations are more gender-differentiated in advanced industrial than in developing and transitional societies. . . . [T]he gender gap in eighth-graders' affinity for math, confidence in math abilities, and interest in a math-related career is significantly smaller in less affluent countries than in rich ("postmaterialist") ones. Clearly, there is more going on than intrinsic male and female preferences.

Questioning STEM's Masculinity

Playing on stereotypes of science as the domain of socially awkward male geniuses, CBS's hit comedy "The Big Bang Theory" stars four nerdy male physicists and one sexy but academically challenged waitress. (Female physicists, when they do show up, are mostly caricatured as gender deviants: sexually unattractive and lacking basic competence in human interaction.) This depiction resonates with popular Western understandings of scientific and technical pursuits as intrinsically masculine.

But representations of scientific and technical fields as *by nature* masculine aren't well-supported by international data. They're also difficult to reconcile with historical evidence pointing to long-term historical shifts in the gender-labeling of some STEM fields. In *The Science Education of American Girls*, Kim Tolley reports that it was *girls* who were overrepresented among students of physics, astronomy, chemistry, and natural science in 19th-century American schools. Middle-class boys dominated the higher-status classical humanities programs thought to require top rational powers and required for university admission. Science education was regarded as excellent preparation for motherhood, social work, and teaching. Sociologist Katharine Donato tells a similar story about the dawn of American computer programming. Considered functionally analogous to clerical work, it was performed mostly by college-educated women with science or math backgrounds. This changed starting in the 1950s, when the occupation became attractive to men as a growing, intellectually demanding, and potentially lucrative field. The sex segregation of American STEM fields—especially engineering, computer science, and the physical sciences—has shown remarkable stability since about 1980.

The gender (and racial) composition of fields is strongly influenced by the economic and social circumstances that prevail at the time of their initial emergence or expansion. But subsequent transformative events, such as acute labor shortages, changing work conditions, and educational restructuring can effect significant shifts in fields' demographic profiles. Tolley, for example, links men's growing dominance of science education in the late 19th and early 20th century to changing university admissions requirements, the rapid growth and professionalization of science and technology occupations, and recurrent ideological backlashes against female employment.

A field's designation as either "male" or "female" is often naturalized through cultural accounts that reference selected gender-conforming aspects of the work. Just as sex segregation across engineering subfields is attributed to physical location in Malaysia (inside work for women, outside work for men), American women's overrepresentation among typists and sewers has been attributed to these occupations' "feminine" task profiles, specifically their requirements for manual dexterity and attention to detail. While the same skills might be construed as essential to the work of surgeons and electricians, explanations for men's dominance of these fields are easily generated with reference to other job requirements that are culturally masculine (technical and spatial skills, for example). Difference-based explanations for sex segregation are readily available because most jobs require diverse skills and aptitudes, some equated with masculinity, some with femininity.

Looking Forward

What then might be done to increase women's presence in STEM fields? One plausible strategy involves changes to the structure of secondary education. Some evidence suggests more girls and women complete degrees in math and science in educational systems where curricular choice is restricted or delayed; *all* students might take mathematics and science throughout their high-school years or the school might use performance-based tracking and course placement. Although such policies are at odds with Western ideals of individual choice and self-expression, they may weaken penetration of gender stereotypes during the impressionable adolescent years.

Of course, the most obvious means of achieving greater integration of STEM is to avoid reinforcing stereotypes about what girls and boys like and what they are good at. Cultural shifts of this sort occur only gradually, but some change can be seen on the horizon. The rise of "geek chic" may be one sign. Aiming to liberate teen-aged girls from the girls-can't-do-math and male-math-nerd stereotypes, television star and self-proclaimed math geek Danica McKellar has written three how-to math books, most recently *Hot X: Algebra Exposed*, presenting math as both feminine and fun. Even Barbie has been updated. In contrast to her math-fearing Teen Talk sister of the early 1990s, the new Computer Engineer Barbie, released in December 2010, comes decked out in a tight t-shirt printed in binary code and equipped with a smart phone and a pink laptop. Of course, one potential pitfall of this math-is-feminine strategy is that it risks swapping one set of stereotypes for another.

So, what gender is science? In short, it depends. When occupations or fields are segregated by sex, most people suspect it reflects fields' inherently masculine or feminine task content. But this presumption is belied by substantial cross-national variability in the gender composition of fields, STEM in particular. Moreover, this variability follows surprising patterns. Whereas most people would expect to find many more female engineers in the U.S. and Sweden than in Colombia and Bulgaria, new data suggest that precisely the opposite is true.

Ironically, the freedom of choice that's so celebrated in affluent Western democracies seems to help construct and give agency to stereotypically gendered "selves." Self-segregation of careers may occur because some believe they're naturally good at gender-conforming activities (attempting to build on their strengths), because they believe that certain fields will be seen as appropriate for people like them ("doing" gender), or because they believe they'll enjoy gender-conforming fields more than gender-nonconforming ones (realizing their "true selves"). It's just that, by encouraging individual self-expression in postmaterialist societies, we may also effectively promote the development and expression of culturally gendered selves.

Recommended Resources

Shelley J. Correll, "Constraints into Preferences: Gender, Status, and Emerging Career Aspirations." *American Sociological Review* (2004), 69:93–113. Presents evidence from experiments on how beliefs about gender influence our own competence and constrain career aspirations.

Paula England, "The Gender Revolution: Uneven and Stalled." *Gender & Society* (2010), 24:149–66. Offers reasons for the persistence of some forms of gender inequality in the United States.

Wendy Faulkner, "Dualisms, Hierarchies and Gender in Engineering." *Social Studies of Science* (2000), 30: 759–92. Explores the cultural linkage of masculinity and technology within the engineering profession.

Sarah Fenstermaker and Candace West (eds.), *Doing Gender, Doing Difference: Inequality, Power, and Institutional Change* (Routledge, 2002). Explores how and why people reproduce gender (and race and class) stereotypes in everyday interactions.

Cecilia L. Ridgeway, *Framed by Gender: How Gender Inequality Persists in the Modern World* (Oxford University Press, 2011). Describes how cultural gender beliefs bias behavior and cognition in gendered directions and how this influence may vary by context.

Yu Xie and Kimberlee A. Shauman, *Women in Science: Career Processes and Outcomes* (Harvard University Press, 2003). Uses data from middle school to mid-career to study the forces that lead fewer American women than men into science and engineering fields.

5

Buying and Selling Gender

Mary Nell Trautner and Catherine G. Valentine

In the video *Adventures in the Gender Trade* (Marenco, 1993), Kate Bornstein, a transgender performance artist and activist, looks into the camera and says, "Once you buy gender, you'll buy anything to keep it." Her observation goes to the heart of deep connections between economic processes and institutionalized patterns of gender difference, opposition, and inequality in contemporary society. Readings in this chapter examine the ways modern marketplace forces such as commercialization, commodification, and consumerism exploit and construct gender. However, before we explore the buying and selling of gender, we want to review briefly the major elements of contemporary American economic life—elements that embody corporate capitalism—which form the framework for the packaging and delivery of gender to consumers.

Defining Corporate Capitalism

Corporate capitalism (also termed *commodity capitalism*, *consumer capitalism*, or *neoliberal capitalism*) is an economic system in which large national and transnational corporations are the dominant forces. The basic goal of corporate capitalism is the same as it was when social scientists such as Karl Marx studied early capitalist economies: converting money into more money (Johnson, 2001). Corporate capitalists invest money in the production of all sorts of goods and services for the purpose of selling at a profit. Capitalism, as Todd Gitlin (2001) observes, requires a consumerist, market-driven way of life.

In today's society, corporate capitalism affects virtually every aspect of social life—most Americans work for a corporate employer, whether a fast-food chain or bank, and virtually everyone buys the products and services of capitalist production (Johnson, 2001; Ritzer, 1999). Those goods and services include things we must have to live (e.g., food and shelter) and, most important for contemporary capitalism's survival and growth, things we have learned to want or desire (e.g., mobile phones, televisions, cruises, fitness fashions, cosmetic surgery), even though we do not need them to live (Ritzer, 1999).

The United States can be fairly characterized as a nation of consumers—people who buy and use a dizzying array of objects and services conceived, designed, and sold to us by corporations. Nearly twenty years ago, George Ritzer (1999), a leading analyst of consumerism, observed that consumption plays such a big role in the lives of contemporary Americans that it has, in many respects, come to define our society, a point that is truer now than ever before. In fact, as Ritzer notes, Americans spend most of their available resources on consumer goods and services. Corporate, consumer capitalism depends on luring people into what he calls the "cathedrals of consumption"—such as superstores, shopping malls, theme parks, fast-food restaurants, and casinos—where we will spend money to buy an array of goods and services.

Our consumption-driven economy counts on customers whose spending habits are relatively unrestrained and who view shopping as pleasurable. Indeed, Americans spend much more today than they did just 40 years ago (Ritzer, 1999). Most of our available resources go to purchasing and consuming "stuff." Americans consume more of everything and more varieties of things than do people in other nations. We are also more likely to go into debt than Americans of earlier generations and people in other nations today. Some social scientists (e.g., Schor, 1998, p. 204) use the term *hyperconsumption* to describe what seems to be a growing American passion for and obsession with consumption.

Ritzer (2011) also argues that the Great Recession, beginning in 2007, was preceded by "the greatest consumer-driven expansion" in U.S. history and resulted in "perhaps the greatest economic setbacks," including high unemployment and foreclosure rates. He and others such as Robert Manning (2011) and Juliet Schor (2010) are persuaded that hyperconsumption and our "business-as-usual" economy (Schor, 2010) are unsustainable, if not outright dysfunctional, and they predict that as these economic patterns move to other economies, such as those in Asia (Ritzer, 2011), we may witness in years to come economic setbacks in those nations similar to the Great Recession in the United States. In addition, scholars of the ethics of consumption (e.g., Adams & Raisborough, 2008; Lewis & Potter, 2010; Shaw & Newhold, 2002) ask if "sustainable consumption" is, in fact, possible and what it would look like given the reality of fast-diminishing resources and rapidly increasing consumption worldwide.

Marketing Gender

Gender is a fundamental element of the modern machinery of marketing. It is an obvious resource from which the creators and distributors of goods and services can draw ideas, images, and messages. The imagery of consumer culture thrives on gender difference and asymmetry. For example, consumer emblems of hyperfemininity and hypermasculinity, such as Barbie and GI Joe, stand in stark physical contrast to each other (Schiebinger, 2000). This is not happenstance. Barbie and GI Joe intentionally reinforce belief in the idea of natural differences between women and men. The exaggerated, gendered appearances of Barbie and GI Joe can be purchased by adult consumers who have the financial resources to pay for new cosmetic surgeries, such as breast and calf implants, that literally inscribe beliefs about physical differences between women and men into their flesh (Sullivan, 2001).

As Suzanna Dunata Walters (2001) observes, turning difference into "an object of barter is perhaps the quintessentially American experience" (p. 289). Indeed, virtually every product and service, including the most functional, can be designed and consumed as masculine or feminine (e.g., deodorants, bicycles, greeting cards, wallpaper, cars, and hairstyles). In a study of gender differences in prices charged for personal care products and services (i.e., women pay more), Megan Duesterhaus, Liz Grauerholz, Rebecca Weichsel, and Nicholas A. Guittar (2011) underscore the fact that "marketers have successfully convinced women and men that the gendered products they sell [e.g., body lotions and deodorants] are in fact different . . . and consumers have 'bought into' this essentialist-based marketing" (p. 187).

Gender-coding of products and services is a common strategy employed by capitalist organizations to sell their wares. It is also integral to the processes by which gender is constructed, because it frames and structures gender practices. As contemporary anthropologists argue, material culture (e.g., weaponry, musical instruments, cloth and clothing, residential buildings) are a significant "medium through which people come to know and understand themselves" and others (Tilley, 2011, p. 348).

To illustrate how consumer culture participates in the construction of gender through one material form, let's look at the gender coding of clothing. Gender archeologist Marie Louise Stig Sørensen (2000) observes that clothing is an ideal medium for the expression of a culture's gender beliefs because it is an extension of the body and an important element in identity and communication. No wonder corporate capitalists have cashed in on the business of fabricating gender through clothing (Sørensen, 2000). Sørensen notes that simple observation of the clothing habits of people reveals a powerful pattern of "dressing gender" (p. 124). Throughout life, she argues, the gender coding of colors, patterns, decorations, fabrics, fastenings, trimmings, and other aspects of clothing create and maintain differences between boys and girls and men and women. Even when clothing designers and manufacturers create what appear to be "unisex" fashions (e.g., tuxedos for women), they incorporate just enough gendered elements (e.g., lacy trim or a revealing neckline) to ensure that the culturally created gender categories—feminine and masculine—are not completely erased. Consider the lengths to which the fashion industry has gone to create clothing that conveys a "serious yet feminine" business appearance for the increasing number of women in management and executive levels of the corporate world (Kimle & Damhorst, 1997). Contemplate the ferocity of the taboo against boys and men wearing skirts and dresses. Breaking the taboo (except on a few occasions, such as Halloween) typically results in negative sanctions. The reading in this chapter by Adie Nelson examines the extent to which even fantasy dress for children ends up conforming to gender stereotypes.

Gender-coded clothing is one example of corporate exploitation of gender to sell all kinds of goods and services, including gender itself. Have we arrived at a moment in history when identities, including gender identity, are largely shaped within the dynamics of consumerism? Will we, as Bornstein observed, buy anything to keep up gender appearances? The readings in this chapter help us answer these questions. They illuminate some of the key ways capitalist, consumer culture makes use of cultural definitions and stereotypes of gender to produce and sell goods and services.

In our "consumers' republic" (Cohen, 2003), the mass media (e.g., television, movies, and magazines) play a central role in delivering potential consumers to advertisers whose job it is to persuade us to buy particular products and services (Kilbourne, 1999; Ritzer, 1999). The advertising industry devotes itself to creating and keeping consumers in the marketplace, and it is very good at what it does. Today's advertisers use sophisticated strategies for hooking consumers. The strategies work because they link our deepest emotions and most beloved ideals to products and services by persuading us that identity and self-worth can be fashioned out of the things we buy (Featherstone, 1991; Zukin, 2004). Advertisers transform gender into a commodity and convince consumers that we can transform ourselves into more masculine men and more feminine women by buying particular products and services. Men are lured into buying cars that will make them feel like hypermasculine machines, and women are sold a wondrous array of cosmetic products and procedures that are supposed to turn them into drop-dead beauties.

Jacqueline Urla and Alan Swedlund (1995) explore the story that Barbie, a well-advertised and wildly popular toy turned icon, tells about femininity in consumer culture. They note that although Barbie's long, thin body and big breasts are remarkably unnatural, she stands as an ideal that has played itself out in the real body trends of *Playboy* magazine centerfolds and beauty pageant contestants. The authors provide evidence that between 1959 and 1978, the average weight and hip size for women centerfolds and beauty contestants decreased steadily. A follow-up study for 1979 to 1988 found the acceleration of this trend with "approximately 69 percent of Playboy centerfolds and 60 percent of Miss America contestants weighing in at 15 percent or more below their expected age and height category" (p. 298). One lesson we might glean from this story is that a toy (Barbie) and real women (centerfolds and beauty contestants) are converging in a culture in which the bonds of beauty norms are narrowing and tightening their grip on both products and persons (Sullivan, 2001). To further illustrate the influence of representations of gender in advertising, Kristen Barber and Tristan Bridges' reading in this chapter takes a close look at news ways that marketers use ideas about masculinity to sell products to men that they might otherwise avoid because of their "feminine" connotation, such as salad and body wash. Likewise the gun industry is finding ways to capitalize on a previously-untapped market, women, by selling a "culture of fear" and linking gun ownership to women's "natural" protective and nurturing instincts. Any analysis of the marketing of femininity and masculinity has to take into account the ways the gendering of products and services is tightly linked to prisms of difference and inequality such as sexuality, race, age, and ability/disability. Consumer culture thrives, for example, on heterosexuality, whiteness, and youthfulness. Automobile advertisers market cars made for heterosexual romance and marriage. Liquor ads feature men and women in love (Kilbourne, 1999). Research on race and gender imagery in the most popular advertising medium, television, confirms the continuing dominance of images of White, affluent young adults. "Virtually all forms of television marketing perpetuate images of White hegemonic masculinity and White feminine romantic fulfillment" (Coltrane & Messineo, 2000, p. 386). In spite of what is called niche marketing or marketing to special audiences such as Latinos, gay men, and older Americans, commercial television imagery continues to rely on stereotypes of race, gender, age, and the like (Coltrane & Messineo, 2000). Stereotypes sell.

Two readings in this chapter address intersections of prisms of difference and inequality in consumer culture. The first, by Kimberly Hoang, is an intriguing analysis of women sex workers in Ho Chi Minh City, Vietnam, who sell impressions of themselves as dark-skinned, poor women to Western businessmen and budget travelers who not only want to purchase sex and intimacy but also want to see themselves as helping poor, Third World women. The workers Hoang studied "racialized their bodies" by using skin darkeners and wearing either simple clothing or ethnic, Orientalizing dresses to cater to the racialized, sexualized, and gendered stereotypes of Western men. The second reading, by Toni Calasanti and Neal King, offers detailed insight into the mass-marketing of "successful aging" products, services, and activities to older men in the United States. They highlight the fact that marketing that targets older people plays on the stigma of aging in American culture and, specifically, on the often desperate attempts of aging men to hang on to youthful manliness.

Can You Buy in Without Selling Out?

The tension between creativity, resistance, and rebellion on the one hand, and the lure and power of commercialization on the other is a focus of much research on consumerism and consumer culture (Quart, 2003; Schor, 2004). Can we produce and consume the gendered products and services of corporate capitalism without wanting and trying to be just like Barbie or Madonna, the Marlboro Man or Brad Pitt? Does corporate, commercial culture consume everything and everyone in its path, including the creators of countercultural forms?

The latter question is important. Consider the fact that "grunge," which began as anti-establishment fashion, became a national trend when companies such as Diesel and Urban Outfitters co-opted and commercialized it (O'Brien, 1999). Then contemplate how commercial culture has cleverly exploited the women's movement by associating serious social issues and problems with trivial or dangerous products. "New Freedom" is a maxi pad. "ERA" is a laundry detergent. Commercial culture is quite successful in enticing artists of all sorts to "sell out." For example, Madonna began her career as a rebel who dared to display a rounded belly. But, over time, she has been "normalized," as reflected in the transformation of her body to better fit celebrity appearance norms (Bordo, 1997).

The culture of the commodity is also successful in mainstreaming the unconventional by turning nonconformity into obedience that answers to Madison Avenue (Harris, 2000). Analysts of the commodification of gayness have been especially sensitive to the potential problems posed by advertising's recent creation of a largely fictional identity of gay as "wealthy White man" with a lifestyle defined by hip fashion (Walters, 2001). What will happen if lesbian and gay male styles are increasingly drawn into mass-mediated, consumer culture? Will those modes of rebellion against the dominance of heterosexism lose their political clout? Will they become mere "symbolic forms of resistance, ineffectual strategies of rebellion" (Harris, 2000, p. xxiii)?

The commodification of modes of rebellion and activist politics is serious business in our corporate, consumer culture, and marketers have been successful in manipulating our feelings of compassion and our sense of justice by persuading many Americans that they can "make a difference" by buying products and services. A thoughtful example of this phenomenon is explored in a book by Gayle Sulik (2011). She focuses on the commodification of the breast cancer movement by the pink ribbon industry and carefully sets out the negative consequences, which include, for example, the sale of products from which little or no money goes to breast cancer research and the sale of products whose manufacturers are linked to the production of toxic substances implicated in breast cancer.

The Global Reach of American Gender Images and Ideals

The global reach of American culture is yet another concern of consumer culture researchers. Transnational corporations are selling American popular culture and consumerism as a way of life in countries around the world (Kilbourne, 1999; Ritzer, 1999). People across the globe are now regularly exposed to American images, icons, and ideals. For example, *Baywatch,* with its array of perfect (albeit cosmetically enhanced) male and female bodies, was seen by more people in the world than any other television show during the years it aired (Kilbourne, 1999). American popular music and film celebrities dominate the world scene. Everyone knows Marilyn Monroe and James Dean, Tom Cruise and Julia Roberts.

You might ask, and quite legitimately, so what? The answer to that question is not a simple one, in part because cultural import–export relations are intricate. As Gitlin (2001) observes, "the cultural gates . . . swing both ways. For example, American rhythm and blues influenced Jamaican ska, which evolved into reggae, which in turn was imported to the United States via Britain" (p. 188). However, researchers have been able to document some troubling consequences of the global advantage of American commercial, consumer culture for the lifeways of people outside the United States. Thus, social scientists (e.g., Collins, 2009; Connell, 1999; Herdt, 1997) are tracing how American categories of sexual orientation are altering the modes of organization and perception of same-gender relations in some non-Western societies that have traditionally been more fluid and tolerant of sexual diversity than the United States or have constructed different, non-Western performances of gay masculinity.

Scientists are also documenting the impact of American mass media images of femininity and masculinity on consumers in far corners of the world. The island country of Fiji is one such place. Researchers discovered that as the young women of Fiji consumed American television on a regular basis, eating disorders such as anorexia nervosa appeared for the first time. The ultra-thin images of girls and women that populate U.S. TV shows and ads have become the measuring stick of femininity in a culture in which an ample, full body was previously the norm for women and men (Goode, 1999). The troubling consequences of the globalization of American consumer culture do not end with these examples. Consider the potential negative impact of idealized images of whiteness in a world in which most people are brown. (See reading by L. Ayu Saraswati in Chapter 6.) Or how about the impact of America's negative images of older women and men on the people of cultures in which the elderly are revered?

Although corporate, capitalist economies provide many people with all the creature comforts they need and more, as well as making consumption entertaining and more accessible, there is a price to pay (Ritzer, 1999). This chapter explores one troubling aspect of corporate, consumer culture—the commodification and commercialization of gender.

A few final questions emerge from our analysis of patterns of gender in relationship to consumer capitalism. How can the individual develop an identity and self-worth *not* contingent on and defined by a whirlwind of products and services? How do we avoid devolving into caricatures of stereotyped images of femininity and masculinity whose needs and desires can be met only by gendered commodities? Is Bornstein correct when she states, "Once you buy gender, you'll buy anything to keep it"? Or can we create and preserve alternative ways of life, even those that undermine the oppression of dominant images and representations?

References

Adams, M., & Raisborough, J. (2008). What can sociology say about fair trade? Class, reflexivity and ethical consumption. *Sociology*, *42*, 1165–1182.

Bordo, S. (1997). Material girl: The effacements of postmodern culture. In R. Lancaster & M. di Leonardo (Eds.), *The gender/sexuality reader* (pp. 335–358). New York: Routledge.

Cohen, L. (2003). *A consumers' republic: The politics of mass consumption in postwar America.* New York: Vintage Books.

Collins, D. (2009). We're there and queer: Homonormative mobility and lived experience among gay expatriates in Manila. *Gender & Society*, *23*(4), 465–493.

Coltrane, S., & Messineo, M. (2000). The perpetuation of subtle prejudice: Race and gender imagery in 1990s television advertising. *Sex Roles*, *42*, 363–389.

Connell, R. W. (1999). Making gendered people: Bodies, identities, sexualities. In M. Ferree, J. Lorber, & B. Hess (Eds.), *Revisioning gender* (pp. 449–471). Thousand Oaks, CA: Sage.

Duesterhaus, M., Grauerholz, L., Weichsel, R., & Guittar, N. A. (2011). The cost of doing femininity: Gendered disparities in pricing of personal care products and services. *Gender Issues*, *28*(4), 175–191.

Featherstone, M. (1991). The body in consumer culture. In M. Featherstone, M. Hepworth, & B. S. Turner (Eds.), *The body: Social process and cultural theory* (pp. 170–196). London: Sage.

Gitlin, T. (2001). *Media unlimited: How the torrent of images and sounds overwhelms our lives.* New York: Henry Holt.

Goode, E. (1999, May 20). Study finds TV alters Fiji girls' view of body. *The New York Times*, p. A17.

Harris, D. (2000). *Cute, quaint, hungry and romantic: The aesthetics of consumerism.* Cambridge, MA: Da Capo Press.

Herdt, G. (1997). *Same sex, different cultures.* Boulder, CO: Westview Press.

Johnson, A. (2001). *Privilege, power, and difference.* Mountain View, CA: Mayfield.

Kilbourne, J. (1999). *Can't buy my love.* New York: Simon & Schuster.

Kimle, P. A., & Damhorst, M. L. (1997). A grounded theory model of the ideal business image for women. *Symbolic Interaction*, *20*(1), 45–68.

Lewis, T., & Potter, E. (Eds.). (2010). *Ethical consumption: A critical introduction*. Abingdon, UK: Routledge.

Manning, R. D. (2011). Crisis in consumption OR American capitalism: A sociological perspective on the consumer-led recession. *Consumers, Commodities & Consumption*, *13*(1). Retrieved from http://csrn.camden.rutgers.edu/newsletters/13-1/manning.htm

Marenco, S. (with Bornstein, K.). (1993). *Adventures in the gender trade: A case for diversity* [Motion picture]. New York: Filmakers Library.

O'Brien, J. (1999). *Social prisms.* Thousand Oaks, CA: Pine Forge Press.

Quart, A. (2003). *Branded: The buying and selling of teenagers.* New York: Basic Books.

Ritzer, G. (1999). *Enchanting a disenchanted world.* Thousand Oaks, CA: Pine Forge Press.

Ritzer, G. (2011). The dinosaurs of consumption. *Consumers, Commodities & Consumption, 13*(1). Retrieved from http://csrn.camden.rutgers.edu/newsletters/13-1/ritzer.htm

Schiebinger, L. (2000). Introduction. In L. Schiebinger (Ed.), *Feminism and the body* (pp. 1–21). New York: Oxford University Press.

Schor, J. (1998). *The overspent American.* New York: Basic Books.

Schor, J. (2004). *Born to buy.* New York: Scribner.

Schor, J. (2010). *Welcome to Plenitude* [Blog]. Retrieved from http://www.julietschor.org/2010/05/welcome-to-plenitude/

Shaw, D., & Newholm, T. (2002). Voluntary simplicity and the ethics of consumption. *Psychology and Marketing, 19,* 167–185.

Sørensen, M. L. S. (2000). *Gender archaeology.* Cambridge, UK: Polity Press.

Sulik, G. (2011). *Pink ribbon blues: How breast cancer culture undermines women's health.* New York: Oxford.

Sullivan, D. A. (2001). *Cosmetic surgery: The cutting edge of commercial medicine in America.* New Brunswick, NJ: Rutgers University Press.

Tilley, C. (2011). Materializing identities: An introduction. *Journal of Material Culture, 16,* 347–357.

Urla, J., & Swedlund, A. C. (1995). The anthropometry of Barbie: Unsettling ideals of the feminine body in popular culture. In J. Terry & J. Urla (Eds.), *Deviant bodies: Critical perspectives on difference in science and popular culture* (pp. 277–313). Bloomington: Indiana University Press.

Walters, S. D. (2001). *All the rage: The story of gay visibility in America.* Chicago: University of Chicago Press.

Zukin, S. (2004). *Point of purchase: How shopping changed American culture.* New York: Routledge.

Introduction to Reading 22

Adie Nelson's article offers a marvelously detailed analysis of one way the modern marketplace reinforces gender stereotypes—the gender coding of children's Halloween costumes. Nelson describes the research process she employed to label costumes as masculine, feminine, or neutral. She provides extensive information about how manufacturers and advertisers use gender markers to steer buyers, in this case parents, toward "gender-appropriate" costume choices for their children. Overall, Nelson's research indicates that gender-neutral costumes, whether ready-to-wear or sewing patterns, are a tiny minority of all the costumes on the market.

1. Many perceive Halloween costumes as encouraging children to engage in fantasy play. How does Nelson's research call this notion into question?
2. Describe some of the key strategies employed by manufacturers to "gender" children's costumes. What strategies do manufacturers use to "gender" adults' costumes? To answer the latter question, look at adult costumes online or in costume shops in malls.
3. How do Halloween costumes help reproduce an active-masculine/passive-feminine dichotomy?

The Pink Dragon Is Female

Halloween Costumes and Gender Markers

Adie Nelson

The celebration of Halloween has become, in contemporary times, a socially orchestrated secular event that brings buyers and sellers into the marketplace for the sale and purchase of treats, ornaments, decorations, and fanciful costumes. Within this setting, the wearing of fancy dress costumes has

Nelson, A. (2000). The pink dragon is female: Halloween costumes and gender markers. *Psychology of Women Quarterly, 24*(2). Reprinted by permission of SAGE Publications, Inc.

such a prominent role that it is common, especially within large cities, for major department stores and large, specialty toy stores to begin displaying their selection of Halloween costumes by mid-August if not earlier. It is also evident that the range of masks and costumes available has broadened greatly beyond those identified by McNeill (1970), and that both children and adults may now select from a wide assortment of ready-made costumes depicting, among other things, animals, objects, superheroes, villains, and celebrities. In addition, major suppliers of commercially available sewing patterns, such as Simplicity and McCall's, now routinely include an assortment of Halloween costumes in their fall catalogues. Within such catalogues, a variety of costumes designed for infants, toddlers, children, adults, and, not infrequently, pampered dogs are featured.

On the surface, the selection and purchase of Halloween costumes for use by children may simply appear to facilitate their participation in the world of fantasy play. At least in theory, asking children what they wish to wear or what they would like to be for Halloween may be seen to encourage them to use their imagination and to engage in the role-taking stage that Mead (1934) identified as play. Yet, it is clear that the commercial marketplace plays a major role in giving expression to children's imagination in their Halloween costuming. Moreover, although it might be facilely assumed that the occasion of Halloween provides a cultural "time out" in which women and men as well as girls and boys have tacit permission to transcend the gendered rules that mark the donning of apparel in everyday life, the androgyny of Halloween costumes may be more apparent than real. If, as our folk wisdom proclaims, "clothes make the man" (or woman), it would be presumptuous to suppose that commercially available children's Halloween costumes and sewing patterns do not reflect both the gendered nature of dress (Eicher & Roach-Higgins, 1992) and the symbolic world of heroes, villains, and fools (Klapp, 1962, 1964). Indeed, the donning of Halloween costumes may demonstrate a "gender display" (Goffman, 1966, p. 250) that is dependent on decisions made by brokering agents to the extent that it is the aftermath of a series of decisions made by commercial firms that market ready-made costumes and sewing patterns that, in turn, are purchased, rented, or sewn by parents or others. . . .

Building on Barnes and Eicher's (1992, p. 1) observation that "dress is one of the most significant markers of gender identity," an examination of children's Halloween costumes provides a unique opportunity to explore the extent to which gender markers are also evident within the fantasy costumes available for Halloween. To the best of my knowledge, no previous research has attempted to analyze these costumes or to examine the ways in which the imaginary vistas explored in children's fantasy dress reproduce and reiterate more conventional messages about gender.

In undertaking this research, my expectations were based on certain assumptions about the perspectives of merchandisers of Halloween costumes for children. It was expected that commercially available costumes and costume patterns would reiterate and reinforce traditional gender stereotypes. Attempting to adopt the marketing perspective of merchandisers, it was anticipated that the target audience would be parents concerned with creating memorable childhood experiences for their children, envisioning them dressed up as archetypal fantasy characters. In the case of sewing patterns, it was expected that the target audience would be primarily mothers who possessed what manufacturers might imagine to be the sewing skills of the traditional homemaker. However, these assumptions about merchandisers are not the subject of the present inquiry. Rather, the present study offers an examination of the potential contribution of marketing to the maintenance of gender stereotypes. In this article, the focus is on the costumes available in the marketplace; elsewhere I examine the interactions between children and their parents in the selection, modification, and wearing of Halloween costumes (Nelson, 1999).

Method

The present research was based on a content analysis of 469 unique children's Halloween ready-made costumes and sewing patterns examined from August 1996 to November 1997 at craft stores, department stores, specialty toy stores, costume rental stores, and fabric stores containing catalogues of sewing patterns. Within retail stores, racks of children's Halloween costumes typically appeared in August and remained in evidence, albeit in dwindling numbers, until early November each year. In department stores, a subsection of the area generally devoted to toys featured such garments; in craft stores and/or toy stores, children's Halloween costumes were typically positioned on long racks in the center of a section devoted to the commercial paraphernalia now associated with the celebration of Halloween (e.g., cardboard witches, "Spook trees," plastic pumpkin containers). Costumes were not segregated by gender within the stores (i.e., there were no separate aisles or sections for boys' and girls' costumes); however, children's costumes were typically positioned separately from those designed for adults. . . .

All costumes were initially coded as (a) masculine, (b) feminine, or (c) neutral depending on whether boys, girls, or both were featured as the models on the packaging that accompanied a ready-to-wear costume or were used to illustrate the completed costume on the cover of a sewing pattern. . . . The pictures accompanying costumes may act as safekeeping devices, which discourage parents from buying "wrong"-sexed costumes. The process of labeling costumes as masculine, feminine, or neutral was facilitated by the fact that these public pictures (Goffman, 1979) commonly employed recognizable genderisms. For example, a full-body costume of a box of crayons could be identified as feminine by the long curled hair of the model and the black patent leather pumps with ribbons she wore. In like fashion, a photograph depicting the finished version of a sewing pattern for a teapot featured the puckish styling of the model in a variant of what Goffman (1979, p. 45) termed "the bashful knee bend" and augmented this subtle cue by having the model wear white pantyhose and Mary-Jane shoes with rosettes at the base of the toes. Although the sex of the model could have been rendered invisible, such feminine gender markers as pointy-toed footwear, party shoes of white and black patent leather, frilly socks, makeup and nail polish, jewelry, and elaborately curled (and typically long and blonde) hair adorned with bows/barrettes/hairbands facilitated this initial stage of costume placement. By and large, female models used to illustrate Halloween costumes conformed to the ideal image of the "Little Miss" beauty pageant winner; they were almost overwhelmingly White, slim, delicate-boned blondes who did not wear glasses. Although male child models were also overwhelmingly White, they were more heterogeneous in height and weight and were more likely to wear glasses or to smile out from the photograph in a bucktooth grin. At the same time, however, masculine gender markers were apparent. Male models were almost uniformly shod in either well-worn running shoes or sturdy-looking brogues, while their hair showed little variation from the traditional little-boy cut of short back and sides.

The use of gender-specific common and proper nouns to designate costumes (e.g., Medieval Maiden, Majorette, Prairie Girl) or gender-associated adjectives that formed part of the costume title (e.g., Tiny Tikes Beauty, Pretty Witch, Beautiful Babe, Pretty Pumpkin Pie) also served to identify feminine costumes. Similarly, the use of the terms "boy," "man," or "male" in the advertised name of the costume (e.g., Pirate Boy, Native American Boy, Dragon Boy) or the noted inclusion of advertising copy that announced "Cool dudes costumes are for boys in sizes" was used to identify masculine costumes. Costumes designated as neutral were those in which both boys and girls were featured in the illustration or photograph that accompanied the costume or sewing pattern or in which it was impossible to detect the sex of the wearer. By and large, illustrations for gender-neutral ads featured boys and girls identically clad and depicted as a twinned couple or, alternatively, showed a single child wearing a full-length animal costume complete with head and "paws," which, in the style of spats, effectively covered the shoes of the model. In addition, gender-neutral costumes were identified by an absence of gender-specific nouns and stereotypically gendered colors.

Following this initial division into three categories, the contents of each were further coded into a modified version of Klapp's (1964) schema of heroes, villains, and fools. In his work, Klapp suggested that this schema represents three dimensions of human behavior. That is, heroes are praised and set up as role models, whereas villains and fools are negative models, with the former representing evil to be feared and/or hated and the latter representing figures of absurdity inviting ridicule. However, although Klapp's categories were based on people in real life, I applied them to the realm of make-believe. For the purposes of this study, the labels refer to types of personas that engender or invite the following emotional responses, in a light-hearted way from audiences: Heroes invite feelings of awe, admiration, and respect, whereas villains elicit feelings of fear and loathing, and fools evoke feelings of laughter and perceptions of cuteness. All of the feelings, however, are mock emotions based on feelings of amusement, which make my categories quite distinct from Klapp's. For example, although heroes invite awe, we do not truly expect somebody dressed as a hero to be held in awe. . . .

For the purposes of this secondary classification of costumes, the category of hero was broadened to include traditional male or female heroes (e.g., Cowboy, Robin Hood, Cinderella, Cleopatra), superheroes possessing supernatural powers (e.g., Superman, Robocop, Xena, the Warrior Princess) as well as characters with high occupational status (e.g.,

Emergency Room Doctor, Judge) and characters who are exemplars of prosocial conformity to traditional masculine and feminine roles (e.g., Team USA Cheerleader, Puritan Lady, Pioneer Boy). The category of villain was broadly defined to include symbolic representations of death (e.g., the Grim Reaper, Death, The Devil, Ghost), monsters (e.g., Wolfman, Frankenstein, The Mummy), and antiheroes (e.g., Convict, Pirate, The Wicked Witch of the West, Catwoman). Fool was a hybrid category, distinguished by costumes whose ostensible function was to amuse rather than to alarm.

Within this category, two subcategories were distinguished. The first subcategory, figures of mirth, referred to costumes of clowns, court jesters, and harlequins. The second, nonhuman/inanimate objects, was composed of costumes representing foodstuffs (e.g., Peapod, Pepperoni Pizza, Chocolate Chip Cookie), animals and insects, and inanimate objects (e.g., Alarm Clock, Bar of Soap, Flower Pot). Where a costume appeared to straddle two categories, an attempt was made to assign it to a category based on the dominant emphasis of its pictorial representation. For example, a costume labeled Black Widow Spider could be classified as either an insect or a villain. If the accompanying illustration featured a broadly smiling child in a costume depicting a fuzzy body and multiple appendages, it was classified as an insect and included in the category of nonhuman/inanimate objects; if the costume featured an individual clad in a black gown, long black wig, ghoulish makeup, and a sinister mien, the costume was classified as a villain. Contents were subsequently reanalyzed in terms of their constituent parts and compared across masculine and feminine categories. In all cases, costumes were coded into the two coding schemes on the basis of a detailed written description of each costume. . . .

Results

The initial placement of the 469 children's Halloween costumes into masculine, feminine, or neutral categories yielded 195 masculine costumes, 233 feminine costumes, and 41 gender-neutral costumes. The scarcity of gender-neutral costumes was notable; costumes that featured both boys and girls in their ads or in which the gender of the anticipated wearer remained (deliberately or inadvertently) ambiguous accounted for only 8.7% of those examined. Gender-neutral costumes were more common in sewing patterns than in ready-to-wear costumes and were most common in costumes designed for newborns and very young infants. In this context, gender-neutral infant costumes largely featured a winsome assortment of baby animals (e.g., Li'l Bunny, Beanie the Pig) or foodstuffs (e.g., Littlest Peapod). By and large, few costumes for older children were presented as gender-neutral; the notable exceptions were costumes for scarecrows and emergency room doctors (with male/female models clad identically in olive-green "scrubs"), ready-made plastic costumes for Lost World/Jurassic Park hunters, a single costume labeled Halfman/Halfwoman, and novel sewing patterns depicting such inanimate objects as a sugar cube, laundry hamper, or treasure chest.

Beginning most obviously with costumes designed for toddlers, gender dichotomization was promoted by gender-distinctive marketing devices employed by the manufacturers of both commercially made costumes and sewing patterns. In relation to sewing patterns for children's Halloween costumes, structurally identical costumes featured alterations through the addition or deletion of decorative trim (e.g., a skirt on a costume for an elephant) or the use of specific colors or costume names, which served to distinguish masculine from feminine costumes. For example, although the number and specific pattern pieces required to construct a particular pattern would not vary, View A featured a girl-modeled Egg or Tomato, whereas View B presented a boy-modeled Baseball or Pincushion. Structurally identical costumes modeled by both boys and girls would be distinguished through the use of distinct colors or patterns of material. Thus, for the peanut M&M costumes, the illustration featured girls clad in red or green and boys clad in blue, brown, or yellow. Similarly, female clowns wore costumes of soft pastel colors and dainty polka dots, but male clowns were garbed in bold primary colors and in material featuring large polka dots or stripes. Illustrations for ready-to-wear costumes were also likely to signal the sex of the intended wearer through the advertising copy: models for feminine costumes, for example, had long curled hair, were made up, and wore patent leather shoes. Only in such costumes as Wrinkly Old Woman, Grandma Hag, Killer Granny, and Nun did identifiably male children model female apparel. . . .

[A]lthough hero costumes constituted a large percentage of both masculine and feminine costumes, masculine costumes contained a higher percentage of villain costumes, and feminine costumes included substantially more fool costumes, particularly those of nonhuman/inanimate objects. It may be imagined that the greater total number of feminine costumes would provide young girls with a broader range of costumes to select from than exists for young boys, but in fact the obverse is true. . . . [W]hen finer distinctions were made within the three generic categories, hero costumes for girls were clustered in a narrow range of roles that, although distinguished by specific names, were functionally equivalent in the image they portray. It would seem that, for girls, glory is concentrated in the narrow realm of beauty queens, princesses, brides, or other exemplars of traditionally passive femininity. The ornate, typically pink, ball-gowned costume of the princess (with or without a synthetic jeweled tiara) was notable, whether the specific costume was labeled Colonial Belle, the Pumpkin Princess, Angel Beauty, Blushing Bride, Georgia Peach, Pretty Mermaid, or Beauty Contest Winner. In contrast, although hero costumes for boys emphasized the warrior theme of masculinity (Doyle, 1989; Rotundo, 1993), with costumes depicting characters associated with battling historical, contemporary, or supernatural Goliaths (e.g., Bronco Rider, Dick Tracy, Sir Lancelot, Hercules, Servo Samurai, Robin the Boy Wonder), these costumes were less singular in the visual images they portrayed and were more likely to depict characters who possessed supernatural powers or skills.

Masculine costumes were also more likely than feminine costumes to depict a wide range of villainous characters (e.g., Captain Hook, Rasputin, Slash), monsters (e.g., Frankenstein, The Wolfman), and, in particular, agents or symbols of death (e.g., Dracula, Executioner, Devil Boy, Grim Reaper). Moreover, costumes for male villains were more likely than those of female villains to be elaborate constructions that were visually repellant; to feature an assortment of scars, mutations, abrasions, and suggested amputations; and to present a wide array of ingenious, macabre, or disturbing visual images. For example, the male-modeled, ready-to-wear Mad Scientist's Experiment costume consisted of a full-body costume of a monkey replete with a half-head mask featuring a gaping incision from which rubber brains dangled. Similarly, costumes for such characters as Jack the Ripper, Serial Killer, Freddy Krueger, or The Midnight Stalker were adorned with the suggestion of bloodstains and embellished with such paraphernalia as plastic knives or slip-on claws.

In marked contrast, the costumes of female villains alternated between relatively simple costumes of witches in pointy hats and capes modeled by young girls, costumes of the few female arch villains drawn from the pages of comic books, and, for older girls, costumes that were variants of the garb donned by the popular TV character Elvira, Mistress of the Dark (i.e., costumes that consisted of a long black wig and a long flowing black gown cut in an empire-style, which, when decorated with gold brocade or other trim at the top of the ribcage, served to create the suggestion of a bosom). The names of costumes for the female villains appeared to emphasize the erotic side of their villainy (e.g., Enchantra, Midnite Madness, Sexy Devil, Bewitched) or to neutralize the malignancy of the character by employing adjectives that emphasized their winsome rather than wicked qualities (e.g., Cute Cuddly Bewitched, Little Skull Girl, Pretty Little Witch).

Within the category of fools, feminine costumes were more likely than masculine costumes to depict nonhuman/inanimate objects (33.1% of feminine costumes vs. 17.4% of masculine costumes). Feminine costumes were more likely than masculine costumes to feature a wide variety of small animals and insects (e.g., Pretty Butterfly, Baby Cricket, Dalmatian Puppy), as well as flowers, foodstuffs (BLT Sandwich, Ice Cream Cone, Lollipop), and dainty, fragile objects such as Tea Pot. For example, a costume for Vase of Flowers was illustrated with a picture of a young girl wearing a cardboard cylinder from her ribcage to her knees on which flowers were painted, while a profusion of pink, white, and yellow flowers emerged from the top of the vase to form a collar of blossoms around her face. Similarly, a costume for Pea Pod featured a young girl wearing a green cylinder to which four green balloons were attached; on the top of her head, the model wore a hat bedecked with green leaves and tendrils in a corkscrew shape. When costumed as animals, boys were likely to be shown modeling larger, more aggressive animals (e.g., Velociraptor, Lion, T-Rex); masculine costumes were unlikely to be marketed with adjectives emphasizing their adorable, "li'l," cute, or cuddly qualities. In general, boys were rarely cast as objects, but when they were, they were overwhelmingly shown as items associated with masculine expertise. For example, a costume for Computer was modeled by a boy whose face was encased in the computer monitor and who wore, around his midtorso, a keyboard held up by suspenders. Another masculine costume depicted a young boy wearing a costume for Paint Can; the lid of the can was crafted in the style of a chef 's hat, and across the cylindrical can worn from midchest to mid-knee was written "Brand X Paint" and, in smaller letters, "Sea Blue." Although rarely depicted as edibles or consumable products, three masculine costumes featured young boys as, variously, Root Beer Mug, Pepperoni Pizza, and Grandma's Pickle Jar.

Discussion

Although the term "fantasy" implies a "play of the mind" or a "queer illusion" (Barnhart, 1967, p. 714), the marketing illustrations for children's Halloween costumes suggest a flight of imagination that remains largely anchored in traditional gender roles, images, and symbols. Indeed, the noninclusive language commonly found in the names of many children's Halloween costumes reverberates throughout many other dimensions of the gendered social life depicted in this fantastical world. For example, the importance of participation in the paid-work world and financial success for men and of physical attractiveness and marriage for women is reinforced through costume names that reference masculine costumes by occupational roles or titles but describe feminine costumes via appearance and/or relationships (e.g., "Policeman" vs. "Beautiful Bride"). Although no adjectives are deemed necessary to describe Policeman, the linguistic prompt contained in Beautiful Bride serves to remind observers that the major achievements for females are getting married and looking lovely. In addition to costume titles that employ such sex-linked common nouns as Flapper, Bobby Soxer, Ballerina, and Pirate Wench, sex-marked suffixes such as the *-ess* (e.g., Pretty Waitress, Stewardess, Gypsy Princess, Sorceress) and *-ette* (e.g., Majorette) also set apart male and female fantasy character costumes. Costumes for suffragettes or female-modeled police officers, astronauts, and fire fighters were conspicuous only by their absence.

Gender stereotyping in children's Halloween costumes also reiterates an active-masculine/passive-feminine dichotomization. The ornamental passivity of Beauty Queen stands in stark contrast to the reification of the masculine action figure, whether he is heroic or villainous. In relation to hero figures, the dearth of female superhero costumes in the sample would seem to reflect the comparative absence of such characters in comic books. Although male superheroes have sprung up almost "faster than a speeding bullet" since the 1933 introduction of *Superman,* the comic book life span of women superheroes has typically been abbreviated, "rarely lasting for more than three appearances" (Robbins, 1996, p. 2). Moreover, the applicability of the term "superhero" to describe these female characters seems at least somewhat dubious. Often their role has been that of the male hero's girlfriend or sidekick "whose purpose was to be rescued by the hero" (Robbins, 1996, p. 3).

In 1941 the creation of *Wonder Woman* (initially known as Amazon Princess Diana) represented a purposeful attempt by her creator, psychologist William Marston, to provide female readers with a same-sex superhero. . . . Nevertheless, over half a decade later, women comic book superheroes remain rare and, when they do appear, are likely to be voluptuous and scantily clad. If, as Robbins (1996, p. 166) argued, the overwhelmingly male comic book audience "expect, in fact demand that any new superheroines exist only as pinup material for their entertainment," it would seem that comic books and their televised versions are unlikely to galvanize the provision of flat-chested female superhero Halloween costumes for prepubescent females in the immediate future.

The relative paucity of feminine villains would also seem to reinforce an active/passive dichotomization on the basis of gender. Although costumes depict male villains as engaged in the commission of a wide assortment of antisocial acts, those for female villains appear more nebulous and are concentrated within the realm of erotic transgressions. Moreover, the depiction of a female villain as a sexual temptress or erotic queen suggests a type of "active passivity" (Salamon, 1983), whereby the act of commission is restricted to wielding her physical attractiveness over (presumably) weak-willed men. The veritable absence of feminine agents or symbols of death may reflect not only the stereotype of women (and girls) as life-giving and nurturing, but also the attendant assumption that femininity and lethal aggressiveness are mutually exclusive.

Building on the Sapir–Whorf hypothesis that the language we speak predisposes us to make particular interpretations of reality (Sapir, 1949; Whorf, 1956) and the assertion that language provides the basis for developing the gender schema identified by Bem (1983), the impact of language and other symbolic representations must be considered consequential. The symbolic representations of gender contained within Halloween costumes may, along with specific costume titles, refurbish stereotypical notions of what women/girls and men/boys are capable of doing even within the realm of their imaginations. Nelson and Robinson (1995) noted that deprecatory terms in the English language often ally women with animals. Whether praised as a "chick," "fox," or "Mother Bear" or condemned as a "bitch," "sow," or an "old nag," the imagery is animal reductionist. They also noted that language likens women to food items (e.g., sugar, tomato, cupcake), with the attendant suggestion that they look "good enough to eat" and are "toothsome morsels." Complementing this, the present study suggests that feminine Halloween costumes also employ images that reduce females to commodities intended for amusement, consumption, and sustenance. A cherry pie, after all, has only a short shelf life before turning stale and unappealing. Although a computer may become obsolete, the image it conveys is that of rationality, of a repository of wisdom, and of scientifically minded wizardry.

In general, the relative absence of gender-neutral costumes is intriguing. Although it must remain speculative, it may be that the manufacturers of ready-to-wear and sewing pattern costumes subscribe to traditional ideas about gender and/or believe that costumes that depart from these ideas are unlikely to find widespread acceptance. Employing a supply–demand logic, it may be that marketing analysis of costume sales confirms their suspicions. Nevertheless, although commercial practices may reflect consumer preferences for gender-specific products rather than biases on the part of merchandisers themselves, packaging that clearly depicts boys or girls—but not both—effectively promotes gendered definitions of products beyond anything that might be culturally inherent in them. This study suggests that gender-aschematic Halloween costumes for children compose only a minority of both ready-to-wear costumes and sewing patterns. It is notable that, when male children were presented modeling female garments, the depicted character was effectively desexed by age (e.g., a wizened, hag-like "grandmother") or by calling (e.g., a nun).

The data for this study speak only to the gender practices of merchandisers marketing costumes and sewing patterns to parents who themselves may be responding to their children's wishes. Beyond this, the findings do not identify precisely whose tastes are represented when these costumes are purchased. It is always possible that, despite the gendered nature of Halloween costumes presented in the illustrations and advertising copy used to market them, parents and children themselves may engage in creative redefinitions of the boundary markers surrounding gender. A child or parent may express and act on a preference for dressing a male in a pink, ready-to-wear butterfly costume or a female as Fred Flintstone and, in so doing, actively defy the symbolic boundaries that gender the Halloween costume. Alternatively, as a strategy of symbolic negotiation, those parents who sew may creatively experiment with recognizable gender markers, deciding, for example, to construct a pink dragon costume for their daughter or a brown butterfly costume for their son. Such amalgams of gender-discordant images may, on the surface, allow both male and female children to experience a broader range of fantastical roles and images. However, like Persian carpets, deliberately flawed to forestall divine wrath, such unorthodox Halloween costumes, in their structure and design, may nevertheless incorporate fibers of traditional gendered images.

References

Barnes, R., & Eicher, J. B. (1992). *Dress and gender: Making and meaning in cultural contexts*. New York: Berg.

Barnhart, C. L. (1967). *The world book dictionary: A–K.* Chicago: Field Enterprises Educational Corporation.

Bem, S. L. (1983). Gender schema theory and its implications for child development: Raising gender-aschematic children in a gender-schematic society. *Signs: Journal of Women in Culture and Society*, *8*, 598–616.

Doyle, J. A. (1989). *The male experience.* Dubuque, IA: Wm. C. Brown.

Eicher, J. B., & Roach-Higgins, M. E. (1992). Definition and classification of dress: Implications for analysis of gender roles. In R. Barnes & J. B. Eicher (Eds.), *Dress and gender: Making and meaning in cultural contexts* (pp. 8–28). New York: Berg.

Goffman, E. (1966). Gender display. *Philosophical Transactions of the Royal Society of London*, *279*, 250.

Goffman, E. (1979). *Gender advertisements.* London: Macmillan.

Klapp, O. (1962). *Heroes, villains and fools.* Englewood Cliffs, NJ: Prentice-Hall.

Klapp, O. (1964). *Symbolic leaders.* Chicago: Aldine.

McNeill, F. M. (1970). *Hallowe'en: Its origins, rites and ceremonies in the Scottish tradition*. Edinburgh: Albyn Press.

Mead, G. H. (1934). *Mind, self and society.* Chicago: University of Chicago Press.

Nelson, E. D. (1999). *Dressing for Halloween, doing gender.* Unpublished manuscript.

Nelson, E. D., & Robinson, B. W. (1995). *Gigolos & Madame's bountiful: Illusions of gender, power and intimacy.* Toronto: University of Toronto Press.

Robbins, T. (1996). *The great women superheroes.* Northampton, MA: Kitchen Sink Press.

Rotundo, E. A. (1993). *American manhood: Transformations in masculinity from the revolution to the modern era.* New York: Basic Books.

Salamon, E. (1983). *The kept women: Mistresses of the '80s.* London: Orbis.

Sapir, E. (1949). *Selected writings of Edward Sapir on language, culture and personality*. Berkeley: University of California Press.

Whorf, B. L. (1956). The relation of habitual thought and behavior to language. In J. B. Carroll (Ed.), *Language, thought, and reality* (pp. 134–159). Cambridge, MA: Technology Press of MIT.

Introduction to Reading 23

Sociologists Kristen Barber and Tristan Bridges analyze "satirical masculinity," a new form of masculinity that has recently appeared in advertisements for traditionally "feminine" products such as yogurt, salad, and body wash. They consider how this form of a "hybrid masculinity" emerged, why they are used to sell to men, and whether such representations are subversive or whether they ultimately reinforce dominant ideologies of masculinity.

1. What is "satirical masculinity," and why do advertisers use it to sell products to men?
2. According to Barber and Bridges, do satirical representations subvert—or reinforce—hegemonic masculinity? How does this process work?
3. What changes in the larger society are related to why consumption has become a new "rite of passage to manhood"?

Marketing Manhood in a "Post-Feminist" Age

Kristen Barber and Tristan Bridges

In recent years, Old Spice has relied on a satirical display of masculinity to rebrand its merchandise. Satire mocks while also revealing a supposed truth about some group, idea, thing, or behavior—here, who men are and who they can and should be. While Mustafa has achieved widespread recognition as the "Old Spice Guy," the satirical masculinity he helped to make famous is used to market an incredibly diverse array of products. Consider Kraft's "Let's Get Zesty" campaign for a line of salad dressings. Like Mustafa, actor Anderson Davis is presented as both a farce and a representative of a masculine ideal. Similarly, Yoplait features square-jawed, furrowed-brow actor Dominic Purcell in marketing its "fluffy," low-calorie Greek yogurt. When he produces a tiny spoon from his pocket, we see that men can maintain masculinity while enjoying a whipped, fruity treat.

All of these advertisements proceed from the assumption that masculinity is naturally at odds with anything even vaguely associated with women, like body wash, salad, and yogurt. Satirical masculinity produces a set of facetious cultural scripts that bridge this divide to create a new consumer base that "mans up" to purchase otherwise "feminine" products.

All of these advertisements present recognizable markers of what sociologist Raewyn Connell calls hegemonic—or culturally celebrated—masculinity, but they also poke fun at those very same markers. The joke plays on the use of contrasting symbols, behavior, and products associated with femininity alongside enactments of hypermasculinity. For instance, in the Yoplait commercials, Purcell's size suggests strength and the ability to dominate others (consistent with his television roles in "Prison Break" and "Legends of Tomorrow"), and his deep voice thrums as he tells us in all seriousness: "It's like a little fluffy cloud in my mouth. Fluffy, fluffy cloud." Kraft's "Let's Get Zesty" uses the tagline, "The only thing better than dressing is undressing," playing on the advertising cliché that "sex sells"—no matter what you're selling. And Mustafa conjures images of intentionally over-the-top romance novel covers. These representations of masculinity are so distant from most men's everyday lives that they are laughable. But scholars who study humor have found that jokes allow for the perpetuation of sexism, as well as racism, homophobia, and classism.

It's worthwhile to ask whether these satirical representations of masculinity are indeed subversive—by making outdated cultural ideals comical—or whether they help to reproduce the very forms of inequality they seem to mock. By analyzing three popular advertising campaigns in the satirical masculinity genre, we connect a cultural phenomenon to the emerging theory of "hybrid masculinities," which considers shifting definitions of manhood in terms of their larger consequences for equality.

Barber, K., & Bridges, T. (2017). Marketing manhood in a "post-feminist" age. *Contexts, 16*(2): 38–43. Reprinted with permission from the authors and the American Sociological Association.

Isaiah Mustafa, shirtless, rises from a bubble bath on a white foam horse to spread the word: "Make Sure Your Man Smells Like a Man." This slogan appears at the bottom of the page, implying there is indeed such a thing as a "manly" scent. He wears foam cowboy boots, chaps, and gloves, as well as a foam hat and sheriff's badge. In his left hand, he displays a red bottle of Old Spice body wash. A woman, wrapped in only a towel, gazes at Mustafa as he whips his foam lasso in the air, sardonically engaging the iconic image of the American cowboy. With a subtle smirk that suggests a casual confidence, the meanings of Mustafa's gender display are diverse. Aimed at men and women alike, this advertisement identifies a potentially deficient man and offers a solution to shore up his masculinity. While the story of transformation through consumption is not new, the jocular exaggeration, wit, and satire seen here are unique features of a popular portrayal of men in advertising today.

A still of Dominic Purcell in Yoplait's "Hunger" television commercial for Yoplait Greek 100 yogurt.

Satirical Masculinity and Hybridity

Satire is part of a larger cultural shift in masculinity and gender relations. Both social theory and empirical scholarship challenge the claim that certain historical periods are marked by masculinity crises, proceeding, as it does, from some idea of a stable masculinity. On the contrary, what we think of as "manly," "macho," or "masculine" varies by society, subculture, and time. But shifts in masculinity do follow a curious pattern: they are reactive rather than anticipatory. Masculinities scholar Michael Kimmel argues that anxiety about what masculinity actually is tends to follow transformations in femininity. Though the idea is counter-intuitive (we don't think of groups in power as being "pushed around" in this way), the historical record bears out Kimmel's point. When women enter into historically "masculine" arenas, like sports or the workplace, they shift the boundaries of femininity. And those are the moments when we get anxious about masculinity, claim that it is "in crisis," and find groups rallying around "solutions" to this suddenly pressing social issue.

Sociologists studying culturally dominant groups are often interested in how these groups retain power and under what circumstances their claims to dominance are challenged. When challenges to inequality fail to produce lasting change, sociologists want to know why. C.J. Pascoe and Tristan Bridges, who theorize hybrid masculinities, have surveyed the masculinities literature and argue that it shows White, straight, and well-to-do men sometimes adopt various elements of "other," more socially marginalized masculinities (and sometimes femininities, too) into their own performances of masculinity. This behavior produces the impression that social change has occurred without any real shifts in privilege.

Inequalities that persist, adapt. The theory of hybrid masculinities suggests that adaption takes place via three interrelated processes. First, hybrid masculinities produce symbolic distance between men and hegemonic forms of masculinity. Michael J. Murphy, for example, shows that public service campaigns like "Real Men Don't Rape" or "My Strength is Not for Hurting" allow some men to position themselves as beyond reproach. Second, hybrid masculinities involve men "strategically borrowing" elements from disadvantaged groups, such as when White men adopt elements of African-American culture. Third, hybrid

Isiah Mustafa as a biker for the Old Spice "Smell Like a Man, Man" campaign.

masculinities fortify boundaries between groups in ways that obscure the inequalities defining those boundaries. So, when straight men play with "gay" culture, it can create the appearance that incredible change has occurred. But if structural inequalities between straight and queer men remain intact, playfulness accomplishes little more than expanding their gender repertoires via a kind of identity tourism.

As a heuristic device, the satirical masculinity in so much contemporary advertising aimed at men is a powerful illustration of hybrid masculinity. These ads show us one way that current forms of inequality persist even as popular and academic critiques of "toxic" masculinity gain attention.

In 2008, Old Spice began a drastic rebranding effort with its "Swagger" advertisements featuring football player Brian Urlacher and musician LL Cool J. The brand has long been associated with older men, partially because it was a trailblazer in personal care products aimed at men; the reliable red packaging with a white sailboat has been available in drugstores since 1938.

As market research companies like Euromonitor International encourage cosmetic companies to appeal to young men and their wallets, though, we have seen a

Anderson Davis in Kraft's "Let's Get Zesty" campaign.

proliferation of commercials and print advertisements that engage satirical masculinity. (Super Bowl commercials have become particularly famous for this.) But satirical humor has not been analyzed much by social scientists, perhaps because its tongue-in-cheek character makes it difficult to take seriously.

Once-popular markers of adulthood such as marriage, children, and breadwinning jobs are less accessible to younger generations of men. In step, retail companies offer consumption as the new rite of passage to manhood. In the case of Old Spice, tackling the femininity of self-care and the association of its products with older men, the company's advertisements play with masculine stereotypes to widen its consumer base. And proceeding from the notion that men perceive themselves as deficient when it comes to masculinity, Old Spice facetiously provides everyday men the potential to feel masculine. The material evidence of masculinity has shifted, from steady paychecks to sudsy products.

Old Spice's "Smell Like a Man, Man" campaign offers young men something they are presumably lacking—manhood—embodied by the cowboy, the biker, and the lumberjack. Culturally salient and rewarded, these representations symbolize power and dominance.

Yet Mustafa's presentation of manhood is complicated. His cowboy is not *all* macho; it could even suggest a queer sensibility. It's not hard to imagine him as a member of The Village People, an exaggerated presentation of masculinity echoing the clone culture among gay men in the 1970s and '80s. The clone pulled from working-class fashions while emphasizing the buff, masculine body. At the same time, the presence of a woman in the ads allows for the potential of dual advertising, a strategy originated by Calvin Klein to produce a more subtle resonance with gay men alongside a more overt pitch to straight men. This

Kraft print advertisement for zesty Italian dressing.

Dominic Purcell eats "fluffy" Yoplait yogurt.

queerness creates a contradiction in masculinity, making it difficult to take seriously the hegemonic markers of power. In this symbolic murkiness, Pascoe and Bridges explain how hybrid masculinities make existing inequalities hard to identify.

Playing on the White, working-class, biker aesthetic in another of the series' ads, Mustafa dons a long, foamy handlebar mustache and mullet. In line with hybrid masculinity, however, this manly, "white trash" look is undermined by the comedic mismatch of Mustafa's serious facial expression. His masculinity is intentionally outlandish, a characteristic of this type of gender play. And if that's not enough, his raised eyebrow and slight smirk underline this playfulness to viewers.

"Oh, hey ladies. Are you in the mood to do something special?" Actor Anderson Davis slowly sucks on a wooden spoon he's just used to toss a salad. "I sure am." The camera pans to a stick of butter that melts under his smoldering gaze. "You know, once you go Italian, you never go back." It's unclear if Davis is talking about himself or the sausages in the skillet. With a backdrop of the Tuscan countryside behind him, Davis steps over to a pot of boiling water. He lifts the lid: "Steamy." His tight white t-shirt clings to his chest, wet from the condensation. The idea is that this will be a romantic dinner and, drawing from larger cultural ideas that women are lucky if their husbands or boyfriends do the cooking, Davis is presented as the perfect man. He's presumably a reflection of just how far feminism has come.

But in the three-part series of Kraft ads starring Davis, the romantic man appears ridiculously cliché. It might be more accurate to call his performance of masculinity a caricature. With his faintly raised eyebrow, Kraft represents Davis as simultaneously—if awkwardly—celebrating and mocking domestic, hetero-romantic masculinity. This simultaneity illustrates change and continuity in relations between men and women—the audience is primed to see it as funny that a "manly-man" is enacting tropes about women's sexual desire (stoked by the fantasy of temporarily surrendering "their" domestic tasks).

In a print advertisement, Kraft depicts a naked Davis on a red and white-checkered picnic blanket. A corner of the blanket keeps him modest by covering his genitals, and a picnic basket overflows next to him. Women, and perhaps men, are invited to imagine themselves at this picnic—curiously absent any salad. The image plays on tropes about hetero-romantic masculinity, and especially on stereotypes about women's desires.

This campaign celebrates gender relationships and sexual practices that feminists have shown perpetuate gender inequality, but with a wink and a nod. They are the exception that proves the rule in that they revive heterosexist assumptions of masculine subjectivity that structure unequal relations between men and women. Again, the humor is in the presentation of women as sexual subjects rather than as the objects of desire.

General Mills, co-owner of Yoplait, declares that "Dominic Purcell is a Man of Yogurt." He is "noted for acting rugged and tough," the company's blog post states, but he can still tuck into a "smooth, creamy, and sometimes fluffy" treat.

Indeed, Purcell is built like a linebacker, and he doesn't smile in any of the Yoplait advertisements or television commercials. "You see this look on my face?" He asks the audience in one commercial, "It's not anger; it's hunger." And apparently, a cup of 100-calorie yogurt will take the edge off. He promises that as he eats it, "a look of satisfaction and contentment [will] blossom across my face." His deep, sober tone remains flat, as does his expression. "Yum. See?" That's the joke: that Purcell has only one mode: stoic, serious, possibly dangerous. Eating yogurt doesn't compromise it a bit.

Like both the Old Spice and Kraft campaigns, Yoplait is attempting to bridge a gender gap in consumption by acknowledging yogurt is culturally coded as "feminine" and potentially emasculating. Successful representations of hybrid masculinities that mitigate the gender of these products require strategic framing to be understood as "masculine" among intended audiences.

Recall that satirical masculinity is used to sell men on products they presumably avoid for fear of what it might say about their gender and sexual identities. In response, all of these advertisements share a common feature: the intentionally excessive displays of masculinity. Through this, Yoplait's depiction of Purcell—like Kraft's use of Anderson and Old Spice's reliance upon Mustafa—appears to simultaneously celebrate and mock his masculinity. After commenting on the "fluffy cloud" quality of the yogurt, the lambo doors wing open on a silver sports car. By contrasting feminine yogurt with the phallic muscle car, this commercial appears to indicate some change, perhaps a softening of masculinity or a calling-out of unrealistic masculine types. But it also continues to celebrate elements of masculinity that feminist scholars argue are in dire need of change.

Satirical masculinity offers viewers the appearance of something progressive by seemingly mocking or deriding configurations of masculinity that have sustained feminist criticism, including hetero-romantic and hard-bodied, action hero masculinities. These images offer a playful, ironic masculinity, and invite us to take pleasure in men who clearly embody idealized forms of masculinity while engaging in "feminine" consumption practices. But this joke really only works if the systems of power are both carefully concealed *and* reinforced. As illustrations of hybrid masculinity, these displays of masculinity underscore stereotypical gender and sexual differences—though in novel ways. The use of satire helps to obscure the full consequences of hybrid masculinities despite them being on full display. It is comical to see Purcell's big hands skillfully wield a tiny spoon with which to eat yogurt, but only because his association with masculine power remains intact.

These ads might also look like sexual objectification, and a great bit of popular discussion assumes this objectification affects men's self-esteem and relationships to their bodies. As gender scholar Susan Bordo argues in her book, *The Male Body*, "I never dreamed that 'equality' would move in the direction of men worrying more about their looks rather than women worrying less." But emergent research on hybrid masculinities finds that looks can be deceiving. This is because sexual objectification is a cultural process that disempowers women on a larger scale, affecting their abilities to be taken seriously in the workplace, for example, or setting the stage for sexual violence and slut-shaming. While displays of satirical masculinity are often sexual, they are not disempowering. These commercials do not wholly undermine the "feminine" nature of these products, nor do they challenge the masculinity of those men hired to sell them. And anyone who finds the joke offensive is implicitly chided for caring too much about something so superficial as an advertisement for body wash. After all, it's just a joke—right?

Introduction to Reading 24

In this reading, Jennifer Dawn Carlson considers the movement of women into a market that has traditionally been associated with masculinity: guns. She asks how women negotiate this gendered space, and whether women's gun ownership challenges gender ideologies of men as protectors and women as nurturers.

1. Carlson talks about gun ownership and gun shooting in terms of "contested ideological space." What does she mean by this?
2. What dominant gender ideologies are used by advertisers, instructors, and women themselves to make sense of women as gun owners and gun users?
3. How do guns impact women's experiences of public space?

Carrying Guns, Contesting Gender

Jennifer Dawn Carlson

I was nervous about adequately concealing my pistol that morning and carrying it into public space, but as a researcher, I was even more curious about what it was like to carry a gun on an everyday basis. Would it feel different to grab a cup of coffee knowing that I was armed with a 9 mm pistol—even if I was the only person who knew? How would my status as a licensed concealed carrier change how gun owners and carriers—disproportionately men—viewed me at the shooting range, in the firearm classroom, or at pro-gun events?

In many respects, gun culture is a man's world. As Scott Melzer argues in his book *Gun Crusaders: The NRA's Culture War*, the gun lobby's success since the 1970s can, at least in part, be tied to the way guns reinforce masculine identity in America. In legal gun use, the language of self-defense laws and doctrine suggest that guns are a man's prerogative: America's "No Duty to Retreat" doctrine has long been defended by the logic that "a man's house is his castle." And in their criminal uses, guns are sometimes wielded to enact gendered violence—especially domestic. According to a 2003 study by Jacquelyn Campbell and colleagues, an abuser's access to guns ranks (after previous abuse and employment status) as one of the top predictors of a woman being killed by that abuser.

Others, however, highlight the protection guns can afford women. According to a series of papers published by Gary Kleck and colleagues that focus on women's victimization during street crime acts of resistance against an attacker—including, but not limited to, brandishing or firing a gun—lessen the likelihood that an attack, such as sexual assault, is "completed." Indeed, many Americans argue women stand to gain as much as, if not more than, men from armed protection. From the NRA to the firearms industry to gun instructors, gun advocates are cozying up to the idea of women owning, learning about, and carrying guns.

Pundits and policymakers on both sides of the gun issue argue whether armed women represent collusion with the patriarchy or the dawn of armed feminism. But politically evocative as it may seem, this binary just doesn't capture the dynamic, contested, and at times pragmatic nature of women's participation in gun culture. At the shooting range, in the gun store, and walking through daily life with a holstered firearm, the women I met during my fieldwork defied a single, straightforward narrative. They engaged in an ever-changing set of *gendered negotiations* through which women's empowerment and masculine protectionism were meshed in sometimes familiar, sometimes surprising ways.

Masculine Terrains

I met Kathy at Gun Sports & More, a shop and shooting range nested deep in the suburbs of Detroit. I had heard about the store because they hold "ladies only" shooting nights and firearms classes for women. A blond, SUV-driving suburban mom who wore a smart polo shirt (the store uniform) and a holstered handgun, Kathy led Gun Sports' efforts with women: she oversaw many of the firearms courses and pressed range owners to open up dedicated range time for women.

In contrast to the male gun carriers I met, who couched their desire to carry in terms of crime and a desire to protect themselves and their families, Kathy offered a different narrative. Sure, she would be capable of willing to defend herself and her kids with a handgun if needed. But, given the low crime rates in the middle-class suburb where she lived, she thought that was unlikely. Kathy's decision to carry a gun was, instead, motivated by her involvement in gun culture as an enthusiast, an instructor, and a competitive shooter. As we chatted, I saw that Kathy simply enjoyed shooting firearms and loved excelling at a "men's" sport.

When she first started competitive shooting, Kathy experienced exclusion and ridicule based on her gender. Then she won a major championship. She was the only woman to compete that year, and she expected to be mocked and dismissed by the other competitors—especially if she placed in the top three. When her name was called for first place in a room full of men, "they all just clapped!"

Scholars like Michael Messner have talked about the gendering of space, arguing that certain arenas like

Carlson, J. D. (2015). Carrying guns, contesting gender. *Contexts, 14*(1): 20–25. Reprinted with permission from the author and the American Sociological Association.

sports and the military have long been "masculine terrain." These social spaces, predicated on women's exclusion, are where bonds among men are forged and links between masculinity and a variety of social attributes—strength, competitiveness, courage—are naturalized. Women who dare to cross the threshold have historically been rejected and ridiculed, told they can't cut it in a man's world. As women enter, these spaces become what Messner calls "contested ideological terrain." The problem isn't their innate abilities, it's how women upset the male monopoly on certain social spaces and social practices.

Betty, who worked in the public sector, was one armed woman who crossed this threshold early in the 1960s, decades before Michigan's concealed carry law was changed. Back then, the law required demonstrating to licensing authorities that you had a legitimate need to carry a gun. Betty argued that her job made her vulnerable, particularly as a woman, and the gun board agreed. However, she recalled being bullied by her male co-workers: "'Look at you, why do you need a gun?' They were just jealous! They told me there's no way someone as small as me could even use a gun!"

Chris told a similar story: after her dad had been the victim of a kidnapping in Detroit's infamous Cass Corridor, she decided she wanted to learn how to use a gun for self-defense. When she called a couple of local gun stores to ask about an impromptu lesson, she "basically got laughed off the phone." She soon found out that in her county, women were being denied concealed pistol licenses because the men on the licensing boards couldn't wrap their heads around the idea of an armed woman capable of self-defense. Chris was angry: "Particularly for women who were in situations where they felt threatened by a former boyfriend, husband, any kind of stalking situation, being denied the ability to defend themselves is absolutely unacceptable."

Committed to teaching women how to shoot and use guns for self-defense, Chris eventually became politically active in the grassroots effort to pass Michigan's 2001 concealed carry law, which would remove the discretion of the licensing board to decide who was worthy of a concealed pistol license. As part of her activism, she was slated to speak at a men's-only hunting lodge. As she described it, she was one of the few women who "had ever crossed that doorway." In fact, it turned out that her invitation was a mistake: "[O]ne of the guys came over to me and said, 'I'm so sorry, Chris, we didn't realize, but with your name, we didn't know if you were a man or a woman, but we assumed you were a man.'" Instead of leaving, Chris decided she wouldn't "drop to that level" and lectured the men: "What about your wife? Your daughter? Your aunt? Your mother? Your grandmother? Your neighbor who doesn't have a husband or boyfriend? When you are not around to defend the homestead, what about them? Don't they deserve the right to take care of themselves and to do it with good training?"

Ten years later, when I started researching gun politics in 2010, I'd hear the same arguments—not just from women, but from pro-gun men, as well.

The Great Equalizer

Today, Chris's arguments about guns as gender equalizers—captured by slogans like "God Created Man and Woman, But Samuel Colt Made Them Equal"—have gained traction. Many of the men I met in Michigan during the course of my research insisted they liked the idea of armed women. And from the looks of shooting ranges, gun shops, and gun magazines, things are changing: at a local shooting range, you're likely to see pink handguns, pink rifles, and even ammunition labeled with a pink bow in support of breast cancer awareness.

Certain guns are feminized: small guns, pink guns, and revolvers have all become "women's guns," and there's now a whole cottage industry—from designer concealed carry bags to bra holsters—geared at making concealed handguns a fashionable, comfortable option for women. Gendering (and sexualizing) the "culture of fear"—to use Barry Glassner's term—gun advertisements often highlight women's vulnerability and promote guns as a "woman's best friend." Glock's "Wrong Girl" commercial depicts a young woman in a tank top and panties alone at home. She hears a disturbance and grabs her Glock. Before she can shoot, the would-be intruder passes out in stock. Was it the sight of the beautiful woman or the Glock? Either way, the threat is neutralized.

There's a good reason the firearms industry wants to make guns appealing to women: they represent a largely untapped consumer base in a market already swimming with 300 million guns. Women mean more gun sales, more NRA memberships. This embrace, as Chris intimated, also fits with dominant gender ideologies stipulating men as protectors: how can a man truly committed to protecting his family deny his wife and children a weapon of self-defense?

But old habits die hard. With this openness comes new narratives that negotiate women's presence in a "man's world." Today, Chris wouldn't be laughed off the phone or stage. Instead, she'd be encouraged:

women are often assured that their gender makes them *great* shooters. Why? Male gun instructors told me women are more docile learners. And if they aren't quite the warriors that men are, they still have a natural, maternal instinct to protect. Their smaller fingers, which apparently make women "naturally" more dexterous, don't hurt either.

I seemed to be a prime example of just how wrong those generalizations were. It took a while for me to understand the feel of the gun at a tactile level, knowing just where the pad of my finger should sit on the trigger or how to roll with the recoil. Like other women I met, I became irritated that a gendered standard had already been set for me and my shooting abilities; it added another layer of stress to conducting fieldwork. And I found that while increasing my firearms proficiency and knowledge made me feel more comfortable and even, like Kathy, a bit "empowered" on the range, it didn't really change how men saw me and my gun. Most often, I was treated as a novice until the man saw how well I shot a .460 Rowland ("My wife would never shoot that!") or how detailed my knowledge of ammunition became ("You know what a 9mm Kurz round is? Most men don't even know that!")

Gender still matters on today's shooting range. There are men insisting that women carry uselessly small, but "gender appropriate" .380s or familiarize themselves with revolvers (semi-automatics might be too mechanically complicated). There are women "surprising" men with their full capability as shooters. But rather than a politics of exclusion, gender works more through a contradictory, contested politics that informally segregates everything from handguns to holsters into gender appropriate categories.

The shooting range is contested ideological terrain. Under new concealed carry laws, the gendering of guns and gun paraphernalia extends into public space, too.

Public Guns

Thanks to the passage of "shall-issue" laws in dozens of US states, guns can become part of everyday life. These laws require licensing authorities to issue a license to carry so long as the applicant meets a predetermined list of criteria. In contrast to Betty and Chris, who had to convince a gun board that they were fit to carry, in most states license applicants today can simply take a firearms course, fill out an application, pay a fee and—assuming their criminal record comes back clean—receive their license in a matter of weeks. In the US, there are roughly 11 million licensed gun carriers, and over 400,000 of them are in Michigan. Though there is no national database of concealed carriers, state-level data—such as Michigan's—suggests that about 1 in 5 of these licensees are women.

For many women, the experience of carrying a gun—even if it is concealed—means experiencing public space differently. Angela, a woman in her 50s, became interested in guns during a camping trip "up north." She explained her first exposure in gendered terms: "You know how the story goes: the girls are talking about sewing, and the guys are talking about hunting. So I overheard one of the guys saying he wanted to go shooting. Nobody had any interest. So I said, 'I will!'" Angela's comfort on the range soon translated into a desire to learn more; she found herself in a concealed pistol licensing class. She didn't intend to ever carry her gun until she became friends with one of the instructors, who offered to go out with her on her first day as a gun carrier. No one could see her gun, of course, but she still felt nervous with it on her hip. What if someone saw? What if it somehow popped out of its holster? Angela told me stepping out of the car, going to the grocery store, or just walking around all felt different. But carrying her gun (eventually, two guns) gave her a degree of confidence: even if her concealed gun was her secret, she'd proved she was capable of carrying it.

Though she was acutely attentive to gender as she described her own turn to guns, Angela didn't use the term "feminist." In fact, only one woman gun carrier I met described herself in explicit feminist terms, jokingly calling herself a "feminazi." Nevertheless, Angela's story evokes Martha McCaughey's argument that self-defense is a form of "physical feminism": teaching one's body to "fight back" provides women with a skillset should they find themselves in a violent confrontation such as an attempted sexual assault. Carrying a gun, as well as the confidence that they can defend themselves, thus transforms how these women move through and experience space. It entails an embodied rejection of dependent femininity. Put differently, Angela's gun is a challenge to what Iris Marion Young calls masculine protectionism: the idea that men have an *exclusive duty* to protect women and children. With a gun at her hip, Angela feels confident she—and she alone—can protect herself.

But just as women's presence on the shooting range is contested ideological terrain, so too are women's holsters: men had a great deal to say about how, what, and when women carried. When I met Cheryl and Matthew, an older, White married couple, Cheryl linked the gun she carried to feelings of empowerment and fearlessness: "It's funny. When you got the gun, you aren't scared." She told me she felt freer in public spaces and less afraid of being victimized when working as a real estate agent, entering unknown spaces. Like Betty and other women, Cheryl's participation in the workforce was tied to her participation in gun culture.

But as I watched Cheryl and Matthew describe their gun carry habits and their politics, I soon realized that Matthew was the "gun nut" of the two. While Matthew emphasized politics, his disdain for the president, and his military service, Cheryl took a no-nonsense attitude—"He's the political one." Cheryl recognized the appeal of a gun, but she also asserted that she didn't need her gun *every* time she stepped outside. Matthew neither understood nor respected her logic. He insisted his wife *always* needed her gun, because "you never know." Motioning to her heavy purse, he argued, "If you can carry all that, you can carry a gun!" Matthew's words followed what I had observed among many other gun-toting men: they were vociferously in support of armed women, but on men's terms. By attempting to override Cheryl's capacity to discern whether and when she needed a gun, he strained to extend *his* duty to protect into her holster. Cheryl seemed unfazed by Matthew's finagling. More pragmatic than political, Cheryl laughed off her husband's insistence. Her embrace of situational self-protection was a subtle negotiation of masculine protectionism.

Double-Barreled

For all of my anticipation walking out the door that first day, carrying a firearm in itself did not transform—for better or for worse—my own gendered sense of self. No doubt, I felt the sense of confidence women had told me about. I also experienced strong encouragement—even pressure—from men who thought it was great that I was carrying a gun and who believed that, as a woman, I would be vulnerable to violent attack without one.

But, truth be told, that first experience of gun carry was anti-climatic. At the end of the day, the gun at my hip was just . . . a gun. It was gun *culture*—and the practices of gun-carrying men and women—that made my 9mm a gendered object.

In this regard, the proliferation of gun carry represents neither the dawn of a new feminism nor the resurgence of the patriarchy—at least not at the individual level of individual armed women, still largely outnumbered by men. Lying somewhere between the domain of "physical feminism" and "masculine protectionism," carrying guns serves—for the women I met—as a way to subtly and not-so-subtly negotiate gender norms around safety, security, and even caring for others. For some women, like Chris and Betty, the gendered world of guns transformed their weapons from tools of self-protection to symbols of women's empowerment. For other women, like Cheryl, guns simply balanced personal, pragmatic needs and men's desire to protect.

The visceral, lived experience of guns is contradictory and contested. A woman's gun can be a tool of embodied empowerment, but it can also be a vehicle of complicity with masculine protectionism. It might even be both, simultaneously. In a complicated, pro-gun country, the gendered meaning of the gun is double-barreled.

Introduction to Reading 25

Kimberly Hoang, a sociologist and Vietnamese American woman, conducted 22 months of participant observation and ethnography in a variety of roles (e.g., hostess and bartender) in a variety of settings (e.g., bars, cafés, malls, streets) in which sex workers and their clients met in Ho Chi Minh City, Vietnam, between 2006 and 2007 and 2009 and 2010. When she began her project she discovered that the women sex workers she studied, all over the age of 18, were not only aware of the conditions of their work but also willing participants in sex work because it offered financial opportunities for themselves and their families that could not be matched by doing factory work or service-sector work. In this article, Hoang focuses on two sex work arenas, one catering to Western budget travelers and the other to Western businessmen. She highlights the complexities of client–worker relationships, in particular the ways sex workers sell impressions of themselves as dark-skinned, poor women to Western men who embrace fantasies of helping poor women living in the third world.

1. Why/how does Hoang challenge the widespread view in Western nations that sex workers in countries such as Vietnam are kidnapped, sold, or forced into sex work?
2. Discuss the ways buying and selling gender, race, and class are incorporated into the sex worker–client relationship in Hoang's study.
3. How do you feel about the ways sex workers "dupe" clients and, in turn, clients play into the "game" as described in this article?

Performing Third World Poverty

Racialized Femininities in Sex Work

Kimberly Hoang

Over the last two decades, the issue of human trafficking has captivated governments, NGOs, researchers, and activists around the world. Images of women in handcuffs and chains circulate through print, television, and online news outlets, perpetuating a view of trafficked women as victims of Third World poverty who are kidnapped, sold, or forced into sex work. As governments, corporations, religious organizations, and celebrities devote millions of dollars to save trafficked women, they neglect to conduct systematic research on the ground to assess this problem.

This article advances the scholarship on sex trafficking by providing a different lens through which to examine the sex industry. Rather than viewing women as victims of global poverty who are kidnapped, forced, or duped into the sex industry, I follow the work of Brennan (2004), Mahdavi (2011), and Parreñas (2011) to analyze the complex social structures that shape the *range of choices* available to women in their relationships with clients, club owners, and brokers. I show how sex workers in Ho Chi Minh City [HCMC], Vietnam, capitalized on media images and NGO narratives that portray sex workers as victims to advance their lives. By *performing Third World poverty*, sex workers who freely entered the sex industry portrayed themselves as victims to procure large remittances from their male clients.

In this article, I raise two primary questions. First, how do sex workers capitalize on global economic restructuring to improve not only their lives but also the lives of their extended families? Second, how do male clients make sense of and respond to sex workers' performances of Third World poverty? To answer these questions, I turn to my ethnographic research in HCMC's sex industry conducted over a total of 22 months in two phases between 2006–2007 and 2009–2010 as well as in-depth interviews with 71 sex workers and 55 clients in two niche markets that catered to Western budget travelers/tourists and Western businessmen who were part of a transnational circuit.[1] I show that just as organizations in the developed world pull on the heartstrings of their donors in the "fight" to end human trafficking, sex workers capitalize on globally circulating narratives to dupe male clients for money and visas by engaging in strategic and racialized performances of third world poverty. This article also looks at how male clients respond, react to, and make meaning out of sex workers' performances as they open their wallets to "save" women from a lifetime of poverty or entrapment in sex work. As such, this article also fills a gap in the literature by paying attention to both sides of client and sex-worker relationships.

Global Sex Work

In recent years, sex work abolitionists examined the lives of women forced into the sex trade (Bales and Soodalter 2009; Bolkovac and Lynn 2011; Mukherjee 2010). Their work fueled a worldwide explosion of representations of Third World women as victims routinely bought and sold to fulfill male sexual desires.

These efforts sensationalized poor Third World women as victims while making heroes of those who studied them and worked to save them. These efforts also prompted serious policy changes regarding the sex industry.

Sex work abolitionists sensationalize this "multi-billion dollar industry" and the "millions of women" kidnapped and forced into sex work without engaging in systematic, in-depth research of the women involved.

In addition, abolitionists assert that in order to end human trafficking, we need to eliminate the demand side of the equation. These researchers depict male sex patrons as predators who brutalize poor Third World women. However, very few studies engage in any kind of systematic research on the men who patronize these establishments. As I have shown elsewhere (Hoang 2011), male patrons of sex work vary a great deal in their behaviors, motivations, and desires. These men are involved in complex, and sometimes intimate, relationships with sex workers, that involve much more than one-time direct sex-for-money exchanges as sex workers engage in a variety of emotional labors that involve a complex intermingling of money and intimacy (Bernstein 2007; Hoang 2010).

A parallel body of research emerged in recent years from sociologists and anthropologists who conducted in-depth ethnographies inside the bars and brothels where sex workers work. These researchers engaged in systematic research, spending a significant amount of time building rapport with sex workers (Brennan 2005), working alongside them (Parreñas 2011; Zheng 2009), and providing complex analyses of the *choices* women made as they entered into sex work. These studies show that, contrary to popular belief, women often knowingly and willingly entered into complex arrangements with brokers, club owners, and entertainment establishments in what Parreñas (2011) refers to as a process of *indentured mobility*.

This article builds on these studies by illustrating how sex workers in HCMC, Vietnam, *perform Third World poverty* to procure large sums of money from Western male clients and how they strategically darkened their skin to embody what I term *racialized femininities.* This article not only points to women's *range* of choices in entering sex work but also highlights how workers strategically capitalize on global economic restructuring to enhance their socioeconomic trajectories. More importantly, I illustrate two sides of the client-worker relationship.

Sex Trafficking: Duped Victims or Hustlers?

In 2006, when I first began preliminary research for this project, I went to Vietnam equipped with information on local NGOs who "helped" the women I thought I would find. I visited a small but growing number of organizations looking to save victims of sex trafficking. However, I was shocked to find that few of the sex workers I met had been duped or sold into the sex trade. Many workers migrated to HCMC from villages to work in factories or the service sector before entering into sex work. During a tour of a local NGO, I was surprised to find that the organization had set up a private business exporting garments made by local sex workers to be sold abroad with "fair trade" stickers and pamphlets with images of women in handcuffs or trapped behind barbed wire fences.

Naively, I approached the workers and asked them how they ended up in this small clothing factory. At first, the women were quiet and refused to acknowledge my presence. After a long moment of awkward silence, Ai-Nhi, a 19-year-old worker, said to me, "I got caught on the street and had to spend several months in the rehabilitation center. This organization came to get me, so I am here now, but I would rather be in the bars." Dumbfounded, I asked her why she would rather be there and she responded:

> I used to work in a factory making $70 US a month. I did that for many years but I was not making enough money to send to my family, so I quit and went to work on the streets . . . One day, I got caught and was sent into the center, and then this woman told me that I could get out early if I went with this organization that would teach me how to sew clothes. I agreed so that I could get out of the center . . . I am here now doing the same thing I was trying to escape. . . . They say they want to help us, but they are only helping themselves. I get paid $30 million VND [$200 US] a month to work here . . . [but] I know that they sell these clothes for a lot more money abroad. . . .

When I asked the women if they were happier in this shop sewing clothes than engaging in sex work, Ha, a 23-year-old worker, told me that she could make a lot more money in the bars where she also had more control over her time. She told me to go to the backpackers' area (*khu tay ba lo*) where I would meet several women who went right back into the bars after leaving the rehabilitation center. She wrote down the name of a bar in that area and told me to talk to the women working there. This was my first introduction to sex work in the backpackers' area, where I would eventually meet several women and their clients who were Western men traveling on a budget.

Performing Third World Poverty

Racialized Bodies

In 2006, both budget travelers and transnational businessmen spent time in the backpackers' area of HCMC. However, by 2009, the area was run down by transient tourists looking to explore Vietnam as cheaply as possible and Western businessmen had carved out a distinct niche in the sex industry. In the backpackers' area, clients could walk into bars where women would immediately greet them, serve them drinks, wipe their faces with wet towels, and provide them with shoulder massages. Men would receive all of these services just by ordering a $2 US beer. The women in this bar engaged in a variety of practices to make their bodies desirable to their male clients. For example, all of the workers darkened their skin with bronzers and skin creams. When I asked the workers why they did so, Xuong, a 26-year-old worker, said to me:

> The men here like darker skin and women who just came up from the village. Girls who just come up from the villages get the most clients because they look fresh. Men like women with dark skin. They will always touch you and say, 'Wow, your skin is so dark and soft.'

Altering their skin color was the most notable strategy these women adopted to racialize their bodies to look like poor women in a Third World country. Those who were darker often received more attention, particularly from white men traveling on a budget.

Sex workers in the backpackers' area also wore very plain clothing consisting of jean skirts or shorts and a plain tank top because they wanted to convey to their clients that they were poor workers. One afternoon while helping Thu, a 22-year-old worker, pack for a trip home, I rummaged through her clothes and noticed that she had a lot of beautiful dresses. When I asked her why she never wore them to work, she explained:

> I have two sets of clothes—one for work and one for the village. I don't wear nice clothes to work because I don't want the men to think that I have money. If they think that I have money, they will not give me any money. I have to make sure that I look like a poor village girl.

Thu was not the only one who strategically dressed down in the bars. All of the sex workers did the same because it allowed them to convey an image that they were poor workers who could not even afford food, let alone nice things. They saved their nicer clothing for their visits to the village because these displays of wealth at home de-stigmatized their work in the bars.

Sex workers who catered to Western businessmen also played into their clients' racialized desires by strategically darkening their skin. However, rather than wearing simple clothing, these workers wore low-cut versions of the traditional Vietnamese *Ao dai*. Women who catered to Western businessmen dressed in ethnic dresses that made them look exotic while allowing workers to appear more expensive than the women in the poor backpackers' area. Lilly, the owner of the bar I studied in this district, reminded her workers on a daily basis that the clients in her bar were respectable businessmen, not cheap backpackers. The women who catered to this higher-paying clientele should therefore dress accordingly. She commissioned a tailor to come to her shop to measure all of her workers and sew their uniforms. Each dress was cut a little differently so that each woman would have a unique look. However, all of the dresses were made of the same faux silk Chinese fabric that closed in around the neck but had an opening that exposed women's cleavage and slits along the skirt to expose the workers' full thigh and bottom.

By darkening their skin and dressing in either plain clothing or exotic, Orientalizing garments, sex workers racialized their bodies to make themselves desirable to their clients. Their outward appearance was crucial to attracting men into the bars. However, workers also had to engage in a variety of emotional and bodily labors (Hoang 2010) to maintain ties with their clients and procure small and large sums of money from them. The next section examines how women enacted *Third World poverty* in ways that extended beyond simple self-presentation in order to attract long-term remittances from their clients.

Performances of Poverty

Regardless of whether women catered to Western men traveling on a budget or Western businessmen, all of the women engaged in various performances of gender (West and Zimmerman 1987) through their embodied practices and through relational work (Zelizer 2005) with their clients. I learned of women's strategies to *perform Third World poverty* through the English lessons that I provided them in the afternoons. I visited *Naughty Girls*, the bar in the backpackers' area, at 2:00 pm, three to four days a week for four months between 2009–2010 to provide the women with free English lessons. Many of the women were excited about the opportunity to work with someone who would help them translate fabricated stories without judging them for lying or duping their clients. During these days, I helped women translate a series of emails, text messages, and key phrases that they wanted to have in their back pockets. They asked me to help them translate phrases like "My motorbike broke down. I have to walk to work. Can you help me buy new motorbike?" and "My father very sick and no one in my family help so I have to work. I am from An Giang village. You go to village before?"

During these lessons, I often asked the women why they lied to their clients and why they were careful not to display too much wealth in front of them. Xuan, a 19-year-old worker, said to me:

> A lot of the men here think that Vietnam is still a poor country. They want to hear that your family is poor and that you have no options so you came here to work. If you make them feel sorry for you as a poor Vietnamese village girl, they will give you a lot more money. We lie to them because it works . . . We tell them that Vietnam is changing and growing so fast and that the price of food and gas has gone up and people from poor rural areas cannot afford to live off of the rice fields anymore.

The act of creating fictive stories about their "rural lives" enabled many women to procure large sums of money from clients through remittances. Xuan's strategic move allowed her to capitalize on Vietnam's changing position in the global economy and the widening inequalities between local rich and local poor. The women I studied were certainly much more financially secure than their family members who worked in the rice fields, textile or manufacturing industries, or even as service workers in HCMC. In an informal interview with Vy, a 22-year-old worker, she said to me:

> Western men in America hear about girls who are sold and forced to sell their bodies, but no one here is forced to do anything. I come to the bar to work and if I want to have sex with a client, I have sex with him. If I don't, then I won't. No one forces me to do anything I don't want to do. People come here trying to give us condoms or save us, but how can they help me when I make more than them?

In that same conversation Chau, a 23-year-old worker interjected and said:

> There are some men who come to Vietnam often and they know what bar life is like . . . There are other men who are new to Vietnam—it is their first trip. Sometimes after sex, I will cry in the room and tell them how hard my life is and how mean the bar owner is. I will tell him that I want to get out and that I need $5000 US to pay for my debt to get out. So they help us.

Sex workers like Vy and Chau were aware of images that circulated across the globe of Asian sex workers as victims forced into a trade that involved horrific forms of violence. Their lives did not match these representations, but they capitalized on the images to procure large sums of money from their clients. Several men willingly provided the $5000 US the workers requested or some portion of that amount to "save" them from their life as a bargirl. In this way, sex workers also helped men feel like superior Western men from strong nations who were engaged in charity projects by helping poor women desperately looking to change their lives in a developing nation.

In their quest to see an authentic Vietnam, from a local's perspective, male clients often asked women if they would take them on tours of their villages and hometowns. Sex workers whose families were based in the city sometimes relied on their connections in the bar to take clients on tours of the Mekong Delta to visit with fake families. For example, in a conversation that I had with Thuy-Linh, she said:

> I am going to Kien Giang tomorrow with one of the guys here because he wants to see my village, but most of my family lives in Saigon now. We moved here about 10 years ago . . . I am taking him to stay with Vi's family so that he will think that I am really poor and maybe give me money to rebuild the house or help my family out.

Several clients expressed a desire to visit villages where they could walk through rice fields, ride bicycles, and bargain for produce in street markets. Therefore, the women in the bars organized tours to visit fake families in nearby villages. In my conversations with the workers about these tours, I was struck by their awareness of their clients' desire to see Vietnam as a developing Third World country. They organized tours that would portray an "authentic" Vietnam removed from signs of global change, modernization, and capitalism. More often than not, sex workers were happy to play into their clients' desires because doing so enabled them to ask for larger sums of money.

In conversations with the clients upon their return from visiting the Mekong Delta, many expressed a sense of deep sadness for the conditions of poverty that the women portrayed to them. For example, after spending three days with Nhi's family in their village, John, a man in his late 50s to early 60s, said to me:

> There are so many things that we in the West take for granted. Roofs over our heads, hot water, shoes . . . When I was with Nhi, I had to shower with buckets of cold water. It was so disgusting because I was brushing my teeth and I didn't realize that the bucket had a bunch of maggots in there. I felt these tiny worms swimming around in my mouth and I had to spit it out. I asked them how much it would cost to put in a proper shower and they said $500 US, so I gave it to them. They were such gracious hosts to me that I wanted to give something back.

While workers certainly employed strategies to embody Third World poverty, they also engaged in performances that highlighted their poverty in relation to the economic situations of their clients. John sympathized with Nhi's life and her conditions of poverty, and he genuinely wanted to help provide her family with a new faucet. Regardless of whether they were to women's true families, these visits to the village allowed workers to capitalize on their client's genuine concern, compassion, and empathy. Men provided women with money to help them escape poverty and transition from a basic standard of living to a comfortable standard of living.

By allowing men to believe that they were fulfilling the provider role, workers performed a femininity that was linked to financial dependence. Clients displayed a class-based masculinity in relation to these women. Several of the men I spoke with expressed a desire to have a traditional marriage where men were the economic providers and women took care of the home. However, because they could not maintain these roles in the US, many hoped to develop these relationships across transnational spaces in less-developed nations. In a long conversation with Jason, a man in his mid-60s from Montana, he told me:

> Men like me probably make women like you very uncomfortable. We come here and get hooked to younger women . . . I grew up at a time in America when women stayed home and took care of the family while men worked. My wife and I were happily married for many years. When she died two years ago, my world fell apart. I didn't know how to cook, or clean, or take care of myself. I was depressed. I needed a wife . . . [or] someone to take care of me. In Asia . . . some women still hold on to those traditional values, and I can afford to take care of a woman on my retirement fund.

It was clear in Jason's mind that while he could not be an economic provider for a woman in the US, he could successfully assert his masculinity across transnational borders because of Vietnam's status as a Third World country. The act of creating fictive stories about their "rural lives" enabled many women to procure large sums of money from clients like Jason through remittances. These scripts enabled men to act as the economic provider and in effect preserve a sense of masculinity lost back at home.

Clients' Scripted Performances

Not all men were as easily duped for their money. By 2009, stories had circulated among male clients of workers lying to them for their money. I sat in the bars listening day after day as men warned each other of the various ways that they might be tricked. However, I was surprised to find that several of the men I spoke with actually preferred to be unaware of workers' desires to consume luxury goods like nicer clothing, expensive cellular phones, and electronics because those items symbolized access to global capital, mobility, status, and most importantly, dignity in their work. For example, Edward, a 38-year-old man from the US said to me:

> We're not as naive as you think we are. We know better. But here is the thing. When you walk into a bar, everyone plays a role. The girls pretend to act excited to see us and flirt with us. We flirt back. They tell us about their poor families and we give them money here and there to help them. We may never know their real stories, but that is not the point. The point is that they make us feel like strong men who can provide for them and they get paid for it.

Some clients developed long-term relationships or friendships with a few workers; those relationships also involved a certain level of duplicity. Howard, a 39-year-old expatriate from Sweden, told me about his experience with duplicitous games:

> I have known this one worker for three years now and I still do not know very much about her. I feel comfortable there. It is predictable. I go in and have the same conversation with her every night. Someone could write a short story about this because it is the same script every single night. I come in and tell her I missed her, that I think she looks beautiful. She tells me about her family and money problems and I pretend to care. Sometimes we have sex, and that is it.

Relationships between clients and sex workers were bound by a set of predictable scripts that allowed clients to feel comfortable. As Howard described, when he walked into a bar, he also engaged in an emotional performance by telling a woman he saw on a regular basis that he missed her, thought she was beautiful, and was concerned about her financial situation.

These games became particularly apparent when new clients came in with friends who were regulars. Often, I listened in as men coached each other on what were appropriate and inappropriate behaviors within the bar. For example, one night while I was working in *Secrets*, a bar that catered to Western expatriates, I listened as Kevin, a regular client, explained the workings of the bar to his friend, Joseph. Kevin said:

> The girls in this bar are really nice. They are very low-key. You can talk and flirt with them with no pressure. If you like them, *you* ask them to meet you afterwards for about $100 US. It is a game. They play it and you play it.

Indeed, clients paid for a service from sex workers who worked as bartenders and hostesses. However, the clients also performed their own scripts in relation to the workers. They informed each other of the unspoken set of norms that guided interactions between workers and clients in the bar. Male clients did not always believe that the workers were truthful, but that was beside the point, because both men and women played a particular role inside the bars. Regardless of their truth, scripted interactions enabled men to maintain a comfortable and predictable relationship with workers based on a fantasy that they were heroic saviors helping women in need.

Conclusion

The issue of human trafficking has become a topic of great interest around the world. Celebrities like Demi Moore, Lucy Liu, and Ricky Martin and fashion designers such as Calvin Klein, Donna Karan, and Diane von Furstenberg have committed resources to raise awareness around the issue of sex trafficking. Presidents George W. Bush and Barack Obama both agree that the US should commit resources to save women from the horrors of forced sexual labors internationally. Through in-depth ethnography, this article explores beneath the surface of this issue to examine the lives of men and women deeply embedded in the sex industry. None of the women I describe in this article were kidnapped, forced, or duped into sex work. Rather, they made conscious choices to sell sex because they saw factory work and service work as more exploitative forms of labor.

This is not to say that trafficking does not occur, or that women are not subject to forms of *indentured mobility* (Parreñas 2011). Rather, I suggest that the story involves greater complexity than the representations of NGOs might suggest. Indeed, women capitalize on images of victimized women and on Vietnam's rapid economic restructuring by *performing Third World poverty* to get money from their clients. Some clients played into these stories because it allowed them to preserve a sense of masculinity lost back at home. While nearly all of the men that I met in 2006 believed the sex workers' lies, by 2009, enough men had been duped that stories of women procuring large sums of money by lying began to circulate among Western clients. Even with this knowledge, clients and sex workers both engaged in scripted practices and discourses related to Third World poverty and women's victimhood. These scripts not only allowed "men to be men," as Nhi put it, but to enable men to be Western men from developed nations by maintaining a fantasy that they are helping poor women living in the Third World. Beyond images of barbed wire and handcuffs are actual men and women engaged in complex relationships. As governments and activists intervene in women's lives, aid efforts must move beyond the sex industry and into other locations such as factories to address the multiple and interconnected sources of women's "exploitation."

Note

1. I have used pseudonyms for all individuals and bars in this paper to protect the anonymity of my research participants.

References

Bales, Kevin and Ron Soodalter. 2009. *The Slave Next Door: Human Trafficking and Slavery in New America Today*. Berkeley: University of California Press.

Bernstein, Elizabeth. 2007. *Temporarily Yours: Intimacy, Authenticity, and the Commerce of Sex*. Chicago: University of Chicago Press.

Bolkovac, Kathryn and Cari Lynn. 2011. *The Whistleblower: Sex Trafficking, Military Contractors, and One Woman's Fight for Justice*. New York: Palgrave Macmillan.

Brennan, Denise. 2004. *What's Love Got to Do with It? Transnational Desires and Sex Tourism in the Dominican Republic*. Durham: Duke University Press.

———. 2005. "Methodological Challenges in Research with Trafficked Persons: Tales from the Field." *International Migration* 43:35–54.

Hoang, Kimberly. 2010. "Economies of Emotion, Familiarity, Fantasy and Desire: Emotional Labor in Ho Chi Minh City's Sex Industry." *Sexualities* 13:255–272.

———. 2011. "'She's Not a Low Class Dirty Girl': Sex Work in Ho Chi Minh City, Vietnam." *Journal of Contemporary Ethnography* 40: 367–396.

Mahdavi, Pardis. 2011. *Gridlock: Labor, Migration, and Human Trafficking in Dubai*. Palo Alto: Stanford University Press.

Parreñas, Rhacel. 2011. *Illicit Flirtations*. Palo Alto: Stanford University Press.

West, Candace and Don Zimmerman. 1987. "Doing Gender." *Gender and Society* 1:125–151.

Zelizer, Viviana. 2005. *The Purchase of Intimacy*. Princeton: Princeton University Press.

Zheng, Tiantian. 2009. *Red lights: The lives of sex workers in postsocialist China*. Chicago: University of Chicago Press.

Introduction to Reading 26

This reading is a good example of the application of intersectional analysis, employing categories of gender, age, and social class. The authors studied a mass-marketed program of "successful aging" that targets old men in an effort to persuade them to spend their money on products and activities that will supposedly make them look and feel youthfully and heterosexually virile and successful. Toni Calasanti and Neal King analyze the ageism of "successful aging" consumer campaigns and their implications for old men's "physical health, unequal access to wealth, heterosexual dominance, and fears of impotence" (from abstract).

1. How does ageism permeate "successful aging" consumer campaigns?
2. Why is it important to examine age relations and their intersections with other inequalities?
3. Discuss the "dirty"/"impotent" double bind and its link to the rise of "successful aging" consumer programs.

Firming the Floppy Penis

Age, Class, and Gender Relations in the Lives of Old Men

Toni Calasanti and Neal King

The rise of a consumer market that targets old people and their desire to remain young brings into sharp relief the problems that old age poses to manhood. This article proposes an expansion of research approaches to the lives of old men so that they may enrich our understandings of masculinities at a time when scientific breakthroughs and high-priced regimens sell visions of manhood renewed. We begin with a brief review of the (relative lack of) research on old men, continue with a look at the mass marketing of "successful aging," and conclude with an overview of the potential rewards that sustained scholarship on the old, and a theorizing of age relations as a dimension of inequality, can offer the studies of men and masculinities.

(Young) Men's Studies

Studies of old men are common in the gerontological literature, but those that theorize masculinity remain

Calasanti, T., & King, N. (2005). Firming the floppy penis: Age, class, and gender relations in the lives of old men. *Men and Masculinities, 8*(1). Reprinted with permission of SAGE Publications, Inc.

rare. As in many academic endeavors, men's experiences have formed the basis for much research, but this androcentric foundation goes largely unexplored because manhood has served as invisible norm rather than as explicit focus of theory. Men's lives have formed the standard for scholarship on retirement, for example, to such an extent that even the *Retirement History Study,* a longitudinal study conducted by the Social Security Administration, excluded married women as primary respondents (Calasanti 1993). In recent years, feminist gerontologists have urged that scholars examine not only women but gender relations as well, and a handful of scholars such as Woodward (1999), Cruikshank (2003), and Davidson (2001) have done so. Despite the proliferation of feminist theorizing, however, most mainstream gerontological studies of women still ignore gender (Hooyman 1999), and research on men lags further. Few studies examine old men *as men* or attend to masculinity as a research topic.

At the same time, profeminist studies of masculinity have studied neither old men nor the age relations that subordinate them. Ageism, often inadvertent, permeates this research, stemming from failures to study the lives of old men, to base questions on old men's accounts of their lives, or to theorize age the way we have theorized relations of gender, race, and class. Mentions of age inequality arise as afterthoughts, usually at the ends of lists of oppressions, but they remain unexamined. As a result, our understanding and concepts of manhood fall short because they assume, as standards of normalcy, men of middle age or younger. Aging scholars' inattention to old *men,* combined with men's studies' lack of concern with *old* men, not only renders old men virtually invisible but also reproduces our own present and future oppression. This article examines a range of popular representations of old men in the context of research about their lives to outline some ways in which the vital work on men and masculinity might benefit by taking age relations into account as a form of inequality that intersects with gender, race, sexuality, and class.

Denial of Aging

Our ageism—both our exclusion of the old and our ignorance of age relations as an inequality affecting us all—surfaces not only in our choices of what (not) to study but also in how we theorize men and masculinities. Listening to the old and theorizing the inequality that subordinates them require that we begin with elementary observations. People treat signs of old age as stigma and avoid notice of them in both personal and professional lives. For instance, we often write or say "older" rather than "old," usually in our attempts to avoid negative labels. But rather than accept this stigma attached to the old and help people to pass as younger than that, we should ask what seems so wrong with that stage of life. In a more aggregate version of this ageism, one theorizes old age as social construction and then suggests that people do not automatically become old at a particular age. One continues to treat "old age" as demeaning and merely seeks to eradicate recognition of it by granting reprieves from inclusion in the group. As well intended as such a theoretical move may be, it exacts a high price. It maintains the stigma rather than examining or removing it. As Andrews (1999) observed, all life cycle stages are social constructions, but "there is not much serious discussion about eliminating infancy, adolescence, or adulthood from the developmental landscape. It is only old age which comes under the scalpel" (302). Emphasis on the socially constructed status of this age category does nothing to eliminate its real-world consequences.

Old age has material dimensions, the consequences of actors both social and biological: bodies *do* age, even if at variable rates, just as groups categorize and apportion resources accordingly. Emphasizing their subjective nature makes age categories no less real. Bodies matter; and the old are not, in fact, just like the middle-aged but only older. They are different, even though cultures and people within them define the differences in divergent ways. We need to consider the social construction of old age in conjunction with the aging of bodies (which, in a vexing irony, we understand only through social constructions).

Successful Aging

A more refined form of ageism attempts to portray old age in a positive light but retains the use of middle age as an implicit standard of goodness and health, in contrast to which the old remain deviant. One may see this ageism in the popular notion that men should "age successfully." From this "anti-aging" perspective, some of the changes that occur with age might seem acceptable—gray hair and even, on occasion, wrinkles—but other age-related changes do not, such as losses of libido, income, or mobility. Aging successfully requires that the old maintain the activities popular among the middle-aged. Successful aging, in effect, requires well-funded resistance to culturally designated markers of old age, including relaxation. Within this paradigm, those signs of seniority remain thoroughly stigmatized.

To be sure, a research focus on men who have aged "successfully" flows from good intentions. Study of successful agers helps us negate stereotypes of the old as "useless," unhappy, and the like. Nevertheless, a theory of the age relations underlying this movement must recognize their interrelations with class, sexual, and racial inequalities. The relevant standards for health and happy lifestyles have been based on leisure activities accessible only to the more well-to-do and middle-aged: tennis, traveling, sipping wine in front of sunsets, and strolls on the beaches of tony resorts that appear in the advertising campaigns for such lifestyles.

The dictate to age successfully by remaining active is both ageist and ignorant of the lives of the working classes. Spurred by the new anti-aging industry, the promotional images of the "active elder" are bound by gender, race, class, and sexuality. The sort of consumption and lifestyles implicated in ads for posh retirement communities with their depiction of "'imagineered' landscapes of consumptions marked by 'compulsively tidy lawns' and populated by 'tanned golfers'" (McHugh 2000, 110) assumes a sort of "active" lifestyle available only to a select group: men whose race and class make them most likely to be able to afford it, and their spouses.

Regimens of successful aging also encourage consumers to define any old person in terms of "what she or he is no longer: a mature productive adult" (McHugh 2000, 104). One strives to remain active to show that one is not really old. In this sense, successful aging means not aging and not being old because our constructions of old age contain no positive content. Signs of old age continue to operate as stigma, even in this currently popular model with its many academic adherents. The successful aging movement disapproves implicitly of much about the lives of the old, pressuring those whose bodies are changing to work hard to preserve their "youth" so that they will not be seen as old. As a result, the old and their bodies have become subject to a kind of disciplinary *activity.* This emphasis on productive activity means that those who are chronically impaired, or who prefer to be contemplative, become "problem" old people, far too comfortable just being "old" (Holstein 1999; Katz 2000).

This underlying bias concerning successful aging and "agelessness" is analogous to what many white feminists have had to learn about race relations, or indeed many men have had to learn about gender relations. Many whites began with the notion that nonwhites were doing fine as long as they acted like whites (just as women in many workplaces were deemed OK to the extent that they acted like men). That actual diversity would benefit our society was news to many, its recognition hard-won by activists of color who championed an awareness of the structuring effects of race relations. Only when we can acknowledge and validate these constructed differences do we join the fights against racism and sexism. The same is true of age relations and the old. We must see the old as legitimately different from the middle-aged, separated by a systematic inequality—built on some set of biological factors—that affects all of our lives. To theorize this complex and ever-changing construction is to understand age relations.

The experience of ageism itself varies by gender and other social inequalities (just as the experience of manhood varies by age and the like). Others have already pointed to the double standard of aging whereby women are seen to be old sooner than men (Calasanti and Slevin 2001). But the experience of ageism varies among different social hierarchies. Women with the appropriate class background, for instance, can afford to use various technologies to "hide" signs of aging bodies (such as gray hair and wrinkles) that will postpone their experiences of ageism. Some women of color, such as African Americans, accept more readily the superficial bodily signs of aging that might bother middle-class white women. Within their communities, signs of aging may confer a status not affirmed in the wider culture (Slevin and Wingrove 1998). By failing to reflect on our own ageism and its sources, we have left age relations and its intersections with such other inequalities unquestioned and misunderstood. We have given lip service to age relations by placing it on a list of oppressions, but we have only begun to theorize them. And so we have left unexplored one of the most important systems shaping manhood.

Examining age relations and its intersections with other inequalities will allow us to address ageism in its deepest form and address the structural inequities that deny power to subgroups of the old. It involves breaking the ethical hold that successful, active aging has on our views of aging. Just as feminists have argued for women's emancipation from stigmatizing pressure to avoid the paths that they might like to take, so too must the old be free to choose ways to be old that suit them without having to feel like slackards or sick people. Old age should include acceptance of inactivity as well as activity, contemplation as well as exertion, and sexual assertiveness as well as a well-earned break. Old people will have achieved greater equality with the young when they feel free not to try to be young, when they need not be "exceptional," and when they can be frail, or flabby, or have "age spots" without feeling ugly. *Old* will have positive content and not be defined mainly by disease, mortality, or the absence of economic value.

Old Men in Popular Culture

The study of masculinity benefits from a look at mass-produced images of old men, because they suggest much about the changing definitions of their problems and the solutions offered. Viewed in context of the experiences of diverse old men as well as the structural constraints on various groups, these popular images illustrate the pressures to be masculine and ways in which men respond to accomplish old manhood. On one hand, the goal of consumer images is to convince others to buy products that will help them better their lives. What is instructive about such images is what they reveal about how people—in this case, aging men—should go about improving their lives (i.e., what it is that they should strive for). On the other hand, images of powerful older men—such as CEOs and politicians—periodically appear in the news media, demonstrating what old men should be striving for in the consumer ads: money, power, and the like. We use mass-produced images of old men, then, to explore the ways that men and masculinities intersect with other systems of inequality—including age relations—to influence various experiences of manhood.

Current Images: New Manhood in Old Age

The recent demographic shift toward an aged population has inspired consumer marketers to address the old with promises of "positive" or successful aging. A massive ad campaign sells anti-aging—the belief that one should deny or defy the signs and even the fact of aging, and treat the looks and recreation of middle-aged as the appropriate standards for beauty, health, and all-around success. As Katz (2001–2002) recently put it, "The ideals of positive aging and anti-ageism have come to be used to promote a widespread anti-aging culture, one that translates their radical appeal into commercial capital" (27). These ads present a paradox for old men, whom ads depict as masculine but unable by virtue of infirmity and retirement to achieve the hegemonic ideals rooted in the lives of the young. Thus, old masculinity is always wanting, ever in need of strenuous affirmation. Even when blessed with the privileges of money and whiteness, old men lack two of hegemonic masculinity's fundamentals: hard-charging careers and robust physical strength. The most current ads promise successful aging with interesting implications for these forms of male privilege.

"Playing Hard"

The first image in this "new masculinity" shows men "playing hard," which differs from previous ads in important ways. It emphasizes activities modeled after the experiences of middle-aged, white, middle-class men. Men pursue leisure but not in terms of grandparenting, reading, or other familial and relaxing pastimes. Instead, they propel themselves into hard play as consumers of expensive sports and travel. Having maintained achievement orientations during their paid-work years, they now intensify their involvement in the expanding consumerist realm, trading production or administration for activity-based consumption. They compete not against other men for salaries and promotions but against their own and nature's incursions into their health as they defy old age to hobble them.

Katz (2001–2002) noted that many ads portray the "older person as an independent, healthy, flexi-retired 'citizen'; who bridges middle age and old age without suffering the time-related constraints of either. In this model . . . 'retirement is not old age'" (29). For instance, McHugh observed that the marketing of sunbelt retirement communities includes the admonition to seniors to busy themselves in the consumption of leisure, to "rush about as if their very lives depended upon it" (McHugh 2000, 112). Similarly, Aetna advertisements selling retirement financial planning show pictures of retired men in exotic places, engaging in such activities as surfing or communing with penguins. Captions offer such invitations as

> Who decided that at the age of 65 it was time to hit the brakes, start acting your age, and smile sweetly as the world spins by? . . . [W]hen you turn 65, the concept of retirement will be the only thing that's old and tired. (*Newsweek*, January 5, 1998, 9)

This active consumer image reinforces a construction of old age that benefits elite men in two ways. First, it favors the young in that the old men pictured do nothing that would entitle them to pay. Instead, they purchase expensive forms of leisure. Readers can infer that old men neither need money nor deserve it. Retired, their roles center around spending their money (implicitly transferring it to the younger generations who do need and deserve it). Such ads affirm younger men's right to a cushion from competition with senior men for salaried positions, power, and status. Second, this active consumer image favors the monied classes by avoiding any mention of old men's financial struggles or (varied) dependence on the state. Indeed, age relations work to heighten economic inequalities, such that the greatest differences in income and wealth appear among the old (Calasanti and Slevin 2001). This polarization of income and wealth creates a demographic situation in which only the most privileged men—white, middle-class or better, and physically similar to middle-aged men—can engage in the recreation marketed.

There, we see an additional benefit to the young of such images of men—the emphasis on the physical abilities that the young are more likely to have. Featherstone and Hepworth (1995) noted that the consumer images of "positive aging" found in publications for those of retirement age or planning retirement ultimately have "serious shortcomings" because they do not counter the ageist meanings that adhere to "other" images of the old, that is, "decay and dependency." In other words, we look more kindly on those old persons engaged in "an extended plateau of active middle age typified in the imagery of positive aging as a period of youthfulness and active consumer lifestyles" (46). In this sense, the new, "positive," and consumer-based view of the old is one steeped in middle-aged, middle-class views and resources. The wide variety of retirement and other magazines—and, more recently, a large and expanding number of Web sites—convey the idea that the body can be "serviced and repaired, and . . . cultivate the hope that the period of active life can be extended and controlled" through the use of a wide range of advertised products (44). This image does not recognize or impute value to those more often viewed to be physically dependent, for example. As a result, those men who are able to achieve this masculine version of "successful aging" appear acceptable within this paradigm, but this new form of acceptance does not mitigate the ways in which we view the old. It denies the physical realities of aging and is thus doomed to failure. Not only are the majority of old men left out of this image of new masculinity for old men, but also the depiction is in itself illusory and transitory. Note the gender inequality in these depictions of aging denied through consumption. Most women participate in the lifestyles of the well-to-do as parts of married couples, dependent on men. Old men may lose status relative to younger men but still maintain privilege in relation to old women.

However hollow such promises of expensive recreation might be for most men, the study of men's physical aggression and self-care suggests that illusions drive many indeed and that men will often sacrifice health and even their lives to accomplish this exaggerated sense of physical superiority to women and resistance to the forces of nature. Researchers of health, violence, and manhood have already documented the harms that men do to themselves. Whether disenfranchised men of color in neighborhoods of concentrated poverty (Franklin 1987; Lee 2000; Staples 1995), athletes desperate to perform as champions (Dworkin and Messner 1999; Klein 1995; White, Young, and McTeer 1995), or ordinary men expressing rage through violence (Harris 2000) and refusing to consult physicians when ill (Courtenay 2000), all manner of men undercut themselves and endanger their lives in the pursuit of their ideals. Harris (2000, 782), for instance, referred to the violence as part of the "doing" of manhood, in line with the sociological theory of gender as accomplishment (Fenstermaker and West 2002). Injury in the pursuit of masculinity extends to social networks, which men more often than women neglect to the point of near isolation and desolation (Courtenay 2000). For those not killed outright, the accumulated damage results in debilitating injury and chronic disease leading to depression (Charmaz 1995; McTeer, White and Young 1995), fatal heart disease (Helgeson 1995), and high rates of suicide born of lonely despair (Stack 2000). The effect of all of this on old manhood is tremendous, with men experiencing higher death rates than women at every age except after age ninety-five (Federal Interagency Forum on Aging-Related Statistics 2000), at which point few men remain alive.

More important to this discussion, however, than the results of such self-abuse on old age are the effects of age relations on this doing of manhood. To be sure, criminal combat and bone-crunching sports decline with age (much earlier in life, actually) such that old men commit few assaults and play little rugby. The increasing fragility of their bodies leads to relatively sedate lifestyles. Nevertheless, the recent anti-aging boom sells the implicit notion that relaxation equals death or at least defeat and that, once he retires, only high-priced recreation keeps a man a man. Age and gender ideals to which any man can be held accountable shift from careerism to consumption, from sport to milder recreation, but maintain notions of performance all the while.

The theoretical gain here lies in recognizing the historical (and very recent) shift to old manhood as a social problem solved through the consumption of market goods. Men throughout history and across the globe appear always to feel defensive about manhood, in danger of losing or being stripped of it (Solomon-Godeau 1995). This theme takes different forms in different periods, however, and in our own appears as the notion that old men lose their hardness if they relax but can buy it back from leisure companies and medical experts.

"Staying Hard"

Given the importance of heterosexuality to hegemonic masculinity, we should consider the ways in which age and gender interact with sexuality, so often equated for men with "the erect phall[us]" (Marsiglio and Greer 1994, 126). Although graceful acceptance by men of their declining sexual desire had previously served as a hallmark of proper aging (Marshall and Katz 2002), current depictions of old men's masculinity focus on virility as expressed in a (hetero)sexuality enabled by medical products. "Staying hard" goes hand-in-hand with playing hard in the construction of age-appropriate gender ideals in this consumer economy.

Examples of the link among continued sexual functioning, manhood, and resistance to aging, in a context of individual responsibility and control, appear throughout the anti-aging industry, which has been growing as a part of our popular culture through the proliferation of Web sites, direct-mail brochures, journal and magazine advertisements, blurbs in academic newsletters, appearances on talk shows and infomercials, self-help paperbacks, and pricey seminars designed to empower the weakening old. For instance, a few passages from *Newsweek* (Cowley 1996) on the movement toward the use of human growth hormone (HGH) and testosterone draw connections among virility, aging, masculinity, individual control, and consumerism.

> Five years ago, on the eve of his 50th birthday, Ron Fortner realized that time was catching up with him. . . . His belly was soft, his energy and libido were lagging and his coronary arteries were ominously clogged up. After his advancing heart disease forced him into a quintuple bypass operation, Fortner decided he wasn't ready to get old. He . . . embarked on a hormone-based regimen designed to restore his youthful vigor. . . . [H]e started injecting himself with human growth hormone. . . . He claims the results were "almost instantaneous." First came a general sense of wellbeing. Then within weeks, his skin grew more supple, his hair more lustrous and his upper body leaner and more chiseled. . . . Awash in all these juices, he says he discovered new reserves of patience and energy, and became a sexual iron man. "My wife would like a word with you," he kids his guru during on-air interviews, "and that word is stop." (Cowley 1996, 68, 70)

Significantly, a yearlong supply of HGH in 1996 ran between $10,000 and $15,000, making it most accessible to elite men.

Another "success story" from the article concerns

> Robert, a 56-year-old consultant who wore a scrotal patch [for testosterone] for two and a half years. . . . Since raising his testosterone level from the bottom to the top of the normal range, Robert has seen his beard thicken, his body odor worsen and his libido explode. "Whether it's mental or physical, you start feeling older when you can't do physical things like you could," he says. "Sexually, I'm more comfortable because I know I'm dependable." His only complaint is that he's always covered with little rings of glue that won't come off without a heavy-duty astringent. (Cowley 1996, 71–72)

Finally, the story concludes by noting that

> as the population of aging males grows, the virility preservation movement is sure to grow with it. "Basically, it's a marketing issue," says epidemiologist John McKinley, director of the New England Research Institute. . . . " The pharmaceutical industry is going to ride this curve all the way to the bank." (Cowley 1996, 75)

Scientific discourse and practice equate, especially for men, sex with "not aging," and propose technology to retain and restore sexual "functionality" (Katz and Marshall 2003). Indeed, as anti-aging guru Dr. Karlis Ullis, author of *Age Right* and *Super T* (for testosterone), proclaims on his Web site, "Good, ethical sex is the best anti-aging medicine we have" (2003). The appearance of such chemical interventions as sildenafil (Viagra) and the widespread advertising campaigns to promote them have also helped to reconstruct old manhood. A recent ad shows an old, white, finely dressed couple dancing a tango, with the man above and the woman leaning back over his leg. The strenuous dance combines with the caption to convey his virility: "Viagra: Let the dance begin" (*Good Housekeeping* April 1999, 79). Here is a man who likes to be on top and has the (newly enhanced) strength to prove it. Still another ad affirms the role of phallic sex in marital bliss. The bold letters next to a black man with visibly graying hair state, "With Viagra, she and I have a lot of catching up to do." And, at the bottom: "Love life again" (*Black Enterprise,* March 2000, 24–25).

Such ideals of virility appear in age-defying ads for active leisure—such as one for Martex towels, which features the caption, "Never, ever throw in the towel." Below this line, three old men stand, towels around their waists, in front of three surfboards that stand erect, stuck in the beach sand. Beneath, one reads that the towels are "for body and soul" (*Oprah*, April 2001, 118). An Aetna financial planning ad shows an old white man paddling in the surf, his erect board standing upward between his legs. The caption reads, "A Rocking Chair Is a Piece of Furniture. Not a State of Mind" (*Newsweek*, October 27, 1997, 15). In the ideal world of these ads, age is a state of mind, one to be conquered through public displays of a phallic, physical prowess. One accomplishes old manhood, then, by at least appearing to try to live up to some of the ideals pictured in these magazines. The resulting widespread doing of old manhood as consumption of the right products and maintenance of the right activities serves in turn to render natural the ideals toward which men strive.

Masculinity and sexual functioning have long been linked to aging in our popular culture, but the nature of this relationship has shifted as age relations have transformed and come under medical authority. Contemporary drug marketers build on an ancient quest but market it in new ways.

By the 1960s, therapists blamed psychological factors for male impotence and suggested that "to cease having sex would hasten aging itself" (Katz and Marshall 2003, 7). They later redefined male impotence as a physiological event—"erectile dysfunction"—to be addressed through such technologies as penile injections and sildenafil (Viagra)—and declared intercourse vital to successful aging (Marshall and Katz 2002; Potts 2000). More recently, advertisers have catered to a popular notion of "male menopause"—an umbrella label for the consequences of the fears of loss that expectations of high performance, in the context of women's rising status, can engender (Featherstone and Hepworth 1985). Marketers have built their depictions of old manhood on these links among sex, success, and masculinity. Sexual functioning now serves as a vehicle for reconstructions of manhood as "ageless," symbolizing the continued physical vigor and attractiveness derived from the experiences of younger men. To the extent that men can demonstrate their virility, they can still be men and stave off old age and the loss of status that accrues to that label.

To be sure, this shift in advertising imagery toward the phallic can work to the benefit of old men, convincing people to take them seriously as men full of potency as well as consumer power. To stop our analysis there, however, leaves unquestioned the ageism on which these assertions rest, the fact that we root these ideals of activity and virility in the experiences of the younger men. The ads avoid sexuality based on attributes other than hard penises and experiences other than heterosexual intercourse, and these are hegemonic sexual symbols of the young. The little research available suggests that orgasm and intercourse recede in importance for some old men, who turn to oral sex and other expressions of love (Wiley and Bortz 1996). But these phallic ads value men only to the extent that they act like younger, heterosexual (and wealthy) men. Their emphases on both playing and staying hard reveal some of the ways in which gender and other inequalities shape old age. Old men are disadvantaged in relation to younger men, no matter how elite they may be.

The renewed emphasis on sexual intercourse among old men also reinforces the gender inequalities embedded in phallic depictions of bodies and sexuality. Historically, women's bodies and sexualities have been of only peripheral interest in part because they did not fit the "scientific" models based on men's physiologies. For example, rejuvenators were uncomfortable touting sex gland surgery for women (one variation promoted grafting the ovaries of chimpanzees to those of female patients) partly because they knew that they could not restore fertility in women. Thus, when they did speak of women, they tended to focus instead on the "mental" fertility that might result. Part of the problem was that women's "losses" in terms of sexuality (i.e., menopause) occurred much earlier in life. Those women were often "young," which confounded the equation of "loss of sexuality" with "old" (Hirshbein 2000).

People continue to define old women's sexuality in relation to old men's, assessing it in terms of penile-vaginal penetration. An old woman, in such popular imagery, remains passive and dependent on her man's continued erection for any pleasure of her own. Research on old women's accounts of their experiences, however, makes clear that these models represent little of what they want from their sex lives. These popular definitions also ignore that many old women have no partners at all. Even if old women "accept" and try to live up to the burden of being sexual and "not old" in male-defined terms, there are not enough old men for them to be partnered (and our age-based norms do not allow them to date younger men).

Finally, the ageism implicit in the demand to emulate the young is self-defeating and ignores the reality that even with technology and unlimited resources, bodies still change. Ultimately, individuals cannot control this; it is a "battle" one cannot win.

Theorizing Age, Class, and Gender Relations

The rewards for the inclusion of a marginalized group into research extend beyond the satisfaction of listening to oft-ignored voices. The study of old manhood stands to enrich our theories of masculinity as social problem, as disciplinary consumer object, as the accomplishment of heterosexuality, and as the "crisis"-torn struggle to achieve or resist the hegemonic ideals spread through our popular culture.

Studying age relations can render insights into ways that we theorize gender. For instance, Judith Kegan Gardiner (2002) suggested that we clarify gender relations by making an analogy to age relations. This would help reconstruct thinking about gender in our popular culture, she argued, because many people already recognize *continuity* in age categories while they still see gender as dichotomous. People already see themselves as *performing* age-appropriate behavior ("acting their ages") while continuing to take for granted the doing of gender (Fenstermaker and West 2002). And popular culture more fully recognizes enduring group *conflicts* (over divisions of resources) between generations than between sexes. Gardiner (2002) suggested that a fuller theorizing of age relations has much to offer the study of men, that scholars may move beyond their polarization of biological and social construction, and that our popular culture may more fully appreciate the power struggles that govern gender relations.

We recommend just this view—of age and gender, race and class, and other dimensions of inequality—as accomplished by social as well as biological actors; as accountable to ever-changing ideals of age- and sex-appropriate behavior; as constructed in the context of a popular culture shaped by consumer marketing and technological change; and as imposing disciplinary regimens in the names of good health, empowerment, beauty, and success.

Taken together, the mass media reviewed above posit ideals of old manhood to which most if not all men find themselves held accountable. To the men fortunate enough to have been wealthy or well paid for their careerism, corporations (often with the support of those gerontologists who implicitly treat old age as a social problem) sell regimens through which those old men may live full lives, working, playing, and staying hard. If careerism kept the attention of these men from their families and leisure lives, constricting their social networks and degrading their physical health, then this high-priced old age serves as a promised payoff. Once retired, those few wealthy enough to do it can enjoy a reward: high-energy time with a spouse and some friends, enjoyment of tourism, surfing, and sex. Men sacrificed much, even their lives, in their pursuits of hegemonic masculine status. Those who survive face a rougher time with old age as a result: few sources of social support and bodies weakened by self-abuse. Thus, the accomplishment of manhood comes to require some response to the invitation to strain toward middle-age activities. Some men reach with all of their strength for the lifestyle ideals broadcast so loudly, whereas many give up for lack of means to compete, and still others deliberately resist. In a cruel irony, the ideals move all the further out of reach of the men who pursued them with such costly vigor in younger years and damaged their health beyond repair. The final push for hegemonic masculinity involves spending money and enjoying health that many old men do not have to pursue the recreation and phallic sex that the ads tell them they need.

Certainly, the study of old men offers striking views of a popular struggle over heterosexuality (although the study of old gay men will surely be as transforming, the near total lack of research on them prevents us from speculating how). Widely held views of old men's sexuality suggest dominance over women as a form of virility. But, as bodies change, outright predation recedes as an issue and impotency moves to the center of concern. A popular (consumer) culture that figures old manhood in terms of *loss* hardly departs from any trend in images of masculinity. Men have always felt that they were losing their manhood, their pride, and their virility, whether because their penises actually softened or because women gained status and so frightened them. But the study of this transition—from the feelings of invincibility that drive the destructiveness of youth to the growing expectation of vulnerability—throws old masculinity into a valuable relief. For instance, theories that center on violence and predation capture little of the realities of old men's lives, just as scholarly emphasis on coercion and harassment of women excludes most of the experiences of old women. For old women, the more important sexual theme may be that of being *cast aside* (Calasanti and Slevin 2001, 195). For old men, *impotence* in its most general sense, leading to many responses ranging from suicidal depression to more graceful acceptance, may be a more productive theme. It serves as both positive and negative ideal in a classic double-bind: old men should, so as not to intrude on the rights of younger men, retreat from the paid labor market; but they should also, so as to age successfully, never stop consuming opportunities to be active. They should, so as not to be "dirty," stop becoming erect; but they should also, so as to age successfully, never lose that erection. Old men fear impotence to the point that many suffer it who otherwise would not. Anxieties drain them at just the moment when expectations of aggressive consumption, of proving themselves younger than they are, reach their heights.

The notion that men accomplish age just as they do gender has much to offer, with its sensitivity to relations of inequality, its moment-to-moment accountability to unreachable but hegemonic ideals, and the perpetually changing nature of such accomplishments. Never have erections been so easily discussed in public, and never has this "dirty"/"impotent" double bind been tighter, than since the rise of this consumer regimen. Nor have old men, before now, lived under such pressure to remain active further into their lengthening life spans. The ideals of manhood that tempted so many to cripple themselves in younger years now loom large enough to shame those who cannot play tennis or waltz the ballrooms of fancy resorts. The study of manhood should take careful notice of the ways in which men do old manhood under such tight constraints. The popular images that we have reviewed provide ideals of old manhood, but they do not necessarily describe the lives of very many old men. Given how little we know of the ways in which old men respond to such ideals, the research task before us seems clear.

Conclusion

Scholars tend to ignore age relations in part because of our own ageism. Most are not yet old, and even if we are, we often deny it (Minichiello, Browne, and Kendig 2000). Most people know little about the old because we seldom talk to them. Family and occupational segregation by age leave the old outside the purview of the work that most young people do.

Resulting in part from such segregation, the study of men, although no more than any other social science and humanist scholarship, has focused on the work, problems, sexuality, and consumption patterns of the young. This neglect of the old results in theories of masculinity that underplay the lengths to which men go to play and stay hard, the long-term effects of their strenuous accomplishment of manhood, and the variety of ways in which men remain masculine once their appetites for self-destruction begin to wane. Research on the old can reveal much about the desperate struggle for hegemonic masculinity and the varied ways in which men begin to redefine manhood. At the same time, it also uncovers the young and middle-aged biases that inhere in typical notions of masculinity that tend to center on accomplishments and power in the productive sphere, for instance. Few researchers have considered the reality of masculinities not directly tied to the fact of or potential for paid labor.

To leave age relations unexplored reinforces the inequality that subordinates the old, an inequality that we unwittingly reproduce for ourselves. Unlike other forms of oppression, in which the privileged rarely become the oppressed, we will all face ageism if we live long enough. As feminists, scientists, and people growing old, we can better develop our sense of interlocking inequalities and the ways in which they shape us, young and old. Our theories and concepts have too often assumed rather than theorized these age relations. The study of men and masculinity and the scholarship on age relations are just beginning to inform each other.

References

Andrews, M. 1999. The seductiveness of agelessness. *Ageing and Society 19*(3): 301–18.

Calasanti, T. M. 1993. Bringing in diversity: Toward an inclusive theory of retirement. *Journal of Aging Studies 7*(2): 133–50.

Calasanti, T. M., and K. F. Slevin. 2001. *Gender, social inequalities, and aging.* Walnut Creek, CA: Alta Mira.

Charmaz, K. 1995. Identity dilemmas of chronically ill men. In *Men's health and illness: Gender, power, and the body,* edited by D. Sabo and D. F. Gordon, 266–91. Thousand Oaks, CA: Sage.

Courtenay, W. H. 2000. Behavioral factors associated with disease, injury, and death among men: Evidence and implications for prevention. *Journal of Men's Studies 9*(1): 81–142.

Cowley, Geoffrey. 1996. Attention: Aging men. *Newsweek,* September 16, 68–75.

Cruikshank, M. 2003. *Learning to be old: Gender, culture, and aging.* Lanham, MD: Rowman & Littlefield.

Davidson, K. 2001. Later life widowhood, selfishness, and new partnership choices: A gendered perspective. *Ageing and Society* 21: 297–317.

Dworkin, S. L., and M. A. Messner. 1999. Just do . . . what? Sport, bodies, gender. In *Revisioning gender,* edited by M. M. Ferree, J. Lorber, and B. B. Hess, 341–61. Thousand Oaks. CA: Sage.

Featherstone, M., and M. Hepworth. 1985. The male menopause: Lifestyle and sexuality. *Maturitas* 7: 235–46.

———. 1995. Images of positive aging: A case study of *Retirement Choice* magazine. In *Images of aging: Cultural representations of later life*, edited by M. Featherstone and A. Wernick, 29–47. London: Routledge.

Federal Interagency Forum on Aging-Related Statistics. 2000. *Older Americans 2000: Key indicators of well-being.* http://www.agingstats.gov.

Fenstermaker, S., and C. West, eds. 2002. *Doing gender, doing difference: Inequality, power, and institutional change*. New York: Routledge.

Franklin, C. 1987. Surviving the institutional decimation of black males: Causes, consequences, and intervention. In *The making of masculinities: The new men's studies,* edited by H. Brod, 155–69. Winchester, MA: Allen and Unwin.

Gardiner, J. K. 2002. Theorizing age and gender: Bly's boys, feminism, and maturity masculinity. In *Masculinity studies & feminist theory: New directions,* edited by J. K. Gardiner, 90–118. New York: University of Columbia Press.

Harris, A. P. 2000. Gender, violence, and criminal justice. *Stanford Law Review* 52: 777–807.

Helgeson, V. S. 1995. Masculinity, men's roles, and coronary heart disease. In *Men's health and illness: Gender, power, and the body,* edited by D. Sabo and D. F. Gordon, 68–104. Thousand Oaks, CA: Sage.

Hirshbein, L. D. 2000. The glandular solution: Sex, masculinity and aging in the 1920s. *Journal of the History of Sexuality 9*(3): 27–304.

Holstein, Martha. 1999. Women and productive aging: Troubling implications. In *Critical gerontology: Perspectives from political and moral economy,* edited by Meredith Minkler and Carroll L. Estes, 359–73. Amityville, NY: Baywood.

Hooyman, N. R. 1999. Research on older women: Where is feminism? *The Gerontologist* 39: 115–18.

Katz, S. 2000. Busy bodies: Activity, aging, and the management of everyday life. *Journal of Aging Studies 14*(2): 135–52.

———. 2001–2002. Growing older without aging? Positive aging, anti-ageism, and anti-aging. *Generations 25*(4): 27–32.

Katz, S., and B. Marshall. 2003. New sex for old: Lifestyle, consumerism, and the ethics of aging well. *Journal of Aging Studies 17*(1): 3–16.

Klein, A. M. 1995. Life's too short to die small: Steroid use among male bodybuilders. In *Men's health and illness: Gender, power, and the body,* edited by D. Sabo and D. F. Gordon, 105–21. Thousand Oaks, CA: Sage.

Lee, M. R. 2000. Concentrated poverty, race, and homicide. *Sociological Quarterly 41*(2): 189–206.

Marshall, B., and S. Katz. 2002. Forever functional: Sexual fitness and the ageing male body. *Body & Society 8*(4): 43–70.

Marsiglio, William, and Richard A. Greer. 1994. A gender analysis of older men's sexuality. In *Older men's lives,* edited by Edward H. Thompson, Jr., 122–40. Thousand Oaks, CA: Sage.

McHugh, K. 2000. The "ageless self"? Emplacement of identities in sun belt retirement communities. *Journal of Aging Studies 14*(1): 103–15.

Minichiello, V., J. Browne, and H. Kendig. 2000. Perceptions and consequences of ageism: Views of older people. *Ageing and Society 20*(3): 253–78.

Potts, A. 2000. The essence of the "hard on": Hegemonic masculinity and the cultural construction of "erectile dysfunction." *Men and Masculinities 3*(1): 85–103.

Slevin, K. F., and C. R. Wingrove. 1998. *From stumbling blocks to stepping stones: The life experiences of fifty professional African American women*. New York: New York University Press.

Solomon-Godeau, A. 1995. Male trouble. In *Constructing masculinities,* edited by M. Berger, B. Wallis, and S. Watson, 69–76. New York: Routledge.

Stack, S. 2000. Suicide: A 15-year review of the sociological literature. *Suicide & Life-Threatening Behavior 30*(2): 145–76.

Staples, R. 1995. Health among Afro-American males. In *Men's health and illness: Gender, power, and the body,* edited by D. Sabo and D. F. Gordon, 121–38. Thousand Oaks, CA: Sage.

Ullis, Karlis. 2003. *Agingprevent.com.* http://www.agingprevent.com/flash/index.html.

White, P. G., K. Young, and W. G. McTeer. 1995. Sport, masculinity, and the injured body. In *Men's health and illness: Gender, power, and the body*, edited by D. Sabo and D. F. Gordon, 158–82. Thousand Oaks, CA: Sage.

Wiley, D., and W. M. Bortz. 1996. Sexuality and aging—usual and successful. *Journal of Gerontology 51A*(3): M142–M146.

Woodward, K., ed. 1999. *Figuring age: Women, bodies, generations.* Bloomington: Indiana University Press.

Topics for Further Examination

- Find articles and websites that offer critiques of gender stereotypes in the mass media, popular culture, and consumer culture. For intriguing insights into sexism in "art," visit www.guerrillagirls.com.

- Google automobile ads. Compare and contrast the gendered "marketing" strategies within and across manufacturers. Select other products and compare, contrast, and critique.

- Check out song lyrics by artists who criticize hegemonic masculinity and emphasized femininity. For example, read Pink's lyrics for the tune titled "Stupid Girls."

6

Tracing Gender's Mark on Bodies, Sexualities, and Emotions

Catherine G. Valentine

This chapter explores the ways gender patterns are woven into three of the most intimate aspects of the self: body, sexuality, and emotion. The readings we have selected make the general sociological argument that there is no body, sexuality, or emotional experience independent of culture. That is, all cultures sculpt bodies, shape sexualities, and produce emotions. One of the most powerful ways a gendered society creates and maintains gender difference and inequality is through its "direct grip" on these intimate domains of our lives (Schiebinger, 2000, p. 2). Gender ideals and norms require work to be done in, and on, the body to make it appropriately feminine or masculine. The same ideals and norms regulate sexual desire and expression, and require different emotional skills and behaviors of women and men.

At first glance, it may seem odd to think about body, sexuality, and emotion as cultural and gendered products. But consider the following questions: Do you diet, lift weights, dehair your legs or face, or use makeup? In public places, do you feel more comfortable sitting with legs splayed or legs crossed at the ankles? Are you conscious of how you feel and move your body through city streets when you are alone at night? How you answer these questions offers insight into the types of gendered body work you do. Now think about the gendering of sexuality in the United States. The mark of gender on sexual desire and expression is clear and deep, tied to gendered body ideals and norms. Whose breasts are eroticized and why? Are women who have many sexual partners viewed in the same way as men who have many sexual partners? Why are many heterosexually identified men afraid of being perceived as homosexual? Why do women shoulder the major responsibility for contraceptive control? Like sexuality, emotions are embodied modes of being. And like sexuality and body, emotions are socially regulated and constructed. They are also deeply gendered.

Consider these questions: Do you associate emotionality with women or men? Is an angry woman taken as seriously as an angry man? Why do we expect women's body language (e.g., their touch and other gestures) to be more affectionate and gentle than men's? What is your reaction to this word pair—tough woman/soft man? The readings in this chapter explore the complex and contradictory ways bodies, sexualities, and emotions are brought into line with society's gender scheme. Two themes unite the readings. First, they demonstrate how the marking of bodily appearance, sexual desire and behavior, and emotional expression as masculine or feminine reinforces Western, U.S. culture's insistence on an oppositional gender binary pattern. Second, the readings show how patterns of gender inequality become etched into bodies, sexualities, and emotions.

Gendered Bodies

All societies require body work of their members (Black & Sharma, 2004; Lorber & Moore, 2007). But not all societies insist on the molding of men's and women's bodies into visibly oppositional and asymmetrical types—for example, strong male bodies and fragile female bodies. Only societies constructed around the belief and practice of gender dualism and hierarchy require the enactment of gender inequality in body work. To illustrate, consider the fact that the height ideal and norm for heterosexual couples in the United States consists of a man who is taller and more robust than his mate (Gieske, 2000). This is not a universal cultural imperative. As Sabine Gieske states, the tall man/short woman pattern was unimportant in 18th-century Europe. In fact, the ideal was created in the Victorian era under the influence of physicians and educators who defined men as naturally bigger and stronger than women. The expectation that men be taller than their female partners persists today, even though the average height gap between men and women is closing (Schiebinger, 2000). The differential height norm is so strong that many contemporary Americans react to the pairing of a short man and a tall woman with shock and even disapproval (Schiebinger, 2000).

How are bodies made feminine and masculine as defined by U.S. culture? It takes work, and lots of it. Well into the 20th century, American gender ideology led men and women alike to perceive women as frail in body and mind. Boys and men were strongly exhorted to develop size, muscle power, physical skills, and the "courage" to beat each other on the playing field and the battlefield, while girls and women were deeply socialized into a world of distorted body image, dangerous dieting, and physical incompetence (Dowling, 2000). Throwing, running, and hitting "like a girl" was a common cultural theme that we now understand to be a consequence of the cultural taboo against girls developing athletic stature and skills. As Collette Dowling (2000) notes, "There is no inherent biological reason for girls not to throw as far, as fast, or as hard as boys do" (p. 64). But there is a cultural reason: the embodiment of the belief in gender difference and inequality. We literally translate the "man is strong, woman is weak" dictum into our bodies. This dictum is so powerful that many people will practice distressed and unhealthy body routines and regimens to try to emulate the images of perfect male and female bodies. The mere threat of seeming masculine or mannish has kept a lot of girls and women from developing their own strength, and the specter of seeming effeminate or sissy has propelled many boys and men into worlds in which their bodies are both weapons and targets of violence (Dowling, 2000; Messner, 1990).

The power of masculinity requirements in the United States to compel some men to risk serious injury and even death has crystallized around the revelation that football players, at every level of the sport, regularly suffer brain injuries from repeat concussions and sub-concussive trauma (amanda, 2012). Research on football and other violent sports (e.g., rugby) shows that men and boys police one another's gender performance on the playing field, enforcing a "pain principle" that is perverse (Sabo, 1994; Vaccaro, Schrock, & McCabe, 2011). But perhaps there is no better example of an institution that molds men's bodies into weapons and targets than the military. The chapter reading by Orna Sasson-Levy examines masculinity requirements in the Israeli military and focuses on the harsh, even violent, individualized bodily and emotional practices that transform Israeli men into combat soldiers who will serve the interests of the state, as well as reinforcing a strong link between hegemonic masculinity and Israeli militarism. The practices examined by Sasson-Levy are typical of militaries globally, including the United States.

However, men's bodies in the United States and similar nations are not held to equally severe ideals and expectations of body work as are women's bodies. Gender inequality is etched into women's flesh in more debilitating ways.

American cultural definitions of femininity equate attractiveness with a white, youthful, slim, fit body that, in its most ideal form, has no visible "flaws"—no hair, pores, discolorations, perspiration marks, body odors, or trace of real bodily functions. In fact, for many women, the body they inhabit must be constantly monitored and managed—dehaired, deodorized, denied food—so that it doesn't offend. At the extreme, girls and women turn their own bodies into fetish objects to which they devote extraordinary amounts of time and money. What are the models of feminine bodily perfection against which girls and women measure themselves and are evaluated by others? The images and representations are all around us in magazine ads, TV commercials and programs, music videos, and toy stores.

Although much media and scholarly attention has been paid to topics such as anorexia and "fear of fat" among thin women, Janna Fikkan and Esther Rothblum (2011) bring together wide-ranging data on what may be a far more troubling issue, "the disproportionate degree of bias experienced by fat women" in the United States (p. 576). Fikkan and Rothblum use the term *fat* because it is descriptive while, as they point out, the term *overweight* implies "unfavorable comparison" to a socially constructed standard and, similarly, the term *obese* is problematic, created by medical practitioners through a biased medical–economic lens (p. 576; see also Kwan, 2010, p. 147). They emphasize two aspects of the oppression of women by impossible ideals of thinness: (1) "The ever thinner cultural ideal means that practically every woman will feel badly about her body," and (2) "because of the pervasiveness and gendered nature of weight-based stigma, a majority of women stand to *suffer significant discrimination*" because they do not and cannot conform to a near-impossible bodily ideal intended to enforce gender inequality (Fikkan & Rothblum, 2011, p. 590). "Weight bias" against women has far-reaching and significant harmful consequences in almost every aspect of women's lives—including, for example, discrimination in hiring, wages, and promotion in the workplace; barriers to good health care; and degrading representation in the mass media (pp. 578–585).

In sharp contrast, the evidence is clear that men pay few if any penalties in any realm of life (e.g., employment and romantic relationships) for being robust and large-bodied (Fikkan & Rothblum, 2011). Samantha Kwan's (2010) study of fat women and men reinforces the findings of Fikkan and Rothblum. As she states, the men in her research sample experienced "very little body consciousness" and rarely spoke of concerns about body management (Kwan, 2010, p. 153). For example, they viewed dining out in pleasurable terms and had no qualms about eating with gusto. Just think of ads for man-sized meals and foods and consider the fact that a comparable "feminine" concept (e.g., woman-sized meal) is either unimaginable or would have the opposite meaning.

Kwan's (2010) study provides us with a useful framework for understanding what scholars such as Joan Chrisler (2011) argue is the threat, even hatred, of women's bodies in patriarchies. Kwan theorizes that a "Western cultural body-hierarchy" has created what she calls body privilege, "an invisible package of unearned assets that thin or normal-sized individuals can take for granted" in everyday life (p. 147). This body-hierarchy is complicated and mediated by prisms such as race and sexual orientation. It appears from as-yet-limited research that lesbians and African American women are more likely to experience greater body satisfaction and protection (via family and community support) from the oppressive ideal of thinness compared with White, heterosexual women (Fikkan & Rothblum, 2011; Kwan, 2010).

The embodiment of women's subordination in gender-stratified societies takes some extreme forms. For example, in highly restrictive patriarchies, women's bodies may be systematically deformed, decimated, and restricted. Foot binding in traditional China is one of the most dramatic examples of an intentionally crippling gender practice. Although foot binding may seem to have no parallels in Western societies, sociologist Fatema Mernissi (2002) challenges the ethnocentric tendency to dismiss practices such as foot binding in China—and the veiling of women in some nations today—as alien or primitive. She does so by revealing the symbolic violence hammered directly on the Western female body by fashion codes and cosmetic industries that place Western women in a state of constant anxiety and insecurity. Mernissi's work suggests the usefulness of comparing practices such as foot binding in China to cosmetic surgery, extreme dieting, and body sculpting among contemporary Western girls and women. Consider the following questions: What do these seemingly different forms of body work have in common? Do they serve similar functions? How do they replicate gender inequality?

The story of the gendering of bodies in the United States and other countries is not only about oppositional and asymmetrical masculinity and femininity. Prisms of difference and inequality come into play. Social class, race, and gender intersect in Erynn Masi de Casanova's chapter reading. She looks at the ways in which women domestic workers felt degraded as "classed, gendered, and sometimes racialized bodies" in her study of embodied dimensions of domestic work in Ecuador's largest city (p. 309). Casanova shows how unequal relations between employers and workers are played out through the degradation of paid domestic workers' bodies. If we focus on the intersections of race and gender in the United States, specifically, the power of appearance norms to engage women's bodies is striking. Consider the following: What's the answer to the question, "Mirror, mirror on the wall, who's the fairest of them all?" (Gillespie, 1998)? You know what it is. The beauty standard is White, blonde, and blue-eyed. It is not Asian American, African American, Latin American, or Native American. In other words, there is a hierarchy of physical attractiveness, and the Marilyn Monroes and Madonnas of the world are at the top. Yes, there are African and Asian American models and celebrities.

But they almost always conform to White appearance norms. When they don't, they tend to be exoticized as the "Other." Just consider the fact that eyelid surgery, nasal implants, and nasal tip refinement procedures are the most common cosmetic surgery procedures undergone by Asian American patients, largely women (Kaw, 1998). The facial features that Asian American women seek to alter, including small, narrow eyes and a flat nose, are those that define them as racially different from White norms (Kaw, 1998). Racial and gender ideologies come together to reinforce an ethnocentric and racist beauty standard that devalues the "given" features of racial and ethnic minority women. The medical system has cashed in on it by promoting a beauty standard that requires the surgical alteration of features that don't fit the ideal.

Gendered Sexualities

Like the body, sexuality is shaped by culture. The sexualization process in a gendered society such as the United States is tightly bound to cultural ideas of masculinity and femininity. In dominant Western culture, a real man and a real woman are assumed to be opposite human types, as expressed in the notion of the "opposite sex." In addition, both are assumed to be heterosexual as captured in the notion that "opposites attract." Conformity to this gendered sexual dichotomy is strictly enforced.

The term *compulsory heterosexuality* refers to the dominance of heterosexual values and the fact that the meanings and practices of nonhetero(sexuality) as well as hetero(sexuality) are shaped by the dominant heterosexual script. For example, "real sex" is generally conceived of as penile–vaginal intercourse. This "coital imperative" limits the control individuals have in determining what counts as sexual activity and frames women's sexuality as passive/receptive and men's as active/penetrative (Gavey, McPhillips, & Doherty, 2001; Katz & Tirone, 2009). That is, the coital imperative defines sex as something men "do" to women privileging men's sexual needs above women's, limiting possibilities for men and women to have truly reciprocal, safe sexual relations. Importantly, the Western obsession with the homosexual/heterosexual identity distinction is relatively new. Created in the 19th century, it became a mechanism by which masculinity and femininity could be further polarized and policed (Connell, 1999). Gay masculinity and lesbian femininity came to be defined as abnormal and threatening to the "natural gender order." Consequently, in the United States, the fear of being thought of as homosexual has had a powerful impact on presentation of self. Men and boys routinely police each other's behavior and mete out punishment for any suggestion of "effeminacy" (Connell, 1999; Pascoe, 2005; Plante, 2006), while women and girls who engage in masculinized activities such as military service and elite sports risk being labeled lesbians.

Contemporary Western gender and sexuality beliefs have spawned stereotypes of lesbians as "manly" women and gay men as "effeminate." The reality is otherwise. For example, research shows that many gay men and lesbians are gender conformists in their expression of sexuality (Kimmel, 2000). Gay men's sexual behavior patterns tend to be masculine—oriented toward pleasure, orgasm, experimentation, and many partners—while lesbian sexuality is feminine in a Western sense, emphasizing sexual intimacy within romantic relationships. However, this account of nonheterosexuality is incomplete. Recent research suggests that compared with heterosexuals, nonheterosexuals have more opportunities to reflect on and experiment with ways of being sexual. Although the meanings attached to sexuality in society at large shape their erotic encounters, they are also freer to challenge the dominant sexual scripts (Weeks, Heaphy, & Donovan, 2001).

The imposition of gender difference and inequality on sexuality in societies such as the United States is also reflected in the sexual double standard. The double standard, which emphasizes and normalizes male pleasure and female restraint, is a widespread product and practice of patriarchy. Although Western sexual attitudes and behaviors have moved away from a strict double standard, the sexual lives of women remain more constrained than those of men. For example, girls and women are still under the control of the good-girl/bad-girl dichotomy, a cultural distinction that serves to pit women against each other and to produce sexual relations between women and men that can be confusing and dissatisfying. Men grow up with expectations to embrace a sexual script by which they gain status from sexual experience (Kimmel, 2000). Women grow up with expectations that being sexually active compromises their value—but, at the same time, that they must be flirtatious and sexy. Doesn't this seem confusing? Imagine the relationship misunderstandings and disappointments that can emerge out of the meeting of these "opposite-sex" sexual scripts.

Erin Hatton and Mary Nell Trautner (2011) offer persuasive evidence that the sexual double standard persists and has intensified in new negative ways for girls and women. They studied more than 40 years of covers of *Rolling Stone* magazine and found that although sexualized images of both men and women have increased, women are more frequently sexualized than men and, most significantly, women are "increasingly likely to be 'hypersexualized,'" but men are not (p. 303). Hatton and Trautner conclude that the dramatic increase in hypersexualized representations of women signifies a dangerous trend toward a narrowing of culturally acceptable ways to do femininity, including the diminishment of other aspects of "emphasized femininity" such as sociability and nurturance (p. 338).

The prisms of difference and inequality alter the experience and expression of gendered sexuality in significant ways in the United States. For example, Trautner (2005) found that social class differences are scripted and represented as sexual differences in exotic dance clubs catering to clientele who come from either the middle class or the working class. Dancers in working-class clubs are more likely to perform stereotypes of bad-girl sexuality, while those who dance in middle-class clubs enact good-girl sexuality. Rebecca Plante (2006), an expert on the sociology of sexuality, analyzes the intersection of race and sexuality in the United States and argues that African Americans "do not enter the discourse and debate about sexualities from the same place where white, middle-class people enter" (p. 231). Black sexuality was historically stereotyped as perverted and predatory, a theme that continues today in pornographic and other representations of African Americans. In a recent study of Black women who work in the pornography industry, Mireille Miller-Young (2010) confirms Plante's observations about the interactions among gender, race, and sexuality. Miller-Young found that Black women are "devalued as hyperaccessible and superdisposable in an industry that simultaneously invests in and ghettoizes fantasies about black sexuality" (p. 219). Offering a nuanced, Black feminist analysis of fatness, Blackness, and women's sexuality, Courtney J. Patterson-Faye, in her chapter reading, shows how Black women in the plus size fashion world are countering "mammy," one of the dominant cultural stereotypes of Black women. Black women in this industry are rewriting gendered and raced sexual scripts that define what is attractive, sexy, and fashionable.

The chapter reading by Breanne Fahs is very helpful in summarizing and sorting out the contradictions that overwhelm women's (and men's) experience of sexuality in the contemporary United States, where living between celebrations of sexual liberation and progress is often cancelled out by alarming rates of sexual violence, body shaming, and other forms of sexual repression. She argues for a more inclusive and critical feminist sex-positive movement that integrates *freedom from* oppressive sexual mandates and requirements for women alongside *freedom to* enjoy and express sexuality in consenting relationships. She states that "women need to be able to deny access to their bodies, say no to sex as they choose, and engage in sexual expression free of oppressive homophobic, sexist, and racist intrusions" (p. 364).

Gendered Emotions

Masculinity and femininity are defined as emotionally opposed in Western culture (Bendelow & Williams, 1998). This opposition expresses itself in both obvious and subtle ways. The obvious opposition is that the emotions that can be expressed—for example, anger and love—and how they are expressed are tied to gender. Boys and men must appear "hard" by hiding or shutting down feelings of vulnerability, such as fear, while girls and women are encouraged to be "soft"—that is, emotionally in touch, vulnerable, and expressive. Consider the impact of learning and enacting different, even oppositional, emotional scripts and feeling rules on intimate relationships. If men are not supposed to be vulnerable, then how can they forge satisfying affectionate bonds with women and other men? Additionally, men who embrace the gender–emotion stereotype of hard masculinity may pay a price in well-being by concealing their own pain, either physical or psychological (Real, 2001; Shields, Garner, Di Leone, & Hadley, 2006; Vaccaro et al., 2011). Also, it is important to recognize the negative consequences of gendered emotionality for girls and women. Although girls and women receive cultural encouragement to be "in touch with" themselves and others emotionally, this has strong associations with weakness and irrationality. Thus, Stephanie Shields and colleagues (2006) observe that the strong cultural association between emotionality and femininity has emphasized "the comparatively ineffectual nature of women's emotion" (p. 63). When women express emotion according to cultural rules, they run the risk of being labeled hypersensitive, temperamental, and irrational. The stereotype of the emotionally erratic and unstable woman has been widely used in efforts to undermine the advancement of women in politics, higher education, the professions, business, and other realms of public life. You know how it goes: "We can't risk having a moody, irritable, irrational woman at the helm."

How do the prisms of difference and inequality interact with gendered emotions in the United States?

Looking through a number of prisms simultaneously, we can see that the dominant definition of hard, stiff-upper-lip masculinity is White, heterosexual, and European (Seidler, 1998). Hegemonic masculinity assigns rationality and reason to privileged adult men. All other people, including women, minorities, and children, are assumed to be more susceptible to negative influence by their bodies and emotions and, as a consequence, less capable of mature, reasoned decision making (Seidler, 1998).

In her chapter reading, L. Ayu Saraswati analyzes yet another way in which gendered emotions intersect with other inequalities. Saraswati argues that some emotions are instruments of conformity that compel women to conform to restrictive appearance ideals and norms. In her research on Indonesian women's decisions to use skin whiteners, Saraswati found that the emotion of *malu* or shame, a gendered emotion, played a central role in pressuring women to cover their skin with whiteners. On the surface, it may seem that Saraswati's research has little meaning for women in the United States, but, just scratch the surface, and the relevance will become clear. Skin whiteners are widely available in the United States and have a long history of use by women of color. As Saraswati emphasizes, the use of skin whiteners by women is linked to the maintenance of gender, racial, and global hierarchies of race, color, and nation. We'd like to conclude this chapter introduction by asking you to think about individual and collective strategies to reject conformity to patterns of gendered body work, sexual expression, and emotionality that demean, disempower, and prove dangerous to women and men. What would body work, emotional life, and sexual desire and experience be like if they were not embedded in and shaped by structures of inequality? How would personal growth, self-expression, and communication with others change if we were not under the sway of compulsory attractiveness, compulsory heterosexuality, the sexual double standard, and gendered emotional requirements? How would your life change? What can we do, as individuals and collectively, to resist and reject the pressure to bring our bodies, sexual experience, and emotional life in line with oppressive and dangerous ideals and norms?

References

amanda. (2012, November 15). Football and brain damage, or how American masculinity ravages men's bodies [Web log post]. Retrieved from http://thesocietypages.org/sociologylens/2012/11/15/football-and-brain-damage-or-how-american-masculinity-ravages-mens-bodies/

Bendelow, G., & Williams, S. J. (1998). Introduction: Emotions in social life. In G. Bendelow & S. Bendelow (Eds.), *Emotions in social life* (pp. xv–xxx). London: Routledge.

Black, P., & Sharma, U. (2004). Men are real, women are "made up." In J. Spade & C. Valentine (Eds.), *The kaleidoscope of gender* (pp. 286–296). Belmont, CA: Wadsworth.

Chrisler, J. C. (2011). Leaks, lumps, and lines: Stigma and women's bodies. *Psychology of Women Quarterly, 35*(2), 202–214.

Connell, R. W. (1999). Making gendered people. In M. Feree, J. Lorber, & B. Hess (Eds.), *Revisioning gender* (pp. 449–471). Thousand Oaks, CA: Sage.

Dowling, C. (2000). *The frailty myth.* New York: Random House.

Fikkan, J. L., & Rothblum, E. D. (2011). Is fat a feminist issue? Exploring the gendered nature of weight bias. *Sex Roles, 66,* 575–592.

Gavey, N., McPhillips, K., & Doherty, M. (2001). "If it's not on it's not on"—or is it? Discursive constraints on women's condom use. *Gender & Society, 15*(6), 917–934.

Gieske, S. (2000). The ideal couple: A question of size? In L. Schiebinger (Ed.), *Feminism and the body* (pp. 375–394). New York: Oxford University Press.

Gillespie, M. A. (1998). Mirror mirror. In R. Weitz (Ed.), *The politics of women's bodies* (pp. 184–188). New York: Oxford University Press.

Hatton, E. & Trautner, M. N. (2001). Equal opportunity objectification: The sexualization of men and women on the cover of *Rolling Stone*. *Sexuality & Culture* (15), 256-278.

Katz, J., & Tirone, V. (2009). Women's sexual compliance with male dating partners: Associations with investment in ideal womanhood and romantic well-being. *Sex Roles, 60,* 347–356.

Kaw, E. (1998). Medicalization of racial features: Asian-American women and cosmetic surgery. In R. Weitz (Ed.), *The politics of women's bodies* (pp. 167–183). New York: Oxford University Press.

Kimmel, M. (2000). *The gendered society.* New York: Oxford University Press.

Kwan, S. (2010). Navigating public spaces: Gender, race, and body privilege in everyday life. *Feminist Formations, 22*(2), 144–166.

Lorber, J., & Moore, L. J. (2007). *Gendered bodies: Feminist perspectives.* Los Angeles: Roxbury.

Mernissi, F. (2002). *Scheherazade goes west.* New York: Washington Square Press.

Messner, M. (1990). When bodies are weapons: masculinity and violence in sport. *International Review for Sociology of Sport, 25,* 203–218.

Miller-Young, M. (2010). Putting hypersexuality to work: Black women and illicit eroticism in pornography. *Sexualities, 13*(2), 219–235.

Pascoe, C. J. (2005). "Dude, you're a fag": Adolescent masculinity and fag discourse. *Sexualities, 8*(3), 329–346.

Plante, R. (2006). *Sexualities in context.* Boulder, CO: Westview Press.

Real, T. (2001). Men's hidden depression. In T. Cohen (Ed.), *Men and masculinity* (pp. 361–368). Belmont, CA: Wadsworth.

Sabo, D. (1994). Pigskin, patriarchy, and pain. In M. Messner & D. Sabo (Eds.), *Sex, violence and power in sports: Rethinking masculinity.* Berkeley, CA: Crossing Press.

Schiebinger, L. (2000). Introduction. In L. Schiebinger (Ed.), *Feminism and the body* (pp. 1–21). New York: Oxford University Press.

Seidler, V. J. (1998). Masculinity, violence and emotional life. In G. Bendelow & S. Williams (Eds.), *Emotions in social life* (pp. 193–210). London: Routledge.

Shields, S., Garner, D., Di Leone, B., & Hadley, A. (2006). Gender and emotion. In J. Stets & J. Turner (Eds.), *Handbook of the sociology of emotions* (pp. 63–83). New York: Springer.

Trautner, M. N. (2005). Doing gender, doing class: The performance of sexuality in exotic dance clubs. *Gender & Society, 19*(6), 771–778.

Vaccaro, C. A., Schrock, D. P., & McCabe, J. M. (2011). Managing emotional manhood: Fighting and fostering fear in mixed marital arts. *Social Psychology Quarterly, 74*(4), 414–437.

Weeks, J., Heaphy, B., & Donovan, C. (2001). *Same sex intimacies: Families of choice and other life experiments.* New York: Routledge.

Introduction to Reading 27

Erynn Masi de Casanova's study of women domestic workers in Guayaquil, Ecuador, examines the embodied dimensions of domestic work and demonstrates how unequal relations between domestic workers and their employers are played out in and on the bodies of women workers through health, food, and appearance. She asks: What can domestic work done by women reveal about the social categorization of bodies? To answer that question, Casanova interviewed fourteen domestic workers, all women, and three employers. She adopted the vantage point of the workers in order to understand their accounts of issues such as the physical demands of their work and the toll it takes on their bodies. Their bodies, Casanova found, are marked as poor and undesirable in relationship to their employers' bodies.

1. What are the racial dynamics of the creation of embodied differences between women domestic workers and their employers?
2. How does embodied inequality manifest itself in dress and physical appearance?
3. How does the author employ both "body as resource" and "body as symbol" theoretical frameworks in the analysis of her data?

Embodied Inequality

The Experience of Domestic Work in Urban Ecuador

Erynn Masi de Casanova

I was cooking and I felt like I was suffocating. I wanted to lie down, because I felt sick . . . I took [spread out] some newspapers and lay on the kitchen floor. I felt like I was dying, that I couldn't get air . . . And the employer gets home and the other employee tells her, "The empleada *[domestic worker] is sick." "Oh, no," she says, "what's my daughter going to eat now, who will cook for her?"*

Cristina, domestic worker, age 40

Cristina has spent much of her life cooking and cleaning in private homes in Guayaquil, Ecuador. Her anecdote highlights domestic work's physical demands, and employers' privileging of their corporeal needs over those of their employees. Based on interviews with domestic workers and employers, I argue that bodies matter for how domestic employees experience their work (apart from often discussed issues of sexual harassment or abuse). Domestic workers' accounts emphasized physical labor and the embodied inequality between employer and employee. Domestic work assumes particular forms in coastal Ecuador, where workers and employers often have similar racial backgrounds, and middle-class people see their position as increasingly precarious under a left-leaning political regime.

What can domestic work reveal about social categorizations of bodies? Domestic employment entails a unique physical proximity of bodies from different class groups, a boundary-threatening situation that must be managed by workers and employers. In this private sphere, bodies can reproduce or challenge class inequality. Although "domestic work constitutes bodily subjectivity in a particular way" (Bahnisch 2000, 59, referencing Gatens 1996, 69), research tends not to place the social meanings of workers' bodies at the center of the analysis. Ecuador, with its long history of paid domestic work and rigid class system (Miles 1998; Roberts 2012)—in which even lower-middle-class families have traditionally employed domestics—is the ideal site for exploring class/work/body. Like many low-prestige jobs, domestic work draws on and propagates social constructions of poor people's bodies as deviant and worthless.

Necessary research on working bodies "progress[es] only by adopting the vantage point of the embodied worker and listening to their accounts of workplace experience 'from the inside'" (Wolkowitz 2006, 183). I take domestic workers' accounts of issues concerning health, food, and appearance/clothing as a starting point, asking: How do bodies matter in domestic work? How does this employment arrangement relate to broader ideas about differently classed bodies in Ecuador?

Theoretical Perspectives

Two complementary theoretical and empirical approaches apply to working bodies. The first, rooted in Marxian theory, views the body as a limited resource, damaged and deformed in exploitative production processes. Marx described factory work's destructive effects on workers' bodies and psyches ([1844] 1978, 74) as the collateral damage of capitalist expansion. Indeed, Labor power, the power of the body, is central to the reproduction and accumulation of capital (Bahnisch 2000, 64). In today's service economies, the "human body continues to be deeply involved in every aspect of paid work" (Wolkowitz 2006, 55). The "body as resource" perspective is often used to describe the types of physical labor identified with men.

The second approach, drawing on Bourdieu's (1990) concept of *habitus* and symbolic interactionism (e.g., Goffman 1959), views the body as symbol. The symbolic body does not communicate unequivocal messages, comprehended similarly by everyone we encounter.[1] However, habitus is observed and interpreted by others according to the "dominant symbolic" (Skeggs 2004a, 87), even if we are unconscious of how social structures produce particular behaviors and (bodily) dispositions. For Bourdieu, habitus "causes an individual agent's practices, without either explicit reason or signifying intent, to be . . . 'sensible' and 'reasonable' to members of the same society" (1977, 79). Bodily aspects of habitus make sense to those able to recognize and classify them.

Dress, appearance, and movement communicate bodies' positions in hierarchies of race, class, gender, and occupation. Embodied habitus is "a statement of social entitlement" (Skeggs 2004b, 22), reproducing class inequalities in, on, and through bodies and becoming a source of conflict or approbation in interactions between people of different classes. Workplaces can be agents of socialization, building or reinforcing habitus, especially when workers begin at young ages, as in domestic employment. Expanding on Bourdieu, Wolkowitz (2006) elaborated the idea of *occupational habitus* related to people's work identities. Domestic work is one setting for the construction of occupational habitus.

The "body as symbol" perspective often focuses empirically on gendered appearance. Research shows how workplaces value certain forms of gender performance (Freeman 2000; Hall, Hockey, and Robinson 2007; McDowell 1997; Nencel 2008; Salzinger 2003). Yet there is scant research on embodied work outside of organizations, except for sex work. Building on the "body as resource" approach, research shows how emotions, signaled through physical cues, are

de Casanova, E. M. (2013). Embodied inequality: The experience of domestic work in urban Ecuador. *Gender & Society, 27*(4): 561–585.

harnessed in gender-segregated service occupations (Hochschild 1983; Kang 2010). Yet Wolkowitz laments the "relative invisibility of the corporeal in the employment-oriented literature on emotion" (2006, 79), and the sociology of the body is just beginning to examine the large portion of an employed person's life that is spent at work.

The term body *work* (Kang 2010; Shilling [1993] 2003) synthesizes the material and symbolic aspects of bodies. Here, drawing on Gimlin, I consider body work: "(iii) the management of embodied emotional experience and display, and (iv) the production or modification of bodies through work" (2007, 353). Consistent with a "body as symbol" frame, employers expect domestic workers to present a certain appearance. As in "body as resource" theories, domestic workers' bodies are produced and transformed through the work.

I propose a more holistic micro-level approach, "embodied inequality." A theoretical framework of embodied inequality bridges the "body as resource" and "body as symbol" approaches and their conceptual offshoots, "body work" and "occupational habitus." In this case, domestic workers' bodies are used as tools, and suffer the physical consequences, yet they must also have an appearance acceptable to their employers and indicative of their socioeconomic position. Unlike academic narratives that privilege only the material or the symbolic, the workers' accounts combine these perspectives for a broader view of embodied inequality.

Paid Domestic Work

. . . This article investigates embodied aspects of domestic work in a thus far unexplored site, urban coastal Ecuador. Taking my cue from workers' interview accounts, I set the body at the center of my inquiry. "Zooming in" on the body, we see how crucial embodiment is to women's understanding and experience of this type of work. Domestic workers described how class relations become embodied and personalized when acted out between individuals in the private sphere. Distinctions between employers' and employees' bodies are not simply symptoms of larger inequalities, played out in rote ways; the very distinctions themselves are created through sometimes hurtful everyday interactions. Although bodily aspects of habitus build up over time so that people are not always conscious of the reasons for their actions (Bourdieu 1977), the domestic workers interviewed point to specific moments when they felt degraded as classed, gendered, and sometimes racialized bodies. We can better understand domestic workers' experience of their work by listening to how they talk about their working bodies.

The Local Context

. . . The female labor market in Ecuador is bifurcated, with a limited set of "good" jobs available to college-educated women, and a set of less desirable jobs, or informal self-employment, to less educated women (Casanova 2011). Women see domestic work as the least appealing employment option because of low pay and potential exploitation by unregulated employers; many prefer other informal work (e.g., selling goods) (Casanova 2011, 164–65).

Guayaquil, Ecuador's largest city, has approximately 2.4 million people living inside the city limits (M.I. Municipalidad de Guayaquil 2012) and a metro area population of around 3 million. Many of Guayaquil's residents live in poverty (87 percent in the early 2000s [Floro and Messier 2006, 234]), and middle-class families commonly employ domestic workers. Employment arrangements vary, from elite families employing entire live-in staffs to lower-middle-class households having someone work a few hours per week. Many families' only claim to middle-class status is the presence of a domestic worker in their household. Long an informal, under the table, contract-free type of employment, domestic work is now the object of increased government scrutiny and public consciousness raising by worker organizations, giving workers and employers the sense that the sands are shifting.

The timing of this study was ideal, as the left-leaning government of President Rafael Correa began enforcing labor laws protecting domestic workers—though not systematically or continuously—in 2009. Domestic workers' issues were frequently featured in news media, and workers' organizations ramped up advocacy and outreach to leverage state support and improve working conditions. It had always been difficult for domestic workers to negotiate living wages and bearable workloads because of the unequal positions of employer and employee and the lack of regulation; public attention around domestics' legal rights in this period made such discussions even more uncomfortable. In 2010, when I began fieldwork, the Ministry of Labor was conducting house-by-house inspections in wealthy neighborhoods to determine the presence of domestic workers in the home and whether they were receiving the government-mandated minimum wage and benefits (e.g., Social Security). Inspections have since ended, though the middle and upper classes still feel threatened by the populist, socialist-identified rhetoric of the current administration. Employers hearing of the inspectors' presence in their neighborhood often "erased" a domestic worker's body by giving her the day off, hiding her in another part of the house, or presenting her as a cousin visiting from the countryside.

The complex racial dynamics of contemporary Ecuador, with most citizens identifying as *mestizo* (of mixed indigenous and European ancestry) and sizable indigenous and Afro-Ecuadorian minorities, are discussed in many studies (Casanova 2004; de la Torre and Striffler 2008; Rahier 1998; Roberts 2012). Throughout Latin America, the high degree of social and economic inequality is often literally written on the body (Bank Muñoz 2008; Casanova 2011; Edmonds 2010; Roberts 2012). Nonwhite appearance and darker skin are generally associated with low class status, and domestic workers are stereotyped as having these characteristics. In Guayaquil, few people self-identify as indigenous or wear traditional dress. Coastal *mestizos/as* have much less contact with their indigenous compatriots than do *mestizos* in the Andean region [sierra]; and the paternalistic *mestizo*-indigenous relations seen in the *sierra* are largely absent on the coast. Thus, it is rare to find indigenous domestic workers in guayaquileños' homes. In fact, in 13 years of conducting research in Guayaquil, I have never encountered a woman who self-identified (or was identified by others) as indigenous laboring in a private home. Guayas province, where Guayaquil is located, has the country's largest concentration of self-identified Afro-Ecuadorians (INEC 2012). While it is more common to find Afro-Ecuadorian women among Guayaquil's domestic workers, there is widespread discrimination against Blacks. My sense from speaking informally with Employers is that they prefer *mestizo* domestic workers; I have also observed more *mestizo* than Black workers in middle-class homes.

Thus, unlike locales where employers and employees are separated by caste (Ray and Qayum 2009) or race (Gill 1990, 1994; Rollins 1985), in Guayaquil it is common for domestic employment arrangements to link women whose official racial classification would be *mestizo.* As in the Philippines (Arnado 2003), this employment relationship is not usually a cross-racial one, meaning that class differences—especially embodied ones—can be more fruitfully explored, as they are not conflated with racial differences. However, identifications of race and class in Ecuador are somewhat fluid and mutually constituting: *mestizos* with higher class status are perceived (and see themselves) as whiter than poorer *mestizos,* regardless of phenotype. Whereas popular conceptions of race in the U.S. rest on ideas about immutable biological differences, in Andean societies, including Ecuador, race is "experienced as alterable, through changes in body and comportment . . . [rather than] genetically determined" (Roberts 2012, 120). People who are *mestizo/a* can become white(r) by complying with middle-class norms of bodily self-presentation, education, and employment—by successfully embodying middle-class habitus. Thus, the "fabrication" of race is not "theoretical" or a priori, but rather, race is enacted and . . . reenacted through a wider range of characteristics than physical appearance as transmuted through genes (Roberts 2012, 114). This racial mutability makes creating embodied differences a priority for employers, who want to visually distinguish themselves from "the help." Compounding the racial uncertainty and jockeying for whiteness is the precarious status of the middle class. In my recent ethnographic research, middle-class people's generally low salaries and their (mostly overblown) sense that they are targeted by government policies that favor the poor have surfaced as important to this group's self-identification. Thus, it is unsurprising to see lines drawn in the sand in the home-as-workplace to remind domestic workers of their (racialized) class position. The most obvious means for marking the bodies of employees is the uniform. The more frequent use of uniforms in Latin America (compared to the United States) underlines the different and perhaps more acute embodied experience of these workers.[2] Strategies of corporeal distance, degradation, and differentiation are best seen as ways of shoring up tenuous claims to privileged statuses such as whiteness, decency [*decencia*], and middle-classness in one local race/class gender context.

Meanwhile, the daily life of domestic workers—as described by participants and as I have observed—remains much the same as in decades past. Most families eat their largest meal at midday, which (in households that can afford it) includes a first course of soup, followed by a protein with rice and vegetables. Cooking can be laborious, often requiring tedious and physically demanding tasks such as grating green plantains by hand or peeling and straining fruits for juice. Rather than using sponge or string mops, domestic workers are usually asked to clean floors with wet rags draped over brooms or long sticks. Most middle-class homes have washing machines; however, employers sometimes require special clothing (like children's school uniforms) to be washed by hand, usually at stone sinks with washboard-like ridges. Even in homes with automatic dryers, clothing is usually hung outside to dry, which is more physically taxing. Other considerations include the dust that accumulates in this urban environment (even in closed-in structures) and the caustic contents of common cleaning products (domestic workers tend not to bring their own products or request specific products). Most live-out domestic workers have long workdays and live in areas distant from public transportation, making for long commutes.

Methods

This article draws on interviews with fourteen current and former domestic workers (all women) and three employers (two women and one man) conducted in Guayaquil, Ecuador, from June through August 2010.[3] Because of the volatility of the domestic labor market, seven of the worker-interviewees were unemployed, though all self-identified as domestic workers and expressed interest in returning to work in private homes (two of these women had been recently hired and were negotiating start dates with new employers). Interviewees ranged in age from 28 to 62. Some had spent most of their working lives in domestic employment, often from adolescence. I asked women about each of their domestic work experiences, beginning with their first and ending with the most recent or current job.[4] Despite the stereotype of domestic workers as rural-to-urban migrants, eight of the fourteen workers interviewed were born in or near Guayaquil. Although I did not ask about racial self-identification, three of the women would likely be identified as Black by most Guayaquileans, and two others appear to have some African ancestry but may not be considered Black. All the women but one were mothers, and most were not legally married. . . .

Findings

Physical Labor in Domestic Work (Body as Resource)

Nearly every worker interviewed stressed the physical demands of housework and child care: "the production or modification of bodies through work" (Gimlin 2007, 353). Cecilia, middle-aged and unemployed, complained that "physical exhaustion" set in over time:

> In the first month, we're wonderful, but in two or three months we're feeling a physical exhaustion (*agotamiento fisico*) that leads to us feeling more and more tired, and we end up getting an illness . . . related to the tiredness, the stress . . . and then the employer begins to complain . . ." What's going on with you? You started off working hard but now you don't clean here, you don't clean over here"—but she doesn't recognize that I am human, too.

Most of the domestic workers interviewed began work at a young age, some at 10 or 12. The toll on the body over time, and the difficulty of getting hired and danger of being fired as an older worker because of perceived or actual physical limitations, were common concerns.

Some domestic tasks are more physically demanding than others. Longtime domestic Patty recounted:

> I have been doing laundry for 22 years, and my hands began to swell up on me . . . there was a horrible pain that grabbed me in my [lower] back . . . and my whole hands were full of fungus, and they bled every time I washed [by hand], they were so irritated . . . it was a rash . . . from the detergent, the soap . . . and the bleach.

After spending a day washing clothes for her employers, Patty would return home and wash her own family's clothes. She consulted a doctor, who recommended she stop hand-washing clothes: "If you keep washing, you are going to die," because the exposure to bleach could cause cancer. Despite the tremendous pain, Patty "went on for some time more for [her] children," whom she helped support. She said, "I was not the same person you see today"—strong, energetic, with a physically commanding presence—because of her overwork. Despite the fact that she no longer washes clothes for a living, I could see the effects of this toil on her hands. For Patty, one of the most troubling aspects of domestic work is that "physically a person gets worn out, deteriorates, doesn't care for herself."

As with much manual labor, domestic work's detrimental physical effects are not well compensated.[5] In interviews, workers listed negative physical consequences: back pain, exposure to hazardous chemicals without protective masks or gloves, and injuries. Former worker Ximena recounted cleaning marble stairs during the final weeks of her pregnancy, when she slipped and fell. The pain she felt afterward, she soon realized, was the beginning of labor. (Perhaps Ximena was fortunate to be working, as two other domestic workers reported being fired when employers learned of their pregnancies.) Patty affirmed, "The work in the home is hard . . . and one arrives home dead, pulverized." Since asking for time off is often not an option, several workers noted that they would simply quit a job when they needed rest, taking a month or two off before returning to work, usually for a new employer.

Workers used embodied metaphors to describe their toil. Belén, employed in one of Guayaquil's wealthiest neighborhoods, referred to her job as earning money "with the sweat of your brow," and Cristina claimed to have "given up even my lungs" to the work; these powerful embodied images signify the bodily sacrifice of domestic work. Such statements recall a tradition of "body as resource" theories inspired by Marx, who saw the body destroyed by the capitalist mode of production. These bodies are modified negatively, and (re)produced as lower-class bodies, through physically demanding tasks. Although their occupational habitus involves executing physically difficult work, the women identify its negative consequences on their bodies.

Employers' Bodies as More Valuable (Body as Symbol)

Workers often discussed distinctions made by employers between employers' and workers' bodies as related to health and health care, food, and clothing. These accounts recall the "body as symbol" perspective: Bodies are seen as carrying different amounts of social worth. Health issues also relate to the physicality of the work and the body as an exhaustible resource. Workers' bodies are so devalued by differential treatment that Cecilia asserted, "The family dog is treated better than the household worker."

Nearly all the workers described myriad health problems, some leaving them temporarily unable to work. Ailments may stem from a variety of causes, including physically demanding work, poor women's precarious health status, and lack of health care access. As former domestic Ana Maria put it, "A [domestic] worker has little time to go to the doctor." Rather than seeing a doctor, she said domestic workers "go to the pharmacy[6] to self-medicate," or consult neighbors and friends. Because most of these women depend on subsidized health services or low-cost clinics, a doctor visit can involve waiting an entire day (or more) to be seen.[7] Paola, a full-time student and former domestic worker, discussed employers' inflexibility with regard to workers' schedules: "We don't have any right to get sick." Paola and other interviewees connected health disparities to class inequality: "We are also human, just like them [employers], except the difference is that they have money, and we don't." When employers get sick, they visit the doctor and take time off from work to recuperate; when domestic employees get sick, they are often unable to do either. When Fátima asked her last employer to help pay for her prescription, she was berated by his adult son, and felt that skin color and class prejudice were at play: "They let me know that because a person is ugly, [and] Black . . . and they [the employers] are white, they can give you a kick in the rear. . . . But why, if we are all human and have feelings?" She compared this employer to a previous one, who, when she fell ill, took her to an expensive private hospital and paid the bill. Workers see the body as a resource they can use, and see the employers' denial of their health needs as a material and symbolic devaluation of their bodies.

Food was another site for drawing lines between upper- and lower-class bodies. Based on my interviews and observations, deep-rooted practices, such as having domestic workers eat separately and offering them different (or less) food, are still common in urban Ecuador. Several women complained that employers denied them food during the workday. Fátima said, "They didn't even give me a piece of warm bread." Others recounted having to eat reheated leftovers when there was plenty of fresh food available, or watching employers throw out food the worker had requested. Domestic workers, said Cecilia, often do not get "decent food."

More offensive to Belén was being forced to use dishes that were just for the help: "From the teacups to the spoons, everything . . . all the utensils were different from theirs." This humiliating experience made her feel insignificant. The symbolism of objects that touch only a worker's body, but never an employer's body, communicates a powerful message of inferiority to the worker. Another worker recalled having to eat outdoors while the employer's family ate inside; typically, domestic workers eat in the kitchen or an adjoining room. Legitimating the study of food as a social and cultural object, Bourdieu ([1984] 2007) focused on divergent eating habits based on class-based tastes, yet here different eating practices are based not on taste but on the exclusion of lower-class bodies from nourishment and desirable foods. By refusing to share eating utensils and eating space with domestic workers, employers shore up embodied class boundaries and prevent even indirect bodily contact.[8] In describing "good" employers, workers often pointed to eating the same food, or eating together at the table, as evidence of kindness. Alternatively, we could see this egalitarian gesture as a distraction from the inherent economic inequality in the employment situation.[9]

Given the incorporation of food-related routines into domestic workers' occupational habitus, it is likely that many workers, unlike those interviewed, do not resist or complain about these practices.

Cristina recalled taking her young daughter to work, so she could "see what I do." Her daughter asked, "*Mami,* why do you make such delicious food here?" Cristina replied, "At home we don't have the same money as they do here." Her daughter complimented Cristina's cooking, saying she wished she could cook such rich foods at home. Cristina responded, "*M'ija* [My daughter], that is why I am working here, because sometime we want to eat well too."

Employers I spoke with connected food and bodies in ways that reaffirmed or obscured class boundaries. Clara, a 30-something woman, claimed to have fired a previous employee in part because of her messy cooking style, saying the food "had a bad flavor [because] her hands were dirty, and her whole appearance was . . . messy." With their current worker, Clara and her husband Alfredo emphasized, they all ate together.

They looked down on those who made domestic workers eat separately, since "we are the same, human beings." The new employee's ability to present a neat appearance made her presence at the dinner table palatable to her employers.

Interviewees also discussed embodied inequality in dress and physical appearance. Domestic workers admitted that there were clothes or other items that they enjoyed or aspired to purchase. As Cristina put it, "Although I don't have money, I like to go to stores, to look and to fantasize." Fátima referred to the popular Ecuadorian saying, "They treat you according to how you look [*a uno como lo ven, lo tratan*]." She used this social practice of judging appearance (Casanova 2011, Chapter 4), which connotes a particular—often racialized—class status, as a justification for always wanting to look good in public. With such statements, workers like Fátima reaffirmed the symbolic importance of physical appearance for women of all class backgrounds. Her employers took notice, and when she requested their help in buying her prescription medicine, they criticized her for having money to do her hair and nails and buy perfume but not to provide for her health, implying that lower-class bodies did not deserve to be made attractive. Her attempt to make a claim on middle-class bodily habitus was thus delegitimized.

Workers often criticized employers' "vanity" or "fashion," referring to the value placed on presenting a socially acceptable middle-class appearance. Elsa had recently left her job over issues of back pay and vacation time. Elsa's employers claimed they couldn't pay her the salary she was owed, yet Elsa noted that any time the mother and daughter were invited to "half a party [*medio reunión*]," they would rush out to get their hair or nails done. Other workers said money that could have been used to pay domestic workers (whose presence in the home is a key symbol of middle-class status) was spent on what they perceived as superficial, temporary, bodily markers of class. Irma, an experienced domestic worker, was overwhelmed by the quantity of clothing, makeup, and accessories in her employer's home, confiding, "I'd go crazy with so much clothing, it's amazing . . . there were suitcases full of purses, full of every kind of makeup." It is harder for low-income women, including domestic workers, to access and "properly" use these symbolic props for middle-class respectability.

Female employers sometimes loan clothing to workers, or give them used clothing. Some workers viewed this as a benefit of their job. At Christmas, Patty's former employer gave her money to buy clothes, and loaned her clothes to attend social events in her rural hometown.[10] Yet she maintained the class line distinguishing worker from employer: Patty was told never to wear her employer's borrowed clothing "in the [employer's] neighborhood," only out of town, and she complied. Marina, a worker whose daughter had also worked in private homes, told me that she accepted used clothes from her former employer because she did not want to waste them; she distributed them among family and friends. Not all workers appreciate hand-me-down clothing. Marina's daughter Francisca frowned on those "who give you the blouse they don't want any more . . . No, that's not a good employer. A good employer says, 'Come on and I'll buy you a new blouse.'" Cast-off clothing, especially if worn or illfitting, marks the lower-class body (Adair 2001), whereas new clothing adds value to the worker's appearance.[11]

The Uniform as Embodiment of Inequality

In popular culture portrayals, it is easy to pick out the domestic worker in a privileged space: She's the one wearing the maid uniform. Uniforms are sold in Guayaquil's department stores and grocery stores, where uniformed domestic workers can be spotted pushing employers' shopping carts or tending to their children. These garments tend to be loose-fitting housecoats, or resemble medical scrubs: The stereotypical, black-and-white version is rare. Uniforms are different enough from everyday street clothing to visually distinguish women as domestics at a glance. Many employers pay for or provide uniforms for their employers, whereas others (to the chagrin of workers) take the cost out of the employee's pay.

I asked both workers and employers[12] about uniforms. The employers interviewed did not require uniforms, describing them as unnecessary and old-fashioned. They mentioned wanting the worker to feel comfortable, especially when caring for young children. Employer Clara, who wore a uniform in her job as a hotel manager, said she had denied her domestic worker's request for a uniform. Perhaps because of her own embodied experience of wearing a uniform as a professional, she saw uniforms not as indicative of low social status, but the opposite: "I personally see the uniform as a differentiation of social strata, so here you see people wearing uniforms who are . . . from a higher socioeconomic level." For Clara, the uniform symbolized middle-class, professional, rather than low, status; it was part of her occupational habitus as a college-educated manager.

Of the workers who expressed opinions about uniforms, seven viewed them negatively and one positively.[13] Plain-spoken Paola, who had to wear a uniform a few times at a previous job, declared it "a humiliation" and "a piece of trash [*una porqueria*]." When a domestic worker goes out in public with her employer, Paola said, "a person can distinguish who is the employee and who is the boss." This is especially relevant when both the employer and employee are *mestiza.* Patty, who had never worn a uniform, described domestic workers' uniforms as "sad dresses." She said that simply because a woman was a domestic she shouldn't have to be "all scruffy . . . with ugly sandals." (In Guayaquil, "ugly" sandals or flip-flops signal lower-class status—middle-class habitus generally reserves flip-flops for beachwear or housewear.) The point of such a display, Patty said, was for an employer to demonstrate that "that person is his/her employee." When the lady of the house goes shopping with her daughter, no one mistakes the daughter for a domestic worker, Patty remarked. Domestic workers generally felt that when they wore uniforms in public, people thought, "Look, there's an *empleada,* a nanny." The uniform clarifies and amplifies the message of the symbolic body.

Francisca, otherwise calm and soft-spoken as we chatted, spoke excitedly when the talk turned to uniforms: "It's like they [uniforms] make you feel like a [derogatory term for maid], like you're less than." While working for a downtown family, she had to wear a uniform only when she went shopping. When asked how she felt, she replied, "Ooooh, like the lowest, because, imagine, in the middle of downtown and dressed like that . . . I felt really very bad [*me senti recontra que mal*]." Francisca eventually told her employer that she didn't like going shopping, and, surprisingly, the employer did not push back. Francisca forfeited an opportunity to escape the confines of the home because of the embarrassment the uniform entailed.

While workers expressed concerns about public perceptions, they also discussed how these stigmatizing clothes looked and felt on the body. Several bemoaned the poor quality of the uniforms required by employers. Francisca described feeling physically uncomfortable in the uniform, which was "like, too hot, different, [when] one is used to wearing her own little clothes." Paola said vehemently:

> Aside from everything else, it's a poorly made uniform. . . . If it were a uniform, a little pantsuit with a little T-shirt, great, fine. Or, why can't I wear pants and a T-shirt like they do in offices? . . . But instead they have to buy the worst . . . fabric because that's how they've treated us, like the lowest of the low [*la última rueda del coche*].

More egregious than bad fabric was bad fit. Patricia complained about having to use worn-out uniforms left behind by previous employees. Women's bodies are all different: "One is fat and the other is thin . . . there should be a uniform that fits one's body." When Belén protested the poor fit of her uniform, her employer replied, "I bought it for your body." Belén disagreed, telling me, "I look like a potato sack with no potatoes, because it fits me so big, and I don't like that." So she wore it around the house, but changed into her own clothes to go out, because the uniform "fit me so ugly." Ugliness is thus associated with both domestic work and lower-class status.

Several workers described households as having just one uniform, to be used by whoever was working in the home. References to a one-size-fits-all, used uniform evoke the image of a garment hanging on a hook in the kitchen, waiting for a domestic worker to literally put on or to embody the role of the help. The uniform stays on the hook and the individual worker (with her unique size, shape, and preferences) changes. This represents the ultimate de-individualization of the worker, in which anybody can be stuffed into or swim around in a generic uniform not chosen for her needs, but to symbolize a social/occupational role. Irma, who generally liked uniforms and whom I often saw wearing medical scrubs-type garments, decided to get her own well-fitting uniforms in order to avoid those provided by employers, which she described as "mistreated" and "badly washed."

Some interviewees interpreted the employer's choice of ill-fitting (usually too large) and unattractive uniforms not just as a demarcation of workers' low status but as an effort to desexualize their bodies, to prevent them from being viewed or targeted sexually by male employers. Patricia commented, "The important thing is that it fits you big, because they don't like it small," yet for workers, such a large garment felt like "a nightgown." When Cristina was searching for words to describe the physical encumbrance posed by overly large uniforms, I interjected:

> Erynn: You have to be able to move to do things . . .
>
> Cristina: Yes, of course, right? So sometimes they give you a long dress, but I say, what's the reason for the long dress? . . . They must be thinking that you're going to steal their husband or something, right?

Cumbersome clothes limit the ability to use the body as a resource, but also have a symbolic dimension: Cristina viewed these modesty requirements as a way of managing the sexual threat posed by the presence of an unrelated female in the home of a married couple. Paola was explicitly instructed by a female employer to "be careful around the son and the husband," and to "wear a bigger pair of shorts, so that you couldn't see my . . ." While we might applaud the employer for wanting to protect her, it is worth noting that she put the onus on Paola to discourage harassment, rather than expecting her male relatives to behave appropriately. Paola agreed with the employer's suggestion, saying "it was obvious" that she should dress more modestly. Domestic workers thus adjust their appearance as employers require, whether or not they internalize stereotypes that characterize them as hypersexual.

Conclusion: Embodied Inequality

The questions driving this research were (1) How do bodies matter in domestic work? and (2) How does this employment arrangement relate to broader social ideas about bodies of different class status? In the accounts presented here, workers described their embodied employment experience in ways that fit both "body as resource" and "body as symbol" theoretical frameworks, pointing to the usefulness of a more comprehensive, holistic approach to embodied inequality.

Bodies matter in all jobs: Even the most cerebral, intellectual tasks are performed by humans who have/are bodies. In discussing their work, domestic workers emphasized the physicality of the tasks and the deleterious effects on their bodies. They described their work as exhausting, accelerating the deterioration of their bodies, and potentially dangerous. These accounts conceive of the body as a limited resource women draw on to do their work, which can be used up or damaged in the process.

Bodies also matter in terms of the symbolic distinctions drawn between "good," middle-class/elite bodies and "bad," lower-class/deviant bodies—between employers' and workers' bodies. Workers face clear boundaries between themselves and employers in relation to health, food consumption, and appearance. Even employers who buck tradition by pursuing more egalitarian relations (e.g., dining with their workers) are aware of the differential values typically placed on differently classed bodies. The uniform, an iconic symbol of domestic service throughout the Americas, was viewed by most of the workers and, interestingly, by the young employers I interviewed, as a superfluous relic of a more oppressive class order. Yet, 70 percent of the domestic workers interviewed had been asked by employers to wear uniforms, so the practice is still alive. Most workers hated being made to wear ill-fitting, cheaply-made uniforms, and several described feeling embarrassed for other women they saw visibly marked as domestics in public spaces. Many who referred to the physical challenges of domestic work also drew on the idea of the body as a malleable symbol of status in describing the embodied aspects of their relations to employers. . . .

These accounts highlight "the management of embodied emotional experience and display" (Gimlin 2007, 353) seen in domestic workers' acquiescence to the bodily regimes of employers in a context of limited employment alternatives. The successful embodiment of modesty (in dress and manner) required by employers is one example of the ways that domestic workers manage their bodies and emotions. Workers' stories of the physical damage and health hazards of the job exemplify "the production or modification of bodies through work" (Gimlin 2007, 353). Bodies are often changed for the worse by engaging in domestic work, and lower-class women's bodies are reproduced as "less than" or "the lowest of the low," to use the workers' terms.

This empirical study can help further theoretical understandings of bodies, work, and class in Latin America and beyond, following the lead of the domestic workers as folk theorists who combine the "body as resource" and "body as symbol" perspectives in a comprehensive view of embodied inequality. Examining one case in depth demonstrates how class-based occupational habitus is created or resisted in employment situations, especially those that begin early in life, like domestic work. Domestics in Guayaquil come from lower-class (sometimes rural) backgrounds, and their bodies are marked as poor and undesirable prior to entering the workforce—in a sense, the lower-class body is already a sort of uniform. Workers become further accustomed to the material and symbolic devaluation of their bodies as they are fed inferior food with separate dishes in separate areas of the home, not permitted to attend to their health problems (some of which emerge from the repetitive physical damage of domestic labor), and denied the objects associated with an acceptable or attractive middle-class feminine appearance. Workers exercise the most resistance in the area of appearance, subverting employers' dress codes, or compensating for the symbolic degradation of their bodies at work by investing in a feminine appearance outside of work. In a local context characterized by racial and class anxiety, and (perceived or actual) political challenges to middle-class status by Ecuador's left-leaning government, employers are invested in holding the line that separates them from employees, and employees are acutely aware of this.

Notes

1. Aside from its symbolic aspects, habitus can be a source of identification and meaning-making for individuals.

2. In her study of domestic work in contemporary South Africa, King identifies the uniform as a "symbolic representation of the regulation of their [workers'] constructed role" (King 2007, 36), For uniforms' role in society, see Joseph (1986).

3. These interviews are part of an ongoing multi-method study of domestic work in Guayaquil, including informal interviews, participant observation with a domestic worker organization, archival research, and analysis of help wanted ads. I met the domestic worker-interviewees through fieldwork at the organization, and employer interviewees were referred by personal contacts.

4. All workers interviewed were connected to the domestic worker advocacy organization where I conducted ethnographic fieldwork. This may bias my sample: Interviewees may be more committed to domestic workers' rights or more politically oriented than other domestic workers.

5. The legal monthly minimum wage in Ecuador, to which domestic workers were legally entitled at the time of the interviews, was $240 (plus benefits: Social Security, overtime, vacation, etc.). Most domestic workers in Guayaquil were paid around $200 per month, with unpaid overtime and no benefits.

6. In Ecuador (as in many developing countries), most drugs are available in pharmacies without a prescription.

7. See Auyero's (2011) excellent ethnography of "poor people's waiting" in government benefits offices in Argentina. Sutton's interviews with domestic workers in Argentina also highlighted long waits to see doctors as a reason that women "neglected signs of illness in their bodies' (Sutton 2010, 56). One domestic worker Sutton interviewed said, "I know I cannot get sick" (ibid.).

8. Historian Shailaja Paik notes the similar embodied exclusion of Datit people in India, particularly in educational settings (personal communication).

9. Thanks to Tamara Mose Brown for highlighting this point (personal communication).

10. Patty offered clothes lending as an example of how a "good employer" behaves.

11. Rollins argues that gifts from employer to domestic, not expected to be reciprocated and often used or worn, highlight employees' inferior status, symbolically defining them as "needy" and "dependent" (1985, 192–94).

12. The employers I interviewed were working professionals in their 30s, just starting their families. Thus, their opinions on uniforms may differ from older or wealthier employers.

13. This worker favored uniforms, yet was critical of the quality and appearance of those typically offered by employers, and thus preferred to buy or make her own (at her expense).

References

Adair, Vivyan C. 2001. Branded with infamy: Inscriptions of poverty and class in the United States. *Signs* 27: 451–71.

Auyero, Javier. 2011. Patients of the state: An ethnographic account of poor people's waiting. *Latin American Research Review* 46:5–29.

Bahnisch, Marc. 2000. Embodied work, divided labour: Subjectivity and the scientific management of the body in Frederick W. Taylor's 1907 "Lecture on Management." *Body & Society* 6:51–68.

Bank Muñoz, Carolina. 2008. *Transnational tortillas: Race, gender, and shop-floor politics in Mexico and the United States.* Ithaca, NY: Cornell University Press.

Bourdieu, Pierre. 1977. *Outline of a theory of practice.* Cambridge, UK: Cambridge University Press.

Bourdieu, Pierre. 1990. *The logic of practice.* Cambridge, UK: Polity.

Bourdieu, Pierre. (1984) 2007. *Distinction: A social critique of the judgment of taste* (R. Nice, trans.). Cambridge, MA: Harvard University Press.

Casanova, Erynn Masi de. 2011. *Making up the difference: Women, beauty, and direct selling in Ecuador.* Austin: University of Texas Press.

De la Torre, Carlos, and Steve Striffler, eds. 2008. *The Ecuador reader: History, culture, politics.* Durham, NC: Duke University Press.

Edmonds, Alexander, 2010. *Pretty modern: Beauty, sex, and plastic surgery in Brazil.* Durham, NC: Duke University Press.

Ehrenreich, Barbara, and Arlie R. Hochschild, eds. 2004. *Global woman: Nannies, maids, and sex workers in the new economy.* New York: Henry Holt.

Estrada, Daniela. 2009. Strides and setbacks for domestic and rural workers. *Inter Press Service.* http://www.ipsnews.net/2009/09/latin-america-strides-and-setbacks-for-domestic-and-rural-workers.

Floro, María, and John Messier, 2006. Tendencias y patrones de crédito entre hogares urbanos pobres en Ecuador. In *La Persistencia de la desigualdad: Género, trabajo y pobreza en América Latina,* edited by Gioconda Herrera. Quito, Ecuador: FLASCO.

Freeman, Carla. 2000. *High tech and high heels in the global economy. Women, work, and pink-collar identities in the Caribbean.* Durham, NC: Duke University Press.

Gatens, Moira. 1996. *Imaginary bodies: Ethics, power, and corporeality.* London: Routledge.

Gill, Lesley. 1990. Painted faces: Conflict and ambiguity in domestic servant employer relations in La Paz, 1930–1988. *Latin American Research Review* 25:119–36.

Gill, Lesley. 1994. *Precarious dependencies: Gender, class and domestic service in Bolivia.* New York: Columbia University Press.

Gimlin, Debra. 2007. What is "body work"? A review of the literature. *Sociology Compass* 1:353–70.

Goffman, Erving. 1959. *The presentation of self in everyday life.* Garden City. NY: Doubleday.

Hall, A., J. Hockey, and V. Robinson. 2007. Occupational cultures and the embodiment of masculinity: Hairdressing, estate agency and firefighting. *Gender, Work, and Organization* 14:534–51.

Hochschild, Arlie R. 1983. *The managed heart: The commercialization of human feeling.* Berkeley: University of California Press.

INEC (Instituto Nacional de Estadística y Censos). 2012. http://www.inec.gob.ec.

Joseph, Nathan. 1986. *Uniforms and nonuniforms: Communication through clothing.* New York: Greenwood.

Kang, Miliann. 2010. *The managed hand: Race, gender; and the body in beauty service work.* Berkeley: University of California Press.

King, Alison Jill. 2007. *Domestic service in post-apartheid South Africa: Deference and disdain.* Burlington, VT: Ashgate.

Marx, Karl. (1844) 1978. Economic and philosophic manuscripts of 1844. In *The Marx-Engels reader,* 2nd edition, edited by Robert C. Tucker. New York: Norton.

McDowell, Linda. 1997. Capital culture: Gender at work in the city. Oxford: Blackwell.

M.I. Municipalidad de Guayaquil. 2010. http://www.guayaquil.gob.ec.

Miles, Ann. 1998. Women's bodies, women's selves: Illness narratives and the "Andean" body. *Body & Society* 4:1–19.

Nencel, Lorraine. 2008. "Que viva la minifalda!": Secretaries, miniskirts and daily practices of sexuality in the public sector in Lima. *Gender, Work, and Organization* 17: 69–90.

Rahier, Jean Muteba. 1998. Blackness, the "racial"/spatial order, migrations, and Miss Ecuador 1995–1995. *American Anthropologist* 100:421–30.

Ray, Raka, and Seemin Qayum. 2009. *Cultures of servitude: Modernity, domesticity, and class in India.* Stanford, CA: Stanford University Press.

Roberts, Elizabeth F. S. 2012. *God's laboratory. Assisted reproduction in the Andes.* Berkeley: University of California Press.

Rollins, Judith. 1985. *Between women: Domestics and their employers.* Philadelphia, PA: Temple University Press.

Salzinger, Leslie. 2003. *Genders in production: Making workers in Mexico's global factories.* Berkeley: University of California Press.

Shilling, Curt. (1993) 2003. *The body and social theory,* 2nd edition. London: Sage.

Skeggs, Beverly. 2004a. Exchange, value, and affect: Bourdieu and the "self." *The Sociological Review* 52:75–95.

Skeggs, Beverly. 2004b. Context and background. Pierre Bourdieu's analysis of class, gender, and sexuality. *The Sociological Review* 52: 19–33.

Sutton, Barbara. 2010. *Bodies in crisis: Culture, violence, and women's resistance in neoliberal Argentina.* New Brunswick, NJ: Rutgers University Press.

Introduction to Reading 28

Orna Sasson-Levy, a sociologist at Bar-Ilan University in Israel, is a specialist in gender and ethnic aspects of the military and militarism. She is the author of *Identities in Uniform: Masculinities and Femininities in the Israeli Military* (2006, Jerusalem: Magnes Press). In this article, she offers extensive analysis of the transformative bodily and emotional practices that have become key to framing male combat service in Israel as an individualistic enterprise in which the soldier perceives his military activities as "masculine self-actualization." Sasson-Levy argues that young men in Israel are willing to kill and be killed not only for the good of the collective but also in the name of individualized hegemonic masculinity. The author's exploration of links between sexist views and nationalist views in Israel has resulted in both praise and condemnation.

1. How does the soldier's body provide the "material infrastructure" of the connection between military service and the state?
2. Discuss the role of self-control and thrill as sources of the soldier's "status of exclusivity and its justification."
3. How does gendering of the soldier's body act as a control mechanism that encourages young men to go to war and risk their lives?

Individual Bodies, Collective State Interests

The Case of Israeli Combat Soldiers

Orna Sasson-Levy

This article seeks to advance a new understanding of the ways in which hegemonic institutions are embodied and reproduced through the construction of extreme masculinities. In particular, I focus on the management of the body and emotions among Israeli combat soldiers as an interface between state and military constructions and individual experience.

Until recently, most writing on the military and gender focused on the problematic of women's military service, while taking men's service for granted. However, Kovitz suggests that research should "shift from problematizing women's service to problematizing that of men" (Kovitz 2003, 1). For Kovitz, the main question is why men agree to go to war and not why women are excluded from it. More specifically, she asks how democratic societies, whose liberal values would seem to contradict the coercive values found in the military, succeed in persuading men that they want to enlist in the army and participate in warfare.

The primary question this article raises, then, concerns how states convince young men to go to war. This is not a new question, and it has been answered before in different ways. First and foremost, states generate nationalist feelings in their subjects by constructing collective identities based on (real or invented) traditions and a common origin, and develop militarized patterns of socialization that prepare their youth to join the military forces (Furman 1999). At the same time, they produce (or maintain) a perception of existential threat, which can be very real or utterly constructed. In both cases, men enlist into the army in response to the call of the state.

The state's call to sign up for military service is supplemented by a promise for equal citizenship. In the West, military service and war have been integral to the definition of citizenship (Janowitz 1994; Tilly 1996). Thus, minority groups sought to perform military service to demonstrate their political loyalty and worthiness as citizens and to enhance their civic standing (Burk 1995; Gill 1997).

These appeals to men to risk their lives for the good of the collective are always gendered. Militaries have been identified as masculine institutions not only because they are populated with men but also because they constitute a major arena for the construction of masculine identities and play a primary role in shaping images of masculinity in society at large (Barrett 1996; Connell 1995; Morgan 1994). Nagel (1998) shows that terms like *honor, cowardice, bravery, heroism, duty,* and *adventure* are hard to distinguish as either nationalistic or masculine. Thus, despite far-reaching political, social, and technological changes, the warrior is

Sasson-Levy, O. (2001). Individual bodies, collective state interests: The case of Israeli combat soldiers. *Men and Masculinities, 10*(3): 296–321. Reprinted with permission of SAGE Publications, Inc.

still a key symbol of masculinity (Morgan 1994, 165), and militaries are often described as "the last bastion of masculinity" (Addelston and Stirratt 1996). The chance to "become a man" is therefore another important enticement into military life.

Another reason for mobilization is economic: When the military in many Western states shifted from mandatory conscription to all-volunteer forces, it was often those in dire economic situations who enlisted. In the United States, for example, military service is perceived as a path for social mobility, and most soldiers sign up for economic reasons (Enloe 1980; Moskos 1993, 86). And, finally, in some countries, though not as many as in the past, men enlist because of coercive state laws requiring mandatory conscription, which is the case in Israel.

Most of these reasons for enlistment are related to the formation of the modern nation state and are thus relevant to Israel as well. However, with the impact of globalization, the configuration of the modern nation-state is rapidly changing (Comaroff 1996), and the link between citizen and state is taking on a new character, a consequence of which is changing mobilization rates. In Israel, the impact of globalization has been a gradual shift from a collective orientation toward a more individualistic one (Horowitz and Lissak 1989; Ram 1999).[1] The roots of this social shift are usually traced to the end of the 1970s, with Israel's signing of a peace accord with Egypt, the growth of the capitalist free market, and the influence of globalizing developments that accelerated processes of privatization. These processes stimulated the decline of communal public spiritedness and the rise of an individualistic orientation that emphasizes self-fulfillment and human and civic rights (Shafir and Peled 1998).

One might expect that the decline of Israeli collectivism would result in the devaluing of the militaristic ethos in general and the status of the warrior in particular. Moreover, the current armed conflict is not perceived as posing an existential threat to Israel's security and does not demand total mobilization. In such a situation of a protracted "low-intensity" conflict, the state may well encounter difficulties in maintaining the dominance of a republican discourse, which exalts the self-sacrifice of the masculine combat soldier.

However, in spite of social and ideological changes, Israel's Jewish community perceives the military as the emblem of pure patriotism and as one of the major symbols of the collective (Kimmerling 1993). In this militaristic culture, the (Jewish) combat soldier has achieved the status of hegemonic masculinity and is identified with good citizenship.

There seems to be a gap, then, between the individualistic trends and the tenacity of the status of the combat soldier. How can we explain the persistence of the warrior's hegemonic status, despite the changes in the relationship between the individual and the collective, the citizen and the nation-state? In this ambivalent social context, how can the state create and maintain "armed masculinity" (Snyder 1999) as a normative social ideal? My aim in this article is to propose other, more subtle ways through which men are lured into fighting at a time when the link between the individual and the state is being transformed.

In the past few years, there has been a growing interest in militarized masculine bodies (Armitage 2003). The research in this field explores how the masculine civilian body is adapted to military use and demonstrates how, with the inculcation of military principles, the body is classified, transformed, and reshaped to meet military and state goals (Ben-Ari 1998; Higate 2003; Peniston-Bird 2003). I suggest that we should look at the connection between these transformative body techniques and the motivation or persuasion question posed by Kovitz (2003), to analyze how specific constructions of militarized bodies lure men into fighting and thus serve the interests of the state. Based on in-depth interviews with combat soldiers, I argue that the construction of the Israeli combat soldier involves two seemingly opposing themes: on one hand, self-control, and on the other, thrill (in Hebrew, *rigush*). While the theme of self-control is characterized by introversion, self-restraint, and self-repression, the theme of thrill accentuates the outward expression of wild, unrestrained feelings, stemming from life-endangering events, adventurous activities, and unique opportunities the military offers for intimacy among men. These interdependent themes accentuate a growing sense of agency and self-actualization, thus allowing, and even promoting, their interpretation through an individualistic frame. The individualistic framing of bodily and emotional practices enables an ambivalent and nonpolitical interpretation of military actions, as it disguises the coercive nature of military service and obscures the collective state interests that are served by the individual male body.

Theoretical Overview: Gendered Body and Emotion Management

Social theories have shifted in the past two decades from emphasizing the cognitive dimension of identity construction to exploring the embodiment of identities (Shilling 1993, 3). The growing interest in the sociology of the body derives both from feminist thought, which highlighted women's bodies as the major political site of patriarchal domination (Bordo 1993), and from Foucault's (1975, 1978, 1980) writings, which enabled research on the "history of the body" and provided the outline for it.

Foucault's innovation lies in analyzing the body as produced by and existing in the discourses and social institutions that govern it. For Foucault (1975), the body is an object of power and a direct locus of social control. His ultimate example of the external construction of the body is the soldier, who, by the late eighteenth century, had "become something that can be made; out of formless clay, an inept body, the machine required can be constructed" (p. 135). The soldier's docile body serves to exemplify the link between daily practices on one hand and the large-scale organization of power on the other (Dreyfus and Rabinow 1982).

While Foucault's writings portrayed the body as a surface to be written on, other sociologists argue that one does not have to accept naturalistic or sociobiological views to acknowledge the body's materiality and corporeality, which always limit its range of possibilities. Frank, for example, proposes to analyze the body as constituted in the intersection of an equilateral triangle, the points of which are institutions, discourses, and corporeality (Frank 1991, 48–49). Similarly, Shilling (1993, 12) conceptualizes the body as "an unfinished biological and social phenomena, which is transformed, within certain limits, as a result of its entry into and participation in society." Foucault himself, in his later writings, attributed more agency to the individual, suggesting that a theory of domination must start with the body dominating itself (Foucault 1978). Social analysis has thus expanded from studying the body as an object of social control and discipline "in order to legitimate different regimes of domination" (Bordo 1993; Foucault 1980) to perceiving the body as participating in its own shaping, as it creates meaning and performs social action (Davis 1997).

As a signifier of social identity, the body participates in the constitution of social inequalities. This is clear when we look at men's and women's bodies. Connell (1987) argued that by neglecting biological similarities and exaggerating biological differences, gendered practices create corporeal differences between men and women where none existed previously. These differences are then used to justify and legitimize the original hierarchal social categories.

Frank (1991) points to two types of bodies that can characterize the combat soldier: the disciplined body (pp. 54–61) and the dominating body (pp. 69–79). When the disciplined body, which "makes itself predictable through its regimentation" (p. 55), directs practices of discipline and control upon others, this body turns into the second type, the dominating body, which is, according to Frank, exclusively male. A prime example of the dominating body can be found in *Male Fantasies,* Theweleit's (1987–1989) rich description and analysis of masculinities constructed in the German Freikorps, an army unit that was formed at the end of World War I (soldiers in this unit later served in the SA and SS). For the German fascist, Theweleit argues, there existed only two types of bodies; the first was the erect, steel-hard, "organized machine" body of the German master. This controlled and emotionally bereft male warrior ideal reacts with revulsion and fear to the second body, the flaccid and fluid female body of the negative Other, lurking inside the male body.

As is obvious from this analysis, bodies and embodiment cannot be discussed in isolation from emotions, which are experienced through the body and shape the way we experience it (Csordas 1993). Similarly to the body, emotions "are treated as material things" (Lutz and White 1986, 407), as a universal aspect of human experience that is least subjected to social construction (Abu-Lughod and Lutz 1990, 1). However, the historical and cultural variability of emotions suggests that "subjective experiences and emotional beliefs are both socially acquired and socially structured" (Thoits 1989, 319; Rosaldo 1984). Emotions can thus be understood as *discursive practices,* which are "created in, rather than shaped by, speech" (Abu-Lughod and Lutz 1990, 10, 12).

As socially constructed *ideological practices,* emotions are always "involved in negotiation over the meaning of events, over rights and morality, over control of resources" (Lutz 1988, 5). Power relations determine what must, can, or cannot be said about the self and emotions; what is believed to be true or false about them; and what can only be said about them by certain individuals and not others. These "feeling rules" (Hochschild 1983) are always gendered; since women are believed to be "more emotional than men," any discourse on emotion is also a discourse on gender (Lutz 1988, 69). The masculine imperative for emotional self-control, which is especially pertinent to combat soldiers, confers men with prestige and locates them in a superior position to women. Emotional norms are thus produced by dominant institutional arrangements and function to sustain them (Thoits 1989, 328). This is particularly accurate when one examines "extreme" gender identities that are constituted by and for the state, identities such as warrior masculinity.

Combat Masculinity and the State

Military service, and especially warfare, is one of the primary means by which Western states established their power within societies (Giddens 1985; Tilly 1985). The link between war and state building is expressed in the republican ethos, which defines the subject's citizenship according to his (but not her) contribution to the state. In militaristic societies, the most significant contribution to the state is participating in the armed forces. This connection between military service and the state is based upon the glorification of militarized masculinity, with the soldier's body providing its material infrastructure.

Following the transition to voluntary professional armies in Western nation-states, the military lost some of its power to shape the meaning of citizenship. In Israel, however, the link between citizenship, military service, and masculinity carries a special meaning (Kimmerling 1993). War and routine conflict management have played a central role in shaping Israel's Jewish community of citizens, a community in which civic virtue is often constructed in terms of military virtue (Helman 1997). In this social context, the Jewish combat soldier has achieved the status of hegemonic masculinity,[2] which has become synonymous with good citizenship (Lomsky-Feder and Ben Ari 1999; Sasson-Levy 2003a). Although combat soldiers make up only 20 percent of the total complement (Cohen 1997, 86), the ideal of warrior masculinity is consensual, transcending ethnic, religious, and class boundaries (though not national ones).

War and military service are represented as enabling the male subject to "become a man" (Morgan 1994; Mosse 1996; Peniston-Bird 2003). Although the military should not be seen as constructing a single embodied masculinity (Barrett 1996; Enloe 1988; Higate 2003; Morgan 1994; Theweleit 1987–1989), theoretical literature on the combat soldier's identity tends to emphasize the single aspect of physical and emotional self-control (Arkin and Dubrofsky 1978; Ben-Ari 1998). War provides the opportunity to nurture the individual's ability to endure pain and even mutilation and to control emotion. Any sign of weakness, vulnerability, or even sensitivity can be interpreted in the military as a sign of homosexuality and, hence, of "failed masculinity" (Petersen 1998, 53).

The Israeli militarized body is fashioned according to a utopian collectivist and gendered ideology, emphasizing the "chosen body" of the healthy, strong, and active Jewish male (Weiss 2002). This "perfect" body, which in reality was the lot of only a select few, was seen as an ideal type, and comparisons with it generated a hierarchy of bodies, defining the boundaries of the collective and its internal stratification (see also Peniston-Bird 2003). The combat soldier—who possesses the perfect body—proves his masculinity through emotional self-control that is attained to cope with stress, anxiety, chaos, and confusion, all of which characterize the battlefield (Ben-Ari 1998, 42–46). As the schema of the battle is also a schema for achieving and reaffirming Jewish Israeli manhood (Ben-Ari 1998, 112), controlling one's emotions is perceived as successfully passing the test of masculinity.

My argument is that alongside the discourse of self-control and discipline, an additional discourse of thrill and excitement enacts and shapes the body and emotions of the warrior in different ways. The discourse of thrill is of critical importance, as it is emphasized by soldiers as a major force in mobilizing their motivation and willingness to go to war. Thrill plays an important role both in the discourse the army uses to lure men into soldiering and in the narratives of soldiers themselves. It appears, then, that even within the military cultural arena, which has been presented as the epitome of hegemonic top-down construction, there are signs of agency, where soldiers actively seek bodily and emotional experiences with which to constitute and assert themselves.

Method

This article is based on interviews with twenty male combat soldiers within one year of their release from army service: one artillery man, three infantry soldiers, three parachutists, four from the engineering corps, three from reconnaissance units, one from the navy, four from the armored forces, and one former combat intelligence officer.

Body practices and emotional management were not the main focus of the interviews. However, these issues were raised by the soldiers so often that I was forced to recognize their centrality in the military experience of the combat soldier. In comparison, noncombat soldiers whom I also interviewed (see Sasson-Levy 2003b) did not talk about body practices in the interviews at all.

It is important to note right from the start that there is no unified and universal version of "Israeli combat masculinity." The masculine model of the pilot, for example, is quite different from that of men in the armored forces, the navy, or the infantry brigades. Moreover, there are even differences between one infantry brigade and another. For instance, the masculinity of the paratroopers' unit in Israel emphasizes rationality, self-control, and self-discipline, while the men of the Golani infantry brigade accentuate resistance to discipline and rebelliousness along with physical capabilities and courage. These different masculine identities indicate that the act of fighting itself does not require one specific gendered ideology and that there is nothing "natural" or essential about the warrior's identity that derives from the act of fighting itself. Due to space limitations, this article cannot show the different nuances of armed masculinity in Israel and can only delineate its general contours.

Bodily and Emotional Self-Control

Soldiering the Body

According to military belief, "masculinity is determined primarily by a healthy body, not a healthy mind" (Arkin and Dobrofsky 1978, 156). Having a masculine body that is healthy, strong, and sturdy is a prerequisite to becoming a combat soldier. However, having the right body is insufficient; one must also be willing to shape and discipline it so that it meets military objectives. Therefore, the tests for combat units (held a few months before enlistment) examine both the physical abilities of adolescent boys and their willingness to stretch these capacities to the limits. Yonatan, who served in the prestigious air force rescue unit, explained, "In basic training, you receive all your credit through physical activity alone. The minute you're physically inactive [if you're injured], you're a leper, you're a leech, you're a parasite, and you're good for nothing."[3]

In other interviews, soldiers repeatedly emphasized the theme of self-control over their bodies. To be more precise, the men did not talk about self-control as much as they described their struggles to achieve control over their bodies. Eldad, a tank commander, explained,

> During long marches, the strain would push people into screaming at each other, really cursing one another; their legs would hurt and they'd feel that this is the end, that they'd reached their limit, and then they would see that they could stretch that limit a little further. This exercise teaches them that the limit is not where they thought it was but always a little further.

Basic training is devoted to forging and strengthening the male body, to taking it to new extremes. At the same time, it inscribes on the body the signs of one's specific combat role. Through the specific body management of each unit, the soldiers shape a new military identity. Simple daily activities such as walking, eating, and sitting assume a new form and meaning. Guy, an infantry soldier in an elite unit, vividly described the unique bodily dimensions of each unit, even during supposedly "passive" activities such as platoon talks:

> We would see the other platoons, sitting relaxed and talking to their commanders. But in my unit, we would have to sit absolutely straight, our gun upright and hands stretched straight in front of the body. The talk could last forty-five minutes, or an hour, but if a fly zooms by your eyes and you take your hand off your rifle to shoo the fly away, you pay. You start running, and running, and running.

Guy's description of his unit's specific style demonstrates how military thinking and culture are transmitted through the body and imprinted on it (Foucault 1975). His quote alludes to the central role of the body in military punishment. As one of the main mechanisms for discipline, punishment is often inflicted directly on the body, through recurring "stretcher hikes," carrying heavy loads, crawling on thorns, doing dozens of push-ups or hundreds of sit-ups, and more. Physical punishment inscribes on the soldier's body the fear of military discipline and the dread of authority, until he internalizes military principles and they become a part of who he is. Thus, formal homogenization processes, such as uniforms and haircuts, are accompanied by violent mechanisms that aim at creating a standardized combatant male body.

Foucault (1975, 138) argued that "the military apparatus explores and studies the soldier's body in order to break it down and rearrange it according to its needs." Indeed, at the end of basic training, if the soldier completes it "in one piece" (and not all do), he has a body capable of things it was not capable of before; he has built the "body of a soldier." Dudi, an officer in the combat engineering corps, said,

> I never understood how it works, but after you're in the army for about a year, all these problems vanish. I don't know why, but I guess the body adjusts somehow. As a commander, I would carry things much heavier than during basic training, and I would say to myself that this is it, you just get used to it. I guess that just as the soles of your feet are getting rough, the skin on your back also gets rough at a certain stage.

The thick skin that Dudi develops "should not be understood metaphorically" (Theweleit 1987–1989, 144). The physical transformation bears institutional implications: as the soldier ceases to experience his body's pain or hunger as his own (Frank 1991, 56), soldiering becomes easier and more tolerable. Now, the soldier can better meet the needs of the state.

Bodily Masculinity Rites

The transformation of a body into a "soldier's body" has an unmistakable gendered meaning. Through the soldiering "body project" (Connell 1995, 50), men's bodies become visibly different from those of women. In this sense, the soldier's embodiment plays a central role in the social construction of polarized gender identities and hierarchal gender regimes (Connell 1987). While intense physical strain prepares soldiers for combat conditions, it also serves as a selection mechanism and a rite of passage into Israeli hegemonic masculinity.

Service in the Israeli military is seen as the primary rite of passage that initiates adolescent boys into full membership in the masculine Zionist civil religion (Aronoff 1999; Ben-Ari 2000; Levy-Schreiber and Mazali 1993). Reuven's description of crucial moments during basic training indicates that the body is the main arena for these militarized rites of passage:

> The procedure is you get into the shooting position; you have to remain firm so that nothing moves you. We would get into shooting position, stand in line, and the officer . . . the bastard . . . he would come and kick us in the knees, the chest, the muscles of the legs, and we're supposed to stand firm, not to move, the muscles all tight . . . all your strength is invested in holding the weapon so that when he kicks you again it won't move you and won't make you shoot off-target.
>
> If I moved . . . did you notice my eye's reflex action? If that would have happened to me then it would have been a sign that I was frightened and that would have meant I was finished. He would have grabbed me by the shoulders and kneed me in the gut. He would beat me to a pulp.

Recall Spencer's (1965, 103) description of rites of passage through public circumcision by the Samburu: "Each youth, placed on view before his male relatives and prospective in-laws, must remain motionless and silent during the cutting. Even an involuntary twitch would be interpreted as a sign of fear." The similarities between the two descriptions of rites of passage—one in an African culture, the other in a seemingly modern western culture—are self-evident. In the same way, parachuting in the Israeli army is viewed as "a test which allows those who pass it to join an exclusive club, to be initiated into an elite group" (Aran 1974, 150). Soldiers who go through military hardships intuitively understand that those who withstand them will achieve elite status. As hardship becomes the mark of masculinity, soldiers not only expect to suffer in basic training but are also proud of it and reiterate this in the interviews, as if to validate their hegemonic status (see also Gill 1997). Moreover, most combat soldiers complete basic training with various physical afflictions, some of which will never disappear. Military casualties are often the most dramatic claim to centrality in Israeli society (Aronoff 1999, 42). Therefore, the soldiers are proud of their wounds and scars, which serve as evidence of their willingness to sacrifice their bodies for the good of the collective.

The combat body project not only creates hierarchal gender differences, however, but also stratifies male bodies. The "chosen body" (Weiss 2002) of the combat soldier depends on the existence of the wrong body, the body that fails to become a combat soldier. The literature often specifies female or homosexual bodies as representative of the "wrong" military body. However, the soldiers I talked to did not compare themselves to women or homosexuals but mostly to other male heterosexual soldiers who had failed to endure the physical training—that is, the fat soldier, the lazy soldier, the "crybaby," or the soldier who is too small. As Robert Connell (1995, 75) notes, masculinity is a relational identity that is often constructed in relation to other masculinities. Michael Kimmel quotes the playwright David Mamet, saying, "Women have, in men's minds, such a low place on the social ladder . . . that it's useless to define yourself in terms of a woman. What man needs is men's approval" (Kimmel 1996, 7).

This self-comparison among men is, for Connell (1995), the reason behind most acts of violence. Most episodes of major violence are transactions among men, used as a means of drawing boundaries and making exclusions (Connell 1995, 83). Violence can become a way of claiming or asserting masculinity in struggles between groups. This is obvious in the attitudes of combat soldiers to those of their peers who do not meet the hegemonic model of masculinity.

Among those who fail the tests, the fat soldier is especially prone to abuse. The overweight soldier represents the opposite of "everything military." Omer, the tank driver, told the following story, which demonstrates the moral threat embodied in the fat soldier's body:

> We wouldn't let soldiers weep during punishments. We had this fat soldier, Danny. He would cry as a matter of principle, whether or not it was difficult. I remember that fatso starting to put on a show as if he was throwing up. I told him, "Hey, watch it, I'm warning you. Shut up," I told him, "You're crying like a little girl."

Danny, the antihero of this story, is doubly wrong—he is fat and cries. The other soldiers are angry with him because they see him as a spoiled, childish liar—all the characteristics that represent for them the opposite of manhood—but also because they worry that his behavior will reflect negatively on them. "It projects on all of us," Omer said. "We wanted to be seen as good guys, the best team." Those who have the wrong body present a threat to the masculinity of the whole group, a threat to social solidarity, and therefore, they constitute a legitimate target for ostracism, ridicule, and abuse. Thus, militarized initiation rites may produce group bonding, solidarity, and coherence (Winslow 1999), but they also create and maintain hierarchical differences among men.

Emotional Control

The perception of emotional control as a signifier of masculinity and hence of higher status (Lutz 1988) is not unique to the Israeli context. Emotional control has been associated with masculinity in Western culture from the early writings of the Greek philosophers through Kant, Descartes, and Hegel (Lloyd 1986; Ortner 1974). In their writings, men's emotional self-control (in other words, the emphasis on reason) signifies superiority (Seidler 1989) and is used to arrange and justify gendered hierarchies. Militarized masculinities seem particularly "obsessed" with emotional control. The Israeli combat soldier refers to emotional control or composure (in Hebrew, *kor ruach*—literally, "coolness of spirit"; Ben-Ari 1998, 45) as a personal and professional masculine achievement, an ideal that should be adopted by all soldiers. Kor ruach, which refers to the ability to act with confidence, poise, and composure under trying circumstances (Ben-Ari 1998, 45), is perceived as a key element of effective performance in day-to-day situations and especially in times of crisis. Alon, a tank commander, said, "Eventually, I became a better commander, more professional, and more understanding. I mean, I became better at commanding with composure: they shoot at us and I remain composed, a soldier tries to commit suicide and I remain composed." Alon describes a process by which he gradually achieves composure. This is important because he equates composure with successfully performing as an officer and considers it a criterion for evaluating competence. Similarly to Alon, other soldiers I talked to also did not perceive emotional self-control as a given masculine characteristic. Rather, they described it as an attribute that one learns, acquires, and perfects as one develops into a more professional soldier (and by extension, "more of a man"). Rami, a combat medic, described how the military creates this composure during the medics' course. Note how he links emotional composure to hypermasculinity:

> [In the medics course,] you come in, play with the dog, it wags its tail and all that, and a minute later, you see him in pieces like in a butcher's shop. It's this kind of act, which says, "I'm macho. I'm not afraid of blood." They want to get you to a certain threshold, so they show you pictures of horrible things. Victims of accidents, smashed limbs, smashed people. They show this to you so that you get accustomed to all this kind of stuff.

While the military applauds emotional self-control and develops mechanisms for teaching and even enforcing it, it would not be accurate to say that the army forbids all emotional expression. Rather, there seems to be an unwritten social and spatial "emotional mapping," similar to Hochschild's (1983, 18) "emotional rules," that dictates which emotions are allowed, when, where, and for what audience. Soldiers are encouraged, for example, to feel motivation and ambition; they can express happiness and pride on the day they complete basic training or more advanced courses; they are allowed to feel homesick (but only to a limited degree); and they are expected to feel desire for women. Most of all, they are supposed to feel affection and camaraderie toward other soldiers in the unit, an emotion that the army creates and exploits (and to which I will return later).

However, the demand for emotional self-control is often loaded with internal contradictions and paradoxical effects. First, acquiring emotional self-control facilitates the soldier's "automatic docility" (Foucault 1975), as it ensures that the soldier will not rebel or be paralyzed by fear in combat. By guaranteeing a high level of obedience and passivity, the imperative of emotional self-discipline provides a solid base for the constitution of the conformist citizen. However, paradoxically, the docility of the "good citizen" is experienced by the soldier as increased self-control and heightened masculinity. Thus, compulsory military service contributes to the disciplining of the citizen and reproduces the association between rational and restrained masculinity (Seidler 1989) and the interests of the state.

The second paradox relates to the fact that although power and control are perceived as central to the definition of *combat masculinity,* in reality, the combat soldier has only a limited degree of autonomy. To become a combat soldier, one must surrender one's autonomy and obey one's commanders for the major part of everyday life. Amos, a paratrooper, expressed this paradox when he explained how he survived basic training: "You simply have to say O.K., you, the army, can do whatever you want with me. I . . . in the end, I will make it. Even if I'm the last one, I will get there and succeed." When Amos says, "You can do whatever you want with me," we learn that the power and control of the combat soldier are paradoxically achieved by surrendering control over the self. Moreover, giving up autonomy is not a passive project. On the contrary, the soldier needs to mobilize all his willpower to get through his seemingly endless training and to gain control over his body and emotions. Obedience is associated here with strenuous effort, hard physical work, and a strong will. Stiehm observed that "patriarchy promises power and benefits to the young men who ultimately prove themselves. But is that proof one of talent, of merit or of morality? No. It is, in fact, proof of submission. It is evidence of obedience. It is a demonstration of compliance and of willingness to risk and sometimes sacrifice" (Stiehm 1989, 227). As a reward for their strenuous efforts to attain physical and emotional self-control, soldiers are given the opportunity to experience physical and emotional thrill, which is regarded as an exclusive experience of combat soldiers.

Thrill

While most of the literature focuses on the theme of self-control, a second, and apparently contradictory, theme of thrill (in Hebrew, *rigush*—or "rush" in American slang) appeared time and again in the soldiers' narratives. Rigush implies intense feelings, both emotional and physical, that derive from extraordinary experiences that are perceived as inaccessible in everyday life. While militarized masculinity demands emotional control, military life nonetheless provides unique opportunities for experiencing the extraordinarily deep feeling of rigush.

The major source of rigush is the risk to one's life involved in being a combat soldier. Israeli culture imparts the heroic notion of self-sacrifice to Israeli males from early childhood. Even at this age, children are exposed to themes of persecution, heroism, and war, which recur throughout the year. Aronoff, for instance, claims that the memorialization of the dead soldier is so central in Israeli culture that it has evolved into a "national cult of the dead" (Aronoff 1999, 43). Although the ethos of self-sacrifice has been modified in the past twenty years (Zerubavel 1995), it still constitutes a major part of the combat soldiers' narratives. Dudi, a platoon commander in the combat engineering corps, explained,

> Do I think the army makes a man out of the boy? Look, it all depends on where you serve. The army turns a child into a man only if he serves in a combat unit. There you have to deal with issues of life and death, you get a sense of proportion concerning what life is all about, and you understand what is real in life. But I don't think the army makes men out of everybody. If you were in the units where I served, or in the infantry, or even in the armored corps or artillery [then it does], but the rest of the units—I can't guarantee it.

For Dudi, masculinity is directly associated with confronting issues of life and death and is achieved through facing the biggest test of all: the willingness to sacrifice one's life for one's country. Self-sacrifice is the sign of the hero, he who has the courage to rise above his basic instinct for life and fight for the good of his imagined community. Serving as a combat soldier provides one of the very few chances in life to receive recognition for heroism, which is a source of honor and thrill in itself. Endangering one's life is seen by soldiers as the ultimate actualization of both masculinity and nationality, and thus, it serves as a criterion in social and military stratification systems (E. Levy 1998; Lomsky-Feder 1998; Weiss 2002). Soldiers perceive life-threatening events as rewards in and of themselves, a source of gratifying rigush. This is what Alon, the tank commander, said about it:

> When we were in Lebanon, they shot at me lots of times. It gives you a feeling of gratification. To my mom, it sounded dumb, but I felt gratified. Because when they shoot at you, you feel the best in the world. Whether you returned fire or not, it doesn't matter. The minute they shoot at you, you already feel satisfaction. Why? Because your life was at risk. . . . It's a special feeling, reserved only for combat soldiers.

The Lebanon war (1982–2000) was perceived by many Israelis as a war of choice rather than a war of self-defense, and it was therefore a bone of contention in Israeli society, a cause for demonstrations and political strife. Alon and his friends, however, interpreted their combat experience in Lebanon as an adventurous and thrilling way to actualize their masculinity and seldom talked about the political meanings and ramifications of their service.

Somewhat different voices were heard when the soldiers talked about serving in the occupied territories, where they were primarily fighting against civilians. There, the soldiers were more aware of the political meanings of their military service and expressed moral doubts. Yet only a tiny minority of soldiers refused to serve in the territories, an act which had the potential to modify the link between masculinity and self-sacrifice in Israel (Helman 1997).

Only three of the combat soldiers whom I interviewed spoke out explicitly against the collective ethos of risking one's life. Eli, from the Nachal infantry brigade, said,

> Look, most of the soldiers that were killed were from *Giva'ati* [another infantry brigade]. In my brigade, people didn't like getting killed. We liked the good life, we liked doing things wisely. If you ask people honestly if they want to fight, without the pretence of the other guys, they'll say no.

Yoram, a soldier in the prestigious bomb squad unit, reached the same conclusion but only during a long trip abroad after his release from the army. Apparently, his long stay outside of Israel enabled his estrangement, lending a new perspective to his role as combat soldier:

> When I was in South America, I became very, very antimilitaristic. I had thoughts I never had before. . . . I asked myself: Do I really want to die for my country? Am I really prepared to die for my country? No way. What kind of twisted thinking is that? I've only got one life, what good will it do if I . . . what good will it do me?

Yoram and Eli represent an alternative perspective. They do not see life-endangering events as exciting and prestigious goals in their own right. By disclaiming the ethos of self-sacrifice, they abrogate the connection between masculinity and risking one's life. However, they reflect the opinions of only a small minority of combat soldiers; furthermore, they continue to serve as combat soldiers when called to reserve duty.[4]

A second source of militarized thrill lies in having control over weapons and technology. Sally Hacker (1989) has pointed out that basic pleasure is gained from the ability to operate complicated technological instruments. This widespread masculine fascination with technology (Morgan 1994, 173) seems to increase when it involves weaponry, which is linked to control, hierarchy, prestige, and power over others.

Rami, a combat medic in the armored corps, described an officer giving the command to open fire:

> You see the armored regiment commander waging a war with his words. . . . It's a crazy male act, wacko machoism. He utters these few phrases, and then he lets them go, as if letting go of their reins, or as if he's lifting the gate at the horse race, and you see forty tanks fire at once. . . . You cannot but be impressed.

Rami associates the enjoyment of warfare technologies with domination and hypermasculinity. Note the sexual undercurrent in his choice of metaphor of "letting go of the horses at the race." Hacker (1989) has pointed out illuminating parallels between technology and eroticism: both inscribe feelings on the body through kinesthetic experience, the pleasure in both is shaped by domination and control, and both are defined predominantly by men and are stratified by gender. When the soldiers express the "near orgasmic excitement of nighttime explosions" (Morgan 1994, 173), technologies of destruction become erotic in and of themselves (Hacker 1989, 46–55). Indeed, describing weapons in terms of sexual metaphors is by now utterly clichéd, but soldiers do it often, especially when talking about the actual act of shooting. Eli, an infantry soldier who was earlier quoted expressing his rejection of the ethos of sacrifice and military values, admitted that he enjoys shooting:

> Why do I like shooting? I don't know, I can't even explain to myself what I love about it. But it's both something physical, and something you can see. It's real, it moves. And there are very close and immediate measures of success. And maybe it's also the feeling, of, ha . . . [sexual] relief.

This "peculiar masculine eroticism of technology" (Hacker 1989, 46) was reiterated by many soldiers when referring to rifles, tanks, or hand grenades, again turning a blind eye to the violent and destructive effects of the technologies of war.

A third source of excitement was the unique feeling of youthful adventure that characterizes combat life. "If you like action, you'll love [military] service in the territories," said Guy.[5] In a similar way, Alon only saw "fun" in the occupied city of Hebron, where forty-two thousand Palestinians are bullied and intimidated by five hundred extremist rightwing Jewish settlers:

> In Hebron, we would go wherever we felt like. We'd raise hell. They'd throw stones, shoot at you, and throw things at you. What a mess. It's just paradise for those who like this kind of thing. It was the most action you could get.

Ofer, an officer from the engineering corps, said explicitly, "I love action. Action itself. Once, I didn't sleep for three days because I couldn't bear the thought that something would happen and I wouldn't be there."

The excitement of dangerous situations is felt in the body itself, in the tone of the muscles, the heartbeat, and the bloodstream. Parachuting, for example, which involves both exhilaration and fear, "brings men into a state of trance, a sort of ecstasy" (Aran 1974, 125, 131). This is a unique emotional "perk" that the military offers to combat soldiers. They feel that they are "alive," their bodies, emotions, and senses all exposed and active. "I was willing to die for the thrill," said a soldier in an interview with a daily newspaper (Alon 1998, 22).

> True, in bungee jumping or car racing, one can also experience an emotional and physiological rush associated with achieving and proving hypermasculinity. But the differences between bungee jumping and combat are not trivial. First, combat is more dangerous as it is connected to real risks to one's life, and second, one does not enter it out of choice—it is always linked to the interests of the state. Its significance, then, is always beyond the individual.

The last source of excitement, albeit an unspoken one, is military homosociality (Sedgwick 1985). Military service provides rare legitimacy for physical and emotional intimacy among men, including homoerotic sensations, without the stigma of homosexuality. Yoram, from the bomb squad, explained,

> It was in the army that all the barriers about physical contact between men came down. In the army, there is always touching. You walk together, shower together, you touch, you live, everything together. I think most people in the army touch each other. I really can't explain it. Maybe it's because it's so tough in the army, so you slap each other and it helps; it feels good.

The soldiers' narratives on men's bonding in the military reveal the contradictions inherent within combat masculinity. While hegemonic masculinity does not allow for emotional expressiveness, it does create specific areas in which physical and emotional intimacy among men is allowed and even encouraged. This intimacy, known as "camaraderie in arms," is a significant motif in Israel's heroic epics and is conveyed to Israeli youth long before they enlist into the army. Camaraderie in arms is described in Israeli culture (through popular songs, canonical prose and poetry, and Memorial Day rituals, for instance) as a lifelong relationship that flourishes despite social or political differences. Research among eighteen-year-old boys found that they rank camaraderie in arms as one of the main motivating factors for enlisting to combat units (E. Levy 1998, 263). The perception of camaraderie as a significant reward for military service (Lieblich 1989) is reflected in the words of Oren Abman, an infantry lieutenant colonel, who was quoted in the newspaper as saying, "The amazing intimacy that is created among men after combat, the affection they express for each other, it may sound strange to you, but it doesn't exist even in sex" (Becker 2000). Abman creates a hierarchy whereby the bond between warriors is superior to any relationship with a woman. Ofer, from the engineering corps, added,

> [It's] freezing cold, everybody is hugging and huddling under the covers with each other. No problem. These are the fun parts of the army that you remember later, and you miss them. It's all so natural, there's nothing sexual in these relationships. Even today, I don't hug everyone I bump into on the street, but if he's an army buddy, then yes, I'll hug him.

The fraternity I told you about—hugging an army buddy but not another guy—it couldn't happen with women.

Ofer proclaims an exclusive link between male intimacy and combat life, but he hastens to reject any homoerotic interpretation of this relationship, thus reinforcing the centrality of heterosexuality in the construction of hegemonic masculinity. Apparently, the social license for male intimacy is only awarded to combat soldiers (and football players), probably because they have already proven their masculinity beyond any doubt (Chapkis 1988). Noncombat soldiers, on the other hand, often talk with envy about warriors' camaraderie, which they missed out on by not serving in combat units.

By marking the uniqueness of the combat soldier, camaraderie in arms creates an exclusive, imagined community of warriors, a community that embodies "the essence of Israeliness" (Helman 1997). This masculine community is based on inner unity and homogeneity and connects the individual with the state. At the same time, the mutual commitment among men serves to draw hierarchal boundaries between those who are entitled to belong to dominant groups and those who will never be able to, namely, women, Palestinian citizens of Israel, and lower-class noncombat soldiers.

Discussion: The Individual Body and the Collective State

As we have seen, the management of the Israeli combat soldier's body and emotions merges two seemingly contradictory themes: self-control and thrill. This combination creates someone who ostensibly has the agency to take charge of his destiny—a man who can control his body and emotions—and dares to take risks and enjoys them. Thus, the combination of self-control and thrill accentuates values of autonomy and self-actualization, which call for an individualistic frame of interpretation.

The individualization of military practices leads soldiers to interpret their military experience in ambivalent and nonpolitical terms. For example, in the interviews, soldiers often framed their growing obedience to military discipline as increasing physical and emotional self-control, which created a strong sense of agency and empowerment. As we saw in the soldiers' quotations, life-endangering events were perceived as unique and rewarding opportunities for self-actualization. The strict surveillance typical of military life was interpreted as their own choice and sometimes even as a privilege, a sign of the intensity of their specific combative "trials of manhood" (Ben-Ari 1998). The authoritarian principles and practices of the military are thus disguised as belonging to, and even as promoting, an individualistic discourse.

The individualization of the soldier's body can be seen as an expression of the effects of globalization on Israeli society. One primary effect of globalization in the cultural sphere is the permeation of consumerist values, which present self-fulfillment as a prime social virtue (Ram 1999; Shafir and Peled 1998). In a consumerist society, with its focus on the individual, cultivation of the masculine body is seen, for the first time in Israeli history, as no less legitimate than the cultivation of the feminine body. Military bodily transformation, which involves strenuous efforts, can be framed in this context as a personal choice that brings prestige. Thus, the soldier's identity is not that of the Spartan, who is willing to sacrifice himself for the good of the collective. Rather, it can be seen as an individual effort to improve oneself, a masculine personal achievement of self-actualization.

When military combat service is framed in an individualistic discourse, it turns out to have a dual, ambivalent nature. This duality in the perception of the military blurs the boundaries between choice and compulsion, and the coercive nature of military service becomes obscured. This dual nature of military experience enables mandatory conscription to be perceived as voluntary and fulfilling (Y. Levy 1993). In fact, only volunteers can serve in the most prestigious, dangerous, and secretive units. Thus, voluntarism becomes a status symbol in itself. At the same time, voluntarism conceals the military disciplining of Israel's body of citizens.

Likewise, the "individual" nexus of self-control and thrill masks the price entailed by combat masculinity and conceals its repressive nature. Current research has pointed at some of the mental and physical injuries suffered by combat soldiers, but since I interviewed only soldiers who served their full terms of three years as combat soldiers, I could not expose the more brutal damages of combat military service, such as physical and emotional disability. Most soldiers who encounter serious emotional difficulties do not stay in combat courses, and at times, they are exempt from service altogether. Thus, the voices of the soldiers who did not conform to the norm are missing, as are the voices of the mentally impaired or those suffering posttraumatic symptoms. It is these injured and distressed soldiers who pay the high price for the image of the ideal of combat masculinity in Israel.

Furthermore, as noted earlier, when soldiers perceive their militaristic activities as masculine self-actualization, they can ignore their moral and political consequences. The individualistic interpretation of military endeavors enables the soldier to overlook the evils that are often carried out in the name of combat masculinity. Thus, the individualistic framing allows for the perpetuation of Israel's militant aggressive policies, in particular the nearly forty-year-long occupation of the West Bank and (until recently) the Gaza Strip.

Therefore, the individualized body and emotion management of the combat soldier serves the symbolic and pragmatic interests of the state, as it reinforces and reproduces the cooperation between hegemonic masculinity and Israeli militarism: Young men are still willing to kill and be killed for the good of their country but now in the name of individualized dominant masculinity. It is the individual body that functions as the instrument of the militaristic state. The gendering of the soldier's body turns out to be a control mechanism that encourages young men to go to war and risk their lives.

Individualism here is harnessed for the good of the collective and is hence an instrument that creates, in a roundabout way, collectivism and obedience to the state. The soldier's autonomy and individualism are a kind of illusion, a façade, because as the body of the soldier is transformed, he becomes part of the state. His body is the material superstructure that links the (male) individual to the state.

I claim that the combat soldier is "marked" as an idealized figure that others cannot emulate. Levy-Schreiber and Ben-Ari (2000), for example, follow Connell (1987, 85) in claiming that through military body practices, state power becomes naturalized, as if it belongs to the order of nature. I argue to the contrary: Through military body practices, the soldier's power becomes unique and visible; the body belongs to the order of nation, signifying the link between manhood and nationhood. Through specific gendered body practices, and their relation with the nation-state, particular male bodies become more significant than others (Petersen 1998, 42). When the body of the adolescent boy turns into the muscular and brave body of the combat soldier, this transformation marks him with an air of heroism and masculinity, the signs of devotion and contribution to the state. His uniform, posture, walk, muscle tone, facial expression, and manner of speech, all signal both self-control and anticipation of the rare thrill of combat. This body provides the combat soldier with a physical presence that many Israelis claim they can identify even from afar. The soldier's body becomes a focus for public identification, a source of national pride, and a locus of sympathy and support. The signs of the nation are inscribed on the body, and the soldier's body becomes the symbol of the nation. This national symbol provides a common, consensual symbol around which a Jewish Israeli imagined community gathers—a symbol that is both masculine and militarized. Nationality, that unseen, imagined quality, receives a public visual expression in the body of the soldier. Thus, the body of the combat soldier is signified by the mark of the nation and serves as a signifier of the gendered nation-state.

Notes

1. Others modify this statement, arguing that the collective ethos has been replaced by two contesting ideologies that crystallized as the major alternatives for future development. The first ideology is a neo-Zionist ethnonationalism (Ram 1999; Shafir and Peled 1998), which characterizes the lower classes and the religious groups. This ideology elevates the exclusionary Jewish collectivity and its motherland, rather than the Israeli state, as a political community defined by common citizenship (Ram 1999, 333–35). The second, competing ideology is a liberal-individual oriented post-Zionist one (Ram 1999; Shafir and Peled 1998). Held by the secular, mostly Ashkenazi middle class and the Israeli Palestinians, this sensibility values individual rights more highly than collective glory. In sharp contrast to neo-Zionism, it is less national, considering the collectivity as a tool for the welfare of the individual (Ram 1999, 333).

2. Evidence of the soldier's dominant status is found in various studies that indicate that young men (and most women) rank the combat soldier at the top of social and military hierarchies (E. Levy 1998; Sasson-Levy 2000). Applications for various prestigious combat units exceed available places by a ratio of eight to one (Cohen 1997, 107). Combat soldiers receive higher salaries during military service and enjoy a range of privileges after their release. For example, they are entitled to academic scholarships and grants that are not available to noncombat soldiers. In the economic and political realms, Israeli ex-colonels still enjoy immense power, as is clear from their growing presence in Israeli governments. In the cultural sphere, the consensual esteem for the combat soldier is reflected in the many commercials that use his image to sell anything from cell phones and medical insurance to laundry detergent and cream cheese. Noncombat soldiers, on the other hand, do not feature in commercials and are rarely represented in the media at all (E. Levy 1998).

3. All names have been changed to protect the soldiers' privacy.

4. Real resistance, however, is not to be found among combat soldiers. Young Israeli men who resist the warrior ethos either make a special effort not to serve in combat roles or refuse to serve in the army altogether, for ideological or personal reasons. In refusing to serve, they express dissension over the essence of the warrior ethos and rejection of the militarized nature of hegemonic Jewish Israeli masculinity. The majority, however, still strongly support compulsory military service in general and the warrior ethos in particular (Ben-Ari 1999; Cohen 1997 and Lomsky-Feder).

5. All the soldiers interviewed used the English word action in their Hebrew speech

References

Abu-Lughod, L., and C. A. Lutz. 1990. Introduction: Emotion, discourse, and the politics of everyday life. In *Language and the politics of emotion,* edited by C. A. Lutz and L. Abu-Lughod, 1–23. Cambridge, UK: Cambridge University Press.

Addelston, J., and M. Stirratt. 1996. The last bastion of masculinity. In *Masculinities in organizations,* edited by C. Cheng, 54–76. Thousand Oaks, CA: Sage.

Alon, M. 1998. Perhaps the I.D.F. prefers its heroes dead? [Hebrew] *Yediot Aharonot,* November 27.

Aran, G. 1974. Parachuting. *American Journal of Sociology* 80(1): 124–53.

Arkin, W., and L. R. Dubrofsky. 1978. Military socialization and masculinity. *Journal of Social Issues* 34(1): 151–69.

Armitage, J. 2003. Militarized bodies: An introduction. *Body and Society* 9(4): 1–12.

Aronoff, M. 1999. Wars as catalysts of political and cultural change. In *The military and militarism in Israeli society,* edited by E. Lomsky-Feder and E. Ben-Ari, 37–57. Albany: State University of New York Press.

Barrett, F. 1996. The organizational construction of hegemonic masculinity: The case of the U.S. navy. *Gender, Work and Organization* 3(3): 129–42.

Becker, A. 2000. The last ones on the ridge [Hebrew]. *Ha'aretz weekend supplement,* June 16, pp. 76–80.

Ben-Ari, E. 1998. *Mastering soldiers.* New York: Berghan.

Bordo, S. 1993. *Unbearable weight.* Berkeley: University of California Press.

Burk, J. 1995. Citizenship status and military service: The quest for inclusion by minorities and conscientious objectors. *Armed Forces and Society* 21(4): 503–29.

Chapkis, W. 1988. Sexuality and militarism. In *Women and the military system,* edited by E. Isaksson, 106–13. New York: St. Martin's.

Cohen, S. 1997. Towards a new portrait of the (new) Israeli soldier. *Israeli Affairs* 3(3/4): 77–117.

Comaroff, J. L. 1996. Ethnicity, nationalism, and the politics of difference in an age of revolution. In *The politics of difference: Ethnic premises in a world of power,* edited by E. N. Wilmsen and P. McAllister. Chicago: University of Chicago Press.

Connell, R. W. 1987. *Gender and power: Society, the person and sexual politics.* Stanford, CA: Stanford University Press.

———. 1995. *Masculinities.* Berkeley: University of California Press.

Csordas, J. T. 1993. Somatic modes of attention. *Cultural Anthropology* 8(2): 135–56.

Davis, K. 1997. Embodying theory: Beyond modernist and postmodernist reading of the body. In *Embodied practices: Feminist perspectives on the body,* edited by K. Davis, 1–23. London: Sage.

Dreyfus, H., and P. Rabinow. 1982. *Michel Foucault: Beyond structuralism and hermeneutics.* Sussex, UK: Harvester.

Enloe, C. 1980. *Ethnic soldiers: State security in divided society.* Athens: University of Georgia Press.

———. 1988. *Does khaki become you?* London: Pandora.

Foucault, M. 1975. *Discipline and punishment: The birth of the prison.* New York: Vintage.

———. 1978. *The history of sexuality.* New York: Vintage.

———. 1980. *Power/knowledge.* New York: Pantheon.

Frank, A. W. 1991. For sociology of the body: An analytical review. In *The body,* edited by M. Featherstone, M. Hepworth, and B. S. Turner, 36–102. London: Sage.

Furman, M. 1999. Army and war: Collective narratives of early childhood in contemporary Israel. In *The military and militarism in Israeli society,* edited by E. Lomski-Feder and E. Ben-Ari, 141–69. New York: State University of New York Press.

Giddens, A. 1985. *The nation-state and violence.* Berkeley: University of California Press.

Gill, L. 1997. Creating citizens, making men: The military and masculinity in Bolivia. *Cultural Anthropology* 12(4): 527–50.

Hacker, S. 1989. *Pleasure, power and technology: Some tales of gender, engineering and the cooperative workplace.* Boston, MA: Unwin Hyman.

Helman, S. 1997. Militarism and the construction of community. *Journal of Political and Military Sociology* 25:305–32.

Higate, P. 2003. *Military masculinities: Identity and the state.* Westport, CT: Praeger.

Hochschild, A. 1983. *The managed heart: The commercialization of human feeling.* Berkeley: University of California Press.

Horowitz, D., and M. Lissak. 1989. *Troubles in utopia: The overburdened polity of Israel.* Albany: State University of New York Press.

Janowitz, M. 1994. Military institutions and citizenship in Western societies. In *Citizenship,* edited by B. Turner and P. Hamilton. London: Routledge.

Kimmel, M. 1996. *Manhood in America.* New York: Free Press.

Kimmerling, B. 1993. Patterns of militarism in Israel. *European Journal of Sociology* 34:196–223.

Kovitz, M. 2003. The roots of military masculinity. In *Military masculinities: Identities and state,* edited by P. Higate, 1–14. Westport, CT: Praeger.

Levy, E. 1998. Heroes and helpmates: Militarism, gender and national belonging in Israel. PhD diss., University of California, Irvine.

Levy, Y. 1993. The role of the military sphere in constructing the social-political order in Israel [Hebrew]. PhD diss., Tel Aviv University.

Levy-Schreiber, E., and E. Ben-Ari. 2000. Body building, character building and nation-building: Gender and military service in Israel. *Studies in Contemporary Judaism* 16:171–90.

Lieblich, A. 1989. *Transition to adulthood during military service: The Israeli case.* Albany: State University of New York Press.

Lloyd, G. 1986. Selfhood, war and masculinity. In *Feminist challenges: Social and political theory,* edited by C. Pateman and E. Gross, 63–76. Boston: Northeastern University Press.

Lomsky-Feder, E. 1998. *As if there was no war: The perception of war in the life stories of Israeli men* [Hebrew]. Jerusalem: Magnes.

Lomsky-Feder, E., and E. Ben-Ari, eds. 1999. *The military and militarism in Israeli society.* Albany: State University of New York Press.

Lutz, C. 1988. *Unnatural emotions.* Chicago: Chicago University Press.

Lutz, C., and G. White. 1986. The anthropology of emotions. *Annual Review of Anthropology* 15:405–36.

Mazali, R. 1993. Military service as initiation rite. *Challenge* IV(4): 36–37.

Morgan, D. H. J. 1994. Theatre of war: Combat, the military and masculinities. In *Theorizing masculinities,* edited by H. Brod and M. Kaufman, 165–83. London: Sage.

Moskos, C. 1993. From citizens' army to social laboratory. *Wilson Quarterly* 17:83–94.

Mosse, G. L. 1996. *The image of man: The creation of modern masculinity.* Oxford, UK: Oxford University Press.

Nagel, J. 1998. Masculinity and nationalism: Gender and sexuality in the making of nations. *Ethnic and Racial Studies* 21(2): 242–70.

Ortner, S. 1974. Is female to male as nature is to culture? In *Women culture and society,* edited by M. Rosaldo and L. Lamphere. Stanford, CA: Stanford University Press.

Peniston-Bird, C. 2003. Classifying the body in the Second World War: British men in and out of uniform. *Body & Society* 9(4): 31–48.

Petersen, A. 1998. *Unmasking the masculine: "Men" and "identity" in a sceptical age.* London: Sage.

Ram, U. 1999. The state and the nation: Contemporary challenges to Zionism in Israel. *Constellations* 6(3): 325–39.

Rosaldo, M. 1984. Toward an anthropology of self and feeling. In *Cultural theory essays on mind, self and emotion,* edited by R. Shweder and R. Le Vine, 137–57. Cambridge, UK: Cambridge University Press.

Sasson-Levy, O. 2000. Constructions of gender identities within the Israeli army. PhD diss., Hebrew University, Jerusalem.

———. 2003a. Feminism and military gender practices: Israeli women soldiers in "masculine" roles. *The Sociological Inquiry* 73(3): 440–65.

———. 2003b. Military, masculinity and citizenship: Tensions and contradictions in the experience of blue-collar soldiers. *Identities: Global Studies in Culture and Power* 10(3): 319–45.

Sedgwick, E. K. 1985. *Between men—English literature and male homosocial desire.* New York: Columbia University Press.

Seidler, V. 1989. *Rediscovering masculinity.* London: Routledge.

Shafir, G., and Y. Peled. 1998. Citizenship and stratification in an ethnic democracy. *Ethnic and Racial Studies* 21(3): 408–28.

Shilling, C. 1993. *The body and social theory.* London: Sage.

Snyder, C. R. 1999. *Citizen-soldier and manly warriors: Military service and gender in the civic republic tradition.* Lanham, MD: Rowman & Littlefield.

Spencer, P. 1965. *The Samburu: A study of gerontocracy in a nomadic tribe.* Berkeley: University of California Press.

Stiehm, J. H. 1989. *Arms and the enlisted woman.* Philadelphia: Temple University Press.

Theweleit, K. 1987–1989. *Male fantasies.* Minneapolis: University of Minnesota Press.

Thoits, P. 1989. The sociology of emotions. *Annual Review of Sociology* 15:317–42.

Tilly, C. 1985. War making and state making as organized crime. In *Bringing the state back in,* edited by P. Evans, D. Rueschemeyer, and T. Skocpol, 169–91. Cambridge, UK: Cambridge University Press.

———. 1996. The emergence of citizenship in France and elsewhere. *International Review of Social History* 40(3): 223–36.

Weiss, M. 2002. *The chosen body: The politics of the body in Israeli society.* Stanford, CA: Stanford University Press.

Winslow, D. 1999. Rites of passage and group bonding in the Canadian airborne. *Armed Forces and Society* 25(3): 429–57.

Zerubavel, Y. 1995. *Recovered roots: Collective memory and the making of Israeli national tradition.* Chicago: University of Chicago Press.

Introduction to Reading 29

Breanne Fahs explores the tensions between *freedom to* and *freedom from* in the feminist sex-positive movement. Sex-positive feminists argue that women's freedom must include sexual freedom to explore and enjoy sex within consenting relationships. The focus of the sex-positive movement, Fahs argues, has been "positive liberty" or women's *freedom to* expand sexual expression and embrace sexual diversity. This focus ignores "negative liberty," defined as women's freedom from oppressive sexual mandates and requirements. Fahs makes a strong argument that the sex-positive movement must integrate *freedom from* as it relates to *freedom to* in order to create a feminist movement that is cohesive and powerful. Thus, women must have freedom from oppressive heterosexist and racist constraints over their sexuality as well as positive liberty. Fahs examines seven examples of contemporary dilemmas (e.g., rape and sexual coercion; same-sex eroticism) women face in their sexual lives demonstrating the ways in which *freedom from* can be incorporated into the feminist sex-positive movement.

1. How does Fahs employ anarchist theories of liberation and freedom to advance her argument that the sex-positive movement must include negative liberty in its theory and activism?
2. Why did sex-positive feminism come to focus on *freedom to* rather than *freedom from?*
3. What role has radical feminism played in forwarding the importance of negative liberty in the movement for women's sexual freedom?
4. Discuss the three goals that Fahs sees as essential to a new vision for sex positivity.

"Freedom to" and "Freedom From"

A New Vision for Sex-Positive Politics

Breanne Fahs

Introduction

Walking through the exhibit hall at the Society for the Scientific Study of Sexuality conference recently, the displays featured a vast array of new possibilities for sexual expression: dildos shaped like tongues, edgy books and journals on bisexuality and polyamory, videos for helping heterosexual women gain comfort with penetrating their (receptive) boyfriends and husbands, and even a pamphlet for an "autoerotic asphyxiation support club." Clearly, the sex-positive movement—inclusive of those who argue against all restrictions on sexuality aside from issues of safety and consent—has made significant advances in how scholars, feminists, practitioners, and the public think about, feel about, and "do" sexuality. For example, women's access to feminist sex toy shops, pornography, blogs, and representations of queer sexuality have increased dramatically in the past several decades (Loe, 1999; Queen, 1997a; Queen, 1997b). Despite the relentless attacks from conservatives in the U.S., people today generally have more expansive options for how they express "normal" sexuality, and they can do so more openly and with more formal and social support. Sex-positive feminists have, in many ways, turned upside-down the notion of the once highly-dichotomous public/private, virgin/whore, and deviant/normal. The sex-positive movement has helped to decriminalize sex work (Jennes, 1993), expand representations in pornography (McElroy, 1995), teach people how to embrace sexuality as normal and healthy (Queen, 1997a; Queen, 1997b), explore "sexual enhancement" devices (Reece et al., 2004), advocate comprehensive sex education (Irvine, 2002; Spencer et al., 2008; Peterson, 2010), and challenge overly simplistic notions of "good" and "bad" sex (Rubin, 1993). From Annie Sprinkle showing her cervix as a "Public Cervix Announcement" (Shrage, 2002), to Carol Queen (1997a) arguing against "whore stigma," to Gayle Rubin (1993) fighting tirelessly against a fixed (and faulty) construction of sex offending, sex positivity has laid the groundwork to depathologize sexuality, particularly for women, sexual minorities, people of color, and sex workers.

Still, the whole scene—clearly intent on a "progress narrative" of sexuality—gives me a strong sense of unease, as many contradictions still overwhelm women's sexual lives. Along with these newfound modes of pleasure-seeking and knowledge-making, women struggle within a plethora of urgent contemporary challenges: alarming rates of sexual violence (Luce, 2010), body shame (Salk and Engeln-Maddox, 2012), eating disorders (Calogero and Pina, 2011), pressure to orgasm (Farvid and Braun, 2006; Fahs, 2011a), distance from

Fahs, B. (2014). "Freedom to" and "freedom from": A new vision for sex-positive politics. *Sexualities, 17*(3): 267–290. Reprinted with permission of SAGE Publications, Inc.

their bodily experiences (Martin, 2001), disempowerment with childbirth (Martin, 2001; Lyerly, 2006), masochistic sexual fantasies (Bivona and Critelli 2009), and a host of other sexual and political crises. When asked why they orgasm, women cite their partner's pleasure as more important than their own (Nicolson and Burr, 2003) and they increasingly associate sexual freedom with consumerism, fashion, and commodities (Hakim, 2010). Women internalize normative pressures to hate their pubic hair and body hair (Riddell et al., 2010; Fahs, 2011b; Bercaw-Pratt, 2012) and hide their menstrual cycles from others (Bobel, 2006; Mandziuk, 2010). Across demographic categories, women suffer immensely from feeling disempowered to speak, explore, and embrace the kinds of sexual lives they most want.

This paper takes this vast contradictory moment—that is, living between celebrations of progress and alarmingly regressive notions of women's sexual "empowerment"—to explore the problems of (uncritical) sex positivity in this post-sexual-revolution age. Drawing from anarchist theories from the past two centuries—particularly the notion that true liberation and freedom must include *both* the freedom *to* do what we want to do AND the freedom *from* oppressive structures and demands—I argue that the sex-positive movement must advance its politics to include a more serious consideration of the freedom from repressive structures (or "negative liberty"). More specifically, by outlining several ways that the *freedom from* and the *freedom to* are currently in conversation in discourses of women's sexuality, I argue that the integration of these two halves could lead to a subtler and more complete understanding of contemporary sexual politics, particularly around tensions that arose during the infamous "sex wars" of the 1980s, thus helping to build a more cohesive and powerful feminist movement as a whole.

Anarchism and Sexuality

. . . While anarchy and political theory may not seem like an intuitive bedfellow for feminists who study sex, the political and social bases of anarchy have much to teach feminists interested in bodies and sexualities as sites of social (in)justice. Anarchists have long espoused important divisions between those interested in individualist versus social anarchism. On the one hand, individualist anarchism, what Isaiah Berlin (1969) termed "negative liberty," argued for freedom from the state and corporate apparatuses. On the other hand, social anarchism advocated for both negative liberty (freedom from the state) and positive liberty (freedom to do what we want to do). . . .

Anarcha-feminists were among the first who refused to conceptualize love, relationships, and domesticity as separate from state politics, calling for an end to "sex slavery" (de Cleyre, 1914), jealousy (Goldman, 1998), oppression through motherhood, marriage, and love (Goldman, 1917; Marso, 2003), and control of love and relationships (Haaland, 1993; Molyneux, 2001). In the 1960s, rifling on the work of anarchists, radical feminists—many of whom came from Marxist backgrounds that prioritized the elimination of class inequalities and the importance of structural equalities—fought for the *freedom to* as much as the *freedom from,* simultaneously advancing ideas about women *gaining* access to certain previously impenetrable spheres (e.g., all-male faculty clubs, all-male jobs) and *blocking access* to others (e.g., all-women consciousness-raising groups, all-women classrooms, all-women music festivals). (While *freedom to* and *freedom from* rarely serve as precise opposites, the ability to willfully *separate* has been considered more a "freedom from" than a "freedom to.") Though radical feminists rarely referred directly to anarchism as their inspiration, these same ideological concepts (negative and positive liberty) informed the center of their political claims about the pathological implications of patriarchy. Radical feminist circles explored the varying modes of power imbalances between men and women (and, later, between women) and attended to distinctions between "power over" (domination and oppression), "power to" (freedom to do and act), and "power with" (collective power to do and act) (Allen, 2008). Far from the oversimplified notion of a singular, rebellious anarchy, they argued for a multi-faceted approach to understanding the workings of power, the state, and social relationships. . . .

Even though anarchist theories of rebellion from the state form an intuitive companion to claims for sexual "freedom," sexuality and anarchy have not typically joined forces (with Greenway, 1997; Passet, 2003; and Heckert and Cleminson, 2011 as notable exceptions). Nevertheless, the two share an important set of common goals, particularly around the *freedom from* oppressive structures, mandates, statutes, and interventions. As Gustav Landauer wrote, "The state is not something which can be destroyed by a revolution, but is a condition, a certain relationship between human beings, a mode of human behavior; we destroy it by contracting other relationships, by behaving differently" (Heckert, 2010). Sexuality, then, allows us to "relate differently" and "behave differently," to reimagine our relationships to sex, love, friendship, and kinship (Heckert, 2011), to forego the boredom and monotony capitalism engenders (CrimethInc, 2012).

If we understand sexuality as a tool of creativity, as a force that reconsiders power, equality, and freedom, it becomes a perfect companion for anarchist sensibilities. Sexuality as *process* rather than *outcome* (as desire rather than merely orgasm, as exchange rather than merely physical release) links up with the political sentiments of anarchy by suggesting that interaction between people—ideally devoid of power imbalances—matters far more than goal-oriented drives toward an *end.* Thus, drawing upon these claims, I argue that merging the *freedom from* and the *freedom to* with sexuality may result in a powerful overhaul of fragmented segments of social and political life: the personal and political, the corporeal and the cognitive, the "sex-positive" and the "radical feminist."

Sex Positives and the "Freedom To"

Though sex-positive feminists as currently constituted would likely not categorize themselves exclusively as fighting for the *freedom to,* the vast majority of their work has centered on the expansion of sexual rights, freedoms, and modes of expression (Queen, 1997a; Queen, 1997b). The movement emerged in response to highly repressive discourses of sexuality, particularly those that they perceived as embracing a state-centered, conservative ideology of "good sex"—as heterosexual, married, monogamous, procreative, non-commercial, in pairs, in a relationship, same generation, in private, no pornography, bodies only, and vanilla—and "bad sex"—as homosexual, unmarried, promiscuous, non-procreative, commercial, alone or in groups, casual, cross-generational, in public, pornography, with manufactured objects, and sadomasochistic (Rubin, 1993). Though the sex wars of the 1980s—where feminists battled about pornography in particular—suggest that sex positivity arose in response to the supposedly pro-censorship notions from radical feminists (though that, too, is highly controversial), in fact sex positivity more clearly rebelled against conservatism, evangelical culture, homophobia, and religious dogma (Duggan and Hunter, 1995). . . .

Sex positives fought fiercely for the *freedom to* have diverse, multiple, expansive, and agentic sexual expression; that is, to dislodge histories of repression, sex positives argued that women (and men) should freely embrace new modes of experiencing and expressing their sexuality. Often taking a sort of libertarian perspective on sexual freedom, some wings of the sex-positive movement value lack of government intervention into sex as their top priority (Weeks, 2002) while other aspects of the sex-positive movement more clearly value education and expansion of sexual knowledge (Irvine, 2002). Sex positives share a concern with external limitations placed on sexual expression and with the moralizing judgments placed upon diverse sexualities. For example, Carol Queen defined sex positivity as a community of people who "don't denigrate, medicalize, or demonize any form of sexual expression except that which is not consensual" (Queen, 1997a, p. 128).

One of the key battlegrounds for sex positives—sex education—has also been (arguably) its most successful terrain. Sex positives fought against abstinence-only sex education in favor of comprehensive sex education that would support not only *more* knowledge about STIs and birth control, but also more expansive ideas about queer sexuality, pleasure, alternative families, and options for abortion (Irvine, 2002). Further, as an offshoot of the push toward *more* sex education, feminist-owned sex toy shops began to open (and thrive) throughout the country (e.g., Good Vibes in San Francisco, Smitten Kitten in Minneapolis, Early to Bed in Chicago, Aphrodite's Toy Box in Atlanta, and several others). These stores state that they promoted safe, fun, non-sexist ways to enjoy sex toys, pornography, and erotica, and have worked hard to navigate the tricky terrain of selling commercial products while nurturing an inclusive, community-based, potentially activist space (Loe, 1999; Attwood, 2005). The conflicts between supposedly promoting sex education while also embracing commercial gains presents tricky territory for feminist critics.

As another key victory for sex-positive feminists, they have fought for heightened awareness about, and advocacy for, queer sexualities. This has included the expansion of trans rights (NCTE, 2003; Currah et al., 2006), more legal and social rights for lesbians and gay men (Bevacqua, 2004; Sycamore, 2004), and acceptance for bisexuality within the queer movement (Garber, 1995). Sex positives have rejected anti-gay-marriage statutes even while exploring marriage as fundamentally flawed (Sycamore, 2004), encouraged better social services for queer youth (McCabe and Rubinson, 2008; Orechia, 2008), and worked to understand sexuality as fluid, flexible, and not beholden to dogmatic and religious doctrines (Diamond, 2009). In doing so, they have embraced a diversity of bodies, expressions, and identities and have critically examined the all-too-narrow construction of the "sexual body" as white, heterosexual, young, female, and passive. Instead, sex positives have moved to recognize the sexualities of those on the "fringe": fat bodies (Johnson and Taylor, 2008), older bodies (Chrisler and

Ghiz, 1993), people of color (Landrine and Russo, 2010; Moore, 2012), gender queer (Branlandingham, 2011), and "alternative" bodies (Hughes, 2009). These *freedom to* victories have certainly helped the feminist movement link up with other movements for social and political justice, particularly as sex positives reject claims of "deviance" for any particular group (Showden, 2012).

Radical Feminists and the "Freedom From"

At the same time that sex positives argued to decriminalize, expand, and embrace sexuality—often constructing pornography as positive, educational, and antirepressive—radical feminists countered these claims by looking at the *freedom from* (MacKinnon, 1989). While radical feminists have often been seen as the *opposite* of, or contrary to, the beliefs of sex positivity, I argue that radical feminists have merely wanted *more* recognition of "negative liberty," or the *freedom from* oppressive structures that most women confront on a daily basis. (Not all sex positives have totally neglected the *freedom from,* but rather, have *deprioritized* the *freedom from* in comparison to other goals and priorities.) Advocating caution about the unconditional access to women that is built into the sex-positive framework, radical feminists essentially said that, without women's *freedom from* patriarchal oppression, women lacked freedom at all. Real sexual freedom, radical feminists claimed, must include the *freedom from* the social mandates to have sex (particularly the enforcement of sex with men) and *freedom from* treatment as sexual objects.

Looking collectively at the work of Teresa de Lauretis (1988), Marilyn Frye (1997), Adrienne Rich (1980), Audre Lorde (2007), Andrea Dworkin (1997), and Catharine MacKinnon (1989), they all share the belief that men's access to women is a *taken-for-granted* assumption often exercised on women's bodies and sexualities. Indeed, all powerful groups demand unlimited access to less powerful groups, while less powerful groups rarely have access to more powerful groups. As Marilyn Frye (1997) said, "Total power is unconditional access; total powerlessness is being unconditionally accessible. The creation and manipulation of power is constituted of the manipulation and control of access" (411). For example, the poor rarely have access to the rich (particularly if demanded by the poor), while the rich almost always have access to the poor (e.g., buying drugs or sex in poor neighborhoods and returning to their "safe" communities). Affluent whites often live in gated communities, in part to deny access to people of color, while middle-class whites live in segregated suburban cul-de-sacs. Men notoriously operate in spheres of power that exclude women (e.g., country clubs, golf circles, "good ol' boys" hiring practices, and so on). At its core, radical feminism argues against the patriarchal assumption that men have the right to access women (and the patriarchal notion that women must internalize this mandate).

Lesbian separatism, particularly *political* lesbian separatism not based specifically or primarily on sexual desire for women (Atkinson, 1974; Densmore, 1973), represents a rebellion against this mandated access. By revealing the assumptions of access embedded within sexuality—namely that men can always access the bodies of women for their sexual "needs"—lesbian separatists make a clear case for the *freedom from.* All-women spaces that excluded men did not merely allow women to physically separate from men, but also to rebel against the constant surveillance of the male gaze and men's *assumptions of access* (Fahs, 2011a). These separatists understood that the "free love" of the 1970s, far from a celebratory moment of progress for women, merely allowed men's sexual access to *more* women and largely ignored women's experiences with "brutalization, rape, submission [and] someone having power over them" (Dunbar, 1969, 56). They also (rightly) noted that women's assertion of their *freedom from* interacting with men provokes dangerous resistances, anger, and hostilities.

Moving forward to the 1980s, scholars like Dworkin (1997) and MacKinnon (1989) made similar claims about sexual access when they argued for the *freedom from* messages and images embedded in pornography that portrayed women as "hot wet fuck tubes" (Dworkin, 1991). Further, they called out the power-imbalanced practices of sexual intercourse and theorized about the dangers of women tolerating their own oppression (e.g., not reporting rape). By calling out sexuality as a site of dangerous power imbalances between men and women, their version of *freedom from* became particularly threatening within and outside of feminist circles. Recall that both MacKinnon and Dworkin received death threats and had to retain security personnel for even *suggesting* that women can and should assert their *freedom from* pornography and power-imbalanced sex (MacKinnon and Dworkin, 1998). Taken together, these examples reveal that the *freedom from* provokes greater threat from men and the culture at large, with increasingly hostile backlashes against those who assert their right to negative liberty. To suggest that women *deny access to men* undercuts core cultural assumptions about gender politics and patriarchal power.

Cycles of Liberation

Looking at these histories together, a clear pattern emerges: Rarely does the feminist movement (or the queer movement, or sexuality studies) adequately address the dialectic between the *freedom from* and the *freedom to.* I am particularly concerned about the degree to which sex positivity neglects the *freedom from,* particularly when the rhetoric of sex positivity often inadvertently allows for the unconditional access to women that many of their projects, goals, and narratives rely upon. That said, I also worry about the ways that negative liberty (ironically) could create new norms that require women to label their (heterosexual, pornographic, masochistic, etc.) desires as necessarily patriarchal (e.g., Showden, 2012 has suggested that sex positivity must become a "politics of *maybe*" rather than a "politics of *yes*"). The relative crisis of women's "sexual freedom"—along with the hazards of *solving* this problem—becomes increasingly clear when examining contemporary dilemmas and quandaries women face in their sexual lives. In the following seven examples, I outline the ways that the *freedom from* has fallen more and more out of focus, even as the *freedom to* achieves small victories.

Orgasm

In my previous work (Fahs, 2011a), I looked at what I consider to be the most dangerous aspect of defining sexual freedom: It is subject to continued appropriations and distortions, and it requires continual reinvention. Sexual freedom has no fixed definition and is not static. Rather, because freedoms are easily co-opted, all definitions of freedom (sexual or otherwise) are transient and transitory, require constant re-evaluation and reassessment, and present an ongoing set of new challenges to each cohort and generation. In the articulation of freedom, co-optation is not only possible, but *probable,* particularly when addressing issues of women's sexuality. As a key example, consider a brief history of women's orgasms. During the sexual revolution, women fought hard for the right to have the clitoris recognized as a site of legitimate pleasure. Until the late 1960s, it was assumed (largely due to the influence of psychoanalysis) that vaginal orgasm represented "maturity" and clitoral orgasm represented "immaturity" (studies by medical doctors—as recent as 2011—still argue this! See Costa and Brody, 2011; and Brody and Weiss, 2011); still more, women were expected to prioritize phallic male pleasure over anything clitoral, often resulting in harsh pressures for women to vaginally orgasm. As Ti-Grace Atkinson (1974) said, "Why *should* women learn to vaginal orgasm? Because that's what men want. How about a facial tic? What's the difference?" (7). During the sexual revolution, the convergence of the queer movement, the women's movement, and the sexual revolution led to a concerted interest in dethroning the vaginal orgasm in favor of the clitoral orgasm (Gerhard, 2000). If women valued and recognized the power of the clitoris, some argued, they could embrace sex with other women, orgasm more easily and efficiently, and enjoy the same sexual pleasures men had always enjoyed (Koedt, 1973).

This celebration of progress on the orgasm front—notably something radical feminists expressed some concerns about even then—had a short-lived period of revolutionary potential. At the tail end of the sexual revolution, radical feminists like Dana Densmore (1973) and Roxanne Dunbar (1969) began to worry that all of this focus on women's clitoral orgasms could lead to an "orgasm frenzy" where women would feel *mandated* to orgasm and men would use clitoral orgasms as yet another tool to oppress women. Dunbar believed that sexual liberation became equated with "the 'freedom' to 'make it' with anyone anytime" (49), while Sheila Jeffreys (1990) claimed retrospectively that "sexual liberation" from the 1960s and 1970s merely substituted one form of oppression for another. Indeed, over the next twenty years, clitoral orgasms evolved from a much-fought-for occurrence to a social mandate between (heterosexual) sex partners. Fast-forward to today and we find some startling data: over half of women have faked orgasms, often regularly (Wiederman, 1997), and women describe faking orgasms for reasons that still ensure male dominance and power: they want to support the egos of their (male) partners; they want to end the encounter (primarily intercourse), often because they feel exhausted; and they imagine that orgasms make them "normal" (Fahs, 2011a; Roberts et al., 1995). This suggests that, though orgasm once served a symbolic role as a tool of liberation (a new *freedom to* personal desire), orgasm eventually reverted to being another tool of patriarchy as orgasm became a marker of male prowess. Given the high rates of faking orgasm (Fahs, 2011a; Roberts et al., 1995) now seen among women, many women may now need the *freedom from* orgasm as a *mandate;* further, the mandate exists not primarily to please women themselves, but to please their (male) partners.

Sexual Satisfaction

As an offshoot of the "orgasm problem," recent research on sexual satisfaction has pointed to some disturbing trends. While popular culture (particularly movies, music, and magazines) generally advocates women's *freedom to* have sexual pleasure and satisfaction, little attention is paid to how women themselves *construct* satisfaction or how women's sexual satisfaction still showcases men's overwhelming sexual power. What does a "satisfied woman" say about her sexual experiences? Studies consistently suggest correlations between sexual satisfaction and intimacy, close relationships (Pinney et al., 1987; Sprecher et al., 1995), emotional closeness (Trompeter et al., 2012), reciprocal feelings of love, and versatile sexual techniques (Haavio-Mannila and Kontula, 1997). These findings underlie the relational dimensions of women's sexual satisfaction (and undermine pop culture's "bodice ripping" stereotypes). In this regard, women may have the *freedom to* sexual pleasure as long as it remains in the stereotyped confines of marriage and romantic love. This may represent a sign of progress, or it may signify the trappings of traditional femininity. Similarly, studies also show that sexually satisfied women typically have better body image (Meltzer and McNulty, 2010), lower rates of eating disorders, lower self-objectification tendencies (Frederickson and Robert, 1997; Calogero and Thompson, 2009), thereby suggesting other modest victories for the association between sexual expression and personal empowerment even while raising questions about women's interpretation of "satisfaction."

When looking more closely at data about women's sexual satisfaction, particularly measures of "deservingness" and "entitlement," women fall far short of full equality with men with regard to seeking pleasure, though differing definitions and assessments of satisfaction make these gender findings increasingly complicated (McClelland, 2010; McClelland, 2011). For example, the same relational dimensions that may help women equate sexual satisfaction with emotional closeness also demand that they prioritize their partner's (especially men's) pleasure over their own. When asked why they want to orgasm, women say that their partners' pleasure matters more than their own and that their partners' pleasure is a conditional factor that determines their own pleasure (Nicholson and Burr, 2003), calling into question *whose* pleasure women value when assessing their own satisfaction. Compared to men, women also more often equate sex with submissiveness, and when they do so, this leads to lower rates of sexual arousal, autonomy, and enjoyment (Sanchez et al., 2006). Further, compared to men, women also describe orgasm as a far less important component of their sexual satisfaction (Kimes, 2002).

Additionally, sexual satisfaction is certainly not static or equally distributed *between* all women, as different demographic groups report vastly different sexual satisfaction. Strong correlations between sexual satisfaction, sexual activity, and social identities have been found, as lower status women (e.g., women of color, less-educated women, working-class women) reported having more *frequent* but less satisfying sex, particularly compared to white, upper-class (typically higher status) women (Fahs and Swank, 2011). That is, women without as much social and political power still have sex, but they have to endure less satisfying sex on a much more regular basis than women with more power. This suggests, most specifically, that lower status women more often lack the *freedom from* unwanted sex or unsatisfying sex even while they have the apparent *freedom to* have frequent sexual activity. Compared to higher status women, lower status women do not have the same social permissions to deny others access to their bodies, and they do not feel as entitled to refuse sex when not satisfied.

Treatment for Sexual Dysfunction

Medicalization has also served as a tool for ensuring women's lack of *freedom from* sex, as women who abstain from sex, refuse sex, or construct themselves as "asexual" or celibate have been labeled as fundamentally dysfunctional by the medical community. For example, to turn a social problem into something supposedly derived from women's inadequacies, a recent study characterized a staggering 43.1% of women as "sexually dysfunctional," even though far fewer women reported subjective distress about such sexual "problems" (Shifren, 2008). When the medical community decides that women have dysfunction if they do not conform to some medically defined prescription of the "normal," they do not account for women's own narratives about their sexual experiences. Rarely do studies account for partner abilities, contextual factors in women's lives, or women's personal narratives about their "dysfunction." As it stands now, women can receive a psychiatric diagnosis of sexual dysfunction (according to the DSM-IV) if they refuse penetration, fail to orgasm, have "inadequate" lubrication or swelling response, refuse to have sex with partners, and feel aversion toward sexuality in general. (These diagnoses are labeled Hypoactive Sexual Desire

Disorder, Sexual Aversion Disorder, Female Sexual Arousal Disorder.) Such diagnoses normalize heterosexuality and penetrative intercourse as the pinnacle of "healthy sex" while setting clear and monolithic standards for sexual normality.

As a more precise example of the dangerous power of medicalization, consider the (especially egregious) recent treatments developed by the medical community to cure vaginal pain disorders like vaginismus. One recent treatment advises doctors to inject Botox into three sites of the vagina in order to allow women to "tolerate" penetrative intercourse (Ghazizadeh and Nikzad, 2004). Another common treatment advises doctors to insert vaginal dilators into women's vaginas in order to stretch out their vaginal opening to allow for penile penetration (Crowley et al., 2009; Grazziotin, 2008; Raina et al., 2007). These treatments ensure that women's vaginas can effectively ingest a penis, thereby constructing "normal" vaginas and "normal" sex as penile-vaginal intercourse. Normal sex becomes that which meets men's sexual needs even if it induces pain in women. Even if women can orgasm through manual or oral stimulation, even if women report sexual violence and abuse histories, and even with reliable statistics that consistently point to penile-vaginal intercourse as an unreliable facilitator of women's orgasms (Hite, 1976), these treatments are considered standard and routine for sexual pain disorders. Women do not have the *freedom from* penile-vaginal intercourse, even if it causes them physical or psychological discomfort and pain. If sociocultural scripts mandate that a "normal" woman has "normal" sex, the medical community will ensure that she complies.

Rape and Sexual Coercion

In an age so often characterized as "empowering" for women—and with so much rhetoric devoted to women's supposed choices about their bodies and sexualities—the occurrence of rape and sexual coercion of women serve as a sobering reminder of patriarchy's widespread influence. In addition to the staggering rates of reported sexual violence both within the U.S. (Elwood et al., 2011) and globally (Koss et al., 1994), women also deal (sometimes on a daily basis) with their lack of *freedom from* sexual harassment, street harassment, pornography, objectification, and coercion. Women typically minimize coercive encounters they have had, often to avoid the stigma and label of "rape victim" (Bondurant, 2001; Kahn et al., 2003). They also protect boyfriends, husbands, family members, and dating partners as "not rapists" by denying or minimizing the coercion that these men enact (Fahs, 2011a), often while endorsing "rape myth" beliefs that women deserve rape or brought it on themselves (Haywood & Swank, 2008). While women may have the *freedom to* experiment with their sexuality in new ways, such experimentation often goes hand-in-hand with coercion, abuses of power, pressure, and lopsided power dynamics.

Women's lack of *freedom from* coercion and harassment also extends into their relationship with space itself. Women construct "outside" as unsafe and "inside" as safe, refuse to walk alone at night (Valentine, 1989), and imagine their (benevolent) boyfriends, husbands, brothers, male acquaintances, and male friends as "protectors," even when these men most often perpetrate sexual violence (Fahs, 2011a). Women's relative lack of freedom to occupy public space, travel alone, protect themselves from violence, or ensure non-coercive sexual exchanges represents a major component of women's sexual consciousness, even if they have not experienced violent rape or clear-cut coercion. The literature on "sexual extortion"—where women engage in sexual acts to avoid domestic violence—also speaks to this continuum between rape and not rape (DeMaris, 1997). Their lack of *freedom from* violence is, for many women, an everyday occurrence that harms their well-being and blocks access to institutions that men use as sources of their power (e.g., better paying jobs, education, public space).

Body Hair as "Personal Choice"

As a reminder of the invisibility of power, women also imagine that they have far more personal freedom when "choosing" how to groom and present their bodies to the outside world and to friends, partners, and families. As a prime example, women, particularly younger women, generally endorse the idea that removing body hair (particularly underarm, leg, and pubic hair) is a "personal choice" that they simply "choose" to do. That said, when women refuse to remove their body hair, they often face intense negative consequences: homophobia, harassment, objectification, partner disapproval, family disapproval, coworker disapproval, threats of job loss, anger and stares from strangers, and internal feelings of discomfort and disgust (Fahs, 2011b). Women who do not remove body hair are labeled by others as "dirty" or "gross" (Toerien and Wilkinson, 2004), and are seen as less sexually attractive, intelligent, happy, and positive compared to hairless women (Basow and Braman, 1998). Women who do not shave their body hair were also judged as less friendly, moral, or relaxed, and as *more* aggressive, unsociable, and dominant compared to women who removed their body hair (Basow and Willis, 2001). While older age, feminist identity, and lesbian identity predicted less negative attitudes toward body hair (Toerien et al., 2005), few women receive full protection from the cultural negativity surrounding women's body hair.

These findings reveal that women overwhelmingly lack the *freedom from* regulating their bodies through depilation. With increasingly vicious rhetoric directed toward their "natural" body hair as inherently dirty, disgusting, and unclean, women spend great energy and time fighting against these stereotypes in an effort to have acceptable bodies, particularly for already stigmatized groups like women of color and working-class women (Fahs, 2011b; Fahs and Delgado, 2011). With trends indicating that increasing numbers of women in the U.S. remove pubic hair, the hairlessness norms seem only to expand, particularly in the last decade (Herbenick et al., 2010). Women may have the *freedom to* groom their pubic hair into triangles, landing strips, or Vajazzled ornaments (thanks to the growth of corporate techniques of hair maintenance, see Bryce, 2012), but they cannot go *au natural* or not shave their bodies without serious social punishments (Toerien et al., 2005).

Same-Sex Eroticism

For several decades, the queer movement has fought to expand legal and social rights for same-sex couples, increase representation of same-sex eroticism, and garner cultural acceptance of LGBT identities as normative and non-"deviant" (Clendinen and Nagourney, 2001; Swank, 2011). In particular, women have more *freedoms to* express their sexual interests in other women and to explore same-sex eroticism more openly, with blatant hostile homophobia diminishing some in the last decades (Loftus, 2001). That said, these supposed "freedoms" have also been appropriated by the patriarchal lens and converted into actions that women undertake to gain acceptance and approval in certain settings: bars, fraternity parties, clubs, bars, and so on (Yost and McCarthy, 2012). Women "making out" and "hooking up" with other women in these settings constitutes an increasingly normative practice, as long as men *watch* women doing this behavior and as long as the women fit stereotypical standards of "sexiness" for male viewers. Following the trends of *Girls Gone Wild,* increasing numbers of women report *pressure* to kiss and have sex with other women in front of men, either by "hooking up" at bars, engaging in threesomes (male partner initiated, only involving multiple women and one man), or allowing men to watch women kiss in public (Fahs, 2009; Yost & McCarthy, 2012).

Such pressures exist regardless of women's sexual identity, as queer women report pressures to "hook up" in front of men just as heterosexual women say their boyfriends and male friends ask them to "pretend" to enjoy same-sex eroticism for their viewing benefit (Fahs, 2009). Women increasingly describe these pressures toward "bisexuality" as compulsory (Fahs, 2009), as 33% of college women had engaged in this behavior, while 69% of college students had observed this behavior (Yost and McCarthy, 2012). Thus, large segments of women lack the relative *freedom from* mandated same-sex eroticism that is performed *in front of* men and for the sake of men's pleasure. The rebellious, transformative qualities of same-sex eroticism have also been distorted to serve the interests of men (and patriarchy) by ensuring that men are physically and psychologically present during these encounters. No wonder, then, that such "performative bisexuality" (Fahs, 2009) has *not,* for the women engaging in these acts, consistently translated into shifts in political consciousness like increased identification with bisexuality, more support for gay marriage laws, and more LGBT activism (Fahs, 2009; Fahs, 2011a).

Sexual Fantasy

As a final example—and likely the one most directly contrasting with sex positivity—women lack the *freedom from* a sexuality that corresponds with mainstream (pornographic?) fantasies. They also lack the freedom from internalized heterosexist beliefs that distorts their imagination about what constitutes exciting sex. Even when women enjoy pornography without internalizing pornographic fantasies *as real,* they still grapple with increasingly narrow definitions of "the sexual." When examining women's sexual fantasies, an incredible amount of internalized passivity, lack of agency, and desire for domination appears (Bivona and Critelli, 2009; Fahs, 2011a). In particular, women report fantasies of men dominating them as their most common sexual fantasies, even when their best lived sexual experiences do not include such content. When women described their most pleasurable sexual encounters, these descriptions did not include dominance and power narratives, yet when women described their sexual fantasies, themes of power, coercion, dominance, and passivity appeared (Fahs, 2011a). In addition to fantasy, women described pressures from their partners to engage in (increasingly rough) anal sex (Stulhofer and Ajduković, 2011), threesomes (Fahs, 2009), forceful encounters (Koss et al., 1994), role playing, and dominance (Bivona and Critelli, 2009), indicating that many women negotiate these themes not only in their minds, but also in their partnered practices.

While women have certainly made advances in their *freedoms to* expand sexual expression and ideas about sexuality, they now face their relative lack of *freedom from* such dynamics both in their imaginations and in their body practices. Pornographic fantasies have entered mainstream consciousness in many ways: body and hair grooming (Fahs, 2011b), new procedures like anal bleaching, increasing desires for "designer vaginas" and labiaplasty surgeries (Braun and Tiefer, 2010), the much-read and discussed *Fifty Shades of Gray* (at the time of writing it held the number one spot on the *New York Times* bestseller list), and pressures for women to conform to men's desires for threesomes, painful anal sex (Stulhofer and Ajduković, 2011), and rougher sex in general. (Indeed, the only recent study about women's experiences with anal sex [Stulhofer and Ajduković, 2011] asks in the title, "Should we take [anal pain] seriously?") Sexual fantasy cannot be dismissed as mere frivolity, as rape fantasies have become increasingly common (Bivona & Critelli, 2009), and men's treatment of women often closely reflects the messages and themes they absorb from pornography (Jensen, 2007).

New Visions for Sex Positivity

The tensions between *freedom to* and *freedom from* in women's sexuality constitute a central dialectic in the study of, and experience of, women's sexuality (Vance, 1984). In order to move toward the ever-elusive "sexual liberation," women need to be able to deny access to their bodies, say no to sex as they choose, and engage in sexual expression free of oppressive homophobic, sexist, and racist intrusions. Women should have, when they choose, the *freedom from* unwanted, mediated versions of their sexuality (e.g., Facebook, Internet intrusions, "sexting"), hetero-sexist constructions of "normal sex," and sexist assumptions about what satisfies and pleases them. If women cannot have *freedom from* these things without social penalty, they therefore lack a key ingredient to their own empowerment.

Those who avoid sex, choose asexuality, embrace celibacy (either temporary or permanent), or otherwise feel disinclined toward sex (perhaps due to personal choice, histories of sexual violence, health issues, hormonal fluctuations, irritation or emotional distance with a partner, and so on) should be considered healthy and *normal* individuals who are making healthy and normal choices. Recent studies have begun to look at "sexual assertiveness," sexual autonomy, and the importance of women's right to refuse sex (Morokoff et al., 1997; Sanchez et al., 2005). The assumption that women must have consistency in their sexual expression, desire, and behavior does not fit with the way sexuality ebbs and flows and responds to circumstances in their lives. Sexual freedom means both the *freedom to* enjoy sexuality, and *freedom from* having to "enjoy" it, just as reproductive freedom means the *freedom to* have children when desired *and* the *freedom from* unwanted pregnancies.

As a new vision for sex positivity, I argue that we need three broadly defined goals, each of which contributes to a larger vision that prioritizes a complex, multifaceted sexual freedom that fuses the political goals of both sex positives and radical feminists: first, *more* critical consciousness about any vision of sexual liberation. Definitive and universal claims about freedom and choice for all women must be met with caution or even downright suspicion. For example, while sex toys can represent a positive aspect of women's lives—they allow for more efficient masturbation for some, exploration for others, and fun and quirkiness for still others—these toys still exist within a capitalistic framework. While sex toys can be empowering, pleasurable, fun, and exciting, they also equate liberated sexuality with *purchasing power,* buying things, and (perhaps) distancing women from their bodies (not to mention that the labor politics around making such toys, where women's labor in developing countries is often exploited in the name of First World pleasure). These debates also overlap with notions of "sexual citizenship" and the ways capitalism shapes not only sexual rights but also desire itself (see Evans, 2002). Also, when women masturbate with sex toys, they learn *not* to touch their vaginas in the same way. When couples "spice things up" with accessories, they often avoid the harder conversations about their goals, desires, and relationship needs. Further, sex toy packaging and marketing fall into the all-too-common associations between acquiring objects and achieving personal happiness (not to mention that sex toy companies often use poor quality plastics and exploit their workers, another decidedly "unsexy" side to the industry). In short, the relentless insistence upon a critical consciousness regarding sexual liberation (and claims of what is "sexually liberating") is a *requirement* if we want to illuminate the complexities of women's sexual freedom.

Second, *more* attention should be paid to how sexual access functions in the lives of lower status people (particularly women). Those with lower socially-inscribed statuses—women, people of color, queer people, working-class and poor people, less educated people (etc.)—are often expected, in numerous areas in their lives, to provide *access* to higher status people. Their bodies are expected to provide certain kinds of labor that serve the interests of high status people (e.g., physical, sexual, emotional labor) (Barton, 2006). Thus, it is especially important that lower-status people not equate sexual liberation with sexual access. These groups should not equate sexual empowerment with providing sexual access to others; rather, their sexual empowerment might derive from the *freedom from* such access to their bodies and (eroticized) labor.

Third, *more* attention must be devoted to the insidious aspects of disempowerment. There is no single definition of "liberated sex." Rather, sexual empowerment is a constantly moving target that requires continual critique, revision, self-reflexiveness, and (re)assessment of our own practices, cultural norms, ideologies, and visions for the self. Even my own vision of better incorporating the *freedom from* into ideas of sexual empowerment carries with it many dangerous trappings that must be cautiously navigated (e.g., not creating new norms and hierarchies of "good" and "bad" sex; forgetting about the "politics of maybe"). Because our culture so often pathologizes "non-normative" sexual behavior, many individuals spend much time and energy defending their sexual choices, behaviors, and lifestyles against conservative, religious, politically regressive individuals and institutions. While this work is much needed and often politically effective, particularly in our current political climate, these defenses *cannot* preclude a *critical assessment* of how our sexual choices still often reflect and perpetuate sexist, classist, racist, and homophobic ideals. We must release our attachment to certain forms of negative liberty that defend us against critical "intruders." In other words, when we insist upon radically examining, critiquing, and unpacking our own sexual lives—even at the cost of unsettling and dislodging the barricades that defend us against intrusions and judgments from the radical right—we move ever closer to a fully realized notion of sexual liberation, sexual empowerment, and sexual equality for all.

References

Allen, A. (2008). Feminist perspectives on power. *Stanford Encyclopedia of Philosophy,* URL (accessed 10 April 2012): http://plato.stanford.edu/entries/feminist-power/

Atkinson, T.-G. (1974). *Amazon Odyssey: The First Collection of Writings by the Political Pioneer of the Women's Movement.* New York: Links Books.

Barton, B. C. (2006). *Stripped: Inside the Lives of Exotic Dancers.* New York: New York University Press.

Bercaw-Pratt, J. L. (2012). The incidence, attitudes, and practices of the removal of pubic hair as a body modification. Journal of *Pediatric & Adolescent Gynecology, 25*(1), 12–14.

Berlin, I. (1969). *Four Essays on Liberty.* New York: Oxford University Press.

Bevacqua, M. (2004). Feminist theory and the question of lesbian and gay marriage. *Feminism & Psychology, 14*(1), 36–40.

Bivona, J., & Critelli, J. (2009). The nature of women's rape fantasies: An analysis of prevalence, frequency, and contents. *Journal of Sex Research, 46*(1), 35–45.

Bobel, C. (2006). "Our revolution has style": Contemporary menstrual product activists "doing feminism" in the third wave. *Sex Roles, 54*(5), 331–345.

Bondurant, B. (2001). University women's acknowledgment of rape: Individual, situational, and social factors. *Violence Against Women, 7*(3), 294–314.

Branlandingam, B. (2011). Hot piece of hipster: Summer genderqueer hair. *The Queer Fat Femme Guide to Life,* URL (accessed 7 March 2012): http://queerfatfemme.com/2011/05/29/hot-piece-of-hipster-summer-genderqueer-hair/

Braun, V., & Tiefer, L. (2010). The "designer vagina" and the pathologisation of female genital diversity: Interventions for change. *Radical Psychology, 8*(1), http://www.radicalpsychology.org/v018–1/brauntiefer.html

Brody, S., & Weiss, P. (2011). Simultaneous penile-vaginal intercourse orgasm is associated with satisfaction (sexual, life, partnership, and mental health). *The Journal of Sexual Medicine, 8*(3), 734–741.

Bryce. (2012). I got vajazzled (and had a camera crew). *The Luxury Spot,* URL (accessed 2 April 2012): http://www.theluxuryspot.com/features/i-got-vajazzled-and-had-a-camera-crew/

Calogero, R. M., & Thompson, J. K. (2009). Potential implications of the objectification of women's bodies for women's sexual satisfaction. *Body Image, 6*(2), 145–148.

Calogero, R. M., & Pina, A. (2011). Body guilt: Preliminary evidence for a further subjective experience of self-objectification. *Psychology of Women Quarterly, 35*(3), 428–440.

Chryslers, J. C., & Ghiz, L. (1993). Body images of older women. *Women & Therapy, 14*(1–2), 67–75.

Clendinen, D., & Nagourney, A. (2001). *Out for Good: The Struggle to Build a Gay Rights Movement in America.* New York: Simon & Schuster.

Costa, M. R., & Brody, S. (2011). Sexual satisfaction, relationship satisfaction, and health are associated with greater frequency of penile-vaginal intercourse. *Archives of Sexual Behavior, 41*(1), 9–10.

CrimethInc Ex-Workers Collective. (2012). Your politics are boring as fuck. CrimethInc, URL (accessed 25 March 2012): http://www.crimethinc.com/texts/selected/asfuck.php

Crowley, T., Goldmeier, D., & Hiller, J. (2009). Diagnosing and managing vaginismus. *BMJ,* 339, 2284.

Currah, P., Juang, R. M., & Minter, S. (2006). *Transgender Rights.* Minneapolis: University of Minnesota Press.

De Cleyre, V. (1914). Sex slavery. *Selected Works of Voltairine de Cleyre.* New York: Mother Earth Publishing Association, pp. 342–358.

De Lauretis, T. (1988). Sexual indifference and lesbian representation. *Theatre Journal*, *40*(2), 155–177.

DeMaris, A. (1997). Elevated sexual activity in violent marriages: Hypersexuality or sexual extortion? *Journal of Sex Research, 34*, 361–373.

Densmore, D. (1973). Independence from the sexual revolution. In A. Koedt, E. Levine, & A. Rapone (Eds.), *Radical Feminism.* New York: Quadrangle Books, pp. 107–118.

Diamond, L. M. (2009). *Sexual Fluidity: Understanding Women's Love and Desire.* Cambridge: Harvard University Press.

Duggan, L., & Hunter, N. D. (1995). *Sex Wars: Sexual Dissent and Political Culture.* New York: Routledge.

Dunbar, R. (1969). Sexual liberation: More of the same thing. *No More Fun and Games, 3,* 49–56.

Dworkin, A. (1997). *Intercourse.* New York: Simon & Schuster.

Dworkin, A. (1991). *Woman hating.* New York: Plume.

Elwood, L. S., Smith, D. W., Resnick, H. S., Gudmundsdottir, B., Amstadter, A., Hanson, R. F., Saunders, B. E., & Kilpatrick, D. G. (2011). Predictors of rape: Findings from the National Survey of Adolescents. *Journal of Traumatic Stress, 24*(2), 166–173.

Evans, D. T. (2002). *Sexual Citizenship: The Material Construction of Sexualities.* New York: Routledge.

Fahs, B. (2011a). *Performing Sex: The Making and Unmaking of Women's Erotic Lives.* Albany, NY: SUNY Press.

Fahs, B. (2011b). Dreaded "Otherness": Heteronormative patrolling in women's body hair rebellions. *Gender & Society, 25*(4), 451–472.

Fahs, B. (2009). Compulsory bisexuality? The challenges of modern sexual fluidity. *Journal of Bisexuality, 9*(3), 431–149.

Fahs, B., & Delgado, D. A. (2011). The specter of excess: Race, class, and gender in women's body hair narratives. In C. Bobel & S. Kwan (Eds.), *Embodied Resistance: Breaking the Rules, Challenging the Norms.* Nashville: Vanderbilt University Press, pp. 13–25.

Fahs, B., & Swank, E. (2011). Social identities as predictors of women's sexual satisfaction and sexual activity. *Archives of Sexual Behavior, 40*(5), 903–914.

Farvid, P., & Braun, V. (2006). "Most of us guys are raring to go anytime, anyplace, anywhere": Male and female sexuality in *Cleo* and *Cosmo. Sex Roles, 55*(5), 295–310.

Frederickson, B. L., & Roberts, T.-A. (1997). Objectification theory: An explanation for women's lived experience and mental health risks. *Psychology of Women Quarterly*, *21*, 173–206.

Frye, M. (1997). Some reflections on separatism and power. In D. T. Meyers (Ed.), *Feminist Social Thought: A Reader.* New York: Routledge, pp. 406–414.

Garber, M. B. (1995). *Vice Versa: Bisexuality and the Eroticism of Everyday Life.* New York: Simon & Schuster.

Gerhard, J. (2000). Revisiting "The myth of the vaginal orgasm": The female orgasm in American sexual thought and second wave feminism. *Feminist Studies*, *26*(2), 449–476.

Ghazizadeh, S., & Nikzad, M. (2004). Botulinum toxin in the treatment of refractory vaginismus. *Obstetrics & Gynecology*, *104*(5), 922–925.

Goldman, E. (1917). Marriage and love. In E. Goldman (Ed.), *Anarchism and Other Essays,* 2nd ed. New York: Dover Publications, pp. 227–239.

Goldman, E. (1998). Jealousy: Causes and a possible cure. In A. K. Shulman (Ed.), *Red Emma speaks: An Emma Goldman reader.* Amherst, NY: Humanity Books, pp. 214–221.

Grazziotin, A. (2008). Dyspareunia and vaginismus: Review of the literature and treatment. *Current Sexual Health Reports, 5*(1), 43–50.

Greenway, J. (1997). Twenty-first century sex. In J. Purkis & J. Bowen (Eds.), *Twenty-first century anarchism: Unorthodox ideas for a new millennium.* London: Continuum Books, pp. 170–180.

Haaland, B. (1993). *Emma Goldman: Sexuality the Impurity of the State.* Montreal: Black Rose Books.

Haavio-Mannila, E., & Kontula, O. (1997). Correlates of increased sexual satisfaction. *Archives of Sexual Behavior*, *26*(4), 399–419.

Hakim, C. (2010). Erotic capital. *European Sociological Review*, *26*(5), 499–518.

Haywood, H., & Swank, E. (2008). Rape myths among Appalachian college students. *Violence and Victims*, *23,* 373–389.

Heckert, J. (2010). Love without borders? Intimacy, identity, and the state of compulsory monogamy. In M. Barker & D. Langdridge (Eds.), *Understanding Non-Monogamies.* New York: Routledge, pp. 255–265.

Heckert, J. (2011). Fantasies of an anarchist sex educator. In J. Heckert & R. Cleminson (Eds.), *Anarchism & Sexuality: Ethics, Relationships, and Power.* London/New York: Routledge, pp. 154–180.

Heckert, J., & Cleminson, R. (2011). Ethics, relationships, and power: An introduction. In J. Heckert & R. Cleminson (Eds.), *Anarchism & Sexuality: Ethics, Relationships, Power.* London: Routledge, pp. 1–22.

Herbenick. D., Schick, V., Reece, M., Sanders, S., & Fortenberry, J. D. (2010). Pubic hair removal among women in the United States: Prevalence, methods, and characteristics. *The Journal of Sexual Medicine, 7*(10), 3322–3330.

Hite, S. (1976). *The Hite Report: A Nationwide Study of Female Sexuality.* New York: Macmillan.

Hughes, B. (2009). Wounded/monstrous/abject: A critique of the disabled body in the sociological imaginary. *Disability & Society*, *24*(4), 399–410.

Irvine, J. M. (2002). *Talk About Sex: The Battles Over Sex Education in the United States.* Berkeley: University of California Press.

Jeffreys, S. (1990). *Anticlimax: A Feminist Perspective on the Sexual Revolution.* New York: New York University Press.

Jenness, V. (1993). *Making It Work: The Prostitute's Rights Movement in Perspective.* New York: Aldine de Gruyter.

Jensen, R. (2007). *Getting Off: Pornography and the End of Masculinity.* Cambridge: South End Press.

Johnston, J., & Taylor, J. (2008). Feminist consumerism and fat activists: A comparative study of grassroots activism and the Dove real beauty campaign. *Signs*, *33*(4), 941–966.

Kahn, A. S., Jackson, J., Kully, C., Badger, K., & Halvorsen, J. (2003). Calling it rape: Differences in experiences of women who do or do not label their sexual assault as rape. *Psychology of Women Quarterly*, *27*(3), 233–242.

Kimes, L. A. (2002). "Was it good for you too?" An exploration of sexual satisfaction. PhD dissertation, University of Kansas.

Koedt, A. (1973). The myth of the vaginal orgasm. In A. Koedt, E. Levine, & Rapone, A. (Eds.), *Radical Feminism.* New York: Quadrangle Books, pp. 199–207.

Koss, M. P., Heise, L., & Russo, N. F. (1994). The global health burden of rape. *Psychology of Women Quarterly*, *18*(4), 509–537.

Landrine, H., & Russo, N. F. (2010). *Handbook of Diversity in Feminist Psychology.* New York: Springer.

Loftus, J. (2001). America's liberalization in attitudes toward homosexuals. *American Sociological Review*, *66*, 762–782.

Loe, M. (1999). Dildos in our toolbox: The production of sexuality at a pro-sex feminist sex toy store. *Berkeley Journal of Sociology*, *43,* 97–136.

Lorde, A. (2007). *Sister Outsider: Essays and Speeches.* Berkeley: Crossing Press.

Luce, H. (2010). Sexual assault of women. *American Family Physician*, *81*(4), 489–495.

Lyerly, A. D. (2006). Shame, gender, birth. *Hypatia*, *21,* 101–118.

MacKinnon, C.A. (1989). *Toward a Feminist Theory of State.* Cambridge: Harvard University Press.

MacKinnon, C. A., & Dworkin, A. (1998). *In Harm's Way: The Pornography Civil Rights Hearings.* Cambridge: Harvard University Press.

Mandziuk, R. M. (2010). "Ending women's greatest hygienic mistake": Modernity and the mortification of menstruation in Kotex advertising, 1921–1926. *Women's Studies Quarterly*, *38*(2), 42–62.

Marso, L. J. (2003). A feminist search for love: Emma Goldman on the politics of marriage, love, sexuality, and the feminine. *Feminist Theory*, *4*(3), 305–320.

Martin, E. (2001). *The Woman in the Body: A Cultural Analysis of Reproduction.* Boston: Beacon Press.

McCabe, P. C., & Rubinson, F. (2008). Committing to social justice: The behavioral intention of school psychology and education trainees to advocate for lesbian, gay, bisexual, and transgendered youth. *School Psychology Review*, *37*(4), 469–486.

McClelland, S. I. (2010). Intimate justice: A critical analysis of sexual satisfaction. *Social and Personality Psychology Compass*, *4*(9), 663–680.

McClelland, S. I. (2011). Who is the "self" in self-reports of sexual satisfaction? Research and policy implications. *Sexuality Research and Social Policy*, *8*(4), 304–320.

McElroy, W. (1995). *XXX: A Woman's Right to Pornography.* New York: St. Martin's Press.

Meltzer, A. L., & McNulty, J. K. (2010). Body image and marital satisfaction: Evidence for the mediating role of sexual frequency and sexual satisfaction. *Journal of Family Psychology*, *24*(2), 156–64.

Molyneux, M. (2001). *Women's Movements in International Perspective: Latin America and Beyond.* New York: Palgrave Macmillan.

Moore, M. R. (2012). Intersectionality and the study of black, sexual minority women. *Gender & Society*, *26*(1), 33–39.

Morokoff, P.J., Quina, K., Harlow, L.L., Whitmire, L., Grimley, D. M., Gibson, P. R., & Burkholder, G. J. (1997). Sexual Assertiveness Scale (SAS) for women: Development and validation. *Journal of Personality and Social Psychology*, *73*(4), 790–804.

National Center for Transgender Equality. (2011). About trans equality. NCTE, URL (accessed 5 April 2012): http://transequality.org/About/about.html

Nicolson, P., & Burr, J. (2003). What is "normal" about women's (hetero)sexual desire and orgasm?: A report of an in-depth interview study. *Social Science & Medicine*, *57*(9), 1735–1745.

Orecchia, A. C. (2008). Working with lesbian, gay, bisexual, transgender, and questioning youth: Role and function of the community counselor. *Graduate Journal of Counseling Psychology*, *1*(1), 66–77.

Passet, J. E. (2003). *Sex Radicals and the Quest for Women's Equality.* Urbana, IL: University of Illinois Press.

Peterson, Z. D. (2010). What is sexual empowerment? A multidimensional and process-oriented approach to adolescent girls' sexual empowerment. *Sex Roles*, *62*(5–6), 307–313.

Pinney, E. M., Gerrard, M., & Denney, N. W. (1987). The Pinney Sexual Satisfaction inventory. *The Journal of Sex Research*, *23*(2), 233–251.

Queen, C. (1997a). Sex radical politics, sex-positive feminist thought, and whore stigma. In J. Nagle (Ed.), *Whores and Other Feminists.* New York: Routledge, pp. 125–135.

Queen, C. (1997b). *Real, Live, Nude Girl: Chronicles of Sex-Positive Culture.* San Francisco: Cleis Press.

Raina, R., Pahlajani, G., Khan, S., Gupta, S., Agarwal, A., & Zippe, C. D. (2007). Female sexual dysfunction: Classification, pathophysiology, and management. *Fertility & Sterility*, *88*(5), 1273–1284.

Reece, M., Herbenick, D. M., & Sherwood-Puzello, C. (2004). Sexual health promotion and adult retail stores. *Journal of Sex Research*, *41*(2), 173–180.

Rich, A. (1980). Compulsory heterosexuality and lesbian existence. *Signs*, *5*(4), 631–660.

Riddell, L., Varto, H., & Hodgson, Z. G. (2010). Smooth talking: The phenomenon of pubic hair removal in women. *Canadian Journal of Human Sexuality*, *19*(3), 121–130.

Roberts, C., Kippax, S., Waldby, C., & Crawford, J. (1995). Faking it: The story of "Ohh!" *Women's Studies International Forum*, *18*(5–6), 523–532.

Rubin, G. (1993). Thinking sex: Notes for a radical theory of the politics of sexuality. In M. A. Barale, H. Abelove, & D. M. Halperin (Eds.), *The Lesbian and Gay Studies Reader.* New York: Routledge, pp. 3–44.

Salk, R. H., & Engeln-Maddox, R. (2012). Fat talk among college women is both contagious and harmful. *Sex Roles*, *66*(9–10), 636–645.

Sanchez, D., Crocker, J., & Boike, K. R. (2005). Doing gender in the bedroom: investing in gender norms and the sexual experience. *Personality and Social Psychology Bulletin*, *31*(10), 1445–1455.

Sanchez, D., Kiefer, A. K., & Ybarra, O. (2006). Sexual submissiveness in women: Costs for sexual autonomy and arousal. *Personality and Social Psychology Bulletin*, *32*(4), 512–524.

Shifren, J. L., Monz, B. U., Russo, P. A., Segreti, A., & Johannes, C. B. (2008). Sexual problems and distress in United States women: Prevalence and correlates. *Obstetrics and Gynecology*, *112*(5), 970–978.

Showden, C. (2012). Theorising maybe: A feminist/queer theory convergence. *Feminist Theory*, *13*(1), 3–25.

Shrage, L. (2002). From reproductive rights to reproductive Barbie: Post-porn modernism and abortion. *Feminist Studies*, *28*(1), 61–93.

Spencer, G., Maxwell, C., & Aggleton, P. (2008). What does "empowerment" mean in school-based sex and relationships education? *Sex Education*, *8*(3), 345–356.

Sprecher, S., Barbee, A., & Schwartz, P. (1995). "Was it good for you, too?" Gender differences in first sexual intercourse experiences. *The Journal of Sex Research*, *32*(1), 3–15.

Stulhofer, A., & Ajduković, D. (2011). Should we take anodyspareunia seriously? A descriptive analysis of pain during receptive anal intercourse in young heterosexual women. *Journal of Sex & Marital Therapy*, *37*(5), 346–358.

Swank, E., & Fahs, B. (2011). Pathways to political activism among Americans who have same-sex sexual contact. *Sexuality Research and Social Policy*, *8*(2), 126–138.

Sycamore, M. B. (2004). *That's Revolting: Queer Strategies for Resisting Assimilation.* Berkeley: Soft Skull Press.

Toerien, M., & Wilkinson, S. (2004). Exploring the depilation norm: A qualitative questionnaire study of women's body hair removal. *Qualitative Research in Psychology*, *1*(1), 69–92.

Toerien, M., Wilkinson, S., & Choi, P. Y. (2005). Body hair removal: The "mundane" production of normative femininity. *Sex Roles*, *52*(5–6), 399–406.

Trompeter, S. E., Bettencourt, R., & Barrett-Connor, E. (2012). Sexual activity and satisfaction in healthy community-dwelling older women. *The American Journal of Medicine*, *125*(1), 37–43.

Valentine, G. (1989). The geography of women's fear. *Area (London 1969)*, *21*(4), 385–390.

Vance, C. (1984). *Pleasure and Danger: Exploring Female Sexuality.* New York: Routledge & Kegan Paul.

Wiederman, M. W. (1997). Pretending orgasm during sexual intercourse: Correlates in a sample of young adult women. *Journal of Sex & Marital Therapy, 23*(2), 131–139.

Yost, M. R., & McCarthy, L. (2012). Girls gone wild? Heterosexual women's same-sex encounters at college parties. *Psychology of Women Quarterly*, *36*(1), 7–24.

Introduction to Reading 30

How is sexuality embodied in a fat Black female body? Sociologist Courtney J. Patterson-Faye offers research-based insight into how to answer this question in her chapter reading. She argues that fat Black women's sexuality can be understood through a "sartor-sexuality" lens which she poses as an alternative to the standard mammy image and cultural script. Patterson-Faye's analysis is based on in-depth interviews with plus size models, bloggers, and designers.

1. Define the term interpersonal scripts. How does the author apply this concept in the readings?
2. How does Patterson-Faye use "sartor-sexuality" and the two subcategories, "silhou sexuality" and "kine-sexuality," to illustrate how fat Black women resist and alter the mammy image and script?
3. The author highlights the public and private enterprises of self-reclamation. What is the meaning of "public and private self-reclamation" in the lives of the Black women who discussed their bodies and sexuality with Patterson-Faye?

"I Like the Way You Move"

Theorizing Fat, Black and Sexy

Courtney J. Patterson-Faye

Whenever fatness and blackness share a theoretical space with sexuality, they more than likely manifest in the corpulent body of 'mammy,' the smiling, docile, asexual figure designated to care for white children and their families. Patricia Hill Collins' (2000) controlling image of mammy asserts that this woman's lack of sexual prowess permits her surrogacy to white children, not only because she can claim them through no sexual act of her own, but also because she—the embodied opposite of westernized beauty ideals—is the impossible partner of white men (Collins, 2000 [1990]: 84), thus causing no threat to Eurocentric frameworks of race, gender and sexuality. Often portrayed as fat, dark and covered from ankle-to-headscarf, mammy's sexuality is further muted by her clothing and her perceived invisibility to white men, and thus all men.

I posit that mammy, her clothing, and her body's movement underneath her clothing were indeed visible, if not hypervisible, to white women and men, and that her proximity to whiteness furnished fat black women, specifically those with bodies deemed unattractive and excessive in size, with peripheral power to subvert previous notions of fatness, blackness and sexuality. By employing a black feminist lens, mapping it onto sexual script theory and analyzing in-depth interviews with plus size models, bloggers and designers, I will theorize a fat black female sexuality by discussing and analyzing how fat black women have: (1) long embodied and rejected the conceptualizations of mammy; (2) consistently regarded sex and sexuality as both public and private enterprises of self-reclamation; and (3) utilized clothing as a means to subscribe to and complicate cultural norms of fat black (a)sexuality.

Theoretical Framing: (De)scripting Mammy

Behind her wide grin, underneath her headscarf, and covered by her long skirt, lies the truth about 'mammy.' She is scripted to be the ultimate caretaker; she not only knows how to rear children and clean households, but she also relinquishes the definitions of femininity and sexuality to be defined by white men (and acted out by white women).[1] However, the mammy that we have come to know and understand, birthed in slavery and developed through waves of popular culture, does not exist in a cultural or historical vacuum. Instead, mammy's complexity has been formed and reformed alongside her long-established identity of corpulence, asexuality and docility.

In fact, what is occurring underneath her apron is just as significant as her donning the service garment. Just as Foucault (I 990 [1978]) outlines, the struggle for power to define one's sexuality exists concomitantly with society's ability to regulate and monitor sexuality.[2]

Current plus size models and their experiences in the fashion industry are a testament to this paradigm. Under the guise of mammy, these women struggle to be recognized by mainstream fashion outlets as beautiful and sexually attractive. However—as I imagine how mammy operated in her own community—they create their own spaces where their respective aesthetics are celebrated, revered, and upheld against mainstream beauty ideals.

Society's depiction, or scripting, of mammy's sexuality tells us that she is no sexual threat to (white) women anywhere because her bigger body simply does not fit into a westernized ideal of beauty, and thus makes her undesirable. However, I know mammy to have a family of her own, complete with a sexual partner and children. These discrepancies are indicative of the interplay between *cultural scripts,* which act as 'instructional guides' that dictate social behavior; *interpersonal scripts,* which are one way for people to collaborate and act as scriptwriters to change dominant cultural scripts; and *sexual scripts,* which are scripts that give meaning to sexual activities and naturalize that meaning through those who engage in said sexual activities (Simon and Gagnon, 1986).[3] The cultural script—mammy as servant—represents how a woman of her stature is supposed to behave. Mammy conceived as a lover or mother of her own children is indicative of an interpersonal script, where mammy and her partner/child are 'scriptwriters' and 'actors' working together to alter the larger, more accepted cultural script. In terms of sexual scripts, mammy and

Patterson-Faye, C. J. (2016). "I like the way you move": Theorizing fat, black and sexy. *Sexualities, 19*(8): 926–944.

her sexual partner would engage in interpersonal scripting to rewrite sexual scripts that give new meaning to their sexual activities.

This article focuses on those interpersonal scripts between fatness and sexuality that not only challenge our conceptualizations of mammy, but also encourage us to move away from discourses that make these two categories mutually exclusive. Historian Thavolia Glymph disrupts the mammy script by arguing that black women working inside of plantation homes were often teenagers with smaller physiques that caused them to be a direct sexual threat to white female slave owners/employers (2008: 53, 119). Jennifer Morgan (2004) does the same by asserting that enslaved African women served as both productive and reproductive laborers, where all women (regardless of size) were expected to tend to the fields/home and birth children at the mercy of their slaveholders (2004: 36, 92). While both authors may reinforce cultural scripts that label black women as overtly sexual because of their African heritage, or the idea that only young, thinner bodies are considered attractive, I recognize their conceptualizations and argue that they speak to mammy's pervasiveness.

Looking to plus size models (and plus size women more generally) we can redirect our attention to how self-representation is especially important for fat black women as they engage in interpersonal scripts.[4] What these women wear and how they wear their clothing represents how they see themselves in relation to their environment and how their environment views them. Like mammy, they may have a 'uniform' of sorts, a collection of clothing that signifies 'what plus size models wear,' such as restrictive garments (formally referred to as 'girdles,' popularly referred to as 'spanx') or dresses with belt-cinched waistlines. Unlike mammy, they do not have to wear clothing that mutes their sexuality. These women, whose bodies are often masculinized and made deviant when juxtaposed to white femaleness and/or thinness (Shaw, 2006: 50), have lived at the intersections of body size, race and gender with full, complete lives that are rife with social, sexual and cultural experiences that continue to be flattened under mammy's existence. I argue that perhaps the mammy script by itself is no longer useful for fat black women, and that we need to find alternative ways through which to examine and explore fat black women's sexuality.

This article offers one of those alternative lenses through which to view fat black female sexuality. Picking apart the idea of mammy's uniform, I argue that fat black women's sexuality can be discussed through what I define as *sartor-sexuality,* which means that how the group defines and executes its sexuality is directly correlated to the relationship between their clothing and their bodies. Moving beyond the idea that sexual attraction can be fixated on the naked body, sartor-sexuality is attributed to the specific struggle that fat black women experience with self-representation. How a fat black woman presents herself to the world can be indicative of her socioeconomic status, self-esteem and self-actualization. Of course, people can treat clothing as a façade through which they can conceal various parts of their lives, but with black women, the idea of concealing (or not concealing) a fat body is as much an act of agency as it is a response to society's dominant cultural scripts. Because of this, I believe sartor-sexuality becomes a critical analytical tool through which to view black female sexuality.

I organize sartor-sexuality into two subcategories: *silhou-sexuality* (the idea that a fat black woman's silhouette is directly related to her sexuality) and *kine-sexuality* (the idea that the movement of fat black women's bodies sends sexual messages to society members). While both of these terms are generalizable and can operate independently of clothing, I argue that for fat black women, and their ability to fashion silhouettes and movement through the clothes that they wear (and do not wear), it would be virtually impossible to define and explore these terms without considering their relationships to clothing. Under silhou-sexuality, I acknowledge that fat black women's bodies have long varied in size and shape. However, culturally speaking, those who find black women attractive typically gravitate toward a low waist-hip ratio (WHR) that accentuates a smaller waist with robust posterior (Freedman et al., 2004). For fat black women who do have a low WHR, this preference may increase their perceived cultural desirability, but—according to Robert Staples (2006: 90)—fat women *without* these dimensions are 'lucky' to receive sexual attention. This attitude is present in both straight and queer communities, rendering black women who have large stomachs and/or small posteriors and/or small hips invisible to a desiring gaze (Weinstein, 2012).

Of course the relationship between WHR and female attractiveness is not a closed concept. Many people find a variety of silhouettes attractive. Large or small sexualized body parts, like breasts or thighs, may not be a factor for some, but dominant narratives within black communities insist that someone with a small waist and a large behind is considerably more desirable than someone lacking these characteristics. For fat black women, silhou-sexuality is the consideration of how their bodily silhouettes directly impact the development of their sexualities. We might also consider within this category some of the ways in which black women tailor and manipulate their clothing and bodies (or not) in accordance with a particular aesthetic. Some women wear corsets, girdles, bras and other types of restrictive garments underneath their clothing to gain a certain silhouette, while others wear clothing that 'fits their body type' so as to appear thinner or more proportionate. However, there exists a group of women who wear little to no restrictive garments and embrace whatever shapes manifest through their clothing from their naked bodies.[5] This manipulation (or lack of manipulation) is critical when thinking about the fat black woman's conceptualization of herself and the way that she wants to present herself to the world.

With kine-sexuality, I focus particularly on the literal movement of fat black women's bodies that sends sexual messages to society members (fat black women included) who decode these movements into sexual and non-sexual behavior. How fat black women's bodies move makes the movement itself imperative to investigating how they develop their sexuality. Because fat bodies make larger bodily movements, can fat women ever escape being sexualized, even when they do not set out to do so? How can they be invisible when this motion, especially for those who choose not to wear restrictive garments, is insistent as well as persistent? Bodily movement, whether voluntary or involuntary, has a considerable impact on how fat black women develop their sexuality and how others develop their ideas of fat black women's sexualities. For example, I saw a model walking down a runway wearing a 'work-appropriate' ensemble that included a blouse whose buttons puckered and exposed her bra as she walked. Although it was not her intention to do so, the model inadvertently drew attention to her breasts, thus giving the audience an opportunity to sexualize her. Had she been modeling lingerie or sleepwear with her bra already exposed, the audience may have still sexualized her, but her (or rather the designer's) intention would have been for them to do so. Kine-sexuality con notes the tensions between intentionality and the idea that fat bodies have been historically sexualized, whether they are in sexual scenarios or not.

I look to fat black women in fashion, namely those in the plus size fashion industry, to understand how they see themselves, (un)clothe their bodies, and write scripts for their own sexualities. The eight women included here, taken from a large data-set that includes 32 interviewees (21 formal interviews, 11 semi-formal) aged 26 to 65, plus size blog content analysis, and ethnographic fieldwork at plus size fashion shows, represent a nucleus of mostly US-born black women, some with continental African or Caribbean parents or grandparents, that help me consider what type of standards exists for plus size models and how these standards affect fat black women. Chosen mainly from their popularity in the plus size blogosphere and audience reaction during the Full Figured Fashion Week (FFFW) in 2013, women were interviewed in person or via skype/telephone about connections between their body size and childhood experiences, development of sexuality, and fashion experience.

Situating their experiences at the center of my analyses, I use black feminist thought to anchor their responses in notions of agency, inequality and self-definition.

Fat Black Female Sexuality, Clothing and Fashion

> There was a person dressed in drag at the show . . . one of the women in my row said, 'Look at that body, I'm almost mad,' referring to the person's silhouette. Everyone here seems to be aware that they have to have their bodies 'snatched' meaning they have to have the right undergarments to smooth out any lumps/bumps and get as close to an hour glass figure as possible.[6]

Whether a body is scantily clad, fully clothed, or naked, it is important to look at how the body is dressed because the way we wear (and do not wear) clothing has always had a significant impact on our social development, especially in terms of power and access (Ash and Wilson, 1993; Crane, 2000). Sartor-sexuality, generally speaking, is a reflection of this phenomenon. Frequently executed and exhibited in the fashion world, sartor-sexuality creates aesthetic boundaries as women walk on runways, appear in magazines and pose for photographers. For fat black women, who are virtually invisible to high fashion designers such as Chanel, Gucci, Louis Vuitton and Prada, and magazines like Vogue and Cosmopolitan, sartor-sexuality is a way in which they can negotiate notions of beauty and empower themselves to make choices about how they present their bodies to the world.

Mammy, in her apron, button-up shirt, long skirt and head tie, may have dressed herself, but she aligned her self-presentation to what was expected of her occupation. What if mammy had dressed 'sexy'? Was it her fatness or her sartorial representation (or both) that rendered her non-threatening to the white male and female adults in the household? Could her protruding bosom underneath her apron be considered sexy? Furthermore, if she did dress 'sexy,' what would that look like and how would it be compared to what was conventionally sexy or fashionable? I argue that a fat black woman's sartor-sexuality is dependent mainly on how she desires to present her body and sexuality to the world. Mammy's type of sartor-sexuality can be considered as a protective mechanism, whereby she may have been able to avoid sexual advances from men and violent attacks from women who were threatened by her embodying a more overtly recognized sexuality. Mammy's sartor-sexuality can also be seen as a way to subvert definitions of sexy. What did she have on underneath her uniform? Does her fatness make her sexier with or without clothing?

I believe the power inherent in choosing a particular type of clothing speaks to the ability of fat black women to create a space for their varied identities. It is true that another person's perception of a fat black woman's body in this clothing impacts her decision-making, but I would hesitate to assert that it is a total motivator. For instance, a fat black Muslim woman wearing an abaya may be adhering to religious beliefs, but if she has large breasts and/or large buttocks, the fabric may very likely outline her body, especially when she moves. This kine-sexual marker shows that while she makes the conscious decision to cover her body, her body's movement is just as important as the body itself when considering how her body can be sexualized. She may be muting her sexuality under religious law, but how we view her body underneath the abaya complicates how we would (or should) label her sexuality.

To grapple with the question: 'Does a fat black woman look sexier with or without clothing?' invites us to have a conversation about who determines what 'sexy' should look like. What makes 'six-pack' abdominal muscles more attractive, or acceptable, than a 'hanging' stomach? Medical literature, religious doctrine and middle and upper-class values have all had a hand in determining what is medically, morally and socially acceptable when it comes to how we present ourselves and our bodies to our communities (Campos, 2006; Crane, 2000; Wilson, 2003). The 'obesity epidemic' that links fatness to chronic disease is laced with moral, personal responsibility narratives that imply that 'overweight and obese' people are responsible for their 'inevitable' health afflictions (Campos, 2006; Kwan, 2009; Rothblum and Solovay, 2009; Saguy and Riley, 2005). This placement of blame ignores the impact of sociocultural realities on overweight and obese peoples, causing even more pronounced stratification between mainstream society and fat black women, especially in terms of healthcare, access to healthy foods and environmental racism.

Whether building their sartor-sexuality through silhou-sexual or kine-sexual ways, fat black women are critically aware of mainstream ideals of blackness, fatness and womanhood, and how the cultural scripts belonging to these social constructs continue to relegate them to the outskirts of society. Because of this, Shaw (2006) asserts that the mere existence of fat black women is an act of resistance. However, there are women who present themselves, knowingly, as sites of resistance. Gabi Gregg, of 'GABIFRESH', is a key player in the plus size blogosphere who acknowledges that the fashion world does not make much room for a woman of her stature. In 2012, Gregg was involved in controversy that included her appearance in a 'fatkini' on her website, and 30 other 'fatshionistas' who posted pictures of themselves in fatkinis on the XOJane website. The pictures received mixed responses, but two main threads dominated the conversation. Some celebrated the women's courage to wear two-pieces and attempted to motivate other women to do the same; whilst others blamed the fat women for encouraging obesity and demanded that they cover up ('Gabi Fresh', 2012; 'Gabi Gregg', 2012; Gregg, 2012a, 2012b).

Gregg responded on the *Today Show* noting that she took the pictures to encourage people to love the bodies that they were in 'right now,' and that she was a strong supporter of healthy living. Gregg may have been referring to the 'health at every size' movement,[7] but I posit that the responses insisting that the women cover up, while couched in references to obesity and unhealthiness, were really ways to attempt to regulate these fat (mostly) black women's bodies. As Foucault (1990[1978]) reminds us, sexuality is built upon regulations and people's response to the regulations. The women consciously donning their two-piece bathing suits and posting them for public consumption were exercising their silhou-sexuality to a very narrow audience. They brought the private, oppressed reality of fat and fatness into a very public arena to resist larger cultural and sexual scripts that demand they cover up and/or lose weight.

These types of regulatory responses are not limited to the women wearing fatkinis. In fact, a number of plus size fashion bloggers have become accustomed to responding to this type of discrimination. Marie Denee of 'The Curvy Fashionista' admitted to experiencing some anger in regards to a 2011 Essence.com article entitled 'Sound-Off: Suffering from a Confidence Overdose?' in which the author not only reprimanded fat women for wearing clothes that did not fit them, but also for not losing weight in order to fit into them. The Essence.com contributor writes:

> But baby, on the flip side, if you are a size 16, no amount of squeezing, pulling, tugging, yanking, or praying is going to make getting into a pair of size 10 shorts look good. Yet I see sisters strutting around all the time in clothes that are clearly from the junior section when they need to be front and center at Ashley Stewart. They're parading around in baby tees with their tummies spilling over their waistbands, ready to verbally assassinate anybody who dares suggest it ain't cute . . . but it's not a good look. ('Sound-Off', 2011)[8]

Is it the movement or silhouette in or outside her clothing that makes this size 16 woman's look unacceptable? Is she aware of how she is seen by others insomuch as she purposefully wears her clothing small as a way to assert her sexuality, no matter the audience member? Or maybe is it that she has lost weight and is struggling to find clothing that she believes fits her body well? Perhaps she has gained weight and is in transition to purchase new clothing. Either way, this author shows how cultural scripts will always attempt to regulate behavior, especially when related to how a fat body is supposed to be dressed.

Fat black women are scrutinized both inside and outside of their clothing, and many of those whom I have interviewed revealed that until they performed the psychological work to accept themselves, they felt sexier covered up and in private. The growth that occurs from their psychological labor most times reverses how fat black women define their bodies and sexualities. In my interactions with plus size fashion models and entrepreneurs, they are clear that how they view and value (their own) fat bodies is not only contingent upon the industry in which they work, the communities where they live and the networks in which they participate, but also upon a bevy of personal experiences intertwined between these three spaces. The next section explores how all of these factors coalesce to determine their respective realities regarding fatness, blackness and sexuality in both public and private spaces.

Fat Black Women's Sexuality and Reality

Plus size fashion is one of the main venues that transports the fat black woman's body between private and public spaces. It not only directly bridges fashion with the fat black female body, but it also gives insight into how plus size models, bloggers, designers and other women can subscribe, reject or integrate dominant cultural scripts of sexuality. Structured and semi-structured interviews with these women in the plus size industry showed me just how complex fat black female sexuality can be. Many of the women acknowledged a need for a sartor-sexuality framework by either sharing how cultural scripts determined how they were made to think about their bodies, using language akin to silhou and/or sartor-sexuality, or refusing outright to believe that fat black women are not sexual beings.

The women, who all varied in shape and size, were very aware of how important. their body size was in their sexual development.

When asked about when she was made aware of her body size, Danielle,[9] a 31-year-old plus size model and blogger stated, 'I've always been aware,' explaining that from a young age, she was not only teased by classmates, but pressured to dress appropriately by family members. She said 'My grandmother taught me how to hide,' and that: 'You don't tuck your shirt in on the inside. No belts,' so she could draw attention away from her body and thus escape scrutiny. However, Danielle indicated that older men became interested in her (at a young age) and that boosted her confidence. She admitted that 'boys my age were never interested in me. They wanted the stick-thin girls.' While Danielle's silhou-sexual development was not exercised overtly through her clothing, it appears that her developing body—arguably her kine-sexual development—still drew men to her. She also mentioned 'If I were smaller, I wouldn't have my husband.'

However, women like Sarah, a 27-year-old entrepreneur and fat-advocate said, 'I felt like my body was objectified . . . my confidence in my own body made me comfortable,' meaning that the attention she got from men made her feel exploited, and her sense of attractiveness stemmed from the way she processed these interactions. Another woman, a stylist named Janelle said, 'I don't need outside validation. God made me this way, so this the way I'ma stay.' Her 'staying the way that God made her' is indicative of religious cultural scripts that discourage body modification. In either case, whether potential partners gave them attention or they had to process their sexual identity through their own sociocultural processes after the attention, these women have used interpersonal scripts to rewrite 'fat as sexy.'

Because of this, most of the women identify as 'plus positive' and have come to love the bodies that they have. Nevertheless, this did not prevent some of the women from deploying obesity epidemic language. A few of the women felt strongly about the obesity epidemic, saying that they did not believe in health at every size and that people should do whatever they can to be medically healthy. Michelle, a petite plus size model and aspiring designer simply stated, 'Fat is fat. Obese is a problem.' Michelle tapped into the notion that obesity denotes disease and deviance, and that while fat is simply 'fat,' there is a limitation on its healthiness.

Building upon the 'fat is fat' quotation, all of the women had very strong opinions about the usage of 'fat' to describe their bodies. A majority of the women preferred alternative terms with positive connotations such as 'plus size' or 'curvy' to describe themselves. The politics surrounding their terms of choice stemmed from negative childhood experiences, awareness of fat fetishization and the perceived linkages between fatness and disease. These women used the word fat when they referred to themselves negatively, especially when they gained weight or did not like how they looked in a particular garment. However, a few of the women had reclaimed the word fat, and as a political statement used the word to describe themselves whether inside or outside a group of fat women. One of the women confessed, 'I'm just fat,' meaning that she did not subscribe to any negative aspects of the word and that those who had issues saying or hearing the word still had 'work to do' in regards to their body size.

Whether they referenced fat in negative or positive ways, all of the women seemed to have a keen understanding that a proportionate silhouette was not only what designers look for when they cast plus size models, but also what potential sexual partners prefer. Michelle confidently asserted, 'Cuteness depends on the silhouette.' Here, Michelle was engaging the core of silhou-sexuality. Some women have hour-glass, proportionate figures and some simply do not.

Casey, a plus size model I met during Full Figured Fashion Week in New York, noted that she thought her proportionate body enhanced her sexuality (and along with her warm personality, it increased her chances of getting hired): 'I love my body. I love that I have big boobs, a small waist . . . I put on a pair of jeans from [Levi's] curvy collection . . . *killa!'*

Portraying and confirming who we are vis-á-vis our clothing choice has always been an integral part of fashion (Arnold, 2001). For black women in particular, appearing 'well put together,' especially in the workplace, is representative of one's socioeconomic status (Gill, 2010; Miller, 2009). Casey declared, 'Clothing says who we are, it's a way to communicate. I'm a lady. I want you to acknowledge that I am a lady. It's how you wear it as well. I wear them, they don't wear me.' This shows that what our bodies do in our clothing is pivotal to how we wear our clothing. By saying 'it's how you wear it,' Casey, like the majority of the respondents, employed kine-sexuality to discuss what is meant to move and what it meant to be restricted.

Fat black sexuality, at least on the runway, did not necessarily mean larger movements. In fact, less movement was preferred. For instance, Michelle declared that she disliked it when 'people don't wear foundations.' Here, she was referring to foundation garments, such as body shapers, that pull a woman's body in so that it appears smoother and smaller, making her clothing look more 'presentable' on her body. Michelle said that people should wear these garments, as well as clothing that fits their body type, to put forward their best sartorial representation. Her usage of such terms, couched in respectability politics, dictated that black women should always present their 'best' selves to stave off negative stereotypes (Higginbotham, 1994). For fat black women, the idea of restricting their bodies through these foundation garments shows how inequality can be replicated in plus size fashion. Can a fat black woman ever put forth her best sartorial representation without foundation garments? What about fat women larger than size 26 who struggle to find foundation garments that fit? Can their clothes ever be flattering?

The types of clothing fat black women wear and what their bodies do underneath their clothing are both vital to understanding their sexual development. Not simply because of self-representation, but also because of the impact that the 'right' type of clothing has on a woman's confidence and sense of self. When asked what type of clothing she liked to wear, Bethany said, 'Anything tight! I'm so serious! I love it! I love everything that accentuates my curves.' Bethany, a plus size model and blogger who had dated both men and women, mentioned that at times she would like her thighs to be smaller, but her breasts 'sit up on their own' and she felt good about herself when her curves were displayed through her clothing. This is a different narrative from her days as an adolescent when she noticed that she was 'always overly developed.' She went on to say, 'I never understood my body, and I guess, the power of what it could do for someone or have on someone. I remember being young and not that I was never called pretty or beautiful, I was always being called sexy by someone.'

Her clothing was not tight then, but people could label her body as sexy without any intention of her own. Through her own psychological work to figure out the power of her body, Bethany has since chosen to wear tight clothing to flatter her curves, boost her confidence and satisfy her self-representation of 'sexy.' Danielle was one of the first models to admit, 'if clothes are flattering, they make me less self-conscious.' She was referring to clothing made to fit her body type, or clothing that made her feel proportionate. Because of the inability to have flattering clothing at the ready, many of the respondents described the difficulties of shopping for fashionable, age-appropriate clothing. Sarah said it was 'hard to find dresses so I started to sketch my own.'

Literally taking her representation into her own hands, Sarah had done what many fat black women, especially those in the plus size fashion world, have done in a world that scripts fatness outside of fashionable. Lynette declares, 'Plus [size] fashion is very much driven by black women. We have huge buying potential and an incredible sense of style and we want better options . . . black women are represented more than anything else.' If four out of every five black women are over-weight or obese in the USA,[10] this is highly significant because it shows that over half of the black population in the USA is made to deal with issues pertaining to intersections of race, gender and body size. It also provides insight into the actions of and beliefs about fat black female sexuality as it pertains to sexual activity and sexual partners. Fat black women, while being labeled deviant, are thriving and writing their own interpersonal scripts with partners who find them sexually attractive. Lastly, it proves that fat black women have economic agency when it comes to self-representation through clothing and/or sexual identities.

Danielle voiced her frustration about not exercising her economic agency when seeing fashionable clothing on the runway that she could not wear: 'I should be able to see what that garment looks like on me.' This is indicative of how many fashion companies, even those who have plus size lines such as Michael Kors, do not wish to market plus size clothing as a part of their image. I argue that companies do this not only to uphold mainstream fashion ideals, but also to apply the power that they have to define what is fashionable to people outside of their fashion boundaries. Brenda, a popular blogger who would not categorize herself as a plus size blogger but has a considerable plus size following said:

> They don't even want press surrounding the fact that they have plus sizes. They think it will affect their image. They want to make sure that their brand is in line with what the industry standards is . . . that the industry standard is that my collection is for the size 4, Upper East Side debutant. That's who we want to showcase in our advertisements and that's who we're gonna do outreach for for press . . . they want the plus size dollar, but they don't be associated with the community.

Brenda's statement about how mainstream clothing designers seem to follow the conventions of straight-size fashion (namely white, thin and upper class) mirrors the ways in which mainstream society has compartmentalized fat black female bodies into images of mammies. Just as white households profited from fat black female bodies only being considered as mammies, these companies are able to profit from conceptualizing fashion in such a manner that excludes plus size black consumerism while still accepting black plus size dollars. Thus, it has become critical for plus size black women to design, purchase and wear clothing that not only gives women choices in representing themselves, but also gives them the confidence to use clothing to define their sexuality. Working collaboratively as 'fat,' 'thick, or 'curvy goddesses' to redefine blackness, fatness and sexuality, these women recognize that the act of wearing clothing is inextricably tied to their identity formation.

Conclusion

Venturing into a theoretical space with fatness, blackness and sexuality requires an alternative and reflective approach that leaves space for unconventional under standings of what these three identities mean. While important in understanding fat black women's sociopolitical positions throughout the world, mammy is no longer the only way to discuss how sexuality is embodied in a fat black female body. Both historical and contemporary black female identities, even the mammies themselves, need other ways in which to define and redefine their sexualities. The use of plus size fashion allows us to view fat bodies alternatively and in a more complete manner. Unlike the dominant fashion world, plus size fashion shows us that fat black women are indeed visible and active in inscribing meaning onto their bodies and experiences. Together, they are designing, choosing and wearing the outfits that help them, their sexual partners and their communities rewrite cultural scripts that define what is attractive, sexy and fashionable.

When Danielle said, 'If I were smaller, I wouldn't have my husband,' she conveyed exactly what I aim to prove: although fat black women have been rendered invisible and somewhat powerless by dominant scripts, there have always been scripts that highlight this group's visibility, proving that they are indeed worthy of (sexual) attention and powerful enough for self-definition. Women like Danielle show that blackness, fatness and sexuality straddle public and private spaces, and that the plus size fashion world reminds us of what Foucault discovered and fat black women have worked so hard to prove—that dominant norms of blackness, fatness and sexuality do not occur and will not exist alone.

Notes

1. In *Black Feminist Thought,* Patricia Hill Collins discusses other *controlling images* to show how pervasive and destructive conceptualizations of black women can be. Known for her sexual prowess, 'Jezebel,' has an insatiable appetite for sex, while 'Sapphire' takes pride in scolding and belittling anyone in her path, especially men, and 'the welfare queen,' sits at home, unemployed and suctioning up any and all kind of governmental assistance. I chose to discuss mammy because she is the only image that has a direct connection to fat and fatness. Some, but not all welfare queens have been associated with being overweight, but race takes precedence over body size with that image, while with mammy, body size is equally considered to race.

2. Michel Foucault's *The History of Sexuality, Volume I* (1990 [1978]) explains that while societies may attempt to regulate sexuality, their constant monitoring and regulation helps to create subversive spaces for people to explore their sexualities, both publicly and privately.

3. Simon and Gagnon (1986) define cultural scenarios as part of larger *cultural scripts* that act as 'instructional guides' to socially determine how people should behave. Just as Foucault discovered, these cultural scripts are not always followed. Thus, Simon and Gagnon offer *interpersonal scripts* and *intrapsychic scripts* (which represent an internal dialogue that affects personal behavior) as means to change dominant scripts. The authors claim that intrapsychic scripts have a greater impact in changing cultural scripts, but because of the nature of my methods, I only examine interpersonal scripts because they are measurable.

4. I employ 'fat' as a term throughout this article to align myself with Fat Studies scholars to reclaim the social and cultural power of the word and move away from medical language such as overweight and obese that signifies deviance. I use 'black' to represent descendants of African peoples all over the world, particularly those born in the USA, the Caribbean and the continent of Africa. 'Woman' is used as a gender term to represent ciswomen from my research, those whose reproductive capabilities align with their societally assigned sex and gender.

5. This type of fashion could be considered 'oppositional dress' as described by Elizabeth Wilson in Adorned *in Dreams* (2003), where people resisting cultural norms refuse to partake in conventional forms of fashion (see also Cohen, 2004). I argue that most times, whatever fat black women wear is in direct opposition to societal norms because their bodies are not categorized as 'normal.'

6. Excerpt from field notes from a plus size fashion show during Full Figured Fashion Week in New York City.

7. The idea that a person can be healthy at any size stems from Linda Bacon's *Health At Every Size: The Surprising Truth About Your Weight* (2008), where the author not only dispels the correlations between obesity and chronic disease, but also indicates the implications of dieting for physical and mental health.

8. Ashley Stewart is a chain of plus size clothing stores marketed toward urban communities.

9. All names of respondents have been changed.

10. The US Department of Human Health and Services' Center for Disease Control and Prevention found in its 2012 study that 80% of African American women were overweight or obese. These are the highest rates compared to any other group in the USA.

References

Arnold R (2001) Fashion, Desire and Anxiety: Image and Morality in the 20th Century. London: IB Tauris.

Ash J and Wilson E (1993) Chic Thrills: A Fashion Reader. Berkeley, CA: University of California Press.

Bacon L (2008) Health at Every Size: The Surprising Truth About Your Weight. Dallas, TX: BenBella Books.

Campos P (2006) The epidemiology of overweight and obesity: Public health crisis or moral panic? International Journal of Epidemiology 35(1): 55–60.

Cohen CJ (2004) Deviance as resistance: A new research agenda for the study of black politics. Du Bois Review l(l): 27–45.

Collins PH (2000 [1990]) Black Feminist Thought: Knowledge, Consciousness, and the Politics of Empowerment. New York, NY: Routledge.

Crane D (2000) Fashion and its Social Agenda: Class, Gender, and Identity in Clothing. Chicago, IL: University of Chicago Press.

Crenshaw K (1989) Demarginalizing the intersection of race and sex: A black feminist critique of antidiscrimination doctrine, feminist theory and antiracist politics. The University of Chicago Legal Forum (1989): 139–168.

Foucault M (1990 [1978]) The History of Sexuality, Volume I: An Introduction. New York, NY: Random House.

Freedman Rachel EK, et al (2004) Ethnic differences in preferences for female weight and waist to hip ratio: A comparison of African-American and White American college and community samples. Eating Behaviors 5(3): 191–198.

'Gabi Fresh' (2012) Gabi Fresh's 'fatkini' gallery on XOJane receives overwhelming applause. Available at: http://www.ibtimes.com/gabi-freshs-fatkini-gallery-xojane-receives-overwhelming-applause-photos-700997 (accessed 18 June 2012).

Gabi Gregg (2012) Gabi Gregg, blogger who started 'fatkini' photos, talks body image on the 'Today' show. Available at: http://www.huffingtonpost.com/2012/05/30/gabi-greggfatkini-today-show_n_l556480.html (accessed 18 June 2012).

Gill TM (2010) Beauty Shop Politics: African American Women's Activism in the Beauty Industry. Urbana, IL: University of Illinois Press.

Glymph T (2008) Out of the House of Bondage: The Transformation of the Plantation Household. Cambridge: Cambridge University Press.

Gregg G (2012a) Fatkini. Available at: http://gabifresh.com /2012/04/fatkini-2012/ (accessed 18 December 2012).

Gregg G (2012b) The XOJane and Gabi Fresh fatkini gallery: 31 hot sexy fat girls in skimpy swimwear. Available at: http://www.xojane.com/fun/gallery/fatkini (accessed 18 June 2012).

Higginbotham EB (1994) Righteous Discontent: The Women's Movement in the Black Baptist Church 1880–1920. Cambridge, MA: Harvard University Press.

Hine DC (1989) Rape and the inner lives of Black women in the middle west. Signs 14(4): 912–920.

Kwan S (2009) Framing the fat body: Contested meanings between government, activists, and industry. Sociological Inquiry 79(1): 25–50.

Miller ML (2009) Slaves to Fashion: Black Dandyism and the Styling of Black Diasporic Identity. Durham, NC: Duke University Press.

Morgan JL (2004) Laboring Women: Reproduction and Gender in New World Slavery. Philadelphia, PA: University of Pennsylvania Press.

Outkast (2003) The Way You Move, on Speakerboxx/The Love Below, CD. Atlanta, Laface/Arista Records.

Rothblum E and Solovay S (eds) (2009) The Fat Studies Reader. New York, NY: New York University Press.

Saguy AC and Riley K (2005) Weighing both sides: Morality, mortality, and framing contests over obesity. Journal of Health Politics, Policy and Law 30(5): 869–921.

Shaw AE (2006) The Embodiment of Disobedience: Fat Black Women's Unruly Political Bodies. Lanham, MD: Lexington Books.

Simon W and Gagnon JH (1986) Sexual scripts: Permanence and change. Archives of Sexual Behavior 15(2): 97–120.

Smith R (2014) Nicki Minaj explains racy 'Anaconda' video. ABCNews.com, 22 August. Available at: http://abcnews.go.com/Entertainment/nicki-minaj-explains-racy-anacondavideo/story?id=25084030 (accessed 12 October 2014).

Sound-Off (2011) Sound-off: Suffering from a confidence overdose? Available at: http://www.essence.com/2011/05/25/black-women-suffering-from-a-confidence/ (accessed 30 April 2013).

Staples R (2006) Exploring Black Sexuality. Lanham, MD: Rowman & Littlefield.

Weinstein R (2012) Fat Sex: The Naked Truth. CreateSpac Independent Publishing.

Wilson E (2003) Adorned in Dreams: Fashion and Modernity. New Brunswick, NJ: Rutgers University Press.

Introduction to Reading 31

L. Ayu Saraswati offers a deep analysis of the meaning of skin-whitening practices among Indonesian women of Indian, Malay, Chinese, European, and Arab backgrounds. For this study, she conducted interviews with 46 women living in two cities, Jakarta and Balikpapan, and from the interviews she developed the proposition that Indonesian women are motivated to use skin whiteners, not by a strong desire to look attractive, but by the desire to avoid shame and embarrassment. Thus, Saraswati argues, the practice of skin whitening is an embodied expression of the "gendered management of affect." Central to this reading is *malu,* an Indonesian term that is similar to shame. However *malu* also signals vulnerability, worry, shyness, and propriety. Importantly, *malu* is a moral and gendered emotion. Saraswati's research demonstrates how women manage feelings of *malu,* or shame, about their skin color by using skin whiteners, and, in so doing, gender, racial, and global hierarchies are reinforced.

1. How do race and nation play a major role in shaping the meaning of skin color among Indonesian women?

2. How does putting on skin-whitening cream result in a "never-ending cycle of *malu*"? Do you think that the use of cosmetics by women in the United States might have a similar outcome?

3. Discuss the links among "being good," "feeling good," "looking good," and "having good" skin color as presented in this reading.

"Malu"

Coloring Shame and Shaming the Color of Beauty in Transnational Indonesia

L. Ayu Saraswati

In Indonesia, the fourth most populous nation in the world, skin-whitening products are ranked highest among all revenue generating products in the cosmetics industry.[1] Unilever Indonesia spent IDR 97 billion ($10.4 million) in 2003 advertising just one of its skin-whitening creams.[2] This sum is larger than the estimated IDR 72 billion spent on advertising anti-dandruff shampoo—the top product in the hair care industry.[3] Indonesia is not anomalous in this regard: Transnational corporations such as Unilever, L'Oreal, and Shiseido have aggressively marketed their skin-whitening creams throughout Asia, Africa, Europe, and the United States.[4] Skin-whitening products are available in Indonesia, the Philippines,[5] Vietnam, Singapore, Malaysia, Japan, China, Korea, Hong Kong, Taiwan, India, Saudi Arabia, Brazil, Peru, Bolivia, Venezuela, Mexico, Malawi, Ivory Coast, the Gambia, Tanzania, Senegal, Mali, Togo, Ghana, Canada, and the United States. Even in countries where they have been banned due to medical or political reasons—South Africa, Zimbabwe, Nigeria, and Kenya, for example—skin-whitening products continue to be circulated underground.[6]

Many skin-whitening products have been deemed medically dangerous[7] because they contain illegal ingredients such as mercury or hydroquinone beyond the allowable 2 percent limit.[8] Mercury can cause black spots, skin irritation, and in high dosages can cause brain and kidney damage, fetal problems, lung failure, and cancer; hydroquinone is known to cause skin irritation, nephropathy (kidney disease), leukemia, hepatocellular adenoma (liver cell adenoma), and ochronosis (adverse pigmentation). And yet, despite warnings that the chemicals in these products may cause harm, women—the target market and primary consumers of these products—continue to use them.

Why are these products so popular even when they are known to be harmful? I am not the first to pose this question. Existing studies on the popularity of skin-lightening creams tend to focus on the political and racial meanings of these products within the context of colonialism and/or transnationalism. . . .

Other studies focus on media representations of skin-lightening creams and, less frequently, reference biological or psychological perspectives. Nancy Etcoff suggests that a preference for lighter-skinned women may reveal the working of a "fecundity detector" whereby possible mates detect women's fecundity by looking at their skin color believing that young ovulating women have lighter skin.[9] She is not oblivious, of course, to the fact that women's skin-whitening practices are also related to racism.[10]

I propose to offer a different approach. Although I shall also situate whitening practices within a transnational context and query their political and racial meanings, I have turned to the users themselves to ask why they use whitening creams and how they understand their meanings. What would we learn if we relied on women's representation of *themselves* as they make sense of skin-whitening practices? In 2015 I pursued this question through in-depth interviews of forty-six Indonesian women; they ranged across many occupations, and their median age was twenty-nine years.[11] This article draws on my analysis of these interviews. Indonesia is a particularly interesting site to carry out such an exploration, with its highly variegated demography: the country claims over three hundred ethnic groups. The two cities where I conducted my interviews, Jakarta and Balikpapan, are the most transnational in their populations: My interviewees included women with Indian, Malay, Chinese, European, and Arab backgrounds.[12] Although the focus of this article is on women living in Indonesia, this article also attends to the ways in which women's experiences of living and traveling abroad have helped shape how they felt (and managed their feelings) about their skin color. Fifteen out of the forty-six women had lived or traveled abroad, and three of them had lived in North America.[13] All of the women interviewed came from lower, middle, and upper-middle classes—the

Saraswati, L. A. (2012). "Malu": Coloring shame and shaming the color of beauty in transnational Indonesia. *Feminist Studies, 38*(1), 113–140.

demographic groups that are most targeted in the marketing of whitening creams.

The interviews led me to a fascinating proposition: that it was the urge to avoid shame and embarrassment rather than an active desire to be attractive that shaped women's decisions to practice (or not practice) skin-whitening routines. I argue that women's practice of skin whitening is a manifestation of "gendered management of affect"—"affect" here drawing from Theresa Brennan's definition of the term as a "physiological shift accompanying a judgment."[14] This gendered management of affect is also important in the maintenance of gender, racial, and global hierarchies. I came to this proposition after noting that the women in my study typically articulated their responses through various "affective vocabularies."[15] One example of affective vocabulary that was often mentioned during the interviews is *malu.* Malu is an Indonesian term that registers not only as a mostly negative affect equivalent to "shame" but also as a positive affect.[16] For the anthropologist Clifford Geertz, malu, or its Balinese equivalent *lek,* can be understood as "a diffuse, usually mild, though in certain situations virtually paralyzing, nervousness before the prospect (and the fact) of social interaction, a chronic, mostly low grade worry that one will not be able to bring it off with the required finesse."[17] In this sense, malu signals one's "vulnerability to interaction."[18] Similarly, social anthropologist Johan Lindquist, drawing from the works of linguist Cliff Goddard and anthropologist Michael Peletz, points out that malu can be translated as "shame, embarrassment, shyness, or restraint and propriety"[19] and is culturally understood as a "moral affect" that is considered "necessary to constrain the individuated self from dangerous and asocial act of impulse, lust, and violence."[20] Lindquist further argues, "it is the experience of malu, or of being identified as someone who should be malu, which becomes an organizing principle for social action and the management of appearances."[21] Hence, malu is an important affective term that works beyond the level of the individual and, as Lindquist points out, has the capacity to structure social encounters and feelings and to link the individual to his or her larger transnational social structures. As such, malu can be regarded both as a negative and a productive affect, a "constraint" or a "stimulus,"[22] rather than having only negative connotations as is often the case in the Western world.[23]

Moreover, malu, as I will tease out throughout this article, is a gendered affect. Men, as Elizabeth F. Collins and Ernaldi Bahar argue, tend to "react aggressively" when managing their feelings of malu.[24] The religious-based violence toward non-heterosexuals that occurred in the late 1990s and early 2000s in Indonesia, for example, can be understood as the behavior of religious men exercising their feelings of malu, according to Tom Boellstorff.[25] Women, however, tend to be "withdrawn or avoidant" when experiencing feelings of malu.[26] This article demonstrates how women tend to manage their feelings of malu about their skin color by becoming withdrawn and avoiding uninvited attention, as well as by performing skin-whitening practices.

In writing about malu or "feelings," there were times when the rich literature of malu proved insufficient to address some of the difficulties I encountered. In such instances, I turned to feminist literature on affect and particularly on shame. For example, many of the women interviewed narrated stories that highlighted specific experiences of being discriminated against because of their skin color, but when I followed up by asking how they felt about being ignored for not having lighter skin or not being considered "beautiful," many simply said "fine" (*biasa saja*) in a tone that suggested it did not matter to them. Comments such as "maybe they were just lucky" or "but we were indeed physically different from them, so [we] just accepted our fate" were also common. Feminist geographer Liz Bondi, who has written that some feelings are "unexpressed and inexpressible,"[27] was helpful here. Feminist philosopher Teresa Brennan provides another, but equally helpful explanation for this phenomenon:

> "Feeling" refers to the sensations that register these stimuli and thence to the senses, but feelings includes something more than sensory information insofar as they suppose a unified interpretation of that information. . . . I define feelings as sensations that have found the right match in words.[28]

Applying Brennan's insight to make sense of the interviewees' comments leads me to suggest that although these women might have felt "sensations," they might not have found the "right match in words" to name them. Particularly because many women are taught to sugar-coat their feelings,[29] they might be at a loss for adequate words to articulate negative emotions.

Additionally, I identify a problem related to the mutability, fluidity, and flexibility of feelings. For example, the same affective vocabulary may mean different things in different times and cultures;[30] indeed, the feelings themselves may change.[31] This means that during the interviews the women might no longer harbor any particular feelings about those moments in the past; they had learned to "accept their fate" and felt "fine." Not being attentive to the fluidity and flexibility of feelings, scholars may fall into the trap of writing about feelings as "fetishized" objects of inquiry detached from their historical context.[32] This article therefore situates malu within a specific context. Hence, rather than simply drawing from theories of affect that are produced and circulated in the United States to analyze women's representation of skin whitening in Indonesia, this article employs malu, an Indonesian term, and places it in conversation with feminist theories of affect and cultural studies of emotions produced elsewhere to make an intervention in the fields of feminist theories of affect and Indonesian theories of malu.

Also, with the disparities between "affective vocabularies" and actual emotions, it became necessary to look for the underlying emotions implicit in the interviewees' statements by paying attention to their body language and tone of voice and by employing the malu perspective to decode the interviews. To understand what I mean by employing a malu perspective, an analogy with a "gender" perspective is helpful. Stories that women tell might not be in and of themselves "gendered," let alone "feminist," stories; yet by employing a gender perspective to decode their stories, it is possible to make gender visible in their stories. Similarly, the interviewees might not necessarily have told stories of malu as such; however, by employing malu as a theoretical-emotional lens to decode the stories, the ways in which malu has been deployed in these women's lives become visible.

In what follows, I provide evidence that the "gendered management of affect" plays a significant role in perpetuating gender, racial, and global hierarchies and in shaping women's decision to practice skin-whitening routines. . . .

Malu in a Transnational Context

What causes a person to feel malu to begin with? Writing from a Western perspective, psychologist Rom Harré argues that embarrassment ("the major affective instrument of conformity"[33]) happens when "[o]ne has become the focus of (an apparently excessive) attention from others whose opinion one values with respect to what one has said or done, or how one appears" and that that person "has become aware that others have taken the sayings, doings or appearances in question to be abnormal."[34] Focusing on the specific context of Southeast Asia, Michelle Rosaldo argues that shame involves "the sanction of tradition, the acknowledgment of authority, the fear of mockery . . . the anxiety associated with inadequate or morally unacceptable performance" and "embracing notions of timidity, embarrassment, awe, obedience, and respect."[35] Particularly relevant to the women interviewed in this study is the fear of mockery as a driving force for them to lighten their skin.

This is best explained by using an example from the interviews: Alya, a middle-class Javanese woman who was born in 1973 in Jakarta and was working in Balikpapan at the time of the interview, shared her childhood experiences of being called *dakko-chan* (from a Japanese term for dark-skinned African dolls). This clearly caused her to feel malu.

> My relatives would twist my name so it had the word "Negro" in it. At school, I was called *dakko-chan* [laughing]. So I was happy when there was someone else in the classroom who was darker than me. That means I wouldn't be the target of their jokes.

There are various layers at work here. First, when she laughed, she turned her head away from me as if wanting to hide. This turning away action is body language that suggests that a person is feeling malu. Even if the assumption that she feels malu can be applied only at the time of the interview, and not necessarily to the past, it is telling that she expressed how happy she was (exposing the implicit feeling of unhappiness at being subjected to the comments) whenever there were darker-skinned people in the classroom because they drew unwanted attention away from her. She recognized that it was her dark skin color that invited the attention. Moreover, these seemingly harmless jokes further expose the racial and color hierarchy in transnational Indonesia within which "Negro" and Japanese *dakko-chan* dolls were represented as "abnormal," and therefore a cause for malu. Interestingly, these racialized concepts that originated in other national locations were the ones producing these Indonesian women's feelings of malu about themselves and their skin color.

The history of racial formation in Indonesia is deeply transnational. Prior to European colonialism, as early as the late ninth century, Aryan Indian domination of high culture and the ruling class had already imprinted its light-skinned beauty standard upon Java, as is evidenced in the oldest surviving Indonesian literatures.[36] In the nineteenth and through the early twentieth century when Dutch colonialism peaked in colonial Indonesia, preference for light skin color was strengthened through specific gendered racial projects. A comment made in 1925 suggested that if men could afford it, they preferred to keep their wives at home because they "like[d] pretty hands and a pretty complexion."[37] At schools, students and teachers highly valued white and fair-skinned European students.[38]

When Japan took over as the new colonial power in Indonesia (1942–1945), they propagated a new Asian beauty ideal; although "Asian" was constructed as the preferred race, white remained the preferred color, while dark skin was configured as an abject form of femininity. In postcolonial Indonesia, particularly since the late 1960s, US popular culture has become one of the strongest influences against which an Indonesian white beauty ideal is articulated. This transnational history of racial formation provides a rich context for understanding how women feel and manage malu as it relates to constructions of gender, skin color, and race in transnational Indonesia.

Based on my interviews, dark-skinned people in today's Indonesia are perceived as "scary," "criminal,"[39] "smelly," "dirty," and "weird-looking." When asked about their perceptions of dark-skinned Indonesians and Africans in Indonesia, almost all of the women interviewed articulated how dark-skinned people are deemed undesirable. . . .

Although dark-skinned people of whatever nationality are often considered undesirable, the opposite is not necessarily the case with light skin color. Not all white skin color is considered desirable in the same way. In the Indonesian context, desiring white skin does not translate to desiring the skin color of those of European descent. This becomes clear in how the interviewed women answered the question, "what is your ideal skin color?" They would often say that they disliked the skin color of *bule* (a slang term used to refer to white foreigners), as well as Chinese white skin color, and that they preferred Indonesian or Japanese white skin color:

> Ami (age 34): "I like Japanese [women's] skin color . . . Japanese [skin] is not pale. Their look is more elegant than the Chinese. I just don't like Chinese white skin."

> Dian (age 29): "I don't like white like bule. . . . Japanese women are so beautiful and so white, they are yellowish. I like Sundanese [western Javanese] white; [it is] transparent."

. . . What is most striking is that in specifying a particular skin color, the women use race, nation, and ethnicity to signify the skin color they prefer. The *quality* of white skin is signified by nationality and race. The whiteness of a particular skin color often becomes undesirable because of the race or nation that signifies it. This is most evident in the case of the Chinese. The long history of discrimination against Chinese people in Indonesia seems to resurface when these women discuss the ideal skin color: Chinese skin color is not preferred. . . . I argue that because it is race and nation that give meaning to skin color, this construction disrupts the otherwise neat racial hierarchy of white at the top and black at the bottom. This allows us to understand that in the Indonesian context, lighter skin color is *not* always better because skin color, rather than being a signifier of race, is signified by both race and nation. Lighter could only be better insofar as the race and nation that signify it is the preferred race and nation. Although there is little discernible difference between Japanese and Chinese skin color, a person may prefer Japanese white skin above Chinese white skin, for example, because of perceptions about race and nation. Interestingly, the presence of people from other countries such as Japan, China, or the United States in Indonesia has helped these women articulate the specific skin color that they prefer. Their construction of an ideal skin color is inevitably linked to national identities; it is typically constructed against people from other nations.

Although it is obvious that they do not desire the skin color of those of European descent, these women are aware that such people are considered superior and receive preferential treatment in Indonesia. This perception of whites as superior may be traced back to the Dutch colonial period.

. . . The interviewees recognized white foreigners as receiving better treatment than Indonesians allowed for the possibility of reading the women's whitening practices as reflecting their desire to be treated well and not necessarily their desire for whiteness as such.

In the interviews, all of the women shared some story to indicate how white foreigners in Indonesia are considered "better" and receive better treatment. Here are just a couple of examples. Both comments came from women married to "non-Indonesian" white-skinned men:

> Andarini (age 33): Why do they always think that [my daughters] are beautiful because of their father? Don't they think the mother is also beautiful? [laughing]. They want to know why my children are so beautiful; and they never say that their mother is beautiful. They always think that there is something else. When they see my husband, they say, "No wonder."

> Widhi (age 41): They always ask me where the father came from. . . . Sometimes people even thought I was their nanny [laughing].

These expressions expose the underlying assumptions that Indonesian women were considered less valuable compared to their white-skinned husbands whose "good genes" were perceived to have made their children beautiful.

As Benedict Anderson has noted, one of the residues of colonial racism is that white Americans and Europeans receive preferential treatment in Indonesia.

Indonesia has a long history of transnational circulation of people coming from India, China, Arab countries, the United States, and Europe that continues to the present day. Based on 2004 data, foreign (and not necessarily white) workers in Indonesia number about 20,000 (6,000 top executives, 11,600 professionals, 1,200 supervisors, 500 technicians, and the rest were "other").[40] More than half of these workers live in Jakarta; only 400 workers live in East Kalimantan (mostly in Balikpapan). Foreign workers are given many privileges in order to encourage them to feel at home and spend their dollars in Indonesia. Examples from the interviews include being seated at different lunch tables (marked as "staff only") with more menu choices compared to Indonesians, getting their requests met faster compared to Indonesian employees, or being given more expensive airline tickets so that they could fly on time when a flight was canceled while Indonesians only got their money back. Here is an example from Ira (age 40):

> When I was staying in B. hotel [in Indonesia] and my relatives came, they were not allowed to swim. They [the staff] came right away and said that it was for members only. But when my bule friend came with her kid [and they were] swimming there, no one asked anything. They weren't members nor staying in the hotel. But they let them (swim). . . . But when my relatives came with their kids, they said "members only." I even asked, "Can I pay you, can we pay you?" "Oh no, it's members only" [they said]. So it seemed that we were dirty, while bule were not.

This example reflects the notion of Indonesia as a good host welcoming white foreigners even over and above its own citizens, to the point of refusing Indonesians and thus making them feel "dirty" and less worthy.

Interestingly, in cases where the interviewed women had moved to another country and had fewer encounters with other Indonesian women who, by way of embarrassing comments, would make them feel bad about their skin color, they felt less pressured to conform to the light-skinned norm. Ira, Andarini, and Amanda, all of whom currently reside in North America, admitted that they are less concerned about whitening practices when they are in the United States. Alya, Lily, Nia, and Lidya, who have lived abroad, in France, England, and the Netherlands, also claimed that they began to value their "tanned" skin once they left Indonesia. The examples in this section suggest that the transnational emulation of people plays a role in producing certain feelings about specific skin colors and helps maintain hierarchies of gender, race, and color in a transnational context.

Coloring Malu

This article argues that malu is deployed as an affective instrument of conformity, particularly in regard to compelling a person to conform to color, race, national, and gender norms. Interestingly, the majority of women in Indonesia do not have light skin, at least not like the light skin that is on view in whitening advertisements. A light-skin norm does not really exist if by "norm" one supposes that everyone's skin is light and one person's dark skin therefore stands out. Rather, the light-skin norm works by way of comparison: Women are compared to others or to themselves in the past. This is why the presence of others who have darker skin, such as in Alva's case, would change one's position within the skin-color hierarchy.

In the interviews, I often heard the women say that they began using skin-whitening creams after comments were addressed to them about their dark skin color. For example, Nina, who was born in 1977 in Kediri and who worked at a fitness center in Balikpapan at the time of the interview, admitted that in a previous job, her office mates often said to her:

> "Goodness (*Aduh*), you have the darkest skin." Although I would tell them "I don't care," inside I felt like saying how dare you say that. But then I always thought about how I could look not too dark. . . . So every time there was going to be a big meeting, I'd make sure that I whitened my skin, at least my hand, legs, and face, so that I would feel *ahem ahem* [making a sound of clearing her throat while smiling].

Here, the "big meeting" became a public space where an "audience" important in the production of shame[41] gathered. It is also a point of social contact that exposes her "vulnerability to interaction,"[42] or to feel malu. In this case, her use of skin-whitening cream could be read as her desire to not stand out, or to conform, because to stand out would invite (undesirable) attention from other office mates. The attention was deemed undesirable because it would embarrass rather than flatter her. Interestingly, her application of skin-whitening cream was intended to *avert* rather than as commonly narrated in whitening ads, to *invite* the gaze of others. The assumption is that to be noticed for having dark skin is to be exposed to the possibility of embarrassing comments. This desire not to attract the gaze of others echoes Frantz Fanon's famous depiction of the colonial subject's plea: "I strive for anonymity, for invisibility. Look, I will accept the lot, as long as no one notices me!"[43]

Nina chose to use lightening cream so that others would think well of her. Another interviewee, Ajeng (age 31), admitted that she used whitening cream so that other people would not comment that her skin was *jorok* (dirty). Women pay attention to these comments because they are "signs" that tell them whether or not they are accepted by their group. . . .

More importantly, because light skin color is not a "norm" in Indonesia but made into an imaginary ideal, comments and practices of whitening routines are crucial in perpetuating a light-skinned norm. . . .

The very act of dropping remarks about each other's skin color normalizes light skin as desirable; it is an example of the constant "surveillance"[44] that women carried out. Women get rewarded by receiving pleasing comments such as "you look beautiful: your skin looks lighter" and get punished by receiving embarrassing comments such as "your skin looks darker; you should . . ." (and one may fill in the blank with various tips that women offer each other).

Offering comments about others' skin color also suggests that malu in the Indonesian context is "collectively shared."[45] One's identity is linked to others' behaviors. Not surprisingly, mothers discipline their daughters, as women-friends discipline each other, as a form of caring about the other, or acting in "solid arity"[46] and preventing the other from feeling malu—by reminding each other, they could all avoid feeling malu.

Out of forty-six interviewees, only eight claimed that they had never tried any skin-whitening product. These eight women, however, admitted that they had seen other women use it and were pressed to try it but never did. However, when asked what they did everyday to care for their skin, almost all admitted that there were some things they did or avoided doing became they did not want their skin to get darker. These various acts, from putting on skin-whitening cream or staying out of the sun, to choosing particular colors and outfits that would make their skin appear lighter, further perpetuate light skin as the "imaginary" norm.

The interviews reveal that skin color mattered for these women because it is one of the sources of their self-confidence. The interviewees reacted with a loss of self-esteem to comments that exposed their "abnormality."[47] In this sense, there was a specifically gendered affect produced. Some of the more common expressions during the interviews were:

> Ina (age 32): "When I had to meet with many people and my skin was dark, I didn't feel confident."
>
> Vanti (age 34): "Generally women feel more confident when their skin is white."
>
> Wati (age 30): "Because I was born dark-brown (*sawo matang*), I didn't have any self confidence. So I whiten my skin to be more confident."

These kinds of statements came up over and over again in the interviews, suggesting the extent to which women's self-confidence is closely linked to their skin color. An interviewee, Ira, an Indonesian American 40-year-old stay-at-home mom, born in West Sumatra and currently living in the United States, admitted that she knew from a very young age that she was ugly and undesirable because of her dark skin color. She said:

> Sometimes people would tell me, "You don't have any sex appeal. Your skin is too dark." Even guys said things like, "Why are you so black?" So they didn't think that they would hurt you, you know. They just blurted it out in front of you, so I got used to it. Even my family, relatives, they would say, "Oh, how come you're so black, so ugly? You're a girl!" . . . I married a bule but not necessarily because they're not Indonesians [and therefore considered better]. But it's the other way around. Indonesian men didn't really think I was beautiful so I had no chance with Indonesian men.

It was these comments about how ugly she was because her skin was dark, so often made to her that she got "used to" hearing them, that made her accept her place in society. She admitted that, contrary to the common assumption that women married bule men because they were of higher status, she came to believe that she *had* to marry a bule man because Indonesian men perceived her as ugly. There are a couple of points here that are noteworthy. First, although she ended up being the one who moved to the United States, it was the presence of white foreign men in Indonesia that provided her with a solution to being perceived as undesirable by Indonesians: She learned that many of these white foreign men chose dark-skinned Indonesian women as their romantic partners. Second, and importantly, most Indonesian men do not actually always marry light-skinned Indonesian women—there are more women in Indonesia with medium tanned skin than light skin. Yet, it was her "feeling" that played a key role here: the comments about her dark skin made her feel malu and hence that she would never be able to attract an Indonesian man. Thus, her marriage to a European American is a result of her *feelings* about her skin color.

In teasing out the gendering of malu based on skin color that impacts a person's self-esteem, it is useful to take a step back and look at how women in general are positioned differently from men. Gendered socialization, as psychologists Tamara Ferguson and Heidi Eyre point out, provides an environment in which women are more likely than men to feel guilt and shame.[48] . . .

There is much evidence in the interviews that women were socialized differently from men. First, from their babyhood, girls, more so than boys, received comments on their skin color and how the "right" color would make them "pretty." Second, men, unlike women, were not asked (by other people in their lives or by popular culture) to put on different kinds of make-up[49] to cover up their otherwise "deficient" selves.[50] Whitening cream advertisements circulating in Indonesia mostly target women. The recent fad of whitening lotion for men is nonetheless marketed in gendered terms—L'Oreal, for example, launched a whitening cream for men called "White Active," highlighting the masculine aspect of men as active beings. Moreover, the interviewees consistently revealed that men are not subjected to the same light-skinned beauty standard as women. Men can have dark skin color because, as the interviewees suggested, dark skinned men are actually perceived as more masculine. Third, fathers, unlike mothers, were not asked to "shame" their daughters about their looks. An interviewee, Titi (age 47), bluntly noted, the most significant person who made her feel bad about herself was her mother. Mothers are responsible for teaching their children about malu and for the moral standard of the family. [51] Taking care of their looks may not be important for men because they are not threatened by other men's beauty—unlike women, who can feel threatened by the beauty of other women because their husbands might take an additional wife or leave their spouse for a more beautiful woman. Representations of second wives as more beautiful are popular in Indonesia and certainly promote such fear.

If light-skinned beauty is desired to keep one's husband, are lesbians then immune from this light-skinned beauty norm? Although I interviewed only two women who identified themselves as lesbians, too few to draw any conclusions, it is interesting that both these women indicated that light-skinned women are also considered more desirable than dark-skinned women in the lesbian communities that they belonged to.

Covering Malu

The desire to conform, to avoid embarrassing comments, or to feel more self-confident are all adequate reasons for using whitening creams; however, I wish to add yet another perspective on malu to further the understanding of these practices. That is that the feeling of malu underpins the very act of covering one's skin with whitening cream. Various scholars have pointed out that the English word "shame" comes from "the Goth word *Scham,* which refers to covering the face,"[52] or from the 'Indo-Germanic root *kam*/*kem* meaning "to cover.'"[53] In addition to "cover," other scholars mention other expressions for feelings of malu, such as to "hide from that other" or to "turn away from the other's gaze."[54] . . .

When interviewees heard embarrassing comments about their dark skin color, they took note of how the skin is an important site upon which their feelings (good or bad) hinge. Managing the skin becomes necessary for the management of affect. The logic here is that if a woman *feels* "good" about her skin color, she will feel more confident in uncovering her skin. Accordingly, when she feels "bad" or ashamed about her skin color, she "resolve[s] the experience of shame"[55] by "hiding" or "covering" her skin to resolve the feeling of malu. In a few cases, women admitted that they would rather stay home than go out when their skin was dark because, in Yasmin's (age 27) words, "I was embarrassed (malu)." Her staying home can be understood as "hiding," or in Collins and Baliar's term, "withdrawing" herself, which is a gendered manifestation of malu. Moreover, malu is understood as an emotion "that describes the failures to live up to the ideals of the nation,"[56] or, in this case, the beauty ideal. Yasmin's feeling of malu can therefore be read as her recognizing how she failed to live up to the light-skinned beauty ideal. She resolved feeling malu by avoiding contacts with others.

In many other cases, however, malu is resolved in covering oneself. . . .

The need to cover oneself often manifests itself in the (psychological) covering of the skin with "whitening" cream that is perceived to have the capacity to free a person from feeling malu. I am extending here a notion that cultural studies scholar Elspeth Probyn articulates: "most experiences of shame make you want to disappear, to hide away and to cover yourself."[57] Volanda's (age 38) succinct response when asked why she would use whitening cream indicated this: "so that when I look in the mirror I am not ashamed (malu) of myself." Here, the shame of the self is articulated by the very action of covering the skin with whitening cream. What I am hinting at is this: just as popular culture often represents women taking a shower after a rape scene to signify the desire to cleanse the body and psyche, women putting skin-lightening cream on their bodies can be read as a psychological covering of their malu about their skin color. As such, whitening practices can be read as manifesting women's managing of that experience of malu.

But why, when these women feel malu, is it the skin that is being covered?

. . . Skin is a site where the self is "exposed" to others. It is both the public and private nature—at times what literally "borders" the private self from others—of the skin that allows the skin to be one of the sites through which others can "read" us, and hence upon which our feelings can be (dis)articulated. I am invoking here both Harré's notion of the "body," taking it to mean specifically the skin as a "legible" surface from which the moral judgment can be read,[58] and Fanon's notion of "epidermalization of inferiority."[59] Because the skin is (constructed to be) telling of who the person is and is a site where one's "inferiority" or one's "shame" materializes, there is a need to "manage" it to reflect what the person wants it to tell others. I am intrigued here by how skin functions similarly to shame in that it "resides on the borderline between self and other."

Certainly, not all parts of the skin are seen by others. That is why, to recall an example mentioned earlier, Nina would whiten "at least" her face, legs, and hands prior to an important meeting. Another interviewee, Pingkan (age 25), explained that the face is most important for her because "usually people notice the face first, not other parts." . . .

Hence, it is the capacity of the skin to expose the self to others and the ways in which others may respond to that exposed skin that make the skin an important site to be "managed." But, taking it further while drawing from Harré and Fanon, I argue that the interviewees' emphasis on face and other visible parts of the skin indicate how the skin, because of its embodiment of private and public self, functions as a site upon which moral judgments are based. Because the skin is exposed for others to see, it grants others permission to judge people based on how "good" their skin (and therefore the person inhabiting that skin) is. . . .

As a marketing executive of a whitening-cream company, who agreed to talk to me on condition of anonymity, explained: "a woman proves to her family and society that she is responsible enough to take care of herself and therefore can be trusted to care for her family by using whitening cream." General expressions of negative feelings, such as "I don't feel good" (*nggak enak*), to illustrate how the interviewees felt when not using these whitening creams, were quite common. Although the line separating "*feeling* good" from a feeling of "*being* good" is not always sharply delineated, what is unmistakable is the equating of "being good," "feeling good," "looking good," and "having good" skin color.

Interestingly, putting on whitening cream turns out to put women in a never-ending cycle of malu: women usually end up feeling more malu for using whitening creams. This is evident from the interviews: when asked the first time around if they had ever tried whitening their skin, many women said they had not. However, as I asked differently phrased questions throughout the interviews, the same women began to tell me which whitening brand they used. Riana (age 24), a waitress in one of Balikpapan's cafes, did not want to be interviewed at first because she said she never used whitening cream and hence had nothing to say, After she changed her mind, she asked me, "I told you that I didn't use whitening cream when you asked me about it the other day. How did you know I used it?" I did not, in fact, know if she had used whitening cream; but I was certain that even if she hadn't, she would still have many things to share with me. Nonetheless, what so many of the interviewees suggested is that the whitening practice itself needs to be covered up, because it is considered in and of itself to be a shameful practice. Herlina, a recent high school graduate who worked at a mall in Balikpapan, commented on her friends:

> They use it but they won't admit it. . . . Maybe they were afraid that they would be mocked for turning white all of a sudden.

Her comment further exposed the feeling of malu that underpins the whitening practice. As Ferguson and Eyre point out, people tend to hide and manage their shame privately because these experiences are painful for them.[60]

However, as women cover their skin with whitening cream, possibly as a psychological "covering" of malu, it leaves larger institutional structures of racism/colorism and (hetero)sexism unarticulated. Instead of feeling the need to "fix" these structures, women displace their feelings onto their skin and therefore feel the need to fix their skin to manage their feelings.

Conclusion

This article uses the optic of malu to expose how people and objects that circulate transnationally to and from Indonesia have helped structure the feeling of malu among Indonesian women, as well as to show that the gendered management of affect is key to understanding women's decision to practice skin-whitening routines. This management of affect in turn maintains racial, color, gender, and global hierarchies.

Notes

1. In Asia alone the market for skin-whitening products is estimated at thirteen billion euros. See http://www.cosmeticsdesign-europe.com/Products-Markets/In-Cosmetics-Asia-focuses-on-skin-lightening-trend (accessed August 25, 2009).

2. Jason W. Clay, *Exploring the Links between International Business and Poverty Reduction: A Case Study of Unilever in Indonesia* (London: Oxfam, 2005), 95.

3. From 2007 data, http://www.asiamedia.ucla.edu/article.asp?parentid=70144 (accessed August 25, 2009).

4. Evelyn Nakano Glenn, "Yearning for Lightness: Transnational Circuits in the Marketing and Consumption of Skin Lighteners," *Gender and Society* 22, no. 3 (June 2008): 281–302.

5. Joanne Rondilla, "Filipinos and the Color Complex: Ideal Asian Beauty," in *Shades of Difference: Why Skin Color Matters,* ed. Evelyn Nakano Glenn (Stanford, CA: Stanford University Press, 2009), 63.

6. Evelyn Nakano Glenn, "Consuming Lightness: Segmented Markets and Global Capital in the Skin-Whitening Trade," in *Shades of Difference,* 171.

7. Not all whitening products are dangerous. Whitening products come in different forms: creams, pills, body and facial soaps, "papaya" soaps, moisturizers, facial cleansers, deodorants, sunblocks, lamb placenta, prescribed drugs, and injections. Most of the interviewees used whitening soaps, lotions, and moisturizers, which they applied twice daily—Pond's and Citra were the two most popular brands among the interviewees. These creams, according to the dermatologist whom I interviewed, are usually not effcient because they contain 2% or less of hydroquinone—to be effcient, creams must contain at least 5% hydroquinone. To buy products with a higher level of hydroquinone, women have to go to dermatologists, beauty salons, or underground markets.

8. This limit is set by the BPOM (Badan Pengawas Obat dan Makanan)—the Indonesian equivalent to the US Food and Drug Administration.

9. Nancy Etcoff, *Survival of the Prettiest: The Science of Beauty* (New York: Doubleday, 1999), 105–106.

10. Ibid., 106.

11. I used snowball and random methods. I chose Jakarta because it is the most developed, most populated, and most transnational city in Indonesia. I also included Balikpapan—a city on the island of Kalimantan—to avoid a "Java-centric" research. Balikpapan has the highest number of foreign workers in Kalimantan.

12. Although the women I interviewed lived in Jakarta and Balikpapan, they came from various cities on the island of Java (Bandung, Brebes, Cianjur, Gombong, Indra-mayu, Jakarta, Kediri, Pekalongan, Purwodadi, Semarang, Sidoarjo, Solo, Sukanagalih, Surabaya), Kalimantan (Balikpapan, Banjarmasin, Pulau Bulu, Tarakan), Sumatra (Medan, Padang, Palembang), Bali, and various other islands from the eastern part of Indonesia (Maluku, Sumba, Ternate); and one was born in Portugal. Their occupations were: engineers, domestic workers, homemakers, sales associates, herb beverage (*jamu*) sellers, owners of small convenience stores, researchers, live-in nannies, preschool teachers, bookstore attendants, gas-station attendants, waitresses, students, foreign language teachers, entrepreneurs, and event organizers. A few occupied managerial level positions at transnational corporations while others were unemployed high school graduates. One of the women was illiterate; several had not graduated from elementary school; many were high school graduates; some had completed four years of college education and even attained master's degrees. Most of the women were heterosexual and two women identified themselves as lesbian. Almost everyone was able-bodied, except for one woman who was partially blind. At the time of the interviews, the women were in their twenties (24 people), thirties (18 people), forties (1 person), fifties (2 people) and eighties (1 person).

13. I put these women's narratives in conversation with each other, rather than grouping them based on the city where they lived. I did so to demonstrate the similarities and linkages of these women's stories, to show that skin-whitening practices are found across various geographical locations among women in Indonesia, and to show how their feelings toward their skin color changed once they moved abroad. As such, I do not essentialize them—I noted and recognized their differences when appropriate.

14. Teresa Brennan, *The Transmission of Affect* (Ithaca, NY: Cornell University Press, 2004), 5. For other definitions of affect see Brian Massumi, *Parables for the Virtual: Movement, Affect, Sensation* (Durham, NC: Duke University Press, 2002); and Patricia Clough and Jean Hayley, eds., *The Affective Turn: Theorizing the Social* (Durham, NC: Duke University Press, 2007).

15. Thomas Hunter, "Indo as Other: Identity, Anxiety and Ambiguity in 'Salah Asoehan,'" *in Clearing a Space: Postcolonial Readings of Modern Indonesian Literature,* ed. Keith Foulcher and Tony Day (Leiden, Holland: KITLV, 2002), 125.

16. See Eve Sedgwick and Adam Frank, *Shame and Its Sisters: A Silvan Tomkins Reader* (Durham, NC: Duke University Press, 1995) on the transformative aspect of shame.

17. Clifford Geertz, *Interpretation of Cultures (New York: Basic Books,* 1973), 402.

18. Ward Keeler, "Shame and Stage Fright in Java," *Ethnos* 11, no. 3 (1983): 158.

19. Johan Lindquist, "Veils and Ecstasy: Negotiating Shame in the Indonesian Borderlands," *Ethnos* 69, no. 4 (2004): 487–508.

20. Michelle Rosaldo, "The Shame of Headhunters and the Autonomy of Self," *Ethnos* 2, no. 3 (1983): 136.

21. Lindquist, "Veils and Ecstasy," 488.

22. Rosaldo, "The Shame of Headhunters," 139.

23. Elizabeth Fuller Collins and Ernaldi Bahar, "To Know Shame: *Malu* and Its Uses in Malay Societies," *Crossroads: An Interdisciplinary Journal of Southeast Asian Studies* 14, no. 1 (2000): 39.

24. Collins and Bahar, "To Know Shame," 48.

25. Tom Boellstorff, *A Coincidence of Desires: Anthropology, Queer Studies, Indonesia* (Durham, NC: Duke University Press, 2007).

26. Collins and Bahar, "To Know Shame," 48.

27. Liz Bondi, "The Place of Emotions," in *Emotional Geographies,* ed. Liz Bondi, Joyce Davidson, and Mick Smith (Aldershot, UK: Ashgate, 2005), 237.

28. Brennan, *The Transmission of Affect,* 5.

29. Mario *Jacoby, Shame and the Origins of Self-Esteem: A Jungian Approach,* trans. Douglas Whitcher (New York: Routledge, 2001), 56.

30. Rom Harré and Grant Gillett, *The Discursive Minds* (Thousand Oaks, CA: Sage, 1994), 160.

31. Bondi, "The Place of Emotions," 237.

32. Sara Ahmed, *The Cultural Politics of Emotion* (Edinburgh: Edinburgh University Press, 2004), 32.

33. Ibid.

34. See L. Ayu Saraswati Prasetyaningsih, *Seeing Beauty, Sensing Race in Transnational Indonesia* (Honolulu: University of Hawai'i Press, Forthcoming).

35. See L. Ayu Saraswati Prasetyaningsih, *Seeing Beauty, Sensing Race in Transnational Indonesia* (Honolulu: University of Hawai'i Press, Forthcoming).

36. Elsbeth Locher-Scholten, *Women and the Colonial State: Essays on Gender and Modernity in the Netherlands Indies 1900–1942* (Amsterdam: Amsterdam University Press, 2000), 50.

37. Paul van der Veur, "Race and Color in Colonial Society: Biographical Sketches by a Eurasian Woman Concerning Pre-World War II Indonesia," *Indonesia* 8 (October 1969): 71.

38. Saraswati, *Seeing Beauty.*

39. In Indonesia, criminals are most often represented as lower class. See James Siegel, A *New Criminal Type in Jakarta: Counter-Revolution Today* (Durham, NC: Duke University Press, 1998).

40. Benedict Anderson, *Mythology and the Tolerance of the Javanese* (Ithaca, NY: Cornell University Southeast Asia Program Publications, 1988), 2.

41. Most of these workers came from Japan (3,500), South Korea (1,900), the United States, Australia, England, India, Canada, Malaysia, and China (about 1,500 each). (http://www.nakertrans.go.id/pusdatinnaker/tka/TKA_WNegara%202004.htm, http://www.nakertrans.go.id/pusdatinnaker/tka/TKA_Jab2004.htm, and http://www.nakertrans.go.id/pusdatinnaker/tka/TKA_Jab%202004.htm.) This number does not include families accompanying the workers, tourists, students, unemployed, and other undocumented workers and foreigners in Indonesia.

42. Gabriele Taylor, *Pride, Shame, and Guilt: Emotions of Self-Assessment* (Oxford: Clarendon, 1985), 53.

43. Keeler, "Shame and Stage Fright," 158.

44. Frantz Fanon, *Black Skin, White Masks,* trans. Charles Markmann (New York: Grove Press, 1967), 116.

45. Ibid., 84.

46. Collins and Bahar, "To Know Shame," 41.

47. To understand how malu is linked to "social harmony and group solidarity," see ibid., 42.

48. In the US context, Sandra Bartky, citing John Rawls, pointed out "shame is an emotion felt upon the loss of self-esteem." See Sandra Bartky, *Femininity and Domination* (New York: Routledge, 1990), 87.

49. Tamara Ferguson and Heidi Eyre, "Engendering Gender Differences in Shame and Guilt: Stereotypes, Socialization, and Situational Pressures," in *Gender and Emotion: Social Psychological Perspectives,* ed. Agneta H. Fischer (Cambridge: Cambridge University Press, 2000), 256.

50. In South Korea men have begun to use whitening cream.

51. Bartky, *Femininity and Domination,* 40.

52. Collins and Bahar, "To Know Shame," 49–50.

53. Probyn, *Blush,* 131.

54. Jacoby, *Shame and the Origins,* 1.

55. Sara Ahmed, "The Politics of Bad Feeling," *Australian Critical Race and Whiteness Studies Association Journal* 1 (2005): 75; and Janice Lindsay-Hartz, Joseph de Rivera, and Michael Mascolo, "Differentiating Guilt and Shame and Their Effects on Motivation," in *Self-Conscious Emotions: The Psychology of Shame, Guilt, Embarrassment, and Pride*, ed. June Tangney and Kurt Fisher (New York: Guilford Press, 1995), 295.

56. Lindsay-Hartz et al., "Differentiating Guilt," 298.

57. Johan Lindquist, *The Anxieties of Mobility: Migration and Tourism in the Indonesian Borderlands* (Honolulu: University of Hawai'i Press, 2009, 14.

58. Probyn, *Blush,* 39.

59. Rom Harré, *Physical Being: A Theory for a Corporeal Psychology* (Oxford: Blackwell, 1991), 142.

60. Fanon, *Black Skin,* 13.

Topics for Further Examination

- Look up research on patterns of cosmetic surgery procedures among women and men in the United States today. Discuss the gender and racial politics of procedures such as skin bleaching, brachioplasty, labiaplasty, and hair transplants.
- There is a complex relationship between hooking-up culture on college campuses and the sexual double standard. To explore this relationship, read online articles by sociologists (e.g., Rachel Allison and Barbara J. Risman; Lisa Wade) who have studied hooking up.
- Compare and contrast the messages about gendered emotions in ads, articles, columns, and other features in popular women's and men's magazines. How are stereotypes of "emotional women" and "cool-headed men" reinforced or challenged?

7

Gender at Work

Mary Nell Trautner with Joan Z. Spade

Throughout this book, we emphasize the social construction of gender, a dominant prism in people's lives. This chapter explores some of the ways the social and economic structures within capitalist societies create gendered opportunities and experiences at work, and how work and gender affect life choices, particularly as they relate to family and parenting. The gendered patterns of work that emerge in capitalist systems are complex, like those of a kaleidoscope. These patterns reflect the interaction of gender with other social prisms such as race, age, sexuality, and social class. Furthermore, gendered patterns at work are intertwined with patterns from other social institutions such as education and family. Readings in this chapter support points made throughout the book. First, women's presence, interests, orientations, and needs tend to be diminished or marginalized within occupational spheres. Second, one can use several of the concepts we have been studying to understand the relationships of men and women at work, including hegemonic masculinity, "doing gender," the commodification of gender, and the idea of separate spaces for men and women.

In this chapter, we explore the construction and maintenance of gender within both paid and unpaid work in the United States. We begin with a discussion of work and gender inequality. The history of gender discrimination in the paid labor market is a long one (Reskin & Padavic, 1999), with considerable social science research that documents gendered practices in workplace organizations. The first reading, by Joan Acker, discusses what she calls "inequality regimes," or the ways that work organizations create and maintain inequality across the intersections of gender, race, and social class. In this piece, she looks beneath the surface to almost invisible institutional practices that maintain unequal opportunities within organizations, such as recruitment or promotion practices (see also Acker, 1999). The second reading in this chapter, by Christine L. Williams, Chandra Muller, and Kristine Kilanski, applies Acker's paradigm to examine the characteristics of gendered organizations that women geoscientists face in the global oil and gas industry.

Consider the various ways the workforce in the United States is gendered. Think about different jobs (e.g., nurse, engineer, teacher, mechanic, domestic worker) and ask yourself if you consider them to be "male" or "female" jobs. Now take a look at Table 7.1, which lists job categories used by the Bureau of Labor Statistics (2018a). You will note that jobs tend to be gender typed; that is, men and women are segregated into particular jobs. The consequences for men and women workers of this continuing occupational gender segregation are significant in the maintenance of gendered identities. Included in Table 7.1 are jobs predominantly held by men (management, architecture and engineering, and construction) and those predominantly held by women (education, health care support, and office and administrative support).

Gender segregation of jobs is linked with pay inequity in the labor force. In 2017, all full-time women workers earned, on average, 81.8% of what all men earned, or median weekly earnings of $770 compared with $941 for men (Bureau of Labor Statistics, 2018b). As you look through Table 7.1, locate those jobs that are the highest paid and determine whether they employ more men or more women. Also, compare

Table 7.1 2017 Median Weekly Salary and Percentages of Men and Women in Selected Occupational Categories by Gender and Race/Ethnic Group[1,2]

		White Women	*White Men*	*Black Women*	*Black Men*	*Hispanic Women*	*Hispanic Men*
		Median Weekly Wages					
Occupational Category	*Total Number Employed in Category (16 years and older) (all numbers in this table are in thousands)*	*% in Category*	*% in Category*	*% in Category*	*% in Category*	*% in Category*	*% in Category*
All occupations	$860	$849	$1,065	$657	$710	$603	$690
	113,272	27.3%	34.7%	6.7%	6.1%	6.8%	10.5%
Management, professional, and related occupations							
Management occupations	$1,392	$1,230	$1,672	$971	$1,143	$970	$1,159
	13,169	30.5%	43.5%	4.5%	4.1%	4.5%	5.7%
Business and financial operations occupations	$1,174	$1,085	$1,440	$950	$1,075	$912	$1,106
	6,245	36.7%	33.2%	6.7%	3.6%	5.9%	4.7%
Computer and mathematical occupations	$1,465	$1,240	$1,538	$1,054	$1,252	$995	$1,300
	4,350	14.9%	45.4%	3.4%	5.2%	1.3%	5.6%
Architecture and engineering occupations	$1,478	$1,332	$1,516	$1,099[3]	$1,356	$1,120[3]	$1,450
	2,936	9.8%	60.9%	1.1%	4.4%	1.2%	7.4%
Life, physical, and social science occupations	$1,286	$1,208	$1,345	$899[3]	$1,405[3]	$1,134	$1,168
	1,238	29.1%	37.4%	3.5%	3.0%	4.4%	5.0%
Community and social services occupations	$900	$897	$1,045	$797	$895	$774	$904
	2,165	38.2%	22.3%	14.9%	6.8%	8.9%	3.2%
Legal occupations	$1,443	$1,231	$1,919	$1,004	$1,652[3]	$843	$1,609
	1,379	42.9%	33.4%	6.0%	1.8%	6.0%	3.7%
Education, training, and library occupations	$1,002	$967	$1,241	$830	$1,033	$871	$1,022
	6,978	52.5%	20.3%	8.6%	3.0%	7.8%	2.5%

Arts, design, entertainment, sports, and media occupations	$1,066	$960	$1,193	$840	$791	$928	$1,075
	1,778	31.4%	42.3%	3.5%	3.5%	4.0%	7.8%
Health care practitioner and technical occupations	$1,124	$1,093	$1,401	$889	$1,072	$957	$1,125
	6,970	51.4%	16.1%	10.4%	2.9%	5.7%	2.6%
Service occupations							
Health care support occupations	$542	$554	$625	$500	$569	$570	$595
	2,500	39.2%	6.0%	26.4%	2.2%	16.4%	3.4%
Protective service occupations	$852	$806	$986	$595	$731	$653	$782
	2,739	9.4%	51.4%	6.0%	14.4%	2.8%	10.1%
Food preparation and serving related occupations	$484	$484	$514	$435	$452	$424	$495
	4,465	22.8%	21.7%	7.1%	8.8%	13.3%	17.8%
Building and grounds cleaning and maintenance occupations	$522	$505	$615	$420	$503	$435	$535
	3,641	10.7%	29.2%	6.8%	9.4%	16.0%	23.3%
Personal care and service occupations	$520	$509	$652	$493	$512	$483	$592
	2,699	34.7%	12.9%	13.6%	5.2%	14.0%	4.9%
Sales and office occupations							
Sales and related occupations	$763	$676	$1,021	$524	$624	$506	$724
	9,953	27.1%	39.6%	5.8%	5.1%	7.5%	8.0%
Office and administrative support occupations	$701	$708	$774	$655	$621	$648	$645
	13,733	45.3%	16.5%	11.2%	4.6%	11.0%	5.1%
Natural resources, construction, and maintenance occupations							
Farming, fishing, and forestry occupations	$539	$540	$689	$645[3]	$415[3]	$444	$512
	962	6.7%	36.1%	1.0%	4.7%	13.7%	36.5%
Construction and extraction	$796	$895	$914	$745[3]	$703	$549[3]	$672
	6,147	1.5%	51.3%	0.3%	7.0%	0.7%	36.8%

(Continued)

Table 7.1 (Continued)

		White Women	*White Men*	*Black Women*	*Black Men*	*Hispanic Women*	*Hispanic Men*
		Median Weekly Wages					
Occupational Category	*Total Number Employed in Category (16 years and older) (all numbers in this table are in thousands)*	*% in Category*	*% in Category*	*% in Category*	*% in Category*	*% in Category*	*% in Category*
Installation, maintenance, and repair occupations	$878	$749	$926	$904[3]	$776	$585[3]	$759
	4,400	1.8%	65.8%	0.7%	8.8%	0.7%	18.3%
Production, transportation, and material moving occupations							
Production occupations	$701	$621	$834	$545	$668	$489	$672
	7,589	12.0%	43.4%	4.9%	9.0%	7.2%	16.0%
Transportation and material moving occupations	$681	$585	$798	$507	$624	$455	$621
	7,236	6.7%	44.5%	3.6%	15.6%	3.6%	19.9%

[1]Data for this table were taken from the Current Population Survey, Table A2: Usual Weekly Earnings of Employed Full-Time Wage and Salary Workers by Intermediate Occupation, Sex, Race, and Hispanic or Latino Ethnicity and Non-Hispanic Ethnicity, Annual Average 2017 (Bureau of Labor Statistics, 2018a).

[2] Data for White women and White men are White, non-Hispanic. However, there may be overlap between Blacks and Hispanics (that is, an individual might identify as Black and Hispanic). The table does not include data for Asian men or women, or individuals who identify as more than one race. Thus, percentages may not add to 100.

[3]These estimates do not meet the Bureau of Labor Statistics standard for statistical reliability (50,000 cases); therefore, they must be used cautiously.

women's to men's salaries across occupational categories. Clearly, a "gender wage gap" is evident in Table 7.1. Even in those job categories predominantly filled by women, men earn more than women. For example, going beyond the data in Table 7.1 and looking specifically at elementary and middle school teachers—a traditionally female job in which women outnumber men 3.62 to 1—the 2017 median weekly earnings for men, regardless of race or ethnicity, is $1,139 compared with $987 for women (a $152/week or $7,904/year average difference). In the occupation that Adia Harvey Wingfield studies in this chapter, registered nurses, women outnumber men 7.96 to 1, but earn $117 less per week ($1,143 compared to men's $1,260, an annual average difference of $6,084; Bureau of Labor Statistics, 2018b). The article by Wingfield discusses the "glass escalator" effect, where men in predominately female jobs earn more and get promoted more easily, even when they are not trying to be promoted or earn raises. But, as she discusses in her article, the glass escalator effect does not have a similar impact for African American men, as shown in Table 7.1. In addition, there is no glass escalator effect for women in traditionally male jobs. For example, looking at the specific occupation of lawyer, men still outnumbered women 1.32 to 1 in 2017 and also out-earned women $2,105 to $1,753 (a $352/week or $18,304/year difference on average; Bureau of Labor Statistics, 2018b); or civil engineers, where men outnumber women 5.44 to 1 and earn $1,524 per week to women's $1,343 per week median wages (a $181/week or $9,412/year difference on average; Bureau of Labor Statistics, 2018b); or physicians and surgeons, where men outnumber women 1.32 to 1 and earn $2,277 per week median wages to women's $1,759 per week (a $518/week or $26,936/year difference on average; Bureau of Labor Statistics, 2018b). These four specific job categories are contained within the larger selected occupational categories in Table 7.1 and provide further understanding of the differences you see there (the Bureau of Labor Statistics only calculates median weekly incomes within specific occupations when there are more than 50,000 persons in that category; therefore, we were limited in the detailed job categories available to us).

The pattern you see does not deny that *some* women are CEOs of corporations, and today we see women workers everywhere, including on construction crews (the article by Amy M. Denissen and Abigail C. Saguy in Chapter 9 describes what it is like for women working in the construction trades). However, although a few women crack what is often called "the glass ceiling," getting into the top executive or hypermasculine jobs is not easy for women and minority group members. The glass ceiling refers to the point at which women and others, including racial minorities, reach a position in their organizations beyond which they cannot continue on an upward trajectory (Purcell, MacArthur, & Samblanet, 2010; see also the articles by Acker, Wingfield, and Williams, Muller, and Kilanski in this chapter). Informal networks generally maintain the impermeability of glass ceilings, with executive women often isolated and left out of "old boys' networks," finding themselves "outsiders on the inside" (Davies-Netzley, 1998, p. 347). Similar internal mechanisms within union and trade-related organizations also keep women and minority group members out, because "knowing" someone often helps one get a job in the higher paid, blue-collar occupations.

Some of these patterns are particularly apparent when we consider the perspectives of "outsiders within." The reading by Kristen Schilt in this chapter studies the workplace experiences of female-to-male (FTM) transsexuals before and after their transition. These workers, having worked as women and now as men, make visible many of the hidden inequalities and structural disadvantages discussed by Acker and others in this chapter.

Gender, Race, and Social Class at Work

When we incorporate the prisms of race/ethnicity, and social class with gender, segregation in the workforce and pay inequality become more complex, as illustrated in the Wingfield article in this chapter. Another look at Table 7.1 indicates that individuals who identify as Hispanic or Latino and African American earn less than White, non-Hispanic men and women, although minority men earn more than White women in some occupational categories. In addition, Hispanic and African American women and men are much less likely to be found in the job categories with higher salaries than their percentages in the labor force would suggest. The continuing discrimination against African Americans, Hispanics, and other ethnic minority groups (as indicated in Table 7.1) shows patterns similar to the discrimination against women, both in the segregation of certain job categories and in the wage gap that exists within these same job categories. These processes operate to keep African American, Hispanic, and other marginalized groups "contained" within a limited number of occupational categories in the labor force.

The inequities of the workplace carry over into retirement (Calasanti & Slevin, 2001). Women and other marginalized groups are at a disadvantage when they retire, because their salaries are lower during their paid work years. Toni M. Calasanti and Kathleen F. Slevin find considerable inequalities in retirement income, which indicate that the inequalities in the labor force have a long-term effect for women and racial/ethnic minorities. They argue that only a small group of the workforce—privileged White men—are able to enjoy their "golden years," and the reasons for this situation are monetary. Likewise, many of the FTMs in Kristen Schilt's study who are not white find themselves unable to capitalize on the gender privilege in the workplace.

Efforts to change inequality in the workplace by combating wage and job discrimination through legislation have included both gender and race. In 1963, Congress passed the Equal Pay Act, prohibiting employment discrimination by sex but not by race. Men and women in the same job, with similar credentials and seniority, could no longer receive different salaries. Although this legislation was an important step, Kim M. Blankenship (1993) cites two weaknesses in it. First, by focusing solely on pay equity, this legislation did not address gender segregation or gender discrimination in the workplace. Thus, it was illegal to discriminate by paying a woman less than a man who held the same job, but gender segregation of the workforce and differential pay across jobs was legal. As Blankenship notes, this legislation saved "men's jobs from women" (p. 220) because employers could continue to segregate their labor force into jobs that were held by men and those held by women and then pay the jobs held by men at a higher rate. Second, this legislation did little to help minority women, as a considerable majority of employed women of color were in occupations such as domestic workers in private households or employees of hotels/motels or restaurants that were not covered by the act (Blankenship, 1993).

In 1964, Congress passed Title VII of the Civil Rights Act. Congress drafted this legislation to address racial discrimination in the labor force. This act prohibited discrimination in "hiring, firing, compensation, classification, promotion, and other conditions of employment on the basis of race, sex, color, religion, or national origin" (Blankenship, 1993, p. 204). Sex-based discrimination was not originally part of this legislation but was added at the last minute, an addition that some argue was to ensure the bill would not pass. However, the Civil Rights Act did pass Congress and women were protected along with the other groups. Unfortunately, the enforcement of gender discrimination legislation was much less enthusiastic than that for race discrimination (Blankenship, 1993).

Blankenship (1993) argues that the end result of these two pieces of legislation to overcome gender and race discrimination was to "protect white men's interests and power in the family" (p. 221), with little concern about practices that kept women and men of color out of higher paying jobs. Sadly, these attempts seem to have had little impact on race and gender discrimination (Sturm & Guinier, 1996). In this chapter, articles by Acker, and Williams, Muller, and Kilanski describe the more subtle ways wage discrimination can take place in higher paying occupations. Take another look at Table 7.1 and think about the ways the different allocations of jobs and wages affect women's and men's lives across race and social class—their ability to be partners in relationships and their ability to provide for themselves and their families.

As you think about the differences that remain in wage inequality, consider what still needs to be accomplished. Pay equity may seem like a simple task to accomplish. After all, now we have laws that should be enforced. However, the process by which most companies determine salaries is quite complex. They rank individual job categories based on the degree of skill needed to complete job-related tasks. Ronnie Steinberg (1990), a sociologist who has studied comparable worth of jobs for almost 40 years, portrays a three-part process for determining wages for individual jobs. First, jobs are evaluated based on certain job characteristics, such as "skill, effort, responsibility, and working conditions" (p. 457). Second, job complexity is determined by applying a "value to different levels of job complexity" (p. 457). Finally, the values determined in the second step help set wage rates for the job. However, care work and other types of work typically performed by women are undervalued in this wage-setting process (England, 2005; Steinberg, 1990).

On the surface, this system of determining salaries seems consistent and "compatible with meritocratic values," where each person receives pay based on the value of what he or she actually does on the job (Steinberg, 1987, p. 467). What is recognized as "skill," however, is a matter of debate and is typically decided by organizational leaders who are predominately White, upper-class men. The gender and racial bias in the system of determining skills is shocking. Steinberg (1990, p. 456) gives an example from the State of Washington in 1972 in which two job categories, legal secretary and heavy equipment operator, were evaluated as "equivalent in job complexity," but the heavy equipment operator was paid $400 more per month than the legal secretary. Although it appears that all wages are determined in the same way based on the types of tasks they do at work, Steinberg (1987, 1990, 1992) and others (including Acker in this chapter) argue that the processes used to set salaries are highly politicized and biased.

Gender Discrimination at Work

One way of interpreting why these gendered differences continue in the workforce is to examine workplaces as gendered institutions, as discussed in the introduction to this book. Acker, and Williams, Muller, and Kilanski—in the first two readings in this chapter—and other researchers examine work as a gendered institution (Acker, 1999, 2012). For example, Patricia Y. Martin (1996) studied managerial styles and evaluations of men and women in two different organizations: universities and a multinational corporation. She found that when promotions were at stake, male managers mobilized hegemonic masculinity to benefit themselves, thus excluding women. Understanding the processes and patterns by which hegemonic masculinity is considered "normal" within organizations is one avenue to understanding how organizations work to maintain sex segregation and pay inequity. These "inequality regimes" disadvantage all but a few, and, as Acker notes in this chapter, things are not likely to get better.

Gender discrimination at work is much more than an outcome of cultural or socialization differences in women's and men's behaviors in the workplace. Corporations have vested interests in exploiting gender labor. The exploitation of labor is a key element in the global as well as the U.S. economy, particularly as companies seek to reduce labor costs. Women in particular are likely targets for large, multinational corporations. In developing nations, companies exploit poor women's desires for freedom for themselves and responsibility to their families. For example, research by McKay (2006) illustrates these points as he describes the gendered assumptions, policies, and practices that multinational corporations bring to their factories in the Philippines and how they reflect the cultures of their home countries and illustrate the various ways "inequality regimes" are created in the workplace.

Looking at some of the top-wage jobs in the United States, Williams, Muller, and Kilanski in this chapter describe some of the subtle and not-so-subtle mechanisms of discrimination for geoscientists in the global gas and oil industry. The "inequality regimes" surrounding career trajectories and compensation patterns make meritocracy a myth and discourage women from trying because it is clear that they are on an uneven playing field. As you read the articles in this chapter, consider those mechanisms and others where gender segregates workplaces and keeps women from advancing into particular jobs.

The Effect of Work on Our Lives

The work we do shapes our identities, affecting our expectations for ourselves and others (Kohn, Slomczynski, & Schoenbach, 1986) and our emotions. It is not just paid work that affects our orientations toward self and others (Spade, 1991) but also work done in the home. In Western societies, work also defines leisure, with leisure related to modernization and the definition of work being "done at specific times, at workplaces, and under work-specific authority" (Roberts, 1999, p. 2). Although the separation of leisure from work is much more likely to be found in developed societies, work is not always detached from leisure, as evidenced by the professionals who carry home a briefcase at the end of the day or the beepers that summon individuals to call their workplaces.

Care Work

Care work is one gendered pattern that restricts women's leisure more so than men's (England, 2005). Women's leisure is often less an escape from work and more a transition to another form of work—domestic work. In an international study using time budgets collected from almost 47,000 people in 10 industrialized countries, Michael Bittman and Judy Wajcman (2000) found that men and women have a similar amount of free time; however, women's free time tends to be more fragmented by demands of housework and caregiving. Another study using time budgets found that women spent 30.9 hours on average performing various different family care tasks such as cooking, cleaning, repairs, yard work, and shopping, while men spent 15.9 hours per week performing such tasks (Robinson & Godbey, 1997, p. 101). Women also reported more stress in the Bittman and Wajcman (2000) study, which the authors attributed to the fact that "fragmented leisure, snatched between work and self-care activities, is less relaxing than unbroken leisure" (p. 185).

Domestic work, while almost invisible and generally devalued, cannot be left out of a discussion of work and leisure (Gerstel, 2000). Care work is devalued—particularly unpaid care work, which rests largely on women's shoulders. Gerstel and others refer to the contribution of women to care work as "the third shift." As a result, domestic labor and caregiving, being unpaid, are done by people least valued in the paid market. The undervaluation of care work carries over to the paid market as well. Look again at Table 7.1 and identify those job categories that encompass care work, such as health support workers and personal care and service workers. Now compare salaries and percentages of men and women in these caregiving jobs. As you start to consider these issues, ask why we undervalue care work—the unpaid care work in the home as well as care work in the workplace? Why are men encouraged not to participate in care work, and why are women the default caregivers? How is it that the work of the home is undervalued, and how is this pattern related to the workplace and the amount of leisure time available to men and women?

Work, Family, and Parenting

Unfortunately things have changed very little since the Bittman and Wajcman (2000) study, as Amy S. Wharton reports in her review of changes in the distribution of domestic labor in this chapter. Wharton calls it a "stalled revolution," as the changes that did occur hit a plateau in the late 1990s. Women's progress slowed in the workforce as well. Wharton sees the problems associated with this uneven workload as related to the institutions of work, family, and gender. With care work perceived as a "feminine" activity, it is not surprising that women's lives are more likely to be focused around, or expected to be focused around, care work activities, whereas the image of the ideal worker is an employee who is totally devoted to his or her job with no other responsibilities that might interfere with those work responsibilities. The reading by Erin Reid in this chapter delves into this "ideal worker image." While researchers have typically found that this expectation disadvantages women, particularly mothers, Reid finds that both men and women experience conflicts in their ability to conform to this ideology, as both men and women have family and care responsibilities and desires. However, men and women are able to cope with these conflicts differently. Men are more likely to hide their deviations from the ideal worker image, whereas women are more likely to make their conflicts known to their employer. Thus, work, family, and parenting become gendered institutions, reinforcing each other in maintaining a gender binary of separate spheres for women and men.

We can illustrate only a few patterns of work in this chapter. The rest you can explore on your own as you take the examples from the readings and apply them to your own life. When you read through the articles in this chapter, consider the consequences of maintaining gendered patterns at work for yourself and your future. While you are at it, consider why these patterns still exist and what these patterns of inequality look like in your life.

References

Acker, J. (1999). Gender and organizations. In J. S. Chafetz (Ed.), *The handbook of the sociology of gender* (pp. 171–194). New York: Kluwer Academic/Plenum.

Acker, J. (2012). Gender and organizations and intersectionality: Problems and possibilities. *Equality, Diversity and Inclusion: An International Journal, 31*(3), 214–224.

Bittman, M., & Wajcman, J. (2000). The rush hour: The character of leisure time and gender equity. *Social Forces, 79*(1), 165–189.

Blankenship, K. M. (1993). Bringing gender and race in: U.S. employment discrimination policy. *Gender & Society, 7*(2), 204–226.

Bureau of Labor Statistics. (2018a). Table A-2: 2017 Median weekly earnings of full-time wage and salary workers by detailed occupation, sex, race and Hispanic or Latino ethnicity and Non-Hispanic ethnicity. Annual Average 2017. Source: *Current Population Survey* [Unpublished data—sent by request].

Bureau of Labor Statistics. (2018b). Table 39: Usual weekly earnings of employed full-time wage and salary workers by intermediate occupation and sex. Annual Average 2017. Source: *Current Population Survey.*

Calasanti, T. M., & Slevin, K. F. (2001). *Gender, social inequalities, and aging.* Walnut Creek, CA: AltaMira Press.

Davies-Netzley, S. A. (1998). Women above the glass ceiling: Perceptions on corporate mobility and strategies for success. *Gender & Society, 12*(3), 339–355.

England, P. (2005). Emerging theories of care work. *Annual Review of Sociology, 31*(1), 381–399.

Gerstel, N. (2000). The third shift: Gender and care work outside the home. *Qualitative Sociology, 23*(4), 467–483.

Kohn, M. L., Slomczynski, K. M., & Schoenbach, C. (1986). Social stratification and the transmission of values in the family: A cross-national assessment. *Sociological Forum, 1,* 73–102.

Martin, P. Y. (1996). Gendering and evaluating dynamics: Men, masculinities, and managements. In D. Collinson & J. Hearn (Eds.), *Men as managers, managers as men* (pp. 186–209). Thousand Oaks, CA: Sage.

McKay, Steven C. (2006). Hard drives and glass ceilings: Gender stratification in high-tech production. *Gender & Society*, 20, 207-235.

Purcell, D., MacArthur, K. R., & Samblanet, S. (2010). Gender and the glass ceiling at work. *Sociology Compass, 4*(9), 705–717.

Reskin, B. F., & Padavic, I. (1999). Sex, race, and ethnic inequality in United States workplaces. In J. S. Chafetz (Ed.), *Handbook of the sociology of gender* (pp. 343–374). New York: Kluwer Academic/Plenum.

Roberts, K. (1999). *Leisure in contemporary society.* Oxon, UK: CABI.

Robinson, J. P., & Godbey, G. (1997). *Time for life: The surprising ways Americans use their time.* University Park: Pennsylvania State University Press.

Spade, J. Z. (1991). Occupational structure and men's and women's parental values. *Journal of Family Issues, 12*(3), 343–360.

Steinberg, R. J. (1987). Radical changes in a liberal world: The mixed success of comparable worth. *Gender & Society, 1*(4), 446–475.

Steinberg, R. J. (1990). Social construction of skill: Gender, power, and comparable worth. *Work and Occupations, 17*(4), 449–482.

Steinberg, R. J. (1992). Gendered instructions: Cultural lag and gender bias in the Hay System of job evaluation. *Work and Occupations, 19*(4), 387–423.

Sturm, S., & Guinier, L. (1996). Race-based remedies: Rethinking the process of classification and evaluation: The future of affirmative action. *California Law Review, 84*(4), 953–1036.

Introduction to Reading 32

Joan Acker draws from her vast research on gender, class, work, and organizations to describe the structure of organizations that maintain gender, class, and race disparities in wages as well as power in organizations. She also explores why inequalities in organizational structures and practices are not likely to change. She describes "inequality regimes," or practices and policies embedded in the organization itself, and shows how they work to create and maintain inequality across gender, race, and class. In this article, Acker provides detailed examples of how organizations maintain the gender inequalities in wages described in Table 7.1 and also why individuals seem powerless to overcome these gender inequalities.

1. Using your own life, think about whether you can identify any "inequality regimes" in the organizations you have worked in.
2. How does Acker's description of inequality regimes explain the data in Table 7.1?
3. What would have to change to reduce "inequality regimes" in the workplace? How might this threaten masculinity?

Inequality Regimes

Gender, Class, and Race in Organizations

Joan Acker

All organizations have inequality regimes, defined as loosely interrelated practices, processes, actions, and meanings that result in and maintain class, gender, and racial inequalities within particular organizations. The ubiquity of inequality is obvious: Managers, executives, leaders, and department heads have much more power and higher pay than secretaries, production workers, students, or even professors. Even organizations that have explicit egalitarian goals develop inequality regimes over time, as considerable research on egalitarian feminist organizations has shown (Ferree and Martin 1995; Scott 2000).

I define inequality in organizations as systematic disparities between participants in power and control over goals, resources, and outcomes; workplace decisions such as how to organize work; opportunities for promotion and interesting work; security in employment and benefits; pay and other monetary rewards; respect; and pleasures in work and work relations. Organizations vary in the degree to which these disparities are present and in how severe they are. Equality rarely exists in control over goals and resources, while pay and other monetary rewards are usually unequal. Other disparities may be less evident, or a

Acker, J. (2006). Inequality regimes: Gender, class, and race in organizations. *Gender & Society, 20*(4): 441–464. Reprinted by permission of SAGE Publications Inc., on behalf of Sociologists for Women in Society.

high degree of equality might exist in particular areas, such as employment security and benefits.

Inequality regimes are highly various in other ways; they also tend to be fluid and changing. These regimes are linked to inequality in the surrounding society, its politics, history, and culture. Particular practices and interpretations develop in different organizations and subunits. One example is from my study of Swedish banks in the late 1980s (Acker 1994). My Swedish colleague and I looked at gender and work processes in six local bank branches. We were investigating the degree to which the branches had adopted a reorganization plan and a more equitable distribution of work tasks and decision-making responsibilities that had been agreed to by both management and the union. We found differences on some dimensions of inequality. One office had almost all women employees and few status and power differences. Most tasks were rotated or shared, and the supervision by the male manager was seen by all working in the branch as supportive and benign. The other offices had clear gender segregation, with men handling the lucrative business accounts and women handling the everyday, private customers. In these offices, very little power and decision making were shared, although there were differences in the degrees to which the employees saw their workplaces as undemocratic. The one branch office that was most successful in redistributing tasks and decision making was the one with women employees and a preexisting participatory ethos.

* * *

What Varies? The Components of Inequality Regimes

Shape and Degree of Inequality

The steepness of hierarchy is one dimension of variation in the shape and degree of inequality. The steepest hierarchies are found in traditional bureaucracies in contrast to the idealized flat organizations with team structures, in which most, or at least some, responsibilities and decision-making authority are distributed among participants. Between these polar types are organizations with varying degrees of hierarchy and shared decision making. Hierarchies are usually gendered and racialized, especially at the top. Top hierarchical class positions are almost always occupied by white men in the United States and European countries. This is particularly true in large and influential organizations. The image of the successful organization and the image of the successful leader share many of the same characteristics, such as strength, aggressiveness, and competitiveness. Some research shows that flat team structures provide professional women more equality and opportunity than hierarchical bureaucracies, but only if the women function like men. One study of engineers in Norway (Kvande and Rasmussen 1994) found that women in a small, collegial engineering firm gained recognition and advancement more easily than in an engineering department in a big bureaucracy. However, the women in the small firm were expected to put in the same long hours as their male colleagues and to put their work first, before family responsibilities. Masculine-stereotyped patterns of on-the-job behavior in team-organized work may mean that women must make adaptations to expectations that interfere with family responsibilities and with which they are uncomfortable. In a study of high-level professional women in a computer development firm, Joanne Martin and Debra Meyerson (1998) found that the women saw the culture of their work group as highly masculine, aggressive, competitive, and self-promoting. The women had invented ways to cope with this work culture, but they felt that they were partly outsiders who did not belong.

Other research (Barker 1993) suggests that team-organized work may not reduce gender inequality. Racial inequality may also be maintained as teams are introduced in the workplace (Vallas 2003). While the organization of teams is often accompanied by drastic reductions of supervisors' roles, the power of higher managerial levels is usually not changed: Class inequalities are only slightly reduced (Morgen, Acker, and Weigt n.d.).

The degree and pattern of segregation by race and gender is another aspect of inequality that varies considerably between organizations. Gender and race segregation of jobs is complex because segregation is hierarchical across jobs at different class levels of an organization, across jobs at the same level, and within jobs (Charles and Grusky 2004). Occupations should be distinguished from jobs: An occupation is a type of work; a job is a particular cluster of tasks in a particular work organization. For example, emergency room nurse is an occupation; an emergency room nurse at San Francisco General Hospital is a job. More statistical data are available about occupations than about jobs, although "job" is the relevant unit for examining segregation in organizations. We know that within the broad level of professional and managerial occupations, there is less gender segregation than 30 years ago, as I have already noted. Desegregation has not progressed so far in other occupations. However, research indicates that "sex segregation at the job level is more extensive than sex segregation at the level of occupations" (Wharton 2005, 97). In addition, even when women and men "are members of the same occupation, they are likely to work in different jobs and firms" (Wharton 2005, 97). Racial segregation also persists, is also complex, and varies by gender.

Jobs and occupations may be internally segregated by both gender and race: What appears to be a reduction in segregation may only be its reconfiguration. Reconfiguration and differentiation have occurred as women have entered previously male-dominated occupations. For example, women doctors are likely to specialize in pediatrics, not surgery, which is still largely a male domain. I found a particularly striking example of the internal gender segregation of a job category in my research on Swedish banks (Acker 1991). Swedish banks all had a single job classification for beginning bank workers: They were called "aspiranter," or those aspiring to a career in banking. This job classification had one description; it was used in banking industry statistics to indicate that this was one job that was not gender segregated. However, in bank branches, young women aspiranters had different tasks than young men. Men's tasks were varied and brought them into contact with different aspects of the business. Men were groomed for managerial jobs. The women worked as tellers or answered telephone inquiries. They had contact only with their immediate supervisors and coworkers in the branch. They were not being groomed for promotion. This was one job with two realities based on gender.

The size of wage differences in organizations also varies. Wage differences often vary with the height of the hierarchy: It is the CEOs of the largest corporations whose salaries far outstrip those of everyone else. In the United States in 2003, the average CEO earned 185 times the earnings of the average worker; the average earnings of CEOs of big corporations were more than 300 times the earnings of the average worker (Mishel, Bernstein, and Boushey 2003). White men tend to earn more than any other gender/race category, although even for white men, the wages of the bottom 60 percent are stagnant. Within most service-sector organizations, both white women and women of color are at the bottom of the wage hierarchy.

The severity of power differences varies. Power differences are fundamental to class, of course, and are linked to hierarchy. Labor unions and professional associations can act to reduce power differences across class hierarchies. However, these organizations have historically been dominated by white men with the consequence that white women and people of color have not had increases in organizational power equal to those of white men. Gender and race are important in determining power differences within organizational class levels. For example, managers are not always equal. In some organizations, women managers work quietly to do the organizational housekeeping, to keep things running, while men managers rise to heroic heights to solve spectacular problems (Ely and Meyerson 2000). In other organizations, women and men manage in the same ways (Wajcman 1998). Women managers and professionals often face gendered contradictions when they attempt to use organizational power in actions similar to those of men. Women enacting power violate conventions of relative subordination to men, risking the label of "witches" or "bitches."

Organizing Processes That Produce Inequality

Organizations vary in the practices and processes that are used to achieve their goals; these practices and processes also produce class, gender, and racial inequalities. Considerable research exists exploring how class or gender inequalities are produced, both formally and informally, as work processes are carried out (Acker 1989, 1990; Burawoy 1979; Cockburn 1985; Willis 1977). Some research also examines the processes that result in continuing racial inequalities. These practices are often guided by textual materials supplied by consultants or developed by managers influenced by information and/or demands from outside the organization. To understand exactly how inequalities are reproduced, it is necessary to examine the details of these textually informed practices.

Organizing the general requirements of work. The general requirements of work in organizations vary among organizations and among organizational levels. In general, work is organized on the image of a white man who is totally dedicated to the work and who has no responsibilities for children or family demands other than earning a living. Eight hours of continuous work away from the living space, arrival on time, total attention to the work, and long hours if requested are all expectations that incorporate the image of the unencumbered worker. Flexibility to bend these expectations is more available to high-level managers, predominantly men, than to lower-level managers (Jacobs and Gerson 2004). Some professionals, such as college professors, seem to have considerable flexibility, although they also work long hours. Lower-level jobs have, on the whole, little flexibility. Some work is organized as part-time, which may help women to combine work and family obligations, but in the

United States, such work usually has no benefits such as health care and often has lower pay than full-time work (Mishel, Bernstein, and Boushey 2003). Because women have more obligations outside of work than do men, this gendered organization of work is important in maintaining gender inequality in organizations and, thus, the unequal distribution of women and men in organizational class hierarchies. Thus, gender, race, and class inequalities are simultaneously created in the fundamental construction of the working day and of work obligations.

Organizing class hierarchies. Techniques also vary for organizing class hierarchies inside work organizations. Bureaucratic, textual techniques for ordering positions and people are constructed to reproduce existing class, gender, and racial inequalities (Acker 1989). I have been unable to find much research on these techniques, but I do have my own observations of such techniques in one large job classification system from my study of comparable worth (Acker 1989). Job classification systems describe job tasks and responsibilities and rank jobs hierarchically. Jobs are then assigned to wage categories with jobs of similar rank in the same wage category. Our study found that the bulk of sex-typed women's jobs, which were in the clerical/secretarial area and included thousands of women workers, were described less clearly and with less specificity than the bulk of sex-typed men's jobs, which were spread over a wide range of areas and levels in the organization. The women's jobs were grouped into four large categories at the bottom of the ranking, assigned to the lowest wage ranges; the men's jobs were in many more categories extending over a much wider range of wage levels. Our new evaluation of the clerical/secretarial categories showed that many different jobs with different tasks and responsibilities, some highly skilled and responsible, had been lumped together. The result was, we argued, an unjustified gender wage gap: Although women's wages were in general lower than those of men, women's skilled jobs were paid much less than men's skilled jobs, reducing even further the average pay for women when compared with the average pay for men. Another component in the reproduction of hierarchy was revealed in discussions with representatives of Hay Associates, the large consulting firm that provided the job evaluation system we used in the comparable worth study. These representatives would not let the job evaluation committees alter the system to compare the responsibilities of managers' jobs with the responsibilities of the jobs of their secretarial assistants. Often, we observed, managers were credited with responsibility for tasks done by their assistants. The assistants did not get credit for these tasks in the job evaluation system, and this contributed to their relatively low wages. But if managers' and assistants' jobs could never be compared, no adjustments for inequities could ever be made. The hierarchy was inviolate in this system.

In the past 30 years, many organizations have removed some layers of middle management and relocated some decision making to lower organizational levels. These changes have been described as getting rid of the inefficiencies of old bureaucracies, reducing hierarchy and inequality, and empowering lower-level employees. This happened in two of the organizations I have studied—Swedish banks in the late 1980s (Acker 1991), discussed above, and the Oregon Department of Adult and Family Services, responsible for administration of Temporary Assistance to Needy Families and welfare reform (Morgen, Acker, and Weigt n.d.). In both cases, the decision-making responsibilities of frontline workers were greatly increased, and their jobs became more demanding and more interesting. In the welfare agency, ordinary workers had increased participation in decisions about their local operations. But the larger hierarchy did not change in either case. The frontline employees were still on the bottom; they had more responsibility, but not higher salaries. And they had no increased control over their job security. In both cases, the workers liked the changes in the content of their jobs, but the hierarchy was still inviolate.

In sum, class hierarchies in organizations, with their embedded gender and racial patterns, are constantly created and renewed through organizing practices. Gender and sometimes race, in the form of restricted opportunities and particular expectations for behavior, are reproduced as different degrees of organizational class hierarchy and are also reproduced in everyday interactions and bureaucratic decision making.

Recruitment and hiring. Recruitment and hiring is a process of finding the worker most suited for a particular position. From the perspectives of employers, the gender and race of existing jobholders at least partially define who is suitable, although prospective coworkers may also do such defining (Enarson 1984). Images of appropriate gendered and racialized bodies influence perceptions and hiring. White bodies are often preferred, as a great deal of research shows (Royster 2003). Female bodies are appropriate for some jobs; male bodies for other jobs.

A distinction should be made between the gendered organization of work and the gender and racial characteristics of the ideal worker. Although work is organized on the model of the unencumbered (white) man, and both women and men are expected to perform according to this model, men are not necessarily the ideal workers for all jobs. The ideal worker for many jobs is a woman, particularly a woman who, employers believe, is compliant, who will accept orders and low wages (Salzinger 2003). This is often a woman of color; immigrant women are sometimes even more desirable (Hossfeld 1994).

Hiring through social networks is one of the ways in which gender and racial inequalities are maintained in organizations. Affirmative action programs altered hiring practices in many organizations, requiring open advertising for positions and selection based on gender- and race-neutral criteria of competence, rather than selection based on an old boy (white) network.

These changes in hiring practices contributed to the increasing proportions of white women and people of color in a variety of occupations. However, criteria of competence do not automatically translate into gender- and race-neutral selection decisions. "Competence" involves judgment: The race and gender of both the applicant and the decision makers can affect that judgment, resulting in decisions that white males are the more competent, more suited to the job than are others. Thus, gender and race as a basis for hiring or a basis for exclusion have not been eliminated in many organizations, as continuing patterns of segregation attest.

Wage setting and supervisory practices. Wage setting and supervision are class practices. They determine the division of surplus between workers and management and control the work process and workers. Gender and race affect assumptions about skill, responsibility, and a fair wage for jobs and workers, helping to produce wage differences (Figart, Mutari, and Power 2002).

Wage setting is often a bureaucratic organizational process, integrated into the processes of creating hierarchy, as I described above. Many different wage-setting systems exist, many of them producing gender and race differences in pay. Differential gender-based evaluations may be embedded in even the most egalitarian-appearing systems. For example, in my study of Swedish banks in the 1980s, a pay gap between women and men was increasing within job categories in spite of gender equality in wage agreements between the union and employers (Acker 1991). Our research revealed that the gap was increasing because the wage agreement allowed a small proportion of negotiated increases to be allocated by local managers to reward particularly high-performing workers. These small increments went primarily to men; over time, the increases produced a growing gender gap. In interviews we learned that male employees were more visible to male managers than were female employees. I suspected that the male managers also felt that a fair wage for men was actually higher than a fair wage for women. I drew two implications from these findings: first, that individualized wage-setting produces inequality, and second, that to understand wage inequality it is necessary to delve into the details of wage-setting systems.

Supervisory practices also vary across organizations. Supervisory relations may be affected by the gender and race of both supervisor and subordinate, in some cases preserving or reproducing gender or race inequalities. For example, above I described how women and men in the same aspiranter job classification in Swedish banks were assigned to different duties by their supervisors. Supervisors probably shape their behaviors with subordinates in terms of race and gender in many other work situations, influencing in subtle ways the existing patterns of inequality. Much of this can be observed in the informal interactions of workplaces.

Informal interactions while "doing the work." A large literature exists on the reproduction of gender in interactions in organizations (Reskin 1998; Ridgeway 1997). The production of racial inequalities in workplace interactions has not been studied so frequently (Vallas 2003), while the reproduction of class relations in the daily life of organizations has been studied in the labor process tradition, as I noted above. The informal interactions and practices in which class, race, and gender inequalities are created in mutually reinforcing processes have not so often been documented, although class processes are usually implicit in studies of gendered or racialized inequalities.

As women and men go about their everyday work, they routinely use gender-, race-, and class-based assumptions about those with whom they interact, as I briefly noted above in regard to wage setting. Body differences provide clues to the appropriate assumptions, followed by appropriate behaviors. What is appropriate varies, of course, in relation to the situation, the organizational culture and history, and the standpoints of the people judging appropriateness. For example, managers may expect a certain class deference or respect for authority that varies with the race and gender of the subordinate; subordinates may assume that their positions require deference and respect but also find these demands demeaning or oppressive. Jennifer Pierce (1995), in a study of two law firms, showed how both gendered and racialized interactions shaped the organizations' class relations: Women paralegals were put in the role of supportive, mothering aides, while men paralegals were cast as junior partners in the firms' business. African American employees, primarily women in secretarial positions, were acutely aware of the ways in which they were routinely categorized and subordinated in interactions with both paralegals and attorneys. The interaction practices that re-create gender and racial inequalities are often subtle and unspoken, thus difficult to document. White men may devalue and exclude white women and people of color by not listening to them in meetings, by not inviting them to join a group going out for a drink after work, by not seeking their opinions on workplace problems. Other practices, such as sexual harassment, are open and obvious to the victim, but not so obvious to others. In some organizations, such as those in the travel and hospitality industry, assumptions about good job performance may be sexualized: Women employees may be expected to behave and dress as sexually attractive women, particularly with male customers (Adkins 1995).

The Visibility of Inequalities

Visibility of inequality, defined as the degree of awareness of inequalities, varies in different organizations. Lack of awareness may be intentional or unintentional. Managers may intentionally hide some forms of inequality, as in the Swedish banks I studied (Acker 1991). Bank workers said that they had been told not to discuss their wages with their coworkers. Most seem to have complied, partly because they had strong feelings that their pay was part of their identity, reflecting their essential worth. Some said they would rather talk about the details of their sex lives than talk about their pay.

Visibility varies with the position of the beholder: "One privilege of the privileged is not to see their privilege." Men tend not to see their gender privilege; whites tend not to see their race privilege; ruling class members tend not to see their class privilege (McIntosh 1995). People in dominant groups generally see inequality as existing somewhere else, not where they are. However, patterns of invisibility/visibility in organizations vary with the basis for the inequality. Gender and gender inequality tend to disappear in organizations or are seen as something that is beside the point of the organization. Researchers examining gender inequality have sometimes experienced this disappearance as they have discussed with managers and workers the ways that organizing practices are gendered (Ely and Meyerson 2000; Korvajärvi 2003). Other research suggests that practices that generate gender inequality are sometimes so fleeting or so minor that they are difficult to see.

Class also tends to be invisible. It is hidden by talk of management, leadership, or supervision among managers and those who write and teach about organizations from a management perspective. Workers in lower-level, nonmanagement positions may be very conscious of inequalities, although they might not identify these inequities as related to class. Race is usually evident, visible, but segregated, denied, and avoided. In two of my organization studies, we have asked questions about race issues in the workplace (Morgen, Acker, and Weigt n.d.). In both of these studies, white workers on the whole could see no problems with race or racism, while workers of color had very different views. The one exception was in an office with a very diverse workforce, located in an area with many minority residents and high poverty rates. Here, jobs were segregated by race, tensions were high, and both white and Black workers were well aware of racial incidents. Another basis of inequality, sexuality, is almost always invisible to the majority who are heterosexual. Heterosexuality is simply assumed, not questioned.

The Legitimacy of Inequalities

The legitimacy of inequalities also varies between organizations. Some organizations, such as cooperatives, professional organizations, or voluntary organizations with democratic goals, may find inequality illegitimate and try to minimize it. In other organizations, such as rigid bureaucracies, inequalities are highly legitimate. Legitimacy of inequality also varies with political and economic conditions. For example, in the United States in the 1960s and 1970s, the civil rights and the women's movements challenged the legitimacy of racial and gender inequalities, sometimes also challenging class inequality. These challenges spurred legislation and social programs to reduce inequality, stimulating a decline in the legitimacy of inequality in many aspects of U.S. life, including work organizations. Organizations became vulnerable to lawsuits for discrimination and took defensive measures that included changes in hiring procedures and education about the illegitimacy of inequality. Inequality remained legitimate in many ways, but that entrenched legitimacy was shaken, I believe, during this period.

Both differences and similarities exist among class, race, and gender processes and among the ways in which they are legitimized. Class is fundamentally about economic inequality. Both gender and race are also defined by inequalities of various kinds, but I believe that gender and racial differences could still conceivably exist without inequality. This is, of course, a debatable question. Class is highly legitimate in U.S. organizations, as class practices, such as paying wages and maintaining supervisory oversight, are basic to organizing work in capitalist economies. Class may be seen as legitimate because it is seen as inevitable at the present time. This has not always been the case for all people in the United States; there have been periods, such as during the depression of the 1930s and during the social movements of the 1960s, when large numbers of people questioned the legitimacy of class subordination.

Gender and race inequality are less legitimate than class. Antidiscrimination and civil rights laws limiting certain gender and race discriminatory practices have existed since the 1950s. Organizations claim to be following those laws in hiring, promotion, and pay. Many organizations have diversity initiatives to attract workforces that reflect their customer publics. No such laws or voluntary measures exist to question the basic legitimacy of class practices, although measures such as the Fair Labor Standards Act could be interpreted as mitigating the most severe damages from those practices. In spite of antidiscrimination and affirmative action laws, gender and race inequalities continue in work organizations. These inequalities are often legitimated through arguments that naturalize the inequality (Glenn 2002). For example, some employers still see women as more suited to child care and less suited to demanding careers than men. Beliefs in biological differences between genders and between racial/ethnic groups, in racial inferiority, and in the superiority of certain masculine traits all legitimate inequality. Belief in market competition and the natural superiority of those who succeed in the contest also naturalizes inequality.

Gender and race processes are more legitimate when embedded in legitimate class processes. For example, the low pay and low status of clerical work is historically and currently produced as both a class and a gender inequality. Most people take this for granted as just part of the way in which work is organized. Legitimacy, along with visibility, may vary with the situation of the observer: Some clerical workers do not see the status and pay of their jobs as fair, while their bosses would find such an assessment bizarre. The advantaged often think their advantage is richly deserved. They see visible inequalities as perfectly legitimate.

High visibility and low legitimacy of inequalities may enhance the possibilities for change. Social movements may contribute to both high visibility and low legitimacy while agitating for change toward greater equality, as I argued above. Labor unions may also be more successful when visibility is high and legitimacy of inequalities is low.

Control and Compliance

Organizational controls are, in the first instance, class controls, directed at maintaining the power of managers, ensuring that employees act to further the organization's goals, and getting workers to accept the system of inequality. Gendered and racialized assumptions and expectations are embedded in the form and content of controls and in the ways in which they are implemented. Controls are made possible by hierarchical organizational power, but they also draw on power derived from hierarchical gender and race relations. They are diverse and complex, and they impede changes in inequality regimes.

Mechanisms for exerting control and achieving compliance with inequality vary. Organization theorists have identified many types of control, including direct controls, unobtrusive or indirect controls, and internalized controls. Direct controls include bureaucratic rules and various punishments for breaking the rules. Rewards are also direct controls. Wages, because they are essential for survival in completely monetized economies, are a powerful form of control (Perrow 2002). Coercion and physical and verbal violence are also direct controls often used in organizations (Hearn and Parkin 2001). Unobtrusive and indirect controls include control through technologies, such as monitoring telephone calls or time spent online or restricting information flows. Selective recruitment of relatively powerless workers can be a form of control (Acker and Van Houten 1974). Recruitment of illegal immigrants who are vulnerable to discovery and deportation and recruitment of women of color who have few employment opportunities and thus will accept low wages are examples of this kind of control, which preserves inequality.

Internalized controls include belief in the legitimacy of bureaucratic structures and rules as well as belief in the legitimacy of male and white privilege. Organizing relations, such as those between a manager and subordinates, may be legitimate, taken for granted as the way things naturally and normally are. Similarly, a belief that there is no point in challenging the fundamental gender, race, and class nature of things is a form of control. These are internalized, often invisible controls. Pleasure in the work is another internalized control, as are fear and self-interest. Interests can be categorized as economic, status, and identity interests, all of which may be produced as organizing takes place. Identities, constituted through gendered and racialized images and experiences, are mutually reproduced along with differences in status and economic advantage. Those with the most powerful and affluent combination of interests are apt to be able to control others with the aim of preserving these interests. But their self-interest becomes a control on their own behavior.

* * *

Globalization, Restructuring, and Change in Inequality Regimes

Organizational restructuring of the past 30 years has contributed to increasing variation in inequality regimes. Restructuring, new technology, and the globalization of production contribute to rising competitive pressures in private-sector organizations and budget woes in public-sector organizations, making challenges to inequality regimes less likely to be undertaken than during the 1960s to the 1980s. The following are some of the ways in which variations in U.S. inequality regimes seem to have increased. These are speculations because, in my view, there is not yet sufficient evidence as to how general or how lasting these changes might be.

The shape and degree of inequality seem to have become more varied. Old, traditional bureaucracies with career ladders still exist. Relatively new organizations, such as Wal-Mart, also have such hierarchical structures. At the same time, in many organizations, certain inequalities are externalized in new segmented organizing forms as both production and services are carried out in other, low-wage countries, often in organizations that are in a formal, legal sense separate organizations. If these production units are seen as part of the core organizations, earnings inequalities are increasing rapidly in many different organizations. But wage inequalities are also increasing within core U.S.-based sectors of organizations.

White working- and middle-class men, as well as white women and all people of color, have been affected by restructuring, downsizing, and the export of jobs to low-wage countries. White men's advantage seems threatened by these changes, but at least one study shows that white men find new employment after layoffs and downsizing more rapidly than people in other gender/race categories and that they find better jobs (Spalter-Roth and Deitch 1999). And a substantial wage gap still exists between women and men. Moreover, white men still dominate local and global organizations. In other words, inequality regimes still seem to place white men in advantaged positions in spite of the erosion of advantages for middle- and lower-level men workers.

Inequalities of power within organizations, particularly in the United States, also seem to be increasing with the present dominance of global corporations and their free market ideology, the decline in the size and influence of labor unions, and the increase in job insecurity as downsizing and reorganization continue. The increase in contingent and temporary workers who have less participation in decisions and less security than regular workers also increases power inequality. Unions still exercise some power, but they exist in only a very small minority of private-sector organizations and a somewhat larger minority of public-sector unions.

Organizing processes that create and re-create inequalities may have become more subtle, but in some cases, they have become more difficult to challenge. For example, the unencumbered male worker as the model for the organization of daily work and the model of the excellent employee seems to have been strengthened. Professionals and managers, in particular, work long hours and often are evaluated on their "face time" at work and their willingness to put work and the organization before family and friends (Hochschild 1997; Jacobs and Gerson 2004). New technology makes it possible to do some jobs anywhere and to be in touch with colleagues and managers at all hours of day and night. Other workers lower in organizational hierarchies are expected to work as the employer demands, overtime or at odd hours. Such often excessive or unpredictable demands are easier to meet for those without daily family responsibilities. Other gendered aspects of organizing processes may be less obvious than before sex and racial discrimination emerged as legal issues. For example, employers can no longer legally exclude young women on the grounds that they may have babies and leave the job, nor can they openly exclude consideration of people of color. But informal exclusion and unspoken denigration are still widespread and still difficult to document and to confront.

The visibility of inequality to those in positions of power does not seem to have changed. However, the legitimacy of inequality in the eyes of those with money and power does seem to have changed: Inequality is more legitimate. In a culture that glorifies individual material success and applauds extreme competitive behavior in pursuit of success, inequality becomes a sign of success for those who win.

Controls that ensure compliance with inequality regimes have also become more effective and perhaps more various. With threats of downsizing and off-shoring, decreasing availability of well-paying jobs for clerical, service, and manual workers, and undermining of union strength and welfare state supports, protections against the loss of a living wage are eroded and employees become more vulnerable to the control of the wage system itself. That is, fear of loss of livelihood controls those who might challenge inequality.

* * *

Conclusion

Greater equality inside organizations is difficult to achieve during a period, such as the early years of the twenty-first century, in which employers are pushing for more inequality in pay, medical care, and retirement benefits and are using various tactics, such as downsizing and outsourcing, to reduce labor costs. Another major impediment to change within inequality regimes is the absence of broad social movements outside organizations agitating for such changes. In spite of all these difficulties, efforts at reducing inequality continue. Government regulatory agencies, the Equal Employment Opportunity Commission in particular, are still enforcing antidiscrimination laws that prohibit discrimination against specific individuals (see www.eeoc.gov/eeoc/statistics/). Resolutions of complaints through the courts may mandate some organizational policy changes, but these seem to be minimal. Campaigns to alter some inequality regimes are under way.

References

Acker, Joan. 1989. *Doing comparable worth: Gender, class and pay equity.* Philadelphia: Temple University Press.

———. 1990. Hierarchies, jobs, and bodies: A theory of gendered organizations. *Gender & Society* 4:139–58.

———. 1991. Thinking about wages: The gendered wage gap in Swedish banks. *Gender & Society* 5:390–407.

———. 1994. The gender regime of Swedish banks. *Scandinavian Journal of Management* 10:117–30.

Acker, Joan, and Donald Van Houten. 1974. Differential recruitment and control: The sex structuring of organizations. *Administrative Science Quarterly* 19:152–63.

Adkins, Lisa. 1995. *Gendered work.* Buckingham, UK: Open University Press.

Barker, James R. 1993. Tightening the iron cage: Concertive control in self-managing teams. *Administrative Science Quarterly* 38:408–37.

Burawoy, Michael. 1979. *Manufacturing consent.* Chicago: University of Chicago Press.

Charles, Maria, and David B. Grusky. 2004. *Occupational ghettos: The worldwide segregation of women and men.* Stanford, CA: Stanford University Press.

Cockburn, Cynthia. 1985. *Machinery of dominance.* London: Pluto.

Ely, Robin J., and Debra E. Meyerson. 2000. Advancing gender equity in organizations: The challenge and importance of maintaining a gender narrative. *Organization* 7:589–608.

Enarson, Elaine. 1984. *Woods-working women: Sexual integration in the U.S. Forest Service.* Tuscaloosa: University of Alabama Press.

Ferree, Myra Max, and Patricia Yancey Martin, eds. 1995. *Feminist organizations.* Philadelphia: Temple University Press.

Figart, D. M., E. Mutari, and M. Power. 2002. *Living wages, equal wages.* London: Routledge.

Glenn, Evelyn Nakano. 2002. *Unequal freedom: How race and gender shaped American citizenship and labor.* Cambridge, MA: Harvard University Press.

Hearn, Jeff, and Wendy Parkin. 2001. *Gender, sexuality and violence in organizations.* London: Sage.

Hochschild, Arlie Russell. 1997. *The time bind: When work becomes home & home becomes work.* New York: Metropolitan Books.

Hossfeld, Karen J. 1994. Hiring immigrant women: Silicon Valley's "simple formula." In *Women of color in U.S. society,* edited by M. B. Zinn and B. T. Dill. Philadelphia: Temple University Press.

Jacobs, Jerry A., and Kathleen Gerson, 2004. *The time divide: Work, family, and gender inequality.* Cambridge, MA: Harvard University Press.

Korvajärvi, Päivi. 2003. "Doing gender"—Theoretical and methodological considerations. In *Where have all the structures gone? Doing gender in organizations, examples from Finland, Norway and Sweden,* edited by E. Gunnarsson, S. Andersson, A. V. Rosell, A. Lehto, and M. Salminen-Karlsson. Stockholm, Sweden: Center for Women's Studies, Stockholm University.

Kvande, Elin, and Bente Rasmussen. 1994. Men in male-dominated organizations and their encounter with women intruders. *Scandinavian Journal of Management* 10:163–74.

Martin, Joanne, and Debra Meyerson. 1998. Women and power: Conformity, resistance, and disorganized coaction. In *Power and influence in organizations,* edited by R. Kramer and M. Neale. Thousand Oaks. CA: Sage.

McIntosh, Peggy. 1995. White privilege and male privilege: A personal account of coming to see correspondences through work in women's studies. In *Race, class, and gender: An anthology,* 2nd ed., edited by M. L. Andersen and P. H. Collins. Belmont, CA: Wadsworth.

Mishel, L., J. Bernstein, and H. Boushey. 2003. *The state of working America 2002/2003.* Ithaca, NY: Cornell University Press.

Morgen, S., J. Acker, and J. Weigt. n.d. *Neo-Liberalism on the ground: Practising welfare reform.*

Perrow, Charles. 2002. *Organizing America.* Princeton, NJ: Princeton University Press.

Pierce, Jennifer L. 1995. *Gender trials: Emotional lives in contemporary law firms.* Berkeley: University of California Press.

Reskin, Barbara. 1998. *The realities of affirmative action in employment.* Washington, DC: American Sociological Association.

Ridgeway, Cecilia. 1997. Interaction and the conservation of gender inequality. *American Sociological Review* 62:218–35.

Royster, Dierdre A. 2003. *Race and the invisible hand: How white networks exclude Black men from blue-collar jobs.* Berkeley: University of California Press.

Salzinger, Leslie. 2003. *Genders in production: Making workers in Mexico's global factories.* Berkeley: University of California Press.

Scott, Ellen. 2000. Everyone against racism: Agency and the production of meaning in the anti racism practices of two feminist organizations. *Theory and Society* 29:785–819.

Spalter-Roth, Roberta, and Cynthia Deitch. 1999. I don't feel right-sized; I feel out-of-work sized. *Work and Occupations* 26:446–82.

Vallas, Steven P. 2003. Why teamwork fails: Obstacles to workplace change in four manufacturing plants. *American Sociological Review* 68: 223–50.

Wajcman, Judy. 1998. *Managing like a man.* Cambridge, UK: Polity.

Wharton, Amy S. 2005. *The sociology of gender.* Oxford, UK: Blackwell.

Willis, Paul. 1977. *Learning to labor.* Farmborough, UK: Saxon House.

Introduction to Reading 33

In this article, Christine L. Williams, Chandra Muller, and Kristine Kilanski use Joan Acker's theory (from the previous reading) to examine the work experiences of women geoscientists in oil and gas companies. The 30 women they interviewed are highly educated (22 had master's degrees, and 8 had PhDs), ranged in age from 30 to 52 (average age 38), and worked in 14 different companies, including large global corporations such as Exxon Mobil, BP, and Shell. They used snowball sampling to locate the women they interviewed by asking women at professional meetings they attended to refer them to other women who held similar jobs. Through this process, they were also able to include three women who had left the industry. In-depth interviews with these women ranged from 1 to 2 hours. They also did observations at three professional meetings and interviewed three men supervisors. Their findings give us an inside look at the job experiences of women in STEM (science, technology, engineering, and mathematics; see also the article by Maria Charles in Chapter 4) and help us understand why women leave these fields.

1. Do men and women "do gender" in these professional fields, thus maintaining a system of inequality?
2. How does the "looser" style of "new management" practices in these powerful global corporations advantage or disadvantage women?
3. Using the findings from this study, explain why women engineers earn less than men.

GENDERED ORGANIZATIONS IN THE NEW ECONOMY

Christine L. Williams, Chandra Muller, and Kristine Kilanski

After making spectacular strides toward gender equality in the twentieth century, women's progress in the workplace shows definite signs of slowing (England 2010). Although women have entered occupations previously closed to them, many jobs remain as gender segregated today as they were in 1950. At both the top and the bottom of the employment pyramid, women continue to lag behind men in terms of pay and authority, despite closing gender gaps in educational attainment and workplace seniority. What accounts for these persistent gender disparities?

To explain gender inequality at work, many sociologists draw on Joan Acker's (1990) theory of gendered organizations. Acker argued that gender inequality is tenacious because it is built into the structure of work organizations. Even the very definition of a "job" contains an implicit preference for male workers (Acker 1990). Employers prefer to hire people with few distractions outside of work who can loyally devote themselves to the organization. This preference excludes many women, given the likelihood that they hold primary care responsibilities for family members. Consequently, for many employers the "ideal worker" is a man (see also Williams 2001).

Acker (1990) further identified five processes that reproduce gender in organizations: the division of labor, cultural symbols, workplace interactions, individual identities, and organizational logic. The latter process—organizational logic—was at the center of Acker's original critique of gendered organizations (Acker 1990) and is the focus of this article. The concept of organizational logic draws attention to how hierarchies are rationalized and legitimized in organizations. It encompasses the logical systems of work rules, job descriptions, pay scales, and job evaluations that govern bureaucratic organizations. Acker describes organizational logic as the taken-for-granted policies and principles that managers use to exercise legitimate control over the workplace. Workers comply because they view these policies and principles as "natural" or normal business practices.

While others had previously identified organizational logic as key to the reproduction of class inequality, Acker's breakthrough identified it as a source of gender inequality as well, even though it appears gender neutral on the surface. . . .

For example, organizations supposedly use logical principles to develop job descriptions and determine pay rates. But Acker argues that managers often draw on gender stereotypes when undertaking these tasks, privileging qualities associated with men and masculinity that then become reified in organizational hierarchies. Through organizational logic, therefore, gender discourses are embedded in organizations, and gender inequality at work results.

A great deal of research supports Acker's theoretical claims (for a review, see Britton and Logan 2008). But in the decades since the article was published, the social organization of work has changed considerably. Starting in the 1970s, organizations began to experience downsizing, restructuring, computerization, and globalization (DiMaggio 2001; Kalleberg 2000; Vallas 2011). Referred to as "work transformation," this general and vast process of change is affecting the structure of work in the United States and around the world. Whereas in the past, many workers looked forward to a lifetime of loyal service to a single employer, workers in the so-called new economy expect to change employers frequently in search of better opportunities and in response to lay-offs, mergers, and downsizing. Organizational logic is changing, too. Under the former system, workers carried out narrow and specific tasks identified by their job descriptions and were evaluated and compensated by managers who controlled the labor process. Today, as corporations shed layers of management, work is increasingly organized into teams composed of workers with diverse skills who work with considerable discretion on time-bounded projects and are judged on results and outcomes, often by peers. Furthermore, in the new economy, standardized career "ladders"—with clearly demarcated rungs that lead to higher-paying and more responsible positions—are being eliminated or replaced by career maps, or "I-deals," which are individualized programs of career development. Networking has become a principal means through which

Williams, C. L., Muller, C., & Kilanski, K. (2012). Gendered organizations in the new economy. *Gender & Society, 26*(4): 549–573. Reprinted by permission of SAGE Publications Inc., on behalf of Sociologists for Women in Society.

workers identify opportunities for advancement both inside and outside their firms (Babcock and Laschever 2003; DiMaggio 2001; Osnowitz 2010; Powell 2001; Rousseau 2005; Vallas 2011).[1]

In this study, we seek to extend Acker's (1990) analysis and critique of gendered organizations by investigating how gender is embedded in the organizational logic of the new economy. Acker's theory explains how gender is embedded in traditional organizations that value and reward worker loyalty and that are characterized by standardized job descriptions, career ladders, and manager-controlled evaluations—features that do not characterize jobs in the new economy. We investigate how organizational logic is gendered when work is precarious, teams instead of managers control the labor process, career maps replace career ladders, and future opportunities are identified primarily through networking.

Geoscientists in the Oil and Gas Industry

To investigate gendered organizations in the new economy, we draw upon our research on women geoscientists in the oil and gas industry. Women geoscientists have increased their numbers radically in recent decades, currently constituting about 45 percent of graduates with master's degrees in geology, the entry-level credential in the field (AGI 2011). Also, according to anecdotal data, women geoscientists are entering professional careers in industry in almost equal numbers as men. Despite these encouraging advances, there is a strong perception that women stall out in mid-career and eventually leave their jobs at the major companies (AAPG 2009). This pattern is not uncommon among women scientists in general (Preston 2004). The glass ceiling is firmly in place in the oil and gas industry, with very few women represented at the executive levels and on boards of directors (*Catalyst* 2011).

The oil and gas industry is an ideal setting to study gendered organizations in the new economy for several reasons. First, it is arguably the most powerful, global, essential, and lucrative industry in the world. In 2007, the largest oil and gas companies made roughly two trillion dollars (U.S.) in combined revenue and 150 billion dollars in profit (Pirog 2008). Despite its critical importance, few sociologists have examined the gender dynamics in this industry (see Miller 2004 for an exception). Second, the industry has a high demand for so-called knowledge workers (scientists and engineers), which is a defining feature of the new economy; one solution to the perceived shortage of these workers has been to increase the numbers of women in these fields (National Academy of Sciences 2010). Third, and most importantly for our analysis, the industry has been in the forefront of implementing the new organizational logic (McKee, Mauthner, and Maclean 2000). Throughout the 80s and 90s, the industry experienced numerous mergers, leading to reorganization and downsizing that exacerbated the vulnerability of its workforce. Consistent with the general process of work transformation, the major corporations have altered the career structure for their professional workforce by institutionalizing career maps and teamwork. The expectation of frequent career moves has enhanced the importance of networking for professional success. These innovations make the oil and gas industry a paradigmatic case for investigating gendered organizations in the twenty-first century.

* * *

Findings

Organizational changes associated with the new economy are reflected in the careers of geoscientists in the oil and gas industry. Gone is the expectation of a lifelong career spent in loyal service to a single employer. Oil and gas companies frequently expand and contract their workforce in response to economic cycles and mergers (Yergin 1993). . . . Job insecurity is described by [one] respondent as both a constant and a "very scary" feature of the oil and gas industry.

The constant threat of layoffs no doubt causes high levels of stress and performance pressures for geoscientists. But how is performance measured? In periods of downsizing and merging, how do individuals survive the periodic cuts and even succeed in the industry?

Given the work geoscientists are hired to do, it would seem that whoever finds the most oil and gas would receive the most rewards. Indeed, after a respondent drilled a successful well, headhunters tried to lure her away from her current company, offering incentives such as stock options. But corporations have good reason to be wary of using this particular metric of productivity, since it may incentivize geologists to overstate their claims, a risky and costly prospect for companies. To protect themselves from this lone wolf phenomenon and insure greater reliability, companies instituted the team structure. This geologist, who experienced both individual- and team-based work, explains the economic stakes:

> When I first started in the mid-80s, I was working an exploration play in northern Louisiana, and the engineer who was going to drill a well for me was based in

> Corpus Christi. I never met him. I would do my maps and put them in the mail because we didn't have electronic submission. We might have a few conference calls before we drilled a million dollar well. That was when it cost $50,000 a day to drill a well. Now a well in the Gulf of Mexico is a million dollars a day. And so, [changing to the team structure] was part of that. You had to be able to get people face-to-face. There was too much on the line from a risk standpoint, and from a financial standpoint.

In the experience of this geologist, teams produce more reliable results than do individuals working alone. With more people involved, she believes that companies get better advice on where to drill and also where not to drill, lessening their economic risks.

Teams are now a standard organizational form for scientists working in industry (Connelly and Middleton 1996). The Bureau of Labor Statistics (2009) identifies the ability to work on teams as an important feature of geoscientists' careers. The women we talked to worked on teams ranging in size from five to 20. Some teams were interdisciplinary, while others were composed of members with a single specialty, all of whom were investigating a particular "play" or geographical area for potential drilling. Individuals' team assignments typically last from three to five years, and many require relocation to a different city, oil field, and/or country. Each team is headed by a supervisor, typically a professional peer working alongside the rest of the team. Supervisors also move around to different teams every few years. The result is a work organization in perpetual flux, with teams forming and disbanding, and team members and supervisors constantly circulating around the country and, indeed, all over the globe.

Even though work is team based and essentially collaborative, careers are still individual. Raises, promotions, and opportunities are allocated to individuals, not to teams (although team members can receive additional bonuses if their collective results contribute to a company's profits). Out of this particular context, oil and gas companies replaced career ladders and standardized job descriptions with career maps—individualized programs for career development. A career map establishes goals and sets expectations that are then used to monitor a worker's productivity and evaluate his or her performance. The supervisor plays a central role in crafting workers' career maps and making sure that they have the tools to achieve their goals. As the primary channel to management, the supervisor identifies high performers on the team, recommends raises and bonuses, and determines the quality of future placements. Thus, individual workers must gain the support of their supervisors in order to further their careers in the industry.

A second major pathway to success in the oil and gas industry is through networking. In many of the large corporations, professionals are assigned mentors for their first three to five years, but by mid-career, we were told, they are basically left on their own to find support and encouragement as well as opportunities for career growth. Networking is viewed by respondents as the principal means to this end. Networks can be internal or external, formal or informal. Through these networks professionals gain exposure for lateral moves (after layoffs) and for leadership opportunities.

The new organizational logic appears gender neutral on the surface. Some have argued that because the new system of teams, career maps, and networking is less rigid than the older system of standardized career ladders and job descriptions, it may be more compatible with women's careers (e.g., Hewlett 2007). In fact, the transition to the new economy has taken place at the same time that major corporations have embraced gender and racial/ethnic diversity (Eisenstein 2009). The giant oil and gas companies tout their efforts to recruit women and minority men. Both Chevron and BP, for example, feature women scientists in recent publicity campaigns. Nevertheless, as we explain in the remainder of this article, these new forms may explain persistent patterns of gender inequality. . . .

Teamwork

In some recent studies, the team structure has been found to attenuate gender inequality in organizations (Kalev 2009; Plankey Videla 2006; Reskin 2002; Smith-Doerr 2004). However, we found that women may be disadvantaged on male-dominated teams. By the very nature of teamwork, the individual's contribution to the final product is obscured. Yet because careers are still individual, members of the team must engage in self-promotion to receive credit and rewards for their personal effort. Our study suggests that women encounter difficulties when promoting their accomplishments and gaining the credibility of their supervisors and other team members. This finding is consistent with experimental studies showing that, in general, women are given disproportionately less credit than men for the success they achieve when they work on teams in male-dominated environments (Heilman and Haynes 2005).

Because female workers are not given the benefit of the doubt in assessments of their work efforts by others, it is especially important that they are willing and able to tout their contributions to team accomplishments. Many of the women we interviewed are conscious of the importance of self-promotion, though they are not always secure in their ability to do it effectively. One geoscientist shared her misgivings about her own presentation skills, as well as her hunch that presentation skills may be more important than scientific ability to get ahead in industry:

> I don't know especially if you have to be as good, or if you have to be just as loud and belligerent as the other people. You definitely/the personality here is, to prove your point, you have to bang the table sometimes. I think women are more reluctant to do that. It's not me to do that.

This woman attributes her reluctance to "bang the table" to her personality, which she suggests is a reflection of an essential gender difference. But the following quote, from the only woman geoscientist in her entire division, indicates that women may be regarded negatively when they promote themselves:

> It's kind of interesting that I feel that I have to fight more to keep promoting what my expertise is. And it keeps getting kind of pushed back. The other people with less expertise in structural geology, they seem to get a little more recognition. Now, they've been working for the company for years. But still, I'm the one that has the expertise in that area. I just don't know how to do it. You don't want to be the one that yells and screams all the time. It's a delicate balance to keep promoting yourself.

Virtually everyone we interviewed talked about the fine line, or "delicate balance," between being assertive and being a "bitch." This perennial dilemma faced by women in the workplace is exacerbated in a team structure that requires workers to engage in assertive self-promotion in order to achieve recognition.

One woman reflected on her experience speaking at a partner meeting, at which she was the only woman, and youngest person, in attendance:

> I had to stand up and tell why I thought the well location should be somewhere and I could absolutely tell that no one was taking me seriously. They didn't care what I had to say—it was very obvious. Part of that I'm sure is being young, part of it was being the first time I had to stand up and tell them that. Because now, after eleven years, I can stand up and I can talk [laughs], but you have to get to that point. You have to know your stuff. I know that I have to cross every "t" and dot every "i," because if I don't, someone is going to pick it apart. There will be some man in the audience that wants to heckle you because he can—and I know that.

As this observation suggests, the difficulties that women encounter with self-promotion may be compounded by age. The following quote also indicates that younger women may face additional hurdles when attempting to bring attention to their accomplishments:

> I think automatically that anything I say is questioned. My supervisor, in my first go-round through the performance, told me I had to speak up—I have to believe what I'm saying, and I can't let them railroad me . . . which, I think he feels is more of an age thing. You get some credibility with age. I'm sure some people think you get more credibility being a guy. [I've got] kind of the short stick on both of those.

Her supervisor admonished her for not being assertive enough. But she perceived that, even when she did speak up, her views were constantly challenged because she was the only woman and the youngest member of the team.

At the professional meetings we attended, we observed that age is often treated as a status group in the industry. For example, when executives discussed "diversity" goals at their companies, they included age as well as gender and race/ethnicity. Layoffs that occurred in the 1980s and late 1990s were reported to have contributed to a large age gap among industry geoscientists (with a virtual absence of workers aged 35–45). Some of the geoscientists that we interviewed believed the age gap contributed to tension within teams. Young geoscientists do not always receive the recognition they seek from the older generation nearing retirement.

However, youth tends to operate differently based on gender and race. Youth can convey certain advantages to men, who may become the protégés of senior men (Roper 1994). In contrast, young women struggle to get noticed in positive ways. Some young women described feeling sexualized by men in their work teams. Others told us that they succeeded only because they fell into the "daughter" role with senior male mentors. Both roles are constraining in the quest for professional credibility. As Ollilainen and Calasanti (2007) have argued, family metaphors can disadvantage women who work on teams by encouraging a gendered division of labor and compelling women to engage in uncompensated emotional labor. Furthermore, in white male-dominated teams, metaphorical family roles may be available only to white women (Bell and Nkomo 2001).

Minority women may be disadvantaged compared to white men and women in additional ways, according to one Asian American woman we interviewed:

> It's all sorts of behaviors and soft skills that they look at for leadership potential. And a lot of the Asian people don't do well in those because we're culturally expected to be modest and we're culturally expected to not stand out. It's OK for us to be introverted or quiet. You actually get respected for being quiet, a man of few words. But at [my oil and gas company], that is not how you get success.

This statement suggests that self-promotion may have different meanings for racial/ethnic minority men and women. Furthermore, other research suggests that those who engage in it may be viewed negat ively by white colleagues and supervisors (Harvey Wingfield 2010).

Interestingly, we observed that women who worked in gender-balanced teams (absent in some companies) felt like they received greater recognition and respect for their contributions. If correct, this observation would confirm theories of tokenism that predict less bias in numerically balanced work groups (Kanter 1977). But how do teams achieve this numerical balance? Supervisors play a key role in determining the composition of the work group. However, as we suggest in the next section, supervisor's discretionary power is not necessarily exercised in the interest of gender equality.

In sum, in order to achieve recognition and rewards for their contributions, individuals working on teams must be willing and able to stand out from the group and advertise their accomplishments. Our findings suggest that this apparently gender neutral requirement can discriminate against women. As other researchers have found (Babcock and Laschever 2003; Bowles, Babcock, and Lai 2007; Broadbridge 2004), self-promotion can have negative meanings and consequences for women in male-dominated environments. When work is organized on the teamwork model, gender inequality is the likely result.

Career Maps

In many companies, career maps have replaced standardized career ladders for highly valued professionals. The purpose of a career map is to chart an individualized course of professional development that incorporates both the company's needs and the personal aspirations of the worker. Sometimes called "I-deals" (Rousseau 2005), these idiosyncratic arrangements often include employees' plans for reduced or flexible hours (e.g., to accommodate family needs) in addition to their career ambitions. Career maps are normally negotiated with supervisors, and they evolve over time.

Respondents were mostly positive about career maps because of the perception that they allow workers to manage their own careers. This was preferable to having, in the words of one geologist, "big brother" determine their futures with a one-size-fits-all set of career expectations (see also Hewlett 2007). However, in practice, the geoscientists we interviewed experienced several problems with career maps, stemming from the perceived ineptitude or gender bias of their supervisors. First, difficulties can arise if the criteria drawn are too vague or subjective. A woman with a PhD in geophysics explained that some workers, and especially new employees, struggled to figure out their job responsibilities. Supervisors sometimes assigned work without explaining the steps necessary or directing new employees to the resources needed to complete their assigned tasks. In fact, it wasn't until right before she left the industry that [one] particular woman felt she understood the "work flow." . . .

Without standardized job descriptions, workers can experience confusion about their job duties. Developing excellent communication skills becomes mandatory in this new context. One geologist attributed her success in the industry to the fact that she has "effectively communicated my career plan to the right people." She said, "Not everyone is so fortunate. . . . I do know of some people who haven't had as much influence on where they have gone. But when I've spoken with them, I really feel like they have not effectively communicated what they wanted to do." From her perspective, it is up to individual workers—not the corporation—to ensure that careers stay on the right track.

A second problem with career maps is that decisions about raises, promotions, and other rewards based on this system can appear arbitrary. This woman shared her confusion and frustration that her husband—who had started his job around the same time she did—had been promoted "a lot faster" than she had:

> And I've seen that, just on the side, watching. . . . I'm like, "OK, what are you doing differently that I need to do to get this going?" He said, "Nothing. I haven't done anything." He is a quiet guy by nature. So he didn't know why he was getting promoted himself. And I thought that was very interesting.

The lack of common job descriptions and career ladders contributes to uncertainty about why some individuals receive recognition and others do not. Because career maps are tailored to the individual—and because most companies prohibit employees from sharing salary information—it is difficult for workers to compare their career progress with others.[2]

Third, geoscientists perceive problems with career maps when supervisors do not actively advocate for them. A 35-year-old geologist working at a major described the importance of supervisors in obtaining good project assignments. . . .

This worker was grateful when a supervisor several levels above her recommended her for a job opening. Even though she didn't end up receiving that job, she felt "fortunate" to have been considered. She wondered aloud, "How do I get that to happen again?"

When opportunities are experienced as a windfall, workers are unsure how to advance themselves. At the same time, workers felt pressured to take any opportunities presented by a supervisor. Turning down more than one assignment was believed to foreclose them from receiving any in the future.

Without a supportive supervisor, careers can flounder. One geologist found herself in a precarious position when her supervisor left the company and another group subsumed her team. The manager of this group was an engineer rather than a geologist, which this respondent saw as a disadvantage. Not only did the person in charge of assigning and judging her work not understand it, he was already responsible for the careers of a large number of people. Without a supervisor advocating for her, this geologist said she felt "unnerved" and stressed out because she didn't know what her next assignment or career move would be.

While all of these issues with supervisors' discretion over career maps can impact both men and women equally, women may be especially disadvantaged if their supervisors harbor gender biases. As we know from previous research, supervisors who harbor biases against women (or in favor of men) can easily derail women's careers, even in the sciences (DiTomaso et al. 2007). Virtually every woman we interviewed encountered an individual supervisor at some point in her career who stymied her advancement. One geoscientist felt her career at a mid-size company was progressing well until she was assigned a new supervisor. The new supervisor would accept her work only if she had it preapproved by a male employee on her team. . . .

Gender bias is also expressed in supervisors' decisions about whom to hire into their teams. Studies suggest that managers favor people who are like themselves, a process known as "homosocial reproduction" (Elliott and Smith 2004; Kanter 1977). Gender differences emerge because women are rarely in a position to make personnel decisions. Even when women are in a supervisory position, their hiring decisions may be scrutinized. One female supervisor hired a woman to her team. When asked if it was controversial to pick a woman, she said that she "got that comment" but was able to defend herself because she had offered the job to a man first. She said, "I wasn't out looking for a female. It turns out we got a female in the group. In this particular case, she is the best fit." Thus, she was put on the defensive for a practice that is common among male supervisors. When gender bias appears to favor women, it is noticed and controversial (a topic we return to in the next section).

Part of developing a career map involves planning for maternity leave and flexible schedules, including part-time. Supervisors often have a great deal of control over these arrangements. One woman said the human resources (HR) department at the major company where she worked "purposefully wrote the rules [regarding flex time] kind of in a gray zone," leaving them open to the interpretation of supervisors. Smaller companies, which often lack formal HR departments, may give supervisors even more discretion than the larger companies do. However, a number of women working at majors gave examples of how supervisory discretion could impact workers' knowledge and ability to take advantage of flexible working options. . . .

This situation captures a paradox at the heart of career maps. On the one hand, they enable greater flexibility in career development, which some argue is in women's best interests (Hewlett 2007). As this geologist attests, "everybody" is unlikely to "want the same thing." On the other hand, if designing a career map that accommodates motherhood depends on having a sympathetic supervisor, potential gender bias is built into the organization. The lack of a "consistent, accepted solution" is frustrating and anxiety producing. . . .

Those we interviewed who had experience working in European offices experienced standardized maternity leave policies that were part of their host country's social welfare system. However, those who worked for European companies in the U.S. faced similarly limited options as those working in U.S. companies, with only supervisor-approved accommodations for maternity leave and part-time schedules available to them.

Because this study was motivated in part to understand women's attrition from the industry, we asked respondents their opinions about why women leave. Many speculated that it was because women tend to "opt out" of the labor force to bear and raise children, which they considered a deeply personal choice. Interestingly, few could cite specific examples. And the three women we talked to who left the industry did not regard children or family as their primary reason for leaving. Nevertheless, we contend that the institution of career maps, which grants supervisors the ability to negotiate family accommodations on a case-by-case basis, may leave mothers without viable and meaningful alternatives. Furthermore, in an industry characterized by constant mergers and downsizing, we suspect that some women may use the framework of "opting out" as a face-saving way to explain a decision to leave prior to an impending layoff. Unfortunately, this framing reinforces the stereotype that women naturally prioritize family over careers and absolves organizations of the responsibility for structuring the workplace in more equitable ways.

In sum, career maps give supervisors a great deal of discretion over individuals' career development. In the absence of accountability or an effective affirmative action program, supervisory discretion can be a breeding ground for gender bias (Reskin and McBrier 2000). Given the difficulty of comparing career progression in this context, patterns of gender and racial disparities may be obscured. Nevertheless, the logic of career maps encourages workers to blame themselves, not the organization, when their careers are stymied.

Networking

Virtually everyone we talked to said that networks are fundamental to achieving professional success. In an industry where layoffs are common and anticipated, workers must rely on their formal and informal networks to survive periodic cuts and to identify new opportunities. Yet, as we know from numerous research studies, networks are highly gendered and racialized (Burt 1998; Loscocco et al. 2009; McGuire 2002; Smith 2007). A geophysicist who worked for several large companies and who now owns her own consulting business explained that many people, and women especially, "work hard as opposed to work smart." Networking, rather than simply doing one's job well, was, she believed, the key to success in the industry. She reflected on the importance of this knowledge to boosting one's career: "If I had known then what I know now, I would be CEO of a company."

In the male-dominated oil and gas industry, not surprisingly, the most powerful networks are almost exclusively male. Often these are organized around golf or hunting (Morgan and Martin 2006). The women we interviewed provided classic accounts of exclusion from these groups.

> The men at upper management were quite comfortable making seat-of-the-pants decisions with each other, and they trusted each other. They had lunch together, they played golf together, they trusted each other. If somebody is going to make a seat-of-the-pants decision, the other guy's going to say "fine." A woman comes in and tries to make a seat-of-the-pants decision, same process, same gut kind of thing, you're not going to be trusted, you're not going to be believed.

Some women perceive that men's networks, sustained through company-sponsored sports and hunting/fishing trips, are not considered networks at all, even though in these spaces men are likely to develop strong relationships of mutual trust (see also DiTomaso et al. 2007). In one egregious case, a woman described how female strippers were positioned at each putting green at an annual company-sponsored golf tournament. While some women have no interest in attending these networking events, others try to fit in because of their critical importance to success in the industry. One independent producer told us that although she doesn't play golf, she makes it a point to "ride in the cart." Another woman tried to join her male colleagues' fantasy football league. Although they were resistant to letting a woman join, she was finally allowed when one man agreed to be her partner (to the others' chagrin).

In response to this exclusion, and in acknowledgment of the importance of networking for career development, some corporations have formed official women's networks. However, these networks have dubious status in corporations and joining may not be in women's best interest. For instance, DiTomaso and colleagues argue that "special mentoring programs for women set up by companies may be a disadvantage for those who use them" (DiTomaso et al. 2007, 198). The women we interviewed concurred, viewing women's corporate-sponsored networks as neither powerful nor especially useful. . . .

One problem [mentioned] was that the company brought together all women from the company, rather than just geoscientists. While she saw value in allowing women to network from across the company, she thought the other women came from "a little bit of a different perspective." Moreover, this type of networking is unlikely to result in future opportunities for a geologist.

At some companies, the women's network is not limited to women, the rationale being that in the interests of "equal opportunity," women should not receive "special treatment." Consequently, when women's groups are formed, they rarely address issues concerning discrimination or inequality. Topics like work-family balance are sometimes addressed, but in a way that does not challenge the structure or policies of the organization. For example, a few years after joining the major at which she works, one respondent and her colleagues started an online "family support network" in order to provide employees with children a chance to connect and give them a place to ask questions and receive advice. This "grass-roots network" received immense support from top managers and has since become institutionalized. . . .

Importantly, this network requires no resources from the employer, nor does it challenge the company's limited support for new parents. Yet the existence of the network makes the company appear to be doing something to promote gender equity.

Furthermore, while some women appreciate this focus on work-family balance, others find it alienating because they do not have children, and feel oppressed by the assumption that they do. For example, one woman spoke of receiving an invitation to a "women in science" session at a local seismic conference. She explained that she was originally excited to hear the experiences of "wicked smart" women scientists talking about how to thrive in a male-dominated environment. Instead she was disappointed that the group focus would be on motherhood. She added, "I don't tend to seek out female-dominated groups because you inadvertently end up sitting next to someone talking about their kids—which is fine. I can hear about your kids for a while. But I don't want to have kids."

On the other hand, some convey more than a hint of cynicism about corporate-sponsored events that highlight the accomplishments of senior women. One woman expressed frustration that corporate diversity events seemed to feature the same senior women retelling their success stories. She explained, "Marilyn is [the company's] poster child. But for every Marilyn there are fifteen women who are not getting what Marilyn gets"—referring to the same opportunities, exposure, and access to powerful networks.

Given the perceived limitations of official women's networks, some women turn to informal networks instead. Unfortunately, these also occupy a highly dubious space in the corporate world. They may be perceived as mere outlets for complaining, venting, or "bitching." A woman who organized a weekend retreat for a group of senior executive women was criticized by detractors for arranging a "ladies' boondoggle," an accusation she felt was "outrageous" because men do equivalent outings all the time.

Not surprisingly, some women are reluctant to disclose their interest in forming or joining a women's group. One woman talked about returning from an AAPG [American Association of Petroleum Geologists] event with the idea of starting a women's mentoring group to mimic those in the larger companies. She and a small group of women had started to organize, but had decided it was in their best interest to keep their intentions secret. This woman expressed palpable fear that if found out, the women involved would suffer negative repercussions since company policy strictly forbids any discussion of salary or contracts among employees. These women knew they were taking a chance by organizing a women's group, so they were planning to hold their meeting 200 miles away in order to avoid detection.

Networking has always been important for professional development. In the new economy, strong networks are needed not only to thrive but to survive periodic downsizing and layoffs. The heightened importance of networking places women geoscientists in a paradoxical position: They are often excluded from powerful men's networks, yet women's formal networks, when they exist, are not powerful and may actually have negative consequences for women's career development. Women's informal networks may be forced to operate under the radar. Because of the centrality of networking, the resulting gender inequality is thus embedded in the organizational logic of the new economy.

Conclusion

The traditional career model, in which a worker spends his or her entire career with one employer, in some, cases climbing a defined career ladder, is on the decline (Vallas 2011). Workers today expect to switch jobs and employers frequently throughout their careers. While some moves are in response to better opportunities, in many cases they are the result of corporate practices, common to some industries, that make workers vulnerable to job loss.

The new career model, created by corporations to reduce their economic risk and responsibility for workers, has several defining features. Under this new model, employees are evaluated based on individualized standards developed in conjunction with their direct supervisors, rather than by a standardized assessment tool. Although workers are evaluated on an individual basis, work is typically performed by self-managed teams. As it is difficult to determine individuals' level of effort, supervisors have a great deal of discretionary power in rewarding employees for a job well done (i.e., giving employees good team placements). The proliferation of career maps may obscure inequality in the pace of career progress.

Given the level of job insecurity, the ability to maintain large networks to identify job opportunities inside and outside of the organization becomes critically important for successful careers.

We examined the careers of geoscientists in the oil and gas industry—an industry at the forefront of implementing these organizational changes—to explore the gendered consequences of these job features. Our research suggests that teams, career maps, and networking reflect gendered organizational logics. To excel at teamwork, individuals must be able to engage in self-promotion, which can be difficult for women in male-dominated environments—even though they are the ones who may need to do it the most. In contexts where supervisors have discretion over careers, gender bias can play a significant role in the allocation of rewards. And networking is gendered in ways that disadvantage women.

These features of work organization are not new, and, in fact, previous research has shown that all three of these elements can be problematic for women (Bowles, Babcock, and Lai 2007; Broadbridge 2004; Burt 1998; Loscocco et al. 2009; McGuire 2002; Ollilainen and Calasanti 2007). This article's contribution has been to connect them to work transformation. Previously, gender inequality has been institutionalized (in part) through the mechanisms of career ladders, job descriptions, and formal evaluations (Acker 1990). In the new economy, these elements of organizational logic have been replaced by teams, career maps, and networking. These have become principal mechanisms through which gender inequality is reproduced in the new economy. . . .

Our findings suggest that addressing workplace gender inequality in the twenty-first century will require focused attention on transforming these job features, or altering their consequences for women. For example standard options for organizing career maps should be made available to workers. In the interest of gender equity, workers should be informed of the I-deals and salaries of their peers. In addition, supervisors should be made accountable to diversity goals, and incentivized to encourage workers to use company flexibility options. While companies should encourage networking activities, all corporate-sponsored events must include women and minority men, and informal male-only social events must somehow be made culturally taboo. These are the sorts of changes that we believe will enhance the careers of women scientists in the new economy.

When Joan Acker (1990) first articulated the organizational logic underlying gendered organizations, she was operating under the assumptions of the traditional career model. Those assumptions no longer apply in many organizations. Organizations are still gendered, but the mechanisms for reproducing gender disparities are different than those in the traditional career path. By exploring women's experiences of work in the new economy, we add an essential but previously missing dimension to the critique of work transformation. By paying close attention to the new organizational logic, we hope that effective policies can be devised to enhance gender equality in the twenty-first century workplace.

Notes

1. These descriptions of "old" and "new" forms of work organizations refer to trends that in actual practice can overlap considerably, so they should be treated as "ideal types" in the Weberian sense.

2. The proliferation of career maps may also make it difficult for human resource departments to detect patterns (and potential disparities) in men's and women's career development.

References

AAPG (American Association of Petroleum Geologists). 2009. Results from the American Association of Petroleum Geologists (AAPG) Professional Women in the Earth Sciences (PROWESS) Survey. American Association of Petroleum Geologists. June 7, 2011, http://www.aapg.org/committees/prowess/AAPG_Jun3.final.pdf.

Acker, Joan. 1990. Hierarchies, jobs, bodies: A theory of gendered organizations. *Gender & Society* 4:139–58.

AGI (American Geological Institute). 2011. Currents 30–35: Minorities, temporary residents, and gender parity in the geosciences. June 27, 2011, http://www.agiweb.org/workforce/webinar-videos/video_currents30–35.html.

Babcock, L., and S. Laschever. 2003. *Women don't ask: Negotiation and the gender divide.* Princeton, NJ: Princeton University Press.

Bell, E. E., and S. Nkomo. 2001. *Our separate ways: Black and white women and the struggle for professional identity.* Boston: Harvard Business School Press.

Bowles, H. R., Babcock, L., and L Lai. 2007. Social incentives for gender differences in the propensity to initiate negotiations: Sometimes it does hurt to ask. *Organizational Behavior and Human Decision Processes* 103:84–103.

Britton, D., and L. Logan. 2008. Gendered organizations: Progress and prospects. *Sociology Compass* 2:107–21.

Broadbridge, A. 2004. It's not what you know, it's who you know. *Journal of Management Development* 23:551–62.

Bureau of Labor Statistics. 2009. Geoscientists and hydrologists. December 13, 2011, http://www.bls.gov/oco/ocos312.htm.

Burt, Ronald S. 1998. The gender of social capital. *Rationality and Society* 10:5–46.

Catalyst. 2011. Women in U.S. mining, quarrying, and oil and gas extraction. September 12, 2011, http://www.catalyst.org/publication/503/women-in-us-mining-quarrying-and-oil-gas-extraction.

Connelly, J. D., and J. C. Middleton. 1996. Personal and professional skills for engineers: One industry's perspective. *Engineering Science and Education Journal* 5:139–42.

DiMaggio, Paul. 2001. *The twenty-first century firm.* Princeton, NJ: Princeton University press.

DiTomaso, N., C. Post, R. Smith, G. Farris, and R. Cordero. 2007. The effects of structural position on allocation and evaluation decisions for scientists and engineers in industrial R&D. *Administrative Science Quarterly* 52:175–207.

Elliott, J., and R. Smith. 2004. Race, gender, and workplace power. *American Sociological Review* 69:365–86.

England, Paula. 2010. The gender revolution: Uneven and stalled. *Gender & Society* 24:149–66.

Harvey Wingfield, Adia. 2010. Are some emotions marked "whites only"? Racialized feeling rules in professional workplaces. *Social Problems* 57:251–68.

Heilman, M.E., and M.C. Haynes. 2005. No credit where credit is due: Attributional rationalization of women's success in male-female teams. *Journal of Applied Psychology* 90:905–16.

Hewlett, Sylvia A. 2007. *Off-ramps and on-ramps: Keeping talented women on the road to success.* Boston, MA: Harvard Business School Publishing.

Kalev, Alexandra. 2009. Cracking the glass cages? Restructuring and ascriptive inequality at work. *American Journal of Sociology* 114:1591–1643.

Kalleberg, Arne. 2000. Nonstandard employment relations. *Annual Review of Sociology* 26:341–65.

Kanter, Rosabeth Moss. 1977. *Men and women of the corporation.* New York: Basic.

Loscocco, K., S. M. Monnat, G. Moore, and K. B. Lauber. 2009. Enterprising women: A comparison of women's and men's small business networks. *Gender & Society* 23:388–411.

McGuire, G. M. 2002. Gender, race, and the shadow structure: A study of informal networks and inequality in a work organization. *Gender & Society* 16:303–22.

McKee, L., N. Mauthner, and C. Maclean. 2000. Family friendly policies and practices in the oil and gas industry: Employer's perspectives. *Work, Employment and Society* 14:557–71.

Miller, Gloria E. 2004. Frontier masculinity in the oil industry: The experience of women engineers. *Gender; Work and Organization* 11:47–73.

Morgan, L., and K. Martin. 2006. Taking women professionals out of the office: The case of women in sales. *Gender & Society* 20:108–28.

National Academy of Sciences. 2010. *Rising above the gathering storm, revisited: Rapidly approaching category 5.* Washington, DC: National Academies Press.

Ollilainen, M., and T. Calasanti. 2007. Metaphors at work: Maintaining the salience of gender in self-managing teams. *Gender & Society* 21:5–27.

Osnowitz, Debra. 2010. *Freelancing expertise: Contract professionals in the new economy.* Ithaca, NY: ILR Press.

Pirog, Robert. 2008. *Oil industry profit review 2007.* Washington, DC: Congressional Research Service. September 1, 2011, assets.opencrs.com/rpts/RL34437_20080404.pdf.

Plankey Videla, Nancy. 2006. Gendered contradictions: Managers and women workers in self-managed teams. *Research in the Sociology of Work* 16:85–116.

Powell, Walter W. 2001. The capitalist firm in the 21st century: Emerging patterns in Western enterprise. In *The twenty-first century firm,* edited by Paul DiMaggio. Princeton, NJ: Princeton University Press.

Preston, Anne E. 2004. *Leaving science. Occupational exit from science careers.* New York: Russell Sage Foundation.

Reskin, Barbara. 2002. Rethinking employment discrimination and its remedies. In *The new economic sociology: Developments in an emerging field,* edited by M. F. Guillen, R. Collins, P. England, and M. Meyer. New York: Russell Sage.

Reskin, B., and D. McBrier. 2000. Why not ascription? Organizations' employment of male and female managers. *American Sociological Review* 65:210–33.

Roper, Michael. 1994. *Masculinity and the British organization man since 1945.* Oxford: Oxford University Press.

Rousseau, Denise. 2005. *I-deals: Idiosyncratic deals employees bargain for themselves.* Armonk, NY: M. E. Sharpe.

Smith, Sandra. 2007. *Lone pursuit: Distrust and defensive individualism among the Black poor.* New York: Russell Sage.

Smith-Doerr, Laurel. 2004. *Women's work: Gender equality versus hierarchy in the life sciences.* Boulder, CO: Lynne Rienner.

Vallas, Steven. 2011. *Work: A critique.* Boston: Polity Books.

Williams, Joan. 2001. *Unbending gender: Why family and work conflict and what to do about it.* New York: Oxford University Press.

Yergin, Daniel. 1993. *The prize: The epic quest for oil, money, and power.* New York: Free Press.

Introduction to Reading 34

Sociologists and others use the term *glass ceiling* to describe the barriers to promotion and advancement that women face in the world of work. At the same time, however, it is argued that men have a glass escalator, particularly men employed in what are traditionally women's jobs. In this article, Adia Harvey Wingfield describes the glass escalator and gives an overview of the research on men in traditionally female occupations. While men make up 8% of all nurses, the percentage of nurses who are Black men is unknown. Therefore, this study helps us understand the intersections of race and gender and how the experiences of Black men differ from those of White men in previous studies. Wingfield's study gives insight into the various ways race and gender intersect to discriminate against Black men in the workplace.

1. How are the experiences of the Black men she studied different from the results of previous studies of White men on the glass escalator?
2. Do Black male nurses do masculinity differently than White male nurses? Why or why not?
3. What forms of discrimination are described in this article? What would you recommend to eradicate such discrimination?

Racializing the Glass Escalator

Reconsidering Men's Experiences With Women's Work

Adia Harvey Wingfield

Sociologists who study work have long noted that jobs are sex segregated and that this segregation creates different occupational experiences for men and women (Charles and Grusky 2004). Jobs predominantly filled by women often require "feminine" traits such as nurturing, caring, and empathy, a fact that means men confront perceptions that they are unsuited for the requirements of these jobs. Rather than having an adverse effect on their occupational experiences, however, these assumptions facilitate men's entry into better paying, higher status positions, creating what Williams (1995) labels a "glass escalator" effect.

The glass escalator model has been an influential paradigm in understanding the experiences of men who do women's work. Researchers have identified this process among men nurses, social workers, paralegals, and librarians and have cited its pervasiveness as evidence of men's consistent advantage in the workplace, such that even in jobs where men are numerical minorities they are likely to enjoy higher wages and faster promotions (Floge and Merrill 1986; Heikes 1991; Pierce 1995; Williams 1989, 1995). Most of these studies implicitly assume a racial homogenization of men workers in women's professions, but this supposition is problematic for several reasons. For one, minority men are not only present but are actually overrepresented in certain areas of reproductive work that have historically been dominated by white women (Duffy 2007). Thus, research that focuses primarily on white men in women's professions ignores a key segment of men who perform this type of labor. Second, and perhaps more important, conclusions based on the experiences of white men tend to overlook the ways that intersections of race and gender create different experiences for different men. While extensive work has documented the fact that white men in women's professions encounter a glass escalator effect that aids their occupational mobility (for an exception, see Snyder and Green 2008), few studies, if any, have considered how this effect is a function not only of gendered advantage but of racial privilege as well.

In this article, I examine the implications of race–gender intersections for minority men employed in a female-dominated, feminized occupation, specifically

Wingfield, A. H. (2009). Racializing the glass escalator: Reconsidering men's experiences with women's work. *Gender & Society, 23*(1): 5–26. Reprinted by permission of SAGE Publications Inc., on behalf of Sociologists for Women in Society.

focusing on Black men in nursing. Their experiences doing "women's work" demonstrate that the glass escalator is a racialized as well as gendered concept.

Theoretical Framework

In her classic study *Men and Women of the Corporation,* Kanter (1977) offers a groundbreaking analysis of group interactions. Focusing on high-ranking women executives who work mostly with men, Kanter argues that those in the extreme numerical minority are tokens who are socially isolated, highly visible, and adversely stereotyped. Tokens have difficulty forming relationships with colleagues and often are excluded from social networks that provide mobility. Because of their low numbers, they are also highly visible as people who are different from the majority, even though they often feel invisible when they are ignored or overlooked in social settings. Tokens are also stereotyped by those in the majority group and frequently face pressure to behave in ways that challenge and undermine these stereotypes. Ultimately, Kanter argues that it is harder for them to blend into the organization and to work effectively and productively, and that they face serious barriers to upward mobility.

Kanter's (1977) arguments have been analyzed and retested in various settings and among many populations. Many studies, particularly of women in male-dominated corporate settings, have supported her findings. Other work has reversed these conclusions, examining the extent to which her conclusions hold when men were the tokens and women the majority group. These studies fundamentally challenged the gender neutrality of the token, finding that men in the minority fare much better than do similarly situated women. In particular, this research suggests that factors such as heightened visibility and polarization do not necessarily disadvantage men who are in the minority. While women tokens find that their visibility hinders their ability to blend in and work productively, men tokens find that their conspicuousness can lead to greater opportunities for leadership and choice assignments (Floge and Merrill 1986; Heikes 1991). Studies in this vein are important because they emphasize organizations—and occupations—as gendered institutions that subsequently create dissimilar experiences for men and women tokens (see Acker 1990).

In her groundbreaking study of men employed in various women's professions, Williams (1995) further develops this analysis of how power relationships shape the ways men tokens experience work in women's professions. Specifically, she introduces the concept of the glass escalator to explain men's experiences as tokens in these areas. Like Floge and Merrill (1986) and Heikes (1991), Williams finds that men tokens do not experience the isolation, visibility, blocked access to social networks, and stereotypes in the same ways that women tokens do. In contrast, Williams argues that even though they are in the minority, processes are in place that actually facilitate their opportunity and advancement. Even in culturally feminized occupations, then, men's advantage is built into the very structure and everyday interactions of these jobs so that men find themselves actually struggling to remain in place. For these men, "despite their intentions, they face invisible pressures to move up in their professions. Like being on a moving escalator, they have to work to stay in place" (Williams 1995, 87).

The glass escalator term thus refers to the "subtle mechanisms in place that enhance [men's] positions in [women's] professions" (Williams 1995, 108). These mechanisms include certain behaviors, attitudes, and beliefs men bring to these professions as well as the types of interactions that often occur between these men and their colleagues, supervisors, and customers. Consequently, even in occupations composed mostly of women, gendered perceptions about men's roles, abilities, and skills privilege them and facilitate their advancement. The glass escalator serves as a conduit that channels men in women's professions into the uppermost levels of the occupational hierarchy. Ultimately, the glass escalator effect suggests that men retain consistent occupational advantages over women, even when women are numerically in the majority (Budig 2002; Williams 1995).

Though this process has now been fairly well established in the literature, there are reasons to question its generalizability to all men. In an early critique of the supposed general neutrality of the token, Zimmer (1988) notes that much research on race comes to precisely the opposite of Kanter's conclusions, finding that as the numbers of minority group members increase (e.g., as they become less likely to be "tokens"), so too do tensions between the majority and minority groups. . . . Reinforcing, while at the same time tempering, the findings of research on men in female-dominated occupations, Zimmer (1988, 71) argues that relationships between tokens and the majority depend on understanding the underlying power relationships between these groups and "the status and power differentials between them." Hence, just as men who are tokens fare better than women, it also follows that the experiences of Blacks and whites as tokens should differ in ways that reflect their positions in hierarchies of status and power. . . .

Relationships With Colleagues and Supervisors

One key aspect of riding the glass escalator involves the warm, collegial welcome men workers often receive from their women colleagues. Often, this reaction is a response to the fact that professions dominated by women are frequently low in salary and status and that greater numbers of men help improve prestige and pay (Heikes 1991). Though some women workers resent the apparent ease with which men enter and advance in women's professions, the generally warm welcome men receive stands in stark contrast to the cold reception, difficulties with mentorship, and blocked access to social networks that women often encounter when they do men's work (Roth 2006; Williams 1992). In addition, unlike women in men's professions, men who do women's work frequently have supervisors of the same sex. Men workers can thus enjoy a gendered bond with their supervisor in the context of a collegial work environment. These factors often converge, facilitating men's access to higher-status positions and producing the glass escalator effect.

The congenial relationship with colleagues and gendered bonds with supervisors are crucial to riding the glass escalator. Women colleagues often take a primary role in casting these men into leadership or supervisory positions. In their study of men and women tokens in a hospital setting, Floge and Merrill (1986) cite cases where women nurses promoted men colleagues to the position of charge nurse, even when the job had already been assigned to a woman. In addition to these close ties with women colleagues, men are also able to capitalize on gendered bonds with (mostly men) supervisors in ways that engender upward mobility. Many men supervisors informally socialize with men workers in women's jobs and are thus able to trade on their personal friendships for upward mobility. Williams (1995) describes a case where a nurse with mediocre performance reviews received a promotion to a more prestigious specialty area because of his friendship with the (male) doctor in charge. According to the literature, building strong relationships with colleagues and supervisors often happens relatively easily for men in women's professions and pays off in their occupational advancement.

For Black men in nursing, however, gendered racism may limit the extent to which they establish bonds with their colleagues and supervisors. The concept of gendered racism suggests that racial stereotypes, images, and beliefs are grounded in gendered ideals (Collins 1990, 2004; Espiritu 2000; Essed 1991; Harvey Wingfield 2007). Gendered racist stereotypes of Black men in particular emphasize the dangerous, threatening attributes associated with Black men and Black masculinity, framing Black men as threats to white women, prone to criminal behavior, and especially violent. Collins (2004) argues that these stereotypes serve to legitimize Black men's treatment in the criminal justice system through methods such as racial profiling and incarceration, but they may also hinder Black men's attempts to enter and advance in various occupational fields.

For Black men nurses, gendered racist images may have particular consequences for their relationships with women colleagues, who may view Black men nurses through the lens of controlling images and gendered racist stereotypes that emphasize the danger they pose to women. This may take on a heightened significance for white women nurses, given stereotypes that suggest that Black men are especially predisposed to raping white women. Rather than experiencing the congenial bonds with colleagues that white men nurses describe, Black men nurses may find themselves facing a much cooler reception from their women coworkers.

Gendered racism may also play into the encounters Black men nurses have with supervisors. In cases where supervisors are white men, Black men nurses may still find that higher-ups treat them in ways that reflect prevailing stereotypes about threatening Black masculinity. Supervisors may feel uneasy about forming close relationships with Black men or may encourage their separation from white women nurses. In addition, broader, less gender-specific racial stereotypes could also shape the experiences Black men nurses have with white men bosses. Whites often perceive Blacks, regardless of gender, as less intelligent, hardworking, ethical, and moral than other racial groups (Feagin 2006). Black men nurses may find that in addition to being influenced by gendered racist stereotypes, supervisors also view them as less capable and qualified for promotion, thus negating or minimizing the glass escalator effect.

Suitability for Nursing and Higher-Status Work

The perception that men are not really suited to do women's work also contributes to the glass escalator effect. In encounters with patients, doctors, and other staff, men nurses frequently confront others who do not expect to see them doing "a woman's job." Sometimes this perception means that patients mistake men nurses for doctors; ultimately, the sense that men do not really belong in nursing contributes to a push "*out* of the most feminine-identified areas and *up* to those regarded as more legitimate for men" (Williams 1995,

104). The sense that men are better suited for more masculine jobs means that men workers are often assumed to be more able and skilled than their women counterparts. As Williams writes (1995, 106), "Masculinity is often associated with competence and mastery," and this implicit definition stays with men even when they work in feminized fields. Thus, part of the perception that men do not belong in these jobs is rooted in the sense that, as men, they are more capable and accomplished than women and thus belong in jobs that reflect this. Consequently, men nurses are mistaken for doctors and are granted more authority and responsibility than their women counterparts, reflecting the idea that, as men, they are inherently more competent (Heikes 1991; Williams 1995).

Black men nurses, however, may not face the presumptions of expertise or the resulting assumption that they belong in higher-status jobs. Black professionals, both men and women, are often assumed to be less capable and less qualified than their white counterparts. In some cases, these negative stereotypes hold even when Black workers outperform white colleagues (Feagin and Sikes 1994). The belief that Blacks are inherently less competent than whites means that, despite advanced education, training, and skill, Black professionals often confront the lingering perception that they are better suited for lower-level service work (Feagin and Sikes 1994). Black men in fact often fare better than white women in blue-collar jobs such as policing and corrections work (Britton 1995), and this may be, in part, because they are viewed as more appropriately suited for these types of positions. . . .

As minority women address issues of both race and gender to negotiate a sense of belonging in masculine settings (Ong 2005), minority men may also face a comparable challenge in feminized fields. They may have to address the unspoken racialization implicit in the assumption that masculinity equals competence. Simultaneously, they may find that the racial stereotype that Blackness equals lower qualifications, standards, and competence clouds the sense that men are inherently more capable and adept in any field, including the feminized ones.

Establishing Distance From Femininity

An additional mechanism of the glass escalator involves establishing distance from women and the femininity associated with their occupations. Because men nurses are employed in a culturally feminized occupation, they develop strategies to disassociate themselves from the femininity associated with their work and retain some of the privilege associated with masculinity. Thus, when men nurses gravitate toward hospital emergency wards rather than obstetrics or pediatrics, or emphasize that they are only in nursing to get into hospital administration, they distance themselves from the femininity of their profession and thereby preserve their status as men despite the fact that they do "women's work." Perhaps more important, these strategies also place men in a prime position to experience the glass escalator effect, as they situate themselves to move upward into higher-status areas in the field.

Creating distance from femininity also helps these men achieve aspects of hegemonic masculinity, which Connell (1989) describes as the predominant and most valued form of masculinity at a given time. Contemporary hegemonic masculine ideals emphasize toughness, strength, aggressiveness, heterosexuality, and, perhaps most important, a clear sense of femininity as different from and subordinate to masculinity (Kimmel 2001; Williams 1995). Thus, when men distance themselves from the feminized aspects of their jobs, they uphold the idea that masculinity and femininity are distinct, separate, and mutually exclusive. When these men seek masculinity by aiming for the better paying or most technological fields, they not only position themselves to move upward into the more acceptable arenas but also reinforce the greater social value placed on masculinity. Establishing distance from femininity therefore allows men to retain the privileges and status of masculinity while simultaneously enabling them to ride the glass escalator.

For Black men, the desire to reject femininity may be compounded by racial inequality. Theorists have argued that as institutional racism blocks access to traditional markers of masculinity such as occupational status and economic stability, Black men may repudiate femininity as a way of accessing the masculinity—and its attendant status—that is denied through other routes (hooks 2004; Neal 2005). Rejecting femininity is a key strategy men use to assert masculinity, and it remains available to Black men even when other means of achieving masculinity are unattainable. Black men nurses may be more likely to distance themselves from their women colleagues and to reject the femininity associated with nursing, particularly if they feel that they experience racial discrimination that renders occupational advancement inaccessible. Yet if they encounter strained relationships with women colleagues and men supervisors because of gendered racism or racialized stereotypes, the efforts to distance themselves from femininity still may not result in the glass escalator effect.

On the other hand, some theorists suggest that minority men may challenge racism by rejecting hegemonic masculine ideals. . . . The results of these studies suggest that Black men nurses may embrace the femininity associated with nursing if it offers a way to combat racism. In these cases, Black men nurses may turn to pediatrics as a way of demonstrating sensitivity and therefore combating stereotypes of Black masculinity, or they may proudly identify as nurses to challenge perceptions that Black men are unsuited for professional, white-collar positions.

Taken together, all of this research suggests that Black men may not enjoy the advantages experienced by their white men colleagues, who ride a glass escalator to success. In this article, I focus on the experiences of Black men nurses to argue that the glass escalator is a racialized as well as a gendered concept that does not offer Black men the same privileges as their white men counterparts. . . .

Findings

Reception From Colleagues and Supervisors

When women welcome men into "their" professions, they often push men into leadership roles that ease their advancement into upper-level positions. Thus, a positive reaction from colleagues is critical to riding the glass escalator. Unlike white men nurses, however, Black men do not describe encountering a warm reception from women colleagues (Heikes 1991). Instead, the men I interviewed find that they often have unpleasant interactions with women coworkers who treat them rather coldly and attempt to keep them at bay. Chris is a 51-year-old oncology nurse who describes one white nurse's attempt to isolate him from other white women nurses as he attempted to get his instructions for that day's shift:

> She turned and ushered me to the door, and said for me to wait out here, a nurse will come out and give you your report. I stared at her hand on my arm, and then at her, and said, "Why? Where do you go to get your reports?" She said, "I get them in there." I said, "Right. Unhand me." I went right back in there, sat down, and started writing down my reports.

Kenny, a 47-year-old nurse with 23 years of nursing experience, describes a similarly and particularly painful experience he had in a previous job where he was the only Black person on staff:

> [The staff] had nothing to do with me, and they didn't even want me to sit at the same area where they were charting in to take a break. They wanted me to sit somewhere else. . . . They wouldn't even sit at a table with me! When I came and sat down, everybody got up and left.

These experiences with colleagues are starkly different from those described by white men in professions dominated by women (see Pierce 1995; Williams 1989). Though the men in these studies sometimes chose to segregate themselves, women never systematically excluded them. Though I have no way of knowing why the women nurses in Chris's and Kenny's workplaces physically segregated themselves, the pervasiveness of gendered racist images that emphasize white women's vulnerability to dangerous Black men may play an important role. For these nurses, their masculinity is not a guarantee that they will be welcomed, much less pushed into leadership roles. As Ryan, a 37-year-old intensive care nurse says, "[Black men] have to go further to prove ourselves. This involves proving our capabilities, *proving to colleagues that you can lead,* be on the forefront" (emphasis added). The warm welcome and subsequent opportunities for leadership cannot be taken for granted. In contrast, these men describe great challenges in forming congenial relationships with coworkers who, they believe, do not truly want them there.

In addition, these men often describe tense, if not blatantly discriminatory, relationships with supervisors. While Williams (1995) suggests that men supervisors can be allies for men in women's professions by facilitating promotions and upward mobility, Black men nurses describe incidents of being overlooked by supervisors when it comes time for promotions. Ryan, who has worked at his current job for 11 years, believes that these barriers block upward mobility within the profession:

> The hardest part is dealing with people who don't understand minority nurses. People with their biases, who don't identify you as ripe for promotion. I know the policy and procedure, I'm familiar with past history. So you can't tell me I can't move forward if others did. [How did you deal with this?] By knowing the chain of command, who my supervisors were. Things were subtle. I just had to be better. I got this mostly from other nurses and supervisors. I was paid to deal with patients, so I could deal with [racism] from them. I'm not paid to deal with this from colleagues.

Kenny offers a similar example. Employed as an orthopedic nurse in a predominantly white environment, he describes great difficulty getting promoted, which he primarily attributes to racial biases:

> It's almost like you have to, um, take your ideas and give them to somebody else and then let them present them for you and you get no credit for it. I've applied for several promotions there and, you know, I didn't get them. . . . When you look around to the, um, the percentage of African Americans who are actually in executive leadership is almost zero percent. Because it's less than one percent of the total population of people that are in leadership, and it's almost like they'll go outside of the system just to try to find a Caucasian to fill a position. Not that I'm not qualified, because I've been master's prepared for 12 years and I'm working on my doctorate.

According to Ryan and Kenny, supervisors' racial biases mean limited opportunities for promotion and upward mobility. This interpretation is consistent with research that suggests that even with stellar performance and solid work histories, Black workers may receive mediocre evaluations from white supervisors that limit their advancement (Feagin 2006; Feagin and Sikes 1994). For Black men nurses, their race may signal to supervisors that they are unworthy of promotion and thus create a different experience with the glass escalator.

Strong relationships with colleagues and supervisors are a key mechanism of the glass escalator effect. For Black men nurses, however, these relationships are experienced differently from those described by their white men colleagues. Black men nurses do not speak of warm and congenial relationships with women nurses or see these relationships as facilitating a move into leadership roles. Nor do they suggest that they share gendered bonds with men supervisors that serve to ease their mobility into higher-status administrative jobs. In contrast, they sense that racial bias makes it difficult to develop ties with coworkers and makes superiors unwilling to promote them. Black men nurses thus experience this aspect of the glass escalator differently from their white men colleagues. They find that relationships with colleagues and supervisors stifle, rather than facilitate, their upward mobility.

Perceptions of Suitability

Like their white counterparts, Black men nurses also experience challenges from clients who are unaccustomed to seeing men in fields typically dominated by women. As with white men nurses, Black men encounter this in surprised or quizzical reactions from patients who seem to expect to be treated by white women nurses. . . .

Yet while patients rarely expect to be treated by men nurses of any race, white men encounter statements and behaviors that suggest patients expect them to be doctors, supervisors, or other higher-status, more masculine positions (Williams 1989, 1995). In part, this expectation accelerates their ride on the glass escalator, helping to push them into the positions for which they are seen as more appropriately suited.

(White) men, by virtue of their masculinity, are assumed to be more competent and capable and thus better situated in (nonfeminized) jobs that are perceived to require greater skill and proficiency. Black men, in contrast, rarely encounter patients (or colleagues and supervisors) who immediately expect that they are doctors or administrators. Instead, many respondents find that even after displaying their credentials, sharing their nursing experience, and, in one case, dispensing care, they are still mistaken for janitors or service workers. Ray's experience is typical:

> I've even given patients their medicines, explained their care to them, and then they'll say to me, "Well, can you send the nurse in?"

Chris describes a somewhat similar encounter of being misidentified by a white woman patient:

> I come [to work] in my white uniform, that's what I wear—being a Black man, I know they won't look at me the same, so I dress the part—I said good evening, my name's Chris, and I'm going to be your nurse. She says to me, "Are you from housekeeping?". . . I've had other cases. I've walked in and had a lady look at me and ask if I'm the janitor. . . .

These negative stereotypes can affect Black men nurses' efforts to treat patients as well. The men I interviewed find that masculinity does not automatically endow them with an aura of competency. In fact, they often describe interactions with white women patients that suggest that their race minimizes whatever assumptions of capability might accompany being men. They describe several cases in which white women patients completely refused treatment. Ray says,

> With older white women, it's tricky sometimes because they will come right out and tell you they don't want you to treat them, or can they see someone else.

Ray frames this as an issue specifically with older white women, though other nurses in the sample described similar issues with white women of all ages. Cyril, a 40-year-old nurse with 17 years of nursing experience, describes a slightly different twist on this story:

> I had a white lady that I had to give a shot, and she was fine with it and I was fine with it. But her husband, when she told him, he said to me, I don't have any problem with you as a Black man, but I don't want you giving her a shot.

While white men nurses report some apprehension about treating women patients, in all likelihood this experience is compounded for Black men (Williams 1989). Historically, interactions between Black men and white women have been fraught with complexity and tension, as Black men have been represented in the cultural imagination as potential rapists and threats to white women's security and safety—and, implicitly, as a threat to white patriarchal stability (Davis 1981; Giddings 1984). In Cyril's case, it may be particularly significant that the Black man is charged with giving a shot and therefore literally penetrating the white wife's body, a fact that may heighten the husband's desire to shield his wife from this interaction. White men nurses may describe hesitation or awkwardness that accompanies treating women patients, but their experiences are not shaped by a pervasive racial imagery that suggests that they are potential threats to their women patients' safety.

This dynamic, described primarily among white women patients and their families, presents a picture of how Black men's interactions with clients are shaped in specifically raced and gendered ways that suggest they are less rather than more capable. These interactions do not send the message that Black men, because they are men, are too competent for nursing and really belong in higher-status jobs. Instead, these men face patients who mistake them for lower-status service workers and encounter white women patients (and their husbands) who simply refuse treatment or are visibly uncomfortable with the prospect. These interactions do not situate Black men nurses in a prime position for upward mobility. Rather, they suggest that the experience of Black men nurses with this particular mechanism of the glass escalator is the manifestation of the expectation that they should be in lower-status positions more appropriate to their race and gender.

Refusal to Reject Femininity

Finally, Black men nurses have a different experience with establishing distance from women and the feminized aspects of their work. Most research shows that as men nurses employ strategies that distance them from femininity (e.g., by emphasizing nursing as a route to higher-status, more masculine jobs), they place themselves in a position for upward mobility and the glass escalator effect (Williams 1992). For Black men nurses, however, this process looks different. Instead of distancing themselves from the femininity associated with nursing, Black men actually embrace some of the more feminized attributes linked to nursing. In particular, they emphasize how much they value and enjoy the way their jobs allow them to be caring and nurturing. Rather than conceptualizing caring as anathema or feminine (and therefore undesirable), Black men nurses speak openly of caring as something positive and enjoyable.

This is consistent with the context of nursing that defines caring as integral to the profession. As nurses, Black men in this line of work experience professional socialization that emphasizes and values caring, and this is reflected in their statements about their work. Significantly, however, rather than repudiating this feminized component of their jobs, they embrace it. Tobias, a 44-year-old oncology nurse with 25 years of experience, asserts,

> The best part about nursing is helping other people, the flexibility of work hours, and the commitment to vulnerable populations, people who are ill.

Simon, a 36-year-old oncology nurse, also talks about the joy he gets from caring for others. He contrasts his experiences to those of white men nurses he knows who prefer specialties that involve less patient care:

> They were going to work with the insurance industries, they were going to work in the ER where it's a touch and go, you're a number literally. I don't get to know your name, I don't get to know that you have four grandkids, I don't get to know that you really want to get out of the hospital by next week because the following week is your birthday, your 80th birthday and it's so important for you. I don't get to know that your cat's name is Sprinkles, and you're concerned about who's feeding the cat now, and if they remembered to turn the TV on during the day so that the cat can watch *The Price Is Right.* They don't get into all that kind of stuff. OK, I actually need to remember the name of your cat so that tomorrow morning when I come, I can ask you about Sprinkles and that will make a world of difference. I'll see light coming to your eyes and the medicines will actually work because your perspective is different.

Like Tobias, Simon speaks with a marked lack of self-consciousness about the joys of adding a personal touch and connecting that personal care to a patient's improvement. For him, caring is important, necessary, and valued, even though others might consider it a feminine trait.

For many of these nurses, willingness to embrace caring is also shaped by issues of race and racism. In their position as nurses, concern for others is connected to fighting the effects of racial inequality. Specifically, caring motivates them to use their role as nurses to address racial health disparities, especially those that disproportionately affect Black men. Chris describes his efforts to minimize health issues among Black men:

> With Black male patients, I have their history, and if they're 50 or over I ask about the prostate exam and a

> colonoscopy. Prostate and colorectal death is so high that that's my personal crusade.

Ryan also speaks to the importance of using his position to address racial imbalances:

> I really take advantage of the opportunities to give back to communities, especially to change the disparities in the African American community. I'm more than just a nurse. As a faculty member at a major university, I have to do community hours, services. Doing health fairs, in-services on research, this makes an impact in some disparities in the African American community. [People in the community] may not have the opportunity to do this otherwise.

As Lamont (2000) indicates in her discussion of the "caring self," concern for others helps Chris and Ryan to use their knowledge and position as nurses to combat racial inequalities in health. Though caring is generally considered a "feminine" attribute, in this context it is connected to challenging racial health disparities. Unlike their white men colleagues, these nurses accept and even embrace certain aspects of femininity rather than rejecting them. They thus reveal yet another aspect of the glass escalator process that differs for Black men. As Black men nurses embrace this "feminine" trait and the avenues it provides for challenging racial inequalities, they may become more comfortable in nursing and embrace the opportunities it offers.

Conclusions

Existing research on the glass escalator cannot explain these men's experiences. As men who do women's work, they should be channeled into positions as charge nurses or nursing administrators and should find themselves virtually pushed into the upper ranks of the nursing profession. But without exception, this is not the experience these Black men nurses describe. Instead of benefiting from the basic mechanisms of the glass escalator, they face tense relationships with colleagues, supervisors' biases in achieving promotion, patient stereotypes that inhibit caregiving, and a sense of comfort with some of the feminized aspects of their jobs. These "glass barriers" suggest that the glass escalator is a racialized concept as well as a gendered one. The main contribution of this study is the finding that race and gender intersect to determine which men will ride the glass escalator. The proposition that men who do women's work encounter undue opportunities and advantages appears to be unequivocally true only if the men in question are white.

* * *

References

Acker, Joan. 1990. Hierarchies, jobs, bodies: A theory of gendered organizations. *Gender & Society* 4:139–58.

Britton, Dana. 1995. *At work in the iron cage.* New York: New York University Press.

Budig, Michelle. 2002. Male advantage and the gender composition of jobs: Who rides the glass escalator? *Social Forces* 49(2): 258–77.

Charles, Maria, and David Grusky. 2004. *Occupational ghettos: The worldwide segregation of women and men.* Palo Alto, CA: Stanford University Press.

Collins, Patricia Hill. 1990. *Black feminist thought.* New York: Routledge.

———. 2004. *Black sexual politics.* New York: Routledge.

Connell, R. W. 1989. *Gender and power.* Sydney, Australia: Allen and Unwin.

Davis, Angela. 1981. *Women, race, and class.* New York: Vintage.

Duffy, Mignon. 2007. Doing the dirty work: Gender, race, and reproductive labor in historical perspective. *Gender & Society* 21:313–36.

Espiritu, Yen Le. 2000. *Asian American women and men: Labor, laws, and love.* Walnut Creek, CA: AltaMira.

Essed, Philomena. 1991. *Understanding everyday racism.* New York: Russell Sage.

Feagin, Joe. 2006. *Systemic racism.* New York: Routledge.

Feagin, Joe, and Melvin Sikes. 1994. *Living with racism.* Boston: Beacon Hill Press.

Floge, Liliane, and Deborah M. Merrill. 1986. Tokenism reconsidered: Male nurses and female physicians in a hospital setting. *Social Forces* 64:925–47.

Giddings, Paula. 1984. *When and where I enter: The impact of Black women on race and sex in America.* New York: HarperCollins.

Harvey Wingfield, Adia. 2007. The modern mammy and the angry Black man: African American professionals' experiences with gendered racism in the workplace. *Race, Gender, and Class* 14(2): 196–212.

Heikes, E. Joel. 1991. When men are the minority: The case of men in nursing. *Sociological Quarterly* 32:389–401.

hooks, bell. 2004. *We real cool.* New York: Routledge.

Kanter, Rosabeth Moss. 1977. *Men and women of the corporation.* New York: Basic Books.

Kimmel, Michael. 2001. Masculinity as homophobia. In *Men and masculinity,* edited by Theodore F. Cohen. Belmont, CA: Wadsworth.

Lamont, Michelle. 2000. *The dignity of working men.* New York: Russell Sage.

Neal, Mark Anthony. 2005. *New Black man.* New York: Routledge.

Ong, Maria. 2005. Body projects of young women of color in physics: Intersections of race, gender, and science. *Social Problems* 52(4): 593–617.

Pierce, Jennifer. 1995. *Gender trials: Emotional lives in contemporary law firms.* Berkeley: University of California Press.

Roth, Louise. 2006. *Selling women short: Gender and money on Wall Street.* Princeton, NJ: Princeton University Press.

Snyder, Karrie Ann, and Adam Isaiah Green. 2008. Revisiting the glass escalator: The case of gender segregation in a female dominated occupation. *Social Problems* 55(2): 271–99.

Williams, Christine. 1989. *Gender differences at work: Women and men in non-traditional occupations.* Berkeley: University of California Press.

———. 1992. The glass escalator: Hidden advantages for men in the "female" professions. *Social Problems* 39(3): 253–67.

———. 1995. *Still a man's world: Men who do women's work.* Berkeley: University of California Press.

Zimmer, Lynn. 1988. Tokenism and women in the workplace: The limits of gender neutral theory. *Social Problems* 35(1): 64–77.

Introduction to Reading 35

Erin Reid tackles a well-known ideology in many workplaces: that of the "ideal worker image." This is an ideology that says the best and most desirable workers are those who are completely devoted to their work, ahead and above all other concerns or aspects of their lives. While the ideal worker expectation has historically been noted as particularly disadvantageous for women (especially mothers), Reid finds that contemporary men and women both experience conflict with the ideal worker image. The difference, she shows, is in how men and women cope with that conflict. Women were more likely to reveal their conflict to their employer by asking for less travel or shorter working hours, while men were more likely to "pass," that is, find under-the-radar ways to not conform but still give the appearance of conforming.

1. What is the difference between an expected professional identity and an experienced professional identity? What was the expected professional identity at AGM?
2. What tools were available to workers who strayed from the ideal worker image?
3. How did men and women combine "passing" and "revealing" across different audiences?
4. What were the consequences for those men and women who "passed" or "revealed"?

Embracing, Passing, Revealing, and the Ideal Worker Image

How People Navigate Expected and Experienced Professional Identities

Erin Reid

Introduction

People today are expected to be wholly devoted to work, such that they attend to their jobs ahead of all else, including family (Blair-Loy 2003), personal needs (Kreiner et al. 2006), and even their health (Michel 2011). These expectations are personified in the ideal worker image: a definition of the most desirable worker as one who is totally committed to, and always available for, his or her work (Acker 1990). Embracing this image is richly rewarded, particularly for people in professional and managerial jobs; in many such workplaces, advancement and prizes accrue to those perceived to best embody this image (Bailyn 2006). Although scholars have focused on the difficulties that women in such jobs experience with these expectations

Reid, E. (2015). Embracing, passing, revealing, and the ideal worker image: How people navigate expected and experienced professional identities. *Organization Science, 26*(4): 997–1017. Reprinted with permission from INFORMS.

(e.g., Blair Loy 2003, Stone 2007), research increasingly suggests that their male colleagues may also find these expectations challenging (Galin sky et al. 2009, Humberd et al. 2014). Thus, many people may encounter a conflict between employer expectations that they be ideal workers and the sort of workers that they believe and prefer themselves to be.

This paper closely examines how people working at a demanding professional service firm navigate tensions between organizational expectations that they be ideal workers—which I conceptualize as an *expected professional identity*—and the sort of workers they believe and prefer themselves to be—their experienced professional identities. I find that people cope with conflict between these two identities by straying from the expected identity and seeking to remain true to their experienced identities. I draw on Goffman's (1963) concepts of passing and revealing, typically used to explain how people manage discredited social identities (Clair et al. 2005, Jones and King 2014, Ragins 2008), to develop a theory about how men and women navigate organizational audiences in ways that disclose or that mask their deviance, and I explore how they are consequently perceived and treated.

Theoretical Background

Professional Identity and the Ideal Worker Image

Identity, and its significance for people's work experiences, is a central concern of contemporary organizational scholarship (Ashforth et al. 2008, Ramarajan 2014, Roberts and Dutton 2009). This study focuses on professional identity (Ibarra 1999, Pratt et al. 2006). Professional identities are role identities, or the "goals, values, beliefs, norms, interaction styles and time horizons that are typically associated with a role" (Ashforth 2001, p. 6). Like most social roles, professional roles are subject to external expectations of incumbents' identities; I focus here on organizational, or employer, expectations and refer to these as *expected professional identities*. People, however, have their own preferences about their identities, and these do not always match those expected of them. I use the term *experienced professional identities* to describe people's beliefs and preferences regarding who they are as professionals. As people form their identities in relation to their past, future, alternative, and possible selves (Ibarra 1999, Markus and Nurius 1986, Obodaru 2012), their statements about their experienced identities may include allusions to these other selves.

Many organizations expect professionals to assume an identity that centers on the ideal worker image, such that they are fully committed to and totally available for their work, with no external commitments that limit this devotion (Acker 1990, Bailyn 2006, Williams et al. 2013). Although professional identities also include profession specific content, this image is central to many professions' expected identities. For example, surgeons, who spend years honing technical skills, are expected to embrace a professional identity that includes always placing "their patients first, over and above any personal commitment" (Kellogg 2011, p. 51). In such jobs, pressures to be ideal workers are often embedded in the very design of work, which routinely spills into evenings and weekends (Moen et al. 2013, Perlow 1998).

This image, and its attendant expectations of devotion, is viewed as a key driver of workplace gender inequality (Bailyn 2006, Correll et al. 2014, Williams 2000), and perhaps consequently, scholars have mostly examined how women, particularly mothers, navigate expectations that they devote themselves to work (e.g., Blair-Loy 2003, Christopher 2012, Webber and Williams 2008). Little work has considered men's experiences in this regard, echoing more general tendencies to frame work—family conflict as a woman's problem (for a review, see Leslie and Manchester 2011). Yet as a core element of an expected professional identity, this image necessarily shapes all workers' experiences, including men's. Moreover, studies increasingly suggest that men also find demands for work devotion challenging (Galinsky et al. 2009, Humberd et al. 2014), suggesting that difficulties with expectations that one assume the identity of an ideal worker are not necessarily restricted to women.

Yet although people apply for jobs in part based on assumptions about incumbents' identities (Barbulescu and Bidwell 2013), many workers are ambivalent about the identities their organization expects them to take on (Collinson 2003, Gagnon and Collinson 2014, Ramarajan and Reid 2013), suggesting that conflict between expected and experienced identities may be relatively common. However, deviance from expected identities may go unrecognized: people's identities do not necessarily match how others perceive them (Gecas 1982). To develop theory about the ways that people may manage incongruence between expected and experienced professional identities, and how this shapes how they are perceived, I turned to Goffman's (1963) concepts of "passing" and "revealing."

Identity Management Strategies: Passing and Revealing

Passing and revealing are ways that people control others' beliefs about who they are. The need to pass or to reveal arises when a person does not belong to a group of people to whom social rewards accrue (Goffman 1963). Some characteristics that disqualify one from membership in a favored group are clearly visible (e.g., skin color) and are managed through methods that "cover" or reduce the salience of the characteristic (Phillips et al. 2009, Rosette and Dumas 2007, Yoshino 2007). Other characteristics, however, are invisible (e.g., sexual preference), and people may choose how to manage them (Clair et al. 2005, Ragins 2008). That is, people may either misrepresent themselves as members of the favored group—thus, passing—or disclose that they are nonmembers—thus revealing. Passing can be intentional, as when a person lies about his or her identity, or accidental, as when others make incorrect assumptions; revealing also occurs across a continuum of intentionality.

Method

I explored these issues through a field study of a consulting firm. The study draws principally on semistructured interview data. I link the findings from the interview data to performance data, turnover data, and participants' stories about each other. Archival data (e.g., human resources (HR) documents) provided contextual information about the firm and industry.

Research Setting

I conducted this study at AGM (a pseudonym), a global consulting firm with a strong U.S. presence. Like many such firms, AGM offered advisory services in multiple areas, such as strategy, marketing, and finance and used small teams to complete projects over a period of weeks to months. Consulting is a notoriously demanding profession: consultants must typically be available for overnight travel to client sites and often work evenings and weekends on short notice. Within AGM, consultants advanced through several levels: associate, junior manager, senior manager, partner, and senior partner.

This setting provided certain advantages for investigating how people navigate tensions between expected and experienced professional identities. First, identity expectations in professional jobs are often strong, and AGM's status as one of the more demanding consulting firms within the industry qualified it as an "extreme" case (Eisenhardt 1989), where pressures to be an ideal worker might be especially acute and hence particularly visible (Pratt et al. 2006). Second, as AGM hired from elite colleges and MBA programs through a complex interview process, its hires were fairly homogeneous in terms of intellect, education level, and social skills. Participants were therefore all likely to be capable of doing the work; this helped to focus the analysis on how they coped with the firm's identity expectations.

Data Collection

Participants. I conducted 115 interviews with people associated with AGM. The core data for this study came from interviews with consultants, I interviewed 70 consultants. I added to this sample by accessing transcripts of 18 interviews conducted by other researchers as part of a study of AGM's culture; these covered topics pertinent to this study (discussed below). This sample included several of AGM's senior partners and senior leaders in the internal HR department. I met about half of these people during meetings at AGM and interviewed two of them during my own data collection. Because of cross-national differences in norms regarding the relationship between work and non work (Uhlmann et al. 2013), I excluded four people employed by non-North American offices. With these duplications and exclusions, the total number of consultants analyzed here is 82.

All consultants held undergraduate or advanced degrees (e.g., MBA, PhD, LL.B.) from elite schools (e.g., Williams, Harvard, Stanford). Twenty-two percent were women, similar to the proportional representation of women at AGM at the time (in 2009, 24%) and similar to or higher than that at competitor firms. Thirteen percent were visible racial minorities (e.g., African American, Southeast Asian).

As the study progressed, I expanded my sampling to include interviews with 27 other people whose experiences might inform the research. These included six employees in non consulting roles, six consultants who had left AGM prior to my study, eight people who worked at competitor firms, and seven of the consultants' spouses. These people were contacted through either random sampling from lists provided by AGM or personal contacts. Table 1 describes participant characteristics.

Interviews

The interview guide included structured questions, which enabled comparisons across people, as well as

Table 1 Participant Characteristics

		Expected vs. experienced identify		
Congruent			*Conflicting*	
Primary identity management strategy[a]	*Total*	*Embracing*	*Passing*	*Revealing*
Consultants	82	35 (43%)	22 (27%)	25 (30%)
Men	64	27 (42%)	20 (31%)	17 (27%)
Women	18	8 (44%)	2 (11%)	8 (44%)
White	71	29 (41%)	19 (27%)	23 (32%)
Visible minority	11	6 (55%)	2 (18%)	3 (27%)
Associate	12		1 (8%)	4 (33%)
Junior Manager	26		7 (27%)	11 (42%)
Senior management	22		7 (32%)	7 (32%)
Partner	13		4 (31%)	2 (15%)
Senior partner	9		3 (33%)	1 (11%)
Additional participants	31			
Nonconsulting Employees	6			
Former AGM consultants	6			
Consultants from competitor firms	8			
Consultants' spouses	7			
Non-North American AGM Consultants	4			
Total participants	113[b]			

[a] "Primary identify management strategy" refers to the strategy people employed in their interactions with senior members of the firm.
[b] Represents 115 interviews in total: 2 participants were interviewed twice.

unstructured questions, which permitted open ended reflection. In the interviews with consultants, I began by asking about people's work histories (e.g., months between promotions), job tasks, work hours, and travel. I then asked about their future goals, the importance of work to their sense of self, recent team projects, and colleagues whom they viewed as successful. Later, I asked about gender dynamics and their nonwork lives. By grounding the interview in the details of people's work, I hoped to limit opportunities for them to misrepresent their experiences. Interviews with other participants followed similar guides tailored to their particular experiences (e.g., I asked former consultants why they had left). The interviews conducted as part of the study of AGM's culture included similar questions about work histories and experiences and perceptions of AGM's success metrics.

Performance Data

I accessed quantitative ratings of consultants' performance for the year preceding the interviews (2009) and for the year of the interviews (2010). As I detail in the findings section, consultants were rated on several dimensions following each project, and these ratings were compiled into one annual rating at the end of each calendar year. The 2009 performance ratings cover 54 of the 60 non-partner consultants in the study (some were unreachable, one refused). Because of departures during 2010 and one promotion to partner, the 2010 data include 43 participants.

Archival Data. I also accessed internal HR documents that described hiring and evaluation practices, newspaper articles about AGM, and reviews of AGM on career websites. These data helped me to better understand AGM and its position in the industry.

Data Analysis

First, to understand the sort of worker that consultants believed AGM favored, I coded for experiences, behaviors, and characteristics that they associated with success at AGM. Once I had stabilized a description of AGM's expected professional identity and how it was communicated, I examined consultants' responses to this expectation. This process revealed that some easily embraced the expected identity, but it also uncovered widespread conflict between this expected identity and people's experienced professional identities. I deduced that people managed this conflict through passing and revealing. I coded the tools that they used in these efforts and their target audiences.

To examine external perceptions, I combined three data sources. First, I used the performance data, which, given the ambiguity of competence and importance of image in professional service work, corresponds well to AGM's perceptions of how well consultants fit its expected professional identity. Second, I quantified the average number of months people reported between promotions. I used these data to rate each person's career progress at AGM as slow, average, or fast. Finally, I examined the transcripts to identify instances where consultants described colleagues' work habits. I identified 78 such accounts of 47 people (some were mentioned several times). Of these, I had interviewed 32. Coding these accounts for consistency with the person's own account revealed that people who had worked directly with the focal individual tended to view their work habits in ways consistent with the person's own account, but that people whose experiences were less direct held less consistent views.

I now report the findings that emerged from this analysis. I begin by describing AGM's expected professional identity, how it was imposed, and its fit with people's experienced identities. I then explain how people strayed in ways that fostered passing or revealing. I close with a discussion of the study's contributions to theories of people's management of their professional identities, the ideal worker image, and passing and revealing in organizations.

Expected and Experienced Professional Identities

Consultants believed that AGM expected them, like ideal workers, to be fully devoted to work: primarily committed to and available for their work at all times and in all places. Although people sometimes associated other attributes with success (e.g., courage, charisma), mention of these attributes was sporadic relative to the near-constant emphasis on commitment and availability that permeated accounts of life at AGM. Tellingly, nearly all senior partners and leaders of AGM's HR group cited commitment and availability as attributes that distinguished successful from unsuccessful consultants. This analysis therefore focuses on AGM's identity expectations regarding commitment and availability

Expected Identity: Committed and Available

Consultants believed that success at AGM required being committed, passionate, and dedicated, such that their work occupied a central place in their lives. "Star" consultants would "give everything they have to the company." Commitment involved loyalty: despite the industry's high turnover, good consultants sought to remain at AGM. Commitment also meant placing work ahead of other life demands. Curtis (Partner, M), for example, had spent Thanksgiving "running a project remotely from the outside deck of [my in-laws'] condominium in Florida." Despite his wife's fury, he believed being a consultant required this commitment:

> I will sometimes have to get calls on Sunday nights. Sometimes, I have to do calls on Saturday mornings. So that the weekend is not sacrosanct. . . . If the client needs me, I will generally take [the call]. And you know when the client needs me to be somewhere, I just have to be there. In the consulting—in the professional services industry, generally—you don't really have the latitude of saying "I can't really be there." And if you can't be there, it's probably because you've got another client meeting at the same time. You know it's tough to say I can't be there because my—my son had a Cub Scout meeting.

The personal sacrifices such commitment entailed were justified by the intense "love" that successful consultants were expected to feel for their work. Suzanne (Junior manager, F) told me that to succeed, "You have to really love client service. I really love my clients. I wake up in the morning and wonder whether my clients are awake, whether they've emailed me, whether I need to do something for them."

Successful consultants were also believed to be fully available for work. Although availability was associated with commitment, the two were not the same: commitment involved dedicating oneself to work ahead of other demands and responsibilities, but availability corresponded to work hours and willingness to travel. People were expected to "work all night, if needed, to get things done" and travelled at "the drop of a hat." The need to be fully available, along with the need to be primarily committed to work, characterized Amos's (Junior manager, M) description of his colleagues:

> You know AGM people, we're on our Black Berries. We're thinking about our work 24/7. I mean, maybe you tune out for a little while here and there, but AGM people work all the time, all the time. I mean, you wake up at night, you're dreaming about it. The first thing you do is you pick up your BlackBerry, you're on it through the morning. You get

> to the office, you're working through the day, you sit at your desk, you know, you're cancelling plans.

Thus, consultants believed that "AGM people" were primarily committed to and fully available for work.

To assess the extent to which consultants' views about the identity of a successful consultant were shared by those who evaluated them, I compared the perceptions of people in client service-based consulting roles (associate through partner) to those of people who led the firm and who controlled recruiting and evaluation (senior partners and leaders of the HR department). Nearly all shared the consultants' beliefs regarding the importance of commitment and availability. For example, Sharon (Partner (HR), F) said,

> The culture at AGM is "give, give, give." The guy you saw leaving my office is leaving AGM, and he came to talk to me and said, "This place is crazy. It's like you're supposed to love this place and give your soul . . . . And when you leave, the norm is to write an email to everyone saying, 'Thank you AGM for all you have given me.' "But no one thanks you. So it's like the message is, we will only love you if you "give, give, give."

These shared beliefs between consultants and those who evaluated them confirms this identity's position as a category that distinguished between favored and un favored consultants (Goffman 1963).

Mechanisms of Identity Control: Structure of Work and Performance Evaluations

AGM pressured people to adopt this identity through the structure of work and the performance evaluation system. Together, these mechanisms encouraged consultants to adopt the expected identity by constructing work demands that seemed to require conforming to this identity while rewarding those who seemed to conform and penalizing those who did not.

Expectations regarding consultants' identities were embedded in AGM's haphazard work structure: crisis situations wherein teams worked late into the night were common, and partners often promised clients new work mid project. Clients often expected travel at short notice: two people arrived for our interview uncertain whether they would travel that day, and several rescheduled interviews because of unanticipated client travel. Kristi's (Junior manager, F) comments about a recent project illustrate the demands that ensued:

> On a recent technology project, the partners were very busy. They would get a document at 10 a.m. and not look at it until 10 p.m. Then, at 11 p.m., I'd have to work on it and get the team online to do the work so they could turn it around for the next day. I ended up working more with the team on the nuts and bolts than I was supposed to. But it was all so last minute.

To satisfy these work demands, one had to be committed to and available for work. Indeed, partners acknowledged that the structure of work demanded a certain sort of person: "Occasionally my teams have to work overnight, you know, around the clock. . . . Some people thrive on 'It's a gold medal game,' and others don't. And I think this job requires that you thrive on 'It's a gold medal game.' You know, it uses every bit of you" (Partner, M).

Performance evaluations served as a second mechanism of identity control. Assessing competence and work quality is difficult in professional service work, and firms consequently may evaluate people based on perceptions of their identities (Alvesson 2001, Rivera 2012). Each year, partners and HR leaders sorted consultants into four performance tiers based on their project performance and "extracurricular" firm service (e.g., recruiting).[1] The highest tier (4) was denoted AGM's "stars." Many acknowledged this evaluation system's subjectivity; an HR document described it as a "highly individualized. . . highly subjective process." AGM officially assessed performance along multiple dimensions, including relational and analytic skills, but members of its HR department stressed the importance of availability and commitment and described these attributes in terms of an expected identity. Keith (Partner (HR), M), the leader of the HR department, described successful consultants in the following terms:

> Consulting is a profession where we hold beliefs regarding what it takes to be a good consultant. . . . Look at Melissa. We hire her because she's willing to be over responsible, highly committed, and we fall into the trap of thinking everyone is always available all the time. . . . I have person A and person B. Re person B, they don't seem that passionate, responsible, committed, [willing to] go the extra mile; if I ask them to do something, they huff around and it feels like work to get it done. Person A, I ask to do something, it gets done immediately; if I have a problem I can call them, and the next day they've taken a crack at it and with a smile on their face. We will use that in appraisal and recruiting.

It is notable that Keith's description of "what it takes to be good consultant" centered on commitment and availability—"highly committed" and "always available all the time"—not expert knowledge and skills.

Because of its effects on apportioning bonuses, recommending promotions, and counseling people out of the firm, the evaluation system, together with the structure of work, was key to how the firm controlled who succeeded and who failed. These control mechanisms loomed large in consultants' minds; they drew on their beliefs about what AGM rewarded, as well as the structure of their work, to argue that one had to conform to the expected identity to succeed, as illustrated in the following quotations:

> To be viewed as successful, you have to take conference calls at 9 p.m. on Sunday evenings. You have to answer your BlackBerry or your emails the second you receive them. You have to put everything on the line for the client and for the partners. And sort of hand over the keys and head down, elbows out. (Junior manager, F)

> The system is incentivized to reward people for a certain set of behaviors. . . . Surprise: the people who have a new family, a new kid, and want to spend time with them may have less time to devote to their job and may not rise as fast as the people who are more single-mindedly devoted to advancing. (Junior manager, M)

Thus, taken together, the structure of work and the performance evaluation system pressured consultants to adopt the expected professional identity.

Congruence or Conflict With People's Experienced Professional Identities

Nearly all consultants were aware of this expected professional identity, but whether they embraced or strayed from it varied according to its fit with their experienced professional identities: the professionals they believed and preferred themselves to be. I first briefly describe those whose experienced identities were congruent, then turn to those whose identities were conflicting.

Many people's experienced professional identities were congruent with the expected professional identity, and they easily embraced this identity (35 consultants, 43% of the sample). They were primarily committed to their work, speaking frequently of their "passion" for their work and "what we're trying to do in the world." Many described being offered good jobs elsewhere but choosing to stay at AGM. Indeed, one year following the interviews, only three of these consultants had left, one of whom was sponsored for an MBA and later rehired into a higher position. They were also fully available: most regularly worked late nights and weekends, more than 70 hours a week, and willingly traveled at a client's "whim." Dave (Senior manager, M) told me, "You know what? At the end of the day, I want to work hard. I like working hard. I want to be successful. I want to make a lot of money. It's important to me. I rationalize it as, you know, trying to provide for my family. So I don't mind so much if I'm at work at 9 p.m."

Table 2 Congruence or Conflict with the Expected Professional Identity

	Total (N)	*Congruence*	*Conflict*
Men	64	27 (42%)	37 (58%)
Women	18	8 (44%)	10 (56%)
Parents	41	17 (41%)	24 (59%)
Non-parents	41	18 (44%)	23 (56%)
Married	55	21 (38%)	34 (62%)
Single	27	14 (52%)	13 (48%)
Total	82	35 (43%)	47 (57%)

Most people (of the sample), however, encountered conflict between the expected professional identity and their experienced professional identities. As noted, scholars typically identify the ideal worker image as chiefly problematic for women, especially mothers, but at AGM, conflict with the expected professional identity was not restricted to these groups. Rather, as shown in Table 2, most people reported conflict with this expected identity.

These people were unwilling to make work their primary life commitment, unwilling to make themselves fully available for their work, or both. Further, their experienced identities centered primarily on attributes that AGM treated as peripheral, which they often directly compared to those attributes considered core to the expected identity. For example, Michael (Junior manager, M) described himself in the following way:

> I've made sure I'm the problem solver. Everything, I mean, even my hobbies usually involve some sort of problem solving. I mean, I enjoy the intellectual part of the job, I enjoy the challenges. . . . But you know, a lot of times our partners can be focused on really needing to delight the client, and so we can never say no to them. . . .

These people's experienced identities thus conflicted with AGM's expected professional identity. Such a conflict is illustrated in Thomas's (Senior manager, M) musings about his future at AGM:

> I am kind of at a crossroads about how much I want to push for partner. I kind of want to do it on my terms, as opposed to assume I have to be like some of the other partners. . . . There's definitely the road warrior model, the guy who's always on the road, who's always walking the halls with clients, he's sending emails on Saturday and Sunday, you know, and he's sending out requests at 6 p.m. expecting something the next day. And I don't want that. . . . I might be more of an outlier than a main stream consultant.

Thus, because of his unwillingness to be a "road warrior," Thomas viewed himself as an "outlier" and was uncertain about his next career steps.

The pressures that organizations' demands for devotion place on people's non work lives are well established (e.g., Kreiner et al. 2009, Perlow 1998), and indeed, many viewed embracing the expected identity as detrimental to their ability to engage meaningfully in their non work lives. However, the data from AGM also show that people's non work lives provoked conflict over their professional identities. To illustrate, Cliff (Junior manager, M) told me,

> [I'm] someone that doesn't work as hard as I should, is a little quicker to say "this is good enough" and pass it along than my peers are. . . . I think that might, if I don't change it soon, [affect] my ability to be really, really successful here. . . . The decision for me is, [do] I get into bed and watch some TV with my fiancée, or sit down and have dinner with her as opposed to wolfing it down and going back to work? I always choose not to work. I think that it makes me a little less likely to be CEO [chief executive officer] of this place one day.

Thus, by not working constantly and maintaining non-work commitments, Cliff perceived himself as not fitting the identity expected of him ("a little quicker to . . . pass it along than my peers") and believed this might limit his success at the firm ("less likely to be CEO").

Straying: Passing and Revealing

People coped with this conflict by straying from the expected identity. They did so by altering the structure of their work—a key means of identity control in this setting. By altering aspects of their work (e.g., client types, client location), people constructed opportunities to remain true to their experienced professional identities. Unlike those who embraced, these people reported working about 60 hours per week or less, having predictable work schedules, and having regular engagement in other aspects of life. For example, Colin (Partner, M) told me, "I work until 5:30 or 6. I go home. I have dinner with my family. I put the kids to bed. Then I'll probably work an hour or two after that if I need to, or if I want to." Most limited weekend work to exceptional circumstances; several minimized travel, and for these people, work did not normally trump other life commitments. Thus, they were both less committed to their work, and less available for it, than the expected identity demanded.

Although some who altered their jobs were penalized, others seemed to pass as having embraced the expected identity. My data show that these differences in how people were perceived and treated originated from information they shared as a result of how they altered their work—personally or asking for help—as well as the information they shared with others. I now elaborate on how the use of different tools enabled people to stray while passing or revealing.

Tools for Straying

Personally Cultivating Necessary Work Conditions: Passing. Some people personally altered the structure of their work in ways that constructed space to enact their experienced selves, thereby straying from the expected identity. People described cultivating local, repeat, or nonprofit clients who required less time and commitment than more typical clients. Some found ways to work on internal firm projects, which reduced travel time and also had more predictable demands. Others worked from home, reducing travel time and creating space for other aspects of life. These efforts bear resemblance to "job crafting": altering the aspects of one's job in ways that reshape work identities (Wrzesniewski and Dutton 2001). However, my findings go further, showing that these efforts to alter the structure of work also permitted people to avoid disclosing their desire to stray from the expected identity and allowed them to pass as having embraced it.

For example, Lloyd (Senior manager, M) viewed himself as an "odd duck" and did not embrace the work devotion he saw in his colleagues ("I'm going to misquote *The Matrix* here, but I feel like the problem is choice . . . the perception of autonomous choice is what makes it palatable. People are more willing to work harder because its perceived to be their choice . . ."). Lloyd strayed from the expected identity: "I skied five days last week. I took calls in the morning and in the evening but I was able to be there for my son when he needed me to be, and I was able to ski five days in a row." He clarified that these were work days, not vacation days: "No, no one knows where I am. . . . Those boundaries are only practical with my local client base. . . . Especially because we're mobile, there are no boundaries." Thus, by using local clients and telecommuting, Lloyd altered the structure of his work in ways that allowed him to stray from the expected professional identity. His statement that "no one knows where I am" indicates that he believed others were unaware of his deceit. Indeed, despite his deviance, senior colleagues viewed him as an incumbent of the expected identity. Cameron (Partner, M), for example, labeled Lloyd a "rising star" who worked "much harder than" he did. This assessment—in combination with Lloyd's star performance rating of 4 and his promotion to partner that year—suggests he had successfully passed in the eyes of senior members of the firm.

Asking for Help in Restructuring Work: Revealing. By contrast, those who requested AGM's help to restructure their work, through informal alterations such as local clients or more formal accommodations such as parental leave, thereby revealed their deviance and were penalized. Doug (Junior manager, M) recounted how he had lost a promotion because, following months in the Middle East, he had requested a U.S.-based project:

> I told the firm, you know, I don't think I can go back to the Middle East again. And if that means I'm going to have to look for something else, I'm going to look for something else. And that was kind of what resulted in the nonpromotion, because they said, "Well, you'll probably get it if you stay out there." . . . Because I'm a brown guy it's easy to think that the Middle East is no big hurdle for me . . . They said, "Well, it's easier for you, you know. You don't drink already." They don't drink in the country I was working in. I said, "Listen, drinking and not drinking is not the hardest thing . . . It's about being away from your family for that long. Right?"

Doug's story later arose during an interview I conducted with Barry (Senior manager, M), who had also worked in the Middle East. Barry told me, "Doug's wife didn't want him to do it, but he did it anyway and that was a much different experience for him . . . . He stayed for about five months and then came back and refused to go back again." Barry identified working in the Middle East as an opportunity that had signaled his personal commitment to AGM and had enabled a recent promotion. Thus, the man who went to the Middle East happily was promoted; the man who publicly cut his stay short because of his non work commitments, thereby revealing his deviance, was denied a promotion.

Accessing formal accommodations also revealed deviance. For example, Michael told me,

> When my daughter was born, one of the things I wanted to do was take off three months and do the full FMLA[2] and be a stay-at-home. Dad. . . . I felt like this was the only time in my career I would be able to do this . . . But the original reaction I actually got inside of AGM was, "Oh no, you can't take three months off."

He settled for six weeks of unpaid leave and worked 80-hour weeks, travelling weekly, for the rest of the year. Yet he found that "people still talked like I was out three months." At his annual review he was told that AGM could not properly evaluate him because the six weeks he had taken off meant he "had this big do nut hole in [his] year." That year, his performance rating fell from a 3 to a 2, and he did not receive a hoped-for promotion. Thus, Michael' deviance was both recognized and penalized. In a subsequent conversation, he reflected, "No one questioned my commitment until I had a family."

Hiding or Sharing Personal Information Passing and Revealing. The personal information that people hid or shared, such as details about how one worked or about how one felt about one's work, also affected whether they passed or revealed. Some deliberately misrepresented themselves as having embraced the expected identity. For example, one afternoon, Venkat (Junior manager, M) told me, "Every one inflates their hours. I would guess I work 50–60 hours a week but would tell others 60. . . . Right now it's about 40, on this particular client." The next morning, I met Robert (Junior manager, M), who had recently begun working with Venkat. Robert, reflecting on his own work ethic, commented, "I could work every night, every weekend, way over deliver, make new work for myself, [but] I'm more laid back than other people on projects Last week when I worked with Venkat, he was a thousand times better than me." He later confirmed that he meant Venkat worked longer hours, suggesting that Venkat had successfully passed to him as fully available.

Others, however, revealed their deviance by telling colleagues about their struggles with AGM's identity expectations. Philippa (Junior manager, F), who found AGM's work structure "difficult for someone like me who's very operational, very structured, [who likes to] have a good plan about where we're going and have flexibility," said that she had disclosed her deviance to colleagues. "I have been very, very open about the fact that I'm unsatisfied. . . ."

As these examples show, how people altered the structure of their work, in tandem with how they controlled their personal information, enabled them to stray from the expected identity while also shaping whether they passed or revealed in their interactions with others. People did not, however, pass or reveal exclusively. Next, I expand on how people combined these efforts across different audiences at the firm.

Integrated Identity Management: Combining Passing and Revealing Across Audiences

People managed their identities differently in their interactions with audiences based on four factors: the status of the audience, the closeness of the relationship, perceived access to the firm's formal accommodations, and the extremity of the conflict that people experienced. These efforts at passing and revealing were interdependent: in the examples I present, people are often revealing to some audiences while simultaneously passing to others. In addition, the perceptions of targeted audiences could spill over to shape other audiences' perceptions through three avenues: labeling, construction of opportunities for passing, and a need to continually negotiate accommodations.

Situational Factors Shaping Passing and Revealing Across Audiences

Audience Status

Consultants typically sought to pass with high-status audiences who had clear power over their chances at the firm, consistent with theoretical ideas about how status distance shapes people's management of stigmatized identities (Phillips et al. 2009). Junior consultants typically focused on passing to high-status audiences within AGM (e.g., partners); more senior consultants, who needed strong client relationships and high sales to these clients, focused their efforts on clients. For example, Veronica (Senior manager, F), who worked only an 80% schedule and had thus revealed her deviance within AGM, still attempted to pass to clients as an always available consultant. She explained to me, "I have full-time day care. . . . [I use my day off] to accommodate client things so that it's not really visible to the clients that I work a reduced schedule."

Closeness of Relationship

People sometimes revealed their deviance to close friends. These were typically people at the same hierarchical level in the organization. For example, Chris (Junior manager, M), describing a recent night with two colleagues, told me, "The three of us had like five pitchers and talked for four hours, just running around in a circle, questioning why we can't imagine doing this demanding of a job for long." People also disclosed to close personal mentors. Although mentors typically occupied higher-status positions, their history of providing professional guidance and the friendship that often (but not always) developed in these relationships seemed to encourage people to reveal their deviance to them. Revealing to these close colleagues seemed to function as a release valve for the tensions that people experienced with straying from the expected identity: being known as their true selves by at least some colleagues may have enabled them to continue passing to others. However, this was not the only consequence of revealing: studies of work/non work boundary management have shown the importance of relationships to one's ability to alter work boundaries (Trefalt 2013, Trefalt and Heaphy 2014). My data similarly show that revealing one's deviance to close colleagues and mentors sometimes led to informal fixes to the structure of work that in turn facilitated straying. Amos, describing a mentor who had become "a buddy of mine," said, "When I had trouble, when I raised my hand and said 'This is BS,' at that time I was under resourced and I was working insane hours. I was hitting obstacles. He'd say, 'Alright, let me take care of that.' I'd get a call two hours later, done, gone, everything." Thus, for Amos, like others, revealing to a close mentor permitted immediate alterations to the structure of work.

Perceived Access to Formal Accommodations

People varied in whether they believed they were entitled to formal accommodations (e.g., parental leave, part time schedule), and these beliefs shaped how readily they sought these options. AGM targeted its accommodations to mothers, and mothers who encountered difficulties with the expected identity tended to gravitate toward requesting formal accommodations. Although some needs, such as maternity leaves, could only be solved formally, other, more chronic issues with the expected identity could possibly be handled through informal accommodations (e.g., personally cultivating non profit clients). Mothers tended simply to seek the organization's help ahead of exploring other, informal means of restructuring their work. For example, Veronica told me,

> I have two kids, so I took two pretty long leaves . . . And then from then on, I've been working an 80% schedule pretty much consistently. . . . And certainly my preference after having my kids was just to be able not to travel. So it's mostly worked out. . . . My preference is accommodated by AGM so far. . . . It's kind of a combination of serendipity and my preference slash AGM being willing to accommodate that preference. . . . Theoretically, [I] would become a partner in four years. . . . I'm assuming it would be a little longer trajectory because I only work four days a week.

Although AGM "accommodated" Veronica's preferences through an 80% schedule and little travel, as her comments regarding her trajectory suggest, use of these tools clearly revealed her deviance. Like Veronica, other mothers gravitated toward official alterations to their schedule. Other people at AGM, however, faced resistance if they requested formal accommodations, or they believed these accommodations were simply not available to them. For example, although AGM was legally required to offer parental leave to fathers, Doug told me that after his son was born, "I was off for a week. There's no paternity leave policy here. But you kind of go to your current case manager and say, Look, I'm going to be off this week. And, they're like, okay. Just pick up the mobile if you get a call from my cell."

Situation-Specific Conflict. Although accommodations and formal alterations of work were typically viewed as accessible to mothers but difficult for others to obtain at AGM, people other than mothers did sometimes seek the firm's help in restructuring their work. I found that people typically did so in order to solve situation-specific problems that resulted from sudden collisions between work demands (e.g., working on an excessively demanding project) and events in their personal lives (e.g., illness of a family member).In such situations, people sought formal accommodations or other sanctioned modifications of work practices (e.g., local client assignments), or they simply told senior colleagues about their problems. Kate (Junior manager, F), following an illness brought on by work stress, began openly questioning and resisting pressures to always "over deliver." In doing so, she outed herself to the partners managing the project:

> One of the partners called me a whiner. He said, "Why are you always whining about this and that?" And I said, "Ok, I don't really understand why you're making us do all this work. The case is already going well. Yeah, we could do all this additional work to over deliver, but at what cost, right?" . . . Literally three people left the case, and two of them have left AGM since. So, not good.

In addition to being labeled a "whiner," Kate was poorly evaluated for her work on this case—each an indication that she had revealed her deviance.

Spread of Perceptions Across Audiences

Thus, people managed their deviance differently across audiences, passing to some, revealing to others. These efforts to pass or to reveal in relation to specific audiences often spilled over to shape other audiences' perceptions.

Passing to High Status Audiences Facilitates Passing More Broadly. Passing to high-status audiences seemed to facilitate passing to equal- or lower-status colleagues as well. For example, Alex (Junior manager, M) worked fewer than 60 hours a week and never travelled overnight, which he managed by focusing on repeat clients and a local industry:

> I've managed to be the junior manager for several cases on one account, which is great. . . . The account happens to be in Connecticut. So I manage it so I go there for day trips, but I almost never spend the night away from home . . . I try to head out by 5 o'clock, get home at 5:30, have dinner, [and] play with my daughter . . . [On weekends,] I try to limit it to, you know, two hours at most, really just catching up on emails.

Alex targeted his efforts to pass at clients: "I know what clients are expecting. So I deliver above that, but I deliver only above that to impress them, not to know that I did . . . everything I could for a particular case." Although Alex thus targeted his efforts at clients, he also passed more generally within AGM: equal-status colleagues viewed him as a star, he received a star performance rating (4) that year, and he had been promoted relatively quickly. Such spillovers in perceptions likely occurred in part because the largely invisible ways that people altered their work to pass to high-status audiences also avoided revealing their deviance more generally. In addition, however, being labeled a star performer by particular, high-status audiences seemed to create a powerful halo effect, such that other audiences also assumed the person was a star. For example, Bill (Senior manager, M) told me,

> My ability here to ascend this hierarchy rapidly is partly about my own abilities and so forth, but it's also partly about the connection that exists between me and my kind of advocates, and the chief advocate is the guy who runs my group. So is he going to value me in the same way as another person who has been flagged by the firm as a star? Probably not.

Once one had been labeled a star, this label was, as one person told me, sticky. Indeed, perceptions that someone had embraced the expected identity could persist even when evidence was presented of the person's deviance. Caroline (Partner, F) said,

> The women say they look up and see women like me and don't want to live my life—they think I work more than I do. If I am client-facing and commercially successful, I must be working all the time. And then they get emails from me at 8 at night and Sunday 5 a.m. What they don't know is that I have taken a half a day off to go on my son's field trip, so I do the work when I fit it in. I try to tell them, but still feel there are misperceptions.

Thus, although Caroline tried to unravel junior colleagues' assumptions, "misperceptions" persisted. Indeed, junior consultants' assumptions about their managers' work habits often seemed more grounded in their managers' reputations than in their actual behavior. For example, Jimmy (Associate, M) assumed that his manager, who was known as a star, worked in ways consistent with the expected identity: "I don't know [how much she worked] because she was never in the office. But it was my impression, I'm sure this is right, that she was working a lot." When pressed, however, he was able to offer no evidence of her work hours aside from this "impression." Thus, the strength of consultants' assumptions that success required embracing the expected identity, passing to the firm's senior partners, and being marked by them as successful enabled passing to the broader audience of the firm.

Revealing to Close Colleagues Facilitates Passing More Broadly

People's choices to reveal to close colleagues tended to result in informal fixes to their work structure that, because of their informality, enabled them to stray from the expected identity while passing to the broader audience in AGM, including high-status audiences. Some, like Wesley (Partner, M), were aware of this spillover effect:

> We kind of have a shared agreement as to what work—life balance is on our team. We basically work really closely with each other to make sure that we can all do that. A lot of us have young kids, and we've designed it so we can do that. We've really designed the whole business[unit] around having intellectual freedom, making a lot of money, [and] having work—life balance. It's pretty rare. And we don't get push back from above because we are squaring that circle—from the managing partners—'cause we are one of the most successful parts of the company. Most of the partners have no idea our hours are that light.

Thus, Wesley acknowledges that he and his colleagues revealed their deviance to each other ("shared agreement"). He identifies the target of their passing behavior as AGM's two managing partners. But as a result, in his account, a broader audience—most of AGM's "partners"—was in fact unaware that people in his unit strayed from the expected identity.

Revealing to High-Status Audiences Entails Revealing More Broadly

Revealing to high-status audiences tended to result in revealing to the broader audience of the firm. This occurred in part because of the visibility of the accommodations people received and the complexity of negotiating them: an extended leave, or an internal assignment, often required negotiations with multiple people over several weeks. In addition, however, formal accommodations typically required ongoing negotiations with clients, teams, and partners that drew continued attention to the person's deviance over time. The following quotations illustrate this dynamic:

> It's hard to stay on the line, doing client service, working part time. You're kind of all in or you're not. We set that expectation for clients. If you're working part time, you'll pay for it. If you're working three days, four days, you will be asked, "Can you really not come in on that day off?" People are wondering, are they in the game or not? (Senior manager, M)

> I worry that [those who go part-time] are getting paid 60% but end up doing 100%. But it's up to the individual to manage this. Some partners are understanding and will remember that someone is 60%, and some will not. So it's up to the individual to "remind" the manager. . . . All in all, it's not good. (Senior manager, F)

Consequences of Passing and Revealing

By managing their identities differently across audiences, people found ways to stray from AGM's expected identity such that they mostly passed in their interactions with senior members of the firm or mostly revealed their deviance to these people. Although, as previously noted, conflict with the expected identity was not restricted to any particular demographic group, men and women seemed to cope with conflict in different ways. Namely, women who strayed from the expected identity were unlikely to engage identity management strategies that enabled passing to senior members of the firm; rather, most (80% of those who strayed) ultimately revealed their deviance to senior members of the firm. The strategies of men who strayed, by contrast, seemed more evenly split between passing (54% of those who strayed) and revealing (45% of those who strayed). The reasons for these differences are likely complex; however, my analyses suggest that one important reason may be that mothers were targeted by AGM's formal accommodation policies and thus tended to gravitate toward these policies. Men, however, were not targeted and instead tended to experiment with informal strategies for straying.

How people were perceived by senior members of the firm in turn influenced the performance evaluation system, a key mechanism through which AGM controlled consultants' identities. As many of the examples I have shown suggest, at AGM, both those who embraced the expected identity and those who successfully passed to senior members of the firm were typically labeled successes and rewarded, whereas those who revealed to senior members were recognized as deviant and penalized. In what follows, I draw on performance and promotion data to further support these assertions.

External Perceptions and the Performance Evaluation System

Embracing Celebrated Successes

The 35 people (42% of men, 44.5% of women) who embraced AGM's expected identity were typically regarded as among AGM's top consultants, described as stars and "superheroes" by their colleagues. They typically received high performance ratings relative to their colleagues (mean rating of 3.0 in 2009 and 3.14 in 2010) (see Table 3).Most reported straightforward career paths, with few stories of disappointments. Three of the 35 were promoted the year after the study, though 2 did not receive hoped for promotions. Partners often occupied internal leadership positions, further signs that they were perceived as having embraced the expected identity.

Passing: Celebrated Successes

The 22 people (31% of men, 11% of women) who strayed yet managed the identities in ways that promoted passing to senior members of the firm were typically perceived as embracing AGM's expected identity and were favorably regarded and highly rewarded. Like those who embraced, others described them in superlative terms, e.g., "stars" and "top senior men." Echoing these perceptions, their performance rankings were slightly higher than those who embraced the expected identity (mean rating of 3.08 in 2009 and 3.13 in 2010)[3] and significantly better than those who revealed their deviance to senior members of the firm. They enjoyed straightforward, even accelerated advancement; one was described by a colleague as "by far the fastest person I've ever seen make partner here." Three were promoted in 2010; none reported being denied a promotion. Some of those who were partners occupied leadership roles within their groups. Thus, AGM did not appear to distinguish between those who embraced and those who passed. In this way, they evaded the performance evaluation system, a key mechanism of identity control.

Revealing: Penalized Deviance

By contrast, the 25 people (27% of men, 44.5% of women) who revealed their deviance to senior

Table 3 Performance Data

	2009		2010	
	N	*Mean (SD)*	*N*	*Mean (SD)*
Strategy				
Embracing (E)	22	3.0 (0.62)	21	3.14 (0.47)
Passing (P)	12	3.08 (0.67)	8	3.13 (0.78)
Revealing (R)	20	2.45 (0.69)	14	2.85 (0.66)
Total	54[a]		43[b]	
Kruskal—Wallis test statistics[c]				
(df), with ties	8.65* (2)		1.5 (3)	
Mann—Whitney test statistics (z)				
E vs P	–0.37		0.06	
E vs R	2.56*		1.31	
P vs R	2.53*		0.78	

[a] Total N is 54, not 60, because a few participants declined to release their data or were unreachable.
[b] Total N is lower in 2010 because of departures from the firm and one promotion to partner.
[c] Grouping variable: Type.

members of the firm were largely recognized as deviant and penalized accordingly. Their performance ratings were significantly lower than those of other consultants (mean rating of 2.45 in 2009 and 2.85 in 2010). When I interviewed them, just one had been recently promoted, and seven reported not receiving anticipated promotions. They complained of being persistently placed on difficult projects with demanding clients, and they had slow career trajectories, both indicators that they were not highly valued by AGM. Eight of the 25 left within a year for other jobs, the highest turnover rate of the sample. A few senior partners revealed their deviance by significantly reducing travel and working far less, without apparent penalty. They may have, after years of embracing the expected identity, accrued enough "idiosyncrasy credits" to openly stray without penalty(Hollander 1958). Overall, however, most who revealed their deviance to senior members of the firm were penalized.

Discussion

I set out to understand how people cope with organizational expectations that they embrace a professional identity that centers on the ideal worker image in light of their experienced professional identities. In the firm I studied, most workers—not simply women and not simply those with families—encountered conflict between these identities, and they responded by straying from the expected identity. I found that this deviance did not in itself beget penalties: rather, some people strayed while still passing as having embraced the expected identity. Moreover, although men and women both experienced conflict, they managed their deviance differently: men tended to pass, whereas women revealed.

The analyses suggest a conceptual model of how people navigate conflict between expected and experienced professional identities people who experienced conflict coped by engaging tools that permitted straying from the expected identity. People's use of these tools to pass or to reveal were shaped by situational factors, and efforts to manage one audience's perceptions sometimes spilled over to shape other audiences' perceptions. Together, people's efforts at passing and revealing across different audiences coalesced to shape the perceptions of senior members of the firm, influencing the performance evaluation system, such that those who passed were highly evaluated and rewarded, whereas those who revealed were penalized. Overall, my findings suggest that people's management of conflict between an expected professional identity and their experienced professional identity is best understood as a layered process involving passing and revealing across audiences. Together, the findings deepen our understanding of and suggest fruitful new directions for scholarship on how men and women can navigate ideal worker images and expected professional identities; they also enrich our understanding of passing and revealing in organizational contexts.

Contributions to Theory

A contribution of this study is to show that the gender inequalities typically associated with the ideal worker image may arise principally from systematic differences in how men and women cope with conflict with this expected identity, rather than from differences in who embraces it. As noted, this image has historically been identified as mostly problematic for women, particularly mothers. Conversely, at AGM, these expectations were experienced as problematic by most workers: men as well as women, parents and non parents, married and single people. Men and women coped with this conflict differently, however: fewer women than men passed; rather, they tended to reveal their deviance. At AGM, an important reason for this divergence seemed to be that its HR accommodations were targeted at mothers, who were consequently more likely to take advantage of these accommodations, which revealed their deviance. Men, not expected to take HR accommodations, instead experimented with less formal, under-the-radar ways of straying from the expected identity.

However, access to accommodations is unlikely to be the only reason why women coped differently than men, and further analysis of gender differences in coping strategies, and the organizational and cultural factors shaping them, would be useful to understand how the ideal worker image contributes to work place inequality. For example, some of the tools for passing required coordination with colleagues or clients; as women typically have different workplace networks than men (Ibarra 1997), they may have been relatively less able to access these tools. Another possible reason is that professional identities are often associated with particular social identities (Ashcraft 2013, Clair et al. 2012, Ramarajan and Reid 2013); in this setting, most consultants were men. Women might have been more focused on managing their status as women in a male dominated role than on finding opportunities to pass. Racial minorities might face similar challenges, as they typically have different workplace networks than their white colleagues (Ibarra 1995) and, like women, often face stereotypes regarding their suitability for a particular job (e.g., Rosette et al. 2008). Overall, for scholars interested in the role of the ideal worker image in inequality, my findings suggest broadening the analytical lens to include all workers' experiences and moving beyond *examining who experiences* conflict to focus on *how* people manage this conflict, and the resources available to them to do so.

Practical Insights

This research also offers important lessons for practice. Society still tends to assume that primarily women, and mainly mothers, experience difficulties with devoting themselves wholly to work. This study shows that problems with demands for work devotion are neither only a mother's issue nor only a women's issue: rather, this conflict is experienced by most workers. It is particularly striking that so many people in this firm experienced this conflict, as AGM, like the consulting industry more generally, was well known to be demanding: people accepted this job with some knowledge of its demands. That so many still experienced conflict with the expected identity underscores a troublesome mismatch between people's preferences and organizations' expectations. The widespread nature of this conflict both heightens the importance for organizations to assess the need for demands for work devotion and suggests that solutions should be targeted at all workers, not simply women.

Conclusion

Overall, this study underscores the continued salience of demands to be an ideal worker in professional work settings and the complex ways these demands shape men and women's work experiences. As the need to pass or reveal is typically associated with highly stigmatized social identities, the fact that many privileged workers who strayed from the expected identity still felt the need to pass is both surprising and speaks to the power of the ideal worker image in defining success in this setting. Yet the very fact that people passed demonstrates that the association between total devotion and success maybe as much a matter of perception as reality.

Notes

1. Partners were excluded from this assessment: their performance was assumed to fall between 3 and 4, and underperforming partners were asked to leave.

2. FMLA stands for the Federal Medical Leave Act of 1993.

3. As noted, men were likely to pass than to reveal, and women were more likely to reveal than to pass. Women often receive poorer evaluations than men in male-type jobs and are held to higher standards for promotion (Lyness and Heilman 2006). To examine whether such differences in men and women's performance evaluations drove the observed difference between the scores of those who passed and those who revealed. I reran the performance data with only men's performance scores. This analysis revealed the same pattern of results and significant differences between people who embraced, passed, and revealed.

References

Acker J (1990) Hierarchies, jobs, bodies: A theory of gendered organizations. *Gender Soc.* 4(2):139–158.

Agresti A, Finlay B (1997) *Statistical Methods for the Social Sciences,* 3rd ed. (Prentice-Hall, Upper Saddle River, NJ).

Alvesson M (2001) Knowledge work: Ambiguity, image and identity. *Human Relations* 54(7):863–886.

Alvesson M, Kärreman D (2007) Unraveling HRM: Identity, ceremony, and control in a management consulting firm. *Organ. Sci.*18(4):711–723.

Alvesson M, Willmott H (2002) Identity regulation as organizational control: Producing the appropriate individual. J. *Management Stud.* 39(5):619–644.

Anteby M (2008a) Identity incentives as an engaging form of control: Revisiting leniencies in an aeronautic plant. *Organ. Sci.*19(2):202–220.

Anteby M (2008b) *Moral Gray Zones: Side Productions, Identity and Regulation in an Aeronautic Plant* (Princeton University Press, Princeton, NJ).

Anteby M (2013) Relaxing the taboo on telling our own stories: Upholding professional distance and personal involvement. *Organ. Sci.* 24(4):1277–1290.

Ashcraft KL (2005) Resistance through consent? Occupational identity, organizational form, and the maintenance of masculinity among commercial airline pilots. *Management Comm. Quart.* 19(1):67–90.

Ashcraft KL (2013) The glass slipper: "Incorporating" occupational identity in management studies. *Acad. Management Rev.* 38(1):6–31.

Ashforth BE (2001) *Role Transitions in Organizational Life: An Identity-Based Perspective* (Lawrence Erlbaum Associates, Mahwah, NJ).

Ashforth BE, Harrison SH, Corley KG (2008) Identification in organizations: An examination of four fundamental questions. *J. Management* 34 (3):325–374.

Bailyn L (2006) Breaking *the Mold: Redesigning Work for Productive and Satisfying Lives* (Cornell University Press, Ithaca, NY).

Barbulescu R, Bidwell M (2013) Do women choose different jobs from men? Mechanisms of application segregation in the market for managerial workers. *Organ. Sci.* 24(3):737–756.

Beatty JE, Joffe R (2006) An overlooked dimension of diversity: The career effects of chronic illness. *Organ. Dynam.* 35(2):182–195.

Becker HS, Carper JW (1956) The development of identification with an occupation. *Amer. J. Sociol.* 61(4):289–298.

Bennett RJ, Robinson SL (2003) The past, present, and future of workplace deviance research. Green berg J, ed. *Organizational Behavior: The State of the Science* (Lawrence Erlbaum Associates, Mahwah, NJ), 247–281.

Blair-Loy M (2003) *Competing Devotions: Career and Family Among Women Executives* (Harvard University Press, Cambridge, MA).

Cech EA, Blair-Loy M (2014) Consequences of flexibility stigma among academic scientists and engineers. *Work Occupations* 41(1):86–110.

Charmaz K (2006) *Constructing Grounded Theory: A Practical Guide Through Qualitative Analysis* (Sage, Thousand Oaks, CA).

Christopher K (2012) Extensive mothering: Employed mothers' constructions of the good mother. *Gender Soc.* 26(1):73–96.

Clair JA, Beatty JE, Maclean TL (2005) Out of sight but not out of mind: Managing invisible social identities in the workplace. *Acad. Management Rev.* 30(1):78–95.

Clair JA, Hundred BK, Caruso HM, Roberts LM (2012) Marginal memberships: Psychological effects of identity ambiguity on professionals who are demographically different from the majority. *Organ. Psych. Rev.* 2(1):71–93.

Collinson DL (2003) Identities and insecurities: Selves at work. *Organization* 10(3):527–547.

Correll SJ, Kelly EL, O'Connor LT, Williams JC (2014) Redesigning, redefining work. *Work Occupations* 41(1):3–17.

Creed WED, Scully M (2000) Songs of ourselves: Employees' deployment of social identity in workplace encounters. J. *Management Inquiry* 9(4):391–412.

De Jordy R (2008) Just passing through: Stigma, passing, and identity decoupling in the workplace. *Group Organ. Management* 33(5):504–531.

Eisenhardt KM (1989) Building theories from case study research. *Acad. Management Rev.* 14(4):532–550.

Ely R, Padavic I (2007) A feminist analysis of organizational research on sex differences. *Acad. Management Rev.* 32(4):1121–1143.

Ely RJ (1995) The power in demography: Women's social constructions of gender identity at work. *Acad. Management J.* 38(3):589–634.

Ely RJ, Meyer son DE (2010) An organizational approach to undoing gender: The unlikely case of offshore oil platforms. *Res. Organ. Behav.* 30:3–34.

Gagnon S, Collinson DL (2014) Rethinking global leadership development programmes: The interrelated significance of power, context and identity. *Organ. Stud.* 35(5):645–670.

Galinsky E, Aumann K, Bond JT (2009) Times are changing: Gender and generation at work and at home. 2008 National Study of the Changing Workforce, Families and Work Institute, New York.

Gecas V (1982) The self-concept. *Annual Rev. Sociol.* 8:1–33.

Goffman E (1963) Stigma: *Notes on the Management of Spoiled Identity* (Simon & Schuster, New York).

Goffman A (2014) On the Run: *Fugitive Life in an American City* (University of Chicago Press, Chicago).

Hall DT (1976) *Careers in Organizations* (Goodyear, Pacific Palisades, CA).

Hewlin PF (2003) And the award for best actor goes to . . . . Facades of conformity in organizational settings. *Acad. Management Rev.* 28(4):633–642.

Hollander EP (1958) Conformity, status, and idiosyncrasy credit. *Psych. Rev.* 65(2):117–127.

Humberd BK, Ladge J, Harrington B (2014) The "new" dad: Navigating fathering identity within organizational contexts. *J. Bus. Psych.*, e Pub ahead of print May 22, http://dx.doi.org/10.1007/s10869-014-9361-x.

Ibarra H (1995) Race, opportunity and diversity of social circles in managerial networks. *Acad. Management J.* 38(3):673–703.

Ibarra H (1997) Paving an alternative route: Gender differences in managerial networks. *Soc. Psych. Quart.* 60(1):91–102.

Ibarra H (1999) Provisional selves: Experimenting with image and identity in professional adaptation. *Admin. Sci. Quart.* 44(4):764–791.

Ibarra H, Barbulescu R (2010) Identity as narrative: Prevalence, effectiveness, and consequences of narrative identity work in macro work role transitions. *Acad. Management Rev.* 35(1):135–154.

Jackall R (1988) *Moral Mazes: The World of Corporate Managers* (Oxford University Press, New York).

Jones KP, King EB (2014) Managing concealable stigmas at work: A review and multilevel model. *J. Management* 40(5):1466–1494.

Kellogg KC (2011) *Challenging Operations: Medical Reform and Resistance in Surgery* (University of Chicago Press, Chicago).

Kreiner GE, Hollens be EC, Sheep ML (2006) Where is the "me" among the "we"? Identity work and the search for optimal balance. *Acad. Management J.* 49(5):1031–1057.

Kreiner GE, Hollensbe EC, Sheep ML (2009) Balancing borders and bridges: Negotiating the work-home interface via boundary work tactics. *Acad. Management J.* 52(4):704–730.

Kunda G (1992) Engineering *Culture: Control and Commitment in a High-Tech Corporation* (Temple University Press, Philadelphia).

Ladge JJ, Clair JA, Green berg D (2012) Cross-domain identity transition during liminal periods: Constructing multiple selves as "professional and mother" during pregnancy. *Acad. Management J.*55(6):1449–1471.

Lamont M, Swidler A (2014) Methodological pluralism and the possibilities and limits of interviewing. *Qualitative Sociol.* 37(2):153–171.

Lepisto DA, Crosina E, Pratt MG (2015) Identity work within and beyond the professions: Toward a theoretical integration and extension. Desilva A, Aparicio M, eds. *International Hand book About Professional Identities. Forthcoming.*

Leslie LM, Manchester CF (2011) Work—family conflict is a social issue, not a women's issue. *Indust. Organ. Psych.* 4(3):414–417.

Lyness KS, Heilman ME (2006) When fit is fundamental: Performance evaluations and promotions of upper-level female and male managers. J. *Appl. Psych.* 91(4):777–785.

Maitlis S (2009) *Who Am I Now? Sense making and Identity in Post traumatic Growth* (Psychology Press, New York).

Manchester CF, Leslie LM, Kramer A (2013) Is the clock still ticking? An evaluation of the consequences of stopping the tenure clock. *ILR Rev.* 66(1):3–31.

Markus H, Nurius P (1986) Possible selves. *Amer. Psychologist* 41(9):954–969.

Michel A (2011) Transcending socialization: A nine-year ethnography of the body's role in organizational control and knowledge workers' transformation. *Admin. Sci. Quart.* 56(3):325–368.

Moen P, Lam J, Ammons SK, Kelly EL (2013) Time work by overworked professionals: Strategies in response to the stress of higher status. *Work Occupations* 40(2):79–114.

Obodaru O (2012) The self not taken: How alternative selves develop and how they influence our professional lives. *Acad. Management Rev.* 37(1):34–57.

Ollier-Malaterre A, Roth bard NP, Berg J (2013) When worlds collide in cyberspace: How boundary work in online social networks impacts professional relationships. *Acad. Management Rev.* 38(4):645–669.

Perlow LA (1998) Boundary control: The social ordering of work and family time in a high-tech corporation. *Admin. Sci. Quart.*43(2):328–357.

Petriglieri G, Petriglieri JL (2010) Identity work spaces: The case of business schools. Acad. Management Learn. Ed. 9(1):44–60.

Phillips KW, Rothbard NP, Dumas TL (2009) To disclose or not to disclose? Status distance and self-disclosure in diverse environments. *Acad. Management Rev.* 34(4):710–732.

Pratt MG (2000) The good, the bad, and ambivalent: Managing identification among Amway distributors. *Admin. Sci. Quart.*45(3):456–493.

Pratt MG, Rockmann KW, Kaufmann JB (2006) Constructing professional identity: The role of work and identity learning cycles in the customization of identity among medical residents. *Acad. Management J.* 49(2):235–262.

Ragins BR (2008) Disclosure disconnects: Antecedents and consequences of disclosing invisible stigmas across life domains. *Acad. Management Rev.* 33(1):194–215.

Ragins BR, Singh R, Cornwell JM (2006) Making the invisible visible: Fear and disclosure of sexual orientation at work. J. *Appl. Psych.* 92(4):1103–1118.

Ramarajan L (2014) Past, present and future research on multiple identities: Toward an intra personal network approach. *Acad. Management Ann.* 8(1):589–659.

Ramarajan L, Reid EM (2013) Shattering the myth of separate worlds: Negotiating non-work identities at work. *Acad. Management Rev.* 38(4):621–644.

Rivera L (2012) Hiring as cultural matching: The case of elite professional service firms. *Amer. Sociol. Rev.* 77(6):999–1022.

Roberts LM (2005) Changing faces: Professional image construction in diverse organizational settings. *Acad. Management Rev.* 30(4):685–711.

Roberts LM, Dutton JE, eds. (2009) *Exploring Positive Identities and Organizations: Building a Theoretical and Research Foundation* (Psychology Press, New York).

Rosette AS, Dumas TL (2007) The hair dilemma: Conform to main stream expectations or emphasize racial identity. *Duke J. Gender Law Policy* 14(1):407–422.

Rosette AS, Leonardelli G, Phillips K (2008) The white standard: Racial bias in leader categorization. *J. Appl. Psych.* 93(4):758–777.

Roth LM (2006) *Selling Women Short: Gender and Money on Wall Street* (Princeton University Press, Princeton, NJ).

Rothbard NP, Phillips KW, Dumas TL (2005) Managing multiple roles: Work-family policies and individuals' desires for segmentation. *Organ. Sci.* 16(3):243–258.

Stone P (2007) Opting Out? Why *Women Really Quit Careers and Head Home* (University of California Press, Berkeley).

Trefalt S (2013) Between you and me: Setting work-non work boundaries in the context of workplace relationships. *Acad. Management J.* 56(6):1775–1801.

Trefalt S, Heaphy E (2014) Process and management expertise: The relational construction of temporal flexibility. *Acad. Management Proc.* 1(Meeting abstract supplement): Abstract 11125.

Turco CJ (2010) Cultural foundations of tokenism: Evidence from the leveraged buyout industry. *Amer. Sociol. Rev.* 75(6):894–913.

Uhlmann EL, Heaphy E, Ashford SJ, Zhu L, Sanchez-Burks J(2013) Acting professional: An exploration of culturally bounded norms against non-work role referencing. *J. Organ. Behav.* 34(6):866–886.

Van Maanen J, Barley SR (1984) Occupational communities: Culture and control in organizations. *Res. Organ. Behav.* 6:287–365.

Van Maanen J, Schein EH (1979) Toward a theory of organizational socialization. *Res. Organ. Behavior.* 1:209–264.

Warren DE (2003) Constructive and destructive deviance in organizations. *Acad. Management Rev.* 28(4): 622–632.

Wayne SJ, Ferris GR (1990) Influence tactics, affect, and exchange equality in supervisor/subordinate interactions: A laboratory experiment and field study. J. *Appl. Psych.* 75(5):487–499.

Webber G, Williams C (2008) Mothers in "good" and "bad" parttime jobs: Different problems, same results. *Gender Soc.* 22(6):752–777.

Williams JC (2000) *Unbending Gender: Why Family and Work Conflict and What to Do About It* (Oxford University Press, New York).

Williams JC, Blair-Loy M, Berdahl JL (2013) Cultural schemas, social class, and the flexibility stigma. *J. Soc. Issues* 69(2):209–234.

Wrzesniewski A, Dutton JE (2001) Crafting a job: Revisioning employees as active crafters of their work. *Acad. Management Rev.* 26(2):179–201.

Yoshino K (2007) *Covering: The Hidden Assault on Our Civil Rights* (Random House, New York).

Introduction to Reading 36

In this reading, Kristen Schilt studies a unique population in order to shed light on the underpinnings of gendered workplace disparities. She interviews a sample of female-to-male (FTM) transsexuals about changes in their workplace interactions and experiences from when they worked as women to when they worked as men, as this "dual" experience has the potential to provide them with an "outsider-within" perspective about gender inequalities at work. Her findings illustrate how structural disadvantages for women are reproduced in workplace interactions, disadvantages that cannot be traced back to individual abilities or skills.

1. What is the "outsider-within" perspective? Why were some transmen not able to develop this perspective? How does being an "outsider within" make workplace disparities visible?
2. In Schilt's study, what were the workplace advantages that FTM workers experienced? Why did some FTM workers not receive gender advantages after transition?
3. What do Schilt's findings suggest about human capital theory as an explanation for gender inequality at work?

Just One of the Guys?

How Transmen Make Gender Visible at Work

Kristen Schilt

Theories of gendered organizations argue that cultural beliefs about gender difference embedded in workplace structures and interactions create and reproduce workplace disparities that disadvantage women and advantage men (Acker 1990; Martin 2003; Williams 1995). As Martin (2003) argues, however, the practices that reproduce gender difference and gender inequality at work are hard to observe. As these gendered practices are citations of established gender norms, men and women in the workplace repeatedly and unreflectively engage in "doing gender" and therefore "doing inequality" (Martin 2003; West and Zimmerman 1987). This repetition of well-worn gender ideologies naturalizes workplace gender inequality, making gendered disparities in achievements appear to be offshoots of "natural" differences between men and women, rather than the products of dynamic gendering and gendered practices (Martin 2003). As the active reproduction of gendered workplace disparities is rendered invisible, gender inequality at work becomes difficult to document empirically and therefore remains resistant to change (Acker 1990; Martin 2003; Williams 1995).

The workplace experiences of female-to-male transsexuals (FTMs), or transmen, offer an opportunity to examine these disparities between men and women at work from a new perspective. Many FTMs enter the workforce as women and, after transition, begin working as men.[1] As men, they have the same skills, education, and abilities they had as women; however, how this "human capital" is perceived often varies drastically once they become men at work. This shift in gender attribution gives them the potential to develop an "outsider-within" perspective (Collins 1986) on men's advantages in the workplace. FTMs can find themselves benefiting from the "patriarchal dividend" (Connell 1995, 79)—the advantages men in general gain from the subordination of women—after they transition. However, not being "born into it"

Schilt, K. (2006). Just one of the guys?: How transmen make gender visible at work. *Gender & Society, 20*(4): 465–490.

gives them the potential to be cognizant of being awarded respect, authority, and prestige they did not have working as women. In addition, the experiences of transmen who fall outside of the hegemonic construction of masculinity, such as FTMs of color, short FTMs, and young FTMs, illuminate how the interplay of gender, race, age, and bodily characteristics can constrain access to gendered workplace advantages for some men (Connell 1995).

In this article, I document the workplace experiences of two groups of FTMs, those who openly transition and remain in the same jobs (open FTMs) and those who find new jobs posttransition as "just men" (stealth FTMs).[2] I argue that the positive and negative changes they experience when they become men can illuminate how gender discrimination and gender advantage are created and maintained through workplace interactions. These experiences also illustrate that masculinity is not a fixed character type that automatically commands privilege but rather that the relationships between competing hegemonic and marginalized masculinities give men differing abilities to access gendered workplace advantages (Connell 1995).

Theories of Workplace Gender Discrimination

Sociological research on the workplace reveals a complex relationship between the gender of an employee and that employee's opportunities for advancement in both authority and pay. While white-collar men and women with equal qualifications can begin their careers in similar positions in the workplace, men tend to advance faster, creating a gendered promotion gap (Padavic and Reskin 2002; Valian 1999). When women are able to advance, they often find themselves barred from attaining access to the highest echelons of the company by the invisible barrier of the "glass ceiling" (Valian 1999). Even in the so-called women's professions, such as nursing and teaching, men outpace women in advancement to positions of authority (Williams 1995). Similar patterns exist among blue-collar professions, as women often are denied sufficient training for advancement in manual trades, passed over for promotion, or subjected to extreme forms of sexual, racial, and gender harassment that result in women's attrition (Byrd 1999; Miller 1997; Yoder and Aniakudo 1997). These studies are part of the large body of scholarly research on gender and work finding that white-and blue-collar workplaces are characterized by gender segregation, with women concentrated in lower-paying jobs with little room for advancement.

Among the theories proposed to account for these workplace disparities between men and women are human capital theory and gender role socialization. Human capital theory posits that labor markets are neutral environments that reward workers for their skills, experience, and productivity. As women workers are more likely to take time off from work for child rearing and family obligations, they end up with less education and work experience than men. Following this logic, gender segregation in the workplace stems from these discrepancies in skills and experience between men and women, not from gender discrimination. However, while these differences can explain some of the disparities in salaries and rank between women and men, they fail to explain why women and men with comparable prestigious degrees and work experience still end up in different places, with women trailing behind men in advancement (Valian 1999; Williams 1995).

A second theory, gender socialization theory, looks at the process by which individuals come to learn, through the family, peers, schools, and the media, what behavior is appropriate and inappropriate for their gender. From this standpoint, women seek out jobs that reinforce "feminine" traits such as caring and nurturing. This would explain the predominance of women in helping professions such as nursing and teaching. As women are socialized to put family obligations first, women workers would also be expected to be concentrated in part-time jobs that allow more flexibility for family schedules but bring in less money. Men, on the other hand, would be expected to seek higher-paying jobs with more authority to reinforce their sense of masculinity. While gender socialization theory may explain some aspects of gender segregation at work, however, it leaves out important structural aspects of the workplace that support segregation, such as the lack of workplace child care services, as well as employers' own gendered stereotypes about which workers are best suited for which types of jobs (Padavic and Reskin 2002; Valian 1999; Williams 1995).

A third theory, gendered organization theory, argues that what is missing from both human capital theory and gender socialization theory is the way in which men's advantages in the workplace are maintained and reproduced in gender expectations that are embedded in organizations and in interactions between employers, employees, and coworkers (Acker 1990;

Martin 2003; Williams 1995). However, it is difficult to study this process of reproduction empirically for several reasons. First, while men and women with similar education and workplace backgrounds can be compared to demonstrate the disparities in where they end up in their careers, it could be argued that differences in achievement between them can be attributed to personal characteristics of the workers rather than to systematic gender discrimination. Second, gendered expectations about which types of jobs women and men are suited for are strengthened by existing occupational segregation; the fact that there are more women nurses and more men doctors comes to be seen as proof that women are better suited for helping professions and men for rational professions. The normalization of these disparities as natural differences obscures the actual operation of men's advantages and therefore makes it hard to document them empirically. Finally, men's advantages in the workplace are not a function of simply one process but rather a complex interplay between many factors, such as gender differences in workplace performance evaluation, gendered beliefs about men's and women's skills and abilities, and differences between family and child care obligations of men and women workers.

The cultural reproduction of these interactional practices that create and maintain gendered workplace disparities often can be rendered more visible, and therefore more able to be challenged, when examined through the perspective of marginalized others (Collins 1986; Martin 1994, 2003; Yoder and Aniakudo 1997). As Yoder and Aniakudo note, "marginalized others offer a unique perspective on the events occurring within a setting because they perceive activities from the vantages of both nearness (being within) and detachment (being outsiders)" (1997, 325–26). This importance of drawing on the experiences of marginalized others derives from Patricia Hill Collins's theoretical development of the "outsider-within" (1986, 1990). Looking historically at the experience of Black women, Collins (1986) argues that they often have become insiders to white society by virtue of being forced, first by slavery and later by racially bounded labor markets, into domestic work for white families. The insider status that results from being immersed in the daily lives of white families carries the ability to demystify power relations by making evident how white society relies on racism and sexism, rather than superior ability or intellect, to gain advantage; however, Black women are not able to become total insiders due to being visibly marked as different. Being a marginalized insider creates a unique perspective, what Collins calls "the outsider-within," that allows them to see "the contradictions between the dominant group's actions and ideologies" (Collins 1990, 12), thus giving a new angle on how the processes of oppression operate. Applying this perspective to the workplace, scholars have documented the production and reproduction of gendered and racialized workplace disparities through the "outsider-within" perspective of Black women police officers (Martin 1994) and Black women firefighters (Yoder and Aniakudo 1997).

In this article, I posit that FTMs' change in gender attribution, from women to men, can provide them with an outsider-within perspective on gendered workplace disparities. Unlike the Black women discussed by Collins, FTMs usually are not visibly marked by their outsider status, as continued use of testosterone typically allows for the development of a masculine social identity indistinguishable from "bio men."[3] However, while both stealth and open FTMs can become social insiders at work, their experience working as women prior to transition means they maintain an internalized sense of being outsiders to the gender schemas that advantage men. This internalized insider/outsider position allows some transmen to see clearly the advantages associated with being men at work while still maintaining a critical view to how this advantage operates and is reproduced and how it disadvantages women. I demonstrate that many of the respondents find themselves receiving more authority, respect, and reward when they gain social identities as men, even though their human capital does not change. This shift in treatment suggests that gender inequality in the workplace is not continually reproduced only because women make different education and workplace choices than men but rather because coworkers and employers often rely on gender stereotypes to evaluate men's and women's achievements and skills.

Method

I conducted in-depth interviews with 29 FTMs in the Southern California area from 2003 to 2005. My criteria for selection were that respondents were assigned female at birth and were currently living and working as men or open transmen. These selection criteria did exclude female-bodied individuals who identified as men but had had not publicly come out as men at work and FTMs who had not held any jobs as men since their transition, as they would not be able to comment about changes in their social interactions that were specific to the workplace. My sample is made up of 18 open FTMs and 11 stealth FTMs.

At the onset of my research, I was unaware of how I would be received as a non-transgender person doing research on transgender workplace experiences, as well as a woman interviewing men. I went into the study being extremely open about my research agenda and my political affiliations with feminist and transgender politics. I carried my openness about my intentions into my interviews, making clear at the beginning that I was happy to answer questions about my research intentions, the ultimate goal of my research, and personal questions about myself. Through this openness, and the acknowledgment that I was there to learn rather than to be an academic "expert," I feel that I gained a rapport with my respondents that bridged the "outsider/insider" divide (Merton 1972).

Generating a random sample of FTMs is not possible as there is not an even dispersal of FTMs throughout Southern California, nor are there transgender-specific neighborhoods from which to sample. I recruited interviewees from transgender activist groups, transgender listservers, and FTM support groups. In addition, I participated for two years in Southern California transgender community events, such as conferences and support group meetings. Attending these community events gave me an opportunity not only to demonstrate long-term political commitment to the transgender community but also to recruit respondents who might not be affiliated with FTM activist groups. All the interviews were conducted in the respondents' offices, in their homes, or at a local café or restaurant. The interviews ranged from one and a half to four hours. All interviews were audio recorded, transcribed, and coded.

Drawing on sociological research that reports longstanding gender differences between men and women in the workplace (Reskin and Hartmann 1986; Reskin and Roos 1990; Valian 1999; Williams 1995), I constructed my interview schedule to focus on possible differences between working as women and working as men. I first gathered a general employment history and then explored the decision to openly transition or to go stealth. At the end of the interviews, I posed the question, "Do you see any differences between working as a woman and working as a man?" All but a few of the respondents immediately answered yes and began to provide examples of both positive and negative differences. About half of the respondents also, at this time, introduced the idea of male privilege, addressing whether they felt they received a gender advantage from transitioning. If the concept of gender advantage was not brought up by respondents, I later introduced the concept of male privilege and then posed the question, saying, "Do you feel that you have received any male privilege at work?" The resulting answers from these two questions are the framework for this article.

In reporting the demographics of my respondents, I have opted to use pseudonyms and general categories of industry to avoid identifying my respondents. Respondents ranged in age from 20 to 48. Rather than attempting to identify when they began their gender transition, a start date often hard to pinpoint as many FTMs feel they have been personally transitioning since childhood or adolescence, I recorded how many years they had been working as men (meaning they were either hired as men or had openly transitioned from female to male and remained in the same job). The average time of working as a man was seven years. Regarding race and ethnicity, the sample was predominantly white (17), with 3 Asians, 1 African American, 3 Latinos, 3 mixed-race individuals, 1 Armenian American, and 1 Italian American. Responses about sexual identity fell into four main categories, heterosexual (9), bisexual (8), queer (6), and gay (3). The remaining 3 respondents identified their sexual identity as celibate/asexual, "dating women," and pansexual. Finally, in terms of region, the sample included a mixture of FTMs living in urban and suburban areas. (See Table 1 for sample characteristics.)

The experience of my respondents represents a part of the Southern California FTM community from 2003 to 2005. As Rubin (2003) has demonstrated, however, FTM communities vary greatly from city to city, meaning these findings may not be representative of the experiences of transmen in Austin, San Francisco, or Atlanta. In addition, California passed statewide gender identity protection for employees in 2003, meaning that the men in my study live in an environment in which they cannot legally be fired for being transgender (although most of my respondents said they would not wish to be a test case for this new law). This legal protection means that California transmen might have very different workplace experiences than men in states without gender identity protection. Finally, anecdotal evidence suggests that there are a large number of transgender individuals who transition and then sever all ties with the transgender community, something known as being "deep stealth."

Table 1 Sample Characteristics

Pseudonym	*Age*	*Race/Ethnicity*	*Sexual Identity*	*Approximate Number of Years Working as Male*	*Industry*	*Status at Work*
Aaron	28	Black/White	Queer	5	Semi-Professional	Open
Brian	42	White	Bisexual	14	Semi-Professional	Stealth
Carl	34	White	Heterosexual	16	Higher Professional	Stealth
Christopher	25	Asian	Pansexual	3	Semi-Professional	Open
Colin	31	White	Queer	1	Lower Professional	Open
Crispin	42	White	Heterosexual	2	Blue-Collar	Stealth
David	30	White	Bisexual	2	Higher Professional	Open
Douglas	38	White	Gay	5	Semi-Professional	Open
Elliott	20	White	Bisexual	1	Retail/Customer Service	Open
Henry	32	White	Gay	5	Lower Professional	Open
Jack	30	Latino	Queer	1	Semi-Professional	Open
Jake	45	White	Queer	9	Higher Professional	Open
Jason	48	White/Italian	Celibate	20	Retail/Customer Service	Stealth
Keith	42	Black	Heterosexual	1	Blue-Collar	Open
Kelly	24	White	Bisexual	2	Semi-Professional	Open
Ken	26	Asian/White	Queer	6 months	Semi-Professional	Open
Paul	44	White	Heterosexual	2	Semi-Professional	Open
Peter	24	White/Armenian	Heterosexual	4	Lower Professional	Stealth
Preston	39	White	Bisexual	2	Blue-Collar	Open
Riley	37	White	Dates women	1	Lower Professional	Open
Robert	23	Asian	Heterosexual	2	Retail/Customer Service	Stealth
Roger	45	White	Bisexual	22	Lower Professional	Stealth
Sam	33	Latino	Heterosexual	15	Blue-Collar	Stealth
Simon	42	White	Bisexual	2	Semi-Professional	Open
Stephen	35	White	Heterosexual	1	Retail/Customer Service	Stealth
Thomas	42	Latino	Queer	13	Higher Professional	Open
Trevor	35	White	Gay/Queer	6	Semi-Professional	Open
Wayne	44	/Latino	Bisexual	22	Higher Professional	Stealth
Winston	40	White	Heterosexual	14	Higher Professional	Stealth

This lack of connection to the transgender community means they are excluded from research on transmen but that their experiences with the workplace may be very different than those of men who are still connected, even slightly, to the FTM community.

Transmen as Outsiders Within at Work

In undergoing a physical gender transition, transmen move from being socially gendered as women to being socially gendered as men (Dozier 2005). This shift in gender attribution gives them the potential to develop an "outsider-within" perspective (Collins 1986) on the sources of men's advantages in the workplace. In other words, while they may find themselves, as men, benefiting from the "patriarchal dividend" (Connell 1995, 79), not being "born into it" can make visible how gendered workplace disparities are created and maintained through interactions. Many of the respondents note that they can see clearly, once they become "just one of the guys," that men succeed in the workplace at higher rates than women because of gender stereotypes that privilege masculinity, not because they have greater skill or ability. For transmen who do see how these cultural beliefs about gender create gendered workplace disparities, there is an accompanying sense that these experiences are visible to them only because of the unique perspective they gain from undergoing a change in gender attribution. Exemplifying this, Preston reports about his views on gender differences at work posttransition: "I swear they let the guys get away with so much stuff! Lazy ass bastards get away with so much stuff and the women who are working hard, they just get ignored. . . . I am really aware of it. And that is one of the reasons that I feel like I have become much more of a feminist since transition. I am just so aware of the difference that my experience has shown me." Carl makes a similar point, discussing his awareness of blatant gender discrimination at a hardware/home construction store where he worked immediately after his transition: "Girls couldn't get their forklift license or it would take them forever. They wouldn't make as much money. It was so pathetic. I would have never seen it if I was a regular guy. I would have just not seen it. . . . I can see things differently because of my perspective. So in some ways I am a lot like a guy because I transitioned younger but still, you can't take away how I was raised for 18 years." These comments illustrate how the outsider-within perspective of many FTMs can translate into a critical perspective on men's advantages at work. The idea that a "regular guy," here meaning a bio man, would not be able to see how women were passed over in favor of men makes clear that for some FTMs, there is an ability to see how gender stereotypes can advantage men at work.

However, just as being a Black woman does not guarantee the development of a Black feminist perspective (Collins 1986), having this critical perspective on gender discrimination in the workplace is not inherent to the FTM experience. Respondents who had held no jobs prior to transition, who were highly gender ambiguous prior to transition, or who worked in short-term, high-turnover retail jobs, such as food service, found it harder to identify gender differences at work. FTMs who transitioned in their late teens often felt that they did not have enough experience working as women to comment on any possible differences between men and women at work. For example, Sam and Robert felt they could not comment on gender differences in the workplace because they had begun living as men at the age of 15 and, therefore, never had been employed as women. In addition, FTMs who reported being very "in-between" in their gender appearance, such as Wayne and Peter, found it hard to comment on gender differences at work, as even when they were hired as women, they were not always sure how customers and coworkers perceived them. They felt unable to speak about the experience of working as a woman because they were perceived either as androgynous or as men.

The kinds of occupations FTMs held prior to transition also play a role in whether they develop this outsider-within perspective at work. Transmen working in blue-collar jobs—jobs that are predominantly staffed by men—felt their experiences working in these jobs as females varied greatly from their experiences working as men. This held true even for those transmen who worked as females in blue-collar jobs in their early teens, showing that age of transition does not always determine the ability to see gender discrimination at work. FTMs working in the "women's professions" also saw a great shift in their treatment once they began working as men. FTMs who transitioned in their late teens and worked in marginal "teenage" jobs, such as fast food, however, often reported little sense of change posttransition, as they felt that most employees were doing the same jobs regardless of gender. As a gendered division of labor often does exist in fast food jobs (Leidner 1993), it may be that these respondents worked in atypical settings, or that they were assigned "men's jobs" because of their masculine appearance.

Transmen in higher professional jobs, too, reported less change in their experiences posttransition, as many of them felt that their workplaces guarded against gender-biased treatment as part of an ethic of professionalism. The experience of these professional respondents obviously runs counter to the large body of scholarly research that documents gender inequality in fields such as academia (Valian 1999), law firms (Pierce 1995), and corporations (Martin 1992). Not having an outsider-within perspective, then, may be unique to these particular transmen, not the result of working in a professional occupation.

Thus, transitioning from female to male can provide individuals with an outsider within perspective on gender discrimination in the workplace. However, this perspective can be limited by the age of transition, appearance, and type of occupation. In addition, as I will discuss at the end of this article, even when the advantages of the patriarchal dividend are seen clearly, many transmen do not benefit from them. In the next section, I will explore in what ways FTMs who expressed having this outsider-within perspective saw their skills and abilities perceived more positively as men. Then, I will explore why not all of my respondents received a gender advantage from transitioning.

Transition and Workplace Gender Advantages[4]

A large body of evidence shows that the performance of workers is evaluated differently depending on gender. Men, particularly white men, are viewed as more competent than women workers (Olian, Schwab, and Haberfeld 1988; Valian 1999). When men succeed, their success is seen as stemming from their abilities while women's success often is attributed to luck (Valian 1999). Men are rewarded more than women for offering ideas and opinions and for taking on leadership roles in group settings (Butler and Geis 1990; Valian 1999). Based on these findings, it would be expected that stealth transmen would see a positive difference in their workplace experience once they have made the transition from female to male, as they enter new jobs as just one of the guys. Open FTMs, on the other hand, might find themselves denied access to these privileges, as they remain in the same jobs in which they were hired as women. Challenging these expectations, two-thirds of my respondents, both open and stealth, report receiving some type of posttransition advantage at work. These advantages fell into four main categories: gaining competency and authority, gaining respect and recognition for hard work, gaining "body privilege," and gaining economic opportunities and status.

Authority and Competency

Illustrating the authority gap that exists between men and women workers (Elliott and Smith 2004; Padavic and Reskin 2002), several of my interviewees reported receiving more respect for their thoughts and opinions posttransition. For example, Henry, who is stealth in a professional workplace, says of his experiences, "I'm right a lot more now. . . . Even with folks I am out to [as a transsexual], there is a sense that I know what I am talking about." Roger, who openly transitioned in a retail environment in the 1980s, discussed customers' assumptions that as a man, he knew more than his boss, who was a woman: "People would come in and they would go straight to me. They would pass her and go straight to me because obviously, as a male, I knew [sarcasm]. And so we would play mind games with them. . . . They would come up and ask me a question, and then I would go over to her and ask her the same question, she would tell me the answer, and I would go back to the customer and tell the customer the answer." Revealing how entrenched these stereotypes about masculinity and authority are, Roger added that none of the customers ever recognized the sarcasm behind his actions. Demonstrating how white men's opinions are seen to carry more authority, Trevor discusses how, posttransition, his ideas are now taken more seriously in group situations—often to the detriment of his women coworkers: "In a professional workshop or a conference kind of setting, a woman would make a comment or an observation and be overlooked and be dissed essentially. I would raise my hand and make the same point in a way that I am trying to reinforce her and it would be like [directed at me], 'That's an excellent point!' I saw this shit in undergrad. So it is not like this was a surprise to me. But it was disconcerting to have happen to me." These last two quotes exemplify the outsider-within experience: Both men are aware of having more authority simply because of being men, an authority that happens at the expense of women coworkers.

Looking at the issue of authority in the women's professions, Paul, who openly transitioned in the field of secondary education, reports a sense of having increased authority as one of the few men in his work environment:

> I did notice [at] some of the meetings I'm required to attend, like school district or parent involvement [meetings], you have lots of women there. And now I feel like there are [many times], mysteriously enough, when I'm picked [to speak]. . . . I think, well, why me, when nobody else has to go to the microphone and talk about their stuff? That I did notice and that [had] never happened before. I mean there was this meeting . . . a little while ago about domestic violence where I appeared to be the only male person between these 30, 40 women and, of course, then everybody wants to hear from me.

Rather than being alienated by his gender tokenism, as women often are in predominantly male workplaces

(Byrd 1999), he is asked to express his opinions and is valued for being the "male" voice at the meetings, a common situation for men in "women's professions" (Williams 1995). The lack of interest paid to him as a woman in the same job demonstrates how women in predominantly female workspaces can encourage their coworkers who are men to take more authority and space in these careers, a situation that can lead to the promotion of men in women's professions (Williams 1995).

Transmen also report a positive change in the evaluation of their abilities and competencies after transition. Thomas, an attorney, relates an episode in which an attorney who worked for an associated law firm commended his boss for firing Susan, here a pseudonym for his female name, because she was incompetent—adding that the "new guy" [i.e., Thomas] was "just delightful." The attorney did not realize that Susan and "the new guy" were the same person with the same abilities, education, and experience. This anecdote is a glaring example of how men are evaluated as more competent than women even when they do the same job in careers that are stereotyped requiring "masculine" skills such as rationality (Pierce 1995; Valian 1999). Stephen, who is stealth in a predominantly male customer-service job, reports, "For some reason just because [the men I work with] assume I have a dick, [they assume] I am going to get the job done right, where, you know, they have to second-guess that when you're a woman. They look at [women] like well, you can't handle this because you know, you don't have the same mentality that we [men] do, so there's this sense of panic . . . and if you are a guy, it's just like, oh, you can handle it." Keith, who openly transitioned in a male-dominated blue-collar job, reports no longer having to "cuddle after sex," meaning that he has been able to drop the emotional labor of niceness women often have to employ to when giving orders at work. Showing how perceptions of behavior can change with transition, Trevor reports, "I think my ideas are taken more seriously [as a man]. I had good leadership skills leaving college and um . . . I think that those work well for me now. . . . Because I'm male, they work better for me. I was 'assertive' before. Now I'm 'take charge.'" Again, while his behavior has not changed, his shift in gender attribution translates into a different kind of evaluation. As a man, being assertive is consistent with gendered expectations for men, meaning his same leadership skills have more worth in the workplace because of his transition. His experience underscores how women who take on leadership roles are evaluated negatively, particularly if their leadership style is perceived as assertive, while men are rewarded for being aggressive leaders (Butler and Geis 1990; Valian 1999).[5]

This change in authority is noticeable only because FTMs often have experienced the reverse: being thought, on the basis of gender alone, to be less competent workers who receive less authority from employers and coworkers. This sense of a shift in authority and perceived competence was particularly marked for FTMs who had worked in blue-collar occupations as women. These transmen report that the stereotype of women's incompetence often translated into difficulty in finding and maintaining employment. For example, Crispin, who had worked as a female construction worker, reports being written up by supervisors for every small infraction, a practice Yoder and Aniakudo (1997, 330) refer to as "pencil whipping." Crispin recounts, "One time I had a field supervisor confront me about simple things, like not dotting i's and using the wrong color ink. . . . Anything he could do, he was just constantly on me. . . . I ended up just leaving." Paul, who was a female truck driver, recounts, "Like they would tell [me], 'Well we never had a female driver. I don't know if this works out.' Blatantly telling you this. And then [I had] to go, 'Well let's see. Let's give it a chance, give it a try. I'll do this three days for free and you see and if it's not working out, well then that's fine and if it works out, maybe you want to reconsider [not hiring me].'" To prove her competency, she ended up working for free, hoping that she would eventually be hired.

Stephen, who was a female forklift operator, described the resistance women operators faced from men when it came to safety precautions for loading pallets:

> [The men] would spot each other, which meant that they would have two guys that would close down the aisle. . . so that no one could go on that aisle while you know you were up there [with your forklift and load] . . . and they wouldn't spot you if you were a female. If you were a guy . . . they got the red vests and the safety cones out and it's like you know—the only thing they didn't have were those little flash-lights for the jets. It would be like God or somebody responding. I would actually have to go around and gather all the dykes from receiving to come out and help and spot me. And I can't tell you how many times I nearly ran over a kid. It was maddening and it was always because [of] gender.

Thus, respondents described situations of being ignored, passed over, purposefully put in harm's way, and assumed to be incompetent when they were working as women. However, these same individuals, as men, find themselves with more authority and with their ideas, abilities, and attributes evaluated more positively in the workforce.

Respect and Recognition

Related to authority and competency is the issue of how much reward workers get for their workplace contributions. According to the transmen I interviewed, an increase in recognition for hard work was one of the positive changes associated with working as a man. Looking at these stories of gaining reward and respect, Preston, who transitioned openly and remained at his blue-collar job, reports that as a female crew supervisor, she was frequently short staffed and unable to access necessary resources yet expected to still carry out the job competently. However, after his transition, he suddenly found himself receiving all the support and materials he required:

> I was not asked to do anything different [after transition]. But the work I did do was made easier for me. [Before transition] there [were] periods of time when I would be told, "Well, I don't have anyone to send over there with you." We were one or two people short of a crew or the trucks weren't available. Or they would send me people who weren't trained. And it got to the point where it was like, why do I have to fight about this? If you don't want your freight, you don't get your freight. And, I swear it was like from one day to the next of me transitioning [to male], I need this, this is what I want and [snaps his fingers]. I have not had to fight about anything.

He adds about his experience, "The last three [performance] reviews that I have had have been the absolute highest that I have ever had. New management team. Me not doing anything different than I ever had. I even went part-time." This comment shows that even though he openly transitioned and remained in the same job, he ultimately finds himself rewarded for doing less work and having to fight less for getting what he needs to effectively do his job. In addition, as a man, he received more positive reviews for his work, demonstrating how men and women can be evaluated differently when doing the same work.

As with authority and competence, this sense of gaining recognition for hard work was particularly noticeable for transmen who had worked as women in blue-collar occupations in which they were the gender minority. This finding is not unexpected, as women are also more likely to be judged negatively when they are in the minority in the workplace, as their statistical minority status seems to suggest that women are unsuited for the job (Valian 1999). For example, Preston, who had spent time in the ROTC as a female cadet, reported feeling that no matter how hard she worked, her achievements were passed over by her men superiors: "On everything that I did, I was the highest. I was the highest-ranking female during the time I was there. . . . I was the most decorated person in ROTC. I had more ribbons, I had more medals, in ROTC and in school. I didn't get anything for that. There was an award every year called Superior Cadet, and guys got it during the time I was there who didn't do nearly what I did. It was those kinds of things [that got to me]." She entered a blue-collar occupation after ROTC and also felt that her workplace contributions, like designing training programs for the staff, were invisible and went unrewarded.

Talking about gender discrimination he faced as a female construction worker, Crispin reports,

> I worked really hard. . . . I had to find myself not sitting ever and taking breaks or lunches because I felt like I had to work more to show my worth. And though I did do that and I produced typically more than three males put together—and that is really a statistic—what it would come down to a lot of times was, "You're single. You don't have a family." That is what they told me. "I've got guys here who have families." And even though my production quality [was high], and the customer was extremely happy with my work . . . I was passed over lots of times. They said it was because I was single and I didn't have a family and they felt bad because they didn't want Joe Blow to lose his job because he had three kids at home. And because I was intelligent and my qualities were very vast, they said, "You can just go get a job anywhere." Which wasn't always the case? A lot of people were—it was still a boy's world and some people were just like, uh-uh, there aren't going to be any women on my job site. And it would be months . . . before I would find gainful employment again.

While she reports eventually winning over many men who did not want women on the worksite, being female excluded her from workplace social interactions, such as camping trips, designed to strengthen male bonding.

These quotes illustrate the hardships that women working in blue-collar jobs often face at work: being passed over for hiring and promotions in favor of less productive male coworkers, having their hard work go unrecognized, and not being completely accepted.[6] Having this experience of being women in an occupation or industry composed mostly of men can create, then, a heightened appreciation of gaining reward and recognition for job performance as men.

Another form of reward that some transmen report receiving posttransition is a type of bodily respect in the form of being freed from unwanted sexual advances or inquiries about sexuality. As Brian recounts about his experience of working as a waitress, that customer service involved "having my boobs grabbed, being called 'honey' and 'babe.'" He noted that as a man, he no longer has to worry about these types of experiences. Jason reported being constantly harassed by men bosses for sexual favors in the past. He added, "When I transitioned . . . it was like a relief! [laughs] . . . I swear to God! I am not saying I was beautiful or sexy but I was always attracting something." He felt that becoming a man meant more personal space and less sexual harassment. Finally, Stephen and Henry reported being "obvious dykes," here meaning visibly masculine women, and added that in blue-collar jobs, they encountered sexualized comments, as well as invasive personal questions about sexuality, from men uncomfortable with their gender presentation, experiences they no longer face posttransition. Transitioning for stealth FTMs can bring with it physical autonomy and respect, as men workers, in general, encounter less touching, groping, and sexualized comments at work than women. Open FTMs, however, are not as able to access this type of privilege, as coworkers often ask invasive questions about their genitals and sexual practices.

Economic Gains

As the last two sections have shown, FTMs can find themselves gaining in authority, respect, and reward in the workplace posttransition. Several FTMs who are stealth also reported a sense that transition had brought with it economic opportunities that would not have been available to them as women, particularly as masculine women.

Carl, who owns his own company, asserts that he could not have followed the same career trajectory if he had not transitioned:

> I have this company that I built, and I have people following me; they trust me, they believe in me, they respect me. There is no way I could have done that as a woman. And I will tell you that as just a fact. That when it comes to business and work, higher levels of management, it is different being a man. I have been on both sides [as a man and a woman], younger obviously, but I will tell you, man, I could have never done what I did [as a female]. You can take the same personality and it wouldn't have happened. I would have never made it.

While he acknowledges that women can be and are business entrepreneurs, he has a sense that his business partners would not have taken his business venture idea seriously if he were a woman or that he might not have had access to the type of social networks that made his business venture possible. Henry feels that he would not have reached the same level in his professional job if he were a woman because he had a nonnormative gender appearance:

> If I was a gender normative woman, probably. But no, as an obvious dyke, I don't think so . . . which is weird to say but I think it's true. It is interesting because I am really aware of having this job that I would not have had if I hadn't transitioned. And [gender expression] was always an issue for me. I wanted to go to law school but I couldn't do it. I couldn't wear the skirts and things females have to wear to practice law. I wouldn't dress in that drag. And so it was very clear that there was a limit to where I was going to go professionally because I was not willing to dress that part. Now I can dress the part and it's not an issue. It's not putting on drag; it's not an issue. I don't love putting on a tie, but I can do it. So this world is open to me that would not have been before just because of clothes. But very little has changed in some ways. I look very different but I still have all the same skills and all the same general thought processes. That is intense for me to consider.

As this response shows, Henry is aware that as an "obvious dyke," meaning here a masculine-appearing woman, he would have the same skills and education level he currently has, but those skills would be devalued due to his nonnormative appearance. Thus, he avoided professional careers that would require a traditionally feminine appearance. As a man, however, he is able to wear clothes similar to those he wore as an "obvious dyke," but they are now considered gender appropriate. Thus, through transitioning, he gains the right to wear men's clothes, which helps him in accessing a professional job.

Wayne also recounts negative workplace experiences in the years prior to his transition due to being extremely ambiguous or "gender blending" (Devor 1987) in his appearance. Working at a restaurant in his early teens, he had the following experience:

> The woman who hired me said, "I will hire you only on the condition that you don't ever come in the front because you make the people uncomfortable." 'Cause we had to wear like these uniforms or something and when I would put the uniform on, she would say, "That makes you look like a guy." But she knew I was not a guy because of my name that she had on the application. She said, "You make the customers uncomfortable." And a couple of times it got really busy, and I would have to come in the front or whatever, and I remember one time

> she found out about it and she said, "I don't care how busy it gets, you don't get to come up front." She said I'd make people lose their appetite.

Once he began hormones and gained a social identity as a man, he found that his work and school experiences became much more positive. He went on to earn a doctoral degree and become a successful professional, an economic opportunity he did not think would be available had he remained highly gender ambiguous.

In my sample, the transmen who openly transitioned faced a different situation in terms of economic gains. While there is an "urban legend" that FTMs immediately are awarded some kind of "male privilege" post-transition (Dozier 2005), I did not find that in my interviews. Reflecting this common belief, however, Trevor and Jake both recount that women colleagues told them, when learning of their transition plans, that they would probably be promoted because they were becoming white men. While both men discounted these comments, both were promoted relatively soon after their transitions. Rather than seeing this as evidence of male privilege, both respondents felt that their promotions were related to their job performance, which, to make clear, is not a point I am questioning. Yet these promotions show that while these two men are not benefiting undeservedly from transition, they also are not disadvantaged.[7] Thus, among the men I interviewed, it is common for both stealth and open FTMs to find their abilities and skills more valued posttransition, showing that human capital can be valued differently depending on the gender of the employee.

Is It Privilege or Something Else?

While these reported increases in competency and authority make visible the "gender schemas" (Valian 1999) that often underlie the evaluation of workers, it is possible that the increases in authority might have a spurious connection to gender transitions. Some transmen enter a different work field after transition, so the observed change might be in the type of occupation they enter rather than a gender-based change. In addition, many transmen seek graduate or postgraduate degrees posttransition, and higher education degrees afford more authority in the workplace. As Table 2 shows, of the transmen I interviewed, many had higher degrees working as men than they did when they worked as women. For some, this is due to transitioning while in college and thus attaining their bachelor's degrees as men. For others, gender transitions seem to be accompanied by a desire to return to school for a higher degree, as evidenced by the increase in master's degrees in the table.

A change in educational attainment does contribute to getting better jobs with increased authority, as men benefit more from increased human capital in the form of educational attainment (Valian 1999). But again, this is an additive effect, as higher education results in greater advantages for men than for women. In addition, gender advantage alone also is apparent in these experiences of increased authority, as transmen report seeing an increase in others' perceptions of their competency outside of the workplace where their education level is unknown. For example, Henry, who found he was "right a lot more" at work, also notes that in daily, nonworkplace interactions, he is assumed, as a man, to know what he is talking about and does not

Table 2 Highest Level of Education Attained

	Stealth FTMs		*Open FTMs*	
Highest Degree Level	*As Female*	*As Male*	*As Female*	*As Male*
High school/GED	7	2	3	2
Associate's degree	2	3	3	3
Bachelor's degree	2	4	7	4
Master's degree	0	1	2	4
Ph.D.	0	1	1	2
J.D.	0	0	1	2
Other	0	0	1	1
Total	11	11	18	18

Note: FTM = female-to-male transsexuals.

have to provide evidence to support his opinions. Demonstrating a similar experience, Crispin, who had many years of experience working in construction as a woman, relates the following story:

> I used to jump into [situations as a woman]. Like at Home Depot, I would hear . . . [men] be so confused, and I would just step over there and say, "Sir, I work in construction and if you don't mind me helping you." And they would be like, "Yeah, yeah, yeah" [i.e., dismissive]. But now I go [as a man] and I've got men and women asking me things and saying, "Thank you so much," like now I have a brain in my head! And I like that a lot because it was just kind of like, "Yeah, whatever." It's really nice.

His experience at Home Depot shows that as a man, he is rewarded for displaying the same knowledge about construction—knowledge gendered as masculine—that he was sanctioned for offering when he was perceived as a woman. As a further example of this increased authority outside of the workplace, several FTMs report a difference in their treatment at the auto shop, as they are not assumed to be easy targets for unnecessary services (though this comes with an added expectation that they will know a great deal about cars). While some transmen report that their "feminine knowledge," such as how to size baby clothes in stores, is discounted when they gain social identities as men, this new recognition of "masculine knowledge" seems to command more social authority than prior feminine knowledge in many cases. These stories show that some transmen gain authority both in and out of the workplace. These findings lend credence to the argument that men can gain a gender advantage, in the form of authority, reward, and respect.

Barriers to Workplace Gender Advantages

Having examined the accounts of transmen who feel that they received increased authority, reward, and recognition from becoming men at work, I will now discuss some of the limitations to accessing workplace gender advantages. About one-third of my sample felt that they did not receive any gender advantage from transition. FTMs who had only recently begun transition or who had transitioned without using hormones ("no ho") all reported seeing little change in their workplace treatment. This group of respondents felt that they were still seen as women by most of their coworkers, evidenced by continual slippage into feminine pronouns, and thus were not treated in accordance with other men in the workplace. Other transmen in this group felt they lacked authority because they were young or looked extremely young after transition. This youthful appearance often is an effect of the beginning stages of transition. FTMs usually begin to pass as men before they start taking testosterone. Successful passing is done via appearance cues, such as hairstyles, clothes, and mannerisms. However, without facial hair or visible stubble, FTMs often are taken to be young boys, a mistake that intensifies with the onset of hormone therapy and the development of peach fuzz that marks the beginning of facial hair growth. Reflecting on how this youthful appearance, which can last several years depending on the effects of hormone therapy, affected his work experience immediately after transition, Thomas reports, "I went from looking 30 to looking 13. People thought I was a new lawyer so I would get treated like I didn't know what was going on." Other FTMs recount being asked if they were interns, or if they were visiting a parent at their workplace, all comments that underscore a lack of authority. This lack of authority associated with looking youthful, however, is a time-bounded effect, as most FTMs on hormones eventually "age into" their male appearance, suggesting that many of these transmen may have the ability to access some gender advantages at some point in their careers.

Body structure was another characteristic some FTMs felt limited their access to increased authority and prestige at work. While testosterone creates an appearance indistinguishable from bio men for many transmen, it does not increase height. Being more than 6 feet tall is part of the cultural construction for successful, hegemonic masculinity. However, several men I interviewed were between 5' 1" and 5' 5", something they felt put them at a disadvantage in relation to other men in their workplaces. Winston, who managed a professional work staff who knew him only as a man, felt that his authority was harder to establish at work because he was short. Being smaller than all of his male employees meant that he was always being looked down on, even when giving orders. Kelly, who worked in special education, felt his height affected the jobs he was assigned: "Some of the boys, especially if they are really aggressive, they do much better with males that are bigger than they are. So I work with the little kids because I am short. I don't get as good of results if I work with [older kids]; a lot of times they are taller than I am." Being a short man, he felt it was harder to establish authority with older boys. These experiences demonstrate the importance of bringing the body back into discussions of masculinity and gender advantage, as being short can constrain men's benefits from the "patriarchal dividend" (Connell 1995).

In addition to height, race/ethnicity can negatively affect FTMs' workplace experiences posttransition. My data suggest that the experiences of FTMs of color is markedly different than that of their white counterparts, as they are becoming not just men but Black men, Latino men, or Asian men, categories that carry their own stereotypes. Christopher felt that he was denied any gender advantage at work not only because he was shorter than all of his men colleagues but also because he was viewed as passive, a stereotype of Asian men (Espiritu 1997). "To the wide world of America, I look like a passive Asian guy. That is what they think when they see me. Oh Asian? Oh passive. . . . People have this impression that Asian guys aren't macho and therefore they aren't really male. Or they are not as male as [a white guy]." Keith articulated how his social interactions changed with his change in gender attribution in this way: "I went from being an obnoxious Black woman to a scary Black man." He felt that he has to be careful expressing anger and frustration at work (and outside of work) because now that he is a Black man, his anger is viewed as more threatening by whites. Reflecting stereotypes that conflate African Americans with criminals, he also notes that in his law enforcement classes, he was continually asked to play the suspect in training exercises. Aaron, one of the only racial minorities at his workplace, also felt that looking like a Black man negatively affected his workplace interactions. He told stories about supervisors repeatedly telling him he was threatening. When he expressed frustration during a staff meeting about a new policy, he was written up for rolling his eyes in an "aggressive" manner. The choice of words such as "threatening" and "aggressive," words often used to describe Black men (Ferguson 2000), suggests that racial identity and stereotypes about Black men were playing a role in his workplace treatment. Examining how race/ethnicity and appearance intersect with gender, then, illustrates that masculinity is not a fixed construct that automatically generated privilege (Connell 1995), but that white, tall men often see greater returns from the patriarchal dividend than short men, young men and men of color.

Conclusion

Sociological studies have documented that the workplace is not a gender-neutral site that equitably rewards workers based on their individual merits (Acker 1990; Martin 2003; Valian 1999; Williams 1995); rather "it is a central site for the creation and reproduction of gender differences and gender inequality" (Williams 1995, 15). Men receive greater workplace advantages than women because of cultural beliefs that associate masculinity with authority, prestige, and instrumentality (Martin 2003; Padavic and Reskin 2002; Rhode 1997; Williams 1995)—characteristics often used to describe ideal "leaders" and "managers" (Valian 1999). Stereotypes about femininity as expressive and emotional, on the other hand, disadvantage women, as they are assumed to be less capable and less likely to succeed than men with equal (or often lesser) qualifications (Valian 1999). These cultural beliefs about gender difference are embedded in workplace structures and interactions, as workers and employers bring gender stereotypes with them to the workplace and, in turn, use these stereotypes to make decisions about hiring, promotions, and rewards (Acker 1990; Martin 2003; Williams 1995). This cultural reproduction of gendered workplace disparities is difficult to disrupt, however, as it operates on the level of ideology and thus is rendered invisible (Martin 2003; Valian 1999; Williams 1995).

In this article, I have suggested that the "outsider-within" (Collins 1986) perspective of many FTMs can offer a more complex understanding of these invisible interactional processes that help maintain gendered workplace disparities. Transmen are in the unique position of having been socially gendered as both women and men (Dozier 2005). Their workplace experiences, then, can make the underpinnings of gender discrimination visible, as well as illuminate the sources of men's workplace advantages. When FTMs undergo a change in gender attribution, their workplace treatment often varies greatly—even when they continue to interact with coworkers who knew them previously as women. Some posttransition FTMs, both stealth and open, find that their coworkers, employers, and customers attribute more authority, respect, and prestige to them. Their experiences make glaringly visible the process through which gender inequality is actively created in informal workplace interactions. These informal workplace interactions, in turn, produce and reproduce structural disadvantages for women, such as the glass ceiling (Valian 1999), and structural advantages for men, such as the glass escalator (Williams 1995).

However, as I have suggested, not all of my respondents gain authority and prestige with transition. FTMs who are white and tall received far more benefits posttransition than short FTMs or FTMs of color. This demonstrates that while hegemonic masculinity is defined against femininity, it is also measured against subordinated forms of masculinity (Connell 1995; Messner 1997). These findings demonstrate the need for using an intersectional approach that takes into consideration the ways in which there are crosscutting relations of power (Calasanti and Slevin 2001; Collins 1990; Crenshaw 1989), as advantage in the workplace is not equally accessible for all men. Further research on FTMs of color can help develop a clearer understanding of the role race plays in the distribution of gendered workplace rewards and advantages.[8]

The experiences of this small group of transmen offer a challenge to rationalizations of workplace inequality. The study provides counterevidence for human capital theories: FTMs who find themselves receiving the benefits associated with being men at work have the same skills and abilities they had as women workers. These skills and abilities, however, are suddenly viewed more positively due to this change in gender attribution. FTMs who may have been labeled "bossy" as women become "go-getting" men who seem more qualified for managerial positions. While FTMs may not benefit at equal levels to bio men, many of them do find themselves receiving an advantage to women in the workplace they did not have prior to transition. This study also challenges gender socialization theories that account for inequality in the workplace. Although all of my respondents were subjected to gender socialization as girls, this background did not impede their success as men. Instead, by undergoing a change in gender attribution, transmen can find that the same behavior, attitudes, or abilities they had as females bring them more reward as men. This shift in treatment suggests that gender inequality in the workplace is not continually reproduced only because women make different education and workplace choices than men but rather because coworkers and employers often rely on gender stereotypes to evaluate men and women's achievements and skills.

It could be argued that because FTMs must overcome so many barriers and obstacles to finally gain a male social identity, they might be likely to overreport positive experiences as a way to shore up their right to be a man. However, I have reasons to doubt that my respondents exaggerated the benefits of being men. Transmen who did find themselves receiving a workplace advantage posttransition were aware that this new conceptualization of their skills and abilities was an arbitrary result of a shift in their gender attribution. This knowledge often undermined their sense of themselves as good workers, making them continually second guess the motivations behind any rewards they receive. In addition, many transmen I interviewed expressed anger and resentment that their increases in authority, respect, and recognition came at the expense of women colleagues. It is important to keep in mind, then, that while many FTMs can identify privileges associated with being men, they often retain a critical eye to how changes in their treatment as men can disadvantage women.

This critical eye, or "outsider-within" (Collins 1986) perspective, has implications for social change in the workplace. For gender equity at work to be achieved, men must take an active role in challenging the subordination of women (Acker 1990; Martin 2003; Rhode 1997; Valian 1999; Williams 1995). However, bio men often cannot see how women are disadvantaged due to their structural privilege (Rhode 1997; Valian 1999). Even when they are aware that men as a group benefit from assumptions about masculinity, men typically still "credit their successes to their competence" (Valian 1999, 284) rather than to gender stereotypes. For many transmen, seeing how they stand to benefit at work to the detriment of women workers creates a sense of increased responsibility to challenge the gender discrimination they can see so clearly. This challenge can take many different forms. For some, it is speaking out when men make derogatory comments about women. For others, it means speaking out about gender discrimination at work or challenging supervisors to promote women who are equally qualified as men. These challenges demonstrate that some transmen are able, at times, to translate their position as social insiders into an educational role, thus working to give women more reward and recognition at these specific work sites. The success of these strategies illustrates that men have the power to challenge workplace gender discrimination and suggests that bio men can learn gender equity strategies from the outsider-within at work.

Notes

1. Throughout this article, I endeavor to use the terms "women" and "men" rather than "male" and "female" to avoid reifying biological categories. It is important to note, though, that while my respondents were all born with female bodies, many of them never identified as women but rather thought of themselves as always men, or as "not women." During their time as female workers, however, they did have social identities as women, as coworkers and employers often were unaware of their personal gender identities. It is this social identity that I am referencing when I refer to them as "working as women," as I am discussing their social interactions in the workplace. In referring to their specific work experiences, however, I use "female" to demonstrate their understanding of their work history. I also do continue to use "female to male" when describing the physical transition process, as this is the most common term employed in the transgender community.

2. I use "stealth," a transgender community term, if the respondent's previous life as female was not known at work. It is important to note that this term is not analogous with "being in the closet," because stealth female-to-male transsexuals (FTMs) do not have "secret" lives as women outside of working as men. It is used to describe two different workplace choices, not offer a value judgment about these choices.

3. "Bio" man is term used by my respondents to mean individuals who are biologically male and live socially as men throughout their lives. It is juxtaposed with "transman" or "FTM."

4. A note on pronoun usage: This article draws from my respondents' experiences working as both women and men. While they now live as men, I use feminine pronouns to refer to their female work histories.

5. This change in how behavior is evaluated can also be negative. Some transmen felt that assertive communication styles they actively fostered to empower themselves as lesbians and feminists had to be unlearned after transition. Because they were suddenly given more space to speak as men, they felt they had to censor themselves or they would be seen as "bossy white men" who talked over women and over women and people of color. These findings are similar to those reported by Dozier (2005).

6. It is important to note that not all FTMs who worked blue-collar jobs as women had this type of experience. One respondent felt that he was able to fit in, as a butch, as "just one of the guys." However, he also did not feel he had an outsider-within perspective because of this experience.

7. Open transitions are not without problems, however. Crispin, a construction worker, found his contract mysteriously not renewed after his announcement. However, he acknowledged that he had many problems with his employers prior to his announcement and had also recently filed a discrimination suit. Aaron, who announced his transition at a small, medical site, left after a few months as he felt that his employer was trying to force him out. He found another job in which he was out as a transman. Crispin unsuccessfully attempted to find work in construction as an out transman. He was later hired, stealth, at a construction job.

8. Sexual identify also is an important aspect of an intersectional analysis. In my study, however, queer and gay transmen worked either in lesbian, gay, bisexual, transgender work sites, or were not out at work. Therefore, it was not possible to examine how being gay or queer affected their work-place experiences.

References

Acker, Joan. 1990. Hierarchies, jobs, bodies: A theory of gendered organizations. *Gender & Society* 4:139–58.

Butler, D., and F. L. Geis. 1990. Nonverbal affect responses to male and female leaders: Implications for leadership evaluation. *Journal of Personality and Social Psychology* 58:48–59.

Byrd, Barbara. 1999. Women in carpentry apprenticeship: A case study. *Labor Studies Journal* 24 (3):3–22.

Calasanti, Toni M., and Kathleen F. Slevin. 2001. *Gender, social inequalities, and aging*. Walnut Creek, CA: Alta Mira Press.

Collins, Patricia Hill. 1986. Learning from the outsider within: The sociological significance of Black feminist thought. *Social Problems* 33 (6): S14–S31.

———. 1990. *Black feminist thought*. New York: Routledge.

Connell, Robert. 1995. *Masculinities*. Berkeley: University of California Press.

Crenshaw, Kimberle. 1989. Demarginalizing the intersection of race and sex: A Black feminist critique of antidiscrimination doctrine, feminist theory, and antiracist politics. *University of Chicago Legal Forum* 1989: 139–67.

Devor, Holly. 1987. Gender blending females: Women and sometimes men. *American Behavioral Scientist* 31 (1): 12–40.

Dozier, Raine. 2005. Beards, breasts, and bodies: Doing sex in a gendered world. *Gender & Society* 19:297–316.

Elliott, James R., and Ryan A. Smith. 2004. Race, gender, and workplace power. *American Sociological Review* 69:365–86.

Espiritu, Yen. 1997. *Asian American women and men*. Thousand Oaks, CA: Sage.

Ferguson, Ann Arnett. 2000. *Bad boys: Public schools in the making of Black masculinity*. Ann Arbor: University of Michigan Press.

Leidner, Robin. 1993. *Fast food, fast talk: Service work and the routinization of everyday life*. Berkeley: University of California Press.

Martin, Patricia Yancy. 1992. *Gender, interaction, and inequality* in organizations. In Gender, interaction, and inequality, edited by Cecelia L. Ridgeway. New York: Springer-Verlag.

———. 2003. "Said and done" versus "saying and doing": Gendering practices, practicing gender at work. *Gender & Society* 17:342–66.

Martin, Susan. 1994. "Outsiders-within" the station house: The impact of race and gender on Black women police officers. *Social Problems* 41:383–400.

Merton, Robert. 1972. Insiders and outsiders: A chapter in the sociology of knowledge. *American Journal of Sociology* 78 (1): 9–47

Messner, Michael. 1997. *The politics of masculinities: Men in movements*. Thousand Oaks, CA: Sage.

Miller, Laura. 1997. Not just weapons of the weak: Gender harassment as a form of protest for army men. *Social Psychology Quarterly* 60 (1): 32–51.

Olian, J. D., D. P. Schwab, and Y. Haberfeld. 1988. The impact of applicant gender compared to qualifications on hiring recommendations: A meta-analysis of experimental studies. *Organizational Behavior and Human Decision Processes* 41:180–95.

Padavic, Irene, and Barbara Reskin. 2002. *Women and men at work*. 2d ed. Thousand Oaks, CA: Pine Forge Press.

Pierce, Jennifer. 1995. *Gender trials: Emotional lives in contemporary law firms.* Berkeley: University of California Press.

Reskin, Barbara, and Heidi Hartmann. 1986. *Women's work, men's work: Sex segregation on the job.* Washington, DC: National Academic Press.

Reskin, Barbara, and Patricia Roos. 1990. *Job queues, gender queues*. Philadelphia: Temple University Press.

Rhode, Deborah L. 1997. *Speaking of sex: The denial of gender inequality*. Cambridge, MA: Harvard University Press.

Rubin, Henry. 2003. *Self-made men: Identity and embodiment among transsexual men*. Nashville, TN: Vanderbilt University Press.

Valian, Virginia. 1999. *Why so slow? The advancement of women.* Cambridge, MA: MIT Press.

West, Candace, and Don Zimmerman. 1987. Doing gender. *Gender & Society* 1:13–37.

Williams, Christine. 1995. *Still a man's world: Men who do "women's"* work. Berkeley: University of California Press.

Yoder, Janice, and Patricia Aniakudo. 1997. Outsider within the firehouse: Subordination and difference in the social interactions of African American women firefighters. *Gender & Society* 11:324–41.

Introduction to Reading 37

In her presidential address to the Pacific Sociological Association, Amy S. Wharton provides an overview of changes in the institutions of work, family, and gender. She frames this address within the 50 years since the passage of the Civil Rights Act. In this piece, she describes the changes that have been made and the patterns that remain problematic. Using her own research on academic institutions, Wharton helps to explain why the social institutions of work, family, and gender are so resistant to change.

1. What was the intent of the Civil Rights Act? In what ways did it succeed? In what ways did it fail?
2. What does she mean by the "stalled revolution"?
3. What is meant by "egalitarian essentialism," and will it usher in further change in equity in institutions?

(Un)Changing Institutions

Work, Family, and Gender in the New Economy

Amy S. Wharton

As sociologists, we are all students of change. In fact, at the most abstract level, change is central to sociological thinking and practice. The study of social life at all levels involves close attention to the reciprocal and interdependent relations between social reproduction and transformation, or between continuity and disruption. Both forces are simultaneously present in the social world—whether at the societal level, the organizational level, in social interaction, or within individuals. In the larger society, change and the forces that produce it receive much more attention than continuity or stability, and this is perhaps not that surprising. However, our agenda in sociology is to capture both the ongoing reproduction of social life and its moments of disturbance or disorder. An interest in exploring those relations as they are expressed in the interconnected realms of work, gender, and family motivates this address.

The timing is right for this discussion. The year 2014 marks the 50th anniversary of the War on Poverty, which was launched by President Lyndon Johnson in his 1964 State of the Union Address. One of the most significant pieces of legislation passed that year was the Civil Rights Act. For those like myself who study workplace inequality, this law's most critical element is the fact that it outlawed discrimination by race, color, religion, national origin, and sex in employment. . . .

Wharton, A. S. (2015). (Un)Changing institutions: Work, family, and gender in the new economy. *Sociological Perspectives, 58*(1): 7–19.

In his book *Inventing Equal Opportunity,* Frank Dobbin (2009:22) notes that the Civil Rights Act was a "broad brush" attempt to forbid discrimination and promote equal opportunity, but it left open exactly what this meant and how it was to be done. Dobbin's argument is germane to this address in a number of important respects. First, the story of civil rights legislation is relevant for underscoring the important role of organizations, particularly work organizations, as a critically important arena where large-scale societal changes are played out. Second, the history of civil rights as told by Dobbin also underscores the messiness of organizational change and the factors that thwart or make it possible for change to occur. Among these factors is the process whereby legislation or other initiatives move from the realm of language to the realm of implementation and practice.

Finally, this history calls attention to the multifaceted and changing societal definitions of gender equality. The civil rights era made equal opportunity central to the meaning of this concept (Burstein, Bricher, and Einwohner 1995), and this emphasis remained predominant over decades of change in women's and men's lives. For example, almost 30 years after the passage of the 1964 Civil Rights Act, President George H. W. Bush signed the Civil Rights Act of 1991 to strengthen laws prohibiting sex discrimination in the workplace, but he vetoed family and medical leave bills (Burstein and Wierzbicki 2000). Today, equal opportunity is viewed as a necessary but not sufficient condition for gender equality, while work-family issues and new narratives about equality and choice have become more central.

To examine these ideas, I begin at the societal level, reviewing progress toward and away from gender equality. Next, I turn to the topic of organizational change. Societal changes are played out in the workplace, but organizations have their own change dynamics. These dynamics are important in understanding why and how organizational change fails. Finally, I use an example drawn from my own research on the academic workplace to examine leaders' gender narratives during a time of organizational change.

Societal Changes in Gender, Work, and Family

The last half-century or more has been a time of fundamental change in gender, work, and family (Goldin 2006). In North America, Western Europe, and indeed throughout the globe, women's participation in the paid labor force rose steadily during the latter half of the twentieth century (Heymann and Earle 2009). In the United States, the increase in women's labor force participation occurred across all educational levels and among almost all racial and ethnic groups. During this time, women made inroads into jobs traditionally dominated by men and they made progress closing the gender earning gap. This pattern was fueled (and reinforced) by women's increasing levels of educational attainment—from primary school to college and to professional and graduate programs (Buchmann and DiPrete 2006). With respect to caregiving and household work, trends suggest a similar pattern of relatively continuous change over the past several decades and across a wide geographic area. Women spend fewer hours working at home, while more men spend more (Geist and Cohen 2011).

Gender attitudes have changed as well. Survey data show relatively consistent movement toward more liberal gender attitudes in the United States between the mid-1970s and 1990s (Cotter, Hermsen, and Vanneman 2011). Majorities of both women and men came to agree that a mother's employment was not damaging to her children, that women's role was not simply to care for the home, and that men did not necessarily make better politicians. North America, Europe, and other developed economies show similar patterns. In fact, attachment to women's and men's "traditional roles" has weakened among both women and men across the globe (Pew Research Global Attitudes Project 2010).

That progress toward gender equality in one area is connected to progress in another is not surprising. Thus, rather than a series of distinct changes, many note a pattern of convergence toward greater gender equality. One form of convergence is cross-national. For example, Claudia Geist and Philip N. Cohen (2011) found that in the last few decades, the amount of housework shouldered by women declined faster in more traditional countries than in those that were less traditional. This created a cross-national convergence of sorts as countries moved at different rates as they converged toward the same outcome: greater equality in the domestic division of labor. Economist Claudia Goldin (2014) conceives of convergence in a slightly different way, referring to "the converging roles of men and women," which she views as among the most important advances in society and the economy in the last century. As evidence for this, she points to the shrinking gap between women and men in labor force participation, hours of paid and unpaid work, labor force experience, occupational attainment, and education.

Uneven Gender Change and the Stalled Revolution

The evidence for twentieth century change (and convergence) in gender, work, and family is thus powerful and compelling. Increasingly, too, is the evidence that progress toward gender equality has gone through a period of deceleration or "stalling," as David A. Cotter, Joan M. Hermsen, and Reeve Vanneman (2004) referred to it in their report for the Russell Sage Foundation (see also England 2010). However, while there is some evidence of a global slowdown in progress toward gender equality, the United States is distinctive in certain respects (Lee 2014).

Cotter et al. (2004) show that the slowdown in the United States occurred across a number of domains. For example, U.S. women's rates of labor force participation leveled off in the late 1990s and have declined from their peak in 1999. This leveling off appears to have occurred across all categories of education, presence of children, and marital status (Lee 2014). With respect to the gender wage gap, the pattern is roughly similar. The wage gap narrowed steadily through the 1970s and 1980s, but progress slowed in the 1990s and early 2000s (Blau and Kahn 2007). During the 10-year period between 2004 and 2013, the gender wage gap barely changed, declining by only 1.7 percent (Institute for Women's Policy Research 2014).

Sociological research has revealed other, more nuanced looks at the stalled progress toward gender equality. In their study of occupational sex and race segregation from the 1960s to the present, Kevin Stainback and Donald Tomaskovic-Devey (2012) find that desegregation slowed considerably after political pressure by the civil rights and (later) the women's movement eased. Similarly, U.S. women's entrance into management positions increased steadily in the second half of the twentieth century, only to slow in the 1990s (P. N. Cohen, Huffman, and Knauer 2009). Although most women do not hold management positions—especially higher level positions—this slowdown has broader relevance. Several studies have shown that the demographic mix of managers shapes many aspects of the work environment, including the behaviors of managers themselves. The percentage of women in management jobs in an organization is positively related to the percentage of women in non-management jobs, and it affects the percentage of new jobs in an organization that are filled by women relative to men (L. E. Cohen and Broschak 2013).

Women in almost all industrialized countries earn a higher proportion of college degrees than men (Buchmann and DiPrete 2006; Charles and Bradley 2009). In the United States, the proportion of degrees received by women surpassed men in the early 1980s, and the gender gap has been growing steadily ever since, as men's college graduation rates decline. Despite their advantage in college graduation rates, other aspects of education show a more complicated picture with respect to movement toward gender parity or equality. In particular, the increase in women's share of college degrees in industrialized countries has been accompanied by a robust pattern of gender segregation by field of study (Charles and Bradley 2009). Paula England (2010) found a similar type of pattern when she looked at trends in doctoral degree attainment. Women's share of doctoral degrees went up fairly steadily over the last several decades (since the 1970s), but there has not been much change in the relative femaleness of different fields. Fields of study that were more female relative to others in the 1970s remain more female than others; fields of study that were less female than others 40 years ago remain less female than others today (England 2010).

Compared with data on employment and education, the evidence with respect to gender attitudes is more equivocal, especially with respect to recent trends. Cotter et al. (2011) show that support for more egalitarian views leveled off somewhat in the mid-1990s, and this leveling occurred among both women and men, of all ethnicities (except Asians) and across all levels of income and education. They found a small "rebound" in more egalitarian attitudes since 2000, but note "a growing but decelerating social liberalism among recent generations" (Cotter et al. 2011:282). However, in more recent analyses, these authors suggest that this rebound has been more robust, as indicated by steady increases since 2006 in popular support for gender equality and women's labor force participation (Cotter, Hermsen, and Vanneman 2014).

The Rise of Egalitarian Essentialism

Although many forces have contributed to the "stalled revolution," the role of cultural factors has received particular attention. Central to these arguments is the claim that a new frame or narrative about gender has gained prominence in politics and popular culture. Sociologists refer to this cultural frame as "egalitarian essentialism" (Cotter et al. 2011:261; see also Charles and Grusky 2004). This frame is distinct from traditional notions of "separate spheres," a dominant perspective in the first half of the twentieth century. It is also distinct from feminist egalitarianism, a frame that emerged from and helped to fuel the feminist movements of the 1960s and 1970s. Egalitarian essentialism is a hybrid, containing an endorsement of the principle of gender equality, while defining equality as the right of individual women to choose what is best for them.

This emphasis on choice aligns with other efforts to describe new "post-feminist" standpoints. The most prominent is "choice feminism," a position described as being "concerned with increasing the number of choices open to women and with decreasing judgments about the choices individual women make" (Kirkpatrick 2010:241). When combined with a belief in essential gender differences, an emphasis on the value of individual choice tends to reinforce the status quo. Maria Charles and Karen Bradley (2009) show how this cultural frame has helped to perpetuate gender segregation in higher education, especially in industrial societies where beliefs in individual self-expression and choice are deeply entrenched. In addition to reinforcing the status quo, these narratives have been critiqued for their political implications. Choice feminism, in particular, has been described as an attempt to represent feminism as non-threatening and "seem appealing to the broadest constituency possible" (Ferguson 2010:248).

In sum, recent history reminds us that that social reproduction and social transformation are inextricably linked. The steady and mostly broad-based progress toward gender equality that marked the last half of the twentieth century has been disrupted or slowed. However, change and stability are relative concepts, and there is room for debate about whether and to what degree gender inequality has increased in recent years. Whether egalitarian essentialism, choice feminism, or similar cultural logics have contributed to this pattern is also in need of further study. Nevertheless, these gender narratives remain alive and well in popular debates about professional women "opting out" of the workforce and have become deeply embedded in work-family debates more generally (Kirkpatrick 2010; Stephens and Levine 2011).

Societal forces, including cultural logics and ideologies, also penetrate organizations, where they are expressed in the perspectives and practices of workers and employers. Organizations have their own change dynamics, however, which shape how cultural narratives are deployed.

Organizational Change and Changing Organizations

Organizational change receives a tremendous amount of attention from researchers. Perhaps one reason for this is that so much of what we understand to be true about organizations emphasizes their immobility or immovability. Rules, routines, and hierarchy are defining features of bureaucratic organizations and help to explain the tremendous inertia (and dysfunction) that is often associated with them (Perrow 1986). Organizations also act to prevent or deflect change. For example, loose coupling is a means by which organizations can create a firewall between outside demands and their normal operations and ways of doing business. Organizations portray themselves to outside constituencies in ways that signal movement, while leaving existing practices and routines untouched (Meyer and Rowan 1977).

The case of work-family policies provides a good example of this process. Many organizations have adopted formal work-family policies around flexibility, parental leaves, and so forth, but implementation often lags (Williams, Blair-Loy, and Berdahl 2013). The policies themselves face resistance or indifference among key organizational gatekeepers, such as managers or supervisors. Meanwhile, workers who may want to use these policies avoid doing so, as they recognize that their employer's commitment is more symbolic than real (Blair-Loy and Wharton 2002; Jacobs and Gerson 2004). The gendered culture of work and its ideal worker norms persist despite even well-intentioned efforts to make work accommodating to parents.

Organizations can face pressures to change from the outside, yet the external environment is more often a source of organizational continuity rather than disruption. Imitation is a major principle of human *and* organizational action (March 1996). Whether seeking solutions to immediate problems, or attempting to chart aspirations for the future, organizations (as well individuals) look not only to their own past performance but also to the past performance of relevant others (March 1996). Imitation contributes to the diffusion of ideas, knowledge, policies, or practice. It not only helps to increase predictability and continuity but also constrains large-scale change and transformation. Thus, when considering some of the basic principles that drive organizations, continuity often wins.

The continuity-change tradeoff is not always resolved in favor of continuity, however. Organizations do change and sometimes change in the direction of greater gender equality. When we look sociologically at these cases, however, the prime movers are often "behind our backs"—unexpected, unanticipated, and difficult to explain. In their study of work on offshore oil rigs, Robin J. Ely and Debra E. Meyerson (2010) identified an unforeseen effect of organizational efforts to enhance safety and performance. Expressions of hegemonic masculinity most often associated with dangerous, predominately male, jobs were significantly reduced. New workplace practices around safety ushered in new kinds of masculine identities and behaviors. In this way, the organization inadvertently "disrupt[ed] the gender status quo through practices that encourage[d] men to let go of conventional masculine scripts" (Ely and Meyerson 2010:28).

In contrast to unplanned or inadvertent transformation, organizations sometimes intentionally seek change. Yet, these experiences sometimes end up validating the most change averse among us. This is because a planned organizational change often goes badly awry (Hannan, Polos, and Carroll 2003). Organizational actors may miscalculate the risks and rewards of change; leaders underestimate how long a change will take and its costs, both monetary and in human terms. Furthermore, as sociologists, we are only too familiar with the unintended consequences of changes, whether planned or unplanned, and sometimes the failure of what seem like self-evident fixes.

Cautionary tales abound. Research by Alexandra Kalev, Frank Dobbin, and Erin Kelly (2006), for example, shows that one of the most ubiquitous approaches to increasing diversity in the workplace—diversity training—has been among the least effective in increasing the racial and gender diversity of managers in U.S. firms. Emilio J. Castilla and Stephen Benard's (2010) study of merit-based reward systems finds that these practices, which are enthusiastically embraced as a means to insure that pay is based on performance—not gender, race, or other considerations—may not be doing what many hoped. Instead, Castilla and Benard have uncovered what they call "the paradox of meritocracy." Organizations that emphasize meritocracy can (under some conditions) unintentionally create conditions that lead to more bias, not less, in the evaluative process.

Another example of well-intentioned organizational change that produces unintended negative consequences derives from the work-family literature. In their 20-nation, cross-national study of the effects of family-friendly policies on women's wages, Hadas Mandel and Moshe Semyonov (2005) found that these policies were associated with a larger gender earning gap, not a more egalitarian earning distribution. The reasons for this are complex, but these researchers suggest that it can be partly attributed to the fact that mothers more so than fathers are likely to take advantage of policies that facilitate work-family integration. This leaves mothers (and women in general) subject to discrimination by employers who penalize them for their work interruptions (such as long maternity leave).

This is not an argument against change efforts or work-family policies but rather another reminder that organizational changes—in the form of practices aimed at reducing inequality and discrimination or to increase work-family flexibility—are much more complicated than they seem. The mechanisms that facilitate change, like those that undermine it, operate at more than one level and sometimes work at cross purposes. For example, formalization is encouraged as a way to reduce bias and discrimination (such as the case of pay for performance or other mechanisms), yet while this may help mitigate the effects of cognitive bias, formalization can introduce biases of its own. Well-intentioned and planned organizational change can be resisted, deflected, or transformed in ways that undermine rather than facilitate desired outcomes.

Continuity and Change in the Academic Workplace

The academic workplace is a useful site for examining the dueling forces of continuity and change and understanding the role that gender narratives play in these dynamics. While bureaucratic organizations of all types may resist change, academic institutions are perceived as especially resistant (Lane 2007; Lucas 2000). Yet, as we have seen, higher education has not been immune from the broader set of forces reshaping gender, work, and family over the last several decades. One particular way these forces have affected the academic workplace is through federally funded initiatives designed to increase the gender diversity of the faculty. Much of this interest derives from concerns about the future of STEM disciplines (i.e., science, technology, engineering, and math) in the academy and the barriers faced by women and underrepresented minorities in these fields (Committee on Women in Science and Engineering 2006).

In 2001, the National Science Foundation created its ADVANCE Program to address these issues. The goals of ADVANCE are to increase the representation of women in academic science and engineering careers, develop ways to promote gender equity in STEM, and increase the diversity of the STEM workforce. This program has not been modest about its investments or intentions. Since 2001, ADVANCE has spent over 130 million dollars to support ADVANCE projects at over 100 colleges and universities and some non-profit (National Science Foundation 2014). The most visible and well-funded ADVANCE award is its Institutional Transformation award. Averaging about 3.5 million dollars, these institutional grants are intended to transform universities in ways that make academe and STEM in particular more accommodating to women and other underrepresented groups.

ADVANCE-funded institutions have pursued many strategies to accomplish this goal (Bilimoria and Liang 2012; Bystydzienski and Bird 2006; Laursen and Rocque 2009). In general, ADVANCE initiatives fall into three broad categories, including those focused on policy reform and creation, departmental or institutional climate, and training of faculty and administrators (Stewart, Malley, and LaVaque-Manty 2007). This investment in institutional change has been fueled by and helped foster an outpouring of sociological research on gender, work, and family in the academe, both within and outside of STEM. This research has included studies of work-family issues in the academy (e.g., Fox, Fonseca, and Bao 2011; King 2008; Mason and Goulden 2004; Misra, Lundquist, and Templer 2012), as well as research on gender inequality in academic life (e.g., Bird 2011; Ecklund, Lincoln, and Tansey 2012; Jacobs and Winslow 2004; Misra et al. 2011; Roos and Gatta 2009; Winslow 2010). Climate, especially departmental climate, has also received significant attention in ADVANCE institutions, and climate studies have become useful diagnostic tools for universities trying to understand the experiences of women and other underrepresented groups (Callister 2006; Maranto and Griffin 2011; Settles et al. 2006).

Leadership and Organizational Change

These studies have helped to explain women's underrepresentation in STEM fields and the barriers that remain to be overcome. Less attention, however, has been paid to the organizational change process itself and particularly the forces that derail or deflect change efforts. My own research examines this issue with a focus on departmental leaders.

Leaders are vitally important to the change process. Frank Dobbin and Alexandra Kalev (2007:280) argue that "In the corporate world, as in academia, programs that establish clear leadership and responsibility for change have produced the greatest gains in diversity." Similarly, Sara I. McClelland and Kathryn J. Holland (2014:3) suggest that leaders' sense of accountability and personal responsibility for diversity initiatives are critical to the success of these efforts. Michael Schwalbe et al. (2000:435) highlight leaders' role in "regulating discourse" through formal or informal mechanisms. By filtering and framing information, leaders shape perceptions of their subordinates (Dragoni 2005). Leaders' beliefs about gender may be especially powerful, given the role of these beliefs in reproducing gender inequality (Ridgeway 2011).

Leadership in academe is multi-layered, but for faculty, the departmental leader is most critical. That institutional transformation in academe requires attention to departmental processes is widely acknowledged, making departmental practices, policies, routines, relationships, and dynamics important topics. Chairs influence all these aspects of departmental life (Bilimoria et al. 2006). In this way, they also shape faculty's satisfaction with their careers, colleagues, and work environment (Bensimon, Ward, and Sanders 2000). Chairs seem to have a particularly important influence on women's work lives in the academy (Settles et al. 2006). Recognition of their role has made departmental leaders a key audience for various types of training opportunities, and climate surveys typically ask faculty about their perceptions of their chair and other leaders. Ironically, however, while we know much about faculty perceptions, chairs' own beliefs are less well understood.

In 2010, I was part of a four-person research team at an ADVANCE institution that set out to investigate departmental leaders' perspectives on their own roles and responsibilities with respect to diversity and organizational change. During the course of this project, graduate student Mychel Estevez and I became attuned to the ways that chairs framed issues of gender and gender inequality, especially as these topics were invoked in the context of the university's broader efforts at improving gender equity and increasing women's representation in STEM fields (see Wharton and Estevez 2014, for a full discussion of this research). Some data from this project, in addition to more recently published research by other scholars, have revealed how leaders' narratives about gender, work, and family can slow or undermine change efforts. Leaders may deflect responsibility for change by emphasizing the choices of others, particularly female faculty, and many fail to act out of a belief that gender change is inevitable and progressive.

* * *

Choice and Change

Choice is personally empowering, connoting independence, freedom, and autonomy. It has many positive consequences for those who have choices or believe themselves to have them (Savani, Stephens, and Markus 2011; Stephens and Levine 2011). This is especially true in American society and, as we have seen, in academe, where the ability to control the conditions of one's work is highly valued. Although having the ability to choose is personally beneficial, it is socially disadvantageous. Experimental research shows that exposure to a choice perspective weakens support for policies designed to advance collectivities or society as a whole (Savani et al. 2011). As Nicole M. Stephens and Cynthia S. Levine (2011:1235) note, Americans' strong embrace of a choice framework helps explain why they "readily dismiss gender barriers as a vestige of the past in the face of evidence to the contrary." Choice fortifies notions of personal responsibility and thereby assigns blame to others for their disadvantages while minimizing the role of external forces or constraints.

Marieke van den Brink and Yvonne Benschop (2012:89) argue that change in the academy is slow because practices and beliefs that perpetuate inequality "may hinder, alter, or transform equality measures." This summarizes the story told here, as good faith and intentional efforts to make change are deflected, rearticulated, and transformed. Leaders perceive work-family issues through the lens of choice, treating these matters as the responsibility of the individual (women) faculty members, and not the institution. This belief in choice also shapes chairs' perceptions of gender inequality more generally. They do not necessarily believe that gender inequality has been eliminated, yet are reluctant to view problems as structural or systemic. The need for change is depoliticized and viewed as inevitable, incremental, and "naturally" occurring over time through generational replacement. Most important, by assigning responsibility for change to others, chairs' willingness, capacity, and resolve to act are substantially weakened.

Conclusion

The passage of the Civil Rights Act and the pursuit of equal opportunity it endorsed were about improving the chances for women and other underrepresented groups to compete in an essentially unchanged workplace. Paul Burstein and colleagues (Bricher 1997; Burstein et al. 1995; Burstein and Wierzbicki 2000) note that what they call the "work-family accommodation" frame was more far reaching politically. This frame contained an implicit critique of the organization of work and drew attention to its impact on women's and men's family responsibilities and commitments. This broader vision of gender equality has yet to gain popular support or a foothold in the political arena. The resurgence of a choice framework—in the form of egalitarian essentialism or choice feminism—has likely played a role in depoliticizing the work-family agenda and undermining the case for change. It has also served as a reminder that narratives about gender are a central ingredient in the broader system of practices that reproduce inequality.

The strong forces of change in the gender system that occurred during the twentieth century were set into motion by numerous forces—including by conscious, political efforts to reduce gender inequality. These changes were not inevitable, nor can they be assumed to be permanent and ongoing. This makes it all the more important that we return our attention to the ways of change. These include the recognition that the forces of continuity and change are simultaneously present in society and the organizations that comprise it, that beliefs and practices that maintain continuity or the status quo restrain and circumvent those that promote equality practices and beliefs, and that many forces tip the balance in favor of continuity.

It is impossible to predict the twists and turns that are in our future. The past decade may look like a small blip 20 years out or may in fact represent a major turning point of some kind. Most of us here are not waiting to see how things turn out or believe (naively) that evolution or generational replacement will by itself pave a way toward greater gender equality. Instead, we seek change—to transform the workplace, to eliminate discrimination and reduce inequality, and to restart the stalled gender revolution. Fulfilling these goals requires us to look carefully at the ways in which inequality practices and beliefs may be undermining our efforts.

References

Bensimon, Estela M., Kelly Ward, and Karla Sanders. 2000. *The Department Chair's Role in Developing New Faculty into Teachers and Scholars.* Bolton, MA: Anker.

Bilimoria, Diana and Xiangfen Liang. 2012. *Gender Equity in Science and Engineering: Advancing Change in Higher Education.* New York: Routledge.

Bilimoria, Diana, Susan R. Perry, Xiangfen Liang, Eleanor P. Stoller, Patricia Higgins, and Cyrus Taylor. 2006. "How Do Female and Male Faculty Members Construct Job Satisfaction? The Role of Perceived Institutional Leadership and Mentoring and Their Mediating Processes." *Journal of Technology Transfer* 31:355–65.

Bird, Sharon. 2011. "Unsettling the Universities' Incongruous, Gendered Bureaucratic Structures: A Case Study Approach." *Gender, Work & Organization* 18:202–30.

Blair-Loy, Mary and Amy S. Wharton. 2002. "Employees' Use of Work-Family Policies and the Workplace Social Context." *Social Forces* 80: 813–845.

Blau, Francine D. and Lawrence M. Kahn. 2007. "The Gender Pay Gap: Have Women Gone as Far as They Can?" *Academy of Management Perspectives* 21:7–23.

Buchmann, Claudia and Thomas A. DiPrete. 2006, "The Growing Female Advantage in College Completion: The Role of Family Background and Academic Achievement." *American Sociological Review* 71:515–41.

Burstein, Paul and Marie Bricher. 1997. "Problem Definition and Public Policy: Congressional Committees Confront Work, Family, and Gender, 1945–1990." *Social Forces* 75:135–69.

Burstein, Paul, Marie Bricher, and Rachel L. Einwohner. 1995. "Policy Alternatives and Political Change: Work, Family, and Gender on the Congressional Agenda, 1945–1990." *American Sociological Review* 60:67–83.

Burstein, Paul and Susan Wierzbicki. 2000. "Public Opinion and Congressional Action on Work, Family, and Gender, 1945–1990." Pp. 31–66 in *Work & Family: Research Informing Policy,* edited by Toby L. Parcel and Daniel B. Cornfield. Thousand Oaks, CA: Sage Publications.

Bystydzienski, Jill and Sharon R. Bird, eds. 2006. *Removing Barriers: Women in Academic Science, Engineering, Technology and Mathematics Careers.* Bloomington, IN: Indiana University Press.

Callister, Ronda R. 2006. "The Impact of Gender and Department Climate on Job Satisfaction and Intentions to Quit for Faculty in Science and Engineering Fields." *Journal of Technology Transfer* 31:367–75.

Castilla, Emilio J. and Stephen Benard. 2010. "The Paradox of Meritocracy in Organizations." *Administrative Science Quarterly* 55:543–76.

Charles, Maria and Karen Bradley. 2009. "Indulging Our Gendered Selves: Sex Segregation by Field of Study in 44 Countries." *American Journal of Sociology* 114:924–76.

Charles, Maria and David B. Grusky. 2004. *Occupational Ghettoes: The Worldwide Segregation of Women and Men.* Stanford, CA: Stanford University Press.

Cohen, Lisa E. and Joseph P. Broschak. 2013. "Whose Jobs Are These? The Impact of the Proportion of Female Managers on the Number of New Management Jobs Filled by Women." *Administrative Science Quarterly* 58:509–41.

Cohen, Philip N., Matt L. Huffman, and Stefanie Knauer. 2009. "Stalled Progress? Gender Segregation and Wage Inequality among Managers, 1980–2000." *Work and Occupations* 36:318–42.

Committee on Women in Science and Engineering. 2006. *To Recruit and Advance: Women Students and Faculty in Science and Engineering.* Washington, DC: National Academy Press.

Cotter, David A., Joan M. Hermsen, and Reeve Vanneman. 2004. *Gender Inequality at Work.* A volume in the series, The American People: Census 2000. New York: Russell Sage Foundation and Population Reference Bureau.

Cotter, David A., Joan M. Hermsen, and Reeve Vanneman. 2011. "The End of the Gender Revolution? Gender Role Attitudes from 1977 to 2008." *American Journal of Sociology* 117:259–89.

Cotter, David A., Joan M. Hermsen, and Reeve Vanneman. 2014. "Back on Track? Stall and Rebound for Gender Equality 1977–2012." Retrieved November 3, 2014 (http://thesocietypages.org/ccf/2014/08/05/gender-revolution-rebound-symposium/).

Dobbin, Frank. 2009. *Inventing Equal Opportunity.* Princeton, NJ: Princeton University Press.

Dobbin, Frank and Alexandra Kalev. 2007. "The Architecture of Inclusion: Evidence from Corporate Diversity Programs." *Harvard Journal of Law & Gender* 30:279–301.

Dragoni, Lisa. 2005. "Understanding the Emergence of State Goal Orientation in Organizational Work Groups: The Role of Leadership and Multilevel Climate Perceptions." *Journal of Applied Psychology* 90:1084–95.

Ecklund, Elaine Howard, Anne E. Lincoln, and Cassandra Tansey. 2012. "Gender Segregation in Elite Academic Science." *Gender & Society* 26:693–717.

Ely, Robin J. and Debra E. Meyerson. 2010. "An Organizational Approach to Undoing Gender: The Case of Offshore Oil Platforms." *Research in Organizational Behavior* 30:3–34.

England, Paula. 2010. "The Gender Revolution: Uneven and Stalled." Gender & Society 24:149–66.

Ferguson, Michaele L. 2010. "Choice Feminism and the Fear of Politics." *Perspectives on Politics* 8:247–70.

Fox, Mary Frank, Carolyn Fonseca, and Jinghui Bao. 2011. "Work and Family Conflict in Academic Science: Patterns and Predictors among Women and Men in Research Universities." *Social Studies of Science* 41:715–35.

Geist, Claudia and Philip N. Cohen. 2011. "Headed toward Equality? Housework Change in Comparative Perspective." *Journal of Marriage and Family* 73: 832–44.

Goldin, Claudia. 2006. "The Quiet Revolution That Transformed Women's Employment, Education, and Family." *AEA Papers and Proceedings* 96 (May): 1–21.

Goldin, Claudia. 2014. "A Grand Gender Convergence: Its Last Chapter." *American Economic Review* 104: 1091–119.

Hannan, Michael T., Laszlo Polos, and Glenn R. Carroll. 2003. "The Fog of Change: Opacity and Asperity in Organizations." *Administrative Science Quarterly* 48:399–432.

Heymann, Jody and Alison Earle. 2009. *Raising the Global Floor: Dismantling the Myth That We Can't Afford Good Working Conditions for Everyone.* Stanford, CA: Stanford University Press.

Institute for Women's Policy Research. 2014. "The Gender Wage Gap 2013: Differences by Race and Ethnicity, No Growth in Real Wages for Women." Fact Sheet.

Retrieved December 15, 2014 (http://www.iwpr.org/publications/pubs/the-gender-wuge-gap-2013-differences-by-race-and-ethnicity-no-growth-in-real-wages-for-women).

Jacobs, Jerry A. and Kathleen Gerson. 2004. *The Time Divide: Work, Family, and Gender Inequality.* Cambridge, MA: Harvard University Press.

Jacobs, Jerry A. and Sarah Winslow. 2004. "Understanding the Academic Life Course, Time Pressures, and Gender Inequality." *Community, Work & Family* 7:143–61.

Kalev, Alexandra, Frank Dobbin, and Erin Kelly. 2006. "Best Practices or Best Guesses? Assessing the Efficacy of Corporate Affirmative Action and Diversity Policies." *American Sociological Review* 71:589–617.

King, Eden B. 2008. "The Effect of Bias on the Advancement of Working Mothers: Disentangling Legitimate Concerns from Inaccurate Stereotypes as Predictors of Advancement in Academe." *Human Relations* 61:1677–711.

Kirkpatrick, Jennet. 2010. "Introduction: Selling Out? Solidarity and Choice in the American Feminist Movement." *Perspectives on Politics* 8:241–45.

Lane, India F. 2007. "Change in Higher Education: Understanding and Responding to Individual and Organizational Resistance." Retrieved December 5, 2014 (http://www.ccas.net/files/ADVANCE/Lane_Change%20in%20higher%20ed.pdf).

Laursen, Sandra and Bill Rocque. 2009. "Faculty Development for Institutional Change: Lessons from an ADVANCE Project." *Change: The Magazine of Higher Learning,* March-April 2009. Retrieved December 5, 2014 (http://www.changemag.org/archives/back%20issues/mnrch-april%202009/full-advance-project.html).

Lee, Jin Y. 2014. "The Plateau in U.S. Women's Labor Force Participation: A Cohort Analysis." *Industrial Relations* 53:46–71.

Lucas, Ann F., ed. 2000. *Leading Academic Change: Essential Roles for Department Chairs.* San Francisco, CA: Jossey-Bass.

Mandel, Hadas and Moshe Semyonov. 2005. "Family Policies, Wage Structures, and Gender Gaps: Sources of Earnings Inequality in 20 Countries." *American Sociological Review* 70:949–67.

Maranto, Cheryl and Andrea E. C. Griffin. 2011. "The Antecedents of a 'Chilly Climate' for Women Faculty in Higher Education." *Human Relations* 64:139–59.

March, James G. 1996. "Continuity and Change in Theories of Organizational Action." *Administrative Science Quarterly* 41:278–87.

Mason, Mary A. and Marc Goulden. 2004. "Marriage and Baby Blues: Redefining Gender Equity in the Academy." *The Annals of the American Academy of Political and Social Science* 596:86–103.

McClelland, Sara I. and Kathryn J. Holland. 2014. "You, Me, or Her: Leaders' Perceptions of Responsibility for Increasing Gender Diversity in STEM Departments." *Psychology of Women Quarterly,* doi: 10.1177/0361684314537997, first published on June 5, 2014.

Meyer, John W. and Brian Rowan. 1977. "Institutionalized Organizations: Formal Structure as Myth and Ceremony." *American Journal of Sociology* 83:340–63.

Misra, Joya, Jennifer H. Lundquist, Elissa Holmes, and Stephanie Agiomavritis. 2011. "The Ivory Ceiling of Service Work." *Academe* 97:22–26.

Misra, Joya, Jennifer H. Lundquist, and Abby Templer. 2012. "Gender, Work Time, and Care Responsibilities among Faculty." *Sociological Forum* 27:300–23.

National Science Foundation. 2014. ADVANCE at a Glance. Retrieved November 3, 2014 (http://www.nsf.gov/crssprgm/advance/).

Perrow, Charles. 1986. *Complex Organizations: A Critical Essay.* New York: McGraw-Hill.

Pew Research Global Attitudes Project. 2010. Gender Equality Universally Embraced, but Inequalities Acknowledged. Retrieved November 3, 2014 (http://www.pewglobal.org/2010/07/01/gender-equality/).

Ridgeway, Cecilia L. 2011. *Framed by Gender: How Gender Inequality Persists in the Modern World.* New York: Oxford University Press.

Roos, Patricia A. and Mary L. Gatta. 2009. "Gender (In)-Equity in the Academy: Subtle Mechanisms and the Production of Inequality." *Research in Social Stratification and Mobility* 27:177–200.

Savani, Krishna, Nicole M. Stephens, and Hazel R. Markus. 2011. "The Unanticipated Interpersonal and Societal Consequences of Choice: Victim Blaming and Reduced Support for the Public Good." *Psychological Science* 22:795–802.

Schwalbe, Michael, Sandra Godwin, Daphne Holden, Douglas Schrock, Shealy Thompson, and Michele Wolkomir. 2000. "Generic Processes in the Reproduction of Inequality: An Interactionist Analysis." *Social Forces* 79:419–52.

Settles, Isis H., Lilia M. Cortina, Janet Malley, and Abigail J. Stewart. 2006. "The Climate for Women in Academic Science: The Good, the Bad, and the Changeable." *Psychology of Women Quarterly* 30:47–58.

Stainback, Kevin and Donald Tomaskovic-Devey. 2012. *Documenting Desegregation: Racial and Gender Segregation in Private-Sector Employment since the Civil Rights Act.* New York: Russell Sage Foundation.

Stephens, Nicole M. and Cynthia S. Levine. 2011. "Opting Out or Denying Discrimination? How the Framework of Free Choice in American Society Influences Perceptions of Gender Inequality." *Psychological Science* 22:1231–236.

Stewart, Abigail, Janet E. Malley, and Danielle LaVaque-Manty (eds.). 2007. *Transforming Science and Engineering: Advancing Academic Women.* Ann Arbor, MI: University of Michigan Press.

Van den Brink, Marieke and Yvonne Benschop. 2012. "Slaying the Seven-Headed Dragon: The Quest for Gender Change in Academia." *Gender, Work & Organization* 19:71–92.

Wharton, Amy S. and Mychel Estevez. 2014. "Department Chairs' Perspectives on Work, Family, and Gender

Pathways for Transformation." *Advances in Gender Research* 19:131–50.
Williams, Joan C., Mary Blair-Loy, and Jennifer L. Berdahl. 2013. "Cultural Schemas, Social Class, and the Flexibility Stigma." *Journal of Social Issues* 69:209–34.
Winslow, Sarah. 2010. "Gender Inequality and Time Allocations among Academic Faculty." *Gender & Society* 24:769–93.

Topics for Further Examination

- Look up the most recent research on women and work done by the Institute for Women's Policy Research (http://www.iwpr.org) and the current activism under way at 9 to 5 National Association of Working Women (https://9to5.org/). Check out workplace policies related to the topics discussed in this chapter, for example, family-work leaves and practices related to workplace discrimination.
- Using the Web, find a list of the top executives in a sample of the largest firms in this country and calculate a gender ratio of women to men. (Hint: Fortune 500 is one such list.)
- Find information on workplace discrimination policies in your state or country. Search the Web to find workplace discrimination policies in another country to compare with those where you live.

8

Gender in Intimate Relationships

Mary Nell Trautner with Joan Z. Spade

Although social institutions and organized activities such as work, religion, education, and leisure provide frameworks for our lives, it is the relationships within these activities that hold our lives together. What surprises many people is that these everyday relationships are patterned. We don't mean "the daily routine" kind of patterns; we are referring to gendered patterns across individuals and relationships. For example, sociologists consider "the family" to be more than just a personal relationship; they view it as a social institution, with relatively fixed roles and responsibilities that meet some basic needs in society such as caring for dependent members and providing emotional support for its members. As you read through this chapter, you will come to realize how social norms influence all gendered relationships, including intimate relationships. This introduction and the readings in this chapter illustrate two key points. First, gendered intimate relationships always evolve, often in response to social changes unrelated to the relationships themselves. Second, gender is embedded in an idealized version of intimacy—the traditional, heterosexual family. As you probably already know, the traditional family is not the reality in the United States and most parts of the world today, as illustrated in readings in Chapters 1 and 3.

Before going on about these details of intimacy, let's stop for a moment and look at relationships in general. The word *relationship* takes on many different meanings in our lives. We can have a relationship with the server at our favorite restaurant because he or she is usually there when we dine out. We have relationships with our friends; some we may have known most of our lives, whereas others we have met more recently. And we have relationships with our family and with people who are like family. Some of these relationships surround us with love, economic support, intimacy, and/or almost constant engagement. All these relationships are shaped by gender. You have already read about relationships at work in Chapter 7; in this chapter, you will learn about how gender shapes more intimate relationships—from friendships to partnering to parenting.

Gendered Relationships

Consider the impact of gender on our relationships. We can have many friends with whom we share affection. In the past, researchers often argued that friendships varied by gender in predictable and somewhat stereotypical ways. That is, they described women's friendships as more intimate, or focused on sharing feelings and private matters, while describing men's friendships as more instrumental or focused around doing things, such as golfing or fishing (Walker, 1994). Some time ago, Francesca M. Cancian (1990) argued that men were more instrumental or task oriented in their love relationships, whereas women expected emotional ties. For example, in Cancian's study, when one man was asked how he expressed his love to his wife, he told the interviewer that he washed her car. A more recent study (Felmlee, Sweet, & Sinclair, 2012) focused on same-gender and cross-gender friendships and found that both had somewhat similar norms, which is an encouraging sign in terms of a breakdown in the gender binary. However, women were more upset than men when friends broke trust and intimacy, for example, when their friends failed to be there for them and stand up for them if they were being attacked. And, men and women both judged a woman who betrayed them more severely than a man (Felmlee et al., 2012).

While these studies are revealing, it is important to remember that other social prisms influence the gendering of our relationships, putting social constraints and expectations on even our most intimate times. For example, Karen Walker (1994) found that while both men and women hold stereotypical views of gendered behavior in friendships, actual friendship patterns were more complex and related to the social class of the individual. Even though working-class men recognized what was gender-appropriate behavior for friendships, they tended to describe their friendships in ways that would be defined as more stereotypically female (disclosure and emotional intimacy), while professional women tended to describe more masculine friendship patterns with other women. These exceptions to stereotypical views of gendered friendships, however, are also patterned; that is, they appear to vary by social class and reflect the constraints of work lives.

A recent study in this chapter examines a new pattern of men's friendships. Stefan Robinson, Adam White, and Eric Anderson analyze "bromances," or intimate friendships among straight-identified men. The "bromance" includes emotional intimacy and non-sexual physicality, and is regarded by many men in their study as more intimate and more honest than their romantic relationships with women. Men in their study say that in their relationships with women, they felt the need to present a false image of themselves to maintain their masculinity, whereas they could "truly" be themselves with their bromance. Readings in this chapter contribute to our understanding of how prisms of class and race—and age—influence gendered patterns in relationships.

Gender and Changing Households

One of the strongest gendered influences on relationships is the expectation that only men and women will fall in love and marry. As we have emphasized throughout this book, American culture assumes idealized intimate relationships to be heterosexual, accompanied by appropriate gendered behaviors and, of course, based in nuclear families. As gender changes, we should expect relationships to change as well. It may surprise you as you read Ellen Lamont's reading on courtship to find that, even as college-educated women have become more egalitarian and independent, they still expect traditional courtship rituals such as having the man ask for and pay for a date or to propose marriage.

You will notice that we did not include the word family in the title of this chapter. We did this because the stereotypical vision of family—mom, dad, two kids, and a dog, all living in a house behind a white picket fence—is only a small percentage of households today and never was the predominant form of family relationships.

As we explore relationships in this chapter, it is important to begin by examining how households actually are patterned. In Table 8.1, we list the various household configurations and the percentage in each category by race and ethnicity in the United States today. These data may surprise you. Relationships in the United States are changing, as indicated by the diversity noted in Table 8.1. Just slightly more than half of all Americans age 18 or older, 51.4%, were married and living with a spouse in 2017 (U.S. Census Bureau, 2017a). These percentages differ for racial and ethnic groups. For example, 31.8% of Black people are married with a spouse present, compared to 54.4% of White people (see Table 8.1). As you look at other relationship statuses on Table 8.1, you will see that clearly, race and ethnicity, and, although not included in this table, social class, influence the patterns of intimate relationships of men and women.

The increase in single and nonfamily households reflects both a trend toward postponing marriage and a longer life expectancy. A White boy born in 1960 had a life expectancy of 68.0 years, compared to a White boy born in 2015, who is projected to live to nearly ten years longer, to 77.1 years. A White girl born in 1960 had a life expectancy of 75.6, and born in 2015, projected to be 81.8 (U.S. Census Bureau, 2017c). These life expectancies are shaped by social factors such as race and social class as well as gender. For example, the projected life expectancy for a Black boy born in 2015 is 71.4 years and for a Black girl, 78.2 years (U.S. Census Bureau, 2017c).

Table 8.1 Marital Status of People 18 Years and Over in the United States, 2017

	Percentage				
	All Races	*White Non-Hispanic*	*Black*	*Hispanic*	*Asian*
Married Spouse Present	51.4	54.4	31.8	47.0	59.5
Married Spouse Absent	1.6	1.1	2.0	2.6	2.9
Widowed	6.1	7.0	6.0	3.4	4.4
Divorced	10.4	11.2	11.5	8.0	4.3
Separated	2.0	1.5	3.6	3.4	0.9
Never Married	28.5	28.3	45.3	35.6	27.6
	Percentages for Men Only				
	All Races	*White Non-Hispanic*	*Black*	*Hispanic*	*Asian*
Married Spouse Present	53.1	55.8	36.2	46.2	59.0
Married Spouse Absent	1.7	1.2	2.1	3.0	3.3
Widowed	2.8	3.2	2.7	1.8	1.3
Divorced	9.1	10.1	10.3	6.7	3.2
Separated	1.8	1.5	3.4	2.6	0.5
Never Married	31.4	26.4	45.2	39.7	32.3
	Percentages for Women Only				
	All Races	*White Non-Hispanic*	*Black*	*Hispanic*	*Asian*
Married Spouse Present	49.8	53.1	28.1	47.7	60.0
Married Spouse Absent	1.4	1.0	1.9	2.2	2.6
Widowed	9.2	10.6	8.8	5.0	7.2
Divorced	11.5	12.4	12.4	9.2	5.3
Separated	2.2	1.6	3.7	4.3	1.2
Never Married	25.8	20.4	45.3	31.5	23.5

Note: Hispanic is an overlapping category. All other racial groups are individuals who filled out their Census form indicating that racial group only and not people who indicated more than one race/ethnicity.

Source: U.S. Census Bureau (2017a).

Increased life expectancy means we have more "time"; therefore, postponing first marriage makes sense. The median age at first marriage has risen for both men and women. In 1980, the median age at first marriage for women was 22 and for men 24.7; however, by 2017 the estimated median age at first marriage increased to 27.4 for women and 29.5 for men (U.S. Census Bureau, 2017d). Also, because we live longer, we may be less inclined to stay in a bad relationship, since it could last for a very long time. The distributions in Table 8.1 are influenced by multiple "prisms" beyond race and ethnicity in the patterning of relationships across groups in the United States.

The "Idealized" Family

The growing diversity of household and family configurations in the United States has reshaped and challenged the rigid gender roles that pattern the ways we enter, confirm, maintain, and envision long-term intimate relationships. Not surprisingly, the rigid gender roles associated with that mythical little home behind the white picket fence, with mom staying at home to care for children and dad heading off to work, is not the reality for most households in the United States. Instead, as noted earlier, we have many different household patterns, with some households headed by a single person, same-sex households, and others unrelated by blood or family bonds living together. Some new household patterns include more single-person households. Others, such as grandparents raising grandchildren or single parents (typically mothers) raising children alone, often arise out of divorce and/or poverty. Whatever the reasons, the idealized, traditional family with its traditional gendered relationships never really was the norm (Coontz, 2000) and certainly is not the norm in American households today.

To illustrate how rare that idealized family is, only 23.4% of all married partner family groups with children under 18 years of age had a stay-at-home mother and less than 1% (0.97% or 209,000 married-couple family groups) had stay-at-home fathers in the United States in 2016 (U.S. Census Bureau, 2017b). When looking only at mothers who are married with spouses present and children under the age of 3, 43% work full-time in the labor force (Bureau of Labor Statistics, 2018). Considering these numbers, it is easy to see that only a small percentage of households fit the "Ozzie and Harriet" model for traditional families in the United States, with mom at home taking care of the children while dad is off at work. Like gender, "the family" is a culturally constructed concept that often bears little resemblance to reality. Simply put, the idealized, traditional family with separate and distinct gender roles does not exist in most people's lives.

Historically, families changed considerably in terms of how they are formed and how they function. While enduring relationships typically involve affection, economic support, and concern for others, marriage vows of commitment are constructed around love in the United States today. In previous generations, most marriage vows promised commitment and love "until death do us part."

However, marriages in the 1800s, even when rooted in love, often were based on economic realities. These 19th-century marriages were likely to evolve into fixed roles for men and women linked to the economy of the time, roles that reinforced gender difference but not necessarily gender inequality. For example, farm families developed patterns that included different, but not always unequal, roles for men and women, with both earning money from different tasks on the farm (Smith, 1987). In the latter half of the 19th century and into the early 1900s, families—particularly immigrant families—worked together to earn enough for survival. Women often worked in the home, doing laundry and/or taking in boarders, and many children worked in the factories (Bose, 2001; Smith, 1987).

What changed, and how did the current idealized roles of men and women within the stereotypical traditional family come to be? Martha May (1982) argues that the father as primary wage earner was a product of early industrialization in the United States. She notes that unions introduced the idea of a family wage in the 1830s to try to give men enough income so their wives and children were not forced to work. In the early 1900s, Henry Ford expanded this idea and developed a plan to pay his male workers $5.00 per day if their wives did not work for money (May, 1982). The Ford Motor Company then hired sociologists to go into workers' homes to make sure that the wives were not working for pay either outside or inside the home (i.e., taking boarders or doing laundry) before paying this family wage to male workers (May, 1982). In fact, very few men actually were paid the higher $5.00 per day wage (May, 1982). You might ask, why was Ford Motor Company so interested in supporting the family with an adequate wage? At that time, factories faced high turnover because work on these first assembly lines was demanding, paid very little, and the job was much more rigid and unpleasant than the farm work that most workers were accustomed to. Ford enacted this policy to reduce this turnover and lessen the threat of unionization. Thus, one reason behind the social construction of the "ideal family" was capitalist motivation to tie men to their jobs for increased profits, not the choices of individual men wanting to control their families (May, 1982).

Workplace policies are very important in ushering in changes in work, family, and gender. Historical and structural changes in the family affect our interpersonal relationships both inside and outside of marriage. For example, Amy S. Wharton in Chapter 7 describes the (un)changing institutions of work, family, and gender, including some changes over the past 50 years and also what she calls a "stalled revolution" that began in the late 1990s.

Same-Sex/Gender Couples

A change that is hidden in Table 8.1 is the growing number of same-sex/gender couples. Beginning with the 2000 Census, the category "unmarried partner" was added to the questionnaire, allowing an estimated count of same-sex/gender couples. In 2010, the Census Bureau estimated that 0.6% of households described themselves as partners of same-sex/gender (Lofquist, Lugaila, O'Connell, & Feliz, 2012). In 2016, the Census Bureau estimated there to be 887,456 same-sex couples in the United States, 54.9% of whom described their relationship as married (Census Bureau, 2017e). A recent report by the Census Bureau (Lewis, Bates, & Streeter, 2015) warns that there are still inconsistencies in how individuals respond to two measures on the American Housing Survey identifying same-sex couples, one asking about the relationship with the second adult in the household (including same-sex and opposite-sex married and unmarried partners) and the other asking about the sex of the other adult in the household. Although these estimates are better than in the past, they still must be treated cautiously.

It is important to note that the identification of same-sex/gender relationships is complicated by the use of terms to define these relationships, all of which are imprecise—sex, gender, and sexuality. Each of these terms refers to a false dichotomy (Lucal, 2008; also see the introduction to Chapter 1), with the expectation that there are two and only two categories in each. However, social relationships and individuals are much more complicated than male/female, man/woman, and heterosexual/homosexual. As Paula C. R. Rust (2000) notes, same-sex may be appropriate in some instances and same-gender in others as we describe one social context versus another. Leila J. Rupp and Verta Taylor show some of this complexity in their reading about college-aged women who identify as straight yet share sexual intimacy with other women. As they say, this kind of sexual fluidity can remind us that "sexuality is gendered and that sexual desire, sexual behavior, and sexual identity do not always match." Given the inadequacy of these categories to describe the complexity of sex, gender, and sexual identities and the shifting terminology used to do so, we use same-sex/gender in our text. However, we honor authors' use of their terminology in readings selected for this book.

The legal parameters for same-sex/gender couples are rapidly changing, beginning when Massachusetts was the first state to allow same-sex/gender marriages in 2004. Other states in the United States followed, but the big change came in June, 2015, when the Supreme Court ruled that same-sex marriage is a right under the U.S. Constitution, thus making it legal in all states in the United States. This followed changes in the Netherlands, Norway, and Canada, who were among the first nations to allow same-sex/gender marriages, with gender-neutral marriage laws that transcend the problematic terms same-sex and opposite-sex and instead extend marriage rights to any two adults. However, many other countries legally recognize same-sex/gender couples, giving them some of the rights of married couples. At the same time, some countries still prohibit same-sex/gender marriages.

Same-sex/gender couples create families and, in doing so, redefine gender, parenthood, and family, as Irene Padavic and Jonniann Butterfield describe in their reading in this chapter. Padavic and Butterfield help us understand the constraints on parenting children within lesbian families as the reactions of others constrain their mothering behaviors. Indeed, parenting relationships, while continuously changing, are clearly socially constructed, and expectations for motherhood and fatherhood have strong impacts on gendered behavior across societies (Christopher, 2012; Gregory & Milner, 2011).

Feminization and Juvenilization of Poverty

Another reality of today's families that differs from the idealized, traditional family is the number of children living in poverty, with only a broken picket fence, if that. In 2016, the official poverty level for a family of four was $24,563. According to the Children's Defense Fund (2017), over 13.2 million children lived below that income level in the United States in 2017, with over 45% of those children living at less than half the poverty level in extreme poverty. The youngest children are the most vulnerable, with one in six children under the age of 6 living in poverty in 2016 (Children's Defense Fund, 2017). This change in household composition challenges idealized gendered relationships expected in families but also relates to social class inequalities tied to gender. The increase in the number of households headed by poor women raising children has been called the feminization or juvenilization of poverty (Bianchi, 1999; McLanahan & Kelly, 1999). Diana Pearce coined the term the feminization of poverty in 1978 (Bianchi, 1999, p. 308), at a time when the number of poor, women-headed families rapidly increased. The rate of women's poverty relative to men's fluctuates over time and is 50% to 60% higher than for men (Bianchi, 1999, p. 311), a reflection of the inequality in wages for men and women discussed in Chapter 7. These rates also reflect racial and ethnic inequalities, with children of color disproportionately represented among the poor. In 2016, the poverty rate for Black children was 30.8%, American Indian/Native Alaskan children 31.0%, and Hispanic children 26.6% (Children's Defense Fund, 2017).

The juvenilization of poverty refers to an increase in the poverty rates for children that began in the early 1980s, whether in single- or two-parent families (Bianchi, 1999). Unfortunately, the juvenilization of poverty continues to increase. From 2000 to 2010, the rate of child poverty increased 36% (Children's Defense Fund, 2014). The feminization and juvenilization of poverty are serious problems, and both are gender as well as race and social class issues.

Government Policies and Family Relationships

Governmental policies play a role in shaping families in the form of tax laws, education policies, health and safety rules, and other legislation. These policies have real gender implications as well as implications for parenting in a modern world. A multitude of policies frame parents' decisions to have children, structure how much time parents have with their children, and shape how children are expected to act in societies, with varying impact across social class (Williams, 2010; see also Wharton in Chapter 7). Early in the 20th century in the United States, health and safety laws increased the age of employment for factory workers (Bose, 2001); these laws are still in effect in terms of dictating what age a child can begin to work for pay. These laws related to the creation of adolescence, with the expectation that children would remain in school throughout high school. Another act of legislation, the dependent deduction on tax returns, was first intended to encourage families to have more children at a time when politicians worried about the declining birth rate in the U.S. At the same time and for the same reason, Canada instituted a policy, the Canada Child Tax Benefit, in which caregivers are given a payment each month for every child under the age of 18. The government makes this payment directly to the caregiver. This law was based on the assumption that, in a two-parent family, the mother assumes responsibility for the children; however, payment may go to the father if a written note is submitted by the mother to indicate that he is the primary caregiver. While only a nominal sum, it was distributed across social classes as an incentive to bear and raise children. In 2007, Canada also instituted the Universal Child Care Benefit, which provides $160 per month for child care for each child under age 6 and $60 for children ages 6 through 17 (Government of Canada, 2015).

Work-family policies vary across nations, and can enforce or challenge gendered assumptions about roles at work and home. Joan C. Williams (2010) describes U.S. policies as "family-hostile." Policies and laws affect all relationships because they often idealize the woman's role in two-parent families, reinforcing hegemonic masculinity and emphasizing femininity, while ignoring other choices and life situations, such as same-sex/gender couples. The impact of policies and practices on relationships in poor families is also considerable, making it difficult to "do" idealized gender because living, in and of itself, is challenging.

Changing Relationships

Marital and parenting relationships have changed considerably over time, as has the way sociologists study the family (Ferree, 2010). In addition to government regulations, the feminization of the workforce has affected men's and women's roles in marriages (Blackwelder, 1997). The fact that most families now have two workers has changed relationships in the home. Kathleen Gerson (2002) described how young people imagine their commitments to marriage and family, balancing that against the autonomy they believe is necessary to succeed in the world of work. Race is another prism that influences families and parenting, as discussed by Dawn Marie Dow in this chapter. She analyzes the particular challenges that African American mothers face in raising their sons to protect them from being perceived as criminals, or "thugs." In addition, race should also be considered as it relates to social class and gender differences in salaries, discussed in Chapter 7.

Gender difference and inequality continue to permeate and frame ever-evolving relationships even though fathers are more involved in household labor, particularly caring for children (Bianchi, Robinson, & Milkie, 2006; see also Wharton in Chapter 7). However, although there may be more equal distribution of household labor today, mothers continue to feel the pressure of caring for children and the household (Bianchi et al., 2006). Thus, the idealized, traditional gendered responsibility for women in the United States to care for children and the household remains, even though most women work outside the home. Jerry A. Jacobs and Kathleen Gerson (2001) argue that the changes in family composition and gender relations have created situations in which members of families, particularly women and most particularly single women, are overworked, with little free time left for themselves or their families. They describe the situation as particularly acute for those couples whose work weeks are 100 hours or more and who tend to be highly educated men and women with prestigious jobs. Readings in this chapter help us understand how patterns at work can influence patterns at home and lead to conflict between commitments to work and those to family discussed in the previous chapter.

The readings in this chapter provide a fuller understanding of how our most intimate relationships are socially constructed around gender. As friends, lovers, parents, and siblings, we are defined in many ways by our gender. Ask yourself what choices you have made or wish to make as you consider how gender influences what you expect in your intimate relationships. It is important to keep in mind that many social factors, including institutional rules and policies as well as social class position, influence the decisions people make in terms of hours worked outside and inside the home (Craig, 2011; Hook, 2010).

References

Bianchi, S. M. (1999). Feminization and juvenilization of poverty: Trends, relative risks, causes, and consequences. *Annual Review of Sociology, 25*, 307–333.

Bianchi, S. M., Robinson, J. P., & Milkie, M. A. (2006). *Changing rhythms of American family life*. New York: Russell Sage Foundation.

Blackwelder, J. K. (1997). *Now hiring: The feminization of work in the United States, 1900–1995*. College Station: Texas A&M University Press.

Bose, C. E. (2001). *Women in 1900: Gateway to the political economy of the 20th century*. Philadelphia: Temple University Press.

Bureau of Labor Statistics. (2018). Economic News Release. Table 6: Employment status of mothers with own children under 3 years old by single year of age of youngest child and marital status, 2016–2017 annual averages. Retrieved from https://www.bls.gov/news.release/famee.t06.htm

Cancian, F. M. (1990). The feminization of love. In C. Carlson (Ed.), *Perspectives on the family: History, class and feminism* (pp. 171–185). Belmont, CA: Wadsworth.

Children's Defense Fund. (2014). *Child poverty. The state of America's children 2014*. Washington, DC: Author.

Children's Defense Fund. (2017). *Child poverty. The state of America's children 2017*. Washington, DC: Author. Retrieved from http://www.childrensdefense.org/library/state-of-americas-children/documents/Child_Poverty.pdf

Christopher, K. (2012). Extensive mothering: Employed mothers' constructions of the good mother. *Gender & Society, 26*(1), 73–96.

Coontz, S. (2000). *The way we never were: American families and the nostalgia trap*. New York: Basic Books.

Craig, L. (2011). How mothers and fathers share childcare: A cross-national time-use comparison. *American Sociological Review, 76*(6), 834–861.

Felmlee, D., Sweet, E., & Sinclair, H. C. (2012). Gender rules: Same- and cross-gender friendship norms. *Sex Roles, 66*, 518–529.

Ferree, M. M. (2010). Filling the glass: Gender perspectives on families. *Journal of Marriage and Family, 72*(3), 420–439.

Gerson, K. (2002). Moral dilemmas, moral strategies, and the transformation of gender: Lessons from two generations of work and family change. *Gender & Society, 16*(1), 8–28.

Government of Canada. (2015). Families and children. Service Canada, Government of Canada website: http://www.servicecanada.gc.ca/eng/audiences/families/

Gregory, A., & Milner, S. (2011). What is "new" about fatherhood: The social construction of fatherhood in France and the UK. *Men and Masculinities, 14*(5), 588–606.

Hook, J. L. (2010). Gender inequality in the welfare state: Sex segregation in housework, 1965–2003. *American Journal of Sociology, 115*(5), 1480–1523.

Jacobs, J. A., & Gerson, K. (2001). Overworked individuals or overworked families? Explaining trends in work, leisure, and family time. *Work and Occupations, 28*(1), 40–63.

Lewis, J. M., Bates, N., and Streeter, M. (2015). Measuring same-sex couples: The what and who of misreporting on relationship and sex (SEHSD Working Paper 2015–12). U. S. Census Bureau.

Lofquist, D., Lugaila, T., O'Connell, M., & Feliz, S. (2012, April). Table 3: Household type by race and Hispanic origin. In Households and Families: 2010 (p. 8). Washington, DC: U.S. Census Bureau. Retrieved from http://www.census.gov/prod/cen2010/briefs/c2010br-14.pdf

Lucal, B. (2008). Building boxes and policing boundaries: (De)constructing intersexuality, transgender, and bisexuality. *Sociology Compass, 2*(2), 519–536.

May, M. (1982). The historical problem of the family wage: The Ford Motor Company and the five dollar day. *Feminist Studies, 8*(2), 399–424.

McLanahan, S. S., & Kelly, E. L. (1999). The feminization of poverty: Past and future. In J. S. Chafetz (Ed.), *Handbook of the sociology of gender* (pp. 127–145). New York: Kluwer Academic/Plenum.

Rust, P. C. R. (2000). *Bisexuality in the United States: A social science reader*. New York: Columbia University Press.

Smith, D. (1987). Women's inequality and the family. In N. Gerstel & H. E. Gross (Eds.), *Families and work* (pp. 23–54). Philadelphia: Temple University Press.

U.S. Census Bureau. (2017a). America's families and living arrangements: 2014. Marital status of people 18 years or older by age, sex, personal earnings, race and Hispanic origin: 2016.

U.S. Census Bureau. (2017b). America's families and living arrangements: 2016: Family groups. Table FG8. Married couple family groups with children under 15 by stay-at-home status of both spouses: 2016. Retrieved from https://www2.census.gov/programs-surveys/demo/tables/families/2016/cps-2016/tabfg8-all.xls

U.S. Census Bureau. (2017c). Births, deaths, marriages, and divorces: Life expectancy. Table 104. Expectation of life at birth, 1960 to 2008, and projections, 2010 to 2020. Retrieved from http://www.census.gov/compendia/statab/cats/births_deaths_marriages_divorces/life_expectancy.html

U.S. Census Bureau. (2017d). Table MS-2. Estimated median age at first marriage, by sex: 1890 to present. Current Population Survey. Retrieved from https://www.census.gov/data/tables/time-series/demo/families/marital.html

U.S. Census Bureau. (2017e). Table 1. Estimates of same-sex couple households in the American Community Survey. Retrieved from: https://www2.census.gov/programs-surveys/demo/tables/same-sex/time-series/ssc-house-characteristics/ssex-hist-tables.xlsx

Walker, K. (1994). Men, women, and friendship: What they say and what they do. *Gender & Society*, *8*(2), 246–265.

Williams, J. C. (2010). The odd disconnect: Our family-hostile public policy. In K. Christensen & B. Schneider (Eds.), *Workplace flexibility: Realigning 20th-century jobs for a 21st century workforce* (pp. 23–54). Ithaca, NY: ILR Press.

Introduction to Reading 38

The world of gender is constantly changing. Women and men are increasingly looking for egalitarian relationships in their lives. However, as noted in this research by Ellen Lamont, women are still looking for traditional gendered patterns in serious relationships. This reading provides some interesting insights in terms of how women reconcile these two opposing behaviors; that is, how does one manage an egalitarian relationship while, at the same time, expecting one's date to pick up the tab for dinner and open the car door? She reports results of interviews with 38 women between the ages of 25 and 40 in the San Francisco area. The interviews with these middle- to upper-middle-class women lasted about 3 hours and should provide some insights into contradictions in our own attitudes and behaviors.

1. How did the respondents in this sample reconcile the discrepancies between what they expect from men in serious dating relationships and their desire for equality?
2. Why did the women's behaviors and acceptance of traditional gender norms change over the course of getting to know someone in a dating relationship?
3. Do you think the social class of these women might affect some of the results reported here? If so, how might findings be different for women lower or higher in social class?

Negotiating Courtship

Reconciling Egalitarian Ideals With Traditional Gender Norms

Ellen Lamont

Courtship conventions delineate distinct gendered behaviors for men and women based on the model of an active, breadwinning male and a passive, dependent female (Bailey 1988; Cate and Lloyd 1992). These norms situate men as the initiators in relationships. Men are responsible for asking women out, paying for dates, determining when the relationship will shift from casual to committed, and proposing marriage, while women are limited to reacting to men's overtures (Bogle 2008; England, Shafer, and Fogarty 2008; Laner and Ventrone 2000; Sassler and Miller 2011). Yet, as women have increased their access to earned income, there has been a rising ideological and behavioral commitment to egalitarian relationships (Bianchi, Robinson, and Milkie 2006; Gerson 2010). College-educated women expect to

Lamont, E. (2014). Negotiating courtship: Reconciling egalitarian ideals with traditional gender norms. *Gender & Society, 28*(2): 189–211.

pursue lucrative and rewarding careers and form peer relationships that provide room for independence and self-development (Coontz 2005; Hamilton and Armstrong 2009). In spite of these destabilizing shifts, traditional gender ideologies remain remarkably resilient, as courtship conventions symbolizing men's dominant, breadwinning status stubbornly persist (Eaton and Rose 2011). These competing sets of behavioral rules create a "moral dilemma" (Gerson 2002) for women as they seek to negotiate romantic relationships in what has been referred to as an "uneven" gender revolution (England 2010), with women's employment opportunities changing more rapidly than gendered patterns in the home. This study looks at how a sample of highly educated women navigate this contradiction, examining how their economic resources and their expressed support for egalitarian relationships intersect with persisting gender norms to shape their contemporary courtship behaviors and attitudes.

Since the courtship period may influence couples' expectations regarding gendered behaviors during marriage (Humble, Zvonkovic, and Walker 2008; Laner and Ventrone 2000), it is important to understand how courtship conventions may impede women's equal status in romantic relationships and where openings for change, and greater equality, may be occurring. While some scholars have speculated that the intransigence of these norms may not be critical to achieving equality (Graf and Schwartz 2011), others posit that they contribute to the power imbalance between men and women (England 2010; Sassier and Miller 2011). Given that the focus of recent research on middle-class women's relationships has been on the college years (Bogle 2008; England, Shafer, and Fogarty 2008; Hamilton and Armstrong 2009), little is known about how women are negotiating conventional scripts in the changing social landscape as they exit college and begin the process of forming long-term partnerships. By looking at how college-educated women navigate courtship and the narratives they use to make sense of their behaviors, this study reveals the contradictory processes of social change among middle-class women.

Symbolic Gendering and the Changing Context of Courtship

Gender scholars have pointed to the ways that gendered meanings continue to influence social relationships, even as the material dimensions that support inequality erode (England 2010; Ridgeway 2011; Tichenor 2005). Ridgeway asserts that gender remains a primary frame that men and women use to define who they are, how they will behave, and how they expect others to behave. Individuals draw on cultural knowledge, or "shared," "common" knowledge that "everybody knows," to coordinate their behavior and facilitate social cohesion (Ridgeway 2011, 35). This knowledge reflects cultural stereotypes about how men and women behave as a result of their sex category and emphasizes the perceived differences between the two groups. As Ridgeway shows, most adults continue to believe that men and women are innately different with either complementary or conflicting needs and desires, especially in heterosexual romantic relationships where sex differences are highlighted. In particular, many studies reveal the existence and pervasiveness of beliefs about men's and women's relative levels of assertiveness versus responsiveness and interest in casual sex versus a committed relationship (Hamilton and Armstrong 2009; Ridgeway 2011). These perceived differences are especially salient during courtship, when people tend to fall back on scripts to ease uncertainty and reassure themselves and others that they conform to normative sexual standards (Eaton and Rose 2011).

Scripts that become culturally hegemonic tend to enshrine the behavior of white, middle-class heterosexuals, who remain the dominant social group (Bailey 1988; Ridgeway 2011). As a result, research on dating and courtship shows remarkable stability and agreement between men and women in support of the traditional, and highly gendered, courtship script (Eaton and Rose 2011; Laner and Ventrone 2000). This script demonstrates men's status as active economic providers and women as more passive dependents, dictating that men initiate, plan, and pay for first dates while women limit themselves to reactive behavior, such as accepting physical contact and being walked to the door (Laner and Ventrone 2000). Relationship progression, including decisions about exclusivity and engagement, also favors the man's desired timetable, with women less empowered to openly seek their desired outcomes (Bogle 2008; England, Shafer, and Fogarty 2008; Sassler and Miller 2011). According to the prevailing narrative, women are looking for commitment, while men are trying to avoid it (Bogle 2008), but women's lack of power to define the terms of courtship holds true even when women do not desire commitment. Hamilton and Armstrong (2009) thus show that women are often coerced into a commitment they do not desire. Alternatively, women who act "too forward" are often passed over for future dates (Bogle 2008) or face relationship destabilization (Sassler and Miller 2011). Even college students who describe themselves as egalitarian engage in these inegalitarian dating patterns, as beliefs about men's assertiveness and women's responsiveness continue to function "unconsciously" (Laner and Ventrone 1998, 475). Drawing on Goffman's (1976) argument that individuals symbolically enact cultural beliefs about men and women through "gender displays," I use the term "symbolic gendering" to refer to the cultural practices associated with courtship. These practices are used to represent what are assumed to be essential, biological differences between women and men, making them appear inevitable while obscuring how they privilege men (Bourdieu 1998).

Yet changes in U.S. women's educational and employment opportunities challenge conventional beliefs about gender difference. As women's wages become increasingly important to attaining middle-class status, romantic relationships based on women's financial dependence appear increasingly less desirable to both men and women (Gerson 2010). Middle-class parents prepare their daughters for professional careers (Lareau 2003), and, with women's college graduation rates now exceeding men's, these women are far more likely to expect career trajectories that mirror those of their male counterparts (Damaske 2011). Among highly educated women, successes in the workplace have undermined the use of gendered self-descriptions, with women more likely to think of themselves as agentic than in the 1960s (Twenge 2001).

The changes have prompted new approaches to relationships. College-educated women expect to pursue career opportunities in young adulthood while delaying marriage until their late twenties or early thirties (Hamilton and Armstrong 2009). The hook-up culture on college campuses allows young adults to engage in sexual encounters outside the context of a relationship, which are often viewed as too time consuming by women hoping to succeed at school and in careers (Hamilton and Armstrong 2009). When relationships *are* formed, women thus expect room for independence and self-development (Cancian 1987; Coontz 2005). The majority of young women seek a relationship in which both partners share work and home responsibilities, while many say they would rather forgo a relationship than be in an unequal one (Gerson 2010). Although women continue to do the majority of housework and child care, men have increased their contributions, while women have decreased theirs (Bianchi, Robinson, and Milkie 2006), and there is some evidence that indicates that this shift is at least partly due to women's increased earnings relative to men's (Bianchi et al. 2000).

As women form stronger attachments to paid labor, conflicts between a commitment to self-development and personal relationships (Hamilton and Armstrong 2009) create a "moral dilemma" with no clear socially sanctioned solutions (Gerson 2002). This dilemma provides an opportunity to challenge conventional gender norms, which no longer represent an obvious, or necessarily viable, pathway. This tension between convention and change can be found in the contrasting findings on hooking up and dating, which highlight the constraints on women's ability to negotiate their desired ends (Bogle 2008; Hamilton and Armstrong 2009), and the literature on relationships, which points to women's efforts to reject inequality in the home and find value in their personal achievements (Graf and Schwartz 2011). These cultural messages are especially likely to clash for college-educated women, who face pressures both to achieve autonomy and to defer to men. In this study, I ask: How does women's commitment to self-development and economic independence intersect with traditional gender norms to shape their courtship narratives and behaviors? To explain how middle-class women negotiate these conflicting norms, I argue that women have disassociated the two scripts, drawing on cultural narratives of choice, individualism, and essentialism to assert that the symbolic gendering in courtship is unrelated to the equality they seek in their married lives as high-achieving professionals. In this manner, however, women inadvertently perpetuate ideologies of gender difference, a basis of inequality in both the household and the workplace.

* * *

Findings

The majority of the women expressed a preference for conventional courtship behaviors and expected men to ask and pay for the first date, confirm the exclusivity of the relationship, and propose marriage. Drawing on the interview data, I illustrate how women justified their support for these conventions in spite of their desire for egalitarian relationships. . . .

Narratives of Gender Difference

Women referred to popular, essentialist beliefs about men's need to be the assertive, dominant partner to explain why they preferred men to ask for dates.

Over a third indicated that they did not ask men on dates because it was in man's nature to like "the thrill of the chase" or it was the man's "role" to do the pursuing. Jenna, 26, a research assistant, said, "It's just partly biological. In animals, the guy always flashes. The male bird always flashes his colors—his feathers or something—to go after what he wants." By locating behavioral differences in biology, women framed these behaviors as natural, inevitable, and legitimate and so did not challenge them (Bourdieu 1998). And because they assumed that men need to be the dominant partners, they argued that women who took that away from men would be considered unappealing partners. Caroline, 31, a marketing director, said, "I feel like men need to feel like they are in control and, if you ask them out, you end up looking desperate and it's a turn-off to them." Anna, 40, a high school teacher, said, "I know that with a man they like to take charge." Although Anna admitted that she, too, really liked to be in control, saying that she liked to ask men out, she attributed this to her personality, rather than to her nature. As a result, women believed they must adjust their behavior to men's natural, unchanging desires.

Only two women made essentialist arguments about women's needs. Instead, like Anna, women discussed their own preferences in terms of their personalities. Indeed, almost one third of the women explained that their unwillingness to ask for a date was due to their personality. Olivia, 26, a lawyer, said she didn't like to approach men "more because I'm shy than out of traditional gender roles or anything like that." Breanna, 36, an internal auditor, said, "I would never approach a guy. I think one is, I'm shy. Two is I've never felt like 'Oh, I'm gorgeous,' so it would be fear of rejection." As Ridgeway (2011) argues, people are more likely to explain the behaviors of others using stereotypes than their own, making gendered self-reports more progressive. Yet, while these women didn't explicitly discuss gendered expectations as a factor, they still attributed to themselves a level of acceptable passivity. Women did not view this passivity as a hindrance to getting dates, a position not possible for men. None of the women acknowledged that men, too, might be shy and afraid of rejection, and none of these women admitted to shyness in their professional lives, indicating that their courtship narratives are potentially unconsciously gendered.

Women's passivity provided assurance of men's interest and protected them from rejection. Amelia, 33, an environmental consultant, said, "I think it's just because I'm old-fashioned that way. I want to know the guy is interested in me." Only 10 of the 38 women had ever asked a man out, and half of them described their actions in less than empowering terms. Their experiences reinforced the cultural stereotype that if a man is actually interested, he *will* pursue, and that women are better off waiting for the man to take the lead. Abby, 33, a postdoctoral fellow, said, "I tended to be the one to approach guys, but those were usually the ones that didn't like me." Heather, 27, an operations technician, said that she had stopped contacting men on an online dating site:

> When I have, they're not interested. There was this one time I saw this guy was looking at my profile two or three times. He was kind of a dick. He was like, "Yeah. Sometimes I just click on people's profiles. It doesn't mean I want to date them." It's funny. The times I've contacted people, I've never ended up meeting them.

Women experienced men's negative reactions as sanctions for transgressing appropriate gendered behavior. Because people know what is expected of their gender and can anticipate these sanctions (Rudman and Fairchild 2004), many women focused on making their interest clear in ways other than suggesting a date. While only a handful of women reported asking a man out on a first date, half reported at least one partnership in which she had pursued a man. Ariana, 30, a doctor, described aggressively pursuing a man she was interested in dating: "I called him and told him to come to a party and that I would make it worth his while. I told him to sit next to me, and took his arm and put it around me. I said, 'Finally, I have all of your attention.' He said, 'You sure do' and kissed me. Then he asked me out." Caroline said, "I approached him at a bar and flirted with him and then invited him [and his friends] to go to a strip club with me and my friends. While we were there, I sat on his lap. . . . But I let him call me the next day. He did and he asked me out on our first date." In this manner, women were able to test the boundaries of appropriate gender behavior without completely challenging them. . . .

Ensuring Men's Commitment

The ubiquity of the hook-up culture on college campuses continued to influence women's understandings of men even after they exited college and entered adulthood. While they may have felt ambivalent about long-term commitment during college (Bogle 2008; Hamilton and Armstrong 2009), as adults in their late twenties and early thirties most of the women were looking for a partner interested in the possibility of marriage. Because many of the women viewed men as still commitment phobic or more interested in casual sex than a relationship, a belief that was reinforced by popular narratives of men's "nature" (Geller 2001), they used an adherence to courtship norms to confirm men's genuine interest in them. Thus, women used the formal date, with its attendant rituals, to distinguish men who were interested in the possibility of a relationship from men who were just looking for a casual sexual encounter.

Since women frequently cited "chivalry" as a sign that the man was respectful, caring, and interested in more than sex, many wanted the man to ask for, plan, and pay for the first few dates. Olivia said, "I tend to like a formal date, like asking me out on a date. Like, 'Would you like to go to dinner?' It just seems like a more clear idea of what's going on. And I also think I like the chivalry of the formal date invitation as opposed to 'Let's just see what happens, maybe we'll hook up.'" Ariana also alluded to chivalry, saying,

> I mean, usually the first time they go out with me I'll offer to pay. I'm like, "Oh, let me split it with you," you know? And it's really honestly a test. I don't want them to say, "Okay." I want them to say, "No, I'll get it," you know? But I usually offer to split it. And then if we go out, like, four times, by the fourth time I'll be like, okay, this is my turn now. Like I want to make sure the guy offers to pay, the guy opens my door, the guy, you know, doesn't just walk ahead of me. Things like that. And that's become more important to me, how gentlemanly they are. I've talked to guys about it before. If you like a chick and you want to impress her, you do everything you can. [If a guy doesn't pay] they just probably don't like you that much.

More than two thirds of the women said that all their first dates were paid for by men. Just like Ariana, many of the women referenced payment of the first date as a test. If the man took them up on their offer to pay or split the check, it was a sign that he wasn't someone they wanted to date, assuming he wasn't "out to impress" and must not be sufficiently interested in them. Only a handful of the women indicated that men's payment was a way to confirm breadwinning ability, perhaps because men who weren't able to take on this responsibility in the long run were screened out before the date even took place, as the majority of the women dated men with similar career opportunities. Instead, men's payment for the first date appears to have taken on new meanings as women have gained their own breadwinning abilities.

These dating conventions became less important to women over the course of their relationships. Most of the women reported, for example, that, once they started dating a man regularly, payment for dates frequently alternated between the two of them. The conventions acquired significance again, however, during moments that were highly scripted and where assumptions about men's commitment to the relationship became salient. Therefore, women expected men to confirm the exclusivity of the relationship and propose marriage as a signal that let women know they were committed to the relationship, as there was a consistently stated belief that men were reluctant to commit. As a result, women initiated fewer than a fifth of the conversations on the exclusivity of the relationship. Consistent with previous findings on gender performance (Sassler and Miller 2011), when women did bring up the exclusivity of the relationship, they tended to do so in an indirect manner, asking questions such as "Where is this relationship going?" or "What are we?" This approach provided women with the ability to initiate the conversation topic, protecting them from a more direct rejection of their desired ends, but gave men the power to confirm or deny an exclusive commitment.

In addition, because many of the women believed that men who were interested in commitment were rare because of essentialist beliefs that men are naturally promiscuous, when men initiated conversations about exclusivity, women sometimes ended up committing to a relationship before they were ready (Hamilton and Armstrong 2009). Nicole, 28, a marketing manager, described how the man she had been dating gave her an ultimatum after he found out she was also having sex with other men: "He said that he wanted me to be his girlfriend and that I couldn't see other people." She agreed because his insistence on commitment "made me see how much he cared about me." Because of the cultural belief that women always want commitment, they worried that their attempts to secure exclusivity could be construed as desperation. In contrast, because of the cultural belief that men avoid commitment, men who did commit were viewed as especially devoted. As a result, women frequently ended up prioritizing men's desires.

Women were even more reluctant to propose marriage. Men proposed in each of the 22 cases analyzed in this study; in addition, all the unmarried women expected men to do the proposing. Again, women discussed men's initiation of the proposal as a sign of commitment to her and to the marriage. Caroline said, "I wanted him to do it since he was really the one who had been slower to be there emotionally. I wanted him to be the one to drive it." Jane, 31, a student who was waiting for her boyfriend to propose, said, "I want to feel adored and I feel like if I was doing the proposing, it was kind of like, 'What, I'm not special enough that you're willing to put yourself out there and be vulnerable for me?'" The act of being chosen remained a powerful draw for the women in this study. To be chosen meant to be considered worthy of love and a lifelong commitment. Rather than view a female-initiated proposal as an expression of valid desire and unwillingness to remain passive, women viewed it as embarrassing reflection of their partner's lack of love or their own desperation. When I asked Ashleigh, 29, a stay-at-home mother, if she would have been willing to propose to her husband, she said, "Never. In my mind, that's not my role. Like, I would feel like he didn't really value me if he wasn't going to propose to me," while Alice, 34, a computer programmer, said, "I think I wouldn't do it because I want to make sure the other person loves me as much, if not more."

Yet the majority of the engagements occurred on a mutually agreed-on timetable. Most of the couples discussed marriage extensively before getting engaged, often going over how they envisioned their lives together, as this was considered pragmatic. After the couple decided when they wanted to get engaged, the man was expected to "surprise" the woman with a proposal. . . .

This approach allowed women to preserve the narrative of the male-initiated proposal, cementing their "chosen" status, while protecting their inclusion in the decision-making process.

Only three women said that they "waited it out" until their partner was ready to propose. In fact, a quarter of the women in the study were aggressive in influencing the timing of the engagement, often giving their boyfriends ultimatums and timelines for proposing. Just as with women who pursued men for first dates, these women were more likely to be highly paid professionals than the women who took a more "hands-off" approach, potentially indicating that their careers empowered them to be more assertive. Alice moved out of state when her boyfriend failed to propose on her timetable and agreed to move back only after he expressed a willingness to propose within the year: "I told him, 'If I move back, there better be a ring on my finger.'" He proposed one year later. . . .

Thus, the women in this sample were almost always able to negotiate desired outcome.[1]

However, because of the often repeated sentiment that proposals initiated by women, whether directly or indirectly, were coercive and indicated a lack of interest by the man, women felt conflicted about issuing ultimatums or otherwise influencing the timing of the engagement, again citing the fear of negative reactions for transgressing gender norms. As a result, while women were willing to play a decisive role in the timing of their marriage proposal, they preferred to keep the illusion of surprise with their peers. Caroline said, "I didn't tell any of my friends what actually happened. I told him to put the ring away so he wouldn't feel like he was backed into a corner. I said he should do it the way he wanted to and that I would say yes. He proposed three weeks later on a boat. That's the story our friends know." In this manner, women's "official" stories of their marriage proposals rarely acknowledged the "backstage work" that took place (Goffman 1959).

The narratives of the women who did not embrace courtship conventions show how support for these norms is rooted in women's desire to secure men's commitment. Only nine women expressed reservations about these conventions and only four uniformly rejected them. These women were not any less likely to express essentialist beliefs about men, but the four who voiced the strongest objections to gendered courtship did not want to get married and have children. Keira, 36, a researcher and recruiter for a tech company, said,

> I never fantasized about the wedding the way my friends did in school. . . . I wanted to see what I could do with my life. Getting married and having kids were probably, if they were even on the list, they were like number 99 and 100 on the list of 100. . . . I think the men I was with knew. It would just be ridiculous if they were on a bended knee offering me a ring.

As a result, they didn't need to rely on courtship conventions to ferret out which men were truly interested in commitment, nor did they have to worry about scaring men off by appearing too interested in commitment, because they weren't.

Yet, even though these women disavowed courtship rituals, they often found themselves engaging in them anyway because the men they dated fell back on these patterns and they "just didn't care enough" to challenge them. Rachel, 26, a vice president of business development, said that she let men ask her out because she "wasn't someone who always needed a boyfriend." Keira always brought money to pay her share on dates, but said her partners were "old-fashioned" and insistent on paying. And while her ex-husband did indeed propose to her, she argued that it was because she "didn't want [marriage] enough," not because she saw it as his role: "He definitely felt more strongly about me than I did about him." Thus, this group of women described their courtship behaviors as the result of men's desire for convention and explained their de facto conformity as the result of their disinterest in the assumed relationship goals of women. Still, consistent with the narratives of the women who wanted men to propose, Keira associated the initiation of the marriage proposal with the strength of her partner's love and commitment. His level of interest in her was reflected in his willingness to ask for long-term commitment.

Competing Narratives Reconciled

Although most of the women supported conventional courtship through symbolic gendering, almost all of them also described their ideal relationship as one in which partners shared breadwinning, housework, and child care relatively equally. Consistent with other findings (Damaske 2011), among the 32 women in the sample who wanted or had children, three quarters reported they had not interrupted or would not interrupt their careers. Caroline broke up with the man she dated during business school when it became clear that they had different visions of their life together. While he wanted a wife who would stay home and raise their children, Caroline planned on staying in the workforce and supporting herself financially, arguing, "I don't want to be in a dynamic where anyone is mooching on anybody or anybody feels entitled to other people's stuff." She finally decided to break up with her boyfriend when she saw the way his father treated his stay-at-home wife. . . .

Not only did Caroline enjoy her work, she also recognized that, without a job, she would have less power to negotiate a fair division of housework, a goal shared by 33 of the women in the sample.[2]

As Gerson (2010) found, if an egalitarian relationship seems unlikely, women are more likely to choose self-reliance than a traditional partnership that poses "the dangers of domesticity." Caroline expressed the fear experienced by many that, without financial independence, she would lose power in her relationship and the ability to leave if necessary. Ariana grew up watching her mother suffer physical and emotional abuse by her father and wanted to avoid the same fate, so she attended medical school to become a surgeon: "My mom was a stay-at-home mom and my dad had a lot of financial control over her and she always emphasized be independent, get a good education so you have your own financial independence and so no man can use that to control you. So that's always been embedded in my brain." After observing their parents' relationships or learning of the challenges faced by women without money, many of the women not only wanted access to their own incomes, but also their own bank accounts. Just over two thirds said that it was important for them to have a separate bank account in order to protect their assets and provide them with greater control over their personal finances. While this was aspirational for the unmarried, more than half of the married women had their own bank accounts, including three of the five women with children. When asked why she kept all her money separate from her husband of 18 years, Anna replied, "I needed to have my own money. I don't need [my husband] looking over my shoulder and telling me what to do," echoing the sentiments of many others. This approach provided women not only with income, but with control over it.

Interestingly, the nine women who planned to leave the workforce for more than a year after having children or who planned to take on a greater share of the household labor in their marriages were no more likely to express support for courtship conventions that symbolized men's dominant, breadwinning status than women who desired an egalitarian partnership. Rather, the women in this study differentiated between the symbolic gendering of courtship and gendered behaviors in the home and workplace, most of them denying a relationship between the two. Thus, while the egalitarian narratives expressed by the majority appear to contradict their commitment to courtship norms, they did not perceive that symbolic gendering would undermine these goals. Seeing symbolic gendering as either a personal preference or a mere convention (Hamilton, Geist, and Powell 2011; Swim and Cohen 1997), it appeared inconsequential to interpersonal power relations and any goals for an egalitarian marriage. . . .

Because stereotypical representations can purportedly provide benefits to women, such as "the belief that women should be protected and taken care of by men," there are fewer incentives to challenge symbolic gendering than to address overt sexism or discrimination in the workplace (Becker and Wright 2011, 63). Aashi, 29, a marketing associate, explained:

> I feel like men and women should be treated equally as far as in their career and their political day-to-day lives, things where they should be treated as equals. But when it comes to biology and manners, it's not like a woman can't open the door, it's not like she can't pay for herself, but when a man does it, it's a nice gesture and it's just . . . It's a nice thing.

This interpretation of equality draws on liberal feminist themes, such as women's legal and economic rights, and is consistent with ideologies of American individualism (England 2010). Focused on women's entry into formerly "male" spheres, the women downplayed how difference narratives in personal relationships contribute to inequality. Caroline, who had broken up with the boyfriend who wanted a gendered division of labor, wasn't as opposed to symbolic gendering: "I am okay with the fact that the gestures [my fiancé] makes are not identical to mine . . . sort of in the same way that I know he'll never carry our child for nine months in his belly, but I trust that he'll do other things to be a great father." Thus, even though women and men were expected to engage in distinct behaviors, this was not viewed as inherently unequal.

In addition, because most women were able to support themselves financially, they did not see gestures that grant men symbolic dominance as a risk to their autonomy or power within relationships. Anna was pleased when her now-husband asked her father for permission to propose: "Not that it would really matter because I'm a pretty independent, liberal-thinking woman, but . . . it is what it is." Breanna said,

> I obviously easily could take over. I am, like I mentioned, independent and self-sufficient. So obviously if I wanted to put my foot on the ground and he didn't want to go my way I could walk away—I'm not dependent on him. I don't need him for anything. But I *choose* not to take that position. . . . I do like a dominant man. . . . [I don't want] them to be submissive in any way. Gross. That would totally turn me off that guy. I even came to see where some women were insisting on paying on the dates to establish their independence. I think it's totally wrong. I mean, I think it's good to be strong and independent, but then to, like, you know . . . kind of . . . force it out there, like "I'm letting you know I'm independent." Like, "I don't need you"—that kind of thing. I don't think that's the best. Even if he's not truly dominant, even if you're his equal, I still think you should let him feel like a man, that kind of thing.

Breanna's argument again reflects the essentialist belief that men need to be in charge in their romantic relationships in order to be happy. She frames men's symbolic dominance as a charade that allows men to "do gender" in spite of women's increased economic independence (West and Zimmerman 1987). But she also states that *she* finds this enactment of gender difference attractive, too, revealing the "cathexis" experienced by many of the women that results from this "social patterning of desire" (Connell 1987, 112). By emphasizing this behavior as chosen, however, she denies an association between her behavior and the reproduction of gender inequality.

As Stone (2007) found, "choice rhetoric" is used by high-achieving women to disavow the constraints women continue to face. A narrative of choice is appealing because it draws on "the language of privilege, feminism, and personal agency" and is, therefore, consistent with how they have constructed understandings of themselves (Stone 2007, 125). By emphasizing symbolic gendering as a choice rather than a requirement, and by reaffirming their autonomy, women were able to take comfort in the courtship conventions that felt "safe" and "right," without sacrificing their sense of an independent, empowered self. As McCall (2011) speculates, the educational and professional attainments of privileged women transgress enough gender boundaries in the workplace and the home to allow them to downplay gender inequality as a relevant social problem.

Conclusion

This study shows how college-educated women's commitment to courtship norms reinforces narratives of gender difference even as they lay claims to gender equality. New norms about gender equality and women's autonomy now compete with more traditional courtship scripts, creating a cultural contradiction for women as they seek to reconcile discrepant sets of behavioral rules. Women's narratives reveal the interactions between the processes that promote both social change and social persistence. They are encountering men on new terms and creating relationships that challenge the assumptions underlying a gendered division of labor. They express comfort asserting financial independence, personal autonomy, and a desire for an egalitarian partnership. The progress they have made, however, has perhaps led them to believe that they can pick and choose between gendered meanings with no consequences. These findings demonstrate that, to ease the conflict between a desire for equality and a persistence of conventional courtship rituals, women conclude that the symbolic gendering of courtship does not contribute to the perpetuation of gender inequality. They construe men's participation in unequal courtship patterns as natural and inevitable and they explain their own participation as a personal choice that is rooted in their personalities and preferences. In this manner, women are able to reaffirm their autonomy and deny the significance of inequality in courtship, demonstrating how narratives of empowerment based on ideologies of individualism can be used to conceal the continuation of male privilege in ways that make individuals feel good about their conformity. Unfortunately, this approach not only limits the options for more privileged women, it also reinforces norms for women whose limited resources provide them with fewer opportunities to challenge gender inequality. When solutions are framed as individual rather than structural, the inability to negotiate a preferred arrangement, often the result of a lack of bargaining power, instead becomes framed as a personal failing, discouraging a broader challenge to the norms that structure relationships.

* * *

Notes

1. This finding is in contrast to Sassler and Miller's (2011) study of working class couples, where women found their efforts to secure desired marriage proposals mostly rebuffed, indicating that well-educated, middle-class women may be better positioned to challenge gender norms successfully.

2. Given space limitations, I do not present findings on how married couples enact gender difference symbolically, although this was discussed by the 17 women who were married or had been married. But activities such as initiating and planning recreation and gifting were more likely to be performed by women, as they took over the bulk of the "relationship work." Household labor remains gendered, as well, and the majority of the married women took their husband's last name.

References

Bailey, Beth. 1988. *From front porch to back seat: Courtship in twentieth-century America.* Baltimore: Johns Hopkins University Press.

Becker, Julia, and Stephen Wright. 2011. Yet another dark side of chivalry: Benevolent sexism undermines and hostile sexism motivates collective action for social change. *Journal of Personal and Social Psychology* 101: 62–77.

Bianchi, Suzanne, Melissa Milkie, Liana Sayer, and John Robinson. 2000. Is anyone doing the housework? Trends in the gender division of household labor. *Social Forces* 79: 191–228.

Bianchi, Suzanne, John Robinson, and Melissa Milkie. 2006. *Changing rhythms of American family life.* New York: Russell Sage Foundation.

Bogle, Kathleen. 2008. *Hooking-up: Sex. dating, and relationships on campus.* New York: New York University Press.

Bourdieu, Pierre. 1998. *Masculine domination.* Stanford, CA: Stanford University Press.

Cancian, Francesca. 1987. *Love in America: Gender and self-development.* Cambridge, UK: Cambridge University Press.

Cate, Rodney, and Sally Lloyd. 1992. *Courtship.* Newbury Park, CA: Sage.

Connell, Raewyn. 1987. *Gender and power.* Stanford, CA: Stanford University Press.

Coontz, Stephanie. 2005. *Marriage, a history: How love conquered marriage.* New York: Penguin.

Damaske, Sarah. 2011. *For the family? How class and gender shape women's work.* New York: Oxford University Press.

Eaton, Asia, and Suzanna Rose. 2011. Has dating become more egalitarian? A 35 year review using sex roles. *Sex Roles* 64: 843–62.

England, Paula. 2010. The gender revolution: Uneven and stalled. *Gender & Society* 24: 149–66.

England, Paula, Emily Fitzgibbons Shafer, and Alison Fogarty. 2008. Hooking up and forming romantic relationships on today's college campuses. In *The gendered society reader,* 3rd ed., edited by Michael Kimmel and Amy Aronson. New York: Oxford University Press.

Geller, Jaclyn. 2001. *Here comes the bride: Women, weddings, and the marriage mystique.* New York: Four Walls Eight Windows.

Gerson, Kathleen. 2002. Moral dilemmas, moral strategies, and the transformation of gender: Lessons from two generations of work and family change. *Gender & Society* 16: 8–28.

Gerson, Kathleen. 2010. *The unfinished revolution: How a new generation is reshaping family, work, and gender in America.* New York: Oxford University Press.

Goffman, Erving. 1959. *The presentation of self in everyday life.* New York: Anchor.

Goffman, Erving. 1976. Gender display. *Studies in the Anthropology of Visual Communication* 3: 69–77.

Graf, Nikki, and Christine Schwartz. 2011. The uneven pace of change in heterosexual romantic relationships: Comment on England. *Gender & Society* 25: 101–07.

Hamilton, Laura, and Elizabeth Armstrong. 2009. Gendered sexuality in young adulthood: Double binds and flawed options. *Gender & Society* 23: 589–616.

Hamilton, Laura, Claudia Geist, and Brian Powell. 2011. Marital name change as a window into gender attitudes. *Gender & Society* 25: 145–75.

Humble, Aine, Anisa Zvonkovic, and Alexis Walker. 2008. "The royal we": Gender ideology, display, and assessment in wedding work. *Journal of Family Issues* 29: 3–25.

Laner, Mary Riege, and Nicole Ventrone. 1998. Egalitarian daters/traditionalist dates. *Journal of Family Issues* 19: 468–77.

Laner, Mary Riege, and Nicole Ventrone. 2000. Dating scripts revisited. *Journal of Family Issues* 21: 488–500.

Lareau, Annette. 2003. *Unequal childhoods: Class, race, and family life.* Berkeley: University of California Press.

McCall, Leslie. 2011. Women and men as class and race actors: Comment on England. *Gender and Society* 25: 94–100.

Ridgeway, Cecilia. 2011. Framed by gender: *How gender inequality persists in the modern world.* New York: Oxford University Press.

Rudman, Laurie, and Kimberly Fairchild. 2004. Reactions to counterstereotypic behavior: The role of backlash in cultural stereotype maintenance. *Journal of Personality and Social Psychology* 87: 157–76.

Sassier, Sharon, and Amanda Miller. 2011. Waiting to be asked: Gender, power, and relationship progression among cohabiting couples. *Journal of Family Issues* 32: 482–506.

Stone, Pamela. 2007. *Opting out? Why women really quit careers and head home.* Berkeley: University of California Press.

Swim, Janet, and Laurie Cohen. 1997. Overt, covert, and subtle sexism: A comparison between the attitudes toward women and modern sexism scales. *Psychology of Women Quarterly* 21: 103–18.

Tichenor, Veronica. 2005. *Earning more and getting less: Why successful wives can't buy equality.* New Brunswick, NJ: Rutgers University Press.

Twenge, Jean. 2001. Changes in women's assertiveness in response to status and roles: A cross-temporal meta-analysis. *Journal of Personality and Social Psychology* 81: 133–45.

West, Candace, and Don Zimmerman. 1987. Doing gender. *Gender & Society* 1: 125–51.

Introduction to Reading 39

In this short reading, Leila J. Rupp and Verta Taylor examine the somewhat-recent phenomenon of college women making out with other women, often at parties. They suggest that this trend is an example of the social construction of sexuality, or the idea that whom we desire, our sexual behaviors and attitudes, and our sexual identity is shaped by society. While the authors mention the same-sex sexual experiences of men only briefly, think about how the social construction of sexuality intersects with the social construction of gender to understand why men and women who identify as heterosexual have such different experiences when it comes to same-sex sexual encounters.

1. What explanations do the authors give for why some college women kiss other women?
2. Under what conditions does "making out" with women impact college women's identity as heterosexual? What does it mean to "cross the line"?
3. Why doesn't "heteroflexibility" give both men and women equal leeway in engaging in same-sex behavior?

Straight Girls Kissing

By Leila J. Rupp and Verta Taylor

The phenomenon of presumably straight girls kissing and making out with other girls at college parties and at bars is everywhere in contemporary popular culture, from Katy Perry's hit song, "I Kissed a Girl," to a Tyra Banks online poll on attitudes toward girls who kiss girls in bars, to AskMen.com's "Top 10: Chick Kissing Scenes." Why *do* girls who aren't lesbians kiss girls?

Some think it's just another example of "girls gone wild," seeking to attract the boys who watch. Others, such as psychologist Lisa Diamond, point to women's "sexual fluidity," suggesting that the behavior could be part of how women shape their sexual identities, even using a heterosexual social scene as a way to transition to a bisexual or lesbian identity.

These speculations touch on a number of issues in the sociology of sexuality. The fact that young women on college campuses are engaging in new kinds of sexual behaviors brings home the fundamental concept of the social construction of sexuality—that whom we desire, what kinds of sexual acts we engage in, and how we identify sexually is profoundly shaped by the societies in which we live. Furthermore, boys enjoying the sight of girls making out recall the feminist notion

Rupp, L. J., & Taylor, V. (2010). Straight girls kissing. *Contexts, 9*(3): 28–32. Reprinted with permission from the authors and the American Sociological Association.

of the "male gaze," calling attention to the power embodied in men as viewers and women as the viewed. The sexual fluidity that is potentially embodied in women's intimate interactions in public reminds us that sexuality is gendered and that sexual desire, sexual behavior, and sexual identity do not always match. That is, men do not, at least in contemporary American culture, experience the same kind of fluidity. Although they may identify as straight *and* have sex with other men, they certainly don't make out at parties for the pleasure of women.

The hookup culture on college campuses facilitates casual sexual interactions (ranging from kissing and making out to oral sex and intercourse) between students who meet at parties or bars. Our campus is no exception. The University of California, Santa Barbara, has a long-standing reputation as a party school (much to the administration's relief, it's declining in those rankings). In a student population of twenty thousand, more than half of the students are female and slightly under half are students of color, primarily Chicano/Latino and Asian American. About a third are first-generation college students. Out of over two thousand female UC Santa Barbara students who responded to sociologist Paula England's online College and Social Life Survey on hooking-up practices on campus, just under one percent identified as homosexual, three percent as bisexual, and nearly two percent as "not sure."

National data on same-sex sexuality shows that far fewer people identify as lesbian or gay than are sexually attracted to the same sex or have engaged in same-sex sexual behavior. Sociologist Edward Laumann and his colleagues, in the National Health and Social Life Survey, found that less than two percent of women identified as lesbian or bisexual, but over eight percent had experienced same-sex desire or engaged in lesbian sex. The opposite is true for men, who are more likely to have had sex with a man than to report finding men attractive. Across time and cultures (and, as sociologist Jane Ward has pointed out, even in the present among white straight-identified men), sex with other men, as long as a man plays the insertive role in a sexual encounter, can bolster, rather than undermine, heterosexuality. Does the same work for women?

The reigning assumption about girls kissing girls in the party scene is that they do it to attract the attention of men. But the concept of sexual fluidity and the lack of fit among desire, behavior, and identity suggest that there may be more going on than meets the male gaze. A series of formal and informal interviews with diverse female college students at our university, conducted by undergraduates as part of a class assignment, supports the sociological scholarship on the complexity of women's sexuality.

The College Party Scene

What is most distinctive about UC Santa Barbara is the adjacent community of Isla Vista, a densely populated area made up of two-thirds students and one-third primarily poor and working-class Mexican American families. House parties, fraternity and sorority parties, dance parties (often with, as one woman student put it, "some sort of slutty theme to them"), and random parties open to anyone who stops by flourish on the weekends. Women students describe Isla Vista as "unrealistic to the rest of the world . . . It's a little wild," "very promiscuous, a lot of experimenting and going crazy," and "like a sovereign nation. . . a space where people feel really comfortable to let down their guards and to kind of let loose." Alcohol flows freely, drugs are available, women sport skimpy clothing, and students engage in a lot of hooking up. One sorority member described parties as featuring "a lot of, you know, sexual dance. And some people, you know, like pretty much are fucking on the dance floor even though they're really not. I feel like they just take it above and beyond." Another student thinks "women have a little bit more freedom here." But despite the unreality of life in Isla Vista, there's no reason to think life here is fundamentally different than on other large campuses.

At Isla Vista parties, the practice of presumably hetero- sexual women kissing and making out with other women is widespread. As one student reported, "It's just normal for most people now, friends make out with each other." The student newspaper sex columnist began her column in October 2008, "I kissed a girl and liked it," recommending "if you're a girl who hasn't quite warmed up to a little experimentation with one of your own, then I suggest you grab a gal and get to it." She posed the "burning question on every male spectator's mind . . . Is it real or is it for show?" As it turns out, students offered three different explanations of why students do this: to get attention from men, to experiment with same-sex activity, and out of same-sex desire.

Getting Attention

Girls kissing other girls can be a turn-on for men in our culture, as the girls who engage in it well know.

A student told us, "It's usually for display for guys who are usually surrounding them and like cheering them on. And it seems to be done in order to like, you know, for the guys, not like for their own pleasure or desire, but to like, I don't know, entertain the guys." Alcohol is usually involved: "It's usually brought on by, I don't know, like shots or drinking, or people kind of saying something to like cheer it on or whatever. And it's usually done in order to turn guys on or to seek male attention in some way." One student who admits to giving her friend what she calls "love pecks" and engaging in some "booby grabbing" says "I think it's mainly for attention definitely. It's usually girls that are super drunk that are trying to get attention from guys or are just really just having fun like when my roommate and I did it at our date party . . . It is alcohol and for show. Not experimentation at all." Another student, who has had her friends kiss her, insists that "they do that for attention . . . kind of like a circle forms around them . . . egging them on or taking pictures." One woman admitted that she puckered up for the attention, but when asked if it had anything to do with experimentation, added "maybe with some people. I think for me it was a little bit, yeah."

Experimentation

Other women agree that experimentation is part of the story. One student who identifies as straight says "I have kissed girls on multiple occasions." One night she and a friend were "hammered, walking down the street, and we're getting really friendly and just started making out and taking pictures," which they then posted on Facebook. "And then the last time, this is a little bit more personal, but was when I actually had a three-some. Which was at a party and obviously didn't happen during the party?" She mentions "bisexual tendencies" as an explanation, in addition to getting attention: "I would actually call it maybe more like experimentation." Another student, who calls herself straight but "bi-curious," says girls do it for attention, but also, "It's a good time for them, something they may not have the courage to express themselves otherwise, if they're in a room alone, it makes them more comfortable with it because other people are receiving pleasure from them." She told us about being drunk at a theme party ("Alice in Fuck-land"): "And me and 'Maria' just started going at it in the kitchen. And this dude, he whispers in my ear, 'Everyone's watching. People can see you.' But me and 'Maria' just like to kiss. I don't think it was like really a spectacle thing, like we weren't teasing anybody. We just like to make out. So we might be an exception to the rule," she giggled.

In another interview, a student described a friend as liking "boys and girls when she's drunk . . . But when she's sober she's starting to like girls." And another student who called herself "technically" bisexual explained that she hates that term because in Isla Vista "it basically means that you make out with girls at parties." Before her first relationship with a woman, she never thought about bisexuality: "The closest I ever came to thinking that was, hey, I'd probably make out with a girl if I was drinking." These stories make clear that experimentation in the heterosexual context of the hookup culture and college party scene provides a safe space for some women to explore non-heterosexual possibilities.

Same-Sex Desire

Some women go beyond just liking to make out and admit to same-sex desire as the motivating factor. One student who defined her sexuality as liking sex with men but feeling "attracted more towards girls than guys" described her coming out process as realizing, "I really like girls and I really like kissing girls." Said another student, "I've always considered myself straight, but since I've been living here I've had several sexual experiences with women. So I guess I would consider myself, like, bisexual at this point." She at first identified as "one of those girls" who makes out at parties, but then admitted that she also had sexual experiences with women in private. At this point she shifted her identification to bisexual: "I may have fallen into that trap of like kissing a girl to impress a guy, but I can't really recollect doing that on purpose. It was more of just my own desire to be with, like to try that with a woman." Another bisexual woman who sometimes makes out with one of her girlfriends in public thinks other women might "only do it in a public setting because they're afraid of that side of their sexuality, because they were told to be heterosexual you know . . . So if they make out, it's only for the show of it, even though they may like it they can't admit that they do."

The ability to kiss and make out with girls in public without having to declare a lesbian or bisexual identity makes it possible for women with same-sex desires to be part of the regular college party scene, and the act of making out in public has the potential to lead to more extensive sexual activity in private. One student described falling in love with her best friend in middle school, but being "too chicken shit to make the first move" because

"I never know if they are queer or not." Her first sexual relationship with a bisexual woman included the woman's boyfriend as well. In this way, the fact that some women have their first same-sex sexual encounter in a threesome with a man is an extension of the safe heterosexual space for exploring same-sex desire.

Heteroflexibility

Obviously, in at least some cases, more is going on here than drunken women making out for the pleasure of men. Sexual fluidity is certainly relevant; in Lisa Diamond's ten-year study of young women who originally identified as lesbian or bisexual, she found a great deal of movement in sexual desire, intimate relationships, and sexual identities. The women moved in all directions, from lesbian to bisexual and heterosexual, bisexual to lesbian and heterosexual, and, notably, from all identities to "unlabeled." From a psychological perspective, Diamond argues for the importance of both biology and culture in shaping women as sexually fluid, with a greater capacity for attractions to both female and male partners than men. Certainly the women who identify as heterosexual but into kissing other women fit her notion of sexual fluidity. Said one straight-identified student, "It's not like they're way different from anyone else. They're just making out."

Mostly, though, students didn't think that making out had any impact on one's identity as heterosexual: "And yeah, I imagine a lot of the girls that you know just casually make out with their girlfriends would consider themselves straight. I consider myself straight." Said another, "I would still think they're straight girls. Unless I saw some, like level of like emotional and like attraction there." A bisexual student, though, thought "they're definitely bi-curious at the least . . . I think that a woman who actually does it for enjoyment and like knows that she likes that and that she desires it again, I would say would be more leaning towards bisexual."

Everybody but Lesbians

So, although girls who kiss girls are not "different from anyone else," if they have an emotional reaction or *really* enjoy it or want to do it again, then they've apparently crossed the line of heterosexuality. Diamond found that lesbians in her study who had been exclusively attracted to and involved with other women were the only group that didn't report changes in their sexual identities. Sociologist Arlene Stein, in her study of lesbian feminist communities in the 1980s, also painted a picture of boundary struggles around the identity "lesbian." Women who developed relationships with men but continued to identify as lesbians were called "ex-lesbians" or "fakers" by those who considered themselves "real lesbians." And while straight college students today can make out with women and call themselves "bi-curious" without challenge to their heterosexual identity, the same kind of flexibility does not extend to lesbians. A straight, bi-curious woman explained that she didn't think "the lesbian community would accept me right off because I like guys too much, you know." And she didn't think she had "enough sexual experience with the women to be considered bisexual." Another student, who described herself as "a free flowing spirit" and has had multiple relationships with straight-identified women, rejected the label "lesbian" because "I like girls" but "guys are still totally attractive to me." She stated that "to be a lesbian meant . . . you'd have to commit yourself to it one hundred percent. Like you'd have to be in it sexually, you'd have to be in it emotionally. And I think if you were you wouldn't have that attraction for men . . . if you were a lesbian."

In contrast to "heteroflexibility," a term much in use by young women, students hold a much more rigid, if unarticulated, notion of lesbian identity. "It's just like it's okay because we're both drunk and we're friends. It's not like we identify as lesbian in any way. . . ." One woman who has kissed her roommate is sure that she can tell the difference between straight women and lesbians: "I haven't ever seen like an actual like lesbian couple enjoying themselves." Another commented, "I mean, it's one thing if you are, if you do identify as gay and that you're expressing something." A bisexual woman is less sure, at first stating that eighty percent of the making out at parties is for men, then hesitating because "that totally excludes the queer community and my own viewing of like women who absolutely love other women, and they show that openly so, I think that it could be either context." At that point she changed the percentage to fifty percent: "Cause I guess I never know if a woman is like preferably into women or if it's more of a social game." A bisexual woman described kissing her girlfriend at a party "and some guy came up and poured beer on us and said something like 'stop kissing her you bitch,'" suggesting that any sign that women are kissing for their own pleasure puts them over the line. She went on to add that "we've gotten plenty of guys staring at us though, when we kiss or whatever, [and] they think that we're doing it for them, or we want them to join or whatever. It gets pretty old."

So there is a lot of leeway for women's same-sex behavior with a straight identity. But it is different than for straight men, who experience their same-sex interactions in a more private space, away from the gaze of women. Straight women can be "barsexual" or "bi-curious" or "mostly straight," but too much physical attraction or emotional investment crosses over the line of heterosexuality. What this suggests is that heterosexual women's options for physical intimacy are expanding, although such activity has little salience for identity, partner choice, or political allegiances. But the line between lesbian and non-lesbian, whether bisexual or straight, remains firmly intact.

Introduction to Reading 40

Whereas much past research on men's friendships has focused on how homophobia, or what Stefan Robinson, Adam White, and Eric Anderson refer to as "homohysteria," limits physical and emotional intimacy among heterosexual men, in this reading, the authors examine a new form of male friendship, the "bromance," and compare these relationships with heterosexual romances. The "bromance" includes emotional intimacy and non-sexual physical intimacy, and is regarded by many men in their study as more intimate and more honest than their relationships with women.

1. What is a "bromance," and in what ways does it differ from a heterosexual romance?
2. The authors say that for contemporary university men, that "self-checking among male friends has diminished." What do they mean by "self-checking," and why has this decline occurred?
3. How do "bromances" impact the quality of heterosexual romantic relationships? How do "bromances" benefit or disadvantage women?

Privileging the Bromance

A Critical Appraisal of Romantic and Bromantic Relationships

Stefan Robinson, Adam White, and Eric Anderson

The concept of twentieth-century friendship between both heterosexual (Ibson 2002) and homosexual (Nardi 1999) men is well examined in the social sciences. While friendship is primarily experienced by individuals as a complex psychological phenomenon (Poplawski 1989), its dimensions, behavioral requisites, and prohibitions are nonetheless socially defined and regulated. However, during much of the twentieth century, investigations of friendship between men have focused on what is missing, by contrast to what exists in women's friendships, namely, emotional and physical intimacy (Lewis 1978; Pleck 1975).

Recent research (Robinson, Anderson, and White 2017) established a framework for defining and characterizing the popular "bromance" term by analyzing heterosexual undergraduate men's perspectives on the subject. That study showed that young men openly pronounce love for their bromances and engage in highly intimate behaviors, both emotionally and physically, which have until recently been socially prohibited in same-sex male friendships. In this article, we examine whether close male friendships have the capacity to rival the intimacy and affection traditionally reserved for romantic, heterosexual relationships. We know relatively little about how romantic and peer relations are similar or different, and this research sets out to examine the similarities and differences between bromances and heterosexual romances, with a specific focus on the nature of intimacy and self-disclosure.

Robinson, S., White, A., & Anderson, E. (2017). Privileging the bromance: A critical appraisal of romantic and bromantic relationships. *Men and Masculinities.*

The Influence of Homohysteria on Homosociality

The level of physical and emotional intimacy expressed between heterosexual young men is dependent on a number of sociohistorical variables (Lipman-Blumen 1976; Sedgwick 1985). For example, homosocial intimacy flourished before the modern era (Deitcher 2001). Exemplifying this, late nineteenth-century and early twentieth-century men not only posed for photography in physically intimate ways, but they wrote endearing letters to one another and even slept in the same beds (Ibson 2002). Tripp (2005) highlights that, for four years, President Abraham Lincoln shared a bed with his intimate male partner, Joshua Speed, and that President George Washington wrote endearing letters to other men.

However, this intimacy that Tripp (2005) describes began to be policed when the awareness of homosexuality grew in the twentieth century, particularly in the 1970s and peaking in the 1980s. In this epoch, straight men began to fear being homosexualized for displaying physical or emotional intimacy. Consequently, this interfered with the development of close male friendships (Morin and Garfinkle 1978). Instead of the nineteenth-century homosociality, late twentieth-century hypermasculine discourse arose in response to the mass cultural awareness of homosexuality among Western populations. This was facilitated by the spread of the HIV/AIDS virus, which brought such cultural visibility that it solidified the notion that homosexuals existed in great numbers (Halkitis 2000)—something made even more salient by large numbers of gay men dying from AIDS-related illnesses. Accordingly, the *General Social Survey* and the *British Social Attitudes Survey* show that cultural homophobia reached unprecedented heights in the mid-1980s to early 1990s (Loftus 2001; Clements and Field 2014).

Sociological research from this era highlights that men began to emotionally distance themselves from other men (Komarovsky 1974; Pleck 1975). Lewis (1978) wrote that men ". . . have not known what it means to love and care for a friend without the shadow of some guilt and fear of peer ridicule" (p. 108). Jourard (1971) showed that self-disclosure—a vital component of emotional intimacy—was lacking between males. Instead, young men knew that they had a friendship with another male when they engaged in activities together like playing sports, drinking, fixing things, or gambling (Seiden and Bart 1975). The difference between the early and later stages of the twentieth century was growing recognition in the latter half that homosexuality exists as a static sexual orientation among a significant portion of the population and corresponding antipathy toward it. Accordingly, Anderson (2009) theorizes that it was the fear of being thought gay that ended the physical and emotional intimacy that heterosexual men once shared, suggesting, alongside Ibson (2002) and Kellner (1991), that by the 1980s, heterosexual men were severely regulated in their behaviors.

Anderson's (2009) inclusive masculinity theory and his concept of homohysteria explain this shift in the physical and emotional dispositions of men before the first half of the twentieth century and the decades of the latter half. McCormack and Anderson (2014) more recently define homohysteria as the fear of being socially perceived as homosexual—something made possible because heterosexuality cannot be definitively proven among straight men in a culture that is both aware and fearful of homosexuality. Subsequently, men are culturally compelled to perform certain overtly heterosexual behaviors and avoid engaging in those that would feminize them. Thus, one way of looking at homohysteria is to suggest that whereas homophobia limits the lives of homosexual men, homohysteria limits the lives of heterosexual men, too (Anderson and McCormack 2016).

The fear of male homosexualization, and its associated femininity, circulated not only within institutions of education (Connell 1989; Mac an Ghaill 1994) but also among other influential institutions including sport (Anderson 2005; Connell 1995), government (Ahmed 2013; Boyle 2008), and the military (Dunivin 1994). During this time, the requirement for men to refrain from emotional vulnerability had filtered into almost all aspects of men's personal lives (Field 1999).

Conversely, matters have been vastly different for women. While Worthen (2014) shows that homohysteria also applies to women, they have been less policed by the structures of homophobia. Hence, women have been able to display a far wider range of emotional behaviors than men (Kring and Gordon 1998; Sprecher and Sedikides 1993; Williams 1985). For example, unlike men who maintained friendship through shared activities, women have maintained friendships through sharing emotions and disclosing secrets (Caldwell and Peplau 1982). This is not to say that certain hypermasculine behaviors did not associate women with lesbianism (Griffin 1998; Worthen 2014), but that women were given more leeway for emotional expression than men. This freedom of expression is more closely associated with young women, as research has shown a decline in the closeness between women who are married (Babchuk and Anderson 1989). Indeed, powerful patriarchal forces have sought to limit the closeness of adult female friendships, as they pose a threat to the strength of male alliances and patriarchal rule (Sedgwick 2015). Accordingly, men have historically denied women seats in cultural institutions (such as sport, the military, academia, religion, and business) and have relegated them to the domestic sphere where they have not had the opportunity to form large friendship networks and lobby for equal treatment (Kahlenber, Thompson, and Wrangham 2008; Flood 2008).

Expressing Love in the 1980s

Writing in the decade of homohysteria, Cancian (1987) said that we have "a feminine conception of love. We identify love with emotional expression and talking about our feelings" (p. 69). Accordingly, Anglo-American male youth of this time would refrain from even using the word love (Williams 1985); they were structured into exceptionally narrow masculine identities that rejected emotionality (Kellner 1991). Instead, they aspired to the muscular, heterosexual, hostile, and patriotic action heroes who filled Hollywood (Pope et al. 2000). Love and intimacy for men had no place in this era as it projected a feminine (read homosexual) image. Swearing, abuse, readiness to fight (Dunning, Murphy, and Williams 2014), and unemotionality (Cancian 1987) were compulsory male characteristics; boys and men did not have the liberty to express fear, weakness, uncertainty, or affection for other men (Plummer 1999).

Pleck (1975) similarly finds men of this era to be restricted in their expression, describing in detail the social confines of male intimacy and emotionality. He found that men were expected to exhibit greater control in their emotional behavior than women, being far removed from their feelings. He goes on to say, "at the same time, men appear to become angry or violent more easily than women and are often rewarded for doing so . . . [having] greater fears about homosexuality than do women" (p. 156). Men were to avoid emotional intimacy with other men, finding and expecting legitimate intimacy and companionship only within the confines of heterosexual relationships with women. There was a great impersonality in male friendships, and women's social exemption to be expressive meant that men were dependent on emotional support almost exclusively from women, while equally possessing disdain for their emotionality.

In this era, the cultural fear of homosexuality, and consequent emphasis on masculine stoicism, led men to depend entirely on women for the little emotional disclosure they were socially permitted. Komarovsky (1974), for example, discovered that undergraduate men were more likely to use women as a confidant, as opposed to men. This is problematized by the fact that the antipathy toward homosexuality, and love of stoicism, was propagated years before men enter university. Restrictive masculinity found its routes in the lives of even very young boys. Exemplifying this, Pollack (1999) showed that fathers of this era would withhold their love and affection from their children, and before boys even reached their teenage years, they could be subjected to abusive and shaming torments from peers and teachers for performing feminine behaviors such as skipping and poetry readings for not being "real boys" (Pollack 1999). The literature consistently documents a cultural zeitgeist of homophobia, hypermasculinity, and emotional abstention among men from the 1970s (Olstad 1975) through the 1990s (Pollack 1999), leaving a generation of men with a life of nonintimate male-to-male connections (Collins and Sroufe 1999; Tognoli 1980).

Inclusive Masculinity Theory

Contrasting twentieth-century literature on adolescent males, in the twenty-first century, young men have a greater social entitlement to express themselves through a diverse spectrum of behaviors and emotions that would have previously socially coded them as gay (Anderson 2014). This occurs without judgment from others (McCormack 2012; Weeks 2007). We use inclusive masculinity theory (Anderson 2009; Anderson and McCormack 2016) to explain this shift in young men's practice of masculinity.

Inclusive masculinity theory is based on the social inclusion of those traditionally marginalized by an orthodox form of masculinity. Inclusive masculinity theorists argue that a substantial shift—a softening—has occurred among male youth and that this can be observed with prominence in institutions such as education (McCormack 2011), sport (Adams 2011; Magrath, Anderson, and Roberts 2015; Murray et al. 2016; White and Robinson 2016), and social media (Morris and Anderson 2015; Scoats 2017). Young men in these forums align themselves away from orthodox tropes of masculinity and are less concerned about whether others perceive them to be gay or straight; masculine or feminine (White and Hobson 2017). Embracing and performing inclusive behaviors have meant that male youth have little fear of homosexualization for the performance of femininity or homosocial behavior (Savin-Williams 2005). Simply put, the decline in cultural homohysteria has relinquished heterosexual men's burden to police their gendered behaviors.

Building on the growing body of work of decreasing homohysteria and the changing nature of adolescent masculinities in the twenty-first century (e.g., McCormack 2012), young men today are now able to have highly intimate homosocial relationships alongside casual friends. Like men of the 1980s, they make friends through sports, drinking, and video games, but unlike men of the 1980s, however, they also shop, dine, vacation, and sleep together (Anderson 2014). They also maintain the opportunity to form deep emotional relationships, based on emotional disclosure with one another (Murray and White 2017). Whereas Bank and Hansford (2000) previously found that male friendships struggle due to emotional restraint, masculine hierarchies, and homophobia, many scholars now suggest that the millennial generation has promoted a culture that is much more inclusive and cohesive (Adams 2011; McCormack 2012; Thurnell-Reid 2012). Significantly, recent research has found that both late adolescent men (Robinson, Anderson, and White 2017) and men in their thirties (Magrath and Scoats forthcoming) believe that the pressures of heterosexual marriage and house buying limit their capacity to maintain such close friendships. Indeed, they recognize the temporal context of university life and that it grants significant social freedoms that will perish as they enter adulthood. Therefore, age in the context of our research is highly pertinent.

Instead of aspiring to the likes of Rambo, both qualitative and quantitative research show that adolescent males today much prefer the feminized charms and homosocial tactility of the members of the boy band One Direction, or popular YouTube vloggers (Morris and Anderson 2015), or the intellect, financial success and charity of Bill Gates (JWT 2013). Young men are also increasingly pursuing interests in the arts, music, and fashion industries (Edwards 2006; McCormack 2012), which illustrates the diversification of male norms. What recent generations of men would have considered a highly feminized notion of masculinity, today's adolescents and young men have greatly expanded upon the gendered and sexual behaviors that are not only permissible but expected of their friends (Adams 2011; McCormack 2012).

In middle to late adolescence, when many young men are at university, there is an increased opportunity and desire among young men to form peer attachment bonds, whereby deep interdependent relationships are developed premised on self-disclosure and intimacy (Collins and Sroufe 1999). This can occur with either romantic partners or same-sex friends (Kobak et al. 2007). This yearning for intimate bonding arises as young men search for independence from their parents and seek new avenues for advice and companionship (Collins and Repinski 1994). While most of these friendships will not be enduring, some will become highly intimate friendships with a small minority of friends providing a "safe haven" for full emotional disclosure (Kobak et al. 2007) that is known to them as a bromance.

The Bromance

Inclusive behaviors have been widely identified in popular television programs and films, where the majority of scholarly attention on bromances has been focused (Boyle and Berridge 2014). While "buddy movies" have existed since the late 1980s (Fuchs 1993), the relationships between the two leading male characters have become more sentient and compassionate than ever before in mainstream cinema. Gill and Hansen-Miller (2011) describe a new type of film called the "Lad Flick" or "Lad Movie," explaining that these films are a hybrid of the buddy movie and "romantic comedy" genre that depict intimate male friendships. Blockbuster films such as *21 Jump Street* (2012), *Due Date* (2010), and *The 40-Year Old Virgin* (2005) have drawn attention to men's capacity to constitute complex and dynamic relationships grounded in male closeness, trust, and homosociality, at least in movies. Indeed, recent research revealed that young men use these types of films as a reference point when defining the bromance and how it operates (Robinson, Anderson, and White 2017). The narratives and relationships played out on screen generally reflect the way we understand our own lives. Thompson (2015) notes that there is disagreement among scholars regarding both the benefit of having such a phenomenon as the bromance and its definition, nonetheless concluding that television and film are "highlighting a subtext of male emotion within bromance[s] that warrants further exploration" (p. 3).

The bromance term was popularized among the mainstream media around 2005, when there was a sharp rise in the amount of on-screen bromances (Boyle and Berridge 2014; DeAngelis 2014; Gill and Hansen-Miller 2011). Of the minimal academic work conducted around the concept, it is exclusively movie and celebrity culture focused. The term was adopted in an attempt to account for the increasingly intimate and emotional affection being displayed between heterosexual men on the silver screen and in celebrity culture (DeAngelis 2014). Accordingly, the vast majority of scholarly attention paid to the bromance revolves around media analysis; work that highlights the changing nature of male friendships in movies and television (DeAngelis 2014).

We recently documented how heterosexual men at a UK university engage with the bromance (Robinson, Anderson, and White 2017). Through interviewing participants about their involvement in bromances, and how their experiences influenced their understanding of the term, data showed that bromances are achieved through certain social freedoms concerning shared interest, emotional intimacy, and non-sexual physical intimacy. Bromances achieved an important level of cultural resonance and meaning to the men, with those involved describing themselves as holding a brotherly or girlfriend-type status with their bromance.

Important for the social creation of same-sex intimacy among heterosexual men, the participants also shared a unanimous definition and understanding of what a bromance is. It was expressed to be an intimate same-sex male friendship based on unrivalled trust and self-disclosure that superseded other friendships. The men publicly expressed "love" for their bromance(s), describing intimate examples of emotionality and physical intimacy that they had shared in such friendships.

The decisive component that characterized these friendships was demonstrated through the emphasis placed on a "lack of boundaries" and seemingly "nothing being off limits." In line with these findings, Hammarén and Johansson (2014) suggest that bromances are same-sex male relations that "emphasize love, exclusive friendship, and intimacy . . . [and] are not premised on competition [or hierarchies]" (p. 6).

The cultural adoption of the bromance term represents an increased recognition that young men are permitted to have more diverse and homosocial masculine identities. Their behavior shows that contrasting to the 1970s (Olstad 1975), 1980s (Askew and Ross 1988), and 1990s (Kimmel 1995, 2004; Pollack 1999) research, young heterosexual men are now able to confide in each other and develop and maintain deep emotional friendships based on intimacy and the expression of once taboo emotional sentimentality.

Method

Over a three-month period, between August 2014 and November 2014, thirty semi-structured interviews into the romantic and bromantic lives of undergraduate males were conducted. To be part of the research, participants needed to identify as heterosexual and be in the second year of their university studies. The limitation to second-year students was to enable the examination of men who had sufficient time to develop friendships with their university peers—having had 18 months to befriend and develop intimate bonds with other men and live in the same house as close male friends.

To assure that the men we interviewed were not strategically presenting positive or overly exaggerated support for gay men and male homosexuality (a prerequisite for inclusive masculinities), eighteen months prior we distributed Herek's (1988) *Attitudes Towards Gays and Lesbians* scale to these students. The survey was administered anonymously upon the students' first day of arrival at the university. Results showed wide support for male homosexuality which meant that all men espoused progay attitudes on arrival at university.

Although men were not selected for race, the virtually exclusive white student body of this British university limited our analysis. The sample was white, with only one exception. The sample was also populated by participants from middle-class backgrounds. We do not therefore conduct a race or class analysis with this research, limiting our findings to white, middle-class, heterosexual, undergraduate men from one university. We identify the demographic of importance to this research as that of age and gender and use this sample to develop and discuss the ways young men perceive the value of their bromances compared to that of their romances.

Participants were asked to discuss their experiences of bromances and the homosocial aspects of their same-sex friendships, before being asked how those intimacies compare to those shared with their romantic partners/girlfriends. The line of questioning intended to tease out what boundaries (or lack of) bromances and romances operate under. Particular focus was given to limits of self-disclosure and emotional intimacy.

Results

Each of the thirty men interviewed for this research described themselves as having at least one bromance and at least one romance, in either the past or present. When describing what he understood a bromance to be, Bruce compared his bromance to a romance, "We are basically like a couple . . . we get called like husband and wife all the time" and Martin agreed, "It's like having a girlfriend, but then not a girlfriend ."Hamish went as far to say "It's your best friend. You are closer to him than anyone. They are like a guy girlfriend." It was clear from the participants that similar behaviors and feelings existed in both bromantic and romantic relationships, which complicate the differentiation of the two.

As the central theme of enquiry, all participants were directly asked to explain what the difference is between a bromance and romance. Most described this as a difficult question. Chris articulated what many suggested:

> That's a very hard question to answer. I feel like I've got to say that there is a difference. But I really don't know. I can't really identify a clear difference. There is a different feeling, but nothing I can particularly describe. Oh! A romance is with a girl and a bromance is with a guy [he laughs].

Participants suggested that the primary difference between a bromance and a romance ultimately hinged on their desire for sex with their romance and not their bromance. Aaron tried to model how the two were different, suggesting that there are three factors to consider, "sexual, emotional and personality. A bromance needs the last two and a girl needs two including sexual." Bob just said, "Sex, really. That's all." Aaron said, "When you have a bromance with a friend, it's motivated by your interest in that person, love and friendship, and not because you want sex." Indeed, sexual desire is often perceived as the traditional missing link between a friendship and a romantic relationship. Beyond the need for sex, we found that for this cohort of men, bromances performed a very similar and often superior function to romances.

Disclosure, Emotionality, and Physical Intimacy

The vast majority of participants suggested that they have a preference for disclosing personal matters and exclusive secrets sharing with their bromance(s), more so than their romances. They were clear that a bromance offers a deep sense of unburdened disclosure and emotionality based on trust and love, in which vulnerabilities can be revealed. Of the men interviewed, twenty-eight of the thirty said that they would prefer to discuss personal matters with a bromance than a romance. They felt able to express and emote in their bromances, and divulge their most personal issues, without ridicule.

Brad said, "There are absolutely things I tell my bromances and not the girlfriend. She expects so much from the relationship and will have a go if I say something out of line, and with Matt we just tell each other everything." Harvey shared a similar sentiment, "There are no boundaries between us [he and his bromance] and what we can say. I couldn't tell her [girlfriend] as much as him cause she might not like me after." When asked to elaborate, Harvey explained, "Well, for example, Tim knows I love listening to Taylor Swift and Beyonce, but I keep that quiet because she would judge me. I feel like I have to be more manly around her."

Bob used a health-related example to clarify his favorability of disclosing to a bromance. "If I found a lump on my testical, I'd talk to [bromance] rather than my girlfriend." The notion of a health concern was named by five other men as well. Hamish spoke about when he was in hospital. "Charlie was there for me all the time when I was recovering. But, when my girlfriend came, I kind of wanted her to leave so that I could have a laugh with Charlie instead of being all serious like she is." George gave a different kind of example. He spoke of his desire to for his girlfriend to finger him (penetrate his anus during sex). However, he thought his girlfriend would think he was gay, and thus ruining the relationship, he had only told his bromance about his sexual desires.

Martin said, "because you can be more truthful about stuff, and because he won't judge me, it's easier to talk to him and I don't have to hold back." Robbie spoke in more general terms, "They [romances] can't do the laddish banter we do. You can piss around and go dirty in your conversations with bro's, but you don't tell girlfriends about that stuff. With guys I know where I stand." Hamish similarly said that, "Bromances are more honest and have more banter cause you say things to the guys you wouldn't' want to say to the girl." Beck believed that this increased disclosure to other men is because, "A girlfriend will judge you and a bromance will never judge you."

Ollie gave an example of where he didn't want his girlfriend to see him emote. "When I was upset about my granddad, I wouldn't let the girlfriend over cause I didn't want her to see me upset. Stewart kept me company, and I think it pissed her off." He elaborated, "Stewart stayed in my bed till the morning." It is this type of behavior that led the men in this study to widely proclaim love for their bromance(s). Ollie was not the only man to be physically intimate with his bromance.

Of the thirty men interviewed, twenty-nine said that they had experienced cuddling with a same-sex friend, and many expressed that it occurred frequently with their bromances. Aaron said that with his bromance, "We hug when we meet, and we sleep in the same bed when we have sleepovers. Everyone knows it, and nobody is bothered by it because they do it as well." Alan shows that there is no shame in this behavior, "There's a great photo of me and Tom on Facebook cuddling" he said. Patrick said, "I think most guys in bromances cuddle, it's a usual in my main friendship group. It's not a sexual thing, either. It shows you care."

Consistent with other research on British undergraduate men (Anderson, Adams, and Rivers 2012), kissing was also widely spoken about as a sign of affection for a bromance. Robbie said, "You see guys kissing and cuddling each other loads. It's never an issue to anyone." Tony said, "I kiss him [bromance] all the time," and Max said that in his bromance, "I hug him and kiss him and tell him I love him." Beck said, "Guys nowadays, in my generation, there is so much kissing between guys because it's showing affection."

Significantly, for men with girlfriends, their bromantic cuddling is known by their girlfriend's and, ostensibly, approved. Joe said, for example, that his girlfriend knows that he has a strong bromance and that he cuddles with him. When asked if his girlfriend is "okay with that," he replied, "My girlfriend is fine with it." Similarly, Tony told us that he has a "fair few bromances" and has also been in a relationship with his girlfriend for seven years. Yet he is still, "quite comfortable touching other people [males]" and that his girlfriend is "not bothered by it." So whereas one might think that heterosexual men with girlfriends might not desire or seek same-sex intimacy, this is not the case. The physical interactions that these men engage in with other men are unanimously expressed to be devoid of sexual desire. However, it remains that they desire to be physically intimate, albeit nonsexually, with their bromances; a behavior traditionally reserved exclusively for romantic relationships.

Hiding From Judgment

There was a clear acknowledgment among the participants that bromances had a heightened capacity for emotional disclosure, beyond that which they felt free to express to romances. Many participants suggested that they limit their emotionality and disclosure to their girlfriends because they feel compelled to self-monitor, perform, and "act differently" to regulate their girlfriend's perception of them. They explained that there was more judgment passed between romances (back and forth) than bromances, and more emotional instability, meaning that it was sometimes easier to refrain from certain disclosures for the sake of conflict limitation and/or receiving negative judgment. Conversely, bromances were expressed to be rather boundless, in terms what can be said without judgment.

Specifically, participants said that they did not want to get in trouble for saying or doing the "wrong thing." As Harvey explained, "You want to project a better self [to your romance] and maintain the standard you had at the beginning of the relationship." Beck explained the other side to this, "With a guy, you don't' have to impress them. You are just so relaxed around each other. Sometimes it feels like I'm always on eggshells with her [girlfriend]."

The fear of persecution for displaying desires for other women in front of their romances was also described as problematic for these men. Many would not talk about or even look at other girls in front of their romances. This is because they believed that they would cause relationship trouble for doing so. Hamish said, "You defiantly have to be more careful. My god, if you talk about another girl, you get your head ripped off" and David agreed. "The first rule is you don't speak about other girls! It just causes trouble." Toby said:

> I feel comfortable talking to my girlfriend about most things. Although, if I look at girls, or mention other girls, she's always suspicious of me. I've said stuff in front of her before and I got in trouble. So, now I know to keep quiet. The guys say I'm under the thumb and they're probably right.

Accordingly, the desire to project a "better self" to romances has caused these men to modify their behavior through attempting to uphold what they consider to be a false image of themselves. Harvey said, "The words I'd use with a friend are different. I found myself talking in a correct manner to my girlfriend." Liam said that with his bromances, he is free to "say more things that won't result in an argument." George adds to this, suggesting, "with a girl you kind of have to think a lot more before you say things, because of the way they will react." Max gave an example, "I don't feel like I can say no to anything she asks. If she wants me to come over and I'm too tired, or busy, I still drop everything and go to avoid arguing." Mark said:

> I don't talk to my girlfriend about the drugs I use, even weed, cause she will definitely have a go. . . . I lie about things to keep her happy, and I nod along with all her plans for our future. I only speak to [bromance] and

> [bromance] about the fact I want to travel for a year and maybe even move abroad.

One participant felt that he needed to artificially maintain a masculine identity; Jay said, "I have to try and keep up this figure of masculinity [around his romance], whereas a guy friend isn't going to care." Under the homosysteric culture of the 1980s and 1990s, men would find themselves strictly monitoring their behavior around other men; however, within this cohort, this self-checking among male friends has diminished.

On the other hand, bromances were described to be more fluid and relaxed. Regi said, "If I don't talk to [bromance] for a while, they are cool. But girls would freak out. . . . In my bromance nothing is off limits." David agreed, "I hold nothing back in my bromance. There are no boundaries like there is with my girlfriend." Jason added, "With the guy. . . you are always on the same wavelength. I don't think girls can relate to me in the same way guys can." Adam said, "With my last girlfriend I would always have to prioritize her over my friends, or it would always end up with me getting an ear full; or no sex [he laughs]." Zani (1993) finds that conflicts in romances are often caused because of jealousy over the amount of time one spends with their same-sex friends, although this does not surmount to the jealousy one is likely to feel if a partner is spending time with someone from the opposite sex (Roth and Parker 2001).

Adam highlights a point made by several others; that sex (or a lack of) can be used as a pawn to reward boyfriends for being loyal and appreciative of their girlfriends; often at the expense of male bonding. Nathan said, "I know if I choose to go out with the guys instead of her, she won't have sex with me for like a week." When speaking about the difference between a bromance and a romance, Henry said, "The only thing with a girl is that it brings sex into it—if you are nice to them for the whole day that is." Other men similarly suggested that to achieve sex with a romance, it often came at a cost. Regi said, "They [girlfriends] always feel like they should take priority, they expect so much from the relationship, and we give them priority because of sex. It's because we can't control our sex drive."

Other participants could relate to this, recognizing that it was problematic: that their desire for sex was not a good thing for the emotional side of their romantic emotional relationship. Some felt that the emotional intimacy they shared with women was artificially enhanced in order to achieve sex and maintain the romance. Harvey said, "In romances, people don't like to truly show their feelings. Other than that, the only difference between romances and bromances is the sexual desire." Sam said, "With women, the sexual stuff is great, but the romantic stuff might not be as honest because you will say whatever for sex." Joe went as far to call this sexual pollution, explaining, "Sex is expected and it interferes with the emotional stuff . . . bromances are stronger because there is no sexual pollution." Beck agreed, "Being able to betruthful with a bromance can be superior to a romance, but you don't get the sexual pleasure." For these men, sex clearly holds power in a heterosexual relationship, leading many to suggest that it can complicate the emotional side of the relationship.

Conflict Resolution

Arguments and conflict resolution were discussed concerning both bromances and romances. When questioned, the participants overwhelmingly stated that arguments with girlfriends were more intense, trivial, and long-lasting in comparison to their bromances. When speaking about arguments with his girlfriend, Harvey said, "She will store up something you did wrong two years ago and recall it, with the exact date and time."

This example was given in similar terms by other participants too. Toby agreed, "With men, it's over. But women are very good at remembering things." Adam said, "I'd say guys are a less emotional about arguments. So if you say something offensive, they wouldn't take it as strong as a female." Hamish said that this makes personal conflict easier to overcome with a bromance. He explained that if he has a problem with a bromance, he handles it by saying, "Stop being a prick." He adds:

> You can't say that to a girlfriend. It will cause all sorts of trouble. . . . Sometimes we [bromance] clash and quite often he says 'ah fair enough, I shouldn't have done that.' Then that's it. It's over. We just move on.

Hamish suggested that men overcome conflict easier because "We are more honest with each other and perhaps more forgiving." Adam said, "Women have culturally been taught to take things more literally and get upset, whereas guys can laugh out their frustrations. So it's easier with guys." Gavin said, "There isn't anything that can go wrong in a bromance. Unless you're specifically a massive tool." Theo argued with his bromance recently, however, it resolved quickly. "We can't hold an argument. We just let it go. We talk about it," he said. Liam summarized a variety of these findings in saying:

> There are just things that you can tell guys that you can't tell a girlfriend: things that if you told to a girlfriend it would start an argument. So you don't tell her. You tell him [the bromance], instead. This allows you to talk about it, to get it out and process it; and it keeps the peace with the girlfriend.

It is for these reasons that this research unmasks perhaps uncomfortable and socially frowned-upon ideas about heterosexual relationships. It questions the quality of relationships for young straight men and women and advocates that, perhaps, heterosexual men widely benefit from long-term intimate (albeit not sexual) relationships with other men. When asked if a long-term heterosexual same-sex relationship was possible, Adam offered, "Yeah, it seems like a logical idea in fact." Harvey felt these relationships would have more success, concluding, "Lovers are temporary, a bromance can last a lifetime."

Discussion

By engaging with a critical appraisal of bromances, alongside traditional heterosexual romances, we illustrate how these thirty millennial men conceptualize and express the value of their close male friendships. The most significant finding in this study concerns the virtually unanimous narrative that these men found it easier to open up and express their feelings to their bromances, more so than their romances. This was suggested for two reasons. First, the most consistent finding concerned the lack of emotional boundaries and limits in bromances. Bromances were described—even if overly idealized—as being judgment free and having a lack of boundaries which allows them to push to margins of traditional masculinity through more physical and emotional behaviors. Conversely, many of the men did identify boundaries in their romances. Often, they could not talk fully about their interests, anxieties, health, and sexual desires, even when emerging adults often idealize romantic partners and exaggerate their supportiveness (Murray, Holmes, and Griffin 1996).

Second, men we interviewed expressed that, with a romance, one was constantly posturing and self-monitoring, not only to achieve desired heterosexual sex but to prevent relationship destruction. The men would restrict what they would say and instead act the part of the adoring boyfriend. The men reasoned that they did this because, in their view, women held onto grudges longer than men and were more unpredictable in their emotional responses, often recalling and reusing historical instances of conflict in later arguments. Indeed, some scholars reason that gendered stereotypes about emotionality inherently affect our display of emotion (Shields 1991), with women continuing to report, from a self-gauge perspective, that they have more frequent and intense emotional experiences and are more sensitive to their feelings (Sprecher and Sedikides 1993; Kring and Gordon 1998). On the contrary, the participants found it much easier to resolve disputes and arguments with their bromances because they found them to be more forgiving. Consequently, they were less guarded in their personal disclosure and identity management with their bromances, despite their romances following a more traditional trajectory.

There was a conclusive determination from the men we interviewed. On balance, they argued that bromantic relationships were more satisfying in their emotional intimacy, compared to their heterosexual romances. They saw social liberty in a bromance that exceeds the disclosure and openness achievable in their romantic relationships. This was articulated to be because of the effort required to maintain a romance, compared to the ease upon which two males can relate in a bromance. This is perhaps why men without girlfriends did not seem altogether longing for one.

We contend that the male preference for emotionality between other men, rather than women, has come about due to a significant cultural shift in the structure of masculinity. In the time that has passed since the 1980s, where a cultural zeitgeist of hegemonic masculinity existed, young men have rapidly come to esteem a more advanced and complex level of emotionality in their same-sex friendships. Men would have previously denounced the presence of intimacy in their friendships (Walker 1994), but they now embrace and speak of it openly. Where men had once reserved secret sharing and exclusive disclosure for women only (Komarovsky 1974), it was clear from our research that these millennial men have now transcended the emotional regulation experienced in the homohysteric era before them, to become highly tactile, inclusive, and caring toward other men.

There are however significant and worrying results here for women. These men perceived women to be the primary regulators of their behavior, and this caused distain for them as a whole in some instances. The men often generalized personal experiences to women as a collective, under an "us and them" binary which associated all women with any negative experiences the men had. The narratives used by the participants undoubtedly reflect the allegiances that they feel toward their own sex, and the nature of their disclosures suggests that some have a limited respect for their past and present girlfriends. Much in the same way that woman are portrayed in contemporary cinema as objects for male gratification (Gill and Hansen-Miller 2011), several of the participants spoke of women they knew in a generally negative way. There was a tendency, as in Hollywood, to deliver sexist perspectives in a humorous and banterous way to deter accusations of sexism, and this is problematic. Mehta and Strough (2009) propose that the strengthening of homosocial bonds contributes toward the devaluing of cross-sex socialization, and it may be that the rise of the bromances may not altogether be liberating and socially positive for women. We believe that the binary approach to questioning (i.e., bromances vs. romances), and the fact that the interviewer was of the same sex as participants, may have subtly influenced the nature of the language used to describe women.

Men in this research highlight that the physical and emotional dimensions of bromances resemble the traditional expectations of romantic companionship, namely, the declarations of love, kissing, cuddling, and exclusive emotional confidence. We show that while one of the fundamental differences between bromances and romantic relationships is sex, these are less rigid due to the progressively inclusive attitudes around same-sex touch. We find that the organization of bromances and romantic relationships is not dissimilar from one another. Under the rubric of inclusive masculinity, young men at this university are embracing their innate desire to search for companionship (Collins and Repinski 1994; Zorn and Gregory 2005), particularly with men, free from social stigma.

These findings are consistent with Anderson's notion of inclusive masculinity (2009) and resonate with other recent findings on young men (Hammarén and Johansson 2014; Zorn and Gregory 2005). It seems that, for the millennial men in this and other studies, they do not hold back on embracing their capacity for emotional versatility; rather, they are free to develop dynamic relationships with other men, offering them "valuable, tangible and socio-emotional support" (Zorn and Gregory 2005, 211).

There are interesting potential implications for domesticity too. These heterosexual millennial men cherish their close male friends so much, so that they may even provide a challenge to the orthodoxy of traditional heterosexual relationships. Given that young men are now experiencing a delayed onset of adulthood and an extended period of adolescence (Arnett 2004), men may choose to cohabit as a functional relationship in the modern era. Just as many of the men in this research share exclusive same-sex houses while at university, they may continue with their bromances and domesticity well beyond university years. Howard (2012) has already found some older men doing this, and Magrath and Scoats (forthcoming) find that men in their thirties have regrets about not maintaining their bromances into later life, with marriage being a key barrier to this. This, again, is another indication that bromances may not altogether benefit cross-sex relations. In other words, because heterosexual sex is now achievable without the need for romantic commitment (Bogle 2008), and because bromances are privileged for these men, the bromance could increasingly become recognized as a genuine lifestyle relationship, whereby two heterosexual men can live together and experience all the benefits of a traditional heterosexual relationship.

References

Adams, Adi. 2011. "'Josh Wears Pink Cleats': Inclusive Masculinity on the Soccer Field." *Journal of Homosexuality* 58:579–96.

Ahmed, Sara. 2013. *The Cultural Politics of Emotion*. New York: Routledge.

Anderson, Eric. 2005. *In the Game: Gay Athletes and the Cult of Masculinity*. New York: University of New York Press.

Anderson, Eric. 2009. *Inclusive Masculinity: The Changing Nature of Masculinities*. New York: Routledge.

Anderson, Eric. 2014. *21st Century Jocks: Sporting Men and Contemporary Heterosexuality*. New York: Macmillan

Anderson, Eric, Adi Adams, and Ian Rivers. 2012. "'I Kiss Them because I Love Them': The Emergence of Heterosexual Men Kissing in British Institutes of Education." *Archives of Sexual Behavior* 41:421–30.

Anderson, Eric, and Mark McCormack. 2015. "Cuddling and Spooning: Heteromasculinity and Homosocial Tactility among Student-athletes." *Men and Masculinities* 18:214–30.

Anderson, Eric, and Mark McCormack. 2016. "Inclusive Masculinity Theory: Overview, Reflection and Refinement." *Journal of Gender Studies*:1–15. doi:10.1080/09589236.2016.1245605.

Anderson, Eric, Mark McCormack, and Harry Lee. 2012. "Male Team Sport Hazing Initiations in a Culture of Decreasing Homohysteria." *Journal of Adolescent Research* 27:427–48.

Arnett, Jeffrey Jensen. 2004. *A Longer Road to Adulthood*. New York: Oxford University Press.

Askew, Sue, and Carol Ross. 1988. *Boys Don't Cry: Boys and Sexism in Education*. Milton Keynes: Open University Press.

Babchuk, Nicholas, and Trudy Anderson. 1989. "Older Widows and Married Women: Their Intimates and Confidants." *International Journal of Aging and Human Development* 28: 21–35.

Bank, Barbara J., and Suzanne L. Hansford. 2000. "Gender and Friendship: Why are Men's Best Same-sex Friendships Less Intimate and Supportive?" *Personal Relationships* 7: 63–78.

Bogle, Kathleen A. 2008. *Hooking Up: Sex, Dating, and Relationships on Campus*. New York: New York University Press.

Boyle, Ellexis. 2008. "Building a Body for Governance: Embodying Power in the Shifting Media Images of Arnold Schwarzenegger." PhD diss., University of British Columbia.

Boyle, Karen, and Susan Berridge. 2014. "I Love You, Man: Gendered Narratives of Friendship in Contemporary Hollywood Comedies." *Feminist Media Studies* 14(3): 353–368.

Caldwell, Mayta A., and Letitia Anne Peplau. 1982. "Sex Differences in Same-sex Friendship." *Sex Roles* 8:721–32.

Cancian, Francesca. 1987. "Love and the Rise of Capitalism." *In Gender in Intimate Relationships*, edited by B. Risman and P. Schwartz, 12–25. Belmont, CA: Wadsworth.

Clements, Ben, and Clive D. Field. 2014. "Public Opinion toward Homosexuality and Gay Rights in Great Britain." *Public Opinion Quarterly* 78:523–47.

Collins, W. Andrew, and L. Alan Sroufe. 1999. "Capacity for Intimate Relationships." *In The Development of Romantic Relationships in Adolescence,* edited by W. B. Furman, Bradford Brown, and Candice Feiring, 125–47. Cambridge: Cambridge University Press.

Collins, W. Andrew, and Daniel J. Repinski. 1994. "Relationships during Adolescence: Continuity and Change in Interpersonal Perspective." In *Personal Relationships during Adolescence,* edited by R. Montemayor, G. R. Adams, and T. P. Gullotta, 7–36. Thousand Oaks, CA: Sage.

Connell, Robert W. 1989. "Cool Guys, Swots and Wimps: The Interplay of Masculinity and Education." *Oxford Review of Education* 15:291–303.

Connell, Robert W. 1995. *Masculinities*. Berkeley: University of California Press.

Deitcher, David. 2001. *Dear Friends: American Photographs of Men Together,* 1840–1918. Ann Arbor: Harry N. Abrams.

DeAngelis, Michael. 2014. *Reading the Bromance: Homosocial Relationships in Film and Television*. Detroit: Wayne State University Press.

Dunivin, Karen O. 1994. "Military Culture: Change and Continuity." *Armed Forces & Society* 20:531–47.

Dunning, Eric, Patrick J. Murphy, and John Williams. 2014. *The Roots of Football Hooliganism (RLE Sports Studies): An Historical and Sociological Study*. New York: Routledge.

Edwards, T. 2006. *Cultures of masculinity*. New York: Routledge.

Field, Tiffany. 1999. "American Adolescents Touch Each Other Less and Are More Aggressive toward Their Peers as Compared with French Adolescents." *Adolescence* 34:753–58.

Flood, Michael. 2008. "Men, Sex, and Homosociality How Bonds between Men Shape Their Sexual Relations with Women." *Men and Masculinities* 10:339–59.

Fuchs, Cynthia. 1993. "The Buddy Politic. Screening the Male: Exploring Masculinities in Hollywood Cinema." *In Screening the Male: Exploring Masculinities in Hollywood Cinema,* edited by I. Rae and S. Cohan, 194–210. New York: Routledge.

Gill, Rosalind, and David Hansen-Miller. 2011. "Lad Flicks: Discursive Reconstructions of Masculinity in Popular Film." In *Feminism at the Movies: Understanding Gender in Contemporary Popular Cinema,* edited by H. Radner and E. Pullar, 36–50. New York: Routledge.

Griffin, Pat. 1998. *Strong Women, Deep Closets: Lesbians and Homophobia in Sport.* Champaign, IL: Human Kinetics.

Halkitis, P. (2000). Masculinity in the age of AIDS: HIV-seropositive gay men and the buff agenda. *Research on men and masculinities series* 12:130–151.

Hammarén, Nils, and Thomas Johansson. 2014. "Homosociality." *SAGE Open* 4: 1–11. Accessed September 29, 2017. https://doi.org/10.1177/2158244013518057.

Herek, Gregory M. 1988. "Heterosexuals' Attitudes toward Lesbians and Gay Men: Correlates and Gender Differences." *Journal of Sex Research* 25:451–77.

Howard, Hillary. 2012. "A Confederacy of Bachelors." *New York Times,* August 3. Accessed September 29, 2017. http://www.nytimes.com/2012/08/05/nyregion/four-men-sharing-rent-and-friendship-for-18-years.html?mcubz=1.

Ibson, John. 2002. *Picturing Men: A Century of Male Relationships in Everyday American Photography.* Chicago: University of Chicago Press.

Jourard, S. 1971. *The Transparent Self.* New York: D. Van Nostrand.

JWT. 2013. Accessed September 29, 2017. www.slideshare.net/jwtintelligence/the-state-of-men.

Kahlenberg, S., M. Thompson, and R. Wrangham. 2008. Female competition over core areas among Kanyawara chimpanzees, Kibale National Park. *International Journal of Primatology* 29(4): 931–947.

Kellner, Douglas. 1991. "Film, Politics, and Ideology: Reflections on Hollywood Film in the Age of Reagan." *Velvet Light Trap* 27:9–24.

Kimmel, Michael. 1995. *Manhood in America*. New York: Free Press.

Kimmel, Michael S. 2004. "Masculinity as Homophobia: Fear, Shame, and Silence in the Construction of Gender Identity." *Race, Class, and Gender in the United States: An Integrated Study* 81–93.

Kobak, Roger, Natalie L. Rosenthal, Kristyn Zajac, and Stephanie D. Madsen. 2007. "Adolescent Attachment Hierarchies and the Search for an Adult Pair-bond." *New Directions for Child and Adolescent Development* 117:57–72.

Komarovsky, Mirra. 1974. "Patterns of Self-disclosure of Male Undergraduates." *Journal of Marriage and the Family* 36:677–86.

Kring, Ann M., and Albert H. Gordon. 1998. "Sex Differences in Emotion: Expression, Experience, and Physiology." *Journal of Personality and Social Psychology* 74: 686–703.

Lewis, Robert A. 1978. "Emotional Intimacy among Men." *Journal of Social Research* 34: 108–21.

Lipman-Blumen, Jean. 1976. "Toward a Homosocial Theory of Sex Roles: An Explanation of the Sex Segregation of Social Institutions." *Signs* 1:15–31.

Loftus, Jeni. 2001. "America's Liberalization in Attitudes toward Homosexuality, 1973 to 1998." *American Sociological Review* 66:762–82.

Mac an Ghaill, Mairtin. 1994. *Making of Men*. Philadelphia: Open University Press.

Magrath, Rory, Eric Anderson, and Steven Roberts. 2015. "On the Door-step of Equality: Attitudes toward Gay Athletes among Academy-level Footballers." *International Review for the Sociology of Sport* 50:804–21.

Magrath, Rory, and Ryan Scoats. Forthcoming. "Young Men's Friendships: Inclusive Masculinities in a Post-university Setting." *Journal of Gender Studies*.

McCormack, Mark. 2011. "Hierarchy without Hegemony: Locating Boys in an Inclusive School Setting." *Sociological Perspectives* 54:83–101.

McCormack, Mark. 2012. *The Declining Significance of Homophobia: How Teenage Boys are Redefining Masculinity and Heterosexuality*. New York: Oxford University Press.

McCormack, Mark, and Eric Anderson. 2014. "The Influence of Declining Homophobia on Men's Gender in the United States: An Argument for the Study of Homohysteria." *Sex Roles* 71:109–20.

Mehta, Clare M., and JoNell Strough. 2009. "Sex segregation in friendships and normative contexts across the life span." *Developmental Review* 29(3): 201–220.

Morin, Stephen F., and Ellen M. Garfinkle. 1978. "Male Homophobia." *Journal of Social Issues* 34:29–47.

Morris, Max, and Eric Anderson. 2015. "'Charlie is so cool like': Authenticity, popularity and inclusive masculinity on YouTube." *Sociology* 49(6): 1200–1217.

Murray, Ashnil, and Adam White. 2017. "Twelve not So Angry Men: Inclusive Masculinities in Australian Contact Sports." *International Review for the Sociology of Sport* 52(5): 536–550. 1012690215609786.

Murray, Ashnil, Adam White, Ryan Scoats, and Eric Anderson. 2016. "Constructing Masculinities in the National Rugby League's Footy Show." *Sociological Research Online* 21:11.

Murray, Sandra L., John G. Holmes, and Dale W. Griffin. 1996. "The benefits of positive illusions: Idealization and the construction of satisfaction in close relationships." *Journal of personality and social psychology* 70(1): 79.

Nardi, Peter M. 1999. *Gay Men's Friendships: Invincible Communities*. Chicago: University of Chicago Press.

Office for National Statistics. 2012. Accessed September 29, 2017. http://www.ons.gov.uk/ons/dcp171778_366530.pdf.

Olstad, Keith. 1975. "Brave New Men: A Basis for Discussion." *In Sex/Male/Gender/Masculine,* edited by J. Petras, Port Washington, NY: Alfred.

Pleck, Joseph. 1975. "Issues for the Men's Movement: Summer, 1975." *Changing Men: A Newsletter for Men Against Sexism* 1–23.

Plummer, David. 1999. *One of the Boys: Masculinity, Homophobia, and Modern Manhood*. New York: Routledge.

Pollack, William. 1999. *Real Boys: Rescuing Our Sons from the Myths of Boyhood.* New York: Macmillan.

Pope, H., K. Phillips, and R. Olivardia. 2000. *The Adonis complex: The secret crisis of male body obsession*. New York: Simon and Schuster.

Poplawski, Paul E. 1989. "Psychological and Qualitative Dimensions of Friendship among Men: An Examination of Intimacy, Sex-role, Loneliness, Control and the Friendship Experience." PhD diss., Temple University.

Robinson, Stefan, Anderson, Eric, and Adam White. 2017. "The Bromance: Undergraduate Male Friendships and the Expansion of Contemporary Homosocial Boundaries." *Sex Roles.* doi:10.1007/s11199–017–0768–5.

Roth, Melanie A., and Jeffrey G. Parker. 2001. "Affective and behavioral responses to friends who neglect their friends for dating partners: Influences of gender, jealousy and perspective." *Journal of Adolescence* 24(3): 281–296.

Savin-Williams, Ritch C. 2005. *The New Gay Teenager*. Cambridge, MA: Harvard University Press.

Scoats, R. 2017. "Inclusive masculinity and Facebook photographs among early emerging adults at a British university." *Journal of Adolescent Research* 32(3): 323–345.

Sedgwick, Eve. 1985. *Between Men: English Literature and Male Homosocial Desire.* New York: Columbia University Press.

Sedgwick, Eve. 2015. *Between Men: English Literature and Male Homosocial Desire.* New York: Columbia University Press.

Seiden, Anne, and Pauline Bart. 1975. "Woman to Woman: Is Sisterhood Powerful." In *Old Family/New Family,* edited by N. Glazer-Malbin, 189–228. New York: D. Van Nostrand.

Shields, Stephanie A. 1991. "Gender in the psychology of emotion: A selective research review." *International Review of Studies on Emotion* 1:227–245.

Sprecher, Susan, and Constantine Sedikides. 1993. "Gender Differences in Perceptions of Emotionality: The Case of Close Heterosexual Relationships." *Sex Roles* 28:511–30.

Thompson, Lauren Jade. 2015. "Reading the Bromance: Homosocial Relationships in Film and Television." *Journal of Gender Studies,* 24(3): 1–3.

Thurnell-Read, Thomas. 2012. "What Happens on Tour: The Premarital Stag Tour, Homosocial Bonding, and Male Friendship." *Men and Masculinities* 15:249–70.

Tognoli, Jerome. 1980. "Male Friendship and Intimacy across the Life Span." *Family Relations* 29: 273–79.

Tripp, Clarence Arthur. 2005. *The Intimate World of Abraham Lincoln*. New York: Free Press.

Walker, Karen. 1994. "Men, Women, and Friendship: What They Say, What They Do." *Gender & Society* 8:246–65.

Weeks, Jeffrey. 2007. *The World We Have Won: The Remaking of Erotic and Intimate Life.* New York: Routledge.

White, Adam, and Michael Hobson. 2017. "Teachers' stories: Physical education teachers' constructions and experiences of masculinity within secondary school physical education." *Sport, Education and Society,* 22(8): 905–918.

White, Adam, and Stefan Robinson. "Boys, Inclusive Masculinities and Injury: Some Research Perspectives." *Boyhood Studies* 9:73–91.

Williams, Dorie Giles. 1985. "Gender, Masculinity-femininity, and Emotional Intimacy in Same-sex Friendship." *Sex Roles* 12:587–600.

Worthen, Meredith G. F. 2014. "The Cultural Significance of Homophobia on Heterosexual Women's Gendered Experiences in the United States: A Commentary." *Sex Roles* 71:141–51.

Zani, Bruna. 1993. "Dating and Interpersonal Relationships in Adolescence." *Adolescence and Its Social Worlds:* 95–119.

Zorn, Theodore E., and Kimberly Weller Gregory. 2005. "Learning the Ropes Together: Assimilation and Friendship Development among First-year Male Medical Students." *Health Communication* 17:211–31.

Introduction to Reading 41

Based on an analysis of interviews with over 60 African American mothers, Dawn Marie Dow discusses how mothers conceptualize the challenges their young sons will face as African American boys and men, and the strategic choices they make as a result. The mothers in Dow's study used a variety of gendered, raced, and classed strategies to actively help their sons navigate negative images of black masculinity.

1. What is a "controlling image"? What challenges to African American mothers envision for their sons?
2. Does middle class status protect sons from the negative experiences related to the "thug" stereotype? Why or why not?
3. What are the primary strategies that mothers in this study used to manage their sons' social interactions and reduce their vulnerability?

The Deadly Challenges of Raising African American Boys

Navigating the Controlling Image of the "Thug"

Dawn Marie Dow

I interviewed Karin, a married mother, in her apartment while she nursed her only child. Karin let out a deep sigh before describing how she felt when she learned the baby's gender:

> I was thrilled [the baby] wasn't a boy. I think it is hard to be a black girl and a black woman in America, but I think it is dangerous and sometimes deadly to be a black boy and black man. Oscar Grant[1] and beyond, there are lots of dangerous interactions with police in urban areas for black men. . . . so I was very nervous because we thought she was a boy. . . . I was relieved when she wasn't. It is terrible, but it is true.

Karin's relief upon learning her child was not a boy underscores how intersections of racial identity, class, and gender influence African American middle- and upper-middle-class mothers' parenting concerns. They

Dow, D. M. (2016). The deadly challenges of raising African American boys: Navigating the controlling image of the "thug." *Gender & Society, 30*(2): 161–188.

are aware their children will likely confront racism, often start addressing racism during their children's infant and toddler years (Feagin and Sikes 1994; Staples and Johnson 1993; Tatum 1992, 2003), and attempt to protect their children from racially charged experiences (Uttal 1999). Responding to these potential experiences of racism, parents believe giving their children the skills to address racism is an essential parenting duty (Feagin and Sikes 1994; hill 2001; Staples and Johnson 1993; Tatum 1992, 2003). Although the participants in this research were middle- and upper-middle-class, and thus had more resources than their lower–income counterparts, they felt limited in their abilities to protect their sons from the harsh realities of being African American boys and men in America.

Research demonstrates that race and gender influence how African Americans are treated by societal institutions, including schools (Eitle and Eitle 2004; Ferguson 2000; Holland 2012; Morris 2005; Pascoe 2007; Pringle, Lyons, and Booker 2010; Strayhorn 2010), law enforcement (Brunson and Miller 2006; Hagan, Shedd, and Payne 2005; Rios 2009), and employment (Bertrand and Mullainathan 2004; Grodsky and Pager 2001; Pager 2003; Wingfield 2009, 2011). African American children also experience gendered racism (Essed 1991). African American boys face harsher discipline in school and are labeled aggressive and violent more often than whites or African American girls (Eitle and Eitle 2004; Ferguson 2000; Morris 2005, 2007; Pascoe 2007). Although African American families engage in bias preparation with their children (McHale et al. 2006), the content of that preparation and how gender and class influence it is often not researched. Anecdotal evidence depicts African American parents as compelled to provide gender– and race–specific guidance to their sons about remaining safe in various social interactions, even within their own, often middle–class, neighborhoods (Graham 2014; Martinez, Elam, and Henry 2015; Washington 2012).

This article examines how African American middle- and upper-middle-class mothers raising young children conceptualize the challenges their sons will face and how they parent them in light of these challenges. I focus on mothers because they are often primarily responsible for socializing young children (Hays 1996), and specifically on middle- and upper-middle-class African American mothers because they typically have more resources to address discrimination than do lower-income mothers. Indeed, one might assume that these mothers' resources would enable them to protect their sons from certain challenges. African American mothers are more likely to engage in the racial socialization of younger children and to prepare children to address experiences of racism than are African American fathers (McHale et al. 2006; Thornton et al. 1990). They are also more likely to be single and, thus, principally responsible for decisions related to their children's educational, social, and cultural resources and experiences. Although there has been substantial public discourse about African American mothers' ability to teach their sons to be men, there has been little systematic analysis of their involvement in these processes (Bush 1999, 2004). Also, cultural stereotypes of uninvolved African American fathers overshadow research demonstrating their more active involvement (Coles and Green 2010; Edin, Tach, and Mincy 2009; Salem, Zimmerman, and Notaro 1998).

Although masculinity is associated with strength, participants' accounts of their parenting practices revealed their belief that the thug image made their sons vulnerable in many social interactions. Participants feared for their sons' physical safety and believed their sons would face harsher treatment and be criminalized by teachers, police officers, and the public because of their racial identity and gender. Their accounts revealed four strategies used to navigate these challenges, which I term *experience*, *environment*, *emotion*, and *image management*.

Raced, Classed, and Gendered Parenting Challenges

Gendered Racism and Controlling Images

Scholars have examined how race, class, and gender influence African Americans' experiences in various settings (Ferguson 2000; Morris 2005, 2007; Wingfield 2007, 2009). African American boys and girls experience different levels of social integration within suburban schools (Holland 2012; Ispa-Landa 2013). Boys are viewed as "cool" and "athletic" by classmates and are provided more opportunities to participate in high-value institutional activities, while girls are viewed as aggressive and unfeminine, and are provided with fewer similar opportunities (Holland 2012; Ispa-Landa 2013). Despite having somewhat positive experiences with peers, boys' encounters with teachers and administrators are fraught, as educators often perceive them as aggressive, violent, and potential criminals (Ferguson 2000; Morris 2005; Pascoe 2007). Compared to whites and African American girls, African American boys are disciplined more severely in school (Welch and Payne 2010), and their in-school discipline is more likely to lead to criminal charges (Brunson and Miller 2006).

African American boys are also more likely to have encounters with law enforcement than are whites or African American girls, and these interactions are more likely to have negative outcomes (Brunson and Miller 2006; Quillian, Pager, and University of Wisconsin-Madison 2000) and become violent (Brunson and Miller 2006). The news provides numerous examples of fatal shootings of unarmed African American teenage boys, often by white police officers and private citizens (Alvarez and Buckley 2013; McKinley 2009; Severson 2013; Yee and Goodman 2013). Initiatives like the White House–sponsored "My Brother's Keeper" are responding to an expansive body of research that demonstrates African American boys face disproportionate challenges to their success from schools, their communities, law enforcement, the workplace, and beyond (Jarrett and Johnson 2014).

Collins (2009) theorizes how controlling images function as racialized and gendered stereotypes that justify the oppression of certain groups and naturalize existing power relations, while forcing oppressed populations to police their own behavior. Scholars studying controlling images examine how these inaccurate depictions of black sexuality, lawfulness, temperament, and financial well–being are used to justify policies that disempower women of color (Collins 2004, 2009; Gilliam 1999; Hancock 2003; Harris-Perry 2011) and impact African Americans' experiences in their workplaces, school settings, and other social contexts (Beauboeuf–Lafontant 2009; Dow 2015; Ong 2005; Wingfield 2007, 2009). These images depict African American men as hypermasculine: revering them as superhuman or reviling them as threats to be contained (Ferber 2007; Noguera 2008). Scholars suggest that African American men enact the thug, a version of subordinate masculinity associated with violence, criminality, and toughness, because they are not permitted to attain hegemonic masculinity (Schrock and Schwalbe 2009). Indeed, African American men who enact alternative versions of manhood that are associated with being educated or middle class confront challenges to their masculinity and racial authenticity (Ford 2011; Harper 2004; Harris III 2008; Noguera 2008; Young 2011).

Expanding on this scholarship, I examine how the thug image influences African American middle- and upper-middle-class mothers' parenting concerns and practices when raising sons. Building on Ford's view that "black manhood refers to imagined constructions of self that allow for more fluid interactions in Black and nonblack, public and private social spaces" (Ford 2011, 42), I argue that this fluidity is not just permitted but required to protect black male bodies and manage their vulnerability in different contexts. Black manhood and double consciousness (Du Bois [1903] 1994) are complementary concepts because each requires individuals to see themselves through the broader society's eyes. These concepts also illuminate how individuals who are associated with privileged identities, such as "man" or "American," confront obstacles that prevent them from benefiting from those identities' privileges.

Emotional Labor and Identity Work

Scholarship on emotional labor and identity work examines how African Americans navigate stereotypes. Hochschild (2003) argues that individuals who perform emotional labor induce or suppress the display of certain feelings to produce specific emotional states in others, thereby contributing to their subordinate position. Studying a predominately white law firm, Pierce (1995) uncovers how men, but not women, garner rewards for expressing a range of negative emotions. Summers-Effler (2002) examines how "feeling rules" become associated with particular positions in society and the members of groups generally occupying those positions. Building on Hochschild's (2003) theories, scholars demonstrate that, fearing they will affirm controlling images, African Americans believe there is a limited range of emotions they can display in the workplace without confronting negative stereotypes, and thus feel less entitled to express discontent or anger (Jackson and Wingfield 2013; Wingfield 2007, 2011, 2013).

Historically, interactions between whites and African Americans have been guided by unspoken rules of conduct that signaled different status positions and maintained and reproduced a social structure that subordinated African Americans through acts of deference (Doyle 1937). These acts included African Americans using formal greetings to signal respect to whites, while whites used less formal greetings to signal their superiority (Doyle 1937). Violations of these rules resulted in frustration, anger, and violence from whites and anxiety, fear, and submission among African Americans (Doyle 1937). Rollins's (1985) research reveals how African American women employed as domestics suppressed their emotions and physical presence in interactions with white female employers. Indeed, adhering to specific feeling rules maintains and reproduces racial, class, and gender hierarchies, even as individuals circumvent them.

As African Americans traverse different economic and social strata that are governed by different rules, scholars identify how they manage the expression of their racial identity and class through code–switching (Anderson 1990), shifting (Jones and Shorter–Gooden 2003), identity work (Carbado and Gulati 2013), and cultural flexibility (Carter 2003, 2006). Carter's (2003, 2006) and Pugh's (2009) research demonstrates that African American children and families, respectively, often necessarily retain some fluency in "low-status" cultural capital, even as they ascend economically. Lacy's (2007) research also suggests that some middle-class African Americans emphasize their racial identity, class identity, or racially infused class identities, depending on social context, to gain acceptance. Although these scholars examine how African American middle-class children and families negotiate race and class, gender is not central to their analysis. This article complicates their scholarship by analyzing how race, class, and gender affect how mothers encourage their sons to express their racial identity and masculinity. Schrock and Schwalbe argue, "learning how to signify a masculine self entails learning how to adjust to audiences and situations and learning how one's other identities bear on the acceptability of a performance" (Schrock and Schwalbe 2009, 282). Mothers play an important part in this gendered, classed, and racialized socialization process (Schrock and Schwalbe 2009).

Methods

This article is based on data from a larger project that examined how African American middle- and upper-middle-class mothers approach work, family, parenting, and child care. Participants were recruited using modified snowball sampling techniques. Study announcements were sent via email to African American and predominately white professional and women's organizations. Announcements were made at church services and in bulletins, and were posted at local businesses and on physical or Internet bulletin boards of community colleges, local unions, and sororities. Announcements were also posted to list servers catering to parents, mothers, or African American mothers. Participants who were interviewed were asked to refer others. Through these methods, 60 participants[2] were recruited to the study, of which 40 were raising sons only or sons and daughters. Aside from the opening quote describing a mother's relief upon learning she was not having a son, this analysis focuses on participants raising sons.

All participants lived in the San Francisco Bay Area and were middle-or upper-middle-class as determined by their education and total family income. Participants attended college for at least two years, and their total annual family incomes ranged from $50,000 to $300,000. Participants' total family incomes were as follows: (1) 27 percent were between $50,000 and $99,000; (2) 23 percent were between $100,000 and $149,000; (3) 23 percent were between $150,000 and $199,000; and (4) 27 percent were between $200,000 and $300,000. The upper end of this income range is high by national standards; however, in the San Francisco Bay Area between 2006 and 2010, the median owner–occupied home value was $637,000 (Bay Area Census 2010). Homeownership is an important marker of middle–class status (Sullivan, Warren, and Westbrook 2000). Participants at the upper end of this income range were among the few who could easily attain that marker. Half of the participants were home-owners and half were renters. Participants' ages spanned from 25 to 49 years. The majority of participants (63 percent) earned advanced degrees such as MD, JD, PhD, or MA, with 27 percent earning college degrees and 10 percent attending some college. Three-fourths of the participants were married or in a domestic partnership, and one-fourth were divorced, never married, or widowed. All participants were raising at least one child who was 10 years old or younger, as this research focused on mothers who are raising young children. Participants' employment status included working full-time or part-time, or not working outside of the home (i.e., stay-at-home mothers).

Using grounded theory (Glaser and Strauss 1967) and the procedures and techniques described by Strauss and Corbin (1998), I transcribed interviews and coded them to identify and differentiate recurring concepts and categories. A key concept that emerged was the controlling image of the "thug," a version of subordinate masculinity identified in masculinity and black feminist scholarship. Some participants used the term "thug" or "thuggish." Others used language that referred to components of the thug, such as criminality, violence, and toughness.

As a middle-class African American mother, I shared traits with my participants. These characteristics, in some ways, positioned me as an insider with participants and facilitated building rapport and their willingness to share information about their lives. This status also required that I refrain from assuming I understood a participant's meanings. I balanced building rapport with guarding against making assumptions by probing for additional clarification when a participant suggested I understood something based on our shared background.

Protecting Sons From Baby Racism and Criminalization

Although participants described parenting concerns that transcended gender and related to fostering other aspects of their children's identity, this article examines their specific concerns about raising sons. Participants' concerns included ensuring the physical safety of their sons in interactions with police officers, educators, and the public, and preventing their sons from being criminalized by these same groups.

Gender, Racial Identity, and Parenting

Generally, middle-class children are thought to live in realms of safety, characterized by good schools, an abundance of educational resources, and protection from harsh treatment from police, teachers, and the public. However, numerous scholars have demonstrated that despite the expansion of the African American middle class, its members face economic, social, residential, and educational opportunities that are substantively different from those of middle-class whites (Feagin and Sikes 1994; Lacy 2007; Pattillo 1999). Middle-class African Americans continue to face discrimination in lending, housing (Massey and Denton 1993; Oliver and Shapiro 1995; Sharkey 2014), and employment (Pager 2003). African American middle-class children often attend schools that are poorly funded, lack adequate infrastructure, and are characterized by lower academic achievement than their white counterparts (Pattillo 1999, 2007). These children are also more likely to grow up in neighborhoods with higher levels of crime and inferior community services as compared to their white counterparts (Oliver and Shapiro 1995; Pattillo 1999). Although participants recognized that their middle-class status afforded them additional resources, they believed that their sons' access to middle-class realms of safety were destabilized and diminished because of their racial identity and gender.

Charlotte, a married mother of four sons, who lived in an elite and predominately white neighborhood, held back tears as she described her fears about how others would respond to them:

> I look at the president. I see how he is treated and it scares me. I want people to look at my sons and see them for the beautiful, intelligent, gifted, wonderful creatures that they are and nothing else. I do not want them to look at my sons and say, "There goes that Black guy," or hold onto their purse.

Similarly, Nia, a married mother of two sons, who lived in an economically diverse, predominantly African American neighborhood, described interactions with other families at local children's activities that she called "baby racism":

> From the time our first son was a baby and we would go [to different children's activities]. Our son would go and hug a kid and a parent would grab their child and be like, "Oh, he's going to attack him!" And it was just, like, "Really? Are you serious?" He was actually going to hug him. You see, like little "baby racism." . . . I have even written to local parents' listservs to ask, "Am I imagining this . . .?" And the response was interesting. Almost all the black mothers wrote in, "You're not imagining this, this is real. You're going to have to spend the rest of your life fighting for your child." And all the white mothers said, "You're imagining it. It's not like that. You're misinterpreting it." And it was like, okay, so I'm not imagining this.

Charlotte and Nia, like other participants, believed that when African American boys participated in activities that were engaged in by predominantly white and middle-class families, their behavior faced greater scrutiny. Race and gender trumped class; poverty and crime were associated with being an African American boy. Participants believed the process of criminalizing their sons' behaviors began at an early age, and was not confined to educational settings but was pervasive. Although participants had no way of knowing how others were thinking about their sons, numerous studies support their belief that African American boys' actions are interpreted differently in a range of settings (Ferguson 2000; Morris 2005; Pascoe 2007).

Participants also saw teachers and educators as potential threats to their sons' development. Karlyn, a single mother of a son and daughter, described her son's experience of being harshly disciplined at school:

> A teacher was yelling at my son because some girls reported that he cheated in Four Square. . . . I had to let her know "don't ever pull my son out of class for a Four Square game again. . . . And don't ever yell at my child unless he has done something horrible." . . . I told the principal, "You know, she may not think she is racist but what would make her yell at a little black boy over a stupid Four Square game?" . . . He said, "Oh my God, I am just so glad that you have the amount of restraint that you did because I would have been really upset." I said, "As the mother of a black son, I am always concerned about how he is treated by people."

Like Karlyn, others relayed stories of educators having disproportionately negative responses to their sons' behavior, describing them as aggressive or scary, when similar behavior in white boys was described as more benign. Karlyn, and others, continuously monitored their sons' schools to ensure they received fair treatment. Ferguson's (2000) and Noguera's (2008) research supports their assessment, identifying a tendency among educators to criminalize the behavior of African American boys. Participants' middle–class status did not protect their sons from these experiences.

Mary, a married mother of a son and daughter, also believed her son faced distinct challenges related to his racial identity, class, and gender and sought out an African American middle- and upper-middle-class mothers' group to get support from mothers who were negotiating similar challenges. Mary described a conversation that regularly occurred in her mothers' group, revealing her worries about adequately preparing her son to navigate interactions with teachers and police officers:

> With our sons, we talk about how can we prepare them or teach them about how to deal with a society, especially in a community like Oakland, where black men are held to a different standard than others, and not necessarily a better one. . . . When you are a black man and you get stopped by the policeman, you can't do the same things a white person would do because they might already have some preconceived notions, and that might get you into a heap more trouble. . . . We talk about our sons who are a little younger and starting kindergarten. What do we have to do to make sure teachers don't have preconceived ideas that stop our sons from learning because they believe little brown boys are rambunctious, or little brown boys are hitting more than Caucasian boys?

It is worth emphasizing that although these participants were middle- and upper-middle-class African American mothers with more resources than lower-income mothers, these resources did not protect their sons from gendered racism. Also, middle-class mothers are depicted as viewing educators as resources (Lareau 2011), but these participants viewed educators as potential threats. They believed their sons' racial identity marked them as poor, uneducated, violent, and criminal, and they would have to actively and continuously challenge that marking and assert their middle–class status in mainstream white society—a version of the politics of respectability (Collins 2004). Some participants attended workshops aimed at helping them teach their sons to safely engage with teachers, police officers, and the public. Like the parents described by Lareau and McNamara (1999), some used race–conscious strategies and others used color–blind strategies to address concerns about gendered racism.

Although most participants believed their sons faced challenges related to the thug, a few did not. These participants attributed their lack of concern to their sons' racially ambiguous appearance. Kera, a married mother of two sons, said, "The way they look, they're like me. They could be damn near anything depending on how they put their hair. . . . I don't think they'll have the full repercussions of being a black man like my brothers or my husband." Kera's comments echo research suggesting that skin color differences impact African Americans' experiences in employment, school, and relationships (Hunter 2007).

Participants also believed their sons faced pressure to perform specific versions of African American masculinity that conformed to existing raced, classed, and gendered hierarchies. Nora, a married mother of a son and daughter, said, "There is a lot of pressure for black boys to assume a more 'thuggish' identity. There aren't enough different identity spaces for black boys in schools . . . and so I want my kids to have choices. And if that's the choice, I might cringe . . . but I would want it to be among a menu of choices." Elements of the thug, such as criminality, aggression, and low academic performance, recurred in participants' accounts as something they and their sons navigated. Scholars (Ong 2005; Wingfield 2007) have identified how African American adults negotiate controlling images, but Nora's comments underscore that these negotiations begin at a young age.

Given these pressures to perform specific versions of African American masculinity associated with poverty and criminality, participants tried to protect their sons from early experiences of subtle and explicit racism because of the potential impact on their identity formation. Sharon, a married mother of a son and daughter, captured a sentiment shared by many participants when she stated,

> Each time a black boy has a racially charged interaction with a police officer, a teacher, or a shop owner, those experiences will gradually start to eat at his self-worth and damage his spirit. He might become so damaged that he starts to believe and enact the person he is expected to be, rather than who he truly is as a person.

Participants believed their sons were bombarded by negative messages about African American manhood from the broader white society and, at times, the African American community. Participants worried about the toll these messages might take on their sons' self-perception as they transitioned to manhood. They steered their sons away from enacting the thug, but also observed an absence of other viable expressions of racially authentic middle-class masculinity.

Strategies to Navigate the Thug

Legal scholar Krieger (1995) argues that the law has a flawed understanding of racial prejudice and that, rather than being an active and explicit set of beliefs, racism operates by shaping our perceptions of behaviors. A loud white boy is viewed as animated and outgoing; a loud black boy is viewed as aggressive and disruptive (Ferguson 2000). Similar to the interracial interactions in the South that Doyle (1937) describes, participants believed that whites expected African American boys to adjust their behavior depending on the racial identity of the person with whom they were interacting. Participants walked a tightrope between preparing their sons to overcome the gendered racism they might confront and ensuring they did not internalize these views or use them as excuses to fail. Christine, who was engaged to be married and the mother of a son, explained that in teaching her son what it means to be an African American man, she wanted to ensure that he did not grow up "with that black man chip on the shoulder. Feeling we are weak. Whites have done something to us and we can't do something because of white people." Christine wanted her son to understand how some viewed him, but she tried to foster a version of double consciousness that emphasized his agency and discouraged him from feeling bitter toward whites, disempowered, or constrained by others' views.

Next, I outline the strategies participants used to navigate the thug image and teach their sons how to modulate their expression of masculinity, race, and class. Participants often preferred one strategy but they may have used other strategies, or a combination of strategies, during different periods of their sons' lives.

Experience and Environment Management

Participants used two explicitly race-, class-, and gender-conscious strategies to manage their sons' regular social interactions: *experience* and *environment management.* Experience management focused on seeking out opportunities for sons to engage in activities to gain fluency in different experiences—both empowering and challenging—of being African American boys and men. *Environment management* focused on monitoring their sons' regular social environment, such as their school or neighborhood, with the aim of excluding sources of discrimination. These environments were often primarily middle-class but diverse in terms of racial identity, religion, and sexual orientation. Participants often used environment management when children were preschool age to avoid early experiences of discrimination. Despite having additional resources, participants navigated a landscape of institutionalized child care, which they believed included racially insensitive providers.

Participants using experience management tried to help their sons acquire what they viewed as an essential life skill: the ability to seamlessly shift from communities that differed by race, class, and gender. Experience management involved shuttling sons to activities, such as Little League baseball, basketball, or music lessons, in a variety of neighborhoods comprising African Americans from different economic backgrounds. Participants also exposed sons to African American culture and history and African American men, including fathers, uncles, cousins, coaches, or friends, whom they believed expressed healthy versions of masculinity. Karlyn said, "I worry about my son because he is not growing up with the kind of 'hood' mentality that me and his father had, but he will have to interact with those people." Karlyn's son was not completely ensconced within the safety of a middle-class community. She believed as her son traveled through his day—to school, riding on buses, walking down the street, going in and out of stores, and interacting with police officers and the public—he would be perceived in a range of different and primarily negative ways. Karlyn believed her son would have to adjust the expression of his masculinity, racial identity, and class to successfully interact with people from that "hood mentality"—a version of subordinate masculinity and people from other racial and class backgrounds. She believed that lacking regular experiences in settings like the one she grew up in put her son at a disadvantage in these situations. Karlyn sought out experiences to help her son learn to navigate a world that she believed viewed him primarily as an African American boy and potential troublemaker, rather than a good middle-class kid. She ensured that her son had regular contact with his father and other African American men. She also regularly discussed examples of clashes between African American men and the police with her son.

Maya, a married mother of four, also used experience management. She described how she and her husband exposed their son to alternative and, in her view, more positive ideals of masculinity:

> With our son, we definitely have a heightened level of concern, especially around public schools, about what it means to be a black male in this society. . . . [My] husband does stuff with him that is very much male socializing stuff. . . . But, it is worrisome to think about sending him into the world where he is such a potential target. . . . I know how to make a kid that does well in school and can navigate academic environments. My husband knows how to help young people—black young people—understand their position, how the world sees them and how they might see themselves in a different and much more positive way.

Through these experiences, out of necessity, participants aimed to help their sons develop a double consciousness—"a sense of always looking at one's self through the eyes of others" (Du Bois [1903] 1994, 5). Maya and her husband did this by teaching their son how others might perceive him while rejecting prevailing images of African American masculinity and crafting alternatives.

Environment management involved managing sons' daily social interactions by excluding specific kinds of exposures. Rachel, a married mother of a son and daughter, said, "My son thinks he is street-smart but he is used to being in an environment in which he is known. No one thinks of my son as a black boy, they think of him as my son, but when he goes out into the real world people will make assumptions about him." Rachel lived in a predominately white neighborhood with few other African American families. She believed her neighbors did not view her family as "the African American family," but simply as a family, and this protected her son from challenges associated with being an African American male in the broader society where he might be assumed to be part of the urban underclass. Charlotte, mentioned earlier, described her efforts to find a neighborhood with the right kind of community:

> When we lived in [a different predominately white suburb], none of the mothers spoke to me. Maybe they would wave but I was really taken aback by how shunned I felt. We were the only black family in the school and no one spoke [to us]. . . . Here [another predominately white area], over the summer, people knew my name and I didn't know their name. . . . There was a feeling of welcome and friendliness from the group. . . . You know, I just worry so much for them. I want them to be accepted, and not judged, and not looked at like a black kid. I want people to look at them as "that is a good young man or a good boy.". . . Maybe if they know my sons and me and my husband, it won't be "Oh, there are the black kids"; it will be "There is us."

Charlotte wanted her sons to have access to better resources and schools, and that translated to living in primarily white neighborhoods. Nonetheless, revealing the diversity in white settings, she looked for white neighborhoods where she believed her sons would not face discrimination. Charlotte hoped to transform her sons from "anonymous" African American boys, assumed to be up to no good, to "the kid next door." Being African American was accompanied by assumptions about lower-class status and criminality that participants sought to overcome. Charlotte's experience underscores how intersections of race, gender, and class are used to value individuals and the challenges her sons confronted to be seen as both African American and "good middle-class kids."

Participants living in economically diverse predominantly African American communities with higher crime rates faced particular challenges when using environment management. Jameela, a single mother of a son, explained, "I live in Richmond because it is more affordable, but I don't see a lot of parents like me. I keep a tight leash on my son because of where we live. I don't want him to get involved with the wrong element." Jameela, and participants living in similar environments, often did not let their sons play with neighborhood children. Her experiences highlight class divisions within African American communities and the intensive peer group monitoring parents engaged in when their residential choices were limited. These children's regular environment did not include their immediate neighborhood but was confined to controlled spaces, including their school, church, or other settings that were diverse, free of racial discrimination, and often primarily middle class.

Experience and environment management both focus on social interactions but with different aims. Experience management aims to inform sons through regular controlled activities about the challenges they may face as African American boys and men and teach them how to modify the expression of their masculinity, class, and racial identity. Environment management aims to reduce or eliminate the challenges of being an African American male so they are not the defining features of their sons' lives. These mothers tried to find or create bias-free environments that would not limit their sons' expression of their masculinity but worried about their sons' treatment outside of these "safe havens."

Image and Emotion Management

Participants also used *image* and *emotion management* to reduce the vulnerability they believed their sons experienced related to the thug image and to prevent them from being associated with poor urban African Americans. These strategies were also explicitly race, gender, and class conscious and focused on their sons' emotional expressions and physical appearance. Sons were encouraged to restrain their expressions of anger, frustration, or excitement lest others view them as aggressive or violent. Participants also counseled their sons to strictly monitor their dress and appearance so they would be viewed not as criminals but as middle-class kids.

Karlyn engaged in something she called "prepping for life" with her son. She said, "I talk to [my son] constantly. We do scenarios and we talk about stuff. I'll pose a situation, like say, if you are ever kidnapped, what do you do? If the police ever pull you over, how do you need to react? So we do scenarios for all of that, it's just prepping for life." It would not be unreasonable for a parent to instruct their child to view police officers as sources of help. What is striking about Karlyn's examples is that she viewed child predators and police officers as equally dangerous to her son. She used emotion management with the hope that preparing her son for these scenarios would give him some agency in his response in the moment.

Some participants looked for places where their sons could safely express "normal boy" behaviors while gaining control over those behaviors. Heather, a divorced mother of a son and two daughters described her plan to help her son control his emotions at school: "I'm hoping to get [my son] into enough relaxation-type yoga classes so he is a little bit calmer when he does go to school. I want to make sure he lets it all out in the play yard and activities after school." Through activities like yoga, karate, and meditation, these participants hoped their sons would learn to restrain their emotions, and that this ability would translate to their interactions with teachers, police officers, peers, and the public. Participants emphasized that there were appropriate times to express feelings and advised their sons to refrain from responding to discrimination in the moment, instead taking their time to determine the best approach. This often meant reframing race-related grievances in nonracial terms so they would be better received by white teachers and administrators. Although masculinity is associated with strength, participants believed their sons were vulnerable and did not have the freedom to exhibit certain feelings or behaviors.

Participants also encouraged their sons to engage in image management to avoid being viewed as thugs. Rebecca, a widow with one son who also raised her nephew in his teenage years, recounted discussions during which she counseled her nephew about how people interpreted his clothing:

> Things like him wearing his hoodie and the assumption that he is up to no good. I tried to explain that to him because he didn't understand. He said, "I am just wearing my hoodie." "But baby, I understand what you are doing, and there is nothing wrong with that, but if you walk through the [poor, primarily African American and high-crime] neighborhood near my school, we see something different." You know, just having to protect him and trying to shelter him from unnecessary stress and trauma. . . . You know, the sagging pants and all the things that teenage boys do that don't necessarily mean they are doing anything wrong. . . . Is it fair? No. Is it reality? Yes.

Rebecca's comments illustrate a parenting paradox. Even as Rebecca challenged the double standards that she believed were used to evaluate her nephew's and son's behavior and appearance, as a practical matter, she felt compelled to educate them about these different standards. At times, she counseled them to adhere to those standards for their own safety. Given the recurring news stories of unarmed African American boys shot by police officers and private citizens, Rebecca's approach seems reasonable. Participants believed their sons might be labeled thugs because of their attire, thus leaving them vulnerable to attacks from others. Participants could not prevent these interactions from happening, but wanted their sons to survive them.

Conclusion

This research was bookended by two shooting deaths of unarmed African American males. The first, Oscar Grant, was shot in the back by Officer Johannes Mehserle while lying face-down on a Bay Area Rapid Transit platform (McLaughlin 2014). The second, Trayvon Martin, was pursued, shot, and killed by George Zimmerman, a neighborhood watch coordinator, while walking home in his father's "safe," middle-class, gated community (Alvarez and Buckley 2013). Despite being a child from that community, it was not safe for Mr. Martin. He was not viewed as a good middle-class kid, but was instead interpreted as a threat. Since these incidents, African American parents are increasingly sharing the concerns they have for their sons' safety. Associated Press writer Jesse Washington (2012) wrote a heart-wrenching but matter-of-fact editorial describing how he advises his son to behave in affluent neighborhoods and in interactions with police and others. These instructions may have damaged his son's spirit but increased his chance of remaining alive. Incidents like these reminded participants that their sons have different experiences with the public than do white boys and men.

Initiatives like My Brother's Keeper focus on heightening African American male youths' agency in their lives, often paying less attention to the societal constraints they face. Some might suggest that recent videos of unarmed African American boys and men being shot by officers are shedding light on those constraints and are compelling the US government to take a closer look at law enforcement's interactions with African American boys and men. These incidents draw attention to contradictions between American ideals and practices, underscoring the fact that solving these challenges is not just a matter of changing behavior or increasing resources. These concerns about safety and vulnerability transcend class and are produced by societal forces.

Although the practices of fathers were not directly examined, it is clear from participants' statements that they helped to execute these strategies. Nonetheless, given that African American fathers' parenting practices at times differ from those of mothers (McHale et al. 2006), future research might directly examine their concerns and strategies. Researchers might also examine how different intersections of race, class, and gender produce different forms of vulnerability and protection.

Existing research suggests that having a male body and access to masculinity confers privileges and protections that serve as a symbolic asset in social interactions. However, my research demonstrates that depending on its racialization, the male body can be a "symbolic liability." The thug image derives its power and strength from intimidation and is used to justify attacks on African American boys' and men's bodies and minds. Participants' additional labor to protect their sons and its raced, classed, and gendered nature is largely invisible to the people it is meant to make more comfortable. Despite having additional resources, participants and their sons were not immune to a social system that required them to police their behaviors, emotions, and appearance to signal to others that they were respectable and safe middle-class African American males. Ironically, by feeling compelled to engage in strategies that encouraged their sons to conform to stricter standards and engage in acts of deference, participants contributed to reproducing a social structure that subordinates African Americans. Their accounts show a continuing need for African Americans to have a double consciousness through which they understand how society views them. Their actions also suggest a tension between individual strategies of survival and strategies that challenge and transform existing gendered, classed, and raced hierarchies.

Notes

1. On New Year's Day 2010, Johannes Mehserle, a white Bay Area Rapid Transit police officer, fatally shot Oscar Grant, an African American teenager, in Oakland. During the incident, Grant was unarmed, lying face-down on the train platform, and had been subdued by several other officers. On July 8, 2010, Mehserle was found guilty of involuntary manslaughter, not the higher charges of second-degree murder or voluntary manslaughter (McLaughlin 2014).

2. Sixty-five mothers were interviewed. Five were excluded because they did not meet the income and educational criteria of the study.

References

Alvarez, Lizzett, and Cara Buckley. 2013. Zimmerman is acquitted in Trayvon Martin killing. *The New York Times*, 13 July.

Anderson, Elijah. 1990. *Streetwise: Race, class, and change in an urban community*. Chicago: University of Chicago Press.

Bay Area Census. 2010. http://www.bayareacensus.ca.gov/counties/alamed-acounty.htm.

Beauboeuf–Lafontant, Tamara. 2009. *Behind the mask of the strong black woman: Voice and the embodiment of a costly performance.* Philadelphia: Temple University Press.

Bertrand, Marianne, and Sendhil Mullainathan. 2004. Are Emily and Greg more employable than Lakisha and Jamal? A field experiment on labor market discrimination. *American Economic Review* 94:991–1013.

Brunson, Rod K., and Jody Miller. 2006. Gender, race, and urban policing: The experience of African American youths. *Gender & Society* 20:531–52.

Bush, Lawson. 1999. *Can black mothers raise our sons?* 1st ed. Chicago: African American Images.

Bush, Lawson. 2004. How black mothers participate in the development of manhood and masculinity: What do we know about black mothers and their sons? *Journal of Negro Education* 73:381–91.

Carbado, Devon W., and Mitu Gulati. 2013. *Acting white? Rethinking race in post–racial America*. New York: Oxford University Press.

Carter, Prudence L. 2003. "Black" cultural capital, status positioning, and schooling conflicts for low-income African American youth. *Social Problems* 50:136–55.

Carter, Prudence L. 2006. Straddling boundaries: Identity, culture, and school. *Sociology of Education* 79:304–28.

Coles, Roberta L., and Charles Green. 2010. *The myth of the missing black father*. New York: Columbia University Press.

Collins, Patricia Hill. 2004. *Black sexual politics: African Americans, gender, and the new racism*. New York: Routledge.

Collins, Patricia Hill. 2009. Black feminist thought: *Knowledge, consciousness, and the politics of empowerment, 2nd ed., Routledge classics*. New York: Routledge.

Dow, Dawn. 2015. Negotiating "The Welfare Queen" and "The Strong Black Woman": African American middle-class mothers' work and family perspectives. *Sociological Perspectives* 58:36–55.

Doyle, Bertram Wilbur. 1937. *The etiquette of race relations in the South: a study in social control*. Chicago: University of Chicago Press.

Du Bois, William Edward Burghardt. (1903) 1994. *The souls of black folks*. New York: Gramercy Books.

Edin, Kathryn, Laura Tach, and Ronald Mincy. 2009. Claiming fatherhood: Race and the dynamics of paternal involvement among unmarried men. *Annals of the American Academy of Political and Social Science* 621:149–77.

Eitle, Tamela McNulty, and David James Eitle. 2004. Inequality, segregation, and the overrepresentation of African Americans in school suspensions. *Sociological Perspectives* 47:269–87.

Essed, Philomena. 1991. *Understanding everyday racism*. New York: Russell Sage.

Feagin, Joseph R., and Melvin P. Sikes. 1994. *Living with racism: The black middle–class experience*. Boston: Beacon Press.

Ferber, Abby L. 2007. The construction of black masculinity: White supremacy now and then. *Journal of Sport & Social issues* 31:11–24.

Ferguson, Ann Arnett. 2000. *Bad boys: Public schools in the making of black masculinity, law, meaning, and violence*. Ann Arbor: University of Michigan Press.

Ford, Kristie A. 2011. Doing fake masculinity, being real men: Present and future constructions of self among black college men. *Symbolic interaction* 34:38–62.

Gilliam, Franklin D., Jr. 1999. The "Welfare Queen" experiment. *Nieman Reports* 53:49–52.

Glaser, Barney G., and Anselm L. Strauss. 1967. *The discovery of grounded theory: Strategies for qualitative research, observations*. Chicago: Aldine.

Graham, Lawrence O. 2014. I taught my black kids that their elite upbringing would protect them from discrimination. I was wrong. *The Washington Post*, 6 November.

Grodsky, Eric, and Devah Pager. 2001. The structure of disadvantage: Individual and occupational determinants of the black-white wage gap. *American Sociological Review* 66:542–67.

Hagan, John, Carla Shedd, and Monique R. Payne. 2005. Race, ethnicity, and youth perceptions of criminal injustice. *American Sociological Review* 70:381–407.

Hancock, Ange-Marie. 2003. Contemporary welfare reform and the public identity of the "Welfare Queen." *Race, Gender & Class* 10:31–59.

Harper, Shaun R. 2004. The measure of a man: Conceptualizations of masculinity among high-achieving African American male college students. *Berkeley Journal of Sociology* 48:89–107.

Harris, Frank, III. 2008. Deconstructing masculinity: A qualitative study of college men's masculine conceptualizations and gender performance. *NASPA Journal* 45:453–74.

Harris-Perry, Melissa V. 2011. *Sister citizen: Shame, stereotypes, and black women in America*. New haven, CT: Yale University Press.

Hays, Sharon. 1996. *The cultural contradictions of motherhood*. New Haven, CT: Yale University Press.

Hill, Shirley A. 2001. Class, race, and gender dimensions of child rearing in African American families. *Journal of Black Studies* 31:494–508.

Hochschild, Arlie Russell. 2003. *The managed heart: Commercialization of human feeling*, 20th anniversary ed. Berkeley: University of California Press.

Holland, Megan M. 2012. Only here for the day: The social integration of minority students at a majority white high school. *Sociology of Education* 85:101–20.

Hunter, Margaret. 2007. The persistent problem of colorism: Skin tone, status, and inequality. *Sociology Compass* 1:237–54.

Ispa-Landa, Simone. 2013. Gender, race, and justifications for group exclusion: Urban black students bused to affluent suburban schools. *Sociology of Education* 86:218–33.

Jackson, Brandon A., and Adia Harvey Wingfield. 2013. Getting angry to get ahead: Black college men, emotional performance, and encouraging respectable masculinity. *Symbolic interaction* 36:275–92.

Jarrett, Valeria, and Broderick Johnson. 2014. My brother's keeper: A new White House initiative to empower boys and young men of color. The White House Blog, 27 Feburary. https://www.whitehouse.gov/blog/2014/02/27/my-brother-s-keeper-new-white-house-initiative-empower-boys-and-young-men-color.

Jones, Charisse, and Kumea Shorter-Gooden. 2003. *Shifting: The double lives of black women in America*. New York: HarperCollins.

Krieger, Linda Hamilton. 1995. The content of our categories: A cognitive bias approach to discrimination and equal employment opportunity. *Stanford Law Review* 47:1161–1248.

Lacy, Karyn R. 2007. Blue–chip black: *Race, class, and status in the new black middle class*. Berkeley: University of California Press.

Lareau, Annette. 2011. *Unequal childhoods: Class, race, and family life*, 2nd ed. Berkeley: University of California Press.

Lareau, Annette, and Horvat Erin McNamara. 1999. Moments of social inclusion and exclusion: Race, class, and cultural capital in family-school relationships. *Sociology of Education* 72:37–53.

Martinez, Michael, Stephanie Elam, and Eric Henry. 2015. Within black families, hard truths told to sons amid Ferguson unrest. *CNN,* 5 February. http://www.cnn.com/2014/08/15/living/parenting-black-sons-ferguson-missouri.

Massey, Douglas S., and Nancy A. Denton. 1993. *American apartheid: Segregation and the making of the underclass*. Cambridge, MA: Harvard University Press.

McHale, Susan M., Ann C. Crouter, Kim Ji–Yeon, Linda M. Burton, Kelly D. Davis, Aryn M. Dotterer, and Dena P. Swanson. 2006. Mothers' and fathers' racial socialization in African American families: Implications for youth. *Child Development* 77:1387–1402.

McKinley, Jesse. 2009. In California, protests after man dies at hands of transit police. *The New York Times*, 8 January.

McLaughlin, Michael. 2014. Ex-transit officer who killed Oscar Grant, unarmed black man, wins lawsuit. *Huffington Post*, 1 February 1. http://www.huffingtonpost.com/2014/07/01/oscar-grant-lawsuit-bart-officer_n_5548719.html.

Morris, Edward W. 2005. "Tuck in that shirt!": Race, class, gender, and discipline in an urban school. *Sociological Perspectives* 48:25–48.

Morris, Edward W. 2007. "Ladies" or "loudies"?: Perceptions and experiences of black girls in classrooms. *Youth & Society* 38:490–515.

Noguera, Pedro. 2008. The trouble with black boys: *And other reflections on race, equity, and the future of public education*. San Francisco: Jossey-Bass.

Oliver, Melvin L., and Thomas M. Shapiro. 1995. *Black wealth/white wealth: A new perspective on racial inequality*. New York: Routledge.

Ong, Maria. 2005. Body projects of young women of color in physics: Intersections of gender, race, and science. *Social Problems* 52:593–617.

Pager, Devah. 2003. The mark of a criminal record. *American Journal of Sociology* 108:937–75.

Pascoe, C. J. 2007. *Dude, you're a fag: Masculinity and sexuality in high school*. Berkeley: University of California Press.

Pattillo, Mary E. 1999. Black picket fences: *Privilege and peril among the black middle class*. Chicago: University of Chicago Press.

Pattillo, Mary E. 2007. *Black on the block: The politics of race and class in the city*. Chicago: University of Chicago Press.

Pierce, Jennifer L. 1995. *Gender trials: Emotional lives in contemporary law firms*. Berkeley: University of California Press.

Pringle, Beverley E., James E. Lyons, and Keonya C. Booker. 2010. Perceptions of teacher expectations by African American high school students. *Journal of Negro Education* 79:33–40.

Pugh, Allison J. 2009. *Longing and belonging: Parents, children, and consumer culture*. Berkeley: University of California Press.

Quillian, Lincoln Grey, Devah Pager, and University of Wisconsin-Madison. 2000. Black neighbors, higher crime? The role of racial stereotypes in evaluations of neighborhood crime. CDE working paper. Madison: Center for Demography and Ecology, University of Wisconsin–Madison.

Rios, Victor M. 2009. The consequences of the criminal justice pipeline on black and Latino masculinity. *Annals of the American Academy of Political and Social Science* 623:150–62.

Rollins, Judith. 1985. *Between women: Domestics and their employers, labor and social change*. Philadelphia: Temple University Press.

Salem, Deborah A., Marc A. Zimmerman, and Paul C. Notaro. 1998. Effects of family structure, family process, and father involvement on psychosocial outcomes among African American adolescents. *Family Relations* 117:331–41.

Schrock, Douglas, and Michael Schwalbe. 2009. Men, masculinity, and manhood acts. *Annual Review of Sociology* 35:277–95.

Severson, Kim. 2013. Asking for help, then killed by an officer's barrage. *The New York Times*, 16 September.

Sharkey, Patrick. 2014. Spatial segmentation and the black middle class. *American Journal of Sociology* 119:903–54.

Staples, Robert, and Leanor Boulin Johnson. 1993. *Black families at the crossroads: Challenges and prospects*. San Francisco: Jossey-Bass.

Strauss, Anselm L., and Juliet M. Corbin. 1998. *Basics of qualitative research: Techniques and procedures for developing grounded theory*. Thousand Oaks, CA: Sage.

Strayhorn, Terrell L. 2010. When race and gender collide: Social and cultural capital's influence on the academic achievement of African American and Latino males. *Review of Higher Education* 33:307–32.

Sullivan, Teresa A., Elizabeth Warren, and Jay Lawrence Westbrook. 2000. *The fragile middle class: Americans in debt*. New Haven, CT: Yale University Press.

Summers-Effler, Erika. 2002. The micro potential for social change: Emotion, consciousness, and social movement formation. *Sociological Theory* 20:41–60.

Tatum, Beverly Daniel. 1992. *Assimilation blues: Black families in a white community*, 1st ed. Northampton, MA: Hazel-Maxwell.

Tatum, Beverly Daniel. 2003. *"Why are all the black kids sitting together in the cafeteria?" And other conversations about race*. New York: Basic Books.

Thornton, Michael C., Linda M. Chatters, Robert Joseph Taylor, and Walter R. Allen. 1990. Sociodemographic and environmental correlates of racial socialization by black parents. *Child Development* 61:401–9.

Uttal, Lynet. 1999. Using kin for child care: Embedment in the socioeconomic networks of extended families. *Journal of Marriage and the Family* 61:845–57.

Washington, Jesse. 2012. Trayvon Martin, my son, and the "black male code." Huffington Post, 24 March 24. http://www.huffingtonpost.com/2012/03/24/trayvon-martin-my-son-and_1_n_1377003.html.

Welch, Kelly, and Allison Ann Payne. 2010. Racial threat and punitive school discipline. *Social Problems* 57:25–48.

Wingfield, Adia Harvey. 2007. The modern mammy and the angry black man: African American professionals' experiences with gendered racism in the workplace. *Race, Gender & Class* 14:196–212.

Wingfield, Adia Harvey. 2009. Racializing the glass escalator: Reconsidering men's experiences with women's work. *Gender & Society* 23:5–26.

Wingfield, Adia Harvey. 2011. *Changing times for black professionals*, 1st edition, Framing 21st-century social issues. New York: Routledge.

Wingfield, Adia Harvey. 2013. *No more invisible man: Race and gender in men's work*. Philadelphia: Temple University Press.

Yee, Vivian, and J. David Goodman. 2013. Teenager is shot and killed by officer on foot patrol in the Bronx. *The New York Times*, 4 August.

Young, Alford A., Jr. 2011. The black masculinities of Barack Obama: Some implications for African American men. *Daedalus* 140:206–14.

Introduction to Reading 42

In this reading, Erin M. Rehel looks at the work/family nexus using a different perspective from that in Chapter 7, looking at the impact of workplace policies on spousal and parenting relationships in the home. She interviewed 50 white-collar fathers along with more than two thirds of their female partners in the United States and Canada. Rehel distinguishes between those men who take "extended" parental leaves of 3 weeks or more and those who take shorter or no parental leaves. She studies these decisions under three different policy frameworks for men who all work for the same organization. The advantage of her research design is that she can look at how fathers make decisions relative to how and for how long they use parental leaves within the same organization but different policy conditions, and she looks at the impact of their decisions on their parenting roles. Although parental leave decisions would seem to be an individual or couple decision, Rehel's research suggests otherwise. This examination of parental leave policies and subsequent attitudes toward such policies within the work organization elaborates on some of the policy and organizational changes Amy S. Wharton, in Chapter 7, suggests are important in changing the institutions of work, family, and gender.

1. What features of these three different policy conditions for family/parental leave seem to be most critical in explaining these men's use of parental leave.
2. How do traditional gender expectations influence decisions even in the face of a policy that allows men to take parental leave upon the birth of their children?
3. What are the consequences of fathers' decisions to take extended parental leaves for parenting and relationships with their spouses and their children across time?

When Dad Stays Home Too

Paternity Leave, Gender, and Parenting

Erin M. Rehel

The transition to parenthood is a time of dramatic change for a couple. New mothers often exit the workforce, for varying lengths of time, to recover from birth and to adjust to their new role (Fox 2009). In the United States and Canada, maternity leave, whether state or employer sponsored, often provides the context for this temporary exit.[1] Far fewer fathers experience even a temporary absence from the workforce at the transition to parenthood. Instead, new fathers typically maintain, or sometimes strengthen, their employment ties in the post-birth period (Glauber 2008; Sanchez and Thomson 1997). As a result, men and women experience structurally different pathways into parenthood, which can

Rehel, E. M. (2014). When dad stays home too: Paternity leave, gender, and parenting. *Gender & Society, 28*(1): 110–132. Reprinted by permission of SAGE Publications Inc., on behalf of Sociologists for Women in Society.

contribute to different understandings and enactments of parenting.

Research on this important life course event consistently demonstrates that the birth of a child results in a gendered division of labor for most heterosexual couples (Cowan and Cowan 1992; Walzer 1998); women take on the bulk of the unpaid labor, particularly child care (Bianchi et al. 2000; Craig and Mullan 2011), even when couples' pre-parenting relationship was relatively egalitarian (Calasanti and Bailey 1991; Shelton 2000). A manager-helper dynamic often develops between new parents: Mothers are primarily responsible for child care and related matters, while fathers serve as helpers when needed and asked (Allen and Hawkins 1999; Coltrane 1996; Ehrensaft 1987; Gerson 1993). Largely overlooked in the literature, however, is what happens when men and women experience the transition to parenthood in structurally *similar* ways. More specifically, do men develop understandings and enactments of parenting that mirror those of women when they, too, exit the workforce temporarily in the immediate post-birth period?

In this article, I argue that when the transition to parenthood is structured for fathers in ways comparable to mothers, fathers come to think about and enact parenting in ways that are similar to mothers. The opportunity to experience the transition to parenthood freed of the demands and constraints of work provides fathers the space to develop a sense of responsibility that is often positioned as a core element of mothering (Fox 2009; Hays 1996; McMahon 1995; Ruddick 1995), while simultaneously gaining mastery of and confidence in parenting tasks. Extended time off for fathers, defined here as greater than three weeks, challenges the popular perception of the naturalness of mothering by highlighting the hands-on, learned nature of parenting (Lamb 2004). By comparing fathers who took extended time off following the birth of a child to fathers who did not, I demonstrate that when fathers do take time off after the birth of a child, they are drawn into the daily realities of responsibility and active parenting much as mothers are. By sampling fathers employed by the same financial services firm but living in three different policy contexts (the Canadian province of Quebec, Canada, and the United States), several important aspects of a father's employment are held constant across policy context. This research design highlights the importance of state-level policy in facilitating leave-taking experiences for new fathers.

Gender and the Transition to Parenthood

Research on parenting consistently finds that heterosexual couples respond to parenthood by adopting a gendered division of paid and unpaid labor (Baxter, Hewitt, and Western 2005). This finding endures, even as men continue to increase their levels of involvement in family life (Bianchi, Robinson, and Milkie 2006; Sayer 2005). Time use data from the United States and Canada have shown a steady increase in the number of hours men spend in both domestic labor and child care (Fisher et al. 2007; Hook 2006), but research also illustrates that men's involvement is somewhat selective (Jump and Haas 1987). Men tend to participate more in "fun" aspects of child care, aspects of domestic labor that suit their tastes and interests (Coltrane 1995), and highly visible, or public, fathering activities (Shows and Gerstel 2009); women continue to do the more quotidian, labor-intensive tasks, such as meal preparation and bathing (Offer and Schneider 2011).

Three theoretical approaches have guided much of the research on why this gendering occurs: relative resources, time availability, and gender ideology (Coltrane 2000; Greenstein 2000). The first two approaches, drawing heavily from economics theory (Hank and Jurges 2007), emphasize rationality in the division of paid and unpaid labor, positioning housework as something undesirable that both men and women attempt to avoid. A relative resources explanation posits that the partner who brings the most resources to a relationship, often in terms of income, has the most power, enabling that partner to opt out of unpaid labor (Lundberg and Pollack 1996). Similarly, a time availability explanation suggests that child care and domestic labor should fall to the person who has the most time available (Greenstein 2000); more specifically, that the partner who is engaged in the most hours of paid labor performs less unpaid labor. The gender ideology approach emphasizes how attitudes around who should do what vis-à-vis paid and unpaid labor shape how these forms of labor are distributed within couples (Bianchi et al. 2000; Davis and Greenstein 2009). Beliefs that certain tasks and responsibilities are appropriate for women or for men explain why women are more likely to take on certain tasks, while men are more likely to do others (Bulanda 2004).

Although these theories provide useful frameworks for thinking about the division of domestic labor, some argue that they are less helpful in thinking about child care (Coltrane 2007; Craig and Mullan 2011). Both men and women now spend more time in child care than any previous period since the 1960s (Bianchi, Robinson, and Milkie 2006). Scholars trace this to emerging ideals around intensive parenting (Craig and Mullan 2011; Hays 1996) and concerted cultivation (Lareau 2003). Despite fathers' increased hours spent in child care, models developed specifically to understand father involvement illustrate a persistent lag in the ways fathers are involved with their children.

Lamb and colleagues provide a useful and popular model for understanding father involvement in child care (Lamb 1987, 2004; Lamb et al. 1985). Unlike previous approaches that enumerated specific (and often very gendered) tasks, this typology identifies broad groupings of ways a *parent* might be involved, specifically engagement, accessibility, and responsibility. This model captures various forms of involvement, from reading and playing (engagement), to meal preparation while a child does homework (accessibility), to planning and orchestrating around the child (responsibility). When framing the available data on father involvement using this typology, we see that fathers have significantly increased their levels of engagement and accessibility but have changed little in terms of responsibility. Responsibility for children is consistently understood as one of the most fundamental elements of good mothering (Christopher 2012; Doucet 2009; Fox and Worts 1999; Ruddick 1995) and continues to be a form of labor, often invisible, that adds to women's share of labor in significant ways.

The days, weeks, or months new mothers spend with their newborns following birth, often in the absence of other adults and free of work obligations, is when what is colloquially referred to as maternal instinct develops (Chodorow 1978; Oakley 1979). During this initial period, women develop a sense of responsibility that comes from being the primary care provider, learning cues, needs, and patterns (Bobel 2002; Miller 2007; Walzer 1998). Fathers, more often than not, do not have this time. In important ways, this period establishes parenting patterns that are both difficult to undo and difficult to discern as they become naturalized over time. Moreover, women's childhood socialization and surrogate caretaking experiences (Coltrane 1989; Lamb 2000) provide them with the opportunity to develop some of the necessary skills for and a sense of confidence in parenting, enabling them to adopt the role of primary caretaker more easily. Together, these experiences contribute to the gendered division of labor when partners become parents.

Although a gendered division of labor is most common among parents, and has strong and meaningful roots in social norms and expectations (West and Zimmerman 1987), several studies of couples that intentionally parent equally illustrate that less gendered ways of apportioning paid and unpaid labor are also possible (Deutsch 1999, 2001; Dienhart 1998; Ehrensaft 1987). Much of this research focuses on whether or not men are capable of being active and nurturing co-parents, rather than simply mother's helpers. This research suggests that when parents share parenting tasks from the beginning, men develop greater confidence and skill in their own parenting, leading to greater father involvement (Coltrane 1996; Lamb 2000). My study expands beyond the focus on the choices of individual fathers who have elected to share parenting, found in these studies, to highlight the role played by structure in enabling men to develop as active co-parents.

Lending further support to the idea that men can parent as fully as women do is the small, but growing, body of research on fathers who are primary care providers for their children, specifically stay-at-home dads (Doucet 2006, 2009; Rochlen et al. 2008) and fathers parenting alone (Coles 2010; Hook and Chalasani 2008; Risman 1987; Ziol-Guest 2009). Overall, research on both groups finds that when fathers are required to be primarily *responsible* for all aspects of child care, they are able to do so. By comparing "reluctant fathers," those who find themselves parenting alone not by their own choice, to single mothers and heterosexual co-parents, Risman (1987) finds that fathers who are situated to parent alone do so in ways quite similar to mothers, revealing much about how structure matters in the enactment of parenting. Much of this research demonstrates men's capacity for "mothering" (Doucet 2006, 2009), reinforcing the idea that parenting is most often learned by doing, or, to borrow from Lamb, "on the job" (Lamb 1987, 2000).

In short, the gendered division of labor that occurs when men and women parent together is far from biologically inevitable. We have evidence that men and women can do "parenting" in the same way, but research shows that this occurs less frequently when men and women parent together than when men parent alone. In fact, the research that most clearly and definitively illustrates fathers parenting as completely as mothers is the research on stay-at-home dads and single fathers, fathers who are structurally situated to parent as women most commonly do. All this research led me to suspect that when men experience the transition to parenting in ways that are structurally similar to women, men develop a sense of responsibility that is seen as characteristic of mothers' parenting but far less common in fathers' parenting. Drawing primarily from interview data with fathers, I demonstrate that fathers who are home during the initial transition to parenthood come to develop a sense of responsibility that permits shared parenting, regardless of the policy context in which they live. . . .

Data and Methods

This article draws on data collected as part of a larger, comparative project examining the influence of social policy on father involvement in parenting in the United States, English Canada, and the French-Canadian province of Quebec. I conducted semi-structured interviews with 50 fathers and 35 of their female partners in Chicago, Toronto, and Montreal. These cities were selected to reflect three of the different family-focused social policy contexts that currently exist in North America. An important component of each of these contexts is the family leave policy.

The Family and Medical Leave Act (FMLA), the federal policy of the United States, provides qualified male and female workers 12 weeks of unpaid leave following the birth or adoption of a child. Department of Labor statistics estimate that approximately 60 percent of the U.S. labor force is covered by this policy, leaving many without access to any type of protected family leave. The Canadian federal plan, Employment Insurance Maternity and Parental Leave Benefits (EI), provides new mothers with 15 weeks of maternity leave, paid at a 55 percent wage replacement level up to a maximum amount. An additional 35 weeks of parental leave is available to either parent, again paid at a wage replacement level of 55 percent. Although parents must meet employment criteria in order to qualify, these requirements are significantly less than those of the FMLA, allowing more workers to qualify. Structured in a similar way to the Canadian federal plan, the Quebec Parental Insurance Plan (QPIP) provides 18 weeks of maternity leave and an additional 32 weeks of shared parental leave. One notable difference from the Canadian plan is the designation of five weeks of nontransferable paternity leave for the father only. Unlike parental leave, this paternity leave is provided on a "use it or lose it" basis: If a father opts not to take it, his female partner cannot add this time to her maternity leave. The Quebec plan also provides a higher average wage replacement level (55–70 percent), higher maximum salary amounts, and lower workforce participation requirements to qualify.

For my study, I recruited fathers from within a single financial services firm with operations in both Canada and the United States. Recruiting within a single firm enabled me to control for some employer-specific structural variation, such as organizational culture around and management support for men's involvement in families, which has been shown to influence father involvement (Russell and Hwang 2004). The firm's interest in providing me access to its employees stems from a strong desire to promote workforce gender equality, a goal consistent with its reputation for workplace diversity initiatives, progressive and generous employee benefits, and corporate social responsibility. With regard to family leave, the firm currently has different policies in Canada and the United States, largely because of significant differences in federal leave policies. Employees in Canada or Quebec who take state-supported leave can have up to six weeks of that leave "topped up" by the firm to 90 percent of an employee's salary, a benefit unavailable to American employees.

* * *

Deciding to Take Leave

To understand the influence of leave-taking on a father's parenting requires first considering the decision to take, or not take, leave. Individual attitudes, structural opportunities or limitations, and maternal desire influence this decision in complicated and nuanced ways. My data suggest that although some men make leave-taking decisions based on personal attitudes about work, family, and parenting, others' decisions are enabled or constrained by policy.

Personal attitudes certainly inform fathers' decisions around taking leave. This is just as much the case for fathers who do take leave as it is for those who do not. Many fathers who took leave, like 43-year-old Montreal father Tony, expressed a clear desire to be a very involved father, right from the start. For Tony, this included taking the five weeks of paternity leave available to him: "I was just so excited, I was so excited for him. I wanted to be around him 100 percent of the time. I wanted to be his whole world, you know?"[2] Chad, a 39-year-old father of one from Chicago, articulated a similar interest in being involved right from birth: "You know, I just wanted to be there from the beginning with our first child. I wasn't actually sure how much time I wanted to be home, so I wanted to be home as long as I could, really." Tony and Chad captured what most leave-taking fathers described: a sense of excitement about becoming a dad and an enthusiasm for being an involved co-parent. These types of explanations for leave-taking hint at self-selection in terms of who takes leave.

Personal attitudes also played a role in the decision of those men who did not take leave. No less enthusiastic about becoming a dad, fathers who did not take leave either did not see a need to be home or simply did not want to be out of work for an extended period of time. When asked if he would have taken extended (paid) leave were that available, 32-year-old Chicago father Mark said,

> I wouldn't have stayed home for two weeks when [our daughter] was born. Even if it was written in stone and that was common practice, I wouldn't have stayed home for two weeks. I was home for a week, plus the weekend, and they didn't need me, so I don't know if anything more generous, I would take advantage of.

Opting out of leave-taking because of a lack of interest indicates perhaps a more traditional orientation toward gender, parenting, and division of labor. For these leave-takers and non-leave-takers alike, personal beliefs and orientations figured prominently in leave-taking behavior. However, structural factors also influenced fathers' decisions around leave-taking in important ways.

The leave-taking fathers I've described were drawn to paternity leave because of a personal orientation toward shared parenting, but other fathers were more extrinsically motivated to take leave, primarily by policy. Here, the case of Quebec's leave policy is illustrative of how policy matters. In 2005, the year prior to the introduction of the current plan, 32 percent of Quebec fathers took leave (Marshall 2008). In 2011, just six years later, 76 percent took leave (Findlay and Kohen 2012). This dramatic rise in the number of fathers taking leave coincides with the introduction of Quebec's new parental leave plan. A plausible explanation for this rapid change is that the policy itself motivated fathers to take leave. With five weeks of nontransferable paternity leave paid at 70 percent of one's salary, the structure of the Quebec policy makes leave accessible to large numbers of new fathers. A type of threshold effect is detectable: The policy reduces barriers to leave-taking, enabling large numbers of men to take leave. As more men take leave, leave-taking becomes normalized. This suggests that new fathers take leave because that is the norm and not *necessarily* because they share the types of attitudes and beliefs articulated by the fathers described earlier.

Explaining why he took five weeks of leave after the births of both of his daughters, 33-year-old Montreal father Allan aptly captured the idea that, for many fathers in Quebec, the existence of the policy served as a very real motivator:

> Because they gave me five weeks and I was, like, "Yeah!" I mean, really? This is Quebec. I pay, like, 40percent tax on everything I earn plus 15 percent on everything I buy, plus the extra on gas and alcohol, and anything good in life they tax it twice as much, okay? And then, every once in a while, you get a social program. And this is one of them. So you just look at it and say, "Yeah, I'm taking it."

What this comment, which iterates a commonly expressed sentiment, reveals is that taking leave is not simply a matter of individual attitudes. Instead, particular types of family leave policies appear to facilitate leave-taking among fathers who might otherwise be disinclined to do so.

Policy also constrained fathers in important ways. Across all three policy contexts, many non-leave-taking fathers described wanting to be home for more time after birth, but pointed to one of three structural limitations that made leave-taking impossible: concerns about reactions from superiors and colleagues; wanting to maximize the weeks of leave available to their partners; and financial limitations.

The most commonly cited reason for not taking more extended time off was a concern for how this would be perceived by supervisors, colleagues, and, sometimes, clients. When asked about the possibility of taking longer than the two weeks he did take, 43-year-old Chicago father Patrick captured a fear expressed by many fathers: "Well, that's kind of a tough question. I probably wouldn't have because of the way it would've been viewed. I mean, honestly—and I've heard executives say this—excuse the language—'I can't fucking believe that guy took a month off after the birth of his baby.' I've heard people say it." Like Patrick, many fathers felt pulled back to work by concerns about how violating the image of the ideal worker would impact their work lives. That this was the only reason given by Quebec fathers who opted not to take paternity leave points to the continued salience of a breadwinner identity among men.

A uniquely Canadian constraint relates to the structure of the federal leave policy. Because the only weeks of leave fathers outside Quebec have access to are the shared weeks of parental leave, a father taking leave reduces the number of weeks a mother can take. Thirty-five-year-old Toronto father Brad, for example, said his wife wanted to be home for the whole year, so he let her take the fifty weeks of combined maternity and parental leave, plus the mandatory two-week waiting period: "I know she loved it and I just would never do that to her. It would've taken weeks off of her year and I just would never have done that."

While fathers like Brad describe not wanting to reduce their partner's leave, others told me that their partners were not open to sharing parental leave, suggesting material gatekeeping when it comes to leave allocation (Allen and Hawkins 1999). When I asked Jack, a 43-year-old Toronto father, why he opted not to take leave, he straightforwardly replied that his wife had said she was taking all the available leave.

Finally, many families felt unable to survive the significant reduction in wages that came with both partners being on leave. Here again we see how much policy can play a role in deciding to take leave: America fathers, who did not have access to any wage replacement, invoked this limitation more than their Canadian counterparts.

There is certainly a degree of selectivity in who does and does not take leave. That some Quebec fathers decide against leave-taking in a policy supportive context, where generous paid leave is available, while American fathers do take leave, despite a lack of significant policy support for this decision, certainly validates the idea that a degree of person preference plays into leave-taking behavior. My data also show, however, that it is not the case that all men who take leave do so because of a predisposition to ideals of co-parenting. Furthermore, it would be inaccurate to assume that all those who do not take leave opt not to because of personal preference or traditional views on gender, parenting, and the division of labor. Whatever the reason for it, leave-taking enables fathers to develop the responsibility necessary for them to actively co-parent along with their partners, as I will now show.

Leave-Taking Men

Unlike women, who often have some "surrogate" parenting experiences prior to becoming mothers (e.g., babysitting, caring for siblings extended family), men's more limited exposure to infant/child care mean they often find themselves engaged in child care without much direct experience prior to becoming fathers (Lamb 2004). Eric, a 45-year-old father of two from Montreal, described his experience: "Like, in my case I had no exposure. I'm the youngest sibling in my family. No exposure to infants, diapers—I was walking into a whole new reality." For Eric, taking five weeks of paternity leave was an eye-opening experience, one that he felt really showed him what infant care entails. . . .

The availability of an extended period of parental leave allows fathers the opportunity to gain a sense of the "concerns" of parenting, many of which are invisible and therefore might go unnoticed by a father who is back at work. Forty-three-year-old Toronto father James had taken eight weeks of leave after the birth of his first child and six weeks following the birth of his second child. Like Eric, he found the experience to be invaluable in gaining a deeper understanding of caretaking: "I think, you know, every spouse should do that because it's an experience that will only help you understand in the long run what the heck your wife is going through." While James's comment reflects a continued connection between women and care work, he is also pointing to the way being on leave provides men with a fuller understanding of parenting that might otherwise be inaccessible to them. While paternal skill-building is also important, it is this fuller understanding of parenting that enables fathers to actively engage in parenting in a self-directed way, rather than relying on their partner's guidance. By sharing more than just tasks, partners become more equal co-parents than when one partner manages and delegates child care and related domestic labor.

For Paul, a 27-year-old first-time father from Montreal, five weeks of paternity leave challenged his previous understanding of what it meant to be home with an infant:

> I had this naive thinking that I'm going to be off and I'm going to be able to catch up on all these things. I'm going to have time to myself, to write music and do this stuff. Oh, my gosh, it was such a slap in the face! All my friends at work who were parents didn't say a thing—they just smirked: "Oh yeah, you're going to do all of that, eh? Have fun with that." It did not happen. Those five weeks went by so fast—we were constantly taking care of [our son]. Really made me realize to what extent taking care of a child is more than a full-time job. You don't get your 15-minute breaks, your half-hour breaks when you want. You don't get time off. You don't have a switch off like you do at work. Really, your attention is always—especially with a newborn—100 percent on him.

Expecting a more leisurely experience while on paternity leave, Paul found his expectations to be at odds with the reality of daily life with an infant. In place of "free time," Paul found his days structured by his new reality as a father. My interview with Paul's wife, Sarah, revealed how this experience continued to inform his parenting after he returned to work, while Sarah was still on maternity leave:

> He's never once told me, "You have it easy," you know, he's never, ever said that. He's always respected that this

> is a job and, I think, the five weeks that they give us paternity leave in the beginning is so fantastic. Because it makes the husband realize what kind of a . . . the responsibility and job, everything that the day entails and I think, you know, Paul experienced that and he's, like, "This is harder than what I do," you know. And he's said that before, because he knows he can take breaks and, you know, he could have his hour lunch and he can just walk away and be hands-free.

Sarah's comment reflects what many mothers whose partners had taken leave said about how leave-taking influenced not only how their partners thought about parenting but also how they enact parenting, even after their leave had ended.

For the men who took advantage of the opportunity to be home for several weeks, this expanded understanding of what it means to care for a child was complemented by the opportunity to develop the skills necessary to share care responsibilities with their partners. Chad, a Chicago father mentioned earlier, pointed to the "24 hours a day" aspect of being on leave as being particularly helpful in learning to parent:

> I think I kind of needed that. Because, especially when she was first born, both of us . . . I mean, your mind is going a hundred miles an hour and you really don't know what to expect when you bring the baby home. And what you're supposed to do. And, just in those first few weeks, I think you learn a lot. Being able to spend 24 hours a day, you know, at home with the child, yeah, I think it helped.

Being able to spend "24 hours a day" is key in extended leave-taking. With this kind of presence, freed of workforce obligations, fathers are able to learn to parent in much the same way as mothers, through continuous hands-on participation (Miller 2007), creating the space for shared parenting.

Claudio, a 41-year-old father of one from Montreal who took a total of six weeks of leave, explained why he thought leave-taking is so important:

> Because or else it becomes a routine, where the mom does the everyday necessities with the child and the dad comes home at night, spends a little time, plays with him, and that's it. But I find that if you're in there, every day with the child, taking care of him, making his meals in the morning, at lunch, putting him to sleep, like, all the little details, you'll become attached just as much as the mom. Then it no longer seems like just the mom who has the *initiative* to look after all these things. It becomes the dad and the mom together.

Informed by his experience of being on leave, Claudio described parenting as a mutually shared endeavor between partners. He went on to describe a "divide and conquer" strategy of parenting common among leave-taking fathers:

> When you have a child, you have to work together. To give an example, say in the evening, depending on who comes home first, we'll eat together, but then I'll go give him his bath and then my partner looks after the kitchen. And the reverse happens: If she goes and does the bath, then I take care of the kitchen and all that.

To use Claudio's language, leave-taking allows fathers to develop "the initiative," which moves a couple beyond a manager-helper style of parenting and toward a co-parenting relationship.

With these parenting skills and newly developed carework capabilities, many fathers who took leave did see themselves as co-parents, capable of all aspects of child care, rather than just helpers. After taking two six-week parental leaves, Jon, a 40-year-old Torontonian father, experienced an internal struggle with having to return to work: "Honestly, I can do everything you can do. Why do I have to go back to work? Then there is the argument, 'Well, I had the kid and I'm the mother.' Yeah, I get that, but I'm a hands-on dad and I can do everything you can do." Jon felt he was capable of all the same dimensions of parenting as his wife; he believed his time off enabled him to gain mastery of the necessary parenting skills and the confidence to parent, two factors that have been shown to enhance father involvement (Lamb 1987; Lamb et al. 1985; Pleck and Masciadrelli 2004).

Non-Leave-Takers

The experiences of fathers who took more than three weeks of parental leave stand in sharp contrast to those of men who took little leave following the birth of their children. The understandings of parenting articulated by fathers who did not take leave provide further support for the claim that extended leave-taking by fathers has the potential to challenge gendered understandings of parenting in significant ways.

In North America, extended parental leave is uncommon among new fathers. Although concerns about how extended time off would be perceived by co-workers, particularly managers and supervisors, were often cited as reasons for not taking leave, fathers just as frequently said they did not take leave because they did not have a sense of their own utility at home during the first few months following birth. Jack, one of the Toronto fathers whose partner would not let him take any of the available weeks of leave, stated:

> That's kind of the time [the first six weeks] when you're the least helpful around the house, from my perspective. Like, if you told me the last six weeks of the [first] year . . . it would make a lot of sense, right? Because at that point your kid is running around, walking, you know, interactive and a lot of work. But the first six weeks they're just sleeping, pooping, and eating. So, I think that's part of it. It's like, okay, so you take six weeks off. You're really just sitting there most of the time. You're not really helping.

In many ways, this understanding of infant care reflects Paul's views *prior to taking leave.* Jack, who took less than a week of leave following each of his children's births, believed there would have been little for him to do had he taken leave: "You're . . . just sitting there most of the time." Likewise, Paul expected that being home with his new son would involve lots of "free time," but instead experienced quite the opposite. In the absence of this experience, Jack retained an understanding of infant care as undemanding and non-labor intensive. Jack's understanding of parenting focused solely on visible material tasks and physical labor, and responsibility, which is often an invisible form of labor, is completely absent from his viewpoint. Instead, a manager-helper dynamic is evident in how Jack articulates his presence as either helpful or not helpful.

Returning to work after a short time off also serves as an impediment to new fathers gaining the mastery and confidence that would enable them to actively co-parent. Mark, who, earlier in this article, said he wouldn't have taken paid leave were it available, felt that his wife "picked it [parenting] up pretty quickly" and thus saw no reason for him to be home for longer than he was. Later in our interview, however, Mark reflected on the limitations of his parenting:

> I need instructions to feed [our daughter]. I could be more, sort of, in tune with, you know, being able to pick up where Leslie dropped off, you know, right away. Like, I sometimes have to think about, what if, you know, something happened to Leslie and it was just me with Haley? You know, would she be in as good hands? And, I think, she would be eventually because, you know, I would . . . I would learn. But, I guess, I just don't know as much about taking care of [our daughter] as Leslie does now.

In my interview with Mark's wife Leslie, I asked if they had a system for dividing up child care and domestic labor. Leslie was quick to answer: "Oh yes, we have a system. It's called I do everything, and Mark does nothing." She paused and then added, "I take that back; he walks the dog." Taken together, Mark and Leslie's comments clearly illustrate the manager-helper dynamic that commonly develops when fathers quickly return to work: While their wives develop the necessary knowledge and parenting skills, their own parenting capabilities are less autonomous and rely more on the direction of their partners. Right from birth, this structurally different experience of parenting produces a gendered division of labor. Mothers' time at home, engaged completely in parenting, naturalizes and erases the hands-on, learned nature of parenting, while fathers' return to work curbs their growth in this area.

Charles, a 41-year-old Chicago father whose wife was expecting their third child at the time of our interview, planned to take off no more than a week when his son arrived. Echoing the sentiments of other fathers who took about a week off, Charles felt that there was no reason for him, as the dad, to be at home for long: "Well, I mean, I'm the dad, so I don't really need to take that many days off, so I'll probably only take, you know, five days, a week, just to help my wife acclimate." As with the other fathers who were at home briefly, Charles did not see why—as the father—he would need to be off for any extended period of time. Unlike those fathers who took an extended leave, non-leave-taking fathers are not intimately involved in the daily realities of family life beyond this initial period of adjustment. This structural reality very much limits their understanding of parenting, making them less able to respond to new, previously unknown needs or tasks than fathers who do stay home.

Conclusion

. . . The findings presented here likely underestimate how paternity leave can help lay the foundation for a co-parenting relationship where mothers and fathers share responsibility. Because all leave-taking fathers in this study took leave concurrently with their partners, they were never solely responsible for their child(ren). Despite this, leave-taking fathers still gained a broader understanding of parenting than fathers who did not take leave. This suggests that if fathers did have the opportunity to be on leave alone, their understandings and enactments of parenting would have even greater depth. Similarly, the comparatively short leaves taken by fathers, even those defined here as extended, meant that mothers spent significantly longer periods fully immersed in parenting. Again, if fathers spent similarly lengthy periods of time fully engaged in parenting, enhanced parenting skills and sense of responsibility would likely develop.

* * *

Notes

1. Within the United States and Canada, access to maternity leave of any kind—paid or unpaid, government sponsored or employer based—is uneven (Hegewisch and Gornick 2011).

2. All participant names have been changed.

References

Allen, Sarah, and Alan Hawkins. 1999. Maternal gatekeeping: Mothers' beliefs and behaviors that inhibit greater father involvement in family work. *Journal of Marriage and Family* 61:199–212.

Baxter, Jennifer, Belinda Hewitt, and Mark Western. 2005. Post-familial families and the domestic division of labor. *Journal of Comparative Family Studies* 36: 583–600.

Bianchi, Suzanne, Melissa Milkie, Liana Sayer, and John Robinson. 2000. Is anyone doing the housework? Trends in the gender division of household labor. *Social Forces* 79:191–228.

Bianchi, Suzanne, John Robinson, and Melissa Milkie. 2006. *Changing rhythms of American family life.* New York: Russell Sage Foundation.

Bobel, Chris. 2002. *The paradox of natural mothering.* Philadelphia, PA: Temple University Press.

Bulanda, Ronald. 2004. Paternal involvement with children: The influence of gender ideology. *Journal of Marriage and Family* 66:40–45.

Calasanti, Toni, and Carol Bailey. 1991. Gender inequality and the division of household labor in the United States and Sweden: A socialist-feminist approach. *Social Problems* 38:34–53.

Chodorow, Nancy. 1978. *The reproduction of mothering: Psychoanalysis and the sociology of gender.* Berkeley: University of California Press.

Christopher, Karen. 2012. Extensive mothering: Employed mothers' constructions of the good mother. *Gender & Society* 26:73–96.

Coles, Roberta. 2010. *Best kept secret: Single black fathers.* New York: Rowman & Littlefield.

Coltrane, Scott. 1989. Household labor and the routine production of gender. *Social Problems* 36:473–90.

Coltrane, Scott. 1995. The future of fatherhood: Social, demographic, and economic influences on men's family involvement. In *Fatherhood: Contemporary theory, research, and social policy,* edited by William Marsiglio. Thousand Oaks, CA: Sage.

Coltrane, Scott. 1996. *Family man: Fatherhood, housework, and gender equity.* New York: Oxford University.

Coltrane, Scott. 2000. Research on household labor: Modeling and measuring the social embeddedness of routine family work. *Journal of Marriage and Family* 62:1208–33.

Coltrane, Scott. 2007. Fatherhood, gender and work-family policies. In *Real Utopias,* edited by Erik Olin Wright. Madison, WI: The Havens Center.

Cowan, Carolyn, and Philip Cowan. 1992. *When partners become parents: The big life change for couples.* New York: Basic Books.

Craig, Lyn, and Killian Mullan. 2011. How mothers and fathers share childcare: A cross-national time use comparison. *American Sociological Review* 76:834–61.

Davis, Shannon N., and Theodore N. Greenstein. 2009. Gender ideology: Components, predictors, and consequences. *Annual Review of Sociology* 35:87–105.

Deutsch, Francine. 1999. *Halving it all: How equally shared parenting works.* Cambridge, MA: Harvard University Press.

Deutsch, Francine. 2001. Equally shared parenting. *Current Directions in Psychological Science* 10:25–28.

Dienhart, Anna. 1998. *Reshaping fatherhood: The social construction of shared parenting.* Thousand Oaks, CA: Sage.

Doucet, Andrea. 2006. *Do men mother? Fathering, care and domestic responsibility.* Toronto: University of Toronto Press.

Doucet, Andrea. 2009. Dad and baby in the first year: Gendered responsibilities and embodiment. *Annals of the American Academy of Political and Social Science* 624:78–98.

Ehrensaft, Diane. 1987. *Parenting together: Men and women sharing the care of their children.* New York: Free Press.

Findlay, Leanne C., and Dafna E. Kohen. 2012. Leave practices of parents after the birth or adoption of young children. *Canadian Social Trends.* Statistics Canada Catalogue no. 11–008-X.

Fisher, Kimberly, Muriel Egerton, Jonathan I. Gershuny, and John P. Robinson. 2007. Gender convergence in the American heritage time use study (AHTUS). *Social Indicators Research* 82:1–33.

Fox, Bonnie. 2009. *When couples become parents: The creation of gender in the transition to parenthood.* Toronto: University of Toronto Press.

Fox, Bonnie, and Diana Worts. 1999. Revisiting the critique of medicalized childbirth: A contribution to the sociology of birth. *Gender & Society* 13:326–46.

Gerson, Kathleen. 1993. *No man's land: Men's changing commitment to family and work.* New York: Basic Books.

Glauber, Rebecca. 2008. Gender and race in families and at work: The fatherhood wage premium. *Gender & Society* 22:8–30.

Greenstein, Theodore. 2000. Economic dependence, gender and the division of labor in the home. *Journal of Marriage and Family* 62:322–35.

Hank, Karstan, and Hendrik Jurges. 2007. Gender and the division of household labor in older couples: A European perspective. *Journal of Family Issues* 28:399–421.

Hays, Sharon. 1996. *The cultural contradictions of motherhood.* New Haven, CT: Yale University Press.

Hegewisch, Ariane, and Janet Gornick. 2011. The impact of work-family policies on women's employment: A review of research from OECD countries. *Community, Work & Family* 14:119–38.

Hook, Jennifer. 2006. Care in context: Men's unpaid work in 20 countries, 1965–2003. *American Sociological Review* 71:639–60.

Hook, Jennifer, and Satvika Chalasani. 2008. Gendered expectations? Reconsidering single fathers' child-care time. *Journal of Marriage and Family* 70:978–90.

Jump, Teresa L., and Linda Haas. 1987. Fathers in transition: Dual-career fathers participating in child care. In *Changing men: New directions in research on men and masculinity,* edited by Michael Kimmel. Newbury Park, CA: Sage.

Lamb, Michael. 1987. *The father's role: Cross-cultural perspectives.* Hillsdale, NJ: Lawrence Erlbaum.

Lamb, Michael. 2000. The history of research on father involvement: An overview. *Marriage and Family Review* 29:23–42.

Lamb, Michael, Ed. 2004. *The role of the father in child development,* 4th ed. Hoboken, NJ: Wiley.

Lamb, Michael E., Joseph H. Pleck, Eric Chamov, and James A. Levine. 1985. Paternal behavior in humans. *American Zoologist* 25:883–94.

Lareau, Annette. 2003. *Unequal childhoods: Class, race, and family life.* Berkeley: University of California Press.

Lundberg, Shelly, and Robert A. Pollack. 1996. Bargaining and distribution in marriage. *Journal of Economic Perspectives* 10:139–58.

Marshall, Katherine. 2008. *Fathers' use of paid parental leave.* Statistics Canada, Catalogue no. 75–001-X.

McMahon, Martha. 1995. *Engendering motherhood: Identity and self-transformation in women's lives.* New York: Guilford.

Miller, Tina. 2007. Is this what motherhood is all about? Weaving experiences and discourse through transition to first-time motherhood. *Gender & Society* 21:337–58.

Oakley, Ann. 1979. *Becoming a mother.* Oxford, UK: Oxford University Press.

Offer, Shira, and Barbara Schneider. 2011. Revisiting the gender gap in time-use patterns: Multitasking and well-being among mothers and fathers in dual-earner families. *American Sociological Review* 76:809–33.

Pleck, Joseph, and Brian Masciadrelli. 2004. Paternal involvement by U.S. residential fathers: Levels, sources, and consequences. In *The role of the father in child development,* edited by Michael Lamb. New York: John Wiley.

Risman, Barbara J. 1987. Intimate relationships from a microstructural perspective: Men who mother. *Gender & Society* 1:6–32.

Rochlen, Aaron, Ryan A. McKelley, Marie-Anne Suizzo, and Vanessa Scaringi. 2008. Predictors of relationship satisfaction, psychological well-being, and life satisfaction among stay-at-home fathers. *Psychology of Men and Masculinity* 9:17–28.

Ruddick, Sara. 1995. *Maternal thinking: Towards a politics of peace,* 2nd ed. Boston: Beacon Press.

Russell, Graeme, and Carl Hwang. 2004. The impact of workplace practices on father involvement. In *The role of the father in child development,* 4th ed., edited by Michael Lamb. New York: John Wiley.

Sanchez, Laura, and Elizabeth Thomson. 1997. Becoming mothers and fathers: Parenthood, gender, and the division of labor. *Gender & Society* 11:747–72.

Sayer, Liana. 2005. Gender, time and inequality. *Social Forces* 84:285–303.

Shelton, B. A. 2000. Understanding the distribution of housework between husbands and wives. In *The ties that bind: Perspectives on marriage and cohabitation,* edited by Michelle Hindin, Arland Thornton, Elizabeth Thomson, Christine Bachrach, and Linda Waite. Boston: Aldine de Gruyter.

Shows, Carla, and Naomi Gerstel. 2009. Fathering, class, and gender: A comparison of physicians and emergency medical technicians. *Gender & Society* 23:161–87.

Walzer, Susan. 1998. *Thinking about the baby: Gender and the transitions into parenthood.* Philadelphia, PA: Temple University Press.

West, Candace, and Don Zimmerman. 1987. Doing gender. *Gender & Society* 1:125–51.

Ziol-Guest, Kathleen. 2009. A single father's shopping bag: Purchasing decisions in single-father families. *Journal of Family Issues* 30:605–22.

Introduction to Reading 43

The social construction of gender is embedded in institutions as well as interactions between individuals, as Irene Padavic and Jonniann Butterfield show in this reading. They interviewed 17 women who were nonbiological and nonlegal co-parents of children in lesbian families. These women, from an area near a medium-sized city in Florida, were identified using a snowball sampling method in which individuals interviewed were asked for names of other women in a similar situation. Motherhood and the expectations surrounding it, particularly the expectation that a biological birth was part of the mothering identity, proved to be a major barrier for the nonbiological co-parents they interviewed. These co-parents struggled with "undoing" the social construction of familial expectations, which links gender and biology to mothering, as they attempted to co-parent their children. The obstacles they faced were not just legal but also social, as individuals these women came in contact with during parenting refused to believe that they were their children's "real" mothers who could make daily decisions for their children. Their findings tell us a great deal about how gendered and fixed in heteronormativity the identity of "mother" is.

1. What were the obstacles that stood in the way of co-parenting for these women?
2. What were the differences between lesbian co-parents who identified themselves as "mothers," "fathers," and "mathers," and what factors were involved in their taking on one of these designations?
3. How does the new category some of these women created, "mathers," allow for a less gendered vision of parenting? What stands in the way of a less gendered vision of parenting without a heterosexual, gendered parenting dichotomy?

Mothers, Fathers, and "Mathers"

Negotiating a Lesbian Co-Parental Identity

Irene Padavic and Jonniann Butterfield

In a society marked by binary categorizations and an ideological preference for a "one mother–one father" family model, lesbian co-parents muddy the waters. Previous research on the identity struggles of lesbian co-parents has focused on their experiences as mothers, but in doing so scholars themselves may have inadvertently reinscribed the heteronormative relationships that many lesbian families seek to dismantle. The assumption that women engaged in parenting want to be "mothers," with all the behavioral prescriptions the role entails, precludes an understanding of how the existence of lesbian families can help unhinge sex from gender. This article argues that to gain a more complete understanding of lesbian families, we must consider how co-parents negotiate a parental identity, rather than presuming that women parents want to mother. This article asks how a nonbiological woman parent determines a parental identity in a system constrained by language that offers only two options (mother or father) and in which the dominant motherhood ideology disqualifies her from achieving the status of good mother. Moreover, it asks how these personal and interpersonal dynamics play out in an institutional context that refuses to legally recognize them as parents.

We employ a social constructionist approach to identity and to gender, which suggests that identities are variable and actively created through interactions (Schwalbe and Mason-Schrock 1996) and that because gender is dynamic, it can be "undone" as well as "done" (Deutsch 2007; West and Zimmerman 1987). As a result of this dynamism, "gender can be openly challenged by non-gendered practices in ordinary interaction, in families, childrearing, language, and organization of space" (Lorber 2000, 88). Lesbian parents have the unique opportunity to experience parenthood and raise children outside the gendered heterosexual context, and by doing so, they can destabilize gendered arrangements (Dalton and Bielby 2000; Weston 1991).

Padavic, I., & Butterfield, J. (2011). Mothers, fathers, and "mathers": Negotiating a lesbian co-parental identity. *Gender & Society, 25*(2): 176–196.

The possibility of undoing gender via such innovative family arrangements is one thing; emotionally creating such "brave new families" (Stacey 1998) is another, and previous research indicates that doing so is not a simple matter, particularly for a nonbiological lesbian parent, who "is denied access to any socially sanctioned parental category" (Gabb 2005, 594). When the state fails to legally recognize the legitimacy of both parents in such families—as is the case in most of the United States (notwithstanding huge civil rights gains in some jurisdictions)—the task of securing a parental identity is especially difficult. Understanding how women parents lacking a legal entitlement to parenthood struggle to forge a parental identity is important because identifying the pitfalls such women face and the successful strategies they devise may provide hope for others seeking to create a society in which gender and sexuality cease to exist as categories that privilege some groups over others.

Barriers to Identity Creation: The Motherhood Hierarchy, Language, and the Legal System

Social expectations present lesbian co-parents with at least three barriers to smoothly constructing a sense of themselves as parents. First, all women become parents in a society that promotes a motherhood ideology validating the identity claims only of mothers who meet certain criteria (Chase and Rogers 2001; Hequembourg and Farrell 1999). Family law, social policies, and cultural representations endorse the married, middle-class, white, heterosexual family as the ideal (Abramovitz 1996; Fineman 1995; Roberts 1997; Thorne 1993), and a "motherhood hierarchy" rewards those who most closely conform to it. The most honored mother is "a heterosexual woman, of legal age, married in a traditional nuclear family, fertile, pregnant by intercourse with her husband, and wants to bear children" (DiLapi 1989, 110). Moreover, much societal and legal reluctance to accept lesbians as good mothers derives from fear that their children will be psychologically harmed or more likely to identify as homosexual (Thompson 2002), despite considerable research evidence to the contrary (Stacey and Biblarz 2001). Media pundits and politicians have also pathologized lesbian parents, portraying them as egocentric and immoral and their relationships as unstable (Hequembourg 2007; Richey 2010). Women who fail to mother in ways congruent with motherhood ideals are subject to "deviance discourses" (Miall and March 2006, 46) and to being labeled as unfit or bad (Arendell 2000). Thus, lesbian parents face a continual struggle to have their parental identity legitimized in a social context that renders it tenuous.

A second barrier is that the language used to identify parents relies on the norm of "one mother–one father," which provides no descriptively accurate label for women who lack a biological or legal tie to the child. Naming is a central—and fraught—component of identity for many lesbian parents. As Gabb (2005, 594) noted, the "materiality of language" is crucial because people come into being only via the power of discourse. Research (Aizley 2006; Sullivan 2004) indicates that co-parents often feel caught in an identity limbo since they neither fit neatly into the "mother" category (because they did not give birth) nor (because they are not men) fit into the only other possibility offered in a binary system, that of "father." While some appreciate this limbo and devise terms such as lesbian dad, dyke daddy, high-femme dad, mamma II, and the Hebrew word for mother, ima (Aizley 2006), most describe facing difficulty being validated by the outside world and having to continuously justify their family structure, including to people in the gay community (Dunne 2000). Outsiders hold a cultural attachment to the notion that there can be only one mother and that fathers are men, and reactions to the presence of two women parents can run from confusion to discomfort to outright rejection. As one co-parent in Dunne's (2000, 24) study put it, "Well if you're not the biological mother, then what the hell are you?"

The third barrier to creating a parental identity is that co-parents often must surmount a formidable hurdle: The legal system. In many jurisdictions, when lesbian couples create families through artificial insemination, only the birth parent has a legal tie to the child; the other parent has no legally recognized standing. Second-parent adoption is a solution for couples residing in a state permitting it (assuming they have the funds to do this), but 82 percent of states do not explicitly allow this (Human Rights Campaign 2010). [Editor's Note: These legal constraints vary from state to state and may be changing with the recognition of same-sex marriage by the Supreme Court in 2015.]

Our analysis seeks to answer the question of how lesbian co-parents, who cannot conform to societal definitions of good mothering because of both their lesbianism and their nonbiological relationship to their child, contend with the problem of creating a parental—not necessarily a maternal—identity. What factors facilitate and impede their search for a parental identity? How might other lesbian parents and social change agents benefit from an understanding of these factors?

* * *

Results

Co-parents in our study reported that developing a parental identity entailed an emotional struggle made worse by legally sanctioned discrimination and interpersonal discrimination. For virtually all, grappling with what it means to be a "mother" in a dichotomous "either mother or father" social order was the starting point. Most co-parents engaged in behaviors to align their sense of themselves as parents with the categories available—mother or father—but a third group created a new, hybrid category that stretched the limits of heteronormative categorizations.

Threats to Parental Identity Stemming From Social and From Legal Discrimination

Most of the women we interviewed described how their sense of themselves as legitimate parents was undermined by forays into the public sphere, and virtually all said that this discomfiture was compounded enormously by their lack of legal rights. Of the 17 co-parents, 10 gave examples of interactions in public that required them to field questions about their relationship to their child and the structure of their household. Outsiders, including doctors, teachers, and other parents, challenged co-parents' claims, and thus their identities, by not understanding or accepting them as parents. Ruth explained,

> Other people are really attached to the idea that there can only be one mom. Every Saturday Margaret and I take Cameron to Play Center, and there are lots of other parents there. Even though we know a lot of the parents there now because it's a thing for us to go and so we have explained our situation, I feel like most of them don't take me seriously. Just last week, one of the mothers said that her kid was having a birthday party and Cameron was invited and said she would see if it was okay with Margaret. I said, "You don't need to ask her. Cameron can go. We don't have any other plans." She told me point-blank that she really thought that she should ask Margaret since she was Cam's mother. I had to walk away.

Gretchen said,

> We started seeing a new doctor, who one of my gay friends told me was gay-friendly. I took our daughter for a vaccination and one of the questions on the form asked what my relationship to the patient was and I put "mather" [a combination of the words mother and father] and then in parentheses I wrote "parent." When we got in front of the doctor, he asked what a mather was. I told him that I am her parent, but don't consider myself a mother or father. Then he asked me if the child was biologically mine and of course I said no, but I was getting defensive. He refused to give my daughter the vaccination because I was not a legal parent or guardian. He then asked to speak to me alone. He actually told me that in his medical opinion, referring to myself as a mather was harmful to our daughter. This put me into a tailspin about whether I was messing up our daughter.

Stories like these were commonplace. Such troubling interactions are similar to those reported by step-parents, whose lack of biological relationship can also provide grounds for challenge from institutions (Mason et al. 2002). Yet the barely muted hostility in the above excerpts (flatly denying a motherhood claim in the one case and making accusations of bad parenting in the other) raises the possibility of anti-lesbian-family sentiment. Although step-parents facing a question about parental status can usually assume benign intent, lesbian co-parents cannot, which may explain reactions of "having to walk away" and going into a "tailspin."

More destructive to the co-parents' sense of identity, however, was the lack of a legal right to their children. Unlike the public interactions that give rise to social discrimination, which they could choose to ignore, the state's position that women co-parents are not legally allowed to act in the role of parent has more encompassing psychological and social ramifications. Almost all (16 of 17) co-parents indicated they struggled with their parental identities because they were not legally recognized parents, which influenced how they thought about themselves as parents and how they felt others perceived them. For many, legal discrimination was a greater hindrance than the lack of biological relatedness for developing a parental identity. . . .

Karen said,

> I knew it would be rough at the beginning to see my partner breastfeed and not have that biological connection, but that can be somewhat compensated by being a legal parent. When you can't establish a legal connection, though, it is really hard to feel like a good parent. Right now, I feel like a nanny or mommy's sidekick.

Thus, the state's lack of acknowledgment undermined their sense of parental identity and compounded any insecurity stemming from the absence of a biological tie. The women felt the lack of a legally recognized status not only in an abstract sense; the feeling was reinforced by institutional structures and policies.

The school system was a key site where their lack of rights was brought home to them. Karen continued,

> And I think that's how my son's teachers view me. I can't sign any of the official paperwork at school. She has to do the official important stuff. I get delegated to bring in cupcakes or whatever. Apparently, they will allow us non-moms to do that. . . .

The lack of legal rights makes the whole outside world, not just the school system or doctor's office, an arena of potential danger. As their status can always be challenged, many co-parents felt the need to arm themselves with documentation to bolster their claims as legitimate parents. Even so, in a state that allows co-parents no real legal parental status, officials may simply ignore these documents, leaving them vulnerable. . . .

The lack of legal recognition coupled with the lack of institutional acknowledgment, even in matters as trivial as signing report cards, undermined these women's sense of themselves as parents. They found it difficult to feel like parents in the face of the institutional cold shoulder, and they also perceived that their lack of a legal relationship delegitimized their parental status to outsiders. In sum, these women's sense of themselves as parents was undermined by contact with people who distrusted them and a legal system that disenfranchised them.

Parental Identity Construction

A question on the interview guide asked about the word the woman used to describe herself as a parent, but it turned out that naming was a centerpiece of women's stories that required little prompting. One group referred to themselves as mothers, another group rejected that label as not fitting their sense of themselves as masculine and identified as fathers, and a third group collectively coined the term mather to denote the amalgam of mother and father characteristics with which they identified.

Mothers. Of the 17 co-parents, 6 were committed to adopting a mother identity despite their lack of biological or legal links to their child. These women said they felt maternal, had longed to be mothers, and wanted to be called mom or mamma, but they encountered many obstacles that undermined their ability to pull this off. One obstacle was that the language they constantly heard was at odds with their identity claims. According to Ruth,

> I'm not the "mother." I'm the "non-legal parent," the "non-birth parent," the "non-adoptive parent." There are so many things I'm not in relation to my child that it feels like a battle to be a parent at all. It's exhausting.

Being defined by what they are not loosens the link to motherhood. Seeing a hyphen or hearing an adjective before the mother word, as frequently happened, made clear their tenuous claim to the status and also underlines the power of language in identity construction.

The mothers expressed feelings of futility, being second best, and not succeeding at being a mother, all of which caused them to question their self-worth and sense of themselves as mothers. These negative feelings were brought on by challenges from many quarters. Karen described her feeling of futility stemming from various reminders of her ambiguous status as a mother:

> I have always wanted a baby. I was the quintessential little girl who wanted to be a mom when she grew up. Well, when I realized I was gay I knew it wasn't going to be easy, but I still wanted a baby. Now I have the baby, but I didn't have the baby. Before she was born, I kept saying, "It won't bother me. I know I'm her mother. I don't need a law to say so." Well, I was wrong. I feel like a mother, I do, but not the mother. It's like I am always getting slapped in the face. "Oh, you can't breastfeed. Oh, you can't sign this paper. Oh, you have to be related to do that." It's exhausting. . . .

Such reminders, whether intentional or not, police the boundaries of who can be accepted in the motherhood ranks and take an emotional toll.

Samantha considered herself a mother but felt second best because she did not give birth to her child and felt pressured to justify her relationship to him:

> When I was in grad school, I would say something about Evan and people would start with the questions. Like, how do you have a kid? You were never pregnant. Did you adopt? And then I would do this whole explanation thing, like, well, my partner actually birthed him, but he's my kid too. This always made me feel really bad, like I wasn't a real parent you know? One day my [academic] advisor said, "Why don't you just say 'Yes, I'm his mother'? Stop explaining and apologizing." And after that I did. But it took me a long time to get to that place.

As these experiences indicate, the extent of society's commitment to the norm of blood relationships defining the link between parents and children is difficult to overestimate (Katz Rothman 2006; Miall and March 2006). In a study of lesbian families in the United Kingdom, Gabb (2005) also found a pervasive sense of second-class citizenship among nonbiological mothers.

The notion that mothers should be feminine and not masculine is another norm that society readily enforces. A woman who used the mother label despite considering herself masculine faced social pressure to fit into a binary schema of feminine mother and masculine father:

> I look butch. So when my partner and I are together, people assume I must be the "dad" because everyone assumes there is a mom and a dad, even in a lesbian couple. I feel people look us over, like, "Okay, who is the butch here?" Even though I am masculine, I still think of myself as a mother. A second-class mother, but still a mother. I guess I don't buy that mothers have to be feminine and dads have to be masculine. But, it's hard because everyone else thinks like that. Sometimes I have doubts about myself as a mother, but I know I am not a father.

Despite people's reaction to her masculine identity, she insisted on the mother label and was willing to stand up to social pressure urging her to reconsider. Since motherhood and femininity go hand in glove in the popular consciousness (Chodorow 1989; Glenn 1994), her decision forced beholders to question automatic associations between motherhood and femininity, thus raising an alternative to the given order of things. Yet maintaining that stance was an ongoing, lonely, "hard" struggle.

Norms about who may and may not be a mother can also penetrate interactions in the intimate realm, and some co-parents felt expendable even in the privacy of their homes. Yolanda explained,

> I think of myself as a mom, but I don't know that anyone else does, even my partner sometimes. I am not identified anywhere as a parent. It goes as far as photographs. I am always the one taking the photograph. So I am not even in very many pictures. It is always "take a picture of us." Like they are the real family.

Taylor similarly felt that her claim of being a mother was contested in the private sphere:

> I even feel like my partner doesn't consider me an equal mother. We had been debating whether or not to get a certain vaccination. One day she came home and told me that she decided to get the vaccination done. I was floored. I couldn't believe that she made that kind of executive decision without me. And this was right after she had read me the riot act because I got Shaun a buzz cut without consulting her. I felt so irrelevant.

For the one-third of interviewees who wanted to think of themselves as mothers, the challenge did not lie in knowing the parental term they desired; as one said, "I know in my heart I'm [a] mom." Rather, the challenge lay in surmounting stumbling blocks in the form of marginalization by norms linking femininity and motherhood, by outsiders, by the legal institution, by other children, and even by their partners, which left their mother identity in question. Women in this category struggled to validate their mother identity in the face of social forces that positioned them as inferior, including a language that positioned them as non-birth mothers, second mothers, other mothers, and so on, diminishing their ability to embrace the mother label as strongly as they wanted and creating a void that is not merely social ("What name should I respond to?") but also personal ("Who am I as a parent?"). Despite challenges and personal doubts, these women nevertheless were living evidence of an alternative family form. As they went about their daily public business—a family headed by two mothers—they transgressed the deeply held belief that families contain one and only one mother and by doing so weakened it.

Fathers. About one-third of the women we interviewed also honored the gender binary, but they did so by inverting it; for example, "I don't look like a mother and I don't feel like a mother, so I must be a father." Women in this sample who identified as fathers all described themselves as masculine or "butch." They had strong and clear ideas about how mothers looked and acted and felt they failed to embody these ideas, leading them to reject the mother label and adopt the alternative. According to one,

> I am butch. How could I possibly be a mom? It's laughable, really! I would rather ride bikes and play ball with the boys, which is something I think fathers typically do. I do some things moms do, I guess, but I don't feel like a mom. I guess it's more a mental thing. And maybe it's a gender thing. I don't want to be a man, but I guess I kind of feel like a guy.

According to another,

> I struggle with it because I know I'm a woman, but I don't look like a mom. I wear work boots and flannels. I drive a truck, not an SUV. And I sure don't act like the moms I knew growing up or the moms I know now. I don't bake cookies. I'm not nurturing in that way. I mean, I am in my own way, but more like a dad, I guess. I don't feel comfortable being called mom, because I don't feel like one. It just doesn't fit me. It makes it difficult, though, because it is really hard for other people to understand that I'm a woman but I feel more like dad. It's confusing.

For these women, a gender identity as masculine precluded using the mother label; indeed, the thought of doing so was "laughable." None tried to keep the mother label and change the meaning of motherhood to include "riding bikes and playing ball" to better fit their personal attributes. Instead, motherhood remained the domain of people like "the moms I knew growing up or the moms I know now." The only other option they saw, and the one they acted on, was to declare themselves fathers.

It was not a perfect fit. A key source of unease was the negative associations many had with heterosexual fathers' behavior, which one described as "not involved with the kid, not affectionate, and a disciplinarian." Mallory had a similar disaffection for the fathering with which she was familiar:

> I chose father, because I am just not a mother. But, it still bugs me. Maybe it's because I am very woman-identified, a lesbian, and I don't like stuff men do, but I don't like how fathers act. I see some at the park when their kid cries, and they do that "toughen up" macho shit. I don't want to be a father like that.

Since their experiences of fathering are based on observations of men, they, as women, had an understandably hard time defining what fathering consists of when done by someone who is not a man. They did not want to practice the fathering style with which they were familiar, nor did they want to be genderless parents (also see Sullivan 2004). As one woman said, "Even though I think of myself as a dad, I am still a woman and a lesbian." Lacking guidance from existing practice about how to father as women, they tried to invent it. One woman chose a new term for "father" partly to facilitate a new practice of fathering:

> My family is Italian, so Kevin calls me Babbo, which is father in Italian. Calling myself Babbo lets me get a little bit away from the actual dad term, because I don't want to be that 1950s dad. I want to be a lesbian dad, a Babbo. I hope I can figure it out as I go along. I'm not girly at all, but I would like to be a girly father, whatever that means.

She chose a term that allowed her freedom from "acting like the fathers I know," but like the other fathers she struggled with the lack of models. None felt they had a clear sense of how to father in a new way, and variations of her "whatever that means" statement were common.

While most assumed the father label or a close variant on their own volition, this was not the case for all. Some felt pressured by their families and friends to conform to the one mother–one father model. Mallory described how her son pushed her to assume the role of his father:

> I am an androgynous person, and once I picked up Jack at the park and heard him say to his friend, "That's my dad." He had never referred to me as that before; he always called me by my name. Another time he said, "When we go into the store, can you lower your voice?" I think it comes from the social idea that you have a mom and a dad. It made me realize that it was kind of selfish not to be mom or dad for his sake. Jack seemed to think of me as dad, and I sure didn't think of myself as mom—he has one of those—so here I am. A dad.

Krista described the tension with her partner that similarly pushed her to call herself a father:

> To be fair, I am not the mother type, but my partner insisted, very firmly, that I was to take the role of the father. For example, she wants to pick out his clothes because she thinks that's something mothers do. One day I tried to dress him and she freaked out. She felt like I was invading her territory. Oh, another example is our baby shower. She referred to it as her baby shower and didn't even want me to come, but I kind of crashed it. One of the presents was a baby book. She told me that since she was the mother I wouldn't be the one putting stuff in the book, so I went out and bought my own. She flew off the handle. She wants to be the mother and wants me to be a father, so clearly that's what I am. Now I just have to figure out what that means exactly.

Important people in their lives did not permit Mallory and Krista the option of choosing nonalignment with the "father" pole of the gender binary, although for one the pressure was gentle and for the other more coercive. In a final example, Veronica also chose the father label under pressure, although her attempt to adapt to the binary choice was more agonized than for others we interviewed:

> People would hold the baby and then pass her back to me and say, "Okay, go back to your Mommy." I think I turned red every time someone said that. The first time I tried to refer to myself as "Mom," I practically had to choke it out. . . . My friends and my girlfriend said, "Just go by 'Daddy.'" But I wasn't down with that either. There is stuff out there like "dyke daddy," but can you imagine the kid calling you that? I can just imagine her telling her teacher, "Yeah, my dyke daddy is picking me up today." Nope. Doesn't work. Poor kid!

She continued,

> My friends and girlfriend said things like, "Well you have to come up with something. The baby needs to understand who you are." I felt accountable, and I couldn't come up with a way to make it right for everyone.

She ended up referring to herself as a father but having the child use her first name, although she was dissatisfied with this solution. As her story makes clear, having a child was a key moment precipitating an upswing in gender coercion and self-policing. Her close companions pressed her—for the sake of the child—to put aside personal discomfort and choose a label, and an imagined internalized teacher forced her to consider the cost to the child of adopting the gender-radical "dyke daddy" label. As she said, unknowingly echoing West and Zimmerman (1987), when it came to parenting, she felt "accountable" to traditional practices and ideology.

These co-parents identified as fathers because their sense of themselves as masculine precluded using the mother label; for some, pressure from people they were close to also entered into the decision. With no language for the category of "not-mother-and-not-quite-father-either," they used the only other option language afforded: father. Their fit with the label was far from perfect, and all admitted to not knowing how to enact the fatherhood role with their children. Despite these problems, by adopting the father label, these women disrupt the prescription that "fathers are men," and thus their very presence in society as fathers helps deconstruct the edifice of the gender binary.

Mathers. Of 17 co-parents, 6 labeled themselves *mathers*. The term was born of an informal support group that had met for about a year before coining it. The group had begun with gatherings of a few friends meeting biweekly as a lesbian co-parent support group, and in short order acquaintances and friends joined. At the suggestion of one member, they advertised the group at a local LGBT center and on a LGBT Internet forum, and over the course of six months the group grew to about 15 people from across the region.

Interviewees clearly articulated the central problem discussed at these support group meetings: They felt like neither mothers nor fathers and lamented the lack of any other categorization. They sought a label mainly to assuage their pervasive worry about what their children would call them (see also Gabb 2005). Many had experimented with "mother" and "father" and were dissatisfied. Much meeting time was devoted to discussion of the various parental labels popular among lesbian parents, but the consensus was that none fit the bill. They sought a label their children could use publicly and privately and that the co-parents themselves felt good about. After a year of what they described as agonizing discussions, a group member suggested the *mather* term as a gender-bending, gender-blending hybridization of mother and father. . . .

Respondents saw the term as a flexible, dynamic word that captured a larger idea about parenting outside the rigid, gendered mother–father dichotomy, and each co-parent could mold the term as she saw fit.

The common experience seemed to be that once they had established a label, they could now more clearly flesh out their familial roles. Jan explained,

> I don't even know if there is anything a mather does that is really that different from what a mother or father does. Being a mather is more a way of thinking, like a way of dealing with feeling uncertain about what you are. . . . As a mather, I can tell my kid, "Hey, families don't have to have moms and dads. They can be whatever." And I don't have to just do "mom things" or "dad things." We do everything together. We talk, paint, wrestle, whatever. When you think of yourself as mom or dad, there are lots of things that go along with that. Like if you're mom then you do certain things, and dads do certain things. Being a mather, I feel like I can cross those lines without penalties. Like I can play ball with my kid and wear my hat backwards, but I can also let myself be vulnerable.

Jan and others invoked fun (talk, paint, wrestle, play ball, wear backward caps, play, laugh) and conveyed a light heartedness that contrasts starkly to the anxiety pervading the group discussions that preceded the term's invention.

The term had a decidedly serious side as well: women who used it were adamant about deploying the term to promote social change. They wanted their children to call them mather, they introduced themselves as their child's mather, and they considered themselves pioneering agents of social change. . . .

Addison relished how using the term made her feel like a change agent:

> Being a mather feels like activism. I see lots of my gay friends feel like they have to choose. But my options are mom or dad, so "Hmm, which do I feel more like?" I get to bust out of those categories. I get to introduce myself to my daughter's teachers and say, "I'm Addison, Jillian's mather." Inevitably, after they get over their shock, they ask me what that means and I get to educate them! I feel like I will make it easier for other parents down the road who don't want to have to be mom or dad. Man, it's about social change.

All the co-parents, not just mathers, promoted social change by bucking strongly held norms, yet it is far easier and more fulfilling to do so with group backing. The contrast is sharp between the angst Lorraine and Addison attribute to their nonmather friends and the sense of empowerment they themselves felt. Some others described how the mather mentality transcended the parental role to affect their sense of self. According to Jan,

> Since I started thinking of myself as a mather, I have changed in lots of ways. I feel more free to express myself without thinking about gender. I stopped thinking about gender when I parent, so I guess it makes sense it would happen in other areas of my life. For example, sometimes now I wear a tie and men's dress shoes to work, which I never would have done before. It's not because I want to be a man; it's because I have always liked the look and now I feel free to do it.

The mathers' stories compared to those of the other groups illustrate the power of language, social support, and a collective identity. The fathers and mothers faced the same challenge as the mathers—feeling like neither a mother nor father—but unlike them lacked a language to redefine themselves and their role. Having a language carved out an ideological space for mathers to redefine their family role, and the realization that they could exist outside the binary reduced their stress significantly. Mathers were the only group that sought validation from significant others outside the privatized home sphere, and by their accounts, the collective act of defiantly creating a new identity and seeking opportunities to educate others about it was empowering. Even so, it was no panacea, as evidenced by the mather sent into a "tailspin" by her doctor's accusation that her choice of label was harming the child. As for effecting social change, like the mothers and fathers, their presence is testimony to the possibility of alternatives to the heteronormative family, and the proliferation of such possibilities increases the likelihood of more appearing.

Discussion and Conclusion

An approach that assumes that because co-parents are women they seek to identify as mothers conceals the complexity of parental identity development for lesbian co-parents. Previous research on the identity struggles of women who parent in the face of a lack of biological ties to their children has focused on their experiences of becoming and being mothers and described such women as members of "dual-mother families," or as "comothers," who negotiate "mothering experiences." This assumption of a link between female sex and mother identity has precluded asking how co-parents negotiate womanhood, intimate relationships, and social expectations to construct a parental identity that may be at variance with the mother identity.

These women faced both external and internal assaults on their sense of themselves as parents. The institutions and people they regularly encountered—play groups, schools, doctors, children's friends, and, perhaps most importantly, the law—explicitly challenged their parental identity claims. Most women faced anguished identity struggles because of these external assaults and because they felt like their lesbian parenting fit neither into the biologically inflected "mother" category nor into the father category, the only other possibility the language offers in a binary gender system. Mathers seemed to have had the most successful resolution of the internal dilemma, but they too paid an emotional toll from facing the constant external challenges from people and institutions unwilling to recognize their consciously blended roles and identities.

What do these results imply for the goal of destabilizing gendered arrangements (Lorber 2005)? On one hand, they confirm the continuing hold on the public and private imagination exerted by the "motherhood institution" (Bernard 1974; Rich 1977). As an institution, motherhood still grants and withholds the material, institutional, and cultural supports that make child rearing easy or difficult (Bernard 1974; Chase and Rogers 2001; Rich 1977). It still shapes what mothers do and how they feel about it, and it privileges women who fit cultural notions about appropriate characteristics of mothers and disfranchises those who do not. As long as rigid prescriptions for gendered behavior are inscribed in the institution—especially when they are backed by law—members of excluded groups will remain in identity limbo. Thus, while planned lesbian families have the potential to help decenter the gender- and power-laden heterosexual nuclear family (e.g., Dalton and Bielby 2000; Weston 1991), this study illustrates that the task is not easy. On the other hand, there is a positive conclusion to be drawn as well. Social change is propelled forward by people, like these lesbian parents, who refuse to live lives consonant with the given order. The mothers were transgressive by embodying a two-mother family, the fathers disrupted the prescription that "fathers are men," and the mathers generated a new family role and included education about it as part of their mission. These women's transgression of the gender binary makes further transgressions more likely and makes a utopian vision of a non-heteronormative family less distant.

The findings point to the necessity of continuing the civil rights struggle, and lesbian parents need not stand alone in the larger struggle to break down parenting ideology and laws. They have common cause with other groups who also suffer from the exclusivity of the good mother category and thus have a stake in degendering the mothering institution. . . .

References

Abramovitz, Mimi. 1996. *Regulating the lives of women: Social welfare policy from colonial times to present.* Boston: South End.

Aizley, Harlyn. 2006. *Confessions of the other mother: Non-biological mothers tell all.* Boston: Beacon.

Arendell, Terry. 2000. Conceiving and investigating motherhood: The decade's scholarship. *Journal of Marriage and Family* 62:1192–1207.

Bernard, Jessie. 1974. *The future of motherhood.* New York: Dial Press.

Chase, Susan E., and Mary F. Rogers. 2001. *Mothers and children: Feminist analyses and personal narratives.* New Brunswick, NJ: Rutgers University Press.

Chodorow, Nancy. 1989. *Feminism and psychoanalytic theory.* New Haven, CT: Yale University Press.

Dalton, S., and D. Bielby. 2000. That's our kind of constellation: Lesbian mothers negotiate institutionalized understandings of gender within the family. *Gender & Society* 14:36–61.

Deutsch, Francine M. 2007. Undoing gender. *Gender & Society* 21:106–27.

DiLapi, E. M. 1989. Lesbian mothers and the motherhood hierarchy. *Journal of Homosexuality* 18:101–21.

Dunne, Gillian. 2000. Opting into motherhood: Lesbians blurring the boundaries and transforming the meanings of parenthood and kinship. *Gender & Society* 14:11–35.

Fineman, Martha. 1995. *The neutered mother, the sexual family, and other twentieth century tragedies.* New York: Routledge.

Gabb, Jacqui. 2005. Lesbian motherhood: Strategies of familial-linguistic management in lesbian parent families. *Sociology* 39:385–603.

Glenn, Evelyn N. 1994. Social constructions of mothering: A thematic overview. In *Mothering: Ideology experience, agency*, edited by E. N. Glenn, G. Chang, and L. R. Forcey. New York: Routledge.

Hequembourg, A. 2007. *Lesbian motherhood: Stories of becoming.* Binghamton. NY: Hawthorne Press.

Hequembourg, A., and M. Farrell. 1999. Lesbian motherhood: Negotiating marginal-mainstream identities. *Gender & Society* 13:540–57.

Human Rights Campaign. 2010. Parenting laws: Second-Parent Adoption. http://www.hrc.org/documents/parenting_laws_maps.pdf (accessed 19 December 2010).

Katz Rothman, B. 2006. Adoption and the culture of genetic determinism. In *Adoptive families in a diverse society*, edited by K. Wegar. New Brunswick, NJ: Rutgers University Press.

Lewin, Ellen. 1993. *Lesbian mothers.* Ithaca, NY: Cornell University Press.

Lorber, Judith. 2000. Using gender to undo gender. *Feminist Theory* 1:75–95.

Lorber, Judith. 2005. *Breaking the bowls: Degendering and feminist change.* New York: Norton.

Mason, M. A., S. Hanison-Jay, G. M. Svare, and N. H. Wolfinger. 2002. Stepparents: De-facto parents or legal strangers? *Journal of Family Issues* 23:507–22.

Miall, C. E., and K. March. 2006. Adoption and public opinion: Implications for social policy and practice in adoption. In *Adoptive families in a diverse society,* edited by K. Wegar. New Brunswick, NJ: Rutgers University Press.

Rich, Adrienne. 1977. *Of woman born: Motherhood as experience and institution.* New York: Bantam.

Richey, Warren. 2010. Florida ban on gay adoption unconstitutional, court declares, Christian Science Monitor, 23 September. www.csmonitor.com/USA/Justice/2010/0923/Florida-ban-on-gay-adoption-unconstitutional-court-rules (accessed 8 January 2011).

Roberts, Dorothy. 1997. *Killing the Black body: Race, reproduction and the meaning of liberty.* New York: Pantheon.

Schwalbe, Michael L., and Douglas Mason-Schrock. 1996. Identity work as group process. *Advances in Group Processes* 13:113–47.

Stacey, Judith. 1998. *Brave new families: Stories of domestic upheaval in late twentieth century America.* Berkeley: University of California Press.

Stacey, Judith, and Timothy J. Biblarz. 2001. (How) does the sexual orientation of parents matter? *American Sociological Review* 66:159–83.

Sullivan, Maureen. 2004. *The family of woman.* Berkeley: University of California Press.

Thompson, Julie. 2002. *Mommy queerest.* Amherst: University of Massachusetts Press.

Thorne, Barrie. 1993. Feminism and the family: Two decades of thought. In *Rethinking the family: Some feminist questions*, 2nd ed., edited by B. Thorne and M. Yalom. New York: Longman.

West, Candace, and Don H. Zimmerman. 1987. Doing gender. *Gender & Society* 1:121–51.

Weston, Kath. 1991. *Families we choose: Lesbians, gays, kinship.* New York: Columbia University Press.

Topics for Further Examination

- Search the Web for "healthy relationships" and examine the first 20 or so results to see the focus for each website and who is sponsoring it. What does this data collection exercise tell you about how we are to view relationships in our society?
- Check out the most recent research on gender and relationships using an academic database. How does this research differ from that which you found on the Web?
- Examine marriage and engagement announcements in your local paper. What race, gender, and sexuality patterns do you find in these short announcements? Do the same thing with a listing of personal ads from a local paper. What does this tell us about expectations for relationships? How likely is it that women will be dependent or men will be dependent in these relationships?

9

Enforcing Gender

Catherine G. Valentine with Joan Z. Spade

Throughout Part II, we have discussed patterns of learning, selling, and doing gender at work and in intimate relationships. In this final chapter of our section on patterns, we look at patterns surrounding the enforcement of gender. Enforcing conformity to the gender binary and to patterns of gender inequality is about more than just *doing* gender; it is about force and its threat as well as subtle and tacit constraints on individual autonomy. Enforcing gender involves a range of social control strategies, such as physical abuse and rape, sexual harassment, gossip and name calling, as well as formal and informal laws and rules created by governments, work organizations, and religions to coerce people to conform to gender norms. Many readings throughout this book are about the enforcement of gender. For example, Chapter 4 readings emphasize the ways in which learning and doing gender fundamentally concern efforts by various societal agents (e.g., parents, teachers, the state) to compel people to obey gender norms, values, and practices. This chapter extends that prior discussion, explicitly focusing on social controls used to enforce gender.

The enforcement of gender can have profound effects on women's and men's choices, opportunities, self-esteem, relationships, and abilities to care for themselves. We argue two main points in this chapter. First, doing gender is not something that is innate (i.e., uniquely biological or psychological) or that we freely choose; rather, there are times that we are forced to do gender in spite of resistance, and many times we do gender because we have deeply learned that toeing the gender line is natural, normal, and desirable. Immersed in systems of gender control, many people come to internalize the dominant gender ideology of their society and community so that even vast gender inequalities and injustices are widely accepted. The chapter readings by Lynzi Armstong and Alexa Dodge illustrate this point by examining the continuing power of rape myths which are held by women as well as men and used by both to blame the girls and women who are victims of assault and harassment. Second, then, there are many occasions whereby the very acts of maintaining gendered identities and expressing those identities in everyday life hurt ourselves and others, either physically, emotionally, or both. Clearly the internalization of the dominant gender ideology helps to explain the indifference towards and complicity of many people in everyday gender violence (e.g., domestic violence and date rape) as well as an array of micro-inequalities and micro-aggressions aimed at those who do not conform (Thapar-Bjorkert, Samelius, & Sanghera, 2016).

Social Control

Enforcing gender conformity is about the physical and emotional control of everyone in the context of (re)producing the larger gender order. Peter L. Berger (1963) describes the processes by which we learn to conform to the norms of society as "a set of concentric rings, each representing a system of social control" (p. 73). At the middle of the concentric rings, Berger places the individual. Social control agents and mechanisms, including family and friends, are in the next ring, and the legal and political systems of a society are in the outer ring. He argues that most social control of behavior occurs in the inner rings, by ourselves and those closest to us, which he described as "broad coercive systems that every individual shares with a vast number of fellow controlees" (p. 75). Paivi Honkatukia and Suvi Keskinen (2018) elaborate on gender and social control in their study of gendered, racialized, and age-related control of young women's clothing and bodies. They distinguish among four interacting dimensions of social control: formal institutional control (e.g., legislation and formal directives); informal institutional control in everyday life (e.g., gossip, shaming, social rewards); normative control in close relationships (e.g., explicit rules and advice asserted by those in authority positions in intimate relationships such as parents); internalized control (e.g., "unconscious, routine, and voluntary adherence to rules based on values, beliefs, emotional ties and commitment") (p. 147). They point out that the contexts of social control vary from public settings (e.g., the streets and schools) to private spaces such as within families and close friendships. The means of control, they note, can be divided into formal and informal types. As mentioned above, formal controls include, for example, religious doctrines and laws while informal controls are "expectations, rewards, or sanctions" created in everyday interaction (p. 147). Take, for example, homophobia and think about how it operates as a control mechanism that cuts across the four interacting dimensions set out by Honkatukia and Keskinen (2018). Homophobia offers a good illustration of a set of beliefs rooted in compulsory heterosexuality and the larger heterosexist social system that are both formalized in religious doctrines and laws and expressed through everyday social control processes including conversation. Consider the powerful words "fag" and "sissy." These words are regularly deployed by boys and men, as well as some girls and women, to enforce heteronormative gender conformity.

The power of words to shape our feelings and actions and to shore up systems of inequality has long been the subject of analysis by scholars.in a variety of disciplines (e.g., anthropology, feminist studies, linguistics, and sociology). Sexism in language is now well documented through a vast body of research (see Introduction) showing the way words are used to punish, ignore, demean, and marginalize girls and women (Thapar-Bjorkert, Samelius, & Sanghera, 2016). In this chapter, the reading by Stef M. Shuster adds to that research. Shuster analyzes the language that dominates ordinary conversation in public and private settings in order to understand how trans people are regularly mis-recognized or mis-gendered. Shuster uses the term "discursive aggression" to capture the emotional impact on trans people of routinized symbolic annihilation via a language system that assumes a gender binary. In a section of the reading called "self-enforcement," Shuster discusses internalized control, examining why trans people may choose not to challenge or correct people who mis-label them in a variety of relationships and settings. Extend that analysis to your own experience of self-policing in accordance with dominant gender norms. How does it feel to silence yourself and what are the pressures you experience to do so?

Gender Violence

Although the enforcement of gender conformity is often accomplished through subtle and tacit processes that result in people unquestioningly conforming to the gender order even when they are disadvantaged by it, other forms of social control are more blatantly coercive, such as domestic violence, sexual abuse including rape, and sexual harassment. We devote the remainder of this chapter introduction to gender violence and its threat as ordinary modes of direct and indirect coercion that keep structures of gender inequality intact. With that as our focus, we want to emphasize that gender violence can be conceived of as a continuum linking domination through various cultural schemes, discourses, and tools (e.g., language as discussed above; gender stereotypes; mass media images) to actual physical violence (Thapar-Bjorkert et. al., 2016).

Domestic Violence/Intimate Partner Violence

Domestic violence or intimate partner violence is a lynchpin of social control in patriarchal societies. Jeff Hearn (2012, pp. 152-3) summarizes decades of research demonstrating that the vast majority of victims are women and girls while men are the major perpetrators especially of ongoing and brutal violence. The typical relationship context for domestic violence is private and heterosexual. While men and boys can be victims of domestic violence imposed by other men and by women, a significant percentage of women's violent acts are in self-defense (Hearn, 2012). Also intimate partner violence in gay, lesbian, bisexual, transgender, and queer communities is acknowledged by researchers and activists but difficult to assess given the dearth of good research. Taylor Brown and Jody Herman (2015) provide a helpful review of existing research on intimate partner violence and sexual abuse among LGBT people.

Typically defined as behavior by an intimate other (e.g., husband, ex-husband, or father) that causes physical, sexual, or psychological harm, domestic violence (World Health Organization, 2012) has been named as a worldwide social problem by the World Health Organization (WHO) and the United Nations (UN). In 2017, the UN estimated that about "35 per cent of women worldwide have experienced either physical and/or sexual violence or non-partner violence at some point in their lifetime. However, some national studies show that up to 70 per cent of women have experienced" such violence. Most of this violence is domestic or intimate partner violence. In addition, up to 38% of murders of women are committed by a male partner according to WHO. Both organizations note that the rate of violence against girls and women is difficult to pin down with precision. Women victims of violence often do not report acts of violence because of fear for themselves and their children, based on threats of additional physical violence or the withdrawal of economic support and/or the outcome of emotional abuse, which leads them to believe they "asked for it." Abusers exert "coercive control" over their victims, making victims fearful for their lives and the lives of loved ones. As such, victims, particularly women, often become psychologically battered and emotionally dependent on their perpetrators (Mahoney, Williams, & West, 2001; Sinozich & Langton, 2014). Not only does domestic violence lower the victim's self-esteem, but it also impedes the victim's ability to leave an abusive situation because abusers often control the victim's freedom of movement and finances (even for women who work for pay outside the home). Abusers also prevent victims from getting the psychological support they need to leave the abusive situation (Mahoney, Williams, & West, 2001).

Furthermore, intimate partner violence is generally normalized. It may be unregulated or legal in some countries (e.g., Armenia, Haiti, Latvia, Pakistan) but even where illegal (e.g., Argentina, Central African Republic, France, Indonesia, Sweden, United States), domestic violence remains common, shored up and legitimized by the pervasive values and practices of gender inequality. These values define masculine domination and violence as normal and typically blame girls and women for being victims of violence (e.g., "She must have done something to deserve it."). Bottom line, gender violence in intimacy functions to preserve gender inequalities and patriarchal relations (Hearn, 2012; Jakobsen, 2014). It is often a means to an end, a way of ensuring men's control of women thus upholding dominant masculine identity norms as well as preserving the ideal and practice of submissive femininity (Hearn, 2012; Jakobsen, 2014). Hegemonic masculinity depends on the sexual, physical, and emotional degradation of women. Boys and men often find that "becoming" a man requires that one show disdain for women. As Amy M. Denissen and Abigail C. Saguy suggest in their reading in this chapter, many men disrespect women, especially in workplaces where maintaining the ideal of masculinity is important. The same patterns and practices we have been discussing throughout this book also explain how men learn to instigate and justify the physical and emotional abuse of girls and women. These patterns also illustrate what encourages men to be violent.

Sexual Violence

Sexual violence is a form of gender violence that is often a component of domestic abuse. Many individuals experience sexual violence, however victimization disproportionately affects young women, which is not surprising given gendered expectations such as the sexual double standard. WHO and the UN name sexual violence against girls and women as a common experience across nations. The definition of sexual violence used by WHO (2012) is "any sexual act, attempt to obtain a sexual act, unwanted sexual comments or advances, or acts to traffic, or otherwise directed, against a person's sexuality using coercion, by any person regardless of their relationship to the victim, in any setting, including but not limited to home and work." Sexual violence takes place in an array of public and private settings and, like domestic violence, laws as well as religious doctrines in some countries still allow some types of sexual violence (e.g., spousal rape in Ghana, India, Singapore, and Sri Lanka). We are going to focus on sexual violence in one social context: U.S. colleges and universities. The academic context has much in common with other organizational and institutional contexts in the United States and analyses of sexual violence in that context offer helpful insights into the gendered dynamics of sexual assault and rape in general.

Based upon a randomly selected U.S. national sample of 12,727 interviews in 2011, the Centers for

Disease Control and Prevention estimate that 19.3% of women and 1.7% of men have been raped in their lifetimes and 43.9% of women and 23.4% of men have been victims of other sexual violence (Breiding et al., 2014). Studies of sexual violence in the collegiate context find that about one in four college women have been raped or were the victims of attempted rape (Association of American Universities, 2015; Bachar & Koss, 2001; Fisher, Cullen, & Turner, 2000). A recent study examined incidence of rape for 483 first-year women at a large northeastern U.S. university (Carey, Durney, Shepardson, & Carey, 2015), with respondents completing a sexual experiences survey every four months during their first year. These researchers asked about both attempted and completed rape and differentiated forcible rape or nonconsensual sex using physical force from incapacitated rape that involved the use of drugs or alcohol. Prior to college, 15.4% of respondents reported attempted or completed forcible rape and 17.5% attempted or completed incapacitated rape. During their first year of college, 9.0% of respondents reported attempted or completed forcible rape and 15.4% attempted or completed incapacitated rape, bringing the lifetime total for these young women who had experienced forcible rape to 21.7% and incapacitated rape to 25.7% (Carey et al., 2015). Taken together, these studies point to a prevalence of sexual assault that is of serious concern.

These estimates of sexual assault of young women, both from the Centers for Disease Control and Prevention and from college campuses, are based on self-reports on surveys; official rates are considerably lower due to victims' unwillingness to report sexual assault. Data from a crime victim's survey found that, while the incidence of sexual assault was similar for women (between the ages of 18 and 24) who were enrolled full-time in a postsecondary institution and those who were not, women enrolled in a postsecondary institution were less likely to report sexual assault (20% compared to 32% of the women not enrolled in a postsecondary institution; Sinozich & Langton, 2014). In a comprehensive review of research from 2000 to 2015, Lisa Fedina., J. L. Holmes, and B. L. Backes (2018), report that campus sexual violence studies have largely focused on White, heterosexual women students. Consequently, there is limited research on sexual violence among students of color and at-risk student populations (e.g., LGBTQ students and students with disabilities). The research that does exist suggests disproportionate rates of victimization among these groups of students.

The chapter reading by Hamilton and Armstrong examines the power of gender beliefs, especially the sexual double standard, to justify college men's mistreatment of college women, particularly women who engage in hook-ups. These beliefs encourage the sexual exploitation of women and sexual aggressiveness by men and are notably widespread in fraternities and men's intercollegiate athletic teams (Martin, 2016). Patricia Yancey Martin (2016) analyzes both social contexts as sites for sexual assaults of women, regardless of the attitudes or beliefs held by individual men towards sexual violence. In other words, not all men in fraternities and intercollegiate athletic programs view sexual violence as acceptable but those men may choose not to intervene in sexual assault in those settings for fear that they will be stigmatized and otherwise punished by their "brothers." In a short essay on "Donald Trump and the Normalization of Rape," Allan Johnson (2016) gets to the point: "[W]hen a society normalizes violence against women, the line between raping and not, between talk and assault is a line you don't have to be recognizably 'bad' to cross. 'Good' men do it all the time, supported by all those other 'good' men who are too afraid or too ambivalent or even too envious to go out of their way to top it, like the fraternity brothers who stand by and watch or take pictures on their cell phones or turn away and pretend it isn't happening." The term "rape culture," defined as beliefs that condone or normalize sexual violence, is commonly used in analyses of sexual violence against girls and women. In her chapter reading, Alexa Dodge applies this concept in her examination of the damaging effects of the digital dissemination of photographs of sexual assaults on the victims and the perpetrators. Cyberbullying, cyberstalking, internet harassment, and the like are new forms of gender violence made possible through information and communications technologies. Researchers such as Dodge are tracking the ways in which the deployment of rape myths, victim blaming, and slut shaming in social media are intensifying rape culture and, consequently, patriarchal processes (Hearn, 2012; Stubbs-Richardson, Rader, & Cosby, 2018). It is important to understand that boys and men are also victims of sexual violence. Perpetrators may be women but more typically boys and men are victimized by men and, depending on the setting, boys. High profile cases of sexual abuse of boys include the Catholic Church scandal and all-boy schools scandals in the United States and United Kingdom. In their chapter reading, Gabrielle Ferrales and co-authors identify key gender dynamics in the horrific violence perpetrated against boys and men during a period of mass conflict and genocide in Darfur, Sudan. "Perpetrators perform masculinity" and, simultaneously, reinforce ethnic dominance over their victims through emasculating forms of violence that include homosexualization, feminization, genital harm, and sex-selective killing (p. 566). The findings of this reading reinforce a growing literature on collective violence and its relationship to men's domination in violent institutions (e.g., the military) and state control of violence (e.g., police). Ferrales and co-authors observe that societies that favor heterosexual masculine identities built upon ideals of strength, control, dominance, aggression, and courage are most likely to produce gender-based violence. However, critical to this understanding of gendered violence, is the fact that although such violence may be a key form of social control in patriarchies, "men and violence are not equivalents" (Hearn, 2012, p. 161). That is men are not violent by nature but instead may enact violence in relationships and settings calling for them to demonstrate that they are "real men."

Sexual Harassment

The women's movements of the 1970s targeted sexual violence and domestic violence for legal action. They also identified sexual harassment as a serious workplace problem for women. Most recently, the #MeToo movement has reignited concern about the depth and breadth of sexual harassment of women, as well as some men, across workplace and educational settings. The definition of sexual harassment includes two types of behaviors (Welsh, 1999). Quid pro quo harassment is that which involves the use of sexual threats or bribery to make employment decisions. Supervisors who threaten to withhold a promotion to someone they supervise unless that person has sex with them engage in quid pro quo harassment. The second form of sexual harassment is termed "hostile environment" harassment, which the U.S. Equal Employment Opportunity Commission defines as consisting of behavior that creates an "intimidating, hostile, or offensive working environment" (Welsh, 1999, p. 170). A hostile environment occurs when a work organization allows a pattern of discriminatory behaviors that make it difficult or uncomfortable for an employee to accomplish their work.

We do not know exactly how prevalent sexual harassment is because researchers have used different definitions of sexual harassment and have had to rely on limited samples. In an effort to get a fuller picture of sexual harassment, Stop Street Harassment (SSH), a nonprofit organization dedicated to conducting research on gender-based harassment, undertook an online survey (Chatterjee, 2018). The survey used a large, nationally representative sample of women and men and a broader definition of sexual harassment. For example, being catcalled and cyber harassment were included in the continuum of harassment experiences tapped by the survey. Also, the survey examined a variety of locations where people experienced harassment (e.g., public spaces as well as workplaces). Bottom line, the results of the survey confirm that sexual harassment is a commonplace across contexts and settings especially for women. The survey also shows that men are most likely to be the perpetrators and it emphasizes the harmful effects of harassment. As widely reported, victims frequently suffer depression and anxiety. Notably, victims rarely confront harassers. Instead they alter their own lives, for example by changing jobs or moving. The consequences of the latter are many including long-term economic costs to women victims of harassment (McLaughin, Uggen, & Blackstone, 2017).

Historian Alice Kessler-Harris (2018) has tracked the gender and racial dynamics of U.S. workplaces that have created sexual and racial harassment. She notes that the workplace in the United States has long been defined as the domain of White men. During the era of rapid industrialization, White male workers and employers sought to marginalize outsiders such as women, immigrants, and formerly enslaved people, through low wages and exclusion from training. When women entered workplaces in large numbers in the 20th century (see chapters 7 and 10), employers and supervisors as well as male coworkers used various strategies to keep women in their place. Those strategies included what we now refer to as sexual harassment: sexual quid pro quos, sexual innuendo, and other predatory behaviors. These behaviors quickly became embedded in work cultures and have persisted in spite of laws defining sexual harassment as a form of sex discrimination and in spite of anti-harassment programs instituted by workplaces. The #MeToo movement has been a sharp reminder that legalistic schemes and approaches are insufficient for reducing harassment. (Dobbin & Kalev, 2018; West, 2018). Similar problems with legal remedies hold true for domestic violence and sexual violence.

Two readings in this chapter examine sexual harassment. Amy M. Denissen and Abigail C. Saguy look at the impact of gendered homophobia on women in the building trades, a work world that is overwhelmingly dominated by men. They argue that the presence of women threatens the perception of this work as masculine. Their interviews with lesbian and straight tradeswomen show how tradeswomen manage and resist men's attempts to marginalize them but, in the end, the individual-level responses of tradeswomen are ineffective in transforming unequal gender relations. Lynzi Armstrong analyzes interviews with street-based sex workers, all women, and a variety of people from agencies linked to sex worker safety in New Zealand, a country that has decriminalized sex work. In spite of decriminalization, street-based sex workers continue to experience intense verbal and even physical harassment in the public settings where they work. Their experience parallels the kinds of sexual harassment many women in general experience on a daily basis in public contexts. However, Armstrong found that the abuse endured by sex workers is leveled by both women and men perpetrators. This is a significant finding, one that illustrates how internalized sexism operates. Together, the readings underscore the fact that gender enforcement aimed at preserving the gender binary and gender inequality will not disappear without lasting systemic change.

In short, there are no quick and easy fixes for gender violence in its various expressions and, as we now know from research across nations and communities, legal remedies are limited in their impact. So where do we go next? Legal scholar Robin West (2018) urges us to continue the basic work of publicizing the widespread practice of harassment and to advocate for harassment-free workplaces for all workers and, of course, for students and employees in schools and colleges. In a short analysis of the work of Tarana Burke, founder of the MeToo movement, Michelle Rodino-Colocino (2018) underscores Burke's emphasis on transformative justice that heals survivors as well as perpetrators as a key process in ending gender violence. Burke argues for activism that digs deep into the systems that maintain patriarchal oppression in order to make systemic interventions that end that oppression. Doing so will entail ending gender enforcement practices embedded in formal and informal institutional controls from language to laws and in public and private contexts from the streets to our homes. In the next and final chapter of this book, we address some of the ways in which the gender binary and gender inequality have been successfully undone through individual acts of resistance, social movements for equality, and other social change forces.

References

Bachar, K., & Koss, M. P. (2001). From prevalence to prevention: Closing the gap between what we know about rape and what we do. In C. M. Renzetti, J. L. Edleson, & R. K. Bergen (Eds.), *Sourcebook on violence against women* (pp. 117–142). Thousand Oaks, CA: Sage.

Berger, P. L. (1963). *An invitation to sociology: A humanistic perspective.* Garden City, NY: Anchor Books.

Breiding, M. J., Smith, S. G., Basile, K. C., Walters, M. L., Chen, J., & Merrick, M. T. (2014, September 5). Prevalence and characteristics of sexual violence, stalking, and intimate partner violence victimization—National Intimate Partner and Sexual Violence Survey, United States, 2011: Surveillance summaries. *Morbidity and Mortality Weekly Report (MMWR)*, *63*(SS08), 1–18. Retrieved from http://www.cdc.gov/mmwr/preview/mmwrhtml/ss6308a1.htm?s_cid=ss6308a1_e

Brown, T. N. T., & Herman, J. L. (2015). Intimate partner violence and sexual abuse among LGBT people: A review of existing research. Retrieved from https://williamsinstitute.law.ucla.edu/research/violence-crime/intimate-partner-violence-and-sexual-abuse-among-lgbt-people/

Bureau of Justice Statistics. (2012). *National Crime Victimization Survey: Victimizations not reported to the police, 2006–2010* (NCJ Report No. 238536). Washington, DC: U.S. Department of Justice. Retrieved from http://www.bjs.gov/content/pub/pdf/vnrp0610.pdf

Carey, K. B., Durney, S. E., Shepardson, R. L., & Carey, M. P. (2015). Incapacitated and forcible rape of college women: Prevalence across the first year. *Journal of Adolescent Health*, *56*(6), 670–680.

Chatterjee, R. (2018). A new survey finds 81 percent of women have experienced sexual harassment. Retrieved from https://www.npr.org/sections/thetwo-way/2018/02/21/587671849/a-new-survey-finds-eighty-percent-of-women-have-experienced-sexual-harassment.

Dobbin, F., & Kalev. A. (2018). Can anti-harassment programs reduce sexual harassment? *ASA Footnotes* (April/May 2018), *46*(2). Retrieved from www.asanet.org/news-events/footnotes/apr-may-2018

Fedina, L., Holmes, J. L., Backes, B. L. (2018). Campus sexual assault: A systematic review of prevalence research from 2000 to 2015. *Trauma, Violence, & Abuse*, *19*(11), 76–93.

Fisher, B. S., Cullen, F. T., & Turner, M. G. (2000). *The sexual victimization of college women.* Washington, DC: Bureau of Justice Statistics. Retrieved from https://www.ncjrs.gov/pdffiles1/nij/182369.pdf

Hand, J. Z., & Sanchez, L. (2000). Badgering or bantering? Gender differences in experience of, and reactions to, sexual harassment among U.S. high school students. *Gender & Society*, *14*(6), 718–746.

Hearn, J. (2012). The sociological significance of deomestic violence: Tensions, paradoxes and implications. *Current Sociology*, *61*(2), 152–170.

Honkatukia, P., & Keskinen, S. (2018). The social control of young women's clothing and bodies: A perspective of differences on racialization and sexualization. *Ethnicities*, *18*(1), 42–161.

Jakobsen, H. (2014). What's gendered about gender-based violence? An empirically grounded theoretical exploration from Tanzania. *Gender & Society*, *28*(4), 537–561.

Johnson, A. (2016). Donald Trump and the normalization of rape. Retrieved from www.agjohnson.us/

Kessler-Harris, A. (2018). The long history of workplace harassment. *Jacobin* (March 23, 2018). Retrieved from https://www.jacobinmag.com/2018/03/metoo-workplace-discrimination-sexual-harassment-feminism.

Mahoney, P., Williams, L. M., & West, C. M. (2001). Violence against women by intimate relationship partners. In C. M. Renzetti, J. L. Edleson, & R. K. Bergen (Eds.), *Sourcebook on violence against women* (pp. 143–178). Thousand Oaks, CA: Sage.

Martin, P. Y. (2016). The rape prone culture of the academic context: Fraternities and athletics. *Gender & Society, 30*(1), 30–43.

McLaughlin, H., Uggen, C., & Blackstone, A. (2017). The economic and career effects of sexual harassment on working women. *Gender & Society, 31*(3), 333–358.

Rodino-Colocino, M. (2018). Me too, #MeToo: Countering cruelty with empathy. *Communication and Critical/Cultural Studies, 15*(1), 96–100.

Sinozich, S., & Langton, L. (2014). *Rape and sexual assault victimization among college-age females, 1995–2013* (NCJ Report No. 248471). Washington, DC: U.S. Department of Justice. Retrieved from http://www.bjs.gov/content/pub/pdf/rsavcaf9513.pdf

Stubbs-Richardson, M., Rader, N. E., & Cosby, A. G. (2018). Tweeting rape culture: Examining portrayals of victim blaming in discussions of sexual assault cases on twitter. *Feminism and Psychology, 28*(1), 90–108.

Thapar-Bjorkert, S., Samelius, L., & Sanghera, G. S. (2016). Exploring symbolic violence in the everyday: Misrecognition, condescension, consent and complicity. *Feminist Review, 112*(1), 144–162.

United Nations. (2017). Facts and figures: Ending violence against women. *UNWOMEN* (August 2017). Retrieved from http://www.unwomen.org/en/what-we-do/ending-violence-against-women/facts-and-figures.

Welsh, S. (1999). Gender and sexual harassment. *Annual Review of Sociology, 25*, 169–190.

West, R. (2018). Manufacturing consent. *The Baffler*, 39, May 2018. Retrieved from https://thebaffler.com/salvos/manufacturing-consent-west

World Health Organization. (2012). Intimate partner violence. Retrieved from apps.who.int/iris/bitstream/10665/77432/1/WHO_RHR_12.36_eng.pdf

World Health Organization. (2017). Violence against women. Retrieved from https://www.who.int/mediacentre/factsheets/fs239/en/

Introduction to Reading 44

Norms that define women as essentially sexually passive were challenged as *Sex and the City* and *Fifty Shades of Grey* hit television and movie screens. However, did they actually change the way the expression of sexuality is used to enforce gender inequality? This article would suggest that although things have changed, sexuality is still gendered. In this reading, Laura Hamilton and Elizabeth A. Armstrong report their findings from a 4-year ethnographic and interview study of women at a midwestern university. The study began in 2004, when the authors, along with seven other researchers, "occupied" a room on a dormitory floor in what students identified as a "party dorm." In addition to observing and hanging out with the women, they also conducted interviews with 41 of the 53 women who lived on this floor during the first year. This particular paper summarizes the women's reports about hooking up and relationships and how both changed over the course of their study. Their focus on social interaction is particularly valuable in understanding how gender is understood and enforced within the inner circles of Peter L. Berger's concentric rings of social control, by ourselves and those closest to us, as described in the introduction to this chapter. In addition, the authors provide an intersectional analysis, examining how interaction and meaning of these college relationships vary across social class for the young women they studied. Their findings help us think about how gender continues to influence the expression of sexuality in a world where permanent relationships are delayed and "sex" is the theme for young women as well as men. They also help us understand that gender enforcement, while not consistent, is patterned within intersections of different prisms, such as gender and social class in this piece.

1. How do the women in this study describe "hooking up," and what factors influence their orientation to this expression of sexuality?
2. How did gender influence the expression of sexuality differently for women from different social classes in this study?
3. Do you think heterosexuality will be less important in the future in enforcing gender difference and inequality than it has been in the past? Why or why not?

Gendered Sexuality in Young Adulthood

Double Binds and Flawed Options

Laura Hamilton and Elizabeth A. Armstrong

As traditional dating has declined on college campuses, hookups—casual sexual encounters often initiated at alcohol-fueled, dance-oriented social events—have become a primary form of intimate heterosexual interaction (England, Shafer, and Fogarty 2007; Paul, McManus, and Hayes 2000). Hookups have attracted attention among social scientists and journalists (Bogle 2008; Glenn and Marquardt 2001; Stepp 2007). To date, however, limitations of both data and theory have obscured the implications for women and the gender system. Most studies examine only the quality of hookups at one point during college and rely, if implicitly, on an individualist, gender-only approach. In contrast, we follow a group of women as they move through college—assessing all of their sexual experiences. We use an interactionist approach and attend to how both gender and class shape college sexuality. . . .

Gender Theory and College Sexuality

Research on Hooking Up

Paul, McManus, and Hayes (2000) and Glenn and Marquardt (2001) were the first to draw attention to the hookup as a distinct social form. As Glenn and Marquardt (2001, 13) explain, most students agree that "a hook up is anything 'ranging from kissing to having sex,' and that it takes place outside the context of commitment." Others have similarly found that *hooking up* refers to a broad range of sexual activity and that this ambiguity is part of the appeal of the term (Bogle 2008). Hookups differ from dates in that individuals typically do not plan to do something together prior to sexual activity. Rather, two people hanging out at a party, bar, or place of residence will begin talking, flirting, and/or dancing. Typically, they have been drinking. At some point, they move to a more private location, where sexual activity occurs (England, Shafer, and Fogarty 2007). While strangers sometimes hook up, more often hookups occur among those who know each other at least slightly (Manning, Giordano, and Longmore 2006).

England has surveyed more than 14,000 students from 19 universities and colleges about their hookup, dating, and relationship experiences. Her Online College Social Life Survey (OCSLS) asks students to report on their recent hookups using "whatever definition of a hookup you and your friends use." Seventy-two percent of both men and women participating in the OCSLS reported at least one hookup by their senior year in college. Of these, roughly 40 percent engaged in three or fewer hookups, 40 percent between four and nine hookups, and 20 percent 10 or more hookups. Only about one-third engaged in intercourse in their most recent hookups, although—among the 80 percent of students who had intercourse by the end of college—67 percent had done so outside of a relationship.

Ongoing sexual relationships without commitment were common and were labeled "repeat," "regular," or "continuing" hookups and sometimes "friends with benefits" (Armstrong, England, and Fogarty 2009; Bogle 2008; Glenn and Marquardt 2001). Ongoing hookups sometimes became committed relationships and vice versa; generally, the distinction revolved around the level of exclusivity and a willingness to refer to each other as "girlfriend/boyfriend" (Armstrong, England, and Fogarty 2009). Thus, hooking up does not imply interest in a relationship, but it does not preclude such interest. Relationships are also common among students. By their senior year, 69 percent of heterosexual students had been in a college relationship of at least six months.

To date, however, scholars have paid more attention to women's experiences with hooking up than

Hamilton, L., & Armstrong, E. A. (2009). Gendered sexuality in young adulthood: Double binds and flawed options. *Gender & Society, 23*(5): 259–616. Reprinted by permission of SAGE Publications, Inc. on behalf of Sociologists for Women in Society.

relationships and focused primarily on ways that hookups may be less enjoyable for women than for men. Glenn and Marquardt (2001, 20) indicate that "hooking up is an activity that women sometimes find rewarding but more often find confusing, hurtful, and awkward." Others similarly suggest that more women than men find hooking up to be a negative experience (Bogle 2008, 173; Owen et al. 2008) and focus on ways that hookups may be harmful to women (Eshbaugh and Gute 2008; Grello, Welsh, and Harper 2006).

This work assumes distinct and durable gender differences at the individual level. Authors draw, if implicitly, from evolutionary psychology, socialization, and psychoanalytic approaches to gender—depicting women as more relationally oriented and men as more sexually adventurous (see Wharton 2005 for a review). For example, despite only asking about hookup experiences, Bogle (2008, 173) describes a "battle of the sexes" in which women want hookups to "evolve into some semblance of a relationship," while men prefer to "hook up with no strings attached" (also see Glenn and Marquardt 2001; Stepp 2007).

The battle of the sexes view implies that if women could simply extract commitment from men rather than participating in hookups, gender inequalities in college sexuality would be alleviated. Yet this research—which often fails to examine relationships—ignores the possibility that women might be the losers in both hookups and relationships. Research suggests that young heterosexual women often suffer the most damage from those with whom they are most intimate: Physical battery, emotional abuse, sexual assault, and stalking occur at high rates in youthful heterosexual relationships (Campbell et al. 2007; Dunn 1999). This suggests that gender inequality in college sexuality is systemic, existing across social forms.

Current research also tends to see hooking up as solely about gender, without fully considering the significance of other dimensions of inequality. Some scholars highlight the importance of the college environment and traditional college students' position in the life course (Bogle 2008; Glenn and Marquardt 2001). However, college is treated primarily as a context for individual sexual behavior rather than as a key location for class reproduction. Analyzing the role of social class in sex and relationships may help to illuminate the appeal of hookups for both college women and men.

Gender Beliefs and Social Interaction

Contemporary gender theory provides us with resources to think about gender inequality in college sexuality differently. Gender scholars have developed and refined the notion of gender as a social structure reproduced at multiple levels of society: Gender is embedded not only in individual selves but also in interaction and organizational arrangements (Connell 1987; Glenn 1999; Risman 2004). This paper focuses on the interactional level, attending to the power of public gender beliefs in organizing college sexual and romantic relations.

Drawing on Sewell's (1992) theory of structure, Ridgeway and Correll (2004, 511) define gender beliefs as the "cultural rules or instructions for enacting the social structure of difference and inequality that we understand to be gender." By believing in gender differences, individuals "see" them in interaction and hold others accountable to this perception. Thus, even if individuals do not internalize gender beliefs, they must still confront them (Ridgeway 2009).

Ridgeway and coauthors (Ridgeway 2000; Ridgeway and Correll 2004) assert that interaction is particularly important to the reproduction of gender inequality because of how frequently men and women interact. They focus on the workplace but suggest that gendered interaction in private life may be intensifying in importance as beliefs about gender difference in workplace competency diminish (Correll, Benard, and Paik 2007; Ridgeway 2000; Ridgeway and Correll 2004). We extend their insights to sexual interaction, as it is in sexuality and reproduction that men and women are believed to be most different. The significance of gender beliefs in sexual interaction may be magnified earlier in the life course, given the amount of time spent in interaction with peers and the greater malleability of selves (Eder, Evans, and Parker 1995). Consequently, the university provides an ideal site for this investigation.

The notion that men and women have distinct sexual interests and needs generates a powerful set of public gender beliefs about women's sexuality. A belief about what women should not do underlies a *sexual double standard:* While men are expected to desire and pursue sexual opportunities regardless of context, women are expected to avoid casual sex—having sex only when in relationships and in love (Crawford and Popp 2003; Risman and Schwartz 2002). Much research on the sexuality of young men focuses on male endorsement of this belief and its consequences (e.g., Bogle 2008; Kimmel 2008; Martin and Hummer 1989). There is an accompanying and equally powerful belief that normal women should always want love, romance, relationships, and marriage—what we refer to as the *relational imperative* (also see Holland and

Eisenhart 1990; Martin 1996; Simon, Eder, and Evans 1992). We argue that these twin beliefs are implicated in the (re)production of gender inequality in college sexuality and are at the heart of women's sexual dilemmas with both hookups and relationships.

An Intersectional Approach

Gender theory has also moved toward an intersectional approach (Collins 1990; Glenn 1999). Most of this work focuses on the lived experiences of marginalized individuals who are situated at the intersection of several systems of oppression (McCall 2005). More recently, scholars have begun to theorize the ways in which systems of inequality are themselves linked (Beisel and Kay 2004; Glenn 1999; McCall 2005). Beisel and Kay (2004) apply Sewell's (1992) theory of structure to intersectionality, arguing that structures intersect when they share resources or guidelines for action (of which gender beliefs would be one example). Using a similar logic, we argue that gender and class intersect in the sexual arena, as these structures both rely on beliefs about how and with whom individuals should be intimate.

Like gender, class structures beliefs about appropriate sexual and romantic conduct. Privileged young Americans, both men and women, are now expected to defer family formation until the mid-twenties or even early-thirties to focus on education and career investment—what we call the *self-development imperative* (Arnett 2004; Rosenfeld 2007). This imperative makes committed relationships less feasible as the sole contexts for premarital sexuality. Like marriage, relationships can be "greedy," siphoning time and energy away from self-development (Gerstel and Sarkisian 2006; Glenn and Marquardt 2001). In contrast, hookups offer sexual pleasure without derailing investment in human capital and are increasingly viewed as part of life-stage appropriate sexual experimentation. Self-protection—both physical and emotional—is central to this logic, suggesting the rise of a strategic approach to sex and relationships (Brooks 2002; Illouz 2005). This approach is reflected in the development of erotic marketplaces offering short-term sexual partners, particularly on college campuses (Collins 2004).

In this case, gender and class behavioral rules are in conflict. Gender beliefs suggest that young women should avoid nonromantic sex and, if possible, be in a committed relationship. Class beliefs suggest that women should delay relationships while pursuing educational goals. Hookups are often less threatening to self-development projects, offering sexual activity in a way that better meshes with the demands of college. We see this as a case wherein structures intersect, but in a contradictory way (Friedland and Alford 1991; Martin 2004; Sewell 1992). This structural contradiction has experiential consequences: Privileged women find themselves caught between contradictory expectations, while less privileged women confront a foreign sexual culture when they enter college.

* * *

The Power of Gender Beliefs . . .

The "Slut" Stigma

Women did not find hookups to be unproblematic. They complained about a pervasive double standard. As one explained, "Guys can have sex with all the girls and *it makes them more of a man,* but if a girl does, then all of a sudden she's a ho, and she's not as quality of a person" ([interview number] 10–1, emphasis added). Another complained, "Guys, they can go around and have sex with a number of girls and they're not called anything" (6–1). Women noted that it was "easy to get a reputation" (11–1) from "hooking up with a bunch of different guys" (8–1) or "being wild and drinking too much" (14–3). Their experiences of being judged were often painful; one woman told us about being called a "slut" two years after the incident because it was so humiliating (42–3).

Fear of stigma constrained women's sexual behavior and perhaps even shaped their preferences. For example, several indicated that they probably would "make out with more guys" but did not because "I don't want to be a slut" (27–2). Others wanted to have intercourse on hookups but instead waited until they had boyfriends. A couple hid their sexual activity until the liaison was "official." One said, "I would not spend the night there [at the fraternity] because that does not look good, but now everyone knows we're boyfriend/girlfriend, so it's like my home now" (15–1). Another woman, who initially seemed to have a deep aversion to hooking up, explained, "I would rather be a virgin for as much as I can than go out and do God knows who." She later revealed a fear of social stigma, noting that when women engage in nonromantic sex, they "get a bad reputation. I know that I wouldn't want that reputation" (11–1). Her comments highlight the feedback between social judgment and internalized preference.

Gender beliefs were also at the root of women's other chief complaint about hookups—the disrespect of women in the hookup scene. The notion that hooking up is okay for men but not for women was embedded in the organization of the Greek system, where most parties occurred: Sorority rules prohibited hosting parties or overnight male visitors, reflecting notions about proper feminine behavior. In contrast, fraternities collected social fees to pay for alcohol and viewed hosting parties as a central activity. This disparity gave fraternity men almost complete control over the most desirable parties on campus—particularly for the underage crowd (Boswell and Spade 1996; Martin and Hummer 1989).

Women reported that fraternity men dictated party transportation, the admittance of guests, party themes such as "CEO and secretary ho," the flow of alcohol, and the movement of guests within the party (Armstrong, Hamilton, and Sweeney 2006). Women often indicated that they engaged in strategies such as "travel[ing] in hordes" (21–1) and not "tak[ing] a drink if I don't know where it came from" (15–1) to feel safer at fraternity parties. Even when open to hooking up, women were not comfortable doing so if they sensed that men were trying to undermine their control of sexual activity (e.g., by pushing them to drink too heavily, barring their exit from private rooms, or refusing them rides home). Women typically opted not to return to party venues they perceived as unsafe. As one noted, "I wouldn't go to [that house] because I heard they do bad things to girls" (14–1). Even those interested in the erotic competition of party scenes tired of it as they realized that the game was rigged.

The sexual double standard also justified the negative treatment of women in the party scene—regardless of whether they chose to hook up. Women explained that men at parties showed a lack of respect for their feelings or interests—treating them solely as "sex objects" (32–1). This disregard extended to hookups. One told us, "The guy gets off and then it's done and that's all he cares about" (12–4). Another complained of her efforts to get a recent hookup to call: "That wasn't me implying I wanted a relationship—that was me implying I wanted respect" (42–2). In her view, casual sex did not mean forgoing all interactional niceties. A third explained, "If you're talking to a boy, you're either going to get into this huge relationship or you are nothing to them" (24–3). This either-or situation often frustrated women who wanted men to treat them well regardless of the level of commitment.

The Relationship Imperative

Women also encountered problematic gender beliefs about men's and women's different levels of interest in relationships. As one noted, women fight the "dumb girl idea"—the notion "that every girl wants a boy to sweep her off her feet and fall in love" (42–2). The expectation that women should want to be in relationships was so pervasive that many found it necessary to justify their single status to us. For example, when asked if she had a boyfriend, one woman with no shortage of admirers apologetically explained, "I know this sounds really pathetic and you probably think I am lying, but there are so many other things going on right now that it's really not something high up on my list. . . . I know that's such a lame-ass excuse, but it's true" (9–3). Another noted that already having a boyfriend was the only "actual, legitimate excuse" to reject men who expressed interest in a relationship (34–3).

Certainly, many women wanted relationships and sought them out. However, women's interest in relationships varied, and almost all experienced periods during which they wanted to be single. Nonetheless, women reported pressure to be in relationships all the time. We found that women, rather than struggling to get into relationships, had to work to avoid them.

The relational imperative was supported by the belief that women's relational opportunities were scarce and should not be wasted. Women described themselves as "lucky" to find a man willing to commit, as "there's not many guys like that in college" (15–1). This belief persisted despite the fact that most women were in relationships most of the time. As one woman noted, "I don't think anyone really wants to be in a serious relationship, but most, well actually all of us, have boyfriends" (13–1). Belief in the myth of scarcity also led women to stay in relationships when they were no longer happy. A woman who was "sick of" her conflict-ridden relationship explained why she could not end it: "I feel like I have to meet somebody else. . . . I go out and they're all these asshole frat guys. . . . That's what stops me. . . . Boys are not datable right now because . . . all they're looking for is freshman girls to hook up with. . . . [So] I'm just stuck. I need to do something about it, but I don't know what" (30–3). It took her another year to extract herself from this relationship. Despite her fears, when she decided she was ready for another relationship, she quickly found a boyfriend.

Women also confronted the belief that all women are relationally insatiable. They often told stories of men who acted entitled to relationships, expected their relational overtures to be accepted, and became angry when rebuffed—sometimes stalking the rejecting woman. As one explained about a friend, "Abby was having issues with this guy who likes her. He was like, 'You have to like me. . . . I'm not gonna take no for an answer. I'm gonna do whatever it takes to date you'" (24–3). Another noted that "last semester, this guy really wanted to date me, and I did not want to date him at all. He flipped out and was like, 'This is ridiculous, I don't deserve this'" (12–3). A third eventually gave in when a man continually rejected her refusals: "I was like, if I go [out with him] . . . maybe he'll stop. Because he wouldn't stop." She planned to act "extremely conservative" as a way to convince him that he did not want to be with her (39–4).

Gender beliefs may also limit women's control over the terms of interaction within relationships. If women are made to feel lucky to have boyfriends, men are placed in a position of power, as presumably women should be grateful when they commit. Women's reports suggest that men attempted to use this power to regulate their participation in college life. One noted, "When I got here my first semester freshman year, I wanted to go out to the parties . . . and he got pissed off about it. . . . He's like, 'Why do you need to do that? Why can't you just stay with me?'" (4–2). Boyfriends sometimes tried to limit the time women spent with their friends and the activities in which they participated. As a woman explained, "There are times when I feel like Steve can get . . . possessive. He'll be like . . . 'I feel like you're always with your friends over me.' He wanted to go out to lunch after our class, and I was like, 'No, I have to come have this interview.' And he got so upset about it" (42–3). Men's control even extended to women's attire. Another told us about her boyfriend, "He is a very controlling person. . . . He's like, 'What are you wearing tonight?' . . . It's like a joke but serious at the same time" (32–4).

Women also became jealous; however, rather than trying to control their boyfriends, they often tried to change themselves. One noted that she would "do anything to make this relationship work." She elaborated, "I was so nervous being with Dan because I knew he had cheated on his [prior] girlfriend . . . [but] I'm getting over it. When I go [to visit him] now . . . I let him go to the bar, whatever. I stayed in his apartment because there was nothing else to do" (39–3). Other women changed the way they dressed, their friends, and where they went in the attempt to keep boyfriends.

When women attempted to end relationships, they often reported that men's efforts to control them escalated. We heard 10 accounts of men using abuse to keep women in relationships. One woman spent months dealing with a boyfriend who accused her of cheating on him. When she tried to break up, he cut his wrist in her apartment (9–2). Another tried to end a relationship but was forced to flee the state when her car windows were broken and her safety was threatened (6–4). Men often drew on romantic repertoires to coerce interaction after relationships had ended. One woman told us that her ex-boyfriend stalked her for months—even showing up at her workplace, showering her with flowers and gifts, and blocking her entry into work until the police arrived (25–2).

Intersectionality: Contradictions Between Class and Gender

Existing research about college sexuality focuses almost exclusively on its gendered nature. We contend that sexuality is shaped simultaneously by multiple intersecting structures. In this section, we examine the sexual and romantic implications of class beliefs about how ambitious young people should conduct themselves during college. Although all of our participants contended with class beliefs that contradicted those of gender, experiences of this structural intersection varied by class location. More privileged women struggled to meet gender and class guidelines for sexual behavior, introducing a difficult set of double binds. Because these class beliefs reflected a privileged path to adulthood, less privileged women found them foreign to their own sexual and romantic logics.

More Privileged Women and the Experience of Double Binds

The Self-Development Imperative and the Relational Double Bind

The four-year university is a classed structural location. One of the primary reasons to attend college is to preserve or enhance economic position. The university culture is thus characterized by the self-development imperative, or the notion that individual achievement and personal growth are paramount. There are also accompanying rules for sex and relationships: Students are expected to postpone marriage and parenthood until after completing an education and establishing a career.

For more privileged women, personal expectations and those of the university culture meshed. Even those who enjoyed relationships experienced phases in college where they preferred to remain single. Almost all privileged women (94 percent) told us at one point that they did not want a boyfriend. One noted, "All my friends here . . . they're like, 'I don't want to deal with [a boyfriend] right now. I want to be on my own'" (37–1). Another eloquently remarked, "I've always looked at college as the only time in your life when you should be a hundred percent selfish. . . . I have the rest of my life to devote to a husband or kids or my job . . . but right now, it's my time" (21–2).

The notion that independence is critical during college reflected class beliefs about the appropriate role for romance that opposed those of gender. During college, relational commitments were supposed to take a backseat to self-development. As an upper-middle-class woman noted, "College is the only time that you don't have obligations to anyone but yourself. . . . I want to get settled down and figure out what I'm doing with my life before [I] dedicate myself to something or someone else" (14–4). Another emphasized the value of investment in human capital: "I've always been someone who wants to have my own money, have my own career so that, you know, 50 percent of marriages fail. . . . If I want to maintain the lifestyle that I've grown up with . . . I have to work. I just don't see myself being someone who marries young and lives off of some boy's money" (42–4). To become self-supporting, many privileged women indicated they needed to postpone marriage. One told us, "I don't want to think about that [marriage]. I want to get secure in a city and in a job. . . . I'm not in any hurry at all. As long as I'm married by 30, I'm good" (13–4). Even those who wanted to be supported by husbands did not expect to find them in college, instead setting their sights on the more accomplished men they expected to meet in urban centers after college.

More privileged women often found committed relationships to be greedy—demanding of time and energy. As one stated, "When it comes to a serious relationship, it's a lot for me to give into that. [What do you feel like you are giving up?] Like my everything. . . . There's just a lot involved in it" (35–3). These women feared that they would be devoured by relationships and sometimes struggled to keep their self-development projects going when they did get involved. As an upper-class woman told us, "It's hard to have a boyfriend and be really excited about it and still not let it consume you" (42–2). This situation was exacerbated by the gender beliefs discussed earlier, as women experienced pressure to fully devote themselves to relationships.

Privileged women reported that committed relationships detracted from what they saw as the main tasks of college. They complained, for example, that relationships made it difficult to meet people. As an upper-middle-class woman who had just ended a relationship described, "I'm happy that I'm able to go out and meet new people. . . . I feel like I'm doing what a college student should be doing. I don't need to be tied down to my high school boyfriend for two years when this is the time to be meeting people" (14–3). A middle-class woman similarly noted that her relationship with her boyfriend made it impossible to make friends on the floor her first year. She explained, "We were together every day. . . . It was the critical time of making friends and meeting people, [and] I wasn't there" (21–2).

Many also complained that committed relationships competed with schoolwork (also see Holland and Eisenhart 1990). An upper-middle-class woman remarked, "[My boyfriend] doesn't understand why I can't pick up and go see him all the time. But I have school. . . . I just want to be a college kid" (18–3). Another told us that her major was not compatible with the demands of a boyfriend. She said, "I wouldn't mind having a boyfriend again, but it's a lot of work. Right now with [my major] and everything . . . I wouldn't have time even to see him" (30–4). She did not plan to consider a relationship until her workload lessened.

With marriage far in the future, more privileged women often worried about college relationships getting too serious too fast. All planned to marry—ideally to men with greater earnings—but were clear about the importance of temporary independence. Consequently, some worked to slow the progression of relationships. One told us, "I won't let myself think that [I love him]. I definitely don't say that. . . . The person he loves is the person he is going to marry. . . . At the age we are at now, I feel like I don't want anything to be more serious than it has to be until it is" (34–3). Eight privileged women even dated men they deemed unsuitable for marriage to ensure autonomy. One noted, "He fits my needs now because I don't want to get married now. I don't want anyone else to influence what I do after I graduate" (33–3). Others planned to end relationships when boyfriends were not on the same page. An upper-middle-class woman explained, "[He] wants to have two kids by the time he's thirty. I'm like, I guess we're not getting married. . . . I'd rather make money and travel first" (43–3).

For more privileged women, contradictory cultural rules created what we call the *relational double bind.* The relational imperative pushed them to participate in committed relationships; however, relationships did not mesh well with the demands of college, as they inhibited classed self-development strategies. Privileged women struggled to be both "good girls" who limited their sexual activity to relationships and "good students" who did not allow relational commitments to derail their educational and career development.

The Appeal of Hookups and the Sexual Double Bind

In contrast, hookups fit well with the self-development imperative of college. They allowed women to be sexual without the demands of relationships. For example, one upper-class woman described hooking up as "fun and nonthreatening." She noted, "So many of us girls, we complain that these guys just want to hook up all the time. I'm going, these guys that I'm attracted to . . . get kind of serious." She saw her last hookup as ideal because "we were physical, and that was it. I never wanted it to go anywhere" (34–2). Many privileged women understood, if implicitly, that hooking up was a delay tactic, allowing sex without participation in serious relationships.

As a sexual solution for the demands of college, hooking up became incorporated into notions of what the college experience should be. When asked which kinds of people hook up the most, one woman noted, "All. . . . The people who came to college to have a good time and party" (14–1). With the help of media, alcohol, and spring break industries, hooking up was so institutionalized that many took it for granted. One upper-middle-class woman said, "It just happens. It's natural" (15–1). They told us that learning about sexuality was something they were supposed to be doing in college. Another described, "I'm glad that I've had my one-night stands and my being in love and having sex. . . . Now I know what it's supposed to feel like when I'm with someone that I want to be with. I feel bad for some of my friends. . . . They're still virgins" (29–1).

High rates of hooking up suggest genuine interest in the activity rather than simply accommodation to men's interests. Particularly early in college, privileged women actively sought hookups. One noted, "You see a lot of people who are like, 'I just want to hook up with someone tonight.' . . . It's always the girls that try to get the guys" (41–1). Data from the OCSLS also suggest that college women like hooking up almost as much as men and are not always searching for something more. Nearly as many women as men (85 percent and 89 percent, respectively) report enjoying the sexual activity of their last hookup "very much" or "somewhat," and less than half of women report interest in a relationship with their most recent hookup.

In private, several privileged women even used the classed logic of hooking up to challenge stereotyped portrayals of gender differences in sexuality. As one noted, "There are girls that want things as much as guys do. There are girls that want things more, and they're like, 'Oh it's been a while [since I had sex].' The girls are no more innocent than the guys. . . . People think girls are jealous of relationships, but they're like, 'What? I want to be single'" (34–1). When asked about the notion that guys want sex and girls want relationships another responded, "I think that is the absolute epitome of bullshit. I know so many girls who honestly go out on a Friday night and they're like, 'I hope I get some ass tonight.' They don't wanna have a boyfriend! They just wanna hook up with someone. And I know boys who want relationships. I think it goes both ways" (42–2). These women drew on gender-neutral understandings of sexuality characteristic of university culture to contradict the notion of women's sexuality as inevitably and naturally relational.

For more privileged women, enjoyment of hookups was tightly linked to the atmosphere in which they occurred. Most were initiated at college parties where alcohol, music, attractive people, sexy outfits, and flirting combined to generate a collective erotic energy. As one woman enthusiastically noted, "Everyone was so excited. It was a big fun party" (15–1). Privileged women often "loved" it when they had an "excuse to just let loose" and "grind" on the dance floor. They reported turning on their "make-out radar" (18–1), explaining that "it's fun to know that a guy's attracted to you and is willing to kiss you" (16–1). The party scene gave them a chance to play with adult sexualities and interact for purely sexual purposes—an experience that one middle-class woman claimed "empowered" her (17–1).

Hookups enabled more privileged women to conduct themselves in accordance with class expectations, but as we demonstrated earlier, the enforcement of gender beliefs placed them at risk of sanction. This conflict gets to the heart of a *sexual double bind:* While hookups protected privileged women from relationships that could derail their ambitions, the double standard gave men greater control over the terms of hooking up, justified the disrespectful treatment of women, supported sexual stigma, and produced feelings of shame.

Less Privileged Women and the Experience of Foreign Sexual Culture

Women's comfort with delaying commitment and participating in the hookup culture was shaped by class location. College culture reflects the beliefs of the more privileged classes. Less privileged women arrived at college with their own orientation to sex and romance, characterized by a faster transition into adulthood. They often attempted to build both relationships and career at the same time. As a result, a third of the participants from less privileged backgrounds often experienced the hookup culture as foreign in ways that made it difficult to persist at the university.

Less privileged women had less exposure to the notion that the college years should be set aside solely for educational and career development. Many did not see serious relationships as incompatible with college life. Four were married or engaged before graduating—a step that others would not take until later. One reminisced, "I thought I'd get married in college. . . . When I was still in high school, I figured by my senior year, I'd be engaged or married or something. . . . I wanted to have kids before I was 25" (25–4). Another spoke of her plans to marry her high school sweetheart: "I'll be 21 and I know he's the one I want to spend the rest of my life with. . . . Really, I don't want to date anybody else" (6–1).

Plans to move into adult roles relatively quickly made less privileged women outsiders among their more privileged peers. One working-class woman saw her friendships dissolve as she revealed her desire to marry and have children in the near future. As one of her former friends described,

> She would always talk about how she couldn't wait to get married and have babies. . . . It was just like, Whoa. I'm 18. . . . Slow down, you know? Then she just crazy dropped out of school and wouldn't contact any of us. . . . The way I see it is that she's from a really small town, and that's what everyone in her town does . . . get married and have babies. That's all she ever wanted to do maybe? . . . I don't know if she was homesick or didn't fit in. (24–4)

This account glosses over the extent to which the working-class woman was pushed out of the university—ostracized by her peers for not acclimating to the self-development imperative and, as noted below, to the campus sexual climate. In fact, 40 percent of less privileged women left the university, compared to 5 percent of more privileged women. In all cases, mismatch between the sexual culture of women's hometowns and that of college was a factor in the decision to leave.

Most of the less privileged women found the hookup culture to be not only foreign but hostile. As the working-class woman described above told us,

> I tried so hard to fit in with what everybody else was doing here. . . . I think one morning I just woke up and realized that this isn't me at all; I don't like the way I am right now. . . . I didn't feel like I was growing up. I felt like I was actually getting younger the way I was trying to act. Growing up to me isn't going out and getting smashed and sleeping around. . . . That to me is immature. (28–1)

She emphasized the value of "growing up" in college. Without the desire to postpone adulthood, less privileged women often could not understand the appeal of hooking up. As a lower-middle-class woman noted, "Who would be interested in just meeting somebody and then doing something that night? And then never talking to them again? . . . I'm supposed to do this; I'm supposed to get drunk every weekend. I'm supposed to go to parties every weekend . . . and I'm supposed to enjoy it like everyone else. But it just doesn't appeal to me" (5–1). She reveals the extent to which hooking up was a normalized part of college life: For those who were not interested in this, college life could be experienced as mystifying, uncomfortable, and alienating.

The self-development imperative was a resource women could use in resisting the gendered pull of relationships. Less privileged women did not have as much access to this resource and were invested in settling down. Thus, they found it hard to resist the pull back home of local boyfriends, who—unlike the college men they had met—seemed interested in marrying and having children soon. One woman noted after transferring to a branch campus, "I think if I hadn't been connected with [my fiancé], I think I would have been more strongly connected to [the college town], and I think I probably would have stayed" (2–4). Another described her hometown boyfriend: "He'll be like, 'I want to see you. Come home.' . . . The stress he was putting me under and me being here my first year. I could not take it" (7–2). The following year, she moved back home. A third explained about her husband, "He wants me at home. . . . He wants to have control over me and . . . to feel like he's the dominant one in the relationship. . . . The fact that I'm going to school and he knows I'm smart and he knows that I'm capable of doing anything that I want . . . it scares him" (6–4). While she eventually ended this relationship, it cost her an additional semester of school.

Women were also pulled back home by the slut stigma, as people there—perhaps out of frustration or jealousy—judged college women for any association with campus sexual culture. For instance, one woman became distraught when a virulent sexual rumor about her circulated around her hometown, especially when it reached her parents. Going home was a way of putting sexual rumors to rest and reaffirming ties that were strained by leaving.

Thus, less privileged women were often caught between two sexual cultures. Staying at the university meant abandoning a familiar logic and adopting a privileged one—investing in human capital while delaying the transition to adulthood. As one explained, attending college led her to revise her "whole plan": "Now I'm like, I don't even need to be getting married yet [or] have kids. . . . All of [my brother's] friends, 17- to 20-year-old girls, have their . . . babies, and I'm like, Oh my God. . . . Now I'll be able to do something else for a couple years before I settle down . . . before I worry about kids" (25–3). These changes in agendas required them to end relationships with men whose life plans diverged from theirs. For some, this also meant cutting ties with hometown friends. One resolute woman, whose friends back home had turned on her, noted, "I'm just sick of it. There's nothing there for me anymore. There's absolutely nothing there" (22–4).

Discussion

The Strengths of an Interactional Approach

Public gender beliefs are a key source of gender inequality in college heterosexual interaction. They undergird a sexual double standard and a relational imperative that justify the disrespect of women who hook up and the disempowerment of women in relationships—reinforcing male dominance across social forms. Most of the women we studied cycled back and forth between hookups and relationships, in part because they found both to be problematic. These findings indicate that an individualist, battle of the sexes explanation not only is inadequate but may contribute to gender inequality by naturalizing problematic notions of gender difference.

We are not, however, claiming that gender differences in stated preferences do not exist. Analysis of the OCSLS finds a small but significant difference between men and women in preferences for relationships as compared to hookups: After the most recent hookup, 47 percent of women compared to 37 percent of men expressed some interest in a relationship. These differences in preferences are consistent with a multilevel perspective that views the internalization of gender as an aspect of gender structure (Risman 2004). As we have shown, the pressure to internalize gender-appropriate preferences is considerable, and the line between personal preferences and the desire to avoid social stigma is fuzzy. However, we believe that widely shared beliefs about gender difference contribute more to gender inequality in college heterosexuality than the substantively small differences in actual preferences.

The Strengths of an Intersectional Approach

An intersectional approach sheds light on the ambivalent and contradictory nature of many college women's sexual desires. Class beliefs associated with the appropriate timing of marriage clash with resilient gender beliefs—creating difficult double binds for the more privileged women who strive to meet both. In the case of the relational double bind, relationships fit with gender beliefs but pose problems for the classed self-development imperative. As for the sexual double bind, hookups provide sexual activity with little cost to career development, but a double standard penalizes women for participating. Less privileged women face an even more complex situation: Much of the appeal of hookups derives from their utility as a delay strategy. Women who do not believe that it is desirable to delay marriage may experience the hookup culture as puzzling and immature.

An intersectional approach also suggests that the way young heterosexuals make decisions about sexuality and relationships underlies the reproduction of social class. These choices are part of women's efforts to, as one privileged participant so eloquently put it, "maintain the lifestyle that I've grown up with." Our participants were not well versed in research demonstrating that college-educated women benefit from their own human capital investments, are more likely to marry than less educated women, and are more likely to have a similarly well-credentialed spouse (DiPrete and Buchmann 2006). Nonetheless, most were aware that completing college and delaying marriage until the mid-to-late twenties made economic sense. Nearly all took access to marriage for granted, instead focusing their attention on when and whom they would marry.

The two-pronged strategy of career investment and delay of family formation has so quickly become naturalized that its historical novelty is now invisible. It is based on the consolidation of class, along with heterosexual, privilege: Heterosexual men and women attempt to maximize their own earning power and that of their spouse—a pattern that is reflected in increased levels of educational homogamy (Schwatz and Mare 2005; Sweeney 2002). Consolidation of privilege is made possible by women's greater parity with men in education and the workforce. In this new marital marketplace, a woman's educational credentials and earning potential are more relevant than her premarital sexual activity, assuming she avoids having a child before marriage. Relationship commitments that block educational and career investments, particularly if they foreclose future opportunities to meet men with elite credentials, are a threat to a woman's upward mobility.

The gender implications of the consolidation of privilege are most visible when contrasted with gender specialization—a marital strategy once assumed to be universal. Marriage was thought to be a system of complementary interdependence in which the man specialized in the market and the woman in domesticity (Becker 1991). Men maximized earning power while women accessed these benefits by marrying those with greater educational or career credentials. Gender specialization does not logically demand chastity of women; however, historically it has often been offered for trade in the marital marketplace. When this occurs, women's sexual reputation and economic welfare are linked. Although this connection has long been attenuated in the United States, it still exists. For example, the term "classy" refers simultaneously to wealth and sexual modesty.

As marriage in the United States has become less guided by gender specialization and more by the consolidation of privilege, gender inequality—at least within the marriages of the privileged—may have decreased. At the same time, class inequality may have intensified. The consolidation of privilege increases economic gaps between the affluent who are married to each other, the less affluent who are also married to each other, and the poor, who are excluded from marriage altogether (also see Edin and Kefalas 2005; England 2004; Schwartz and Mare 2005; Sweeney 2002). The hookup culture may contribute in a small way to the intensification of class inequality by facilitating the delay necessary for the consolidation of privilege. . . .

References

Armstrong, Elizabeth A., Paula England, and Alison C. K. Fogarty. 2009. Orgasm in college hookups and relationships. In *Families as they really are,* edited by B. Risman. New York: Norton.

Armstrong, Elizabeth A., Laura Hamilton, and Brian Sweeney. 2006. Sexual assault on campus: A multilevel, integrative approach to party rape. *Social Problems* 53:483–99.

Arnett, Jeffrey Jensen. 2004. *Emerging adulthood: The winding road from the late teens through the twenties.* New York: Oxford.

Becker, Gary S. 1991. *A treatise on the family.* Cambridge, MA: Harvard University Press.

Beisel, Nicola, and Tamara Kay. 2004. Abortion, race, and gender in nineteenth century America. *American Sociological Review* 69:498–518.

Bogle, Kathleen A. 2008. *Hooking up: Sex, dating, and relationships on campus.* New York: New York University Press.

Boswell, A. Ayres, and Joan Z. Spade. 1996. Fraternities and collegiate rape culture: Why are some fraternities more dangerous places for women? *Gender & Society* 10:133–47.

Brooks, David. 2002. Making it: Love and success at America's finest universities. *The Weekly Standard,* December 23.

Campbell, Jacquelyn C., Nancy Glass, Phyllis W. Sharps, Kathryn Laughon, and Tina Bloom. 2007. Intimate partner homicide. *Trauma, Violence & Abuse* 8:246–69.

Collins, Patricia Hill. 1990. *Black feminist thought: Knowledge, consciousness and the politics of empowerment.* Boston: Unwin Hyman.

Collins, Randall. 2004. *Interaction ritual chains.* Princeton, NJ: Princeton University Press.

Connell, R. W. 1987. *Gender and power: Society, the person, and sexual politics.* Stanford, CA: Stanford University Press.

Correll, Shelley J., Stephen Benard, and In Paik. 2007. Getting a job: Is there a motherhood penalty? *American Journal of Sociology* 112:1297–1338.

Crawford, Mary, and Danielle Popp. 2003. Sexual double standards: A review and methodological critique of two decades of research. *Journal of Sex Research* 40:13–26.

DiPrete, Thomas A., and Claudia Buchmann. 2006. Gender-specific trends in the value of education and the emerging gender gap in college completion. *Demography* 43:1–24.

Dunn, Jennifer L. 1999. What love has to do with it: The cultural construction of emotion and sorority women's responses to forcible interaction. *Social Problems* 46:440–59.

Eder, Donna, Catherine Colleen Evans, and Stephen Parker. 1995. *School talk: Gender and adolescent culture.* New Brunswick, NJ: Rutgers University Press.

Edin, Kathryn, and Maria Kefalas. 2005. *Promises I can keep: Why poor women put motherhood before marriage.* Berkeley: University of California Press.

England, Paula. 2004. More mercenary mate selection? Comment on Sweeney and Cancian (2004) and Press (2004). *Journal of Marriage and the Family* 66: 1034–37.

England, Paula, Emily Fitzgibbons Shafer, and Alison C. K. Fogarty. 2007. Hooking up and forming romantic relationships on today's college campuses. In *The gendered society reader,* edited by M. Kimmel. New York: Oxford University Press.

Eshbaugh, Elaine M., and Gary Gute. 2008. Hookups and sexual regret among college women. *Journal of Social Psychology* 148:77–89.

Friedland, Roger, and Robert R. Alford. 1991. Bringing society back in: Symbols. practices, and institutional contradictions. In *The new institutionalism in organizational analysis,* edited by W. W. Powell and P. J. DiMaggio, 232–63. Chicago: University of Chicago Press.

Gerstel, Naomi, and Natalia Sarkisian. 2006. Marriage: The good, the bad, and the greedy. *Contexts* 5:16–21.

Glenn, Evelyn Nakano. 1999. The social construction and institutionalization of gender and race: An integrative framework. In *Revisioning gender,* edited by M. M. Ferree, J. Lorber, and B. B. Hess, 3–43. Thousand Oaks, CA: Sage.

Glenn, Norval, and Elizabeth Marquardt. 2001. *Hooking up, hanging out, and hoping for Mr. Right: College women on mating and dating today.* New York: Institute for American Values.

Grello, Catherine M., Deborah P. Welsh, and Melinda M. Harper. 2006. No strings attached: The nature of casual sex in college students. *Journal of Sex Research* 43:255–67.

Holland, Dorothy C., and Margaret A. Eisenhart. 1990. *Educated in romance: Women, achievement, and college culture.* Chicago: University of Chicago Press.

Illouz, Eva. 2005. *Cold intimacies: The making of emotional capitalism.* Cambridge, UK: Polity.

Kimmel, Michael. 2008. *Guyland: The perilous world where boys become men.* New York: Harper Collins.

Manning, Wendy D., Peggy C. Giordano, and Monica A. Longmore. 2006. Hooking up: The relationship contexts of "nonrelationship" sex. *Journal of Adolescent Research* 21:459–83.

Martin, Karin. 1996. *Puberty, sexuality, and the self: Boys and girls at adolescence.* New York: Routledge.

Martin, Patricia Yancey. 2004. Gender as a social institution. *Social Forces* 82:1249–73.

Martin, Patricia Yancey, and Robert A. Hummer. 1989. Fraternities and rape on campus. *Gender & Society* 3:457–73.

McCall, Leslie. 2005. The complexity of intersectionality. *Signs: Journal of Women in Culture and Society* 30:1771–1800.

Owen, Jesse J., Galena K. Rhoades, Scott M. Stanley, and Frank D. Finchaln 2008. "Hooking up" among college students: Demographic and psychosocial correlates. *Archives of Sexual Behavior,* http://www.springerlink.com/content/44j645v7v38013u4/fulltext.html.

Paul, Elizabeth L., Brian McManus and, Allison Hayes. 2000. "Hookups": Characteristics and correlates of college students' spontaneous and anonymous sexual experiences. *Journal of Sex Research* 37:76–88.

Ridgeway, Cecilia L. 2000. Limiting inequality through interaction: The end(s) of gender. *Contemporary Sociology* 29:110–20.

Ridgeway, Cecilia L. 2009. Framed before we know it: How gender shapes social relations. *Gender & Society* 23:145–60.

Ridgeway, Cecilia L., and Shelley J. Correll. 2004. Unpacking the gender system: A theoretical perspective on gender beliefs and social relations. *Gender & Society* 18:510–31.

Risman, Barbara, and Pepper Schwartz. 2002. After the sexual revolution: Gender politics in teen dating. *Contexts* 1:16–24.

Risman, Barbara J. 2004. Gender as a social structure: Theory wrestling with activism. *Gender & Society* 18:429–50.

Rosenfeld, Michael J. 2007. *The age of independence: Interracial unions, same sex unions and the changing American family.* Cambridge, MA: Harvard University Press.

Schwartz, Christine R., and Robert D. Mare. 2005. Trends in educational assortative marriage from 1940 to 2003. *Demography* 42:621–46.

Sewell, William H. 1992. A theory of structure: Duality, agency, and transformation. *American Journal of Sociology* 98:1–29.

Simon, Robin W., Donna Eder, and Cathy Evans. 1992. The development of feeling norms underlying romantic love among adolescent females. *Social Psychology Quarterly* 55:29–46.

Stepp, Laura Sessions. 2007. *Unhooked: How young women pursue sex, delay love, and lose at both.* New York: Riverhead.

Sweeney, Megan M. 2002. Two decades of family change: The shifting economic foundations of marriage. *American Sociological Review* 67:132–47.

Wharton, Amy S. 2005. *The sociology of gender: An introduction to theory and research.* Malden, MA: Blackwell.

Introduction to Reading 45

The policing of women's bodies and sexuality has been extended and intensified in the new world of social media. In this reading, social scientist Alexa Dodge examines several high profile cases of cyberbullying following sexual assaults. In each case the sexual violence experienced by the victim was exacerbated by the digital sharing of photographs of the assault. As Dodge points out, "the damaging effects . . . of photographing sexual assault should be self-evident" as should digital dissemination of those photographs. But that was not the case. Dodge uses Judith Butler's concept of *digitalization of evil* to explain how digitally shared photos of girls being sexually assaulted and abused come to be perceived in a way that blames the victim and reinforces rape culture.

1. Define rape culture. Discuss the author's application of this concept in this reading.
2. How does online sexual violence both extend the trauma of sexual assault and become a form of sexual violence itself?
3. Define "digitalization of evil." Discuss the usefulness of this concept in understanding how "normal" people who shared and commented on the photographs of sexual assaults of women did not hide their identity as bystanders who participated in blaming the victims for the violence they suffered.

Digitizing Rape Culture

Online Sexual Violence and the Power of the Digital Photograph

Alexa Dodge

In 2013, the tragic suicide of a 17-year-old Nova Scotian named Rehtaeh Parsons made headlines across Canada and internationally. When she was 15 years old, Parsons attended a party where she was allegedly sexually assaulted by multiple boys. During the alleged assault, a photo was taken of one of the boys penetrating Parsons from behind while she was vomiting with her head stuck out of a window (*CBC News,* 2013; *Maritime Noon,* 2013b). This photo was then widely circulated on social media networks, the bullying and so-called 'cyberbullying' that Parsons experienced as a result would eventually lead to her taking her own life. Although no formal sexual assault charges have been laid in this case. Parson's case has been widely accepted as an example of the digital dissemination of photographs of sexual violence.[1] Parson's case has been compared to similar incidents in the United States that involved photographs of sexual assault being disseminated and used as the catalyst for cyberbullying over social media (Shariff and DeMartini, 2015: 290). Some have even referred to Parsons' case as 'Canada's Steubenville' (*Day 6,* 2013; *Maritime Noon,* 2013a). The Steubenville case involved a teenage girl in Steubenville, Ohio, known anonymously as Jane Doe, being sexually assaulted by multiple boys and having photographs of the event circulated on social media. The comments made on Twitter in response to the photos of her sexual assault were disturbing, and are only one example of the nature of the bullying and cyber-bullying that Doe endured in the aftermath of this tragedy (*Day 6,* 2013). Others have also drawn parallels between the Parsons' case and the case of Audrie Pott in California (*Day 6,* 2013; *Maritime Noon,* 2013a). Pott was also 15-years-old when three boys sexually assaulted her at a party, wrote lewd comments on her body with marker, photographed the incident and circulated at least one photo on social media networks. Pott also committed suicide following the bullying and cyberbullying she experienced in response to this photo (*Maritime Noon,* 2013a). In each of these cases, these girls were left not only to deal with the effects of being sexually assaulted, but were also left to grapple With the humiliation of their assaults being shared over and over again on social media networks and the resulting bullying and cyberbullying of peers who echoed sentiments that the girls were 'sluts' and therefore deserved this treatment.

The sexual assault and cyberbullying cases of Rehtaeh Parsons, Jane Doe and Audrie Pott all bring up disturbing questions about the ways that sexual violence can be normalized and condoned within western society and the role that new media increasingly plays in the perpetration of this violence. These cases represent the ways that new media can be seen to exacerbate issues surrounding sexual violence by creating digital spaces wherein the perpetuation and legitimization of sexual violence takes on new qualities. For instance, how does the digital capturing and dissemination of photographs of sexual violence intensify victims' experiences? How does the framing of these images on social media influence the ways that they are read and responded to? How are these photographs variously interpreted and reinterpreted through social

Dodge, A. (2015). Digitizing rape culture: Online sexual violence and the power of the digital photograph. *Crime, Media, Culture, 12*(1): 65–82.

media, the law and feminist activism? This paper will explore these questions using the concept of rape culture to investigate how our cultural norms allow for the perpetuation of online sexual violence and for the misinterpretation of photographic evidence of sexual violence. I will use Judith Butler's theory on photography, torture and framing to argue that these cases are an example of what Butler refers to as the digitalization of evil (2007: 961).

Defining "Cyberbullying" and Rape Culture in the Context of Online Sexual Violence

The onslaught of recent cases involving intersections between sexual and social media have been popularly referred to by both the media and government as cases of 'cyberbullying'. Nova scotia's provincial anti-bullying website describes cyberbullying as incidences when 'someone uses technology (social networking sites, e-mails, text messages, and the sharing of images) to bully someone else. Cyberbullying is unique because hurtful messages or pictures can quickly and anonymously be shared' (Province of Nova Scotia, 2014). Based on definitions such as these, I would like to problematize the use of the term cyberbullying in relation to instances of sexual violence over social media. Some of the critiques around the use of this term to refer to cases of harassment involving sexual violence include concerns that the term cyberbullying: conflates multiple complex issues, trivializes these issues and wrongfully frames these issues as politically neutral (Fairbairn, 2014). The term bullying is often associated with relatively benign schoolyard teasing; thus, the use of the term cyberbullying may not properly represent what is going on in cases such as Parsons', cases that may be more accurately explained as instances of online sexual violence, digital sexual assault or as the proliferation of digital evidence of sexual assault. A recent report by Fairbairn et al. titled Sexual violence and social media: Building a framework for prevention discusses the issue of how to frame sexual violence that occurs over social media. As one participant in the report's survey asserts, 'it's important to create spaces for people to have conversations about sexual violence over social media/digitally, and to see it framed AS sexual violence' (Fairbairn et al., 2013: 35). Therefore, although the term cyberbullying will be used in this paper to refer to instances *of* harassment over social media, I will seek to highlight the extreme forms that this harassment can take and to discuss how cyberbullying can manifest as a digital form of sexual violence. For instance, cyberbullying that takes the form *of* the dissemination of photographs of sexual assault will be referred to as instances of online sexual violence. It is important to recognize that not only can online sexual violence be used as a way of extending the trauma of sexual assault, but that it is also a form of sexual violence in and of itself. As Fairbairn et al. explain, in the context of sexual violence and social media, it is particularly important to think about sexual violence on a continuum that involves emotional, psychological, and verbal violence as well as physical violence' (2013: 13). This paper will attempt to convey the complexities of this continuum and the breadth of violence that has been included under the term cyberbullying.

Another term that is often used, and misused, in relation to cases of online sexual violence is that of rape culture. The term rape culture refers to multiple pervasive issues that allow rape and sexual assault to be excused, legitimized and viewed as inevitable (Smith, 2004: 174). In the influential book transforming a Rape Culture, rape culture is described as:

> A complex set of beliefs that encourage male sexual aggression and supports violence against women. [. . .] A rape culture condones physical and emotional terrorism against women and presents it as the norm. [. . .] In a rape culture, both men and women assume that sexual violence is a fact of life, as inevitable as death or taxes. (Buchwald et al., 2005: xi)

The term was originally coined by second-wave feminists in the 1970s (Smith, 2004: 174) and has become more popularly used and understood since the 2011 SlutWalk movement (Shah, 2012: n.p.). Recently, attention has started to be given to the particular ways that rape culture is manifested and experienced online (Fairbairn, 2015: 234; Shariff and DeMartini, 2015: 281). For instance, multiple chapters of the recently released collection *eGirls, eCitizens* speak to the way that online sexual violence both redeploys long-existing manifestations of rape culture and alters/intensifies them due to 'the rapid pace at which various forms of expression, including offensive and demeaning photographs and images, can be distributed and shared online' (sheriff and DeMartini, 2015: 281). I will continue the emerging discussion regarding the ways that online sexual violence both reinscribes long-standing beliefs that support rape culture and changes the way that sexual violence and rape culture are experienced and perpetuated.

Some critics have misunderstood the concept of rape culture and have wrongly conceived of it as a way of blaming cultural factors rather than individual perpetrators for acts of sexual violence (Marcotte, 2014). To clarity, discussions of rape culture are not meant to invisibilize the actions of perpetrators, but rather to help describe the ways in which 'rapists are given a social license to operate by people who make excuses for sexual predators and blame the victims for their own rapes' (Marcotte, 2014). As Stuart Hall explains, culture is about 'shared meanings' (1997: 1). In the cases of Parsons, Doe and Pott, rape culture is manifested in the fact that many of those who shared and commented on these photographs of sexual violence shared a reading of these photographs as something other than evidence of sexual assault. The shared meanings that many of these youth drew from these photos—namely that the girls pictured in them were sluts that deserved having their bodies violated and/or having their humiliation used for entertainment—speaks to the workings of rape culture. As Hall explains:

> Things 'in themselves' rarely if ever have any one, single, fixed and unchanging meaning. [. . .] It is by our use of things, and what we say, think and feel about them—how we represent them—that we give them a meaning. In part, we give objects, people and events meaning by the frameworks of interpretation which we bring to them. (1997: 3)

With the importance of framing in mind, this paper will investigate how photographs of sexual violence are shared in a context of rape culture that frames them as something other than evidence of sexual assault or a digital form of sexual violence.

The Power of the Digital Photograph: Temporal Extension and Ubiquity

It is important to first understand the impact that photographs of sexual assault can have on the victims who are pictured in them. The cases of Parsons, Doe and Pott represent a growing list of sexual assault cases that include the additional element of digital photography and social media sharing. In all three of these cases, photographs of the sexual assault allowed the traumatic event to be memorialized by those on social media; with each share, like, tweet or comment, the sexual assault was reiterated over and over again. The proliferation of these photographs, and the disturbing victim blaming, slut-shaming and refrains of 'lol'[2] that often accompanied their proliferation (Goddard, 2013), allowed these sexual assaults to be temporally extended and to continue to haunt their victims in a very tangible way. As Butler asserts:

> Photographing is a kind of action that is not always anterior to the event. not always posterior to the event. The photograph is a kind of promise that the event will continue is that very continuation of the event, producing an equivocation at the level of the temporality of the event; did those actions happen then; do they continue to happen? Does the photograph continue the event in the present? (2007: 959)

These photographs, and the resulting bullying and cyberbullying by peers, continue to recreate and extend the trauma of these sexual assaults. In the cases of Parsons and Pott, the trauma caused by the permanency of these photos and the cyberbullying experienced as a result of their dissemination, in addition to the sexual assault itself, led to their suicides (Maritime Noon, 2013a, 2013b). As the lawyer for the Pott family stated:

> The bullies savagely took advantage of her while she had no ability to defend herself, and then what happened afterwards may have been worse. They rubbed her *nose* in it. effectively, by spreading around at least one photograph of the assault taking place and various taunting messages about what happened. (*Maritime Noon,* 2013a)

Likewise, reports from Parsons' mother and father about their daughter's trauma focus on their daughter's reaction to the spread of a photograph of the assault and the bullying and cyberbullying she experienced as a result (CBC News, 2013; Maritime Noon, 2013b). The Chronicle Herald summarized this point in the aftermath of Parsons' suicide saying, 'it was a week after 17-year-old Rehtaeh Parsons' death, and one thing that had tormented her was the same thing, after her death that people wouldn't let go: a photo' (The Chronicle Herald, 2013). The online dissemination of photographs of sexual assault not only extends the experience of the sexual assault, but can be seen as an act of sexual violence in and of itself, the impact of which is sometimes described as even more traumatic than the original assault (Bluett-Boyd et al., 2013: 39; Fairbairn et al., 2013: 49).

Normalizing Sexual Violence: Rape Culture and the Digitalization of Evil

The damaging effects, for both victims and perpetrators, of photographing sexual assault should be self-evident. However, in the cases of all three girls, these photographs seem to have been taken and disseminated without regard for the possible consequences. In the Steubenville case, as documented by Goddard.[3] it appears that the teenagers who shared and commented on these photographs did not perceive the images as evidence of sexual assault that should be reported to authorities (ABC News, 2013), but rather read these photos as proof that the girls pictured in them were 'sluts' and 'whores' who deserved to be shamed and laughed at (Goddard, 2013; Maritime Noon, 2013b). For example, a photograph of Doe 'appearing lifeless and being carried like an animal' was responded to on Twitter by her peers calling her a whore, and one boy tweeting 'she looks dead lmao,' among other troubling comments (Goddard, 2013). These reactions not only extend the violence committed against Doe, but are also dehumanizing and misogynistic forms of harassment in and of themselves. Reactions such as these leave a disturbing question lingering in their wake: How is it that photographs of an unconscious girl being sexually assaulted and abused are perceived in a way that allow her to be read as a 'slut' who is 'asking for it' and who, therefore, deserves to be shamed by the continued circulation of these images?

Butler's 'Torture and the ethics of photography' may offer some clues as to how such photographs come to be taken in the first place, and how they are able to be read as entertainment or as evidence against the victim's character rather than as evidence of sexual assault. In this article, Butler argues that our ability to respond ethically to photographs of human suffering is influenced by the way a photograph is presented to us and how this presentation, or 'framing', is influenced by broader norms that affect our ability to perceive individuals as human and to recognize their suffering (2007: 951). Therefore, in the context of photographs of sexual assault, this would mean that rape culture, and the myths that enforce it such as stereotypes about masculinity and female sexuality, Influences the way that these photographs are perceived. For instance, one of the myths that bolsters rape culture is the 'purity myth'. In 'Purely rape: The myth of sexual purity and how it reinforces rape culture', Jessica Valenti explains that the purity myth is an invention that allows sexual violence against women to be excused and invisibilized by positing that 'sexuality defines how "good" women are, and that women's moral compasses are inextricable from their bodies [. . .]. Under the purity myth, any sexuality that deviates from a strict (generally straight, maledefined) norm is punishable by violence' (2008: n.p.). This myth is seen at work in the cases of Parsons, Doe and Pott as those who disseminated the photographs of sexual assault framed them with dialogue that referred to the young women pictured in them as sluts and whores (Goddard, 2013; *Maritime Noon,* 2013b). As documented by screenshots like Goddard's and a slew of reports from news media *(ABC News,* 2013; Goddard, 2013; Macur and Schweber, 2012), many of the peers and social media users Who shared and commented on these photos echoed these sentiments and, thereby, reiterated one of the core beliefs that informs rape culture, the common belief that women who express their sexuality (i.e. 'sluts') 'deserve whatever they get' (Friedman and Valenti, 2008: n.p.). In relation to this, it is important to recognize the prevalence of victim blaming and slut-shaming reactions in cases such as those of Parsons, Doe and Pott. For instance, reports from Parsons and Doe's mothers assert that harassment towards their daughters was widespread and that even many close friends turned against them.

Butler's discussion of photographs taken within the Abu Ghraib prison can further help to explain the framing and reading of photographs of sexual assault. Butler uses the Abu Ghraib photographs to interrogate the ways in which the taking and distributing of photographs of humiliation, rape and murder were normalized within this context (2007: 956). She asserts that the photographs of torture taken within the Abu Ghraib prison speak to the fact that 'the perspective on the so-called enemy was not idiosyncratic, but shared, so widely shared, it seems, that there was hardly a thought that something might be amiss here' (Butler, 2007: 958)[4] argue that something similar was at play when photographs were taken and circulated of the sexual assaults of three teenage girls. In these three cases, it was not a single, morally corrupt boy who committed sexual assault, photographed the abuse and spread the images online. Rather, it was multiple boys who sexually-assaulted, multiple people who stood by and said nothing as someone grabbed the camera to document the event, and multiple people who shared the photos and engaged in bullying and cyberbullying in the aftermath. In each of these cases, a great deal of people felt that there was no need to question the treatment of these teen girls because they were perceived to be 'sluts' and, therefore, deserving of such treatment (*Maritime Noon,* 2013b). Thus, within the context of rape culture, the acts of creating and disseminating digital photographs of the sexual assaults committed against Parsons, Doe and Pott are able to be read by many as entertainment and as legitimate forms of humiliation committed against deserving girls.

To explain how something so unethical as torture, including rape and murder, could be photographed so willingly; even excitedly, Butler takes up Hannah Arendt's theory of the banality of evil. Butler reimagines this theory for her purposes as the digitalization of evil explaining that the soldiers and security personnel at Abu Ghraib are not caught torturing, but rather *pose* for the camera, wait for it to document the moment, and digitally disseminate the resulting photographs; therefore, they do not recognize the moment as morally corrupt (2007: 960–961). This digitalization of evil is demonstrated in the photo circulated of Parsons, as the perpetrator in the photo does not even attempt to hide his identity, but rather turns toward the camera, smiling and posing with a thumbs-up gesture (*The Chronicle Herald,* 2013).

In all three of the cases in question, the people involved did not even attempt to hide their identity when appearing in, sharing and commenting on photographs of sexual assault. The perpetrators and their peers shared and commented on these photos using their own social media accounts and cell phones. In the case of Audrie Pott, the boys involved texted photographs of Pott's assault (*CBC News,* 2014), they did not worry that there could be consequences and that these photographs would be directly traceable to their cellphones and social media accounts. Likewise, Alexandria Goddard, a blogger and activist that was highly involved in reporting on the Steubenville rape case, reflects on how astonishingly brazen bystanders were when tweeting about their presence at, and therefore complicity in, the rape and assault of Jane Doe:

> By the time I reached party-goer Michael Nodianos' Twitter account, I was horrified. I could not believe some of the things this young man was saying about this young girl. Things like 'some people deserve to be peed on #whoareyou' and 'you don't sleep through a wang[5] in the but-thole'. (2013)

Goddard comments that the remarks made by bystanders to this crime caused her to ask herself the ultimate question: If so many kids were tweeting about this rape, 'WHY DIDN'T ANYONE DO ANYTHING TO STOP IT?' (2013). The banality/digitalization of evil allows us to begin to answer Goddard's question. No one did anything[5] because this behavior was normalized and legitimized by a context of rape culture. It is this normalization that allowed many people, as well as some media outlets, to openly sympathize with the rapists in the Steubenville case (Q, 2013). These sympathizers believed that the young men who raped and abused Doe did not deserve punishment because they were just *normal* boys, they were not *evil* people. As Rachel Giese, senior editor at *The Walrus* and a journalist who has been following the Steubenville case, explained in a radio interview:

> I think it gets at the way in which we still believe that people who do bad things are something apart from the rest of us, that they are something so strange and out there and evil [. . .] and I'm sure that these two boys are not evil boys. (0, 2013)

Thus, in line with Arendt's banality of evil, we cannot simply demonize and exceptionalize these rapists, we need to look at the cultural context[6] that allowed these 'normal boys' to rape, and allowed others to stand by, to share photos of the rape and to blame the victim. As Giese elaborates, the real tragedy 'is not that these boys got punished, the tragedy is that any boy might do this, that it could happen to any girl, that this is not something so separate from all of us, but something that lots of people are capable of' (Q, 2013). Therefore, we cannot just hold these boys responsible,[7] but must also pay attention to the pervasiveness of rape culture and the ubiquity of acts of sexism that allow the perpetration of sexual violence to become banal.

Framing the Photograph: Rape Culture and Responsibilization

In regard to the photographs of torture from Abu Ghraib, Butler posits, 'the photos are not only shown, but named; both the way that they are shown, the way they are framed and the words used to describe what is shown work together to produce an interpretive matrix for what is seen' (2007: 957). This kind of interpretative matrix can be seen at work in the presentation of photographs of sexual assault. In the cases of Parsons, Doe and Pott, the comments and tweets that reportedly accompanied the photographs of sexual assault asserted that the victims were 'whores' and 'sluts' and, thereby, utilized myths of rape culture to frame these photographs within narratives of victim blaming. The myths being utilized in this framing include stereotypes about women that construct them as naturally passive and lacking in sexual agency, while, at the same time, constructing their bodies as inherently sexual and tempting (Friedman and Valenti, 2008: n.p.). Myths such as these lead to the cultural belief that women should always be on the defensive, and that if they are sexually assaulted they must have in some way incited it by wearing clothes that revealed too much of their tempting bodies, by drinking too much, by entering spaces that they 'should' have read as unsafe, or by asserting their sexuality and agency.

These myths allow the actions of men who sexually assault to be legitimized, and allow women to be responsibilized for the crimes committed against them. These myths do not hold rapists responsible, but rather send a message to women that it is their responsibility to avoid sexual violence. It is these kinds of myths that inform problematic readings of photographs that may seem to clearly depict acts of violence. Due to the ways that patriarchy, sexism and rape culture frame what is seen, photographs depicting sexual violence are not uniformly read as evidence of sexual assault.

Butler's article 'Endangered/endangering: Schematic racism and white paranoia' discusses the way that photographs and videos that may appear to be objective evidence of violence can be misinterpreted due to the cultural context that frames their reading. Butler uses the video footage of the Rodney King beating to exemplify the powerful influence that cultural context and framing have in affecting the way that videos and photographs of violence are read. Butler explains this powerful influence saying:

> [When describing the Rodney King video], without hesitation, I wrote, 'the video shows a man being brutally beaten.' And yet, it appears that the jury in Simi Valley claimed that what they 'saw' was a body threatening the police, and saw in those blows the reasonable actions of police officers in self-defense. From these two interpretations emerges, then, a contest within the visual field, a crisis in the certainty of what is visible, one that is produced through the saturation and schematization of that field with the inverted projections of white paranoia. (2004: 205)

Butler elaborates, 'this is not a simple seeing, an act of direct perception, but the racial production of the visible, the workings of racial constraints on what it means to "see"' (2004: 206). In a similar way, the photographs of sexual violence being committed against Parsons, Doe and Pott were read through the lens of rape culture and sexism. Just as a racist lens allowed King's beaten body to be read as a body that was threatening and deserving of violence, the young women in these photographs were read through a sexist lens that allowed them to be read as 'sluts' whose bodies were innately tempting and deserving of violation.[8] This lens allowed fellow students and community members to read these photos with the belief that the boys pictured in them were just 'boys being boys', while the victims were read as promiscuous girls who were getting what they deserved because they should have dressed differently, should have drank less, should have stayed home, and so on. Alison Young's work on visual criminology further supports this malleability of photographic meaning:

> The image and the social world cannot be distinguished from each other[,] [. . .] meaning derives from the affective nature of the spectator's encounter with the image. [. . .] affect refers not so much to the emotional landscape of law or criminal justice but rather to the ways in which subject positions such as 'victim', 'criminal' and 'judge' are corporeally registered in the spectator at the moment in which she undergoes the experience of spectatorship, such that, in the process of making meaning out of an event, affect and intellectual categorization are irrevocably intertwined. (2014: 161–162)

Not through the lens of the camera alone, but through the lens of rape culture, these photographs became tools for victim blaming and slut-shaming, rather than evidence of sexual violence.

Photographs as Evidence and the Law's Ability to See

The suicides of two Canadian teens, Rehtaeh Parsons and Amanda Todd, following incidents involving online sexual violence and cyberbullying were major catalysts for the Canadian government's implementation of anti-cyberbullying legislation and campaigns at both the federal and provincial levels. For instance, the Protecting Canadians from Online Crime Act has been marketed as legislation designed to combat cyberbullying (Southey, 2013).[9] This legislation directly addresses the issue of cyberbullying (or what is more akin to online sexual violence) in the form of non-consensually sharing of 'intimate images' over social media and other technology. The legislation makes it a criminal offence to distribute an 'intimate image' of a person without their consent, and thus validates the considerable damage that sharing such images can have on victims (Parliament of Canada, 2013). Although this new legislation may prove to be useful in better addressing the non-consensual sharing of 'intimate images', the law, like the everyday viewer, is still left to grapple with interpreting the actions captured in these photographs. In the case of images of alleged sexual violence, the ability of the photograph to provide evidence of sexual assault still remains ambiguous. Therefore, even when the law is able to reinterpret such photographs as evidence, this visual evidence cannot necessarily be expected to outweigh culture stereotypes that influence the law's 'capacity for vision' (Biber, 2014: 5).

As a growing number of researches in the area of visual criminology have revealed, legal readings of photographs are influenced by histories, biases and framings (Biber, 2014; Young, 2014). In Captive Images: Race, Crime, Photography, Katherine Biber asserts:

> For law, photographs purport to tell the truth; photographs are evidence. The criminal courts pore over photographs, hoping to crack open the secret hidden within the image. Ignoring over one century of scholarship on photography, never attempting to formulate a jurisprudence of the visual, the law looks at photographs as if there was nothing impeding its capacity to see. (2014: 5)

Biber thus echoes Butler's argument that the ability to read photographs is influenced by our preexisting stereotypes. For example, for both Biber and Butler our histories of racism influence the law's 'capacity for vision' (Biber, 2014: 5). This assertion ultimately seeks to undermine the assumption that the photograph provides us with a dear story, that the photograph can necessarily tell us the 'truth'. Thus, this work allows us to question the assumption that 'the rules of evidence are capable of controlling the fantasies that compete when we look into a photograph' (Biber, 2014:11). And, as Michael Salter asserts, the fantasies that compete in regard to women's allegations of sexual violence are particularly strong: 'Cultural mythologies around gender and sexuality have proven to be particularly durable, blunting the impacts of feminist-inspired law reform (Stubbs, 2003)' (2013: 227).

Therefore, although the Canadian government has enacted legal responses to the non-consensual dissemination of 'intimate images', this response may not effectively address the histories of sexism and the context of rape culture that allowed these images to be misread in the first place. This is exemplified by Parsons' case because, as noted earlier, her reported sexual assault has not been deemed legally provable despite the existence of photographic evidence. The following section of this paper provides a brief overview of how the voices of online feminist activists are creating narratives that seek to right such injustices by creating a culture that believes survivors and combats rape culture. These activists seek to challenge rape culture by radically transforming the ways we think and talk about sexual violence, sexism, female sexuality, male sexuality and more. By harnessing social media, blogs, online campaigns and other technologies as tools for activism, these activists are creating anti-hegemonic spaces and narratives that question the dominant discourse around sexual violence. This grassroots activism may have the potential to work from the ground up to create a cultural shift that would allow for images of sexual violence to be more readily interpreted as evidence of sexual assault both in and outside of the court of law.

Reframing the Image: Speaking Out Against Rape Culture and Victim Blaming

While social media, and the internet more generally, often acts as a forum for victim blaming, slut-shaming, sexual violence and the reiteration of gendered stereotypes, these same social media tools have also allowed feminist activists and their allies to organize and have their voices heard. Through this organizing and speaking out, feminist activists have begun to craft new narratives about sexual violence and the various stereotypes that perpetuate it. They are utilizing online platforms to create activism against rape culture and to provide solidarity for those experiencing sexual and gender-based violence. This grassroots activism has created positive change by challenging gender-based hate and victim blaming both online and off. Websites such as The Everyday Sexism Project, Take Back the Tech! and Hollaback! are all examples of the extra-legal ways that online feminist activists are using online spaces to speak out against sexual violence. Although non-consensual intimate images, images of sexual assault and instances of genderbased hate continue to proliferate the online world, these activists are fighting back and giving a voice to those who are often denied one. The work of online feminist activists is geared towards changing the culture of me Internet and beyond by fighting sexual violence and gender-based hate in its many forms and in all of its complexity.

Multiple online activists were directly involved in demanding justice and countering victim blaming narratives in the cases of both Parsons and Doe (Day 6, 2013). As discussed earlier, Goddard was one of these activists, as she was active in retaining evidence and demanding justice in the Steubenville case from very early on. Goddard found, and captured screenshots of, tweets from perpetrators and bystanders posted throughout the night of the sexual assault. She explains that these tweets included jokes about raping and urinating on Doe's unconscious body (Goddard, 2013). Goddard shared these tweets on her blog as a way to bring attention to the horrific nature of these comments and to counter victim-blaming narratives that were being presented by Doe's peers, members of the Steubenville community and some media outlets. Thus, Goddard did what many on the night of the rape did not: she read these photographs and tweets as evidence of sexual assault and used her blog to frame them as such. As Salter asserts, 'the clear potential of online counter-publics is that they may offer girls and women the support and validation that can elude them in off-line contexts and perhaps offset the advantages that the public sphere affords men accused of sexual violence' (2013: 238).

Not only are online activists speaking out in response to specific cases of sexual violence and gender-based hate, many activists and organizations are also creating dialogues that address the overarching cultural factors that lead to sexism and allow acts of sexual violence to be read as 'banal' and 'everyday'. Online feminist activism goes beyond holding individual perpetrators accountable and is able to create a space to discuss the larger social contexts, beliefs and stereotypes that add to the pervasive nature of sexual violence. As Filipovic posits in her contribution to the powerful collection of sex-positive essays Yes Means Yes!: Visions of Female Sexual Power and a World Without Rape, 'eradicating rape may very well be impossible. But as long as we continue to view it as a crime committed by an individual against another individual. absent of any social context, we will have little success in combating it' (2008: n.p.). Many online activists have taken up an approach that challenges rape culture and, thereby, seeks to create a world wherein, for instance, photographs of sexual violence are not read as entertainment or as evidence of a victim's character. Butler discusses the importance of this kind of undermining of the deeply held cultural beliefs and stereotypes that result in injustice. In regard to the Rodney King video, Butler comments:

> To think that the video 'speaks for itself' is, of course, for many of us, obviously true. But if the field of the visible is racially contested terrain, then it will be politically imperative to read such videos aggressively and to repeat and publicize such readings, if only to further an antiracist hegemony over the visual field. (2004: 206)

In the context of photographs of sexual assault, this same aggressive reading is needed to counter *sexist* hegemony over the visual field. This is what is being done by the multiple blogs, online campaigns and petitions by activists that are working to undermine the stereotypes and beliefs that sustain rape culture.

For example, The Everyday Sexism Project, an online blog that asks users to share the sexism they experience on a day-to-day basis, seeks to reveal the prevalence of sexism and thereby create a narrative that takes these issues more seriously. This website asks women to contribute to a forum with stories about the sexism that they experience every day whether these experiences are 'serious or minor, outrageously offensive or so niggling and normalised that you don't even feel able to protest' (The Everyday Sexism Project, 2014). They explain, 'by sharing your story you're showing the world that sexism does exist, it is faced by women everyday and it is a valid problem to discuss' (The Everyday Sexism Project, 2014). This website's mandate directly responds to the concept of the banality of evil discussed above. The discussions of the *everyday* nature of sexism that unfold on their forum illuminate the process by which acts of violence against women become banal and, thereby, acceptable and/or excusable. This is just one example of the feminist work that is being done online to expose the prevalence of sexism and rape culture and to challenge the dominant narratives around these issues.

When online feminist activists reject narratives of victim-blaming and female responsibilization, new narratives about the causes and effects of gender-based violence begin to gain traction. Some of these new narratives involve creating greater understanding about the meaning and importance of consent and undermining myths about masculinity, femininity and sexuality that support rape culture. The work that online feminist activists are doing helps to question the norms of rape culture that allow sexism and sexual violence to remain pervasive. The long-term effect this work will have on combating online sexual violence is yet to be seen, but there is certainly potential for these activists to create change in ways that the law has proven unable to address.

Conclusion

In the introduction to *Yes Means Yes! Visions of Female Sexual Power and* a *World Without Rape* readers are asked to 'imagine a world where women enjoy sex on their own terms and aren't shamed for it. Imagine a world where men treat their sexual partners as collaborators, not conquests. Imagine a world where rape is rare and punished swiftly' (Friedman and Valenti, 2008: n.p.). The cases of Rehtaeh Parsons, Jane Doe and Audrie Pott speak to the fact that this world is still, unfortunately, only an imagined one. Cases of sexual violence over social media have left us with all too real visual and textual evidence that women's bodies are often still seen as objects to be conquered and that the immorality of this conquering is still up for debate. As Butler comments in regard to the photographs from Abu Ghraib, 'We probably need to accept that the photograph neither tortures nor redeems, but can be instrumentalized in radically different directions, depending on how it is discursively framed, and through what media presentation the matter of its reality is presented' (2007: 964). Although it may seem that the reality of certain images can be objectively defined, I have argued that deeply held cultural beliefs and stereotypes can cause the meaning of a photograph to drastically warp in the eye of the reader. In the cases of Parsons. Doe and Pott, and in many similar cases that have been exposed since, photographs of sexual violence were distorted by framings and the context of rape culture in a way that allowed them to be read as humorous or as legitimate acts of punishment against deserving young women. These distorted readings allowed these images of sexual assault to be widely shared online, and thereby allowed their impact to be broadened and the trauma of the original sexual assault to be temporally extended. Thus, the power that the photograph holds must be recognized, and misreadings of photographs such as those taken of Parsons, Doe and Pott must be drowned out by assertive and repeated readings of these photographs as evidence of sexual violence.

Notes

1. Although there have been no formal sexual assault charges laid in the case of Rehtaeh Parsons, it is widely asserted that she was too intoxicated to have the capacity to give consent. This is evidenced by the fact that there is a photograph of her vomiting out of a window while being penetrated. Additionally, I would argue that the non-consensual creation of sexually explicit and humiliating photographs of her during this penetration is a form of sexual violence in and of itself. Therefore, I will be referring to Parsons' case as one that deals with the intersections between social media and sexual assault. Although sexual assault has been deemed not legally provable in this case, it is important to believe victims of sexual violence and to represent all aspects of Parsons' experience of sexual violence. For more details on the child pornography charges laid in this case, and speculation about why sexual assault charges were not laid, see 'Rehtaeh Parsons case prompts scrutiny of police, prosecution procedures', The Chronicle Herald, 22 December 2013, http:/thechronicleherald.ca/metro/1176083-rehtaeh-parsons-caseprompts-scrutiny-of-police-prosecution-procedures; and 'Probation for second man who pleaded guilty in Rehtaeh Parsons case', CTV News, 15 January 2015, http://www.ctvnews.ca/canada/probation-for-second-man-who-pleaded-guilty-in-rehtaeh-parsons-case-1.2189927.

2. A popular online initialism that stands for 'laugh out loud'.

3. When the Steubenville case became a media story many young people began deleting any social media statuses, comments, videos and pictures that may have tied them to the case. Therefore, I must rely heavily on the screen captions taken by Goddard to understand what took place on social media the night that Jane Doe was sexually assaulted.

4. A slang term for penis.

5. There has been very little evidence reported in regard to whether or not, or to what extent, any bystanders attempted to intervene during the sexual assaults and initial acts of online sexual violence committed against Doe. Researchers are left to rely on evidence such as the screenshots taken by Goddard (which do not show evidence of bystander intervention) and some related reports from witnesses, families of the victims and court transcripts when attempting to understand the reactions of bystanders. At least one witness for the prosecution in the Steubenville case claims to have attempted to intervene, however this witness also captured and shared a photo of Doe's assault. This witness (offender Trent Mays' best friend) testified saying, 'I tried to tell Trent to stop it, you know, I told him, "Just wait — wait till she wakes up if you're going to do any of this stuff. Don't do anything you're going to regret"' (Macur and Schweber, 2012). The witness then took a photograph of Mays and Richmond sexually assaulting Doe and, although he testified in court that he took the photograph because he wanted Doe to know what happened to her, he never showed Doe the photograph and deleted it from his phone after showing several other people (Macur and Schweber, 2012). In general, media reports have asserted that bystanders did not intervene. For instance, Steubenville police chief William McCafferty commented on Doe's case saying, 'the thing I found most disturbing about this is that there were other people around when this was going on. [. . .] Nobody had the morals to say, "Hey, stop it, that isn't right."' (Macur and Schweber, 2012).

6. For further discussion of the factors that influence male sexual aggression, especially in groups and in the context of sports teams, see Michael Kimmel's chapter 'Men, masculinity, and the rape culture' in Transforming a Rape Culture (2005).

7. Although it will not be discussed in this paper, it is important to question the ways that these boys are held responsible and to trouble narratives that assert imprisonment as an effective way to 'fix' these boys and to address rape culture. See Mia McKenzie's blog post on *Black Girl Dangerous* titled 'On rape, cages, and the Steubenville verdict' for a thought provoking discussion of the prison industrial complex and pursuit of justice in relation to the Steubenville case, http://www.blackgirldangerous.org/2013/03/20133171g5wckiks8gpa0iahe4zc46go4wawsu/.

8. This example also points to the fact that sexism is not the only cultural belief that allows violence to be excused and legitimized. Racism, ableism, homeophobia and many other prejudiced beliefs also allow violence to be conceived of as legitimate. Therefore, all of these beliefs must be interrogated in the same way that this paper interrogates the workings of rape culture. Rape culture should not be used to invisibilize other forms of violence, but as just one way of addressing the systematic sterotypes that allow violence to be legitimized. All cultural norms that allow the abuse and control of one person over another to be justified are unacceptable and must be undermined in the same way that rape culture attempts to undermine sexist, misogynistic and patriarchal beliefs.

9. Although it is outside of the scope of this paper, it is necessary to address the fact there has been a considerable outcry of concern in reaction to the Protecting Canadians from Online Crime Act. This outcry is due to concerns that the Conservative government is using rhetoric about cyberbullying as a guise to conceal the fact that this legislation includes provisions to allow government and police greater surveillance power over citizens (see Southey, 2013).

References

ABC News (2013) Steubenville: Alter the party's over. *20/20, ABC News,* 22 March. Available at: http://abcnews.go.com/2020/video/steubenville-partys-18795344 (accessed 15 June 2015).

Biber K (2014) The hooded bandit. In: *Captive Images: Race, Crime, Photography.* Oxon: Routledge, pp.1–26.

Bluett-Boyd N, Fileborn B, Quadara A. et al. (2013) The role of emerging communication technologies in experiences of sexual violence: A new legal frontier? Report, Australian Institute of Family Studies, Melbourne, February.

Buchwald E, Fletcher P and Roth M (2005) Preamble. In: Buchwald E, Fletcher P and Roth M (eds) *Transforming a Rape Culture* (revised edition). Minneapolis, MN: Milkweed Editions, pp.XI.

Butler J (2004} Endangered/endangering: Schematic racism and white paranoia. In: Salih Sand Butler J (eds) *The Judith Butler Reader.* Malden: Blackwell Publishing, pp.204–211.

Butler J (2007) Torture and the ethics of photography. *Environment and Planning D: Society and Space* 25: 951–966.

CBC News (2013) Rape, bullying led to N.S. teen's death, says mom. CBC News, 12 April. Available at: http://www.cbc.ca/news/canada/nova-scotia/rape-bullying-led-to-n-s-teen-s-death-says-mom-1.137 0780 (accessed 10 December 2013).

CBC News (2014) Audrie Pott case: 3 boys admit to assault, sharing photos. *CBC News,* 16 January. Available at: http://lwww.cbc.*cai*news/world/audrie-pott-case-3-boys-admit-to-assault-sharing-photos-1.2498612 (accessed 20 January 2014).

Day 6 (2013) Justice and social media, Rehtaeh and Steubenville. *Day 6, CBC Radio,* 12 April. Available at: http://www.cbc.ca/player/News/Canada/NS/Rehtaeh+Parsons/ID/2376487098/ (accessed 15 January 2014).

Fairbairn J (2014) Sexual violence & social media, At: *Carleton University's sexual assault awareness week,* Ottawa, ON, Canada, 13 -February 2014.

Fairbairn J (2015) Rape threats and revenge porn: Defining sexual violence in the digital age. In: Bailey J and Steeves V (eds) *eGirls, eCitizens.* Ottawa: University of Ottawa Press, pp.229–252.

Fairbairn J, Bivens R and Dawson M (2013) Sexual violence and social media: Building a framework for prevention. Report, OCTEVAW & Crime Prevention Ottawa, Ottawa, August.

Filipovic J (2008) Offensive feminism: The Conservative gender norms that perpetuate rape culture, and how feminists can fight back. In: Friedman J and Valenti J (eds) *Yes Means Yes!: Visions of Female Sexual Power and* a *World Without Rape.* Berkeley, CA: Seal Press.

Friedman J (2008) In defense of going wild or: How I stopped worrying and learned to love pleasure (and how you can, too). In: Friedman J and Valenti J (eds) *Yes Means Yes!: Visios of Female Sexual Power and a World Without Rape.* Berkeley, CA: Seal Press.

Friedman J and Valenti J (2008) Introduction. In: Friedman J and Valenti J (eds) *Yes Means Yes!: Visions of Female Sexual Power and a World Without Rape.* Berkeley CA: Seal Press.

Hall S (1997) Introduction. In: *Representation: Cultural Representations and Signifying Practices.* London, UK: Sage Publications, pp.1–12.

Kimmel M (2005) Men, masculinity, and the rape culture. In: Buchwald E, Fletcher P and Roth M (eds) *Transforming a Rape Culture: Revised Edition.* Minneapolis, MN: Milkweed Editions, pp. 140–157.

Macur J and Schweber N (2012) Rape case unfolds on web and splits city. *The New York Times,* 16 December. Available at: http://www.nytimes.com/2012/12/17/sports/high-school-football-rape-case-unfolds-online-and-divides-steubenvflle-ohio.html?pagewanted=all (accessed 15 June 2015),

Marcotte A (2014) RAINN denounces, doesn't understand the concept of 'rape culture'. *Slate,* 18 March. Available at: http://www.slate.com/blogs/xx_factor/2014/03/18/rainn_attacks_the_phrase_rape_culture_in_its_recommendations_to_the_white.html?wpisrc=burger_bar (accessed 25 March 2014).

Maritime Noon (2013a) Parallels between Rehtaeh Parsons' story and case in California. *Maritime Noon, CBC*

Radio, 12 April. Available at: http://www.cbc.ca/player/News/Canada/NS/Rehtaeh+Parsons/ID/2376364632/ (accessed 5 January 2014).

Maritime Noon (2013b) Rehtaeh Parsons' mother. *Maritime Noon, CBC Radio,* 9 April. Available at: http://www.cbc.ca/player/News/Canada/NS/ehtaeh+Parsons AD/2374375198/ (accessed 5 February 2014).

Parliament of Canada (2013) Bill C-13. Available at: http://www.parl.gc.ca/HousePublications/Publication.aspx?Language=E&Mode=1&DocldG311444&Col=1&File4 (accessed 15 March 2014).

Province of Nova Scotia (2014) What is bullying and cyberbullying? Available at: http://antibullying.novascotia.ca/ (accessed 20 January 2014).

Q (2013) What the Steubenville rape case coverage got wrong. Q, *CBC Radio,* 21 March, Available at: http://www.cbc.ca/radio/qlblog/2013/03/21/the-mediahas-come-under-fire/ (accessed 25 January 2014).

Salter M (2013) Justice and revenge in online counter-publics: Emerging responses to sexual violence in the age of social media. *Crime, Media, Culture* 9(3): 225–242.

Shah B (2012) Thoughts on rape culture. Available at: http://binashah.blogspot.ca/2012/10/thoughts-onrape-culture.html (accessed 10 July 2015).

Shariff S and DeMartini A (2015) Defining the legal lines: eGirls and intimate images. In: Bailey J and Steeves V (eds) eGirls, eCitizens. Ottawa: University of Ottawa Press, pp.281–306.

Smith M (2004) E*ncyclopedia of Rape.* Westport, CT: Greenwood Press.

SoutheyT (2013) Bill C-13 is about a lot more than cyberbullying. *The Globe and Mail,* 6 December. Available at: http://www.theglobeandmail.com/globe-debate/columnists/maybe-one-day-revenge-porn-will-be-have-no-power/artide15804000/ (accessed 8 March 2014).

The Chronicle Herald (2013) Rehtaeh Parsons case prompts scrutiny of police, prosecution procedures. *The Chronicle Herald,* 22 December. Available at: http://thechronicleherald.ca/metro/1176083-rehtaeh-parsons-case-prompts-scrutiny-of-police-prosecution-procedures (accessed 12 January 2014).

The Everyday Sexism Project (2014) The everyday sexism project. Available at: http://everydaysexism.com/ (accessed 27 January 2014).

Valenti J (2008) Purely rape: The myth of sexual purity and how it reinforces rape culture. In: Friedman J and Valenti J (eds) *Yes Means Yes! Visions of Female Sexual Power and a World Without Rape.* Berkeley, CA: Seal Press.

Xojane (2013) Goddard A. I am the blogger who allegedly 'complicated' the Steubenville gang rape case-and I wouldn't change a thing. Available at http://www.xojane.com/issues/steubenville-rape-verdict-alexandria-goddard (accessed 10 february 2014).

Young A (2014) From object to encounter: Aesthetic politics and visual criminology. *Theoretical Criminology* 18(2): 1 59–175.

Introduction to Reading 46

The authors of this reading analyzed interview data from the Atrocities Documentation Survey to examine the mechanisms by which gender-based violence was perpetrated against Darfuri men and boys during a protracted period of mass conflict in Sudan in 2003 and 2004. Their analysis contributes to a better understanding of the gendered nature of genocidal violence. As the authors emphasize, the violence against Darfuri men and boys enacted masculinity in accordance with hegemonic gender norms in Sudan.

1. What is the difference between primary victimization and proximate victimization? Why is it important to include proximate victimization in studies of gendered genocidal violence?

2. Define emasculation of the ethnic other. What are the four mechanisms identified by the authors as comprising emasculation in Darfur. How were those mechanisms shaped by gender and ethnic power dynamics?

3. How does this study challenge the binary of men as perpetrators and women as victims in the context of mass violence?

4. What is "gendered" about genocide? What is the gender–genocide nexus?

Gender-Based Violence Against Men and Boys in Darfur

The Gender–Genocide Nexus

Gabrielle Ferrales, Hollie Nyseth Brehm, and Suzy McElrath

Although scholarship on gender-based violence has emphasized violence committed by men against women, scholars have begun to consider how violence against men can also be gendered (e.g., Carpenter 2006; Jones 2006; Sivakurnaran 2007 Zarkov 2001). Following this line .of inquiry, we identity mechanisms through which gender-based violence in Darfur emasculates men and boys.[1] In doing so, we illustrate that rape not only occurred in Darfur but that it was one form of gender-based violence perpetrated against men. We draw upon narratives from 1,136 Darfuri refugees to analyze patterns of gender-based violence against men and boys and demonstrate how genocidal violence is gendered. In line with an interactionist approach (Connell and Messerschmidt 2005; Jakobsen 2014; Ridgeway 2009; West and Zimmerman 1987, 2009), we argue that this gender-based violence reflects a hegemonic ideal of the Sudanese man and communicates an emasculating message to individual victims and targeted social groups. Perpetrators perform masculinity through violence—"doing gender" (West and Zimmerman 1987) by reaffirming their own hegemonic dominance while simultaneously proclaiming power over ethnic victim groups. In essence, gender-based violence enacts, reinforces, and creates meaning on multiple levels to assert a dominant social order—a process we term the *gender–genocide nexus.*

Gender-Based Violence During Mass Atrocity

While gender-based violence during mass conflict has occurred for centuries, it has only recently garnered scholarly attention. This scholarship frequently employs the terms *rape, sexual violence,* and *gender-based violence* synonymously or narrowly focuses on rape, typically defined as penetration of the body (e.g., Rome Statute 2002 [1998]). Yet, rape is but one form of gender-based violence perpetrated during mass conflict (Carpenter 2006). Gender-based violence constitutes an extensive range of physical and psychological actions, including acts of penetration, sexual assault, genital mutilation, forced pregnancy, culturally inappropriate actions that sexually harass and humiliate, as well as nonsexual acts perpetrated on the basis of gender, such as sex-selective killing. Though scholars have debated how gender facilitates and patterns violence, Jakobsen (2014) notes that what is "gendered" about gender-based violence in any context remains woefully undertheorized. In her own work, she argues that gender is salient in domestic violence in Tanzania, where tile "good beating" of wives is prescribed in the performance of hegemonic masculinity and femininity. This interactionist approach illuminates how violence may be "*based on* gender, while at another level violence may *affect* gender," cycling from micro to macro-institutional levels in a matrix of mutually reinforcing processes (543). Extending this analysis to mass atrocities permits a related question: What is gendered about genocidc?[2]

Gender and Violence in Darfur

Darfur was once a sultanate encompassing forty tribes (O'Fahey 1980). Tribes often self-identified as African or Arab, with African tribes[3]—such as the Fur, Masaleit, and Zaghawa—and Arab tribes—such as the Rizeigat and the Beni Halba—coexisting and intermarrying (Flint and de Waal 2005). Most residents practiced Islam, and many people spoke Arabic and traditional languages. The Fur Sultanate became part of Sudan in 1916, and subsequent decades were spent under Anglo-Egyptian rule. Political instability and periods of widespread drought followed Sudanese independence in 1956 (Collins 2008). The new state struggled to meet these challenges while simultaneously engaging in nation-building efforts (Doornbos 1988; Straus 2015). Nation-building continued into the

Ferrales, G., Brehm, H. N., & Mcelrath, S. (2016). Gender-based violence against men and boys in Darfur: The gender–genocide nexus. *Gender & Society, 30*(4): 565–589.

1990s, as President Omar al-Bashir—who took control of the country through a coup in 1989 and remains in power today—implemented policies of Arabization and Islamization. These ideologies and related practices privileged individuals viewed as Arab (Doornbos 1988; Flint and de Waal 2005; Fluehr-Lobban 1990) and often marginalized those viewed as African, including many residents of Darfur.

The process of constructing a national identity directly engages the construction of gender (Charrad 2001; Kandiyoti 1991; Kim, Puri, and KimPuri 2005; Yuval-Davis and Anthias 1989), and Sudan is no exception (Hale 1996; Nageeb 2004; Tonnessen 2007). There, the gendered project of shaping a national identity became closely tied to an "Islamist moral discourse" (Willemse 2007a, 437) and Sudanese identity became aligned with notions of ideal Muslim women and men (Hale 1996). For instance, Sudanese women were portrayed as carriers of Sudanese culture and morality, and mothers were to be engaged in the home. Heteronormative gender and sexual identities are consequently highly regulated in Sudan (Willemse 2007a). Related policies, such as those restricting women's labor opportunities (Hale 1996), reinforced an image of the ideal Muslim man as the financial provider and ultimate guardian of the family (Al-Ahmadi 2003; Willemse 2007a, 2007b). In line with this, homosexuality is seen as an inferior identity and a crime punishable by death upon the third offense (Government of Sudan 1991; S. Martin 2007; Onyango and Hampanda 2011; UN OCHA 2008). Vigilantes have targeted suspected gay men, and state actors have publicly flogged men for wearing women's clothes and makeup (Hartenstein 2010; U.S. State Department 2013).

These norms are prevalent throughout the country, including in Darfur, where men are positioned as protectors who should not flee from peril or stand helpless in moments of danger (Mohamed 2004; Mora 1998; Oladosu 2009). Yet, many prescribed ideals are incompatible with the realities of socioeconomic conditions in Darfur, where a dearth of economic opportunities stems from the state's sustained neglect of the region and a national economic downturn that began in the 1970s (Willemse 2007b, 2009). Among the Fur, a dominant tribe in Darfur, the economic situation led men to migrate domestically or internationally, and thereby kept many from marrying or providing for and protecting their family. State actors constructed these behaviors as familial desertions which, according to Willemse (2007b, 2009), spurred a widespread crisis in masculinity and contributed to subsequent violence.

Widespread neglect of the Darfur region is also widely cited as a factor that contributed to conflict in the region. Darfuri leaders began voicing discontent about decades of systematic marginalization during the 1990s. Sporadic violence culminated in Darfuri rebel attacks on Sudanese military barracks in early 2003 (Tanner and Tubiana 2007). In response, the government of Sudan unleashed an unprecedented campaign of terror on Darfur's civilians, and Sudanese soldiers and government-trained militias known as the Janjaweed began attacking villages (Flint 2009). These attacks have targeted Fur, Zaghawa, Masaleit, and other "African" civilians. For instance, predominantly "African" villages in Darfur have been obliterated, while neighboring "Arab" villages have been left intact. Racial epithets accompanying attacks—such as "this is the last day for blacks" or "we will kill all the black-skinned people"—support these assertions (Hagan, Rymond-Richmond, and Parker 2005). Since the violence seeks to destroy certain groups, it constitutes genocide by both legal and scholarly definitions (see Daly 2010; Kiernan 2007; Luban 2006; Straus 2015; Totten and Markusen 2006). And while scholars have documented much of this genocidal violence, its gendered nature has garnered less attention (though see Kaiser and Hagan 2014).

Methodology

We thus analyze data from the Atrocities Documentation Survey (ADS) to examine gender-based violence against men and boys in Darfur. This project, commissioned by the U.S. State Department, documented violence in Darfur in 2003 and 2004 by interviewing 1,136 Darfuri refugees in Eastern Chad. Chad was the ideal location for the interviews because of its large refugee population (U.S. Department of State 2004), and the ADS team conducted interviews in ten United Nations High Commissioner for Refugee (UNHCR)—run camps and nine informal refugee settlements (Howard 2006).

Three teams of researchers conducted interviews using a multistage, systematic random sampling design to obtain a sample mirroring camp ethnic compositions. Interviews took place in private, with the respondent, interviewer, and a translator present. Questions were primarily open-ended, allowing for detailed narratives in order to document victimization (U.S. Department of State 2004). Each interview is assigned an identification number, which we use to identify respondents below. Respondents, all over the age of 18,[4] self-identified predominantly as Zaghawa (46 percent) or Masaleit (30 percent). Fewer self-identified as Fur (8 percent) or members of other "African" groups (16 percent), and slightly more than half were women (56 percent). To be clear, this is not a random sample of all refugees or civilian victims of the violence, as most respondents fled from within 50 miles of the Sudan–Chad border.

As with any interview data regarding mass atrocity, ours also reflect survivorship bias, as only those who survived were able to share their stories. Furthermore, gender-based violence is likely underreported due to stigma (Abdullah-Khan 2008; Carlson 2008; Javaid 2014; Mullaney 2007) and the criminalization of some forms of violence, such as homosexuality. One man, for example, described being detained with approximately 80 other men. The ADS interviewer suspected from the man's body language and nervousness that he had suffered sexual abuse while detained, though the interviewee denied physical harm. After the interview, however, the man disclosed that "he suffered a 'man beating'" but could not talk about it "because it was too humiliating" (7). Men also may not report sexual violence because of their own nonrecognition as victims (Weiss 2010). In fact, the Sudanese Criminal Code excludes sexual violence against men—as-well as anal penetration and the insertion of objects—from the definition of rape (Government of Sudan 1991).

To capture forms of gender-based violence, we inductively constructed a qualitative coding scheme. Our final 70 codes included violent acts, including sex-selective killing, rape (oral or anal penetration with body parts or objects), sexual assault (sexual contact without penetration), and violence targeting the body. Codes also captured situational characteristics, such as the presence of witnesses, the number of perpetrators, and location. Following the interview guide, we included violence against the respondent as well as violence the respondent witnessed and/or heard about. The majority of responses indicated direct victimization and witnessed violence. Notably, the inclusion of hearsay and witness statements eases (though does not eliminate) some concerns about underreporting, as respondents were often willing to share what happened to *others* despite likely reluctance to disclose personal victimization. This allows us to gain a more complete picture of the violence, including accounts of violence against those who did not survive.

Emasculation in Darfur

The ADS data illustrate that many refugees in Chad experienced both primary and proximate gender-based violence. *Primary* victimization is perpetrated directly against an individual and includes actions like rape, genital harm, or sex-selective killings. While it is difficult to quantify these acts given the inclusion of witness/hearsay, the high number of deaths, and likely under reporting, it is clear that primary victimization of men was neither uncommon nor localized. We documented approximately 30 instances of rape and more than 40 instances of genital harm, while the vast majority of the sample reported sex-selective killing. Notably, these forms of violence also have been reported in other studies (Gingerich and Leaning 2004; S. Martin 2007; Onyango and Hampanda 2011). *Proximate* victimization, or witnessing violence perpetrated against others, often accompanied primary victimization. The vast majority of respondents witnessed violence, frequently against family members. Proximate violence is rarely prosecuted and sometimes not even considered violence, yet it can also be gendered and is thus key to a more expansive conceptualization of gender-based violence (Carpenter 2006; AT. Goldstein 1993).

Taken together, the totality of the violence perpetrated against men and boys forms the basis of our analysis. Both primary and proximate victimization in Darfur emasculated the ethnic other—in this case, the targeted "African" tribes. Deriving from *emasculare,* Latin for "diminutive male" (Taylor 2000), emasculation refers to any practice that "diminishes the potency of men in the family or society more generally" (Ross 2002, 311). Following recent scholarship (Fang 2004, 6), we employ three interrelated meanings of emasculation: to castrate, to deprive of strength or vigor, or to possess unsuitable feminine qualities.

Drawing on Sivakumaran (2007), we find that emasculation in Darfur occurs through homosexualization, feminization, and genital harm.[4] Additionally, we identify a fourth mechanism—sex-selective killing. These mechanisms are neither mutually exclusive nor exhaustive but rather are *complementary.* They simultaneously influence and are influenced by both gender and ethnic power dynamics. To be clear, we are not able to assess the motives of individual perpetrators, and thus we do not know whether each person perpetrating the violence intended their actions to be emasculating. Likewise, we do not know if each man who was victimized felt emasculated. Nevertheless, we contend that the sum of individual actions emasculates the *social group.* These acts communicate the impotence of the targeted group and divest group members of their power dominance, and collective masculine identities, as we illustrate below.

Homosexualization

> Four men were raped in the village. . . . These men were then shot and killed. . . . After they killed the men, they raped them anally with sticks (287).

This 21-year-old Masaleit woman recounted a key method of emasculation: rape. Rape preceded and followed murder, and groups of soldiers and Janjaweed typically used penile penetration or objects, such as sticks and gun barrels, against groups of Darfuri men. For example, a West Darfuri man witnessed government soldiers and Janjaweed rape five men with sticks. He explained, "They tied up arms and legs and [threw] them to the ground and raped them. All burned to death in fire" (488). Much research establishes that rape is an act of power, dominance, and an assertion of strength and manhood (Weiss 2010). Rape can also be homosexualizing, which likewise can be emasculating. Similar to U.S. constructs of masculinity (Messner 2003), the Sudanese heterosexual man has long been conceptnalized as dominant over women, homosexual men, and others (Willemse 2007a). The homosexual man is considered weak and less masculine—a status that carries potentially lethal consequences (e.g., Jones 2006; J. Goldstein 2001; Seifert 1994; Sivakumaran 2007)—as reflected in the Sudanese Criminal Code. Given this context, rape functions as an actual and symbolic means of masculinized dominance, or "doing difference" (West and Fenstermaker 1995), between competing ethnic or national groups (Vojdik 2014). As manhood is intimately tied to ethnic identity (Zarkov 2001), perpetrators do not just rape men—they rape *ethnic* men. In this sense, the Darfuri victim not only becomes "a lesser man but . . . his ethnicity is lesser" (Zarkov 2001, 78). In tum, these acts demarcate group difference, malign the ethnic outgroup, and publicly communicate hegemony, power, and control over the collective. Homosexualization thus communicatess dominance and demarcates group difference.

Meanwhile, the individual rapist and his group are empowered. As Price (2001, 212) noted regarding militarized expressions of violence in Bosnia, "I AM only to the extent that you are not. . . . Your absence marks, verifies my presence and your pain becomes my power." To be clear, the perpetrator likely does not identity as homosexual[5] but rather uses penetration to impose and elevate his dominant heterosexual status (Given 2010; Lewis 2009; Segal 1990; Zarkov 2001). Group participation in a mutually shared crime bolsters the individual attacker's masculinity while simultaneously strengthening communal solidarity among the group (Alison 2007; Cohen 2013a, 2013b; Given 2010), sealing "allegiance in atrocity" (Morrow 1993, 48).[6] The penetration of men with objects postmortem also desecrates the ethnic male body and his family by violating sacred spiritual norms. In Darfur, cultural norms proscribe a highly ritualized treatment of the deceased: the body is washed, wrapped in white cloth, and buried by family members without delay (Totten 2011). Rape post-mortem thus violates the deceased and his family, who are unable to complete these culturally prescribed practices (Komar 2008).

Overall, homosexualization by rape varied by age and status, and elderly Darfuri leaders appear to have been particularly targeted for rape and other forms of violence. For example, a Masaleit woman stated, "Near us, they raped 10 old men using sticks and barrels of the guns" (336). She also noted, "The imam was raped and then taken to the police station." This represents the emasculation of one of the community's most honored members. By contrast, there were no instances in our data of rape of boys under the age of 15, which may suggest that they are not yet viewed as hegemonic adult men and sexual beings (see Jones 2006).

Finally, while we have focused on rape, other acts of violence also homosexualized Darfuti men. For instance, perpetrators often excised victims' penises and inserted them into the victims' mouths. A Fur woman reported, "I saw a young boy and his father dismembered while still alive. They cut off their penises and put them in their mouths" (615). Another woman recalled how she observed the torture of seven men who were dismembered alive. Perpetrators pulled their teeth out, cut off their tongues, severed their penises, and then put them in their mouths (620). Here, genital harm emasculated the victim directly, via castration, and indirectly, via homosexualization.

Feminization

> I have four wives . . . two were raped by the Janjaweed. . . . I saw this start but then had to run and hide (786).

This 50-year-old Zaghawa man's account illuminates how feminization was also used as a related form of emasculation whereby targeted men were demeaned or devalorized on the basis of sex and gender associations. By forcibly imposing attributes and behaviors culturally associated with women, feminization constructs and maintains hierarchies of masculinities (Hooper 2001; MacKinnon 1997; Sjoberg 2015). Hegemonic definitions of masculinity require men to be strong, self-sufficient, impenetrable, dominant, and in control (Kinnnel and Messner 1989). As noted above, the inability to protect oneself, one's family, and others represents a salient transgression against masculinity in Darfur, lowering the status of the individual and the enemy group by inverting gendered constructions of the protector and the protected.[7] In essence, the feminized man (and his group) is "unmanned" and rendered weak and defenseless, contravening markers of manhood.

While feminization may be inflicted through numerous forms of violence, including homosexualization and genital harm (discussed below), we focus on demonstrative violence. This involves publicly displaying bodies to instill fear, communicate threat, and serve as an emblem of group conquest and emasculation. Instances of public victimization were replete in the data. A Fur woman recounted witnessing a man beaten, whipped, and conspicuously hung from a tree (23). Another woman shared how she witnessed two men publicly beaten for an hour with sticks (43). Others described how boys were dragged "behind running horses until they were dead" (563) or behind trucks and paraded through the village (559), illustrating their inability to defend themselves. Indeed, some suggest that defeat in mass conflict can be feminizing, as hegemonic masculinity is associated with victory (Jones 2006).

Men were also prominently shot, slaughtered, and had their throats cut, which respondents described as "extra humiliation" (7; 625; 637). Family members were also forced to watch. A newlywed woman recounted how perpetrators made her watch them kill her husband, take his clothes, and cut his body open (615). Lasting marks from violence were also significant. A Fur man. described how his brother was tortured and beaten, noting he "had so many marks on his body he looked like he was branded" (19). The scars become a perennial symbol of his emasculation and the inability of the targeted group to defend themselves. As respondents explained, these marks humiliated men, symbolically castrating the victim and anathematizing the enemy. Many men likewise reported witnessing their wives, women relatives, and others being raped and sexually assaulted while they were rendered powerless to stop it, which directly attacks the Darfuri man's duty as protector (Mohamed 2004; Morn 1998). One man recounted, "I saw ladies in the village [as I lay wounded] being raped right in front of everyone, even their fathers and their children. . . . We could do nothing, nothing. We had no way to fight" (258). A 36-year-old Zaghawa man similarly recounted, "I ran away because I couldn't stand to see the women hurt in [the] family. . . . The men gathered in [the] yard to try to defend [them]. The soldiers shot them. The men had nothing to protect the village" (629). Another noted how his wife was beaten and whipped until she had "slashes all over her body" (552)—leaving a visible symbol of her husband's impotence. Perpetrators also verbally mocked men while victimizing women. A 30 year "old Masaleit man noted that when he saw four girls being kidnapped, the soldiers shouted, "Come get your girls if you can . . ." (257). Respondents likewise recalled perpetrators laughing while raping women, which likely served to taunt the men present and further underscore their inability to protect their wives, family, and property—in effect leaving them powerless. This demonstrative violence targets one of the basic attributes of hegemonic Darfuri masculinity—his ability to protect his family and community—and simultaneously forcibly imposes on men gender associations culturally associated with women.

Genital Harm

> For seven days, I was detained and tortured by government soldiers. I was made to lie on my back with my hands tied behind my back, ankles tied and they would stomp on my thighs and kick me in the genitals ([and I have had] sexual problems ever since) (5).

This excerpt from a 52 year-old Fur man detained with more than 30 other Darfuri men reveals how genital harm was yet another method of emasculation. Genital harm often prevents procreation, which, when targeting a group, is a recognized crime of genocide (Genocide Convention 1948). Furthermore, because genital harm diminishes hegemonic masculinity—which is equated with virility—this form of sexual victimization emasculates symbolically as well as physically. Genital targeting was neither exceptional nor localized, and it frequently preceded death. For instance, a Masaleit woman described how five men bled to death following castration (259). Another woman recounted speaking with a man who had his "genitals cut off" (261). Others reported how sexual organs were severed during dismemberment

Although there were several reports of injury to the testicles, genital harm often involved the pronounced targeting of the penis, signifying the elimination of a source of power. This form of emasculation has occurred throughout history within societies that construct male bodies as dominant; for example, Persian armies often displayed plates of conquered soldiers' penises when celebrating the defeat of the enemy (Vojdik 2014). In these cases, a victim's body represents the "corporeal embodiment" of the enemy, and the excise of the male organ functions as an extreme form of emasculation and symbolically represents the emasculation of an entire group (Sivakumaran 2007; Zarkov 2001). Likewise, as previously noted, penises were also forced into the mouths of victims during or after death, symbolically silencing the enemy group. These expressions of violence emphasize the association between men's sexual organs and hegemonic masculinity. More broadly, this and other forms of genital harm constitute a salient form of emasculation that targets physical and cultural virility at both the individual and collective level.

Sex-Selective Killing

> I saw five pregnant women have their stomachs ripped open. The soldiers removed the fetuses. If they were male, they destroyed the fetus by smashing it on the ground (615).

This excerpt from a 40-year old Fur woman illustrates how male fetuses were targeted for sex-selective killing.[8] Similarly, a Masaleit woman described how she witnessed attackers check the sex of infants. If they were male infants, they would take the baby by his feet and "slam it against ground until it dies" (489). In line with other sources (Jones 2004), numerous respondents reported that men were also targeted for death. A Fur woman described how soldiers screamed while raping her, "I am Omar al-Bashir and I have orders to take everything, kill the men, and capture the women" (624). Another noted, "We have killed all your men—now we come for the women and the cattle" (250). These and many other statements indicate that sex-selective killing was widespread, with interviewees reporting finding numerous male corpses (e,g,, 7; 19; 24), As noted above, men were also mutilated and butchered "like animals" (625). A Zaghawa woman described:

> They shot him in the body twice but he did not fall. Several Arab militias grabbed him, held him down, first they cut off one arm, then the other. . . . He fell unconscious as they cut off one leg. He was dead by the time they cut off the second leg (256).

Another woman remembered a man whose "arms [and] legs had been cut off with a knife or machete, and chest cut open with heart pulled out" (503). Witnesses also described how perpetrators "burned people to make them cry" (136). These sex-selective patterns, rooted in assumptions about the duties of men during violence, constitute gender-based violence (Carpenter 2006). As one respondent noted, "I also saw the bodies of about 25 young boys—it seemed they were targeting the men and boys because I heard them say 'a puppy can become a dog'" (24), reflecting gendered associations between masculinity, aggression, and violence. The mass annihilation of men from enemy groups appropriates the power of men to provide, protect, and defend. In Darfur, targeted men and boys were literally and figuratively "unmanned" (Zarkov 2001). Moreover, as men are conceived as the bearers of ethnicity in Sudan (Daly 2010; de Waal2009), targeting men and boys destroys existing and potential group members. It also symbolically communicates the group's collective incapacity in the context of mass atrocity.

Conceptualizing the Gender–Genocide Nexus

Analyzing primary and proximate forms of gender-based violence in Darfur, we have identified four mechanisms of emasculation of Darfuri men: homosexualization, feminization, genital harm, and sex-selective killing. By identifying how gendered power relations operate in the context of mass violence, this article extends interactionist scholarship on the links between violence and social constructions of gender (Alison 2007; Vojdik 2014). It also challenges the binary of men as perpetrators and women as victims[9] and begins to answer the question: What is gendered about genocide? Specifically, we conceptualize how dominant norms regarding gender influence forms of mass violence, suggesting that gender-based violence establishes, enforces, and reproduces gendered hierarchies within a broader social system where both body and gender become "highly salient organizing principles of interaction" (Messerschmidt 2002, 209). Ethnicity, age, sexuality, and other identities also pattern violence, illustrating the importance of intersectionality (Crenshaw 1991). Uncovering the relationship between gender and mass violence necessitates attention to how multiple levels interconnect, a process that we term the *gender–genocide nexus.* Specifically, the gendered patterning of violence can be traced, in part, to patriarchal and heteronormative state-supported ideologies and gender constructs that position men as familial protectors and guardians. In Darfur, a state-led ideology also targeted particular groups, such that ethnicity and other social attributes interacted with gender to pattern violence.

Primary and proximate forms of gender-based mass violence produce difference between groups along gender constructs that link heteronormativity, power, and ethnicity with the collective goal of eradicating the enemy group, Darfuri men were systematically denied the attributes of dominant heterosexual masculinity and demarcated as outgroup members through at least four mechanisms of emasculation. In this way, perpetration of violence constitutes a form of doing gender, where gender is salient in the manifestations of violence and the resulting subordination and attempted destruction of the targeted enemy group (Jakobsen 2014; P. Y. Martin 2003). Violence in this context is *based* on gender and simultaneously *affects* gender in a mutually reinforcing process (Jakobsen 2014, 543; see Anderson 2005) whereby a newly emerging social order excluding outgroup members may reinforce or exacerbate state ideologies and the persistence of patriarchal heteronormative ideals (hence the double arrow). An interactionist perspective elucidates how micro- and macro-level victimization occurs through mutual reinforcement (Ridgeway 2009, Ridgeway and Correll 2004). Violence can operate on multiple levels, as individual perpetrators and victims and collective ethnic groups assume divergent but interconnected roles of emasculators and emasculated. Crucially, violence perpetrated against an individual during mass conflict can be emblematic of victimization against the community. In this sense, repertoires of collective violence function to materially and symbolically demarcate group boundaries (Tilly 2003; Wood 2009).

We contend that these processes occur regardless of the intent behind individual actions. Conceptualizing gender as situated action means recognizing variation in agency, yet also recognizing we cannot fully capture differences in individual perpetrator's motives (Miller 2002) that are not reflected in the ADS data. Likewise, our data do not capture whether each victim felt emasculated. Yet, even if an individual victim is consciously unaware of feelings of emasculation (or each perpetrator feels emboldened through his actions), it is nonetheless clear that *doing gender* via repertoires of collective violence reproduces social structure, consolidates power, and weakens collectives along gender and ethnic lines (Miller 2002). The sum of individual social actions nested within the structural context reproduces inequality

While we analyze this process during a specific episode of genocide, this general model may inform the study of other mass violence. Explicating variation in the extent and form of gender-based violence across conflicts requires comparative analysis (Wood 2006), but we can nevertheless suggest several neither mutually exclusive nor exhaustive propositions regarding the factors that may influence comparatively more gender-based violence during conflict. First, gender-based violence and processes of emasculation may be more prevalent when states and other powerful actors sponsor a systematic campaign that explicitly targets sex. Second, cultural norms nested within social systems that privilege hegemonic masculinity make gendered forms of violence and processes of emasculation more likely (Enloe 2000; Vojdik 2014), just as norms may prohibit such violence in other contexts (Wood 2006). Third, conditions that exacerbate threats to masculinity, which can emerge from severe imbalances between hegemonic expectations and the opportunities to achieve them (Willemse 2009), may lead to violence as a mechanism for achieving masculinity (Kimmel and Messner 1989; Kimmel and Mahler 2003; Schrock and Padavic 2007). Lastly, gender-based violence against men may be more likely when manhood and ethnicity are intertwined. Overall, gender-based violence against men is not aberrant or confined to mass conflict but is prevalent in social systems that construct men as heterosexual and dominant (Vojdik 2014). Gendered identities—typically masculine identities that emphasize strength, aggression, and courage—are privileged in settings ranging from the United States (e.g., Copes and Hochstetler 2003) to Hong Kong (e.g., Kong 2009) to Sudan (e.g., Willemse 2009). Given this burgeoning line of literature linking myriad forms of crime with hegemonic masculinity, the patterns and processes that we identify may also inform violence in times of relative peace.

Notes

1. Although our analysis includes both men and boys, we reference men for brevity. We note when patterns of victimization differ by age.

2. Genocide is defined in international law as "the intent to destroy, in whole or in part, a national, ethnical, racial, or religious group" (Genocide Convention 1948). Mass atrocity includes genocide along with other forms of violence (e.g., war).

3. While race and ethnicity are social constructs, race is typically constructed by outsiders, while ethnic identify is often self-defined (Cornell and Hartmann 2004). Darfuris self-identify is often self-defined (Cornell and Hartmann 2004). Darfuris self-identified with tribes, though colonial authorities also determined the race of race of each tribe, categorizing some as Black and others as Arab (Mamdani 2010, 150). We conceptualize Darfuri tribes as racialized ethnicities but refer to ethnicity in line with self-identification during interviews. We use the word tribe because Darfuris use this word (see also Mamdani 2010).

4. We use "genital harm" instead of "prevention of procreation" because we identify other types of genital harm

that do not prevent procreation but nonetheless can still influence emasculation.

5. The ADS data contain no evidence of women perpetrators.

6. Previous work suggests that the proportion of rapes carried out by multiple perpetrators during mass violence is significantly higher than during peacetime (Da Silva, Harkins, and Woodharns 2013; Wood 2013).

7. Emasculation is not only associated with victimization in mass conflict but can occur anywhere hegemonic masculinity is pervasive.

8. Though female fetuses were killed, respondents noted that perpetrators targeted male fetuses.

9. Designations of perpetrator/victim are fluid (Fujii 2009).

References

Abdullah-Khan, Noreen. 2008. *Male rape: The emergence of a social and legal issue.* New York: Palgrave Macmillan.

Al-Ahmadi, Hala Abdel Magid. 2003. *Globalizations, Islamism and gender: Women's political organisations* in *the Sudan.* PhD diss., Radboud University Nijmegen, Nijmegen.

Alison, Miranda. 2007. Wartime sexual violence: Women's human rights and questions of masculinity. *Review of International Studies* 33:75–90.

Anderson, Kristin L. 2005. Theorizing gender in intimate partner violence research. *Sex Roles* 52:853–65.

Carlson, Melanie. 2008. I'd rather go along and be considered a man: Masculinity and bystander intervention, *Journal of Men's Studies* 16 (1):3–17.

Carpenter, R. Charli. 2006. Recognizing gender-based violence against civilian men and boys in conflict situations. *Security Dialogue* 37 (1):83–103.

Charrad, Mounira. 2001. *States and women's rights: The making of postcolonial Tunisia, Algeria, and Morocco.* Berkeley: University of California Press.

Cohen, Dara K. 2013a. Explaining rape during civil war: Cross-national evidence (1980–2009). *American Science Review* 107 (3):461–77.

Cohen, Dara K. 2013b. Female combatants and the perpetration of violence: Wartime rape in the Sierra Leone civil war. *World Politics* 65:383–415.

Collins, Robett O. 2008. *A history of modern Sudan.* Cambridge, UK: Cambridge University Press.

Connell, R., and James W. Messerschmidt. 2005. Hegemonic masculinity: Rethinking the concept. *Gender & Society* 19:829–59.

Convention on the Prevention and Punishment of the Crime of Genocide ("Genocide convention"), *78 U.N. Treaty Series (UNTS) 277,* adopted by the General Assembly, Dec. 8, 1948, entered into force, Jan. 12, 1951.

Copes, Reith, and Andy Hochstetler. 2003. Situational constructions of masculinity among male street thieves. *Journal of Contemporary Ethnography* 32:279–304.

Cornell, Stephen Ellicott, and Douglas Hartmatm. 2004. Conceptual confusions and divides: Race, ethnicity, and the study of immigration. In *Not just black and white,* edited by Nancy Foner and George M. Fredrickson. New York: Russell Sage Foundation.

Crenshaw, Kimberlé. 1991. Mapping the margins: Intersectionality, identity, politics, and violence against women of color. *Stanford Law Review* 43:1241–99.

Daly, Mattin W. 2010. *Darfur's sorrow: A history of destruction and genocide.* Cambridge, UK: Cambridge University Press.

Da Silva, Teresa, Leigh Harkins, and Jessica Woodhams. 2013. Multiple perpetrator of rape: Au international phenomenon. In *Handbook on the study of multiple perpetrator: A multidisciplinary response to an international problem,* edited by Miranda A. H. Horvalth and Jessica Woodhams. New York: Routledge.

de Waal, Alex. 2009. Who are the Darfuriatns? Arab and African identities, violence, and external engagement. In *Darfur and the crisis of governance* in *Sudan: A critical reader,* edited by Salah M. Hassan and Carina E. Ray. Ithaca, NY: Cornell University Press.

Doornbos, Paul. 1988. On becoming Sudanese. In *Sudan: State, capital, and transformation,* edited by Tony Barnett and Abbas Abdelkarims. New York: Croom Helm

Enloe, Cynthia. 2000. *Maneuvers: The international politics of militarizing women's lives.* Berkeley: University of California Press.

Fang, Jincai. 2004. *The crisis of emasculation and the restoration of patriarchy in the fiction of Chinese contemporary male writers Zhang Xianliang and Jia Pingwa.* PhD diss., University of British Columbia, Vancouver, BC.

Flint, Julie. 2009. *Beyond Janjaweed: Understanding the militias of Darfur.* Geneva, Switzerland: Small Arms Survey.

Flint, Julie, and Alex de Waal. 2005. *Darfur: A short history of a long war.* New York: Zed Books.

Fluehr-Lobban, Carolyn. 1990. Islamization in Sudan: A critical assessment. *Middle East Journal* 44 (4):610–23.

Fujii, Lee Ann. 2009. *Killing neighbors: Webs of violence in Rwanda. Ithaca*, NY: Cornell University Press.

Gingerich, Tara, and Jennifer Leaning. 2004. *The use of rape as a weapon of war in the conflict in Darfur.* Boston, MA: Program on Humanitarian Crises and Human Rights, Francoise-Xavier Bagnoud Center for Health and Human Rights, Harvard School of Public Health.

Goldstein, Anne Tierney. 1993. *Recognizing forced impregnation as a war crime.* New York: Center for Reproductive Law and Policy.

Goldstein, Joshua. 2001. *War and gender: How gender shapes the war system and vice versa.* Cambridge, UK: Cambridge University Press.

Government of Sudan. 1991. Criminal Code of 1991. Article 148.

Hagan, John, Wenona Rymond-Richmond, and Patricia Parker. 2005. The criminology of genocide: The rape and death of Darfur. *Criminology* 43:525–61.

Hale, Sondra. 1996. *Gender politics in Sudan: Islamism, socialism, and the state.*
Boulder, CO: Westview.
Hartenstein, Meena. 2010. 19 Sudanese men flogged for being caught crossdressing, dancing in women's clothes and make-up. *Daily News,* 4 August.
Hooper, Charlotte. 2001. *Manly states: Masculinity, international relations, and gender politics.* New York: Columbia University Press.
Howard, Jonathan, 2006, Survey methodology and the Darfur genocide. In *Genocide in Darfur: Investigating the atrocities in the Sudan*, edited by Samuel Totten and Eric Markusen. New York: Routledge.
Jakobsen, Hilde. 2014. What is gendered about gender-based violence? An empirically grounded theoretical exploration in Tanzania. *Gender & Society* 28:537–61.
Javaid, Aliraza, 2014. Feminism, masculinity, and male rape: Bring male rape "out of the closet." *Journal of Gender Studies* 12 (6):1–11.
Jones, Adam. 2004. Case study: Military conscription/impressment. *Gendercide Watch.* http://www.gendercide.org/case_conscription.html.
Jones, Adam. 2006. Straight as a rule: Heteronormativity, gendercide, and the noncombatant male. *Men & Masculinities* 8:451–69.
Kaiser, Joshua, and John Hagan. 2015. Gendered genocide: The socially destructive process of genocidal rape, murder, and forced displacement in Darfur. *Law & Society Review* 49 (1):69–107.
Kandiyoti, Deniz. 1991. *Women, Islam and the state: Women in the political economy.* Philadelphia, PA: Temple University Press.
Kiernan, Ben. 2007. *Blood and soil: A world history of genocide and extermination from Sparta to Darfur.* New Haven, CT: Yale University Press.
Kim, Hyun Sook, Jyoti Puri, and H. J. Kim-Puri. 2005. Conceptualizing gendersexuality-state-nation: An introduction. *Gender & Society* 19 (2):137–59.
Kimmel, Michael S., and Michael A. Messner, eds. 1989. *Men's lives.* New York: Macmillan.
Kimmel, Michael S., and Matthew Mahler. 2003. Adolescent masculinity, homophobia, and violence: random school shootings, 1982–2001. *American Behavioral Scientist* 46(10):1439–1458.
Komar, Debra. 2008. Patterns of mortuary practice associated with genocide: Implications for archaeological research. *Current Anthropology* 49 (1):123–33.
Kong, Travis S. K. 2009. More than a sex machine: Accomplishing masculinity among Chinese male sex workers in the Hong Kong sex industry. *Deviant Behavior* 20:715–45.
Lewis, Dustin. 2009. Unrecognized victims: Sexual violence against men in conflict settings under international law. *Wisconsin International Law Journal* 27:1–50.
Luban, David J. 2006. Calling genocide by its rightful name: Lemkin's word, Darfur, and the UN report *Chicago Journal of lnternational Law* 7 (1):303–20.
MacKinnon, Catharine. 1997. Oncale v. Sundowner Offshore Services, Inc., 96–568, Amici Curiae Brief in Support of Petitioner, 8. *UCLA Women's Law Journal* 9:18–19.
Mamdani, Mahmood. 2010. *Saviors and survivors: Darfur, politics, and the war on terror.* New York: Random House.
Martin, Patricia Yancey. 2003. "Said and done" versus "saying and doing." *Gender & Society* 17:342–66.
Martin, Sarah. 2007. *Ending sexual violence in Darfur: An advocacy agenda.* Washington, DC: Refugees International.
Messerschmidt, James W. 2002. Men, masculinities, and crime. In *Handbook on men and masculinities,* edited by Michael Kimmel, R. W. Connell, and Jeff Hearn. Thousand Oaks, CA: Sage.
Miller, Jody. 2002. The strengths and limits of "doing gender" for understanding street crime. *Theoretical Criminology* 6 (4):433–60.
Mohamed, Adam Azzain. 2004. From instigating violence to building peace: The changing role of women in Darfur region of western Sudan. *African Journal on Conflict Resolution* 1:11–26.
Moro, Leben Nelson. 1998. *A study of some ethnic groups* in *Sudan.* Cairo: Sudan Cultural Digest Project.
Morrow, Lance. 1993. Unspeakable: Rape and war. *Time,* 22 February.
Mullaney, Jamie L. 2007. Telling it like a man: Masculinities and battering men's accounts of their violence. *Men and Masculinities* 10 (2):222–47.
Nageeb, Salma Ahmed. 2004. *New spaces and old frontiers: Women, social space, and Islamization in Sudan.* Lanham, MD: Lexington Books.
O'Fahey, Rex S. 1980. *State and society* in *Dar Fur.* New York: St. Martin's.
Oladosu, Afis A. 2009. Al-Rujūlah: Male and masculinities in modern Sudanese narrative discourse. *Hawwa* 7 (3):249–70.
Onyango, Monica A., and Karen Hampanda. 2011. Social constructions of masculinity and male survivors of wartime sexual violence. *International Journal of Sexual Health* 23:237–47.
Price, Lisa S. 2001. Finding the man in the soldier-rapist: Some reflections on comprehension and accountability. *Women's Studies International Forum* 24 (2):211–27.
Ridgeway, Cecilia L. 2009. Framed before we know it: How gender shapes social relations. *Gender & Society* 23:145–60.
Ridgeway, Cecilia L., and Shelley J. Correll. 2004. Unpacking the gender system: A theoretical perspective on gender beliefs and social relations. *Gender & Society* 18:510–31.
Rome Statute of the International Criminal Court. 1998/2002. UN General Assembly.
Ross, Marlon B. 2002. Protest, masculine strategies of black. In *Masculinity studies & feminist theory: New directions,* edited by Judith Kegan Gardiner. New York: Columbia University Press.

Schrock, Douglas P., and Irene Padavic. 2007. Negotiating hegemonic masculinity in a batterer intervention program. *Gender & Society* 21 (4):625–49.

Segal, Lyune. 1990. *Slow motion: Changing masculinities, changing men.* New Brunswick, NJ: Rutgers University Press.

Seifert, Ruth. 1994. War and rape: A preliminary analysis. In *Mass rape: The war against women* in *Bosnia-Herzegovina,* edited by Alexandra Stiglmayer. Lincoln: University of Nebraska Press.

Sivakumaran, Sandesh. 2007. Sexual violence against men in armed conflict. *European Journal of International Law* 18 (2):253–76.

Sjoberg, Laura. 2015. Seeing sex, gender, and sexuality in international security. *International Journal* 70 (3): 434–53.

Straus, Scott. 2015. *The making and unmaking of nations: War, leadership, and genocide* in *modern Africa.* Ithaca, NY: Cornell University Press.

Tanner, Victor, and Jérôme Tubiana. 2007. *Divided they fall: The fragmentation of Darfur's rebel groups.* Geneva, Switzerland: Small Arms Survey.

Taylor, Gary. 2000. *Castration: An abbreviated history of Western manhood.* London: Routledge.

Totten, Samuel. 2011. *An oral and documentary history of the Darfur genocide, volume 1.* Santa Barbara, CA: Praeger Security International.

Totten, Samuel, and Eric Markusen. 2006. *Genocide in Darfur: Investigating the atrocities in Sudan.* New York: Routledge.

Tønnessen, Liv. 2007. Competing perceptions of women's civil rights in Sudan. Chr. Michelsen Institute, Bergen.

UN OCHA. 2008. Discussion Paper 2: The nature, scope, and motivation for sexual violence against men and boys in armed conflict. *UN OCHA Research* Meeting. 26 June.

U.S. Department of State. 2004. *Documenting atrocities in Darfur.* Washington, DC: Bureau of Democracy, Human Rights, and Labor and the Bureau of Intelligence and Research.

U.S. Department of State. 2013. *Sudan 203 human rights report.* http://www.state.gov/documents/organization/220376.pdf.

Vojdik, Valorie K. 2014. Sexual violence against men and women in war: A masculinities approach. *Nevada Law Journal* 14 (3):923–52.

Weiss, Karen G. 2010. Male sexual victimization: Examining men's experiences of rape and sexual assault. *Men and Masculinities* 12 (3):275–98.

West, Candace, and Don H. Zimmerman. 1987. Doing gender. *Gender & Society* 1:125–51.

West, Candace, and Sarah Fenstermaker. 1995. Doing difference. *Gender & Society* 9 (1):8–37.

West, Candace, and Don H. Zimmerman. 2009. Accounting for doing gender. *Gender & Society* 23:112–22.

Willemse, Karin. 2007a. *One foot in heaven: Narratives on gender and Islam in Darfur, West-Sudan.* Boston: Brill.

Willemse, Karin. 2007b. "In my father's house": Gender, Islam and the construction of a gendered public sphere in Darfur, Sudan. *Journal for Islamic Studies* 27:73–115.

Willemse, Karin. 2009. Masculinity and the construction of a Sudanese national identity. *In Darfur and the crisis of governance in Sudan: A critical reader*, edited by Salah M. Hassan and Carina E. Ray. Ithaca, NY: Cornell University Press.

Wood, Elisabeth Jean. 2006. Variation in sexual violence during war. *Politics & Society* 34 (3):307–41.

Wood, Elisabeth Jean. 2009. Armed groups and sexual violence: When is wartime rape rare? *Politics and Society* 37 (1):131–61.

Wood, Elisabeth Jean. 2013. Multiple perpetrator of rape during war. In *Handbook on the study of multiple perpetrator: A multidisciplinary response to an international problem,* edited by Miranda A. H. Horvalth and Jessica Woodhams. New York: Routledge.

Yuval-Davis, Nira, and Floya Anthias. 1989. *Woman, nation state.* New York: St.Martin's.

Zarkov, Dubravka. 2001. The body of the other man: Sexual violence and the construction of masculinity, sexuality and ethnicity in Croatian media. In *Victims, perpetrators or actors? Gender, armed conflict and political violence,* edited by Caroline O. N. Moser and Fiona C. Clark. London, UK: Zed.

Introduction to Reading 47

Amy M. Denissen and Abigail C. Saguy interviewed women who worked as tradeswomen in the building trades to get a better understanding of how women are treated in a predominantly male workplace. The ability of these women to do their jobs within the building trade was complicated by the enforcement of gender by both the males and the females. While the males enforced masculinity and emphasized the masculine traits that were needed for the job, they also denounced femininity as weak and not belonging on the job sites. However, females also wished to be accepted for their strength and ability to do their jobs. Interestingly, females who were lesbians were more likely to be accepted by the men, while often rejected by other women. Thus, heterosexuality is also used to maintain gendered assumptions on these work sites. These interviews with 35 women, who were located by contacting local unions, apprenticeship training programs, tradeswomen conferences, and through referrals from others interviewed, provide an interesting view of how gender and heteronormativity can be enforced and manipulated on the job.

1. How is both gender and heteronormativity used by the men and the women in this reading, and by doing so, does it maintain male dominance at the workplace?
2. Why do these women distrust each other and fail to work together for change in the workplace?
3. Think about other workplaces you have been in. Was gender reinforced there as well? Imagine how the same dialogues might happen in other male-dominated workplaces, particularly in powerful, upper-level occupations.

Gendered Homophobia and the Contradictions of Workplace Discrimination for Women in the Building Trades

Amy M. Denissen and Abigail C. Saguy

The effects of double binds, in which femininity and competence are seen as mutually exclusive, are well documented in male-dominated workplaces (Jamieson 1995; Valian 1998). Previous research shows that women resist double binds in part by "finding a variety of ways to do gender" (Pierce 1995, 13–14) that trouble boundaries of gender difference. Women may directly challenge gender dualities by, for instance, demanding respectful recognition as women while performing masculinity (Denissen 2010b). They may also invoke shared identities based on race, class, occupational hierarchy, or culture to deemphasize gender difference (Denissen 2010b; Janssens, Cappellen, and Zanoni 2006). Women workers thereby participate in "gender maneuvering" (Schippers 2002; see also Finley 2010), or the manipulation of gender rules to redefine the relationship between femininity and masculinity. . . .

Drawing on interviews with a diverse sample of lesbian and straight women in the construction trades, such as electricians and sheet metal workers, of which women comprise less than 2 percent of the workforce nationwide (Bilginsoy 2009), this article extends our understanding of gender maneuvering by exploring how the meaning of race, body size, and seniority impact the constraints tradeswomen face and the cultural resources available to them for resisting gender boundaries. We argue that the presence of women in male-dominated jobs threatens the perception of this work as inherently masculine (Collinson 2010; Epstein 1992; Paap 2006). We further argue that branding all tradeswomen lesbians, and thus—in the popular imagination—as not fully women, can partly be understood as an attempt to neutralize this threat. While the lesbian label (whether or not women personally identify as such) offers some degree of acceptance and freedom from performing emphasized femininity, it can place demands on tradeswomen to perform a subordinate blue-collar masculinity that may include participating in a misogynistic work culture (Connell 1987; West and Zimmerman 1987).

Moreover, the presence of lesbians (and sexually autonomous straight women whose sexuality is not directed toward tradesmen) threatens heteronormativity and men's sexual subordination of women, or what Ingraham calls "patriarchal heterosexuality" (Ingraham 1994). By sexually objectifying tradeswomen, tradesmen, in effect, attempt to neutralize this threat. While tradeswomen, in turn, are sometimes able to deploy femininity to manage men's conduct

---

Denissen, A. M., & Saguy, A. C. (2014). Gendered homophobia and the contradictions of workplace discrimination for women in the building trades. *Gender & Society, 28*(3): 381–403. Reprinted by permission of SAGE Publications, Inc. on behalf of Sociologists for Women in Society.

and gain some measure of acceptance as women, it often comes at the cost of their perceived professional competence and sexual autonomy and—in the case of lesbians—sexual identity.

Those who refuse to be sexually objectified may subsequently find themselves the target of open hostility. Certain women—including lesbians and those who present as butch, large, or Black—may be less able to access emphasized femininity as a resource and thus more subject to open hostility. We show that tradeswomen navigate among imperfect strategies and engage in complex risk assessments (McDermott 2006). Extending Denissen (2010b), we highlight how tradeswomen reflexively manipulate gender meanings, adding a new emphasis on the intersection between sexuality, gender representation, race, and body size. Ultimately, however, we argue that individual strategies are insufficient and show how tradesmen deploy the stigma of lesbianism to discourage solidarity and collective action among tradeswomen. We consider the implications of these findings within the larger debate about the efficacy of interactional forms of resistance for challenging patriarchy and the dominant gender order.

Gender and Sexuality in Male-Dominated Occupations

Previous work shows that men working in male-dominated blue-collar occupations accentuate their manliness by distinguishing their work from women's work (Epstein 1992; Schrock and Schwalbe 2009) and how managers manipulate gender ideology to control workers (Collinson 2010; Epstein 1992; Paap 2006). For instance, in a coal miner's protest about being asked to lift too much weight, the foreman asked, "What's the matter? Aren't you man enough?" (Epstein 1992, 243). By encouraging workers to identify with their gender and, also, their race, national, and class identities, employers divide workers and distract them from working conditions in order to enhance labor control (Hossfeld 1990). Generalizing from Ramirez (2010), many "macho" masculinities can be understood as working-class men's "compensatory reactions" to subordination when other sources of masculine identity are blocked (Zinn 1982) or become insecure because of declining wages, job security, union power, and social regard (Paap 2006). When men derive psychic and social rewards and managers derive economic benefits from these identifications, both groups can be expected to resist the entrance of women workers, which undermines the sense that it is, in fact, "men's work" (Epstein 1992).

In addition, tradeswomen are at a structural disadvantage as tokens (Kanter 1977) in "doubly male dominated" workgroups that "create a work culture that is an extension of male culture" and where the "numerical dominance of the workplace by men heightens the visibility of, and hostility toward, women workers" (Gruber 1998, 303). Institutional factors further intensify tradeswomen's visibility and vulnerability. For instance, the decentralization of production in the construction industry means that workers regularly change job sites, entering into new work relationships. As a result, tradeswomen prove themselves without the full benefit of their prior accomplishments. When a tradeswoman's reputation precedes her, it is often a liability, as in the case of tradeswomen managing the "sexual harassment lady" (Denissen 2010a) or "looking for a lawsuit" (Paap 2006) label that is sometimes attached to women who complain.

Moreover, despite the autonomy that construction workers enjoy (Applebaum 1999), tradeswomen's success and safety require good relations with tradesmen because (1) the apprentice model creates dependence on journeymen for training, (2) the work requires the cooperation of various trades to achieve tasks, and (3) workers must rely on each other for their physical safety (Applebaum 1999). While tradeswomen often emphasize the crucial role that supportive tradesmen play in their careers, their dependence on tradesmen also presents challenges.

Male homosexuality is also widely viewed as a threat to masculinity. It is common, in the male-dominated trades and elsewhere, for men to distance themselves—through homophobic jokes and the use of derogatory terms like "gay" and "faggot"—from homosexuality as a way of affirming their masculinity (Seidman 2010). C.J. Pascoe describes as "gendered homophobia" high school boys' use of the terms "gay" and especially "fag" to police behavior considered insufficiently masculine on the part of other boys (Pascoe 2005). Pascoe argues that fag discourse is targeted specifically at boys, rather than girls, and is as much about policing masculinity as sexuality (Pascoe 2005).

Yet, research suggests that, in male-dominated occupations, both men and women—straight and gay—are targets of sexist and anti-gay harassment (Paap 2006). Men tease other men who exhibit behavior deemed feminine and tell their female coworkers to eschew makeup and to work "like a man" (Denissen 2010b, 1056). In male-dominated contexts, where simply occupying a trade as a woman is associated with other forms of perceived gender inversion, including same-sex desire (Paap 2006), men direct anti-gay harassment at straight women and lesbians alike (Frank 2001). In fact, in this context, the presumption of heterosexuality, or heteronormativity (Ingraham 1994), may be suspended.

Indeed, to the extent to which lesbians are perceived as not fully women, their presence may be less threatening, than that of straight women, to the idea of male-dominated occupations as "men's work" (Paap 2006). Moreover, lesbians are positioned differently than are gay men within the hierarchical gender system that privileges both masculinity and heterosexuality (Schilt and Westbrook 2009). Whereas gay men are devalued both because of their sexuality and because they are perceived as feminine, lesbians (and those perceived as lesbians) may derive benefits in some contexts from their perceived masculinity, while having to negotiate a devalued sexual identity.

This insight helps make sense of studies showing that open lesbians are sometimes more accepted as coworkers in male-dominated work contexts, compared to straight women (Miller, Forest, and Jurik 2003; Myers, Forest, and Miller 2004; Paap 2006). For example, studies find that male police officers better accept lesbian, compared to gay men, coworkers (Miller, Forest, and Jurik 2003, 369), and that lesbians' sexual orientation offers a waiver from social pressures to enact emphasized femininity (Burke 1994). In some cases, heterosexual men's interest in lesbian sexuality may facilitate lesbians' inclusion in workplace banter (Frank 2001).

The experiences of butch, gender-blending women, and transmen further suggest that people may not always be censured for adopting the socially respected traits of masculinity (Devor 1987; Schilt and Westbrook 2009). We use the term "butch/dyke" to refer to performances of masculinity by women, what Halberstam (1998) calls "female masculinities." We use the term "gender-blending" to refer to women who combine interactional strategies that are alternatively coded as feminine or masculine (Devor 1987; Lucal 1999; Moore 2011). Butch and gender-blending women may be lesbian or straight and may sometimes be taken for men, but—unlike transmen—they do not identify as men.

Schilt and Westbrook find that, in nonsexual interactions, transmen are able to establish a male gender identity on the basis of gendered appearance and demeanor, even when they do not possess male genitalia (Schilt and Westbrook 2009). Male coworkers accept transmen—or at least tall, white transmen—as "just one of the guys," based on visible cues of masculinity (e.g., facial hair), even when they know that they were formerly women (Schilt 2011). Yet, transmen who have not had hormonal therapy and therefore do not appear to be men do not receive such social advantages (Schilt 2011).

Women who do "female masculinities" (Halberstam 1998) may similarly receive some forms of patriarchal dividends. For example, Kazyak's (2012) study of rural gays and lesbians shows that female masculinity may be normative in rural settings. However, to the extent that women clearly identify as women, they are unlikely to be granted the full status of "honorary men" (Schilt 2011). Moreover, they may find that inclusion prompts subjection to the rough and demeaning talk that characterizes many male interactions (Denissen 2010b).

If lesbians are perceived as less threatening to notions of "men's work," their visibility threatens the dictates of compulsory heterosexuality (Rich 1993) and, more broadly, the subordination of women's sexuality to men's desire (MacKinnon 1982; Pateman 1988). Men's efforts to sexually objectify women coworkers can be understood as an attempt to restore this gender-sexual order. In response, women skillfully mix performances of femininity and masculinity to resist being depicted as occupationally incompetent or sexually deviant and to assert their sexual autonomy (Denissen 2010b). Yet, resistance to sexual objectification may elicit more overt hostility from male coworkers.

In response to homophobia, lesbian tradeswomen engage in interactional strategies that vary by perceived risk and other contextual factors. These fall along a continuum from "passing" (Goffman 1963) or "playing it straight" (Sullivan 2001), in which they conceal their sexual orientation, and "covering," in which they prevent this identity from "looming large" (Goffman 1963) to fully "coming out" or "telling it like it is" (Sullivan 2001). Most engage in hybrid strategies, such as "speaking half-truths to power" or adopting an "open closet door policy" (Reimann 2001), in which they carefully manage disclosure by selectively revealing their sexual orientation based on specific context. In addition to sexual orientation, we expect that race, gender presentation, and body size inform which interactional strategies are both possible and preferred (Crenshaw 1989; Fikkan and Rothblum 2011; Moore 2011; Saguy 2012).

* * *

Negotiating Threats to the Masculine Definition of the Work

Tradeswomen report that homophobic comments, jokes, and graffiti are pervasive and that tradesmen regularly use terms like "gay" and "faggot" to publicly establish heteromasculine identities and to reinforce the masculine definition of the work. For example, Monique says her male coworkers "pick on each other, [saying things] like: 'The electricians are faggots,' 'The carpenters are faggots,' 'Because he walks a certain way, he's gay.'" In this example, tradesmen use homophobic comments to assert dominance over "rival" groups of men (such as men from other trades) and to regulate the gender and sexual behavior of men. Yet, unlike the high school boys studied by Pascoe who claim not to direct fag discourse at boys known to be gay (Pascoe 2005), tradesmen unapologetically use homophobic slurs to repudiate both homosexuality and femininity (in men). This was not lost on the tradeswomen interviewed, who attributed the fact that they did not know any openly gay men to their sense that the trades are dangerous for openly gay men.

Similarly, the presence of women on job sites threatens the definition of the construction trades as "men's work." One way that tradesmen make sense of tradeswomen's presence and neutralize this threat is to label them lesbians or likely lesbians. Lynne, an Asian American lesbian, explains, "People think if you're a tradeswoman, you're a lesbian. You want to do a man's job, so you want to be a man, so you're a lesbian." Stephanie, a straight white woman, says, "People think I'm gay a lot of the time because . . . I don't look real feminine." Holly, another straight white woman, says a fellow apprentice "never discussed her love life at work, and she [then] mentioned having a boyfriend. Everybody looked at her like 'You have a boyfriend?' They thought she was gay." Imagining tradeswomen as lesbians, that is, not fully female, preserves the idea of the trades as "men's work."

This opens up the possibility that straight tradeswomen may be perceived as more of a threat to the masculine definition of the work than lesbian tradeswomen. Indeed, Loretta, a Black lesbian, says that her male coworkers do not "want any women at all," but that "somebody like me is safer for them because they can ignore me like a guy they don't like." . . .

Loretta notes that while some tradesmen resent the presence of all women in the trades, straight or lesbian, that she, as a lesbian, is "not really a chick" and her presence does not limit tradesmen's freedom to perform masculinity as they please. This may be especially true for lesbians like Loretta who present as butch. Indeed, Vicky, a lesbian tradeswoman who describes herself as "a bit girlier" notes that tradesmen are more likely to treat a woman who "doesn't look as feminine on the outside" as "one of the guys," while they are more likely to "watch their potty mouth" around more "girly"-presenting lesbians. We also find some evidence that butch lesbians are somewhat less likely to be targeted by sexual advances.

A few tradeswomen claim that, as lesbians, they are fully accepted as "one of the guys." For example, Toni, a white lesbian, who describes herself as someone who "used to be extremely feminine" but no longer bothers because "it required too much maintenance," describes how she is incorporated into the men's sex talk:

> [My coworker] tells his girlfriend, "She's like one of the guys, you know, I can tell her anything." That's how most of the guys think of me anyways. They just talk about whatever they want to. It's, like, [I'll tell the men,] "You should do this [sexual maneuver] or you should try that [sexual position]." [And, later they'll tell me,] "Oh, that worked! Thanks a lot, Toni." So it's all good.

For Toni, offering advice on women's sexuality is a "good" form of inclusion because it takes place within a supportive working relationship with coworkers.

At the same time, finding acceptance as "one of the guys" can be fraught with danger. Lori, the Jewish butch lesbian introduced earlier, describes a lunchtime interaction she had as an apprentice, when she was especially vulnerable:

> They're sitting around talking about the Mike Tyson case when he sexually assaulted this woman. For me, rape is no joking matter. So here's nine of 'em, a foreman, journeymen, apprentices, and one shop steward, and I'm the only woman in this discussion. They're all sitting there talking about it and joking about it, and I'm, like, "Whoa. I'm feeling really, really violent." So I said, "The next person who says anything, I'm gonna get really violent." They all shut the fuck up. Then there was another situation where they were talking about wife beating. I got mad, but sometimes it's not worth it 'cause it's, like, "Oh, she's got no sense of humor." So then I just don't eat lunch with them anymore. . . .

While lesbians may be more likely than straight women to be accepted as "one of the guys," and while this can provide some camaraderie and acceptance as a serious worker, they rarely experience full acceptance. Rather, tradeswomen typically emphasize that acceptance as one of the guys is incomplete and conditional. Many tradeswomen say their male coworkers hold them to an exaggerated standard of masculinity, making them carry heavier materials and do dirtier and more dangerous work, in order to prove they can work "like a guy." Further, as we discuss ahead, acceptance as one of the guys in some contexts does not exempt them from the ideals of emphasized femininity in others.

Managing Perceived Threats to the Heterogendered Order

While the presumption that tradeswomen are likely lesbians neutralizes threats to the masculine definition of the work, it threatens heteronormativity and the sexual and economic subordination of women to men. In response, tradesmen sometimes direct gendered homophobic comments at lesbian tradeswomen. In other instances, they sexually objectify (lesbian and straight) tradeswomen. We examine tradeswomen's accounts of this behavior and how they respond to it.

Keeping Them Guessing, Keeping It Private, and Other Responses to Gendered Homophobia. Just as they use fag discourse to police gender noncomformity among men, so tradesmen use the lesbian label to control the gendered conduct of tradeswomen. For example, Elena (a Latina heterosexual) says tradesmen single out lesbian tradeswomen as deviant "freaks": "The guys talk about them really bad, like, she's trying to play a man role, she likes it rough, men can't satisfy her, she must be freaky and have freaky needs." Lauren, a white lesbian who describes herself as tomboyish but not butch, says that she has heard her coworkers make disparaging comments about "hardcore dyke lesbians." She recounts how one tradesman exclaimed, "Damn, I'm working with this guy and next thing I know she turns around and, shit, she's got tits!" When Lauren asked him if she was a good worker, he responded, "I don't know, I couldn't work with her."

Racial minority status and body size can intersect with sexual identity and gender presentation to heighten stigmatization and otherness. Loretta, the Black butch lesbian, is large and has a shaved head. An electrician, a trade that historically has had among the lowest number of minority workers (Bilginsoy 2005), Loretta describes job sites as "bastions of white male supremacy." She notes that, in recent years, an influx of Latino workers has heightened racial tensions and that the prevailing message conveyed to women, "queers," and people of color is "You shouldn't be here." She tells of hearing tradesmen say, "Now they're letting animals in the trade." When asked to whom they were referring, Loretta exclaims, "Me! Or my crew-member who [was] a person of color." Loretta speaks of how she is threatening, not just as a woman, but as a large, Black, butch lesbian woman with an aggressive personality, a composite that "messes with the whole expectation of what your gender, what your behavior's supposed to be."

Loretta says that tradesmen sometimes "picked on" her about her large size, saying things like, "You're fat" or that her size "ain't cool for chicks." Similarly, Lori, who describes herself as a "big butch dyke" (a "three-part package"), says that her coworkers' negative comments about her size are gendered: "They'll accommodate a big guy where they won't accommodate a big woman."

Sometimes the label "lesbian" is decoupled from women's own sexual identity, as when tradesmen target tradeswomen for gendered homophobia because their appearance or behavior does not conform to tradesmen's gender expectations. For instance, Cheryl, a white heterosexual, explains how one of her coworkers "was mad because I'd showed him up that day," performing better than he in a workplace task. He asked her, "What's the matter with you? Are you one of those lesbian women, you know, and you're not interested in me?" In this example, Cheryl's coworker accuses her of being a lesbian, and thus unfeminine, because she outperforms him. He thereby conflates her occupational competence and sexual orientation, considering both as signs of gender nonconformity.

In response to gendered homophobia, lesbian tradeswomen engage in complex risk assessments and employ a variety of disclosure options. For example, Anna, a Latina lesbian who describes herself as tomboyish and "not real girly but not real butchy," remarks about a coworker, "I've heard him make comments about fags and queers and I didn't want to go there. When he said, 'Are you married?' I said, 'No' and I didn't say I have a partner." Here, Anna speaks a "half-truth to power" (Sullivan 2001). It is true that she is not married, but she conceals the full truth—that she has a same-sex romantic partner—from this coworker because his homophobic comments make that revelation feel unsafe. Further, Anna says that in situations that feel safer, she selectively discloses her sexual identity. . . .

Racial minority status often heightens othering and perceived risk, further limiting tradeswomen's disclosure options. For example, Lori, a self-described Jewish butch lesbian, says she did not disclose her sexual orientation on one job site early in her career because she had heard "a bunch of sexist, racist, and homophobic speech" that made disclosure feel unsafe. While her coworkers were specifically "targeting the Hispanics," their behavior "really frightened" her "because they had swastikas and Nazi and KKK-type talk." Yet, later in her career and on less racist job sites, she developed a strategy of singling out one man with whom she would be more open:

> What I do generally is I'd make allies with one dude who I felt was more open-minded or we have a connection. I would be honest with him about who I was. As long as I had one person I could be myself with, then I felt okay.

> Now I'm pretty much out. I decided that I'm out in the union as a whole, but I pick and choose how much I say.

Several of the respondents similarly spoke about becoming more open, but still guarded, regarding their sexuality as they gained more occupational seniority.

Sometimes tradeswomen conceal their sexual identity not simply out of fear of retaliation but also to resist the salience of their sexual identity in workplace interactions. We call this strategy "keeping it private." Vanesa, a white lesbian, explains that she brought her best friend, rather than her girlfriend, to union picnics both because she wants to keep her "personal life private" and also because she hopes to "keep away from the stigma" and does not "want a guy not to teach me because of who I am." Anita, a Native American lesbian, similarly evokes a concern with both privacy and homophobia, explaining that she was not initially out because "it's nobody's business, and then going into a man's field I figured it's probably not a good idea to advertise." Yet, she says that "if it came up, I didn't deny it," akin to what others have labeled an "open closet door policy" (Reimann 2001). Similarly, Lauren, a white lesbian, says, "There's some guys that don't know. Maybe that's my way of blending in without any confrontation. I like to get in there, get my job done, and get out. I've had a couple of guys ask me, and if they got the balls to ask me, I'll tell them."

Gina, a large, black, straight, married woman, evokes a "keep them guessing" strategy that entails sending mixed messages about sexual identity as part of an attempt to "break that stereotype":

> I had them so fooled there were people that didn't have any idea what my sexual orientation was. If somebody questioned me, [I'd say,] "I'm gay, leave me alone, I'm a lesbian." Or [I'd say,] "I'm single," or "I have two kids," or "I have a husband." People would be running around, [saying,] "No, she told me she was gay." Or "Gina, you're not gay, I met your husband." So you'd keep them guessing because the point was that your sexual orientation didn't matter.

While keeping the men guessing may function partially as an expression of solidarity with lesbian tradeswomen, a sort of reverse passing intended to challenge stereotypes about lesbians, Gina herself says it is also a way of resisting the salience of women's sexual identities at work.

Similarly, Alex, a white lesbian, talks about mixing displays of subordinated feminine heterosexuality with more stereotypically masculine behavior in order to resist homophobia and sexism. She explains that while she used to be mistaken for a man because she "looked completely androgynous," she has grown her hair since joining the trades because short hair "would be such a red flag" that she is a "dyke" or is "so manly":

> I'd rather act feminine and friendly and cute than get harassed, ignored, or treated worse. But at the same time it's like I have to be careful that I don't act overly feminine because they'll think I can't work. Sometimes I'll say something that will totally throw them for a spin [or] make them raise an eyebrow because I'll say it in a masculine way. I'll say something that's really clear, concise, and to the point, and they don't expect that of me. They think I'm a bubbly person; they stereotype me as a female.

Alex is managing a classic double bind where she is held accountable to conflicting expectations for gendered conduct. She is aware that her coworkers may mark (raised eyebrow) and sanction (harassment, isolation) masculine conduct. Alex says she flirts with men and acts "feminine" in an effort to forestall certain forms of harassment and exclusion, but fears that overdoing it may detract from her perceived competence. She performs an intricate gender maneuvering in trying to strike a balance by varying heterofeminine displays with more assertive (masculine) actions to transgress dualistic sexual and gender boundaries. While white respondents, straight and gay, were more likely to speak of incorporating displays of emphasized femininity into their gender maneuvering, Black, butch, and large tradeswomen were more likely to emphasize their ability to "hold their own" with the heaviest, dirtiest, and most dangerous tasks.

Turning the Tables: Resistance to Compulsory Heterosexuality. Another way that tradesmen neutralize the threat of lesbian/female autonomy is by recasting them as objects of men's sexual desire. Some lesbian tradeswomen say tradesmen embrace them through the heterosexual male fantasy of having 'fancy sex' with multiple women. For instance, when asked if she ever was directly targeted by homophobia, Anna, the Latina lesbian introduced earlier, responds, "No, because I'm a female. Some guys say, 'I don't care about the women. I think that's great! That's fancy for me! I just can't stand the guys.'" Yet, Toni, a white lesbian, suggests that this form of acceptance has its costs: "They'll make innuendos like 'You should hook up with her and then hook up with me later.' They know I'm not interested in them. They just continue to do it because they know it bugs me." In this instance, Toni's coworkers impose heterosexual expectations and meanings onto her and intentionally "bug" her. By redefining lesbian relationships as serving male heterosexual desire, tradesmen neutralize the perceived threat of lesbian desire to heterosexism.

Out lesbian tradeswomen use various strategies to resist their coworkers' efforts to heterosexualize them and, sometimes, to reaffirm their sexual identity as lesbians. Jan, a lesbian of white and Native American descent, who is slim and has long blonde hair, and says she "doesn't go out of her way to be feminine" but "doesn't seem butch to the guys," complains about how she has to tell her coworkers that she is not "free porn." Others speak of resisting traditional gender dynamics by showing a sexually assertive interest in their coworker's women partners. Anna, the Latina lesbian, explains:

> [My coworkers] accept me for who I am. [He'll say,] "That's cool, girl. Can we get some?" [laughs] I'll be, like, "Can I get some of yours? I'll let you talk to my girl if you let me talk to your wife." And he'll be, like, "Fuck you." Guys are cool with me. [They'll say,] "How's your girl? She's pretty hot." I go, "Yeah, thank you. So is your wife." [Laughs.]

While Anna describes her interactions with her coworkers as playful and respectful ("They accept me for who I am"), she also experiences counterresistance from her coworker ("Fuck you"). Indeed, it seems that she gets respect, in large part, because she can give as good as she gets, using masculine displays of dominance to neutralize efforts to sexually dominate her. We call this strategy "turning the tables."

Similarly, Lynne, an Asian American lesbian whom we would describe as gender-blending but who is sometimes mistaken for a man, explains how she responded to a coworker who constantly asked her if he could watch her have sex with another woman:

> I said, "Why don't you talk to your girlfriend about it? Bring me a picture; I want to see what she looks like." He got all defensive: "Who, wait, what'd you mean? I don't have a picture. She ain't going for that shit." He backed off that whole line of conversation after that.

Like Anna, Lynne successfully wards off her coworker's efforts to sexualize her by turning the tables and sexualizing his girlfriend. While this interaction seems to have been successful in curtailing demands to watch Lynne have sex with other women, later in the interview Lynne says this incident led to a strained working relationship with this particular coworker.

Moreover, tradeswomen are not equally able to resist their coworker's efforts to sexualize them. Julia, a Latina lesbian apprentice who described herself as "looking like a little dude," describes an extreme case in which a coworker attempted to sexually force himself on her. . . .

This tradesman disregards Julia's identity as a lesbian, as well as her resistance to his sexual advances, trying to force himself on her. He responds to her defiance with threats to "get her," culminating with a sexual assault on the job site. Fearing for her job, she initially refused to report the incident but ultimately did so, upon the urging of the superintendent and the coworker who witnessed the assault. Julia says she never saw the assailant again.

Other tradeswomen also report being targeted with overt hostility and violence after refusing to engage in sexual banter or feminine displays. Some of the more egregious examples include having electrical wires turned on while they were working on them, having tools dropped on them, or finding feces in their hard hat. These sorts of incidents highlight the risks and limitations of individual-level resistance.

How Gendered Homophobia Limits Collective Resistance. While individual strategies have subversive potential, successful "contestation of gender hierarchy is fundamentally a collective process" (Connell 2009, 109). With typically few allies at work, one might expect tradeswomen to seek each other out for safety and support. Yet many of our respondents say they avoid other women both on and off the construction site. In some cases, this stems from their own homophobia, but it is more often described as an effort to protect themselves from homophobic stigma and sexist stereotypes. Vanesa, a white lesbian, explains, "Women will tell me they don't want to be seen with other women or belonging to a women's group because a lot of the guys say, '[If] you women want to be just like us men so much, then why do you have this little women's group?'" Some tradesmen pressure her and other tradeswomen to avoid associating with other women. Vanesa further describes how tradesmen reframe women's efforts to support each other as attempts to gain special privileges. For example, her foreman remarked, after seeing her in a tradeswomen's convention T-shirt, "I don't think there should be separate organizations, you guys need to be treated the same."

Loretta, the Black butch lesbian, says that she "would never hang out with the girls" and that "the girls on the crew wouldn't want to hang out with me, because they wouldn't want the other guys to think that they were gay. Because of that guilt by association thing it's, like, 'Well, if we're nice to you, they might think we're like you.'" Loretta's comments speak to how lesbian stigma is attached not only to joining women's associations but also to socializing with other women on the job. Similarly, Lori, a Jewish butch lesbian, says, "I wanted to start a lesbian tradeswomen group but not even the lesbians want to start it with me." Moreover, she says, "Sometimes even other women in the trades are afraid to be seen with me because I'm an out lesbian. Like it'll spill off on them and the guys will see it." At a conference for women in the trades, the women became particularly animated when they heard that tradesmen were referring to the conference as a "big lesbian orgy" in what seemed like an attempt to discredit the conference and keep both straight and lesbian tradeswomen away.

As Lori describes it, tradesmen effectively use the specter of lesbianism to stymie gender solidarity and political activism: "Sometimes there's solidarity, sometimes not, because the lesbians think they have to align themselves with the men for power and that means turning against other women or a more out lesbian. They'll be more closeted or they're afraid to be seen as lesbian whether they're lesbian or not." In distancing themselves from other women in order to protect themselves from the gendered homophobia of their coworkers, both straight and lesbian tradeswomen are made more vulnerable as they become isolated from each other. Yet, there is also evidence of resistance and change. For example, the tradeswomen conference has grown steadily over time from a state to an international event and active tradeswomen's groups have formed online, demonstrating organizational success despite these challenges.

Conclusion

Drawing on interviews with a diverse sample of lesbian and straight women in the construction trades, this article examines how the cultural meanings of sexual identity, gender presentation, race, and, more tentatively, body size and seniority, inform how men seek to control tradeswomen and how the latter respond to these efforts. We show that labeling tradeswomen as lesbians, and thus—in the popular imagination—as not fully women, both makes sense of their presence and reaffirms the perception of the trades as "men's work." Some lesbian tradeswomen report being more accepted than their straight women coworkers and claim that the lesbian label offers them some freedom from performing emphasized femininity. This acceptance is limited, however, and can place them in uncomfortable situations where they are expected to perform misogynist versions of masculinity. Moreover, while lesbians may be less threatening to the notion of the trades as men's work, their presence threatens heteronormativity and assumptions about the sexual subordination of women. We explain how tradesmen's efforts to sexually objectify tradeswomen can be understood as attempts to neutralize threats to heteronormativity and male privilege.

We demonstrate that in response to these constraints, tradeswomen use gender maneuvering (Schippers 2002) to combine performances of femininity and masculinity, to gain some measure of acceptance as women, and to maintain their perceived competence as workers. While tradeswomen strategically draw upon multiple strategies, we further show how the meanings attributed to tradeswomen's sexuality, gender presentation, race, body size, and seniority influence their preferred strategies. . . .

While individual tradeswomen are creative and sometimes successful in their efforts to resist men's attempts to marginalize and exclude them, our study suggests that individual responses may not be enough to produce widespread or lasting change. Tradeswomen's efforts to organize, however, are stymied by insinuations of lesbianism. Thus, gendered homophobia plays a crucial role in isolating and dividing tradeswomen, undermining their efforts to create solidarity, engage in collective resistance, and bring about institutional change. The risks of associating with lesbians and other women may be greatest for women of color and other especially vulnerable populations, a question that merits additional research.

We show how contradictions in the dominant heterogender order constrain tradeswomen, while opening up possibilities for—and even necessitating—more reflexive, varied, and strategic forms of gender and sexual practices (Denissen 2010b). Since gendered expectations of tradeswomen are intrinsically contradictory (e.g., sufficiently masculine to be deemed competent but sufficiently feminine to be socially acceptable), tradeswomen must constantly vary the way they "do gender" (West and Zimmerman 1987). Earlier work shows that exclusion of women in the building trades is reproduced despite women's resistance at the level of interaction and identity construction (Denissen 2010b). This article sheds light on one key mechanism whereby women's strategic agency is limited: the isolation of tradeswomen from other women. Thus, while individual tradeswomen strategically maneuver among gender and sexual meanings in ways that transgress heterogender boundaries and trouble the heterogender order, they face greater counterresistance when they collectively organize. This study expands on previous research that documents how race, class, and gender identities can be used to divide and control workers (Hossfeld 1990) by showing how tradesmen use gendered homophobia as a means of dividing and subordinating women workers.

These findings speak to debates about the extent to which individual-level resistance disrupts patriarchy or, alternatively, unwittingly reinforces the dominant gender order (Devor 1987; Ridgeway and Correll 2004). According to Finley (2010), transformations in gender relations are more likely in women-controlled than male-dominated spaces. Finley argues that women's networks are crucial for transforming the dominant gender order and that women in male-dominated settings are too isolated from other women to be effective. Our findings regarding women's isolation from each other and limits to collective resistance are consistent with Finley's argument. . . .

References

Applebaum, Herbert. 1999. *Construction workers, U.S.A.* Westport, CT: Greenwood.

Bilginsoy, Cihan. 2005. Registered apprentices and apprenticeship programs in the U.S. construction industry between 1989 and 2003: An examination of the AIMS, RAIS, and California Apprenticeship Agency databases. Department of Economics Working Paper Series (Working Paper #2005–09), University of Utah, Salt Lake City.

Bilginsoy, Cihan. 2009. *Wage structure and unionization in the U.S. construction sector.* Salt Lake City, UT: University of Utah, Department of Economics.

Burke, Marc. 1994. Homosexuality as deviance: The case of the gay police officer. *British Journal of Criminology* 34:192–203.

Collinson, David L. 2010. *Managing the shopfloor.* Boston, MA: De Gruyter.

Connell, R. W. 1987. *Gender and power: Society, the person, and sexual politics.* Stanford, CA: Stanford University Press.

Connell, Raewyn. 2009. "Doing gender" in transsexual and political retrospect. *Gender & Society* 23(1): 94–98.

Crenshaw, Kimberle. 1989. Demarginalizing the intersection of race and sex: A Black feminist critique of antidiscrimination doctrine, feminist theory and antiracist policies. *University of Chicago Legal Forum* 139–67.

Denissen, Amy M. 2010a. Crossing the line: How women in the building trades interpret and respond to sexual conduct at work. *Journal of Contemporary Ethnography* 39:298–327.

Denissen, Amy M. 2010b. The right tools for the job: Constructing gender meanings and identities in the male-dominated building trades. *Human Relations* 63:1051–69.

Devor, Holly. 1987. Gender blending females: Women and sometimes men. *American Behavioral Scientist* 31:12–40.

Epstein, Cynthia Fuchs. 1992. Tinkerbells and pinups. In *Cultivating differences: Symbolic boundaries and the making of inequality,* edited by Michele Lamont and Marcel Foumier. Chicago: University of Chicago Press.

Fikkan, Janna L., and Esther Rothblum. 2011. Is fat a feminist issue? Exploring the gendered nature of weight bias. *Sex Roles* 66:575–92.

Finley, Nancy J. 2010. Skating femininity: Gender maneuvering in women's roller derby. *Journal of Contemporary Ethnography* 39:359–87.

Frank, Miriam. 2001. Hard hats and homophobia: Lesbians in the building trades. *New Labor Forum* 8:25–36.

Goffman, Erving. 1963. *Stigma: Notes on the management of a spoiled identity,* New York: Prentice Hall.

Gruber, James E. 1998. The impact of male work environments and organizational policies on women's experiences of sexual harassment. *Gender & Society* 12:301–20.

Halberstam, Judith. 1998. *Female masculinity.* Durham, NC: Duke University Press.

Hossfeld, Karen J. 1990. "Their logic against them": Contradictions in sex, race, and class in Silicon Valley. In *Women workers and global restructuring,* edited by Kathryn B. Ward. Ithaca, NY: Cornell University Press.

Ingraham, Chrys. 1994. The heterosexual imaginary: Feminist sociology and theories of gender. *Sociological Theory* 12:203–19.

Jamieson, Kathleen H. 1995. *Beyond the double bind: Women and leadership.* New York: Oxford University Press.

Janssens, Maddy, Tineke Cappellen, and Patrizia Zanoni. 2006. Successful female expatriates as agents: Positioning oneself through gender, hierarchy, and culture. *Journal of World Business* 41:133–48.

Kanter, Rosabeth M. 1977. *Men and women of the corporation.* New York: Basic Books.

Kazyak, Emily. 2012. Midwest or lesbian? Gender, rurality, and sexuality. *Gender & Society* 26:825–48.

Lucal, Betsy. 1999. What it means to be gendered me: Life on the boundaries of a dichotomous gender system. *Gender & Society* 13:781–97.

MacKinnon, Catharine. 1982. Feminism, Marxism, method and the state: An agenda for theory. *Signs* 7:533–44.

McDermott, E. 2006. Surviving in dangerous places: Lesbian identity performances in the workplace, social class and psychological health. *Feminism and Psychology* 16:193–211.

Miller, Susan L., Kay B. Forest, and Nancy C. Jurik. 2003. Diversity in blue: Lesbian and gay police officers in a masculine occupation. *Men and Masculinities* 5:355–85.

Moore, Mignon. 2011. *Invisible families: Gay identities, relationships, and motherhood among Black women.* Berkeley: University of California Press.

Myers, Kristen A., Kay B. Forest, and Susan L. Miller. 2004, Officer friendly and the tough cop: Gays and lesbians navigate homophobia and policing. *Journal of Homosexuality* 47:17–37.

Paap, Kris. 2006. *Working construction: Why white working-class men put themselves—and the labor movement—in harm's way.* Ithaca, NY: Cornell University Press.

Pascoe, C. J. 2005. "Dude, you're a fag": Adolescent masculinity and the fag discourse. *Sexualities* 8:329–46.

Pateman, Carol. 1988. *The sexual contract.* Stanford, CA: Stanford University Press.

Pierce, Jennifer L. 1995. *Gender trials: Emotional lives in contemporary law firms.* Berkeley: University of California Press.

Ramirez, Hernan. 2010. Masculinity in the workplace: The case of Mexican immigrant gardeners. *Men and Masculinities* 14:97–116.

Reimann, Renate. 2001. Lesbian mothers at work. In *Queer families, queer politics,* edited by Mary Bernstein and Renate Reimann. New York: Columbia University Press.

Rich, Adrienne. 1993. Compulsory heterosexuality and lesbian existence. In *The lesbian and gay studies reader,* edited by Henry Abelove, Michele Aina Barale, and David Halperin. New York: Routledge.

Ridgeway, Cecilia L., and Shelley J. Correll. 2004. Unpacking the gender system: A theoretical perspective on gender beliefs and social relations. *Gender & Society* 18:510–31.

Saguy, Abigail C. 2012. Why fat is a feminist issue. *Sex Roles* 68:600–7.

Schilt, Kristen. 2011. *Just one of the guys? Transgender men and the persistence of workplace gender inequality.* Chicago: University of Chicago.

Schilt, Kristen, and Laurel Westbrook. 2009. Doing gender, doing heteronormativity: "Gender normals," transgender people, and the social maintenance of heterosexuality. *Gender & Society* 23:440–64.

Schippers, Mimi. 2002. *Rockin'out of the box: Gender maneuvering in alternative hard rock.* New Brunswick: Rutgers University Press.

Schrock, Douglas, and Michael Schwalbe. 2009. Men, masculinity, and manhood acts. *Annual Review of Sociology* 35:277–95.

Seidman, Steven. 2010. *The social construction of sexuality,* 2nd ed. New York: Norton.

Sullivan, Maureen. 2001. Alma mater: Family "outings" and the making of the Modern Other Mother (MOM). In *Queer families, queer politics,* edited by Mary Bernstein and Renate Reimann. New York: Columbia University Press.

Valian, Virginia. 1998. *Why so slow? The advancement of women.* Cambridge: MIT Press.

West, Candace, and Don H. Zimmerman. 1987. Doing gender. *Gender and Society* 1:125–51.

Zinn, Maxine Baca. 1982. Chicano men and masculinity. *Journal of Ethnic Studies* 10:29–44.

Introduction to Reading 48

New Zealand's Prostitution Reform Act (PRA) of 2003 decriminalized sex work in that country. Many researchers and activists as well as global organizations have argued that decriminalization would create a safer environment for sex workers. However, the research carried out by social scientist Lynzi Armstrong makes it clear that sex workers in New Zealand continue to be victims of violence who are subjected to frequent street harassment. In this reading Armstrong analyzes her interviews with street-based sex workers and "informants" from agencies connected to sex worker safety. Her findings reveal the limits of decriminalization in the context of continuing societal views of sex workers as "bad women" and deserving victims.

1. Why are street-based sex workers viewed as undesirable and deviant people? How is this view of sex workers tightly linked to the patriarchal, sexual double standard and the Madonna/whore dichotomy?
2. Why do women who are not sex workers participate in the harassment of sex workers? How does the concept of "internalized misogyny" help to explain this fact?
3. How do sex workers respond to sexual harassment and why might younger workers be less tolerant of harassment?
4. Armstrong says that dismantling structures of gender inequality is essential to creating a safe workplace for women who sell sex. Discuss.

"Who's the Slut, Who's the Whore?"

Street Harassment in the Workplace Among Female Sex Workers in New Zealand

Lynzi Armstrong

This article interrogates the ongoing marginalization and victimization of women working on the streets as sex workers in the decriminalized New Zealand context by focusing on one type of victimization that has not previously been explored in depth: street harassment. The article focuses on answering the following questions: How do female sex workers perceive and manage street harassment they experience? And, what is the significance of these experiences in the context of decriminalization? In doing so, this article attempts to shed new light on how street sex workers experience working in a decriminalized environment, and to begin a discussion on the limitations of decriminalization and what more needs to change to improve the treatment of sex workers in the open space.

Method

This article explores the findings of qualitative research that formed the basis of a 3-year study conducted in Wellington and Christchurch between 2008 and 2011. The research was focused on exploring the strategies employed by female street-based sex workers to manage risks of violence while working on the streets. In-depth semi-structured interviews were conducted with 28 women with experience working on the streets in Wellington and Christchurch. Further interviews were undertaken with 17 key informants from agencies with an interest in sex worker safety, such as the New Zealand Prostitutes Collective (NZPC), police, community leaders, and staff from youth- and social-service-based organizations. Researcher observation represented the third strand of data collection and involved documenting several hundred hours spent in drop-in centers and accompanying outreach workers on the street. Interviews with sex workers were focused on entry into sex work, perceptions of vulnerability, and experiences of violence, the shift to decriminalization, and strategies to manage risks. Interviews with key informants focused on perceptions of risk for street-based sex workers, strategies to support sex worker safety, the impact of the-Prostitution Reform Act, and changes for the future.

The sex workers interviewed were recruited using a snowball approach drawing on networks within the NZPC and at the evening drop-in center run by the Salvation Army in Christchurch. The first five interviews with sex workers, along with two key informant interviews, were conducted in Wellington where there is a relatively low population of street-based sex workers. The rest of the fieldwork took place in Christchurch, primarily as the city has the highest population of street workers in New Zealand as a proportion of all sex workers, 26% compared with 13% in Wellington and 11% in Auckland (Abel et al., 2007). The majority of interviews were conducted in the NZPC and Salvation Army drop-in centers, whereas two interviews were conducted in participant's homes.

The demographic characteristics of the sex worker participants varied considerably. The youngest woman interviewed was 17 and the oldest was 57. The age of entry into sex work also varied and while the average age was 20, one woman described becoming involved in street prostitution at the age of 12 and another had started working in the sex industry when she was 45. Thirteen of the women identified as New Zealand European, 14 as Māori, and 1 as Cook Island Māori. The women described disparate histories of sex work, and the length of time working on the street varied between 2 months and 22 years. Seventeen of the women had children, although not all were full-time caregivers. Ten of the women reported doing no other work outside of the sex industry in the course of their lives, while the remaining 18 had worked in other occupations previous or concurrent to working in the sex industry. Five women were studying either full- or part-time at the time of the research and four women reported that they were involved in voluntary work. Five of the women had started working after the PRA was passed, while the

Armstong, L. (2015). "Who's the slut, who's the whore?": Street harassment in the workplace among female sex workers in New Zealand. *Feminist Criminology, 11*(3): 285–303.

remaining 23 worked in the sex industry before decriminalization. The research therefore incorporates a range of perspectives from women who had started selling sexual services at very different times in their lives and in variable circumstances. All of the women were invited to choose a pseudonym to protect their privacy, while ensuring that they could identify themselves in the research. A few women preferred that a name be chosen on their behalf, and this request was respected.

Limitations

This research focuses on the experiences of cisgender women who work on the streets as sex workers in New Zealand. In New Zealand, the street sex worker population is predominantly female; however, transgender women also make up a moderate proportion of the population (Abel et al, 2007). Male street sex workers are virtually unheard of in New Zealand, and they also form a minority of the street sex work population in comparable countries such as the United Kingdom (Sanders, O'Neill, & Pitcher, 2009). The focus on just cisgender women in this article is a limitation as it does not provide insights into the ways in which sex workers of all gender identities experience harassment. It is possible that the harassment experienced by transgender women, for instance, may differ from the harassment outlined by these women. However, arguably this harassment is all part of the same picture, and greater understanding of the incidence and significance of street harassment in the lives of these cisgender women can lend itself to greater understandings of how street harassment affects the lives of all street sex workers.

Unpacking the Nature of Street Harassment in Street Sex Work

Street harassment is a broad term that is understood to encompass a wide range of behaviors. Much of the literature on street harassment has focused on the experience of street harassment among women in general, focusing on behaviors, such as catcalls, whistling, leering, being followed, and indecent exposure, that women experience while doing everyday mundane activities in the public space (Macmillan, Nierobisz, & Welsh, 2000). While it is possible for men to be victims too, there is an overwhelming agreement that street harassment is a gendered experience, with women being the most likely victims and men the most likely perpetrators (Fileborn, 2013). A number of studies have found experiences of street harassment among women to be widespread, with estimates ranging from 25% to 87% of women having experienced it (Johnson & Bennett, 2015; Macmillan et al., 2000). Street harassment is therefore something that affects *all* women. However, sex workers are arguably particularly vulnerable to being targeted with street harassment Richardson and May (1999) note that in the context of violence, sexual minorities are considered deserving victims, especially if they bring private and taboo behavior into the public domain. Indeed, the Madonna/whore dichotomy dictates that women who digress from expectations of "good" and "virtuous" female behavior are considered to bring violence on themselves (Russell, 1984; Stanko, 1985). As women who publically sell sex, it is likely that some people consider street-based sex workers to be "fair game" when it comes to street harassment Given that women are harassed in the public space when engaging in very mundane activities such as walking to the shops or waiting for a bus, it is perhaps then not surprising that women are also harassed when working on the streets as sex workers. While street harassment is a hazard for *all* women, women who work as street sex workers are vulnerable to being specifically targeted with harassment In line with this, a large amount of research has found that street harassment, particularly verbal abuse, toward sex workers is common (Abel, 2010; Day et al., 2001; Miller & Schwartz, 1995; Morgan Thomas, 2009; Nixon et al., 2002; Pitcher, Campbell, Hubbard, O'Neill, & Scoular, 2006; Sharpe, 1998). In agreement with this finding, all of the women who took part in this research had experienced harassment while working on the streets, and verbal abuse was an almost nightly occurrence.

Verbal Abuse

The findings of this research suggest similarities and differences in how street-based sex workers experience street harassment compared with other women, and this is evident in the verbal abuse that was described by the women. Shannon for instance noted that the harassment she experienced was often sexual in nature and involved gendered and sexualized comments such as "'*Oh show us your tits' and all that and call you a slut or they'll throw five cents at you.*" However, the harassment described by the women interviewed for this research was far broader than sexualized comments and included suggestions of more overt violence. One woman, Amy, recalled a passerby making light of the recent murder of a sex worker in the city, noting,

> These guys the other night there must have been about six of them and I know men when they get together have

> that sort of mob mentality . . . They were like 50/55 and all six of them had a different comment like "Oh well you should be in the river" . . . like one of them said, "Shouldn't you be in the river?" . . . They all said something different and I was thinking "you fucking filthy assholes" you know? . . . But we get that all the time. (Amy, Christchurch)

The areas in which the women worked in Wellington and Christchurch are known as areas in which street sex workers are working. Most of the women working in these areas dressed for the purposes of attracting clients and stood on corners, meaning that they could be easily identified as sex workers. They could do this easily as sex work is decriminalized, and they did not have to fear being identified by the authorities. While this was beneficial in enabling them to attract clients and to manage risks of violence from potential clients through screening processes (Armstrong, 2014), their visibility also meant that they can be easily identified by passersby who wanted to harass them. The verbal abuse described above by Amy indicates that this harassment was specifically targeted at her not just as a woman but as a woman who sells sex. Thus, it can be argued that the verbal abuse directed at sex workers on the street is targeted at them specifically because they are sex workers, and can have darker undercurrents that underline their vulnerability to being targeted with more overt violence while working in the open space.

Another interesting factor of the verbal abuse described by the women interviewed is that women were often the perpetrators of this harassment It is commonly understood that men are the perpetrators of street harassment toward women in the open space. However, several of the women interviewed recalled experiencing verbal abuse delivered to them by other women. Shannon recalled, "I remember one night I was out there and there was these four girls in a car . . . driving up and down abusing the fucking shit out of me." Similarly, Sydney noted, "The worst is when a girl goes past within a car of guys and she's yelling out 'slut' and 'whore'."

The verbal abuse that was described by the women was therefore not only sexualized in nature but was also underpinned by internalized misogyny as both men and women were perpetrators. In verbally abusing sex workers, the female perpetrators were able to position themselves above the sex working women by publically scolding them for stepping outside of the boundaries of acceptable female behavior. In doing so, they could safely locate themselves within the category of "good" women and be reassured that they were not like "those" women. Therefore, while all women are vulnerable to experiencing street harassment perpetrated by men, these findings suggest that women who sell sex are targeted with verbal abuse by a broader range of perpetrators.

Physical Abuse on the Street

All of the women without exception had encountered physical abuse from passersby. Most commonly, this involved eggs, bottles, and other objects being thrown at the women from cars, typically accompanied by verbal abuse. However, a few women reported being shot at with pellet guns, and one woman had been physically and sexually assaulted by a male passerby. Recalling this incident, Amy explained,

> I said "hello" to him and he didn't say "hello" and I was like "oh that's weird" because normally people, you know, say "hello." And, yeah it was weird . . . He stopped . . . And then he crossed the road and was probably quite a bit away from me, still quite a good distance. And I turned around and asked if he was alright because it was like he was lost or something. And he just didn't answer me. And I turned around and he had me just like that. (Amy)

While this degree of physical violence was unusual, all of the women described instances in which they had objects thrown at them by passersby. Key informants also demonstrated an acute awareness of these sorts of incidents. One Police key informant noted,

> Oh I think they get harassed a lot. I mean I've seen people throw bottles at them, like get a drink bottle and throw water . . . I've heard people yelling out to them. (Key informant, Police)

The experiences of street harassment described by the women therefore extended beyond the more common forms of street harassment that is directed toward women in general, and included sexualized verbal abuse perpetrated by both women and men, sometimes with particularly sinister undertones, and physical abuse that ranged from the throwing of objects to physical and sexual assault.

Responding to Street Harassment

Research on women's experiences of street harassment has found that this type of harassment is often minimized, considered "normal" and lower level than other forms of sexual violence (Fileborn, 2013). In agreement with this finding, the majority of the women interviewed described ignoring the verbal abuse they encountered or responding to it liberally. Street harassment was a frequent occurrence and was something that all of the women expected to experience while working on the street. Emotionally reacting to the abuse was considered futile when it was evident that some people were deliberately coming to the street for the purposes of harassing sex working women. Bianca, for instance, described how she responded to verbal abuse. She explained,

> They're drunk, you know, and they're smart and when they get smart to me I go "Well you should be at home with your mummy" [Laughs] . . . They know there's sex workers [in the area]. And I think to myself, "Well, why do they drive around there when there's other ways they could drive around to get to where they want?" But they just drive around just to get smart. Because they're bored and there's no one else they can pick on. But then I just say, "Well, have a lovely night!" [laughs] (Bianca)

For the most part, the women ignored the verbal abuse they experienced because they felt that responding to it would result in increased stress, would impact negatively on their work, and would be unlikely to resolve the situation. Deltah noted,

> The ones that scream and yell at you at the side of the road, well I don't listen to that. You just hear a whole lot of yelling and you just block it out because at the end of the day if you let it get to you then you've just let them ruin your whole night of work I just let it go in one ear and out the other. (Deltah)

Key informants also overwhelmingly understood that the women accepted this type of abuse as inevitable and ignored it, despite the fact they should not have to tolerate this behavior. One key informant noted, "I guess a lot of the girls out there build a wall to it and they ignore it . . . It shouldn't be like that."

While most of the women described ignoring the abuse, they did not passively accept this victimization. According to Goffman (1990), members of stigmatized groups internalize negative portrayals of their identity and are silenced by a fear of judgment, which is termed "felt stigma." However, consistent with other research findings, it appeared that the women resisted internalizing these underlying messages. Instead, they reflected on the ideas implicit in the insults directed at them (Abel & Fitzgerald, 2010; Sallmann, 2010). Reflecting on the harassment that was directed at her by other women, Hollie explained,

> I find it a crack up because, you know, the girls . . . They're just jealous because we can get our money out there and do what we do and make more money than them. And they just yell out and abuse us because they can't do it. You know, they'd have a big cry about it [doing sex work] . . . It's like "Ok well if you're going to yell out that then you get out here and see how you go." You know, "You go and give it away for free—what do you get a couple of drinks out of it?" (Hollie)

Similarly, Sydney noted,

> The worst is when a girl goes past within a car of guys and she's yelling out "slut" and ""whore" and you think "you're the one with all these guys and you're probably rooting them without protection." So who's the slut, who's the whore? (Sydney)

A few women described actively challenging the verbal abuse by defiantly replying to it and refusing to buy into content that was intended to stigmatize and disrespect them (Armstrong, 2010). Vixen, for instance, recalled responding to the calling out of "*whore*" with the sarcastic reply, "*Oh thanks for telling me, I forgot!*" Similarly, Catherine described how she responded to verbal abuse while working on the street:

> Usually I don't react because that just encourages them. But sometimes . . . when they say something really stupid then I will yell out to them . . . Well you know I had someone call out "you fucking slut" and I was like "prostitute actually." "Why don't you get a real job," "well I'm at university actually" and they don't like that at all . . . And the fact is, even if we are out there because we want to have some fun or because we've got a drugs habit or whatever, it's our choice and they have no right to make that judgment. (Catherine)

However, while a few women did describe responding to frequent verbal abuse in some situations, it was clear that the perpetrators were generally unlikely to engage in any direct confrontation. Shannon recounted an experience where she was repeatedly verbally abused by a group of young women in a car:

> I say to them "what fucking wimps get out of the car and come over here." . . . Well the lights at Tuam Street went red and the traffic was backed up and they were right on my corner and I went up to the car and you want to see how fast they locked their doors. You know, just mouth and too scared . . . at the end of the day they're wimps, they're only doing it to show off. (Shannon)

This example suggests that while the perpetrators of verbal abuse wished to engage with sex workers from a distance, this enthusiasm did not extend to any direct or meaningful engagement. This reinforces the extent to which women who sell sex are "othered" and the significance of this othering in legitimizing the harassment of sex workers on the street. While the people in the car felt able to harass Shannon, they were only willing to do so at a distance. This suggests that the harassment of streetbased sex workers is not only opportunistic, but it is also motivated by fear and curiosity, an "us and them" mentality through which women who sell sex are viewed simultaneously as both scary and interesting to these abusive passersby. Stigma is clearly at the heart of this harassment. Several of the women indicated that they knew a lack of understanding was integral to the harassment directed at them. Holly noted,

> Oh yeah . . . you get shit every time man. You get eggs thrown at you, like glass bottles and stuff thrown at you ay . . . they don't know me they just think I'm some slut that works on the corner . . . they just drive off ay they don't really come back. Too scared and stuff you know knowing "Oh fuck I shouldn't have done that." There are idiots out there because they don't realize you're out there to do it for yourself. Like they don't realize how hard it is because their dad and mum's fucking rich and got jobs and they get what they want. But try doing something for yourself ay, and getting it your own way. Yeah. (Zoe, Christchurch)

The experience of verbal abuse, despite its frequency, was generally minimized by the women as unimportant. However, there was less tolerance of the incidence of physical abuse among the women interviewed. Deltah noted, "I don't like the ones that throw things out the windows 'cause I've been hit by a few eggs and a lot of bottles. . . that's horrible ay." Similarly, Kay explained that while she would not generally respond to verbal abuse, she felt that physical abuse needed to be addressed: "I don't comment unless they're getting really smart and throwing shit . . . because that's not cool, throwing things at girls . . . They could seriously injure someone." However, in agreement with the findings of previous research, the women were unlikely to report what was considered "lower level" physical abuse, such as having bottles thrown at them, to the police (Abel, 2010; Campbell & Kinnell, 2000; Dalla, Xia, & Kennedy, 2003; Delacoste & Alexander, 1988; Downe, 1999; Harris, Nilan, & Kirby, 2011; Kinnell, 2006; Shannon et al., 2009; Silbert & Pines, 1981, 1982; Woodward, Fischer, Najman, & Dunne, 2004). Deltah explained that while she would record the vehicle's details, she was reluctant to report these incidents to the police, feeling that this would have a detrimental impact on her work:

> Yeah I normally just take their license plate numbers if I can see them . . . Sometimes I won't go to the police because it just kills my night and I end up sitting at the police station when I could be out there making money. I just come home and get changed and go back out and make money instead of just going to the cop shop because there's not very much they can do. (Deltah)

Although only a few of the women said that they would report the individuals who threw objects at them at work, in Christchurch the police were already well aware that sex workers were frequently abused by some members of the public. Indeed, all of the key informants commented on the extent and nature of abuse that was experienced by sex workers on the street. One police key informant reported making an effort to address this behavior when possible:

> Oh I think they get harassed a lot. I mean I've seen people throw bottles at them, like get a drink bottle and throw water at them and I've stopped and had a go at people about that. And I think things like that they just put up with it. They don't really report that type of thing . . . I think they almost see that that's sort of part of their occupation . . . I mean I have actually seen, I've heard people yelling out to them and I mean I always stop and talk to people doing that sort of thing. (Key informant, Police)

It was evident that some police were genuinely committed to supporting streetbased sex workers to challenge the abuse and harassment they experienced while working on the street. One woman, Claire, recalled reporting an experience she had while working on the street:

> A car went past me and they pulled out a rifle . . . Yeah and shot it. And I happened to be on the side and it just went past me—the bullet. And I had a friend with me who witnessed it as well. And the cameras and stuff had seen it . . . I reported it to [youth workers] . . . And they took me down to the police station and made some complaints which I'd never done before. And . . . it just felt good, I mean having them idiots going past and doing that. I mean they're probably one out of many who do that. And it's not only guns, it's bottles too . . . Everyone's got opinions on the police but yeah I did actually have a little respect for them after that. I mean at least those little boys can't go around shooting people. (Claire, Christchurch)

This example highlights how having a supportive experience of reporting abuse can have a significant impact not only in addressing harassment on the street but also in increasing confidence in the police among some sex workers. It appeared that this was particularly important for younger people working on the street who tended to have less confidence in the extent to which the police could assist them in the event of any kind of abuse or harassment. While older sex workers either ignored verbal abuse and the throwing of objects, or reported license plates to the police, among some younger people there was a tendency to retaliate with violence toward abusive passersby. One key informant explained,

> I think a lot of the older workers get a lot more hardened to it and they can ignore it a lot more. The young ones get really volatile. Sometimes they'll throw things back or so it can end up turning into just a big bloody mess especially when often you might have one person working but behind the fence there's 15 of their friends. So you can imagine what can sometimes end up happening and the original perpetrator of the stupid behavior ends up getting absolutely hammered. (Key informant, Youth)

The stronger reactions of younger workers could also reflect lower levels of tolerance among a new generation of post-law reform street workers to the harassment they experience. This emphasizes further the great importance of police continuing to build relationships with street sex workers and demonstrating a willingness to respond to the harassment they experience.

Discussion

The PRA was passed in 2003 in the spirit of improving the health and safety of sex workers. The inclusion of street-based sex work as part of the legislation was unique as street workers have historically been stigmatized and criminalized to a greater extent than indoor sex workers. The decriminalization of street-based sex work transformed the selling of sex on the streets from an illegal and clandestine activity to a form of labor that individuals can freely engage in. However, the experiences documented in this article suggest that this transformation has not been entirely borne out in the experiences of street-based sex workers. The sex workers who were interviewed for this research recounted numerous experiences of harassment, which included verbal and physical abuse perpetrated by passersby. This finding in itself is not surprising for a number of reasons. One reason is the fact that street harassment is commonly experienced by *all* women. Another is that numerous studies throughout the world have documented the street harassment that sex workers experience in particular. However, the important contribution made by this article is in discussing the significance of this continued harassment in the decriminalized context, what might be done about it, and how this harassment fits into the broader picture of violence against women throughout western society.

Many of the women interviewed for this research expressed an ambivalent attitude toward the frequent harassment they experienced while working on the street. In relation to verbal abuse, there was a sense that these experiences were to be expected, normalized as inevitable background noise that could not be silenced. At the same time, the women resisted the messages implicit in the harassment directed at them. Several women demonstrated an acute awareness that the stigma attached to streetbased sex workers was at the heart of the harassment they experienced. These findings are significant in demonstrating how stigma continues to prevail, even when sex work is decriminalized. In no other occupation would daily verbal abuse and having things thrown at you be considered an occupational hazard. The decriminalization of sex work had affected the women in contradictory ways. On the one hand, working in a decriminalized context meant that they could be more visible while working on the street as there was no risk of arrest. This made it easier to manage risks of violence from potential clients as the women could wait on the street for as long as needed before deciding to get in a car (Armstrong, 2014). On the other hand, this increased visibility arguably made them more vulnerable to being harassed by passersby, who could easily pick them out as they waited on their corner for potential clients. What this suggests is that providing an environment that is conducive to the safety and wellbeing of *all* sex workers requires much more than a law change that removes the criminality from their work. The success of the PRA is limited by social conditions in which it operates. The PRA provided legal change to improve the lives of street-based sex workers by enabling them to work without the fear of arrest. What is now needed is social change, so that street workers are accepted as part of the society, are respected in the open space, and can work in an environment where abuse and harassment are the exception rather than the rule.

The harassment that the women described shared commonalities with the street harassment that women report experiencing in their day-to-day lives (harassment that they too may experience when they are not working on the street). For instance, it was often very sexualized in nature and based on disparaging ideas about women and sexuality. However, it also differed in that it was dealt out by both male and female perpetrators. The content of the harassment sometimes also had a somewhat darker undercurrent that hinted at their vulnerability to more severe violence. Liz Kelly's concept of a continuum of sexual violence provides a useful framework in which to locate these experiences. Kelly (1987) argues that the concept of a continuum is useful in highlighting the fact that all women experience sexual violence at some point in their lives, and that more common everyday abuses that women encounter can be linked to the less common experiences that are classed in law as crimes. The same can be said for the harassment that women experience while working on the streets and the harassment all women can experience when going about their everyday lives. The harassment meted out to the women interviewed represents, to some extent, a magnified version of the more common forms of street harassment that all women can experience. Dominant ideas about women and sexuality that continue to be prevalent support a common view that women who do street-based sex work are lesser than other women and are deserving of violence through the very act of publically soliciting for sex. While some women may be more likely to experience harassment that is characterized by more subtle, everyday sexism in the form of wolf whistles and catcalls, the harassment that street-based sex workers experience can be seen as raw, uncensored misogyny, with no ambiguity over the intentions of the deliverer. It can be argued that these women were targeted with harassment that is specifically directed at them as sex workers, by perpetrators who endorse negative views of women and sexuality. However, the harassment they experience forms part of the same picture of violence against all women, as all forms of sexual violence are underpinned by the same structures and the same social and cultural attitudes (Kissling, 1991).

When the harassment that these women experienced is considered in the context of violence against all women in general, the fact that women were also perpetrators of this harassment is significant. The type of harassment described by the women most often involved verbal insults such as "slut" and "whore," and reflects internalized misogyny among groups of women. This is further reflected in the ways in which some of the sex working women responded to these insults, turning it back on the woman who delivered it by questioning their assumed identity as "good" women. Sydney for instance felt that it was ironic that she was called a slut by women who she surmised may have casual unprotected sex without payment with male acquaintances, causing her to ponder, "*Who's the slut, who's the whore?*" In their research on young women's use of alcohol and social media, Hutton, Griffin, Lyons, Niland, and McCreanor (2015) found that some young women placed themselves in opposition to other women who they labeled as drunk, slutty, and out of control, which the authors termed "positioned othering." This concept of positioned othering is useful in conceptualizing the ways in which the women were harassed by other women while working on the street, and how they in turn responded to this harassment. The harassment they experienced and their responses to it can both be considered a form of positioned othering in which the women doing the harassing disassociated themselves from the sex workers, presumably to avoid the label of the slut or the whore themselves. In this situation, Sydney disassociated herself from the women who harassed her by speculating that their behavior was in fact less respectable than hers as a professional sex worker. What is evident in this exchange is that negative ideas of women and sexuality are so ingrained and are sometimes held and reinforced by women themselves. The street harassment that sex workers in particular experience is therefore deeply complex and relates to factors far broader than the laws surrounding sex work.

Tackling the street harassment that street sex workers experience therefore relies on dismantling the structures that support the subjugation of all women. It also relies on women being part of the change themselves by supporting other women rather than tearing those down who do not adhere to gender norms. What this requires though is the transformation of these norms so that women's worth is not measured against their sexual behavior, and all women are afforded the same respect.

In conclusion, the decriminalization of street-based sex work provided a symbolic shift in the societal positioning of women who sell sex. However, their status is still largely determined by social norms that control the behavior of all women and punish those who do not conform. Creating streets that can be a safe workplace for women who sell sex therefore requires far more dramatic change than a change in the law, rather a fundamental shift in societal views of women and sexuality in general.

Declaration of Conflicting Interests

The author(s) declared no potential conflicts of interest with respect to the research, authorship, and/or publication of this article.

References

Abel, G. (2010). *Decriminalisation: A harm minimisation and human rights approach to regulating sex work* (Unpublished doctoral dissertation). University of Otago, Dunedin, New Zealand.

Abel, G., & Fitzgerald, L. (2010). Decriminalisation and Stigma. In G. Abel., L. Fitzgerald., & C. Healy (Eds.), *Taking the crime out of sex work: New Zealand sex workers' fight for decriminalisation* (pp. 239–259). Bristol, UK: Policy Press.

Abel, G., Fitzgerald, L., & Brunton, C. (2007). *The impact of the Prostitution Reform Act on the health and safety practices of sex workers.* Christchurch, New Zealand: Department of Public Health and General Practice, University of Otago.

Armstrong, L. (2010). Out of the shadows (and into a bit of light): Decriminalisation, human rights and street-based sex work in New Zealand. In K. Hardy, S. Kingston, & T. Sanders (Eds.), *New sociologies of sex work* (pp. 39–55). Farnham, UK: Ashgate.

Armstrong, L. (2011). *Managing risks of violence in decriminalised street-based sex work: A feminist (sex worker rights) perspective* (Unpublished doctoral dissertation). Victoria University of Wellington, Wellington, New Zealand.

Armstrong, L. (2014). Screening clients in a decriminalised street-based sex industry: Insights into the experiences of New Zealand sex workers. *Australian & New Zealand Journal of* Criminology, 47, 207–222.

Benson, C., & Matthews, R. (2000). Police and prostitution: Vice squads in Britain. In R. Weitzer (Ed.), *Sex for sale: Prostitution, pornography, and the sex industry* (pp. 245–264). New York, NY: Routledge.

Brents, B., & Sanders, T. (2010). Mainstreaming the sex industry: Economic inclusion and social ambivalence. *Journal of Law and Society, 37, 40–60.*

Brooks-Gordon, B. (2008). State violence towards sex workers: Police power should be reduced and sex workers' autonomy and status raised. *British Medical Journal, 337,* a908.

Campbell, R., & Kinnell, H. (2000). "We shouldn't have to put up with this": Street sex work and violence. *Criminal Justice Matters, 42,* 12–13.

Dalla, R. L., Xia, Y., & Kennedy, H. (2003). "You just give them what they want and pray they don't kill you": Street-level sex workers' reports of victimization, personal resources, and coping strategies. *Violence Against Women, 9,* 1367–1394.

Dally, J. (2013, November 13). Prostitute copped frustration. *stuff.co.nz.* Retrieved from http://www.stuff.co.nz/national/9392726/Prostitute-copped-frustration

Day, S., Ward, H., & Boynton, P. M. (2001). Violence towards female prostitutes. *British Medical Journal, 323,* 230

Day, S., & Ward, H. (2007). British policy makes sex workers vulnerable. *British Medical Journal, 334,* 187.

Delacoste, F., & Alexander, P. (1988). *Sex work: Writings by women in the sex industry.* London, England: Virago Press.

Downe, P. J. (1999). Laughing when it hurts: Humor and violence in the lives of Costa Rican prostitutes. *Women's Studies International Forum, 22,* 63–78.

Fileborn, B. (2013). *Conceptual understandings and prevalence of sexual harassment and street harassment* (ACSSA Resource Sheet). Melbourne: Australian Centre for the Study of Sexual Assault.

Goffman, E. (1990). Stigma*: Notes on the management of a spoiled identity*. London, England: Penguin Books.

Goodyear, M., & Cusick, L. (2007). Protection of sex workers. *British Medical Journal, 334,* 52–53.

Harris, M., Nilan, P., & Kirby, E. (2011). Risk and risk management for Australian sex workers. *Qualitative Health Research, 21,* 386–398.

Hubbard, P. (2004). Cleansing the metropolis: Sex work and the politics of zero tolerance. *Urban Studies,* 41, 1687–1702.

Hubbard, P. (2006). Out of touch and out of time? The contemporary policing of sex work. In R. Campbell & M. O'Neill (Eds.), *Sex work now* (pp. 1–32). Cullompton, UK: Willan.

Hubbard, P., & Sanders, T. (2003). Making space for sex work: Female street prostitution and the production of urban space. *International Journal of Urban and Regional Research, 27,* 73–87.

Hutton, F., Griffin, C., Lyons, A., Niland, P., & McCreanor, T. (2015). *"Tragic girls" and "crack whores": Alcohol, femininity and facebook.* Manuscript submitted for publication.

Jeffrey, L.A., & Sullivan, B. (2009). Canadian sex work policy for the 21st century: Enhancing rights and safety, lessons from Australia. *Canadian Political Science Review, 3,* 57–76.

Johnson, M., &.Bennett, E. (2015). *Everyday sexism; Australian women's experiences of street harassment.* Canberra, Australian Capital Territory, Australia: The Australia Institute.

Kelly, L. (1987). The continuum of sexual violence. In J. Hanmer & M. Maynard (Eds.), *Women, violence and social control* (pp. 46-60). Basingstoke, UK: Macmillan.

Kinnell, H. (2008). *Violence and sex work in Britain.* Cullompton, UK: Willan.

Kissling, E. (1991). Sexual harassment: The language of sexual terrorism. *Discourse & Society, 2,* 451–460.

Loren, A. (2013, January 29). Prostitution bill delayed again. *Manukau Courier*. Retrieved from http://www.stuff.co.nz/auckland/local-news/manukau-courier/8234780/Prostitution-billdelayed-again

Manukau City Council (2010). *Manukau City Council (Regulation of prostitution in specified places) bil.* Retrieved 19 May 2013 from http://www.legislation.govt.nz/bill/local/2010/0197/latest/versions.aspx*l.*

Macmillan, R., Nierobisz, A., & Welsh, S. (2000). Experiencing the streets: Harassment and perceptions of safety among women. *Journal of Research in Crime & Delinquency, 37,* 306–322.

Miller, J., & Schwartz, M.D. (1995). Rape myths and violence against street prostitutes. *Deviant Behavior, 16,* 1–23.

Morgan Thomas. R. (2009). From "toleration" to zero tolerance: A view from the ground in Scotland. In J. Phoenix (Ed.), *Regulating sex for sale: Prostitution policy reform in the UK* (pp. 137-158). Bristol, UK: Policy Press.

Mossman, E., & Mayhew, P. (2007). *Central government aims and local government responses: The Prostitution Reform Act* 2003. Wellington, New Zealand: Ministry of Justice.

Nixon, K., Tutty, L., Dowoe, P., Gorkoff, K., & Ursel, J. (2002). The everyday occurrence: Violence in the lives of girls exploited through prostitution. *Violence Against Women, 8,* 1016–1104.

Pitcher, J., Campbell, R., Hubbard, P., O'Neill, M., & Scoular, J. (2006). *Living and working in areas of street sex work: From conflict to coexistence.* Bristol, UK: Joseph Rowntree Foundation.

Pyett, P., & Warr, D. (1999). Women at risk in sex work: strategies for survival. *Journal of Sociology, 35(2),* 183–197.

Richardson, D., & May, H. (1999). Deserving victims?: Sexual status and the social construction of violence. *The Sociological Review, 47,* 308–331.

Rosen, E., & Venkatesh, S. A. (2008). A "perversion" of choice: Sex work offers just enough in Chicago's urban ghetto. *Journal of Contemporary Ethnography,* 37, 417–441.

Russell, D. (1984). *Sexual exploitation: Rape, child sexual abuse and workplace harassment.* London, England: SAGE.

Sagar, T. (2007). Tackling on-street sex work: Anti-social behaviour orders, sex workers and inclusive inter-agency initiatives. *Criminology and Criminal Justice,* 7(2), 153–168.

Sallmann, J. (2010). Living with stigma: Women's experiences of prostitution and substance use. *Ajjilia, 25,* 146–159.

Sanders, T., O'Neill, M., & Pitcher, J. (2009). *Prostitution, policy and politics.* London, England: SAGE.

Scrambler, G. (1997). Conspicuous and inconspicuous sex work: The neglect of the ordinary and the mundane. In G. Scrambler & A. Scrambler (Eds.), *Rethinking prostitution, purchasing sex in the 1990s* (pp. 105–120). London, England: Routledge.

Shannon, K., Kerr, T., Strathdee, S., Shoveller, J., Montaner, J., & Tyndall, M. (2009). Prevalence and structural correlates of gender based violence among a prospective cohort of female sex workers. *British Medical Journal, 339,* b2939.

Sharpe, K. (1998). Red light, blue light: *Prostitutes, punters and the police.* Aldershot, UK: Ashgate.

Silbert, M. H., & Pines, A.M. (1981). Occupational hazards of street prostitutes. *Criminal Justice and Behavior*, 8, 395–399.

Stanko, E. (1985). Intimate intrusions: *Women's experiences of male violence, rape, child sexual abuse and sexual harassment.* London, England: Routledge.

Sullivan, B. (2008). Working in.the sex industry in Australia: The reorganisation of sex work in Queensland in the wake of law reform. *Labour and Industry,* 18, 73–92.

West, J. (2000). Prostitution: Collectives and the politics of regulation. *Gender, Work & Organization,* 7, 106–118.

Woodward, C., Fischer, J., Najman, J. M., & Dunne, M. (2004). *Selling sex in Queensland.* Brisbane, Queensland, Australia: Prostitution Licensing Authority.

Wotton, R. (2005). The relationship between street-based sex workers and the police in the effectiveness of HIV prevention strategies. *Research for Sex Work, 8,* 11–13.

Introduction to Reading 49

As discussed in the Chapter Introduction, the enforcement of conformity to the sex/gender/sexuality binary takes many forms and includes mechanisms that are subtle as well as blatantly coercive. In this reading, sociologist Stef M. Shuster addresses one of the most subtle and pervasive modes of gender policing—everyday gendered language. Shuster, a transidentified person, interviewed forty transgender people and reports on their experiences with language as a covert and overt vehicle through which binary gender expectations are conveyed and enforced. Talk, as Shuster says, "conveys our deeply held beliefs and assumptions" that people can or should be easily sorted into two and only two gender categories. In the vignettes presented in this reading, trans people discuss the ways in which they are expected to conform to incorrect binary gender determinations through the mechanisms of covert and overt discursive aggression.

1. Define the term "discursive aggression."
2. What is "self-enforcement accountability?" How does it operate to enforce conformity to binary gender?
3. What is "other-enforcement accountability?" How does this form of accountability illustrate how discursive aggression is built into systems of power?
4. Return to the reading by Lucal in Chapter One and explore the links between Lucal's experiences and those of the trans people reported on in this reading.

Punctuating Accountability

How Discursive Aggression Regulates Transgender People

Stef M. Shuster

As social life is organized by the assumption of a dichotomous gender system that structures social relations, trans people move through everyday life fraught with tension over the possibility of experiencing aggressive interactions based on the guiding expectation of a "two-and-only-two" (Lucal 1999) gender system.[1] The impetus to categorize people is so pervasive that individuals do not, as suggested by Lucal, "even know how to interact with such persons who are neither a woman nor man, and have no way of addressing such a person that does not rely on making an assumption about the person's gender" (1999, 782). One mechanism for assigning a gender to an individual, with limited information about that individual's identity, is through cultural schemas, or frameworks of information that enable quick cognitive processes (Hollander, Renfrow, and Howard 2011). But for trans people, existing schemas do not accommodate the reality that not every individual identifies as a cisgender woman or man.

Another mechanism for how gendered expectations shape social encounters and may detrimentally impact trans individuals is in the concept of gender determination processes (Westbrook and Schilt 2014), which illustrate how gender is an exchange in interactions. People present information about their gender ("doing gender"), while others interpret this information and place them in gender categories ("determining gender"). In interaction, "accountability structures" shape how gender determination processes play out. While typically used in the existing scholarship to describe a manner of holding others accountable to expectations, Hollander (2013) demonstrates that accountability processes also shape how individuals align themselves to cultural expectations for behavior in interaction. Missing from these accounts is the central role of power in shaping expectations for how interactions should unfold (Dennis and Martin 2005; Fine 1993).

Recent work in the sociology of trans studies has attended to the reification of inequality in medical providers' decision making (shuster 2016); the workplace (Schilt 2010); concerns for safety upon transitioning (Abelson 2014); and redefining the family in cisgender-transgender partnerships (Pfeffer 2012). Many of these studies are implicitly guided by the tenets of symbolic interactionism in exploring the everyday experiences of trans people, and the ways in which trans people (and relational partners) make meaning of their identities at the micro-level of interaction. Curiously, the explicit role of language and discourse in shaping everyday experience has been absent. Yet language is a mechanism through which individuals give meaning to social life (Hollander and Abelson 2014) and a crucial component of people's identities and experiences in social interaction.

This article is guided by the questions, How is power mobilized in everyday encounters to reinforce gendered expectations and behavior, and by whom? By bridging insights from scholars of language and cognition and those from gender accountability processes, I focus on how language and talk uphold social order and regulate gender in interaction. Talk and embodied communication both figure centrally in the lives of trans people and reveal important cues about gender expectations. I introduce "discursive aggression" as a concept to describe how communicative acts are used in social interaction to hold people accountable to social and cultural-based expectations (i.e., other-enforcement), and how individuals hold themselves accountable in anticipating the unfolding of interactions (i.e., self-enforcement).

Based on 40 interviews, I present five vignettes that exemplify the common themes that trans people narrated in experiencing discursive aggression in everyday life. For self-enforcement accountability, these include: acknowledging the collective power of an

Shuster, S. M. (2017). Punctuating accountability: How discursive aggression regulates transgender people. *Gender & Society, 31*(4): 481–502.
Reprinted by permission of SAGE Publications, Inc. on behalf of Sociologists for Women in Society.

audience; subverting one's needs for others; and acquiescing to the power structure when an individual holds less power. For other-enforcement accountability, these include having the entitlement to ask invasive questions, and using a dominant status to define a subordinated identity. Discursive aggression contributes to the scholarship on language, cognition, and accountability processes by explicitly placing power at the center of understanding micro-inequalities in everyday life, while adding to the body of scholarship in the sociology of trans studies.

Cognition, Accountability, and Discourse

Theories of social cognition take up the matter of how we think about our social worlds. These theories describe cultural schemas that enable quick cognitive processing, particularly in moments of uncertainty or ambiguity in interaction (Hollander, Renfrow, and Howard 2011). Cultural schemas about gender perpetuate the assumption that all people are either cisgender women or men, with no other permutation of identification readily available in our collective imaginings. Within this limited structure of meaning-making, trans people map onto existing schemas in uneasy ways as they negotiate social life continuously facing the possibility that their gender identification may not be correctly perceived by others. In these instances of friction between the doing and determining of gender (Westbrook and Schilt 2014), the social order begins to break down as the coherence of everyday life is destabilized.

Cognitive theories related to schemas offer the tools to analyze how social structure affects the everyday behaviors of individuals, the role of categories in shaping our perceptions of other people, and expectations for how interactions should unfold. Meanwhile, symbolic interactionist theories exemplify how everyday interactions come to constitute social life. It is through thought, after all, that individuals interpret symbols (Hollander, Renfrow, and Howard 2011). For example, a prevalent symbolic ordering of social life is based on gender categories that are assumed to be stable and natural (West and Zimmerman 1987). But as demonstrated in the work of Lucal (1999) and others whose gender displays provoke uncertainty in interaction, gender determinations can shift for an individual depending on gendered cues, interpretations, and social context. When gender determinations are found to be incorrect, or schemas do not help individuals resolve interactional uncertainty, efforts are made to restore order. Thus, successful social relations require all participants to present, monitor, and interpret gender displays (West and Zimmerman 1987).

Taken collectively, social interaction becomes coherent through mutually shared understandings of symbols, rules that govern behavior, and expectations rooted in social structure. Individuals give meaning to interactions by defining the situation and resolving contradictions and inconsistencies between actors' individual understandings and interpretations of a situation by creating a common perspective that facilitates the accomplishment of a mutual goal (Hollander, Renfrow, and Howard 2011, 9). While cognitive theories have tended to assume that individuals will strive to maintain order in social life, and symbolic interactionism assumes that individuals have shared understandings of symbolic displays, power is an overarching feature of everyday life. How those with more power may feel entitled to define the situation and, because of their social positioning, regulate others to uphold the social order remains an under-examined feature of maintaining and (re)producing inequality (Cook 2006; Fine 1993; Hollander, Renfrow, and Howard 2011).

Gender, Talk, and Social Order

Communicative acts are an important means through which individuals relay expectations and restore or maintain order in interaction. Gender expectations are reflected and inflected within language (Butler 2011). Phrases such as "working mother" or "career woman," for example, shape the experiences of women by defining them in relation to restricted gender roles that further perpetuate gender-based inequality (Eder, Evans, and Parker 1995). These subtle linguistic indicators are part and parcel of the "cultural routines" (Goffman 1959) that make interactions predictable and possible. Like schemas, cultural routines minimize ambiguity in how individuals are perceived as gendered, while restricting the range of assumed genders to either women or men. There are few widely available references in language that do not presume all people are cisgender and that an individual's gender is recognizable by sensory-based information alone (Friedman 2013). As linguistic structures narrowly define the range of acceptable gender presentations, trans people often are not received in social life in the gender-specific ways they desire to be recognized. In this interactive process, language both reproduces gender norms and regulates interaction (Butler 2011; Speer 2005).

When cultural routines begin to break down and cause people to be uncertain about another's gender, it is a "violation of social rules, and signals that something is amiss in interaction" (Lucal 1999, 791). To maintain or restore social order in interaction, participants may engage in boundary maintenance by regulating others through talk. For example, Pascoe (2007) demonstrated that cisgender boys in high school used a "fag discourse" to substantiate their own masculinity by calling that of others into question. The epithet "fag" sanctioned those who were not masculine (enough) or not acting in accordance with the normative standards of masculinity. While Pascoe's (2007) work explored how language and cultural expectations regarding gender play out in the lives of boys and in the institutional context of schools, the central point remains clear: language and talk are crucial foundations for the way we think about, process, and analyze information and people. Talk conveys our deeply held beliefs and assumptions about social order.

Accountability Processes Regulate Interaction Through Discursive Aggression

The existing scholarship on gendered accountability processes has added an important dimension to theories of how gender, culture, and expectations shape interactions in everyday life. Working within West and Zimmerman's (1987) now-classic research on accountability processes and "doing gender," many scholars have emphasized how people are held accountable to gender expectations, or to the enforcement of those expectations by others. As Hollander (2013) suggested, this focus on otherenforcement glosses over how accountability also functions through self-enforcement. That is, how an individual approaches social interaction is shaped by our knowledge and assumptions of how other people think we should behave in social life (Hollander 2013, 4).

Yet, while the growing body of scholarship on accountability processes, gender determinations, and schemas has offered important theoretical accountings of how interactional dynamics are shaped by broader cultural structures, power is a missing though vital part of interactional exchanges (Collins 2008; Cook 2006; Giddens 1984). Holding someone accountable to expectations is claiming power to define the situation and the people who make up the interaction. In everyday life experiences, it becomes clear that for trans people inhabiting a subordinated identity often translates to having less power in interactional exchanges to resist other-enforcement or redefine the situation. Discursive aggression enables an examination of how power is maintained in interaction through talk. These communicative acts are used to hold people accountable to social and cultural-based expectations while reinforcing the inequality in everyday life. There are penalties involved in not doing gender correctly, or in failing to meet the expectations of an individual in interaction. Discursive aggression is one way for individuals to restore order or coherence in social life when expectations for how people ought to be are not realized. Furthermore, these are encounters marked by people socially regulating those who disrupt or create ambiguity in the expectations that govern interactions. As those in subordinated positions are aware of the cultural expectations that bear down in everyday life (Cook 2006), the imagined possibility of discursive aggression also provokes self-enforcement in interaction.

Methods

I conducted 40 in-depth interviews with trans people in a midwestern metropolitan area (referred to throughout as "Metromidwest"). I interviewed transsexuals, people who embrace "transgender" as a catch-all identity label, non-binary identified trans people, and people who at one time identified as trans but no longer wish to be known as trans, living presently in social life in their transitioned gender. All identity labels and pronouns used in this paper correspond with how participants self-identified in the interview. All names that appear are pseudonyms.[2]

To find interviewees, I used a snowball sampling method, which began with personal networks.

As a trans-identified person, I was aware that trans people are mistrustful of researchers because of historically biased research. I addressed this concern by being transparent about my own background and identities. To further establish rapport, I was open to answering questions interviewees had about my lived experiences as a trans person. Interviews ranged in length from one to three hours.

In focusing explicitly on dominant language systems, and rooted in gendered expectations for how interactions should unfold, I found that all of the respondents in my sample—regardless of gender identity, sexual identity, class, race, and age—described frequent experiences with discursive aggression.

Five major subthemes emerged from coding the data. The percentage of respondents who narrated experiences is presented in parentheses. Three subthemes for self-enforcement accountability included acknowledging the collective power of an audience (60%), subverting one's needs for others (85%), and acquiescing to the power structure when an individual holds less power (45%). For other-enforcement accountability, two themes emerged: having the entitlement to ask invasive questions (75%) and using a dominant status to define a subordinated identity (80%). I selected five vignettes as exemplars of each of these themes. Through composing vignettes, I prioritized the voice of participants in the presentation of data (Sprague 2016), helped exemplify the ways that participants narrated accountability processes, and enabled theory-building activities from the patterns that emerged from the interviews.

Self- and Other-Enforcement Accountability

How individuals use language and talk in interaction signals expectations for how interactions should unfold. The regulating force of language becomes particularly salient in interactions where trans people are not referred to or identified in the gender-specific ways they wish to be known. I present five vignettes to explore these interactional moments and, as recalled from the experiences of trans people, to show how language and gender expectations materialize as discursive aggression in everyday life and regulate trans people through both self- and otherenforcement accountability processes.

Self-Enforcement

Acknowledging the collective power of an audience.

One rainy fall afternoon I met in a local pub with Alex, a 28-year-old Latino trans man, to talk about his experiences living in Metromidwest. Alex shared that since physically transitioning at the age of 22, he had, over time, become disconnected from family, friends, and partners, and that he felt isolated from the broader trans community as a trans person of color. Part of this isolation, he reflected, might also stem from his wish to be received in everyday life as "simply a man" and not have his trans background widely known or "read" by others. Thus, Alex is incredibly sensitive to people who misread his gender because it has a double layer of feeling; as he described, "I am not being read by others as a man, while simultaneously being clocked as a trans person."[3] I asked if he could provide an example of being mis-gendered. Without pause he shared,

> One day I was standing in line to get my food and the guy working behind the counter was going down the line for people waiting and quickly calling out, "Sir, ma'am, sir, ma'am" to get everyone's food ordered. He is about to get to me and I just had this feeling. I knew he was going to fuck up. And he's like "Ma'am can I take your order?" I was mortified, and I was in front of so many people. Like, I'm just going to pretend that didn't happen. I'm not going to think about it, it didn't happen. You know, it's something you learn to do just to survive.

In this particular interaction, Alex had anticipated that the food service worker would mis-gender him. But also importantly, this story occurred in a public space with many other people waiting around for their food to be called out. In imagining how other people overheard the mis-gendering, Alex engaged in a self-enforcement accountability process by pretending that it did not happen. Not wanting to call attention to the mistake and draw the interest of others waiting in line with him, Alex perceived that there were few options to redefine the situation and correct the worker without calling further attention to himself that might have outed him as a trans person. The collective power of an audience in public spaces to sustain the interaction and keep social order intact translated to Alex fading back into line and holding himself accountable to the mis-gendering by not acknowledging the mistake.

Using references like "sir" or "ma'am" is a common way to confer respect upon an individual. These references are built upon the assumption that attributing gender is uncomplicated, and that by visual cues alone, one can determine the gender of a person. But, as the above excerpt suggests, when accompanied by incorrect gender attributions what is intended to be a respectful term of reference can become a form of discursive aggression. While the worker likely had no idea that he incorrectly gendered Alex, what matters is that Alex noticed it and was mortified that it had happened in front of so many people. In these instances, for people whose gender identity does not seamlessly translate to observable cues, language structures interaction in a way that gendered terms of respect become discursive aggressions when an incorrect gender is attributed to an individual.

Cultural schemas, and the language structure that helps individuals make sense of others, narrow the range of gender determinations to a binary. While Alex is a trans man and experiences the burden of being mis-gendered as a woman, people who identify with genders outside of the binary related similar experiences in social life. The difference between binary and non-binary people's experiences in interactions such as the one that Alex shared is that, while the collective power of an audience helps to maintain order in interactions by remaining silent when people are mis-gendered, the assumption of a binary gender system is reified through language and talk, since no commonly used references of respect exist that would correctly recognize non-binary people's identities.

Subverting one's needs for others.

Many interviewees turned to their relationships with family members as sites of conflict laden with aggression. Upon transitioning, familial labels become fraught with tension around gendered signifiers and embedded in interactional power struggles whereby trans people may assert their wish to be referred with different labels (e.g., aunt or dad) while family members may assert their own needs for comfort, stability, and maintaining the illusion that nothing has changed in the family. Given the lower power position that trans people inhabit compared to their cisgender counterparts, trans people may subvert their needs to accommodate others.

Lauren, a 73-year-old White trans woman, reflected on the changing dynamics within her family as her gender shifted more towards a feminine presentation. Lauren shared how prior to marrying her wife in the early 1960s she had identified as a woman but kept her trans identity a secret out of fears related to the social stigma and isolation that many trans people experienced during that time. Over the years, Lauren and her wife had several children. Once the children grew up, Lauren came out to her wife as a trans woman. At home Lauren would present as a woman, and in public life she presented as a man. She continued to keep her identity a secret from her children, because she believed it would help keep the family intact. After living for decades in what she described as a "double life," Lauren decided to tell her children about her gender identity so that she could "stop hiding" from them. Part of this urge, Lauren further explained, was that she felt a growing distance between herself and her daughter.

Upon coming out to her daughter, she shared that her daughter's response was to "put her arms around me and say, 'Dad, don't worry. I love you no matter what.'" Lauren smiled in the retelling of this story. She continued, "It felt incredible. So, so good. Like I could finally be myself around her." At this moment in the interview there was a long pause, and Lauren's body language shifted to a slight huddle. After several minutes of silence, Lauren looked up, teary eyed, and quietly shared that while it felt really good in the moment, like "a release of some sort," her daughter's use of the term "dad" was incredibly difficult for her to hear. She went on to describe how it felt like "a dismissiveness of her identity," as the word "dad" indicated that her daughter still thought of her as a man.

For some trans people like Lauren, it takes years to come out to family members because these intimate relational contexts involve people who have known each other for a long time. While Lauren shared that it felt affirming for her daughter to reassure her of her love, using a gendered signifier such as "dad" brought with it complicated feelings. In this instance, gender and heteronormativity combine to uphold the social order. Lauren's daughter supports her as a trans person but subtly communicates the expectation that Lauren will remain "dad" regardless of her coming out as a trans woman.

Lauren's story also demonstrates the double-burden that trans people may face in negotiating gendered language, schemas, and accountability processes. On one hand, Lauren is the parental figure and would be assumed to have more power in the situation. On the other hand, this narrative involves a grown child who has autonomy as an adult. Lauren engaged in self-enforcement by subverting her needs to be recognized as a woman in deference to her family member's needs to maintain the comfort of thinking of Lauren as "dad." Lauren withheld challenging this assumption, as she faced the risk of losing her family altogether. It is the perception of the possible response from her family if she "push[ed] them too far" that created a power imbalance in the negotiation of self-identification and gender determinations by a family member. Engaging in significant emotional labor on behalf of her family members, Lauren iterates a process where she had already asked "so much" of her family in accepting that she is a trans woman that she could not ask for anything more, nor could she assert her right to be referred to as "mom."

These interactions uphold the differential power afforded her daughter in using the familial label with which she felt more comfortable, while undercutting Lauren's agency. Further, the lack of gender-neutral familial labels, or discomfort of family members to change existing labels, signifies how language systems structure interactions to produce discursive aggression. These labels indicate how dominant notions of "family" are aligned with cultural expectations surrounding not only who composes families, but also how to name and make meaning of trans family members in ways that appropriately recognize their gender identities.

Acquiescing to the power structure, when you hold less power.

Samuel, an 18-year-old White trans guy, grew up in a middle-class family in a small suburban area outside of the city. He shared that he came from a fairly supportive family but still had some difficulty in high school coming out as a trans person, because there was not a cohesive trans community in the suburbs and his friendships were with heterosexual cisgender boys. In spite of his family's support, Samuel never quite felt like he fit in because, he said, "Most of the guys just thought of me as some cool chick who was kind of butch and liked to play video games."

Heading off to college, Samuel was excited to enroll in women's studies courses to enable him to have greater access to information about gender diversity and situate him in classes that were more affirming of "trans kids like him." Yet, in contrast to what he imagined his experiences as a first year trans college student would be like in "those kinds of classes," he recounted the daily forms of discursive aggression he experienced around pronouns of reference:

> My teachers are kind of forgetful about my pronouns. In one of my classes, this teacher just cannot remember that I go by male pronouns. This was kind of disappointing for me because this is a women's studies class and I kind of expected that this teacher would be more aware. But she uses female pronouns pretty constantly in reference to me. I have talked to her, e-mailed her, and my friend who is in my class even corrects her sometimes. But {pause} it's probably just because she has a lot of students. I recognize my voice is higher. Yeah {pause} I think it's mostly my voice or my body type, which is like more female. I just assume that they are not doing it maliciously, they're just forgetful.

Pronouns reflect moments when an individual makes a quick inference about another person's gender within existing cultural schemas. Across the interviews, trans people identified pronouns as a common source of daily struggle. Yet many trans people led with the assumption that the person enacting aggression was not intending to harm them. As demonstrated in Samuel's account of consistently being mispronounced in class, there was a tension in how he narrated his experiences. Listening closely to what Samuel shared, he first described how he went through extensive effort to remind his professor of the correct pronouns of reference, as she continued to mispronoun him in class and in front of his peers. He concluded the story by placing the responsibility of the discursive aggression on himself because of his high voice, holding himself accountable to the gender display rule of "male until proven otherwise" because his voice discounted his masculine gender presentation. The professor's behavior was excused away because of Samuel's assumption that she had no malicious intent. But Samuel was left with few ways to correct her, leaving the regulating behavior intact and assumptions about his gender unchecked because the professor continued to use an auditory cue (voice) to determine Samuel's gender.

In contrast to Alex and Lauren, who silenced their needs because of the anticipation of how others might respond, Samuel asserted his agency to the professor by correcting her on multiple occasions. But in the end, and with an unbalanced power relation in student-teacher interactions, he could not force his professor to change her behavior. In this interactional exchange, the professor has more power to define the situation and can dismiss Samuel's assertion to be recognized as a man. For Samuel to continue to challenge the professor's definition of the situation would subject him to the possibility of failing the class, receiving a reprimand, or being further scrutinized by his peers.

Discursive aggression that continues to be enacted by the same people reflects how encounters are wrapped into the gender system and gender expectations that shape communication patterns. These recurring encounters are reflective of how power contributes to self-enforcement by trans people who inhabit lower-power positions in interaction. The burden of responsibility to maintain social order shifts from the person who—often unintentionally—enacts discursive aggression onto the person who experiences it. When these interventions are met with dismissiveness or inaction on the part of the people who initiate the aggression, it signals a disregard for trans people's agency when self advocating a claim to a gender identity. Samuel, like Alex and Lauren, engaged in emotional labor by acquiescing to the power imbalance in the interaction and putting in significant work to make himself feel OK in spite of the professor's refusal to change her behavior and communication patterns

Other-Enforcement

In the vignettes that have been presented so far, language was a covert vehicle through which gender expectations were communicated, and discursive aggression was employed as a communicative act to help uphold the social order. Overt discursive aggression, however, comes to take on a different quality of character. Culled from the same cultural norms and expectations governing gender and interactions as covert discursive aggression, more overt communicative acts demonstrate how discursive aggression is enmeshed in systems of power that regulate and shape social life for marginalized groups in other-enforcement accountability mechanisms.

Having the entitlement to ask invasive questions.

Vic, a 23-year-old Latino trans man, shared an experience of overt discursive aggression while going to the dentist. During the patient intake process, Vic included testosterone on his list of current medications. Upon returning the forms to the front desk, the dental staff questioned why he was taking testosterone. He offered that he was transitioning from female to male and used testosterone to facilitate the transition. In response to this disclosure, for several minutes the dental staff continued to persist in their questions related to testosterone and Vic's gender presentation, eventually sending him back to the waiting area so that they could confer with the office manager. After several minutes, the office staff called Vic to the front desk and shared that they would change his medical records to mark him as female in his file.

What is striking about Vic's story is that within this setting the office staff did not ask Vic what sex marker appeared on his insurance card or government-issued identification card, nor did they ask him his own preference for how he might be identified in his records. In service exchanges, the entitlement to ask questions beyond the purview of professional concern or the service offered signals a power imbalance that Vic had little recourse to address. Beyond certain dental procedures such as X-rays and medications that might present physical harm to a fetus during pregnancy, most dental practices and procedures are not differentiated by one's sex assignment at birth. But here, Vic's requests were denied under the auspices of staff "needing to know" for health-related reasons, which further upholds the power embedded in medical expertise in patient-provider interactions. Instead of maintaining his records to reflect the ways in which he wanted to be acknowledged in the office as a man, his wishes were ignored. His records were changed to reflect his sex assignment at birth, rather than Vic's expressed gender identity. In reflecting upon his experiences in the dental office, and other medical settings, Vic shared:

> It feels like there is a lack of education in the wider medical community. When my trans status comes up, people try to be polite. But they still interrogate me on why I mark myself as male. That is what it says on my driver's license, that is how I present in the world. It becomes complicated when doctors have to do treatments related to my sex assigned at birth. But even at the dentist's office?

There were alternate actions that the dental staff may have enacted, such as affirming Vic's gender identity as a man but making a note in the file to ask him about his pregnancy status for subsequent visits. Mandating that the paperwork associated with Vic corresponded with how they thought his gender should be identified, and by verbally confronting Vic, the office staff suggested that they did not believe Vic had authority over his gender or sex, and that they were justified in exerting power over him.

This scenario also highlights the entitlement with which some people in authority positions believe they have to question a person's self-defined gender and to hold others accountable to dominant narratives surrounding sex and gender. This example from the dentist's office, as well as countless other examples from trans people narrating challenges associated in everyday life, speaks to the far-reaching ways that institutional settings perpetuate inequality and restricts (trans) people. The social patterning of gender expectations, and the gender binary contained in language, uphold assumptions regarding the correspondence of sex and gender categories. While people who work within institutional settings are shaped by the same language systems and cultural expectations as those who move through them, there are ways they can choose to circumvent linguistic barriers by changing the systems, record-keeping, and filing processes to avoid exerting an expert status while disregarding people's right to selfdetermine and identify their own gender and sex.

Using dominant status to define a subordinated identity.

Claiming expertise over another person's gender does not exclusively occur in institutional contexts, such as health care settings that ate characterized by power imbalances. Many trans people shared stories about moving through social spaces and having other people dismiss or challenge their claims to a specific gender identity. Similar to Vic, it is these moments in interactions that trans people interviewed narrated as being incredibly difficult to address. For some, choosing not to respond to discursive aggression when it occurs is a matter of protecting one's emotional, physical, or bodily integrity. For others, there is a lose-lose situation where, because of the very structure of language that assumes the gender binary, there is little room to negotiate identity-challenges.

Jax, a 22-year-old American Indian genderfucked person, used the nonbinary pronouns of ze and zir. Ze expresses zir gender in a variety of ways, depending on the social situation. Zir identity as genderfucked, ze explained, is characterized as "a deliberate performance, a playfulness, and also a political stance on both the shifting terrains and construction of gender." Jax shared that ze faces resistance from many people in social life in honoring zir pronouns and gender playfulness. As one who inhabits a non-binary identity, Jax negotiates language structures that cannot accommodate zir identity, pronouns, or embodied meaning-making. Thus, a lasting problem in how language and gender norms come to shape the experiences of trans people is rooted in the structure of language itself, which reinforces a two-gender system.

Jax shared that sometimes even within trans spaces—sites where one may reasonably expect not to have to deal with transphobia and gender policing—people are weary of building alliances or community with zir. Engaging in respectability politics, or in a boundary maintenance strategy to exclude some members of a community who are perceived to make others "look bad," some trans people regulate others who do not fit the dominant narrative that defines "transgender" as being equated with transitioning from one intelligible gender binary category to the other. Within this logic, "real" trans people are those who physically transition.

Yet the respectability politics and gender policing extends beyond trans-on-trans regulation. Sometimes, even cisgender allies engage in these behaviors. At one social gathering that was for trans people and allies, Jax was approached by a cisgender partner of a trans woman. According to Jax, she told ze was not "really a trans person because I was not taking hormones or transitioning. And then she told me that I shouldn't even be there because I wasn't trans. I looked around and trans people were watching this cis person telling me that I did not belong there. No one said anything!" Jax's claim to a trans identity and community space was dismissed because ze was not transitioning in the ways typically discussed within socio-medical realms. Furthermore, trans bystanders did not come to the aid of Jax, upholding the perceived right of a cisgender person to regulate Jax's claim to trans community spaces.

Non-binary people will continue to face daily slights as they negotiate both a structure of language that relies on a binary to organize gender and a normative accounting of who inhabits a transgender identity. Like the dental staff working with Vic who changed his records to reflect what they presumed his gender should be, the cisgender partner from Jax's story exerted her cisgender privilege, taking on an "expert" status in deciding who counts as transgender by virtue of having a trans partner. That the cisgender person felt emboldened to use her dominant status to define a subordinated identity and dismiss Jax from the space implies that she believed others would support her assertion of authority to define the space as one that held no room for Jax.

Conclusions

This paper examined how in interaction talk is used to enforce expectations and regulate gender. Given the cultural expectation that all people fit within the binary gender system (Lucal 1999), trans people must negotiate social life and expectations that do not accommodate them. As my work demonstrates, part of the difficulty that trans people face in everyday life is rooted in cognitive processes used to maintain and restore social order. Schemas help individuals filter through the chaos of social life by enabling cognitive shortcuts to make meaning of vast amounts of information (Hollander, Renfrow, and Howard 2011). Schemas also shape expectations as to how interaction should unfold. But, in relying on cognitive shortcuts to make meaning of others, the people who do not fit within existing schemas will continue to be miscategorized misrepresented and to face significant barriers in everyday life.

My work adds a conceptual tool to the existing scholarship on microinequalities. Through talk, discursive aggression regulates trans people in everyday social settings and produces for them the feeling that they are not received in the ways they wish to be known, that they are made invisible, and that their self-authorship in naming and claiming a gender identity is questioned. In these interactional exchanges, trans people have limited options to respond to discursive aggression. Some, like Samuel, may at first attempt to address the mistakes that are made. But most of the interviewees eventually acquiesced to the power structure. In being aware of others' expectations for how interactions should unfold, trans people may engage in self-silencing to uphold the social order

Building on the work of Hollander (2013) and others, accountability processes help explain the perseverance of inequality in social life. For other-enforcement accountability processes, trans people's agency to assert their identities and claims to social life are undermined. In enforcing expectations, such as a two-gender binary system, trans people are held accountable to implausible standards of conforming to cisgendercentric understandings of how trans people should behave or identify, or they have their autonomy dismissed outright. Further, there are limited possibilities for responding to incorrect gender determinations in public spaces, as those with more power in the interaction can enforce expectations about how the interaction should unfold. Those with more power regulate "proper" gender expression and work to realign or redefine ambiguity to cohere with cultural expectations. In relation to self-enforcement accountability, trans people anticipate the responses of public and private interactions and silence themselves to maintain social order. Rather than to assert their rights to claim an identity or to inhabit social spaces, at moments it may be a matter of safety (emotional and/or physical) and bodily integrity to choose not to respond to discursive aggression. This particular dimension of accountability processes is crucial to understanding how power may be invoked in interaction, and how subordinated groups put in significant work to help others "save face" by not correcting mistakes, engage in emotional labor to prioritize the needs of family members and friends over their own needs, and are boxed in by restrictive cultural expectations.

Notes

1. I use the term "transgender" as an inclusive term to refer to people whose gender does not correspond with the gender assigned at birth. "Cisgender" refers to people whose gender corresponds with gender assigned at birth. "Transgender" and "trans" are used interchangeably in this paper.
2. This project was approved by the Institutional Review board at the University of Iowa.
3. Being "clocked" refers to moments in interaction when a trans person is read in social life as trans person.

References

Abelson, Miriam. 2014. Dangerous privilege: Trans men, masculinities, and changing perceptions of safety. *Sociological Forum* 29 (3): 549–70.

Butler, Judith. 2011. *Bodies that matter: On the discursive limits of "sex."* NewYork: Routledge.

Collins, Randall. 2008. *Violence: A micro-sociological theory*. Princeton, NJ: Princeton University Press.

Cook, Karen J 2006. Doing difference and accountability in restorative justice conferences. *Theoretical Criminology* 10:107–24.

Dennis, Alex, and Peter J. Martin. 2005. Symbolic interactionism and the concept of power. *British Journal of Sociology* 56 (2): 191–213.

Eder, Donna, Catherine Colleen Evans, and Stephen Parker. 1995. *School talk: Gender and adolescent culture*. New Brunswick, NJ: Rutgers University Press.

Fine, Gary Alan. 1993. The sad demise, mysterious disappearance, and glorious triumph of symbolic interactionism. *Annual Review of Sociology* 19:61–87.

Friedman, Asia. 2013. *Blind to sameness: Sexpectations and the social construction of male and female bodies*. Chicago: University of Chicago Press.

Giddens, Anthony. 1984. *The constitution of society: Outline of the theory of structuration*. Berkeley: University of California Press.

Goffman, Erving. 1959. *The presentation of self in everyday life*. Garden City, NY: Doubleday.

Hollander, Jocelyn A. 2013. "I demand more of people": Accountability, interaction, and gender change. *Gender & Society* 27:5–29.

Hollander, Jocelyn A., and Miriam Abelson. 2014. Language and talk. In *Handbook of the social psychology of inequality*, edited by Jane McLeod, Edward Lawler, and Michael Schwalbe. New York: Springer.

Hollander, Jocelyn A., Daniel G. Renfrow, and Judith A. Howard. 2011. *Gendered situations, gendered selves: A gender lens on social psychology.* Lanham, MD: Rowman & Littlefield.

Lucal, Betsey. 1999. What it means to be gendered me: Life on the boundaries of a dichotomous gender system. *Gender & Society* 13:781–97.

Pascoe, C. J. 2007. *Dude, you're a fag: Masculinity and sexuality in high school.* Berkeley: University of California Press.

Pfeffer, Carla A. 2012. Normative resistance and inventive pragmatism: Negotiating structure and agency in transgender families. *Gender & Society* 26 (4): 574–602.

Schilt, Kristen. 2010. *Just one of the guys?: Transgender men and the persistence of gender inequality.* Chicago: University of Chicago Press.

shuster, stef. 2016. Uncertain expertise and the limitations of clinical guidelines in transgender healthcare. *Journal of Health and Social Behavior* 57 (3):319–32.

Speer, Susan. 2005. The interactional organization of the gender atttibution process. *Sociology* 39 (I): 67–87.

Sprague, Joey. 2016. *Feminist methodologies for critical researchers: Bridging differences,* 2nd ed. Lanham, MD: Rowman & Littlefield.

West, Candace, and Don Zimmerman. 1987. Doing Gender. *Gender & Society* 1:125–51.

Westbrook, Laurel, and Kristen Schilt. 2014. Doing gender, determining gender: Transgender people, gender panics, and the maintenance of the sex/gender/ sexuality system. *Gender & Society* 28 (1): 32–57.

Topics for Further Examination

- Jackson Katz is well-known for his pro-feminist scholarship and activism. Katz's work includes calling on men to take responsibility for violence against women including sexual harassment and domestic violence. Listen to his Ted Talk and summarize his major arguments for the critical involvement of men in stopping violence against women. See Jackson Katz: Violence against women—it's a men's issue - TED Talks.
- Read "The Insidious Symbolism of Boy and Girl Bikes" by sociologist Lisa Wade (https://thesocietypages.org/socimages/2016/08/26/insisting-on-boys-and-girls-bikes/). Consider how the gendering of bikes reinforces the gender binary and inequality. Next, consider the many gendered objects that comprise your life and play a similar role in enforcing conformity to a gender binary.
- Homophobia is understood by sociologists who study gender to be a key mechanism in the policing of masculinity among men and boys. C. J. Pascoe's (2007) work on the use of the fag epithet as a gendered slur used by adolescent boys to ensure that they were seen as masculine by other boys. How does this behavior persist among men in college and in other settings? How does it reinforce masculine power, control, and domination?

PART III

POSSIBILITIES

10

Nothing Is Forever

Catherine G. Valentine

The title of this chapter represents the principle that change is inevitable. Like the ever-evolving patterns of the kaleidoscope, change is inherent in all life's patterns. Anything can be changed and everything does change, from the cells in our bodies to global politics. There is no permanent pattern, no one way of experiencing or doing anything that lasts forever. This fact of life can be scary, but it can also be energizing. The mystery of life, like the wonder of the kaleidoscope, rests in not knowing precisely what will come next.

The readings in this chapter address the changing terrain of gender. If one takes only a snapshot of life, it may appear as though current gender arrangements are relatively fixed. However, an expanded view of gender, over time and across cultures, reveals the well-researched fact that gender meanings and practices are as dynamic as any other aspect of life. Patterns of gender continuously undergo change, and they do so at every level of experience, from the individual to the global. Michael Schwalbe (2001) observes that there is both chance and pattern in the lives of individuals and in the bigger arena of social institutions. He makes the point that no matter how many rules there might be and no matter how much we know about a particular person or situation, "social life remains a swirl of contingencies out of which can emerge events that no one expects" (p. 127). As a result, life, including its gendered dimensions, is full of possibilities.

Social constructionist theory is especially helpful in understanding the inevitability of change in the gender order. Recall that social constructionist research reveals the processes by which people create and maintain the institution of gender. It underscores the fact that gender is a human invention, not a biological absolute. Particular gender patterns keep going only as long as people share the same ideas about gender and keep doing masculinity and femininity in a routine, predictable fashion (Carrera, DePalma, & Lameiras, 2012; Johnson, 1997; Schwalbe, 2001). Given that humans create gender, gender patterns can be altered by people who, individually and collectively, choose to invent and negotiate new ways of thinking about and doing gender.

At the micro level of daily interaction, individuals participate in destabilizing the binary, oppositional sex/gender/sexuality order. They do so by choosing to bend conventional gender rules or changing the rules altogether by undoing or redoing gender (Bobel & Kwan, 2011; Deutsch, 2007; Lorber, 1994; West & Zimmerman, 2009). For example, women and men are creating new forms of partnership based on shared care work and housework roles. Other individuals purposefully transgress the boundaries of sexual and gender identities by mixing appearance cues via makeup, clothing, hairstyle, and other modes of self-presentation (Bobel & Kwan, 2011; Lorber, 1994). Chris Bobel and Samantha Kwan's (2011) research illustrates a variety of ways people employ their bodies in acts of gender resistance and related forms of resistance such as counter-heteronormativity and counter-homonormativity. For example, within the U.S. gay community, big men or "bears" are masculine-presenting, fat, hairy, gay and bisexual men who, with their admirers ("chasers"), have formed alternative spaces where they can interact in comfort. In addition, they display their bodies with pride and challenge the appearance norms of dominant gay society and the heterosexual world (Pyle & Klein, 2011). Research on embodied resistance, such as bear culture, points to the powerful social fact that "humans can be at once rule-bound and wonderfully inventive agents of social change. We can enact the mandates—trudging along, submitting and rationalizing—but we can also assert ourselves and break away" (Bobel & Kwan, 2011, p. 2). When we do the latter, we engage our potential to alter toxic social patterns such as gender inequality.

Trends

At the macro level of the gender order, change comes about through large-scale forces and processes, both planned and unplanned. Trends are unplanned changes in patterns that are sustained over time. For example, Peter Kivisto (2011) states that the Industrial Revolution is a trend, marking the transition from agricultural to industrial economies. This so-called revolution involves complex economic, technological, and related changes, such as urbanization, that have profoundly altered the fabric of social life over time. Consider the impact of industrialization on gender in work and family life in the United States. Prior to industrialization, women's labor was essential to agricultural life. Women, men, and children worked side by side to grow crops, make clothing, raise animals, and otherwise contribute to the family economy (Lorber, 2001). That is, work and family were closely intertwined and the distinction between home and workplace did not exist (Wharton, 2005).

As the Industrial Revolution got under way, productive or waged work moved from the home into factories and other specialist work sites, and work came to be defined as valuable only if it resulted in a paycheck. Although essential work was still done at home, it typically did not produce income. The negative outcome was that household labor was transformed into an invisible and devalued activity. Work and family came to be defined as distinct, firmly gendered domains of life, especially in the White middle class. Women and children were relegated to the home and "good" women were expected to be full-time housewives and mothers, while men were ordained to follow wage work in the capitalist market, embrace the breadwinner role, and participate in the political arena (Godwin & Risman, 2001; Wharton, 2005).

The profound changes in gender relations and the organization of work and family wrought by the Industrial Revolution continue to be a source of conflict for many women and men in the United States today. For example, although most heterosexual married women with children work outside the home, the doctrine of natural separate spheres—unpaid household work for women and paid work for men—continues to operate as an ideal against which "working women" who have children are often negatively evaluated.

Industrialization continues as a force for social change, one that is amplified and altered by processes of globalization. The term *globalization* refers to the increasing interconnectedness of social, cultural, political, and economic activities worldwide (Held, McGrew, Goldblatt, & Perraton, 1999; Connell, 2005). Transnational forces such as geopolitical conflicts, global markets, transnational corporations, transnational media, and the migration of labor now strongly influence what happens in specific countries and locales (Connell, 2000). For example, the international trading system—dominated by nations such as the United States—encompasses almost every country in the world, while films and television programs, especially those produced in the West, circulate the globe (Barber, 2002).

Raewyn Connell (2000) argues that globalization has created a worldwide gender order. This world gender order has several interacting dimensions: (1) a gender division of labor in a "global factory" in which poor women and children provide cheap labor for transnational corporations owned by businessmen from the major economic powers, (2) the marginalization of women in international politics, and (3) the dominance of Western gender symbolism in transnational media.

However, despite the order Connell posits, globalization is not monolithic. There are countervailing forces challenging the homogenizing and hegemonic aspects of globalization. For example, indigenous cultures interact with global cultures to produce new cultural forms of art and music. In addition, globalization has spawned transnational social movements such as the reproductive justice movement, environmental justice movement, and domestic workers movement, which address worldwide problems of Western hegemony, global inequality, and human rights (Connell, 2000).

Social Movements

Large-scale change may also come about in a planned fashion. Social movements are prime examples of change that people deliberately and purposefully create. They are conscious, organized, collective efforts to work toward cultural and institutional change and share distinctive features, including organization, consciousness, noninstitutionalized strategies (such as boycotts and protest marches), and prolonged duration (Kuumba, 2001). The United States has a long history of people joining together in organizations and movements to bring about justice and equality. The labor union movement; socialist movement; civil rights movement; and gay, lesbian, bisexual, transgender, intersex movement have been among the important vehicles for change that might not otherwise have happened.

One of the most durable and flexible social movements is feminism (Ferree & Mueller, 2004). Consider the fact that the feminist movement has already lasted for more than two centuries. At the opening of the 19th century, feminism emerged in the United States and Europe. By the early 20th century, feminist organizations appeared in urban centers around the world. By the turn of the 21st century, feminism had grown into a transnational movement in which groups work at local and global levels to address militarism, global capitalism, racism, poverty, violence against women, economic autonomy for women, and other issues of justice, human rights, and peace (Shaw & Lee, 2001).

One of the major challenges facing feminist movements today is creating an intersectional feminism (see Chapter 2) that enables activists to work together across differences and inequalities among women rooted in cultural, national, religious, and other social differences and inequalities.(Hewitt, 2011; Ryan, 2001). Simply put, "gender is but one strand of oppression among many" (Motta, Flesher Fominaya, Eschie, & Cox, 2011, p.13), and alliances need to be forged across intramovement differences (Hewitt, 2011; Motta et al., 2011). This problem takes many shapes. Not only is it a matter, for instance, of class-privileged women and poor women or White women and women of color forming working alliances, but it is also a matter of women and men being attuned to pitfalls in thinking and organizing across individual and collective differences and inequalities (Motta et al., 2011). The #MeToo movement and the #TimesUp initiative provide us with an immediate example of collective organizing across multiple borders, both intra-national and international. Tarana Burke, founder of that movement, has emphasized that #MeToo is not just a movement for famous White cisgendered women. It is global in its reach and as Burke says "[s]exual violence knows no race, class, or gender, but the response to it does" (Austin-Evelyn, 2018, np).

The chapter reading on reproductive justice (Briggs et al. in this chapter) is a layered and rich examination of the global and American terrain of reproductive politics. This article highlights the limitations of what is largely a Western, White, middle-class feminism focused on individual choice (choice feminism), and it does so through the lens of the transnational reproductive justice movement. The closing words of Rosalind Petchesky, reproductive justice scholar and activist, underscore the significance of cross-border, cross-boundary feminisms. She calls for feminists to think deeply about systemic change and to build broad coalitions with other social justice and antiracist movements. Rachel Thwaites offers an in-depth critique of choice feminism in her chapter reading. Like Briggs et al., she calls for feminists to avoid "uncritical unthinkingness" (p. 67) and to embrace political engagement that is visionary. Two other readings in this chapter examine effective strategies for building bridges across intersectionalities and between feminist organizations. Courtney Martin and Vanessa Valenti assess the critical role of online feminism in the 21st century. They demonstrate how feminist blogs have become a vehicle for consciousness raising and activism that makes a difference. Online feminists and traditional feminist organizations have become allies across boundaries and borders and, together, have successfully altered the genderscape. Martin and Valenti offer clear accounts of the radicalizing force of online feminism. In a second reading, Cathy J. Cohen and Sarah J. Jackson discuss the important role of digital media in allowing marginalized people an activist platform. They also highlight the key ways in which feminism has contributed to the movement for Black lives (Black Lives Matter) with an emphasis on the centering of gay, lesbian, trans, and queer people as members and leaders.

The Complexity of Change

Not only is social change pervasive at micro and macro levels of life and a function of both planned and unplanned processes, it is also uneven and complex (Ridgeway, 2009). Change doesn't unfold in a linear, predictable fashion, and it may be dramatically visible or may take us by surprise. Consider the passage of the 19th Amendment to the U.S. Constitution in 1920, which guaranteed women the right to vote. This one historic moment uplifted the public status of women and did so in a visible fashion. But more often, change consists of alterations in the fabric of gender relations that are not immediately visible to us, both in their determinants and their consequences.

For instance, we now know that a complex set of factors facilitated the entry of large numbers of single and married women into the paid workforce and higher education in the second half of the 20th century. Those factors included very broad economic, political, and technological developments that transformed the United States into an urban, industrial capitalist nation (Stone & McKee, 1998). Yet no one predicted the extent of change in gender attitudes and relations that would follow the entry of women into the workforce. It is only "after the fact" that the implications have been identified and assessed. For example, heterosexual marital relationships in the United States have moved toward greater equity in response to the reality that most married, heterosexual women are not dependent on their husbands' earnings. As married women have increasingly embraced paid work, their spouses have increasingly reconceptualized and rearranged their priorities so they can devote more attention to parenthood (Goldscheider & Rogers, 2001).

In his chapter reading, Michael Messner explores the complexities of recent change in gender relations in the United States by looking at men's planned responses to feminism, ranging from anti-feminist to pro-feminist. His analysis focuses on three large trends (e.g., deindustrialization and the rise of the neoliberal state) of the 1980s and 1990s that have generated particular forms of men's anti-feminist and pro-feminist activism. Messner ends his reading by emphasizing the critical role of large-scale changes created by social movements, i.e., feminist movements, alongside changes in the political-economy in generating significant shifts in the gender order.

Prisms of Gender and Change

Returning briefly to the metaphor of the kaleidoscope, let us recall that the prism of gender interacts with a complex array of social prisms of difference and inequality, such as race and sexual orientation. The prisms produce ever-changing patterns at micro and macro levels of life. Our metaphor points to yet another important principle of dynamic gender arrangements. We can link gender change to alterations in other structural dimensions of society, such as race, class, and age. For example, as Americans have moved toward greater consciousness and enactment of gender equality, they have also come to greater consciousness about the roles that heterosexism (i.e., the institutionalization of heterosexuality as the only legitimate form of sexual expression) and homophobia (i.e., the fear and hatred of homosexuality) play in reinforcing rigid gender stereotypes and relationships (see Chapter 6). It has become clear to many seeking gender justice that the justice sought after cannot be achieved without eliminating homophobia and the heterosexist framework of social institutions such as family and work.

Additionally, gender transformation in the United States is inextricably tied to movements for racial equality. This is true both historically and today. The first wave of feminism was an outgrowth of the antislavery movement, and the politics of racial justice led to the second wave of feminism (Freedman, 2002). Racism, as well as ageism, classism, and other forms of oppression, had to be addressed by feminists, because the struggle to achieve equal worth for women had to include all women and men. Anything less would mean failure.

The Inevitability of Change

Collectively, the articles in this chapter invite the reader to ask, "Why should I care about or get involved in promoting change in the gender status quo?" That is a good question. After all, why should one go to the trouble of departing from the standard package of gender practices and relationships? Change requires effort and entails risk. On the other hand, the cost of "going with the flow" can be high. There are no safe places to hide from change. Even if we choose "not to rock the boat" by closing ourselves off to inner and outer awareness, change will find us. There are two reasons for this fact of life. First, we cannot live in society without affecting others and in turn being affected by them. Each individual life intertwines with the lives of many other people, and our words and actions have consequences, both helpful and harmful. Every step we take and every choice we make affect the quality of life for a multitude of people. If we choose to wear blinders to our connections with others, we run the risk of inadvertently diminishing their chances, and our own, of living fulfilling lives (Schwalbe, 2001). For example, when a person tells a demeaning joke about women, he or she may intend no harm; however, the (unintended) consequences are harmful. The joke reinforces negative stereotypes, and telling the joke gives other people permission to be disrespectful to women (Schwalbe, 2001).

Second, we can't escape broad, societal changes in gender relations. By definition, institutional- and societal-level change wraps its arms around us all. Think about the widespread impact of laws such as the Equal Pay Act and Title VII, outlawing discrimination against women and people of color, or consider how sexual harassment legislation has redefined and altered relationships in a wide array of organizational settings. Reflect on the enormous impact of the large numbers of women who have entered the workforce since the latter half of the 20th century. The cumulative effect of the sheer numbers of women in the workforce has been revolutionary in its impact on gender relations in family, work, education, law, and other institutions and societal structures.

Given the inevitability of change in gender practices and relationships, it makes good sense to cultivate awareness of who we are and what our responsibilities to one another are. Without awareness, we cannot exercise control over our actions and their impact on others. Social forces shape us, but those forces change. Every transformation in societal patterns reverberates through our lives. Developing the "social literacy" to make sense of the changing links between our personal experience and the dynamics of social patterns can aid us in making informed, responsible choices (O'Brien, 1999; Schwalbe, 2001).

References

Austin-Evelyn, K. (2018). From #MeToo to a movement: Building an intersectional, global framework for transformative change. *Ms. Magazine*. Retrieved June 2, 2018 (http://msmagazine.com/blog/2018/05/03/metoo-movement-building-intersectional=).

Barber, B. R. (2002). Jihad vs. McWorld. In G. Ritzer (Ed.), *McDonaldization: The reader* (pp. 191–198). Thousand Oaks, CA: Pine Forge Press.

Bobel, C., & Kwan, S. (2011). Introduction. In C. Bobel & S. Kwan (Eds.), *Embodied resistance: Challenging the norms, breaking the rules* (pp. 1–10). Nashville, TN: Vanderbilt University Press.

Carrera, M. V., DePalma, R., & Lameiras, M. (2012). Sex/gender identity: Moving beyond fixed and "natural" categories. *Sexualities, 15*(8), 995–1016.

Chowdhury, E. H. (2009). Locating global feminisms elsewhere: Braiding US women of color and transnational feminisms. *Cultural Dynamics, 21*(1), 51–78.

Connell, R. W. (2000). *The men and the boys.* Berkeley: University of California Press.

Connell, R. W. (2005). Change among the gatekeepers: Men, masculinities, and gender equality in the global arena. *Signs: Journal of Women in Culture and Society, 3*(30), 1801–1825.

Deutsch, F. (2007). Undoing gender. *Gender & Society, 21,* 106–127.

Ferree, M., & Mueller, C. (2004). Feminism and the women's movement: A global perspective. In D. Snow, S. Soule, & H. Kriesi (Eds.), *The Blackwell companion to social movements* (pp. 576–607). Malden, MA: Blackwell.

Freedman, E. (2002). *No turning back: The history of feminism and the future of women.* New York: Ballantine Books.

Godwin, F. K., & Risman, B. J. (2001). Twentieth-century changes in economic work and family. In D. Vannoy (Ed.), *Gender mosaics* (pp. 134–144). Los Angeles: Roxbury.

Goldscheider, F. K., & Rogers, M. L. (2001). Gender and demographic reality. In D. Vannoy (Ed.), *Gender mosaics* (pp. 124–133). Los Angeles: Roxbury.

Held, D., McGrew, A., Goldblatt, D., & Perraton, J. (1999). *Global transformations.* Stanford, CA: Stanford University Press.

Hewitt, L. (2011). Framing across differences, building solidarities: Lessons from women's rights activism in transnational spaces. *Interface, 3*(2), 65–99.

Johnson, A. (1997). *The gender knot.* Philadelphia: Temple University Press.

Kivisto, P. (2011). *Key ideas in sociology* (3rd ed.).Thousand Oaks, CA: Sage.

Kuumba, M. B. (2001). *Gender and social movements.* Walnut Creek, CA: AltaMira Press.

Lorber, J. (1994). *Paradoxes of gender.* New Haven, CT: Yale University Press.

Lorber, J. (2001). *Gender inequality: Feminist theories and politics.* Los Angeles: Roxbury.

Motta, S., Flesher Fominaya, C., Eschie, C., & Cox, L. (2011). Feminism, women's movements and women in movements. *Interface, 3*(2), 1–32.

O'Brien, J. (1999). *Social prisms.* Thousand Oaks, CA: Pine Forge Press.

Pyle, N., & Klein, N. L. (2011). Fat. Hairy. Sexy: Contesting standards of beauty and sexuality in the gay community. In C. Bobel & S. Kwan (Eds.), *Embodied resistance: Challenging the norms, breaking the rules* (pp. 78–87). Nashville, TN: Vanderbilt University Press.

Ridgeway, C. (2009). Framed before we know it: How gender shapes social relations. *Gender & Society, 23*(2), 145–160.

Ryan, B. (2001). *Identity politics in the women's movement.* New York: New York University Press.

Schwalbe, M. (2001). *The sociologically examined life.* Mountain View, CA: Mayfield.

Shaw, S. M., & Lee, J. (2001). *Women's voices, feminist visions.* Mountain View, CA: Mayfield.

Stone, L., & McKee, N. P. (1998). *Gender and culture in America.* Upper Saddle River, NJ: Prentice Hall.

West, C., & Zimmerman, D. (2009). Accounting for undoing gender. *Gender & Society, 23*(1), 112–122.

Wharton, A. S. (2005). *The sociology of gender.* Malden, MA: Blackwell.

Introduction to Reading 50

Reproductive justice for all women has been a centerpiece of the feminist movement in the United States since the 1980s. The reproductive justice framework was created by women of color to address the multiple facets of reproductive oppression and to move beyond the limitations of a reproductive choice framework. It is a positive approach that links sexuality, health, and human rights to social justice issues and movements such as immigrants' rights, environmental justice, and population control. Reproductive justice theory asserts that every woman has the human right to decide if and when she will have a baby and the conditions under which she will give birth; decide if she will not have a baby and her options for preventing or ending a pregnancy; and parent children in safe, healthy environments. The conversation in this reading is among scholars and activists who have made major contributions to feminist research on reproductive politics and justice. They discuss reproductive justice in light of the legacies of *Roe v. Wade* and related issues and challenges.

1. What is *Roe v. Wade* and, according to the discussants in this reading, what are its negative and positive outcomes?
2. Why have women of color been at the forefront of the reproductive justice movement?
3. What role has racism played in the anti-abortion movement in the United States?
4. How has the United States had significant impact on women's reproductive health issues in other parts of the world?

Roundtable

Reproductive Technologies and Reproductive Justice

Laura Briggs, Faye Ginsburg, Elena R. Gutiérrez, Rosalind Petchesky, Rayna Rapp, Andrea Smith, and Chikako Takeshita

To commemorate the fortieth anniversary of *Roe v. Wade,* and to invite conversation about the broader global and American landscapes of reproductive politics, the *Frontiers'* editors convened a roundtable of scholars and activists who have made major contributions to feminist research in the field. Beginning with a question about the legacies of the *Roe* decision, we also asked our contributors to reflect on other landmarks in the history of struggles for reproductive justice and to share their perspectives on ongoing challenges. . . . The conversation that appears here is based on the contributors' written comments and was put together in this format by Mytheli Sreenivas.

The year 2013 is the fortieth anniversary of Roe v. Wade. What do you think are the most important legacies of this Supreme Court decision for contemporary women's reproductive issues, both in the United States and globally?

SMITH: The legacy of the *Roe v. Wade* decision was to narrow the agenda of reproductive justice to abortion rights. While abortion rights are important, they are only one aspect of a larger reproductive justice agenda. *Roe v. Wade* framed the right to abortion through the right to privacy rather than through the lens of gender equality. This framework easily lent itself to a more libertarian framework around freedom

Briggs, L., Ginsburg, F., Gutiérrez, E. R., Petchesky, R., Rapp, R., Smith, A., & Takeshita, C. (2013). Roundtable: Reproductive technologies and reproductive justice. *Frontiers: A Journal of Women Studies, 34*(3): 102–125. Reprinted with permission from the University of Nebraska Press.

from government intervention. However, this framework was limited in terms of the responsibility of the government to ensure all have equal access to abortion services. Hence, the Hyde Amendment, which prohibits Medicaid funding for abortion except in cases of rape, incest, or if the life of the mother is endangered, was not deemed inconsistent with *Roe v. Wade.* Thus, even today mainstream reproductive rights groups do not address issues like dangerous contraceptives in communities of color, repealing the Hyde Amendment, environmental racism as it impacts the reproductive systems of indigenous women and women of color, poverty as it affects women's ability to access reproductive health services, and so on. Reproductive justice has become equated with the right of some women who can afford it to have abortion.

RAPP AND GINSBURG: *Roe v. Wade* stands as two things: a beacon of the successful U.S. feminist struggle for reproductive rights and an icon of its limitations. As other contributors to this roundtable point out, in the last forty years we have witnessed a shift toward a more encompassing reproductive justice agenda across a broad range of issues linked to women's health. Increasingly, this agenda identifies and struggles against the intersecting and stratified divides through which women experience their reproductive "choices." These include divides that are based on race or ethnicity, class, age, and rural or urban access in the United States and the continued high maternal mortality rates of pregnant women in developing countries, where abortion too often still remains illegal and hence murderously dangerous. Such structural barriers disproportionally affect women from poor and historically discriminated-against communities here as well as abroad. Collectively, feminist activists, advocates, and scholars have taught us to situate our analyses of reproduction in local, national, and global contexts, taking account of the many structural barriers that constrain the real-life choices of women in the actual settings where they live.

GUTIÉRREZ: As when *Roe v. Wade* became law in 1973, the politics of abortion today remain a cornerstone in the health and social disparities that exist for women living in the United States. Forty years after the Supreme Court decided that women have a legal right to have an abortion, most continue to face limited access to pregnancy termination procedures (medical and nonmedical) as well as many other reproductive health care services, including prenatal care, fertility technologies, and pap smears. Although the existence of *Roe v. Wade* has certainly increased the availability of legal abortion services and was responsible for irrevocably bringing reproductive politics into public conversation over the past forty years, the impact of the law has been significantly limited almost since its inception. Most important, the subsequent passage and persistence of the Hyde

Amendment in 1977 was essential to establishing a government-regulated reproductive divide for women in the United States that has since only widened. This legislation, in addition to the declining access to services in many states and increasingly restrictive circumstances nationwide, makes it very difficult and often impossible for low-income women to pay for a legal abortion or experience any semblance of actual reproductive choice as it is popularly conceived. In response local abortion funds, which are almost all grassroots, community-based efforts dependent upon private donations, have grown over the United States to assist women who may need financial assistance to pay for pregnancy termination procedures that they cannot afford.

The Guttmacher Institute recently reported that during 2011 and 2012 more abortion restrictions were enacted in U.S. states than in any other previous years. The year 2011 marked a record high, with ninety-two pieces of legislation being passed throughout the country.[1] These types of measures disproportionately impact women who live in poverty, as they are more likely to have to terminate a pregnancy because of an inability to parent another child due to financial constraints. Forty-two percent of women having abortions are poor, and women of color are more likely to live below income than white women. Thus, low-income women of color are those most impacted by dwindling access to abortion services.

Increased attention to these disparate circumstances has developed from and contributed to steadily growing advocacy movements within women-of-color communities that insist that true reproductive "choice" necessitates an intersectional approach to understanding the many factors that impact women's reproductive options and a more comprehensive rubric of reproductive justice. This means that access to all types of reproductive health care, not only abortion services, is a matter not only of gender equity but of racial, class, sexual, and embodied justice.

PETCHESKY: On first glance the most remarkable thing about approaching the fortieth anniversary of *Roe v. Wade* is a landscape of apparent stagnation. Instead of celebrating how far we've come from a moment when the struggle for abortion rights for women and girls seemed blessedly to have been won in the courts, we encounter a never-ending battle over four decades to counter right-wing strategies that make abortion the stand-in for feminism. And feminism here is clearly coded as antifamily, antichildren, pro-sex—especially for young unmarried women. In writing about abortion politics in 1990—in terms that remain depressingly relevant today—I argued "that abortion is the fulcrum of a much broader ideological struggle in which the very meanings of the family, the state, motherhood, and young women's sexuality are contested." And I linked that confluence of meanings directly to an insidious racism that underlies the anti-abortion movement, in which "birth control and abortion services, widely available without age or marital restrictions, have helped to make the young, white woman's sexuality visible, thereby undermining historical race and class stereotypes of 'nice girls' and 'bad girls.'" In a racist society this makes contraception and abortion doubly dangerous.[2]

So today we have national and state funding attacks on Planned Parenthood centers and Title X, plus systematic efforts to keep abortion and contraception out of health care reform plans. (If fire-bombings of clinics have subsided, picketing and harassment of providers and patients have not; and the last murder of an abortion doctor, George Tiller, occurred as recently as 2009.) The assault on contraception has ratcheted up the anti-abortion movement from saving fetuses to sanctifying embryos. In 2012 electoral politics we have the first lady and the aspiring first lady, and a bevy of other politicians, appealing to (white) women voters by proclaiming their allegiance to familialism, momism, and stand-by-your-manism. We have a Republican Party platform that seeks to take us back to a time when abortion was criminal in absolutely every circumstance. And we have a new and more sophisticated twist on the racist themes of the so-called pro-life campaigns through billboards that attempt to convince African Americans that abortion rights are a form of "genocide"—when, in fact, four times as many women of color as white women suffer from unwanted pregnancies due to lack of access to safe, affordable reproductive health care, including contraceptive services and supplies. In turn this lack of access comes largely from the structural racism and class divisions that permeate our society, resulting in exclusions from jobs that provide health insurance with contraceptive coverage; restrictions on Medicaid funding in many states (Texas in the lead) for not only abortion but also contraception, breast exams, and other routine gynecological services; and lives more burdened with sexual violence, single motherhood, and poverty.[3]

Given this bleak landscape, what then can we count as the legacies of *Roe v. Wade* after forty years? On the positive side I still believe the idea of a right to personal ownership over one's body, its sexuality and reproductive capacity, contained in *Roe*'s "privacy" doctrine is powerful and potentially transformative. There's no question that this idea has exploded in thousands of directions and locales, both in the United States (e.g., informing the Supreme Court's 2003 ruling against sodomy statutes in *Lawrence v. Texas*) and among social movements across the globe for freedom of sexual and gender expression. At the same time *Roe*'s very strength was also its weakness: Restricting abortion decisions to the "personal," the "private," also meant severing their deep connection to issues of economic and social justice. Taking "a woman and her doctor" out of the context of all the social and structural reasons why abortion access is one link in a huge chain of conditions necessary for personal well-being made it easier to demonize women seeking abortions as "evil" at worst or hapless victims at best and providers as purveyors of genocide. But the U.S. legal system, with its emphasis on individualism and property rights, doesn't lend itself well to fights for social justice. So here, I think, we might find a lesson from *Roe v. Wade,* a lesson we have to learn again and again: Litigation is only one tool available to movements for social change. It may shift standards in the courts, but it takes a mobilized social movement to change values, images, and power relations.

TAKESHITA: The political right's relentless attempt to undermine the 1973 Supreme Court decision, regrettably, is one of the most significant legacies of *Roe v. Wade.* Forty years after legalization, American women seeking abortion still face harassment from anti-abortion protestors and encounter obstacles set up by state laws. As of October 2013 seventeen states mandate that women be given counseling that includes anxiety-provoking information that is scientifically unsupported, such as the purported link between abortion and breast cancer, the ability of a fetus to feel pain, and long-term mental health consequences for women. Twenty-six states require a woman seeking an abortion to wait twenty-four hours or more between when she receives counseling and the procedure is performed. Thirty-nine states require parental consent or notification in order for a minor to receive abortion.[4] The right wing has been unrelentingly attacking Planned Parenthood, engineering constitutional amendments to overturn *Roe v. Wade,* and is now attempting to challenge President Obama's Affordable Care Act, which requires new health plans to fully cover birth control for women.

While the recent presidential election drew considerable attention to the rights of American women to receive reproductive health care, there is little awareness of the negative impact that the backlash against *Roe v. Wade* has also had on women abroad. After unsuccessfully attempting to challenge *Roe v. Wade* during the 1970s, anti-abortion leaders turned their effort overseas, gradually cutting aid to family planning programs in developing countries from the United States Agency for International Development (USAID). The Reagan administration instituted the Mexico City Policy in 1984, which denied funding from USAID to foreign nongovernmental organizations (NGOs) that provide abortion counseling, referral, and/or services using non-U.S. funds. Known also as the Global Gag Rule because it prevents health care providers from not only performing but also making references to abortion, this executive order has resulted in diminishing much-needed reproductive health care for women in the global South by forcing clinics to close and curtailing contraceptive supplies from USAID in certain areas. The Global Gag Rule also made it difficult for governments fearful of jeopardizing their relationships with USAID to openly discuss abortion-law reforms in their countries. While it had no effect on reducing global abortion rates, the Global Gag Rule most likely drove numerous women to back-alley abortions performed by untrained people. Sadly, unsafe abortions still account for thousands of maternal deaths and injuries worldwide. Although the Global Gag Rule has been rescinded and funding has been restored every time a Democratic president has taken office, overseas programs critical to women's health have perpetually been "held hostage to the ping-pong game of U.S. partisan politics."[5]

The United States has had significant influence on women's reproductive health issues in the global South for decades. Initially American leaders aggressively urged foreign governments to control population growth and prioritized funding family planning programs over providing aid for other development projects. Global population control was a signature imperialist project of the U.S. during the 1960s and 1970s. Over time, however, American aid arguably also helped meet contraceptive needs of women in countries where resources are scarce. Conservative lawmakers' attempts to curb foreign aid for reproductive health care amounts to another form of American tyranny over the reproductive lives of women in the global South. Given this history, we might say that the adversarial legacy of *Roe v. Wade,* namely American antiwomen's rights activists' attempt to restrain abortion, contraception, and women's sexuality in general, has an impact on women beyond the United States.

BRIGGS: As Chikako Takeshita suggests, the transnationalization of the U.S. culture wars is an important legacy of the post-*Roe* decades. Here I will speak only of the United States and Latin America, as that is the context I know best, although the geographic expansion of Evangelical Christian "values" concerns has been crucial to the (significantly condom-free, homophobic) expansion of HIV education in many nations in Africa.[6] As journalist Michelle Goldberg has argued, the "population control" regime of the 1950s, 1960s, and 1970s was replaced with a fight about abortion and women's rights, decisively after the Cairo Conference of 1994, but beginning in the 1980s, with the election of Ronald Reagan and the growing political power of Evangelical Christian actors in the United States and Latin America. This often occurred in relationship to anticommunism. As Lynn Morgan and Liz Roberts have argued, the same people who so hated the regime of human rights during the anticommunist civil wars in Latin America turned decisively to a regime of reproductive governance that, in their phrase, emphasized the "rights of the unborn" and proliferated a weak notion of rights and responsibilities. Since 1998 abortion has been banned altogether, with no exception even to save the life of the mother, in El Salvador, Nicaragua, the Dominican Republic, and some Mexican states. Costa Rica has banned in vitro fertilization in deference to the Catholic Church.[7]

What additional historical landmarks do you think are important for understanding issues related to reproductive rights, technologies, and justice? For example, how significant has been the policy shift from population control to women's reproductive health, as developed at the UN Conference on Population and Development in Cairo in 1994? What has been the impact of changing reproductive technologies on feminist understandings of reproduction? In other words, if we decenter *Roe v. Wade,* how might our conception of reproductive rights, technologies, and justice change?

PETCHESKY: Thinking historically, let's remember that the campaigns for safe, legal abortion in both the United States and Europe in the early 1970s sparked a profusion of women's health movements—in Latin America and the Caribbean, Asia, Africa, and the Pacific, as well as the "West"—whose aims were far broader than the legalistic and individualistic prism of *Roe v. Wade.* By the mid-1980s the International Women and Health Meetings, which originated in Italy in 1975 and reconvened every three years for over two decades thereafter, were bringing together activists from all these regions and were forging strategies to secure not only safe, legal abortion but also access to safe childbearing; an end to maternal mortality and morbidity; an end to sexual abuse and violence; and effective challenges to racist population policies, poverty, and global economic injustices. Most important, women from the global South not only were building their own context-specific strategies and organizations but also were in the leadership of many of these transnational efforts.[8]

We need to situate the International Conference on Population and Development in Cairo in 1994 and its historic, if still limited, Program of Action in this long trajectory. Cairo was not a beginning but rather a kind of culmination, a (nonbinding) codification at the level of intergovernmental policy making of the visions, aspirations, and energetic campaigns of women's health activists for the previous twenty years. The Program of Action itself was disappointing in terms of its weak resource allocations and its reaffirmation of neoliberal, market-based approaches to "development," to say nothing of its deference to national laws on the matter of abortion and its heteronormative assumptions about sexuality and gender. But it remains to this day a powerful statement of a vision of reproductive health rights that embraces a multitude of intersecting needs. These include (a) a comprehensive definition of reproductive health as a human right encompassing all aspects of obstetric and gynecological care (including prevention and treatment of infertility, HIV, other STDS and gynecological cancers), as well as primary health care; (b) adolescent rights to all these forms of care, as well as full and accurate sexuality education; (c) the legitimacy of "diverse family forms" and the need for government policies that benefit all families; (d) "gender equality, equity and empowerment of women" as not only indispensable to development but also "a highly important end in itself"; and (e) a view of "gender" that includes men (despite regrettable silence on transgender and intersex lives) through demands for "shared male responsibility" around pregnancy, child care, household labor, and sexual health.[9]

With Cairo this expansive definition of reproductive health and its firm link to "internationally recognized human rights" became embedded in the discursive frameworks of UN agencies, donors, health providers, and a wide range of activists and advocates. But for most people in most countries this shift in discourse has still not translated into real-life programs and policies.

SMITH: I would dispute the assumption that Cairo shifted the discourse from population to reproductive health. At Cairo the population paradigm remained. It was simply described in more benevolent language. The impact of Cairo was that people know to use different language, but the assumption that the cause of the world's problems is poor people's ability to reproduce has not fundamentally changed. Dangerous contraceptives are still promoted in third world communities and communities of color in the United States.

I think we not only need to decenter *Roe v. Wade,* but we must decenter the framework of reproductive "choice." This paradigm rests on essentially individualist, consumerist notions of "free" choice that do not take into consideration all the social, economic, and political conditions that frame the so-called choices that women are forced to make. Consequently, pro-choice advocates narrow their advocacy around legislation that affects the one choice of whether or not to have an abortion without addressing all the conditions that give rise to a woman having to make this decision in the first place.

The consequence of the "choice" paradigm is that its advocates often take positions that are oppressive to women from marginalized communities. For instance, this paradigm often makes it difficult to develop nuanced positions on the use of abortion when the fetus is determined to have abnormalities. Focusing solely on the woman's choice to have or not have this child does not address the larger context of able-bodied supremacy that sees children with disabilities as having lives not worth living and provides inadequate resources to women who may otherwise want to have them. Thus, it is important to assess the intersection of expanded reproductive technologies with ail structures of domination.

TAKESHITA: However mundane it may seem today, contraceptive development was an important landmark in the history of reproductive rights and justice movements. Since *Roe v. Wade* had not yet passed when oral contraceptives and IUDs became available during the early 1960s, these new technologies of fertility control seemed like a godsend for American women who were desperate to avoid pregnancy. Many suffered health problems, however, from these "scientific" methods, which were initially conceived by their developers as a tool to prevent "global population explosion." In a rush to disseminate contraceptives that can easily be applied to the masses, potential dangers of the Pill and the IUD were overlooked or downplayed. Unaware of the risks, American doctors did not carefully screen contraceptive users for contraindications, nor did they take very seriously women's complaints of side effects, which sometimes led to severe chronic injuries and life-threatening conditions. The plight of women in the global South and women of color in the United States, who were specifically targeted by long-acting contraceptives such as the IUD, Norplant, and Depo-Provera, was even graver, while underreported. Despite their failings, we cannot deny modern contraceptives have had significant impact on women's ability to manage their reproductive lives. Fortunately, feminist activists' advocacy for women's agency, safer birth-control technologies, and men to share family planning responsibilities has successfully redirected contraceptive research and development away from long-acting "imposable" methods preferred by population control advocates to women-controlled methods such as contraceptive patches, vaginal rings, and gels, as well as to birth control methods for men.[10] In the face of the political right's attack on reproductive rights, contraceptive developers, despite their problematic eugenicist past, have become close allies of pro-choice feminists in supporting women's access to various contraceptive options.

As Rosalind Petchesky points out, the Program of Action of the United Nations Conference on Population and Development in Cairo in 1994 has shifted the language of international population policies and programming that focused on reducing birth rates to one that privileges women-centered reproductive health care and women's empowerment. Together with these changes, the framing of contraceptive technologies has also shifted. During the 1960s and 1970s contraceptive technologies were openly discussed as population control tools. "Women's unmet need" was later used as a synonym for lack of birth control in regions of high fertility. Recently, with the installation of the women's empowerment discourse, contraceptives are increasingly being framed as something that women in the global South desire. Including "imposable" methods that have problematically targeted underprivileged women of color, contraceptive technologies are becoming politically neutralized as they are cast as "choices" and women are transformed from family-planning service recipients to pseudo-consumers in individualist and neoliberalist terms. As Andrea Smith notes, this kind of universalist framing tends to assume that all women's basic reproductive health care needs are fulfilled and that they are "free" to make decisions, which neglect the realities of the majority of women. Problematizing the politics of contraceptive technologies is one way to decenter *Roe v. Wade* and bring intersectionality to the forefront.

GUTIÉRREZ: Since 1973 activists and scholars have demonstrated and documented how reproductive politics in the United States are shaped by gender and patriarchy but also by white supremacy, heteronormativity, ableism, and classism. Following the publication of Dorothy Roberts's path-breaking treatise *Killing the Black Body: Race, Reproduction and the Meaning of Liberty* (1998), a growing literature has documented and theorized the historical patterns and contemporary dynamics of how stratified reproduction plays out for women in the United States, as well as how women of color have resisted repeated episodes of reproductive coercion upon their communities often in the name of population control. This scholarship includes documentation of the racial politics of reproduction, including the usage of women of color for contraceptive trials (i.e., the IUD, the birth control pill, and foam), as well as the preponderance of controlling images and ideologies that circumscribe the reproductive circumstances women experience (i.e., that African American women are teenage moms, that Mexican immigrant women have too many children, or that Asian women are tiger moms). This body of literature enables a much deeper understanding of how women's relation to reproductive expression is impacted by their social location.

For example, the many episodes of coercive sterilization that have occurred in various communities over the years fundamentally challenge the common assumption that simply because abortion is legally available in the United States, all women here equally experience reproductive "choice." The histories of low-income, African American, Native American, Puerto Rican, and Mexican-origin women's forced sterilization demonstrate that racist and classist rhetoric about overpopulation, care of the irresponsible "poor," economic development, and environmental sustainability are used to justify both official and unofficial policies that limit women's autonomy in their reproductive experience. Members of each of these groups of women faced intentional, strategic, and successful efforts to permanently end their childbearing without their knowledge or consent, some when they were actually in the hospital to deliver a child.

Puerto Rican women were first and perhaps most greatly affected by coercive sterilization, with over one-third of those living on the island being sterilized during the 1940s and 1950s, when the surgery was first being practiced as a means of encouraging women's employment and decreasing "overpopulation" on the island. Although institutionalized programs to sterilize women no longer exist, anthropologist Iris Lopez has demonstrated that these high rates of sterilization continue not only among women who live on the island of Puerto Rico but also for those who live in Puerto Rican communities in New York.

In later years, when the surgery became readily available on the continental United States and the procedure was 90-percent paid for by the availability of federal funds, Native American women were sterilized without their knowledge in Indian Health Service clinics, and low-income women in the South and in urban centers were sterilized without their knowledge at great rates, often at publicly funded hospitals. As I have argued elsewhere, different racial logics were utilized to justify the targeting of various communities of women, although always based in economic difference.[11] Some doctors believed that they were doing women a favor, by limiting the number of children that they could have, or that they were ridding society of a welfare burden. Beyond bringing more attention to the reproductive abuses that have occurred in the United States, documentation of these histories has assisted organized calls for reparative measures and resistance to contemporary ideological discourses that pose the reproduction of women of color and poor women as a threat to the health and well-being of society. Over the past thirty years a parallel effort to mainstream reproductive rights organizing, led primarily by women of color, has evolved into a distinct coalition-based reproductive justice movement that calls for a broad advocacy agenda—one that goes beyond a focus on a woman's "choice" to have an abortion.

RAPP AND GINSBURG: One of the structural barriers that constrains women's real-life reproductive choices but that often goes unexamined is the impact of fetal disability on attitudes and practices surrounding abortion decisions. As the journalist Amy Harmon pointed out in a 2007 *New York Times* article: "Seventy percent of Americans said they believe that women should be able to obtain a legal abortion if there is a strong chance of a serious defect in the baby, according to a 2006 poll conducted by the National Opinion Research Center."[12] What does this portend for a more democratic inclusion of people with disabilities, even as we support the rights of pregnant women to make their own decisions to continue or end any given pregnancy? As many feminist disability activists have pointed out, there is insufficient dialogue between their concerns and those of reproductive justice activists.

Like the women's health movement, the disability rights movement is both national and international; like feminist approaches to reproductive justice it struggles to frame positions that are both ethical and activist. Ranging from philosophical to pragmatist, feminist disability activists and scholars have identified a range of issues that affect both mainstream discrimination against children and adults with disabilities and the rights and possibilities for people with disabilities to express their own sexual and reproductive aspirations. We need an ongoing conversation that takes the disability lens as a framework for thinking about reproductive choices. Why is fetally diagnosed disability so routinely considered cause for abortion? To what extent might the experiences of people living with disabilities provide a more robust social fund of knowledge that could better provide a truly informed reproductive decision as to whether or not to terminate a diagnosed pregnancy?

Some of the context that makes this wider conversation both possible and necessary emerges from recent legal changes catalyzed by disability rights activism. The passage and adjudication of the Americans with Disabilities Act (1990) has changed the cultural milieu in which, for example, the almost automatic sterilization of people with disabilities was considered an unexceptional, even ethical practice. Other aspects of our changed context include the improved health care for infants, children, and young adults with disabilities, from neonatal intensive care units to seizure-controlling medications to increasingly sophisticated prosthetics. All enable fuller social participation—in particular in education—over the life cycle, potentially including reproduction. At the international level a vigorous disability rights movement has brought these issues of survival and inclusion into the UN Convention on the Rights of Persons with Disabilities, which has had the highest number of signatories of any human rights framework to date. . . .

What do you think are important ongoing challenges in struggles for reproductive justice? How do you characterize the relationship between scholarship and activism on reproduction?

RAPP AND GINSBURG: The feminist reproductive health agenda has embraced the inclusion of disability rights in its broad agenda. Yet the actualization of such a commitment is anything but clear. How can feminists justify support for the abortion decisions women make and for a disability-inclusion perspective? The profound segregation and discrimination against disability prevents fully grounded knowledge of what it means to live with an impairment from entering into those deliberations. We want to underscore the significance of continued feminist conversations on this topic across movements and coalitions, which always take place in the context of changing reproductive technologies. As feminist activists we ignore "medical advances" at our peril. . . .

SMITH: The most important scholarly interventions are happening among indigenous and women-of-color reproductive justice organizations. They are decentering *Roe v. Wade* by not assuming that if the decision is overturned, reproductive justice organizations cannot provide reproductive health services themselves. These groups are focusing not simply on influencing law and policy but on building a reproductive justice movement in which people begin to take charge of their reproductive health. Such groups do everything from teaching midwifery to doing community gardening as a way to address the relationship between the environment and reproductive health. These interventions also situate reproductive justice within a broader framework for social justice. Some of these interventions are not necessarily found in books but are circulating through the Web or other social networks in a way that directly influences grassroots organizing.

TAKESHITA: Recently I invited an artist and activist, Heather Ault, for an event that marks the fortieth anniversary of *Roe v. Wade* at the University of California, Riverside (UCR). Her project *4000 Years for Choice* is a series of fifty prints that combine image and text that represent a method of contraception and abortion from the past or a historical figure's comment on birth control.[13] Her intention is to generate new visuals and languages for the pro-choice movement that are more powerful than the iconic wire coat hanger and more inviting than the combative and defensive terms used in its call to action such as *fight, struggle, defend, attack, and threat.* Through her artwork Ault underscores how fundamental birth control has been to the history of humanity and suggests terms such as *cherish, embrace, nourish, trust, and unite* to tell stories of "women's reproductive empowerment, wisdom, and self-care that dates back a millennium."[14] She writes that regardless of an audience's position on reproductive rights, history should serve as testament to "women's deeply ingrained desire to control pregnancies for the good of ourselves, our relationships, and our families."[15]

However disheartened and tired we are of the political and ideological deadlock that "pro-choice" feminists are stuck in, countering the "pro-life" movement still must remain one of our major concerns considering its incessant endeavor to narrow women's access to reproductive health care. While securing legal rights to birth control, as contributors to this roundtable point out, is far from enough to address reproductive justice, we cannot ignore the legal challenges by private employers to mandatory contraceptive health care coverage for employees and conservative lawmakers' constant effort to pass anti-choice legislations. Certainly feminist organizations such as Planned Parenthood and NARAL (the National Abortion and Reproductive Rights Action League) are working tirelessly to "win" this "war on women." But is there anything feminist scholarship can do to intervene in the impasse? I myself have often felt that no dialogue seems possible when both sides believe that their opponents are utterly immoral or irrational and share little in common. Yet there may still be room for feminist scholarship and grassroots movements to disrupt the dichotomous thinking, which is what these opposing positions on abortion have come to be, and foreground the complex gray areas of the debate.

Ault attempts to transcend the reproductive-politics gridlock by proposing alternative visuals and languages for the pro-choice movement to overcome the stigma of abortion. Using photos of (presumably) aborted fetuses, anti-abortionists have successfully created a visual narrative of pregnancy termination as equivalent to murder and women who choose to have an abortion as evil, selfish, and horrifying. Unfortunately, symbols that have been available to the pro-choice side, namely photographs of women who lost their lives in blotched abortions and a wire coat hanger with a line drawn across it, have not been capable of evoking as strong an emotional response as the opponent's gory images. With this in mind Ault's *4000 Years for Choice* aims to provide alternative beautiful images for the pro-choice movement along with historical evidence for the universality of the desire and action to control birth. Ault's work clearly moved the audience at UCR. Curious students from all walks of campus came through the exhibit throughout the day. Some brought friends and talked to each other while viewing the artwork. Some asked the artist questions. There were as many male students taking photos of the artwork as female students.

Spending most of the day at Ault's art exhibit and talk prompted me to respond to the third question of this roundtable. First, I suggest that it might be constructive for feminist scholars to reexamine what personal stories might accomplish. There may be something to be learned from the open-mike session of the national conference for March of Life, during which women confess their abortions and express sadness, anger, and regret. These intimate and emotional disclosures bond the speaker and audience with empathy and reinforce the group's conviction that abortion must be eradicated. We might think about telling personal stories of women who made the difficult decision to have an abortion that they ultimately do not regret or of women who have decided not to discontinue their unintended pregnancies despite their pro-choice convictions. Real-life stories that represent the "gray" areas of abortion narratives might be instrumental in complicating the "pro-life versus pro-choice" divide and opening up new dialogues. They might also help establish positive images of pro-choice women that contradict the "selfish baby killers" and "angry feminists" as they have been painted.

Second, echoing the suggestion that Ault made during the course of the day, I want to suggest that there be more collaboration among artists, activists, and scholars. One thing those of us scholars who have access to college students and facilities can do would be to bring artists and activists to campus in order to expose more young people to the human side of the reproductive-choice debate that is currently swamped by political wrestling devoid of discussion that may move us toward conciliation.

GUTIÉRREZ: A reproductive-justice approach to abortion demands that all women must have not only the real choice of terminating a pregnancy but the opportunity to have the true option to deliver a child to full term if that is what she wants, *free from economic constraints.* Thus, issues such as access to education, health care, civil rights, and social services all must be considered under the rubric of reproductive choice in addition to the spectrum of reproductive health services, and until they are available to all women who reside in the United States, reproductive justice has not been achieved.

In addition to the fundamental right to have children, more recent activist developments increasingly demand that all women have the right not only to have however many children they want whenever they want but also to deliver those children in whatever manner that they want. This commitment to providing the kind of childbirth experience that any woman wants has certainly grown in the past ten to fifteen years; alongside a quickly growing and diversifying midwifery communities are doula practitioners who are focused on providing women with various aspects of practical support that might be necessary for anyone who may be having a child. These efforts, which build upon the work of the women's health movement of the 1980s, are most often associated with natural childbirth, but doulas often work with women who deliver in the hospital and provide additional support for a woman before, during, and/or after childbirth, in the form of emotional and physical assistance.

In recent years the large majority of those in this practice are birth and postpartum doulas who are professionally trained in supporting a woman while she goes through the process of pregnancy. It is becoming more common for doulas to also provide support services for pregnancy termination, offering the equivalent emotional and physical support around this experience as during the birth of a child. Abortion doula services include emotional support during the procedure, after care, child care, and practical factors such as preparation. Although such practices are only beginning to become more readily available, they are certainly an important development in the abortion rights movement given the increasingly deleterious circumstances that most women must face when they must go through the procedure. . . .

BRIGGS: The deep and intensifying care-work gap, in the United States and transnationally, would in my mind be a candidate for the most important reproductive-rights issue in the present moment. Not only did the Pill and abortion make heterosexual sex possible for those of us who did not want to be committed to children and a partner before we were twenty, but they also have made it possible for educational institutions, corporations, and the labor market in general to expect many of those in their twenties to delay childbearing for much of that decade of our lifespans, too. In the 1970s feminists joked that they needed a wife so they could work and still do care work and the reproductive labor of the household; now companies, universities, and government demand that we all act like the husbands of that decade—as if we never had a problem because public school is 180 days and the working year is 260, never had a sick child, never were made systematically crazy by the mismatch between school running from 8:30 to 2:30 (when it's not an early-release day) and work from 8:00 to 5:00 at a minimum. And this is to say nothing of the perfectly normal exceptional circumstances—you can't even get this much coverage from public programs if your child is under six years old or has disabilities or behavior problems that kick them into the "special" programs that run, say, from 8:00 to noon. That problem is multiplied for the growing percentage of women who work more than one job, and the racist right is trying to ensure that immigrant women have no recourse even to the public programs there are, by trying to ban immigrant children from school—and while direct efforts have failed, terror that their children's visible presence will get the family deported has succeeded in places like Alabama and Arizona, at least for some period. While this is a fierce tragedy for children, it is also worth noting that it is a huge added care-work burden for families.

As a growing and important body of scholarship on care work and reproductive labor, particularly in sociology, has noted, in the United States and many comparable postindustrial economies, the care-work gap is increasingly being filled by immigrant nannies and, for the elderly and those with disabilities, home-care workers.[16] As many have noted, the work of immigrants in filling the care gap in one place—for middle-class people in the United States, usually but not necessarily white—merely creates one in another place, as nannies are often mothers themselves, and many leave children behind in home countries as a different kind of transnationalization of care labor. Just as it is cheaper for U.S. families to substitute immigrant labor for their own in caring for children, as declining real wages force middle-class households to have two people working for money, so too is it cheaper for immigrant nannies to "outsource" the raising of their children to family members in the global South, with their wages often making the difference between having funds for school fees and even bare survival or not.[17]

The care-work gap drives a host of other issues—including to some extent infant mortality rates, the rising use of reproductive technology (contemporary infertility being largely an artifact of the delayed childbearing demanded by women who seek to be in professional sectors of the labor market), surrogacy, and adoption. Adoption, as the transfer of children from the impoverished to those who are middle class, from the global South to the United States, Europe, Canada, and a handful of other places, and from the young to older parents, should trouble us more than it does, as a place where violence, coercion, and power meet (usually single) mothers and their children.[18] The idea of adoption also hovers over the abortion debate in the Americas, as its more desirable other for liberals and conservatives alike; for many conservative commentators it also represents a preferable alternative to single mothers raising their children. Mothers, on the contrary, seem to have voted with their feet that they prefer either alternative to renting out their bodies for nine months, giving rise to sharply declining rates of placing children in adoption in the United States once single mothers can support their children or get an abortion and "unwed mothers' homes" losing their coercive power, except perhaps for some young Evangelical Christian girls and women.[19]

For the most part, though, scholarship on the care-work gap is not brought together with concerns about abortion and adoption, nor do most historians, sociologists, or political scientists who study the dynamics of the abortion debate in the United States think of it as part of a process across the Americas that is linked to anticommunist civil wars. Among scholars there is no field of "reproductive politics," and so the scholarship on abortion or surrogacy is unrelated to work on the medicalization of pregnancy, poverty, or infant mortality, although they are critical to understanding each other. There are reproductive economies just as there are domestic economies, and the globalization of production, finance, and labor did not take place without a concomitant global adjustment of reproduction. I would argue that we need a scholarship and a reproductive-justice movement that are about reproductive politics writ large—from neoliberalism to reproductive governance to welfare reform, from infant mortality and racial and geographic health disparities to reproductive technologies.

PETCHESKY: It's too easy to blame failures to secure reproductive justice for all women on the continued barrage of attacks from the right, who still see "reproductive health and rights" as code for rampant abortion and promiscuous sex; or the retreat of global health providers and policy makers into vertical, single-issue programs that favor HIV treatment or family planning and eschew a comprehensive, rights-based approach to health and sexuality as too complicated and too costly in resource-scarce societies. But responsibility also lies with social, including feminist, movements whose campaigns are still siloed and "issue" oriented rather than foregrounding the deep connections between reproductive and sexual health and rights and social transformations that challenge global capitalist priorities—the structures that keep resources scarce for the many and plentiful for the few. In the United States, I think, the most radical voices of the past decade on reproductive issues have been those of women of color and particularly the work of SisterSong in promoting a concept of *reproductive justice* that insists on the direct links between access to all aspects of reproductive health care, including safe abortion, and addressing poverty, racism, gender-based violence, community development, education, and labor conditions. To make this vision a reality, feminists need to be thinking deeply about systemic change—and building the broad coalitions with Occupy and other social-justice and antiracist movements to make it happen.

Notes

1. "Laws Affecting Reproductive Health and Rights: 2012 State Policy Review," Guttmacher Institute, http://www.guttmacher.org/statecenter/updates/2012/statetrends42012.html (accessed Mar. 13, 2013).

2. Rosalind P. Petchesky, *Abortion and Woman's Choice: The State, Sexuality, and Reproductive Freedom,* rev. ed. (Lebanon, NH: Northeastern University Press, 1990), xi, xviii.

3. See Carole Joffe, "Abortion Patients and the 'Two Americas' of Reproductive Health," chap. 6 of her superb *Dispatches from the Abortion Wars* (Boston: Beacon Press), 2009.

4. For additional information on state abortion laws see "An Overview of Abortion Laws," State Policies in Brief, Guttmacher Institute, http.//www.guttmacher.org/statecenter/spibs/index.html (accessed Jan. 2013).

5. For additional information on the impact of the Global Gag Rule see the website of Population Action International, http//populationaction.org/topics/global-gag-rule/ (accessed Aug. 2012).

6. Sheryl Gay Stolberg, "In Global Battle on AIDS, Bush Creates Legacy," *New York Times,* Jan. 5, 2008, http://www.nytimes.com/2008/01/05/washington.05aids.html.

7. Lynn M. Morgan and Elizabeth F. S. Roberts, "Reproductive Governance in Latin America," *Anthropology and Medicine* 19, no. 2 (Aug. 2012): 241–54; Michelle Goldberg, *The Means of Reproduction: Sex, Power, and the Future of the World* (New York: Penguin Press, 2009).

8. See Rosalind P. Petchesky, *Global Prescriptions: Engendering Health and Human Rights* (London: Zed Books, 2003), chap.1; S. Corrêa, R. Petchesky, and R. Parker, *Sexuality, Health and Human Rights* (London and New York: Routledge, 2008), chaps. 2, 8.

9. Petchesky, *Global Prescriptions,* 44–45.

10. The women-centered methods and male contraceptive methods are a trend I have seen in the research trajectories of the Population Council in New York, an organization that has played a significant role in the development of IUDs and implants, as well as the establishment of family planning programs in the global South. Over the years the organization has expanded its mission to overall reproductive health, including HIV/AIDS. Its current projects on reproductive technology development can be found on its website: http://www.popcouncil.org/topics/reprotech.asp#/Projects (accessed Sept. 2012). Adele Clarke coined the term *impossible* contraceptives to represent methods that last for a long time once administered and are difficult for users to discontinue at will. These include the IUD, implants, and injectables.

11. Elena Gutiérrez, "Policing 'Pregnant Pilgrims': Welfare, Health Care and the Control of Mexican-Origin Women's Fertility," in *Women, Health and Nation: The U.S and Canada since 1945*, ed. Molly Ladd-Taylor, Gina Feldberg, Kathryn McPherson, and Alison Li (Toronto: McGill-Queens University Press, 2003), 379–403; Elena Gutiérrez, *Fertile Matters: The Politics of Mexican-Origin Women's Reproduction* (Austin: University of Texas Press, 2008).

12. Amy Harmon, "Genetic Testing + Abortion =???" *New York Times,* May 13, 2007, http://www.nytimes.com/2007/05/13/weekinreview/13harm.html?_r=0.

13. Information on Heather Ault and her work is available at www.4000yearsforchoice.com (accessed Feb. 2013).

14. "About 4000 years for choice," http://www.4000yearsforchoice.com/pages/about-the-project (accessed Feb. 2013).

15. "About *4000 Years for Choice.*"

16. For different reasons immigrant maids and nannies are also crucial to other kinds of ferociously transnationalized economies like the United Arab Emirates and Singapore.

17. See e.g., Pierrette Hondagneu-Sotelo and Ernestine Avila, "I'm Here, but I'm There: The meanings of Latina Transitional Motherland," *Gender and Society II* no. 5(1997): 548–71; Rhacel Salazar Parreñas, *Children of Global Migration: Transitional Families and Gendered Woes* (Stanford: Stanford University Press, 2005).

18. That at least is the argument I make in *Somebody's Children: The Politics of Transracial and Transnational Adoption* (Durham: Duke University Press, 2012).

19. See Rickie Solinger, *Wake up Little Susie: Single Pregnancy and Race before* Roe v. Wade (New York: Routledge, 1992), and *Beggars and Choosers: How the Politics of Choice Shapes Adoption, Abortion, and Welfare in the United States* (New York: Hill and Wang, 2001). For a first-hand account of the coercive power of Evangelical Christian families and unwed mothers' homes in the present, see Ruth Graham's thoughtful book, with Sara Dormon, *I'm Pregnant . . . Now What?* (Ventura CA: Regal Books, 2004).

Introduction to Reading 51

Courtney E. Martin and Vanessa Valenti discuss the critical role of online feminism in the 21st century. They argue that feminist blogs are the consciousness-raising groups of this era and that online technologies have produced thousands of activists, writers, bloggers, and tweeters around the world who are able to engage one another across boundaries and borders. Online feminists and feminist organizations have become allies in movements for gender equality. Martin and Valenti identify the ways in which online feminism is able to reach beyond traditional feminist institutions (e.g., relationship-building and engaging young people) as well as the challenges facing online feminist organizing (e.g., funding and sustainable infrastructure).

1. Why are feminist blogs defined as "the consciousness-raising groups of the 21st century"?
2. What is culture jamming, and how are online feminists using this tool?
3. How does online feminist activism push media stakeholders to be more accurate and less harmful?

#FemFuture

Online Revolution

Courtney E. Martin and Vanessa Valenti

Part One: A New Landscape

What Is Online Feminism?

Online feminist work has become a new engine for contemporary feminism. No other form of activism in history has empowered one individual to prompt tens of thousands to take action on a singular issue—within minutes. Its influence is colossal and its potential is even greater. Feminists today, young and old, use the Internet to share their stories and analysis, raise awareness and organize collective actions, and discuss difficult issues.

The beginnings of online feminism were primarily in the form of online forums, newsgroups, journals and blogs developed in response to the need for a public platform where young women could voice their opinions about the state of the world around them. Many created websites and online zines early on; Heather Corinna began Scarlet Letters in the late 1990s, the first site online to specifically address and explore women's sexuality, and soon after, Scarleteen.com, an online resource for teen and young adult sex education. Viva La Feminista's Veronica Arreola took the feminist ideas discussed on the listserv of online organization Women Leaders Online to create a website at Geocities.com that discussed sports, pop culture and feminist politics. Later on, she developed the first pro-choice webring. Jennifer Pozner developed Women in Media and News in order to create a space for feminist media analysis and increase women's voices in public debate.[1]

Women were creating powerful spaces for themselves online, helping to build the next frontier of the feminist movement. These forums began as simple websites and developed into communities of hundreds of thousands of people who needed a platform to express themselves. They found it on the Internet.

This is why so many identify feminist blogs as "the consciousness-raising groups of the 21st century." The very functionality of blogs—the self-publishing platforms and commenting community—allow people to connect with each other, creating an intentional space to share personal opinions, experiences of injustice, and ideas, all with a feminist lens. Consciousness-raising groups were said to be the "backbone" of second-wave feminism; now, instead of a living room of 8–10 women, it's an online network of thousands.

As years went by, social technologies began to evolve into a robust diverse field of web-based tools and platforms. YouTube allowed for vlogging, or "video blogging"; Twitter and Tumblr, or "microblogging," allowed for easier and even more immediate sharing capabilities. Today, this evolution of online technologies has produced thousands of activists, writers, bloggers, and tweeters across the globe who live and breathe this movement, engaging their audience every day in the name of equality.

In a study conducted in 2011, the Pew Research Center's Internet and American Life Project crowned young women between the ages of 18 and 29 years old as "the power users of social networking."[2] Eighty-nine percent of women use social networks and 93 percent of young people between the ages of 18 and 29 are online. Over 584 million people log in to Facebook alone on an average day.

In this rapidly shifting technological age, it shouldn't be a surprise that the next generation of social movement-building in the United States is largely online.

A New Channel for Activism

The typical image of feminist activism has been pretty clear historically: women marching down the street, or protesting at a rally. From suffrage through the second wave, collective chants and painted protest signs had been defining markers of feminist action. And today, offline organizing continues to be a major tactic to galvanize the masses. Waves of protests across South Asia early this year following the death of a 23-year-old gang rape victim in Delhi are just one powerful reminder of the impact that a collective group of people can make on the ground.

Martin, C. E., & Valenti, V. (2012). #FemFuture: Online revolution. *New feminist solutions series*, Vol. 8. Barnard College Center for Research on Women.

The marchers that filled the streets of Delhi, however, weren't just using their feet; they were also using their tweets. #Delhibraveheart—a hashtag, which essentially serves as a filter for a particular theme or meme—was added to millions of impassioned laments for the victim. Government leaders, who initially had an anemic reaction, were compelled to respond to the young people taking up space in the streets, but also those setting the Internet on fire with their rage. This dual approach is just the latest example that demonstrates how feminist activism has expanded to the online sphere.

The rapid innovation and creativity that characterize online activist work are game-changers in the contemporary art of making change. Compared to the weeks or months of prep time it takes to gather thousands of people in one place for a rally or march, online feminists can mobilize thousands within minutes. Whether you're signing an online petition, participating in a Twitter campaign against harmful legislation, or blogging about a news article, technological tools have made it infinitely easier for people invested in social justice to play their part.

Another striking development in online organizing today is the role of citizen-produced media in online activist work. On feminist blogs, for example, writers post commentary about the day's news with a feminist lens, highlighting and amplifying social justice work that is off the mainstream media's radar, and often linking this analysis to action that readers can take. This widespread, collective effort creates the necessary consciousness and a broad range of content that organizations like Hollaback!, Color of Chance, Move On, UltraViolet, and the Applied Research Center draw on as they share articles, connect with others, sign petitions and pledges, and use online tools to mobilize on-the-ground action. Users can then be instantly contacted to request action in the future. Media-making essentially allows activists to become experts in the issues that we care about, and makers—not just advocates—of the change we want to see and be in the world.

As decade-long activists, we have lived and understand the power of boots on the ground. The feminist movement will continue to make strides through lobbying, on-the-ground organizing, and creating meaningful discourse through academia, but online feminism now offers a new entry point for feminist activism.

A Vibrant New Movement

We are currently living in the most hostile legislative environment to reproductive rights in this country in the last forty years. In 2011, we reached a record number of state restrictions on abortion. Contraception coverage is being attacked, access to basic health care services through providers like Planned Parenthood is threatened, and decisions about one's reproductive health are increasingly criminalized. The feminist movement continues to push back against each hurdle thrown at us. The days of proactive work and creating legislation for equality, of securing our rights rather than defending them, has seemed far beyond our capacity when there is so much responsive action to take.

Yet when millions of women and men can tweet their demand for accountability from corporations, governments, and media, we have an opportunity to shift this paradigm. For over a decade, online feminist activists have been working on feminist causes, but it has never been so visible. Now, feminist organizations, media, and corporate stakeholders, and national leaders are beginning to recognize how the power of social media and online organizing is reanimating the feminist movement.

Online Feminism, a Radicalizing Force

For years, online feminists have served as powerful allies for feminist organizations. We liveblog at conferences, tweet calls to action, and translate the sometimes jargon-laden organizational press release into catchy hashtags, nudging people to look twice before they skip to a funny cat video. As we mapped the movement and the role online feminism plays within it at our convening, we were all struck by the hours and hours of labor made visible.

The good news is that most major women's organizations get it. They recognize that online media is a powerful tool to create change, and have begun to leverage online tools in their work. For example, The National Domestic Workers Alliance, an organization that advocates for the rights and support of domestic workers, created a social media campaign, #bethehelp, around the nomination of the Hollywood film *The Help* for an Academy Award. As controversial as the film's portrayal of African American domestic workers was, The National Domestic Workers Alliance recognized that it was a rare moment within mainstream media where domestic workers were in the spotlight, and they didn't shy away from seizing the day for their own radical purposes. The #bethehelp campaign was helped along by individuals joining in and popularizing the trend.

Traditional feminist organizations and online feminists are becoming more and more symbiotic in this way. Meanwhile, independent online feminists continue to invent new methods of action and catalyze new discussions that are pushing institutional feminism forward. In 2012, when the Susan B. Komen Foundation threatened to withdraw funding from Planned Parenthood because they provide abortions along with many women's and reproductive health services, Planned Parenthood had to respond to the Komen Foundation through formal channels in a professionally appropriate tone. Individual online activists were beholden to no such conventions. Digital strategist Deanna Zandt's Tumblr, Planned Parenthood Saved Me, featured hundreds of women from across the country sharing their stories of how Planned Parenthood's health care has saved their lives. Those stories were a large force behind what compelled Komen to change direction.

Young feminists have been at the helm of online activism for the last several years. "We can't move too quickly over the important cultural (and deeply political) feminist work that younger women are leading, largely online," said Erin Matson, the former Action Vice President of the National Organization for Women, in an intergenerational dialogue at *In These Times.* "All this work is rapidly building into a platform that has the power to force big policy changes, and that's exciting."

Ties between organizations and online feminists have become stronger over the years and have sometimes provided resources for bloggers: organizations may contract bloggers to livetweet at their annual conference, pay for campaign ads on their blogs, or hire online influencers as consultants to assist with communications strategy.

But critical gaps remain between institutional feminism and online feminism. As Jensine Larsen of World Pulse pointed out at the convening, each has expertise that the other can benefit from: nonprofit organizations often have the infrastructure (physical space, resources, womanpower) that online feminists crave, and online feminists often deploy the communications innovations that nonprofit organizations struggle to generate while already stretched thin trying to achieve their larger missions.

Thus far, we've been exchanging our resources in piecemeal, inadequate ways. It's time to come up with a sustainable strategy that serves all of us and strengthens the movement in the process. More meaningful collaborations between two of the most powerful sectors of the feminist movement could create huge impact.

Creating a New Pipeline of Feminist Leadership

. . . Leadership development online can provide a means of resisting the hierarchical, insular, monocultural structure of traditional institutions. There has been a lot of debate about whether the World Wide Web itself can provide new tools for democracy, movement-building and alternative models of leadership. The Internet is not inherently egalitarian; after all, it was first created by the military and can be used in ways that directly reinforce patriarchy and structural violence.

But you don't have to spend years making copies, learning a special language, or knowing people who know people to become a leading voice in online feminism; you just have to have something unique to say and the technological skills needed to amplify that story or idea online. This landscape allows for decentralized movements of multiple voices, communities, and identities.

In fact, many feminist blogs were born out of young women's frustration with entry-level jobs at nonprofits where the mission may have been feminist, but the labor distribution made them feel invisible and, too often, exploited. While more traditional feminist institutions—advocacy organizations, cultural institutions, foundations, etc.—develop initiatives designed to "engage Millennials," they often overlook the young women in their own offices, underutilized and anxious to start flexing their leadership muscles, not to mention the hundreds of thousands of young women and men who are engaging with feminism online every single day. These organizations must stop operating on the "if we build it, they will come" assumption, and start going to meet young people where they are—whether bored and underutilized in their own offices and/or channeling their energy into active online spaces.

Making the Personal Political

The capacity for storytelling and relationship-building online allows young women—so many of them living in small pockets of conservative middle America—to feel less alone, to feel like they're part of a community. This is one example of the hundreds of emails that feminist blogs receive on a regular basis: *I just wanted to say a quick "thank you." I have been reading this blog for about a year and a half, and it has provided me with strength to live through some situations that I know I would have never gotten from anyone or anything else in my life. You have given me hope that it might get better and I just wanted to let you know.*

This kind of connectivity can be life-saving. So many young women find feminism, not in their classrooms or even controversial novels, but in online blogs like F-Bomb, a site by and for teen girls about women's rights. Marinated in the voices and ideas of young feminists that share their sensibility, they are made to feel a part of something bigger than themselves—even as that connection is forged through the most intimate of stories.

In one of the most popular posts ever at the Crunk Feminist Collective blog, University of Alabama professor and blogger Dr. Robin Turner wrote in "Twenty things I want to say to my twentysomething self":

> *You are strong (your capacity of strength is so much wider than you think) . . .*
>
> *but being a strongblackwoman is not a necessity or responsibility in your life. Your frailties and vulnerabilities make you human, not weak.*
>
> *You are a storyteller and people will need your stories. Don't stop writing them down.*

It is these kinds of poetic descriptions that transcend some of the more tangible ways in which online spaces like the Crunk Feminist Collective serve to mobilize young feminists. It's not just about organizing on local issues or taking action on federal policy; it's also about healing, reclamation, solidarity, beauty, and wisdom.

Providing an Entry Point

Decades of stigmatization have resulted in a toxic perception of what a feminist is. But that stigma is beginning to dissipate among young people as they see feminism in action online. Here, feminism is cool again.

At the end of 2011, *New York Magazine* journalist Emily Nussbaum highlighted the ways that feminist blogs use popular culture:

> *Instead of viewing pop culture as toxic propaganda, bloggers embraced it as a shared language, a complex code to be solved together, and not coincidentally, something fun. In an age of search engines, it was a powerful magnet: Again and again, bloggers described pop culture posts to me as a "gateway drug" for young women—an isolated teenager in rural Mississippi would Google "Beyonce" or "Real Housewives," then get drawn into threads about abortion.*

Letting young women know that they can be feminists and care about pop culture gives them social permission to care about equality. Tavi Gevinson is one striking example; she began blogging about fashion and feminism on her blog, Style Rookie, when she was 11-years-old. Five years later, she is the Editor-in-Chief of *Rookie Magazine,* the premiere indie online magazine for teenage girls. In *Rookie,* one piece is about how to create the perfect Fourth of July manicure; the next is a guide to protecting your civil liberties.

Another weapon feminist bloggers and writers use is humor, countering the long-held, wildly inaccurate stereotype that feminists have no funny bones.

Convincing the public that feminism can actually be fun through humorous quips on blog posts has evolved into savvy online campaigns that catch like wildfire. One recent example was the Tumblr blog BWinders Full of Women, created after Mitt Romney's controversial remarks in the 2012 presidential debate about getting binders full of women for possible hires when he was Massachusetts Governor. The Tumblr included snapshots of women dressed up as binders full of women for Halloween screenshots with witty captions, mock campaign ads, etc.

Demonstrating the serious side of cultural entrepreneurship like this, the creator of the Tumblr, Veronica De Souza, wrote in her last post on the site:

> *Now that the election is over, I think this whole thing is done. I never thought it would get this big, or that anyone would ask me to talk about memes on CNN or that this would help me find a job. I am so thankful for everything.*

What De Souza and her peers are doing is essentially "culture jamming"—disrupting mainstream political and cultural narratives using crowd-sourced creativity and playfulness. Latoya Peterson of Racialicious spoke to this in a 2011 interview with *Persephone Magazine:*

> *In a way, using pop culture to deconstruct oppressive structures in society is culture jamming. We are, in many ways, creating a distortion in the smoothly packaged ideas being sold to us. Pop culture is about selling lifestyles, selling ideas; it normalizes certain elements of our culture and erases others. Why do so many people have the idea that we are all vaguely middle class? Because that's what's represented in our media environment.*

"Culture jamming" has historically been used as a tool to shape advertisements and consumer culture into public critiques. Online activists and bloggers use media like memes to transform popular culture into a tool for social change. The result? Young people online are transformed from passive pop-culture consumers to engagers and makers.

Humor, pop culture, fashion, and the punchy, sassy writing, tweeting, and memes that online feminists deploy have become the most effective way to engage young people about the seriousness of injustice, using new Internet culture to speak back to pop culture.

Reclaiming the Frame

Working within a media landscape drenched in reality shows and rape jokes is no easy feat for any feminist. With women comprising only 22%[3] of thought leadership in most mainstream media forums and only 3% of clout positions,[4] it's no surprise that pop culture and legacy news can be such sexist, racist, and homophobic environments. In this context, online feminism continues to constitute an alternative space, where feminist values are suffused in every point and click, *and* to influence legacy media.

The immediacy and viral nature of blogs and Twitter have fundamentally changed how we consume the news. Wherein the past relationship between the media and the public consisted of a top-down flow of information, the Internet has allowed the public to participate in and influence the larger public conversation.

This has resulted in a lateral relationship between the public and news media, to the point where online engagement influences the news of the day. Case in point: when a 31-year-old woman died in an Irish hospital after doctors refused to perform a termination of her pregnancy despite the fact that she was already experiencing a miscarriage, feminist blogs were instrumental in spreading the story. RH Reality Check covered the story on November 13, 2012, and by the next day Jill Filipovic of Feministe wrote about it for the *Guardian,* with a number of other outlets following.

These days, feminist blogs and Twitter accounts can often be a source of both breaking developments and overlooked stories for mainstream media outlets. "Paying attention to feminist media through Twitter is essential," said Jamil Smith, segment and digital producer at MSNBC's *The Melissa Harris-Perry Show.*

> *I originally started using the service as an RSS feed of sorts, and that's how some of the first voices I discovered—Jennifer Pozner, Jessica Valenti, Jill Filipovic, and many more—opened up a new source of political perspective, analysis, and leadership in the media for me. Personally and professionally, I owe them all an enormous debt.*

Melissa Harris-Perry herself rose to prominence in part because of her longtime online presence, including her blog, The Kitchen Table, where she discussed a variety of issues with friend and fellow Princeton Professor Yolanda Pierce.

Online feminism is not only bringing attention to the media gender gap through online activism, but also beginning to fill that space with a new generation of media influencers. Zerlina Maxwell, law student and contributor to Feministing and TheGrio, was, for example, recently featured in *The New York Times* as a political voice on Twitter to follow during election season. She strategically used Twitter to get noticed by mainstream outlets:

> *I picked a handful of folks [on Twitter] I admire that I looked up to and followed everyone they followed. Some of those producers, editors and thought leaders followed me back. Then I started tweeting at media folks if I agreed or disagreed. Figured if they saw my name they wouldn't forget it, and that's exactly what happened . . . Twitter shrinks the world and makes everyone accessible.*

Zerlina is now a blogger for *The New York Daily News,* a columnist for *Ebony Magazine,* and a regular commentator on Fox News & Friends, providing a feminist analysis that was largely absent in these spaces. And her story—one of a law student from New Jersey turned mainstream media commentator—speaks to one of the most remarkable things about online activism: It's bringing feminist analysis and voices into the mainstream. "Mainstreaming the voices of feminist media, particularly at national outlets like MSNBC, is essential given the demands of our news consumers," says Jamil Smith. "As a more technologically sophisticated populace devours its daily news diet from a number of different sources, we not only need to provide spaces for women and men in the feminist movement to contribute to the dialogue and analysis we present on our air—but if we fail to do so, we'll be the ones left behind."

Holding Powerbrokers Accountable

The new lateral relationship between the online public and the media has also created possibilities for a stronger culture of accountability. Sexism, transphobia, and nationalism in mainstream media are far too commonplace, but online responses to these biases are helping to push media stakeholders to be more accurate and less harmful. Online activism has convinced *The New York Times,* for example, to publicly acknowledge victim-blaming content in their articles and reexamine their coverage of transgender people.

Another powerful example: In the summer of 2012, The Applied Research Center and *Colorlines* launched a campaign, "Drop the I-Word," calling on news publications to stop referring to undocumented immigrants as "illegals," "illegal alien" and "illegal immigrant." After a multi-media action strategy, including an online pledge and toolkit, a Twitter campaign and widespread blog coverage, mainstream media picked up the initiative. Announcements followed by those renouncing the usage of the term "illegal immigrant," like *The Miami Herald,* Fox News Latino, ABC News and *The Huffington Post,* as well as those who continue to use it, including *The New York Times.* Today, the Drop the I-Word campaign continues to influence media and individuals in their efforts to create better public representation of undocumented immigrant communities. (And still sends letters to *The New York Times* in response to their continued use of the word.)

Feminists can mobilize online in response to politician and corporate actions as well. When the news broke that Representative Todd Akin told KTVI-TV that pregnancy from rape is rare because "if it's a legitimate rape, the female body has ways to try to shut that whole thing down," feminists responded immediately. "Todd Akin" quickly became a trending topic on Twitter. Tumblrs, social media campaigns and Internet memes followed suit, calling Akin unelectable. Thousands took to Akin's Facebook page urging him to withdraw from the race. While he didn't take his constituents' and colleagues' advice, social media no doubt played a role in his loss on Election Day.

Akin's story was covered all over the country, but stories of movement toward accountability are happening in different pockets of the online community on a regular basis, demonstrating the need for positive and pro-active communication. In June 2011, Vanessa Valenti wrote a blog post critiquing New York Senator Kirsten Gillibrand for saying that the women's movement was "stalled," because too many women were "not engaged" and "don't want their voices to be heard" on MSNBC's *Morning Joe.* She was on discussing her new initiative, "Off the Sidelines." As a very engaged participant in the women's movement and co-founder of a blog whose success has been built on the hundreds of thousands of voices seeking to be heard, Vanessa felt a responsibility to disagree:

> *I'm not saying there's anything wrong with trying to mobilize women who aren't politically active, because of course there are folks out there who aren't. But how can you say they don't want their voices heard when you're the one speaking for them? Because that is one of the biggest lessons we here at Feministing have learned—young women do want their voices heard, they just need a platform to do it. We're here, we're engaged, and we sure as hell don't have a stalled movement. Our hundreds of thousands of readers every month at this blog alone is proof of that.*

The next morning, the Senator called her personally to discuss her remarks. It was a powerful moment for her, and an honor that Senator Gillibrand had, in fact, heard and valued a young woman's voice. The Senator's staff and Vanessa now speak regularly about issues facing women and various strategies for engaging and empowering them.

Creating Space for Radical Learning

The feminist movement isn't without its complicated history. Combating racism, homophobia, classism, and other forms of oppression within feminist communities is a decades-old struggle that is far from over. But the Internet has allowed for a more open space of accountability and learning, helping to push mainstream feminism to be less monolithic.

Professor and theorist Kimberle Crenshaw coined the term "intersectionality" in 1989 as a recognition of the intersecting and overlapping identities that women hold, contributing to varied experiences of oppression. Intersectionality is today a well-known and often-discussed theory of practice within the online feminist world.

A lot of feminist dialogue online has focused on recognizing the complex ways that privilege shapes our approach to work and community. Andrea Plaid of Racialicious spoke at the convening of the unaccounted for labor of constantly educating people with white privilege about racial justice issues. She said: "What we need is more white allies [to challenge racism online] . . . continue to come get your people, without excuse."

One powerful example of this dialogue is the wave of online conversations among women of color online that emerged from the increased attention to "SlutWalk" marches in the mainstream media. "SlutWalks" began in Toronto following a police officer's statement that women should avoid acting like "sluts" as an act of rape prevention. Women around the world protested the idea that women's safety should be tied to their appearance, but the choice by some to reclaim the word "slut" as a rallying cry was not universally embraced. Many felt that the word held a different valence for women of color than for white women and that the experiences of women of color were not being included or respected by protest organizers. This was amplified when a picture was shared of an offensive protest sign that a white woman was holding at New York City's SlutWalk, quoting John Lennon and Yoko Ono's song, "Woman is the N—r of the World." The incident sparked emboldened, necessary conversations about racism within the feminist movement, and the women of color who felt that the movement didn't identify with many of their lived experiences. The author of the QueerBlackFeminist blog wrote:

> *I don't think the intent of the organizers of Slutwalk has ever been to trivialize rape, I firmly believe that Nonetheless, intent is of dire importance at this time. Or the ignorance of the real differences and experience of "womanhood," and the intersections of race, class, gender, sex, sexuality and violence that structure the lives of women of color will continue to be a dividing line in feminist movement.*
>
> *I am hopeful that we will keep these conversations, these critiques, open.*

As people continue to hold one another accountable, as open and honest dialogue persists, as blogs written by diverse voices establish wider and wider audiences, the way we approach feminist activism and leadership is changing.

Part II: Challenges and Opportunities

The Urgency of Now

Nonprofit Organizing in the Margins

Online organizing is a relatively new field of work and as such, we are still struggling to establish sustainable infrastructure. For nonprofits like UltraViolet, Hollaback!, and SPARK Movement, who manage to raise some operating support, funding isn't coming fast enough (or enough, period). "We've had to hustle really hard for every dollar, in part because most foundations just don't have a portfolio that we can fit into," says Hollaback! co-founder and Executive Director Emily May. Although Hollaback! has 250 leaders organizing online and on land across the world, the organization has only two full-time members of staff to support their organizers.

Currently, no women's foundations have initiatives specifically dedicated to online feminist work. There are those who have portfolios committed to funding "nontraditional feminist work," like the incredible FRIDA, a fund that supports young feminists in youth-led organizing. But no major foundation or women's fund has intentionally and explicitly developed a portfolio for online feminist organizations and initiatives. . . .

The Band-Aid Business Structure

Feminist blogs and for-profit online organizations each have our own story of struggle behind why we haven't been able to develop a sustainable infrastructure, but there's one problem that's common: Most of us have what we call "the band-aid business structure"; we operate as LLCs or sole proprietorships relying on third-party advertising or random fundraising drives to pay for server costs and other technical fees. Otherwise, many of us work full-time elsewhere, or rely on social media consulting or speaking engagements as temporary sources of income to supplement our free labor.

Feminist blogs are the least sustained entities within online feminism. Daily website and editorial maintenance generally requires at least one person to be behind a highly active blog every day, as well as general management of media inquiries, organizational partnerships, and advertisements. But ad revenue, even earned by the highest trafficked blogs, can't begin to cover this work. Feministing, which has a readership of over half a million every month, made just $30,000 in 2011, so imagine the number of readers needed to support a movement with this model.

Unfortunately, many of the more lucrative revenue models that organizations have adopted elsewhere online come at a cost and raise difficult ethical questions for feminist blogs. While large sites like Jezebel make money on amassing lucrative page views, the mission of feminist blogs is to send people *away* from our sites to take action, not trap them there. Many of us don't want corporate sponsorship from companies that are antithetical to our mission and don't want to sell our devoted readers' emails to third-party companies. Even one of the biggest success stories of an online social justice organization, Change.org, recently changed its policy to accept right-wing and conservative petitions for the sake of sustaining and expanding the company.

Crowdfunding

Since one of the most powerful things about online feminism is its community, online activists often reach out to their constituencies for funding, whether it be to help pay for their website server costs or a specific project they're trying to jump start. This strategy has been termed "crowdfunding," where money is raised online in a collaborative effort for an organization, individual, or project.

Feminist video blogger Anita Sarkeesian started a campaign on the popular crowdfunding website Kickstarter to raise $6,000 to create videos about depictions of women in video games for her Feminist Frequency series. After being attacked by misogynist trolls who vandalized her Wikipedia page, hacked her website, and sent her death and rape threats, her online fans rallied to support her and her project, ultimately raising nearly $160,000.

Queer Nigerian Afrofeminist writer and media activist Spectra raised over $10,000 for her social media and communications training for African women's and LGBT organizations.[5] Miriam Zoila Pérez raised over $4,000 for her self-authored "The Radical Doula Guide," a booklet inspired by her blog Radical Doula that addresses the political context of supporting people during pregnancy and childbirth. These examples, and so many more, demonstrate that crowdfunding can be an effective way for individuals, collectives, or institutions to raise money while circumventing the need for nonprofit status, which many online feminists don't have.

It's far from a systematic or sustainable solution, however. Chances are that none of the activists mentioned above will be going to their communities for support again anytime soon, assuming they've essentially "rapped" their networks.

Crowdfunding is great for discrete projects—a video series, a training, a book—but doesn't lend itself well to creating the infrastructure needed to sustain online organizations. Additionally, the larger a project is, the harder it is to reach the fundraising goal: Only 38% of $10,000 projects on Kickstarter, the most popular crowdfunding site, reach their goal, and that rate drops drastically as the goal gets higher.[6]

Membership and Subscription Model

In counter-point to the foundation and crowdfunding strategies to raise funds, a membership model allows supporters to give funds to online activists that are not project directed. This allows these organizations the freedom to be flexible in their activities with the stability of a constant income stream.

MoveOn, one of the most well-known progressive online nonprofits, follows this membership model; the average donation to MoveOn.org Political Action, according to their site, is less than $20. Women, Action, & the Media (WAM!) is also funded through this system, offering discounts and benefits in exchange for a $45 fee.

A variation on membership is the subscription model common to content producers. A number of media outlets provide enhanced online access for an annual fee, such as *Bitch Magazine.*

But many online platforms balk at the idea of sequestering their content and providing it only to paying customers. And anyone relying on ongoing donations runs a risk that those donations could disappear at any time. Because most membership models are built on a monthly or yearly system, long-term planning can be extremely difficult.

What's at Stake

Online feminism may continue to grow and evolve, but whether it will reach the potential it needs to sustain itself—and make the real, transformative impact the world needs—is yet to be seen. An unfunded online feminist movement isn't merely a threat to the livelihood of these hard-working activists, but a threat to the larger feminist movement itself. Without greater support, the online feminist movement faces a number of risks.

We will remain reactive and myopic. . . . In spite of the powerful successes of online feminists, our stories of impact have a disappointing common trend: They're almost all reactive and short-term. The lack of infrastructure and sustainability for the online feminist movement makes it nearly impossible to think about more meaningful, long-term strategizing. More than ever, we need to create effective proactive campaigns and policies to prevent sexist encroachments in the first place, rather than being in a perpetual state of pushback.

There will be an incredibly high burnout rate. In April 2012, one of the largest global feminist blogs online, Gender Across Borders, ceased operating. After three years of collaboration with international organizations and companies, offering over 30,000 readers feminist analysis and global activism opportunities every month, founder Emily Fillingham and the team of editors decided it was just too difficult to maintain. "Unfortunately, none of us could afford to keep it up. We made a lot of progress in just a few years, but it still wasn't enough to earn us any funding," said editor Colleen Hodgetts. And they're not the first; dozens of underpaid, overworked and exhausted online activists have left the movement, their voices lost and the mix—as a result—much less rich.

An unfunded movement further privileges the privileged. . . . If we don't support this work, the most privileged in the online movement—those who already have the resources and time to blog every day, and do organizing work for free—will have the most amplified voices. Women of color and other groups are already overlooked for adequate media attention and already struggle disproportionately in this culture of scarcity. If feminist movements don't create supportive spaces, the leadership pipeline will grow smaller and more insular, and fewer voices will get promoted.

Anti-feminists will leverage the Internet. Misogyny, both blatant and covert, is rampant online. Online harassment and threats are a daily experience for online activists, and young women and girls are increasingly bombarded with vitriolic and harmful messaging on the very same forums we use for activism. Radically anti-feminist commentator Ann Coulter has over 300,000 followers on Twitter—four times the number of followers as Planned Parenthood. Pinterest—the social networking site of 17 million visits per day[7]—has become immersed in diet tips and images of Victoria Secret models.[8] Anti-feminist video bloggers outnumber feminists in search results for "feminist" on YouTube. Not only is it up to us to build our influence and challenge the sexism and bigotry that exists online, but also to continue to provide safe spaces for young people to engage with one another in healthy and empowering ways.

We'll repeat the same mistakes.

> "If the gender identities were different, it would be a different conversation. How do we combat all of the things in our socialization that teach us that we don't deserve sustainability? We have to embrace the entitlement of saying—'No, I deserve these things, and I need them and I'm not going to wait for someone to hand them to me.'"
>
> —Miriam Zoila Pérez, Radical Doula

The "psychology of deprivation" we speak of is not a new phenomenon for feminist activists. We acknowledge that historically, the feminist movement has not valued its own labor. It has largely depended on unpaid work, slowly evolving into exhaustion and eventual burnout. We believe this is a huge part of what's been holding the movement back from creating the real policy and structural change it needs. We pass this model down to the young women and girls who look up to us: that it's necessary to work for free, and to risk our physical and mental health, and our relationships in order to make change. We convince them that these martyr-like sacrifices are "heroic" and "inspiring," when, in fact, we know they've only been harmful to our well-being and to the movement. We must create a new culture of work, a vibrant and valued feminist economy that could resolve an issue that's existed for waves before us—and create a more hopeful legacy for the generations to come.

What's Needed Now?

To avoid these pitfalls and embrace the opportunities ahead, the online community will have to be strategic and partner with a range of their feminist allies—advocacy and nonprofit organizations, philanthropists and entrepreneurs, corporate leadership with a feminist sensibility, educators, community organizers, artists, and youth—among so many others. It is time to strengthen the connective tissue between those who are most savvy and connected online, and those pushing feminist agendas in our courtrooms, classrooms, boardrooms, and beyond. The results could be profound.

We need to create more spaces and times where strategy and collaboration are prioritized, supported, and expected, and where feminists of all ages—but especially the young and online—have a chance to do the profound work of dreaming together.

Notes

1. Corrected 4/12/13: Based on feedback, this paragraph has been corrected to better reflect the understanding of the work by those mentioned.
2. Madden, Mary and Kathryn Zickuhur, "65% of Online Adults Use Social Networking Sites." Pew Internet and American Life Project, 2011, http://pewinternet.org/Reports/2011/Social-Networking-Sites.aspx
3. http://in.gov/iew/files/benchmark_wom_leadership.pdf
4. http://annenbergpublicpolicycenter.org/Downloads/Information_And_Society/20010314_Progress_and_Women/200110321_Progress_women_report.pdf
5. Corrected 04/12/13: Spectra was originally incorrectly referred to as an "online activist."
6. Mitroff, Sarah. "4 Keys to a Winning Kickstarter Campaign." *Wired Magazine*, 2012. www.wired.com/business/2012/07/kickstarter/
7. http://mbaonline.com/a-day-in-the-internet/
8. http://buzzfeed.com/arnyodell//how-pinterest-is-killing-feminism

Introduction to Reading 52

Choice feminism, also known as neoliberal feminism, is a contemporary form of popular feminism that encourages women to uncritically value individualism and consumerism over collective action against social inequalities. In this reading, Rachel Thwaites offers a critique of choice feminism. She analyzes a subset of data—discussions among women at feminist bride websites—gleaned from a larger study on what British women do with their last names upon marriage to men. Thwaites is especially concerned with the strong tendency among choice feminists to silence dissent in the name of being supportive of other women no matter the life choices they make.

1. How does the choice narrative shut down "politically engaged debate and different viewpoints"? Why is this dangerous to feminist movements for gender justice?
2. Thwaites notes the surnaming decisions in heterosexual marriages are particularly open to the choice narrative. Discuss.
3. Thwaites ends this reading by stating that "[un]thinkingness is not feminist." What does she mean by that statement?

Making a Choice or Taking a Stand? Choice Feminism, Political Engagement and the Contemporary Feminist Movement

Rachel Thwaites

The popular (non-academic), positive image of feminism is of a movement which protests for more choice for women in their daily lives: more opportunities, more freedom, less restraint and less-constricted roles to play. This is an image that could be associated with any 'wave' of feminism, but has perhaps become most connected with the 'third wave,' beginning in the 1990s. In this popular idea and narrative of feminism, 'choice' is the most significant word. Women can *choose* to work or stay at home, choose to marry or not, have children or not; choices are to be made freely as the world becomes a more equal place. This kind of feminism sounds inspiring, welcoming and positive. Claire Snyder has argued that third-wave feminism, from which choice feminism grew, is intended to be more inclusive and diverse (2008:180). However, as liberatory and tolerant as choice feminism initially sounds, it has drawbacks which get to the heart of the question of what feminism is for (Thornton, 2010).[1] These drawbacks are highlighted by feminists who critique the narratives of choice feminism. Indeed, some have been critical of the genuine inclusiveness of third-wave feminism (Springer, 2002), casting doubt upon the idea that the choice narrative allows everyone to follow their own desires and wishes within the modern feminist movement.

In this article, I examine choice feminism and how it influences discussions of decision-making in women's lives, through a set of empirical evidence taken from a small study of feminist bridal websites. This will focus on the discussions around name changing and retaining on marriage, and how 'choice' becomes a part of maintaining the neoliberal status quo. As academic feminists debate the narrative of choice and its negative and anti-equality connections to neoliberalism, popular feminism continues to chart a course of celebrating choice and using it as a means to live a feminist life. The divide between these feminist narratives needs to be bridged to have a more open discussion about what feminism means and how choice fits into it. Without this, we run the risk of seeing it become increasingly difficult for feminism to make political statements for women, and of feminism becoming something which does not translate across academic and popular lines. In this article, I engage with academic critiques of choice feminism, before looking at the set of empirical data and discussing what this means for how contemporary feminism is understood by those identifying as feminist. Finally, I demonstrate why choice feminism needs to be challenged if the movement is to remain politically engaged and useful to creating change.

Critiquing Choice

Choice feminism is often associated with authors like Jennifer Baumgardner and Amy Richards (2000),

Thwaites, R. (2016). Making a choice or taking a stand? Choice feminism, political engagement and the contemporary feminist movement. *Feminist Thoery, 18*(1): 55–68.

who understand every decision a woman makes as potentially feminist, if given thought and made with a political consciousness. Authors such as Natasha Walter have also written highly popular books based on choice feminism (Walter, 1999; although Walter has since distanced herself from this stance—see: Walter, 2010).[2] It was Linda Hirshman (2006) who coined the phrase 'choice feminism' and gave it a pejorative slant, criticising the lack of political thought that went into these choices. In her typically polemical style, Hirshman wrote that '[a] movement that stands for everything ultimately stands for nothing' (2006: 2). Michaele Ferguson is another strong opponent of choice feminism, arguing that judgement is needed to truly live a feminist politics and that all choices are not equal (2010: 251). She argues clearly that judgements may be difficult to make—especially judgements of the lives of loved ones—but that this does not mean they should not be made (Ferguson, 2010: 249). She argues instead that the only way to genuinely improve our world is to make these judgements and to not allow fears of upsetting or alienating people to prevent us from actively engaging in politics in the everyday: 'if we suspend judgement in the context of our personal relationships, we seem to be failing in courage as feminists—for feminism is precisely about reimagining and reworking the personal' (Ferguson, 2010: 249). It is this point about judgement which is so clitical to discussions of choice feminism and its worth. When we allow every choice to be equal, there is no capacity to argue against one form of action and decision-making over another.

Choice feminism certainly opens up a number of critical questions around whether feminism's main focus should be on the individual and their decisions or on the collective and the best decisions for all. Jannet Kirkpatrick states that choice feminists are interested in getting away from the negative judgement of feminism, and in remembering that only each individual woman can really know her own circumstances and reasons for acting as she does (2010: 242). There can be no 'standard feminist' actions, but only individual choices based on what is best for that person and her life; the worst thing a feminist could do is restrain her fellows in making these choices (Kirkpatrick, 2010: 242). For choice feminists, Kirkpatrick argues (2010: 243), feminism is always here, and the movement's gains are simply part of the fabric of life. Instead of seeing feminism as an ongoing battle with the possibility of regression and of the restriction of rights and hard-won freedoms, choice feminists see it as 'in the water' (Kirkpatrick, 2010: 242). Hirschman has claimed that this harms the feminist movement itself, and has argued that those holding this viewpoint should be taken to task (see: Hirschman, 2006). Lori Marso also points out that having diverse desires is a part of politics, and that debate is key to maintaining a political stance (Marso, 2010: 263). The fact, then, that people have different viewpoints, desires and challenges in their life does not mean the demise of the movement: 'we can retain feminist community while also retaining diversity' (Marso, 2010: 264). Embracing the fact there are differences between women and that intersections create difference does not also mean that all commonalities are washed away, or that women cannot strive to understand one another's situations and work together through these differences. As Susan Friedman argues, we have to make political statements about 'women' to make any political progress, and we can do this without pretending there is a united sisterhood (1993: 250). There have always been different and competing desires within feminism, as within most political movements.[3] Being a woman does set up socially constructed ways of relating to the world and of the world relating to yon (Marso, 2010: 266). It is recognising these and speaking across commonalities which women can do. Instead of falling on an individualistic choice rhetoric, which removes the chance to debate, critique and *do* politics, we can work together through challenge, discussion and considered judgement.

This academic feminist criticism of the feminist choice narrative suggests a divide between popular conceptions of feminism and academic ones. The danger of not bridging the gap between these discussions will become clear below. I am not suggesting that non-academic feminists are unable to think without academic feminists—or indeed that academic feminists never turn to and use popular narratives themselves—but that there is a strong choice rhetoric in popular feminism, which helps to make living a feminist life-politics easier in a complex world; however, this also allows feminists to refuse to take responsibility for difficult judgements, in Ferguson's terms, and it is this which is critiqued by academic feminists. Choice within feminism is a very difficult concept to reconcile with the wider emancipatory, communal project and I am not providing a definitive answer to it here. However, there are more or less traditional choices which impact on unequally gendered relations and are linked to patriarchal pasts and futures. Instead of justifying these choices and silencing any critique of them, feminists should be prepared to discuss and critique the context within which they make their life decisions. Of course feminists will make a variety of choices about their lives, and not always in the best interests of women as a whole or of the wider feminist movement, but the reality of this fact should not be hidden away behind the word 'choice'. Instead, feminists have a responsibility to look carefully at their own and others' decisions and the reasons for which they make them; in so doing, feminists will be able to critically examine and debate what it means to be feminist, and what courses of action are better than others for women as a whole. It is important to take responsibility for one's actions, but equally significant to interrogate and challenge the structural and institutional factors which contribute to them, instead of allowing an unhelpful individualism to take hold. This will ensure feminism is not a movement of individual blame, but a collective force for change.

Details of the Study: Methodology and Analysis

The wider study within which this data was situated looked at what British women do with their last names when they marry, and their sense of identity in connection with this decision. I administered an online survey, which 102 women completed, and conducted sixteen in-depth interviews to capture participant experiences of name changing or retaining, their narratives of self and their thoughts and feelings around the norm of name changing more generally, as well as how their name connected with family, sexuality, feminism, ideas of tradition and love (see: Thwaites, forthcoming 2017 for more on the wider study). In this article, I focus on a smaller section of work that I did within my wider study: looking at the articles and following comment discussions on feminist bride websites in relation to name changing. Though this was not the main focus of my study, and hence a small selection of feminist bride websites was examined, the discussions which occurred on these platforms are indicative of wider popular feminist discussions of choice and bring useful insight through use of the words of women themselves. These websites are spaces which are accepted by users to be populated by feminists, and so allow people to openly discuss their feminist life politics without too much fear of rejection or ridicule; they can truthfully share their life decisions and the complexities of coming to them. This makes them a good source of information on how women relate to feminism in their everyday lives.

The websites I surveyed were: *The Feminist Bride* (www.TheFeministBride.com), which was set up in 2011; *Feminist Wedding* (www.feministwedding.com), which was set up in 2009; and *A Practical Wedding* (www.APracticalWedding.com), which was set up in 2008 (I also looked at *Feminist Bride* (www.feministbride. com) and *The Offbeat Bride* (www.offbeatbride.com) for background reading, but my comments come from the first three websites). The *Feminist Bride* describes itself as a source for 'modern brides' to investigate 'substantive questions' around weddings and wedding traditions, to enjoy their wedding but feel it is a space of equality. The editor is Katrina Majkut, who used to work in investments before moving towards writing, via industry analysis of the wedding industry; she describes herself as a 'proud feminist'. The website has a magazine style, but offers lectures on wedding traditions and information on rights in marriage. The editor/author believes there are certain wedding traditions not worth following (name changing is one of them), but she wants the space to be for debate. *Feminist Wedding* is run by Casey, and is a blog of her thoughts on weddings. She provides little information about herself, but follows a number of feminist blogs and has given her blog the tagline 'tradition disrupted. A feminist perspective on weddings'. Her blog revisits naming a few times, describing some of the pitfalls of doing so, such as losing track of people, and is generally highly negative about the practice. *A Practical Wedding* is a website for 'modern wedding planning', and allows anyone to submit posts—moderated by and discussed with editors—about wedding planning. The website is not specifically about feminist weddings but includes posts on breaking with tradition and how to have a more 'modern' wedding, which can include references to more equal weddings. Judging by the entries that are specifically feminist, and the comments, this website does attract women who define themselves as feminist as at least part of its readership. None of these websites provide any explicit definition of feminism.

As these websites are openly public, no ethical approval was sought to use the words of the women, however I will synthesise opinions given in comments rather than single people out, except for the authors of articles who would appear to have given their consent to their words being used in the public domain by presenting their words and videos alongside their name, in a similar way to news journalists. I searched these websites for articles relating to names and decisions about them when marrying, looking specifically at what a feminist community was saying about this decision. Through reading these articles and studying the comments and discussion that followed, I was able to draw out recurring themes. The most significant theme of all was that of choice, and how it should be a woman's free decision what she does with her name. There were some added caveats, such as having carefully thought about the decision beforehand, and in this way they echoed the writings of Baumgardner and Richards (2000), which will be discussed in more detail below.

The women writing on these websites attempt to question the norms of marriage and the amount of money involved in the wedding industry, but there are conflicting and conflicted discussions of whether or not a feminist can change her name. Meg Keene, writing for *A Practical Wedding,* offers up the viewpoint that a name can be changed to a husband's after a period of thought and when that decision feels 'right' to the woman (Keene, 2012). She argues that any decision can be a feminist one if enough time is given to considering it, rather than acting unthinkingly. Keene reasons that sharing a name builds a sense of being a team or unit.[4] Keene's argument aligns very closely in this sense with Baumgardner and Richards (2000) about what being a choice feminist means. The thinking that could be done, and its connection with feminism, is rather less clear, however: is Keene thinking about the historical and contemporary meanings of name changing, about the aesthetics of her name or about the thoughts and feelings of family, friends and society? Each of these areas of thought I have suggested have gendered norms of action which can lead one to act in a particular way; no decision is completely freely made. The underlying position in this article and in the following discussion in the comments section, which is very much in favour of Keene's philosophy, is that any choice should be supported since only the woman in question can really understand her circumstances. The thinking that is done can be of any kind, as long as it occurs. The *thinking* is the feminist act: yet what this thinking really 'looks' like or how one justifies a decision that upholds unequal gender relations (Thwaites, forthcoming 2017) is not explicated. The thinking that is mentioned begins to look more like a lack of thought about one's decision in terms of the feminist movement and women in general, and more about individual needs and desires. And when following a norm can make life easier—which name changing certainly can (Thwaites, forthcoming 2017)—it is understandable that that norm might be followed. However, saying this is a feminist act is far more difficult to accept.

More unusually, *The Feminist Bride* has an entry by Katrina Majkut which advocates *not* changing names on marriage (Majkut, 2012). She argues that there are historical and political reasons not to follow this pattern. The discussions that follow this entry are initially in favour of Majkut's philosophy, but quickly turn to disagreement by stating that all choices should be upheld. The choice narrative becomes the dominant one in this comment thread, and anyone suggesting that choice is not an acceptable answer to the question of name changing would have found it very difficult to be heard. Not upholding every woman's decision, even if that decision follows a traditionally unequal path, is to suggest you are not supportive of your fellow feminists and are therefore not a 'good' feminist. In this sense, the choice narrative actually shuts down politically engaged debate and different viewpoints by encouraging everyone to act as they wish to, and providing a justification with which it is extremely hard to argue. On each of these three websites it becomes clear that there is no obvious answer for feminists as to the best thing to do for themselves as individuals, as well as for women more widely. There is no agreement over the real significance of names, and in this situation—not wanting to blame a woman for her choices or to suggest she is not a proper feminist—the individualised narrative of choice becomes incredibly significant. When it is most difficult to decide 'what to do for the best', choice can appear to reconcile this difficulty with a notion of feminist politics, despite its depoliticising effect.

On *Feminist Wedding,* visitors are asked to fill in a short quiz about name changing created by the site editors in response to debates triggered by the question of what to do with one's name on marriage. The results show that most women who answered felt there was considerable pressure on women to change names, that they were irritated that it is seen as only a woman's problem to grapple with and that they were in the main going to keep their own name, but that they expected backlash for this decision. The question that caused trouble was whether women who change names are making an anti-feminist choice. Despite clear thoughts on the other questions with an easy majority one way, 30 per cent of the women thought it was an anti-feminist choice, 47 per cent thought not and the remainder were unsure (results examined on 28 January 2013). The comments on the websites reveal an equally mixed viewpoint on this question. One *Feminist Bride* author found some support for her viewpoint in the comment discussion that women should not change names, but the idea of choice as more important than following a specific 'standard feminist' route quickly crept into the discussion of her article and became the dominant standpoint for those commenting. One *Practical Wedding* article has a long comments section following it in which feminists argue this point, showing how controversial the issue can be. Ultimately though, most agree that feminism is about offering women choice, and that all decisions should be supported. The feminist websites, helping women make decisions about living a feminist life-politics, ultimately come to the conclusion that choice feminism is the only way to deal with some of the incredible complexities which arise from living as a feminist in a non-feminist world. Naming decisions are one example of this choice feminism, but it is just one example of how judgement and debate over individual actions that actually impact other women's decisions and selfhoods are silenced in favour of choice and individualism.

Standards of Feminism: What Does Being a Feminist Mean?

The silencing of other thoughts and opinions, reducing everything to uncritical 'choice,' forces debate away from politics and into the realm of 'unthinkingness' (Shils, 1971), as everyone must follow the choice majority. In fact, to not follow this majority is to be laid open to accusations of not being a 'good' feminist as one is not supporting women. The complexity surrounding what it means to be a feminist comes to the fore in these discussions, with naming choices a clear case of confusion over definition and lack of comfort in judging others, even when faced with the reality that a 'choice' may be detrimental to the woman herself. Naming decisions are particularly open to choice feminist rhetoric. To expound a complex argument to a loved one about their own name can feel like too much, that there are other, bigger battles, and that this is too intimate a decision into which to bring politics. It is also very difficult to discuss the patriarchal basis of name changing, which still exists in subtler forms than in previous centuries (see: Thwaites, 2013; forthcoming 2017), when most women change their name. For feminists in particular, this discussion is tricky: feminists are meant to understand the patriarchal basis of these kinds of traditions and hence to fight against them, so when a feminist friend changes her name, any explanation other than free choice can be too difficult to face. However, feminists find themselves complicit in non-feminist and anti-feminist decision-making at times—we are all embedded within our society and have desires to follow, and we often find satisfaction in the norms which make people understandable and acceptable within that society.

The name change is perhaps such a particularly difficult and controversial decision for feminists to make because of the association that keeping a name has with feminism. Certain actions and beliefs are associated with feminism (Western feminism in this case, as the naming issue was important to British and American feminism particularly); therefore being critical of marriage, aware of the historical subjugation of women and retaining one's name, are a part of this association. This has been discussed briefly above, but can be seen in my own research (see: Thwaites, 2013; forthcoming 2017) and in other smaller studies into feminists and naming decisions (see: Mills, 2003). A vocabulary of what it means to be a feminist today is created, and name retaining can be seen as a 'standard' move. In fact, in my research I found that the only time women who changed their name had to explicitly justify themselves to others was when they had to justify themselves to their feminist community (see: Thwaites, 2013; forthcoming 2017). The difficulty that arises from having to articulate one's seemingly non-feminist action within the context of a feminist community compels many women towards choice feminism, both to justify themselves and to help make more comfortable those they see having to justify themselves. Justification depends on context, but choice feminism provides an easy route away from this uncomfortable moment and provides women with a sense of empowerment and agency in all their decision-making.

The words 'I made my own choice' are, after all, very hard to argue with. It also shifts the possibility of being a 'bad' feminist onto those who are more inclined to debate, and judgement of other choices becomes 'nagging,' 'judgmental' and not supportive of one's fellow feminists: in short, being labelled with words and phrases most feminists try to avoid as part of the clichéd, negative picture of feminism. Choice is a strongly disciplining narrative (Thornton, 2010: 96); it encourages one to make specific decisions and side with specific ways of being a feminist in order to remain open to everyone's desires and actions, even if those may in fact seem detrimental to women as a wider grouping, or even to just that woman herself. Choice is therefore not as free as it initially sounds: if choice feminism was just about women determining their lives it would be unproblematic, but the academic critiques of this part of the movement are compelling. Making women more comfortable removes the need to interrogate deeply the motives behind following a traditional path connected with patriarchy; it also aligns feminism with neoliberalism and consumerism in a way which should make us all wary, and puts critique and judgement firmly into the 'bad feminist' box. It is important that academic and popular feminism inform one another, and that there are not barriers to communicating critiques of the movement; it is only by doing so that we can move forward in our goal of creating political change. Popular and academic feminism should not be separate entities but parts of a wider whole, yet the critiques of choice which academics debate are not represented on the websites studied, suggesting something of a gap between the two narratives. In order to unite popular and academic feminism, choice feminism must be loudly challenged.

Challenging Choice Feminism

Choice is an important idea in discussions of late modernity and individualisation. Anthony Giddens' idea of the 'reflexive project of the self' (1996: 5) suggests that there are now more choices and options available to people than ever before; life trajectories are not bounded as they once were by traditions and rigid rule structures. However, this argument appears simplistic when thinking about how significant the past remains to decision-making and possibilities for action. Individualisation theorists pit the past against the present (see: Giddens, 1996; Beck and Beck-Gernsheim, 2010; Bauman, 2011): the past is presented as stable and unchanging, whereas the present is in constant flux. This reductionist view of the past is in part, Vanessa May argues, because the past is viewed through its structures, whereas the present is viewed via personal lives (2011: 365): structures then easily appear unchanging, while our personal lives appear to be moving fluidly. However, separating structures and personal life paints an unrealistic picture of the world, as structure and personal life—or society and self—are 'interdependent and permeable, each affected by the other' (May, 2011: 365–366). The past, then, as Matthew Adams argues, influences us all through its 'codes of practice'—structures and norms influencing personal life (2003: 227).

Such 'codes of practice' become traditions and guide the decisions we make. Traditions can become so embedded that they are 'unthought,' and become so taken-for-granted that they need little or no justification for being followed. The choice narrative ignores not only the very important place of unthinkingness within norms and traditions, but also that some seemingly freely made decisions are so influenced by societal practice and opinion that they cannot be considered truly free, in Giddens' sense. As Steven Lukes argues, some powerful social norms may be so ingrained that conceiving of other possibilities for action is practically impossible (2005: 113). Choices remain limited by the past, by resources, by the society in which a person lives, including the influences of the gender order and the ethical and moral standards of the day. The idea that women should be able to do whatever they want and what makes them happy is a part of this narrative of reflexive choice; it is also a part of the pervasive neoliberal rhetoric which is so significant to modern capitalist societies. Though neoliberalism, globalisation and the impacts these systems have on the world are frequently presented by politicians and the mainstream media as inevitable forces—unstoppable in their linear progression—they are actually the consequences of human decisions and policies (Heron, 2008: 95). These decisions have created vast inequalities but, as Taitu Heron argues, the 'role of international and political economic structures and interests as co-determinants to poverty and continuing inequality is not recognized' (2008: 95). Instead of investing in social equality and welfare, states look to solve problems through the market—and inequalities widen.

The idea that we can improve our lives through consuming is an important force within neoliberal capitalism. The consumer makes choices based on the idea of the consumer's *right* to choose (Craven, 2007). The consumer should be given a full range of choices and decide which is best for them. Yet our choices are bounded, as I have argued. Unthinkingness prevents certain thoughts or decisions for different courses of action from even arising. A person may look as if they have made a totally free choice when in fact the powerful cultural norms at work can prevent 'an agent or agents' desires, purposes or interests [. . .] [from being fulfilled] or even from being formulated' (Lukes, 2005: 113). Particular courses of action that are viewed as worthy or valuable, and can even confer a certain status, will shape action, encouraging people to follow the norm to make their lives intelligible to others and viewed societally as valuable. Changing a name is one such norm, which is rarely questioned. The unequal nature of this naming practice—women being expected to change names in a way men are not—continues to be deeply embedded in conceptions of the self and maintenance of inequalities between men and women. It is incredibly difficult not to follow these norms, but when they perpetuate gendered inequalities feminists must take the time to interrogate them rather than finding ways of justifying them.

Upholding all women's decisions may seem like a feminist action in not belittling or talking down to other women, but it remains that feminists should be critical of the taken-for-granted norms that are unequally gendered, and that they should remain alert to the wider patriarchal context in which all decisions are made. Choices are not entirely free, but the rhetoric of their being so is clearly highly important and influences societies across the globe. Feminism has been involved with using the rhetoric of choice to attempt to improve the position of women. Christa Craven points out that feminists called women making decisions about their reproductive rights 'consumers' to attempt to get away from the generally paternalistic relationship with male doctors that women entered into on becoming pregnant: the female patient versus the male doctor (2007: 701–2). However liberating this narrative was intended to be, the use of 'choice' by feminists within a neoliberal capitalist society must be constantly critiqued. This neoliberal rhetoric of 'choice' is often invested in maintaining the status quo by removing the agency of the less powerful and enhancing that of the established powerful elite (Heron, 2008: 95), who actually have the resources and means to make a wider range of choices than those who are disadvantaged. The less powerful are then blamed by this rhetoric for not taking responsibility for themselves to make their lives better and more prosperous.

This last point highlights the more dangerous side of individualism which creeps into these discussions, and to which feminism can anchor itself when critical discussion takes a back seat to embracing every decision. Though the idea of everyone choosing to follow their own desires and achieve their own goals sounds freeing, it is, as Marso contends, an argument which ignores how choices and actions impact upon other people (2010: 264). As she writes, '[f]or feminism to retain its political vision as a force for social justice, we must continue the difficult conversations concerning how acting on our diverse desires impacts the lives of others' (Marso, 2010: 264). This negative individualism also encourages us as a feminist community to forget the differences in access to resources which are available to us, and which influence our ability to make choices. Furthermore it encourages us to blame ourselves as individuals when things go wrong, rather than to look critically at social norms and structures, and purports that any choice made freely by the individual cannot harm them, hence being unhappy or undermined by your choice can only be your own problem. These are worrying statements to make in connection with feminism, a movement which should recognise systematic inequalities and work towards an equal society. Criticality is central to ensuring these important aims do not get lost in a narrative of neoliberal 'choice'.

The power of the imagination is significant to thinking critically and opening up other possibilities: again, what courses of action we follow are in part influenced by whether or not we can even imagine them (Lukes, 2005: 113). Michel De Certeau argues that 'the *thinkable* [. . .] is identified with what one can *do'* (1988: 190; emphasis in original). In other words, if we can think it we believe we can do it. Contrary to this, if we cannot think it we cannot do it—it takes being able to imagine an action first before it becomes a reality. The unthinkingness that sometimes surrounds decision-making, and the justifications we provide for those decisions, should not go by accepted, but should be taken as a call to all feminists to challenge themselves and others to be more creative and imaginative in their thinking. Name changing is the prevalent norm; name retaining is not so well articulated, and with fewer examples and bureaucracy often discouraging it (see: Thwaites, 2013), it can be harder to imagine as a possibility which is genuinely workable. In seeing beyond the traditional, and in imagining a better and more equal society, we open up the possibilities for action and change. Unthinkingness is not feminist, and this is something for us all to bear in mind.

Notes

1. As Thornton in her 2010 article, when looking at the context of 'an incident involving the representation of women's breasts on the cover of an Australian law school student magazine, which included short articles on sexed crime', de-politicising feminist activity and using 'irony' and 'humour' as excuses to objectify women, under the guise of third-wave feminism, empowerment and choice, can have detrimental effects for people's understanding of what feminism's aims are and actually for the position of women within that context.

2. Natasha Walter writes in her 2010 book, *Living Dolls*, that she feels she was overconfident in *The New Feminism* about the gains women had made. She writes that she failed to understand how pervasive and detrimental the sexualisation and objectification of girls and women really was, and that it has only increased and intensified over the 2000s.

3. I see this as a sign of vibrancy and passion, responding to complex concerns.

4. In this way she reflects many of the women who changed their name in my wider study, who wished to be obviously a unit, team or family. Though not discussed in detail on these sites—the correctness of the decision itself rather than the reasons for the decision being the main topic of debate—women in my wider study who changed names gave a number of reasons for doing so, which are given here to provide some context: display of love and commitment to the marriage, a sense of adulthood achieved through marriage

and the status of being a 'wife', and being obviously and intelligibly a family by sharing a name with one's husband but also importantly with one's children (see: Thwaites, 2013; and forthcoming 2017 for more on the reasons women gave for changing names).

References

Adams, Matthew (2003) 'The Reflexive Self and Culture'. *British Journal of Sociology,* 54(2): 221–238.

Bauman, Zygmunt (2011) *Liquid Modernity.* Cambridge: Polity Press.

Baumgardner, Jennifer and Amy Richards (2000) *Manifesta: Young Women, Feminism, and the Future.* New York: Farrar, Straus, and Giroux.

Beck, Ulrich and Elizabeth Beck-Gernsheim (2010) *Individualization.* London: SAGE.

Craven, Christa (2007) A "Consumer's Right" to Choose a Midwife: Shifting Meanings for Reproductive Rights under Neoliberalism'. *American Anthropologist,* 109(4): 701–712.

De Certeau, Michel (1988) *Tile Practice of Everyday Life.* Berkeley, CA: University of California Press.

Ferguson, Michaele L (2010) 'Choice Feminism and the Fear of Politics'. *Perspectives on Politics,* 8(1): 247–253.

Friedman, Susan Standford (1993) 'Relational Epistemology and the Question of Anglo-American Feminist Criticism'. *Tulsa Studies in Women's Literature,* 12(2): 247–261.

Giddens, Anthony (1996) *Modernity and Self-Identity.* Cambridge: Polity Press.

Heron, Taitu (2008) 'Globalization, Neoliberalism and the Exercise of Human Agency'. *International Journal of Politics, Culture, and Society,* 20(1/4): 85–101.

Hirshman, Linda R (2006) *Get to Work: A Manifesto for Women of the World.* New York: Viking.

Keene, Meg (2012) 'What Should We Call Me? Changing My Name as a Feminist Choice', *A Practical Wedding,* 12 September. Available at: http:/fapracticalwedding .com/2012/09/ changing-your-name-to-your-partners-last-name-as-a-feminist-choice/ (accessed 30 September 2014).

Kirkpatrick, Jannet (2010) 'Introduction: Selling Out? Solidarity and Choice in the American Feminist Movement'. *Perspectives on Politics,* 8(I): 241–245.

Lukes, Steven (2005) *Power: A Radical View.* Basingstoke: Palgrave Macmillan.

Majkut, Katrina (2012) 'Why Do Bride's [sic] Take Their Husband's Name?'. *The Feminist Bride,* 18 May. Available at: http://thefeministbride.com/history-of-name-change/ (accessed 30 September 2014).

Marso, Lori J. (2010) 'Feminism's Quest for Common Desires'. *Perspectives on Politics,* 8(1): 263–269.

May, Vanessa (2011) 'Self, Belonging and Social Change'. *Sociology,* 45(3): 363–378.

Shils, Edward (1971) 'Tradition'. *Comparative Studies in Society and History*, 13 (2): 122–159

Springer, Kimberly (2002) 'Third Wave Black Feminism?' *Signs,* 27(4): 1059–1082.

Thornton, Margaret (2010) "Post-Feminism" in the Legal Academy'. *Feminist Review,* 95: 92–98.

Thwaites, Rachel (2013) 'The Making of Selfhood: Naming Decisions on Marriage'. *Families, Relationships, and Societies,* 2(3): 425–439.

Thwaites, Rachel (forthcoming 2017) *Changing Names and Gendering Identity: Social Organisation in Contemporary Britain*. Abingdon: Routledge.

Walter, Natasha (1999) *The New Feminism.* London: Virago.

Walter, Natasha (2010) *Living Dolls: The Return of Sexism.* London: Virago.

Introduction to Reading 53

The 2013 hashtag #BlackLivesMatter (BLM) marked the beginning of a powerful global network and movement organized to intervene in state violence inflicted on Black communities. Notably, BLM has always centered women, queer, and trans people in its social justice organizing and work. The merger of BLM with feminist movements is discussed in depth in the conversation between Sarah J. Jackson and Cathy J. Cohen set out in this reading. Cohen and Jackson discuss the importance of intersectional analysis and practice in BLM. In addition, Cohen emphasizes the core role of young, Black, often queer women in the collective leadership of BLM.

1. According to Cohen, what are the three contributions of feminism to racial justice movements?
2. Why must cis and trans women be at the center of the membership and leadership of Black liberation movements?
3. Why do Jackson and Cohen discuss the Sarah Bland case at length? What conclusions do they reach?
4. What does a "radical trans analysis" contribute to liberation movements?

Ask a Feminist

A Conversation With Cathy J. Cohen on Black Lives Matter, Feminism, and Contemporary Activism

Cathy J. Cohen and Sarah J. Jackson

In the following conversation, Cathy J. Cohen and I discuss the potentials for feminist theory in racial justice movements, the unique ways in which race and gender intersect in state violence, challenges for feminist academics of color engaged in activism, and the shape of the #BlackLivesMatter movement. You can follow us on Twitter at @cathyjcohen and @sjjphd, respectively.

Sarah Jackson *(SJ):* I'd like to begin by asking what role you see feminism, and feminist scholarship in particular, playing in today's racial justice movements, as well as what you think scholars can learn from activists and vice versa.

Cathy Cohen *(CC):* That's a really big and important question. At its most basic level I think that feminism at the very least—and maybe most importantly—makes us stop and ask about the role that women, and here I mean both cis and trans women, are playing at this particular time in the multiple movements that are emerging, and particularly, at least for the work that I'm doing and thinking about, in what people are calling the "black lives movement."

But I think that feminism does a number of different things in relation to racial justice movements today. I am especially thinking about the role of black feminism. I'll give you three things that I think it does: first, it makes us think differently about or, hopefully, expand where we look for victims and resisters to state violence. It says that while there's a traditional or normative model of who we think about as the victim of state violence, which is often a hetero-sexual man in a confrontation with police, we know that state oppression manifests not only in that model but in lots of different places. It happens through the denial of state welfare assistance, and it happens in the ways we militarize the public schools that primarily black, Latino, and poor kids attend. These are different forms of state violence. And I think feminism fundamentally makes us ask the question, when we confront the traditional model, what are the other examples of state violence or state oppression that we need to be paying attention to? Of course, feminism has us intervene in traditional sites of state violence by asking the very basic question about the status of women. So if we're looking at campaigns that are mobilizing against direct police violence and only mention men as targets, feminism would have us ask, where are the women? The Say Her Name campaign (AAPF 2015), the work that the BYP100 is doing in Chicago around the Rekia Boyd case (BYPl00 2015), and, some would argue, even the Sandra Bland case (Alter 2015) are examples of attention to direct police violence where we can say, "Well, wait a minute, this is also happening to women."

SJ: And there was just the Charnesia Corley case in Texas, where the police did a vaginal search during her traffic stop (Lohr 2015).

CC: Yes, and we can go down the list of incidents that people will recognize as police violence. But again I think what feminism says is that we have to expand how we understand state oppression and more specifically state violence. We have to not only be attentive to what are now recognizable forms of state violence but also move beyond the "traditional" models of state violence to mobilize for justice for broader communities of people.

Second, I think feminism is informing the movement for black lives in terms of how it's structured and its leadership. There's some important feminist work that tells us that there are different forms of leadership that we should be paying attention to. Whether it is Belinda Robnett's (1997) work on the civil rights movement and bridge leaders or the exceptional work that Barbara Ransby (2003) has done thinking about Ella Baker and more democratic forms of radical leadership, I think many of the young leaders in the Black Lives Matter Movement recognize that the male charismatic leader, or the singular charismatic leader, is not the form of leadership that they adhere to or they are going to put forth. In fact, many of these new organizations are led by young black women who identify as queer and who promote the idea, as Barbara

Cohen, C. J., & Jackson, S. J. (2016). Ask a feminist: A conversation with Cathy J. Cohen on Black Lives Matter, feminism, and contemporary activism. *Signs, 41*(4): 775–792. Reprinted with permission from The University of Chicago Press.

Ransby has noted, that far from this movement being led by one person or having no leaders, it is a leaderful movement with cis and trans women taking positions of power. So the organizations that are part of a network of groups working under the broad framework of the Black Lives Matter Movement look different and structure their leadership differently than organizations significant to the civil rights movement in part because of feminist teaching, feminist scholarship, especially black feminist teaching and scholarship, and the fact that many of these young activists have been in the classroom learning about these alternative forms of organizing and leadership.

Third, I think that feminism should also require us to think broadly and radically about what we are fighting for—the outcomes we seek to the oppression that we face. Radical black feminists, in particular, have argued that while immediate policy changes can be part of what we fight for, the structural transformation of the lived condition of marginal communities has to guide our struggle. For example, if you take the work of someone like Beth Richie (2012), who is an exceptional black feminist scholar focused on issues of incarceration and violence against women, she has written that when most think about violence against women, the traditional response in the mainstream domestic violence movement has been to involve the state, specifically passing laws against domestic violence and involving the police. But this response doesn't take into account the fact that the state is the oppressor in many communities of color and poor communities. So involving the state or turning to police intervention cannot be the place we land when building a movement to address the structural conditions that foster or contribute to violence against women. Similarly, feminist organizations like INCITE! and the BYP100 have argued that working against the prison industrial complex must be focused on improving and transforming the conditions under which incarcerated folk exist, but we must also question the existence of prisons and seriously contemplate a broader structural approach that promotes a prison abolition agenda.[1] The point is that feminism should turn our attention to structure, not just as a limiting factor but also as a jumping off point as we imagine a broader liberation agenda.

SJ: Yes, absolutely. I think all of those points needed to be made. And something that you just said made me think about how, as a black woman scholar, I have been very interested to see how the contemporary movement is redefining what questions are being asked, who's in leadership, etc. It's also been interesting to see what role academics think they should or shouldn't play in the movement versus what role activists think academics should and shouldn't play. These expectations sometimes seem to butt heads, but there is also a space for academics who are activists, and many of us—particularly those of us who are women of color—are already deeply engaged in the activism we're studying. The academy as The Academy is, like any other American institution, often an oppressive force and contradictory in nature to antiracist and feminist activism, but those of us within the academy who as individuals subscribe to antiracist and feminist politics can have authentic ties to activism. So I wonder if you could say a little bit about what possibilities and roles you see for academics in relation to these movements.

CC: This is a hard one. I can think of at least four or five examples of organizations that have either been started by academics or are deeply in conversation with academics that are helping to fuel the Black Lives Matter movement. For instance I could talk about the BYP100, which grew out of the Black Youth Project, which I head. We held a convening of one hundred young black activists on the same weekend in which the George Zimmern verdict was announced. Out of disgust with the verdict and as a result of die skills that these organizers and activists have, they created this new organization called the BYPl00, which has really taken off and is doing incredible work (BYPl00 2013). And I've been honored to be in conversation with the leadership of that organization. We are getting ready to do a special issue with the journal *Souls* that will focus on black youth activism and will include both academics and activists. This is one model of an activist academic supporting the work of activists rooted outside the academy.

You could also look at Salamishah Tillet's work with A Long Walk Home; you could look at Barbara Ransby's work with the Social Justice Initiative and Ella's Daughters; you could look at academics like Angela Davis or Beth Richie, who are both involved in INCITE!, or Kimberlé Crenshaw's work with the African American Policy Forum and the Say Her Name campaign.[2] We can go down a very long list of feminist academics doing work with the Black Lives Matter movement organizations or doing their own political work—and I apologize to people I didn't include. I would argue that the black feminists I just named and many more have been engaged in activism as much as they have been engaged in the academy. Particularly for women of color, there is an understanding that you may never be fully embraced in the academy, and what this understanding does is give you a kind of freedom to pursue the work that will transform institutions of oppression, including the academy. So I think women of color have always endured a kind of love-hate relationship with the academy, with our positioning allowing us to be insiders-outsiders. At our best moments this insider-outsider status has allowed us to be deeply engaged and accountable to communities outside the academy: the communities from which we come, the communities that we call home, and the communities that many of us study and promote. So it's not surprising, I would argue, in particular for women of color and feminists of color, that this idea of bridging what we do inside the academy and outside the academy is almost second nature—that's what we have always done. And for most of our careers this duel status has been the thing that people have often used to dismiss our work. Now, with the emergence of liberation movements in communities of color, people who are not politically engaged in the same way are excited by it. But, sadly, I believe this too will pass and there will be another generation of young people, young scholars, young feminists, doing political work in and outside the academy who will again be discredited for doing such work but who will understand that their calling is not only about getting tenure but also leveraging the privilege and resources that can come with a life in the academy to resist and transform oppressive institutions that too often include the academy.

SJ: Thank you so much. Honestly that was a very self-indulgent question because this is something that I think a lot about, personally.

CC: Can I say one more thing? Something else that interests me about this moment and the intersection of the academy and organizing outside the academy—and I've said this in other venues—is that we're seeing the emergence of a leadership that has often been in our classrooms. Many of the young leaders that I've interacted with have been in African American studies classes, have been in ethnic studies classes, have been in feminist, gender, and women's studies classes—these might even be their majors. So our entry into the academy may not have transformed that institution, but it has informed the thinking of a generation of young people who are now leading these movements, and I think it's being manifested in decisions about how they structure their organizations, or what leadership looks like to them, and who they understand to be members of their communities—for example, they are conceiving of the black community in ways that are perhaps much more expansive than what we've seen before. I think that's another way in which the academy and activism are productively intersecting.

SJ: That's so true. Speaking of the leadership, we know that the hashtag #BlackLivesMatter was created by queer black women (Garza 2014), which probably can't be said enough, and in some media outlets there's been a lot of buzz and there have been several major newspapers or online magazines that have written pieces about the new civil rights movement (Demby 2014; Eligon 2015) and the new face of racial justice activism. They have suggested that the inclusion of queer folks and women, or queer women, is maybe a new thing. But we know that these folks have been central to activist movements all along—we know who threw the first rock at Stonewall—and we know that this is part of the story that has actually been erased (see Adsit 2015). This may be something that we both know the answer to, but do you think that the inclusivity of this moment is new? You have suggested just now that in some ways it may be, but perhaps in some ways it isn't. So I'd like you to talk a little bit about that, and also discuss how intersectional politics in particular might be evidenced in today's movement and how this is similar to or different from a previous moment.

CC: You've got a lot packed into these questions! On the question of whether it's new: not completely, but partly. There have always been radical black women or radical women engaged in mobilization, organizing, and leadership: we know that is not new. We can read the books on Fannie Lou Hamer (Mills 1994); we can read the books on Ella Baker; we can read Anna Julia Cooper. We can go through that individualized history to say that women have been a part of the leadership; they've been a part of the strategizing. And I would argue that trans women have also had a long history of organizing and resistance that far too often is not held up, ranging from the Compton's Cafeteria Riot, to Stonewall, to the mobilization around CeCe McDonald and beyond. So we know that the women, women of color and queer folk, have always been central to our struggles. That part is not new. I think it's new information for some people, but if you think about the history and know the history, it's not new.

I do think what's new is the ways in which, at this moment in the Black Lives Matter movement, young, black, often queer women are not just doing the work but are part of a collective leadership. The fact that they are visible and vocal, not just in one organization but across a number of organizations, shaping the direction of this movement—this is something that's new. And they are leading not specifically women's organizations but also what many of us recognize as black liberation organizations. While the inclusion and in some cases leadership of women, queer, and LGBT folks in our movements is not new, these individuals, like Bayard Rustin, haven't always had the opportunity to be a visible or foregrounded part of the black struggle, so in that sense I think that it's new.

Another thing that's new at this moment is the recognition and articulation of a kind of black queer nationalist politics that is informing the politics of many organizations involved in the black lives movement. I should say that when I say "black lives movement," I'm talking about the multiple organizations that are engaged in the national or international mobilization that is commonly known as Black Lives Matter, not just the one organization of the same name. To assert the significance of queer bodies as part of the black community is new and important. That is a struggle that has existed for some time and really took root when black activists were responding to HIV and AIDS in black communities (see Cohen 1999), where we were trying to say that there were lesbian and gay and bisexual and trans folks who were part of our communities and their issues, in this case their struggle against HIV/AIDS, had to be addressed by both national organizations and the state but also by indigenous black organizations. Now activists are saying that not only are gay, lesbian, trans, and queer folk part of our communities, but they are part of the leadership, and they—for example, cis and trans women—have to be at the center of how we think about black liberation. The centering of cis and trans women and lesbians and gay men as members and leaders of our communities—that to me is significant and new. It is an expansion of our understanding of black communities that I don't think we have seen articulated in the past. The way in which people are also pushing back by using a queer lens to challenge the static nature of categories and identities is also important. This perspective is not always articulated or productively struggled with, but I do think there's the possibility for these organizations to borrow from a queer lens, we might say, and to think about the ways in which different bodies are marginalized and made to be queer in the eyes of the state as well as in their own communities. This idea of queering marginalized bodies, in particular those of color, was an argument I tried to articulate in my article "Punks, Bulldaggers, and Welfare Queens" (Cohen 1997). This lens allows for and promotes different types of allegiances, not only racialized allegiances but also allegiances based on the positionality of people relative to the state, which queers us all or produces a bond of unity needed for the type of mobilization that we're beginning to see. For me, that is what's new and significant about the forms of leadership and organizational structures that we're seeing in this moment.

The other new development we could talk about is the constant articulation of the intersectional nature of both the oppression that people feel and the type of liberation and resistance that they want to mount; even though the framework for the moment is appropriately focused on police violence, people also understand the ways in which race and class and gender intersect. If we look at a campaign like Fight for $15, which is about securing a living wage—often for women of color, who find themselves doing more work for less money—that's a kind of intersectional approach to understanding the significance and uniqueness of the positionality of black women while also saying that their positionality can speak to the condition of black people more generally.[3] I don't think we have often seen movements say that the common thread of blackness is not just the male body, or the presumed cis male body, but in fact that cis and trans black women can represent the intersectional positionality and oppression that black communities face.

And finally, I think we're in a technological context that allows people who have traditionally been silenced or made invisible to have a voice or at least to have their voices and issues amplified. When we think about how the Black Lives Matter movement started with a hashtag, I wouldn't want to discount the significance of technology as an important tool for organizing today. I would not say that technology is a driving force, but it has been a critical tool in terms of democratizing the voices who can be a part of this movement.

SJ: That's actually a really good bridge to my next question. You already mentioned Sandra Bland, and it seems to me that the Bland case was the first time in recent memory that the mainstream media, not just the black media or the alternative media, actually covered an instance of state violence against a black woman as a major news story. It seemed to be a tipping point, in that this case was pushed into visibility by the current moment, and it suddenly became possible to realize, "Oh, it's not just black men!" Now of course Bland in particular was a middle-class cis black woman, so in many ways she was also an "easier" victim in terms of the media's framework, but I wonder if you could talk a little bit about the fact that we know that there are countless incidences of violence against trans women of color, against immigrant women of color, against cis black women, that we don't see. Could you talk about whether you think we are at a tipping point in the visibility of these other types of stories? Or what other work might need to be done to fully integrate those conversations?

CC: That's a great question. I don't think I have a great answer. This feels confusing to me in the sense that, even given all the things you said, there are reasons why we might expect that the Bland case would resonate. One has to do with what we were talking about before: I think we now have a generation of new leaders, especially among the Black Lives Matter network, who are going to say that in fact the deaths of black women matter and who are going to try to hold up names and the experiences of cis and trans black women, trying to make other organizations and even the mainstream media pay attention to their lives. Today we have that kind of pushing and the organizational capacity to try to make that case. Of course, I think the role of digital media is also important here. There's a way in which, using the tools of new media, activists can circumvent what might be thought to be the mainstream or dominant press and try to build the story themselves, using platforms such as their organization's website, Twitter, YouTube, and Facebook.

I think Bland's case also resonated nationally because it seemed so extreme. While the country may be prepared to accept the killing of black men and marginalized black women at the hands of the state, the idea that someone like Sandra Bland would end up dead in police custody seemed extreme and unthinkable to many in this country. Bland, while a black woman, was also a middle-class, college-educated, light-skinned black woman who was engaged in an act that was not deemed violent. In fact, she was doing something that everyone has done, which is changing lanes without signaling. So everybody can say, "Damn, I've done that." So I think it was the perceived extreme nature of the case, the insignificance of the infraction (and I don't even know if changing lanes qualifies as an "infraction"), the way the officer's outrageous behavior escalates, the fact that Bland was in town to accept a job (thus fulfilling the American norm of self-dependence), the ability to see so much of the interaction unfold on video, and the capacity of organizations to push and to organize around the story with a leadership that understood how important it was to bring attention to the police violence black women face, all of this results in the Sandra Bland death becoming a national story. I think it's one of those moments where many factors came together.

But I'm not sure I'm prepared to say that the Bland case, as important as it was, marks a tipping point. When I see the same type of mobilization nationally around the Rekia Boyd case or the Marlene Pinnock case or the Dajerria Becton case that we have rightfully seen around the cases of Trayvon Martin, Michael Brown, and Eric Garner, then I'll say we have reached a tipping point. I'm not sure if it was this confluence of characteristics that allowed people to pay attention to the Bland case specifically or if it's a longer trend, but I think the pieces are at least in place to help put forward more and more cases of women, trans and cis, who are the targets of state abuse, often police abuse. I just hope people will pay attention.

Finally, I also think there has been a form of training that has gone on over the past year. We have trained people to pay attention to cases of police abuse, which has helped with people's recognition of the Bland case. Sadly, now that people have seen so many cases of black people being killed by the police, it is a familiar story that is recognizable and people are able to say, "Oh, that happened again, and this time it was to a woman." They are able to process the Bland case because far too many cases prior to the Bland death trained the public to pay attention. I don't know if this training process helps your point about having reached a tipping point, but at least people are more likely to believe, listen to, and pay attention to these types of stories.

SJ: Thank you for that. Were you at the Movement for Black Lives Convening in Cleveland?[4]

CC: I was not, but I heard many reports about it.

SJ: Yes, so did I. At this convening, there was a moment in the middle of the meeting when a deliberate effort was made to reaffirm the value of the inclusion of black trans lives. I wonder if you could speak a bit about the significance of trans inclusivity in both contemporary black activist and other activist spaces, as well as in scholarly spaces and in the work that we're doing in feminist scholarship.

CC: I think that it's important for any movement concerned with the liberation in particular of black people to be thinking about those individuals who are most marginalized nationally and in our own communities. And those are often poor people, trans and cis women, as well as LGB folks. So while we have made some progress in having activists and scholars detail the role that racism, sexism, and class play in structuring and truncating the lives of all black people, we still have much work to do to explain and challenge how heterosexism and heteronormativity work to limit the lives of black people. Similarly, I think there's still a lot of work to be done in support of trans members of our community, acknowledging their struggles, vulnerability, and growing resistance.

It's important for those of us who profess to be concerned with the liberation of black people, who say we love black people, to include and make central to our work those most marginal in our communities because improving their lives should help move us all down the road toward liberation. In this case it means modeling trans inclusivity, making visible and central the struggles of trans folks in our analysis and movements. The hesitation to making trans struggles central to our work, at least among many black feminists of my generation, those who came to feminism when second-wave feminism was dominant (if we even want to adhere to that wave structure), is that some of the most visible or prominent trans spokes-people seem to affirm a strict, conservative, and essentialist gender binary. I think people often confuse the more public and visible articulations of some trans celebrities with a universal trans politics. I believe that we have to pay attention to the nuances in trans politics. Just as LGBT politics is different from queer politics and women's politics is different from feminist politics, we have to acknowledge that an essentialist trans politics is very different from a radical feminist trans politics. Thus, concern emerges when one hears the language of what seems like a call to biological essentialism—such as when Caitlyn Jenner says "I am a woman. . . . My brain is much more female than it is male."[5] I think that essentialist propaganda drives me and my feminist friends crazy. But when we hear the informed and radical stories of trans folks who refuse to live on the binary or someone like Janet Mock (2014) who, while choosing a more traditional gender performance, talks incisively about the ways being a trans woman is related to the positionality of cis black women, who also are at the bottom of the racial order and gender hierarchy, the connection between the lives and struggles of black cis and trans women are made clear, highlighting the larger structures that we're both fighting against. A radical trans analysis makes clear that our struggle is not so that everyone can find their essentialist selves, but instead this movement is about breaking down systems of oppression based on gender and class and race and sexuality that limit the ability of people to have full and happy lives—from having good jobs, to having the kind of intimate partners that they want, to experiencing joy, to having agency, to having control over their bodies and sexuality.

Getting us to that point probably means more dialogue, more struggle, as Bernice Johnson Reagan (1983) would say, more being uncomfortable around the coalition and intersection of work that we want to do. I sometimes worry that we've accepted a dialogue around trans politics that we haven't struggled with enough. I think there is a way that both in the academy and outside of the academy, we haven't figured out where the space is to really have this dialogue and come to an understanding around the kind of trans feminist politics that I think is radical and transformative and not primarily essentialist.

SJ: Would you say that that's the case both for feminists working within the academy and for activists outside of it?

CC: I think there is a way that, yes, both in the academy and outside of the academy, we haven't figured out where the space is to really have this dialogue and come to an understanding around the kind of trans feminist politics that I think is radical and transformative and not primarily essentialist. Because of trans-exclusionary radical feminists, I think it's been hard to find a space where people feel like they can actually have that dialogue and move to a more informed position. I'm not sure that we've done enough, either in the academy or outside the academy, but we've probably done more in the academy, actually.

SJ: This is something that I'm also still thinking about. And I have been talking a lot about this recently because Suzanna Walters's (2014) book basically makes the same argument about the LGB folks who say, "I was born this way; it's in my genes," and how this is actually counter to a radical feminist liberatory theory that would say, "actually we shouldn't justify treating people like human beings by any sort of genetic or biological criteria." So I like how you distinguished that narrative that has arisen around trans identity as being an essential thing as opposed to a radical trans identity that is rooted in a more feminist liberatory ideology. It seems like a really good way to think about it.

CC: I'm still struggling to figure it out as well. To me it's the difference or the tension between a women's rights politics and a feminist politics, or an LGBT politics and a queer politics. What I'm looking for and have found, I hope, is a radical trans feminist politics that is thinking about and rooted in the transformation of institutions that would oppress and limit people's understanding and performance of gender. I am, thus, interested in a politic that is rooted in and investigates life in contrast to binaries and static categories. That's the kind of feminist trans politics that I'm committed to, and that's the inclusion I want in the Black Lives Matter movement. People may not like this, but without an intentional politics, I don't see trans as inherently radical. I think there are many instances where marginal individuals are inserted into traditional institutions or movements and they do something to change the dynamics but they don't necessarily change these spaces and entities in a radical way that is open and more equitable. I'm interested in trans feminist politics in the same way that I'm committed to a black feminist politics that is tied to a transformative liberatory agenda.

SJ: That makes perfect sense. There also has been a lot of talk about this idea that there's a generational divide among racial justice activists. The media can't seem to get enough of comparing contemporary activists to civil rights movement activists, which is problematic in and of itself because the contextual differences are endless, as well as because everyone's public memory of the civil rights movement has been so sanitized. But there is this idea that there is a generational divide, and I think we have seen this divide in some spaces. For example, there was that moment in 2014 when Al Sharpton got shouted down (see Demby 2014). People were saying, "We didn't ask you to be here to talk about this." Maybe I'm wrong about this, but I think in some ways these generational differences—I don't like the word "conflict"—are analogous to the waves of thought in feminism, and maybe this current moment of activism looks more like the fourth wave of feminist thought. Or maybe it doesn't? I wonder if you have any thoughts about that.

CC: Here's the problem for me with the generational framework, and I guess we could say it mirrors my concerns about the wave frame-work that other people have already articulated: when you start comparing waves or generations, you homogenize those waves and generations. So I shudder when I hear you say Sharpton might speak for my generation. I'm like, "Oh my God, no! That's just scary!" I think that's what happens when we start to say that there are certain individuals that represent a generation. One major problem with that approach is that the political/ideological tensions between individuals within the generation are made invisible. As I see it, if there is any divide, it's more an ideological one: it is the divide between those who understand themselves to be feminist and who insist that cis and trans women are just as important to our conversation and movement as cis men versus those that don't share this understanding. To me, that's the divide. It's a divide that's built around our larger vision of black liberation and who is a central part of black communities; it's a divide between the structure of more traditional civil rights organizations and the organizations mobilizing today; it's a divide between those who support new models of collective leadership in our protest organizations versus those who want one charismatic leader to point to. That, I think, is the divide, and this speaks to Sharpton also. He comes out of a tradition, an ideological positioning, that would lead him to seek the role of the male charismatic leader and believe that there should be one person and one organization leading the way. I think the divide is less about generation, even though I think that's the easiest place for people to point, but I think if you dig a little deeper, beyond generation, you will see different political and ideological commitments driving the differences in the approach to social movements or in opinions about the Black Lives Matter movement. To me, this divide maps onto the issue of the waves model in feminist studies because, as you know, black feminists have contested the wave theory and questioned when waves start and end and who gets included. In the waves model you see the flattening of differences and dissension within waves as a way to bring into stark contrast those differences that exist between waves.

SJ: It's interesting because now that you say that, I think it does map: it seems like the divide within racial justice activism around what leaders look like and whose story should be included maps more onto the conflicts—both inside and outside of academia—between upper- or middle-class white feminists and feminists of color and working-class feminists, who have always had this ideological head butting, where feminists of color, for example, have been saying, "Your feminism isn't necessarily inclusive; you're helping prop up the carceral state, and so on." And that's interesting because it's less about generations and more about where one's standpoint is coming from in the first place. That's a really interesting way to look at it.

CC: But—to go back to the topic of trans inclusivity—I think another interesting aspect is that many of the trans women that we are talking about often come from poor backgrounds rather than middle-class backgrounds. They've been in completely vulnerable positions in terms of their lived experience. To me, that's a different positioning than a middle-class cis black feminist, so I think the question is, are we prepared to look at all of these differences in stand-point, as Patricia Hill Collins might talk about, and say that they're all significant? The divide that we saw, for example, in second-wave feminism, between black feminists or feminists of color, Latina feminists, Native feminists, and white feminists—that's an important one. But we know that we've complicated that account and that many people in the academy are in the academy because they have access to resources and some form of class mobility. So while folks of color can challenge and open up new possibilities in terms of thinking about feminism, we also want to pay attention to the class positioning of those women of color who are speaking in the academy, often to the exclusion of poor women or trans women, who often have a very different positionality with regard to both feminism and their lived day-to-day experiences.

Notes

1. See INCITE! (2006) and http://www.incite-national.org
2. See http://www.alongwalkhome.org/;http://www.uic.edu/depts/oaa/sji/;http://ellasdaughters.blogspot.com/;http://www.incite-national.org/;http://www.aapf.org/
3. See http://fightfor15.org/
4. See http://movementforblacklives.org/
5. Interview with Diane Sawyer, 20/20, ABC, April 24, 2015, http://abc.go.com/shows/2020/listing/2015-04/24-bruce-jenner-the-interview.

References

AAPF (African American Policy Forum). 2015. "#SayHerName: Resisting Police Brutality against Black Women." Report, July 16. http://www.aapf.org/sayhernamereport/.

Adsit, Lexi. 2015. "Stonewall Is in Our Blood: How the New Film's White Gay Hero Is the Latest Effort to Erase Trans Activists of Color." *Salon,* August 10. http://www.salon.com/2015/08/10/stonewall_is_in_our_blood_how_the_new_films_white_gay_hero_is_the_latest_effort_to_erase_trans_activists_of_color/.

Alter, Charlotte. 2015. "Sandra Bland's Not the First Woman to Experience Police Violence." *Salon,* July 22. http://time.com/3965032/sandra-bland-arrest-video-police-violence/.

Brown, Elaine. 2015. *A Taste of Power: Black Woman's Story.* New York: Anchor.

BYP100 (Black Youth Project 100). 2013. "#BYP100 Responds to George Zimmerman Verdict." YouTube video, July 15. https://www.youtube.com/watch?v=DUxKJXK5WAc.

———. 2015. "National Day of Action for Black women and Girls: #Justicefor Rekia #SayHerName #BlackWomenMatter." Black Youth Project 100, May 20. http://byp100.org/justice-for-rekia/.

Cohen, Cathy J. 1997. "Punks, Bulldaggers, and Welfare Queens: The Radical Potential of Queer Politics!" GLQ 3(4):437–65.

———. 1999. *The Boundaries of Blackness: AIDS and the Breakdown of Black Politics.* Chicago: University of Chicago Press.

———. 2010. *Democracy Remixed: Black Youth and the Future of American Politics.* Oxford: Oxford University Press.

Cohen, Cathy J., Kathleen B. Jones, and Joan C. Tronto, eds. 1997. *Women Transforming Politics: An Alternative Reader.* New York: New York University Press.

Demby, Gene. 2014. "The Birth of a New Civil Rights Movement." *Politico,* December 31. http://www.politico.com/magazine/story/2014/12/fcrguson-new-civil-rights-movement-113906.

Eligon, John. 2015. "One Slogan, Many Methods: Black Lives Matter Enters Politics." *New York Times,* November 18. http://www.nytimes.com/2015/ll/19/us/one-slogan-many-methods-black-lives-matter-enters-politics.html? smprod=nytcore-iphone&smid=nytcore-iphone-share&_r=1.

Garza, Alicia. 2014. "A Herstory of the #BlackLivesMatter Movement." *Feminist Wire,* October 7. http://www.rhefeministwire.com/2014/10/blacklivesmatter-2/.

INCITE! Women of Color against Violence. 2006. *Color of Violence: The Incite! Anthology.* Cambridge, *MA:* South End.

Jackson, Sarah J. 2014. *Black Celebrity, Racial Politics, and the Press: Framing Dissent.* New York: Routledge.

Lohr, David. 2015. "Woman Says Gas Station Strip Search Was Like Sexual Assault." *Huffington Post,* August 10. http://www.huffingtonpost.com/entry/tcxas-suip-search-public_55c8f940c4b0923cl2bdb903.

Mills, Kay. 1994. *This Little Light of Mine: The Life of Fannie Lou Hamer.* Lexington: University Press of Kentucky.

Mock, Janet. 2014. *Redefining Realness: My Path to Womanhood, Identity Love and So Much More.* New York: Atria Paperback.

Ransby, Barbara. 2003. *Ella Baker and the Black Freedom Movement: A Radical Democratic Vision.* Chapel Hill: University of North Carolina Press.

Reagon, Bernice Johnson. 1983. "Coalition Politics: Turning the Century." *In Home Girls: A Black Feminist Anthology*, ed. Barbara Smith, 356–68. New York: Kitchen Table.

Richie, Beth. 2012. *Arrested justice: Black Women, Violence, and America's Prison, Nation.* New York: New York University Press.

Robnett, Belinda. 1997. *How Long? How Long? African-American Women in the Struggle for Civil Rights.* New York: Oxford University Press.

Walters, Suzanna Danuta. 2014. *The Tolerance Trap: How God, Genes, and Good Intentions Are Sabotaging Gay Equality.* New York: New York University Press.

Introduction to Reading 54

How do men respond to feminist movements and to transformations in gender roles and relations that bring about greater gender equality? Michael Messner, author of this reading, offers an answer to this question by examining a range of U.S. men's organized responses to feminism and gender change, from anti-feminist to pro-feminist. He ends on an optimistic note, arguing that the increase in the numbers of men of color doing anti-violence work signals the expansion of an intersectional social justice framework for feminist work.

1. What were the politics of the 1970s men's liberation movement that split the movement into two factions, men's rights organizations and pro-feminist organizations?
2. Discuss the three "substantial and interrelated social changes" that have had a major impact on feminist movements and men's organized responses to feminism.
3. Who is Warren Farrell and what strategies has he developed to keep the men's rights movement afloat?
4. How is the infusion of men of color into anti-violence work having a positive impact on the anti-violence field of study and activism?

Forks in the Road of Men's Gender Politics

Men's Rights vs. Feminist Allies

Michael A. Messner

Introduction

For more than a century, men have responded to feminist movements in the US and in other western jurisdictions in varying ways, ranging from outright hostility, to sarcastic ridicule, to indifference, to grudging sympathy, to enthusiastic support (Kimmel 1987; Messner 1997). In this article I argue that large-scale social changes—those shaped by social movements, changing cultural beliefs, and shifts in political economy—create moments of historical gender formation that in turn shape, constrain and enable certain forms of men's gender politics. In particular, I trace the two most politically engaged tails of a continuum of gender politics—anti-feminist men's rights groups and pro-feminist men allies—with an eye to understanding how moments of historical gender formation shape men's gender politics.

The 1970s and the present moment generated possibilities for men's gender politics: forks in the road, as it were. The image of historical forks in the road implies choices for men's responses to feminism, but not an unlimited range of 'free' choices. Rather, feminist challenges and shifts in the gender order confront men with a limited field of structured options: stop dead in your tracks, befuddled; attempt a U-turn and retreat toward an idealized past of male entitlement; turn right and join a backlash against feminism; or bend left and actively support feminism. Adapted from Omi and Winant's (1986) theory of racial formation, I introduce *historical gender formation*, a concept that provides a more nuanced view of the dynamics of gender politics than the dualistic image of a fork in the road. Central to the theory of racial formation is the idea that the grassroots racial justice movements of the 1950s through the 1970s wrested concessions from the state, altered the ways in which racial categories were defined, and created new foundations upon which subsequent racial tensions and politics arose. Similarly, the women's movements of the 1960s and 1970s wrested concessions from the state, challenged and partially transformed cultural values about sex and gender, and succeeded in bringing about substantial

Messner, M. A. (2016). Forks in the road of men's gender politics: Men's rights vs feminist allies. *International Journal for Crime, Justice and Social Democracy, 5*(2): 6–20.

reforms in various social institutions. Thus, men's engagements with gender politics today take place in a very different context—one partly transformed by feminism—than they did in the 1970s. I will demonstrate that the 1970s and the present are two moments of gender formation that create different limits and possibilities for men's engagements, both for and against feminism.

1970s Gender Formation: The Women's Movement and Men's Liberation

By the early 1970's, following several years of organizing, the women's liberation movement had exploded on to the social scene. In the United States, the most visible feminist activism took place 'in the streets': small local consciousness-raising groups, grassroots groups linked by word-of-mouth and hand-printed newsletters, a sprouting of local rape-crisis centers and women's shelters run by volunteers in private homes or low-rent storefronts, all punctuated by mass public demonstrations for women's rights (Allen 1970; Stansell 2010). In other words, in relation to male-dominated institutions like the state, the economy, military, religion or medicine, feminism in the US was mostly on the outside looking in (with academia, where feminists gained an earlier foothold, a partial exception). The 1970s, then, was a time of deeply entrenched gender inequality across all institutions, against which a grassroots women's movement was organizing on many fronts, characteristically in alliance with gay rights and other social justice movements.

By the early 1970s a few US men—many of them veterans of the new left, anti-war and student movements—responded to the re-emergence of feminism in the 1960s by organizing men's consciousness-raising groups and networks, and asking a potentially subversive question: what does feminism have to do with us (Men's Consciousness-Raising Group 1971)? Some leaders promoted the idea of a 'men's liberation movement' that would work symmetrically with the women's liberation movement to bring about progressive personal and social change (Farrell 1974; Nichols 1975). They reasoned that a men's liberation program that emphasized potential gains *for men* might draw more interest than one that positioned men as oppressors whose only morally correct action was guilty self-flagellation. The language of sex roles, emerging at that time as the dominant discourse of liberal feminism—just one of multiple feminist positions that emerged in the wake of the 1960s rebirth of feminism—was an ideal means through which to package feminism for men in a way that lessened the guilt and maximized the potential gain that men might expect from 'liberation' (Messner 1998). The 'female sex role' had clearly oppressed women, men's liberationists argued, and 'the male sex role' also harmed men.

Leaders posited men's liberation as the logical flipside of women's liberation, but they walked a tightrope from the start They acknowledged that sexism had oppressed women and privileged men; it was pretty hard to ignore that 59 per cent wage gap, the obvious lack of women in political and corporate leadership positions, or the ubiquitous violence against women. But they sought to attract men to feminism by stressing how the 'male sex role' was 'impoverished', 'unhealthy', even 'lethal' for men's health, emotional lives and relationships (Jourard 1974). Thus, from the outset, there was tension in men's liberation's attempt to focus simultaneously on men's institutional power over women *and* on the 'costs of masculinity' to men. Savvy men's liberation leaders sought to connect these seemingly contradictory positions by demonstrating that it was in fact men's attempts to secure access to the institutional privileges of masculinity that enforced boys' and men's emotional stoicism, lack of empathy for self and others, physical risk-taking, and unhealthy daily practices like smoking and drinking. Progressive men's liberationists drew from the works of psychologist Joseph Pleck (1977) who argued that while women were *oppressed* by the female sex role, men were *privileged and simultaneously dehumanized* by the male sex role. The social change corollary to this was the assertion that, when men committed themselves to bringing about full equality for women, this would create the conditions for the full humanization of men, including healthier and longer lives and more satisfying relationships with intimate partners, friends, and children.

It did not take long before serious slippage began to occur with men's liberationists' attempts to navigate the tension between emphasizing men's privileges and the costs of masculinity. Less politically progressive leaders began to assert a false symmetry, viewing men and women as differently but equally oppressed by sex roles (Farrell 1974; Goldberg 1976). This assertion generated critical distrust from politically radical women, and vigorous debate from more politically radical men in the movement. By the mid-to-late 1970's, men's liberation had split directly along this fissure. On the one hand, men's rights organizations stressed the costs of narrow conceptions of masculinity to men, and either downplayed or angrily disputed feminist claims that patriarchy benefited men at women's expense. On the other hand, a profeminist (sometimes called 'anti-sexist') men's movement emphasized the primary importance of joining with women to do away with men's institutionalized privileges. Patriarchy may dehumanize men, profeminists continued to insist, but the costs that men pay for adherence to narrow conceptions of masculinity are linked to the promise of patriarchal power and privilege.

In short, men's liberation had premised itself upon a liberal language of symmetrical sex roles, which contributed both to its promise as a movement and to its eventual demise. Following the fissuring of men's liberation, the men's rights movement continued to deploy a narrowly conservative language of sex roles. Now severed from its progressive roots, a more reactionary tendency within the men's rights movement unleashed overtly anti-feminist and sometimes outright misogynist discourse and actions (Baumli 1985). Meanwhile, the emergent profeminist men's movement largely rejected the language of sex roles, adopting instead a radical language of gender relations that facilitated an activist focus on ending men's institutional privileges and men's violence against women (Messner 1997, 1998).

By the mid-1970s the women's movement had altered the political context in ways that made A men's rights movement possible, if not inevitable. The men's rights movement was not simply a kneejerk backlash against feminism; it was a movement that co-opted the liberal feminist language of symmetrical sex roles and then turned this language back on itself. Men's liberationist-turned men's rights advocate Warren Farrell (1974), for instance, borrowed Betty Friedan's (1963) idea that a 'feminine mystique' oppressed women, arguing that men were trapped in a 'masculine mystique' that narrowly positioned them as breadwinners and protectors. In response to feminist criticisms of the effects on women of being constructed as 'sex objects', Farrell posited an equally negative effect on men in being constructed as 'success objects'. Herb Goldberg's 1976 book *The Hazards of Being Male* asserted that male privilege is a 'myth'. Men actually have it worse than women, Goldberg argued, due to the fact that the male role is far more rigid than the female role, and because women have created a movement through which they can now transcend the limits of culturally-imposed femininity. Men's rights organizations broke from the men's liberation movement's gender symmetry and began to articulate a distinct discourse of overt and angry anti-feminist backlash. By the late 1970's and early 1980's, men's rights advocates were claiming that men are the true victims of prostitution, pornography, dating rituals, sexist media conventions, divorce settlements, false rape accusations, sexual harassment, and domestic violence (Baumli 1985). And in subsequent decades, the beating heart of the men's rights movement has been organizations that focus—largely through the Internet—on fighting for fathers' rights, especially in legal cases involving divorce and child custody (Dragiewicz 2008; Menzies 2007).

Shifting Gender Formations

In the 1980s and into the 1990s the radical power of feminism fractured under a broadside of anti-feminist backlash (Faludi 1991), and fragmented internally from corrosive disputes among feminists around issues of race and class inequalities, and divisive schisms that centered on sex work and pornography (Echols 2002). Some key political efforts by US feminists such as the Equal Rights Amendment (ERA) had failed, and feminism was less visible as a mass movement. However in 1989 sociologist Verta Taylor argued that the US feminist movement had not disappeared; rather, this was a time of 'movement abeyance', when activists in submerged networks continued to fight for equality, sustaining below-the-radar efforts that created the possibility for future political mobilizations. At the same time, in Canada, Australia and other jurisdictions where women's policy machineries were established, feminist networking and activism went 'mainstream', as did states' commitment to gender mainstreaming globally (see Bacchi and Eveline 2003; Franzway, Court and Connell 1989).

But there was something more happening in the 1980s and 1990s US gender politics than 'movement abeyance', Feminist momentum from the 1970s and networks of feminist activists combined with three substantial and interrelated social changes: the institutionalization and professionalization of feminism; the emergence of a widespread postfeminist cultural sensibility; and shifts in the political economy, including deindustrialization and the rise of a neoliberal state that slashes taxes for corporations and the rich, cuts public welfare and education, and celebrates individualism and the primacy of the market. These three changes created the current moment of gender formation that makes possible a range of men's engagements with gender politics, including men's rights organizing and profeminist men's activism that take substantially different forms than they did in the 1970s.

Professionally Institutionalized Feminism

The mass feminist movement was in decline in the 1980s and 1990s United States, but this was also a time of successful and highly visible feminist institutional reform, including the building of large feminist advocacy organizations like the National Organization for Women (NOW) and the National Abortion Rights Action League (NARAL), the institutionalization of women's and gender studies in universities, and the stabilization of myriad community and campus-based rape crisis and domestic violence centers (Martin 1990). Thus, as was occurring in Canada, Australia and elsewhere, feminists reformed police practices and legal responses to rape and domestic violence; workplaces incorporated sexual harassment trainings; and schools revised sexist curricula and expanded opportunities for girls' sports. These reforms were accompanied by the creation and expansion of professional sub-fields and occupational niches that focused on women's issues in social work, law and psychology.

The institutionalization of feminism created new challenges for feminists, not the least of which was what Markowitz and Tice (2002) called 'the paradoxes of professionalization'. On the one hand, professionalization created the conditions for sustaining feminist reform efforts on many fronts, including the creation of career paths for feminists in law, academia, medicine, social work and other professions (Staggenborg 1988). But on the other hand it led to a diversion of activist energies away from radical social change efforts toward finding sustainable funding sources for service provision, and also ushered in different organizational processes, with bureaucratic hierarchies displacing earlier feminist commitments to democratic decision-making processes.

US feminists also managed to wrest significant concessions from the state, including the 1974 passage of Title IX (federal law related to gender equity in schools), and the 1994 *Violence Against Women Act*, which altered the landscape for feminist work against gender-based violence. Even given the fact that state support for women's issues in the US remained minimal, Kristen Bumiller (2013) argues that such 'feminist collaboration with the state' threatens to water down, or even sever, the language and grass-roots politics of feminism. Moreover, with the continuing decline of the welfare state and the conco mitant expansion of neoliberalism, what Ruth Gilmore (2007) calls 'the non-profit industrial complex' emerged as a sort of 'shadow state' (Walch 1990), funded by an exploding number of foundations, and advancing professionalized public health-oriented approaches to issues like violence against women.

The rise of professionally institutionalized feminism, in short, broadened and stabilized the field of feminist action, while simultaneously thinning its political depth, threatening even to make feminist language and analysis disappear altogether: university women's studies programs become 'gender studies programs'; 'violence against women' morphs to 'gender-based violence'; and feminist organizations created by and for women become mixed-gender organizations whose historical roots are easily forgotten in the crush of day-to-day struggles to measure and document the effectiveness of service provisions, needed to win continued funding from foundations or the state (Messner, Greenberg and Peretz 2015). Professionally institutionalized feminism was also accompanied by a widespread shift in cultural values about gender: namely, the emergence of a postfeminist sensibility.

Postfeminism

As movement feminism receded from public view in the 1990s, a new and controversial 'postfeminist' discourse emerged. Feminist scholars have explored and debated the claim that a whole generation of younger people express a postfeminist worldview. On the one side, drawing from public opinion data, sociologists Hall and Rodriguez (2003) found little support for claims of widespread adherence to postfeminism, which they defined as including anti-feminist beliefs. But on the other side most scholars have drawn a distinction between postfeminism and opposition to feminism. Postfeminist narratives normally include an appreciation for feminist accomplishments, coupled with a belief that the work of feminism is in the past, and thus that feminist collective action is no longer necessary (Butler 2013). Sociologist Jo Reger (2012) argues that younger women and men for the most part agree with feminist positions on equal opportunities for women and men, but tend to experience feminism as both 'everywhere and nowhere'. The 'everywhere' refers both to feminism's professional institutionalization and to the ways that liberal feminist values have permeated popular culture in much the same way that fluoride invisibly permeates public drinking water (in fact, Reger (2012) refers to today's youth as 'generation fluoride'). However, feminism today is also experienced as 'nowhere': young people do not see an in-the-streets mass feminist movement, nor do they see any reason for one. The continuing work of professional feminism is, to most young people, as invisible as the cavity-prevention work of fluoride in our public waters.

As with any widely shared generational sensibility, postfeminism contains its own contradictions and limits. Most of the younger women in the supposedly 'postfeminist' generation studied by Aronson (2003) appreciated the accomplishments of the feminist movement and many recognized the existence of continued gender inequalities. However, Aronson described roughly half of them as 'fence-sitters', passive supporters of feminist goals, thus leaving open the question of how, or under what conditions, postfeminist consciousness might convert to feminist identification and political action. Pomerantz and her colleagues (2013) show that postfeminist discourse makes it hard for school girls to name sexism when it happens, yet they argue that postfeminist narratives have a built-in instability, especially when they run up against the lived reality of continuing sexist constraints on girls. For instance, girls and young women (often supported by their fathers and mothers) can become instant gender-equity activists when they discover that they are not being given equal athletic opportunities in schools or colleges, or when college women survivors of sexual assault learn that their own institutions are neither supporting them nor holding perpetrators accountable. In other words, when groups of girls and women bump up against sexist institutional constraints, it is possible for an individualist postfeminist sensibility to convert to collective feminist actions. And when it does, such feminist action is often given form and facilitated by existing institutionalized professional feminism—for instance, campus rape crisis centers or women's law centers.

This feminist optimism, however, faces an uphill struggle against the regressive tendencies built into a postfeminist sensibility that is coterminous with a larger political shift to neoliberal celebrations of individual market consumption choices as drivers for progress. And postfeminism is perfectly consistent with—indeed is shaped by and helps to naturalize—the eclipse of feminist language and politics within the professionalized non-profit industrial complex. Postfeminism also works in tandem with shifts in the political economy, including the nearly four-decade-long trend of deindustrialization that accompanied the ascendance of neoliberalism and that has disproportionately rendered poor and blue-collar young men redundant, a shift eventually referred to by some as 'the decline of men'.

Deindustrialization and 'The Decline of Men'

The early 1980s recession accelerated a continuing deindustrialization of the American labor force that resulted in the elimination of millions of unionized jobs, rising levels of structural unemployment, and the growth of low-paid non-unionized service sector jobs (Wilson 1989). As in the UK under Margaret Thatcher, in Australia under John Howard, and in Canada under Stephen Harper, the policies of Reaganomics facilitated this economic restructuring by tugging the nation away from a New Deal/Great Society welfare state toward a state based on neoliberal ideas that celebrated individualism and the primacy of the market, while slashing taxes on rich individuals and corporations and cutting support for welfare and education. These shifts continued in subsequent decades, resulting in the dramatic growth of a super-rich class of people, a shrinking middle class and a growing proportion of working poor in the population. The economic restructuring that accelerated from the 1980s and 1990s had multiple and devastating effects; however, for my purposes here, I want to focus on how the neoliberalization of the economy was especially devastating for families headed by blue collar male wage earners. As women flowed into the labor market by the millions—as much out of necessity as for reasons sparked by ideals of feminist empowerment—the more educated ones poured into a growing field of professional occupations, while the greater mass of women filled an expanding array of low-paid pink collar and service sector jobs (Charles and Grusky 2004). While professional class men continued to fare reasonably well in this economic restructuring. blue collar and poor men—disproportionately men of color—faced an increasingly bleak field of economic opportunity (Wilson 1996).

As the 1990s came to a close, Connell (1995) documented how the crumbling structural foundation for the male breadwinner role had escalated the gender insecurities of young working class men. Deteriorating public schools, declining hope for decent jobs in inner cities, and the expansion of prisons combined to create—especially for younger generations of black and brown boys and young men—contexts conducive to skyrocketing school dropout rates, neighborhood gang activity, illegal commerce in the informal economy, and high levels of domestic abuse and other forms of violence against women (DeKeseredy, Shahid and Schwartz 2003; Flores and Hondagneu-Sotelo 2013; Rios 2011). And right at the time, when the broader culture is trumpeting the arrival of 'involved fatherhood', the constraints on young poor and working class men make the achievement of the middle class ideal of an involved breadwinning father increasingly unreachable.

Men and Gender Politics Today

The three trends I have outlined—professionally institutionalized feminism, the emergent culture of postfeminism, and a post-industrial political economy characterized by deindustrialization and neoliberal state policies—together help to constitute the present moment of gender formation. Next, I will sketch how these three trends together make possible particular forms of men's antifeminist and profeminist actions.

Possibilities for Antifeminist Men's Rights Activism

By the 2000s, shifts in the political economy, combined with the increased visibility of women in higher education, popular culture and politics, and the growing public awareness of the institutionalization of women's rights, sparked journalistic and political hand-wringing about a supposed 'war against boys' in public schools and a widespread 'decline of males' in the public sphere (Sommers 2001; Tiger 2000). These escalating public concerns about boys and men created fertile ground for a resurgent men's rights movement.

But it is unlikely that we will see a widely popular or even marginally successful frontal attack on feminism from men's rights groups. This is in part because the same postfeminist sensibility that views feminism as a movement of the past is likely also to view aggressively anti-feminist men's rights activism as atavistically misogynistic. As values favoring public equality for women are increasingly institutionalized and defined not in a language of politicized feminism but more in a common sense language of equity and fairness, this shift also contracts the possibilities for anti-feminism. In a sense, I am suggesting that the institutional deck is stacked against overt anti-feminist backlash, be it frontal attacks on Title IX in schools, or men's rights groups' challenges to the state's (still minimal) support for women's shelters. Today, contingent on the specifics of national political developments, institutionalized feminism continues to influence legal and other decision making, albeit in a context that is often marked by deep controversy (Brodie 2008). Arguably, institutionalized feminism in the US now occupies a legal high ground, notwithstanding one that is still sometimes contested.

Rather than overt anti-feminist backlash, I argue that what is more likely to gain traction today in the US is a 'kinder, gentler' form of men's rights discourse and organizing, such as that now characterized by Warren Farrell, often considered the 'godfather' of the men's rights movement. Farrell's analysis in the 1980s and early 1990s drifted from the liberal feminist symmetry of men's liberation to asserting in his book *The Myth of Male Power* (1993) that there are many ways in which men are victimized by women's less visible forms of power. For instance, in response to women's attempts to stop sexual harassment in workplaces, Farrell claimed that in fact it was male employers who were disempowered and victimized by their secretaries' 'miniskirt power, cleavage power, and flirtation power'. In his more recent public speeches, however, Farrell appears to have returned to a less combative language of gender symmetry, reminiscent of his mid-1970s perspective (Farrell 2014). While it may seem on the surface that men are privileged with higher status and higher paying jobs, Farrell asserts, men pay a huge price for accepting the increased responsibilities that come with these jobs. Women just don't choose to enter higher paying careers, he claims, and they are smart to reject the stress; their lives are better for it.

Farrell's strategy is to raise sympathies for men, not to engage in anti-feminist polemics. In a postfeminist context, this more moderate men's rights discourse is likely to ring true as reasonable, as common sense. In this worldview, the women's movement succeeded in improving women's lives, and the logical flip-side is this: in the absence of a symmetrical men's movement to improve men's lives, men suffer harm. While anti-feminist vitriol continues to mark men's rights discourse on the Internet (Dragiewicz 2008, 2011), the emergent 'moderate' voice of the men's rights movement does not directly attack feminism or disparage women. Rather, it maneuvers in the postfeminist interstices between the 'everywhere' and the 'nowhere' of feminism. The means to improve men's lives are articulated to the general public in a depoliticized and individualized 'equality language' (Behre 2015) that resonates in a postfeminist and neoliberal context where present-day feminism seems to be 'nowhere'. Meanwhile, leaders such as Farrell (2014) are apparently coming to realize that they need not rant to a men's rights audience that feminism is 'everywhere', privileging women and holding men down. This is what these men already know; it is the fluoride in their ideological waters.

As a result, men's rights rhetoric that contains an *implicit anti-feminism* is likely to resonate with men who feel insecure or embattled. And I would speculate that moderate men's rights leaders' focus on individual choice and their implicit antifeminism resonates best with educated middle class white men who do not want to appear to be backwards misogynists. A central aspect of privileged men's gender strategies in recent years, after all, is to present one's self as an educated modern man who is supportive of gender equality. And this is achieved partly by projecting atavistic sexism on to less educated men, poor men, immigrant men and men of color (Dekeseredy, Shahid and Schwartz 2003; Hondagneu-Sotelo and Messner 1994).

In this context, is there a potential for less educated poor and working class men to constitute a sort of lump en anti-feminist army for men's rights? After all, declining economic opportunities for working class men to achieve a traditional conception of the male breadwinner role, combined with the perception that the law favors mothers over fathers in divorce and custody settlements might seem to create a perfect storm for the creation of an army of angry working class fathers ready to join men's rights organizations (Kimmel 2013). Thus far, this has hardly been the case. Most leaders of the men's rights movement are not poor and working class men; rather, they are men with the educational and financial resources needed to form organizations, create websites or hire attorneys. But just as the multi-billionaire Koch brothers' well-financed right wing anti-statism appeals to many lower-middle class whites, the men's rights movement's anti-feminist backlash rhetoric could possibly appeal to men with less education and less resources, men who may have a powerful father hunger but feel that their 'rights' have been denied them by controlling mothers, and especially by the state.

Indeed, in their study of poor working class fathers in Philadelphia and Camden, sociologists Edin and Nelson (2014) found that the men they studied frequently express 'a profound, abiding mistrust of women'; they think 'the system' that enforces child support automatically and unfairly favors mothers, who themselves are gatekeepers who keep the men away from their own children. These men feel as though they have few rights, while both mothers and 'the system' treat them as though they are 'just a paycheck' (when many of these fathers have no regular paycheck). In fact, this discourse is precisely what is commonly disseminated on men's rights Internet sites (Dragiewicz 2008, 2011; Mann 2005; Menzies 2007). The common feminist retort to fathers' rights claims have in the past been something like this: 'When you share fully in the *responsibilities* of birth control, and then also share equally the *responsibilities* of child support and childcare before divorce, *then* you can share parental rights afterwards'. But the stories of poor fathers reported by Edin and Nelson (2014) illustrate the inadequacy of this rejoinder. These are men who desire deeply to be foundational and present in their children's lives, but who face seemingly insurmountable institutional barriers to achieving and sustaining this parental ideal. To date, there is very little evidence that masses of poor fathers are joining as foot soldiers in anti-feminist collective action. But if current industrial nations continue to lack the will to address the many ways in which a huge strata of young men are being treated as dispensable by the economy and the criminal justice system, it is possible that some of these men will find resonance with Internet-based anti-feminist men's rights discourse that blames women and the liberal state for men's woes.

Possibilities for Men's Profeminist Activism

The recent institutionalization and professionalization of feminism has included a modest expansion of opportunities for men professionals to work on gender issues in social work, academia, law and other fields. This expanding base of men's professional action in gender fields carries both promise and risk. Especially given the recent explosions of public awareness of sexual assault and domestic violence in academia, the military and men's sports, this moment of opportunity and risk is nowhere more apparent than within the array of professional fields that confront violence against women. In 2014, even the President of the United States—not usually a platform for feminist calls for action—called on men to take an active role in ending violence against women.

Feminist women toiled for the past half-century to transform public awareness about violence against women, and to create and sustain rape crisis centers and domestic violence shelters. For the most part, feminist women welcome male allies who step up to prevent future acts of sexual and domestic violence. But some feminists in the anti-rape and anti-domestic violence community are also cautious about the ways in which male allies still benefit from male privilege that works to the detriment of women professional colleagues (Messner, Greenberg and Peretz 2015). And in a context of postfeminism, long-time feminist activists fear that, just as the field of gender-based violence prevention has expanded to include more men, the politics underlying anti-violence actions have thinned, severing action from feminist historical and political roots (Greenberg and Messner 2014).

In short, I suggest that while feminists continue to strategize vigilantly against eruptions of misogynist anti-feminist backlash, a less obvious but perhaps greater challenge springs from the ways in which the very conditions of historical gender formation that facilitate men's movement into professionalized 'gender work' also threaten to eclipse feminism altogether. In particular, widespread ideologies of postfeminism, coupled with depoliticized and marketized anti-violence initiatives, threaten to further erase feminist women's organizational leadership as well as the feminist analysis that underlies anti-violence work. Today's anti-violence workers commonly refer to 'the movement' not as an eruption of mass activism, but as a network of likeminded anti-violence professionals, and they talk of 'politics' not in terms of activism aimed to bring about structural transformations, but as strategies designed to keep their organizations funded (Messner, Greenberg and Peretz 2015). Much of the violence prevention curricula deployed in schools and communities today has jettisoned the feminist idea that violence against women springs from men's over-conformity with dominant conceptions of masculinity, instead deploying a pragmatic (and more individualistic) strategy of teaching boys and men to make 'healthy choices'—a discourse that, not incidentally, is shaped so that it can be subjected to 'metrics' that document program effectiveness in support of continuing requests for state or foundation funds.

What are the forces that potentially counter the depoliticization of anti-rape and anti-domestic violence work? One—though this is likely to be temporary—is the continued presence of older feminist women in the field, who mentor younger cohorts of professional women and men in ways that keep feminist analysis and goals at the center of the work. A second source of change, potentially more transformative, lies in the recent growth of diversity among men in the anti-violence field. As the field has expanded in the US, the opportunities for men to work in internships and paid jobs in rape and domestic violence prevention, state and foundation funders have increasingly targeted violence prevention efforts to communities of boys and men considered to be 'at risk' due to poverty, crumbling schools, and high rates of gang violence and drug use. There is a widely held perception in the field today that boys of color from poor communities will be more open to learning from young men from their own communities, who look and talk more like they do. This in turn has created a demand for a more racially diverse influx of young men into violence prevention work.

The growing number of young African American and Latino men entering anti-violence work is infusing a much-needed intersectional perspective into professional anti-violence work. Intersectionality—the perspective that takes the simultaneity of gender, race, class and other forms of 'intersecting' inequalities as its conceptual core—has long been central in academic feminism (Collins 1990; Crenshaw 1991). Indeed, it could be argued that, within academic feminism, intersectionality has for some years been a paradigmatic theoretical perspective and research approach (McCall 2005). But the radical insights of feminist intersectionality risk being diluted or even lost in professionally institutionalized violence prevention efforts.

Men of color's movement into professionalized anti-violence work brings to the field not so much a background in academic intersectionality but, rather, an experience-based organic intersectionality, different in two ways from the experiences of most white middle-class men in the field. First, young men of color frequently begin with a commitment to addressing boys' vulnerabilities to various forms of violence—in the home, in the street, and from police. These young men often began working with boys around gang and substance abuse issues, in college internships and then paid jobs in non-profit organizations. In that work, they discovered the links between young men's vulnerabilities to multiple forms of violence with their experiences with rape and domestic violence. In short, it was through doing 'race and class' work with young men that many of these anti-violence workers 'discovered' gender. This in turn created the possibility for an analysis of violence that does not always start with gender as necessarily being foundational (as it so often does with white middle class men who enter the anti-violence field), instead developing into an intersectional understanding of violence, grounded organically in the everyday experiences of race, class, and gender as interlocking processes (Messner, Greenberg and Peretz 2015).

This organically intersectional analysis underlies a second difference between young men of color and white middle class men in the anti-violence field. Young men of color tend to view the now-standard curricula deployed in school- and community-based violence prevention efforts as flat, one-dimensional and thus inadequate. Instead, these young men are innovating and even departing from the standard curricula, developing approaches that draw, for instance, from 'theatre of the oppressed', radical community education pedagogies that plumb the everyday life experiences of boys in order to 'make it real'. Men of color's emergent organically intersectional pedagogies frequently also circle back to academic feminist intersectionality, discovering there a ready resource for understanding connections between violence against women with other forms of 'gender based violence'—like sexual abuse and homophobic bullying of boys and transgender youth—as well as with forms of violence that may not be so obviously (or at least primarily) about gender—such as gang violence or police violence.

The progressive potential of the rise of organic intersectionality in the anti-violence field is twofold. First, it has direct appeal to young boys in poor communities because they can see their stake in working for change in their schools and communities. Second, it can re-infuse a powerful dose of radical social justice-oriented politics back into a professionalized anti-violence field that in recent years has seen a severe thinning of its politics, and a near-evaporation of its ability to address connections between gender-based violence with broader social justice issues like poverty, warfare, and cuts in public support for schools and families.

References

Allen P (1970) Free Space: A Perspective on the Small Group in Women's Liberation. New York: Times Change Press.

Aronson P (2003) Feminists or 'postfeminists'? Young women's attitudes toward feminism and gender relations. Gender and Society 17(6): 903–922.

Bacchi C and Eveline J (2003) Mainstreaming and neoliberalism: A contested relationship. Policy and Society: journal of Public, Foreign and Global Policy 22(2): 98–118. DOI: 10.1016/51449–4035(03)70021–6.

Baumli F (ed.) (1985) Men Freeing Men: Exploding the Myth of the Traditional Male. Jersey City: New Atlantis Press.

Behre KA (2015) Digging beneath the equality language: The influence of the fathers' rights movement on intimate partner violence public policy debates and family law reform. William & Mary journal of Women and the Law 21(3): 525–602.

Brodie J (2008) We are all equal now: Contemporary gender politics in Canada. Feminist Theory 9(2): 145–164. DOI: 10.1177/1464700108090408.

Bumiller K (2013) Feminist collaboration with the state in response to sexual violence: Lessons from the American experience. In Tripp AM, Marx Ferree M and Ewing C (eds) Gender, Violence, and Human Security: Critical Feminist Perspectives: 191–213. New York: New York University Press.

Butler J (2013) For white girls only? Postfeminism and the politics of inclusion. Feminist Formations 25(1): 35–58.

Charles M and Grusky DB (2004) Occupational Ghettos: The Worldwide Segregation of Women and Men. Stanford: Stanford University Press.

Collins PH (1990) Black Feminist Thought: Knowledge, Consciousness, and the Politics of Empowerment. Boston: Unwin Hyman.

Connell R (1995) Masculinities. Berkeley and Los Angeles: University of California Press.

Crenshaw KW (1991) Mapping the margins: Intersectionality, identity, politics, and violence against women of color. Stanford Law Review 43(6): 1241–1299.

Dekeseredy W, Shahid A and Schwartz MD (2003) Under Siege: Poverty and Crime in a Public Housing Community. Lanham, Maryland: Lexington Books.

Dragiewicz M (2008) Patriarchy reasserted: Fathers' rights and anti-VAWA activism. Feminist Criminology 3(2): 121–144. DOI: 10.1177/1557085108316731.

Dragiewicz M (2011) Equality with a Vengeance: Men's Rights Groups, Battered Women, and Antifeminist Backlash. Boston: Northeastern University Press.

Echols A (1984 [2002]) The taming of the ID: Feminist sexual politics, 1968–1983. In Echols A (ed.) Shaky Ground: The Sixties and its Aftershocks: 108–128. New York: Columbia University Press.

Edin K and Nelson TJ (2013) Doing the Best I Can: Fatherhood in the Inner City. Berkeley: University of California Press.

Faludi S (1991) Backlash: The Undeclared War against American women. New York: Crown.

Farrell W (1974) The Liberated Man. New York: Random House.

Farrell W (1993) The Myth of Male Power: Why Men are the Disposable Sex. New York: Simon and Schuster.

Farrell W (2014) International Conference on Men's Issues - Day 2 Excerpt - Warren Farrell. Available at https://www.youtube.com/watch? v=V5PMS6VkJkY (accessed 10 November 2015).

Flores EO and Hondagneu-Sotelo P (2013) Chicano gang members in recovery: The public talk of negotiating Chicano masculinities. Social Problems 60(4): 1–15.

Franzway S, Court D, and Connell RW 1989. Staking a Claim: Feminism, Bureaucracy and the State. Sydney: Allyn and Unwin.

Friedan B (1963) The Feminine Mystique. New York: Norton.

Gilmore RW (2007) In the shadow of the shadow state. In INCITE! Women of Color Against Violence (ed.) The Revolution Will Not be Funded: Beyond the Non-profit Industrial Complex: 41–52. Cambridge, Massachusetts: South End Press.

Goldberg H (1976) The Hazards of Being Male: Surviving the Myth of Masculine Privilege. New York: Signet

Greenberg MA and Messner MA (2014) Before prevention: The trajectory and tensions of feminist anti-violence. In Texler Segal M and Demos V (eds) Gendered Perspectives on Conflict and Violence (Part B): 225–250. Emerald Group Publishing.

Hall EJ and Rodriguez MS (2003) The myth of postfeminism. Gender & Society 17(6): 878–902. DOI: 10.1177/0891243203257639.

Hondagneu-Sotelo P and Messner MA (1994) Gender displays and men's power: The 'New Man' and the Mexican immigrant man. In Brod H and Kaufman M (eds) Theorizing Masculinities: 200–218. Sage Publications.

Jourard SM (1974) Some lethal aspects of the male role. In Pleck JH and Sawyer J (eds) Men and Masculinity: 21–29. Englewood Cliffs, New Jersey: Prentice-Hall.

Kimmel MS (1987) Men's responses to feminism at the turn of the century. Gender& Society 1(3): 261–283.

Kimmel MS (2013) Angry White Men: American Masculinity at the End of an Era. New York: Nation Books.

Markowitz L and Tice KW (2002) Paradoxes of professionalization: Parallel dilemmas in women's organizations in the Americas. Gender& Society 16(6): 941–958.

Mann RM (2005) Fathers' rights, feminism, and Canadian divorce law reform, 1998–2003. Studies in Law Politics and Society 35: 31–68. DOI: 10.1016/S1059–4337(04)35002–7

Martin PY (1990) Rethinking feminist organizations. Gender & Society4(2):182–206.DOI:10.1177/089124390004002004.

McCall L (2005) The complexity of intersectionality. Signs 30(3): 1771–1800. DOI: 10.1086/426800

Men's Consciousness-Raising Group (1971) Unbecoming Men. Washington, New Jersey: Times Change Press.

Menzies R (2007) Virtual backlash: Representations of men's 'rights' and feminist 'wrongs' in cyberspace. In Chunn DE, Boyd SB and Lessard H (eds) Reaction and Resistance: Feminism, Law, and Social Change: 65–97. Vancouver: UBC Press.

Messner MA (1997) Politics of Masculinities: Men in Movements. Lanham, Maryland: Altamira Press.

Messner MA (1998) The limits of 'the male sex role': An analysis of the men's liberation and men's rights movements' discourse. Gender & Society 12(3): 255–276. DOI: 10.1177f0891243298012003002

Messner MA, Greenberg MA and Peretz T (2015) Some Men: Feminist Allies and the Movement to End Violence against Women. New York: Oxford University Press.

Nichols J (1975) Men's Liberation: A New Definition of Masculinity. New York: Penguin.

Omi M and Winant H (1986) Racial Formation in the United States: From the 1960s to the 1980s New York: Routledge and Kegan Paul Inc.

Pomerantz S, Raby Rand Stefanik A (2013) Girls run the world? Caught between sexism and postfeminism in school. Gender & Society 27(2): 185–207. DOI: 10.1177/0891243212473199.

Pleck JH (1977) Men's power with women, other men, and in society: A men's movement analysis. In Hiller DV and Sheets R (eds) Women and Men: The Consequences of Power: 417–433. Cincinnati, Ohio: Office of Women's Studies, University of Cincinnati.

Reger J (2012) Everywhere and Nowhere: Contemporary Feminism in the United States. New York and Oxford: Oxford University Press.

Rios VM (2011) Punished: Policing the Lives of Black and Latino Boys. New York and London: New York University Press.

Sommers CH (2001) The War against Boys: How Misguided Feminism is Harming Our Young Men. New York: Simon and Schuster.

Staggenborg S (1988) The consequences of professionalization and formalization in the prochoice movement American Sociological Review 53(4): 585–605.

Stansell C (2010) The Feminist Promise: 1792 to the Present New York: Random House.

Taylor V (1989) Social movement continuity: The women's movement in abeyance. American Sociological Review 54(5): 761–775.

Tiger L (2000) The Decline of Males: The First Look at an Unexpected New World for Men and Women. New York: St Martin's Press.

Wilson WJ (1989) The Truly Disadvantaged. Chicago, Illinois: University of Chicago Press.

Wilson WJ (1996) When Work Disappears: The World of the New Urban Poor. New York: Knopf.

Wolch J (1990) The Shadow State: The Government and the Voluntary Sector in Transition. New York: The Foundation Center.

Topics for Further Examination

- Listen to the podcast "The Long History of #MeToo (a History Talk podcast)" produced by the Department of History at Ohio State (January 31, 2018). Identify the key lessons about social change for gender equality that are discussed in this podcast.
- Browse feminist websites on women's organizations and gender issues such as Institute for Women's

Policy Research (http://www.iwpr.org), Feminist Majority Foundation (http://www.feminist.org), Black Girl Dangerous (bgdblog.org), Feministing (feministing.com), The F-Word Blog (www.thef-word.org.uk), Autostraddle (https://www.autostraddle.com), Crunk Feminist Collective (www.crunkfeministcollective.com), Women Watch (http://www.un.org/womenwatch/), and Women's International League for Peace and Freedom (http://www.wilpf.org).

- Sociologist and activist, Allan Johnson, offered useful advice about how each of us can work on behalf of dismantling oppressive systems. For example, he urged us to start by acknowledging oppression and learning as much as we can about how we participate in systems of privilege. He also asked us to take "little risks." What little risks might you take to advance gender equality in the worlds in which you live and work?

About the Editors

Catherine (Kay) G. Valentine is Professor Emerita of sociology at Nazareth College in Rochester, New York. She received her PhD from Syracuse University and her BA from the State University of New York at Albany. Kay taught a wide range of courses, such as sociology of gender, senior seminar in sociology, sociology of bodies and emotions, sociology of consumerism, and human sexuality. Her publications include articles on teaching sociology, on women's bodies and emotions, on gender and qualitative research, and on the sociology of art museums. Kay is coeditor of *Letting Go: Feminist and Social Justice Insight and Activism* (Vanderbilt University Press, 2015). She is the founding director of women's studies at Nazareth College and a longtime member of Sociologists for Women in Society and the American Sociological Association. She has also served as president of the New York State Sociological Association. Kay's life partner and travel companion, Paul J. Burgett, passed away in August 2018. He was University of Rochester vice president and professor of music.

Mary Nell Trautner is Associate Professor of sociology at the University at Buffalo, SUNY. She earned her PhD and MA from the University of Arizona and her BA from Southwestern University. She is an award-winning teacher of courses in sociology of gender, criminology, sociology of law, and social problems. Mary Nell is an author of a forthcoming active learning textbook on social problems, and has published a variety of articles on topics ranging from physical appearance bias and images of women in popular media to day labor workers' legal knowledge and environmental pollution. She is active in the American Sociological Association and Law & Society Association. In her free time, she enjoys travel and scuba diving.

Joan Z. Spade is Professor Emerita of sociology at The College at Brockport, State University of New York. She received her PhD from the University at Buffalo, State University of New York; her MA from the University of Rochester; and her BA from the State University of New York at Geneseo. In addition to courses on gender, Joan taught courses on education, family, research methods, and statistics. She published articles on rape culture in college fraternities and on work and family, including women's and men's orientations toward work. She has also coedited two books on education and published articles on education, including research on tracking, and gender and education. Joan was active in Sociologists for Women in Society, Eastern Sociological Society, and the American Sociological Association. In addition to visiting children and grandchildren with her significant other, she enjoys RVing, music and the arts, travel, and being outdoors.